Native and Modified Starches for Food Applications

Applications of Starches in Food and Packaging

Native and Modified Starches for Food Applications

Edited by

Pothiyappan Karthik
Department of Food Technology, Faculty of Engineering,
Karpagam Academy of Higher Education (Deemed to be University),
Coimbatore, Tamil Nadu, India

Tanmay Sarkar
West Bengal State Council of Technical Education,
Government of West Bengal,
Malda, West Bengal, India

Wing-Fu Lai
School of Food Science and Nutrition,
University of Leeds, United Kingdom

Series Editor
Sneh Punia Bangar

ELSEVIER
ACADEMIC PRESS
An imprint of Elsevier

Academic Press is an imprint of Elsevier
125 London Wall, London EC2Y 5AS, United Kingdom
50 Hampshire Street, 5th Floor, Cambridge, MA 02139, United States

ISBN: 978-0-443-27431-2

For information on all Academic Press publications visit our website at https://www.elsevier.com/books-and-journals

Publisher: Mara Conner
Acquisitions Editor: Palak Gupta
Editorial Project Manager: Ritika Dagar
Production Project Manager: M. Balakrishnan
Cover Designer: Gopalakrishnan Venkatraman

Typeset by TNQ Tech

Contents

22. Nanostructured packaging: Innovations in starch-based nanocomposite films and coatings for extended food shelf life

Rachna Gupta, Kanchan Goswami, Aashish Pratap Singh and Vivek Kumar Gupta

23. Starch based nanocomposite materials for enhanced quality and safety of food products during storage

Abhishek Biswal, P. Poojapushkarni, Aravinth Sithivinayagam, V. Sneha, Vinoth Kumar Vaidyanathan, Periyar Selvam Sellamuthu and Emmanuel Rotimi Sadiku

24. Disruptive technologies: Impact of artificial intelligence and blockchain on starch supply chain management

Abubakar Kabir Adam, Dragan Pamucar and Tithli Sadhu

Contributors

Jincy Abraham, Department of Food Processing Technology, PSG College of Arts & Science, Coimbatore, Tamil Nadu, India

Abubakar Kabir Adam, Department of Food Science and Technology, School of Agriculture, SR University, Warangal, Telangana, India

Faraz Ahmad, Department of Integrative Biology, School of Bio Sciences and Technology, Vellore Institute of Technology, Vellore, Tamil Nadu, India

Irene Andressa, Federal University of Viçosa, Viçosa, Brazil

Erandi Arvizu-Hernandez, Tecnologico de Monterrey, School of Engineering and Sciences, Institute for Obesity Research, Monterrey, Mexico; Tecnologico de Monterrey, School of Engineering and Sciences, Zapopan, Mexico

Oluwaseun Peter Bamidele, Food Science and Technology, University of Venda, Thohoyandou, South Africa

Khalid Bashir, Department of Food Technology, Jamia Hamdard, New Delhi, India

Abhishek Biswal, Department of Food Process Engineering, Postharvest Technology Research Lab, School of Bioengineering, SRM Institute of Science and Technology, Chennai, Tamil Nadu, India

Esra Capanoglu, Istanbul Technical University, Department of Food Engineering, Faculty of Chemical and Metallurgical Engineering, Maslak, Turkey

Paulo Henrique Costa Paiva, Agricultural Research Company of Minas Gerais (EPAMIG), Cândido Tostes Dairy Institute, Juiz de Fora, Brazil

P.R. Delna, Department of Food Technology, Faculty of Engineering, Karpagam Academy of Higher Education, Coimbatore, Tamil Nadu, India; Centre for Food Nanotechnology, Karpagam Academy of Higher Education, Coimbatore, Tamil Nadu, India

R. Devadharshini, School of Biosciences and Technology, Faculty of Engineering and Technology, SRM Institute of Science and Technology, Tiruchirappalli, Tamil Nadu, India

Gangaraju Divyashri, Bioproducts Research, Iom Bioworks Pvt. Ltd., Bangalore, Karnataka, India

Rajaraman Eri, School of Science, STEM College, RMIT University, Melbourne, VIC, Australia

Anayansi Escalante-Aburto, Tecnologico de Monterrey, Institute for Obesity Research, School of Engineering and Sciences, Monterrey, Mexico

Sony George, TKM Institute of Technology, Department of Food Technology, Kollam, Kerala, India

Manasi Ghosh, Department of Food Technology, Faculty of Engineering, Karpagam Academy of Higher Education, Coimbatore, Tamil Nadu, India; Centre for Food Nanotechnology, Karpagam Academy of Higher Education, Coimbatore, Tamil Nadu, India

Ulaganathan Gnana Moorthy Eswaran, School of Biosciences and Technology, Faculty of Engineering and Technology, SRM Institute of Science and Technology, Tiruchirappalli, Tamil Nadu, India

Rosa Leonor González-Díaz, Tecnologico de Monterrey, School of Engineering and Sciences, Monterrey, Mexico

Kanchan Goswami, Human Nutrition, Krishi Vigyan Kendra, Jaipur-I, Home Science, Jaipur, Rajasthan, India

Rachna Gupta, Department of Food Technology, School of Biosciences and Technology, Galgotias University, Greater Noida, Uttar Pradesh, India

Shafiul Haque, Jazan University, Research and Scientific Studies Unit, College of Nursing and Health Sciences, Jazan, Saudi Arabia

Harini Hutti, Iom Bioworks Pvt. Ltd., Centre for Cellular and Molecular Platforms (C-CAMP), GKVK Campus, Bengaluru, Karnataka, India

Snehal R. Jadhav, Consumer Analytical Safety Sensory (CASS) Food Research Centre, School of Exercise and Nutrition Sciences, Deakin University, Melbourne, VIC, Australia

Yellutla Jahnavi, Department of Biotechnology, M S Ramaiah Institute of Technology, Bangalore, Karnataka, India

G.C. Jeevitha, Vellore Institute of Technology, Department of Bio-Sciences, Vellore, Tamil Nadu, India

Safreena Kabeer, Department of Food Technology, Faculty of Engineering, Karpagam Academy of Higher Education, Coimbatore, Tamil Nadu, India

Murat Karamese, Kafkas University, Faculty of Medicine, Department of Medical Microbiology, Kars, Turkey

Pothiyappan Karthik, Department of Food Technology, Faculty of Engineering, Karpagam Academy of Higher Education, Coimbatore, Tamil Nadu, India; Centre for Food Nanotechnology, Karpagam Academy of Higher Education, Coimbatore, Tamil Nadu, India

Damanjeet Kaur, School of Health, Deakin University, Melbourne, VIC, Australia

P. Kiran Kumar, Department of Chemistry, Seshadripuram Institute of Technology, Mysuru, India

Naveen Krishna, Integrated Bioprocessing Laboratory, Department of Biotechnology, School of Bioengineering, SRM Institute of Science and Technology, Chennai, Tamil Nadu, India

Kalaiselvi Krishnamoorthy, Centre for Herbal Pharmacology and Environmental Sustainability, Chettinad Hospital and Research Institute, Chettinad Academy of Research and Education, Kelambakkam, Tamil Nadu, India

Vivek Kumar Gupta, Department of Biological Engineering, Biohybrid Systems Research Centre (BSRC), Inha University, Incheon, Republic of South Korea

Rosa L. González-Díaz, Tecnologico de Monterrey, School of Engineering and Sciences, Zapopan, Mexico

Ivan Luzardo-Ocampo, Tecnologico de Monterrey, Institute for Obesity Research, School of Engineering and Sciences, Monterrey, Mexico; Tecnologico de Monterrey, School of Engineering and Sciences, Zapopan, Mexico

Lalitha Prasanna Madduri Venkata, Iom Bioworks Pvt. Ltd., Centre for Cellular and Molecular Platforms (C-CAMP), GKVK Campus, Bengaluru, Karnataka, India

Sahar Akhavan Mahdavi, Department of Food Materials and Process Design Engineering, Gorgan University of Agricultural Science and Natural Resources, Gorgan, Iran

E.D. Mahima, Ramaiah Institute of Technology, Bengaluru, Karnataka, India

Shubasree Manickavel, Department of Biotechnology, Vel Tech Rangarajan Dr. Sagunthala R&D Institute of Science and Technology, Chennai, Tamil Nadu, India

M. Maria Leena, School of Biosciences and Technology, Faculty of Engineering and Technology, SRM Institute of Science and Technology, Tiruchirappalli, Tamil Nadu, India

Mpho Edward Mashau, Department of Food Science and Technology, University of Venda, Thohoyandou, South Africa

Shreya Mechri, Ramaiah Institute of Technology, Bengaluru, Karnataka, India

Aanchal Mittal, Vellore Institute of Technology, Department of Biotechnology, Vellore, Tamil Nadu, India

Vaishnavi Mohanadas Nair, Technical University of Munich, Department of Chemical Engineering, Bavaria, Germany

Girilal Morkattu, Saintgits College of Engineering, Department of Food Technology, Kottayam, Kerala, India

T.S. Murali Krishna, Department of Biotechnology, Vel Tech Rangarajan Dr. Sagunthala R&D Institute of Science and Technology, Chennai, Tamil Nadu, India

Hafsa Nadeem, Department of Biotechnology, M S Ramaiah Institute of Technology, Bangalore, Karnataka, India

Gbeminiyi Olamiti, Department of Food Science and Technology, University of Venda, Thohoyandou, South Africa

Flaviana Coelho Pacheco, Federal University of Viçosa, Department of Food Science and Technology, Vicosa, Brazil

Ana Flávia Coelho Pacheco, Agricultural Research Company of Minas Gerais (EPAMIG), Cândido Tostes Dairy Institute, Juiz de Fora, Brazil

Sakthi Priyanka Palanivel, Department of Biotechnology, Vel Tech Rangarajan Dr. Sagunthala R&D Institute of Science and Technology, Chennai, Tamil Nadu, India

Dragan Pamucar, University of Belgrade, Faculty of Organizational Sciences, Department of Operations Research and Statistics, Belgrade, Serbia; UNEC Applied Artificial Intelligence Research Center, Azerbaijan State University of Economics (UNEC), Baku, Azerbaijan

Anjaly Paul, McGill University, Department of Bioresource Engineering, Sainte-Anne-de-Bellevue, QC, Canada

Néstor Ponce-García, Facultad de Ciencias Agrícolas, UAEMéx, Toluca, Mexico

P. Poojapushkarni, Department of Food Process Engineering, Postharvest Technology Research Lab, School of Bioengineering, SRM Institute of Science and Technology, Chennai, Tamil Nadu, India

R. Poorvasruthi, School of Biosciences and Technology, Faculty of Engineering and Technology, SRM Institute of Science and Technology, Tiruchirappalli, Tamil Nadu, India

Aashish Pratap Singh, Master of Business Admistration, Narsee Monjee Institute of Management Studies (NMIMS), Mumbai, India

Rakesh Raghunathan, International Rice Research Institute, ISARC, Centre of Excellence in Rice Value Addition (CERVA), Varanasi, Uttar Pradesh, India

R. Rajam, Department of Food Technology, Kalasalingam Academy of Research and Education, Virudhunagar, Tamil Nadu, India

Shonisani Eugenie Ramashia, Department of Food Science and Technology, University of Venda, Thohoyandou, South Africa

Yashwanth Ramesh, Department of Biotechnology, M S Ramaiah Institute of Technology, Bangalore, Karnataka, India

Sandeep Singh Rana, School of Bio Sciences & Technology (SBST), Vellore Institute of Technology, Vellore, India

Miguel Rebollo-Hernanz, Department of Agricultural Chemistry and Food Science, Faculty of Science, Universidad Autónoma de Madrid, Madrid, Spain; Department of Production and Characterization of Novel Foods, Institute of Food Science Research, CIAL (UAM-CSIC), Madrid, Spain

Jeny A. Rincón-Aguirre, Tecnologico de Monterrey, School of Engineering and Sciences, Monterrey, Mexico

Tithli Sadhu, Department of Food Science and Technology, School of Agriculture, SR University, Warangal, Telangana, India; Department of Biochemistry, School of Agriculture, SR University, Warangal, Telangana, India

Parnasree Sadhukhan, Department of Food Science & Nutrition Management, J.D. Birla Institute, Kolkata, West Bengal, India

Emmanuel Rotimi Sadiku, Tshwane University of Technology, Department of Chemical, Metallurgical and Materials Engineering, Pretoria, South Africa; Institute of NanoEngineering Research (INER) and Department of Chemical, Metallurgical and Materials Engineering, Tshwane University of Technology, Pretoria West Campus, Pretoria, Republic of South Africa

Ravi Sai Nandhini, Department of Biotechnology, Vel Tech Rangarajan Dr. Sagunthala R&D Institute of Science and Technology, Chennai, Tamil Nadu, India

Ravi Sanjana, Department of Food Technology, Faculty of Engineering, Karpagam Academy of Higher Education, Coimbatore, Tamil Nadu, India

R. Saran, School of Biosciences and Technology, Faculty of Engineering and Technology, SRM Institute of Science and Technology, Tiruchirappalli, Tamil Nadu, India

A. Saravana Raj, Vel Tech High Tech Dr.Rangarajan Dr. Sakunthala Engineering College, Department of Chemical Engineering, Chennai, Tamil Nadu, India

Periyar Selvam Sellamuthu, Department of Food Process Engineering, Postharvest Technology Research Lab, School of Bioengineering, SRM Institute of Science and Technology, Chennai, Tamil Nadu, India

Priyanka Balaji Sethupathy, School of Bio Sciences & Technology (SBST), Vellore Institute of Technology, Vellore, India

Rohan M. Shah, Department of Chemistry and Biotechnology, School of Science, Computing and Emerging Technologies, Iverson Health Innovation Research Institute, Swinburne University of Technology, Melbourne, VIC, Australia

Ashwina Shankar, Integrated Bioprocessing Laboratory, Department of Biotechnology, School of Bioengineering, SRM Institute of Science and Technology, Chennai, Tamil Nadu, India

Sohel Rana Sheakh, School of Biosciences and Technology, Faculty of Engineering and Technology, SRM Institute of Science and Technology, Tiruchirappalli, Tamil Nadu, India

Aravinth Sithivinayagam, Integrated Bioprocessing Laboratory, Department of Biotechnology, School of Bioengineering, SRM Institute of Science and Technology, Chennai, Tamil Nadu, India

V. Sneha, Integrated Bioprocessing Laboratory, Department of Biotechnology, School of Bioengineering, SRM Institute of Science and Technology, Chennai, Tamil Nadu, India

P. Subramanian, Riddet Institute, Massey University, Palmerston North, New Zealand

Sujithra Sureshkumar, Department of Food Technology, Faculty of Engineering, Karpagam Academy of Higher Education, Coimbatore, Tamil Nadu, India

M. N. Tejaswini, School of Biosciences and Technology, Faculty of Engineering and Technology, SRM Institute of Science and Technology, Tiruchirappalli, Tamil Nadu, India

P. Thivya, Department of Food Process Engineering, National Institute of Technology (NIT), Rourkela, Odisha, India

Krishna Priyanka Tippareddy, Department of Food Technology, Faculty of Engineering, Karpagam Academy of Higher Education, Coimbatore, Tamil Nadu, India; Centre for Food Nanotechnology, Karpagam Academy of Higher Education, Coimbatore, Tamil Nadu, India

Likitha V., Ramaiah Institute of Technology, Bengaluru, Karnataka, India

Vinoth Kumar Vaidyanathan, Integrated Bioprocessing Laboratory, Department of Biotechnology, School of Bioengineering, SRM Institute of Science and Technology, Chennai, Tamil Nadu, India

Swetha Vasudevan, Vellore Institute of Technology, Department of Bio-Sciences, Vellore, Tamil Nadu, India

Olivia Wright, School of Human Movement and Nutrition Sciences, Queensland University, Brisbane, QLD, Australia

Preface

Starch is one of the most useful and versatile constituents of the human diet, not only as a source of energy, but also as a key functional component in different kinds of food products. Since the beginning of staple products, which are made of cereals, to the new-fangled researched processes and the convenience foods, starch still defines a texture, stability, greatness, and other factors that make a food marketable. Over the past decades, the scientific capabilities of both native and modified starches have advanced at an accelerated rate, owing to the innovations in food chemistry, processing technologies, and the general increase in the demand toward more healthy, sustainable, and custom-made food. The idea behind this book, *Native and Modified Starches for Food Applications*, is that it is intended to be a complete resource, that is, one covering both the fundamental science together with the practical use in industry application.

The book is divided into three sections to bring the reader to the foundations of the topic and up to the advanced innovations. Part I preliminarily gives the scientific background of investigating the sources, classification, structural properties, physicochemical behavior, sensory dimensions, and the emergent processing and modification method of starches. The purpose of these chapters is to provide the readers with a clear conceptual and analytical framework that they would need to have in order to comprehend how starch functionality can be applied to suit a certain product need. It focuses not only on conventional, but also innovative modification methods, such as the alterations of physical, chemical, enzymatic, and biotechnological methods, that are transforming the world of ingredients.

Part II exchanges theory with practice by demonstrating the working of native and modified starches in various food systems. Their use as fat replacers, texturizing agents, emulsifiers, stabilizers, biopolymers, and delivery matrices is discussed and other specialized applications of the topic including gluten-free formulations, nutraceuticals, meat and seafood restructuring, dairy and frozen products, extruded foods, and age-specific nutrition are also covered.

Part III covers the future of starch science and technology. The examples of emerging starch research directions include nanotechnology, plant-based solutions, nanocomposite packaging, and smart storage. Together with this, the market and consumer behavior are discussed to highlight the economic and social forces that shape ingredient choice and the positioning of the product. Combined, these views demonstrate that starch is not a fixed consumer product, but a changing platform upon which technological advancement and food infrastructures inspire it to work toward sustainability.

This book will help satisfy the needs of food scientists, nutritionists, researchers, students, product developers, academicians, and industry professionals who will find the book intensive (in depth) and extensive (in breadth) in terms of starch applications. The editors have managed to unite multidisciplinary knowledge and actual case investigations to produce a reference that assists in developing innovation, knowledgeable decision-making, and future research directions. Finally, the aim of this work is to show that both the natural and modified forms of starch continue to play a role in the progress of modern food science and technology.

Pothiyappan Karthik
Tanmay Sarkar
Wing-Fu Lai

Part I

Introduction to starches in food

Chapter 1

Introduction to starches in food: Sources and classification

Kalaiselvi Krishnamoorthy[1], Ravi Sanjana[2] and Pothiyappan Karthik[2, 3]

[1]*Centre for Herbal Pharmacology and Environmental Sustainability, Chettinad Hospital and Research Institute, Chettinad Academy of Research and Education, Kelambakkam, Tamil Nadu, India;* [2]*Department of Food Technology, Faculty of Engineering, Karpagam Academy of Higher Education, Coimbatore, Tamil Nadu, India;* [3]*Centre for Food Nanotechnology, Karpagam Academy of Higher Education, Coimbatore, Tamil Nadu, India*

1. Introduction

Starch is an important component of human diets. They consume starch from a variety of sources, which directly contributes to the major health benefits. Also, it serves as a essential source of carbohydrates. At the same time, starch provides the vital nutrients for plant growth and development, so it is called the energy hub of metabolism (Kumar et al., 2012; Zhang et al., 2025; Yadav et al., 2024a). Strach acts as an important substance for the isolation of many products, like glucose-based syrups and maltodextrins. Apart from this, it's involved in food production, in food technology-based industries, starch is majorly used for its nontoxic nature and biodegradability (Hofman et al., 2016; Perez & Agama-Acevedo, 2017). Furthermore, cosmetic productions, cleansing agents, medicine preparations, thin paper production, fabrics, and industrial chemicals preparations, starches are mainly used. Starch serves several important purposes, including thickening, emulsifying, and gelling (Goswami & Mahanta, 2021; Olivato, 2024; Yadav et al., 2024b). For the finished product to have the appropriate qualities, starch utilization must be optimized. It serves as a thickening ingredient in soups and sauces, and because of its stability in texture changes, potato, tapioca, and waxy maize starches are often chosen as important sources (Sheldrake. 2009).

Starch is produced naturally as semicrystalline granules involving two polymers named amylose (AM) and amylopectin (AP) (Table 1.1). AP is a branched-chain polysaccharide composed of glucose units connected linearly by α $(1 \rightarrow 4)$ glycosidic linkages, with branches formed via α $(1 \rightarrow 6)$ linkages. Whereas, AM is a unbranched chain polysaccharide composed of -D-glucose monomers linked by α $(1 \rightarrow 4)$ glycosidic bonds. The composition of AP and AM varies depending on the plant source (Chakraborty et al., 2020). Among all other types of bio-based polymers, starch is one of the most affordable sources. The major challenge in the food industry is providing healthy food products with proper food security. Specifically in Asia, demand for food has risen because of the population increases, and to manage the food demand, some alternative methods of food production are necessary in the current decade. To fulfill the requirement, starch could be helpful and act as a good source of carbohydrates (Falua et al., 2022; Niranjana Prabhu & Prashantha, 2018).

Specifically, starch-based foods are more quickly digestible and produce more glucose in the bloodstream. Factors like starch structural composition, food processing method, and enzymatic breakdown in digestion influence these properties (Bai et al., 2021). Starch is deposited in various plant organs with semicrystalline granules. AP is a highly branched polymer, whereas AM is a comparatively long and less branched polymer (Cao et al., 2022). AM's exclusive molecular structure allows it to react with other substances to modify the physicochemical properties of starch. A, B, and C chains are the three types of AP chains. It was confirmed that AM was more prone to annealing treatment and responsible for the annealing of starch than AP (Li et al., 2025; Yu et al., 2025). Consequently, the current chapter provides key details regarding various starch sources, classifications based on their botanical origin and digestibility. Additionally, the modification techniques to improve the starch properties and their wide range of applications are discussed.

Native and Modified Starches for Food Applications. https://doi.org/10.1016/B978-0-443-27431-2.00007-9

TABLE 1.1 Structural and physicochemical distinctions between amylose (AM) and amylopectin (AP) and their properties.

Component	Properties	Description	References
Amylose (AM) 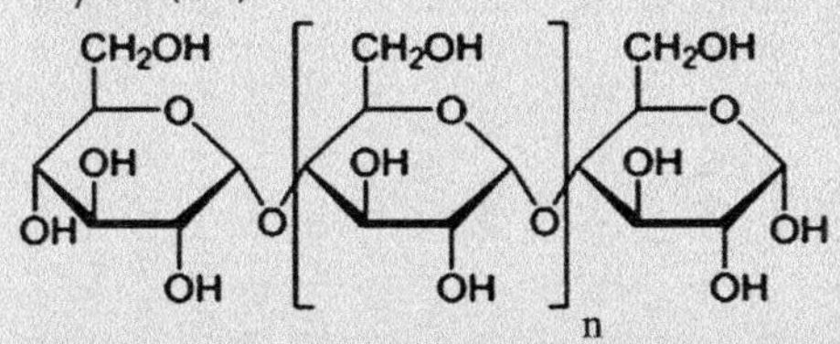	**Structure**	Linear or slightly branched polymer of α-D-glucose units linked by α-(1→4) glycosidic bonds	Li et al. (2025)
	Solubility	Less soluble in water due to its linear structure, which facilitates tight packing	Chakraborty et al. (2020)
	Gelation	Forms a firm gel upon cooling after gelatinization	Yu et al. (2025)
	Function	Contributes to the formation of resistant starch, which has health benefits like improved digestion and lower glycemic response	Falua et al. (2022)
Amylopectin (AP) 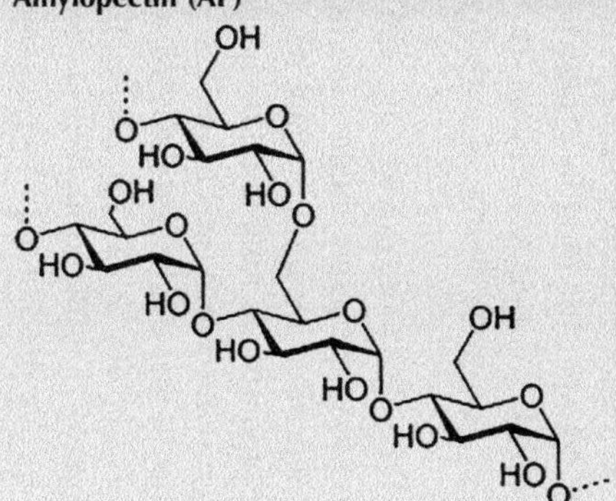	**Structure**	Highly branched polymer of α-D-glucose, containing both α-(1→4) glycosidic bonds (linear regions) and α-(1→6) glycosidic bonds (branch points)	Li et al. (2025)
	Solubility	More soluble in water due to its branched structure, which prevents tight packing	Cao et al. (2022)
	Gelation	Produces a more viscous solution but does not form a firm gel like amylose	Yu et al. (2025)
	Function	Provides the majority of the energy in starch, as it is more rapidly digested compared to amylose	Kong et al. (2008)

2. Sources of starch

Millions of people throughout the world depend on plant sources that contain starch for energy and vital carbohydrates. It is a polysaccharide that is widely distributed in many plant sources. Besides, starch is used in the manufacturing of biofuel, textiles, medications, and food (Rashwan et al., 2024). For human and animal diets, starch plays a major role as carbohydrates. The upcoming section explains the primary sources of starch, categorized into cereal grains, tubers, legumes, fruits, and unconventional sources (Fig. 1.1).

2.1 Cereal grains as starch sources

The richest sources of starch are cereal grains, which provide essential proteins, fibers, vitamins, and minerals. Starch, a multifaceted carbohydrate, is the key energy reserve in cereals and delivers an important portion of daily caloric consumption worldwide. In cereal grains, widely cultivated crop varieties are wheat, corn, rice, amaranth, rye, buckwheat, barley, and millet. These are concerning as major crops provide rich starch sources because of their potential bioactive compounds and added value (Zahid et al., 2025; Zhang et al., 2025). Cereal grains are a major source of human foods and they can provide various nutrients like carbohydrates, vitamins, minerals, and essential amino acids. Therefore, some of the important cereal variants are explained below.

Wheat is a key cereal crop variety that provides calories and proteins. It is generally used for bread, cakes, cookies, pasta, and noodles. Wheat grains are composed of water, carbohydrates, proteins, and lipids. From wheat grains, wheat starch is produced; it contains about 75% grain and 80% flour. One of the highest starch-content-rich cereal crops is corn, which consists of 85% starch at dry weight (Kumar et al., 2011; Kumar et al., 2024). Corn starches are generally used for the production of thickening agents, biodegradable plastics, corn oil, and sweeteners. Corn Starch is used as a major material for bioethanol production due to its simple sugar hydrolysis (Tabasum et al., 2019). Corn starch-based by-products are generally used for humans and animals; apart from this, they act as a biological fermentation and energy source for the microorganisms (Svihus et al., 2005).

Rice is a standard food in the world; it belongs to the Poaceae family, mainly cultivated in the Asia and Africa. Rice is rich in fiber, carbohydrates, vitamins, minerals, and proteins. Rice starch is a form of carbohydrate isolated from rice grains and it has a hierarchical structure (Rostamabadi et al., 2024). Compared to all starch-based crops, rice has more energy and it has a high glycemic index (GI) of 64%-93%. Rice is considered hypoallergenic due to its absence of gliadin proteins generally associated with celiac diseases and is majorly involved in gluten-free product formulations (Zhong et al., 2009). Apart from food sources, it's used in pharmaceutical and industrial applications because of its hydrophilicity, whiteness, and smoothness (Shoukat et al., 2025).

Barley is one of the oldest cultivars of grain varieties, it belongs to the Poaceae family, and it can adapt to cold, drought, and saline soil types. This is the fourth most cultivated cereal after wheat, rice, and maize (Zhu, 2017). Generally, this was used for humans as food sources, animal feed, and health supplements. It is rich in dietary fiber, vitamins, proteins, carbohydrates, and minerals. The major uses of barley are baking and as a thickening agent in soups, beer, and whiskey products. In terms of health benefits, it can improve gut health, cholesterol management, and diabetic

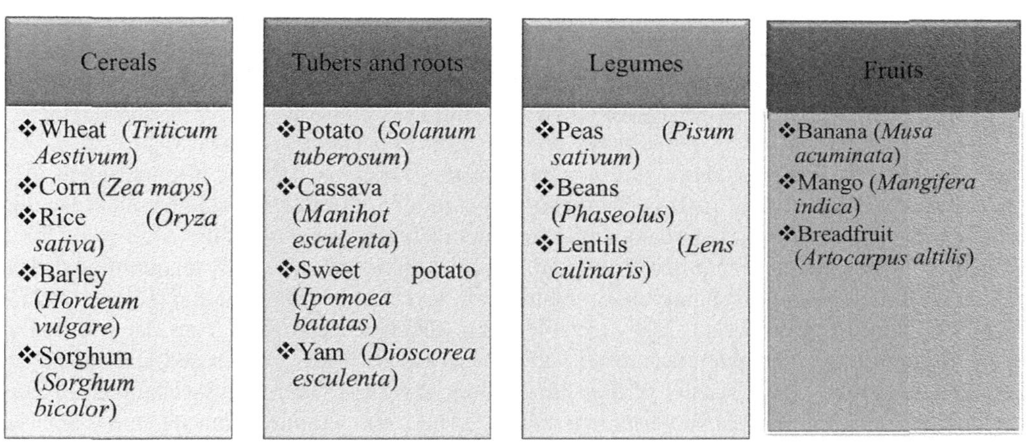

FIGURE 1.1 General sources of plant-based starches and their scientific name.

management, and it is rich in antioxidants. Barley starch is a carbohydrate polymer extracted from barley grains. It is mainly used for gelatinization, baking, and bioplastic manufacturing. It has good water absorption and swelling behavior, and notable enzymatic digestibility (Källman et al., 2015; Punia, 2020).

Sorghum is cultivated and consumed by selected countries such as India, Ethiopia, Argentina, Nigeria, and Mexico (Zhu, 2014). Sorghum is called the king of millets due to its nutraceutical potential activities. It has higher nutritional values, and it is rich in starch. Sorghum showed a lower GI compared to wheat (Dabija et al., 2021). It is an affordable cereal starch variety compared to the other cereals, and is rich in polyphenols, resistant to climate changes, and has noticeable activity, such as being rich in antioxidants and antidiabetic activity (Khoddami et al., 2023; Vanamala et al., 2018).

2.2 Tubers and roots as starch sources

In the human diet, tubers and roots play a significant role. There are different types of tubers and roots are cultivated worldwide, but some of the varieties are only utilized in the food industry, like potato, cassava, sweet potato, yams, and taro (Dorantes-Fuertes et al., 2024; Vilpoux et al., 2019). Extraction of starch from tubers and roots is a long and essential process, and it is consumed at the global level. The major tubers and root-based starch varieties are described in this section. There are several techniques for extracting starch from various sources, particularly roots and tubers (Kringel et al., 2020; Nabeshima et al., 2020; Wickramasinghe et al., 2009).

One of the rich sources of tuber variety starch is the potato, which belongs to the Solanaceae *family*, it's a widely consumed plant source worldwide. Rich in vitamins, carbohydrates, minerals, and notable bioactive compounds, potatoes are present in potatoes (Barbaś et al., 2023; Girigoswami et al., 2024). Potato starch is extracted from the raw potato after the harvest, it is washed, peeled, and cut into small pieces, then crushed to use of blunder to produce the pulp. Afterward, it was filtered and washed multiple times to separate the starch (Mishra et al., 2020; Yassaroh et al., 2019).

The Cassava plant is a perennial variety; this plant has the unique capability of a solo plant that can produce both male and female flowers due to its monecious nature. Nigeria is the major production place for the cassava plant in the world. After Nigeria, Brazil, and Thailand cultivating the cassava plants (Teixeira et al., 2012). In India, the southern regions of Tamil Nadu and Kerala mainly cultivate cassava varieties. It is used for food sources, therapeutic benefits, industrial applications, and animal feed. Based on the research, cassava plant leaf stems and extracts have wound-healing activity and anticancer properties (Ceballos et al., 2015; Scaria et al., 2024). From cassava, tuber starch was prepared, like potato starch preparations; therefore, it involved cleaning, peeling, grinding, sieving, sedimentation, and drying (Dorantes-Fuertes et al., 2024). In the food industry, this cassava starch is mainly used for food thickening and gluten-free baking products (Horstmann et al., 2017).

Sweet potato is a dicotyledonous plant variety belonging to the Convolvulaceae family that originated in Latin America (Zhu et al., 2011). More than 600 species have been identified by biologists,. most of the species have potent anti-inflammatory, antidiabetic, anticancer, antioxidant, antibacterial, and hepatoprotective activities. This is one of the most noticeable food crop varieties because of its extreme stress tolerance and high yield. It's a root vegetable with a sweet taste, vibrant color and nutritional benefits rich in carbohydrates, fiber, vitamins, and antioxidants (Al-Maqtari et al., 2024). Strach from the sweet potato has unique physical and chemical behaviors compared to other plant-based starches. The starch granules of sweet potatoes are found in different sizes and shapes. Like round to polygonal, semioval, oval shapes. Specifically, these starches are used for starch-based food manufacturers like noodles, jellies, cakes, flavoring reagents, alcoholic beverages, sweeteners, and vermicelli production (Lu et al., 2020). This type of starch is suitable for resistant starch formulation, which could be helpful in decreasing the postprandial blood glucose level, managing the risk of diabetes and obesity (Huang et al., 2015; Lyu et al., 2021).

Yam is a *Dioscorea* genus, a monocotyledonous type of crop, it's majorly cultivated in Africa, Asia, and South America. More than 600 species of varieties are available in this *Dioscorea* genus, it's consumed as, either boiled, steamed, fried, or baked (Otegbayo et al., 2014). Yam varieties naturally have the predominant pharmacological activities like antimicrobial, anti-inflammatory, antioxidant, anticancer, antidiabetic, and antiobesity (Mali et al., 2002). Apart from these, vital bioactive compounds are identified from yam names of diosgenin, phenols, allantoin, and flavonoid. In that diosgenin plays an important role in managing obesity, fatty liver, and chronic inflammation (Huang et al., 2006; Li et al., 2023). Specifically in China, they used yams as a vegetable and medicine a decade ago. Yam starch is traditionally used in food products, with applications in pharmaceuticals and biodegradable films. Yam-based starch extraction is a little complicated process due to the high viscosity of the slurry produced by nonstarch polysaccharides. Because of its many advantageous qualities, including low pH and viscosity stability at high temperatures, yam starch has been suggested as an alternate source. However, yam starch gel's rapid retrogradation is unfavorable when used in food systems (Obidiegwu et al., 2020; Shrivastava et al., 2024; Zang et al., 2024).

2.3 Legumes as starch sources

Legumes come under the Fabaceae family. 20,000 specious are come under this family. In that, herbs, shrubs, and trees are included. These legumes were used as a staple food for humans from prehistoric times (Hoover & Zhou, 2003; Keskin et al., 2022; Wani et al., 2016). It provides starch sources along with rich protein content, carbohydrates, proteins, vitamins, and dietary fibers. Legumes have a low GI, and starches derived from legumes are widely used for various applications. Some of the major types of legume-based starches are described below. Pea is an annual legume belonging to the Fabaceae family. This plant is widely cultivated for the seeds, which are consumed fresh, dried, and processed into a variety of food products (Ratnayake et al., 2002). From peas, seed starch was synthesized and utilized for gluten-free product manufacture, animal feed, pharmaceuticals, and biodegradable plastic production. In the food industry, it is specifically used for its thickening capacity to prepare soups and sauces (Wang et al., 2012). It can act as a stabilizer and binding agent in dairy-based products and plant-based alternative preparations. Its lower GI is used for blood glucose level maintenance and gut health maintenance. It can act as a gluten-free alternative; it is also suitable for gluten intolerance and celiac disease (Gao et al., 2022; Wang et al., 2013).

In beans, all the varieties are rich in carbohydrates. Specifically useful for digestion and blood glucose monitoring, it has rich starch content due to its potential physiochemical properties and its use in the food and pharmaceutical industries (Punia et al., 2020). In the food industry, bean-based starches are used for thickening agents, noodles and pasta making, gluten-free baked products and meat alternative foods (Simsek et al., 2012). Due to its prebiotic contents, it is good for the gut function and mechanism. Lentil starch is mainly used in the human diet due to its potent nutritional benefits, weight management, and diabetic management. Besides, lentil-based starch is used in food manufacturing to thicken soups and sauces. Resistant starch, which is abundant in lentil starch, improves intestinal health, blood sugar regulation, and digestion.

2.4 Fruits as starch sources

Fruits are a rich source of minerals, fiber, vitamins, and enriched antioxidants, based on their botanical properties, fruits are generally classified. Fruit juices, pulps, and seeds are mainly used in the food industry as value-added products like jams, chocolates, ice creams, etc. Some major fruit varieties are discussed below, which are vitally used for starch production (Kupervaser et al., 2023; Lan et al., 2025).

Bananas come under the Musaceae family, which is a tropical plant variety with significant economic, medical, and nutritional benefits. The origin of the banana is Southeast Asia, it's a kind of perennial herb with a pseudostem (Singh et al., 2025). Generally, banana fruits are elongated and contain soft pulp with tiny seeds, specifically in wild varieties. In nutritional aspects, bananas are rich in potassium, carbohydrates, fiber, and vitamins like vitamins A, C, D, and B6. Specifically, it contains resistant starch, which would support gut health and proper digestion. In industries, bananas are used for paper production, textiles manufacturing, and biodegradable packaging. For starch production, unripe bananas are used; they contain almost 80% of starch. In the starch isolation process, peeling, slicing, homogenization, centrifuging, and drying processes are involved (Munir et al., 2024; Ranjha et al., 2022; Rawat et al., 2025). The isolated starch is specially used for dietary supplements, functional foods, and bakery products. Banana-derived starch is good for gut health and gluten-free product production (Kaur et al., 2020; Zhang et al., 2005).

Mango fruit is originated in South Asia, it's a kind of tropical and subtropical plant variety. In various sizes, colors, and shapes, varieties of mangoes are available. It's rich in antioxidants, fiber, potassium, and magnesium and it can boost immunity and maintain heart health. For food purposes, mango is used for jams, pickles, and juice preparations. In the medical field, particularly in Ayurveda treatments, it's used as an immunity booster, skin health maintenance, and to improve digestion (Wang et al., 2020). Apart from this, in the cosmetic field, mango butter is used for hair care and skin care product production. Mango starch is ideal for pharmaceuticals, coatings, and beverage productions (Bangar et al., 2021). Breadfruit belongs to the Amoraceae family fruits are round, originating in the Pacific islands and South Asia. It's used for biodegradable product development and animal feed. Breadfruit contains 20%-27% starch, which is a kind of resistant starch. In the food industry, it is used for noodles and bakery products. It acts as a sustainable packaging and dietary fiber. Breadfruit starch can manage the blood glucose level and weight management (Adebowale et al., 2017; Mehta et al., 2023). Overall, banana, mango, and breadfruit starches offer unique industrial, nutritional, and functional qualities. Due to their distinct physicochemical properties, these starches are frequently utilized in food processing, medicines, and biodegradable materials.

Using agricultural waste and food industry by-products as possible sources of starch offers a creative and sustainable way to fulfill expanding industrial and dietary needs. Traditionally, essential crops such as maize, wheat, rice, and potatoes are used to extract starch. However, this technique increases the strain on freshwater and arable land resources and fuels arguments about food versus feed. An environmentally beneficial substitute is provided by using underutilized waste materials, such as fruit peels, seeds, and kernels from the food processing industry or agricultural wastes like cassava peels, banana peels, mango seeds, jackfruit seeds, and even maize husks. These by-products often contain plenty of starch and other useful biomolecules. As an example, jackfruit seeds and banana peels have demonstrated good starch yields with advantageous physicochemical qualities, but mango seeds can contain up to 60% starch. Sustainable packaging, biofilms, medications, and even food compositions as thickening or stabilizing agents can all benefit from the extraction of starch from such waste products. It encourages sustainable industrial practices and a circular economy in the agri-food industry. Additionally, creating scalable and reasonably priced extraction techniques can help these alternative starch sources become more widely used on a commercial scale (Hofman et al., 2016; Perez & Agama-Acevedo, 2017). However, by preserving edible crops for direct consumption, using agricultural leftovers and food industry by-products as sources of starch reduces waste, promotes food security, and is consistent with global sustainability goals. Unlocking the full potential of these unconventional starch sources requires more study and innovation in extraction technology, quality improvement, and application development.

The different sources of starches influence the classification of the starch, such as rapidly digestible, slowly digestible and resistant starch. The rapidly digestible starches are mainly present in simple carbohydrates like bread, potatoes, etc. While the slowly digestible starches are found in complex carbohydrates. The upcoming section provides the details about the characteristics and classification of the starch.

3. Classification of starch

Based on the digestion process, starches are separated into three main groups: rapidly digestible starch (RDS), slowly digestible starch (SDS), and resistant starch (RS). This classification provides significant insight into the varying metabolic responses associated with each starch type, especially regarding blood glucose regulation and energy release. Each starch category has distinct structural and functional features that influence its digestion and absorption behavior.

3.1 RDS

Starch that bypass absorption in the small intestine gets carried to the colon, where it is fermented by gut microorganisms. Because RDS is rapidly absorbed and broken down, blood glucose levels unexpectedly rise, putting more strain on the body to maintain glucose homeostasis (Kim et al., 2024). RDS makes up the bulk of dietary starch. Starchy foods are cooked with moist heat, and a lot of RDS include bread and potatoes. Enzymatic digestion quickly transforms RDS, a kind of starch, into glucose molecules. Significant amounts of RDS in meals will cause a rapid release of glucose into the blood, increasing the ratio of insulin to blood glucose. Starch digestibility is significantly impacted by food processing factors (Birt et al., 2013; Han et al., 2023).

3.2 SDS

SDS, on the other hand, breaks down more slowly as it releases glucose in the small intestine, reducing the health concerns associated with plasma glucose changes and contributing to stable blood glucose levels. A food's glycemic index, which is based on the increase in blood glucose that occurs after consuming a meal, indicates how many carbohydrates it contains (Miao et al., 2015). Because low GI foods' carbohydrates are digested and absorbed more slowly, blood glucose and insulin levels increase more slowly. Furthermore, they are becoming more and more linked to health advantages. Thus, low GI meals have been linked to a lower risk of diabetes, heart disease, and even some types of cancer. SDS has important nutritional roles in the management and prevention of several diseases. Elevated insulin and glucose plasma concentrations after a glucose test are associated with cardiovascular risk and noninsulin-dependent diabetes (Englyst et al., 2018). Slower digestion and absorption of carbohydrates are advantageous for healthy individuals as well as for the dietary treatment of metabolic illnesses like diabetes and hyperlipidemia because they have good effects on many physiological markers. As a result, SDS is increasingly popular as a new functional material (Mishra et al., 2008).

3.3 RS

As a new dietary fiber and prebiotic, RS has received a lot of consideration in recent years. Studies have indicated that RS has a favorable effect on the control of postprandial blood glucose. Developed nations like the US, Europe, and Australia consume 2-4 times as much RS as those in developing nations like China and India, which consume a lot of starchy foods.

In comparison to the other two starches, RS can only ferment in the colon and cannot be broken down in the small intestine. RDS is the term for starch fractions that can be broken down within 20 min of incubation. RS is a combination of several elements that may be divided into five categories: RS1, RS2, RS3, RS4, and RS5. Human metabolism and colon health have been shown to benefit from RS, which has been shown to improve insulin and glycemic responses, regulate lipid metabolism, alter gut microbiota like probiotics, and prevent colon cancer. Though RS cannot be broken down in the small intestine, probiotic bacteria can promote fermentation in the large intestine, which can result in the production of alcohols, organic acids, and short-chain fatty acids (SCFA) (Chen et al., 2024). Gel development, tiny particle size, bland flavor, white color, swelling index, enhanced viscosity, and water-holding capacity are some of the special physico-chemical advantages of RS. As a result, RS is commonly used in many meals, including brownies and cakes. Because RS3 can tolerate high temperatures, it is frequently utilized in fried battered meals. Additionally, RS can enhance the color, taste, oiliness, and crispness of baked or fried meals (Farooq & Yu, 2024).

The above classification of the starch based on the digestibility will have an impact on its absorption rate and thus plays a major role in determining the nutritional and functional properties of the starch. Thus, starch modification techniques, as discussed below, are applied to enhance the characteristic features of the starch and improve its applicability.

4. Properties in starches

Starch is a vital molecule that finds application in various sectors such as paper boards, chemical manufacturers, textiles, and the food and service sector. The varying levels of AM and AP alter the attributes of starch that decide its applicability (Kong et al., 2008). Starch, in its native form, is rarely used as they may be exhibiting poor qualities like retrogradation, syneresis, low viscosity, gelling property, etc. Thus, starch modification is implemented to enhance the functional and structural characteristics of the starch molecules. They are also easily decomposed at high temperatures and are highly unstable to pH and shear stress (He et al., 2023). To overcome these attributes and to contribute in food system as stabilizer, binding agent, thickener, etc., and nonfood applications such as biodegradable packs, thermoplastics, and adhesives; diverse modification approaches are carried out, including physical, chemical, enzymatic, thermal, nonthermal and dual modifications (Suri & Singh, 2023).

4.1 Gelatinization

Starch molecules, when exposed to moisture at higher temperatures, swell up and the amylose molecules leach out, disrupting the granular structure. Further on cooling, new hydrogen bonds are made, leading to a paste-like consistency and increased solubility (Chakraborty et al., 2022). Gelatinization character of the starch makes it eligible to be applied as a thickener and binding agent in foods. Throughout the gelatinization process, a heating phase, holding phase and a cooling phase are observed, which has influence on various parameters such as setback viscosity, peak viscosity, pasting temperature, etc. The values of each of these determine the application of the gelatinized starch to a particular food product. For example, if the peak viscosity is higher, its water absorption capacity is better with an enhanced swelling ratio. Similarly, if the pasting temperature is higher, the cooking time also extends proportionally (Zhang & Lim, 2020). Multiple factors, such as its source of origin, type of proteins, moisture, pH, processing technique, etc., directly affect the gelatinization. Koch and Jane (2000) reported that the bigger the granule size, the higher will be gelatinizing temperatures. Alternatively, as the amylopectin content increases with shorter branches, the gelatinizing temperatures are reduced.

4.2 Retrogradation

Retrogradation is characterized by reassociation of the AM and AP, which were disrupted during the heating stage of starch. It is to be noted that the original structure will not be retained, a different structural arrangement will be adopted. Retrogradation is a time and temperature dependent process. A waxy starch will be retrograded to a soft gel whereas a nonwaxy starch will be retrograded into a firm and hard gel (Tang & Copeland, 2006). A three-stage process of starch retrogradation was investigated by Chang et al. (2021). Initially, the starch molecules are shifted into the aqueous system, followed by rearrangement of the water molecules and concluding with recrystallization of the previously gelatinized starch (Chang et al., 2021). These organizations of molecules can be regulated by a few thermal analyses, including thermogravimetry and differential thermal analysis. They measure the temperature difference and loss in the sample mass to evaluate the range of retrogradation. Rheological analysis measures the pasting viscosity and the textural attributes such as hardness, cohesiveness, elasticity, etc. Microscopic and spectroscopic analyses measure the morphology of the retrograded starch granules. Water activity, storage conditions, and additives (carbohydrate, proteins, lipids, and salts) influence the rate of retrogradation (Wang et al., 2015). The AM content and the granule size are also crucial in the starch retrogradation. Smaller granule size and higher AM content enable quick retrogradation (Tang, 2004). Retrogradation is directly related to staling of the foods due to recrystallization of the starch molecules. Since retrogradation is associated with the spoilage of starchy foods, inhibition of the same is of high interest among researchers.

4.3 Swelling property

The starch molecules have the ability to absorb water and swell at elevated temperatures. At these temperatures, the hydrogen bond is displaced by water molecules. Physical treatments on the starch molecules cause some holes which allow better hydration of the starch, leading to improved swelling property (Uthumporn et al., 2009). The swelling capacity of the starch is allied with the viscosity. Starch with high AP absorbs more water, leading to increased swelling and thus improved viscosity and vice versa (Song & Jane, 2000). Similarly, the waxy starch sources swell better with a small granule size.

4.4 Modification techniques

4.4.1 Physical modifications

These modifications of starch imply alternation of the physical characteristics of starch molecules, like the structural morphology and 3D structure. This will have an impact on both structure like the size of the molecule, and functional properties, like the solubility of starch, swelling properties, gelling and pasting character, and water absorption capacity (Nawaz et al., 2020). This technique is considered safer and eco-friendly than any other modification technique and thus considered as "Green Technology" (Suri & Singh, 2023). Physical modification is broadly classified as Thermal and nonthermal treatments, as discussed below (Fig. 1.2).

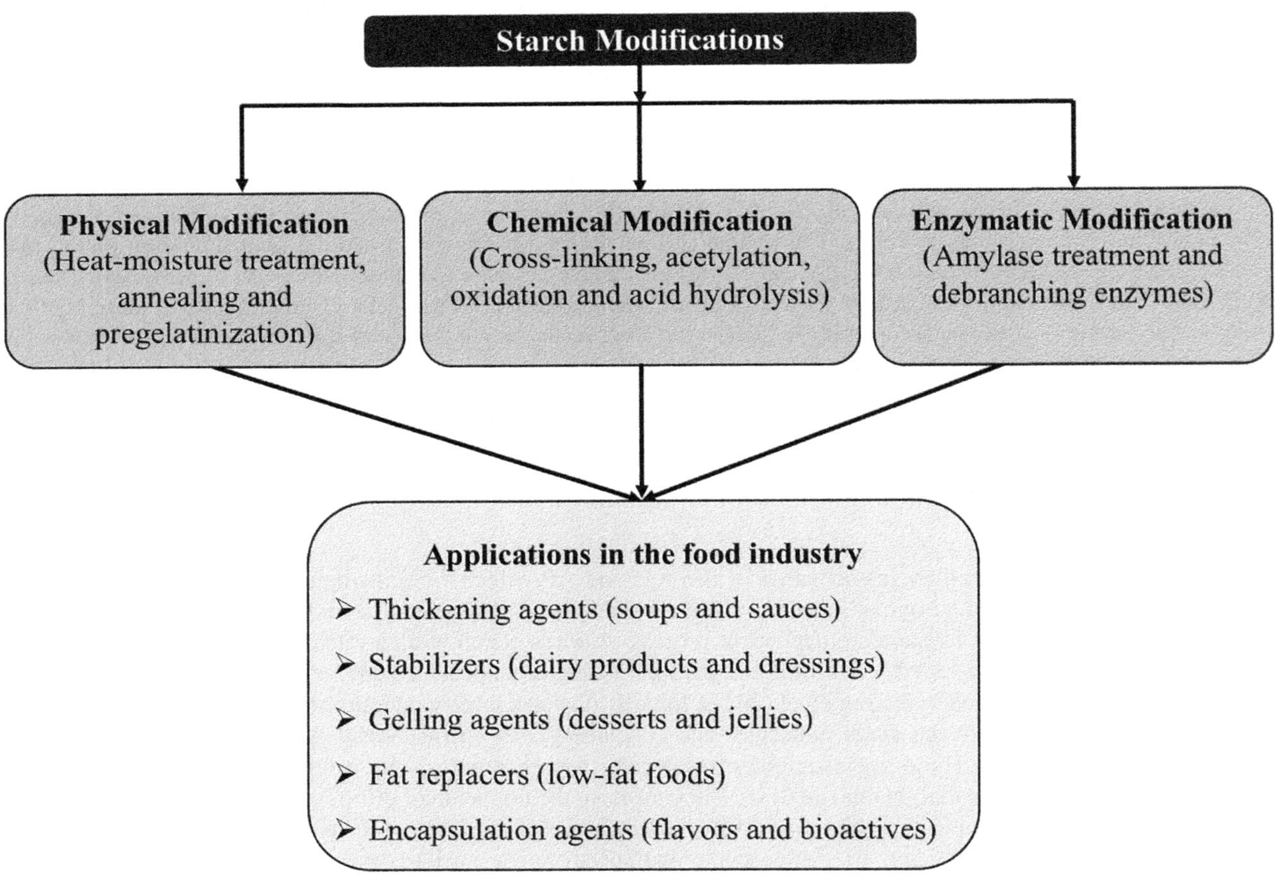

FIGURE 1.2 Functional properties of the starch and its related application in the food industry.

4.4.1.1 Thermal treatments

Pregelatinization of starch involves cooking it completely till it is gelatinized and further drying it using any method like oven drying, spray drying, extrusion drying, and/or drum drying. The hydrogen bond of the starch breaks and further granule fragments, yielding an improved viscosity, swelling, and gelling property, making them easily soluble in water (He et al., 2023). These improved property benefits the modified starch to be utilized in Ready-To-Eat foods, baby foods and as thickeners in soups, as it requires less or no cooking time. Alcázar-Alay and Meireles (2015) mention that drying, cooking method and the sources of starch influence the functionality of the pregelatinized starch. A morphological comparison of the pregelatinsed starch with extrusion drying and drum drying was carried out by Yan and Zhengbiao (2009). Extrusion drying gave the starch irregular holes and shapes, whereas the drum drying technique gave a jagged laminar structure, losing the entire granular structure. This is due to the breakage of H_2 bonds, which damaged the crystallinity of the starch.

The hydrothermal technology is another thermal and physical modification technique that alters the starch properties by keeping the granule as such without breaking it. Thus, the temperature is maintained below the temperature of gelatinization (Suri & Singh, 2023). Annealing (ANN) and Heat Moisture Treatment (HMT) are the best examples of hydrothermal techniques. The major difference between these two methods is the moisture content at which the process occurs. Annealing requires 40%–75% of moisture content, while HMT requires 10%–35% of moisture (Compart et al., 2023). Chung et al. (2009) observed the effects of hydrothermal modification techniques on lentils, peas, and beans. HMT-exposed starch was then allowed for annealing procedure, and it showed reduced swelling property and the leaching of AM.

4.4.1.2 Nonthermal treatments

The nonthermal modification involves modifying the starch properties without heat involvement. It can be as simple as grinding and milling to a complex process as high-pressure treatments, ultrasonication, radiation methods and pulse electric field technology. The major advantage of nonthermal treatments over thermal treatments is that the use of heat treatments could change the sensory properties of the food and cause nutritional loss. Grinding of the starch leads to surface damage and breaking of the crystalline structure, altering the starch morphology and properties (Kaur et al., 2022). Milling involves the application of force and stress, along with friction, to modify the starch attributes. Studies by Zhang et al. (2021) showed that wet ball milling was more efficient than dry ball milling. Huang et al. (2007) concluded that the solubility of starch in cool water increased with addition in ball milling time due to H_2 bond breakage. Alternatively, over-grinding and milling for a longer period of time causes clustering of starch. The swelling property was seen best at the 15th minute and reduced with prolonged timing (Soe et al., 2019). These physical methods can be considered "green technology" as they are eco-friendly with improved starch-modified properties.

In the High Hydrostatic Pressure (HPP) technique, gelatinization of starch is done at high pressures. The pressure range, temperature, time and the starch type and source influence the effect of HPP on modified attributes. This treatment also reduces the crystallinity of the starch, leading to better water absorption capacity and improved swelling property (H. Liu et al., 2016). An ideal temperature of 55–75°C should be maintained to see positive impacts; if this increases to 85–95°C, the desired modified properties do not reflect well (Dhull et al., 2022). Ultrasonication is another novel nonthermal technique that modifies the physical properties of the starch. The ultrasonication treatment deformed the starch, leading to noncrystalline amorphous structure with better solubility (Satmalawati et al., 2020).

Radiation, a similar technology, uses irradiation to alter the starch characteristics. Improved water solubility and swelling with reduced viscosity and weight can be achieved through irradiation. Several small dosages of rays can reduce energy utilization and enhance the starch characteristics (He et al., 2023). Cold plasma treatment is a novel technology that uses ionized plasma gas at lower temperatures to produce ions, atoms and radicals that react with and modify the starch. They degrade the starch crystallinity through holes and surface cracks and increase the swelling properties. These can be combined with physical, chemical and enzymatic modification treatments (Compart et al., 2023).

4.4.2 Chemical modifications

The starch molecules with various hydroxyl groups enable improvement of starch properties and functionality through chemical modification. It involves adding or hindering a functional group to alter the chemical activities of the starch molecules through oxidation, succinylation, crosslinking, esterification, etc. The starch oxidation process proceeds with oxidants such as potassium permanganate and hydrogen peroxide. Starch is depolymerized to break down the crystalline structure. The produced carboxyl and carbonyl groups and other factors, such as pH and time helps determine the level of oxidation (Wojeicchowski et al., 2018). Studies suggest that the water absorption, pasting temperature, and swelling capacity increased in the starches due to oxidation. Ozone, as a contemporary oxidizing agent, is being studied in recent investigations (Hu et al., 2021). It is a green and eco-friendly option, better than other oxidizing agents, as it does not give off residues. It is used to oxidize a variety of starch molecules from corn, tapioca, potato, buckwheat, etc. Oxidized starch is used in 3D printing of food (Maniglia et al., 2019).

Other than oxidation, esterification with acid yields esters (inorganic and organic), another chemical modification technique which can be used due to the presence of numerous hydroxyl groups. It majorly occurs at C2, C3, and C6 of the starch skeleton. The improved starch properties include water absorption, swelling capability and viscosity (Compart et al., 2023). Succinylation, a type of esterification performed in an alkaline atmosphere, is widely studied for its better properties. The succinyl group added to the starch will decrease the retrogradation and oil absorption while enhancing water binding activity (Bushra et al., 2012).

Cross-linking, unlike other modification techniques, reduces the water absorption and solubility due to the formation of more bonds. It is also to be noted that starch that is cross-linked has elevated transition temperatures and enthalpy. The degree of cross-linking and its efficiency are dependent on the concentration of the cross-linking agent, pH, time, etc. This technique is mainly used in food processing and water treatment industries (Compart et al., 2023). When compared with other chemical modifications, oxidation shows better solubility properties (Suri & Singh, 2023). Other major chemical modifications such as acetylation and acid hydrolysis also modify the characteristics of starch and finds its utilization in the food industry as thickeners, stabilizers, binder, and texturizer in salads, sauces, baked goods, etc.

4.4.3 Enzymatic modifications

Enzymatic modification is followed by altering the starch properties through the action of enzymes on the starch molecules. Four major kinds of enzymes are used in the food systems for this purpose. They are exoamylase (β-amylase), endoamylase (α-amylase), transferases (amylomaltase) and debranching enzymes (isoamylase) (Compart et al., 2023). Exoamylase works on the glucose, on the external surface of AM and AP. β-amylase and amyloglucosidase are used in the production of sugar syrups like maltose, glucose and fructose syrups. α-amylase helps in the removal of haze from fruit juices and beer. Endo and exoamylase form pores and cracks on the starch surface, which enhances the solubility of the starch molecule (Wang et al., 2020). The endoamylase acts on the 1,4 bond to decrease the length of the chain. The transferase acts on α-1,4 glycosidic bond while the debranching enzyme acts on the α-1,6 glucosidic bond to produce linear, and straight glucan molecules (Van Der Maarel et al., 2002). Pullulanase, a debranching enzyme used on cassava starch, enhanced the shelf-life of dairy products like yogurt by modifying the water activity (Lopez-Ochoa et al., 2022). Amylomaltase, a transferase enzyme, helped in improving the creaminess as a fat substitute in yogurt (Alting et al., 2008).

Overall, the modified starch promises to have wide applicability with improved functional and structural properties compared to the native starch. Physical modification mostly involves a green technology to alter the starch by gelatinization, ANN, HMT, etc. While the chemical methods are cost-effective and easy to perform. But they might leave residues which pose a threat to human consumption and as a threat to the ecosystem. Enzymatic modification techniques are promising to modify specific properties, but some of the sensitive to the environment, which might hinder the efficiency of the enzymatically modified starch. Dual modification can be incorporated to overcome the limitations of using one modification technique. These improved characteristics of the starch widen the scope of starch application in the food systems as thickeners in soups and jellies, food binding agents and as texturizers. The upcoming section briefly discusses the application of modified starch in food products along with its nutritional aspects.

5. Applications of starch

Starch is an elementary source of carbohydrate present in predominant food products like bread, pasta, baked goods, etc., as a primary contributor of energy. Other than being used directly as food, starch's functional characteristics, such as pasting and gelling properties, enable them to be best utilized as food additives in the form of thickeners, texturizers, and stabilizers. Different sources of starches from tapioca, sago, potato, corn, wheat, rice, etc. are used in various foods as fillers, gelling agents, edible films, sources of bio-ethanol, excipients in pharmaceutics, etc. (Yazid et al., 2018). Hydrolysis of starch would break the linear chains and yield glucose syrup, which is used as sweeteners. This sugar syrup can also undergo alcoholic fermentation with yeast to produce ethanol. Mass production of ethanol can contribute to the biofuel sector. Other by-products of starch processing include polyols, organic acids and amino acids. Polyols such as sorbitol, xylitol, and mannitol are the products of hydrogenation and hydrolysis of starch (sugar). Fermentation of sugars produces a variety of organic acids, such as lactic and citric acid, that are used in food products as an additive. Similarly, starch hydrolysis produces amino acids like lysine, cysteine, etc. These are utilized as supplements for enrichment of animal nutrition (Ogori & Alimi, 2022). Another use of medical manifestation is for patients with dysphagia who face swallowing disorders due to the hardness of the food. Their diet should be made of soft foods that require thickeners. A similar thickening property is also required for infant foods (Moret-Tatay et al., 2015).

5.1 Starch in food products

Modified and processed starches are added to instant mixes of soups and desserts. Instant mixes are usually a combination of the dried form of the ingredients made through extrusion, spray-drying, freeze-drying, drum-drying, etc. The moisture content is maintained at 5% and less for shelf-life and storage purposes. When water is added to rehydrate, most times it forms lumps of starch, which can be avoided through the addition of dispersing substances such as maltodextrins, dextrose and sucrose. In snack products like puffed and extruded, fried and baked foods, starch is generally used as an adhesive, texture, and color modifier. Extruded products are widely made of potato, tapioca and maize starch, which are modified to enhance their textural properties. In fried snack products, the outer coating is usually made of starch. In case of meat analogue, modified starch is used to mimic the texture and improve the texture (Rosicka-Kaczmarek et al., 2017). Starch plays various roles in the food manufacturing and processing industries. It can be directly used as the raw material in the manufacturing of starch-based foods. As an additive, in its capability as an emulsifier, stabilizer and texturizer. Thus, the starch is utilized in different food products like baked goods, puddings, desserts, soups and sauces.

5.2 Starch in bio-films

Recent advancements in the food packaging industry have led to much research and innovations in bio-films with starch. They are abundantly found, cost-effective and most importantly, biodegradable as Generally Recognized As Safe (GRAS). It is to be noted that starch sources containing large proportions of AM have possibilities of forming films. The addition of additives can improve the mechanical strength and antimicrobial property (Zhu, 2025). Commonly available sources of starches such as rice, potato, tapioca and corn are used to make the edible coating and bio-films; likewise, underutilised starches such as tubers, beans and jackfruit are also being explored in recent studies (Nguyen et al., 2022). Otero-Herrera et al. (2025) explored sweet potato in the manufacturing of edible films for candies. This prototype showed good stability of the product for 21 days. Calcium carbonate used in this composition improved the physical and mechanical attributes of the film. $CaCO_3$ has the barrier capacity to O_2, CO_2 and water vapor. Shuprajhaa et al. (2025) investigated the production of bioplastics with starch from bananas. Plasticizers and additives such as glycerol, polyvinyl alcohol and carboxymethyl cellulose are used to improve the tensile strength and other mechanical attributes such as flexibility and stability of the film. Starch-based bio-films are becoming popular due to consumer demand for sustainable choices. Additionally, through various modification techniques, these properties can be enhanced or reduced to widen the starch's applicability.

6. Nutritional aspects of starch

Starch is not just a source of energy; it has several other health benefits. It is a natural, renewable, and biodegradable polymer produced by many plants as an energy storage material. Starches can be used in a variety of ways due to their structural and functional diversity. Starch's functional qualities can be altered and improved by a variety of physical, chemical, and enzymatic procedures to enable its usage for a range of medicinal applications. Starch is an essential component of the human diet, particularly for maintaining metabolism, brain function, and general physical activity. It is a polysaccharide made up of glucose units (Ojogbo et al., 2020). For digestion, blood sugar homeostasis, gut health, and chronic disease prevention. When starch is consumed, digestive enzymes break it down into glucose, which is subsequently taken into the circulation and utilized by cells as fuel. The brain, in particular, relies largely on glucose generated from starch to perform cognitive tasks and retain memories.

Starch is a very important component of human nutrition for the maintenance of metabolism, brain function, and overall physical activity. It is a polysaccharide composed of glucose units. Starch is digested into glucose by enzymes in the gut, absorbed into the bloodstream, and utilized by cells as fuel (Roder et al., 2005; Zhuang et al., 2024). RS is one of the most beneficial components of starch to human health. In contrast to ordinary starch, RS does not undergo small intestine digestion and enters the colon undigested, where it functions similarly to dietary fiber. It works as a prebiotic, supporting the growth of healthy intestinal flora. SCFAs, including butyrate, are produced during this fermentation process and promote colon health, lower inflammation, and maybe even prevent colorectal cancer (Bojarczuk et al., 2022). Preparation and cookery also affect the nutritional quality of starch. For example, retrogradation is the process responsible for the hardening of cooked starches like rice and potatoes during refrigeration (Pellegrini et al., 2020). The nutritional benefit is highlighted by its effect on gut bacteria, blood sugar control, gastrointestinal health, and prevention of chronic diseases. Therefore, well-being and health benefits can be greatly improved by making a priority to consume foods that contain RS and whole, unprocessed carbohydrates.

7. Conclusion

Starch's quantity, adaptability, and effectiveness make it a fundamental component of both human nutrition and food technology. There are types of natural sources of starch, and their structure has been described in this chapter, with a focus on the functions of amylose and amylopectin. The categorization of starches into native, modified, and resistant kinds emphasizes their uses in food systems, ranging from texture alteration to dietary fiber enrichment. Understanding these distinctions is crucial for optimizing starch choices throughout product development and improving food quality, stability, and nutritional value. To advance food science and meet contemporary dietary demands, a in-depth understanding of starch sources and classification will be essential. Hence, this knowledge will support future researches on starch and its innovation.

Acknowledgments

The author (PK) wishes to thank the Faculty of Engineering, Karpagam Academy of Higher Education, for their support and help.

References

Adebowale, O. J., Salaam, H. A., Komolafe, O. M., Adebiyi, T. A., & Ilesanmi, I. O. (2017). Quality characteristics of noodles produced from wheat flour and modified starch of African breadfruit (Artocarpus altilis) blends. *Journal of Culinary Science & Technology, 15*(1), 75–88.

Al-Maqtari, Q. A., Li, B., He, H. J., Mahdi, A. A., Al-Ansi, W., & Saeed, A. (2024). An overview of the isolation, modification, physicochemical properties, and applications of sweet potato starch. *Food and Bioprocess Technology, 17*(1), 1–32.

Alcázar-Alay, S. C., & Meireles, M.a. A. (2015). Physicochemical properties, modifications and applications of starches from different botanical sources. *Food Science and Technology, 35*(2), 215–236.

Alting, A. C., Van De Velde, N. F., Kanning, M. W., Burgering, M., Mulleners, L., Sein, A., & Buwalda, P. (2008). Improved creaminess of low-fat yoghurt: The impact of amylomaltase-treated starch domains. *Food Hydrocolloids, 23*(3), 980–987.

Bai, Y., Li, X., Ji, H., Wang, Y., Zheng, D., Wang, Y., & Jin, Z. (2021). A review of the design and architecture of starch-based dietary foods. *Engineering, 7*(5), 663–673.

Bangar, S. P., Kumar, M., & Whiteside, W. S. (2021). Mango seed starch: A sustainable and eco-friendly alternative to increasing industrial requirements. *International Journal of Biological Macromolecules, 183*, 1807–1817.

Barbaś, P. I. O. T. R., Noaema, A. H., & Sawicka, B. A. R. A. (2023). Potato (Solanum tuberosum L.) as a rich source of nutrients and bioactive compounds: A review. *Journal of Cell and Tissue Research, 23*(2), 7337–7355.

Birt, D. F., Boylston, T., Hendrich, S., Jane, J. L., Hollis, J., Li, L., & Whitley, E. M. (2013). Resistant starch: Promise for improving human health. *Advances in Nutrition, 4*(6), 587–601.

Bojarczuk, A., Skąpska, S., Khaneghah, A. M., & Marszałek, K. (2022). Health benefits of resistant starch: A review of the literature. *Journal of Functional Foods, 93*, Article 105094.

Bushra, M., Yun, X. X., Pan, S. Y., Hydamaka, A., Hua, M. W., & Feng, W. L. (2012). Effect of oxidation and esterification on functional properties of mung bean (Vigna radiata (L.) wilczek) starch. *European Food Research and Technology, 236*(1), 119–128.

Cao, P., Wu, G., Yao, Z., Wang, Z., Li, E., Yu, S., & Li, S. (2022). Effects of amylose and amylopectin molecular structures on starch electrospinning. *Carbohydrate Polymers, 296*, Article 119959.

Ceballos, H., Kawuki, R. S., Gracen, V. E., Yencho, G. C., & Hershey, C. H. (2015). Conventional breeding, marker-assisted selection, genomic selection and inbreeding in clonally propagated crops: A case study for cassava. *Theoretical and Applied Genetics, 128*, 1647–1667.

Chakraborty, I., Pooja, N., Mal, S. S., Paul, U. C., Rahman, M. H., & Mazumder, N. (2022). An insight into the gelatinization properties influencing the modified starches used in food industry: A review. *Food and Bioprocess Technology, 15*(6), 1195–1223.

Chakraborty, I., Pallen, S., Shetty, Y., Roy, N., & Mazumder, N. (2020). Advanced microscopy techniques for revealing molecular structure of starch granules. *Biophysical Reviews, 12*, 105–122.

Chang, Q., Zheng, B., Zhang, Y., & Zeng, H. (2021). A comprehensive review of the factors influencing the formation of retrograded starch. *International Journal of Biological Macromolecules, 186*, 163–173.

Chen, Z., Liang, N., Zhang, H., Li, H., Guo, J., Zhang, Y., & Shi, N. (2024). Resistant starch and the gut microbiome: Exploring beneficial interactions and dietary impacts. *Food Chemistry X, 21*, Article 101118.

Chung, H., Liu, Q., & Hoover, R. (2009). Effect of single and dual hydrothermal treatments on the crystalline structure, thermal properties, and nutritional fractions of pea, lentil, and navy bean starches. *Food Research International, 43*(2), 501–508.

Compart, J., Singh, A., Fettke, J., & Apriyanto, A. (2023). Customizing starch properties: A review of starch modifications and their applications. *Polymers, 15*(16), 3491.

Dabija, A., Ciocan, M. E., Chetrariu, A., & Codină, G. G. (2021). Maize and sorghum as raw materials for brewing, a review. *Applied Sciences, 11*(7), 3139.

Dhull, S. B., Chandak, A., Collins, M. N., Bangar, S. P., Chawla, P., & Singh, A. (2022). Lotus seed starch: A novel functional ingredient with promising properties and applications in Food—A review. *Starch - Stärke, 74*(9–10).

Dorantes-Fuertes, M. G., López-Méndez, M. C., Martínez-Castellanos, G., Meléndez-Armenta, R.Á., & Jiménez-Martínez, H. E. (2024). Starch extraction methods in tubers and roots: A systematic review. *Agronomy, 14*(4), 865.

Englyst, K., Goux, A., Meynier, A., Quigley, M., Englyst, H., Brack, O., & Vinoy, S. (2018). Inter-laboratory validation of the starch digestibility method for determination of rapidly digestible and slowly digestible starch. *Food Chemistry, 245*, 1183–1189.

Falua, K. J., Pokharel, A., Babaei-Ghazvini, A., Ai, Y., & Acharya, B. (2022). Valorization of starch to biobased materials: A review. *Polymers, 14*(11), 2215.

Farooq, M. A., & Yu, J. (2024). Recent advances in physical processing techniques to enhance the resistant starch content in foods: A review. *Foods, 13*(17), 2770.

Gao, L., Wu, Y., Wan, C., Wang, P., Yang, P., Gao, X., & Gao, J. (2022). Structural and physicochemical properties of pea starch affected by germination treatment. *Food Hydrocolloids, 124*, Article 107303.

Girigoswami, A., Deepika, B., Pandurangan, A. K., & Girigoswami, K. (2024). Preparation of titanium dioxide nanoparticles from solanum tuberosum peel extract and its applications. *Artificial Cells, Nanomedicine, and Biotechnology, 52*(1), 59–68.

Goswami, B., & Mahanta, D. (2021). Starch and its derivatives: Properties and applications. *Polysaccharides: Properties and Applications, 18*, 253–281.

Han, J., Wu, J., Liu, X., Shi, J., & Xu, J. (2023). Physiological effects of resistant starch and its applications in food: A review. *Food Production, Processing and Nutrition, 5*(1), 48.

He, R., Li, S., Zhao, G., Zhai, L., Qin, P., & Yang, L. (2023). Starch modification with molecular transformation, physicochemical characteristics, and industrial usability: A state-of-the-art review. *Polymers, 15*(13), 2935.

Hofman, D. L., Van Buul, V. J., & Brouns, F. J. (2016). Nutrition, health, and regulatory aspects of digestible maltodextrins. *Critical Reviews in Food Science and Nutrition, 56*(12), 2091–2100.

Hoover, R., & Zhou, Y. (2003). In vitro and in vivo hydrolysis of legume starches by α-amylase and resistant starch formation in Legumes—A review. *Carbohydrate Polymers, 54*(4), 401–417.

Horstmann, S. W., Lynch, K. M., & Arendt, E. K. (2017). Starch characteristics linked to gluten-free products. *Foods, 6*(4), 29.

Hu, J., Li, X., Cheng, Z., Fan, X., Ma, Z., Hu, X., Wu, G., & Xing, Y. (2021). Modified tartary buckwheat (Fagopyrum tataricum gaertn.) starch by gaseous ozone: Structural, physicochemical and in vitro digestible properties. *Food Hydrocolloids, 125*, Article 107365.

Huang, C. C., Lin, M. C., & Wang, C. C. (2006). Changes in morphological, thermal and pasting properties of yam (Dioscorea alata) starch during growth. *Carbohydrate Polymers, 64*(4), 524–531.

Huang, Z., Xie, X., Chen, Y., Lu, J., & Tong, Z. (2007). Ball-milling treatment effect on physicochemical properties and features for cassava and maize starches. *Comptes Rendus Chimie, 11*(1–2), 73–79.

Huang, T. T., Zhou, D. N., Jin, Z. Y., Xu, X. M., & Chen, H. Q. (2015). Effect of debranching and heat-moisture treatments on structural characteristics and digestibility of sweet potato starch. *Food Chemistry, 187*, 218–224.

Källman, A., Vamadevan, V., Bertoft, E., Koch, K., Seetharaman, K., Åman, P., & Andersson, R. (2015). Thermal properties of barley starch and its relation to starch characteristics. *International Journal of Biological Macromolecules, 81*, 692–700.

Kaur, J., Ahmed, N., Thagunna, B., & Kumar, K. (2022). Starch modification technique: An overview. *Journal of Postharvest Technology, 10*(4), 75–98.

Kaur, L., Dhull, S. B., Kumar, P., & Singh, A. (2020). Banana starch: Properties, description, and modified variations-a review. *International Journal of Biological Macromolecules, 165*, 2096–2102.

Keskin, S. O., Ali, T. M., Ahmed, J., Shaikh, M., Siddiq, M., & Uebersax, M. A. (2022). Physico-chemical and functional properties of legume protein, starch, and dietary Fiber—A review. *Legume Science, 4*(1), Article e117.

Khoddami, A., Messina, V., Vadabalija Venkata, K., Farahnaky, A., Blanchard, C. L., & Roberts, T. H. (2023). Sorghum in foods: Functionality and potential in innovative products. *Critical Reviews in Food Science and Nutrition, 63*(9), 1170–1186.

Kim, M. K., Park, J., & Kim, D. M. (2024). Resistant starch and type 2 diabetes mellitus: Clinical perspective. *Journal of Diabetes Investigation, 15*(4), 395–401.

Koch, K., & Jane, J. (2000). Morphological changes of granules of different starches by surface gelatinization with calcium chloride. *Cereal Chemistry, 77*(2), 115–120.

Kong, X., Bertoft, E., Bao, J., & Corke, H. (2008). Molecular structure of amylopectin from amaranth starch and its effect on physicochemical properties. *International Journal of Biological Macromolecules, 43*(4), 377–382.

Kringel, D. H., El Halal, S. L. M., Zavareze, E. D. R., & Dias, A. R. G. (2020). Methods for the extraction of roots, tubers, pulses, pseudocereals, and other unconventional starches sources: A review. *Starch Staerke, 72*(11–12), Article 1900234.

Kumar, V., Sinha, A. K., Makkar, H. P., De Boeck, G., & Becker, K. (2012). Dietary roles of non-starch polysachharides in human nutrition: A review. *Critical Reviews in Food Science and Nutrition, 52*(10), 899–935.

Kumar, A., Swarnalatha, A. P., Shwetha, J., Saravanan, P., Hatamleh, A. A., Al-Dosary, M. A., ... Ravindran, B. (2024). Pyrolysis behaviour and synergistic effect in co-pyrolysis of wheat straw and polyethylene terephthalate: A study on product distribution and oil characterization. *Heliyon, 10*(17).

Kumar, P., Yadava, R. K., Gollen, B., Kumar, S., Verma, R. K., & Yadav, S. (2011). Nutritional contents and medicinal properties of wheat: A review. *Life Sciences and Medicine Research, 22*(1), 1–10.

Kupervaser, M. G., Traffano-Schiffo, M. V., Dellamea, M. L., Flores, S. K., & Sosa, C. A. (2023). Trends in starch-based edible films and coatings enriched with tropical fruits extracts: A review. *Food Hydrocolloids for Health, 4*, Article 100138.

Lan, T., Wang, J., Li, Y., Wang, Y., Sun, X., Fang, Y., & Ma, T. (2025). Tropical fruit-derived starch: An innovative strategy for high-value nutritional processing of agricultural solid waste. *Food Frontiers, 6*(1), 142–162.

Li, Y., Ji, S., Xu, T., Zhong, Y., Xu, M., Liu, Y., & Lu, B. (2023). Chinese yam (dioscorea): Nutritional value, beneficial effects, and food and pharmaceutical applications. *Trends in Food Science & Technology, 134*, 29–40.

Li, S., Wang, Z., Pan, Y., Sun, C., Li, E., & Gilbert, R. G. (2025). Effects of amylose and amylopectin molecular structures on the emulsification performance of starch nanoparticles. *International Journal of Biological Macromolecules, 308*, Article 142717.

Liu, H., Wang, L., Cao, R., Fan, H., & Wang, M. (2016). In vitro digestibility and changes in physicochemical and structural properties of common buckwheat starch affected by high hydrostatic pressure. *Carbohydrate Polymers, 144*, 1–8.

Lopez-Ochoa, J. D., Cadena-Chamorro, E., Ciro-Velasquez, H., & Rodríguez-Sandoval, E. (2022). Enzymatically modified cassava starch as a stabilizer for fermented dairy beverages. *Starch - Stärke, 74*(7–8).

Lu, P., Li, X., Janaswamy, S., Chi, C., Chen, L., Wu, Y., & Liang, Y. (2020). Insights on the structure and digestibility of sweet potato starch: Effect of postharvest storage of sweet potato roots. *International Journal of Biological Macromolecules, 145*, 694–700.

Lyu, R., Ahmed, S., Fan, W., Yang, J., Wu, X., Zhou, W., & Wang, H. (2021). Engineering properties of sweet potato starch for industrial applications by biotechnological techniques including genome editing. *International Journal of Molecular Sciences, 22*(17), 9533.

Mali, S., Grossmann, M. V. E., Garcia, M. A., Martino, M. N., & Zaritzky, N. E. (2002). Microstructural characterization of yam starch films. *Carbohydrate Polymers, 50*(4), 379–386.

Maniglia, B. C., Lima, D. C., Matta, M. D., Junior, Le-Bail, P., Le-Bail, A., & Augusto, P. E. (2019). Hydrogels based on ozonated cassava starch: Effect of ozone processing and gelatinization conditions on enhancing 3D-printing applications. *International Journal of Biological Macromolecules, 138*, 1087–1097.

Mehta, K. A., Quek, Y. C. R., & Henry, C. J. (2023). Breadfruit (artocarpus altilis): Processing, nutritional quality, and food applications. *Frontiers in Nutrition, 10*, Article 1156155.

Miao, M., Jiang, B., Cui, S. W., Zhang, T., & Jin, Z. (2015). Slowly digestible Starch—A review. *Critical Reviews in Food Science and Nutrition, 55*(12), 1642–1657.

Mishra, S., Monro, J., & Hedderley, D. (2008). Effect of processing on slowly digestible starch and resistant starch in potato. *Starch Staerke, 60*(9), 500–507.

Mishra, T., Raigond, P., Thakur, N., Dutt, S., & Singh, B. (2020). Recent updates on healthy phytoconstituents in potato: A nutritional depository. *Potato Research, 63*, 323–343.

Moret-Tatay, A., Rodríguez-García, J., Martí-Bonmatí, E., Hernando, I., & Hernández, M. J. (2015). Commercial thickeners used by patients with dysphagia: Rheological and structural behaviour in different food matrices. *Food Hydrocolloids, 51*, 318–326.

Munir, H., Alam, H., Nadeem, M. T., Almalki, R. S., Arshad, M. S., & Suleria, H. A. R. (2024). Green banana resistant starch: A promising potential as functional ingredient against certain maladies. *Food Science and Nutrition, 12*(6), 3787–3805.

Nabeshima, E. H., Moro, T. M., Campelo, P. H., Sant'Ana, A. S., & Clerici, M. T. P. (2020). Tubers and roots as a source of prebiotic fibers. *Advances in Food & Nutrition Research, 94*, 267–293.

Nawaz, H., Waheed, R., Nawaz, M., & Shahwar, D. (2020). *Physical and chemical modifications in starch structure and reactivity*. USA: IntechOpen eBooks.

Nguyen, T. K., That, N. T. T., Nguyen, N. T., & Nguyen, H. T. (2022). Development of starch-based bioplastic from jackfruit seed. *Advances in Polymer Technology, 2022*, 1–9.

Niranjana Prabhu, T., & Prashantha, K. (2018). A review on present status and future challenges of starch based polymer films and their composites in food packaging applications. *Polymer Composites, 39*(7), 2499–2522.

Obidiegwu, J. E., Lyons, J. B., & Chilaka, C. A. (2020). The dioscorea genus (Yam)—An appraisal of nutritional and therapeutic potentials. *Foods, 9*(9), 1304.

Ogori, A. F., & Alimi, T. (2022). Starch chemistry and application. *J Food Technol Pres, 6*(6), 130.

Ojogbo, E., Ogunsona, E. O., & Mekonnen, T. H. (2020). Chemical and physical modifications of starch for renewable polymeric materials. *Materials Today Sustainability, 7*, Article 100028.

Olivato, J. B. (2024). Starch: A natural, safe, and multifunctional ingredient for cosmetic formulations. In *Starch industries: Processes and innovative products in food and non-food uses* (pp. 255–269). Academic Press.

Otegbayo, B., Oguniyan, D., & Akinwumi, O. (2014). Physicochemical and functional characterization of yam starch for potential industrial applications. *Starch Staerke, 66*(3–4), 235–250.

Otero-Herrera, A., Fuentes-Gaviria, L., Pérez-Cervera, C., & Andrade-Pizarro, R. (2025). Development of edible films based on sweet potato (Ipomoea batatas) starch and their application in candy packaging. *International Journal of Biological Macromolecules, 299*, Article 140031.

Pellegrini, N., Vittadini, E., & Fogliano, V. (2020). Designing food structure to slow Down digestion in starch-rich products. *Current Opinion in Food Science, 32*, 50–57.

Perez, L. A. B., & Agama-Acevedo, E. (2017). Starch. In *Starch-based materials in food packaging* (pp. 1–18). Academic Press.

Punia, S. (2020). Barley starch modifications: Physical, chemical and enzymatic-a review. *International Journal of Biological Macromolecules, 144*, 578–585.

Punia, S., Dhull, S. B., Sandhu, K. S., Kaur, M., & Purewal, S. S. (2020). Kidney bean (Phaseolus vulgaris) starch: A review. *Legume Science, 2*(3), e52.

Ranjha, M. M. A. N., Irfan, S., Nadeem, M., & Mahmood, S. (2022). A comprehensive review on nutritional value, medicinal uses, and processing of banana. *Food Reviews International, 38*(2), 199–225.

Rashwan, A. K., Younis, H. A., Abdelshafy, A. M., Osman, A. I., Eletmany, M. R., Hafouda, M. A., & Chen, W. (2024). Plant starch extraction, modification, and green applications: A review. *Environmental Chemistry Letters, 22*(5), 2483–2530.

Ratnayake, W. S., Hoover, R., & Warkentin, T. (2002). Pea starch: Composition, structure and Properties—A review. *Starch Staerke, 54*(6), 217–234.

Rawat, M., Chaudhary, P., Tulika, K., Singh, A., Mishra, S., & Janmeda, P. (2025). Pharmacognostical standardization, antioxidant capability, and phytocompound separation from fruit peels. *Environment Conservation Journal, 26*(3).

Roder, N., Ellis, P. R., & Butterworth, P. J. (2005). Starch molecular and nutritional properties: A review. *Advances in Molecular Medicine, 1*(1), 5–14.

Rosicka-Kaczmarek, J., Kwaśniewska-Karolak, I., Nebesny, E., & Komisarczyk, A. (2017). *The functionality of wheat starch.* USA: Elsevier eBooks.

Rostamabadi, H., Yildirim-Yalcin, M., Demirkesen, I., Toker, O. S., Colussia, R., do Nascimentob, L.Á., & Falsafi, S. R. (2024). Improving physicochemical and nutritional attributes of rice starch through green modification techniques. *Food Chemistry, 458*, Article 140212.

Satmalawati, E. M., Pranoto, Y., Marseno, D. W., & Marsono, Y. (2020). The physicochemical characteristics of cassava starch modified by ultrasonication. *IOP Conference Series: Materials Science and Engineering, 980*(1), Article 012030.

Scaria, S. S., Balasubramanian, B., Meyyazhagan, A., Gangwar, J., Jaison, J. P., Kurian, J. T., & Joseph, K. S. (2024). Cassava (Manihot esculenta Crantz)—A potential source of phytochemicals, food, and nutrition—An updated review. *eFood, 5*(1), Article e127.

Sheldrake, P. (2009). Starch. *Food stabilisers, thickeners and gelling agents, 17*, 293–324.

Shoukat, R., Cappai, M., Pilia, L., & Pia, G. (2025). Rice starch chemistry, functional properties, and industrial applications: A review. *Polymers, 17*(1), 110.

Shrivastava, A., Gupta, R. K., & Srivastav, P. P. (2024). Exploring novel frontiers of advancements in purple yam (Dioscorea alata L.) starch extraction, modification, characterization, applications in food and other industries. *Measurement: Food, 15*, Article 100196.

Shuprajhaa, T., Paramasivam, S. K., Pushpavalli, S., Anandakumar, S., & Naik, R. (2025). Influence of additives on the development, mechanical, functional characteristics and biodegradability of banana starch-based bio plastic films. *International Journal of Biological Macromolecules, 295*, Article 139544.

Simsek, S., Ovando-Martínez, M., Whitney, K., & Bello-Pérez, L. A. (2012). Effect of acetylation, oxidation and annealing on physicochemical properties of bean starch. *Food Chemistry, 134*(4), 1796–1803.

Singh, L. J., Prakash, R. O., Mahadevakumar, S., Midigesi, A. K., Naik, M. C., Dhole, P. A., & Ekka, G. A. (2025). History and emended description of an endemic musa L. of Andaman and Nicobar Islands, India: M. Indandamanensis LJ Singh (Musaceae). *Botany Letters, 172*, 1–17.

Soe, M. T., Chitropas, P., Pongjanyakul, T., Limpongsa, E., & Jaipakdee, N. (2019). Thai glutinous rice starch modified by ball milling and its application as a mucoadhesive polymer. *Carbohydrate Polymers, 232*, Article 115812.

Song, Y., & Jane, J. (2000). Characterization of barley starches of waxy, normal, and high amylose varieties. *Carbohydrate Polymers, 41*(4), 365–377.

Suri, S., & Singh, A. (2023). Modification of starch by novel and traditional ways: Influence on the structure and functional properties. *Sustainable Food Technology, 1*(3), 348–362.

Svihus, B., Uhlen, A. K., & Harstad, O. M. (2005). Effect of starch granule structure, associated components and processing on nutritive value of cereal starch: A review. *Animal Feed Science and Technology, 122*(3–4), 303–320.

Tabasum, S., Younas, M., Zaeem, M. A., Majeed, I., Majeed, M., Noreen, A., & Zia, K. M. (2019). A review on blending of corn starch with natural and synthetic polymers, and inorganic nanoparticles with mathematical modeling. *International Journal of Biological Macromolecules, 122*, 969–996.

Tang, H. (2004). Relationship between functionality and structure in barley starches. *Carbohydrate Polymers, 57*(2), 145–152.

Tang, M., & Copeland, L. (2006). Analysis of complexes between lipids and wheat starch. *Carbohydrate Polymers, 67*(1), 80–85. https://doi.org/10.1016/j.carbpol.2006.04.016

Teixeira, E. D. M., Curvelo, A. A., Corrêa, A. C., Marconcini, J. M., Glenn, G. M., & Mattoso, L. H. (2012). Properties of thermoplastic starch from cassava bagasse and cassava starch and their blends with poly (lactic acid). *Industrial Crops and Products, 37*(1), 61–68.

Uthumporn, U., Zaidul, I., & Karim, A. (2009). Hydrolysis of granular starch at sub-gelatinization temperature using a mixture of amylolytic enzymes. *Food and Bioproducts Processing, 88*(1), 47–54.

Van Der Maarel, M. J., Van Der Veen, B., Uitdehaag, J. C., Leemhuis, H., & Dijkhuizen, L. (2002). Properties and applications of starch-converting enzymes of the α-amylase family. *Journal of Biotechnology, 94*(2), 137–155.

Vanamala, J. K., Massey, A. R., Pinnamaneni, S. R., Reddivari, L., & Reardon, K. F. (2018). Grain and sweet sorghum (Sorghum bicolor L. moench) serves as a novel source of bioactive compounds for human health. *Critical Reviews in Food Science and Nutrition, 58*(17), 2867–2881.

Vilpoux, O. F., Brito, V. H., & Cereda, M. P. (2019). Starch extracted from corms, roots, rhizomes, and tubers for food application. In *Starches for food application* (pp. 103–165). Academic Press.

Wang, S., Blazek, J., Gilbert, E., & Copeland, L. (2012). New insights on the mechanism of acid degradation of pea starch. *Carbohydrate Polymers, 87*(3), 1941–1949.

Wang, S., Jin, F., & Yu, J. (2013). Pea starch annealing: New insights. *Food and Bioprocess Technology, 6*, 3564–3575.

Wang, S., Li, C., Copeland, L., Niu, Q., & Wang, S. (2015). Starch retrogradation: A comprehensive review. *Comprehensive Reviews in Food Science and Food Safety, 14*(5), 568–585.

Wang, P., Luo, Y., Huang, J., Gao, S., Zhu, G., Dang, Z., & Chen, Y. (2020). The genome evolution and domestication of tropical fruit mango. *Genome Biology, 21*, 1–17.

Wang, S., Wang, J., Liu, Y., & Liu, X. (2020). *Starch modification and application*. Singapore: Springer eBooks.

Wani, I. A., Sogi, D. S., Hamdani, A. M., Gani, A., Bhat, N. A., & Shah, A. (2016). Isolation, composition, and physicochemical properties of starch from legumes: A review. *Starch Staerke, 68*(9–10), 834–845.

Wickramasinghe, H. A. M., Takigawa, S., Matsuura-Endo, C., Yamauchi, H., & Noda, T. (2009). Comparative analysis of starch properties of different root and tuber crops of Sri Lanka. *Food Chemistry, 112*(1), 98–103.

Wojeicchowski, J. P., De Andrade De Siqueira, G. L., Lacerda, L. G., Schnitzler, E., & Demiate, I. M. (2018). Physicochemical, structural and thermal properties of oxidized, acetylated and dual-modified common bean (Phaseolus vulgaris L.) starch. *Food Science and Technology, 38*(2), 318–327.

Yadav, P., Prabahar, A. E., & Verma, A. (2024a). Extraction of Starch from Different Natural Sources and their Characterization by Advanced Spectroscopic Techniques. *Frontiers in Health Informatics, 13*(3).

Yadav, P., Prabahar, A. E., & Verma, A. (2024b). Chemical Modification and Pharmaceutical Evaluation of Starch obtained from Edible Natural Sources. *South Eastern. European Journal of Public Health*, 303–310. https://doi.org/10.70135/seejph.vi.2664.

Yan, H., & Zhengbiao, G. (2009). Morphology of modified starches prepared by different methods. *Food Research International, 43*(3), 767–772.

Yassaroh, Y., Woortman, A. J., & Loos, K. (2019). A new way to improve physicochemical properties of potato starch. *Carbohydrate Polymers, 204*, 1–8.

Yazid, N. S. M., Abdullah, N., Muhammad, N., & Matias-Peralta, H. M. (2018). Application of starch and starch-based products in food industry. *Journal of Science and Technology, 10*(2).

Yu, X., Huo, J., Wang, K., Duan, Y., Ma, X., Ma, J., & Xiao, Z. (2025). Different interaction behaviors of rice glutelin with amylose and amylopectin within starch under the extrusion environment. *Food Chemistry, 480*, Article 143974.

Zahid, M. K., Ahmad, D., Amin, R., & Bao, J. (2025). Sorghum starch: Composition, structure, functionality, and strategies for its improvement. *Comprehensive Reviews in Food Science and Food Safety, 24*(1), Article e70101.

Zang, Z., Gong, X., Cao, L., Ni, H., & Chang, H. (2024). Resistant starch from yam: Preparation, nutrition, properties and applications in the food sector. *International Journal of Biological Macromolecules, 273*, Article 133087.

Zhang, C., & Lim, S. (2020). Physical modification of various starches by partial gelatinization and freeze-thawing with xanthan gum. *Food Hydrocolloids, 111*, Article 106210.

Zhang, J., Liu, Y., Wang, P., Zhao, Y., Zhu, Y., & Xiao, X. (2025). The effect of protein–starch interaction on the structure and properties of starch, and its application in flour products. *Foods, 14*(5), 778.

Zhang, P., Whistler, R. L., BeMiller, J. N., & Hamaker, B. R. (2005). Banana starch: Production, physicochemical properties, and Digestibility—A review. *Carbohydrate Polymers, 59*(4), 443–458.

Zhang, B., Yuan, Z., Qiao, D., Zhao, S., Lin, Q., & Xie, F. (2021). Wet ball milling of indica rice starch effectively modifies its multilevel structures and pasting behavior. *ACS Food Science & Technology, 1*(4), 636–643.

Zhong, F., Li, Y., Ibáñez, A. M., Oh, M. H., McKenzie, K. S., & Shoemaker, C. (2009). The effect of rice variety and starch isolation method on the pasting and rheological properties of rice starch pastes. *Food Hydrocolloids, 23*(2), 406–414.

Zhu, F. (2014). Structure, physicochemical properties, modifications, and uses of sorghum starch. *Comprehensive Reviews in Food Science and Food Safety, 13*(4), 597–610.

Zhu, F. (2017). Barley starch: Composition, structure, properties, and modifications. *Comprehensive Reviews in Food Science and Food Safety, 16*(4), 558–579.

Zhu, F. (2025). Starch based films and coatings for food packaging: Interactions with phenolic compounds. *Food Research International, 204*, Article 115758.

Zhu, F., Yang, X., Cai, Y. Z., Bertoft, E., & Corke, H. (2011). Physicochemical properties of sweetpotato starch. *Starch Staerke, 63*(5), 249–259.

Zhuang, J., Zhu, J., Cheung, P. C., & Li, C. (2024). The physical and chemical interactions between starch and dietary fiber: Their impact on the physicochemical and nutritional properties of starch. *Trends in Food Science & Technology, 149*, Article 104566.

Chapter 2

Structure, physicochemical, and functional properties of native and modified starches

Pothiyappan Karthik[1,2], Krishna Priyanka Tippareddy[1,2], Manasi Ghosh[1,2], P.R. Delna[1,2], Kalaiselvi Krishnamoorthy[3] and Safreena Kabeer[1]

[1]*Department of Food Technology, Faculty of Engineering, Karpagam Academy of Higher Education, Coimbatore, Tamil Nadu, India;* [2]*Centre for Food Nanotechnology, Karpagam Academy of Higher Education, Coimbatore, Tamil Nadu, India;* [3]*Centre for Herbal Pharmacology and Environmental Sustainability, Chettinad Hospital and Research Institute, Chettinad Academy of Research and Education, Kelambakkam, Tamil Nadu, India*

1. Introduction

Starch is a polysaccharide which is found abundantly in cereals and tubers such as wheat, rice, maize, potato, tapioca, cassava, etc (Ağagündüz et al., 2023; Rapaille & Vanhemelrijck, 1992). Due to their versatile functional properties, economically inexpensive, and nontoxic nature, they are widely used in the food industry sector (Compart et al., 2023). Amylose and amylopectin are the two molecules that make up the starch. Amylose is a primarily linear polymer composed of glucose units linked by α-(1 $\rightarrow$ 4) glycosidic bonds. In contrast, amylopectin is a highly branched polymer of glucose, consisting of linear chains connected by α-(1 $\rightarrow$ 4) glycosidic bonds with branch points formed by α-(1 $\rightarrow$ 6) glycosidic linkages. These branches typically occur every 24 to 30 glucose units. The composition of amylopectin and amylose varies according to the plant sources (Rapaille & Vanhemelrijck, 1992; Chakraborty et al., 2022). Starch possesses functional characteristics such as swelling of starch granules, solubility, gelatinization, retrogradation, syneresis, and rheological properties (Wang et al., 2014). Different industries, such as the textile, pharmaceutical, and food industries, use starch for various purposes. Most food industries avoid using starch in its raw form due to numerous limitations. Native, unprocessed starch does not dissolve in cold water and exhibits high retrogradation, poor solubility, high viscosity, and low acid stability, all of which limit its industrial applications. These limitations led to further modification of the properties of starches for better behavioral characteristics; thus, resulting in the development of modified starch (Abbas et al., 2010; Calvin, 2016; Chakraborty et al., 2022). Each industry has its distinct techniques for producing starch according to the equipment availability. In general, the starch manufacturing process involves the separation of protein and fiber and the extraction of the raw starch, followed by refinement and drying. Physical techniques include crushing, milling, or grinding, which form a starch slurry that is subsequently isolated from other processed raw material components, including fibers, proteins, sugars, and salts. Following this, several chemical changes have been made, either to change its functionality or to obtain sweeteners through the hydrolysis process for further applications. The purified starch is produced by the application of different methods to separate it from the remaining components. For instance, maize starch is derived from maize kernels through a wet milling process, thus resulting in the separation of fiber, oil, and protein. Thereby, it makes no difference what type of extraction process is applied if the insoluble starch is recovered as intact or undamaged granules, known as native starch, which is either dried and washed or remains as a slurry for further processing into modified starches (Chen et al., 2018).

The objective of this chapter is to provide an overview of starch modification that has been physically, chemically, or enzymatically changed to enhance its functional properties. In addition, it explored the applications of native and modified

Native and Modified Starches for Food Applications. https://doi.org/10.1016/B978-0-443-27431-2.00015-8

starches towards food processing, pharmaceutical and biotechnological sectors. Furthermore, it also highlights the Challenges, limitations and future directions of current starch modifications, emerging applications and innovations.

2. Structure of starch

Starch is produced naturally as semicrystalline granules involving two polymers named amylose and amylopectin (Table 2.1). Generally, starch granules are not amorphous; depending on their chemical configurations and the proportion of amylose to amylopectin, they display different levels of crystallinity. The crystallinity of starch is divided into three primary types: A, B, and C types. Cereal varieties come under the A type of starch because of their easy digestion process. Tuber varieties are classified into B types due to their increased water absorption and slow digestion capacity. Legumes cover the C types of starch, which provide good digestion and water retention.

3. Native starch physicochemical properties

Starch is extensively used in food and nonfood industries owing to its unique physicochemical and functional properties. Starch is a ubiquitous polysaccharide composed of two distinct polymer units: amylose and amylopectin. The crystalline regions of starch are primarily attributed to amylopectin, while the amorphous regions are mainly associated with amylose. According to Alcázar-Alay et al. (2015), depending upon the ratio of amylopectin and amylose and the strength of α-glycosidic bonds, the physicochemical properties of starch (thermal properties, moisture, and water absorption properties, viscosity, and flow behavior) change in species (Fig. 2.1).

3.1 Thermal properties

Native starch exhibits a semicrystalline nature, which imparts unique thermal properties. According to some studies (Subroto et al., 2022; Xu et al., 2014), gelatinization temperature, melting enthalpy, and glass transition temperature are the key parameters used here. Gelatinization temperature (Tg) is the temperature at which water-insoluble starch granules absorb water upon heating above 52 °C, disrupting their granular structure and leading to swelling. This temperature is directly proportional to the stability of starch granules, indicating stronger bonds between starch molecules. As noted by Subroto et al. (2022), gelatinization temperature enables more flexible interactions between starch granules. Gelatinization

TABLE 2.1 Comprehensive characteristics of amylose and amylopectin.

Property	Amylose	Amylopectin
Structure	Linear and helical	Highly branched
Function	Gel formation	Viscous and thickening
Bonds	α-(1 → 4) glycosidic	α-(1 → 4) and α-(1 → 6) glycosidic
Digestibility	Slowly digestible	Rapidly digestible
Solubility	Less soluble	More soluble

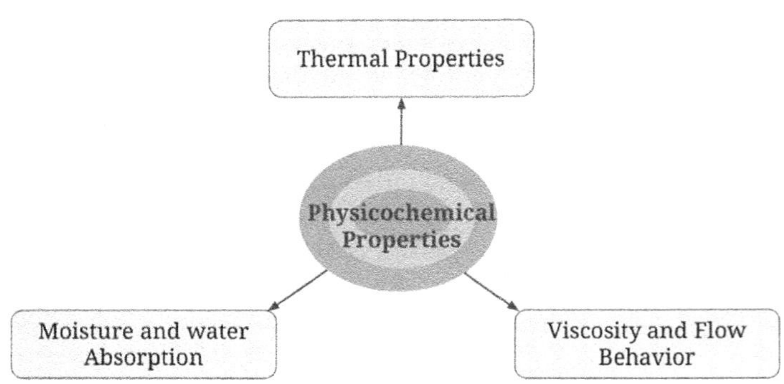

FIGURE 2.1 Pictorial representation of the physicochemical properties of native starch.

temperature can be triggered by factors such as moisture content, pH, and heating rate. Increased gelatinization temperature indicates an increase in the double helical structure of starch (Mojo-Quisani et al., 2024).

The melting enthalpy (ΔH) of starch refers to the temperature at which the crystalline structure of native starch is completely distorted. As the temperature escalates, the crystalline structure of starch breaks down, leading to a forfeiture of its native properties. According to Xu et al. (2014), high melting enthalpy can indicate greater crystallinity, higher intermolecular bonding, tight packing, and a lower presence of plasticizers. The glass transition temperature (Tg') is a critical parameter that marks the temperature at which the amorphous regions of starch undergo a transition, becoming more flexible and prone to chemical changes. This temperature is essential in understanding the behavior of starch in various applications. Liu et al. (2010) stated that the glass transition temperature can be assessed by a differential scanning calorimeter and dynamic mechanical analysis. The thermal properties of native starch can be triggered by some factors, such as moisture content, heating rate, and types of starch. Water availability can affect starch's thermal processing, leading to thermal damage (Sinaki & Scanlon, 2016). Understanding the thermal activities of native starch is mandatory for the evolution performance in various ways, including the food industry, pharmaceutical applications, biodegradable packaging, etc. Native starches like corn, barley, rice, potato, and tapioca exhibit different glass transitions, melting enthalpy, and glass transition states. According to Xu et al. (2014), corn starch has a higher melting enthalpy than other starches, which indicates a higher crystalline nature.

3.2 Moisture and water absorption properties

Starch has been incorporated in many culinary products, such as sauces, soups, desserts, frozen batter, etc. These products are usually experienced in a series of different conditions, which are the reason behind their spoilage. One of the most important factors for these are water and moisture absorption capacity of native starch (Srichuwong et al., 2012). Native starch is hydrophilic and exhibits unique moisture and water absorption properties. Moisture content is the raw amount of water that is present in the starch expressed as a percentage. Moisture is found in two ways in starch: one as bound moisture that strongly accommodates starch granules, and another in free moisture, which is loosely attached to the starch granules. Water absorption refers to the absorb and retain capability of starch granules. Hydrophilicity, porosity, and crystallinity are the reasons behind water absorption. Native starch increases its hydrophilic tendency due to heat treatment. The amorphous region of starch gets expanded because of the disruption of the hydrogen bond between the amorphous region and the crystalline region of starch (Sindhu et al., 2019). Freeze-thaw stability is another important aspect regarding water absorption and moisture content. It refers to the capability of starch to avoid the loss of water when starch experiences thermal cycles between freezing and inclusive conditions (Sofi et al., 2013). Freeze-thaw stability of native starch depends on the amylose content and chain extent of the amylopectin; however, their relationship needs to be further explored (Srichuwong et al., 2012).

3.3 Viscosity and flow behavior

Most of the native starches (e.g., Rice, barley, potato, and tapioca) are non-Newtonian fluids because they change their viscosity under many circumstances. Native starch shows complex viscosity and flow behavior properties, which can be influenced by molecular weight, chemical structure, and concentration of molecules. Viscosity refers to the fluid-resistant behavior of starch, its mainly depends on two factors: shear-thinning behavior and non-Newtonian behavior. Viscosity decreases gradually depending on the increased shear rate. Additionally, rice and tapioca starch lose their viscosity by increasing shear rate, showing shear-thinning behavior, but corn starch behaves more likely as a Newtonian fluid. Likewise, Li et al. (2020) stated that viscosity increases as the molecular chain or molecular concentration increases in starch granules; additionally, as the temperature increases, viscosity starts decreasing. Additionally, viscosity is sometimes governed by the viscosity of the suspending medium, the size of the granule, shape, volume fraction, and the thermodynamic condition of the system (Sinaki & Scanlon, 2016).

In terms of viscosity, pasting properties play an important role. In a standard viscosity graph, there are four distinct temperatures (Peak viscosity, breakdown viscosity, final viscosity, and setback viscosity). The temperature at which viscosity increases sharply is the pasting temperature. Peak viscosity denotes the ideal concept about the water-holding capacity of the starch; however, final viscosity denotes the chilling phase where starch turns into a viscous paste after cooking or cooling. Amylose is the main reason behind the final viscosity. In general, amylose content is inversely proportional to peak viscosity and gelatinization temperature. Final viscosity or retrogradation can be measured by gel turbidity, light scattering, or Fourier Transform Infrared spectroscopy (FTIR) (Sofi et al., 2013). Light transmittance is another interesting property that describes how starch paste behaves when light passes through it. It is calculated at

640 nm. Amylopectin is covalently attached to phosphate monoesters, which increases the clarity of starch paste; however, amylose attached with phospholipids due to fractions makes the solution opaque. Increased light transmittance indicates a more transparent paste (Mahto & Das, 2014). Viscosity reduces as the shear rate rises over time, but as the shear rate lowers, it again returns to its original form, which leads to pseudoplastic behavior. Moreover, viscosity starts decreasing over time when the shear rate remains constant, which leads to thixotropic behavior. Sinaki and Scanlon (2016) stated that viscosity increases with the increase of shear rate, like corn starch viscosity increases with shear stress, but potato starch viscosity decreases with the increase of shear stress. Flow behavior and viscosity are measured by a rheometer and a viscometer. Likewise, Kim et al. (2012) suggested in their study that the flow behavior of corn starch is generally less, which confirms the shear-thickening behaviour. This confirms that concentration, temperature, shear rate, and types of starch are the main factors behind the changes in viscosity and flow behavior of native starch.

4. Functional properties of starch

Attention toward starch in the food industry increased due to its unique functional properties. Starch can act as both a thickening agent, binding, and a gelling agent, through the porous nature of starch granules as well as water absorption ability (Fig. 2.2). Starch can be used as a thickening agent in culinary applications such as soups, gravies, sauces, creams, and even salad dressing (Himashree et al., 2022). Native starches like rice, corn, barley, and wheat have significant gelling, properties and thickening, and binding properties.

4.1 Gel formation

Starch is insoluble in water; however, during the heat treatment, starch undergoes a gelatinization or gel formation process where starch starts to absorb water and loses its crystalline structure and becomes amorphous (Wang et al., 2023). According to Sethi et al., (2022), gelatinization is a bilateral endothermic process. Initially, above 52 °C temperature starch granules swell because of hydrogen bond breakage in the amorphous region, then water associates as a plasticizer, which affects hydration and swelling of the amorphous region. More heating causes the loss of crystallinity and absorption of high water. Increased water absorption leads to increased swelling of the starch granules, which leads to more randomness. After the rupture of starch granules, amylose starts to release from hydrogen bonds, which leads to paste formation; additionally, it increases the solution's viscosity. The gelatinization temperature of starch depends on the moisture, pH, and starch source. High gelatinization temperature amylopectin and amylose granules started entangling with each other, which results in gel formation (Oladebeye et al., 2023). Additionally, the swelling power is an indicator of water-holding capacity and gelatinization temperature. Likewise, gelatinization can change in species specifically depending on their structural composition (such as amylopectin and amylose ratio, lipid, proteins, enzymes, and phosphorus content), the molecular composition of amylopectin starch granular structure, the size distribution of starch granules and morphology (Wani et al., 2012).

According to Yazid et al. (2018), starch-like wheat and potato exhibit high gelling properties. Similarly, Sinaki and Scanlon (2016) stated that potato starch has better gelatinization ability than corn starch; moreover, corn starch acts more as a Newtonian fluid. Likewise, Ali et al. (2016) stated that rice starch is also applied abundantly as a gelling agent in the food industry. The gelatinization process includes the melting of crystalline nature, loss of birefringence, and solubilization of starch due to temperature difference. Starch molecules can return to their double helical structure or amorphous and crystalline nature as the temperature gradually decreases.

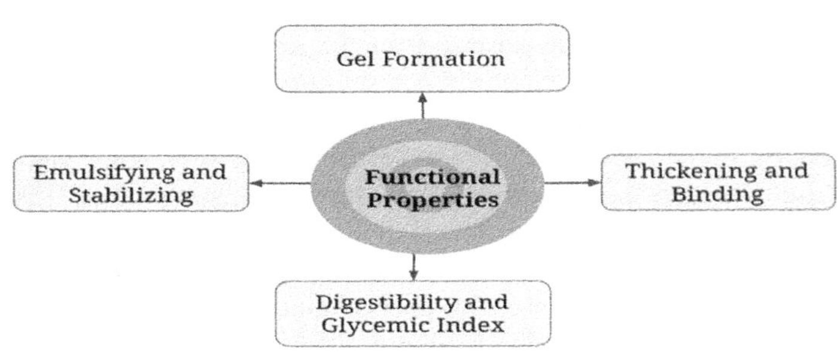

FIGURE 2.2 Illustration of functional properties of starch.

4.2 Thickening and binding properties

Most of the food industry's favourite thickening and binding agent choice is starch due to its versatile functional properties, which enable it to work both as a thickener and a binder. A thickening agent increases viscosity in the solution, whereas a binding agent implies the holding ability of molecules. According to Yazid et al. (2018) native starches such as corn, tapioca, and potatoes exhibit high binding and thickening properties. Similarly, Kumar et al. (2024) stated that millet starch is used as a thickening agent. Thickening and binding properties depend on two mechanisms: hydrogen bonding and chain entanglement. Starch granules form hydrogen bonds with the water; in cold temperatures, these starch granules are insoluble in water (Xie et al., 2023). Additionally, in the presence of temperature, the hydrogen bond starts breaking, which results in water absorption that leads to swelling of the granules. Increased heat causes the entanglement of amorphous and crystalline regions (Oladebeye et al., 2023). Thickening properties and binding properties can be triggered by the origin of starch molecules, such as moisture content and temperature. Yazid et al. (2018) stated in their study that native starches, rice, corn, tapioca, potato, and barley, are highly demanding thickener and binder agents used for culinary purposes. They also noted that temperature is the key to these properties. By maintaining temperature, we can use starch more usefully as a thickener. Additionally, they also observed the structure and starch source as very important indicators.

4.3 Digestibility and glycemic index

The degree to which dietary starch is digestible is measured using different standards. The major one is the glycemic index (GI). The digestibility of starch is divided into three categories. They are rapidly digestible starch (RDS), slowly digestible starch (SDS), and resistant starch (RS). In the *in vitro* study, the Englyst test proved that RDS and SDS are digestible in the human small intestine at different rates. Compared to SDS, ingestion of RDS leads to sudden variations in the blood glucose level in humans. Therefore, SDS plays a major role in the health sector (e.g., diabetic management, stable glucose metabolism, and better satiety). RS amounts to the nondigested starch by enzymes present in the small intestine, which are stored in the large intestine (Giuberti & Gallo, 2018). A study was performed to determine the *in vitro* starch digestibility using selective ingredients like barley, oats, gluten, and guar gum and the glycemic index of traditional foods. The addition of barley and oats at 9 g/100 g and gluten and guar gum at 5 g/100 g decreased *in vitro* GI and resulted in a rise in SDS and RS. The study summarized that GI and digestibility are correlated. The invention of food products with low GI improves people with hyperglycemia and glucose intolerance (Giri et al., 2017). The rate of digestibility of starch was evaluated in cereals and starchy feeds (pig feed). The chemical composition and starch fraction could affect the predicted glycemic index. The principal component analysis revealed that the predicted GI is positively influenced by RDS and negatively impacted by SDS and RS (Giuberti et al., 2012). Ratnaningsih et al. (2020) evaluated legume starch since it is rich in amylose, resistant starch, and low glycemic index. Physical modification (autoclaving) affects the physicochemical properties, estimated glycemic index, and starch digestibility. The study suggests that the cowpea starch autoclaving-cooling process decreases the GI, and this can be provided to people with degenerative diseases. In general digestibility of starch and GI plays a crucial role in human health (e.g., blood glucose level, diabetes, etc.). The RDS causes an increase in the blood glucose level, whereas SDS stabilizes the glucose level. Implementation of low-GI foods in the diet helps to manage glucose-related issues. Certain ingredients, such as barley and oats, control the GI level and improve glucose intolerance and hyperglycaemia.

4.4 Emulsifying and stabilizing properties

Two fluids are dispersed in a colloid called an emulsion i.e., oil-in-water (OW) and water-in-oil (WO). To produce an emulsion, a high shear or high pressure techniques are required. Natural polysaccharides stabilize the emulsions. The molecular structure of starch components (amylose and amylopectin) are altered by octenyl succinic anhydride (OSA), which is used as a stabilizer in emulsions (Agama et al., 2017; Hu et al., 2019; Mu et al., 2021). A study was conducted on the stabilization of oil-in-water emulsion using nonchemically modified gelatinized starch. During the process, nonchemically modified starch was able to stabilize oil-in-water emulsion under high shear (Kasprzak et al., 2018). Among native starches, rice starch is the top starch as an emulsifier. While the concentration of the rice starch increases, the droplet size decreases, which leads to improved stability (Li et al., 2013). Lauroyl chloride was used to esterify hydrolyzed rice starch. Laurate starches demonstrated desirable qualities over hydrolyzed rice starch, such as increased solubility, which is linked to improved emulsion stabilizing capabilities (García et al., 2018). A study was performed with native and annealed oat starch as a substitute for fat in mayonnaise. The study summarized that this starch reduces the amount of oil in mayonnaise, boosting stability and offering a clean-label product alongside items that are healthier and low in calories (Werlang et al., 2021). A study was carried out by Xu et al. (2020) on the characteristics of starch-based pickering emulsion from an interface perspective. For the interface formation, the kinetics of particles of starch in the oil-water contact is important. In the continuous phase of stabilization, free nonadsorbed starch, either clumped or in clusters, is present. One of the most important requirements for the stability and usefulness of starch-based pickering emulsions is a strong oil-water interface. Hydrolyzed rice starch stabilizes the oil-in-water emulsions; rice starch acts as a stable emulsifier. By lowering the oil amount and enhancing the quality of mayonnaise, starch acts as a fat substitute. The above studies summarize that emulsions are important in food products where starches play a major role in stabilizing them.

5. Modified starch: method of modification and physicochemical properties

Modified starch is a food ingredient and a derivative of starch, mainly made by treating the native starch with physical, chemical, and enzymatic treatments to enhance the functional characteristics of the native starch and to broaden its applications in the food industry (Ağagündüz et al., 2023).

5.1 Type of modifications

Modified starch is developed by treating the native starch with physical, chemical, and enzymatic treatments to change its behavioral and functional characteristics. Starch can be modified to attain specific properties in two ways: (i) In *planta* modification or genetic modification, and (ii) Modification after harvesting the plant. This includes physical, chemical, and enzymatic treatments (Compart et al., 2023) (Fig. 2.3).

5.2 In *planta* modification

In *planta* starch modification is defined as altering the starch properties within the plant itself, further eliminating the requirement of postharvest processing treatments such as physical, chemical, and enzymatic treatments (Chen et al., 2021). This can be achieved through genetic engineering and mutagenesis, which is highly efficient and allows precise control of starch characteristics at the molecular level. Multiple isoforms of starch-related enzymes are involved in starch synthesis, and each has distinct functions. Thus, selectively modifying these isoforms can alter the starch composition, granule morphology, and physicochemical properties to meet industrial needs. Along with that, extracting starch-modifying related enzymes from bacteria and fungi and introducing them to the plants will expand several potential

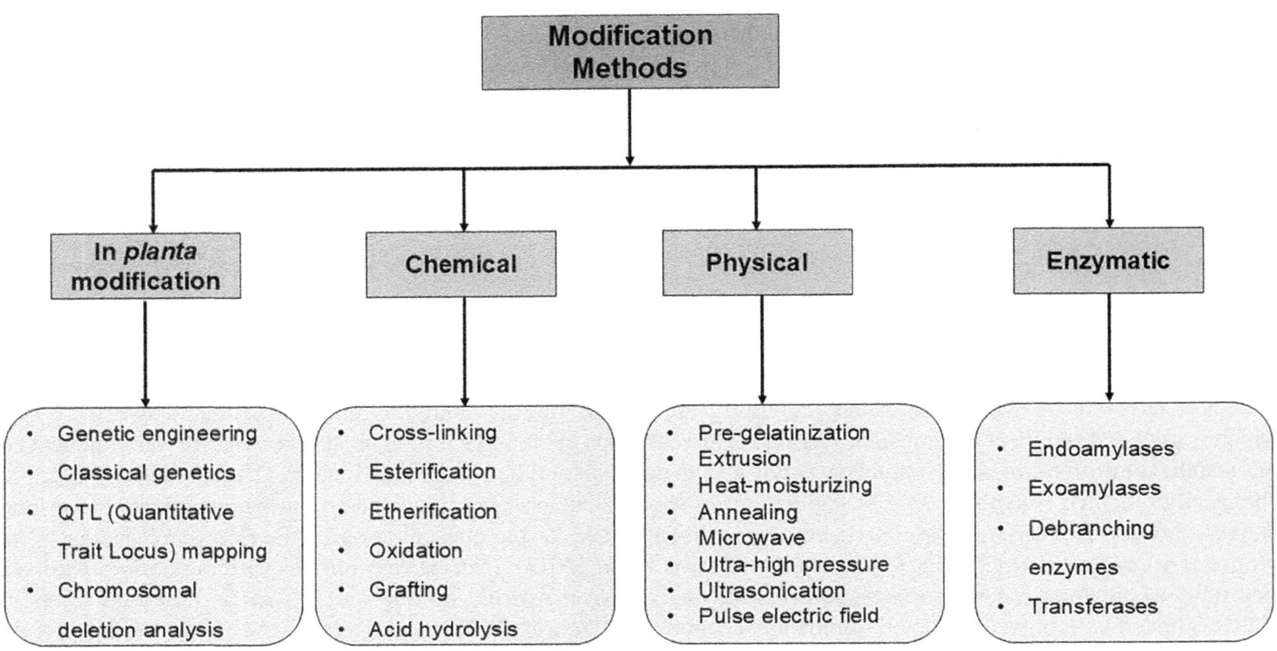

FIGURE 2.3 Represents the method of modifications and its techniques.

applications by enhancing starch structure and functionality (Apriyanto et al., 2022). Modification of starch can be accomplished via genetic engineering, classical genetics, Quantitative Trait Locus (QTL) mapping, and chromosomal deletion analysis. Chia et al. (2020) studied starch granule formation in wheat and barley, which possess a bimodal distribution consisting of large A-type granules that develop early, and smaller B-type granules that develop later. Hence, the BGC1 (B-Granule Content 1) gene plays a significant role in B-type granule formation that affects gelatinization, viscosity, and digestibility properties, which is further used in industrial applications. Peng et al. (2014) discovered that the Floury endosperm 6 (FLO6) gene, present in rice, was identified through map-based cloning. This FLO6 is linked with isoamylase, which influences the starch composition and granule formation, with applications in food, pharmaceuticals, and industry. Saito et al. (2018) identified that FLO6 in mutant barley led to the fractured starch granules, which altered the starch properties. Toyosawa et al. (2016) outlined that integrating Starch SynthaseIIIa (SSIIIa) and Starch SynthaseIVb (SSIVb) into mutant rice resulted in spherical starch granules, which show a positive impact on gelatinization temperature, pasting behavior thus resulting in quick cooking, better digestibility and modified textural properties in food Miura et al. (2018) created a japonica rice mutant EM204 that lacks the SSIIa enzyme, which lowers the gelatinization temperature and increases short-chain amylopectin. This EM204 variant has higher visible amylose content, thus making it more resistant to retrogradation. The mechanism by which phosphate groups bind to starch molecules at the C6 and C3 hydroxyl groups during their synthesis in plastids is known as phosphorylation, and it occurs naturally in plants. Phosphorylation is essential for starch metabolism and shows a significant effect on properties such as solubility and gelatinization. Cereal has little starch-bound phosphate, compared to potato starch. Hence, Carciofi et al. (2011) increased phosphorylation in cereal starch by integrating potato starch phosphorylating enzyme, which is known as Glucan Water Dikinase, StGWD, into barley endosperm, increasing starch-bound phosphate levels, thus leading to use in many industrial applications as thickening agents, improved freeze-thaw stability and enhanced digestibility.

Therefore, in *planta* starch modification offers specific control over starch properties by adjusting key genes and enzymes, thus eliminating the need for postharvest treatments. Targeting genes (specifically modifying, editing, or regulating the activity of certain genes) like BGC1, FLO6, SSIIa, and StGWD results in modifying starch properties such as composition, improving gelatinization, digestibility, and industrial functionality. These alterations pave the way for starch applications in food, pharmaceuticals, and other industries.

5.3 Synthesis of modified starch after harvesting

The modification outside the plant system involves three types: physical, chemical, and enzymatic, which are described in detail below.

5.3.1 Chemical modification of starch

Chemical modification of starch is conducted to improve its functional characteristics, which are naturally limited in the native starch. Native starch possesses low resistance to shear forces (it readily breaks down easily under mechanical stress), limited thermal stability (it degrades at high temperatures), undergoes thermal decomposition (it can decompose during processing), and a tendency to undergo retrogradation (it forms gels and water oozes from the gel when it is cooled down which is undesirable for many food applications). Therefore, chemically modifying starch enhances its structural integrity and functional characteristics, making it appropriate for specific food applications by improving texture, maintaining stability, and enhancing shelf life in the food products (Chen et al., 2018). The chemical modification includes various treatments such as cross-linking, esterification, etherification, oxidation, grafting, and acid hydrolysis (Compart et al., 2023). Table 2.2 depict the techniques, description/function, and application of chemically modified starch.

5.3.2 Cross-linking

One of the main factors preventing the development of starch-based products is the hydrophilic nature of starch. To overcome this limitation, chemical modification is applied to produce water-resistant materials. One promising approach in this treatment involves converting hydrophilic hydroxyl groups to esters by reacting them with carboxylic acids; however, this process is costly. Consequently, cross-linking of starch has gained importance and cost-effective modification techniques. By creating intermolecular connections, the act of cross-linking strengthens polysaccharide molecules. Bifunctional or multifunctional substances, including monosodium phosphate, sodium trimetaphosphate, sodium tripolyphosphate, a combination of adipic and acetic anhydride, and a combination of succinic anhydride and vinyl acetate, are used to accomplish this alteration. These chemicals help polysaccharide molecules create ether or ester bonds, which increase the granules' resistance to swelling, high temperature, high shear, and acidic environments. For example, cross-

TABLE 2.2 Represents the techniques, description/function, and application of chemically modified starch.

Chemical modification method	Reagents used	Applications/properties enhanced	References
Cross-linking	Sodium trimetaphosphate, sodium tripolyphosphate, adipic and acetic anhydride mixtures and succinic anhydride and vinyl acetate mixtures	Improves swelling resistance, thermal and shear stability, used as viscosifiers, texturizers, and chelating agents in bioplastics	Compart et al. (2023)
Esterification	Organic acids, inorganic acid salts and organic acid anhydrides or chlorides	Enhances hydrophobicity, thermal stability, reduces gelatinization and retrogradation, used in frozen foods, emulsions, and encapsulation	Zdybel et al. (2020); Punia et al. (2024); (Kim et al., 2022); Chen et al. (2018)
Etherification	Monochloroacetic acid, propylene oxide, or ethylene oxide	Improves solubility, acts as stabilizers, thickeners, and a resistant starch; widely used as food additives	(Boldrini, 2023; Wang et al., 2020); Gałkowska et al. (2023)
Oxidation	Sodium hypochlorite, hydrogen peroxide	Provides low viscosity, film forming, stabilizing, emulsifying properties; used in paper, textile, and food (puddings, whipped cream, frozen foods)	Vanier et al. (2017)
Grafting	Grafts synthetic polymer chains onto starch backbone via radical polymerization	Increases viscosity, thermal stability, biodegradability, film-forming; important in food and pharma industries	(Çankaya, 2016); Noordergraaf et al. (2024); (Jyothi, 2010)
Acid hydrolysis	Breaks glycosidic bonds using dilute acids (HCl, H_2SO_4) below gelatinization temperature	Improves solubility, lowers gelatinization temperature, enhances dietary fiber, reduces retrogradation; used in desserts, plant-based meat alternatives	Halim et al. (2024)

linked starches are employed in bioplastics, serve as chelation agents and viscosity, and texturize dairy products (Girigoswami et al., 2021; Karthik et al., 2019; Manigandan et al., 2018).

5.3.3 Esterification

Esterification is another treatment in which the hydroxyl (-OH) groups present in the glucose units of starch are replaced by the acyl group, resulting in the formation of starch esters. This transformation enhances key properties such as hydrophobicity and thermal stability with the help of reagents such as organic acids, salts of inorganic acids, as well as anhydrides or chlorides of organic acids (Punia et al., 2024; Zdybel et al., 2020). Esterified starch exhibits improved characteristics such as water resistance, heat stability, and moisture repellence (Kim et al., 2022). For instance, when starch undergoes esterification with anhydrides or vinyl acetate, it produces starch acetate, which demonstrates a lower gelatinization temperature, reduced retrogradation, minimal gel formation tendency, and superior paste clarity. Because of these characteristics, it can be used as an emulsion stabilizer and for encapsulation in frozen and chilled foods (Chen et al., 2018).

5.3.4 Etherification

Etherification of starch is explained as the replacement of hydroxyl groups on glucose units with functional groups such as carboxymethyl, hydroxypropyl, and other substituent groups, forming an ether linkage. This reaction is carried out using reagents such as monochloroacetic acid, propylene oxide, or ethylene oxide (Boldrini, 2023; Wang et al., 2020). The resulting derivatives are hydroxypropyl starch (HPS), hydroxyethyl starch (HES), and carboxymethyl starch (CMS). The food industry uses these modified starches as additives as they possess functional properties such as stabilizers, resistant starch, and thickening properties (Gałkowska et al., 2023; Karthik et al., 2019).

5.3.5 Oxidation

Oxidation of starch includes the alteration of hydroxyl groups in starch molecules into carbonyl and carboxyl groups using oxidizing chemicals like hydrogen peroxide and sodium hypochlorite. This reaction leads to the depolymerization of starch chains, producing modified starch with specific properties. Oxidized starches have low viscosities and high stabilizing, clarifying, film, and binding properties that are useful in many industries, especially in the paper, textile, and food. In food industries, these are applied in puddings, cream puddings, whipped cream, and dough mixtures. Oxidized starches are prominently used in frozen foods as they possess a property of stability in frozen temperatures. Additionally, oxidized starches are also used in emulsifiers, conditioners for bread, and gum arabic replacements (Vanier et al., 2017).

5.3.6 Grafting

Grafting is a specialized technique used to modify the polymers. In the case of starch, this process involves introducing newly synthesized chains onto the existing starch backbone, thereby either enhancing its existing properties or introducing entirely new functionalities (Noordergraaf et al., 2024; Çankaya, 2016). Grafting can be done through three methods, such as activation: in an acidic medium, ceric ammonium nitrate $(NH_4)_2(Ce(NO_3)_6)$ react with hydroxyl (-OH) groups present in starch molecules, resulting in the generation of starch molecules. This process is followed by initiation, where these starch radicals react with an acrylic monomer, triggering the formation of a polymer chain that is attached to the starch backbone. Followed by propagation, where the monomers are added sequentially to the growing radical, resulting in the formation of long-grafted polymer chains. The last method is termination, in which two radicals combine or through other termination mechanisms, resulting in a starch graft copolymer. Starch graft copolymers consist of synthetic polymer chains chemically bonded to the starch backbone. The grafted copolymers exhibit several encouraging characteristics, including high viscosity, water absorption capacity, biodegradability, thermal stability, and strong film-forming capabilities. Owing to these properties, grafted starch copolymers have significant importance in the food and pharma industries (Jyothi, 2010).

5.3.7 Acid hydrolysis

A common and well-known chemical method for creating modified starch is acid hydrolysis. This process yields improved solubility, lower gelatinization temperature, and less pasting viscosity. Acid hydrolysis includes breaking down the glycosidic bonds found in the starch molecules using diluted mineral acids such as hydrochloric acid (HCL) and sulfuric acid (H_2SO_4) while maintaining the temperatures below the gelatinization point. A study by Halim et al. (2024) investigated the acid hydrolysis of quinoa flour, faba bean flour, and native cassava starch. Before undergoing hydrolysis, alkaline steeping was performed to derive native starch from faba bean and quinoa flour. This process involved adding diluted HCL as a reagent, followed by centrifugation and extracting the starch, then milling the starch, and storing it in airtight containers. Acid hydrolysis of these quinoa, cassava, and faba bean native starches led to the enhancement of overall dietary fiber content, resistance to shearing and heating, increased starch solubility, reduced retrogradation, improved gelling properties, which can be mainly used in desserts and plant-based meat alternatives.

5.4 Physical modification of starch

The granular size or molecular packing structure of starch can be altered or controlled by physical manipulation. Without adding any harmful chemicals, starch can be physically modified to get properties that are similar to those of chemical treatments. Starch can be physically modified in a sustainable, economical, and easy way. Thermal and nonthermal treatments are two types of physical modification techniques. Pregelatinization, extrusion, heat-moisturizing, annealing, and microwave treatments are examples of thermal treatments. Ultrahigh pressure, ultrasonication, and pulsed electric fields are examples of nonthermal treatments (Compart et al., 2023). Table 2.3 depict the techniques, description/function, and application of physically modified starch.

5.4.1 Pregelatinization

Pregelatinized starch (PGS) can be synthesized through several procedures, such as spray, drum drying, and extrusion cooking. These starches are mainly utilized as thickeners in instant foods due to their rapid dispersion in water and high viscosity (Lee & Yoo, 2023). Among these, extrusion cooking and drum drying have shown a vital improvement in the textural plus functional properties of pregelatinized starch. The process includes mixing native starch and water to form a starch slurry, which is either spread onto a heated drum or fed into an extruder for high-temperature short-time (HTST) processing. This is followed by scraping/drying and milling of the starch to further improve the solubility in cold water and swelling power of the starch (Wang et al., 2022).

5.4.2 Heat moisture treatment & annealing

Heat moisture treatment (HMT) and annealing (ANN) are hydrothermal treatments applied to starch to improve its functional properties without breaking down its granular structure. These methods control the starch-to-moisture ratio,

TABLE 2.3 Represents the techniques, description/function, and application of physically modified starch.

Physical modification methods	Function	Applications	References
Pregelatinization	Improves cold water solubility, swelling power, and viscosity.	Used as thickeners for instant foods.	(Lee & Yoo, 2023); Ma et al. (2022)
Heat moisture treatment	Modifying pasting properties stabilizes starch and prevents breakdown during heating	Acts as stabilizers, and thickeners in sauces, baked products, and in biodegradable packaging	Zavareze & Dias (2011)
Annealing	Reduces solubility, swelling power, and viscosity. Increases gelatinization temperature and digestibility. Alters hydrogen bonds and molecular vibrations.	Improves starch functionality for food processing.	Braşoveanu and Nemţanu (2014)
Microwave treatment	Reduces solubility, swelling power, and viscosity. Increases gelatinization temperature and digestibility. Alters hydrogen bonds and molecular vibrations.	Improves starch functionality for food processing.	Braşoveanu and Nemţanu (2014)
Pulsed electric field	Decreases gelatinization temperature and crystallinity. Reduces viscosity, improving texture and processing efficiency.	Enhances food processing efficiency and texture.	Han et al. (2009)
Ultra-high pressure	Disrupts hydrogen bonds, alters granule morphology, increases swelling and solubility, and reduces retrogradation. Forms gelatinization at lower temperatures.	Improves starch functionality for food processing and packaging.	Castro et al. (2020)
Ultrasonication	Creates cracks on starch granules, enhancing water intake. Alters gelatinization, pasting behavior, and crystalline structure. Softens grains and stimulates the production of antioxidants and bioactive compounds.	Improves palatability, cooking ease, and nutritional value of rice starch/grains.	Bonto et al. (2021)

temperature, and heating time that influence starch's structural and physicochemical characteristics. Heat-moisture treatment entails heating starch under low moisture conditions and at a high temperature (above the gelatinization temperature). Due to the limited water availability, starch does not undergo full gelatinization but undertakes structural changes, leading to modification of starch's pasting properties, thereby improving thermal stability. Annealing treatment includes soaking starch in lots of water and heating at a temperature above the glass transition but below gelatinization temperature, allowing starch molecules to reorganize themselves, resulting in crystallinity, swelling, and solubility. In food industries, HMT and ANN-treated starches act as stabilizers, and thickeners in making sauces and baked products and are used in biodegradable packaging (Zavareze & Dias, 2011).

5.4.3 Microwave treatment

It's a nonionizing radiation that is used to change starch properties such as water absorption, solubility, swelling power, and viscosity. This treatment induces structural transformations in starch, which reduces solubility, swelling power, and crude protein content in starch and decreases the viscosity, increasing the gelatinization temperature and digestibility by altering the structure of starch (making the surface porous or creating cracks). Analytical techniques such as FTIR and Raman spectroscopy reveal that this treatment leads to changes in the structure of starch granules by modifying the hydrogen bonds and molecular vibration patterns (Braşoveanu & Nemţanu, 2014).

5.4.4 Nonthermal treatments

Nonthermal treatments, including pulsed electric field, ultrasonication, and ultrahigh-pressure are applied to the food without the application of heat, which can negatively impact food quality by degrading heat-sensitive nutrients and flavors. Hence, there is a demand for nonthermal, environmentally friendly technologies to enhance food processing.

5.4.4.1 Pulsed electric field

Pulsed electric field (PEF) treatment is a nonthermal processing technique that involves generating high-voltage short-term electric pulses to the food, which is placed between two electrodes in a treatment chamber. This treatment causes microbial inactivation, derivation of bioactive compounds, improves winemaking, enhances drying and freezing, and enzyme inactivation (Wang et al., 2018). Application of pulsed electric field to corn starch-water suspension led to a reduction in gelatinization temperature, thus making it easily gelatinize, decrease in crystallinity in the starch granule. reduced viscosity which ultimately enhances texture, stability and processing efficiency (Han et al., 2009).

5.4.4.2 Ultra-high pressure

Highpressure treatment breaks molecular linkages of hydrogen bonds within and between the starch molecules, leading to changes in starch granule morphology. The degree of modification depends on factors such as pressure level, solvent type (in which the starch is treated in medium, e.g., water), treatment time, and starch type and origin. Applying high pressure can increase the solubility and swelling of starch based on the temperature and pressure applied. Along with this, high pressure on starch leads to the formation of gelatinization at lower temperatures, alters the pasting properties such as viscosity of starch, alters the rheological behavior (viscous and elastic behavior), reduces retrogradation, and changes the crystalline structure (Castro et al., 2020).

5.4.4.3 Ultrasonication

Ultrasonication refers to generating acoustic cavitation in an aqueous medium, which further creates physical forces (shock waves, shear forces and microjets) that change the organization of starch and grain by creating cracks on the surface of the starch granules, which enhance the intake of water; thus, alter the gelatinization and pasting behavior. As per Bonto et al., (2021), ultrasonication alters the texture of rice starch and grains by softening them and making them more palatable and easier to cook, and changes in crystalline structure show a positive impact on temperature, viscosity, and retrogradation. One significant impact of treating rice starch/grains with high pressure is that it stimulates the production of health-related metabolites such as antioxidants and bioactive compounds, which enhance the value of rice.

5.5 Enzymatic treatment

The starch processing industry has seen rapid growth, expansion and technical advancements in producing starch-based derivatives such as maltodextrin, modified starches, and glucose/fructose syrups. The evolution of processing techniques, ranging from acid hydrolysis to enzymatic treatment, has played a prominent role in this progress. There are four types of enzymatic treatments with different enzymes: endoamylase, exoamylase, transferases, and debranching enzymes (Compart et al., 2023). Table 2.4 depicts the techniques, description/function, and application of enzymatically modified starch.

TABLE 2.4 Represents the techniques, description/function, and application of enzymatically modified starch.

Enzyme category	Effects on starch	Applications	References
Endoamylases	Creates shorter starch fragments and dextrins by rupturing starch granules. Shortens the chain.	Utilized in manufacturing of modified starches and maltodextrins.	Van Der Maarel et al. (2002)
Exoamylases	Spits starch molecules from the ends. Creating cracks on granule surfaces. Produces β-dextrins, maltose, and glucose. Increases solubility, reduces swelling capacity, and lowers pasting viscosity	Utilized in the manufacturing of modified starches and glucose/fructose syrup.	Pandey et al. (2000); Van Der Maarel et al. (2002)
Debranching enzymes	To linearize the glucans, break the branch points in amylopectin	Enhances starch processing in industries.	Van Der Maarel et al. (2002); (Prongjit et al., 2022; Xia et al., 2021)
Transferases	Forms cyclodextrins and amylose-like polymers. Enhances starch properties, texture, and stability.	Improves texture and stability of processed foods.	Compart et al. (2023); (Van Der Maarel et al., 2002)

Endoamylases hydrolyze the (1.4)-glycosidic linkages of amylose and amylopectin in an unpredictable sequence. This technique reduces the whole chain length of starch molecules. One example of endoamylase is α-amylase. Depending upon the nature and concentration of the enzyme, α-amylases create holes in the surface granules that lead to the production of shorter starch fragments and dextrins (van der Maarel et al., 2002). By removing maltose units from the nonreducing ends of starch chains β-amylases produce β-dextrins. Amyloglucosidases and α-glucosidases produce maltose and glucose by breaking down starch chains. Enzymatically modified starch is more soluble, has a lower propensity to swell when heated, and has a lower pasting viscosity than native starch, making it more appropriate for a variety of uses (Pandey et al., 2000). Whereas debranching enzymes such as isoamylase, and pullulanase, hydrolyzes the α-1.6 glucosidic bonds in glycogen, starch, and pullulan, which is a polysaccharide that is similar to starch. Both enzymes specifically target the branch points in amylopectin, which is a branched polymer and break the α-1.6 bonds and linearize the branched structure of amylopectin, producing linear glucans, improving starch utilization and production efficiency in industries (Van Der Maarel et al., 2002; Xia et al., 2021; Prongjit et al., 2022). On the other hand, transferases such as amylomaltase, cyclodextrin glycosyltransferase and amylosucrase break and reform α-1.4 bonds, enabling the formation of new starch structures such as cyclodextrins and amylose-like polymers. These transferase enzymes can be used to enhance the characteristics of starch as well as to enhance the texture and stability of processed foods (Compart et al., 2023; Van Der Maarel et al., 2002).

6. Physicochemical properties of modified starch

The properties of the native starch are limited solubility in cold water and low swelling power (Gupta et al., 2024). Compared with native starch, modified starch shows an expanded spectrum of physicochemical properties. Each modification increases the functional properties that are suitable for different industries. Solubility, light transmittance value, pasting properties, swelling power, rheological properties, thermal properties like gelatinization and retrogradation, starch digestibility, starch granule crystal structure, and freeze-thaw stability are among the physicochemical characteristics of modified starches (Carciofi et al., 2011; Hutabarat & Stevensen, 2023).

7. Applications of native and modified starches

Starch contains two main glucans, amylose and amylopectin. Many industries, including food, pharmaceutical and cosmetic sectors, use starch as a basic material. Native and modified starch has various applications in the food industry, which include gelling agents, thickeners, encapsulating agents, emulsifiers, etc. Similarly, starch is used in paper, textile and pharmaceutical industries (Ai & Jane, 2015). Native starch is an insoluble white powder that has no taste or odor. Plant sources store energy in the form of native starch. Starch that has undergone physical, chemical, or enzymatic modification to change one or more original properties is known as modified starch (Bashir & Aggarwal, 2019). Corn starch is the principal source of starch, with cassava, potato, and wheat coming in second and third. A 70% of starch is produced from corn all over the world. The second most abundant source of starch is potatoes, produced from roots and tubers (Bertolini, 2009). Table 2.5 represents the applications of modified starch.

7.1 Type of starch

7.1.1 Corn starch

Corn starch is the major starch used in industries due to its consistency and gelling properties. It is used mainly in the production of food items like cakes, bread, etc. The starch is isolated from the corn and is used in baked goods. A study done by Yousif et al. (2012) stated that physically modifying (pregelatinization) the corn starch to prepare noodles instead of wheat flour. Several properties (solubility, swelling power, and the ability to bind water and oil) were tested after modification by the pregelatinization process. A study done by Hong et al. (2015) chemically modified the cornstarch to make them a thickener and O/W emulsion stabilizer in the food sector by treating in propionic anhydrous and NaOH as catalysts. With a greater degree of substitution, the starch's swelling power decreases. The cereal has a sugar coating to extend its shelf life. Since people are concerned with their health, an alternative to sugar coating was introduced (Luckett & Wang, 2012). In this study, cereal was coated with Hylon VII extracted from corn, which contains 70% amylose content, which helped to replace the sugar coating. Abbas et al. (2010) summarized that crosslinked corn starch produced stable, stronger, and bigger loaf volume in bread. This can be used as an alternative to bread dough since it improves the rheological properties and quality of bread.

TABLE 2.5 Applications of modified starch.

Type of starch	Method of modification	Treatment	Parameters	Purpose/applications	References
Corn starch	Chemical	Propionic anhydrous (NaOH as catalyst)	12 h, 70 °C water bath	Thickeners, O/W emulsion stabilizers in food, cosmetic, and pharmaceutical industries	Hong et al. (2015)
		Crosslinked corn starch	-	Preparation of dough for bakery foods	Abbas et al. (2010)
	Enzymatic	-	Drying at 50 °C	Functions as a coating material	Luckett and Wang (2012)
	Physical	Pregelatinization	60 °C for 5 min	Substitute for wheat flour noodles	Yousif et al. (2012)
Tapioca starch	chemical	3 wt% microcrystalline cellulose	-	Biodegradable film	Othman et al. (2019)
		4% octenyl succinic anhydride	-	Fat replacements	Balic et al. (2017)
	Physical	HMT	Autoclave at 120 °C for 60 min, 30% moisture	Production of nanocrystals	Fonseca et al. (2021)
	Enzymatic	Thermostable amylose	-	Tablet filler in direct compressor	Charoenthai et al. (2018)
	Radiation	Gamma irradiation	-	Active film for packaging meat	Kanatt (2020)
Potato starch	Chemical	Microwave & ultrasound	-	Emulsifier and surfactants	Kaur et al. (2012)
		Curing	Mechanical, wear resistance	The coating that separates the emulsion	Wang et al. (2020)
Rice starch	Physical	Freezing and thawing	Temperature	Preparation of breads	Tao et al. (2015)
	Enzymatic	4-α-glucanotransferase		Low-fat mayonnaise	Abbas et al. (2010)

7.1.2 *Tapioca starch*

Tapioca starch is utilized as a biodegradable film to develop a sustainable environment by incorporating microcrystalline cellulose (MCC) as a filler at an ideal concentration of 3 wt.%. This MCC improved the thermal property, mechanical property (tensile stress, elongation capacity, and flexibility), and barrier property (water barrier and oxygen barrier) (Othman et al., 2019). Similarly, tapioca and native wheat starches were used to substitute oil in the bread recipe. A 4% octenyl succinic anhydride (OSA) modified starch enhanced its pasting, gel hardness, and sticky mixing qualities (Balic et al., 2017). Hydrothermal processes such as HMT improve the properties of starch. Tapioca starch was autoclaved at 120 °C for 30 min and used to produce nano crystals, which played a significant role in the food sector (Fonseca et al., 2021). Charoenthai et al. (2018) modified the tapioca starch enzymatically (thermo-stable alpha-amylase) to produce crystalline starches in the direct compression process, and this modified starch is used as a tablet filler.

7.1.3 *Potato starch*

Potato starch displayed its properties as an emulsifier and surfactant through chemical modification (microwave and ultrasound irradiation) by esterification of carboxymethyl with octenylsuccinic anhydride. This reduced the esterification time (Kaur et al., 2012; Wang et al., 2020). A coating was prepared using potato starch and polydimethylsiloxane (PDMS) that applies to the separation of oil-water mixtures and emulsions. The chemical bond between PDMS and starch enhances the mechanical and wear resistance.

7.1.4 Rice starch

A significant part of the food industry is frozen technology. Freezing transforms the product into a crystal, extending its shelf life. This reduced the gelatinization temperature range, enthalpy, and pasting viscosities; therefore, breads were prepared from rice starch (Tao et al., 2015). Rice starch was combined with xanthan gum and modified enzymatically using 4-α-glucanotransferase (4αGTase) for the making of reduced-fat mayonnaise (Abbas et al., 2010).

7.2 Native starch uses in the food sector

Native corn starch is the starch in its purest form. It was collected, and various analyses (Proximate analysis, thermal properties, rheological properties and water binding capacity) were completed. This study demonstrated the use of native starch in the food business (Shaari et al., 2017). Starch noodles were prepared from a variety of native starches (corn starch, pea starch and mung bean starch). Dropping, cutting, and extrusion techniques were implemented to produce the noodles using starch (Tan et al., 2009). The starch of cassava varieties (Abrabopa, AGRA, and Bankye hemaa) was incorporated into the yogurt during preparation time to increase the syneresis of the yogurt. The physicochemical, rheological, and sensory properties of the developed product were analyzed, and 7% of the abrabopa cassava variety proved the best result (Agyemang et al., 2020). Native tapioca starch is widely used in tablets and capsules due to its properties (less friability, tensile strength, extended disintegration time, and poor drug dissolution rate). This is used as a binder in the wet granulation process in the form of paste (Charoenthai et al., 2018). A study was carried out using different levels of pectin (0.3% and 0.6%) in the native tapioca starch, and the formulation of thickeners in fruit filling pastries was attained. This was an alternative to modified starches (Agudelo et al., 2014). Native starch has properties like gel-forming agents, thickening agents, and stabilizing food products. A study was carried out by modifying the consistency of sago and tapioca starch in the form of a thickener. This helped the patients with dysphagia who experience difficulty in swallowing foods (Yazid et al., 2018).

7.3 Pharmaceutical and biotechnological applications

Starch is commonly used in pharmaceutical manufacturing due to its low processing, biocompatibility, and nontoxicity. The modified starch plays a versatile role in pharmaceuticals. The modification can be done by heating, gelatinization, extrusion, etc. (Balic et al., 2017). To target the cancer cell, the biocompatibility of starch as a carrier was investigated by Alexiou et al., (2006). Iron oxide nanoparticle was prepared using phosphate-modified starch, followed by mixing mitoxantrone. On the other hand, in a study by Szepes et al., (2008) reported that an isostatic ultrahigh-pressure technique was used in potato and corn starch hydrogel, where corn starch exhibited a constant delivery of active components, while the hydrogel of potato starch showed rapid release.

7.4 Personal care and cosmetic products

7.4.1 Gel formulation

Starches are polysaccharides that are mainly used due to their natural approach and budget-friendly properties. Starches can be an alternative in the cosmetic field due to their ability to increase the stability of emulsions through gelatinization properties. A study was carried out on gel formulation using tapioca and corn starch. A total of 10% of tapioca and corn starch was mixed with 6% glycerine. The developed formulation can be used as a substitute for rheological additives (polymers and silicon) (Infante et al., 2024). An antagonistic study was performed by isolating starch from the medicinal plant *Curcuma phaeocaulis Val* (CPV), *Paeonia suffruticosa Andr* (PSA), and *Paeonia lactiflora Pall* (PLP). The PSA starch contains an equal amount of amylose content as potato and corn starch. The three traditional Chinese medicines have a higher water-binding capacity compared to corn and potato starch. Due to this specific property, it's widely used in the cosmetic industry (Xia et al., 2013).

7.4.2 Powder formulation

A variety of unripe banana fruit was used in the experiment to extract the starch. The swelling power and percentage of solubility in water of the banana starch were found to be optimal. This can be adopted as a powder formulation to provide a smooth feel and better-spreading property (Thanyapanich et al., 2021). A study was carried out using corn starch as an alternative to talc powder by testing the flow and performance effects. Ajayi and Amin (2021) found that a combination of fumed silica and corn starch can be used as an alternative to talc powder. Corn starch at lower concentrations, lower than 5%, can be an alternative to sustainable talc alternatives. A comparative study was performed using Manihot starch and Maydis starch for the powder formulation with natural orange sweet potato. The powder prepared from manihot starch showed better performance than maydis starch (Monika et al., 2023). The starch was isolated from wild yam (*Dioscorea hispida Dennst.)*, and its cosmetic application is studied. The isolated starch was white in fine powder. *D. hispida* was used in a deep powder formulation. The study summarized that starch isolated from wild yams shows stability and is used as talcum powder (Sangkanu et al., 2025).

7.4.3 Thickening agents

Thickeners are a major material in the cosmetic industry since they are used in emulsions and cleansing products. Viscosity has a significant impact on thickening. To increase viscosity, 0.5 wt.% modified starch (2-hydroxyalkyl carboxymethyl) was added to the selective aqueous surfactant, and thickening can be attained (Lehmann et al., 2008). Starch-based hydrogel matrices are suitable for a sustainable alternative to create personal care products. For lipstick (20% starch), a high mineral content and ionic ratio are sufficient (Fig. 2.4) (Torres et al., 2021).

8. Challenges and future directions

The demand for starch is growing to meet food consumption desires, but farmers are shifting toward nonfood, industrial applications to access new markets and achieve sustainability goals. Native starch has boundaries for many industrial uses, requiring modification through chemical, enzymatic, or physical processes (Zarski et al., 2024). Despite significant growth in this field, numerous challenges remain. Starch is an important raw material that addresses universal nutritional needs (Zia-ud-Din et al., 2017). The necessity for both functional and traditional starch-based foods and derivatives is growing. Nonfood-based industries are also turning to starch due to issues with petrochemical raw materials and postproduction waste recycling, which causes major issues to its renewability, biodegradability, and wide availability (Sun et al., 2023). Consequently, efforts to enhance the multifunctional properties of starches are ongoing.

One foremost problems is the lack of interdisciplinary collaboration, which limits the ability to address complex and complicated problems efficiently. Furthermore, there is a gap between theoretical advancements and their practical implementation, which can slow down the progress in solving real-world problems. As the new technology continues to evolve, important challenges such as ethical considerations, particularly concerning privacy, bias, and accountability, persist in growing (Li et al., 2025).

8.1 Limitation of current starch modifications

Starch is a valuable, commercially imperative polymer with abundant applications in food and industrial industries. Native starches, on the other hand, have limited applicability, making industrial use difficult. As a result, starch modification improves the beneficial properties of native starches while eliminating their limitations. Even if current starch modifications improve qualities for a variety of applications, they still have drawbacks, such as the possibility of processing

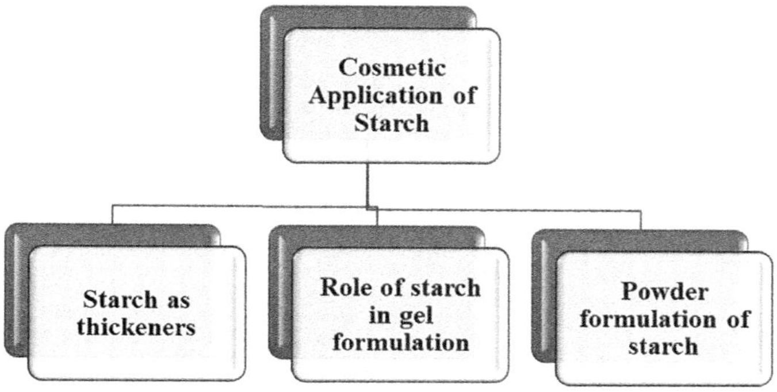

FIGURE 2.4 Application of starch in cosmetic products.

degradation, high cost, and the need for additional study into long-term stability and environmental impact. Starch modifications can result in new polymers with a variety of functional and value-added qualities that meet industry needs (Zarski et al., 2021). Table 2.6 summarizes the common advantages and disadvantages of starch modification.

8.2 Future trends in starch modification

Promising application research has been conducted for starch-based delivery systems in domains such as food packaging materials, health and agricultural products. However, some problems and difficulties still need to be resolved (Thirumalai et al., 2023). Growing environmental concerns and the need for novel food and industrial applications are driving future developments in starch modification, indicating a rise in demand for sustainable, functional, and specialized starches (Sun et al., 2023). The market for modified starch value was USD 18,779.0 million in 2024 and is expected to reach USD 28,885.3 million in 2034. Over the forecast period (2024–2034), the market is anticipated to exhibit a CAGR of 4.4%. Because of the strong demand from the food and pharmaceutical industries, Asia–Pacific region will drive the market's expansion (Future Market Insights Global and Consulting Pvt. Ltd., 2024). With many emerging techniques in starch modification, the most promising approaches have been physical, enzymatic, and genetic alteration.

8.3 Emerging applications and innovations

Starch modifications are enclosed by different new emerging applications, such as developing biodegradable packaging with improved qualities, producing functional food ingredients with better texture and digestibility, and employing modified starches as feedstocks for 3D printing. These advances are mostly due to methods like enzyme alterations, high-pressure processing, and cold plasma (Yudhistira et al., 2024). Starch was improved to expand its suitability for industrial uses and address the drawbacks of native starches (Kaur et al., 2012). Native starches are susceptible to syneresis, gelling, and retrogradation when cooked. Modifying starch improves paste clarity, gloss, texture, film formation, and adhesion, while it reduces the process of retrogradation, gelling, and syneresis (Velazquez et al., 2025). Modification of starch has led to the development of new technologies and market trends. Starch plays a dynamic role throughout a food product's lifecycle, providing high functional derivatives that offer several advantages by enhancing product aesthetics, shortening formulations, decreasing production costs, refining consistency, diminishing batch rejections, and enhancing shelf-life (Kaur et al., 2012; Rajagopal et al., 2022).

9. Conclusion

In food technology, starch plays a major role because of its physicochemical properties, like gelatinization, solubility, pasting nature, and digestibility. At the same time, the limitations of native starch, like low solubility and instability, can't fulfill the requirements. On the other hand, modified starches have gained importance as essential functional ingredients in several industries. Through chemical, physical, and enzymatic methods, the native starch is altered into modified starches

TABLE 2.6 Merits and demerits of the commonly used starch modifications and techniques.

Modification types	Advantages	Disadvantages	References
Inorganic and organic solvents	Low production cost, easily recyclable and wide range of applications in temperature	Toxicity, highly inflammable, high temperature and longer time	Ptak et al. (2014)
Ionic liquids	Nonvolatility, flammability and multifunctionality	Difficult purification and recycling	Ptak et al. (2020)
supercritical CO_2	Easy recycling nontoxicity, flammability and volatility multifunctionality	high cost of production and purification difficulties	Adak and Banerjee (2016)
Microwave radiation	Reduction of reaction time.	High apparatus cost	Ptak et al. (2014)
Ultrasonication			Muljana et al. (2010)

that have enhanced stability, improved texture, and increased resistance to processing conditions. All these modifications to the native starch address the limitations, such as retrogradation, syneresis, and limited solubility, making modified starches a prominent additive in modern food industrial applications. As the technology is increasing, the application of different techniques to modify starch, such as heat-moisture treatment, cross-linking, classic genetics, QTL mapping, genetic engineering, and enzymatic hydrolysis, has expanded the functional properties of starch. Each modification method exhibits unique properties that are useful for different industries. However, the integration of novel and environmentally friendly techniques, such as genetic engineering and green chemistry, is showing a way for sustainable starch-based solutions. Despite numerous advantages, there are some limitations which include safety, cost-effectiveness, clean-label, and natural ingredients as per consumer preferences. Along with that, further studies focus on improving the nutritional and functional properties of starch. Overall, modified starch plays an essential role in different sectors by offering versatile functionalities. Thus, technological advancements pave the way to refine the modification process, which can deliver a clean label, costeffectiveness, and a safer product.

References

Abbas, K. A., Khalil, S. K., & Hussin, A. S. M. (2010). Modified starches and their usages in selected food products: A review study. *Journal of Agricultural Science, 2*(2), 90.

Adak, S., & Banerjee, R. (2016). A green approach for starch modification: Esterification by lipase and novel imidazolium surfactant. *Carbohydrate Polymers, 150*, 359–368.

Ağagündüz, D., Özata-Uyar, G., Kocaadam-Bozkurt, B., Özturan-Şirin, A., Capasso, R., Al-Assaf, S., & Özoğul, F. (2023). A comprehensive review on food hydrocolloids as gut modulators in the food matrix and nutrition: The hydrocolloid-gut-health axis. *Food Hydrocolloids, 145*, Article 109068. https://doi.org/10.1016/j.foodhyd.2023.109068

Agama-Acevedo, E., & Bello-Perez, L. A. (2017). Starch as an emulsions stability: The case of octenyl succinic anhydride (OSA) starch. *Current Opinion in Food Science, 13*, 78–83.

Agudelo, A., Varela, P., Sanz, T., & Fiszman, S. M. (2014). Native tapioca starch as a potential thickener for fruit fillings. Evaluation of mixed models containing low-methoxyl pectin. *Food Hydrocolloids, 35*, 297–304. https://doi.org/10.1016/j.foodhyd.2013.06.004

Agyemang, P. N., T Akonor, P., Tortoe, C., T Johnsona, P.-N., & Manu-Aduening, J. (2020). Effect of the use of starches of three new Ghanaian cassava varieties as a thickener on the physicochemical, rheological and sensory properties of yoghurt. *Scientific African, 9*, Article e00521. https://doi.org/10.1016/j.sciaf.2020.e00521

Ai, Y., & Jane, J.-lin (2015). Gelatinization and rheological properties of starch. *Starch - Stärke, 67*(3–4), 213–224. https://doi.org/10.1002/star.201400201

Ajayi, O. M., & Amin, S. (2021). Flow and performance effects of talc alternatives on powder cosmetic formulations. *International Journal of Cosmetic Science, 43*(5), 588–600.

Alcázar-Alay, S. C., & Meireles, M. A. A. (2015). Physicochemical properties, modifications and applications of starches from different botanical sources. *Food Science and Technology, 35*, 215–236.

Alexiou, C., Schmid, R. J., Jurgons, R., Kremer, M., Wanner, G., Bergemann, C., Huenges, E., Nawroth, T., Arnold, W., & Parak, F. G. (2006). Targeting cancer cells: Magnetic nanoparticles as drug carriers. *European Biophysics Journal, 35*(5), 446–450. https://doi.org/10.1007/s00249-006-0042-1

Ali, A., Wani, T. A., Wani, I. A., & Masoodi, F. A. (2016). Comparative study of the physico-chemical properties of rice and corn starches grown in Indian temperate climate. *Journal of the Saudi Society of Agricultural Sciences, 15*(1), 75–82.

Apriyanto, A., Compart, J., & Fettke, J. (2022). A review of starch, a unique biopolymer–Structure, metabolism and in planta modifications. *Plant Science, 318*, Article 111223.

Balic, R., Miljkovic, T., Ozsisli, B., & Simsek, S. (2017). Utilization of modified wheat and tapioca starches as fat replacements in bread formulation. *Journal of Food Processing and Preservation, 41*(1), Article e12888. https://doi.org/10.1111/jfpp.12888

Bashir, K., & Aggarwal, M. (2019). Physicochemical, structural and functional properties of native and irradiated starch: A review. *Journal of Food Science and Technology, 56*(2), 513–523. https://doi.org/10.1007/s13197-018-3530-2

Bertolini, A. (2009). *Starches: Characterization, properties, and applications*. CRC Press.

Boldrini, D. E. (2023). Starch-based materials for drug delivery in the gastrointestinal tract-A review. *Carbohydrate Polymers, 320*, Article 121258. https://doi.org/10.1016/j.carbpol.2023.121258

Bonto, A. P., Tiozon, R. N., Sreenivasulu, N., & Camacho, D. H. (2021). Impact of ultrasonic treatment on rice starch and grain functional properties: A review. *Ultrasonics Sonochemistry, 71*. https://doi.org/10.1016/j.ultsonch.2020.105383

Braşoveanu, M., & Nemţanu, M. R. (2014). Behaviour of starch exposed to microwave radiation treatment. *Starch - Stärke, 66*(1–2), 3–14. https://doi.org/10.1002/star.201200191

Çankaya, N. (2016). Synthesis of graft copolymers onto starch and its semiconducting properties. *Results in Physics, 6*, 538–542. https://doi.org/10.1016/j.rinp.2016.08.010

Calvin, O. (2016). Starch and modified starch in bread making: A review. *African Journal of Food Science, 10*(12), 344–351. https://doi.org/10.5897/AJFS2016.1481

Carciofi, M., Shaik, S. S., Jensen, S. L., Blennow, A., Svensson, J. T., Vincze, E., & Hebelstrup, K. H. (2011). Hyperphosphorylation of cereal starch. *Journal of Cereal Science, 54*(3), 339–346. https://doi.org/10.1016/j.jcs.2011.06.013

Castro, L. M. G., Alexandre, E. M. C., Saraiva, J. A., & Pintado, M. (2020). Impact of high pressure on starch properties: A review. *Food Hydrocolloids, 106*. https://doi.org/10.1016/j.foodhyd.2020.105877

Chakraborty, I., Mal, S. S., Paul, U. C., Rahman, M. H., & Mazumder, N. (2022). An insight into the gelatinization properties influencing the modified starches used in food industry: A review. *Food and Bioprocess Technology, 15*(6), 1195–1223.

Charoenthai, N., Sanga-ngam, T., & Puttipipatkhachorn, S. (2018). Use of modified tapioca starches as pharmaceutical excipients. *Pharmaceutical Sciences Asia, 45*(4).

Chen, J., Hawkins, E., & Seung, D. (2021). Towards targeted starch modification in plants. *Current Opinion in Plant Biology, 60*, Article 102013.

Chen, Y. F., Kaur, L., & Singh, J. (2018). Chemical modification of starch. In *Starch in food* (pp. 283–321). Woodhead Publishing.

Chia, T., Chirico, M., King, R., Ramirez-Gonzalez, R., Saccomanno, B., Seung, D., Simmonds, J., Trick, M., Uauy, C., Verhoeven, T., Kay, T., & Lunn, J. (2020). A carbohydrate-binding protein, B-GRANULE CONTENT 1, influences starch granule size distribution in a dose-dependent manner in polyploid wheat. *Journal of Experimental Botany, 71*(1), 105–115. https://doi.org/10.1093/jxb/erz405

Compart, J., Singh, A., Fettke, J., Apriyanto, A., & Starch Properties, C. (2023). A review of starch modifications and their applications. *Polymers, 15*(16), 3491. https://doi.org/10.3390/polym15163491

Fonseca, L. M., Halal, S. L. M. E., Dias, A. R. G., & Zavareze, E.d. R. (2021). Physical modification of starch by heat-moisture treatment and annealing and their applications: A review. *Carbohydrate Polymers, 274*. https://doi.org/10.1016/j.carbpol.2021.118665

Future Market Insights Global and Consulting Pvt Ltd. (2024). *Modified starch market set to nearly double, projected revenue of USD 28.9 billion by 2034, fuelled by plant-based and gluten-free expansions*. Future Market Insights, Inc.. https://finance.yahoo.com/news/modified-starch-market-set-nearly-103000874.html?guccounter=1

Gałkowska, D., Kapuśniak, K., & Juszczak, L. (2023). Chemically modified starches as food additives. *Molecules, 28*(22), 7543.

García-Tejeda, Y. V., Leal-Castañeda, E. J., Espinosa-Solis, V., & Barrera-Figueroa, V. (2018). Synthesis and characterization of rice starch laurate as food-grade emulsifier for canola oil-in-water emulsions. *Carbohydrate Polymers, 194*, 177–183.

Giri, S., Banerji, A., Lele, S. S., & Ananthanarayan, L. (2017). Starch digestibility and glycaemic index of selected Indian traditional foods: Effects of added ingredients. *International Journal of Food Properties, 20*(Suppl. 1), Article S290. https://doi.org/10.1080/10942912.2017.1295387

Girigoswami, A., Ghosh, M. M., Pallavi, P., Ramesh, S., & Girigoswami, K. (2021). Nanotechnology in detection of food toxins–focus on the dairy products. *Biointerface Research in Applied Chemistry, 11*(6), 14155–14172.

Giuberti, G., & Gallo, A. (2018). Reducing the glycaemic index and increasing the slowly digestible starch content in gluten-free cereal-based foods: A review. *International Journal of Food Science and Technology, 53*(1), 50–60.

Giuberti, G., Gallo, A., Cerioli, C., & Masoero, F. (2012). In vitro starch digestion and predicted glycemic index of cereal grains commonly utilized in pig nutrition. *Animal Feed Science and Technology, 174*(3–4), 163–173. https://doi.org/10.1016/j.anifeedsci.2012.03.006

Gupta, R. K., Guha, P., & Srivastav, P. P. (2024). Effect of cold plasma treatment and plasma-activated water on physicochemical and structural properties of starch: A green and novel approach for environmental sustainability. *Plasma Processes and Polymers, 21*(4), Article 2300204.

Halim, A., Torley, P. J., Farahnaky, A., & Majzoobi, M. (2024). Investigating the effects of acid hydrolysis on physicochemical properties of quinoa and faba bean starches as compared to cassava starch. *Foods, 13*(23), 3885. https://doi.org/10.3390/foods13233885

Han, Z., Zeng, X.a., Zhang, B.s., & Yu, S.j. (2009). Effects of pulsed electric fields (PEF) treatment on the properties of corn starch. *Journal of Food Engineering, 93*(3), 318–323. https://doi.org/10.1016/j.jfoodeng.2009.01.040

Himashree, P., Sengar, A. S., & Sunil, C. K. (2022). Food thickening agents: Sources, chemistry, properties and applications-A review. *International Journal of Gastronomy and Food Science, 27*, Article 100468.

Hong, L. F., Cheng, L. H., Lee, C. Y., & Peh, K. K. (2015). Characterisation of physicochemical properties of propionylated corn starch and its application as stabiliser. *Food Technology and Biotechnology, 53*(3), 278–285. https://doi.org/10.17113/ftb.53.03.15.3907

Hu, X., Karthik, P., & Chen, J. (2019). Manipulating oral behaviour of food emulsions using different emulsifiers. *International Journal of Food Science and Technology, 54*(7), 2408–2415.

Hutabarat, D. J. C., & Stevensen, J. (2023). Physicochemical properties of enzymatically modified starch: A review. In *IOP conference series: Earth and environmental science*. IOP Publishing, Article 012093. Vol. 1169, No. 1.

Infante, V. H. P., Calixto, L. S., & Campos, P. M. B. G. M. (2024). Application of tapioca and corn starches as an alternative for synthetic polymers in cosmetic products. *Brazilian Journal of Pharmaceutical Sciences, 60*, Article e20124.

Jyothi, A. N. (2010). India starch graft copolymers: Novel applications in industry. *Composite Interfaces, 17*(2–3), 165–174. https://doi.org/10.1163/092764410X490581

Kanatt, S. R. (2020). Irradiation as a tool for modifying tapioca starch and development of an active food packaging film with irradiated starch. *Radiation Physics and Chemistry, 173*, Article 108873.

Karthik, P., Ettelaie, R., & Chen, J. (2019). Oral behaviour of emulsions stabilized by mixed monolayer. *Food Research International, 125*, Article 108603.

Kasprzak, M. M., Macnaughtan, W., Harding, S., Wilde, P., & Wolf, B. (2018). Stabilisation of oil-in-water emulsions with non-chemical modified gelatinised starch. *Food Hydrocolloids, 81*, 409–418. https://doi.org/10.1016/j.foodhyd.2018.03.002

Kaur, B., Ariffin, F., Bhat, R., & Karim, A. A. (2012). Progress in starch modification in the last decade. *Food Hydrocolloids, 26*(2), 398–404. https://doi.org/10.1016/j.foodhyd.2011.02.016

Kim, Y., Jung, C., & Darie-Nita, R. (2022). Reaction mechanisms applied to starch modification for biodegradable plastics: Etherification and esterification. *International Journal of Polymer Science, 2022*, 1–10. https://doi.org/10.1155/2022/2941406

Kim, Y., Yoo, S. H., Park, K. H., Shim, J. H., & Lee, S. (2012). Functional characterization of native starches through thermal and rheological analysis. *Journal of the Korean Society for Applied Biological Chemistry, 55*(3), 413–416. https://doi.org/10.1007/s13765-012-2002-z

Kumar, N., Singh, R., Upadhyay, A., & Giri, B. S. (2024). Application and functional properties of millet starch: Wet milling extraction process and different modification approaches. *Heliyon, 10*(3).

Lee, H., & Yoo, B. (2023). Particle agglomeration and properties of pregelatinized potato starch powder. *Gels, 9*(2), 93. https://doi.org/10.3390/gels9020093

Lehmann, A., Volkert, B., Fischer, S., Schrader, A., & Nerenz, H. (2008). Starch based thickening agents for personal care and surfactant systems. *Colloids and Surfaces A: Physicochemical and Engineering Aspects, 331*(1–2), 150–154. https://doi.org/10.1016/j.colsurfa.2008.09.055

Li, C., Li, Y., Sun, P., & Yang, C. (2013). Pickering emulsions stabilized by native starch granules. *Colloids and Surfaces A: Physicochemical and Engineering Aspects, 431*, 142–149. https://doi.org/10.1016/j.colsurfa.2013.04.025

Li, Y., Shabani, K. I., Liu, H., Guo, Q., & Liu, X. (2020). Structural, physicochemical and rheological properties of a novel native starch obtained from rhizoma gastrodiae. *Food Structure, 25*, Article 100148. https://doi.org/10.1016/j.foostr.2020.100148

Li, R., True, A. D., Sha, L., & Xiong, Y. L. (2025). Structure–functionality relationship and modification strategies of oat protein: Challenges and opportunities. *Comprehensive Reviews in Food Science and Food Safety, 24*(1), Article e70091.

Liu, P., Yu, L., Wang, X., Li, D., Chen, L., & Li, X. (2010). Glass transition temperature of starches with different amylose/amylopectin ratios. *Journal of Cereal Science, 51*(3), 388–391. https://doi.org/10.1016/j.jcs.2010.02.007

Luckett, C. R., & Wang, Y. J. (2012). Application of enzyme-treated corn starches in breakfast cereal coating. *Journal of Food Science, 77*(8), C901–C906.

Ma, H., Liu, M., Liang, Y., Zheng, X., Sun, L., Dang, W., & Liu, C. (2022). Research progress on properties of pre-gelatinized starch and its application in wheat flour products. *Grain and Oil Science and Technology, 5*(2), 87–97.

Mahto, R., & Das, M. (2014). Effect of gamma irradiation on the physico-mechanical and chemical properties of potato (Solanum tuberosum L.), Cv.'Kufri sindhuri', in non-refrigerated storage conditions. *Postharvest Biology and Technology, 92*, 37–45.

Manigandan, V., Karthik, R., Ramachandran, S., & Rajagopal, S. (2018). Chitosan applications in food industry. In *Biopolymers for food design* (pp. 469–491). Academic Press.

Marc, J., C van der Maarel, E., van der Veen, B., Uitdehaag, J.C.M., Leemhuis, H., & Dijkhuizen, L. (2002). Properties and applications of starch-converting enzymes of the α-amylase family. *Journal of Biotechnology*, *94*(2), 137-155. https://doi.org/10.1016/s0168-1656(01)00407-2.

Miura, S., Crofts, N., Saito, Y., Hosaka, Y., Oitome, N. F., Watanabe, T., Kumamaru, T., & Fujita, N. (2018). Starch synthase IIa-deficient mutant rice line produces endosperm starch with lower gelatinization temperature than japonica rice cultivars. *Frontiers in Plant Science, 9*. https://doi.org/10.3389/fpls.2018.00645

Mojo-Quisani, A., Licona-Pacco, K., Choque-Quispe, D., Calla-Florez, M., Ligarda-Samanez, C. A., Mamani-Condori, R., Florez-Huaracha, K., & Huamaní-Melendez, V. J. (2024). Physicochemical properties of starch of four varieties of native potatoes. *Heliyon, 10*(16). https://doi.org/10.1016/j.heliyon.2024.e35809

Monika, K. M., Dewi, P. I. C., & Pasha, N. L. S. P. (2023). The evaluation of loose powder formulation of manihot starch and maydis starch with natural Orange sweet potato. *Ad-Dawaa'Journal of Pharmaceutical Sciences, 31*, 173–181.

Mu, M., Karthik, P., Chen, J., Holmes, M., & Ettelaie, R. (2021). Effect of amylose and amylopectin content on the colloidal behaviour of emulsions stabilised by OSA-Modified starch. *Food Hydrocolloids, 111*, 106363.

Muljana, H., Picchioni, F., Heeres, H. J., & Janssen, L. P. B. M. (2010). Process-product studies on starch acetylation reactions in pressurised carbon dioxide. *Starch - Stärke, 62*(11), 566–576. https://doi.org/10.1002/star.201000021

Noordergraaf, I. W., Witono, J. R., & Heeres, H. J. (2024). Grafting starch with acrylic acid and Fenton's initiator: The selectivity challenge. *Polymers, 16*(2), 255.

Oladebeye, A. O., Oladebeye, A. A., & Arawande, J. O. (2023). Physicochemical properties of wild yam (Dioscorea villosa) starch. *International Journal of Food Science, 2023*(1), Article 8868218.

Othman, S. H., Majid, N. A., Tawakkal, I. S. M. A., Basha, R. K., Nordin, N., & Shapi'i, R. A. (2019). Tapioca starch films reinforced with microcrystalline cellulose for potential food packaging application. *Food Science and Technology, 39*(3), 605–612. https://doi.org/10.1590/fst.36017

Pandey, A., Nigam, P., Soccol, C. R., Soccol, V. T., Singh, D., & Mohan, R. (2000). Advances in microbial amylases. *Biotechnology and Applied Biochemistry, 31*(2), 135–152.

Peng, C., Wang, Y., Liu, F., Ren, Y., Zhou, K., Jia, L., Zheng, M., Zhao, S., Zhang, L., Wang, C., Jiang, L., Zhang, X., Guo, X., Bao, Y., & Wan, J. (2014). FLOURY ENDOSPERM 6 encodes a CBM 48 domain-containing protein involved in compound granule formation and starch synthesis in rice endosperm. *The Plant Journal, 77*(6), 917–930. https://doi.org/10.1111/tpj.12444

Prongjit, D., Lekakarn, H., Bunterngsook, B., Aiewviriyasakul, K., Sritusnee, W., Arunrattanamook, N., & Champreda, V. (2022). In-depth characterization of debranching type I pullulanase from Priestia koreensis HL12 as potential biocatalyst for starch saccharification and modification. *Catalysts, 12*(9), 1014. https://doi.org/10.3390/catal12091014

Ptak, S., Roczkowska, M., Zarski, A., & Kapusniak, J. (2014). Esterification of starch with fatty acids. *New opportunities and challenges, 93*, 472–479.

Ptak, S., Zarski, A., & Kapusniak, J. (2020). The importance of ionic liquids in the modification of starch and processing of starch-based materials. *Materials, 13*(20), 4479.

Punia Bangar, S., Sunooj, K. V., Navaf, M., Phimolsiripol, Y., & Whiteside, W. S. (2024). Recent advancements in cross-linked starches for food applications-a review. *International Journal of Food Properties, 27*(1), 411–430.

Rajagopal, S., Murugan, R., Aavula, T., Aji, A., & Sakthivel, V. (2022). Extraction of chitin and chitosan from marine waste and its applications on food and agriculture products. *Asian Journal of Biological Sciences, 15*(5-6), 271–279.

Rapaille, A., & Vanhemelrijck, J. (1992). Modified starches. In *Thickening and gelling agents for food* (pp. 171–201). Boston, MA: Springer US.

Ratnaningsih, N., Harmayani, E., & Marsono, Y. (2020). Physicochemical properties, in vitro starch digestibility, and estimated glycemic index of resistant starch from cowpea (Vigna unguiculata) starch by autoclaving-cooling cycles. *International Journal of Biological Macromolecules, 142*, 191–200.

Saito, M., Tanaka, T., Sato, K., Vrinten, P., & Nakamura, T. (2018). A single nucleotide polymorphism in the "fra" gene results in fractured starch granules in barley. *Theoretical and Applied Genetics, 131*(2), 353–364. https://doi.org/10.1007/s00122-017-3006-1

Sangkanu, S., Khanansuk, J., Phoopha, S., Udomuksorn, W., Phupan, T., Puntarat, J., Tungsukruthai, S., & Dej-adisai, S. (2025). Utility assessment of isolated starch and extract from Thai yam (Dioscorea hispida dennst.) for cosmetic via in vitro and in vivo studies. *Life, 15*(2), 151. https://doi.org/10.3390/life15020151

Sethi, S., Choudhary, P., Nath, P., & Chauhan, O. P. (2022). *Starch gelatinization and modification advances in food chemistry: Food components, processing and preservation*. India: Springer Nature. https://doi.org/10.1007/978-981-19-4796-4_3

Shaari, N. A., Sulaiman, R., & Cheok, C. Y. (2017). Rheological properties of native and modified corn starches in the presence of hydrocolloids. *International Food Research Journal, 24*(5), 2082–2089.

Sinaki, N. Y., & Scanlon, M. G. (2016). Flow behavior of native corn and potato starch granules in aqueous suspensions. *Food Biophysics, 11*(4), 345–353.

Sindhu, R., Devi, A., & Khatkar, B. S. (2019). Physicochemical, thermal and structural properties of heat moisture treated common buckwheat starches. *Journal of Food Science and Technology, 56*(5), 2480–2489.

Sofi, B. A., Ahmed Wani, I., Ahmad Masoodi, F., Saba, I., & Muzaffar, S. (2013). Effect of gamma irradiation on physicochemical properties of broad bean (Vicia faba L.) starch. *LWT - Food Science and Technology, 54*(1), 63–72. https://doi.org/10.1016/j.lwt.2013.05.021

Srichuwong, S., Isono, N., Jiang, H., Mishima, T., & Hisamatsu, M. (2012). Freeze–thaw stability of starches from different botanical sources: Correlation with structural features. *Carbohydrate Polymers, 87*(2), 1275–1279. https://doi.org/10.1016/j.carbpol.2011.09.004

Subroto, E., Mahani, Indiarto, R., Yarlina, V. P., & Izzati, A. N. (2022). A mini review of physicochemical properties of starch and flour by using hydrothermal treatment. *Polymers, 14*(24). https://doi.org/10.3390/polym14245447

Sun, C., Wei, Z., Xue, C., & Yang, L. (2023). Development, application and future trends of starch-based delivery systems for nutraceuticals: A review. *Carbohydrate Polymers, 308*, Article 120675. https://doi.org/10.1016/j.carbpol.2023.120675

Szepes, A., Makai, Z., Blümer, C., Mäder, K., Kása, P., & Szabó-Révész, P. (2008). Characterization and drug delivery behaviour of starch-based hydrogels prepared via isostatic ultrahigh pressure. *Carbohydrate Polymers, 72*(4), 571–578. https://doi.org/10.1016/j.carbpol.2007.09.028

Tan, H. Z., Li, Z. G., & Tan, B. (2009). Starch noodles: History, classification, materials, processing, structure, nutrition, quality evaluating and improving. *Food Research International, 42*(5–6), 551–576.

Tao, H., Yan, J., Zhao, J., Tian, Y., Jin, Z., Xu, X., & Wu, K. (2015). Effect of multiple freezing/thawing cycles on the structural and functional properties of waxy rice starch. *PLoS One, 10*(5), Article e0127138. https://doi.org/10.1371/journal.pone.0127138

Thanyapanich, N., Jimtaisong, A., & Rawdkuen, S. (2021). Functional properties of banana starch (musa spp.) and its utilization in cosmetics. *Molecules, 26*(12), 3637.

Thirumalai, A., Harini, K., Pallavi, P., Gowtham, P., Girigoswami, K., & Girigoswami, A. (2023). Nanotechnology driven improvement of smart food packaging. *Materials Research Innovations, 27*(4), 223–232.

Torres, M. D., Fradinho, P., Raymundo, A., Sousa, I., Falqué, E., & Domínguez, H. (2021). The key role of thermal waters in the development of innovative gelled starch-based matrices. *Food Hydrocolloids, 117*, Article 106697. https://doi.org/10.1016/j.foodhyd.2021.106697

Toyosawa, Y., Kawagoe, Y., Matsushima, R., Crofts, N., Ogawa, M., Fukuda, M., Kumamaru, T., Okazaki, Y., Kusano, M., Saito, K., Toyooka, K., Sato, M., Ai, Y., Jane, J.-L., Nakamura, Y., & Fujita, N. (2016). Deficiency of starch synthase IIIa and IVb alters starch granule morphology from polyhedral to spherical in rice endosperm. *Plant Physiology, 170*(3), 1255–1270. https://doi.org/10.1104/pp.15.01232

Van Der Maarel, Van der Veen, B., Uitdehaag, J. C., Leemhuis, H., & Dijkhuizen, L. (2002). Properties and applications of starch-converting enzymes of the α-amylase family. *Journal of Biotechnology, 94*(2), 137–155.

Vanier, N. L., El Halal, S. L. M., Dias, A. R. G., & da Rosa Zavareze, E. (2017). Molecular structure, functionality and applications of oxidized starches: A review. *Food Chemistry, 221*, 1546–1559. https://doi.org/10.1016/j.foodchem.2016.10.138

Velazquez, G., Mendez-Montealvo, G., Flores-Silva, P. C., & Soler, A. (2025). Multi-scale structures, functional properties, and applications of starch modified by dry heat treatment. *Biopolymers, 116*(2). https://doi.org/10.1002/bip.70000

Wang, X., Huang, L., Zhang, C., Deng, Y., Xie, P., Liu, L., & Cheng, J. (2020). Research advances in chemical modifications of starch for hydrophobicity and its applications: A review. *Carbohydrate Polymers, 240*, Article 116292. https://doi.org/10.1016/j.carbpol.2020.116292

Wang, R., Li, M., Brennan, M. A., Dhital, S., Kulasiri, D., Brennan, C. S., & Guo, B. (2023). Complexation of starch and phenolic compounds during food processing and impacts on the release of phenolic compounds. *Comprehensive Reviews in Food Science and Food Safety, 22*(4), 3185–3211. https://doi.org/10.1111/1541-4337.13180

Wang, Q., Li, Y., Sun, D. W., & Zhu, Z. (2018). Enhancing food processing by pulsed and high voltage electric fields: Principles and applications. *Critical Reviews in Food Science and Nutrition, 58*(13), 2285–2298. https://doi.org/10.1080/10408398.2018.1434609

Wang, S., Li, C., Yu, J., Copeland, L., & Wang, S. (2014). Phase transition and swelling behaviour of different starch granules over a wide range of water content. *LWT-Food Science and Technology, 59*(2), 597–604.

Wang, X., Liu, S., & Ai, Y. (2022). Gelation mechanisms of granular and non-granular starches with variations in molecular structures. *Food Hydrocolloids, 129*, Article 107658.

Wani, A. A., Singh, P., Shah, M. A., Schweiggert-Weisz, U., Gul, K., & Wani, I. A. (2012). Rice starch diversity: Effects on structural, morphological, thermal, and physicochemical properties-a review. *Comprehensive Reviews in Food Science and Food Safety, 11*(5), 417–436. https://doi.org/10.1111/j.1541-4337.2012.00193.x

Werlang, S., Bonfante, C., Oro, T., Biduski, B., Bertolin, T. E., & Gutkoski, L. C. (2021). Native and annealed oat starches as a fat replacer in mayonnaise. *Journal of Food Processing and Preservation, 45*(3). https://doi.org/10.1111/jfpp.15211

Xia, Y., Gao, W., Wang, H., Jiang, Q., Li, X., Huang, L., & Xiao, P. (2013). Characterization of tradition chinese medicine (TCM) starch for potential cosmetics industry application. *Starch - Stärke, 65*(5–6), 367–373. https://doi.org/10.1002/star.201200153

Xia, W., Zhang, K., Su, L., & Wu, J. (2021). Microbial starch debranching enzymes: Developments and applications. *Biotechnology Advances, 50*, Article 107786. https://doi.org/10.1016/j.biotechadv.2021.107786

Xie, S., Chen, H., Jiang, X., Zhou, B., Guo, Z., Zeng, H., & Zhang, Y. (2023). Structural and physicochemical properties of a Chinese yam starch–tea polyphenol complex prepared using autoclave-assisted pullulanase treatment. *Foods, 12*(20), 3763. https://doi.org/10.3390/foods12203763

Xu, Y., Sismour, E. N., Grizzard, C., Thomas, M., Pestov, D., Huba, Z., Wang, T., & Bhardwaj, H. L. (2014). United States morphological, structural, and thermal properties of starch nanocrystals affected by different botanic origins. *Cereal Chemistry, 91*(4), 383–388. https://doi.org/10.1094/CCHEM-10-13-0222-R

Xu, T., Yang, J., Hua, S., Hong, Y., Gu, Z., Cheng, L., Li, Z., & Li, C. (2020). Characteristics of starch-based pickering emulsions from the interface perspective. *Trends in Food Science & Technology, 105*, 334–346. https://doi.org/10.1016/j.tifs.2020.09.026

Yazid, N. S. M., Abdullah, N., Muhammad, N., & Matias-Peralta, H. M. (2018). Application of starch and starch-based products in food industry. *Journal of Science and Technology, 10*(2).

Yousif, E. I., Gadallah, M. E., & Sorour, A. M. (2012). Physico-chemical and rheological properties of modified corn starches and its effect on noodle quality. *Annals of Agricultural Science, 57*(1), 19–27.

Yudhistira, B., Husnayain, N., Punthi, F., Gavahian, M., Chang, C. K., & Hsieh, C. W. (2024). Progress in the application of emerging technology for the improvement of starch-based active packaging properties: A review. *ACS Food Science & Technology, 4*(9), 1997–2012. https://doi.org/10.1021/acsfoodscitech.4c00260

Zarski, A., Bajer, K., & Kapuśniak, J. (2021). Review of the Most important methods of improving the processing properties of starch toward non-food applications. *Polymers, 13*(5), 832.

Zarski, A., Kapusniak, K., Ptak, S., Rudlicka, M., Coseri, S., & Kapusniak, J. (2024). Functionalization methods of starch and its derivatives: From old limitations to new possibilities. *Polymers, 16*(5), 597. https://doi.org/10.3390/polym16050597

Zavareze, E. D. R., & Dias, A. R. G. (2011). Impact of heat-moisture treatment and annealing in starches: A review. *Carbohydrate Polymers, 83*(2), 317–328. https://doi.org/10.1016/j.carbpol.2010.08.064

Zdybel, E., Zięba, T., Tomaszewska-Ciosk, E., & Rymowicz, W. (2020). Effect of the esterification of starch with a mixture of carboxylic acids from yarrowia lipolitica fermentation broth on its selected properties. *Polymers, 12*(6), 1383. https://doi.org/10.3390/polym12061383

Zia-ud-Din, M., Xiong, H., & Fei, P. (2017). Physical and chemical modification of starches: A review. *Critical Reviews in Food Science and Nutrition, 57*(12), 2691–2705.

Chapter 3

Sensory profiling and rheological characterization of starch-based food ingredients

Néstor Ponce-García[1], Ivan Luzardo-Ocampo[2, 3], Anayansi Escalante-Aburto[2, 3] and Rosa Leonor González-Díaz[3]

[1]*Facultad de Ciencias Agrícolas, UAEMéx, Toluca, Mexico;* [2]*Tecnologico de Monterrey, Institute for Obesity Research, Monterrey, Mexico;* [3]*Tecnologico de Monterrey, School of Engineering and Sciences, Monterrey, Mexico*

1. Introduction

As versatile and widely used carbohydrate polymers, starches play a crucial role in the food industry, particularly in developing texture, mouthfeel, and stability in starch-based foods. Sensory profiling methodologies offer insights into consumer perceptions of these attributes, enabling a deeper understanding of how native and modified starches contribute to overall product quality. Rheological characterization of starches, encompassing both native and modified varieties, serves as a critical tool for evaluating their structural and functional properties under different processing conditions. The interplay between rheological parameters and sensory attributes highlights the importance of linking measurable physical properties to consumer preferences.

Rheology also profoundly influences the processing efficiency of starches, shaping factors such as flow behavior, gelatinization, and viscosity, which directly impact production scalability and cost-effectiveness. Moreover, innovations in starch-based ingredients have opened new pathways for enhancing texture and product stability, enabling food scientists to address emerging trends and demands for clean-label and functional food products. Strategies for improving texture and mouthfeel through tailored use of native and modified starches emphasize the potential for customizing formulations to achieve superior sensory experiences.

Integrating sensory and rheological analysis is a decisive driving innovation, allowing for the design of starch-based systems that meet both technical and consumer demands. Future directions in this field will likely focus on advancing characterization techniques, exploring novel starch sources, and developing eco-friendly, high-performance starch derivatives that align with sustainability goals while ensuring superior sensory and functional qualities.

2. Sensory profiling methodologies in starch-based foods

Sensory profiling is an essential component in the development and evaluation of starch-based food products. Several methodologies, ranging from traditional approaches to modern rapid techniques, are employed to assess their sensory characteristics. According to (Mathobo et al., 2021), a food product achieving an overall acceptability score above five indicates good quality. This evaluation process is essential for identifying the attributes that define the overall quality of food products, providing valuable insights for refining formulations to align with consumer preferences.

There are several methodologies available for conducting sensory analysis of foods, including traditional approaches such as conventional descriptive analysis (DA) and more recent rapid techniques, such as check-all-that-apply (CATA), rate-all-that-apply (RATA), flash Profile (FP), and free listing (FL) (Kim et al., 2023; Miraballes et al., 2018) explore the application of the rapid descriptive sensory method, specifically the Pivot Profile (PP) and its variant, incorporating CATA questions, in the context of food product development. This study contributes to the understanding of sensory profiling

Native and Modified Starches for Food Applications. https://doi.org/10.1016/B978-0-443-27431-2.00002-X

methodologies for starch-based foods by illustrating how these techniques are used to characterize the sensory properties of a whey-based beverage formulated with modified cassava starch.

In specific situations, the combination of two or more techniques can generate more comprehensive and precise data and information. To address the need to quantitatively complement the data obtained through temporal dominance of sensations (TDS) analysis and reduce the analysis time required to achieve a more complete sensory profile (Crepalde et al., 2021) proposed the integration of an optimized descriptive profile (ODP) to also analyze the samples using TDS, intending to integrate both sensory methodologies. This ODP-TDS integration proved effective, as it optimized the time required to obtain a detailed, quantitative profile of the products.

Likewise (Sharma & Duizer, 2019), contribute significantly to the field of sensory profiling in starch-based food systems by examining the dynamic textural properties of pureed foods formulated with hydrocolloids such as starch and xanthan gum. The study underscores the importance of employing TDS and Temporal Check-All-That-Apply (TCATA) to more effectively capture the evolving perception of texture during consumption. By comparing various formulations and distinguishing specific textural attributes (e.g., grainy vs. smooth), the research offers a deeper understanding of how hydrocolloids modulate the sensory experience of starch-based purees. These insights are particularly relevant for developing specialized food products designed to meet dietary requirements, such as those for individuals with dysphagia. Furthermore, the study reflects a methodological shift from conventional static sensory evaluations to dynamic approaches, providing food scientists with robust data to enhance the sensory quality and consumer acceptability of starch-containing foods.

However, extrinsic factors, including processing methods and ingredient interactions, are key determinants of sensory perception. For example, the interaction between starches and proteins within food systems significantly impacts the texture and sensory attributes of the final product (Lamy et al., 2021) explore the influence of salivary protein composition on the sensory perception of starch-based foods, emphasizing the critical role of salivary α-amylase in the tasting and oral processing of these products. Their findings directly connect to sensory profiling methodologies by illustrating how individual differences in enzymatic activity and protein levels in saliva can modulate taste sensitivity and overall sensory evaluation outcomes. These insights underscore the importance of accounting for interindividual variability in salivary composition when designing and interpreting sensory assessments of starch-based food products.

Gaining insights into these interactions is crucial for optimizing formulations to deliver favorable sensory characteristics (Jia et al., 2022; Zhou, Keast, & Roura, 2023) discussed identifying starchy flavor attributes and characterizing distinctive sensory profiles in various starches, specifically corn, potato, rice, and waxy rice. This study presents sensory profiling methodologies that are essential for characterizing the sensory attributes of starch-based foods, offering evidence-driven perspectives on flavor development and consumer perception within this context.

(Fahmy et al., 2023) examine the complex interplay between texture, deformation behavior, and sensory profiling in starch-based foods, particularly in 3D printing technology. The study highlights the critical role of texture modulation in starch-based closed-cell foams, emphasizing how variations in design parameters, such as bubble size and distribution, can profoundly impact both the mechanical properties and the sensory qualities of these food products. Those authors conclude that a thorough understanding of stress-strain behavior and deformation under different design conditions is essential for tailoring the sensory experience to consumer preferences. Likewise (Escalante-Aburto et al., 2021) described that some native and modified starches (from waxy maize, potato, among others) could be successfully used in 3D food printing for producing starch-based bioinks. The addition of starch to fruits and vegetable purees improves the texture of foods according to personalized nutrition, especially for patients with dysphagia.

(Ari Akin et al., 2019) published an interesting study that applied sensory profiling methodologies to evaluate the quality of gluten-free sorghum bread formulated with various starches and hydrocolloids. The researchers developed an extensive lexicon consisting of 28 sensory attributes related to appearance, aroma, flavor, and texture/mouthfeel, showcasing a thorough approach to sensory analysis. Their results indicate that specific combinations of starch and hydrocolloids can notably affect the sensory characteristics of the bread, offering valuable insights for optimizing starch-based food formulations. This study serves as an exemplary case of the effective use of sensory profiling methodologies in assessing the sensory qualities of starch-based products.

On the other hand, various studies have explored the impact of both native and modified starches on sensory characteristics, providing insights into texture, flavor, and overall acceptability (Table 3.1).

As described in Table 3.1, modified and native starches significantly enhance sensory attributes across diverse food applications. Modified starches consistently improve texture-related properties such as consistency, thickness, mouthfeel, and crispiness, as observed in white sauces, baked goods, and meat products. These modifications can often be achieved without compromising sensory acceptability, as demonstrated in low-fat meat products and baked goods. Additionally, the studies emphasize the importance of understanding starch retrogradation and storage effects, as these factors significantly

TABLE 3.1 Studies on sensory profiling methodologies in native and modified starch-based foods.

Food matrix	Key findings	References
White sauces	Modified corn and pearl millet starches (hydroxypropylation and succinylation) improved sensory attributes such as consistency, thickness, and acceptability.	Shaikh et al. (2020)
	Chemical modifications of starches (rice, potato, waxy corn, corn, and modified waxy corn) enhanced mouthfeel and thickness, altering sensory properties.	Arocas et al. (2010)
Low-fat chicken mortadella	Native (cassava and corn) and modified starches (waxy maize) maintained consumer preference without affecting sensory acceptability.	Prestes et al. (2015)
Cookies	High-amylose maize starch (HAMS) replaced up to 20% of flour without altering the sensory profile.	Giuberti et al. (2016)
Gluten-free sorghum bread	The descriptive sensory analysis highlighted the importance of starches and hydrocolloids (potato starch plus xanthan gum, rice starch plus xanthan gum, and tapioca starch plus hydroxypropyl methylcellulose) in consumer acceptance.	Ari Akin et al. (2019)
Retort pouched buffalo meat nuggets	Corn starch increased gumminess and chewiness, positively influencing texture scores compared to wheat semolina, wheat flour, and tapioca starch.	Devadason et al. (2010)
Meat-based snacks	Wheat flour starch improved sensory attributes such as crispiness than native tapioca and modified pregelatinized tapioca starch	Nimitkeatkai et al. (2022)
Starchy foods	Starch retrogradation process affects sensory properties and texture changes over time.	Wang et al. (2015)
Rainbow trout fish burgers	Native and modified starches (chayotextle and potato mix) enhanced water-holding capacity and improved pH stability and texture profile throughout storage.	Acosta-Pérez et al. (2019)
Fermented soy beverage	Enhancing the solubility and heat tolerance of modified cassava starches has significant implications for improving texture and stability during food processing and storage.	Rodriguez-Ruiz, Rodriguez-Sandoval and Hernández (2025)

influence sensory quality and consumer preferences over time. Nevertheless, nutritional profile (caloric and carbohydrate content) is also an important issue to be addressed in further studies.

3. Rheological characterization of native and modified starches

The rheological properties of native and modified starches are essential for understanding their behavior during processing and consumption. These properties directly influence critical attributes such as texture, viscosity, and stability, underscoring the importance of rheological characterization in the design and optimization of starch-based products.

Native starches exhibit distinct rheological behaviors compared to their modified counterparts, primarily due to structural and compositional changes introduced by chemical, physical, or enzymatic treatments (Singh & Kaur, 2017) describe native starches as demonstrating non-Newtonian behavior. Still, they often suffer from notable limitations, including low resistance to shear stress and a pronounced tendency toward retrogradation, which can constrain their industrial applications (Sulaimon et al., 2020).

In contrast, modified (chemical, physical, or enzymatic) starches display superior rheological properties, such as increased viscosity and enhanced stability against retrogradation (Hebeish et al., 2013). Techniques like esterification and crosslinking significantly enhance these properties, as highlighted by (Rożnowski et al., 2021). Furthermore, modified starches are particularly effective in reducing the hardness of starch-based foods while improving elasticity. This enhancement in texture is primarily attributed to a reduced degree of retrogradation, facilitated by incorporating bulky substituent groups, including esterified acetyl and etherified hydroxypropyl groups (Arlai et al., 2022).

Additionally, processing conditions such as heat and moisture play an important role in modifying the rheological and textural properties of starches. As (Molavi & Razavi, 2018) emphasize, these factors significantly impact the final characteristics of both native and modified starches, determining their suitability for specific industrial applications. Furthermore, physical modifications, such as dry heating, can also alter the rheological properties of starches, leading to variations in breakdown viscosity and stability. For instance, dry heat-treated starches have been observed to demonstrate lower breakdown viscosity compared to native starches, due to enhanced granule cohesion resulting from the modification process (Rao et al., 2022).

In summary, starch modification through various physical and chemical means significantly alters its rheological properties, often resulting in improved functionalities essential for food processing and other industrial applications. The nature of these modifications and their impact on starches' behavior underscore the importance of rheological characterization in optimizing starch-based products.

3.1 Influence of rheological characterization

Rheological evaluation offers valuable insights into the functional performance of starches. Analyzing the flow behavior of materials through rheological methods allows for optimizing processing conditions, leading to enhanced product quality and increased production efficiency. Fig. 3.1 highlights key focus areas for rheological analysis in products made from native and modified starches. The rheological characterization of these starches provides critical data for optimizing their application in the food and industrial sectors. By understanding their functional properties, manufacturers can develop innovative, high-quality starch-based products that meet diverse market demands.

3.1.1 Processing efficiency

The influence of rheological characterization on the processing efficiency of starches is integral to their application across various industries, particularly in food processing. The rheological properties of starches, including their viscosity and texture profile, play a crucial role in determining their performance during processing and in final product quality. Rheological tests provide insights into how starch behaves under shear and during thermal treatments, which is fundamental for optimizing processing conditions. One notable effect of rheological analysis on processing efficiency is the pasting behaviors of native versus modified starches. Hydrothermal treatments, such as heat, moisture treatment, or cross-linking, can significantly alter the rheological properties of starch, impacting its function as a thickener or stabilizer in food products.

Understanding flow behavior under various processing conditions enables significant operational efficiency and product consistency improvements. Insights gained from rheological studies empower engineers to design equipment tailored to the specific flow characteristics of the materials, enhancing processing performance and minimizing waste (Stefanes de Santana Varela, 2017). For example, according to (Navaf et al., 2022), modified starches subjected to dual

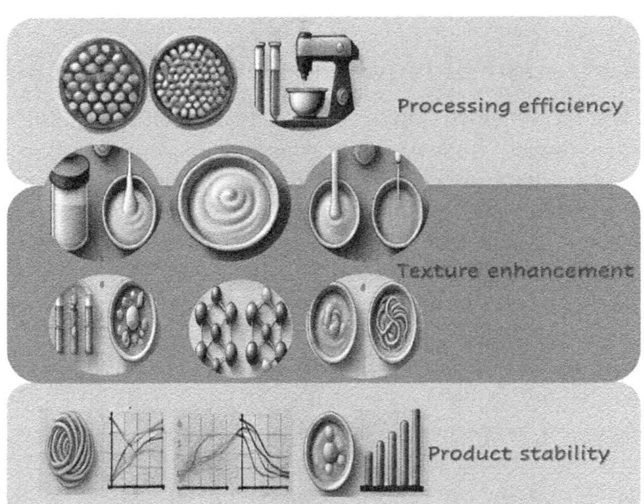

FIGURE 3.1 Areas of rheological interest in products based on native and modified starches. *Made with ChatGPT.*

modifications exhibit higher retrogradation tendencies and enhanced structural complexity, influencing texture and stability in processed foods. The increased retrogradation facilitates a more stable gel formation, which is particularly advantageous in applications requiring prolonged shelf life and product consistency. In the same context (Xie et al., 2012), highlighted that understanding the rheological properties of starch polymer melts during processing is essential for analyzing melt microstructure and optimizing processing parameters.

On the other hand (Cioroiu et al., 2017), pointed out that the viscosity of starch pastes directly impacts the energy needed for mixing and pumping, influencing overall processing efficiency. Starches with higher viscosity may demand greater energy input, affecting production costs and sustainability. Similarly (Erni et al., 2007), established that the presence of hydrophobically modified starches influences interfacial dynamics in emulsions, enhancing both emulsion stability and product rheology. Furthermore (Brandner et al., 2022), highlighted the importance of rheological properties in processing efficiency by evaluating starch–gluten interactions in dough systems. Their findings demonstrated that such interactions can significantly affect dough elasticity and extensibility.

Moreover, comprehending rheological behavior is essential for fine-tuning production processes, as it allows for precise adjustments to formulations and processing parameters to achieve the desired product attributes. Understanding the rheological behavior of both native and modified starches is essential for optimizing processing performance and final product quality. Rheological analyses provide valuable information that supports more effective formulation strategies and processing parameters, ultimately enhancing product functionality and meeting consumer expectations.

3.1.2 *Texture enhancement and product stability*

The texture and stability of food play a crucial role both at the sensory and instrumental levels, serving as a key parameter for assessing the quality and acceptance of a product, intending to ensure consumer satisfaction, and optimizing industrial production and control processes. Adding native or modified starches to food products is crucial in shaping their final texture and overall quality characteristics.

The food industry uses starch-based products as texturizing and gelling agents. In the case of native starches, the amylose-to-amylopectin ratio, granule structure, and additional factors, such as macromolecular interactions, significantly influence the textural and rheological properties (Ačkar et al., 2010; Wang et al., 2015) pointed out that native starch often fails to meet the functional requirements of applications. To overcome challenges such as undesirable changes in texture, resistance to syneresis, appearance, and stability caused by retrogradation or structural degradation during processing and storage, starches are commonly modified to enhance their performance.

In this sense, modified starches are essential for developing innovative food products, as they significantly impact both processing performance and the quality attributes of the final product. Esterification with octenyl succinic anhydride (OSA) introduces notable changes to starch properties, including enhanced swelling capacity, improved paste viscosity, emulsion stability, pH stability, optimized gel texture, and higher resistance to retrogradation (Ovando-Martinez et al., 2017). Additionally, OSA-modified starches exhibit increased emulsion stability due to better droplet coverage, which improves mouthfeel and reduces segregation in emulsified products (Marefati & Rayner, 2020). Thus, specific modifications of native starches (physical, enzymatic, chemical) can improve food stability.

3.2 Rheological characterization of native and modified starches

Native starches are sourced from corn, wheat, potato, and cassava. Their rheological properties are influenced by factors such as the amylose-to-amylopectin ratio, granule size, and the specific plant origin. In general, studies on the rheological behavior of native starches have focused on evaluating parameters such as viscosity development, pasting behavior, gelation, and flow properties. The specific goals of the study determine the selection of rheological tests. These rheological analyses offer crucial insights into the functional properties of native starches and their potential uses in food systems.

Modifications like acetylation, treatment with octenyl succinic anhydride (OSA), and carboxymethylation can notably change the structural and rheological properties of starches (Won et al., 2017). The alteration of starches, whether by physical or chemical processes, can substantially modify their rheological properties (Molavi & Razavi, 2018) showed that acorn starch modified through hydrothermal treatment demonstrated distinct rheological behaviors compared to its native counterpart, suggesting that such modifications can improve starch's ability to stabilize food components (Páramo-Calderón et al., 2016). investigated the impact of cross-linking on the rheological properties of starch, highlighting the necessity of gaining a more thorough understanding of how structural modifications affect rheological behavior.

Future studies should prioritize combining rheological analysis with advanced modeling approaches to forecast starch behavior under different processing conditions (Punia Bangar et al., 2021) found that the swelling power and solubility of modified pearl millet starch had a notable impact on its rheological properties, particularly influencing its thickening behavior. Equally (Kiprop et al., 2021), observed that modified starches can also exhibit reduced syneresis, thereby stabilizing emulsions and contributing to a more desirable consumer experience by improving mouthfeel and overall texture.

In general, modified starches exhibit improved yield stress, viscosity, and flow behavior compared to native starches and frequently employ power law and Herschel-Bulkley to characterize their flow behavior (Almeida et al., 2021; Zambelli et al., 2018). Table 3.2 presents the standard rheological tests used to characterize native and modified starches.

Instrumental rheological measurements provide valuable insights for predicting and interpreting sensory perceptions, thereby playing a pivotal role in developing and optimizing food products (Franzol et al., 2021).

Recently (Escalante-Aburto et al., 2025), evaluated the rheological properties of nixtamalized flours as printable inks using additive manufacturing (3D printing). The study revealed that hydrated masas exhibit pseudoplastic behavior with predominant elastic properties. Rheological characteristics and printability varied significantly by flour type. Viscosity was linked to retrogradation, and strong correlations were found between crowding/packing factors and particle distribution. The Einstein-Roscoe equation supported these findings and aligned with the viscoelastic Maxwell model; a rheological mathematical approach used to explain viscoelasticity in these materials (Ponce-García et al., 2017).

The selection of native or modified starches depends on the specific needs of the application. Native starches are well-suited for more straightforward applications with less demanding processing conditions, whereas modified starches provide superior performance for more complex formulations and challenging processing environments. Table 3.3 shows some of the most significant differences in the rheological properties between native and modified starches.

4. Association between sensory attributes and rheological parameters

The relationship between sensory attributes (consumer-perceived) and rheological parameters (instrumental measurement of flow and strain) is a vital area of study in food science, as it bridges the gap between objective instrumental measurements and subjective consumer perceptions. Sensory characteristics such as texture, mouthfeel, and consistency are closely tied to the rheological properties of food, which are analyzed using scientific techniques. A thorough understanding of these connections allows manufacturers to create products that align with consumer preferences while meeting quality and processing standards. Fig. 3.2 illustrates some key interactions between sensory attributes and food rheological parameters.

TABLE 3.2 Principal rheological tests applied to characterize native and modified starches.

Application	Key findings	Principal equipment	References
Differential scanning calorimetry	Gelatinization temperature; enthalpy	DSC	Gryszkin et al. (2017)
Pasting and gelatinization properties.	Pasting and peak viscosity; breakdown; setback; texture profile analysis.	RVA (rapid visco analyzer) Universal texture analyzer	Wu et al. (2016).
Dynamic oscillatory	Storage modulus (G′) and loss modulus (G″); tan δ (loss factor).	Rheometer	Kumar et al. (2023)
Steady shear	Viscosity (as a function of shear rate); yield stress; flow behavior using models (herschel-bulkley and power law).	Viscosimeters Rheometer	(Molavi & Razavi, 2018; Zambelli et al., 2018).
Others	Stress-relaxation and creep; viscosity as a function of temperature, pH and cooking time.	Universal texture analyzer Viscosimeters	(Huang & Lai, 2010; Zambelli et al., 2018).

TABLE 3.3 Comparative analysis of native and modified starches.

Property	Native starch	Modified starch
Viscosity	Limited by thermal/shear stress	Tunable for specific applications
Gelatinization stability	Susceptible to breakdown	Highly stable
Retrogradation	Pronounced in high amylose	Minimized through modification

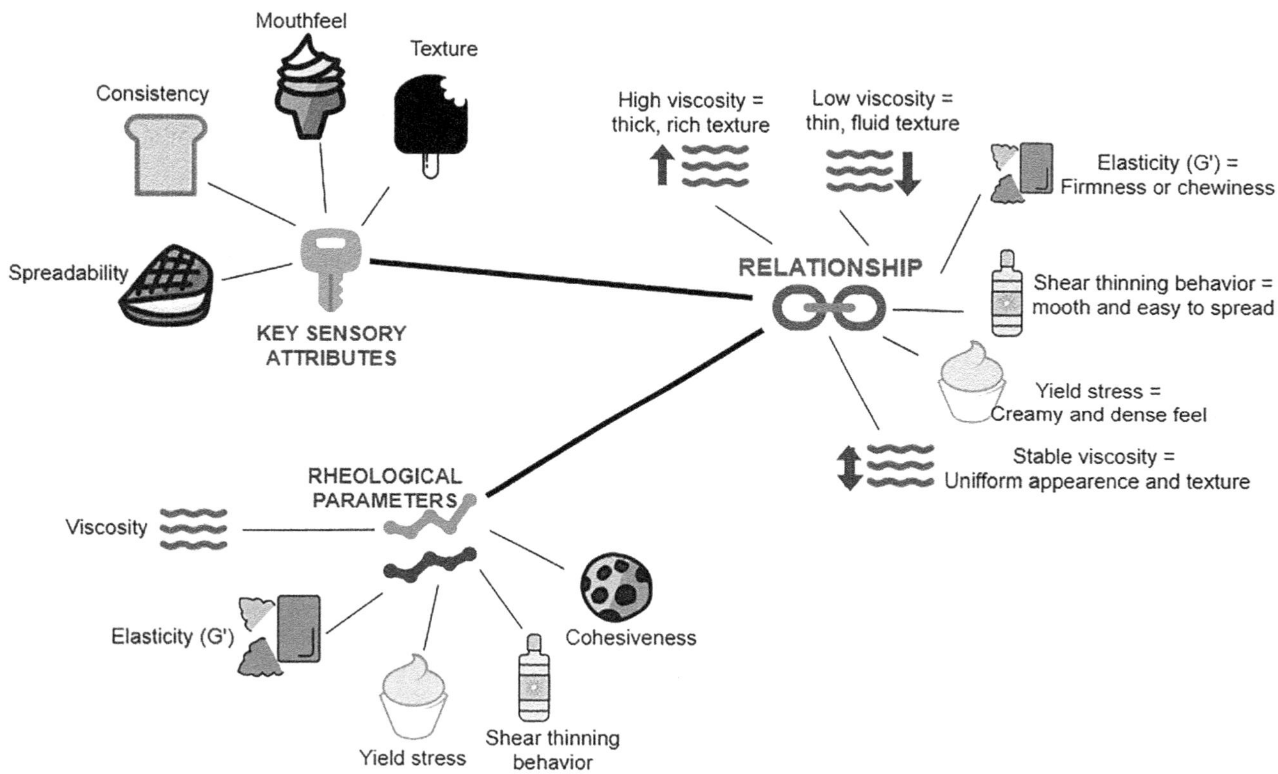

FIGURE 3.2 Key interaction between some sensory attributes and rheological parameters in foods.

(Cheng et al., 2023) thoroughly explained the direct relationship between sensory attributes and rheological parameters in starch-based foods. The authors investigate the impact of various food ingredients and additives on the digestibility of maize starch processed through extrusion and roller-drying, an essential factor in understanding the textural characteristics and overall sensory perception of starch-based products.

The study particularly emphasizes that the physicochemical properties of starch, including its rheological behavior, play a critical role in determining the texture, flavor, and digestibility of the final product. This connection is significant, as it suggests that changes in processing methods, such as extrusion and roller-drying, not only alter rheological parameters (e.g., viscosity and flow behavior) but also influence sensory perceptions like mouthfeel and texture. The findings underscore the importance of understanding these interactions to develop starch-based foods that align with specific sensory expectations, reinforcing the link between rheology and sensory attributes.

(Lukic et al., 2012) highlighted that specific sensory attributes are closely associated with rheological and textural properties, enabling the prediction of sensory characteristics through instrumental analyses. Comparably (Yarovaya et al., 2022), demonstrated strong correlations between rheological parameters, such as yield stress and oscillatory measurements, and sensory attributes, including shape integrity, firmness, glossiness, spreadability, and tackiness (adhesive ability).

The relationship between sensory attributes and rheological parameters in food systems is well-documented (Guo et al., 2020) investigate the interplay between granular swelling and the rheological behavior of various crop starches, presenting empirical evidence on the impact of treatments such as controlled far-infrared heating. The study emphasizes that while granular swelling is a contributing factor, the rheological properties are also strongly dependent on the botanical origin of the starch, underscoring the complexity of these interactions. Those authors demonstrated meaningful insights into how these rheological characteristics correlate with sensory attributes in starch-based foods, directly supporting efforts to explore the relationship between sensory perception and rheological behavior in such products (Kumari & Sit, 2024) contribute to the understanding of the relationship between sensory attributes and rheological properties in starch-based foods by analyzing the physicochemical and rheological characteristics of flours and starches derived from various pulses, including pigeon pea and cowpea. The study identifies critical rheological behaviors, such as water absorption capacity, swelling power, and flow characteristics, that are closely linked to sensory traits like smoothness and viscosity in food products. Notably, the exploration of non-Newtonian fluid behavior and viscosity under shear stress provides valuable insights into how these rheological parameters influence the mouthfeel and texture of food items.

The connection between sensory attributes and rheological parameters in starch-based hydrogels, with particular emphasis on how variations in starch concentration and the incorporation of resistant starch type 2 (RS2) particles influence the textural characteristics of these hydrogels was explored by (Apostolidis et al., 2021), who found that higher concentrations of native starch (NS) resulted in increased hardness, cohesiveness, and gumminess, which are key sensory attributes that significantly influence consumer acceptability (Hirao et al., 2018) offer valuable insights into the relationship between sensory attributes and rheological properties in starch-based foods, with a particular emphasis on sago starch. Their study demonstrates that sago starch possesses advantageous rheological traits (elasticity, softness, and flexibility) that are critical to the texture and mouthfeel of food products. These properties significantly shape consumer sensory perceptions, as reflected in the favorable evaluations of foods in which sago starch is used as a substitute. The findings reinforce the notion that strong rheological performance, including lower adhesiveness, contributes to enhanced physical qualities, which are fundamental to achieving appealing taste and texture in food formulations. Similarly, in starch-pectin-sucrose gels (Lotufo Haddad et al., 2019), reported correlations between sensory properties, such as cohesiveness, consistency, firmness, and spreadability, and their rheological counterparts, including cohesiveness, gumminess, and elasticity.

For instance, in mayonnaise, sensory attributes such as stickiness, firmness, creaminess, thickness, smoothness, and mouth-coating have been shown to interact with rheological parameters like viscosity, yield stress, consistency, shear work, adhesion, firmness, stickiness, and melting point (Agyei Amponsah et al., 2021; Schädle et al., 2022). Additionally in mayonnaise (León, Soto, López, et al., 2019), examined the properties of oxidized cassava starch as a fat replacer in mayonnaise formulations, notably addressing the link between rheological parameters and sensory attributes, detail how the oxidation process enhances the emulsion's stability and alters its rheological properties, which are critical for maintaining the desired texture and consistency. The study shows that the interactions between the starch and protein components can predict the sensory attributes related to texture, thereby establishing a clear connection between the rheological data and the sensory experience of the product.

On the other hand, (Kumar et al., 2023), explored the relationship between sensory attributes and rheological parameters in starch-based foods by examining the role of starch retrogradation. The study delves into how the retrogradation of amylose and amylopectin in gelatinized starch influences both texture and sensory quality. While retrogradation can enhance certain nutritional benefits, it may also lead to unfavorable textural changes that negatively impact sensory characteristics such as mouthfeel and overall product quality during storage. By clarifying the mechanisms through which retrogradation alters the rheological behavior of starch, the authors offer valuable insights for developing formulation strategies aimed at enhancing sensory appeal while minimizing the adverse effects associated with retrogradation.

5. Innovation in starch-based ingredients

Recent advancements in the field of starch-based ingredients have led to significant innovations aimed at improving the functional properties of starch in food systems. As consumer preferences increasingly shift toward healthier, more sustainable, and clean-label products, the food industry has responded by enhancing the properties of starch to meet these evolving demands. Modern research has expanded the applications of starch, moving beyond its traditional role as a thickening agent to include functions such as gelling, emulsification, and fat replacement.

Notable innovations include the development of functional native starches with improved thermal stability, enhanced textural properties, and greater resistance to retrogradation. These starches offer increased versatility in both processed and non-processed food formulations, allowing for the creation of products with improved mouthfeel, consistency, and shelf life. Moreover, there has been a concerted effort to incorporate organic and non-GMO starches into food formulations, responding to the growing consumer demand for clean-label ingredients.

Further research has also focused on using starch-based ingredients for sustainable applications, such as biodegradable packaging materials and plant-based food products. The integration of starches into gluten-free, low-calorie, and plant-based formulations has further broadened their role in modern food science, highlighting their potential in meeting nutritional and sensory demands without compromising on quality.

These innovations underscore the critical role of starch in contemporary food science, offering new avenues for improving both the nutritional and sensory properties of food products while addressing environmental concerns. Continued research into the modification and optimization of starch-based ingredients promises to further enhance their functionality and applicability across a range of food systems.

5.1 Current trends and advancements in starch modification and application

(Awolu & Olofinlae, 2016) analyzed recent advancements in starch modification, focusing on water yam starch and its applications in food. They explored various chemical modification techniques (acetylation, oxidation, and acid thinning) and evaluated their effects on the starch's physicochemical, functional, and pasting properties. These modifications lead to changes in the starch's composition, including reduced fat, protein, and amylose content, while enhancing resistant starch levels and oil absorption capacity. Such improvements are essential for meeting the food industry's demand for healthier, more versatile ingredients. The study highlights the significance of starch modifications in enhancing gelation and retrogradation resistance, especially for products like yogurt. Additionally, it underscores the importance of starch source selection and processing conditions, which affect properties such as solubility and swelling power, crucial for various food formulations.

In general, starch is modified to improve its properties and optimize its performance during processing. A notable trend is the adoption of physical modification techniques, such as annealing and heat-moisture treatment (HMT), which modify the starch without altering its structure or causing granule disruption (Mathobo et al., 2021). Annealing is a type of hydrothermal process that treats starch with an excess of moisture. At the same time, heat-moisture treatment (HMT) involves adding controlled moisture for starch modification (Adebowale et al., 2005). For instance, HMT increases the thermal stability and viscosity of starch, making it particularly suitable for food applications that require optimal texture and stability (Mathobo et al., 2021). HMT significantly increases the hardness of starch gels, which enhances the texture of food products. The increased rigidity of starch granules, resulting from reduced swelling power, contributes to a firmer gel structure, which is especially beneficial in products like sauces and puddings (Molavi & Razavi, 2018). The ability to modify starch properties through innovative techniques, such as acid methanol treatment and hydrothermal treatment, underscores the continuous advancement of starch in food applications. This highlights the growing importance of customizing starch to align with health trends and consumer preferences for low-glycemic index (GI) foods (Ng et al., 2018).

Likewise (Rahman et al., 2020), highlighted that another notable physical modification is high hydrostatic pressure (HHP), used to modify starches, improving viscosity and functional properties while preserving the integrity of the starch granules. In contrast, chemical modifications such as acetylation and carboxymethylation are commonly used to enhance starch properties. For example, acetylated starches show improved swelling and viscosity, making them ideal for use as thickeners and stabilizers in food formulations. Carboxymethyl groups have been found to improve the solubility and moisture retention of starch, thereby helping to extend the shelf life of food products (Patricia et al., 2023; Zdybel et al., 2021).

Additionally, the adoption of innovative techniques, such as ultrasound-assisted modification, is gaining traction as a practical approach to enhance starch properties. This offers control over the modification process, efficient and effective modification, yielding starches with customized functionalities suitable for drug delivery among others (Kumar et al., 2024). Similarly, another emerging trend in producing chemically modified starches is the use of ionic liquids (ILs). These versatile compounds play multiple roles in starch chemistry, including dissolution, gelatinization, modification, and plasticization, while also regard as environmentally friendly alternatives to traditional organic solvents. However, the currently high costs of ILs may limit their widespread commercial applications (Ren et al., 2020).

(Nakkala et al., 2022) described the latest advancements in starch modification, specifically focusing on the chemical modifications of turmeric starch through oxidation, succinylation, and phosphorylation. The study uses reagents such as sodium hypochlorite for oxidation and octenyl succinic anhydride for succinylation, which significantly alter the starch's physicochemical and functional properties. These modifications aim to enhance starch performance in various food applications, particularly where texture and stability are critical. The results show that while the starch's hydration properties mainly remained unchanged, its swelling power varied; oxidized starch exhibited reduced swelling, while phosphorylated starch demonstrated enhanced swelling capabilities.

Thus, advancements in modified starches involve a blend of physical and chemical modification methods that enhance their functional properties for various applications. Research in this field continues to expand the potential uses of modified starches, establishing them as key components in various fields, industrial areas, and materials.

5.2 Strategies for improving texture and mouthfeel through innovative native and modified starch use

Modified starches have gained growing importance in improving the texture and mouthfeel of a wide variety of food products. Native starches often face challenges such as limited texture stability, pronounced retrogradation tendencies, and suboptimal mouthfeel, which can negatively impact the overall sensory quality of food. However, progress in starch modification methods has led to the creation of starches with enhanced functional properties, significantly expanding their use in the food industry.

(Singh et al., 2003) assert that starch is increasingly important in defining the textural properties of food and has a broad range of industrial applications, including its use as a thickener, colloidal stabilizer, gel enhancer, bulking agent, and water retention aid. A key approach to enhancing the texture and mouthfeel of starch-based products is the incorporation of hydrocolloids. Hydrocolloids, such as gellan gum and xanthan gum, are commonly added to starch formulations to improve their rheological properties, enhancing texture and minimizing syneresis. For example, hydrocolloids in modified corn starch gels significantly increased their stability during frozen storage, a critical factor in maintaining texture over time. This stabilization is especially advantageous for products subjected to freeze-thaw cycles, as it helps retain the desired texture and mouthfeel (Lin et al., 2023).

Blending different types of starch, such as high amylose corn starch and waxy cassava starch, has been shown to produce synergistic effects on textural properties (Wang, Wu, et al., 2020) found that blends of these starches exhibited changes in hardness, resilience, and adhesiveness compared to the individual starches, suggesting that the interaction between different starch types can enhance textural characteristics.

Chemically, it has been reported that the incorporation of cross-linked starches has been demonstrated to improve the stability and texture of baked goods by reducing the effects of starch retrogradation, which can cause undesirable textural alterations over time (Wang et al., 2024). Similarly, the acetylation of pigeon pea starch enhances viscosity and mouthfeel in yogurt applications, resulting in a creamier texture and decreased syneresis (Olagunju et al., 2020).

Heat-moisture treatment (HMT), a physical modification technique, alters the internal structure of starch granules, thereby improving stability and providing better texture control, which is essential for various food applications. Research has shown that HMT effectively enhances the textural properties of starch gels, significantly increasing their hardness, cohesiveness, and chewiness, resulting in a more desirable mouthfeel compared to native starch gels (Deng et al., 2022).

Among the innovative strategies explored are enzymatic modifications, which enhance the functional properties of starch by altering its molecular structure. The study by (Marta et al., 2022) highlights strategies for improving texture and mouthfeel through the innovative use of native and modified starches, specifically exploring the properties of banana starch and its modifications. They point out that native starches often lack certain properties required for specific food applications, necessitating various modification methods to enhance their functionality. The study details how physical, chemical, and enzymatic modifications can significantly alter the textural properties of starches, thereby improving their application in food products. Furthermore, enzymatic modifications enable precise molecular-level alterations, creating starches with specific textural characteristics needed for certain applications.

In this regard (Choton et al., 2024), highlighted that enzymatic modification enhances key attributes such as gel strength, viscosity, and emulsification, which are crucial for achieving the desired mouthfeel and texture in food products. Enzymatically modified starch can produce items with greater stability against retrogradation, resulting in superior textural quality and prolonged mouthfeel satisfaction. Moreover, the reference to the versatility of starch in various formulations implies that experimenting with combinations of native and modified starch can lead to enhanced textural properties sought after in diverse food applications. This strategic approach allows food scientists to optimize product formulations, utilizing native and modified starches to achieve superior sensory attributes that align with consumer expectations for quality and mouthfeel.

Recently (Supreethee & Vashishth, 2025), explored strategies to enhance texture and mouthfeel in food products by using native and modified starches. The authors provide a comprehensive review of various modification techniques applied to palm tree starch, including physical, chemical, and enzymatic methods, to improve its functional properties. These improvements are essential for customizing texture and mouthfeel, enabling manufacturers to achieve the desired sensory characteristics in food formulations. Additionally, discusses the use of modified starches in the development of biodegradable packaging films, highlighting an innovative application that enhances food quality and addresses environmental concerns, aligning with consumer demand for sustainable products.

5.3 The role of sensory and rheological analysis in driving innovation

The development of innovative starch-based products requires a comprehensive understanding of both the sensory and rheological properties of native and modified starches. Sensory analysis assesses attributes such as texture, mouthfeel, and overall consumer acceptability. These analyses offer valuable insights into how starch modifications impact the physical properties of food products, affecting their sensory characteristics and overall quality, and are key factors influencing consumer acceptance and satisfaction. Rheological analysis provides quantitative data on flow behavior, deformation, and structural integrity under processing conditions. It enables researchers and food technologists to quantify the viscosity, elasticity, and texture of starch-based systems, key parameters for effective product formulation. Combining these analytical methods is essential for driving product innovation in the food and materials industries.

Rheological analysis provides crucial insights into gelatinization, viscoelastic behavior, and shear-thinning properties that determine the processability and functional performance of starch-based systems. Native starches are generally characterized by pronounced shear-thinning behavior, which can be significantly altered through physical or chemical modifications such as acetylation or pre-gelatinization (Ai & Jane, 2015; Dhull et al., 2021). These modifications induce changes in molecular interactions and granule morphology, as evidenced by microscopic observations (Lawal et al., 2015). Advanced rheological techniques, such as steady shear and oscillatory testing, enable the quantification of key parameters, including consistency index, yield stress, and storage modulus.

These parameters are essential for predicting material behavior under industrial processing conditions (Molavi & Razavi, 2018; Sharma et al., 2023). Moreover, the dynamic rheological properties of starch gels, including storage (G') and loss moduli (G''), provide valuable insights into their viscoelastic behavior, which is vital for predicting product performance during processing and storage (Butt et al., 2020).

Furthermore, rheological characterization offers a direct link between processing conditions and product stability, supporting the formulation of starch-based products with optimized texture, mouthfeel, and overall performance. Sensory analysis complements rheological studies by offering qualitative insights into consumer preferences related to texture and mouthfeel. Different findings highlight the significance of selecting the appropriate starch and modification method to achieve the desired sensory characteristics that improve consumer satisfaction. Additionally, the interaction between starch and other ingredients, such as hydrocolloids, can substantially impact the textural properties of food products.

According to (Ahmt et al., 2004), sensory evaluations including texture profiling combined with advanced chemometric techniques, including radar plots, have demonstrated that alterations in starch structure have a direct impact on textural attributes and overall sensory appeal in food systems. Genetically modified potato starches with extended amylopectin chains and elevated phosphorus content produced more cohesive and gel-like textures, as confirmed through integrated sensory and rheological assessments. Furthermore, the combination of sensory analysis with rheological characterization offers a comprehensive perspective on how starch modifications influence the functional properties of final products. This integrative approach is especially critical in complex biopolymer matrices, where the interactions between starch and other macromolecules (such as proteins or non-starch polysaccharides) must be precisely balanced to achieve desirable sensory qualities and processability (Elgadir et al., 2012).

Incorporating hydrocolloids can enhance the mouthfeel and moisture retention of starch-based products, thereby improving their overall sensory appeal (Borries Medrano et al., 2017). The rheological properties of these mixtures can be fine-tuned to achieve the desired texture, which is critical for products such as sauces, dressings, and baked goods. Advances in starch modification techniques, such as acetylation and cross-linking, have been utilized to control retrogradation, thereby improving the stability and texture of starch-based products (Hunt et al., 2010). Managing retrogradation enhances sensory attributes and extends the shelf life of these products, making them more appealing to consumers. The synergy between sensory and rheological analyses extends beyond traditional food applications. For example, in the case of native waxy starches, comparative rheological assessments are being explored as potential substitutes for chemically modified starches in texture formation, which is increasingly important for clean-label product development (Lewandowicz et al., 2024).

6. Future directions for native and modified starches

The future of native and modified starch research relies on a convergence of innovative modification techniques, nanoscale interventions, bioactive compound integration, and advanced multi-scale characterization methods. These approaches will be critical in overcoming current limitations of native starches, enabling the creation of tailored starch-based materials that are functionally robust, sustainable, and versatile enough to meet the demands of emerging industrial and consumer applications. The potential applications of native and modified starches are vast and continue to expand as research and technology progress. Future directions for these starches will likely focus on several key areas.

6.1 Improved modification techniques

Developing native and modified starches through advanced modification techniques presents significant potential, particularly given the increasing demand for functional food ingredients and sustainable materials. Exploring innovative modification methods is expected to enhance starch properties, broadening their applicability across diverse industrial sectors. Furthermore, as the need for tailored starch functionalities continues to grow, novel approaches to starch modification are likely to emerge.

6.1.1 Microwave and ultrasound processing

Microwave processing has emerged as a promising technique for starch modification, offering rapid heating that can significantly alter functional properties, such as enhancing solubility and gel formation (Li et al., 2023). Similarly, ultrasound has attracted attention for its efficiency and capacity to improve the physical properties of starch without relying on chemical agents (Samarakoon, 2020). Future research could explore using microwave and ultrasound treatments to achieve starches with enhanced and superior functional properties (Zhou, Wang, et al., 2023) provided evidence that ultrasound and microwave modifications lead to improvements in solubility and digestibility of starch, which could pave the way for novel industrial applications that require precise control over starch functionalities.

6.1.2 Extrusion and texture modification

Extrusion is a commonly utilized processing method for modifying starches to enhance their functional properties. This thermomechanical technique promotes starch gelatinization, resulting in improved texture and digestibility. It serves as a highly adaptable processing technology, wherein thermal, mechanical, and chemical forces converge to drive starch gelatinization, molecular fragmentation, and structural reorganization processes that collectively influence the texture and functional performance of starch-based materials (Gumul et al., 2014; Qiu et al., 2024).

Recent advances highlight the potential of *reactive extrusion*, wherein chemical crosslinking agents are introduced in situ during processing, promoting enhanced network formation within the starch matrix. This innovation significantly improves gel strength and textural stability, critical attributes for ensuring consistency, mouthfeel, and structural integrity in food applications. Optimizing reactive extrusion conditions allows for efficient in-process crosslinking while preserving critical granular characteristics, thereby minimizing the need for post-extrusion treatments (Mendes et al., 2023).

(Ren et al., 2022) highlighted that extrusion parameters determine the physicochemical properties and digestibility of starches in pet foods, indicating that optimized extrusion can enhance the nutritional quality of starch-based products. Additionally, incorporating high-amylose starch into gluten-free formulations has been found to improve texture and increase the content of slowly digestible starch, offering significant health benefits (Giuberti & Gallo, 2018).

From a process engineering standpoint, extrusion parameters such as screw configuration, screw speed, and the incorporation of additives (e.g., stearic acid) have been shown to modulate key phenomena, including amylose–lipid complexation and starch granule fragmentation (Panyoo & Emmambux, 2019). Tailoring screw design enables control over shear intensity and residence time, directly influencing the extent of starch gelatinization and retrogradation, factors closely linked to the final texture of the product.

Beyond gelatinization, extrusion also alters the phase behavior and crystallinity of starch, yielding significant changes in its rheological behavior and, consequently, in the mechanical and sensory properties of the final product (Gumul et al., 2014; Qiu et al., 2024). These modifications typically result in reduced shear viscosity and altered setback viscosity, key indicators of textural behavior in semi-solid food matrices. When followed by additional physical treatments such as annealing or heat-moisture conditioning, these processes enable the fine-tuning of starch functionality to meet specific industrial requirements (Qiu et al., 2024).

The integration of extrusion with advanced analytical tools, such as FTIR spectroscopy and texture profile analysis, provides deeper insights into how processing parameters influence molecular interactions and macro-structural development within starch systems. This interdisciplinary approach supports the precise modulation of starch properties, yielding enhanced textural consistency and improved functional performance in applications ranging from food systems to biodegradable bioplastics (Mendes et al., 2023; Panyoo & Emmambux, 2019; Qiu et al., 2024). In conclusion, innovations in extrusion-based starch modification, particularly those involving reactive processing and controlled texture design, offer powerful strategies for engineering starches with customized properties.

6.1.3 Precipitation and graft copolymerization

Precipitation and graft copolymerization are increasingly gaining attention. These methods allow for the development of starch-based materials with improved mechanical, thermal, and functional characteristics, making them suitable for different industrial uses. Precipitation is a commonly employed technique in starch modification to obtain uniform, purified starch derivatives that serve as ideal substrates for further chemical modification. In these processes, precipitation is used for purification (removing unreacted monomers and low-molecular-weight byproducts) and inducing morphological changes that enhance reactivity, particularly for developing starch-based super-absorbents and hydrogels. This process involves introducing a precipitating agent into a starch solution, forming a gel-like structure capable of absorbing substantial amounts of water. Additionally, precipitation can promote the rearrangement of crystalline structures within the starch, yielding materials with improved mechanical properties (León, Soto, González, et al., 2019).

(Nawaz et al., 2020) emphasized how various physical and chemical modifications can alter starch's granule size, crystallinity, and molecular chain arrangement, making it more amenable to subsequent grafting reactions. Improved precipitation procedures afford better control over the starch's surface properties and degree of polymerization, directly influencing the efficiency and outcome of graft copolymerization processes.

On the other hand, graft copolymerization is a promising route to functionalize starch by covalently bonding synthetic polymer chains onto the natural starch backbone. This method allows for the incorporation of desired chemical functionalities with minimal disruption to the inherent biodegradability and renewability of starch, enhancing its suitability for various applications. Precipitation and graft copolymerization in starch modification extend beyond food science to include areas such as biodegradable packaging, agricultural products, and water treatment systems.

(Weerapoprasit & Prachayawarakorn, 2019) demonstrated that grafting polymethacrylamide onto cassava starch improved the mechanical properties and hydrophilicity of thermoplastic starch films. By introducing bulky side groups, grafting disrupts hydrogen bonding between starch chains, preventing recrystallization and increasing the material's flexibility.

Precipitation techniques can be effectively combined with graft copolymerization to develop hybrid materials that capitalize on the advantages of both methods. For instance, integrating grafted polymers into the precipitated starch matrix can improve the material's overall properties, such as biodegradability and hydrophilicity (Weerapoprasit & Prachayawarakorn, 2019; Yahya et al., 2012) showed that mechanical dispersion using precipitation techniques, combined with subsequent graft copolymerization with polybutylacrylate, effectively reduces the hydrophilic nature of sago starch. The resulting graft copolymers exhibit improved compatibility with hydrophobic matrices, as confirmed by scanning electron microscopy (SEM), Fourier transform infrared spectroscopy (FTIR), and X-ray diffraction (XRD) analyses.

Recent research has focused on refining key reaction parameters, such as monomer concentration, type of initiator, reaction duration, and temperature to enhance grafting efficiency and precisely control the molecular architecture of modified starches(Alfuhaid et al., 2023) systematically analyzed the variables influencing grafting efficiency during the incorporation of maleic anhydride-diethylenetriamine (MA-DETA) onto starch. In a complementary effort (Kumar et al., 2017), provided a comprehensive review of graft copolymerization strategies employing free radical initiation mechanisms, which enable high grafting yields while preserving portions of the native starch backbone, ensuring a balance between intrinsic starch characteristics and improved functionalities.

Moreover, a synergistic effect has been observed when graft copolymerization techniques are combined with advanced precipitation methods, facilitating the development of starch-based materials with customized properties (Lele et al., 2018) demonstrated that integrating precipitation purification with controlled grafting processes results in starch derivatives exhibiting superior adhesiveness, water retention capacity, and biodegradability. These integrative strategies offer precise modulation of the starch's physicochemical characteristics, paving the way for the design of tailor-made materials suited to specific industrial demands.

6.2 Health-oriented applications

Health-oriented applications of native and modified starch garner growing interest as modern consumer demands prioritize nutritional quality and the functional and therapeutic attributes of food ingredients. In this context, various modification techniques aim to enhance inherent properties such as resistant starch (RS) content, digestibility, and the capacity to carry bioactive compounds without compromising natural health benefits Fig. 3.3.

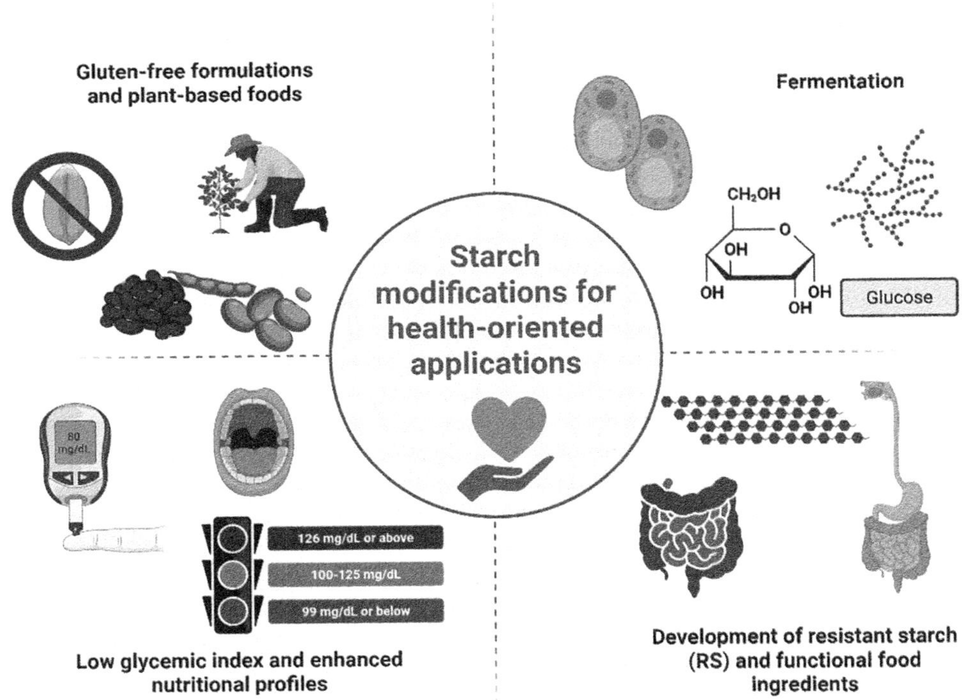

FIGURE 3.3 Starch modification approaches for health-oriented applications, including enhancement of nutritional quality and therapeutic attributes.

6.2.1 Fermentation processes

Fermentation processes have emerged as a promising approach to starch modification. They offer the dual benefits of improving the functional properties of starches and providing a more sustainable and eco-friendlier alternative to conventional chemical methods. Diverse fermentation techniques, especially those utilizing lactic acid bacteria (LAB) and other microbial strains, have been investigated to enhance the physicochemical attributes of starches, resulting in products with improved digestibility, desirable flavors, texture, and nutritional quality. LABs hydrolyze starch into simpler sugars, increasing the availability of fermentable carbohydrates.

(Songré-Ouattara et al., 2010) demonstrated that fermenting a pearl millet-soybean slurry with selected LAB resulted in higher lactic acid production, reflecting effective starch hydrolysis and an enhanced nutritional profile in the resulting gruels. Furthermore, fermenting native starches can increase their resistant starch content, significantly benefiting gut health. This is particularly important for developing functional foods that support healthy gut microbiota (Wronkowska & Soral-Śmietana, 2012) observed that the growth and acidifying activity of *Bifidobacterium* sp. were greater in fermented modified starches compared to native starches, highlighting the potential of fermentation to enhance starch functionality for probiotic applications. This enhanced fermentability leads to improved breakdown of the starch structure into resistant starch (RS), thereby facilitating the production of short-chain fatty acids (SCFAs) during in vitro fermentation. These SCFAs, including butyrate, propionate, and acetate, are crucial for maintaining colonic integrity and modulating metabolic health (Wang et al., 2023).

A key strategy involves modifying traditional fermentation parameters by incorporating specific substrates and controlling environmental conditions to favor the proliferation of beneficial microorganisms. For example, the addition of glucose and yeast extract during the fermentation of sour cassava starch has been shown to significantly increase medium acidity and lower pH, which is associated with enhanced starch modification and improved expansion properties (Jorge A et al., 2018).

On the other hand, enzymatically modified starches, particularly those used as stabilizers in fermented dairy beverages, exhibit modified setback properties and altered amylose content, which contribute to a distinct fermentation profile. The enzymatic treatment before fermentation can modify the molecular structure, yielding a starch that is more suitable as a carrier for probiotics and capable of enhancing the texture and stability of the final fermented product (Lopez Ochoa et al.,

2022). These modifications may promote more rapid and efficient fermentation, resulting in improved nutritional composition while reducing potential gastrointestinal discomfort through the controlled production of SCFAs.

6.2.2 Development of resistant starch (RS) and functional food ingredients

One of the most promising applications of modified starches is the production of resistant starch (RS), which has garnered attention for its health benefits. RS can act as a prebiotic, promoting gut health and improving digestive function. Various modification techniques, including heat-moisture treatment and enzymatic processes, have been demonstrated to effectively increase the resistant starch content in food products (Barretti et al., 2022; Sulistyawati et al., 2024) emphasized that starch modification enhances its potential for health-related applications by transforming it into resistant starch, significantly promoting energy balance and increasing satiety. Modified starches are gaining recognition as functional food ingredients with the potential to deliver health benefits beyond basic nutrition. Improvements have been observed with novel RS types, such as Type-4 RS, which undergo chemical modifications rendering them less digestible in the small intestine and more fermentable in the colon. In vitro studies have demonstrated that these RS variants exhibit a distinct fermentation kinetics profile characterized by gradual SCFA production and a sustained decrease in pH, attributes that are beneficial for gut microbiota modulation (Erickson et al., 2018). Additionally, in vitro fermentation studies using pig fecal inoculum have provided insights into the dynamic production of SCFAs from various RS sources, facilitating the design of fermentation processes tailored to maximize health benefits (Giuberti et al., 2013).

According to (Gałkowska et al., 2023), chemically modified starches can act as carriers for bioactive compounds, improving their stability and bioavailability. This functionality is especially valuable in developing functional foods targeting specific health issues, such as obesity and diabetes. Encapsulating nutrients and bioactive compounds within starch matrices enhances their delivery and effectiveness in supporting overall health.

6.2.3 Low-glycemic index and enhanced nutritional profiles

Modifying starches to develop low glycemic index (GI) foods has become a significant area of interest. Low-GI foods are digested and absorbed more slowly, leading to a gradual increase in blood glucose levels, which is particularly advantageous for individuals with diabetes (Khatun et al., 2019). An emerging and effective approach for improving the nutritional quality of starch-based ingredients involves enhancing resistant starch content through targeted physical processing techniques. Resistant starch is widely acknowledged for its beneficial roles in glycemic control, weight regulation, and colonic health, as it escapes digestion in the small intestine and undergoes fermentation in the colon, producing health-promoting short-chain fatty acids (Farooq & Yu, 2024).

Recent findings have demonstrated that specific physical modifications, such as retrogradation manipulation and controlled thermal processing, can significantly elevate RS concentrations within food matrices. These treatments result in starches with heightened resistance to enzymatic hydrolysis, thereby reducing the glycemic index and enhancing their prebiotic potential (Zailani et al., 2023).

(Taylor et al., 2015) emphasized that processing cereal-based foods, particularly high-amylose starch and targeted thermal treatments, can substantially decrease starch availability and reduce glycemic indices. This is particularly relevant for formulating products that cater to individuals with diabetes or those seeking to manage their weight. In agreement (Dhull et al., 2021), highlight that physical and chemical modifications can influence starch digestibility, resulting in products that provide a more stable and controlled glucose release. For example (Srinang et al., 2023), proposed that the development of rice varieties with low amylose content could offer improved cooking quality and reduced digestibility.

Starch modification can also improve the nutritional profiles of food products, making them more suitable for health-oriented applications. Thus, chemical modifications introducing functional groups can enhance starch solubility and digestibility, promoting better nutrient absorption (Widhyahrini, 2023). Furthermore, starches derived from pulses have been explored for their unique functional properties and health benefits, such as lowering cholesterol levels and improving gut health (Ren et al., 2021). The utilization of underexploited starch sources, such as proso millet starch, offers additional health benefits due to its distinctive nutritional composition. Proso millet is inherently low in glycemic index and high in fiber, and its modified forms have shown promising potential as functional ingredients in diverse food formulations (Kumar et al., 2023). The isolation and subsequent modification of proso millet starch have paved the way for its application as a natural food additive that promotes better glycemic control and enhances overall nutritional quality. In parallel, chemically modified starches are emerging as versatile food additives with tailored functionalities, including encapsulating bioactive ingredients and serving as fat replacers, further contributing to health-oriented applications (Gałkowska et al., 2023). Such advances enrich the nutritional profile of starch-based ingredients and support the prevention of chronic diseases, positioning these modifications as pivotal in the formulation of functional and health-focused food products.

6.2.4 Gluten-free formulations and plant-based foods

Starches are fundamental ingredients in food formulation, and their use in novel food products, such as gluten-free and plant-based foods, is expected to increase. A major area of focus will be the development of starches that can mimic the texture and functionality of gluten or other proteins. Starches are crucial in gluten-free formulations, offering the structure and texture typically provided by gluten in conventional bread and baked goods (Nugraheni et al., 2017) underscored the benefits of gluten-free flour enriched with resistant starch type 3 (RS3) from *Canna edulis*, which has the potential to impact glucose and lipid regulation in health contexts positively. The addition of resistant starch not only improves the nutritional value but also contributes to the desired texture in gluten-free products.

(Kang et al., 2015) highlighted that the shape and size of rice starch granules significantly influence the properties of gluten-free flour and the quality of gluten-free bread. The rheological properties of starches, such as viscosity and gel formation, are essential for achieving the proper consistency and mouthfeel in gluten-free products, which often lack the elasticity typically provided by gluten (Camelo-Méndez et al., 2018) discovered that adding whole blue maize flour to gluten-free pasta formulations enhanced antioxidant capacity and lowered in vitro starch digestibility, emphasizing the potential of functional ingredients to improve the health benefits of gluten-free products.

(Ari Akin & Miller, 2017) demonstrated that the sensory characteristics of gluten-free sorghum bread, combined with different starches and hydrocolloids, can significantly influence consumer preferences. Creating appealing textures and flavors in gluten-free products is crucial for satisfying consumer expectations and fostering broader acceptance of these foods. Incorporating starches into gluten-free and plant-based foods can also enhance their nutritional value. For example, adding resistant starch can improve insulin sensitivity and contribute to better glycemic control, as noted by (Nugraheni et al., 2017), which is especially important for individuals with diabetes or those aiming to manage their blood sugar levels.

Furthermore, using starches from nutrient-rich sources, such as red potatoes, can boost the overall nutritional profile of gluten-free breads (Gumul et al., 2017) found that incorporating freeze-dried red potatoes could increase the protein content and introduce bioactive components, such as polyphenols, into gluten-free formulations. This strategy improves nutritional value and provides functional benefits to the final products. Future research should explore new starch sources and modification techniques to enhance the functionality of gluten-free and plant-based foods. Additionally, studying the interactions between starches and other ingredients will be essential for optimizing formulations and achieving desirable sensory qualities.

Plant-based applications often require tailored starch functionality to offset the absence of gluten, which inherently contributes to dough structure and elasticity. Improved modification techniques, such as octenyl succinic anhydride (OSA) modification, have been shown to enhance the rheological and gas-retaining properties of doughs by reinforcing the interactions among swollen starch granules and accompanying hydrocolloids (Korus et al., 2021). Complementary strategies include incorporating dietary fibers and plant proteins into modified starch matrices. For instance, adding psyllium fiber to maize-based gluten-free bread has resulted in linked starch granules and enhanced dough stability, as confirmed by scanning electron microscopy and rheological assessments (Aprodu & Banu, 2015).

Such approaches improve the physical properties of gluten-free baked goods and increase their nutritional value by elevating the fiber content and potentially reducing the postprandial glycemic response. Moreover, plant-based formulations that utilize native flours from Andean crops or pseudocereals (e.g., buckwheat and sorghum) often benefit from an optimized mixture design that strategically combines rice, maize, and cassava starches to mimic the textural functions of gluten while promoting a more balanced nutrient profile (Horstmann et al., 2017; Vidaurre-Ruiz et al., 2022).

6.3 Innovative applications in non-food industries

The use of native and modified starches in non-food industries is growing rapidly, propelled by their diverse properties and functionalities. Beyond their importance in food formulations, starches also serve key roles in industrial sectors, such as pharmaceuticals, textiles, packaging, and bioplastics. The diverse and innovative applications of native and modified starches in non-food industries underscore their importance as sustainable and multifunctional materials. Ongoing research focused on enhancing the properties of starch and its derivatives is expected to broaden their potential uses across various sectors, solidifying their position as essential alternatives to traditional materials.

6.3.1 Starch in the pharmaceutical industry

The pharmaceutical industry has long acknowledged the versatility of starch, both in its native and modified forms, as a multifunctional excipient. Native starches are widely incorporated into pharmaceutical formulations as fillers, binders, and disintegrants due to their biocompatibility, natural abundance, and inherent biodegradability. However, they often present compressibility, flowability, and stability limitations under various processing conditions (Garba Mohammed, 2017). These challenges have driven extensive research into physical, chemical, and enzymatic modification techniques to enhance starch functionality and expand its utility in drug formulation (Adewale Adetunji, 2020).

Chemically modified starches have demonstrated superior performance as direct compression excipients, yielding tablets with greater hardness and lower friability than those formulated with native starches (Hong et al., 2014). Additionally, physical modifications like pre-gelatinization and retrogradation have enhanced the flow and compaction properties of starches, which are critical for ensuring uniform and efficient tableting (Achor et al., 2015). Optimizing these attributes contributes to improved manufacturing efficiency and the production of dosage forms that consistently meet rigorous quality standards, including stability and content uniformity (Chaksmithanont et al., 2024).

Moreover, native and modified starches are increasingly employed as matrix formers and disintegrants in advanced drug delivery platforms. Their swelling and erosion behaviors can be precisely tuned to modulate the release profiles of active pharmaceutical ingredients (APIs) (Garcia et al., 2020). Modification processes that alter granule morphology and hydration capacity are essential for achieving sustained or targeted drug release, particularly in extended-release systems where a careful balance between enzymatic degradation and structural integrity is required to regulate release kinetics (Adewale Adetunji, 2020; Hong et al., 2014).

(Cisse et al., 2023) emphasized that acetylation-modified starches can increase resistance to retrogradation and syneresis, making them ideal excipients in tablet formulations and improving the release profile of active pharmaceutical ingredients. This growing emphasis on starch-based excipients underscores their promise as cost-effective, renewable, and adaptable materials capable of addressing the complex formulation challenges in contemporary pharmaceutical development (Lawal, 2019).

6.3.2 Starch in bioplastics, packaging, and coating agents

Applying starch in biodegradable plastics represents a promising research area, offering solutions to the increasing concerns about plastic waste. Starch-based bioplastics can be developed using various modification methods, such as graft copolymerization and blending with other biopolymers (Bensaad et al., 2023) explored the potential of modified sweet potato starch for creating biodegradable films, which could provide sustainable alternatives to traditional plastics. Including starch in packaging materials enhances biodegradability and improves the films' mechanical properties and barrier functions, making them suitable for food and non-food applications. Modified starches offer a sustainable alternative for creating packaging materials that are both environmentally friendly and safe for food contact (Amaraweera et al., 2021).

Starches are progressively used as coating agents across various industries, such as textiles and paper, due to their film-forming and adhesive properties, making them ideal for surface coating applications (Cisse et al., 2023) observed that acetylation-modified starches can enhance resistance to retrogradation and syneresis, making them well-suited for coating agents in the paper industry. Modifying starches to improve their performance in coating applications is essential for creating more sustainable and efficient products. The production of starch-based films and coatings supports efforts to reduce plastic waste while preserving food quality and safety, meeting the growing consumer demand for eco-friendly products.

6.3.3 Starch in agriculture

The incorporation of native and modified starches into agricultural systems has garnered growing scholarly attention due to their potential to enhance nutrient efficiency, improve soil moisture retention, and support environmentally sustainable practices. Native starches, valued for their natural abundance, biodegradability, and renewability, offer a promising bio-based alternative to conventional synthetic polymers used in agrochemical formulations and soil conditioners. However, their inherent physicochemical limitations, such as rapid water absorption and insufficient mechanical stability, often constrain their direct applicability. Consequently, various modification strategies have been employed to tailor their properties and extend their utility in targeted agricultural applications (Gamage et al., 2022; Vega et al., 2025).

In agriculture, starches have been used as biodegradable films to enhance soil moisture retention and inhibit weed growth. Starch-based films can improve crop yields while reducing the environmental impact associated with conventional plastic mulches. Studies have demonstrated that biodegradable starch films effectively suppress weed growth and promote soil health, making them a valuable tool for sustainable agricultural practices (Sinhmar et al., 2023).

For example, hydrogel systems composed of modified starch and synthetic polymers have demonstrated considerable efficacy in providing controlled nutrient release, aligning nutrient availability with crop uptake dynamics while minimizing leaching and associated environmental contamination (Gerling et al., 2025; Li et al., 2014). Beyond their role in fertilizer encapsulation, modified starches have also been employed in developing films and hydrogels designed to enhance soil water retention. Notably, the incorporation of lignin into starch-based matrices has been shown to improve water repellency and regulate moisture absorption properties that are essential for designing hydrophilic coatings suited to agricultural applications (Sarwono et al., 2014). Additionally, starch-derived superabsorbent polymers have emerged as sustainable agrochemical carriers, offering a porous structure that facilitates the gradual release of nutrients while conserving soil moisture under fluctuating environmental conditions (León, Soto, González, et al., 2019). These advances not only enhance the functional attributes of starch but also align with broader goals of sustainable water resource management in agriculture (Sarwono et al., 2014).

Moreover, the targeted application of starch-based delivery systems holds promise for addressing persistent challenges related to nitrogen pollution in agroecosystems. Traditional fertilization practices often suffer from poor nutrient use efficiency, resulting in nutrient losses and environmental degradation. Engineered starch polymers capable of encapsulating nitrogenous fertilizers have been developed to synchronize nutrient release with plant uptake, thereby reducing the incidence of nitrogen runoff and contributing to improved crop productivity (Ibrahim et al., 2020; Zou et al., 2015). Enhancements in the rheological behavior and viscosity stability of starch-containing coating formulations further support uniform nutrient distribution, while simultaneously mitigating ecological risks associated with conventional fertilizer use (Ibrahim et al., 2020; Zou et al., 2015).

6.3.4 Starch for textile applications

The application of native and modified starches in the textile industry has significantly developed in recent decades. Native starch, widely valued for its renewability, biodegradability, and low production cost, has traditionally been employed as a sizing agent and thickener in textile processing. Nevertheless, its inherent drawbacks, such as suboptimal adhesion, limited water solubility, and inadequate film-forming capacity, have restricted its effectiveness in high-performance textile applications (Grace & Liew, 2016; He et al., 2023).

In contrast, modified starches produced through chemical, physical, or enzymatic techniques exhibit improved functional attributes, including enhanced viscosity control, superior fiber adhesion, and greater stability under processing conditions. These modifications enable the tailoring of starch rheology to meet specific textile requirements and enhance its compatibility with various finishing agents, thereby optimizing the efficacy of textile treatments (Ding et al., 2014).

In practical textile operations, modified starches have found widespread application in both sizing and desizing processes, as well as in the development of multifunctional finishing agents. For example, the fabrication of zinc coneflower–porous starch composites illustrates the potential of chemical modification to impart textiles with enhanced antimicrobial activity, improved tensile strength, and increased durability during laundering, characteristics essential for extending textile lifespan (Mehrotra et al., 2020). Enzymatic approaches, particularly those utilizing amylases, have been successfully integrated into desizing procedures, facilitating the gentle removal of starch-based sizes while preserving fabric integrity (Kavitha et al., 2020). Furthermore, specialized derivatives such as phosphorylated starch and cyclodextrin-modified starch have been developed to augment fiber adhesion during finishing, regulate moisture retention, and support controlled dye release in textile dyeing applications (El-Sayed and Hassabo, 2021; Ramadan and Sitohy, 2020). These advanced modifications result in starch-based materials with superior physicochemical properties, aligning with the high-performance standards of contemporary textile manufacturing.

6.3.5 Starch in construction and building materials

Native and modified starches have garnered increasing interest in construction and building applications as sustainable, environmentally benign alternatives to conventional binders and composite additives. While native starches are abundant, biodegradable, and derived from renewable resources, their practical application in construction is often constrained by intrinsic limitations such as high-water sensitivity and relatively low mechanical strength. To address these shortcomings, various modification strategies have been employed to enhance starch's functional properties for use in structural materials (Ai & Jane, 2015; Ashogbon & Akintayo, 2014).

A notable example is provided by (Vitola et al., 2024), who developed a modified potato starch binder by incorporating sodium metasilicate and glycerol. Their results indicated that the inclusion of 80% sodium metasilicate and 33%

glycerol (by starch weight) significantly improved the binder's flexibility while preserving its compressive strength. This formulation enhanced the workability of the binder and facilitated the creation of bio-composites with superior mechanical performance, positioning them as viable alternatives to traditional materials in sustainable construction practices.

(Mansour et al., 2020) introduced the concept of "starch sandstones," wherein maize starch, combined with sand and water and subjected to thermal treatment, formed a novel biodegradable composite material. This innovation exploits the natural binding capability of starch while offering a substantially reduced environmental footprint compared to conventional cementitious systems. In the context of lightweight construction (Fahmi et al., 2023), reported the development of a microwave-cured NaCl–starch binder, formulated with food-grade corn starch. Their findings demonstrated the production of hydrophobic hybrid materials with low bulk density, suggesting their suitability for applications where weight reduction and enhanced durability are critical.

Starches are being investigated for their potential applications in construction materials, especially as additives in cement and concrete formulations. The addition of starch can enhance the workability and water retention of cement mixtures, resulting in improved performance. Research suggests that modified starches can function as effective plasticizers, reducing the water-to-cement ratio while maintaining the desired consistency (Wang, Wu, et al., 2020).

Collectively, these studies underscore the pivotal role of both native and modified starches in advancing sustainable building technologies. While native starch provides a renewable and biodegradable foundation, chemical and physical modifications are crucial for tailoring properties such as mechanical strength, water resistance, and long-term durability. The development of starch-based binders and composites represents a promising pathway toward eco-friendly, cost-effective, and high-performance construction materials capable of addressing the environmental challenges posed by traditional building practices.

7. Conclusion

Recent research on sensory profiling methodologies in starch-based foods has significantly advanced our understanding of how starch properties influence sensory perception and overall acceptability. These tools are particularly valuable in linking sensory experiences to the functional properties of starches. Rheological characterization of native and modified starches remains a cornerstone in understanding their functional properties and optimizing their applications in food and non-food industries. The association between sensory attributes and rheological parameters in native and modified starches has gained significant attention in recent research due to its critical role in aligning food product functionality with consumer preferences. Innovation in starch-based ingredients has advanced significantly, driven by recent research focusing on functional modifications and their applications in diverse food systems. Studies have demonstrated the potential of modified starches to enhance texture, stability, and shelf life, thereby meeting the demands for clean-label and health-oriented products. To stay competitive, researchers and industries must prioritize collaborations that integrate advanced characterization techniques, sustainable processing technologies, and consumer-driven sensory profiling. By aligning innovation with environmental responsibility and market demands, native and modified starches will continue to play a transformative role in the future of food science and product development.

AI disclosure

During the preparation of this work the author(s) used ChatGPT in order to elaborate Fig. 1 for representing the interaction about rheology and starches in the food industry. After using this tool/service, the author(s) reviewed and edited the content as needed and take(s) full responsibility for the content of the publication.

References

Ačkar, Đ., Babić, J., Šubarić, D., Kopjar, M., & Miličević, B. (2010). Isolation of starch from two wheat varieties and their modification with epichlorohydrin. *Carbohydrate Polymers, 81*(1), 76–82. https://doi.org/10.1016/j.carbpol.2010.01.058

Achor, M., Oyeniyi, J., Musa, M., & Gwarzo, M. (2015). Physicochemical properties of cassava starch retrograded in alcohol. *Journal of Applied Pharmaceutical Science, 5*(10), 126–131. https://doi.org/10.7324/JAPS.2015.501021

Acosta-Pérez, V. J., Zamudio-Flores, P. B., Ornelas-Paz, J. D. J., Palma-Rodríguez, H. M., Juárez-Muñoz, J., & Vargas-Torres, A. (2019). Effects of native and modified starches on the physicochemical and textural properties of rainbow trout (*Oncorhynchus mykiss*) fish burgers. *CyTA – Journal of Food, 17*(1), 207–213. https://doi.org/10.1080/19476337.2019.1574903

Adebowale, K. O., Afolabi, T. A., & Olu-Owolabi, B. I. (2005). Hydrothermal treatments of finger millet (Eleusine coracana) starch. *Food Hydrocolloids, 19*(6), 974–983. https://doi.org/10.1016/j.foodhyd.2004.12.007

Adewale Adetunji, O. (2020). *Chemically modified starches as excipients in pharmaceutical dosage forms*. USA: IntechOpen. https://doi.org/10.5772/intechopen.88210

Agyei Amponsah, J., Macakova, L., DeKock, H. L., & Emmambux, M. N. (2021). Effect of substituting sunflower oil with starch-based fat replacers on sensory profile, tribology, and rheology of reduced-fat mayonnaise-type emulsions. *Starch - Stärke, 73*(3–4). https://doi.org/10.1002/star.202000092

Ahmt, T., Wischmann, B., Blennow, A., Madsen, F., Bandsholm, O., & Thomsen, J. (2004). Sensory and rheological properties of transgenically and chemically modified starch ingredients as evaluated in a food product model. *Food, 48*(2), 149–155. https://doi.org/10.1002/food.200300407

Ai, Y., & Jane, J.-lin (2015). Gelatinization and rheological properties of starch. *Starch - Stärke, 67*(3–4), 213–224. https://doi.org/10.1002/star.201400201

Alfuhaid, L., Al-Abbad, E., Alshammari, S., Alotaibi, A., Malek, N., & Al-Ghamdi, A. (2023). Preparation and characterization of a renewable starch-g-(MA-DETA) copolymer and its adjustment for dye removal applications. *Polymers, 15*(5). https://doi.org/10.3390/polym15051197

Almeida, R. L. J., dos Santos Pereira, T., Almeida, R. D., Santiago, Â. M., de Lima Marsiglia, W. I. M., Nabeshima, E. H., de Sousa Conrado, L., & de Gusmão, R. P. (2021). Rheological and technological characterization of red rice modified starch and jaboticaba peel powder mixtures. *Nature Research, Brazil Scientific Reports, 11*(1). https://doi.org/10.1038/s41598-021-88627-4

Amaraweera, S. M., Gunathilake, C., Gunawardene, O. H. P., Fernando, N. M. L., Wanninayaka, D. B., Dassanayake, R. S., Rajapaksha, S. M., Manamperi, A., Fernando, C. A. N., Kulatunga, A. K., & Manipura, A. (2021). Development of starch-based materials using current modification techniques and their applications: A review. *Molecules, 26*(22). https://doi.org/10.3390/molecules26226880

Apostolidis, E., Kioupis, D., Kakali, G., Stoforos, N. G., & Mandala, I. (2021). Effect of starch concentration and resistant starch filler addition on the physical properties of starch hydrogels. *Journal of Food Science, 86*(12), 5340–5352. https://doi.org/10.1111/1750-3841.15954

Aprodu, I., & Banu, I. (2015). Influence of dietary fiber, water, and glucose oxidase on rheological and baking properties of maize based gluten-free bread. *Food Science and Biotechnology, 24*(4), 1301–1307. https://doi.org/10.1007/s10068-015-0167-z

Ari Akin, P., & Miller, R. A. (2017). Starch–hydrocolloid interaction in chemically leavened gluten-free sorghum bread. *Cereal Chemistry, 94*(5), 897–902. https://doi.org/10.1094/cchem-05-17-0094-r

Ari Akin, P., Miller, R., Jaffe, T., Koppel, K., & Ehmke, L. (2019). Sensory profile and quality of chemically leavened gluten-free sorghum bread containing different starches and hydrocolloids. *Journal of the Science of Food and Agriculture, 99*(9), 4391–4396. https://doi.org/10.1002/jsfa.9673

Arlai, A., Tananuwong, K., & Yan, Z.-F. (2022). Storage stability of chilled and frozen starch gels as affected by blended starch formulation, sucrose syrup, and coconut milk. *International Journal of Food Science, 2022*, 1–13. https://doi.org/10.1155/2022/9454229

Arocas, A., Sanz, T., Salvador, A., Varela, P., & Fiszman, S. M. (2010). Sensory properties determined by starch type in white sauces: Effects of freeze/thaw and hydrocolloid addition. *Journal of Food Science, 75*(2), Article S132. https://doi.org/10.1111/j.1750-3841.2009.01501.x

Ashogbon, A. O., & Akintayo, E. T. (2014). Recent trend in the physical and chemical modification of starches from different botanical sources: A review. *Starch - Stärke, 66*(1–2), 41–57. https://doi.org/10.1002/star.201300106

Awolu, O. O., & Olofinlae, S. J. (2016). Physico-chemical, functional and pasting properties of native and chemically modified water yam (Dioscorea alata) starch and production of water yam starch-based yoghurt. *Starch - Stärke, 68*(7–8), 719–726. https://doi.org/10.1002/star.201500302

Barretti, B. R. V., de Almeida, V. S., Ito, V. C., Silva, B. M., Filho, M. A.d. S. C., Sydney, E. B., Demiate, I. M., & Lacerda, L. G. (2022). Combination of organic acids and heat-moisture treatment on the normal and waxy corn starch: Thermal, structural, pasting properties, and digestibility investigation. *Food Science and Technology, 42*. https://doi.org/10.1590/fst.33120

Bensaad, D. E., Saleh, M., Ismail, K., Lee, Y., & George, O. (2023). Chemical modifications of starch; a prospective for sweet potato starch. *Jordan Journal of Agricultural Sciences, 18*(4), 293–308. https://doi.org/10.35516/jjas.v18i4.802

Borries Medrano, E.von, Jaime-Fonseca, M. R., & Aguilar-Méndez, M.Á. (2017). *Starch-galactomannans mixtures: Rheological and viscosity behavior in aqueous systems for food modeling*. IntechOpen. https://doi.org/10.5772/intechopen.68915

Brandner, S., Becker, T., & Jekle, M. (2022). Gluten–starch interface characteristics and wheat dough rheology—Insights from hybrid artificial systems. *Journal of Food Science, 87*(4), 1375–1385. https://doi.org/10.1111/1750-3841.16115

Butt, N. A., Ali, T. M., & Hasnain, A. (2020). Development of rice starch-based casein and fat mimetics and its application in imitation mozzarella cheese. *Journal of Food Processing and Preservation, 44*(12). https://doi.org/10.1111/jfpp.14928

Camelo-Méndez, G. A., Flores-Silva, P. C., Agama-Acevedo, E., Tovar, J., & Bello-Pérez, L. A. (2018). Incorporation of whole blue maize flour increases antioxidant capacity and reduces in vitro starch digestibility of gluten-free pasta. *Starch/Staerke, 70*(1–2). https://doi.org/10.1002/star.201700126

Chaksmithanont, P., Bangsitthideth, K., Arunprasert, K., Patrojanasophon, P., & Pornpitchanarong, C. (2024). Statistical-based optimization of modified Mangifera indica fruit starch as substituent for pharmaceutical tableting excipient. *Polymers, 16*(18), 2653. https://doi.org/10.3390/polym16182653

Cheng, Y., Chen, Q., Wang, Z., Zeng, M., Qin, F., Chen, J., & He, Z. (2023). Effects of different food ingredients and additives on the digestibility of extruded and roller-dried maize starch and its application in low glycemic index nutritional formula powder. *Journal of the Science of Food and Agriculture, 103*(13), 6483–6490. https://doi.org/10.1002/jsfa.12726

Choton, S., Bandral, J. D., Singh, J., Bhat, A., Sood, M., Gupta, N., Reshi, M., & Kaur, D. (2024). Enzymatic modification of starch: A review. *Saudi Journal of Medical and Pharmaceutical Sciences, 10*(01), 1–8. https://doi.org/10.36348/sjmps.2024.v10i01.001

Cioroiu, D. R., Parvulescu, O. C., Koncsag, C. I., Dobre, T., & Raducanu, C. (2017). Rheological characterization of algal suspensions for bioethanol processing. *Revista de Chimie, 68*(10), 2311–2316. https://doi.org/10.37358/rc.17.10.5875

Cisse, M., Michel Koffi, D., & Aya Akaffou, F. (2023). Functional properties of acetylation-modified starches of three purple maize cultivars from Côte d'ivoire. *European Journal of Nutrition & Food Safety, 15*(10), 63–72. https://doi.org/10.9734/ejnfs/2023/v15i101346

Crepalde, L. T., Vidigal, M. C. T. R., Carneiro, J.de D. S., & Minim, V. P. R. (2021). Integration of the optimized descriptive profile and temporal dominance of sensations methodologies. *Journal of Sensory Studies, 36*(3). https://doi.org/10.1111/joss.12651

Deng, C., Menbhnk, O., & Luo, Y. (2022). Optimization of heat-moisture treatment on potato starch and study on its physicochemical properties. *Technology Audit and Production Reserves, 3*(65), 43–49. https://doi.org/10.15587/2706-5448.2022.260149

Devadason, I. P., Anjaneyulu, A. S. R., & Babji, Y. (2010). Effect of different binders on the physico-chemical, textural, histological, and sensory qualities of retort pouched Buffalo meat nuggets. *Journal of Food Science, 75*(1), S31. https://doi.org/10.1111/j.1750-3841.2009.01399.x

Dhull, S. B., Punia, S., Kumar, M., Singh, S., & Singh, P. (2021). Effect of different modifications (physical and chemical) on morphological, pasting, and rheological properties of Black rice (Oryza sativa L. indica) starch: A comparative study. *Starch - Stärke, 73*(1–2). https://doi.org/10.1002/star.202000098

Ding, C.-L., Ju, B.-Z., & Zhang, S.-F. (2014). Preparation and properties of novel starch derivatives containing the sulfonic acid of the group and carboxymethyl. *Asian Journal of Chemistry, 26*(11), 3166–3170. https://doi.org/10.14233/ajchem.2014.15869

El-Sayed, E., & Ahmed, H. (2021). Cyclodextrin usage in textile industry. *Journal of Textiles, Coloration and Polymer Science, 0*(0). https://doi.org/10.21608/jtcps.2021.76238.1059, 0–0.

Elgadir, M. A., Akanda, M. J. H., Ferdosh, S., Mehrnoush, A., Karim, A. A., Noda, T., & Sarker, M. Z. I. (2012). Mixed biopolymer systems based on starch. *Molecules, 17*(1), 584–597. https://doi.org/10.3390/molecules17010584Malaysia

Erickson, J., Carlson, J., Stewart, M., & Slavin, J. (2018). Fermentability of novel type-4 resistant starches in in vitro system. *Foods, 7*(2), 18. https://doi.org/10.3390/foods7020018

Erni, P., Windhab, E. J., Gunde, R., Graber, M., Pfister, B., Parker, A., & Fischer, P. (2007). Interfacial rheology of surface-active biopolymers: Acacia Senegal gum versus hydrophobically modified starch. *Biomacromolecules, 8*(11), 3458–3466. https://doi.org/10.1021/bm700578z

Escalante-Aburto, A., Ponce-García, N., Dios Figueroa Cárdenas, J., Maldonado-Rosas, R., Rodríguez-Fernández, I., & Tejada-Ortigoza, V. (2025). Assessing the viability of producing nixtamalized foods through additive manufacturing and rheological analysis. *International Journal of Food Science and Technology, 60*. https://doi.org/10.1093/ijfood/vvaf089

Escalante-Aburto, A., Trujillo-de Santiago, G., Álvarez, M. M., & Chuck-Hernández, C. (2021). Advances and prospective applications of 3D food printing for health improvement and personalized nutrition. *Comprehensive Reviews in Food Science and Food Safety, 20*(6), 5722–5741. https://doi.org/10.1111/1541-4337.12849

Fahmi, A., Zavaragh, S. R., Hanafi, M. R., Rahimpour, H., Zinatloo-Ajabshir, S., & Asghari, A. (2023). Facile preparation, characterization, and investigation of mechanical strength of starchy NaCl-binder as a lightweight construction material. *Scientific Reports, 13*(1). https://doi.org/10.1038/s41598-023-46536-8

Fahmy, A. R., Jekle, M., & Becker, T. (2023). Texture modulation of starch-based closed-cell foams using 3D printing: Deformation behavior beyond the elastic regime. *Journal of Texture Studies, 54*(1), 153–169. https://doi.org/10.1111/jtxs.12729

Farooq, M. A., & Yu, J. (2024). Recent advances in physical processing techniques to enhance the resistant starch content in foods: A review. *Foods, 13*(17), 2770. https://doi.org/10.3390/foods13172770

Franzol, A., Banin, T. M., Brazil, T. R., & Rezende, M. C. (2021). Rheological analyses and artificial neural network as optimization tools to predict the sensory perception of cosmetic emulsions. *Materials Research, 24*(6). https://doi.org/10.1590/1980-5373-MR-2021-0252

Gałkowska, D., Kapuśniak, K., & Juszczak, L. (2023). Chemically modified starches as food additives. *Molecules, 28*(22), 7543. https://doi.org/10.3390/molecules28227543

Gamage, A., Liyanapathiranage, A., Manamperi, A., Gunathilake, C., Mani, S., Merah, O., & Madhujith, T. (2022). Applications of starch biopolymers for a sustainable modern agriculture. *Sustainability, 14*(10), 6085. https://doi.org/10.3390/su14106085

Garba Mohammed, K. (2017). Modified starch and its potentials as excipient in pharmaceutical formulations. *Novel Approaches in Drug Designing & Development, 1*(1). https://doi.org/10.19080/napdd.2017.01.555551

Garcia, M. A. V. T., Garcia, C. F., & Faraco, A. A. G. (2020). Pharmaceutical and biomedical applications of native and modified starch: A review. *Starch/Staerke, 72*(7–8). https://doi.org/10.1002/star.201900270

Gerling, E., Schneider, D. E., Gryczak, M., Petzhold, C. L., & Daitx, T. S. (2025). Hydrogel based on starch and poly(ethylene terephthalate) recycling products for the slow release of urea. *Polymers for Advanced Technologies, 36*(2). https://doi.org/10.1002/pat.70105

Giuberti, G., & Gallo, A. (2018). Reducing the glycaemic index and increasing the slowly digestible starch content in gluten-free cereal-based foods: A review. *International Journal of Food Science and Technology, 53*(1), 50–60. https://doi.org/10.1111/ijfs.13552

Giuberti, G., Gallo, A., Fortunati, P., & Rossi, F. (2016). Influence of high-amylose maize starch addition on in vitro starch digestibility and sensory characteristics of cookies. *Starch - Stärke, 68*(5–6), 469–475. https://doi.org/10.1002/star.201500228

Giuberti, G., Gallo, A., Moschini, M., & Masoero, F. (2013). In vitro production of short-chain fatty acids from resistant starch by pig faecal inoculum. *Animal, 7*(9), 1446–1453. https://doi.org/10.1017/s1751731113001092

Grace, S., & Liew, K. C. (2016). Trans tech publications ltd Malaysia hydrolyzation of edible starches: Their preparations and properties. *Materials Science Forum, 875*, 63–76. https://doi.org/10.4028/www.scientific.net/MSF.875.63

Gryszkin, A., Zięba, T., & Kapelko-Żeberska, M. (2017). Properties of corn starch subjected hydrothermal modification. *International Agrophysics, 31*(1), 53–60. https://doi.org/10.1515/intag-2016-0031

Gumul, D., Berski, W., Ivanišová, E., Gambuś, H., Kačániová, M., Harangozo, L., & Tokár, M. (2017). Characteristics of starch breads enriched with red potatoes. *Potravinarstvo Slovak Journal of Food Sciences, 11*(1), 162–166. https://doi.org/10.5219/720

Gumul, D., Krystyjan, M., Buksa, K., Ziobro, R., & Zięba, T. (2014). The influence of oxidation, extrusion and oxidation/extrusion on physico-chemical properties of potato starch. *Starch - Stärke, 66*(1–2), 190–198. https://doi.org/10.1002/star.201300069

Guo, B., Wu, J., Hu, X., Luo, S., Wang, H., Xu, S., Huang, Z., & Liu, C. (2020). Effects of controlled far-infrared treatment on granular swelling and rheological properties of crop starches. *Starch - Stärke, 72*(3–4). https://doi.org/10.1002/star.201900251

He, R., Li, S., Zhao, G., Zhai, L., Qin, P., & Yang, L. (2023). Starch modification with molecular transformation, physicochemical characteristics, and industrial usability: A state-of-the-art review. *Polymers, 15*(13), 2935. https://doi.org/10.3390/polym15132935

Hebeish, A., Abdul, R. M., Ramadan, M. A., & Saleh, A. (2013). New route for novel polycarboxylic starch hybrid. *Materials Sciences and Applications, 04*(11), 695–703. https://doi.org/10.4236/msa.2013.411087

Hirao, K., Kondo, T., Kainuma, K., & Takahashi, S. (2018). *Starch properties and uses as food for human health and welfare sago palm: Multiple contributions to food security and sustainable livelihoods*. Singapore, Japan: Springer. https://doi.org/10.1007/978-981-10-5269-9_21

Hong, Y., Liu, G., & Gu, Z. (2014). Recent advances of starch-based excipients used in extended-release tablets: a review. *Drug Delivery, 23*(1), 12–20. https://doi.org/10.3109/10717544.2014.913324

Horstmann, S. W., Lynch, K. M., & Arendt, E. K. (2017). Starch characteristics linked to gluten-free products. *Foods, 6*(4), 1–21. https://doi.org/10.3390/foods6040029

Huang, Y.-C., & Lai, H.-M. (2010). Noodle quality affected by different cereal starches. *Journal of Food Engineering, 97*(2), 135–143. https://doi.org/10.1016/j.jfoodeng.2009.10.002

Hunt, A., Getty, K. J., & Park, J. W. (2010). Development of temperature tolerant surimi gels using starch–protein interactions. *Journal of Food Quality, 33*(s1), 119–136. https://doi.org/10.1111/j.1745-4557.2010.00303.x

Ibrahim, K. A., Naz, M. Y., Shukrullah, S., Sulaiman, S. A., Ghaffar, A., & AbdEl-Salam, N. M. (2020). Nitrogen pollution impact and remediation through low cost starch based biodegradable polymers. *Scientific Reports, 10*(1). https://doi.org/10.1038/s41598-020-62793-3

Jia, S., Yu, B., Zhao, H., Tao, H., Liu, P., & Cui, B. (2022). Physicochemical properties and in vitro digestibility of dual-modified starch by cross-linking and annealing. *Starch - Stärke, 74*(1–2). https://doi.org/10.1002/star.202100102

Jorge A, F., Cadena, E. M., Rincon, C. C., & Salcedo, J. G. (2018). Modification of the fermentation process for sour cassava starch with expansion properties. *Advance Journal of Food Science and Technology, 16*(SPL), 55–61. https://doi.org/10.19026/ajfst.16.5937

Kang, T. Y., Sohn, K. H., Yoon, M. R., Lee, J. S., & Ko, S. (2015). Effect of the shape of rice starch granules on flour characteristics and gluten-free bread quality. *International Journal of Food Science and Technology, 50*(8), 1743–1749. https://doi.org/10.1111/ijfs.12835

Kavitha, R., Prabhu*, S. V., Melese Belew, B., Solomon, E., & Work Mekonen, S. (2020). RSM based process modeling and optimization for amylase enzyme production and its desizing potential. *International Journal of Innovative Technology and Exploring Engineering, 9*(7), 382–391. https://doi.org/10.35940/ijitee.f4149.059720

Khatun, A., Waters, D. L. E., & Liu, L. (2019). A review of rice starch digestibility: Effect of composition and heat-moisture processing. *Starch - Stärke, 71*(9–10). https://doi.org/10.1002/star.201900090

Kim, D. E., Kwak, H. S., Lim, M., & Lee, Y. (2023). Comparison of check-all-that-apply (CATA), rate-all-that-apply (RATA), flash profile, free listing, and conventional descriptive analysis for the sensory profiling of sweet pumpkin porridge. *Foods, 12*(19), 3556. https://doi.org/10.3390/foods12193556

Kiprop, V. J., Omwamba, M. N., & Mahungu, S. M. (2021). Influence of gum Arabic from *Acacia Senegal* Var. *kerensis* on the modifications of pasting and textural properties of cassava and corn starches. *Food and Nutrition Sciences, 12*(11), 1098–1115. https://doi.org/10.4236/fns.2021.1211081

Korus, J., Ziobro, R., Witczak, T., Kapusniak (Jochym), K., & Juszczak, L. (2021). Effect of octenyl succinic anhydride (OSA) modified starches on the rheological properties of dough and characteristic of the gluten-free bread. *Molecules, 26*(8), 2197. https://doi.org/10.3390/molecules26082197

Kumar, V., Kumarasamy, V., Bhatt, P., Dixit, R., Kumar, M., Shukla, C. P., Subramaniyan, V., & Kumar, S. (2024). Ultrasound assisted techniques for starch modification to develop novel drug delivery systems: A comprehensive study. *Journal of Bioactive and Compatible Polymers, 39*(4), 279–297. https://doi.org/10.1177/08839115241249143

Kumar, D., Pandey, J., Raj, V., & Kumar, P. (2017). A review on the modification of polysaccharide through graft copolymerization for various potential applications. *The Open Medicinal Chemistry Journal, 11*(1), 109–126. https://doi.org/10.2174/1874104501711010109

Kumar, S. R., Tangsrianugul, N., & Suphantharika, M. (2023). A review on isolation, characterization, modification, and applications of proso millet starch. *Foods, 12*(12), 2413. https://doi.org/10.3390/foods12122413

Kumari, B., & Sit, N. (2024). Physicochemical, rheological properties, and in-vitro starch digestibility of flours and starches from pigeon pea, cowpea, pinto bean, and navy bean. *Starch - Stärke, 76*(9–10). https://doi.org/10.1002/star.202300303

Lamy, E., Santos, V., Barrambana, S., Simões, C., Carreira, L., Infante, P., & Silva, F. C.e (2021). Saliva protein composition relates with interindividual variations in bread sensory ratings. *Starch - Stärke, 73*(1–2). https://doi.org/10.1002/star.202000052

Lawal, M. V. (2019). Modified starches as direct compression excipients – Effect of physical and chemical modifications on tablet properties: A review. *Starch - Stärke, 71*(1–2). https://doi.org/10.1002/star.201800040

Lawal, M. V., Odeniyi, M. A., & Itiola, O. A. (2015). Material and rheological properties of native, acetylated, and pregelatinized forms of corn, cassava, and sweet potato starches. *Starch/Staerke, 67*, 964–975. https://doi.org/10.1002/star.201500044

León, O., Soto, D., González, J., Piña, C., Muñoz-Bonilla, A., & Fernandez-García, M. (2019). Environmentally friendly fertilizers based on starch superabsorbents. *Materials, 12*(21), 3493. https://doi.org/10.3390/ma12213493

León, O., Soto, D., López, D., Bonilla, A. M.-, & García, M. F.- (2019). Fat-replacer properties of oxidized cassava starch using hydrogen peroxide/sodium bicarbonate redox system in mayonnaise formulation and its stability. *Starch - Stärke, 71*(9–10). https://doi.org/10.1002/star.201900112

Lele, V. V., Kumari, S., & Niju, H. (2018). Syntheses, characterization and applications of graft copolymers of sago starch – A review. *Starch - Stärke, 70*(7–8). https://doi.org/10.1002/star.201700133

Lewandowicz, J., Le Thanh-Blicharz, J., & Szwengiel, A. (2024). Insight into rheological properties and structure of native waxy starches: Cluster analysis grouping. *Molecules, 29*(11), 2669. https://doi.org/10.3390/molecules29112669

Li, G., Ge, X., Guo, C., & Liu, B. (2023). Effect of ultrasonic treatment on structure and physicochemical properties of pea starch. *Foods, 12*(13), 2620. https://doi.org/10.3390/foods12132620

Li, J., Hu, Y. S., Pu, L. M., Guo, G. S., & Niu, H. Y. (2014). Trans tech publications ltd China preparation of an environmental friendly slow release nitrogen fertilizer. *Advanced Materials Research, 1015*, 346–349. https://doi.org/10.4028/www.scientific.net/AMR.1015.346

Lin, H.-T. V., Tsai, J.-S., Liao, H.-H., & Sung, W.-C. (2023). The effect of hydrocolloids on penetration tests and syneresis of binary gum gels and modified corn starch–gum gels. *Gels, 9*(8), 605. https://doi.org/10.3390/gels9080605

Lopez Ochoa, J. D., Cadena-Chamorro, E., Ciro-Velasquez, H., & Rodríguez-Sandoval, E. (2022). Enzymatically modified cassava starch as a stabilizer for fermented dairy beverages. *Starch - Stärke, 74*(7–8). https://doi.org/10.1002/star.202100242

Lotufo Haddad, A. M., Ribotta, P., Armada, M., & Goldner, M. C. (2019). Textural, pasting, and rheological behavior of starch-pectin-sucrose gels: Relation with sensory perception. *Starch - Stärke, 71*(9–10). https://doi.org/10.1002/star.201800286

Lukic, M., Jaksic, I., Krstonosic, V., Cekic, N., & Savic, S. (2012). A combined approach in characterization of an effective w/o hand cream: The influence of emollient on textural, sensorial and in vivo skin performance. *International Journal of Cosmetic Science, 34*(2), 140–149. https://doi.org/10.1111/j.1468-2494.2011.00693.x

Mansour, G., Zoumaki, M., Tzetzis, D., Mansour, G., Tzetzis, D., & Kyratsis, P. (2020). Starch sandstones in building bio-materials. *MATEC Web of Conferences, 318*, Article 01046. https://doi.org/10.1051/matecconf/202031801046

Marefati, A., & Rayner, M. (2020). Starch granule stabilized pickering emulsions: An 8-year stability study. *Journal of the Science of Food and Agriculture, 100*(6), 2807–2811. https://doi.org/10.1002/jsfa.10289

Marta, H., Cahyana, Y., Djali, M., & Pramafisi, G. (2022). The properties, modification, and application of banana starch. *Polymers, 14*(15), 3092. https://doi.org/10.3390/polym14153092

Mathobo, V. M., Silungwe, H., Ramashia, S. E., & Anyasi, T. A. (2021). Effects of heat-moisture treatment on the thermal, functional properties and composition of cereal, legume and tuber Starches—A review. *Journal of Food Science and Technology, 58*(2), 412–426. https://doi.org/10.1007/s13197-020-04520-4

Mehrotra, N., Venkatraman, U., Chaitra, V., Kaviyarasi, N. S., & Pradeep, N. (2020). Synthesis of zinc coneflower-porous starch composite for textile finishing. *SN Applied Sciences, 2*(3). https://doi.org/10.1007/s42452-020-2229-y

Mendes, M. P., Raniero, G. Z., Demiate, I. M., & Monteiro, A. R. G. (2023). Effects of reactive extrusion on cross-linking of corn starch. *International Food Research Journal, 30*(4), 1040–1048. https://doi.org/10.47836/ifrj.30.4.20

Miraballes, M., Hodos, N., & Gámbaro, A. (2018). Application of a pivot profile variant using CATA questions in the development of a whey-based fermented beverage. *Beverages, 4*(1), 11. https://doi.org/10.3390/beverages4010011

Molavi, H., & Razavi, S. M. A. (2018). Dynamic rheological and textural properties of acorn (Quercus brantii lindle.) starch: Effect of single and dual hydrothermal modifications. *Starch - Stärke, 70*(11–12). https://doi.org/10.1002/star.201800086

Nakkala, K., Godiyal, S., Kumar Ettaboina, S., & Laddha, K. S. (2022). Chemical modifications of turmeric starch by oxidation, phosphorylation, and succinylation. *Starch - Stärke, 74*(9–10). https://doi.org/10.1002/star.202200053

Navaf, M., Sunooj, K. V., Krishna, N. U., Aaliya, B., Sudheesh, C., Akhila, P. P., Sabu, S., Sasidharan, A., Mir, S. A., & George, J. (2022). Effect of different hydrothermal treatments on pasting, textural, and rheological properties of single and dual modified Corypha umbraculifera L. starch. *Starch/Staerke, 74*(3–4). https://doi.org/10.1002/star.202100236

Nawaz, H., Waheed, R., Nawaz, M., & Shahwar, D. (2020). *Physical and chemical modifications in starch structure and reactivity*. USA: IntechOpen. https://doi.org/10.5772/intechopen.88870

Ng, J. Q., Siew, C. K., Mamat, H., Matanjun, P., & Lee, J. S. (2018). Effect of acid methanol treatment and heat moisture treatment on in vitro digestibility and estimated glycemic index of raw and gelatinized sago (Metroxylon sagu) starch. *Starch/Staerke, 70*(9–10). https://doi.org/10.1002/star.201700198

Nimitkeatkai, H., Pasada, K., & Jarerat, A. (2022). Incorporation of tapioca starch and wheat flour on physicochemical properties and sensory attributes of meat-based snacks from beef scraps. *Foods, 11*(7), 1034. https://doi.org/10.3390/foods11071034

Nugraheni, M., Lastariwati, B., & Purwanti, S. (2017). The potential of gluten free flour enriched with resistant starch type 3 from Canna edulis for the management profile of glucose, lipids and short chain fatty acid in healthy mice. In *Proceedings of the 1st international conference on social, applied science and technology in home economics*. https://doi.org/10.2991/iconhomecs-17.2018.29

Olagunju, A., Omoba, O., Enujiugha, V., Alashi, A., & Aluko, R. (2020). Technological properties of acetylated pigeon pea starch and its stabilized set-type yoghurt. *Foods, 9*(7). https://doi.org/10.3390/foods9070957

Ovando-Martinez, M., Whitney, K., Ozsisli, B., & Simsek, S. (2017). Physicochemical properties of octenyl succinic esters of cereal, tuber and root starches. *Journal of Food Processing and Preservation, 41*(1), Article e12872. https://doi.org/10.1111/jfpp.12872

Páramo-Calderón, D. E., Carrillo-Ahumada, J., Juárez-Arellano, E. A., Bello-Pérez, L. A., Aparicio-Saguilán, A., & Alvarez-Ramirez, J. (2016). Effect of cross-linking on the physicochemical, functional and digestibility properties of starch from macho (Musa paradisiaca L.) and roatan (musa sapientum L.) banana varieties. *Starch/Staerke, 68*(7–8), 584–592. https://doi.org/10.1002/star.201500200

Panyoo, A. E., & Emmambux, M. N. (2019). Effects of screw configuration, screw speed, and stearic acid addition on the functional properties and structural characteristics of maize starch extrudates. *Starch - Stärke, 71*(5–6), Article 1800149. https://doi.org/10.1002/star.201800149

Patricia, J., Nma, Y. B., Ugoanyanwu, F. O., Amaechi, D., Ekpe, I. P., et al. (2023). The effect of carboxymethylation on morphological and physicochemical properties of Faro 61 rice starch. *Biomedical Journal of Scientific & Technical Research, 51*(4). https://doi.org/10.26717/bjst

Ponce-García, N., Ramírez-Wong, B., Torres-Chávez, P. I., Figueroa-Cárdenas, J.d. D., Serna-Saldívar, S. O., Cortez-Rocha, M. O., & Escalante-Aburto, A. (2017). Evaluation of visco-elastic properties of conditioned wheat kernels and their doughs using a compression test under small strain. *Journal of the Science of Food and Agriculture, 97*(4), 1235–1243. https://doi.org/10.1002/jsfa.7855

Prestes, R. C., Silva, L. B., Torri, A. M. P., Kubota, E. H., Rosa, C. S., Roman, S. S., Kempka, A. P., & Demiate, I. M. (2015). Sensory and physicochemical evaluation of low-fat chicken mortadella with added native and modified starches. *Journal of Food Science and Technology, 52*(7), 4360–4368. https://doi.org/10.1007/s13197-014-1496-2

Punia Bangar, S., Nehra, M., Siroha, A. K., Petrů, M., Ilyas, R. A., Devi, U., & Devi, P. (2021). Development and characterization of physical modified pearl millet starch-based films. *Foods, 10*(7). https://doi.org/10.3390/foods10071609

Qiu, C., Hu, H., Chen, B., Lin, Q., Ji, H., & Jin, Z. (2024). Research progress on the physicochemical properties of starch-based foods by extrusion processing. *Foods, 13*(22), 3677. https://doi.org/10.3390/foods13223677

Rahman, M. H., Mu, T. H., Zhang, M., Ma, M. M., & Sun, H. N. (2020). Comparative study of the effects of high hydrostatic pressure on physicochemical, thermal, and structural properties of maize, potato, and sweet potato starches. *Journal of Food Processing and Preservation, 44*(11). https://doi.org/10.1111/jfpp.14852

Ramadan, M. F., Sitohy, M. Z., & Starches, P. (2020). Preparation, properties, functionality, and techno-applications. *Starch - Stärke, 72*(5–6). https://doi.org/10.1002/star.201900302

Rao, H., Sindhu, R., & Panwar, S. (2022). Morphology and functionality of dry heat-treated and oxidized quinoa starches. *Journal of Food Processing and Preservation, 46*(7). https://doi.org/10.1111/jfpp.16672

Ren, Y., Quilliam, C., Weber, L. P., Warkentin, T. D., Tulbek, M., & Ai, Y. (2022). Effects of pulse crop types and extrusion parameters on the physicochemical properties, in vitro and in vivo starch digestibility of pet foods. *Cereal Chemistry, 99*(3), 625–639. https://doi.org/10.1002/cche.10524

Ren, F., Wang, J., Xie, F., Zan, K., Wang, S., & Wang, S. (2020). Applications of ionic liquids in starch chemistry: A review. *Green Chemistry, 22*(7), 2162–2183. https://doi.org/10.1039/c9gc03738a

Ren, Y., Yuan, T. Z., Chigwedere, C. M., & Ai, Y. (2021). A current review of structure, functional properties, and industrial applications of pulse starches for value-added utilization. *Comprehensive Reviews in Food Science and Food Safety, 20*(3), 3061–3092. https://doi.org/10.1111/1541-4337.12735

Rożnowski, J., Juszczak, L., Szwaja, B., & Przetaczek-Rożnowska, I. (2021). Effect of esterification conditions on the physicochemical properties of phosphorylated potato starch. *Polymers, 13*(15), 2548. https://doi.org/10.3390/polym13152548

Rodriguez-Ruiz, J. D., Rodriguez-Sandoval, E., & Soledad Hernández, M. (2025). Influence of modified cassava starches on the rheological and structural properties of a fermented soy beverage during storage. *Starch - Stärke, 77*(2). https://doi.org/10.1002/star.202400003

Samarakoon, E. R. J. (2020). Impact of physical modifications on starch nutritional fractions: Rapidly digestible starch, slowly digestible starch and resistant starch. *Journal of Food Bioactives, 12*, 106–121. https://doi.org/10.31665/jfb.2020.12249

Sarwono, A., Man, Z., Bustam, M. A., & Azizli, K. A. (2014). Water uptake behavior of lignin modified starch film. *Applied Mechanics and Materials, 699*, 204–209. https://doi.org/10.4028/www.scientific.net/amm.699.204

Schädle, C. N., Bader-Mittermaier, S., & Sanahuja, S. (2022). Characterization of reduced-fat mayonnaise and comparison of sensory perception, rheological, tribological, and textural analyses. *Foods, 11*(6), 806. https://doi.org/10.3390/foods11060806

Shaikh, M., Ali, T. M., & Hasnain, A. (2020). Comparative study on the application of chemically modified corn and pearl millet starches in white sauce. *Journal of Food Processing and Preservation, 44*(4). https://doi.org/10.1111/jfpp.14393

Sharma, A., Bist, Y., Chakkingal, F., Kumar, Y., & Chandra Saxena, D. (2023). Preparation, characterization, and utilization of crosslinked and dual-modified banana starch for stabilizing pickering emulsion. *Starch - Stärke, 75*(11–12). https://doi.org/10.1002/star.202300084

Sharma, M., & Duizer, L. (2019). Characterizing the dynamic textural properties of hydrocolloids in pureed Foods—A comparison between TDS and TCATA. *Foods, 8*(6), 184. https://doi.org/10.3390/foods8060184

Singh, S., & Kaur, M. (2017). Steady and dynamic shear rheology of starches from different oat cultivars in relation to their physicochemical and structural properties. *International Journal of Food Properties, 20*(12), 3282–3294. https://doi.org/10.1080/10942912.2017.1286504

Singh, N., Singh, J., Kaur, L., Sodhi, N. S., & Gill, B. S. (2003). Morphological, thermal and rheological properties of starches from different botanical sources. *Food Chemistry, 81*(2), 219–231. https://doi.org/10.1016/S0308-8146(02)00416-8

Sinhmar, A., Kumar Pathera, A., Sharma, S., Nehra, M., Thory, R., & Nain, V. (2023). Impact of various modification methods on physicochemical and functional properties of starch: A review. *Starch - Stärke, 75*(1–2). https://doi.org/10.1002/star.202200117

Songré-Ouattara, L. T., Mouquet-Rivier, C., Humblot, C., Rochette, I., Diawara, B., & Guyot, J. P. (2010). Ability of selected lactic acid bacteria to ferment a pearl millet-soybean slurry to produce gruels for complementary foods for young children. *Journal of Food Science, 75*(5), Article M261. https://doi.org/10.1111/j.1750-3841.2010.01640.x

Srinang, P., Khotasena, S., Sanitchon, J., Chankaew, S., Jogloy, S., & Monkham, T. (2023). New source of rice with a low amylose content and slow in vitro digestion for improved health benefits. *Agronomy, 13*(10), 2622. https://doi.org/10.3390/agronomy13102622

Stefanes de Santana Varela, M. (2017). Determination and modeling of thermophysical and transport properties of tropical pulps. *European Journal of Biophysics, 5*(5), 79. https://doi.org/10.11648/j.ejb.20170505.11

Sulaimon, A. A., Akintola, S. A., Mohd Johari, M. A. B., & Isehunwa, S. O. (2020). Evaluation of drilling muds enhanced with modified starch for HPHT well applications. *Journal of Petroleum Exploration and Production Technology, 11*. https://doi.org/10.1007/s13202-020-01026-9

Sulistyawati, E. Y. E., Mustofa, A., Rismaya, R., & Suhartatik, N. (2024). Physicochemical and functional properties of vermicelli made of modified white rice, brown rice, and Black rice. *International Journal of Advanced Science, Engineering and Information Technology, 14*(3), 1042–1048. https://doi.org/10.18517/ijaseit.14.3.19628

Supreethee, S., & Vashishth, R. (2025). Unlocking the potential of palm tree starch: A review on modification techniques and its application. *International Journal of Food Science and Technology, 60*(1). https://doi.org/10.1093/ijfood/vvaf009

Taylor, J. R. N., Emmambux, M. N., & Kruger, J. (2015). Developments in modulating glycaemic response in starchy cereal foods. *Starch - Stärke, 67* (1–2), 79–89. https://doi.org/10.1002/star.201400192

Vega, Y. A., Arrieta, Álvaro A., Acosta-Humánez, M. F., Consuegra, A., Palencia, M., Combatt, E. M., & Jarma Orozco, A. (2025). Starch and its potential applications as a biomaterial for hydrogels with agricultural purposes: A mini-review. *Journal of Science with Technological Applications, 18*, 1–10. https://doi.org/10.34294/j.jsta.25.18.105

Vidaurre-Ruiz, J., Vargas, R. J. Y., Alcázar-Alay, S. C., Encina-Zelada, C. R., Cabezas, D. M., Correa, M. J., & Repo-Carrasco-Valencia, R. (2022). Andean crops: kañiwa and tarwi flours used for the development of vegan gluten-free muffins. *Journal of the Science of Food and Agriculture, 102* (15), 7282–7292. https://doi.org/10.1002/jsfa.12094

Vitola, L., Pundiene, I., Pranckeviciene, J., & Bajare, D. (2024). Innovative hemp shive-based bio-composites: Part I: Modification of potato starch binder by sodium meta-silicate and glycerol. *Materials, 17*(19), 4911. https://doi.org/10.3390/ma17194911

Wang, S., Li, C., Copeland, Les, Niu, Q., Wang, S., & Retrogradation, S. (2015). A comprehensive review. *Comprehensive Reviews in Food Science and Food Safety, 14*(5), 568–585. https://doi.org/10.1111/1541-4337.12143

Wang, D., Tang, Z., Chen, R., Zhong, X., Du, X., Li, Y.-X., & Zhao, Z. (2024). Physicochemical properties of esterified/crosslinked quinoa starches and their influence on bread quality. *Journal of the Science of Food and Agriculture, 104*(7), 3834–3841. https://doi.org/10.1002/jsfa.13265

Wang, Z., Wang, S., Xu, Q., Qi, K., Li, F., Lin, L., Xu, Y., & Wei, Y. (2023). Synthesis and functions of resistant starch. *Advances in Nutrition, 14*(5), 1131–1144. https://doi.org/10.1016/j.advnut.2023.06.001

Wang, M., Wu, Y., Liu, Y., & Ouyang, J. (2020). Effect of ultrasonic and microwave dual-treatment on the physicochemical properties of chestnut starch. *Polymers, 12*(8), 1718. https://doi.org/10.3390/polym12081718

Wang, H., Zhu, Q., Wu, T., & Zhang, M. (2020). Glass transition temperature, rheological, and gelatinization properties of high amylose corn starch and waxy cassava starch blends. *Journal of Food Processing and Preservation, 44*(9). https://doi.org/10.1111/jfpp.14682

Weerapoprasit, C., & Prachayawarakorn, J. (2019). Effects of polymethacrylamide-grafted branch on mechanical performances, hydrophilicity, and biodegradability of thermoplastic starch film. *Starch - Stärke, 71*(11–12). https://doi.org/10.1002/star.201900068

Widhyahrini, K. (2023). Synthesis study of amylum-carboxymethyl and amylum-sulfonate to increase their solubility in water. *Jurnal Kartika Kimia, 5* (2). https://doi.org/10.26874/jkk.v5i2.156

Won, C., Jin, Y.i., Kim, M., Lee, Y., & Chang, Y. H. (2017). Structural and rheological properties of potato starch affected by degree of substitution by octenyl succinic anhydride. *International Journal of Food Properties, 20*(12), 3076–3089. https://doi.org/10.1080/10942912.2016.1272610

Wronkowska, M., & Soral-Śmietana, M. (2012). Fermentation of native wheat, potato, and pea starches, and their preparations by bifidobacterium - Changes in resistant starch content. *Czech Journal of Food Sciences, 30*(1), 9–14. https://doi.org/10.17221/18/2011-cjfs

Wu, K., Dai, S., Gan, R., Corke, H., & Zhu, F. (2016). Thermal and rheological properties of mung bean starch blends with potato, sweet potato, rice, and sorghum starches. *Food and Bioprocess Technology, 9*(8), 1408–1421. https://doi.org/10.1007/s11947-016-1730-1

Xie, F., Halley, P. J., & Avérous, L. (2012). Rheology to understand and optimize processibility, structures and properties of starch polymeric materials. *Progress in Polymer Science, 37*(4), 595–623. https://doi.org/10.1016/j.progpolymsci.2011.07.002

Yahya, S. R., Azura, A. R., & Azahari, B. (2012). Preparations and characterization of sago starch dispersion and modification. *Advanced Materials Research, 620*, 395–399. https://doi.org/10.4028/www.scientific.net/amr.620.395

Yarovaya, L., Waranuch, N., Wisuitiprot, W., & Khunkitti, W. (2022). Correlation between sensory and instrumental characterization of developed sunscreens containing grape seed extract and a commercial product. *International Journal of Cosmetic Science, 44*(5), 569–587. https://doi.org/10.1111/ics.12807

Zailani, M. A., Kamilah, H., Husaini, A., Awang Seruji, A. Z. R., & Sarbini, S. R. (2023). Starch modifications via physical treatments and the potential in improving resistant starch content. *Starch/Staerke, 75*(1–2). https://doi.org/10.1002/star.202200146

Zambelli, R. A., Galvão, A. M. M. T., De Mendonça, L. G., De Souza Leão, M. V., Carneiro, S. V., Lima, A. C. S., & Melo, C. A. L. (2018). Effect of different levels of acetic, citric and lactic acid in the cassava starch modification on physical, rheological, thermal and microstructural properties. *Food Science and Technology Research, 24*(4), 747–754. https://doi.org/10.3136/fstr.24.747

Zdybel, E., Wilczak, A., Kapelko-Żeberska, M., Tomaszewska-Ciosk, E., Gryszkin, A., Gawrońska, A., & Zięba, T. (2021). Physicochemical properties and digestion resistance of acetylated starch obtained from annealed starch. *Polymers, 13*(23), 4141. https://doi.org/10.3390/polym13234141

Zhou, M., Keast, R., & Roura, E. (2023). Identification of starchy flavour attributes and characterization of distinctive sensory profiles of corn, potato, rice and waxy rice starches. *Proceedings of the Nutrition Society, 82*(OCE2). https://doi.org/10.1017/s0029665123001878

Zhou, Y., Wang, M., Wang, L., Liu, L., Wu, Y., & Ouyang, J. (2023). Comparison of the effect of ultrasound and microwave on the functional properties and in vitro digestibility of normal maize starch and potato starch. *Journal of Food Process Engineering, 46*(2). https://doi.org/10.1111/jfpe.14222

Zou, H., Yao, L., Dang, X., Yu, N., Zhang, Y. L., Zhang, Y. L., & Dong, J. (2015). Solubility characteristics and slow-release mechanism of nitrogen from organic-inorganic compound coated urea. *International Journal of Photoenergy, 2015*, 1–6. https://doi.org/10.1155/2015/705471

Further reading

Chiravi, K., Khan, M. A., Wadikar, D. D., Mahesh, C., & Semwal, A. D. (2023). Starch retrogradation: Role of food ingredients and extrinsic factors. *Food Science and Engineering, 28*, 204–215. https://doi.org/10.37256/fse.4220232939

Chapter 4

Emerging processing techniques for starches: Gelatinization, modification and extrusion

Flaviana Coelho Pacheco[1], Irene Andressa[2], Ana Flávia Coelho Pacheco[3] and Paulo Henrique Costa Paiva[3]
[1]Federal University of Viçosa, Department of Food Science and Technology, Vicosa, Brazil; [2]Federal University of Viçosa, Viçosa, Brazil;
[3]Agricultural Research Company of Minas Gerais (EPAMIG), Cândido Tostes Dairy Institute, Juiz de Fora, Brazil

1. Introduction

Starch is the main source of carbohydrates in several foods, such as cereal grains, tubers, legumes, fruits, leaves and roots. It is composed mainly of amylose (25%–30%) and amylopectin (70%–75%) (Makroo et al., 2021). Amylose is a linear polymer of D-glucose with α(1≥4) linkages, while amylopectin is a highly branched short polymer of glucose units linked by α-(1–4) glycosidic bonds with α-(1–6)-glycosidic bonds interconnected to glucose units (Parker & Ring, 2001). In addition, starch granules contain small amounts of lipids, proteins and mineral salts, such as phosphorus, which influence their functionality (Alcázar-Alay & Meireles, 2015). Starch is used as a thickener, binder, encapsulating agent, stabilizer, and gelling agent in various food products (Singh & Adedeji, 2017). However, starch extracted from plants is known as native starch and cannot be used directly in food applications due to low shear, low thermal stability, and high retrograde rates (Almeida et al., 2022). Therefore, to overcome these limitations, starch is subjected to biological, chemical, or physical transformations, and is now called modified starch (Raghunathan et al., 2021).

Starch modification is related to the improvement of its nutritional quality, physicochemical characteristics of the amylose and amylopectin chains, and functional and textural properties (Kumari & Sit, 2023). These improvements are achieved through changes in chain length, formation of branching points, debranching, and disproportionation (Park et al., 2008). Gelatinization, chemical, physical and enzymatic modification, as well as extrusion processes stand out as key methods in this context.

The chemical modification of starch is a widely applied method, since it alters its molecular and techno-functional properties, such as amylose content, strength, crystallinity and stability, in addition to reducing characteristics such as swelling power and solubility (Lee & Chang, 2019; Zhang et al., 2021). These processes introduce new functional groups, modifying the granular structure of starch and allowing the development of various products, such as gels, nanocrystals and films (Sharma et al., 2021). However, modification through chemical processes negatively impacts the environment, since it generates effluents that are harmful to the environment and health. Therefore, to reduce these impacts, the use of safer compounds, such as organic acids, and more sustainable alternatives, such as physical and enzymatic modification, are sought.

Gelatinization occurs when starch granules are heated in the presence of water, leading to the rupture of their crystalline structure, increased viscosity, and solubilization (Ma et al., 2022). Physical modification is a simple, accessible, and environmentally friendly process and can be categorized into thermal and non-thermal modification processes (Kumari & Sit, 2023). In this case, emerging technologies such as ultrasound, microwave, cold plasma, ohmic heating, high hydrostatic pressure, and high-pressure homogenization are being investigated to improve the control and efficiency of the degree of gelatinization (Kumari & Sit, 2023). These non-thermal physical modification technologies offer

Native and Modified Starches for Food Applications. https://doi.org/10.1016/B978-0-443-27431-2.00027-4

advantages such as fast processing, minimal food alteration, preservation of color, texture, nutrients, and uniformity. Their effects vary depending on the intensity, time, and pressure applied (Kumari & Sit, 2023).

Enzymatic modification is an innovative and sustainable approach to altering the structure of starch, generating fewer by-products and being considered a green technology (Bangar et al., 2022). It modifies the length of the chains, the molecular mass and the ratio between amylose and amylopectin. The resulting starch presents more stable gels, less retrogradation, higher viscosity and gelatinization temperature, but with lower swelling power, solubility and digestibility (Christensen et al., 2023).

Extrusion, in turn, is an indispensable technique in starch processing, in which it is subjected to high pressures, temperatures and shear forces. In addition to improving the solubility, gelatinization and digestibility of starch, extrusion allows the formation of unique textures and the incorporation of functional ingredients, making the process highly versatile for food applications (Qiu et al., 2024).

This chapter discusses recent advances in starch modification to enhance its functionalities in industrial food applications, focusing on alternative techniques that increase the quality and yield of modified starch. In addition, the review presents the industrial applications and performance of native and modified starches, offering a critical analysis of how the botanical origin of starch can influence its properties and identifying opportunities to expand its use in industrial contexts, considering the economic challenges of the market.

2. Starches and their food applications

Plant sources of starch can be classified into conventional and unconventional. Conventional sources include cereals, tubers, and legumes such as maize, potatoes, wheat, cassava, and rice, while non-conventional sources include rhizomes, fruits, stems, and bulbs (Tagliapietra et al., 2021). Although conventional sources are widely used around the world, unconventional sources present themselves as promising alternatives for the development of new products, due to their superior technological properties.

Starch is a polymer composed of glucose units linked in two main ways: amylose, which represents 20%–30% of starch, and which has a linear structure with α1-4 bonds; and amylopectin, which accounts for 70%–80% of starch, which has a branched structure with about 5% α1-6 bonds (Jenkins & Donald, 1995). Variations in the proportions of amylose and amylopectin directly influence the granular structure, physicochemical properties and quality of the final products (Van Hung et al., 2006).

The properties of starch, such as swelling, solubility, gelatinization, crystallinity, paste behavior, amylose-amylopectin ratio, composition, granule structure, and molecular weight, are, in fact, strongly influenced by the size of starch granules. Additionally, amylose content can vary based on the botanical source and starch genotype, with factors such as environmental conditions and extraction methods also impacting this variation. This knowledge is critical to understanding the differences in starch characteristics and performance in different industrial applications (Kaur et al., 2011).

In the food industry, starch can be used both in its native, unprocessed form (extracted directly from the plant) and in its processed or modified form (Mathobo et al., 2021). In general, native starch is not widely employed in food industries due to its low thermal and shear resistance, in addition to its tendency to retrograde, which is the main cause of aging in food products (Wang et al., 2020). On the other hand, modified starch is widely used in the food industry as a fat substitute, thickener, stabilizer, or texture modifier (Fig. 4.1) (Chen et al., 2020).

Starch is widely used as a natural thickener in products such as soups and sauces, with its modification improving interaction with non-starchy hydrocarbons and rheological attributes (Mahmood et al., 2017). In the meat industry, it is a multifunctional ingredient, acting as a binder, emulsifier, fat substitute, gelling agent, and water retention agent. Modified starch gel, due to restricted swelling, has higher density, holding capacity, and shear strength, reducing cooking and thawing losses (Kaur et al., 2019).

In dairy products, starch is used as a thickener and stabilizer, improving texture, reducing syneresis, and increasing viscosity and solids in yogurts (Saleh et al., 2020). Swelling starch granules can be advantageous for increasing viscosity in foods that require thickening agents (Jia et al., 2023), but in products such as noodles, a low swelling power is preferable to maintain firm texture and structural integrity (Compart et al., 2023).

The rheological properties of starch, such as gel strength, flow behavior, and gelatinization kinetics, are influenced by factors such as density, granule size, temperature, concentration, and extraction conditions. These measurements are important for understanding the relationship between the thermal properties and molecular structure of starch, as well as aiding in the development of starch-based food products and films. Various factors, such as concentration, temperature, and shear rate, also affect these properties.

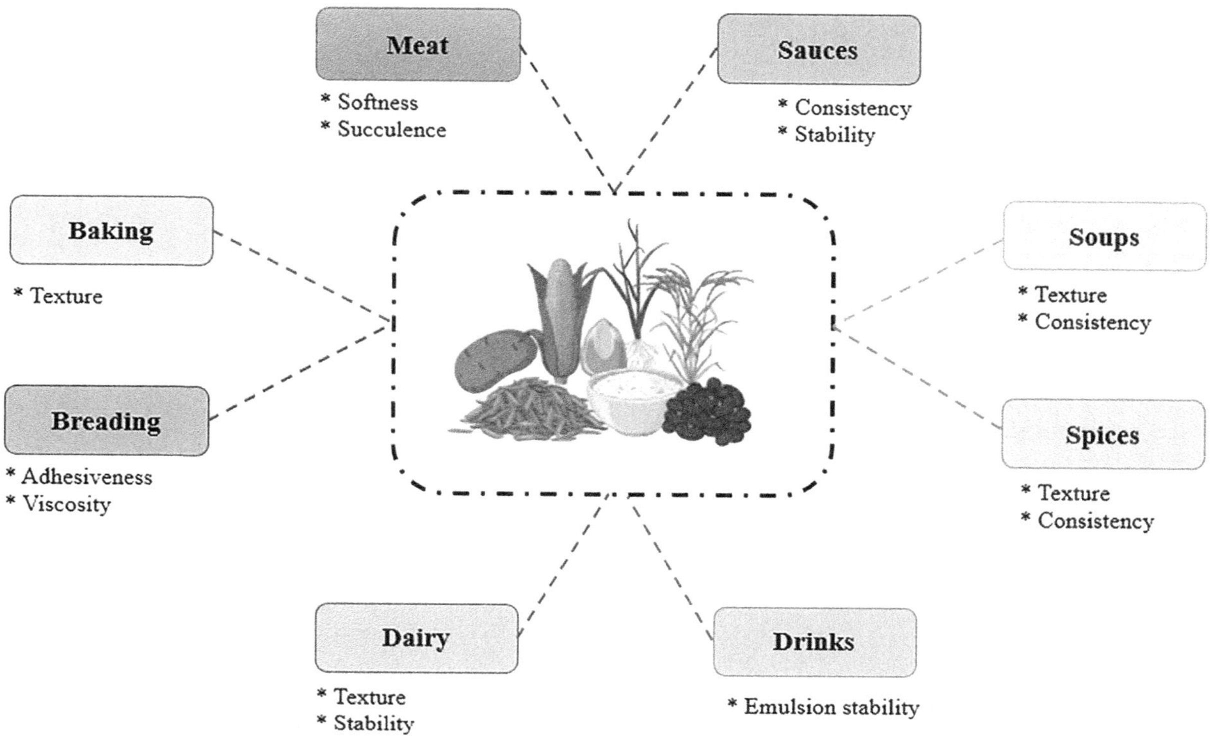

FIGURE 4.1 Examples of food applications using modified starch.

3. Gelatinization of starches: Principles and applications

The gelatinization process of starch granules occurs with hydration and swelling at high temperatures, resulting in the disorder of crystalline structures and the loss of birefringence (Cai et al., 2014). This process begins in the hilum of the granules, since the central region around the hilum is considered the least organized (Chen et al., 2011). Figs. 4.2 and 4.3 illustrate the changes in viscosity during hydration and heating over time.

The starch granules, in their native state, have an organized crystal structure with amylose and amylopectin aligned radially around the hilum, exhibiting the "Maltese cross" pattern under polarized light (Cai & Wei, 2013). During the gelatinization process (hydration and heating), extensive swelling of the starch granules occurs, resulting in the rupture of the crystal structure and leaching of the amylose molecules from the swollen granules, as well as increased viscosity (Wang et al., 2018). Crystallinity disruption and swelling vary depending on the position of the hilum and the type of crystalline polymorphism. For starches with an eccentric hilum, the breakage occurs at the surface of the granule, while for starches with a central hilum, it begins in the center of the granule. With increasing temperature, these effects intensify, resulting in greater loss of crystallinity and increased swelling and viscosity (Cai et al., 2014; Cai & Wei, 2013).

The gelatinizing properties of starches play a key role in the food industry. Differential scanning calorimetry (DSC) is an efficient technique for analyzing the thermal changes, gelatinization, and structural organization of starch granules. Among the key parameters measured, gelatinization temperature (GT) and gelatinization enthalpy (ΔH gel) stand out. These properties are influenced by several factors, such as botanical origin, the proportion of amylose and amylopectin, the shape and size of the granules, crystallinity, and moisture content (Table 4.1) (Govindaraju et al., 2022). According to a study by Hong et al. (2020), changes in the gelatinization properties of starch are closely related to particle size, with smaller granules having lower thermal stability. This is because the smaller granules have a larger surface area, which facilitates interaction with water and speeds up the gelatinization process. In addition, greater exposure of the crystalline surfaces in smaller granules can result in faster structural disorganization, decreasing thermal resistance (Hong et al., 2020).

Salts, along with other food ingredients like sugars, proteins, lipids, and polysaccharides, significantly influence the starch gelatinization process (Yeoh et al., 2020). Starch is commonly gelatinized in the presence of these ingredients, Table 4.1 illustrates the effects of the ingredients on the gelatinization properties of starch.

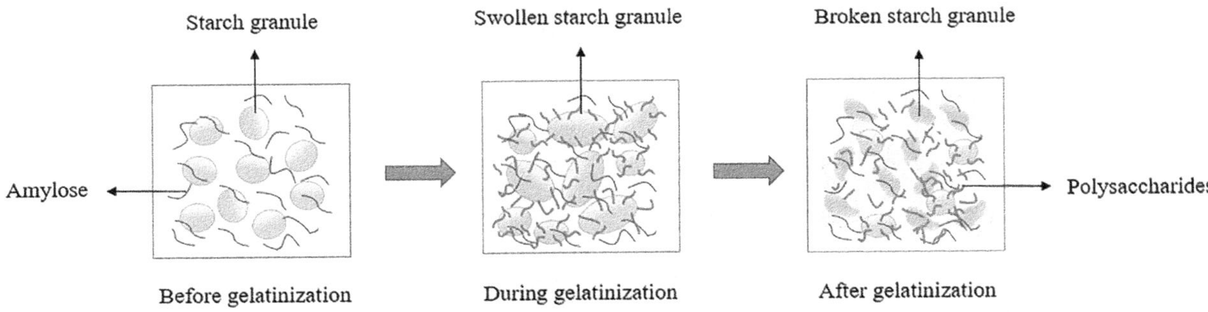

FIGURE 4.2 Modification of the starch structure during heating.

FIGURE 4.3 Changes in starch granules during the gelatinization process.

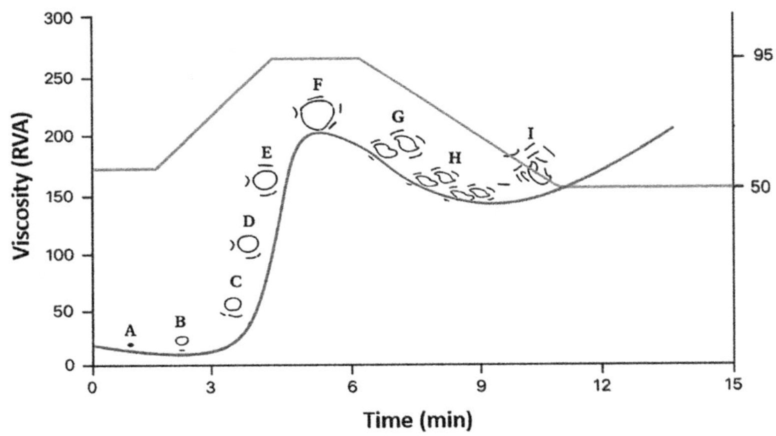

The degree of gelatinization of starch plays a key role in modifying its structural organization, directly impacting the physicochemical, sensory, and gastrointestinal properties of foods that have starch as the main component (Ding et al., 2021). Changes in the degree of gelatinization can influence characteristics such as texture, flavor, and digestibility, being a crucial parameter in the development of foods with specific attributes and better consumer acceptance (Sun et al., 2019).

For example, starch gelatinization is an important transformation in rice grains during the parboiling process, significantly impacting their properties. Optimizing the degree of gelatinization can reduce the grain breakage rate and increase rice yield (Neutzling et al., 2023). However, excessive gelatinization can result in wasted energy, undermining the sustainability of the process (Taghinezhad et al., 2016).

The quality of the noodles is influenced by cooking losses and textural characteristics such as hardness and tenderness. In the case of gluten-free noodles, the formation of starch gels is crucial to their properties. Optimizing the degree of gelatinization reduces cooking losses and increases hardness, due to the formation of a stronger gel network (Sun et al., 2019). However, an excessive degree of gelatinization can increase cooking losses, as starch degradation facilitates water dissolution during cooking (Sun et al., 2019).

In general, as the degree of gelatinization increases, the structure of the starch granules is altered, with a reduction in crystallinity, particle size and molecular weight, in addition to the formation of starch-lipid complexes. As a result, the viscosity of the paste decreases and digestibility increases, although these effects vary depending on the starch source and processing conditions (Yan et al., 2024).

4. Starch modification: Methods and functionality

There are three main types of starch modifications: chemical, physical, and enzymatic (Fig. 4.4). The purpose of these modifications is to alter the three hydroxyl groups present in the starch copolymer, adapting its functional properties for different applications. The degree of modification depends on the genetic origin of the starch and the reaction conditions

TABLE 4.1 Effects of starch characteristics, process conditions, and ingredient addition on starch gelatinization properties.

Conditions	Effects on starch gelatinization	References
Starch characteristics		
Amylose-amylopectin ratio	More orderly structure of amylopectin in starch results in greater thermal stability and changes in the melting enthalpy of starches	Chakraborty et al. (2022)
Granular shape and size	Smaller starch granules result in lower thermal stability	Hong et al. (2020)
Process conditions		
Temperature	Higher temperatures increases swelling levels and starch gelatinization temperature	Liu et al. (2017)
pH	Alkaline pH increases the swelling capacity of starch granules	Shukri & Shi (2017)
Moisture content	In excess water, starch melts as a single peak at high gelatinization temperature (Tp_1). In limited water, partial melting occurs, generating a second peak at higher temperatures (Tp_2).	Chakraborty et al. (2022)
Makings		
Salts	Increase or decrease in gelatinization temperature and starch enthalpy	Zhang et al. (2020)
Sugar	Increases the gelatinization temperature and alters the enthalpy of gelatinization	Li (2022)
Proteins	Increased starch gelatinization temperature	Chakraborty et al. (2022)
Lipids	Inhibit granule swelling and increase starch gelatinization temperatures	Li (2022)
Non-starchy polysaccharides	Delay starch gelatinization and inhibit swelling of starch granules	Li (2022)

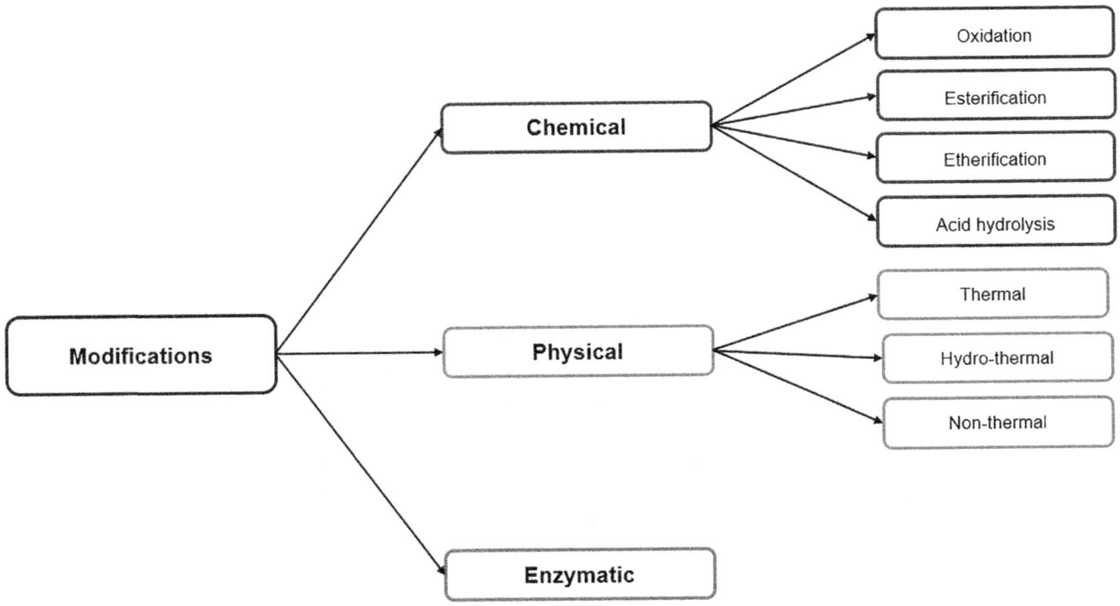

FIGURE 4.4 Three methods of starch modification.

(Pu et al., 2011). Such modifications affect properties such as gelatinization, swelling, solubility, and retrogradation (Tharanathan, 2005), making modified starch more suitable for food applications, compared to native starch.

4.1 Chemical modification of starch

Chemical modification is the most widely explored modification method due to the non-destructive nature of a select few processes and potential increases in the functionality of theModified starch. Chemical modification utilizes chemical reagents that can be classified as either monofunctional or bifunctional reagents based on their chemical properties (Wolf et al., 1999). Monofunctional reagents alter gelatinization and adhesion properties, while bifunctional reagents promote cross-linking by reinforcing granules and adjusting starch solubility and swelling (Xiao, 2013). The three main chemical modifications to starch and their variations are summarized in Fig. 4.4.

Chemical modification by oxidation involves the reaction of starch with oxidizing agents, such as hydrogen peroxide, under controlled conditions, resulting in starch with greater paste clarity, lower viscosity, and greater adhesion, being used as a binder and crispy coating in foods (Chen et al., 2015). Chemical modification by esterification converts the hydroxyl groups of starch into alkyl or aryl derivatives, which improves solubility, swelling, and freezing stability, and is useful as a thickener and stabilizer in refrigerated or frozen foods (Amini et al., 2017). Chemical modification by etherification replaces hydroxyl groups with groups such as carboxymethyl or hydroxypropyl, increasing viscosity and thermal stability, being applied in sauces and puddings, as well as acting as a flavor encapsulant (Clasen et al., 2018). Chemical modification by acid hydrolysis, on the other hand, uses mineral acids to break down amorphous regions of starch granules, improving gelation and gel strength, with applications in jams, gummies, and slow-digesting products (Majzoobi & Farahnaky, 2021).

Conventional chemical modification processes for starch, which involve the use of agents such as hypochlorite, acetates, phosphates, and acids, result in the generation of large volumes of effluent that require recycling. These effluents can have a negative impact on the environment due to their composition. In addition, the presence of residues of these compounds in the product can limit their applications, especially in the pharmaceutical and food industries, which are increasingly demanding ingredients with *clean label* labels, that is, more natural and free of artificial additives.

4.2 Physical modification of starch

Physical modifications of starch are low-cost and more environmentally friendly alternatives compared to chemical and enzymatic processes. These techniques have been shown to be effective in improving starch properties, including changes in molecular structure, packaging arrangements, and crystallinity, without the introduction of chemical reagents (Oliveira et al., 2018). Among the physical methods for starch modification, they are explored, including thermal and non-thermal physical treatments, highlighting ozone, pulsed electric field (PEF), ultrasound (US), microwave, high pressure homogenization and high hydrostatic pressure (HPP) (Fig. 4.5, Table 4.2). Table 4.2 presents the emerging technologies mentioned above in the modification of different starch sources and their main results.

4.3 Ozone

Ozone (O_3), a highly reactive and unstable gas, has been shown to be an environmentally friendly and non-thermal alternative in the modification of starches. Its application can occur both in aqueous and gaseous phases, being more efficient in the presence of water. The effectiveness of the ozonation process is influenced by several factors, including ozone concentration, temperature, pH, and reactor characteristics (Khadre & Yousef, 2001). Among the most common effects of ozonation are the increase in the carbonyl and carboxyl groups, the depolymerization of amylopectin and amylose molecules, and the reduction of the viscosity of starch pastes (Castanha et al., 2017; Lima et al., 2021).

In the study by Lima et al. (2021), the modification of cassava starch was analyzed in terms of its molecules and granules. A reduction in large and branched molecules was observed, with an increase in smaller molecules after ozone treatment at a concentration of approximately 43.5 mg O_3 /L. These structural changes significantly impacted the paste properties, gel texture, water absorption, and starch solubility indexes. In another study, Castanha et al. (2017) investigated the modification of potato starch with ozone (47 mg O_3 /L) and found an increase in carbonyl, carboxyl and reducing sugar levels, as well as a reduction in pH, amylose content and molecular size. Similarly, the study by He et al. (2023) revealed the conversion of hydroxyl groups to carbonyl and carboxyl in sweet potato starch, reinforcing the action of ozone as an oxidizing agent capable of altering the chemical composition and molecular structure of starch. In addition, the authors highlighted the increase in water solubility and clarity of the paste, as well as reductions in water absorption capacity, viscosity and viscoelasticity in sweet potato starch as a result of ozone modification.

While the effects of ozone on starch structure are well documented, each starch source responds differently to treatment, which opens up room for further study. The combination of ozone with other technologies, which is still little explored, can offer promising results, especially in more complex food systems. In the future, the expansion of the industrial application of ozone for starch modification may offer new opportunities in food processing and other industries.

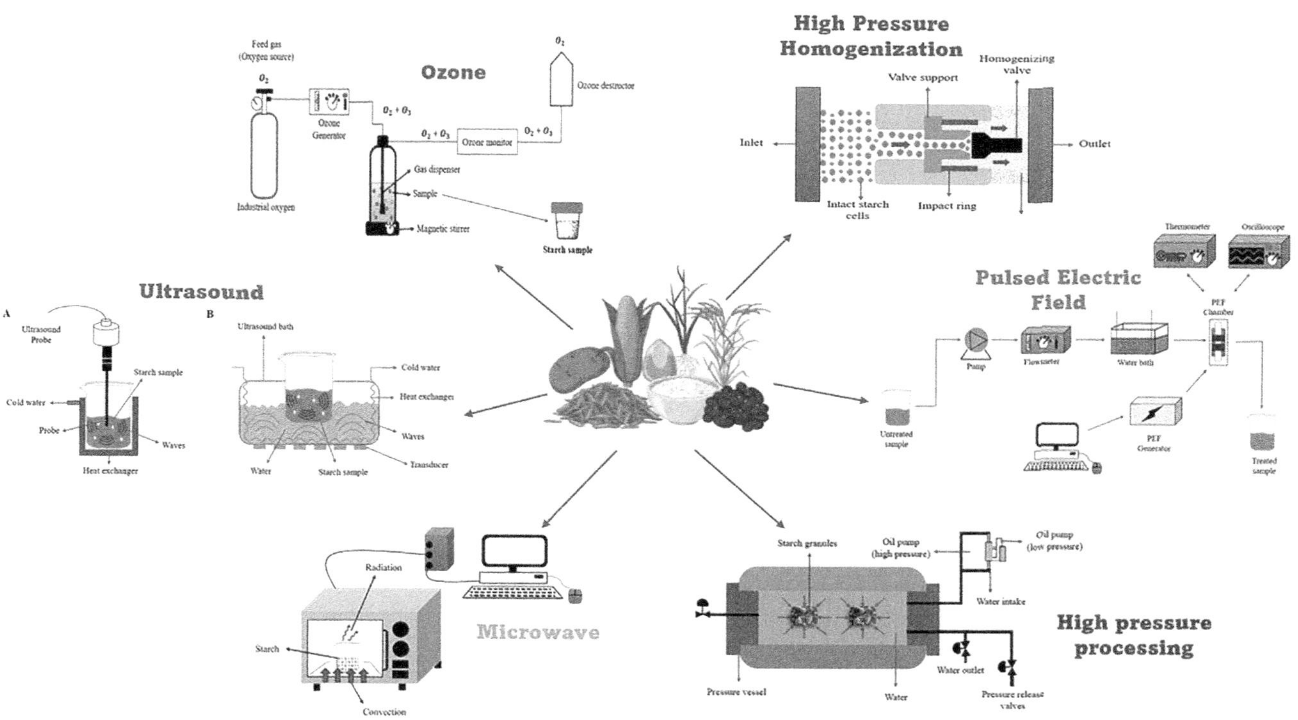

FIGURE 4.5 Physical methods for starch modification.

TABLE 4.2 Application of "clean label" technologies to improve starch performance.

Starch source	Conditions	Observed effects	References
Ozone			
Potato starch	47 mg O_3 /L for 15, 30, 45 and 60 min	Ozonation modified potato starch, increasing oxidative groups and reducing amylose and molecular size, without changing the crystallinity, but changing the morphology and properties of the gel and paste.	Castanha et al. (2017)
Arracacha starch	47 mg $O_3 \cdot$ /L for 15 and 30 min	Ozonation modified the starch structure, reducing the molecular size and forming new functional groups, increased reducing sugars, and improved the transparency of the pastes and the strength of the hydrogels	Lima et al. (2020)
Sweet potato starch	0.75, 1.50, 2.25 and 3.00 g for 15, 30, 45 and 60 min	The ozonation of sweet potato starch promoted molecular changes that increased the solubility and clarity of the paste, reduced viscosity and water absorption capacity, and improved the functionality of the starch, with effects dependent on the treatment time.	He et al. (2023)
Waxy rice starch	1 L/min for 0, 0.5, 1, 1.5, 2 h	Ozone treatment increased the molecular order and improved the surface properties of waxy rice starch nanocrystals without changing the crystal type, making them more suitable for applications in food science.	Yang et al. (2024)
Cassava, corn and potatoes	46 mg $O_3 \cdot$ /L for 15 and 30 min	Ozonation did not affect the morphology of the granules, but reduced the viscosity and the G* modulus, with variable effects on the strength of the gels according to the concentration and source of the starch.	Castanha et al. (2020)
Cornstarch	20 g/h for 10, 20 and 30 min	Ultrasonic and ozone modification improved the solubility and swelling of starch, as well as increased the strength, elasticity, transparency, and water vapor barrier of the produced films.	Huang et al. (2024)
Quinoa starch	17.4 mg O_3/L for 10, 20 and 30 min	Ozonation of quinoa starch reduced amylose and crystallinity, while increasing carbonyl/carboxyl groups, clarity, oil absorption, viscosity and gelatinization temperature, proving to be an efficient and safe method of modification.	Ramos et al. (2024)
Ultrasound			
Corn and cassava starch	40 kHz for 10 and 20 min at 26°C	Ultrasound modified the structure of corn and cassava starches, reducing enthalpy and crystallinity, without changing the gelatinization temperature. This indicates its potential as a non-thermal technique to alter starch properties.	Rahaman et al. (2021)
White millet starch	33 kHz for 30 min at 50°C	The physical modification of starch increased the amylose content to 31%–33%, the enthalpy to 59.18 mJ/g and the syneresis to 66%, in addition to increasing WAC, SP and crystallinity, with the formation of pores and cracks.	Amarnath et al. (2023)

Pearl millet starch	33 kHz for 45 min at 50°C	Sonication increased amylose, swelling, thermal and rheological stability of millet starch, as well as improved films, indicating potential for packaging.	Punia et al. (2021)
Millet and sorghum starches	750 W, 20 kHz for 10, 15 and 20 min	Ultrasound on millet and sorghum starches increased water absorption, swelling, clarity, stability and digestibility, without changing the color. It generated pores on the surface and slightly reduced crystallinity, improving the strength and elasticity of the gel.	Sharma et al. (2023)
Quinoa and corn starch	50%, 20 kHz for 2, 4, 8, 14 and 22 h	Ultrasound increased swelling in the short term and degraded amylopectin with long treatment on quinoa starch; In corn starch, there was a balance between leaching and damage.	Wei et al. (2023)
Wheat and tapioca starch	400 W, 24 kHz.	Ultrasound improved gelatinization and altered starch structure, viscosity, and entropy, reducing energy dissipation and increasing bubble pressure and temperature.	Roohi et al. (2024)
Native cornstarch	40%, 20 kHz for 3–24 min	Ultrasound disorganizes the structure of corn starch primarily by thermal effect, altering its physical properties and facilitating enzymatic hydrolysis to produce porous starches, while crystallinity remains almost intact	Parra et al. (2025)
Microwave			
Quinoa starch	4.8 W/g, 9.6 W/g, 14.4 W/g, 90 s	Heat and humidity (MHT) treatment on quinoa increased crystallinity, granule size, and amylose, reducing digestibility by converting fast starch into slow, resistant digestion, especially at 25% moisture.	Liu et al. (2023)
Cornstarch	2450 MHz, 720 W	The combination of lysine and microwave increased the slowly digestible starch and crystallinity of corn starch, reducing viscosity and enthalpy, as well as causing aggregation and lysine coating, altering its physicochemical properties and digestibility.	Zhou et al. (2022)
Rice starch	100 W, 300 W	A study with constant power microwaves showed that rapid heating caused greater damage and reduced crystallinity in rice starch, being effective in modifying its molecular structure with high repeatability.	Han et al. (2021)
Bambara peanut starch	700 W, 2450 MHz for 0, 10, 30, or 60 s	Reduced crystallinity, viscosity and swelling capacity, raising the gelatinization temperature and decreasing the enthalpy	Oyeyinka et al. (2019)
Fava bean starch	2450 MHz and 1100 W	Treated starch generated a material with low viscosity, high solubility and good potential for food coatings and release of functional compounds.	González-Mendoza et al. (2022)
High Hydrostatic Pressure (HPP)			
Bean starch	600 MPa at 30°C for 30 min	The treatments caused changes in the structure, such as pores on the surface, reduced crystallinity, and thermal and functional changes.	Almeida et al. (2024b)

Continued

TABLE 4.2 Application of "clean label" technologies to improve starch performance.—cont'd

Starch source	Conditions	Observed effects	References
Ozone			
Cornstarch		The modified starch reduced syneresis and improved the viscosity of the puree. In addition, the starch showed greater stability and maintained better physical and nutritional properties for 14 days	Dominguez-Ayala et al. (2023)
Rice starch-alginate	500 MPa at 20 ± 3°C for 20 min	HHP-treated starch-alginate gels with amylose showed adequate texture and fluidity compatible with foods for dysphagia	Rivero-Ramos et al. (2024)
Amorphous granular starch	550 MPa for 30 min	The treatment maintained the granular shape and was more stable, with lower viscosity and solubility	Ye and Baik (2024)
Quinoa starch	100, 200, 300 and 400 MPa for 30 min	The treatment improved quinoa starch modification, increasing the efficiency of the chemical reaction and producing more stable emulsions.	Yin et al. (2025)
White finger millet	200,400 and 600 MPa for 10 min	Relative crystallinity decreased, while hydration and gelling properties increased. In addition, the clarity of the paste was reduced and the gelatinization profile changed.	Sherin et al. (2024)
Pulsed electric field (PEF)			
Wheat and cassava starches	20 kV–100 A, 25 ± 2°C	The treatment reduced the viscosity and increased the firmness of the gels of both starches. In wheat starch, it caused fractures in the granules, depolymerization, reduction of crystallinity and gelatinization enthalpy	Maniglia et al. (2021)
Potato starch	0, 2, 3, 4, 5 and 6 kV/cm for 50 min	The treatment reduced the gelatinization temperature and viscosity of the paste, increased the efficiency of the reaction, and decreased the processing time	Chen et al. (2021)
Cornstarch	0, 3, 6, 9 and 12 kV/cm for 15 min	The treatment increased particle size, solubility, aldehyde group content, and reduced viscosity, making the process more efficient than the traditional method.	Li et al. (2023a)
Cornstarch	0, 4, 6, 8, 10, 12 and 14 kV/cm for 10 min	The treatment increased the carboxyl content, caused surface changes, reduced gelatinization temperature and viscosity, and improved thermal stability	Li et al. (2023a)
High-pressure homogenization			
Cornstarch	140, 200 and 250 MPa for 1, 2, 3 and 4 cycles	The treatment reduced the size of the granules of corn starch with high amylose content, altered its surface and increased the proportion of amylose, maintaining the B-type crystal structure.	Apostolidis and Mandala (2020)
Chestnut starch	20, 40, 60, 80 and 100 MPa for 1 cycle and 100 MPa for 3, 5 and 7 cycles	The treatment improved the properties of chestnut starch, increasing its solubility, swelling power and transparency, while reducing aging and crystallinity, as well as decreasing particle size.	Gao et al. (2022)

Cornstarch	30, 60, 90 and 120 MPa	The treatment reduces the molecular size of starch by degrading long amylose chains into smaller chains and alters the distribution of the amylopectin chains	Li et al. (2020)
Double modification			
Japonica rice starch	Ozone: 0.00042 g for 0, 30, 60 and 90 min Enzyme: α-amylase	Ozone increased the enzymatic susceptibility of rice starch, altered the structure of the granules and reduced the viscosity and rheological properties, facilitating the action of α-amylase.	Almeida et al. (2025a)
Bean starch	PEF: 30 kV/cm, 600 Hz for 60s Ozone: 50 g/h, for 60 and 120 min	The combination of ozone and pulsed electric fields increased oxidation, modified the structure, and increased the solubility of bean starch, reducing its rheological properties and indicating potential for food applications.	Almeida et al. (2024a)
Maize starch	Ozone: 42 mg O_3/L for 15 min Ultrasound: 24–26°C, 25 kHz, 72 W/L for 8 h	Ozone altered the molecular structure and properties of starch. The combination of ultrasound and ozone improved the clarity of the paste, indicating that ultrasound enhances the action of ozone.	Castanha et al. (2019)
Cassava starch	Ultrasound: 200 W for 30 min Enzyme: glucoamylase	Ultrasound and enzymes reduced the size and crystallinity of cassava starch, increased the solubility, hydrolysis and adsorption of water and oil, and raised the gelatinization temperature, indicating greater potential for adsorption.	Liu et al. (2023)
Lotus root starch	Microwave: 300 W, 700 W Cold plasma: 0.8A, 40 V for 60, 90, and 120 s	Combined microwave and cold plasma treatments on lotus root starch modified the structure, reduced crystallinity and resistant starch, and altered its physicochemical properties, without affecting the chemical groups.	Sun et al. (2023)
Quinoa starch	HHP: 400 MPa for 10 min PEF: 10 and 30 kV /cm	HHP and PEF changed the molecular structure, increasing the granule size and gelatinization temperature, improved texture and stability.	Almeida et al. (2025b)

4.4 Ultrasound

Ultrasound, which uses high-frequency acoustic waves (>1 W/cm^2), promotes physical changes in materials and is considered a green technology, with increasing application in the modification of starches (Cárcel et al., 2012). This process occurs mainly in liquid starch suspensions, being driven by acoustic cavitation, compression and expansion cycles that form and implode bubbles, generating high temperatures, pressures and reactive radicals (BeMiller & Huber, 2015). Other effects, such as microjets and shear, also contribute to the changes (Miano et al., 2017). Several factors influence the efficiency of ultrasound, such as potency, frequency, pH, temperature, and starch characteristics (Falsafi et al., 2019). However, the lack of standardization of parameters and the wear and tear of equipment, such as probe tips, in addition to the heat generated, have resulted in inconsistencies in the data, limiting its reliable application (Chatel, 2018).

Wei et al. (2023) evaluated the effect of ultrasound on the structural and physicochemical properties of different starches, identifying distinct behaviors between small and large granule starches. For small granules, an increase in swelling power and a reduction in relative crystallinity have been reported, accompanied by changes in peak areas in prolonged treatments. On the other hand, the starches of large granules showed more severe damage on the surface and significant morphological changes. These results highlight the influence of granule size on the response to ultrasonic treatment. In another study, Amarnath et al. (2023), evaluated ultrasound physical modification in white millet starch. The results showed that US increased the water absorption capacity, the power of swelling and generated cracks in the granules of white millet starch (WFM), in addition to altering its textural and functional properties.

Thus, ultrasound demonstrates great potential as a technology for starch modification, standing out for its efficiency and versatility. However, it is essential to invest in studies that improve the standardization of process parameters and address challenges related to scalability and wear and tear of equipment, ensuring its efficient application on an industrial scale.

4.5 Microwave

Microwaves are high-frequency electromagnetic waves, with frequencies ranging between 300 MHz and 300 GHz and wavelengths between 1 m and 1 mm (Li et al., 2023). Microwave heating is widely used in the food industry, as a physical modification technique, and is increasingly applied in starch modification due to its advantageous characteristics, such as lower environmental impact, time savings, high efficiency, uniform energy distribution, good penetration capacity, and reduction in nutrient loss (de la Hoz & Loupy, 2013; Braşoveanu & Nemţanu, 2014). The electric field associated with microwave radiation promotes the continuous reorganization of polar molecules and ions, which results in friction and collisions with neighboring molecules, generating heat through electromagnetic induction (El Khaled et al., 2018).

Several studies have explored the modification of starch through the use of microwaves, aiming to improve its properties to meet different industrial demands (Balakumaran et al., 2023; Han et al., 2021; Zailani et al., 2021). In a research conducted by Zailani et al. (2021), the modification of sago starch was promoted through heating using microwave radiation. The results showed that the treatment induced irregularly shaped cracks in the starch granules, which resulted in a reduction in the moisture content, solubility and transparency of the starch slurry. On the other hand, the process also led to an increase in swelling power, water retention capacity, amylose content, and syneresis. Another study investigated the modification of white millet starch utilizing microwave radiation (Balakumaran et al., 2023). The results indicated that the treatment promoted an increase in the hydration properties of the starch gel. In addition, there was a reduction in bonding temperature, accompanied by changes in indentation viscosity, and a significant decrease in relative crystallinity.

4.6 High hydrostatic pressure (HPP)

High Hydrostatic Pressure (HPP) is a non-thermal technology widely used in the food industry. This technique is characterized by being isostatic and adiabatic, which ensures an even and instantaneous distribution of pressure (Yu et al., 2022). During processing, the applied pressure, which can vary between 100 and 1000 MPa at room temperature, is maintained for a predefined period. After this interval, the system is depressurized and the sample is removed (Yan et al., 2022). In starch, high hydrostatic pressure can promote changes in the functional properties of the Polysaccharidesaltering their secondary and tertiary structures (Taddei et al., 2018).

Several studies have investigated the influence of PPH on starch properties. For example, Almeida et al. (2023) analyzed the modification of rice starch subjected to high hydrostatic pressure under different conditions: 200, 400, and 600 MPa for 30 min. The results showed that HPP caused disorder in the granular structures of starch, and pressures above

400 MPa induced changes in the diffraction pattern. In addition, the increased pressure resulted in improvements in water, oil, and milk absorption values, especially at 600 MPa. In another study, Dominguez-Ayala et al. (2023) investigated changes in the crystal structure of starches modified by high hydrostatic pressure, applying pressures of 300 and 600 MPa for 5 min at 10°C. The authors observed that the treatment at 300 MPa promoted the aggregation of the starch granules, maintaining their granular structure and resulting in higher viscosity. On the other hand, at 600 MPa, the granules were destroyed, presenting a collapsed structure that contributed to the reduction of viscosity.

The results demonstrate that changes in starch properties depend directly on the pressure applied. In addition, the type and origin of starch play a significant role in the modifications promoted by HPP. For this reason, more studies are needed to clarify these aspects and deepen the understanding of the subject. Overall, the effects of HPP on starches can contribute to the modification or enhancement of starch's nutritional and functional characteristics, benefiting both new product development and consumer health.

4.7 Pulsed electric field (PEF)

The pulsed electric field (PEF) is a non-thermal technique widely used in the food industry for the extraction of bioactive compounds, enzymatic modification, microbial inactivation, and changes in the physicochemical properties of various materials (Prabhu et al., 2020). This technology uses short pulses of high-intensity electric fields, ranging from 1 to 80 kV/cm, with an extremely short duration, which prevents excessive heating and unwanted electrolyte reactions (Zhu, 2018). In the case of starch, PEF can damage its surface, reduce crystallinity, modify digestibility, and increase the stability of emulsions (Maniglia et al., 2021). The effects of this technology depend on parameters such as the strength of the electric field and the energy applied. In general, the process involves the application of high-voltage pulses to products positioned between two electrodes, minimizing heating and the formation of unwanted compounds, making it a promising approach for the controlled modification of food properties (Maniglia et al., 2021).

The effect of the pulsed electric field on starch modification has been widely reported in scientific studies. Maniglia et al. (2021), when investigating the impact of PEF on starch, observed that the technology reduced the apparent peak viscosity and resulted in firmer gels. In addition, depolymerization, reduction of relative crystallinity, and enthalpy of gelatinization were reported, with no changes in functional groups. These modifications have enabled innovative applications, such as 3D printing with smoother surfaces and differentiated textures. In two other studies, Li et al. (2023a, 2023b) conducted complementary studies on the effect of PEF on pretreated corn starch. In the first study, they investigated the changes in the structure and physicochemical properties of starch, finding an increase in particle size, greater solubility, an increase in the content of aldehyde groups and a reduction in viscosity. In the second study, they evaluated the impacts of PEF on the microstructure and thermal properties of starch. The results showed the formation of dents and cracks in the granules, reduction of the gelatinization peak temperature and viscosity, as well as an improvement in the thermal stability of the starch paste. These findings highlight the potential of PEF to modify starch properties in a controlled manner, opening up new possibilities for industrial applications.

4.8 High-pressure homogenization

Homogenization is a technology that uses a high-pressure pump to force emulsions through small orifices, applying a pressure of approximately 30 MPa (Ma et al., 2024). This process has the main objective of breaking up the fat globules in dairy products, increasing the stability of the product. However, due to the positive results achieved, significant technological advances have been made over time, resulting in the development of equipment capable of operating at pressures in excess of 400 MPa (Ma et al., 2024). Thus, a distinction is made between standard homogenization for pressures between 0 and 50 MPa, high-pressure homogenization (HPH) for pressures between 50 and 300 MPa, and ultra-high-pressure (UHPH) homogenization for pressures equal to or greater than 400 MPa (Mesa et al., 2020).

In the food sector, high-pressure homogenization (HPH) has been used to add value to foods by offering new approaches to starch modification. Although HPH affects the crystal structure of starch, changes in amorphous regions, including short-range ordered structures and local organization, are not yet completely understood, representing a developing research area (Apostolidis & Mandala, 2020). In the study conducted by Guo et al. (2021), the structural and functional properties of chestnut starch were evaluated with high-pressure homogenization treatment. The results showed that the treatment promoted significant improvements in chestnut starch, including a higher swelling index, transparency and anti-aging properties. With increasing pressure and the number of cycles, the destruction of the amorphous and crystalline zones of starch occurs, accompanied by variations in the functional groups and morphology of the particles. In addition, the size and crystallinity of the particles progressively decrease. In another study, Apostolidis and Mandala

(2020), evaluated starch modification using high-pressure homogenization treatment. According to the results, the treatment with a high-pressure homogenizer demonstrated greater efficacy in reducing particle size, had reduced crystallinity, higher amylose content, and a morphology characterized by the presence of pores, deep grooves, and aggregates under high-pressure conditions.

4.9 Enzymatic modification of starch

Physical and enzymatic modification of starches is widely recognized as an environmentally friendly method (Zhang & Bao, 2023), falling under the category of "Clean Label ingredients" because it is free of artificial and synthetic additives (Park & Kim, 2021). Among these methods, enzymatic modification has gained prominence, as it makes it possible to obtain specific and desirable functional characteristics (Punia, 2020).

One of the main advantages of using enzymes is their specificity and selectivity, which minimize the formation of unwanted by-products. In addition, enzymatic reactions occur under mild conditions and require less activation energy, facilitating the conversion of the substrate into the final product. From an environmental and consumer health perspective, enzymes stand out as a safer and more sustainable alternative (Park et al., 2018). Despite these advantages, one of the challenges of applying enzymes is the high cost, which can significantly increase the price of the process (Lan et al., 2016).

The enzymes used in starch modification are classified into glycosyl hydrolases and transglycosylases. Glycosyl hydrolases perform hydrolytic cleavage of glycosidic bonds, generating sugar hemiacetals and free aglycones, while transglycosylases catalyze the conversion of glycosides into other glycosidic compounds (Park & Kim, 2021). Examples of these enzymes include α-amylase, β-amylase, amylosucrase, cyclodextrin glycosyltransferase, and branching and debranching enzymes such as isoamylase and pululanase (Park et al., 2018).

Enzymatic modification involves debranching action, phosphate substitution, and disproportion by changing the length of the starch chain (Park et al., 2018). These changes in starch chain length result in the formation of low molecular weight linear short chain fractions of amylopectin, providing increased resistant starch content and increasing starch stability during freezing and thawing (Zhang & Bao, 2023). Table 4.3 presents the main effects observed on starch through enzymatic modification.

Enzymatic modification of starch promotes the formation of short linear chains of low molecular weight from amylopectin, increasing the resistant starch content and improving stability against freezing and thawing cycles, due to the reduction of syneresis in the gel network (Woo et al., 2021). These starches exhibit lower paste viscosity, differentiated rheological behavior, greater elasticity and digestibility (Kim et al., 2017; Singla et al., 2020). The action of debranching enzymes generates short-chain glucans, capable of easily recrystallizing and realigning by hydrophobic interactions and hydrogen bonds, increasing water retention, forming hydrogels and promoting chain aggregation (Liu et al., 2015).

Furthermore, enzymatic hydrolysis induces the formation of pores and cavities in the granules, creating a porous structure that favors liquid absorption, gelatinization, and control of the release of encapsulated compounds (Purcell et al., 2014). These porous starches also have a higher surface area and adsorption capacity, being useful as adsorbent materials (Wu et al., 2020). During cooling, amylose leaching from these starches facilitates the formation of gels, resulting in greater transparency, better paste clarity, and greater stability to freeze-thaw cycles (Punia et al., 2021). Finally, the crystallinity of pullulanase-treated starches is reduced, with the transition from a type A double helix structure to a more amorphous type V single helix structure (Liu et al., 2015).

Enzyme treatment can be performed using one enzyme or a combination of more than one enzyme, as shown in Table 4.2. In general, double or compound enzymatic modifications of various starches are more common in the literature than simple enzymatic modifications, in addition, the most commonly used enzymes for dual modifications are α-amylase (AA) and amyloglucosidase (GA) (Hj. Latip et al., 2021).

4.10 Effect of enzymatic modification on starch properties

Enzymatic modification of starch promotes significant changes in its physicochemical, structural and functional properties, making it more suitable for various industrial applications, especially in the food industry. Unlike chemical methods, enzymatic modification is considered a "green" and "clean label" approach, as it operates under mild temperature and pH conditions, with high specificity and less generation of unwanted by-products (Punia Bangar et al., 2021).

Enzymatic hydrolysis of starch involves different enzymes with specific functions that alter the structure and functional properties of starch. Hydrolases such as α-amylase, β-amylase and glucoamylase act by reducing molecular size and increasing the production of simple sugars. Debranching enzymes, such as pullulanase and isoamylase, promote the

TABLE 4.3 Enzymatic modification of starch.

Starch type	Enzyme	Findings	References
Simple modification			
Potatoes and corn	β-amylase (BA)	Formation of cracks and holes and increased crystallinity in potato and corn starch	Das and Kayastha (2019)
Waxy corn	amylossucrase (AS)	Increased proportion of branched chains of average amylopectin length and relative crystallinity.	Yoo et al. (2018)
Rice	β-amylase (BA)	Increased degree of gelatinization, moisture content, resistant starch and amorphous regions in starch; reduction of the proportion of crystalline regions.	Ye et al. (2018)
Double modification			
Oats	β-amylase (BA) and transglucosidase (TG)	Increase in resistant starch content, decrease in viscosity, formation of discontinuous compact fibrous structure with many surface cracks, and decrease in crystallinity.	Shah et al. (2018)
Potato	Branching enzyme (TE) and transglucosidase (TG)	Increased α-1,6-glycosidic bond ratio and short chain amounts, leading to decreased degree of crystallinity, viscosity, gelatinization and enthalpy temperature, and a notable increase in solubility.	Guo et al. (2019)
Rice	α-amylase (AA) and amyloglucosity (GA)	The results show a strong relationship between crystallinity, gelatinization temperatures, amylose content, and starch thermal properties.	Lacerda et al. (2019)
Wheat	α-amylase (AA) and pullulanase (PUL)	Increased pore number, pore volume and specific surface area of porous starches; increased adsorption capacities and degree of branching; decrease in amylose levels and amylopectin branched chain lengths.	Guo et al. (2021)
Corn	α-amylase:glucoamylase	Increasing the enzymatic hydrolysis time generated microporous starches with greater capacity to retain water and improve the gluten structure, reducing water loss in frozen doughs and improving their anti-freeze properties.	Lei et al. (2025)
Wheat starch and dough	glucoamylase and α-amylase	The modifications altered the structure and physicochemical properties of the starch, notably improving the viscoelasticity of the dough and reducing water mobility, offering a new strategy to enhance the storage and processing performance of the dough.	Gong et al. (2025)
Triple modification			
Rice	β-amylase (BA), transglucosidase (TG), and pullulanase (PUL)	Granule structure has become dense; increased proportion of short chain, amylose content and thermal stability; higher swelling and solubility indexes, and lower apparent viscosity.	Li et al. (2020)
Sweet potato	Branching enzyme, β-amylase (BA) and transglucosidase (TG)	Increased relative crystallinity during hydrolysis and solubility; decrease in viscosity, storage modulus (G′) and loss modulus (G″) of starch slurries	Guo (2018)

linearization of chains, affecting crystallinity and retrogradation. Transferases, such as CGTase, amylosucrase and α-glucanotransferase, promote complex structural rearrangements, increasing branching or generating cyclodextrins. These modifications are essential for industrial applications, such as improving texture, controlling retrogradation and increasing stability in freeze-thaw cycles. Table 4.4 presents a summary of the main enzymes, their function and effect on starch.

Enzymatic modification significantly alters the physicochemical, structural, and functional properties of starch, making it more suitable for industrial applications, especially in the food sector. This eco-friendly and "clean label" approach operates under mild conditions with high substrate specificity and minimal by-product formation (Punia Bangar et al., 2021). Enzymatic hydrolysis reduces paste viscosity by breaking α-1,4 glycosidic bonds, producing low-molecular-weight dextrins (Kim et al., 2017; Singla et al., 2020). It also enhances the elastic behavior and rheological properties of starch due to the formation of short glucan chains, which can readily realign and form stable hydrogel networks (Liu et al., 2015).

Additionally, enzymatic treatments improve freeze-thaw stability by increasing resistant starch content and preventing syneresis through controlled modification of amylopectin branches (Woo et al., 2021; Zhang & Bao, 2023). Enzyme action, especially from α-amylase and debranching enzymes like pullulanase, generates porous starch structures with higher water absorption capacity, aiding gelatinization and controlled release applications (Purcell et al., 2014; Wu et al., 2020). Moreover, enzymatic treatments can reduce starch crystallinity, shifting it from A-type to more amorphous V-type structures, enhancing solubility and paste clarity (Liu et al., 2015). The production of resistant starch is also promoted, offering potential for developing low glycemic index foods (Guo et al., 2019; Zhang et al., 2019).

4.11 Starch extrusion

Extrusion is a processing technology widely used for the modification of starches and the production of cereal-based foods with low moisture content. This process involves an intense combination of heat, shear and pressure, which leads to gelatinization, degradation and molecular reorganization of starch, resulting in products with desirable functional properties for the food industry. During the process, starch granules undergo gelatinization, while amylopectin molecules are fragmented due to the combination of high pressure, temperature and mechanical forces applied. This technique promotes a physical modification of starch, resulting in significant improvements in the physicochemical and nutritional properties and digestibility of extruded products (Mohamed, 2023). According to Qiu et al. (2024), extrusion induces gelatinization of starch by breaking hydrogen bonds and the crystalline structure, allowing starch molecules to reorganize into amorphous or partially crystalline structures. In general, the α-1,6 glycosidic bonds are broken, promoting the debranching of amylopectin and the fragmentation of starch into low molecular weight molecules and short chains. This process also converts insoluble dietary fibers into soluble ones (Liu et al., 2021).

In addition, the extruder allows chemical modifications to the starch, such as esterification and crosslinking, enabling adjustments in digestibility and obtaining extrudates with specific physical and chemical properties. This can improve digestibility, reduce the glycemic index and increase the thermal stability of the products. In addition, extrusion allows the incorporation of other food components, such as proteins, lipids and polyphenols, generating complexes that confer new functional properties to the starch.

In a study developed by Zhang et al. (2022), improved extrusion cooking technology was used to modify high-amylose jackfruit starch. The results showed that the extruded samples presented higher gelatinization enthalpy, crystallinity and molecular order with increasing moisture content, in addition to changes in the microstructure, changing from thicker to more compact surfaces. These findings highlight the impact of moisture on starch modification, especially on the development of resistant starch. In another study, Liu et al. (2021) evaluated the effect of the combination of extrusion and recrystallization treatment on the properties of corn and potato starch. As a result, the authors observed that, after extrusion and recrystallization, both showed a significant reduction in molecular weight and an increase in amylose content. Samples with 60% moisture content presented a type B crystalline pattern, also exhibiting type V peaks in vitro, with slow-digesting and resistant starch contents up to 3.5 and 6.5 times higher than non-extruded corn starch. In general, extrusion improves starch defragmentation and protein denaturation, facilitating interactions between macromolecules through covalent or non-covalent bonds.

The process increases the levels of amylose and resistant starch, essential for self-assembly and increased crystallinity. In addition, it promotes the formation of starch-lipid complexes and their interaction with proteins, generating supramolecular structures. However, due to the variability in the compositions of starch, proteins and lipids, influenced by genetic and environmental factors, the interactions remain complex and require further studies to optimize processed foods with desirable functional properties (Mohamed, 2023).

TABLE 4.4 Main enzymes used for starch hydrolysis and their effects on starch properties.

Enzyme	Classification	Main function	Effects on starch
α-Amylase	Endo-hydrolase (GH13)	Breaks internal α-1,4 linkages (random action)	Reduces viscosity, produces dextrins, increases solubility
Glucoamylase (GA)	Exo-hydrolase	Releases glucose from non-reducing ends	Increases reducing sugars, lowers molecular size
Maltogenic α-Amylase (MA)	Endo-exo hydrolase (GH13)	Produces maltose and short oligosaccharides	Improves freeze-thaw stability, reduces retrogradation
β-Amylase (BA)	Exo-hydrolase	Releases maltose from non-reducing ends	Increases short chains, reduces retrogradation
Pullulanase	Debranching enzyme	Cleaves α-1,6 linkages	Produces linear chains, reduces crystallinity, controls retrogradation
Isoamylase	Debranching enzyme	Cleaves α-1,6 linkages	Produces more linear structures, reduces viscosity
Cyclodextrin Glycosyltransferase (CGTase)	Transferase	Converts starch into cyclodextrins	Alters structure, creates hydrophobic cavities, useful for encapsulation
Amylosucrase (AS)	Transferase	Transfers glucose from sucrose to starch	Extends side chain length, alters crystallization pattern
α-Glucanotransferase (Amylomaltase)	Transferase	Rearranges glucose units within the molecule	Increases branching, improves resistance to retrogradation
Branching Enzyme (BE)	Transferase	Creates new α-1,6 linkages (increases branching)	Increases branching density, reduces viscosity and retrogradation

Table 4.5 presents a summary of the main effects of extrusion on starch, highlighting the modified parameters, the observed effects and the respective references that support these observations.

5. Challenges and future perspectives in starch processing for the food industry

Double modified starches, resulting from the combination of simple modification methods, have complex physico-chemical and functional properties, essential for various applications in the food and non-food industry. Dual modification of starch may involve combining chemical and physical methods or chemical and enzymatic methods, or with extrusion. They are widely used as emulsifiers, thickeners, and binders, as well as applied in the treatment of diseases, such as diabetes and obesity, due to their high content of resistant starch (Zia-ud-Din et al., 2017). Modifications such as acid hydrolysis and enzymatic treatments increase the surface area of starches, improving their adsorption capacity and biodegradable film-forming properties, suitable for packaging (Cabrera Canales et al., 2020). In addition, the dual modification of starch improves viscosity and thermal stability, being useful in the frozen food industry and in the production of gels for pharmaceutical applications (Chen et al., 2020). The characteristics of solubility, low retrogradation and high thermal stability are fundamental for the production of food and packaging with desirable properties.

The non-thermal physical modification of starch, such as high pressure, ultrasonication, pulsed electric field, among others mentioned above, has gained prominence due to its safety, efficiency and sustainability, without generating chemical waste. These methods alter the starch structure, affecting granule morphology, crystallinity, and molecular chains, resulting in changes in properties such as solubility, swelling power, gelatinization temperatures, and in vitro digestibility. On the other hand, the enzymatic approach has several benefits over chemical and physical modifications. However, enzymes must be safe, as plant-derived enzymes can be used as safe food components. Although it is an interesting method, due to the scarcity of sustainable sources and environmental unpredictability during food and starch processing, its applications are limited.

Despite the potential of these techniques, there is a need for further development, especially in the investigation of the effects of different treatments on different types of starch and in the conditions of process optimization. Some treatments can generate excessive heat, which requires strategies to control the temperature and avoid undesirable effects. In addition, non-thermal physical modification has shown promising results in assisted chemical reactions such as cross-linking and acetylation, but more research is needed to improve its efficiency. These modifications can be applied to food and materials, including the creation of starch-based hydrogels with improved mechanical properties.

6. Applications of modified starch in the food industry

Starch modification is a strategy widely used in the food industry to improve the functional properties of these natural polymers, making them more suitable for various applications. Modification techniques, whether physical or enzymatic, allow adjusting characteristics such as solubility, viscosity, thermal stability and resistance to retrogradation, expanding the use of starch in food products. Below, we will discuss in detail the results observed in Table 4.6, complementing them with information from the scientific literature on starch modifications and their applications in food production.

Among the physical modifications, high hydrostatic pressure has emerged as a promising non-thermal technique for altering the structure of starch without compromising the nutritional quality of foods. The treatment of wheat, corn and waxy corn starches with 600 MPa for 20 min at 25°C has been shown to improve the functionality of these polymers in oil/water emulsions with reduced fat content. In this context, modified waxy corn starch acted as an effective fat mimetic, allowing the formulation of emulsions with up to 50% less fat, without compromising texture and stability. These effects are attributed to the partial disorganization of the amorphous and crystalline regions of the starch granules, increasing their water retention capacity and their functionality as a thickener and stabilizer (Heydari & Razavi, 2022).

Another widely explored physical technique is ultrasound, which is based on the application of high-intensity acoustic waves to generate cavitation in a liquid medium, causing structural changes in the starch granules. In the case of black wheat starch, ultrasound treatment at different intensities (40%–60%) for 30 min promoted an increase in amylose concentration, as well as in solubility and swelling capacity. These effects resulted in doughs with better baking properties and texture. In addition, the application of ultrasound to rice flour used for baking contributed to a bread with a softer crumb and a lower tendency to harden during storage. This is directly related to the change in the molecular structure of the starch, favoring more efficient interactions with the dough matrix and greater moisture retention (Panghal et al., 2024).

Advancing in emerging techniques, the pulsed electric field has been studied as an alternative to modify starches without the use of heat. The treatment of different flours with PEF (2.5 kV/cm, 1000 Hz) significantly improved the appearance and elasticity of raw pasta doughs, in addition to increasing the cooking yield and reducing the hardness of the

TABLE 4.5 Main effects of extrusion on starch properties.

Evaluated parameter	Effect of extrusion	References
Gelatinization	Increases with higher moisture and temperature; reduction in gelatinization enthalpy (ΔH)	Cheng et al., 2020; Liu et al., 2017
Water absorption index (WAI)	Increases with higher temperature and moderate moisture	Liu et al. (2017)
Water solubility index (WSI)	Increases with greater shear and temperature	Liu et al. (2017); Ye et al. (2018)
Retrogradation	Reduced due to disruption of crystalline structure and hindered molecular reordering	Wang et al. (2020); Zhang et al. (2019)
Digestibility	Formation of resistant starch (RS) through complexation with lipids and other compounds	Jia et al. (2023); Liu et al. (2017)
Microstructure	Granule damage, increased surface roughness, and structural breakdown	Zhang et al. (2021), Gao et al. (2022)

TABLE 4.6 Food applications using modified starches.

Modification process	Starch	Food system	Findings	References
Physical modification				
High hydrostatic pressure (600 MPa, 20 min, 25°C)	Wheat, maize, and waxy maize	Low-fat O/A emulsions	High-pressure-treated waxy corn starch acts as an effective fat mimetic in O/A emulsions with up to 50% less fat.	Heydari and Razavi (2022)
Ultrasound (40%, 50% and 60%, 30 min, 3 min cycle, 30°C)	Black wheat	Dough	Starch modification increased the amylose concentration, solubility, and swelling ability of the modified starch, promoting better cooking properties and dough texture.	Panghal et al. (2024)
Ultrasound (80% on-off pulse, 180 W, 24 kHz, 20°C, 2, 5, 10, and 20 min)	Rice flour	Bread	The modification of starch provided a bread with a crumb with less tendency to harden	Vela et al. (2023)
Pulsed electric field (2.5 kV/cm, 1000 Hz)	Different flours	Fresh noodles	Modifying the starch improved the appearance and elasticity of their raw noodles, as well as increased cooking yield and decreased hardness	Hong et al. (2020)
Microwave (1, 2, 3, 4, 5 kW, 20 s)	Wheat flour	Fresh Wet Noodles	Starch modification improved the quality of the starch, such as starch stability, extensibility, and viscosity	Zhang et al. (2024)
Microwave (600 W, 700 W and 800 W, 8 min, 20 s cycles)	Rice flour	Rice flour bread	The modification of the starch gave the bread a higher specific volume, increased pores of the gas cells, softer crumb, lower hardness and delayed aging	Md Yunos et al. (2024)
High pressure homogenization (20 MPa, 30 MPa, 50 MPa and 70 MPa)	Wheat flour	Bread	The modification of starch provided a yellowing bread, with greater lightness, less hardness and increased cohesiveness and elasticity values	Sert et al. (2021)
High-pressure ultrasound and homogenization	Native cornstarch powder	Starch-based nanoemulsion	The modification of starch provided stable nanoemulsion to heat and freeze-thaw treatments, showed better oxidative stability, as well as physical stability during storage at 6 and 25°C.	Chutia and Mahanta (2021)
Enzymatic modification				
Thermostable (TS-αA) and mesophilic (MSαA) α-amylases	Wheat	Extruded noodles	Modified starch MSαA (1.6%) proved to be more effective than TSα A in the production of noodles by extrusion, significantly reducing cooking time.	Li et al. (2019)
Cyclodextrin glucanotransferase and debranching enzymes	Corn	Frozen pasta and bread	Dough with 5% modified starch retained more water (19%) and the bread was less hard and with less retrogradation (37%).	Woo et al. (2021)
4-α-glucanotransferase (4αGTase)	Rice and corn	A/O/A Emulsions	4αGTase-modified corn starches form stiffer gels and perform better in emulsions, with slower salt release, greater stability, and less osmotic swelling.	Kim et al. (2017)
Not informed	Sweet potato starch	Ice cream	The modification of the starch provided a reduction in the caloric content of the ice cream by 20%.	Kale et al. (2020)

4-α-glucanotransferase (4αGTase)	Rice	Starch-filled hydrogels	Hydrogels with rice starch treated with 4αGT significantly improved curcumin stability.	Kang et al. (2021)
4-α-glucanotransferase (4αGTase)	Potato starch	Thermoreversible starch gels	Starches treated with 4αGT showed sol-gel reversibility.	Jeong et al. (2020)
Glucoamylase and α-amylase	Wheat	Wheat pasta	The modification with the presence of enzymes improved the viscoelasticity and reduced the fluidity of the water in the dough, optimizing its stability and processability.	Gong et al. (2025)

final product. These effects are explained by the molecular reorganization induced by electrical pulses, which affect both starch and proteins present in the food matrix. Studies suggest that CEP can promote the permeabilization of cell membranes and alter the structure of biopolymers without causing thermal damage, which makes it ideal for fresh and minimally processed foods (Hong et al., 2020).

Similarly, the use of microwaves as a physical modification tool has proven effective in improving the functional characteristics of starch. Thermal treatment with microwaves (between 1 and 5 kW for 20 s or in controlled cycles of 8 min) of wheat and rice flours provided improvements in different products. In the production of fresh pasta, microwave-modified starch contributed to greater dough stability, extensibility and viscosity. In rice flour breads, the technique resulted in greater specific volume, greater crumb porosity, reduced hardness and delayed product aging. These benefits derive from the partial gelatinization of starch and the redistribution of moisture promoted by rapid and uniform heating, favoring the formation of a more cohesive network in the structure of the baked product (Zhang et al., 2024).

Another physical approach that deserves to be highlighted is high-pressure homogenization, used mainly to modify the structure of particles in dispersions. The application of HPH to wheat flour at pressures ranging from 20 to 70 MPa has been shown to significantly improve the quality of the bread produced. Lighter and more uniform coloration, greater lightness, lower hardness, and increased cohesiveness and elasticity values of the dough were observed. These changes are attributed to the reduction in the size of starch and protein particles, promoting a more homogeneous matrix with better gas retention during fermentation and baking (Sert et al., 2021).

The combination of techniques, such as ultrasound and HPH, has also been explored to enhance the desired effects. For example, the combined use of these techniques on native cornstarch allowed the production of heat-, freeze- and thaw-stable nanoemulsions with excellent oxidative and physical stability during storage at different temperatures. The generated nanoemulsions have reduced particle size and uniform distribution, which is essential for products that require effective emulsification, such as functional beverages, sauces and dairy desserts (Chutia & Mahanta, 2021).

In addition to physical modifications, enzymatic techniques have been widely used to alter the structure of starch in a precise and selective manner, with advantages such as mild processing conditions and reaction specificity. Among the main biocatalysts used are α-amylases, which promote the hydrolysis of α-1,4 bonds, reducing the molecular weight of starch. Comparing thermostable (TS-αA) and mesophilic (MS-αA) α-amylases applied to wheat starch for the production of extruded pasta, it was observed that the starch treated with MS-αA (1.6%) was more efficient, resulting in a significant reduction in cooking time and better product quality. The controlled enzymatic action favors the formation of an intermediate structure that improves gelatinization and dough texture (Li et al., 2019).

Other notable enzymes include cyclodextrin glucanotransferase (CGTase) and debranching enzymes, which act to restructure starch chains, promoting greater water retention and reduced retrogradation. In food systems such as frozen pasta and bread, the addition of 5% modified starch resulted in a 19% increase in water retention and significantly less hard bread, with a 37% reduction in retrogradation. These properties are highly desirable in frozen products, as they provide greater stability during storage and better quality after reheating (Woo et al., 2021).

Among the enzymes with promising applications, 4-α-glucanotransferase (4αGTase) also stands out, which has been used to modify corn, rice and potato starches, with interesting results in different contexts. In the case of multiple emulsions (W/O/W), cornstarch treated with 4αGTase formed stiffer gels, with better performance in terms of slow salt release, greater stability and lower osmotic swelling. This is especially important for products with controlled release of nutrients or active compounds, such as functional foods and nutraceuticals (Kim et al., 2017).

The same 4αGTase treatment applied to sweet potato starch allowed a 20% reduction in the caloric value of ice cream formulations, without compromising texture or stability. In products such as ice cream, where replacing fat or sugar can impair creaminess and body, the use of enzymatically modified starches is an effective solution. Rice starch treated with 4αGTase was applied in the production of hydrogels loaded with curcumin, resulting in systems with high stability of curcumin, a bioactive compound of functional and pharmaceutical interest. Finally, the treatment of potato starch with 4αGTase resulted in thermoreversible starch gels, i.e., with the capacity for reversible transition between the sol and gel states, which is useful in systems that require structural changes during preparation or consumption (Kale et al., 2020).

In addition, the combination of glucoamylase and α-amylase applied to wheat doughs has been shown to significantly improve viscoelasticity and reduce water fluidity, promoting greater dough stability and processability. This translates into products that are easier to shape and have better behavior during baking, especially in industrial baking (Gong et al., 2025).

In summary, the different starch modification techniques, both physical and enzymatic, have proven effective in improving their technological properties and expanding their applications in food production. Each technique confers distinct characteristics to the starch, enabling its use in emulsions, doughs, breads, noodles, ice cream, hydrogels and other food systems. The choice of modification method should consider the type of final product desired, the processing conditions, the costs involved and the nutritional and sensory requirements. With the advancement of starch modification and characterization technologies, new possibilities continue to emerge, paving the way for more functional, healthier, stable foods with better sensory quality.

7. Conclusions

Starch modification can be achieved through enzymatic, physical, and chemical methods, each with distinct advantages and limitations. Enzymatic methods, although more commonly applied in the pharmaceutical and food industries due to their specificity, are limited in polymer applications because of high production costs. Physical methods, while more environmentally friendly and cost-effective, often compromise the starch structure, restricting their industrial use. In contrast, chemical methods are widely used in industry for their effectiveness in modifying starch structure, enabling the development of specific functional properties. These modifications impact crystallinity, morphology, molecular weight, and functional attributes such as shelf life, foam stability, air barrier, and moisture control. The global modified starch market is expected to expand, particularly in the food and beverage sector, where modified starch serves as a gluten-free emulsifier and stabilizer. This growth is driven by rising demand for processed foods, bio-based adhesives, and starch-derived products, as well as increasing interest in natural sweeteners and the incorporation of modified starch in cosmetics, paper, and pharmaceuticals.

References

Alcázar-Alay, S. C., & Meireles, M. A. A. (2015). Physicochemical properties, modifications and applications of starches from different botanical sources. *Food Science and Technology, 35*, 215–236.

Almeida, R. L. J., Monteiro, S. S., Santos, N. C., Rios, N. S., & dos Santos, E. S. (2024). Starch modification and its application in Pickering emulsion stabilization: A review. *Journal of Food Measurement and Characterization, 18*(6), 4984–5003.

Almeida, R. L. J., Santos, N. C., de Medeiros, W. R. D., de Queiroga, A. P. R., Rios, N. S., & dos Santos, E. S. (2025). Dual modification of red rice starch: Combining hydrothermal pretreatment with acid hydrolysis for the production of nanomaterials. *Chemical Engineering and Processing-Process Intensification, 213*, 110292.

Almeida, R. L. J., Santos, N. C., da Silva Pedro, M., de Souza Ferreira, I. L., da Silva Eduardo, R., Muniz, C. E. S, da Costa, G. A., et al. (2024). Combined effects of high hydrostatic pressure and pulsed electric fields on quinoa starch: Analysis of microstructure, morphology, thermal, and pasting properties. *Food Chemistry, 460*, 140826. https://doi.org/10.1016/j.foodchem.2024.140826

Almeida, R. L. J., Santos, N. C., de Brito, A. C. O., Leite, A. C. N., Morais, J. R., et al. (2025). Dual modification of starch: Synergistic effects of ozonation and pulsed electric fields on structural, rheological, and functional attributes. *Food Chemistry, 464*, 141718.

Almeida, R. L. J., Santos, N. C., Feitoza, J. V. F., da Silva, G. M., de Sousa Muniz, C. E., da Silva Eduardo, R., & de Almeida Mota, M. M. (2022). Effect of heat-moisture treatment on the thermal, structural and morphological properties of Quinoa starch. *Carbohydrate Polymer Technologies and Applications, 3*, Article 100192.

Almeida, R. L. J., Santos, N. C., Feitoza, J. V. F., dos Santos Pereira, T., da Silva Eduardo, R., de Almeida Mota, M. M., & de Figueiredo, M. J. (2023). Evaluation of the technological properties of rice starch modified by high hydrostatic pressure (HHP). *Innovative Food Science & Emerging Technologies, 83*, Article 103241.

Amarnath, M. S., Muhammed, A., Antony, A. K., Malini, B., & Sunil, C. K. (2023). White finger millet starch: Physical modification (annealing and ultrasound), and its impact on physicochemical, functional, thermal and structural properties. *Food and Humanity, 1*, 599–606.

Amini, H. W., Masruri, M., & Ulfa, S. M. (2017). Study on esterification reaction of starch isolated from cassava (Manihot Esculenta) with acetic acid and Isopropyl myristate using ultrasonicator. *The Journal of Pure and Applied Chemistry Research, 6*(2), 93.

Apostolidis, E., & Mandala, I. (2020). Modification of resistant starch nanoparticles using high-pressure homogenization treatment. *Food Hydrocolloids, 103*, Article 105677.

Balakumaran, M., Nath, K. G., Giridharan, B., Dhinesh, K., Dharunbalaji, A. K., Malini, B., & Sunil, C. K. (2023). White finger millet starch: Hydrothermal and microwave modification and its characterisation. *International Journal of Biological Macromolecules, 242*, Article 124619.

Bangar, S. P., Ashogbon, A. O., Singh, A., Chaudhary, V., & Whiteside, W. S. (2022). Enzymatic modification of starch: A green approach for starch applications. *Carbohydrate Polymers, 287*, Article 119265.

BeMiller, J. N., & Huber, K. C. (2015). Physical modification of food starch functionalities. *Annual Review of Food Science and Technology, 6*(1), 19–69.

Braşoveanu, M., & Nemţanu, M. R. (2014). Behaviour of starch exposed to microwave radiation treatment. *Starch-Stärke, 66*(1–2), 3–14.

Cárcel, J. A., García-Pérez, J. V., Benedito, J., & Mulet, A. (2012). Food process innovation through new technologies: Use of ultrasound. *Journal of Food Engineering, 110*(2), 200–207.

Cabrera Canales, Z. E., Rodríguez Marín, M. L., Gómez Aldapa, C. A., Méndez Montealvo, G., Chávez Gutiérrez, M., & Velazquez, G. (2020). Effect of dual chemical modification on the properties of biodegradable films from achira starch. *Journal of Applied Polymer Science, 137*(45), 49411.

Cai, C., & Wei, C. (January 30, 2013). In situ observation of crystallinity disruption patterns during starch gelatinization. *Carbohydrate Polymers, 92*(1), 469–478.

Cai, C., Zhao, L., Huang, J., Chen, Y., & Wei, C. (February 15, 2014). Morphology, structure and gelatinization properties of heterogeneous starch granules from high-amylose maize. *Carbohydrate Polymers, 102*, 606–614.

Carrillo Parra, E. J., Palavecino, P., Ribotta, P. D., & Penci, M. C. (2025). Effect of ultrasound treatment on structural and physical properties of native maize starch. *Plant Foods for Human Nutrition, 80*(1), 1–8.

Castanha, N., da Matta Junior, M. D., & Augusto, P. E. D. (2017). Potato starch modification using the ozone technology. *Food Hydrocolloids, 66*, 343–356.

Castanha, N., Lima, D. C., Junior, M. D. M., Campanella, O. H., & Augusto, P. E. D. (2019). Combining ozone and ultrasound technologies to modify maize starch. *International Journal of Biological Macromolecules, 139*, 63–74.

Castanha, N., Miano, A. C., Jones, O. G., Reuhs, B. L., Campanella, O. H., & Augusto, P. E. (2020). Starch modification by ozone: Correlating molecular structure and gel properties in different starch sources. *Food Hydrocolloids, 108*, Article 106027.

Chakraborty, I., Pooja, N., Mal, S. S., Paul, U. C., Rahman, M. H., & Mazumder, N. (2022). An insight into the gelatinization properties influencing the modified starches used in food industry: A review. *Food and Bioprocess Technology, 15*(6), 1195–1223.

Chatel, G. (2018). How sonochemistry contributes to green chemistry? *Ultrasonics Sonochemistry, 40*, 117–122.

Chen, Y., She, Y., Zhang, R., Wang, J., Zhang, X., & Gou, X. (2020). Use of starch-based fat replacers in foods as a strategy to reduce dietary intake of fat and risk of metabolic diseases. *Food Science and Nutrition, 8*(1), 16–22.

Chen, B. R., Wen, Q. H., Zeng, X. A., Abdul, R., Roobab, U., & Xu, F. Y. (2021). Pulsed electric field assisted modification of octenyl succinylated potato starch and its influence on pasting properties. *Carbohydrate Polymers, 254*, Article 117294.

Chen, P., Yu, L., Simon, G. P., Liu, X., Dean, K., & Chen, L. (February 1, 2011). Internal structures and phase-transitions of starch granules during gelatinization. *Carbohydrate Polymers, 83*(4), 1975–1983.

Chen, Q., Yu, H., Wang, L., Abdin ul, Z., Chen, Y., Wang, J., & Chen, X. (2015). Recent progress in chemical modification of starch and its applications. *RSC Advances, 5*(83), 67459–67474.

Cheng, W., Gao, L., Wu, D., Gao, C., Meng, L., Feng, X., & Tiang, X. (2020). Effect of improved extrusion cooking technology on structure, physiochemical and nutritional characteristics of physically modified buckwheat flour: Its potential use as food ingredients. *LWT, 133*, 109872.

Christensen, S. J., Madsen, M. S., Zinck, S. S., Hedberg, C., Sørensen, O. B., Svensson, B., & Meyer, A. S. (2023). Enzymatic potato starch modification and structure-function analysis of six diverse GH77 4-alpha-glucanotransferases. *International Journal of Biological Macromolecules, 224*, 105–114.

Chutia, H., & Mahanta, C. L. (2021). Properties of starch nanoparticle obtained by ultrasonication and high pressure homogenization for developing carotenoids-enriched powder and pickering nanoemulsion. *Innovative Food Science & Emerging Technologies, 74*, Article 102822.

Clasen, S. H., Müller, C. M., Parize, A. L., & Pires, A. T. (2018). Synthesis and characterization of cassava starch with maleic acid derivatives by etherification reaction. *Carbohydrate Polymers, 180*, 348–353.

Compart, J., Singh, A., Fettke, J., & Apriyanto, A. (2023). Customizing starch properties: A review of starch modifications and their applications. *Polymers, 15*(16), 3491.

Das, R., & Kayastha, A. M. (2019). Enzymatic hydrolysis of native granular starches by a new β-amylase from peanut (Arachis hypogaea). *Food Chemistry, 276*, 583–590.

de la Hoz, A., & Loupy, A. (2013). *Microwaves in organic synthesis. John Wiley & Sons.*

Ding, X. L., Wang, L. J., Li, T. T., Wang, F., Quan, Z. Y., Zhou, M., & Qian, J. Y. (2021). Pre-gelatinisation of rice flour and its effect on the properties of gluten free rice bread and its batter. *Foods, 10*(11), 2648.

Dominguez-Ayala, J. E., Ayala-Ayala, M. T., Velazquez, G., Espinosa-Arbelaez, D. G., & Mendez-Montealvo, G. (2023). Crystal structure changes of native and retrograded starches modified by high hydrostatic pressure: Physical dual modification. *Food Hydrocolloids, 140*, 108630.

El Khaled, D., Novas, N., Gazquez, J. A., & Manzano-Agugliaro, F. (2018). Microwave dielectric heating: Applications on metals processing. *Renewable and Sustainable Energy Reviews, 82*, 2880–2892.

Falsafi, S. R., Maghsoudlou, Y., Rostamabadi, H., Rostamabadi, M. M., Hamedi, H., & Hosseini, S. M. H. (2019). Preparation of physically modified oat starch with different sonication treatments. *Food Hydrocolloids, 89*, 311–320.

Gao, S., Zhang, Y., Wang, J., Jiang, J., Zhai, X., Wang, W., & Hou, H. (2022). Influence of starch content on the physicochemical and antimicrobial properties of starch/PBAT/ε-polylysine hydrochloride blown films. *Food Packaging and Shelf Life, 34*, 101005.

Gong, J., Xu, W., Zhang, C., Zhu, Q., Qin, X., Zhang, H., & Liu, G. (2025). Effects of esterification and enzymatic modification on the properties of wheat starch and dough. *Food Hydrocolloids, 158*, Article 110509.

González-Mendoza, M. E., Martínez-Bustos, F., Castaño-Tostado, E., & Amaya-Llano, S. L. (2022). Effect of microwave irradiation on acid hydrolysis of Faba bean starch: Physicochemical changes of the starch granules. *Molecules, 27*(11), 3528.

Govindaraju, I., Zhuo, G. Y., Chakraborty, I., Melanthota, S. K., Mal, S. S., Sarmah, B., & Mazumder, N. (2022). Investigation of structural and physico-chemical properties of rice starch with varied amylose content: A combined microscopy, spectroscopy, and thermal study. *Food Hydrocolloids, 122*, Article 107093.

Guo, L. (2018). Sweet potato starch modified by branching enzyme, β-amylase and transglucosidase. *Food Hydrocolloids, 83*, 182–189.

Guo, L., Deng, Y., Lu, L., Zou, F., & Cui, B. (2019). Synergistic effects of branching enzyme and transglucosidase on the modification of potato starch granules. *International Journal of Biological Macromolecules, 130*, 499–507.

Guo, L., Yuan, Y., Li, J., Tan, C., Janaswamy, S., Lu, L., & Cui, B. (2021). Comparison of functional properties of porous starches produced with different enzyme combinations. *International Journal of Biological Macromolecules, 174*, 110–119.

Han, Z., Li, Y., Luo, D. H., Zhao, Q., Cheng, J. H., & Wang, J. H. (2021). Structural variations of rice starch affected by constant power microwave treatment. *Food Chemistry, 359*, Article 129887.

He, Y., Ye, F., Tao, J., Zhang, Z., & Zhao, G. (2023). Ozone exposure tunes the physicochemical properties of sweet potato starch by modifying its molecular structure. *International Journal of Biological Macromolecules, 236*, Article 124002.

Heydari, A., & Razavi, S. M. A. (2022). Impact of HHP-treated starches on LAOS, tribology, and steady shear behavior of reduced-fat O/W emulsions. *Innovative Food Science & Emerging Technologies, 78*, Article 103006.

Hj. Latip, D. N., Samsudin, H., Utra, U., & Alias, A. K. (2021). Modification methods toward the production of porous starch: A review. *Critical Reviews in Food Science and Nutrition, 61*(17), 2841–2862.

Hong, J., An, D., Liu, C., Li, L., Han, Z., Guan, E., & Bian, K. (2020). Rheological, textural, and digestible properties of fresh noodles: Influence of starch esterified by conventional and pulsed electric field-assisted dual technique with full range of amylose content. *Journal of Food Processing and Preservation, 44*(8), Article e14567.

Huang, X., Chen, L., & Liu, Y. (2024). Effects of ultrasonic and ozone modification on the morphology, mechanical, thermal and barrier properties of corn starch films. *Food Hydrocolloids, 147*, Article 109376.

Jenkins, P. J., & Donald, A. M. (1995). The influence of amylose on starch granule structure. *International Journal of Biological Macromolecules, 17*(6), 315–321.

Jeong, D. W., Jeong, H. M., Shin, Y. J., Woo, S. H., & Shim, J. H. (2020). Properties of recombinant 4-α-glucanotransferase from Bifidobacterium longum subsp. longum JCM 1217 and its application. *Food Science and Biotechnology, 29*, 667–674.

Jia, R., Cui, C., Gao, L., Qin, Y., Ji, N., Dai, L., & Sun, Q. (2023). A review of starch swelling behavior: Its mechanism, determination methods, influencing factors, and influence on food quality. *Carbohydrate Polymers*, Article 121260.

Kale, R. V., Sontakke, M. D., Raut, G. S., & Chavan, V. R. (2020). Use of enzyme modified sweet potato starch in formulation of ice cream. *International Journal of Chemical Studies, 8*(4), 3002–3008.

Kang, J., Kim, Y. H., Choi, S. J., Rho, S. J., & Kim, Y. R. (2021). Improving the stability and curcumin retention rate of curcumin-loaded filled hydrogel prepared using 4αGTase-treated rice starch. *Foods, 10*(1), 150.

Kaur, M., Oberoi, D. P., Sogi, D. S., & Gill, B. S. (August 2011). Physicochemical, morphological and pasting properties of acid treated starches from different botanical sources. *Journal of Food Science and Technology, 48*, 460–465.

Kaur, R., Sharma, M., Ji, D., Xu, M., & Agyei, D. (2019). Structural features, modification, and functionalities of beta-glucan. *Fibers, 8*(1), 1.

Khadre, M. A., & Yousef, A. E. (2001). Sporicidal action of ozone and hydrogen peroxide: A comparative study. *International Journal of Food Microbiology, 71*(2–3), 131–138.

Kim, Y. L., Mun, S., Rho, S. J., Do, H. V., & Kim, Y. R. (2017). Influence of physicochemical properties of enzymatically modified starch gel on the encapsulation efficiency of W/O/W emulsion containing NaCl. *Food and Bioprocess Technology, 10*, 77–88.

Kumari, B., & Sit, N. (2023). Comprehensive review on single and dual modification of starch: Methods, properties and applications. *International Journal of Biological Macromolecules*, Article 126952.

Lacerda, L. D., Leite, D. C., & da Silveira, N. P. (2019). Relationships between enzymatic hydrolysis conditions and properties of rice porous starches. *Journal of Cereal Science, 89*, Article 102819.

Lan, X., Xie, S., Wu, J., Xie, F., Liu, X., & Wang, Z. (2016). Thermal and enzymatic degradation induced ultrastructure changes in canna starch: Further insights into short-range and long-range structural orders. *Food Hydrocolloids, 58*, 335–342.

Lee, Y. K., & Chang, Y. H. (2019). Structural and in vitro digestibility properties of esterified maca starch with citric acid and its application as an oil-in-water (O/W) pickering emulsion stabilizer. *International Journal of Biological Macromolecules, 134*, 798–806.

Lei, M., Yuan, H., Jia, R., Huang, Z., Yang, Y., Liang, Q., & Pan, Z. (2025). Effects of different enzymatic hydrolysis times on the structures and properties of corn microporous starch particles and their applications in frozen dough. *Food Hydrocolloids, 164*, Article 111178.

Li, C. (2022). Recent progress in understanding starch gelatinization-An important property determining food quality. *Carbohydrate Polymers, 293*, 119735.

Li, H., Gui, Y., Li, J., Zhu, Y., Cui, B., & Guo, L. (2020). Modification of rice starch using a combination of autoclaving and triple enzyme treatment: Structural, physicochemical and digestibility properties. *International Journal of Biological Macromolecules, 144*, 500–508.

Li, J., Rashed, M. M., Deng, L., Jin, Z., & Jiao, A. (2019). Thermostable and mesophilic α-amylase: Effects on wheat starch physicochemical properties and their applications in extruded noodles. *Journal of Cereal Science, 87*, 248–257.

Li, Y., Wang, J. H., Han, Y., Yue, F. H., Zeng, X. A., Chen, B. R., & Han, Z. (2023). The effects of pulsed electric fields treatment on the structure and physicochemical properties of dialdehyde starch. *Food Chemistry, 408*, Article 135231.

Li, Y., Wang, J. H., Wang, E. C., Tang, Z. S., Han, Y., Luo, X. E., et al. (2023). The microstructure and thermal properties of pulsed electric field pretreated oxidized starch. *International Journal of Biological Macromolecules, 235*, 123721.

Li, Y., Wang, J. H., Wang, E. C., Tang, Z. S., Han, Y., Luo, X. E., & Han, Z. (2023b). The microstructure and thermal properties of pulsed electric field pretreated oxidized starch. *International Journal of Biological Macromolecules, 235*, Article 123721.

Li, C., Wu, A., Yu, W., Hu, Y., Li, E., Zhang, C., & Liu, Q. (2020). Parameterizing starch chain-length distributions for structure-property relations. *Carbohydrate Polymers, 241*, 116390.

Lima, D. C., Castanha, N., Maniglia, B. C., Matta Junior, M. D., La Fuente, C. I. A., & Augusto, P. E. D. (2021). Ozone processing of cassava starch. *Ozone: Science & Engineering, 43*(1), 60–77.

Lima, D. C., Villar, J., Castanha, N., Maniglia, B. C., Junior, M. D. M., & Augusto, P. E. D. (2020). Ozone modification of arracacha starch: Effect on structure and functional properties. *Food Hydrocolloids, 108*, Article 106066.

Liu, G., Gu, Z., Hong, Y., Cheng, L., & Li, C. (2017). Structure, functionality and applications of debranched starch: A review. *Trends in Food Science and Technology, 63*, 70–79.

Liu, G., Hong, Y., Gu, Z., Li, Z., Cheng, L., & Li, C. (2015). Preparation and characterization of pullulanase debranched starches and their properties for drug controlled-release. *RSC Advances, 5*(117), 97066–97075.

Liu, Q., Jiao, A., Yang, Y., Wang, Y., Li, J., Xu, E., & Jin, Z. (2021). The combined effects of extrusion and recrystallization treatments on the structural and physicochemical properties and digestibility of corn and potato starch. *LWT, 151*, Article 112238.

Liu, G., Zhang, R., Huo, S., Li, J., Wang, M., Wang, W., & Zheng, J. (2023). Insights into the changes of structure and digestibility of microwave and heat treated moisture quinoa starch. *International Journal of Biological Macromolecules, 246*, Article 125681.

Ma, H., Liu, M., Liang, Y., Zheng, X., Sun, L., Dang, W., & Liu, C. (2022). Research progress on properties of pre-gelatinized starch and its application in wheat flour products. *Grain & Oil Science and Technology, 5*(2), 87–97.

Ma, W., Tang, J., Cheng, H., Tain, J., Wu, Z., & Zhou, J. (2024). High-resistant starch based on amylopectin cluster via extrusion: From the perspective of chain-length distribution and structural formation. *Foods, 13*(16), 2532.

Mahmood, K., Kamilah, H., Shang, P. L., Sulaiman, S., & Ariffin, F. (2017). A review: Interaction of starch/non-starch hydrocolloid blending and the recent food applications. *Food Bioscience, 19*, 110–120.

Majzoobi, M., & Farahnaky, A. (2021). Granular cold-water swelling starch; properties, preparation and applications, a review. *Food Hydrocolloids, 111*, Article 106393.

Makroo, H. A., Naqash, S., Saxena, J., Sharma, S., Majid, D., & Dar, B. N. (2021). Recovery and characteristics of starches from unconventional sources and their potential applications: A review. *Applied Food Research, 1*(1), Article 100001.

Maniglia, B. C., Pataro, G., Ferrari, G., Augusto, P. E. D., Le-Bail, P., & Le-Bail, A. (2021). Pulsed electric fields (PEF) treatment to enhance starch 3D printing application: Effect on structure, properties, and functionality of wheat and cassava starches. *Innovative Food Science & Emerging Technologies, 68*, Article 102602.

Mathobo, V. M., Silungwe, H., Ramashia, S. E., & Anyasi, T. A. (2021). Effects of heat-moisture treatment on the thermal, functional properties and composition of cereal, legume and tuber starches—a review. *Journal of Food Science and Technology, 58*(2), 412–426.

Md Yunos, N. S. H., Hafid, H. S., Omar, F. N., Mohammed, M. A. P., Wakisaka, M., Mustapha, N. A., & Samsu Baharuddin, A. (2024). Microwave-assisted treatment for the improvement of rice flour properties and rice flour bread quality. *International Journal of Food Science and Technology, 59*(12), 9157–9169.

Mesa, J., Hinestroza-Córdoba, L. I., Barrera, C., Seguí, L., Betoret, E., & Betoret, N. (2020). High homogenization pressures to improve food quality, functionality and sustainability. *Molecules, 25*(14), 3305.

Miano, A. C., Rojas, M. L., & Augusto, P. E. (2017). Other mass transfer unit operations enhanced by ultrasound. In *Ultrasound: Advances for food processing and preservation* (pp. 369–389). Academic Press.

Mohamed, I. O. (2023). Interaction of starch with some food macromolecules during the extrusion process and its effect on modulating physicochemical and digestible properties. The review. *Carbohydrate Polymer Technologies and Applications, 5*, Article 100294.

Neutzling, H. P., Timm, N. D. S., Cañizares, L. D. C. C., Schwab, M. P., Ferreira, C. D., Ramos, A. H., & de Oliveira, M. (2023). Effects of infrared radiation heating temperature and time on starch gelatinization and technological, thermal, and pasting properties of parboiled rice. *Cereal Chemistry, 100*(3), 734–744.

Oliveira, A. V., Da Silva, A. P. M., Barros, M. O., de sá, M., Souza Filho, M., Rosa, M. F., & Azeredo, H. M. (2018). Nanocomposite films from mango kernel or corn starch with starch nanocrystals. *Starch-Stärke, 70*(11–12), 1800028.

Oyeyinka, S. A., Umaru, E., Olatunde, S. J., & Joseph, J. K. (2019). Effect of short microwave heating time on physicochemical and functional properties of Bambara groundnut starch. *Food Bioscience, 28*, 36–41.

Panghal, A., Kumar, R., Bishnoi, P., Rana, D., & Chhikara, N. (2024). Impact of ultrasonication on physicochemical, morphological, thermal, pasting, and pasta quality attributes of black wheat starch. *International Journal of Food Science and Technology, 59*(7), 4470–4478.

Park, S., & Kim, Y. R. (2021). Clean label starch: Production, physicochemical characteristics, and industrial applications. *Food Science and Biotechnology, 30*, 1–17.

Park, S. H., Na, Y., Kim, J., Kang, S. D., & Park, K. H. (2018). Properties and applications of starch modifying enzymes for use in the baking industry. *Food Science and Biotechnology, 27*, 299–312.

Park, K. H., Park, J. H., Lee, S., Yoo, S. H., & Kim, J. W. (2008). Enzymatic modification of starch for food industry. *Carbohydrate-active Enzymes*, 157–183.

Parker, R., & Ring, S. G. (2001). Aspects of the physical chemistry of starch. *Journal of Cereal Science, 34*(1), 1–17.

Prabhu, M., Levkov, K., Levin, O., Vitkin, E., Israel, A., Chemodanov, A., & Golberg, A. (2020). Energy efficient dewatering of far offshore grown green macroalgae Ulva sp. biomass with pulsed electric fields and mechanical press. *Bioresource Technology, 295*, Article 122229.

Pu, H., Chen, L., Li, L., & Li, X. (2011). Multi-scale structural and digestion resistibility changes of high-amylose corn starch after hydrothermal-pressure treatment at different gelatinizing temperatures. *Food Research International, 53*(1), 456–463.

Punia, S. (2020). Barley starch modifications: Physical, chemical and enzymatic-A review. *International Journal of Biological Macromolecules, 144*, 578–585.

Punia Bangar, S., Nehra, M., Siroha, A. K., Petrů, M., Ilyas, R. A., Devi, U., & Devi, P. (2021). Development and characterization of physical modified pearl millet starch-based films. *Foods, 10*(7), 1609.

Punia, S., Kumar, M., Siroha, A. K., Kennedy, J. F., Dhull, S. B., & Whiteside, W. S. (2021). Pearl millet grain as an emerging source of starch: A review on its structure, physicochemical properties, functionalization, and industrial applications. *Carbohydrate Polymers, 260*, Article 117776.

Purcell, S., Wang, Y. J., & Seo, H. S. (2014). Enzyme-modified starch as an oil delivery system for bake-only chicken nuggets. *Journal of Food Science, 79*(5), C802–C809.

Qiu, C., Hu, H., Chen, B., Lin, Q., Ji, H., & Jin, Z. (2024). Research progress on the physicochemical properties of starch-based foods by extrusion processing. *Foods, 13*(22), 3677.

Raghunathan, R., Pandiselvam, R., Kothakota, A., & Khaneghah, A. M. (2021). The application of emerging non-thermal technologies for the modification of cereal starches. *LWT, 138*, 110795.

Rahaman, A., Kumari, A., Zeng, X. A., Farooq, M. A., Siddique, R., Khalifa, I., & Manzoor, M. F. (2021). Ultrasound based modification and structural-functional analysis of corn and cassava starch. *Ultrasonics Sonochemistry, 80*, Article 105795.

Ramos, G. V. C., Branco, I. G., da Matta Junior, M. D., Ferreira, J. A., Ditchfield, C., do Amaral Sobral, P. J., & Moraes, I. C. F. (2024). Structural and functional properties of ozone-treated quinoa starch. *Polymer, 308*, Article 127314.

Rivero-Ramos, P., Valdez, M. I., Sanz, T., Garzón, R., Rosell, C. M., Benlloch-Tinoco, M., & Rodrigo, D. (2024). Rice starch-alginate systems gelatinised by high hydrostatic pressure (HHP) as dysphagia-oriented matrices. *Food Hydrocolloids, 151*, Article 109793.

Roohi, R., Abedi, E., Hashemi, S. M. B., & Akbari, M. (2024). Effect of ultrasound geometry on the production efficiency of damaged starch: Determining rheology parameters, and non-isothermal reaction kinetics. *Ultrasonics Sonochemistry, 106*, Article 106882.

Saleh, A., Mohamed, A. A., Alamri, M. S., Hussain, S., Qasem, A. A., & Ibraheem, M. A. (2020). Effect of different starches on the rheological, sensory and storage attributes of non-fat set yogurt. *Foods, 9*(1), 61.

Sert, D., Mercan, E., Tongur, A., & Kara, Ü. (2021). Production of bread from doughs composed of high-pressure homogenisation treated flour slurries: Effects on physicochemical, crumb grain and textural characteristics. *Journal of Food Measurement and Characterization, 15*(4), 3052–3059.

Shah, A., Masoodi, F. A., Gani, A., & Ashwar, B. (2018). Dual enzyme modified oat starch: Structural characterisation, rheological properties, and digestibility in simulated GI tract. *International Journal of Biological Macromolecules, 106*, 140–147.

Sharma, D., Afzal, S., & Singh, N. K. (2021). Nanopriming with phytosynthesized zinc oxide nanoparticles for promoting germination and starch metabolism in rice seeds. *Journal of Biotechnology, 336*, 64–75.

Sharma, S., Thakur, K., Sharma, R., & Bobade, H. (2023). Molecular morphology & interactions, functional properties, rheology and in vitro digestibility of ultrasonically modified pearl millet and sorghum starches. *International Journal of Biological Macromolecules, 253*, Article 127476.

Sherin, A. J., Sunil, C. K., Chidanand, D. V., & Venkatachalapathy, N. (2024). Structural, physicochemical and functional properties of high-pressure modified white finger millet starch. *International Journal of Biological Macromolecules, 261*, Article 129919.

Shukri, R., & Shi, Y. C. (2017). Structure and pasting properties of alkaline-treated phosphorylated cross-linked waxy maize starches. *Food Chemistry, 214*, 90–95.

Singh, M., & Adedeji, A. A. (2017). Characterization of hydrothermal and acid modified proso millet starch. *LWT-Food Science and Technology, 79*, 21–26.

Singla, D., Singh, A., Dhull, S. B., Kumar, P., Malik, T., & Kumar, P. (2020). Taro starch: Isolation, morphology, modification and novel applications concern-A review. *International Journal of Biological Macromolecules, 163*, 1283–1290.

Sun, X., Sun, Z., Saleh, A. S., Lu, Y., Zhang, X., Ge, X., & Li, W. (2023). Effects of various microwave intensities collaborated with different cold plasma duration time on structural, physicochemical, and digestive properties of lotus root starch. *Food Chemistry, 405*, Article 134837.

Sun, X., Yu, C., Fu, M., Wu, D., Gao, C., Feng, X., & Tang, X. (2019). Extruded whole buckwheat noodles: Effects of processing variables on the degree of starch gelatinization, changes of nutritional components, cooking characteristics and in vitro starch digestibility. *Food & Function, 10*(10), 6362–6373.

Taddei, R., Giacometti, F., Bardasi, L., Bonilauri, P., Ramini, M., Fontana, M. C., et al. (2018). Effect of production process and high-pressure processing on viability of *Listeria innocua* in traditional Italian dry-cured coppa. *Italian Journal of Food Safety, 9*(2), 9133.

Taghinezhad, E., Khoshtaghaza, M. H., Minaei, S., Suzuki, T., & Brenner, T. (2016). Relationship between degree of starch gelatinization and quality attributes of parboiled rice during steaming. *Rice Science, 23*(6), 339–344.

Tagliapietra, B. L., Felisberto, M. H. F., Sanches, E. A., Campelo, P. H., & Clerici, M. T. P. S. (2021). Non-conventional starch sources. *Current Opinion in Food Science, 39*, 93–102.

Tharanathan, R. N. (2005). Starch—value addition by modification. *Critical Reviews in Food Science and Nutrition, 45*(5), 371–384.

Van Hung, P., Maeda, T., & Morita, N. (2006). Waxy and high-amylose wheat starches and flours—characteristics, functionality and application. *Trends in Food Science and Technology, 17*(8), 448–456.

Vela, A. J., Villanueva, M., & Ronda, F. (2023). Physical modification caused by acoustic cavitation improves rice flour bread-making performance. *LWT, 183*, Article 114950.

Wang, S., Chao, C., Xiang, F., Zhang, X., Wang, S., & Copeland, L. (February 14, 2018). New insights into gelatinization mechanisms of cereal endosperm starches. *Scientific Reports, 8*(1), 3011.

Wang, C., Xue, Y., Yousaf, L., Hu, J., & Shen, Q. (2020). Effects of high hydrostatic pressure on the ordered structure, including double helices and V-type single helices of rice starch. *International Journal of Biological Macromolecules, 144*, 1034–1042.

Wei, Y., Li, G., & Zhu, F. (2023). Impact of long-term ultrasound treatment on structural and physicochemical properties of starches differing in granule size. *Carbohydrate Polymers, 320*, Article 121195.

Wolf, B. W., Bauer, L. L., & Fahey, G. C. (1999). Effects of chemical modification on in vitro rate and extent of food starch digestion: An attempt to discover a slowly digested starch. *Journal of Agricultural and Food Chemistry, 47*(10), 4178–4183.

Woo, S. H., Kim, J. S., Jeong, H. M., Shin, Y. J., Hong, J. S., Choi, H. D., & Shim, J. H. (2021). Development of freeze-thaw stable starch through enzymatic modification. *Foods, 10*(10), 2269.

Wu, W., Jiao, A., Xu, E., Chen, Y., & Jin, Z. (2020). Effects of extrusion technology combined with enzymatic hydrolysis on the structural and physicochemical properties of porous corn starch. *Food and Bioprocess Technology, 13*, 442–451.

Xiao, C. (2013). Current advances of chemical and physical starch-based hydrogels. *Starch-Stärke, 65*(1-2), 82–88.

Yan, G., Cui, Y., Lia, D., Ding, Y., Han, J., Wang, S., & Zheng, H. (2022). The characteristics of soybean protein isolate obtained by synergistic modification of high hydrostatic pressure and phospholipids as a promising replacement of milk in ice cream. *LWT, 160*, Article 113223.

Yan, X., McClements, D. J., Luo, S., Liu, C., & Ye, J. (2024). Recent advances in the impact of gelatinization degree on starch: Structure, properties and applications. *Carbohydrate Polymers*, Article 122273.

Yang, B., Fang, X., Chen, L., Du, M., Din, Z. U., Wang, Y., & Ding, W. (2024). Ozone modification of waxy rice starch nanocrystals: Effects on the multi-scale structural and surface properties. *International Journal of Biological Macromolecules, 278*, Article 134500.

Ye, S. J., & Baik, M. Y. (2024). Physicochemical properties of amorphous granular starch (AGS) prepared by non-thermal gelatinization by high hydrostatic pressure (HHP) and spray drying. *International Journal of Biological Macromolecules, 260*, Article 129508.

Ye, J., Liu, C., Luo, S., Hu, X., & McClements, D. J. (2018). Modification of the digestibility of extruded rice starch by enzyme treatment (β-amylolysis): An in vitro study. *Food Research International, 111*, 590–596.

Yeoh, S. Y., Lubowa, M., Tan, T. C., Murad, M., & Easa, A. M. (2020). The use of salt-coating to improve textural, mechanical, cooking and sensory properties of air-dried yellow alkaline noodles. *Food Chemistry, 333*, 127425.

Yin, Y., Zhuang, Y., Sun, L., Gu, Y., Zhang, G., Fan, X., & Ding, Y. (2025). How does high hydrostatic pressure treatment improve the esterification of quinoa (Chenopodium quinoa Willd.) starch? *Food Chemistry, 463*, Article 141166.

Yoo, H. J., Kim, H. R., Choi, S. J., Park, C. S., & Moon, T. W. (2018). Characterisation of low-digestible starch fractions isolated from amylosucrase-modified waxy corn starch. *International Journal of Food Science and Technology, 53*(3), 557–563.

Yu, T., Zhang, X., Feng, R., Wang, C., Wang, X., & Wang, Y. (2022). Comparison of the effects of high hydrostatic pressure and pasteurization on quality of milk during storage. *Foods, 11*(18), 2837.

Zailani, M. A., Kamilah, H., Husaini, A., & Sarbini, S. R. (2021). Physicochemical properties of microwave heated sago (Metroxylon sago) starch. *CyTA-Journal of Food, 19*(1), 596–605.

Zhang, Z., & Bao, J. (2023). Recent advances in modification approaches, health benefits, and food applications of resistant starch. *Starch-Stärke, 75* (9–10), Article 2100141.

Zhang, Y., Chen, C., Chen, Y., & Chen, Y. (2019). Effect of rice protein on the water mobility, water migration and microstructure of rice starch during retrogradation. *Food Hydrocolloids, 91*, 136–142.

Zhang, J., Li, X., Ren, X., An, Y., Song, X., Zhao, Y., & Zhang, W. (2024). Changes in physicochemical characteristics of wheat flour and quality of fresh wet noodles induced by microwave treatment. *Grain & Oil Science and Technology, 7*(3), 177–185.

Zhang, X., Ma, Q., Liu, X., Zhang, D., Ma, L., Luo, D. L., & Liu, X. (2020). Effect of microwave irradiation on the pasting, thermal, and rheological properties of cassava starch–sugar mixtures. *Journal of Food Process Engineering, 43*(8), Article e13431.

Zhang, J., Ran, C., Jiang, X., & Dou, J. (2021). Impact of octenyl succinic anhydride (OSA) esterification on microstructure and physicochemical properties of sorghum starch. *LWT, 152*, Article 112320.

Zhang, Y., Wang, Q., Zhang, Y., Wu, G., Tan, L., & Zhang, Z. (2022). Effects of moisture content on digestible fragments and molecular structures of high amylose jackfruit starch prepared by improved extrusion cooking technology. *Food Hydrocolloids, 133*, Article 108023.

Zhou, Y., Wang, M., Li, W., Liu, G., Wang, W., Zhi, W., & Zheng, J. (2022). Effects of dual modification of lysine and microwave on corn starch: In vitro digestibility and physicochemical properties. *International Journal of Biological Macromolecules, 220*, 426–434.

Zhu, F. (2018). Modifications of starch by electric field based techniques. *Trends in Food Science & Technology, 75*, 158–169.

Zia-ud-Din, Xiong, H., & Fei, P. (2017). Physical and chemical modification of starches: A review. *Critical Reviews in Food Science and Nutrition, 57* (12), 2691–2705.

Chapter 5

Strategies and novel approaches in the modification of starches: Physical, chemical, enzymatic, and biotechnological methods

Ravi Sai Nandhini[1], P. Kiran Kumar[2], Krishna Priyanka Tippareddy[3, 4], T.S. Murali Krishna[1] and Pothiyappan Karthik[3, 4]

[1]*Department of Biotechnology, Vel Tech Rangarajan Dr. Sagunthala R&D Institute of Science and Technology, Chennai, Tamil Nadu, India;* [2]*Department of Chemistry, Seshadripuram Institute of Technology, Mysuru, India;* [3]*Department of Food Technology, Faculty of Engineering, Karpagam Academy of Higher Education, Coimbatore, Tamil Nadu, India;* [4]*Centre for Food Nanotechnology, Karpagam Academy of Higher Education, Coimbatore, Tamil Nadu, India*

1. Introduction

Starch is a ubiquitous, naturally occurring polysaccharide that serves as the principal energy reserve in plants and the cornerstone of human nutrition and industrial applications. Its presence spans a vast array of botanical sources, including cereal grains such as maize, wheat, and rice, tubers notably potato and cassava, roots like arrowroot, legumes, and even unconventional sources such as banana, mango, and various rhizomes (Rashwan et al., 2024). The granules of starch, ranging from 1 to 100 microns in size, are highly species-specific in their morphology, which in turn influences their physicochemical and functional properties (He & Wei, 2017). This diversity in source and structure supports the broad utility and adaptability of starch in food, pharmaceutical, and other industrial sectors.

Starch is composed of two primary glucose polymers: amylose and amylopectin. Amylose is predominantly linear, consisting of α-1,4-linked D-glucose units, and typically constitutes 20%–30% of starch in most plants. Its linearity allows for the formation of helical structures, which are less soluble in water and contribute to the formation of firm gels upon cooling. Amylopectin, on the other hand, is a highly branched molecule, with α-1,4-linked glucose chains and α-1,6-branch points occurring every 20–30 glucose units. This branching imparts a soluble structure, accounting for 70%–80% of most starches, and is responsible for the semicrystalline nature of starch granules (Compart et al., 2023). The ratio and structure of these two polymers vary significantly among botanical sources, influencing properties such as gelatinization temperature, viscosity, and digestibility. The granule architecture of starch features amorphous and semi-crystalline growth rings, typically 100–400 nm thick. These rings are composed of clusters of amylopectin side chains, which determines the crystalline polymorph of the starch: A-type common in cereals, B-type tubers and high-amylose starches, or C-type legumes and some roots (Bertoft, 2017). The crystalline structure is not a structural curiosity; it profoundly affects the functional behavior of starch during processing and digestion (Cheng et al., 2024).

The diversity of starch sources extends beyond the conventional. While maize, wheat, potato, and cassava dominate global production, there is growing interest in underutilized and unconventional sources such as banana, mango, jackfruit, litchi, longan, and various rhizomes. These sources often exhibit unique granule morphologies, amylose/amylopectin ratios, and crystalline types, which can be harnessed for specialized applications (Zhu, 2020). For example, banana starch is notable for its high resistant starch content and B-type crystallinity, making it valuable for dietary fiber enrichment and glycemic control. Similarly, starches from mango and jackfruit kernels display A-type crystallinity and high amylose content, offering potential for novel food and pharmaceutical formulations. The extraction and processing of starch are

Native and Modified Starches for Food Applications. https://doi.org/10.1016/B978-0-443-27431-2.00011-0

adapted to its source and intended application. Wet milling is the predominant method for cereals and tubers, while enzymatic and alkaline treatments are employed to enhance purity and modify functional properties. The choice of extraction method can influence granule integrity, degree of damage, and subsequent performance in food or industrial systems (Dorantes-Fuertes et al., 2024). Advances in green chemistry and biotechnological approaches are increasingly being integrated to improve yield, reduce environmental impact, and enable the production of specialty starches. In food systems, starch acts as a thickener, stabilizer, gelling agent, and texturizer, with its behavior governed by the interplay of amylose and amylopectin, granule size, and crystalline structure. In pharmaceuticals, starch serves as a diluent, binder, disintegrant, glidant, and lubricant in tablet and capsule formulations, with its small granule size and hydrophilicity (Egharevba, 2019). The emergence of starch-based nanomaterials and films are new way for controlled drug delivery, edible coatings, and biodegradable packaging, leveraging the biocompatibility and film-forming properties of both native and modified starches (Pelissari et al., 2019).

Modification of starch is a central theme in expanding its utility. Physical methods such as heat-moisture treatment, pregelatinization, and high-pressure processing alter granule structure and functionality without chemical residues. Chemical modifications, including cross-linking, esterification, and etherification, introduce new functional groups, enhancing stability, viscosity, and resistance to retrogradation (Subroto et al., 2023). Enzymatic approaches exploit the specificity of amylases, branching enzymes, and debranching enzymes to tailor molecular architecture, producing derivatives such as cyclodextrins and resistant starches with unique properties (Park et al., 2017). Biotechnological innovations, including microbial fermentation, genetic engineering, and CRISPR/Cas-mediated genome editing, are enabling the development of starches with amylose/amylopectin ratios, branching patterns, and digestibility profiles directly at the source (Sun, Jiao, et al., 2017). Resistant starch fractions, which bypass digestion in the small intestine and are fermented in the colon, have been linked to improved glycemic control, enhanced satiety, and beneficial modulation of the gut microbiota (Birt et al., 2013). The formation of resistant starch is influenced by granule structure and processing conditions, with high-amylose and B-type starches being effective. Advances in enzymatic and biotechnological modification are facilitating the production of resistant starch-enriched foods for the prevention and management of chronic diseases such as diabetes and obesity (Tian et al., 2024). The global starch industry is a dynamic and rapidly evolving sector, driven by the expanding demand for sustainable, biodegradable, and functional materials. China and the United States are the largest producers, with the European Union also playing a significant role. The integration of unconventional starch sources, green extraction technologies, and advanced modification methods is reshaping the landscape, enabling the development of high-value products for food, pharmaceutical, and other industrial applications (Hu et al., 2025; Karthik et al., 2019). This chapter explores the complex relationship between starch structure and function, highlighting the functions of amylose and amylopectin in influencing physicochemical behavior. It investigates the variety of starch sources, their extraction techniques, and their useful applications in the industrial, food, and pharmaceutical sectors. Starch-modifying techniques based on physical, chemical, enzymatic, and biotechnological methods that increase its application. The advantages for health and sustainable innovation are studied with regard to emerging trends, including genome-edited starches, biodegradable polymers, and resistant starches.

2. Physical modification of starch

Natural starch from corn, rice, or potatoes does not have the ability to mix well in cold water, and it has low water absorption. Hence, to increase starch applications in different foods such as instant noodles, baby food, frozen dough, this native starch should undergo modification to ease the application. There are different kinds of modifications present in the industry. In all this, Physical modification gained popularity globally because it is affordable, ensures safety, and delivers effective results without producing toxicity, which can be a concern with chemical methods. Physical modification of starch consists of two treatments: thermal treatment and nonthermal treatment. Thermal treatment includes methods such as pregelatinization and heat-moisture treatment (HMT), and annealing. Non-thermal treatments include high-pressure processing, ultrasound, and cold plasma treatments (Bonto et al., 2021). Fig. 5.1 depicts the different types of physical modification of starch.

2.1 Pregelatinization treatment

Pregelatinization is a physical modification method and a thermal process that is safe and chemical-free. Pregelatinization starch is made through three common methods: (i) Spray Drying, (ii) Extrusion, and (iii) Drum Drying. In spray drying methods, starch slurry is sprayed into hot air, ensuring quick drying. With this method, the starch gains better cold-water solubility, and it possesses less damage to the granules, eventually increasing viscosity to a moderate level. The

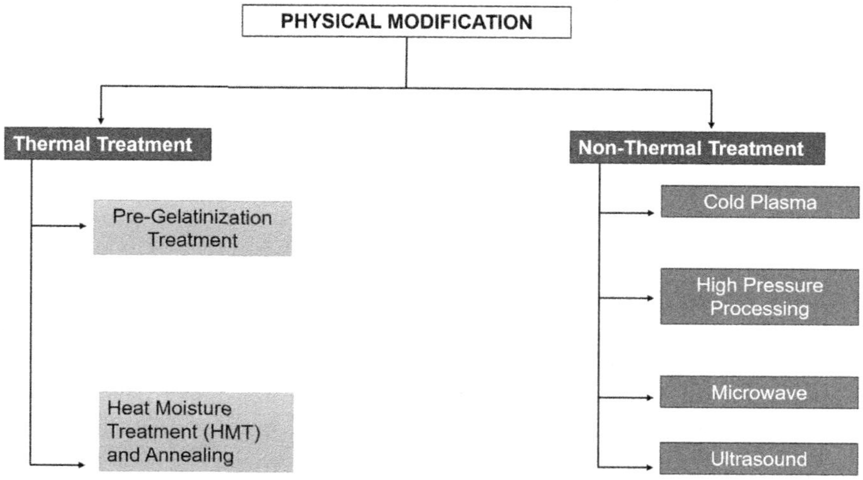

FIGURE 5.1 Types of physical modification of starch.

morphology of spray-dried pregelatinized modified starch is mostly spherical and has a slightly wrinkled surface. When it comes to melting behavior, spray-dried pregelatinized modified starch shows a gelatinization peak. Extrusion method includes a combination of application of high heat, high shear, and low moisture, leading starch granules to break down completely, further forming a porous and honeycomb-like structure (Compart et al., 2023). By this method, the native starch is modified and has cold water solubility, and its viscosity is greatly increased. The pregelatinized starch from the extrusion method is good for encapsulation or flavor holding. The morphology of extruded pregelatinized modified starch has broken granules and has a porous sponge like structure. When it comes to hydration, extruded starch has the highest water absorption due to its more open structure. While extruded starch shows no clear gelatinization peak, which means it has already been fully gelatinized. In the drum drying method, starch slurry is spread on a heated drum. When the slurry is heated, it will be dried in the form of a film, followed by scraping and grinding it into fine powder. Due to the application of severe heat, the starch structure breaks down heavily, hence it becomes modified starch, which has high cold-water viscosity, high solubility, but has lower hot-paste viscosity. The morphology shows that the drum-dried starch has flat flakes and no intact granules. By different methods, the pregelatinization starch is formed, and it changes starch properties such as morphology, structure, hydration, melting behavior, and pasting properties. One challenge of this method is that it sometimes creates irregular flake like particles. With all these methods, pregelatinized starch possesses all the positive applications, but has one downside, which is that it causes clumps. To prevent agglomeration, the process is applied to pregelatinized starch, resulting in larger and more uniform particles. This enhances their flow and mixing by converting fine particles into bigger, loose clusters. At the end, each method affects starch differently in shape, structure and functional properties (Ma et al., 2022). This pregelatinized modified starch is used in the production of buckwheat noodles, where it acts as a binding agent, provides tensile strength, and forms the dough sheet. Pregelatinized starch absorbs water fast and forms a gel layer that protects yeast, resulting in better fermentation, less yeast damage during freeing and saves proofing time and baking time for frozen dough. In breads, it shows moisture retention and helps slow down aging (Lee & Yoo, 2023).

2.2 Heat-moisture treatment & annealing

Heat-Moisture Treatment (HMT) and Annealing are both hydrothermal treatments. HMT includes heating starch at high temperatures (90–120°C) with low water content (10%–30%). This method does not totally cook or gelatinize the starch but rearranges its internal structure. Heat partially disrupts the crystalline areas and double helices inside the starch granule, which promotes new interactions between starch chains, amylose-amylose and amylose-amylopectin. It causes the rearrangement of molecules inside the granule. The swelling power decreased as the treated starch granules don't swell as much in hot water, as the molecules are more tightly packed in crystals and stronger internal bonds. Regarding pasting properties possessing higher gelatinization temperatures have lower peak viscosity and have better stability during heating and stirring, and less paste stays thicker under stress (Zavareze & Dias, 2011). The modified starch from this treatment forms firmer gels and increases resistant starch, which is good for gut health, promotes digestion, and eventually lowers blood sugar spikes. The method raises the temperature needed for starch to gelatinize, makes starch granules swell less and leak less amylose, enhances heat and shear stability, while also increasing resistant starch, which supports gut health. Additionally, it reduces digestibility, promoting a slower and more sustained glucose release. Annealing is another hydrothermal treatment, in which starch is soaked in excess water over 60% at moderate temperatures. Temperature is above the glass transition (Tg) but below the gelatinization temperature. When water penetrates the starch granule, molecules become more mobile, crystalline regions become more ordered with better-packed double helices. Annealing improves crystal perfection, stable viscosity and less retrogradation. Both of these treatments keep starch natural, eco-friendly and safe for food use and allow customizing starch properties for specific industrial or health purposes (Zavareze & Dias, 2011).

2.3 Cold plasma and high-pressure processing

Plasma is a fourth state of matter, and cold plasma is a partially ionized gas with active particles that stays at room temperature; hence, it is safe for heat-sensitive foods. This ionized gas is created using electric or magnetic fields at low or atmospheric pressure. Cold plasma utilizes ionized gas containing active particles to break down bonds, cross-link chains, or add functional groups (Okyere et al., 2022). When this plasma is projected on starch, it bombards starch with ions, radicals, and UV, which further breaks down starch bonds or reorganises into depolymerization or, cross-linking or changing the crystallinity nature of the starch molecules that brings modification in molecular weight, granule shape, crystalline structure, gelatinization temperature and energy needed to cook, pasting properties and rheology. Plasma etching increases the surface roughness of starch granules, creating holes and cracks that enhance water absorption and reduce cooking time (Gupta et al., 2024). Recent innovation shows that application of plasma-activated water, which holds reactive oxygen or nitrogen species that can modify starch in a manner of increasing crystallinity, makes it more resistant to digestion. During industrial applications, cold plasma can be used in the form of low-pressure plasma; this method requires precise control and costly vacuum equipment, whereas atmospheric plasma is a cheaper alternative that operates continuously, making it more suitable for food industries. Regarding the usage of gases, noble gases are used, such as argon and helium, which need less energy but produce fewer reactive species, which can be avoided by mixing gases. To improve the effects of modification of starch, a combination of plasma with other treatments can help ensure the safety for food labeling and approval. When cold plasma was applied to rice starch at different time variations (2, 6 and 10 min), they noticed slight surface changes in the starch granules, and amylose content generally increased with cold plasma as more short double helices formed. The crystallinity has slightly decreased, and the treated starch is more soluble than native starch at low temperatures (50–60°C). When cold plasma-treated rice starch is heated, the viscosity is reduced. Resistant starch has been increased, which possesses slower digestion, which is good for blood sugar control. When the cold plasma treatment is subjected to longer durations, it reduces resistant starch as it leads to the breakdown of too much starch, making it easier to digest (Sun et al., 2022). The cold plasma-treated starch possesses various applications in the food industry as thickeners stabilizers, and is used in sauces, gluten-free baking, and dessert gels. In the packaging industry, it is used in biodegradable films, coating, that block moisture and oxygen. In the pharmaceutical industry, this starch is used in drug carriers, wound dressings and in the textile industry in fabric finishing and antimicrobial coatings. In the biomedical industry, this modified starch is used for controlled-release drugs (Gupta et al., 2024).

High-pressure processing works based on two principles, the Isostatic Principle and Le Chatelier Principle. The isostatic principle is defined as the pressure being applied equally in all directions, irrespective of the shape and size of the food. This way ensures uniform treatment. The Le Chatelier principle is defined as when a system is in balance, and it gets disturbed; it will adjust itself to reduce that disturbance and reach a new balance. To modify the starch, starch is mixed with water between 4% and 33% concentration, and it is poured and sealed in plastic bags, and the air is removed. These bags are placed in special chambers that are filled with liquid, commonly water. Now, at room temperature, a very high pressure is applied (100–600 MPa) for a given time, which is 2–30 min. After this treatment, the pressure is reduced to normal quickly, which causes the breakage of noncovalent bonds, which changes the molecular structure, eventually improving its properties, such as either gelatinizing or swelling differently (Castro et al., 2020). The treated starch is sent for air drying purposes, which might cause slight retrogradation and recrystallization of starch, or to cause less retrogradation, freeze drying helps better, which keeps the new structure better. Hence, this treatment explains that the modification of starch can be done without the application of chemicals or heat by using high pressure. By this treatment, the starch content might be decreased slightly or remain the same. For example, when chestnut flour is treated at 400–600 MPa, there is no change in the total starch content noticed, whereas this treatment can damage starch granules mainly at high pressure. This is noticed in quinoa, where this treatment gave 15%–17% of damaged starch when it is modified. Apart from this limitation, the modified starch from this treatment consists of resistant starch and digestible starch. Resistant starch was noted in lentils 4.47%–6.80%, when it was treated at 400–600 MPa, and digestible starch was noticed in algarrobo seed and apple when both were treated at 500 MPa for a longer duration of time. When native starch undergoes this treatment, it typically shows increased water-holding capacity and improved solubility at lower temperatures. Regarding gelatinization, the high-pressure-treated starch needs less energy to cook the starch, hence lowering gelatinization. The texture of this high-pressure-treated starch has less hardness, gumminess and chewiness in starch gels, eventually making them softer and possessing better stability during storage. Regarding pasting properties, high-pressure-treated starch usually shows higher pasting temperatures and has lower viscosity at very high temperatures (Castro et al., 2020). High-pressure processing of regular potato starch is typically carried out at 600 MPa for 2–3 min at around 20 °C. When examined through NMR (nuclear magnetic resonance), FT-IR spectroscopy, DSC (differential scanning calorimetry) and SEM (scanning electron microscopy) declared that there is a presence of amorphous regions inside starch, and the starch structure broke down partly, gel like structure, crystallinity decreased, resulting in lower gelatinization temperatures. The outer layer stays tougher while the inside becomes gel like, which is useful for making new food textures or low-fat products (Błaszczak et al., 2005).

2.4 Ultrasound and microwave treatment

Ultrasonication is a nonthermal treatment that utilizes ultrasound waves of more than 20 kHz, causing certain physical effects such as cavitation, which means formation and collapse of microscopic bubbles, that generate high shear force, high local temperature and pressure, and free radicals (Zhu, 2015). These effects crack starch granules, create surface pores, break long polymer chains, and change functional properties. The ultrasound treatment on corn starch was conducted to study the effects on structure, rheology, thermal properties, turbidity, and swelling power of corn starch suspensions. They have applied 24 kHz frequency ultrasound, with two types such as ultrasound probe that has high power (100–400 W) and an ultrasound bath, which has lower power (2 W/cm^2). The treatment lasted between 15 and 30 min, with samples labeled from K1 (untreated, 12.97 kJ/kg) to K9 (strongest ultrasound, 7.47 kJ/kg). In this study, they noticed that ultrasound breaks corn starch chains, which leads to a decrease in thickness that further leads to less resistance to flow. When the ultrasound was applied to K1 to K9 samples with the mentioned energy above, it was declared that less energy is needed as the structure has already been damaged when ultrasound has been applied lead to the disrupted starch crystals, hence enthalpy dropped drastically when energy increased from K1 to K2. Turbidity has decreased, swelling power (g/g), increased with ultrasound and temperature (K1 at 70 °C/15 min: 11.05 g/g and K7 at 70°C/15 min: 15.87 g/g). As the ultrasound frequency has increased, it has led to the disruption of starch granules, which opens up space for water to enter, further breaking crystalline bonds, which eventually leads to higher water absorption. The morphology of starch granules when undergoing ultrasound treatment shows that there is visible rupture on the surface, breakage, and granule shrinkage have been noticed in K4 to K9 samples. At the highest frequency (K8 to K9 samples), the starch granules started agglomeration (Jambrak et al., 2010).

The ultrasound treatment involves placing potato, wheat, corn, and rice starches in two different media to examine the effects on the structure and functional properties. The morphological studies are performed using SEM, which shows that potato and wheat starches undergone more damage, and the granules which are suspended in water during ultrasound treatment have more damage than the granules suspended in ethanol and TEM prove that large cavities have been noticed in many granules, and after the ultrasound treatment, the cavities increased, the edges became rough and damages are noticed on the granules, especially in potato and wheat starches. Water absorption and fat absorption both increased for all starches after treatment (Sujka & Jamroz, 2013). Absorption of water and fat has mainly been enhanced in the starch granules that are suspended in water than in ethanol (Chen et al., 2024). Overall, the modified starch by ultrasound treatment has higher clarity, better thickening and gelling ability, and better absorption, which can be used in different food applications.

Different treatments are introduced to modify starch to enhance its properties for industrial applications. Microwave treatment is one such treatment used to modify starch, as it is a safe, fast, and energy-efficient physical modification method. When microwave treatment is applied to starch, the water molecules present in starch vibrate due to the internal heating created by microwave radiation. This vibration between starch molecules causes pressure, leading to moisture loss, granule expansion, and structural damage (Oyeyinka et al., 2021). Microwave processing alters the structure and behavior of potato starch. Potato starch consists of high amylose content, and when it is cooked, the starch looks firm, which might be a problem for making instant foods such as instant noodles or instant mashed potatoes. Hence, microwave treatment has been applied to alter the starch structure so that it can be used in different applications. The potato starch was extracted, dried, and treated with microwave for certain specific durations. The potato starch granules are smooth, round, or elliptical in shape. When a microwave is applied for 5–10 s, the granules show minor cracks and surface damage, and viscosity has increased. After 15–20 s, the starch granules ruptured, collapsed, and surface damage was noticed, which further affected viscosity. Viscosity has been dropped because of too much breakdown of starch chains. Hence, Xie et al., (2025) reported that the microwave treatment is applied at 10–15 s, which is an optimal duration that improves functional properties for food processing industries. This microwave-treated starch is applied in pharmaceutical industries in drug tablets to ensure sustained drug release, can be used in low-calorie foods, thermally stable food products, and in instant noodles (Ramírez-Brewer et al., 2023).

3. Chemical modification of starch

Starch is the second most abundant and essential form of biomass on Earth, serving as a primary energy source in human diets. It has been found in a wide range of foods, plant tissues, stems, roots, leaves, and fruit (Zia-ud-Din et al., 2017). Starch shows properties such as nontoxic, biodegradable and biocompatible nature that makes many researchers gain interest in this field (Kenawy et al., 2014). The Native starch, in its unmodified form, lacks the physical and chemical characteristics necessary to meet the demands of many applications in industrial processes. To address these limitations, starch is often altered using physical, chemical, or enzymatic techniques that improve functionality and adaptability for targeted applications. Among these, chemical modification plays a crucial role, as it introduces new functional groups onto the starch structure, thereby providing specific desired properties. The abundance of hydroxyl groups in the starch offers numerous reactive sites, making it particularly amenable to chemical modifications.

Native starch, in its unmodified form, lacks the physical and chemical characteristics necessary to meet the demands of many industrial processes. To address these limitations, starch is often altered using physical, chemical, or enzymatic techniques that improve its functionality and adaptability for targeted applications. Among these, chemical modification plays a crucial role, as it introduces new functional groups onto the starch molecule, thereby conferring specific desired properties. The abundance of hydroxyl groups in the starch structure offers numerous reactive sites, making it particularly amenable to such chemical transformations. Research into chemically modifying starch began as early as the 1940s and has since evolved into a well-established area of study (Villa Zabala, 2020). The functional groups, such as carboxyl, acetyl and hydroxypropyl groups, are generally added. Generally, the chemical modification can be performed by cross-linking, esterification and etherification, cationic and anionic modifications and dual or multiple chemical modifications as illustrated in Fig. 5.2. Also, the conditions and different methods have a great effect on the properties of the modified starch (Chen et al., 2015).

3.1 Cross-linking of starch

The Cross-linking is a chemical process of forming chemical bonds linking one polymer chain to another (Malik et al., 2022). Starch contains both primary and secondary (6-OH and 2-OH, 3-OH) hydroxyl groups, which can be crosslinked with multifunctional reagents to obtain crosslinked structures. These reagents can create ether or ester bonds between nearby starch chains, not only within a single chain but also between adjacent ones. Due to this close arrangement within the molecules, a small number of crosslinking agents is enough to form stable intermolecular links between starch molecules Fig. 5.3 (Yao et al., 2022). Starch crosslinking provides strong bonding between the molecules, restricting the movement of the polymer chains; thus, the crosslinked starch increases its resistance to temperatures, acidity, and wear, rendering it exceptionally useful in various applications commonly used in industries, food, paper, textiles, and adhesives (Pillai & Thomas, 2023).

Cross-linking in starch forms the covalent bonds between polymer chains, replacing the native bonds, thereby enhancing its stability and functionality. These structural chemical modifications can be done through treating the native starch with various reagents to integrate an ether or ester linkage by substituting a hydroxyl group with phosphate, acetyl, hydroxypropyl and citryl groups either individually or in combination (Li et al., 2019; Okoli et al., 2019). The food-grade crosslinking agents, such as phosphoryl chloride ($POCl_3$), Sodium Tri Metaphosphate (STMP) epichlorohydrin (EPI), are commonly used (Egharevba, 2019). Various studies have reported that the cross-linked starch shows greater properties. The modified banana starch functional and chemical characteristics of cross-linked starch varied by multiple factors, including temperature, pH, reaction time, starch source, and concentration of cross-linking agents. The study demonstrated that the dual modification through cross-linking and esterification significantly enhanced the resistant starch, increasing from 21.49% to 29.14% (Raigond et al., 2015). The optimal conditions for preparing cross-linked resistant starch vary between corn starch and wheat starch. Wheat starch achieves the best results at 38°C and pH 12, whereas corn starch requires a higher temperature of 70°C under the same pH condition (Kahraman et al., 2015).

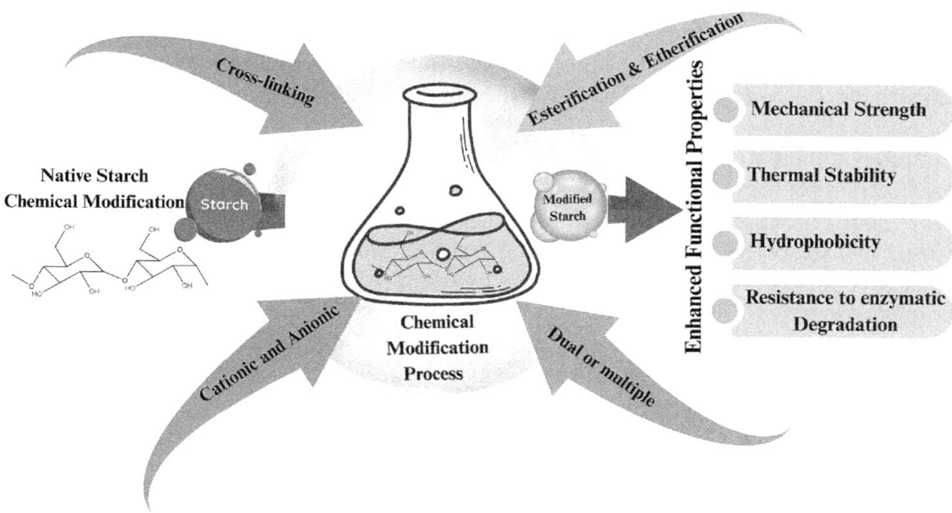

FIGURE 5.2 Overview of key chemical modification methods and functional properties of modified starch.

A study showed that wheat and corn starch were used as starch sources, and the starch sources were blended with polyvinyl alcohol, along with malic acid and succinic acid as cross-linking agents. The films crosslinked with succinic acid exhibited lower swelling capacity and water vapor permeability compared to those with malic acid. Furthermore, the films also confirmed their biodegradability (Thessrimuang & Prachayawarakorn, 2019). The cross-linking modification of starch using chemical treatment involving sodium acetate and adipic acid was carried out at 75°C for 1 h. The study showed notable changes in the ash content; the swelling powder had increased in moisture, thus indicating an alteration in the starch structural and functional characteristics. These changes suggest that the crosslinked condition enhances the thermal stability, which was observed, making it more suitable for industrial applications (Akpa et al., 2012). The cross-linking of corn starch films using citric acid under various pH and curing treatments. The study found that the cross-linking was most effective at pH, while at higher pH levels, there was a reduced efficiency. Interestingly, the thermal curing decreased cross-linking, and surface treatment with citric acid-alcohol showed no significant improvement (Gerezgiher & Szabó, 2022). Overall, the obtained starch was effectively enhanced with water resistance by forming ester bonds and reducing hydroxyl groups.

3.2 Esterification and etherification

The chemical modification technique of esterification and etherification plays a significant role in modifying the native starch. The esterification of starch is the conversion of the three available hydroxyl groups in starch to an alkyl or aryl

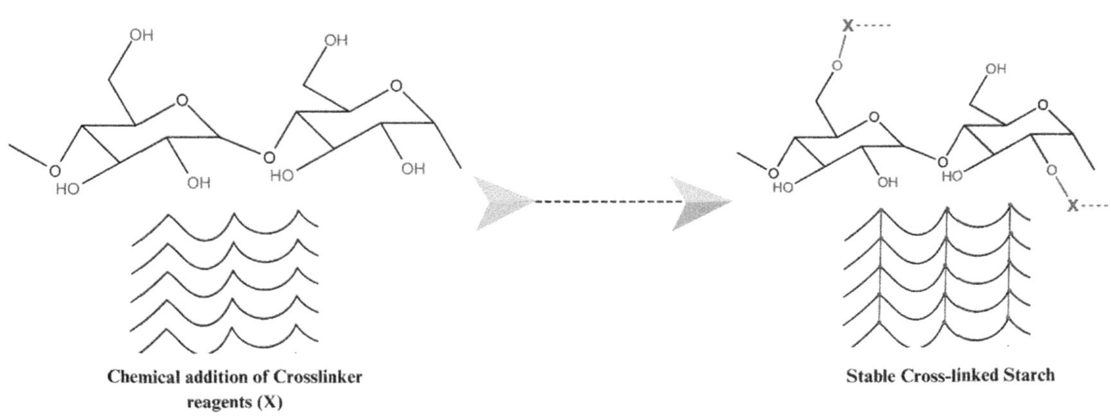

FIGURE 5.3 Crosslinking agents form stable intermolecular links between starch molecules.

derivative (Masina et al., 2017). These methods primarily involve the substitution of hydrophilic hydroxyl groups with hydrophobic groups, enhancing the starch properties. In the esterification process, reagents such as acetic anhydride or vinyl acetate are used to add an acetyl group to the starch structure. This substitution reduces the interaction with water molecules and the starch, thus increasing the hydrophobic properties, thermal resistance, and making the starch more suitable for food and other industrial applications (Kushwaha et al., 2023). Acetylation is a type of esterification that enhances starch functionality by improving the starch thickening properties, reducing gelatinization temperature, reducing the rate of thermal degradation, and increasing the viscosity, making the starch more suitable for diverse food processing applications (Xie et al., 2025). The modification of starch through esterification can significantly enhance certain functional properties, including increased hydrophobicity, improved swelling behavior, and a reduced tendency for thermal degradation (Singh et al., 2007). These factors are especially beneficial for food and industrial applications where stability and hydrophobicity play an important role.

Etherification, on the other hand, typically involves the use of propylene oxide in an alkali medium to introduce hydroxypropyl groups in the starch moiety, resulting in hydroxypropylated starches (Raigond et al., 2015). These hydroxypropyl groups interfere with hydrogen bonds within the starch molecules, increasing the amorphous nature of the starch. These changes improve the starch functional flexibility but also reduce its digestibility, beneficial for applications for lower glycemic response (Kushwaha et al., 2023). Esterification and etherification are widely adopted chemical modification techniques used to tailor the structural and functional properties of starch. These processes work by introducing specific substituent groups that alter the molecular architecture of the starch, thereby enhancing the modified starch's thermal resistance, hydrophobic behavior, and protection against enzymatic breakdown.

In etherification, functional groups such as carboxymethyl and hydroxyethyl are incorporated, forming ether linkages within the starch chain, which significantly improve the viscosity and thermal stability (Clasen et al., 2018). On the other side, esterification reactions typically require an alkali medium with sodium hydroxide frequently employed as a catalyst to accelerate the substitution process on the starch structure (Heinze et al., 2004). A notable study introduced a novel thermosensitive and hydrophobic polymer, PyHES, via esterification, where the inclusion of propynyl groups contributed to its hydrophobic characteristics, thus expanding its potential for specialized applications (Jong & Ju, 2017). An etherified starch shows better thermal stability and rheological performance. These properties vary depending on the functional groups introduced through etherification (Chen et al., 2015). A notable example is carboxymethyl starch, a modified starch that shows a negatively charged ether starch, showing a strong hydrophilic nature and solubility in cold water (Spychaj et al., 2013). Due to their enhanced functional properties, these modified starches find a wide range of applications in diverse fields, including pharmaceuticals, where they support drug delivery systems and environmental applications like wastewater treatment. In the food industry, they serve as effective stabilizers in products such as pudding and fruit pie filling, and are employed in food technology to encapsulate and retain flavors (Egharevba, 2019).

3.3 Cationic and anionic modifications

The key strategic method employed in starch modification is ionic modification, which involves the creation of cationic and anionic derivatives by introducing charged functional groups onto the starch backbone. These modifications enhance intermolecular interactions, thereby improving the starch physiochemical characteristics and expanding its range of industrial applications. In particular, cationic starches containing quaternary ammonium groups are typically produced through an etherification reaction using agents such as 3-chloro-2-hydroxypropyl tri-methylammonium chloride (CHPTC) or EPTAC. The incorporation of these positively charged moieties facilitates strong electrostatic bonding with negatively charged surfaces, significantly enhancing the starch performance in various formulations (Chemelli et al., 2020). Another example of cationic starch modification involves the esterification of starch using 3-carboxy-propyl-trimethyl-ammonium chloride, facilitated by a prior activation step with N, N'-carbonyl diimidazole. Cationic starch derivatives are known for achieving a relatively high degree of substitution and have shown promising potential for antibacterial applications (Pfeifer et al., 2017). In the emulsion polymerization system, various types of surfactants, including nonionic, anionic, and cationic, are employed to improve both polymer yield and stability. Among the different starch modification techniques, cationic modification has gained increasing attention, particularly in the paper industry. These starches are highly effective as dry strength agents, as they enhance fiber-to-fiber bonding by interacting with naturally anionic starch fiber matrix surfaces (Hubbe, 2006). The cationic chemical modification process can be performed under diverse conditions, including homogeneous solutions, aqueous suspensions, water-organic solvent mixtures, or even in dry state reactions (Kavaliauskaite et al., 2008). One of the key reasons for the widespread adoption of cationic starch is its cost-efficiency and practical applicability. The performance of these starches, especially in the papermaking process, is largely influenced by the amount of starch that adheres to fiber surfaces (Eriksson et al., 2005). Cationic starches are widely utilized in the paper-making industry due to their ability to significantly improve the dry strength of paper sheets. They are also employed as emulsion stabilizers in conjunction with internal synthetic sizing agents such as alkyl ketene dimer (AKD) and alkenyl succinic anhydride (ASA), and further, they serve as effective surface sizing agents to improve the paper quality and printability. The cationic starch derived from potato peel waste exhibited effective flocculation performance in livestock wastewater, achieving suspended solid removal, showing effectiveness of 72 and 162.9 mg TSS/mg for swine and dairy wastewater, respectively (Haleem et al., 2023). The improvement of strength and water barrier properties (up to 51.5%) of pulp molded product showed more stability in its shape and lower weight, providing great application for packaging materials (Qin et al., 2022).

Anionic starches are generally inorganic starch esters that are commonly synthesized through chemical reactions involving functional groups such as phosphates, xanthates, carboxylates and sulfates (Peng et al., 2016). By using phosphorylation methods, typically performed using sodium tripolyphosphate under controlled pH and temperature conditions. This process introduces phosphate moieties into the starch, thereby enhancing charge density and other properties. Synthesized potato starch sulfate under optimized reaction conditions, producing the derivatives with therapeutic applications. The synthesis process typically involves the addition of sulfate groups on the starch using sulphonating agents under controlled acidic or thermal conditions (Cui et al., 2009). These have gained considerable attention due to their properties having higher viscosity, glueyness, and the starch obtained is of high transparency; hence, these modified starches have been significantly suitable for use as a thickening agent, stabilizer, adhesive, and excipient in various industrial and pharmaceutical applications (Chen et al., 2015). These cationic and anionic modifications not only improve the functional properties of starch but also provide new paths for its application in biomedical, food, and cosmetic formulations.

3.4 Dual or multiple chemical modifications

Single chemical modification of starch enhances properties like thermal, viscosity, and strength and has supporting applications in nanocrystals, hydrogel, biomedical, pharma, and biodegradable films. Their use has sometimes been limited by poor thermal stability, weak gel integrity and reduced film strength under certain conditions (Zhou et al., 2021). To address the inherent limitations associated with the single starch modification technique, scientists are increasingly exploring dual modification techniques. These combined approaches have shown promising improvements in starch functionality, enhancing its suitability for a wide range of industrial uses. Notably, dual modifications contribute to more effective processing, clearer gel formation, and shorter processing duration (Kong et al., 2016). The effectiveness of dual modification of starch is significantly impacted by its botanical origin and specific conditions under which the reaction occurs, including the duration of the process (Lee et al., 2015). Additionally, the order in which modifications are applied, along with the type of starch selected, plays a critical role in determining the final structure and functional properties of the dual-modified starch (Wang et al., 2022). For instance, starch that undergoes ultrasound treatment before heat-moisture treatment (HMT) tends to exhibit greater peak viscosity and enhanced thermal stability compared to starch treated postultrasound (Han et al., 2021).

Dual modifications of starch can be categorized into different types, based on the methodologies employed: dual chemical, dual physical, dual enzymatic and dual heterogeneous. Each involves the combination of processes that synergistically alter the starch structure and enhance its functional properties. In dual chemical modifications, several combinations have been extensively investigated to tailor starch for specific industrial applications. Notable examples include crosslinking coupled with acetylation (Lee et al., 2015), crosslinking with hydroxypropylation (HPT) (Zhao et al., 2015), oxidation followed by dual chemical routes, including crosslinking/phosphorylation (Sukhija et al., 2016; Sun, Jiao, et al., 2017). Other significant dual chemical routes include crosslinking/phosphorylation (Iacovou et al., 2017), HPT/Crosslinking (Hazarika & Sit, 2016), and acid hydrolysis followed by succinylation (Mehboob et al., 2015), which have demonstrated promising properties of thermal, mechanical and functional characteristics of modified starch derivatives. The initial modification of starch typically acts as a preparatory step that alters the structural integrity of the starch granules, including changes to the surface morphology and internal bonding of the starch chains. This creates a more favorable environment for the subsequent modification. The dual modification largely influenced, whereas the first modification plays a supportive role in enhancing the effectiveness (Ashogbon, 2021). This type of dual chemical modification significantly improves the starch properties to form gels and encapsulating stability. In particular, the dual modifications have enabled the development of stable hydrogels, scratch-resistant starches and tougher films (Lima et al., 2021). A dual chemical-modified biodegradable starch film developed from rice starch. In their approach, rice starch was first hydroxypropylated using propylene oxide, followed by crosslinking with sodium tri-metaphosphate (STMP). The study showed notable enhancements in the starch films, including improvements in tensile strength, flexibility, and water-solubility properties (Woggum et al., 2014).

Chemically modified starch can be produced through various techniques; these modifications alter the starch crystallinity, molecular structure, and physical properties, thereby enhancing its performance in the field of packaging, pharmaceuticals, agriculture and wastewater treatment, as illustrated in Table 5.1. The chemical modification also improved shelf-life, moisture resistance, and showed excellent foaming capacity. Despite its advantages, challenges like water sensitivity and difficulty in processing technology remain. The growing demand in food, adhesives, cosmetics, and paper sectors is expected to drive future research and market growth.

4. Enzymatic modification of starch

4.1 Enzyme types

The enzymatic modification of starch is shaped by the specificity and synergistic action of starch-degrading and modifying enzymes, which enable starch structure and properties for industrial, nutritional, and pharmaceutical applications. Starch, a semicrystalline biopolymer composed primarily of amylose and amylopectin, is naturally resistant to many forms of processing due to its dense granule architecture and variable branching patterns (Bertoft, 2017). The application of enzymes, including α-amylase, β-amylase, glucoamylase, pullulanase, isoamylase, and cyclodextrin glycosyltransferase, allows for targeted hydrolysis, branching, or debranching reactions that alter the amylose-to-amylopectin ratio, chain length distribution, and degree of branching within the starch granule. Each enzyme type acts at a specific position along the starch polymer, and their selection, combination, and sequence of application determine the extent and nature of structural changes, influencing key physicochemical attributes such as crystallinity, gelatinization behavior, viscosity, and digestibility. α-Amylase is an endo-acting enzyme that randomly cleaves internal α-1,4-glycosidic bonds in both amylose and amylopectin, resulting in a rapid reduction in molecular weight and viscosity, and producing a mixture of linear and branched oligosaccharides known as dextrins (de Souza & e Magalhães, 2010; Mu et al., 2021). Thermostable α-amylases, derived from *Bacillus licheniformis*, are widely utilized in high-temperature industrial applications, facilitating efficient starch breakdown in food, brewing, and bioethanol production (Fincan et al., 2021). β-Amylase, is an exoacting enzyme that cleaves α-1,4-glycosidic bonds from the nonreducing ends of starch chains, releasing maltose units. It cannot bypass

TABLE 5.1 Chemical modification techniques and their practical applications.

Modification technique	Description with applications	References
Crosslinking	Casein-based protein films, crosslinked with plant-derived tannic acid (TA)	Picchio et al. (2018)
Crosslinking	Sodium alginate from seaweed and pectin from pineapple rind were crosslinked using natural acids to form biodegradable films.	Singh et al. (2020)
Etherification	Sodium carboxymethyl cellulose was synthesized via etherification under dry conditions. The effect of etherification showed better mechanical properties of the bioplastic for the fruit packaging.	Syarifuddin et al., (2022)
Esterification	Galactomannan was chemically modified through the esterification method to produce hydrophobic, high-strength films used for packaging.	Liu et al. (2021)
Dual modification	Etherified, followed by the oxidation of cassava starch with PVA. The results showed enhanced toughness and flexibility in films.	Li et al. (2020)
Crosslinking	Cross-linking using sodium acetate, ammonium phosphate, sodium acetate with adipic acid and sodium acetate with fumaric acid. The modified starch showed stable viscosity, making it more resistant to shear and retrogradation, and can be used for industrial applications.	Akpa et al. (2012)
Etherification	Esterification using acetic anhydride to produce films with hydrophobic, mechanical and oxygen barrier properties for the development of starch-based biodegradable films for controlled release of fertilizers for mitigating nitrogen pollution in agriculture.	Ibrahim et al. (2020)

α-1,6-branch points, resulting in the accumulation of β-limit dextrins. This enzyme is important for the production of high-maltose syrups and for modifying the chain length distribution of amylopectin, which in turn affects the digestibility and textural properties of starch-based foods. Glucoamylase (amyloglucosidase) is another exoacting enzyme that hydrolyzes both α-1,4- and, more slowly, α-1,6-glycosidic bonds from the nonreducing ends of starch chains, ultimately releasing glucose. It is predominantly produced by fungi and is crucial for the complete saccharification of starch to glucose, especially in the production of glucose syrups and bioethanol (Roy & Gupta, 2004). The enzyme's ability to act on branch points, at a slower rate, distinguishes it from β-amylase and enables the exhaustive hydrolysis of starch polymers. Debranching enzymes such as pullulanase (EC 3.2.1.41) and isoamylase (EC 3.2.1.68) specifically hydrolyze α-1,6-glycosidic bonds in amylopectin and related oligosaccharides, linearizing the branched structure and facilitating the production of resistant starch (RS) and slow-digesting starch fractions. Pullulanase is used in combination with glucoamylase or β-amylase to enhance the yield of linear oligosaccharides and facilitate the production of high-glucose or high-maltose syrups, while isoamylase's preference for larger branched substrates makes it valuable for producing linear amylose and for structural studies of starch (Niu et al., 2022; Wang et al., 2023).

Cyclodextrin glycosyltransferase (CGTase) (EC 2.4.1.19) catalyzes the intramolecular trans glycosylation of starch, producing cyclic oligosaccharides known as cyclodextrins (CDs). These cyclic molecules, composed of six (α-CD), seven (β-CD), or eight (γ-CD) glucose units, have a hydrophobic cavity and a hydrophilic exterior, enabling them to form inclusion complexes with a wide range of guest molecules (Sonnendecker & Zimmermann, 2019; Thombre & Kanekar, 2016). CGTase is a multifunctional enzyme, also capable of disproportionation, coupling, and weak hydrolysis reactions. The enzyme product specificity α-, β-, or γ-cyclodextrin is determined by its active site architecture and can be modulated through protein engineering. Branching enzymes (EC 2.4.1.18) introduce α-1,6-branch points into linear glucan chains, increasing the degree of branching and modifying the physicochemical properties of starch. These enzymes are instrumental in the biosynthesis of amylopectin and in the production of highly branched, slow-digesting starches (Yang et al., 2023). The synergistic use of multiple enzymes can dramatically enhance the modification effect on starch. Dual modification of maize starch with β-amylase and α-glucosidase has been shown to decrease amylose content, increase branching degree, and produce a denser, more crystalline structure with improved thermal stability and slow digestibility. Such synergistic systems allow for the fine-tuning of starch structure, enabling the production of starches with tailored functional properties for specific industrial or nutritional applications (Jia et al., 2024). The branching degree of maize starch can be increased from 4.95% to 7.10% after bienzymatic modification, while the amylose content can be reduced from 28.77% to 18.60%, and the amylopectin content can be elevated from 70.79% to 81.71%. These changes are accompanied by an increase in relative crystallinity from 20.41% to 30.20% and a more compact starch structure, as revealed by SEM, XRD, and Fourier transform infrared spectroscopy. The dual enzyme modification also affects the thermal stability, pasting, light transmittance, solubility, and swelling of maize starch, making it possible to produce starches with specific amylose/amylopectin ratios and functional properties (Li et al., 2019).

The specificity and exclusivity of enzyme modification minimize the generation of by-products, allowing for the efficient production of final products with desired properties. The use of α-D-glucan branching enzyme and cyclodextrin glucosyltransferase in combination will produce starch with a higher content of slow-digestible and resistant starch compared to natural maize starch and single enzyme-treated starch (Kelly et al., 2009). Similarly, the modification of cross-linked tapioca starch with two α-amylases has been used to improve gel strength, while the use of cyclodextranase and CGTase has been employed to alter the molecular structure, oligosaccharide composition, and digestibility of glutinous maize starch (Jia et al., 2024). The effectiveness of enzymatic modification is also influenced by the botanical source of the starch. Starches with A-type crystallinity, such as those from maize and wheat, are generally more resistant to hydrolysis, while those with B-type crystallinity, such as potato and banana starch, are more susceptible to enzymatic attack. This is due to differences in the packing density of the crystalline regions and the accessibility of the amorphous lamellae to enzymatic action. Amylose-deficient starches and those with a higher proportion of short external amylopectin chains are amenable to enzymatic modification, as they allow for more efficient hydrolysis and restructuring of the granule architecture (Li et al., 2018). These structural differences are reflected in the functional properties of the modified starches, with amylose-deficient and A-type starches exhibiting more specific hydrolysis of amorphous lamellae and a greater increase in crystallinity after enzymatic treatment.

4.2 Cyclodextrin production

Cyclodextrin production represents industrially significant applications of enzymatic starch modification, leveraging the catalytic activity of CGTases to transform linear and branched starch molecules into cyclic oligosaccharides with unique inclusion properties. Cyclodextrins (CDs), primarily α-, β-, and γ-CDs, comprising six, seven, and eight glucose units, are characterized by a hydrophobic central cavity and a hydrophilic exterior, enabling them to encapsulate a wide variety of molecules. This property supports their extensive use in pharmaceuticals, food technology, and environmental applications, where they function as molecular containers for flavors, vitamins, drugs, and even pollutants. The process of cyclodextrin formation is an example of biocatalytic precision: CGTases, typically produced by Bacillus and Paenibacillus species, catalyze intramolecular transglycosylation reactions that cleave α-1,4-glycosidic bonds in starch and reform them to yield cyclic structures. The product distribution is determined by the enzyme active site, the source and structural characteristics of the starch substrate, and the reaction conditions, including pH, temperature, and substrate concentration. (Han et al., 2014).

The efficiency and specificity of cyclodextrin production are strongly influenced by the physicochemical properties of the starch substrate. Amaranth starch has been shown to yield higher total cyclodextrins compared to maize starch due to its exceptionally high amylopectin content, small granule size (1–3 μm), and high dispersibility, which collectively enhance enzyme accessibility and catalytic turnover. In a comparative study, amaranth starch yielded 22% total cyclodextrins, while maize starch yielded only 14.5% under identical conditions, highlighting the critical role of substrate selection in optimizing cyclodextrin output (Jia et al., 2024). The CGTase from *Paenibacillus macerans* CCM 2012, when acting on amaranth starch, produced a distinct profile of α-, β-, and γ-CDs, as confirmed by chromatographic analysis. Beyond substrate selection and enzyme engineering, process optimization plays a crucial role in maximizing cyclodextrin yield and purity (Arce-Vázquez et al., 2016). High substrate concentrations (7.5%–10% w/v), precise control of pH and temperature, and the use of immobilized enzyme systems enhance catalytic efficiency and facilitate enzyme reuse. Immobilization of CGTase on supports such as magnetic nanoparticles, chitosan beads, or DEAE-cellulose not only stabilizes the enzyme but also enables its recovery and repeated use in batch or continuous processes, reducing operational costs and environmental impact. β-Cyclodextrin, with its relatively low water solubility, can be readily crystallized from the reaction mixture, while α- and γ-cyclodextrins often require more sophisticated separation techniques such as complexation with organic solvents or chromatographic methods (Ogunbadejo & Al-Zuhair, 2021).

In the pharmaceutical sector, CDs are used to enhance the solubility, stability, and bioavailability of poorly water-soluble drugs, as well as to cover unpleasant tastes and odors. Their ability to form inclusion complexes extends to the encapsulation of volatile flavors and fragrances in the food and cosmetic industries, as stabilizers and controlled-release agents. In environmental technology, CDs are employed for the removal of hydrophobic pollutants from water and soil, capitalizing on their capacity to trap organic molecules within their central cavity (Ogunbadejo & Al-Zuhair, 2021). The versatility of cyclodextrins has also spurred research into their use as carriers for vitamins, antioxidants, and functional lipids, broadening their utility in nutraceuticals and functional foods. The integration of recombinant DNA technology has also facilitated the large-scale production of CGTase in heterologous hosts such as *Escherichia coli* or *Pichia pastoris*, streamlining enzyme supply for industrial processes. In addition, the use of raw starch-digesting CGTases has eliminated the need for pregelatinization, further reducing energy consumption and process complexity. The immobilization of CGTase not only enhances enzyme stability and reusability but also enables continuous processing, reducing water and energy consumption (Ibrahim et al., 2013). The use of renewable and biodegradable starch sources, coupled with the elimination of toxic solvents and reagents, aligns cyclodextrin production with the goals of circular bioeconomy and environmental stewardship. Furthermore, advances in process engineering, such as the development of membrane reactors, integrated fermentation-hydrolysis systems, and high-throughput screening of enzyme variants, are poised to further improve the efficiency and scalability of cyclodextrin synthesis.

4.3 Enzyme immobilization for sustainable modification

Enzyme immobilization has emerged as a transformative strategy for enhancing the sustainability, efficiency, and economic viability of enzymatic starch modification processes. The immobilization of enzymes attaches within solid supports to address critical challenges associated with enzyme stability, reusability, and process scalability, which are significant in industrial starch processing, where cost, operational robustness, and environmental impact (Abdella et al., 2020). By facilitating continuous use of biocatalysts, immobilization not only reduces enzyme consumption and production costs but also minimizes waste generation. The immobilization techniques, including adsorption, covalent binding, entrapment, encapsulation, and cross-linking, have enabled the enzyme for starch modification applications, from the production of resistant starch and cyclodextrins to the design of slow-digesting and functional starch derivatives. Fig. 5.4 illustrates the different techniques used for enzyme immobilization. Adsorption involves the physical attachment of enzymes onto carriers such as DEAE-cellulose, silica, or activated carbon through noncovalent interactions. While this approach is cost-effective and preserves enzyme activity, it is susceptible to enzyme leaching, especially under fluctuating pH or ionic strength. Covalent binding forms stable chemical linkages between the enzyme and functionalized supports such as glutaraldehyde-crosslinked chitosan, polyacrylamide, or magnetic nanoparticles, significantly enhancing enzyme stability

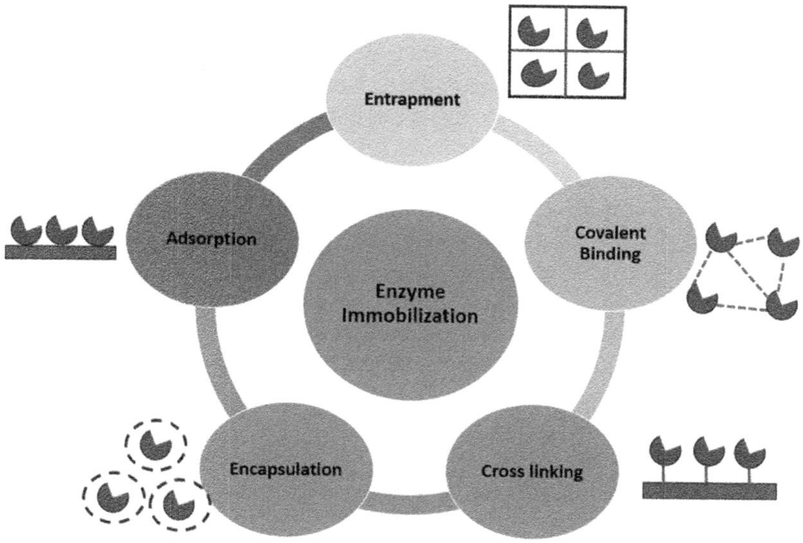

FIGURE 5.4 Different techniques are used for enzyme immobilization.

and preventing leaching process (Datta et al., 2013). Entrapment and encapsulation techniques immobilize enzymes within polymeric matrices like calcium alginate beads, carrageenan, or sol-gel-derived silica, providing a protective microenvironment that shields the enzyme from denaturation and facilitates gentle recovery and reuse. Cross-linking agents like glutaraldehyde, in the presence of protein feeders such as bovine serum albumin, create robust enzyme aggregates with high mechanical stability and resistance to harsh processing conditions (Mostaraddi et al., 2023).

Immobilized α-amylase has been shown to retain over 80% of its initial activity after 15 reaction cycles, while immobilized CGTase on magnetic nanoparticles or chitosan beads can achieve up to 72% β-cyclodextrin yield over multiple cycles (Zusfahair et al., 2020). Multi-enzyme coimmobilization systems such as α-amylase, pullulanase, β-amylase and α-glucosidase complexes enable one-pot conversion of starch to maltose, glucose, or cyclodextrins, streamlining production and reducing the need for intermediate purification steps. These systems can be integrated into batch, packed-bed, or fluidized-bed reactors, supporting both batchwise and continuous operation (Xu et al., 2020). By enabling enzyme reuse, immobilization reduces the overall demand for enzyme production and decreases the frequency of enzyme replenishment, leading to lower operating expenses. The reduction in enzyme leaching and waste generation also minimizes the environmental footprint of starch processing, supporting regulatory compliance and sustainability goals (Maghraby et al., 2023). Moreover, immobilized enzymes exhibit enhanced tolerance to extremes of pH, temperature, and substrate concentration, broadening their applicability across diverse processing environments and enabling the use of less refined or unconventional starch sources.

Immobilized debranching enzymes have been used to increase the content of short linear chains in amylopectin, thereby enhancing gel strength and resistant starch content in the final product. Immobilized CGTase systems facilitate the selective production of specific cyclodextrin types, improving product purity and yield while simplifying downstream purification (Xia et al., 2021). Mesoporous silica and graphene oxide provide high surface area and tunable pore sizes, allowing for high enzyme loading and efficient substrate diffusion. Magnetic nanoparticles enable rapid separation of the immobilized enzyme using external magnetic fields, facilitating process automation and integration with continuous production systems (Kausar et al., 2024). The integration of immobilized enzyme reactors can be coupled with upstream fermentation or downstream purification units, enabling the valorization of agricultural residues and the production of high-value starch derivatives with minimal waste. The use of renewable, biodegradable, and recyclable support materials such as chitosan, alginate, or plant-derived nanocellulose further enhances the environmental profile of immobilized enzyme systems (Bié et al., 2022). The design of stimuli-responsive supports that modulate enzyme activity in response to temperature, pH, or substrate concentration holds promise for dynamic process control and adaptive manufacturing (Claaßen et al., 2019). The coimmobilization of enzyme cascades mimicking natural metabolic pathways can enable the simultaneous transformation of starch into a spectrum of value-added products, from oligosaccharides and cyclodextrins to bioactive peptides and biodegradable polymers.

5. Biotechnological approaches to starch

Biotechnological approaches have revolutionized the field of starch research and industrial application, offering sustainable strategies for modifying starch structure, biosynthesis, and functionality through conventional physical, chemical, or enzymatic methods. Starch is not only a primary carbohydrate reserve in plants but also a critical raw material for the food, pharmaceutical, textile, paper, and biofuel industries. The growing demand for specialty starches with physicochemical and functional properties has driven the exploration of advanced biotechnological tools, including microbial fermentation, genetic engineering of microbes, and genome editing technologies such as CRISPR/Cas systems, to create novel starches with improved or entirely new attributes. The starch modification lies in the deep understanding of starch biosynthesis and granule architecture, based on the complex interplay of core enzymes, including ADP-glucose pyrophosphorylase (AGPase), starch synthases (SSs), starch branching enzymes (SBEs), debranching enzymes (DBEs), and regulatory proteins (Sun, Jiao, et al., 2017). In crops such as rice, maize, potato, and sweet potato, the manipulation of genes encoding these enzymes has been shown to alter the amylose-to-amylopectin ratio, branching pattern, granule morphology, and crystallinity of starch. The suppression of granule-bound starch synthase I (GBSSI) leads to amylose-free or waxy starches, while the overexpression of SBEs and DBEs can produce high-amylose or highly branched starches with unique digestibility and processing characteristics (Sang et al., 2025). The recent elucidation of regulatory pathways, such as the LAZY gene family in rice, which modulates starch biosynthesis in gravity-sensing tissues and affects plant architecture, further expands the toolkit for targeted starch modification (Guo et al., 2020).

Engineered microbial strains, such as *Bacillus* (Li et al., 2022), *Escherichia coli*, *Saccharomyces cerevisiae*, and *Yarrowia lipolytica*, have been developed to express amylolytic enzymes for the direct utilization of raw starch, bypassing the need for energy-intensive gelatinization steps. These microbes can be further optimized through metabolic engineering to enhance substrate uptake, product yield, and process robustness. Further, the introduction of thermostable α-amylase, glucoamylase, and pullulanase genes into industrial yeast strains enables efficient simultaneous saccharification and fermentation of starch to ethanol, reducing process costs and improving sustainability.

Synthetic biology approaches allow for the de novo construction of metabolic pathways for the production of rare starches, such as cyclodextrins, resistant starch, and modified amylopectin. The expression of plant-derived starch biosynthetic enzymes in microbial hosts enables the in vitro assembly of starch granules with controlled structure and composition. The integration of omics technologies, computational modeling, and high-throughput screening accelerates the design-build-test-learn cycle, facilitating the rapid development of microbial cell factories for starch modification. CRISPR-based tools enable targeted knockout, knock-in, or modulation of genes involved in starch metabolism, allowing for the creation of crops with customized starch profiles. CRISPR-mediated editing of SBEIIb in rice and maize has resulted in high-amylose starches with increased resistant starch content, offering health benefits such as reduced glycemic response and enhanced gut health (Ma et al., 2024). Multiplex genome editing enables the simultaneous manipulation of multiple genes, facilitating complex trait engineering and the creation of novel starch phenotypes. The use of base editors and prime editors further expands the range of possible modifications, enabling precise nucleotide changes without introducing double-strand breaks. The engineering of sweet potato starch through CRISPR/Cas9 has enabled the production of starches with altered gelatinization temperature, branching pattern, and granule morphology, expanding their utility in food and industrial applications.

5.1 Microbial fermentation for starch transformation

Microbial fermentation has become a sustainable and versatile platform for converting starch into value-added products. This approach leverages the metabolic capabilities of bacteria, yeasts, and filamentous fungi to hydrolyze, modify, and utilize starch, either as a carbon source for microbial growth or as a substrate for the biosynthesis of specialty compounds. The microbial fermentation lies in its ability to operate under mild conditions, its compatibility with renewable feedstocks, and its adaptability to both conventional and unconventional starch sources, including tubers, cereals, and roots. As global demand for functional foods, bio-based chemicals, and green energy intensifies, microbial fermentation is increasingly integrated into starch biorefineries, enabling the production of organic acids, bioethanol, bioplastics, cyclodextrins, and prebiotic oligosaccharides. The process of microbial starch transformation begins with the enzymatic or acid hydrolysis of starch to release fermentable sugars, such as glucose and maltose. The genetically engineered strains of *Saccharomyces cerevisiae*, *Yarrowia lipolytica*, and *Bacillus subtilis* have been equipped with amylolytic enzymes such as α-amylase, glucoamylase, and pullulanase to facilitate the direct conversion of starch to ethanol, lactic acid, or other target molecules (de Souza & e Magalhães, 2010). The simultaneous saccharification and fermentation process, wherein enzymatic hydrolysis and microbial fermentation occur in a single reactor. This strategy not only improves substrate utilization and product yield but also minimizes contamination risks and operational complexity.

Lactic acid bacteria (LAB), such as *Lactobacillus plantarum* and *Lactobacillus amylovorus*, are employed for the fermentation of starch-rich substrates to produce lactic acid, a precursor for biodegradable plastics and a precursor in food preservation (Akoetey & Morawicki, 2018). These bacteria possess robust amylolytic systems in a variety of starch-containing matrices, including sweet potato, cassava, and maize. Similarly, *Clostridium* species are harnessed for the production of butanol and other solvents via acetone-butanol-ethanol (ABE) fermentation, capitalizing on their ability to degrade complex carbohydrates under anaerobic conditions. The use of filamentous fungi, such as *Aspergillus niger* and *Rhizopus oryzae*, extends the range of fermentative products to include citric acid, fumaric acid, and various enzymes of starch-derived bioproducts. Tubers such as cassava, sweet potato, and colocasia, as well as minor crops like arrowroot and elephant foot yam, provide abundant and cost-effective starch feedstocks, particularly in tropical and subtropical regions. Cassava starch, with its high purity and low lipid and protein content, is well-suited for fermentation, yielding high concentrations of ethanol and lactic acid with minimal byproduct formation. Sweet potato starch, on the other hand, is rich in fermentable sugars and micronutrients, supporting the growth of both bacteria and yeasts. The adaptability of microbial systems to these diverse substrates is further enhanced by strain selection, adaptive laboratory evolution, and metabolic engineering, which enable the adaptation of microbial metabolism to specific feedstock characteristics.

The efficiency of microbial starch fermentation is influenced by a multitude of factors, including substrate composition, particle size, water content, nutrient availability, and fermentation conditions. The presence of inhibitors such as cyanogenic glycosides in cassava or phenolic compounds in certain tubers can impede microbial growth and product formation; pretreatment steps such as detoxification or enzymatic hydrolysis may be required to overcome these challenges. The integration of fermentation with downstream processing, such as distillation for ethanol recovery or membrane filtration for lactic acid purification, further enhances the starch biorefineries (Cheng et al., 1991). In addition to bulk chemicals, resistant starch, a form of dietary fiber that escapes digestion in the small intestine and is fermented by gut microbiota in the colon, can be generated by the fermentation of high-amylose starches. The short-chain fatty acids, such as butyrate, confer health benefits including improved gut health, reduced inflammation, and enhanced mineral absorption. Similarly, the microbial conversion of starch into prebiotic oligosaccharides, such as isomalto oligosaccharides and fructo oligosaccharides, supports the growth of beneficial gut bacteria and contributes to the development of functional foods with targeted health effects (Iram et al., 2023). Secondary metabolites, such as antibiotics, pigments, and bioactive peptides, can also be synthesized from starch-derived substrates, expanding the scope of biotechnological innovation. The cofermentation of starch by LAB and yeasts can enhance flavor development and nutritional quality in fermented foods, while mixed cultures of bacteria and fungi can improve the efficiency of bioplastic or biofuel production.

5.2 Genetically engineered microbes

The core of microbial engineering for starch modification is the introduction or optimization of genes encoding amylolytic and starch-modifying enzymes, such as α-amylase, β-amylase, glucoamylase, pullulanase, isoamylase, CGTase, and branching or debranching enzymes. By expressing these enzymes in microbial hosts, most commonly *Escherichia coli, Bacillus subtilis, Saccharomyces cerevisiae, and Yarrowia lipolytica* directly hydrolyze raw starch, synthesize specialty oligosaccharides, or modify starches with altered amylose/amylopectin ratios and branching patterns (Ong et al., 2008). The use of recombinant DNA technology allows for the fine-tuning of enzyme expression levels, secretion pathways, and posttranslational modifications, maximizing catalytic efficiency and product yield. Engineered Bacillus strains expressing thermostable α-amylase and pullulanase can efficiently liquefy and saccharify starch at elevated temperatures, streamlining the production of glucose syrups, bioethanol, or organic acids. Similarly, the coexpression of CGTase and branching enzymes in *E. coli* enables the one-pot conversion of starch to cyclodextrins, with product specificity modulated by enzyme engineering and process conditions (Liu et al., 2012).

The integration of genetically engineered microbes into starch processing workflows offers several advantages over conventional methods. First, engineered strains can be tailored to utilize a wide range of starch substrates, including unconventional or low-cost sources such as cassava, sweet potato, colocasia, and even agricultural residues. This flexibility supports the diversification of feedstocks and enhances the sustainability of starch biorefineries. Second, the ability to coexpress multiple enzymes or entire metabolic pathways within a single microbial chassis enables complex, multi-step transformations that would be difficult or inefficient to achieve with free enzymes or chemical catalysts (Wang et al., 2018). Engineered yeast strains capable of simultaneous saccharification and fermentation can convert raw starch directly to ethanol, lactic acid, or other bioproducts in a single reactor, reducing process complexity, contamination risk, and energy consumption. Third, advances in synthetic biology have enabled the construction of synthetic operons, regulatory circuits, and dynamic control systems that optimize enzyme expression.

Recent research has demonstrated the potential of genetically engineered microbes to produce starch derivatives with novel structures and functionalities. The coexpression of α-glucan branching enzyme and CGTase in *E. coli* has been shown to generate starches with increased branching density and slow-digesting or resistant starch fractions, which are valued for their health benefits and functional properties in food applications (Pfister et al., 2016). In another study, *Yarrowia lipolytica* was engineered to express plant-derived starch synthases, enabling the in vitro assembly of amylopectin granules with controlled chain length and branching pattern (Yook et al., 2025). These designer starches exhibit unique physicochemical properties, such as altered gelatinization temperature, viscosity, and digestibility, expanding their potential uses in food, pharmaceutical, and materials science. The overexpression of chaperones, proteases, or secretion factors can enhance the folding, stability, and export of recombinant enzymes, while metabolic engineering can redirect carbon flux toward desired products and minimize the accumulation of inhibitory byproducts. Adaptive laboratory evolution and high-throughput screening are increasingly used to identify strains with improved performance under industrial conditions, such as high substrate concentrations, elevated temperatures, or the presence of inhibitors (Liu et al., 2014). The use of CRISPR/Cas genome editing has further accelerated the development of optimized microbial cell factories, enabling precise, multiplexed modifications of metabolic pathways, regulatory networks, and stress response genes. In single-strain engineering, the use of microbial consortia has new opportunities for synergistic starch transformation. By combining strains with complementary enzymatic activities, it is possible to achieve more efficient substrate utilization, product diversification, and process resilience. A consortium of Bacillus and Lactobacillus strains can convert starch to lactic acid with high yield and productivity, while mixed cultures of yeast and filamentous fungi can produce a range of organic acids, enzymes, and bioactive compounds from starch-rich feedstocks (Fossi et al., 2014). The design and optimization of such consortia require microbial interaction, metabolic cross-feeding, and controlling population composition and function.

5.3 CRISPR and synthetic biology in starch biosynthesis modification

The CRISPR/Cas genome editing and the maturation of synthetic biology have revolutionized the landscape of starch biosynthesis modification, offering a level of precision, efficiency, and creativity previously unattainable through traditional breeding, mutagenesis, or even classical genetic engineering. These next-generation tools enable targeted manipulation of genes and regulatory networks that govern starch structure, composition, granule morphology, and functional properties in both plants and microbes. The impact of these technologies is profound, not only for fundamental research into the molecular mechanisms of starch biosynthesis but also for the rational design of crops and microbial systems tailored to meet the evolving demands of food, pharmaceutical, and industrial sectors.

CRISPR/Cas9 and its derivatives have become the method of choice for targeted genome editing in plants. By designing guide RNAs that direct the Cas9 nuclease to specific loci, researchers can induce double-strand breaks and harness the cell's native repair machinery to introduce insertions, deletions, or precise nucleotide substitutions. This technology has been rapidly adopted for starch biosynthesis modification in major crops such as rice, maize, wheat, and potato. For example, the targeted knockout of starch branching enzyme IIb (SBEIIb) in rice and maize has resulted in high-amylose or even amylose-only starches, which exhibit increased resistant starch content and altered gelatinization properties (Wang et al., 2021). In wheat, multiplex CRISPR editing of GBSSI, SSIIa, and SBEIIa has enabled the generation of lines with a broad range of amylose content, from waxy to high-amylose phenotypes, for the development of specialty starches with digestibility and processing characteristics (Ma et al., 2024). The use of base editors and prime editors further expands the toolkit, allowing for precise point mutations without double-strand breaks, reducing off-target effects and enhancing the efficiency of trait engineering.

In starch biotechnology, synthetic biology has been harnessed to construct novel biosynthetic routes for the production of designer starches, cyclodextrins, and functional oligosaccharides in microbial hosts such as *Escherichia coli*, *Saccharomyces cerevisiae*, and *Yarrowia lipolytica* (Yook et al., 2025). By introducing plant-derived starch synthases, branching enzymes, and debranching enzymes into these microbes, the in vitro assembly of amylopectin granules with controlled chain length and branching pattern. The modularity of synthetic biology allows for the combinatorial optimization of enzyme expression, localization, and activity, facilitating the production of starches (Ramli et al., 2013). The coexpression of CGTase and branching enzymes in *E. coli* has enabled the one-pot synthesis of cyclodextrins with adjustable size distribution, while synthetic operons and dynamic regulatory systems have been used to optimize the flux through starch biosynthesis pathways in yeast and filamentous fungi. The characterization of novel regulatory genes such as LAZY3 and LAZY2 in rice has revealed their roles in starch granule formation within gravity-sensing tissues, with direct implications for plant architecture and yield (Cai et al., 2023). The targeted editing of OsPK2, encoding a plastidic pyruvate kinase, has provided new insights into the interplay between primary metabolism, starch synthesis, and grain filling in rice. The ability to introduce or modify multiple genes simultaneously enables the rapid domestication of underutilized species, the enhancement of starch yield and quality, and the starch properties for specific industrial applications. The engineering of sweet potato and cassava for altered starch structure and digestibility holds the ability for the development of functional foods and speciality ingredients in tropical and subtropical regions.

6. Challenges and future directions

Significant obstacles still exist, considering the important developments in starch modification via physical, chemical, enzymatic, and biotechnological methods. The stability and reliability in engineered starch products are further complicated by the inherent complexity and source-dependent variability of starch, including its granular structure, chain-length distribution, and amylose/amylopectin ratio (Maniglia et al., 2020). Chemical processes such as oxidation, esterification, and cross-linking increase processability and stability; however, they require hazardous chemicals and produce potentially risky byproducts that raise safety, regulatory, and environmental issues, especially in food and pharmaceutical applications. Although enzymatic methods provide moderate conditions and great selectivity, scaling up is still difficult because of the intricacy of regulating hydrolysis or remodeling events, costs, and enzyme stability. Physical treatments such as heat-moisture conditioning, ultrasonication, and high-pressure techniques can increase a material's viscosity or resistant starch content, but they frequently result in uneven and poorly managed structural changes that reduce their functional accuracy (Tappiban et al., 2024).

In the field of biotechnology, precise engineering of the genes involved in starch production in microbes and crops has been made possible by CRISPR/Cas-based genome editing. In barley, rice, and cassava, multiplex editing of branching enzymes and starch synthases has resulted in high levels of waxy or resistant starch properties (Yang, Bui, et al., 2024). Nevertheless, there are a number of drawbacks to field-deployable genome editing. Many nonmodel crops still struggle with steady regeneration, effective transformation techniques, and avoiding off-target mutations. The trade-off between functionality and agronomic performance is demonstrated by high-fidelity Cas variants, off-target effects with CRISPR and regulatory risks across regions (Chen et al., 2024). Although engineered strains that can biosynthesize and break down starch are increasingly utilized in microbial platforms, challenges with strain optimization, pathway balancing, and scale-up still exist. Although consistent granule formation and yields are still limited, the successful integration of thermostable amylases, glucoamylases, and pullulanases in *S. cerevisiae*, *E. coli*, *Bacillus*, and *Y. lipolytica* offers potential for *in-situ* starch processing (Yang, Bui, et al., 2024).

New techniques such as pressure-moisture treatments and ultrasonication-assisted production of resistant starch have shown effectiveness when paired with microbiological and enzymatic processes (Tappiban et al., 2024). Target discovery, modification scheme optimization, and functional impact prediction will be improved by combining multi-omics with systems biology and machine learning. Next-generation genome editing techniques, including base editors and prime editors, provide more precise control with fewer side effects and could circumvent GMO regulatory barriers (Hebelstrup et al., 2015). Together with cis-genic and TILLING techniques, RNP or transient delivery systems for CRISPR facilitate the development of nontransgenic, modified starch crops that might gain wide regulatory approval. While using nano-fillers and better cross-linking techniques, the design of starch-based bioplastics and active packaging keeps improving, but large-scale adoption still requires cost and environmental effect reductions (Yudhistira et al., 2024).

7. Conclusion

The improvements in physical, chemical, enzymatic, and biotechnological methods have allowed starch modification to develop into a multidisciplinary domain. Despite its enormous economic potential, natural starch must be structurally and functionally modified to meet the demands of modern applications due to its inherent constraints. Through CRISPR/Cas genome editing and microbial engineering, starches with improved qualities such as higher resistant starch content, changed gelatinization behavior, and enhanced processing stability have been developed. In order to create environmentally friendly and functionally superior starches, future approaches should concentrate on combining green chemistry, systems biology, omics-guided design, and synthetic biology. There is potential for increasing the use of starch in the food, pharmaceutical, and material industries through the investigation of underutilized starch sources, process optimization, and intelligent bioprocessing. The next generation of starch-based technologies in a global bioeconomy will ultimately be defined by the convergence of sustainability, accuracy, and utility.

References

Abdella, M. A. A., El-Sherbiny, G. M., El-Shamy, A. R., Atalla, S. M. M., & Ahmed, S. A. (2020). Statistical optimization of chemical modification of chitosan-magnetic nano-particles beads to promote Bacillus subtilis MK1 α-amylase immobilization and its application. *Bulletin of the National Research Centre, 44*(1). https://doi.org/10.1186/s42269-020-00301-3

Akoetey, W., & Morawicki, R. (2018). The effect of adaptation of Lactobacillus amylovorus to increasing concentrations of sweet potato starch on the production of lactic acid for its potential use in the treatment of cannery waste. *Journal of Environmental Science and Health Part B Pesticides Food Contaminants and Agricultural Wastes, 53*(12), 802–809. https://doi.org/10.1080/03601234.2018.1505076

Akpa, J., Gunorubon, & Dagde, K. (2012). Modification of cassava starch for industrial uses. *International Journal of Engineering and Technology, 2.*

Arce-Vázquez, M. B., Ponce-Alquicira, E., Delgado-Fornué, E., Pedroza-Islas, R., Díaz-Godínez, G., & Soriano-Santos, J. (2016). Integral use of amaranth starch to obtain Cyclodextrin glycosyltransferase, by Bacillus megaterium, to produce β-cyclodextrin. *Frontiers in Microbiology, 7.* https://doi.org/10.3389/fmicb.2016.01513

Ashogbon, A. O. (2021). Dual modification of various starches: Synthesis, properties and applications. *Food Chemistry, 342*, Article 128325. https://doi.org/10.1016/j.foodchem.2020.128325

Błaszczak, W., Valverde, S., & Fornal, J. (2005). Effect of high pressure on the structure of potato starch. *Carbohydrate Polymers, 59*(3), 377–383.

Bertoft, E. (2017). Understanding starch structure: Recent progress. *Agronomy, 7*(3), 56. https://doi.org/10.3390/agronomy7030056

Bié, J., Sepodes, B., Fernandes, P. C. B., & Ribeiro, M. H. L. (2022). Enzyme immobilization and co-immobilization: Main framework, advances and some applications. *Processes, 10*(3). https://doi.org/10.3390/pr10030494

Birt, D. F., Boylston, T., Hendrich, S., Jane, J. L., Hollis, J., Li, L., McClelland, J., Moore, S., Phillips, G. J., Rowling, M., Schalinske, K., Paul Scott, M., & Whitley, E. M. (2013). Resistant starch: Promise for improving human health. *Advances in Nutrition, 4*(6), 587–601. https://doi.org/10.3945/an.113.004325

Bonto, A. P., Tiozon, R. N., Sreenivasulu, N., & Camacho, D. H. (2021). Impact of ultrasonic treatment on rice starch and grain functional properties: A review. *Ultrasonics Sonochemistry, 71.* https://doi.org/10.1016/j.ultsonch.2020.105383

Cai, Y., Huang, L., Song, Y., Yuan, Y., Xu, S., Wang, X., Liang, Y., Zhou, J., Liu, G., Li, J., Wang, W., & Wang, Y. (2023). LAZY3 interacts with LAZY2 to regulate tiller angle by modulating shoot gravity perception in rice. *Plant Biotechnology Journal, 21*(6), 1217–1228. https://doi.org/10.1111/pbi.14031

Castro, L. M. G., Alexandre, E. M. C., Saraiva, J. A., & Pintado, M. (2020). Impact of high pressure on starch properties: A review. *Food Hydrocolloids, 106.* https://doi.org/10.1016/j.foodhyd.2020.105877

Chemelli, A., Gomernik, F., Thaler, F., Huber, A., Ulrich, H., Bauer, W., & Spirk, S. (2020). Cationic starches in paper-based applications—A review on analytical methods. *Carbohydrate Polymers, 235*, Article 115964. https://doi.org/10.1016/j.carbpol.2020.115964

Chen, F., Chen, L., Yan, Z., Xu, J., Feng, L., He, N., Guo, M., Zhao, J., Chen, Z., Chen, H., Yao, G., & Liu, C. (2024). Recent advances of CRISPR-Based genome editing for enhancing staple crops. *Frontiers in Plant Science, 15.* https://doi.org/10.3389/fpls.2024.1478398

Chen, Q., Yu, H., Wang, L., Ul Abdin, Z., Chen, Y., Wang, J., Zhou, W., Yang, X., Khan, R. U., Zhang, H., & Chen, X. (2015). Recent progress in chemical modification of starch and its applications. *RSC Advances, 5*(83), 67459–67474. https://doi.org/10.1039/c5ra10849g

Cheng, P., Mueller, R. E., Jaeger, S., Bajpai, R., & Iannotti, E. L. (1991). Lactic acid production from enzyme-thinned corn starch usingLactobacillus amylovorus. *Journal of Industrial Microbiology, 7*(1), 27–34. https://doi.org/10.1007/bf01575599

Cheng, F., Ren, Y., Warkentin, T. D., & Ai, Y. (2024). Heat-moisture treatment to modify structure and functionality and reduce digestibility of wrinkled and round pea starches. *Carbohydrate Polymers, 324*, 121506.

Claaßen, C., Gerlach, T., & Rother, D. (2019). Stimulus-responsive regulation of enzyme activity for one-step and multi-step syntheses. *Advanced Synthesis & Catalysis, 361*(11), 2387–2401.

Clasen, S. H., Müller, C. M. O., Parize, A. L., & Pires, A. T. N. (2018). Synthesis and characterization of cassava starch with maleic acid derivatives by etherification reaction. *Carbohydrate Polymers, 180*, 348–353. https://doi.org/10.1016/j.carbpol.2017.10.016

Compart, J., Singh, A., Fettke, J., Apriyanto, A., & Starch Properties, C. (2023). A review of starch modifications and their applications. *Polymers, 15*(16), 3491. https://doi.org/10.3390/polym15163491

Cui, D., Liu, M., Wu, L., & Bi, Y. (2009). Synthesis of potato starch sulfate and optimization of the reaction conditions. *International Journal of Biological Macromolecules, 44*(3), 294–299. https://doi.org/10.1016/j.ijbiomac.2009.01.003

Datta, S., Christena, L. R., & Rajaram, Y. R. S. (2013). Enzyme immobilization: An overview on techniques and support materials. *3 Biotech, 3*(1), 1–9.

de Souza, P. M., & e Magalhães, P.d. O. (2010). Application of microbial α-amylase in industry - A review. *Brazilian Journal of Microbiology, 41*(4), 850–861. https://doi.org/10.1590/s1517-83822010000400004

Dorantes-Fuertes, M. G., López-Méndez, M. C., Martínez-Castellanos, G., Meléndez-Armenta, R.Á., & Jiménez-Martínez, H. E. (2024). Starch extraction methods in tubers and roots: A systematic review. *Agronomy, 14*(4). https://doi.org/10.3390/agronomy14040865

H.O. Egharevba, Chemical properties of starch and its application in the food industry chemical properties of starch. (2019).

Eriksson, M., Pettersson, G., & Wågberg, L. (2005). Application of polymeric multilayers of starch onto wood fibres to enhance strength properties of paper. *Nordic Pulp and Paper Research Journal, 20*, 270–276.

Fincan, S. A., Özdemir, S., Karakaya, A., Enez, B., Mustafov, S. D., Ulutaş, M. S., & Şen, F. (2021). Purification and characterization of thermostable α-amylase produced from Bacillus licheniformis So-B3 and its potential in hydrolyzing raw starch. *Life Sciences, 264.* https://doi.org/10.1016/j.lfs.2020.118639

Fossi, B. T., Tavea, F., Fontem, L. A., Ndjouenkeu, R., & Wanji, S. (2014). Microbial interactions for enhancement of α-amylase production by bacillus amyloliquefaciens 04BBA15 and Lactobacillus fermentum 04BBA19. *Biotechnology Reports, 4*(1), 99–106. https://doi.org/10.1016/j.btre.2014.09.004

Gerezgiher, A. G., & Szabó, T. (2022). Crosslinking of starch using citric acid. *Journal of Physics: Conference Series, 2315*(1), Article 012036. https://doi.org/10.1088/1742-6596/2315/1/012036

Guo, D., Ling, X., Zhou, X., Li, X., Wang, J., Qiu, S., Yang, Y., & Zhang, B. (2020). Evaluation of the quality of a high-resistant starch and low-glutelin rice (Oryza sativa L.) generated through CRISPR/Cas9-mediated targeted mutagenesis. *Journal of Agricultural and Food Chemistry, 68*(36), 9733–9742. https://doi.org/10.1021/acs.jafc.0c02995

Gupta, R. K., Guha, P., & Srivastav, P. P. (2024). Physical action of nonthermal cold plasma technology for starch modification. *Food Physics, 1*, Article 100011.

Haleem, N., Osabutey, A., Albert, K., Zhang, C., Min, K., Anderson, G., & Yang, X. (2023). Flocculation of livestock wastewater using cationic starch prepared from potato peels. *Environmental Sciences: Water Research & Technology, 9*(6), 1690–1700. https://doi.org/10.1039/d2ew00794k

Han, L., Cao, S., Yu, Y., Xu, X., Cao, X., & Chen, W. (2021). Modification in physicochemical, structural and digestive properties of pea starch during heat-moisture process assisted by pre- and post-treatment of ultrasound. *Food Chemistry, 360*, Article 129929. https://doi.org/10.1016/j.foodchem.2021.129929

Han, R., Li, J., Shin, H. D., Chen, R. R., Du, G., Liu, L., & Chen, J. (2014). Recent advances in discovery, heterologous expression, and molecular engineering of cyclodextrin glycosyltransferase for versatile applications. *Biotechnology Advances, 32*(2), 415–428.

Hazarika, B. J., & Sit, N. (2016). Effect of dual modification with hydroxypropylation and cross-linking on physicochemical properties of taro starch. *Carbohydrate Polymers, 140*, 269–278. https://doi.org/10.1016/j.carbpol.2015.12.055

He, W., & Wei, C. (2017). Progress in C-type starches from different plant sources. *Food Hydrocolloids, 73*, 162–175. https://doi.org/10.1016/j.foodhyd.2017.07.003

Hebelstrup, K. H., Sagnelli, D., & Blennow, A. (2015). The future of starch bioengineering: GM microorganisms or GM plants? *Frontiers in Plant Science, 6*.

Heinze, T., Liebert, T., Heinze, U., & Schwikal, K. (2004). Starch derivatives of high degree of functionalization 9: Carboxymethyl starches. *Cellulose, 11*(2), 239–245. https://doi.org/10.1023/B:CELL.0000025386.68486.a4

Hu, X., Jiang, H., Liu, Z., Gao, M., Liu, G., Tian, S., & Zeng, F. (2025). The global potato-processing industry: A review of production, products, quality and sustainability. *Foods, 14*(10), 1758. https://doi.org/10.3390/foods14101758

Hubbe, M. A. (2006). Bonding between cellulosic fibers in the absence and presence of dry-strength agents–A review. *BioResources, 1*(2), 281–318.

Iacovou, M., Lim, J., Maningat, C. C., Bogatyrev, A., Ly, E., Dhital, S., Gidley, M. J., Shi, Y.-C., Muir, J., & Seib, P. A. (2017). In vivo digestibility of cross-linked phosphorylated (RS4) wheat starch in ileostomy subjects. *Bioactive Carbohydrates and Dietary Fibre, 12*, 25–36. https://doi.org/10.1016/j.bcdf.2017.08.002

Ibrahim, A. S. S., Al-Salamah, A. A., El-Toni, A. M., El-Tayeb, M. A., & Elbadawi, Y. B. (2013). Immobilization of cyclodextrin glucanotransferase on aminopropyl-functionalized silica-coated superparamagnetic nanoparticles. *Electronic Journal of Biotechnology, 16*(6). https://doi.org/10.2225/vol16-issue6-fulltext-8

Ibrahim, K. A., Naz, M. Y., Shukrullah, S., Sulaiman, S. A., Ghaffar, A., & AbdEl-Salam, N. M. (2020). Nitrogen pollution impact and remediation through low cost starch based biodegradable polymers. *Scientific Reports, 10*(1). https://doi.org/10.1038/s41598-020-62793-3

Iram, A., Ali, O., Turhan, I., & Demirci, A. (2023). Production of value-added products as food ingredients via microbial fermentation. *Processes, 11*(6), 1715. https://doi.org/10.3390/pr11061715

Jambrak, A. R., Herceg, Z., Šubarić, D., Babić, J., Brnčić, M., Brnčić, S. R., Bosiljkov, T., Čvek, D., Tripalo, B., & Gelo, J. (2010). Ultrasound effect on physical properties of corn starch. *Carbohydrate Polymers, 79*(1), 91–100. https://doi.org/10.1016/j.carbpol.2009.07.051

Jia, X., Xu, J., Cui, Y., Ben, D., Wu, C., Zhang, J., Sun, M., Liu, S., Zhu, T., Liu, J., Lin, K., & Zheng, M. (2024). Effect of modification by β-amylase and α-glucosidase on the structural and physicochemical properties of maize starch. *Foods, 13*(23), 3763. https://doi.org/10.3390/foods13233763

Jong, K., & Ju, B. (2017). Thermo-responsive behavior of propynyl-containing hydroxyethyl starch. *Colloid and Polymer Science, 295*(2), 307–315. https://doi.org/10.1007/s00396-016-4009-1

Kahraman, K., Koksel, H., & Ng, P. K. W. (2015). Optimisation of the reaction conditions for the production of cross-linked starch with high resistant starch content. *Food Chemistry, 174*, 173–179.

Karthik, P., Ettelaie, R., & Chen, J. (2019). Oral behaviour of emulsions stabilized by mixed monolayer. *Food Research International, 125*, Article 108603.

Kausar, S., Yousaf, M., Mir, S., Awwad, N. S., Alturaifi, H. A., & Riaz, F. (2024). Mesoporous materials: Synthesis and electrochemical applications. *Electrochemistry Communications, 169*. https://doi.org/10.1016/j.elecom.2024.107836

Kavaliauskaite, R., Klimaviciute, R., & Zemaitaitis, A. (2008). Factors influencing production of cationic starches. *Carbohydrate Polymers, 73*(4), 665–675.

Kelly, R. M., Dijkhuizen, L., & Leemhuis, H. (2009). The evolution of Cyclodextrin glucanotransferase product specificity. *Applied Microbiology and Biotechnology, 84*(1), 119–133.

Kenawy, E. R., Kamoun, E. A., Mohy Eldin, M. S., & El-Meligy, M. A. (2014). Physically crosslinked poly(vinyl alcohol)-hydroxyethyl starch blend hydrogel membranes: Synthesis and characterization for biomedical applications. *Arabian Journal of Chemistry, 7*(3), 372–380. https://doi.org/10.1016/j.arabjc.2013.05.026

Kong, X., Zhou, X., Sui, Z., & Bao, J. (2016). Effects of gamma irradiation on physicochemical properties of native and acetylated wheat starches. *International Journal of Biological Macromolecules, 91*, 1141–1150. https://doi.org/10.1016/j.ijbiomac.2016.06.072

Kushwaha, R., Kaur, S., & Kaur, D. (2023). Potential of jackfruit (Artocarpus heterophyllus lam.) seed starch as an alternative to the commercial starch source – A review. *Food Reviews International, 39*(5), 2635–2654. https://doi.org/10.1080/87559129.2021.1963979

Lee, S. J., Hong, J. Y., Lee, E. J., Chung, H. J., & Lim, S. T. (2015). Impact of single and dual modifications on physicochemical properties of japonica and indica rice starches. *Carbohydrate Polymers, 122*, 77–83. https://doi.org/10.1016/j.carbpol.2015.01.009

Lee, H., & Yoo, B. (2023). Particle agglomeration and properties of pregelatinized potato starch powder. *Gels, 9*(2), 93. https://doi.org/10.3390/gels9020093

Li, Y., Li, C., Gu, Z., Cheng, L., Hong, Y., & Li, Z. (2019). Digestion properties of corn starch modified by α-D-glucan branching enzyme and Cyclodextrin glycosyltransferase. *Food Hydrocolloids, 89*, 534–541. https://doi.org/10.1016/j.foodhyd.2018.11.025

Li, Y., Ren, J., Liu, J., Lu, S., Wang, Y., Liu, B., Li, C., & Li, Z. (2018). Modification by α-d-glucan branching enzyme lowers the in vitro digestibility of starch from different sources. *International Journal of Biological Macromolecules, 107*(Pt B), 1758–1764. https://doi.org/10.1016/j.ijbiomac.2017.10.049

Li, W., Wu, L., Xu, Z., & Liu, Z. (2020). Adhesion-to-fibers and film properties of etherified–oxidized cassava starch/polyvinyl alcohol blends. *Iranian Polymer Journal (English Edition), 29*(4), 331–339. https://doi.org/10.1007/s13726-020-00798-5

Li, H., Yao, D., Pan, Y., Chen, X., Fang, Z., & Xiao, Y. (2022). Enhanced extracellular raw starch-degrading α-amylase production in Bacillus subtilis by promoter engineering and translation initiation efficiency optimization. *Microbial Cell Factories, 21*(1). https://doi.org/10.1186/s12934-022-01855-9

Lima, D. C., Maniglia, B. C., Matta Junior, M. D., Le-Bail, P., Le-Bail, A., & Augusto, P. E. D. (2021). Dual-process of starch modification: Combining ozone and dry heating treatments to modify cassava starch structure and functionality. *International Journal of Biological Macromolecules, 167*, 894–905. https://doi.org/10.1016/j.ijbiomac.2020.11.046

Liu, W., Gu, J., Huang, C., Lai, C., Ling, Z., & Yong, Q. (2021). Fabrication of hydrophobic and high-strength packaging films based on the esterification modification of galactomannan. *International Journal of Biological Macromolecules, 167*, 1221–1229. https://doi.org/10.1016/j.ijbiomac.2020.11.076

Liu, H., Li, J., Du, G., Zhou, J., & Chen, J. (2012). Enhanced production of α-cyclodextrin glycosyltransferase in *Escherichia coli* by systematic codon usage optimization. *Journal of Industrial Microbiology and Biotechnology, 39*(12), 1841–1849. https://doi.org/10.1007/s10295-012-1185-y

Liu, Z., Liu, L., Österlund, T., Hou, J., Huang, M., Fagerberg, L., Petranovic, D., Uhlén, M., Nielsen, J., & Brakhage, A. A. (2014). Improved production of a heterologous amylase in Saccharomyces cerevisiae by inverse metabolic engineering. *Applied and Environmental Microbiology, 80*(17), 5542–5550. https://doi.org/10.1128/aem.00712-14

Ma, H., Mei, L., Liang, Y., Zheng, X., Sun, Le, Dang, W., Li, J., Li, L., & Liu, C. (2022). Research progress on properties of pre-gelatinized starch and its application in wheat flour products. *Grain & Oil Science and Technology, 5*(2), 87–97. https://doi.org/10.1016/j.gaost.2022.01.001

Ma, M., Sun, S., Zhu, J., Qi, X., Li, G., Hu, J., Xie, C., & Liu, C. (2024). Engineering high amylose and resistant starch in maize by CRISPR/Cas9-mediated editing of starch branching enzymes. *The Crop Journal, 12*(4), 1252–1258.

Maghraby, Y. R., El-Shabasy, R. M., Ibrahim, A. H., & Azzazy, H. M. E. S. (2023). Enzyme immobilization technologies and industrial applications. *American Chemical Society, Egypt ACS Omega, 8*(6), 5184–5196. https://doi.org/10.1021/acsomega.2c07560

Malik, M. K., Kumar, V., Sharma, P. P., Singh, J., Fuloria, S., Subrimanyan, V., Fuloria, N. K., & Kumar, P. (2022). Improvement in digestion resistibility of mandua starch (*Eleusine coracana*) after cross-linking with epichlorohydrin. *ACS Omega, 7*(31), 27334–27346.

Maniglia, B., Castanha, N., Le-Bail, P., Le-Bail, A., & Augusto, P. (2020). Starch modification through environmentally friendly alternatives: A review. *Critical Reviews in Food Science and Nutrition*, 61.

Masina, N., Choonara, Y. E., Kumar, P., du Toit, L. C., Govender, M., Indermun, S., & Pillay, V. (2017). A review of the chemical modification techniques of starch. *Carbohydrate Polymers, 157*, 1226–1236. https://doi.org/10.1016/j.carbpol.2016.09.094

Mehboob, S., Ali, T. M., Alam, F., & Hasnain, A. (2015). Dual modification of native white sorghum (Sorghum bicolor) starch via acid hydrolysis and succinylation. *LWT - Food Science and Technology, 64*(1), 459–467. https://doi.org/10.1016/j.lwt.2015.05.012

Mostaraddi, S., Pazhang, M., Ebadi-Nahari, M., & Najavand, S. (2023). The relationship between the cross-linker on chitosan-coated magnetic nanoparticles and the properties of immobilized papain. *Molecular Biotechnology, 65*(11), 1809–1823. https://doi.org/10.1007/s12033-023-00687-1

Mu, M., Karthik, P., Chen, J., Holmes, M., & Ettelaie, R. (2021). Effect of amylose and amylopectin content on the colloidal behaviour of emulsions stabilised by OSA-modified starch. *Food Hydrocolloids, 111*, Article 106363. https://doi.org/10.1016/j.foodhyd.2020.106363

Niu, D., Cong, H., Zhang, Y., Mchunu, N. P., & Wang, Z. X. (2022). Pullulanase with high temperature and low pH optima improved starch saccharification efficiency. *Scientific Reports, 12*(1). https://doi.org/10.1038/s41598-022-26410-9

Ogunbadejo, B., & Al-Zuhair, S. (2021). MOFs as potential matrices in Cyclodextrin glycosyltransferase immobilization. *Molecules, 26*(3), 680. https://doi.org/10.3390/molecules26030680

Okoli, C. P., Adewuyi, G. O., Zhang, Q., Zhu, G., Wang, C., & Guo, Q. (2019). Aqueous scavenging of polycyclic aromatic hydrocarbons using epichlorohydrin, 1,6-hexamethylene diisocyanate and 4,4-methylene diphenyl diisocyanate modified starch: Pollution remediation approach. *Arabian Journal of Chemistry, 12*(8), 2760–2773.

Okyere, A. Y., Rajendran, S., & Annor, G. A. (2022). Cold plasma technologies: Their effect on starch properties and industrial scale-up for starch modification. *Current Research in Food Science, 5*, 451–463.

Ong, R. M., Goh, K. M., Mahadi, N. M., Hassan, O., Rahman, R. N. Z. R. A., & Illias, R. M. (2008). Cloning, extracellular expression and characterization of a predominant β-CGTase from bacillus sp. G1 in *E. coli*. *Journal of Industrial Microbiology and Biotechnology, 35*(12), 1705–1714. https://doi.org/10.1007/s10295-008-0462-2

Oyeyinka, S. A., Akintayo, O. A., Adebo, O. A., Kayitesi, E., & Njobeh, P. B. (2021). A review on the physicochemical properties of starches modified by microwave alone and in combination with other methods. *International Journal of Biological Macromolecules, 176*, 87–95. https://doi.org/10.1016/j.ijbiomac.2021.02.066

Park, S. H., Na, Y., Kim, J., Kang, S. D., & Park, K.-H. (2017). Properties and applications of starch modifying enzymes for use in the baking industry. *Food Science and Biotechnology, 27*(2), 299–312.

Pelissari, F., Ferreira, D., Louzada, L., Santos, F., Corrêa, A., Moreira, F., & Mattoso, L. (2019). Starch-based edible films and coatings: An eco-friendly alternative for food packaging (pp. 359–420).

Peng, H., Zhong, S., Lin, Q., Yao, X., Liang, Z., Yang, M., Yin, G., Liu, Q., & He, H. (2016). Removal of both cationic and anionic contaminants by amphoteric starch. *Carbohydrate Polymers, 138*, 210–214. https://doi.org/10.1016/j.carbpol.2015.11.064

Pfeifer, A., Hampe, R., & Heinze, T. (2017). Synthesis and characterization of novel water-soluble and bactericidic cationic starch esters. *Starch - Stärke, 69*(9–10), Article 1700029.

Pfister, B., Sánchez-Ferrer, A., Diaz, A., Lu, K., Otto, C., Holler, M., Shaik, F. R., Meier, F., Mezzenga, R., & Zeeman, S. C. (2016). Recreating the synthesis of starch granules in yeast. *eLife, 5*, Article e15552.

Picchio, M. L., Linck, Y. G., Monti, G. A., Gugliotta, L. M., Minari, R. J., & Alvarez Igarzabal, C. I. (2018). Casein films crosslinked by tannic acid for food packaging applications. *Food Hydrocolloids, 84*, 424–434. https://doi.org/10.1016/j.foodhyd.2018.06.028

Pillai, R. R., & Thomas, V. (2023). Plasma surface engineering of natural and sustainable polymeric derivatives and their potential applications. *Polymers, 15*(2). https://doi.org/10.3390/polym15020400

Qin, C., Li, J., Wang, W., & Li, W. (2022). Improving mechanical strength and water barrier properties of pulp molded product by wet-end added polyamide epichlorohydrin/cationic starch. *ACS Omega, 7*(26), 22173–22180. https://doi.org/10.1021/acsomega.1c07369

Raigond, P., Ezekiel, R., & Raigond, B. (2015). Resistant starch in food: A review. *Journal of the Science of Food and Agriculture, 95*(10), 1968–1978.

Ramírez-Brewer, D., Quintana, S. E., & García-Zapateiro, L. A. (2023). Effect of microwave treatment on technological, physicochemical, rheological and microstructural properties of mango (*Mangifera indica*) kernel starch variety tommy and sugar. *Lebensmittel-Wissenschaft & Technologie, 187*, Article 115311.

Ramli, N., Abd-Aziz, S., Alitheen, N. B., Hassan, M. A., & Maeda, T. (2013). Improvement of Cyclodextrin glycosyltransferase gene expression in Escherichia coli by insertion of regulatory sequences involved in the promotion of RNA transcription. *Molecular Biotechnology, 54*(3), 961–968. https://doi.org/10.1007/s12033-013-9647-7

Rashwan, A. K., Younis, H. A., Abdelshafy, A. M., Osman, A. I., Eletmany, M. R., Hafouda, M. A., & Chen, W. (2024). Plant starch extraction, modification, and green applications: A review. *Environmental Chemistry Letters, 22*(5), 2483–2530. https://doi.org/10.1007/s10311-024-01753-z

Roy, I., & Gupta, M. N. (2004). Hydrolysis of starch by a mixture of glucoamylase and pullulanase entrapped individually in calcium alginate beads. *Enzyme and Microbial Technology, 34*(1), 26–32. https://doi.org/10.1016/j.enzmictec.2003.07.001

Sang, S., Sun, X., Ma, T., Zhang, Y., Yao, G., Wang, X., Tan, X., Feng, L., Li, J., Ji, S., & Cheng, H. (2025). Efficient promoter editing of the SBEIIb gene enables fine-tuning of the resistant starch content in rice. *International Journal of Biological Macromolecules, 290*, Article 138904. https://doi.org/10.1016/j.ijbiomac.2024.138904

Singh, P., Baisthakur, P., & Yemul, O. S. (2020). Synthesis, characterization and application of crosslinked alginate as green packaging material. *Heliyon, 6*(1), Article e03026.

Singh, J., Kaur, L., & McCarthy, O. J. (2007). Factors influencing the physico-chemical, morphological, thermal and rheological properties of some chemically modified starches for food applications—A review. *Food Hydrocolloids, 21*(1), 1–22.

Sonnendecker, C., & Zimmermann, W. (2019). Change of the product specificity of a cyclodextrin glucanotransferase by semi-rational mutagenesis to synthesize large-ring cyclodextrins. *Catalysts, 9*(3), 242. https://doi.org/10.3390/catal9030242

Spychaj, T., Wilpiszewska, K., & Zdanowicz, M. (2013). Medium and high substituted carboxymethyl starch: Synthesis, characterization and application. *Starch - Stärke, 65*(1–2), 22–33.

Subroto, E., Cahyana, Y., Indiarto, R., & Rahmah, T. A. (2023). Modification of starches and flours by acetylation and its dual modifications: A review of impact on physicochemical properties and their applications. *Polymers, 15*(14). https://doi.org/10.3390/polym15142990

Sujka, M., & Jamroz, J. (2013). Ultrasound-treated starch: SEM and TEM imaging, and functional behaviour. *Food Hydrocolloids, 31*(2), 413–419. https://doi.org/10.1016/j.foodhyd.2012.11.027

Sukhija, S., Singh, S., & Riar, C. S. (2016). Effect of oxidation, cross-linking and dual modification on physicochemical, crystallinity, morphological, pasting and thermal characteristics of elephant foot yam (*Amorphophallus paeoniifolius*) starch. *Food Hydrocolloids, 55*, 56–64.

Sun, Y., Jiao, G., Liu, Z., Zhang, X., Li, J., Guo, X., Du, W., Du, J., Francis, F., Zhao, Y., & Xia, L. (2017). Generation of high-amylose rice through CRISPR/Cas9-Mediated targeted mutagenesis of starch branching enzymes. *Frontiers in Plant Science, 8*, 298.

Sun, X., Saleh, A. S. M., Sun, Z., Ge, X., Shen, H., Zhang, Q., Yu, X., Yuan, L., & Li, W. (2022). Modification of multi-scale structure, physicochemical properties, and digestibility of rice starch via microwave and cold plasma treatments. *Lebensmittel-Wissenschaft & Technologie, 153*. https://doi.org/10.1016/j.lwt.2021.112483

Syarifuddin, S. H., Jannah, M., Ahmad, A., Taba, P., Fauziah, S., & Sondari, D. (2022). The effect of etherification agent on the mechanical properties of sodium carboxymethyl cellulose-based bioplastic and its application as fruit packaging. *Egyptian Journal of Chemistry, 65*(11), 483–488.

Tappiban, P., Sraphet, S., Srisawad, N., Ahmed, S., Bao, J., & Triwitayakorn, K. (2024). Cutting-edge progress in green technologies for resistant starch type 3 and type 5 preparation: An updated review. *Food Chemistry X, 23*, Article 101669. https://doi.org/10.1016/j.fochx.2024.101669

Thessrimuang, N., & Prachayawarakorn, J. (2019). Characterization and properties of high amylose mung bean starch biodegradable films cross-linked with malic acid or succinic acid. *Journal of Polymers and the Environment, 27*(2), 234–244. https://doi.org/10.1007/s10924-018-1340-2

Thombre, R. S., & Kanekar, P. P. (2016). *Cyclodextrin glycosyl transferase (CGTase): An overview of their production and biotechnological applications industrial biotechnology: Sustainable production and bioresource utilization*. India: CRC Press.

Tian, Y., Petersen, B. L., Liu, X., Li, H., Kirkensgaard, J. J. K., Enemark-Rasmussen, K., Khakimov, B., Hebelstrup, K. H., Zhong, Y., & Blennow, A. (2024). Characterization of different high amylose starch granules. Part II: Structure evolution during digestion and distinct digestion mechanisms. *Food Hydrocolloids, 149*. https://doi.org/10.1016/j.foodhyd.2023.109593

Villa Zabala, C. C. (2020). *An overview on starch structure and chemical nature*. Springer Science and Business Media LLC. https://doi.org/10.1007/978-3-030-42542-5_2

Wang, L., Chen, S., & Wu, J. (2018). Cyclodextrin enhanced the soluble expression of Bacillus clarkii γ-CGTase in *Escherichia coli*. *BMC Biotechnology, 18*(1), 72.

Wang, H. C., Hsu, Y. C., Wu, Y. P., Yeh, S. Y., & Ku, M. S. B. (2021). *Production of high amylose and resistant starch rice through targeted mutagenesis of starch branching enzyme iib by Crispr/cas9*. Taiwan: Research Square. https://doi.org/10.21203/rs.3.rs-585478/v1

Wang, Z., Mhaske, P., Farahnaky, A., Kasapis, S., Majzoobi, M., & starch, C. (2022). Chemical modification and its impact on functional properties and digestibility, a review. *Food Hydrocolloids, 129*, Article 107542. https://doi.org/10.1016/j.foodhyd.2022.107542

Wang, L., Zhu, S., Chen, Y., Karthik, P., & Chen, J. (2023). Fabrication and characterization of O/W emulsion stabilized by octenyl succinic anhydride (OSA) modified resistant starch. *Food Hydrocolloids, 141*, Article 108750. https://doi.org/10.1016/j.foodhyd.2023.108750

Woggum, T., Sirivongpaisal, P., & Wittaya, T. (2014). Properties and characteristics of dual-modified rice starch based biodegradable films. *International Journal of Biological Macromolecules, 67*, 490–502.

Xia, W., Zhang, K., Su, L., & Wu, J. (2021). Microbial starch debranching enzymes: Developments and applications. *Biotechnology Advances, 50*, Article 107786. https://doi.org/10.1016/j.biotechadv.2021.107786

Xie, Q., Fan, B., Tao, R., Wang, F., & Sun, Y. (2025). A comprehensive review of starch: Structure, properties, chemical modification, and application in food preservation. *Food Frontiers, 6*, 2643–8429. https://doi.org/10.1002/fft2.70040

Xu, K., Chen, X., Zheng, R., & Zheng, Y. (2020). Immobilization of multi-enzymes on support materials for efficient biocatalysis. *Frontiers in Bioengineering and Biotechnology, 8*. https://doi.org/10.3389/fbioe.2020.00660

Yang, X., Bui, T. A., Mei, H., Aksoy, Y. A., Deng, F., Hutvagner, G., & Deng, W. (2024). Exploring the potential and challenges of CRISPR delivery and therapeutics for genetic disease treatment. *Advanced Functional Materials, 34*(38). https://doi.org/10.1002/adfm.202402630

Yang, T., Hu, Q., Liu, Y., Xu, R., Wang, D., Chang, Z., Jin, M., & Huang, J. (2023). Biochemical characteristics and potential application of a thermostable starch branching enzyme from Bacillus licheniformis. *AMB Express, 13*(1). https://doi.org/10.1186/s13568-023-01511-4

Yang, Q., Ral, J., Wei, Y., Zheng, Y., Li, Z., & Jiang, Q. (2024). Genome editing of five starch synthesis genes produces highly resistant starch and dietary fibre in barley grains. *Plant Biotechnology Journal, 22*(7), 2051–2053.

Yao, S., Wang, B.-J., & Weng, Y.-M. (2022). Preparation and characterization of mung bean starch edible films using citric acid as cross-linking agent. *Food Packaging and Shelf Life, 32*, Article 100845.

Yook, S., Deewan, A., Ziolkowski, L., Lane, S., Tohidifar, P., Cheng, M. H., Singh, V., Stasiewicz, M. J., Rao, C. V., & Jin, Y. S. (2025). Engineering and evolution of Yarrowia lipolytica for producing lipids from lignocellulosic hydrolysates. *Bioresource Technology, 416*. https://doi.org/10.1016/j.biortech.2024.131806

Yudhistira, B., Husnayain, N., Punthi, F., Gavahian, M., Chang, C. K., & Hsieh, C. W. (2024). Progress in the application of emerging technology for the improvement of starch-based active packaging properties: A review. *ACS Food Science & Technology, 4*(9), 1997–2012. https://doi.org/10.1021/acsfoodscitech.4c00260

Zavareze, E. D. R., & Dias, A. R. G. (2011). Impact of heat-moisture treatment and annealing in starches: A review. *Carbohydrate Polymers, 83*(2), 317–328. https://doi.org/10.1016/j.carbpol.2010.08.064

Zhao, J., Chen, Z., Jin, Z., De Waard, P., Buwalda, P., Gruppen, H., & Schols, H. A. (2015). Level and position of substituents in cross-linked and hydroxypropylated sweet potato starches using nuclear magnetic resonance spectroscopy. *Carbohydrate Polymers, 131*, 424–431. https://doi.org/10.1016/j.carbpol.2015.06.005

Zhou, L., He, X., Ji, N., Lei, D., Yang, L., Yang, J., Xiong, L., & Sun, Q. (2021). Preparation and characterization of waxy maize starch nanoparticles via hydrochloric acid vapor hydrolysis combined with ultrasonication treatment. *Ultrasonics Sonochemistry, 80*, Article 105836. https://doi.org/10.1016/j.ultsonch.2021.105836

Zhu, F. (2015). Impact of ultrasound on structure, physicochemical properties, modifications, and applications of starch. *Trends in Food Science & Technology, 43*(1), 1–17. https://doi.org/10.1016/j.tifs.2014.12.008

Zhu, F. (2020). Underutilized and unconventional starches: Why should we care? *Trends in Food Science & Technology, 100*, 363–373. https://doi.org/10.1016/j.tifs.2020.04.018

Zia-ud-Din, M., Xiong, H., & Fei, P. (2017). Physical and chemical modification of starches: A review. *Critical Reviews in Food Science and Nutrition, 57*(12), 2691–2705.

Zusfahair, Ningsih, D., Kartika, D., Kurniasih, M., Nofiani, R., & Fatoni, A. (2020). Improved reuse and affinity of enzyme using immobilized amylase on alginate matrix. *Journal of Physics: Conference Series, 1494*, Article 012028.

Part II

Starch application in food processing

Chapter 6

Starch as a fat replacer: Strategies for healthier food formulations

Gangaraju Divyashri[1], Yellutla Jahnavi[2], Hafsa Nadeem[2] and Yashwanth Ramesh[2]
[1]*Bioproducts Research, Iom Bioworks Pvt. Ltd., Bangalore, Karnataka, India;* [2]*Department of Biotechnology, M S Ramaiah Institute of Technology, Bangalore, Karnataka, India*

1. Introduction

1.1 Overview of fat replacement in consumable food

Over the last 10 years, the field of fat replacers has made significant progress as a result of consumer demand for low-fat foods. A substance that switches all or some of the fat in food while maintaining the same physiological properties is known as a fat replacer. Fat replacers can be of any kind, for example, protein, sugar, or lipid-based. They reduce the extent of fat utilized in the creation of food items and show fat-like properties. Fat replacers display these characteristics by contributing texture and acting as gelling agents, stabilizers, emulsifiers, and thickening agents. It is essential to select the appropriate type of fat substitute (Peng & Yao, 2017). The utilization of fat replacers should not be indefinite, as it may lead to adverse effects. Fat replacers are most helpful when they help in controlling calories and encourage the intake of meals rich in essential nutrients. Accumulative evidences suggest that cutting back on fat intake can help monitor body weight, mitigate cardiovascular disease risks, hypertension, and type-2 diabetes (Syan et al., 2024). Fat replacers are broadly categorized into fat substitutes and fat mimetics (Marangoni et al., 2020). Fat substitutes are the actual ingredients that take the place of fat in food products, which include emulsifiers (help in blending substances to achieve the creamy texture of fat) and structured lipids (fats that have been chemically or physically modified to change their digestion or absorption properties). They can be modified to replicate the mouthfeel of natural fats and their texture while providing potentially healthier effects or fewer calories. Fat mimetics, on the other hand, are ingredients that are structurally distinct and unique but mimic the physical and textural properties of fats (Chen & Yang, 2023, pp. 277–311). Fat mimetics include microparticulated proteins, proteins processed into tiny particles that disperse in water and give a creamy texture without the fat (they are utilized in low-fat dairy products like frozen yoghurt and cheddar to offer mouthfeel) and microparticulated carbohydrate (tiny particles of carbohydrates that assist in accomplishing the creamy texture of fats in various food products).

1.2 Role of starch as a fat replacer

Starch is increasingly utilized as a fat replacer in food products, where it assumes a significant part in reducing the overall fat content while still maintaining desirable sensory and textural qualities. Starch substitution is advantageous from a nutritional and functional perspective. Modified starches can be viewed as successful fat replacers as they retain comparable sizes and morphology of fat globules. Native or modified starch has distinct properties that include bulking, imparting a creamy texture, viscous flow and finally adding form and structure similar to fats (Demiate et al., 2024, pp. 115–154). However, due to the fact that native starch may be unsuitable for the processing conditions like freezing, acidic pH and high heat, the food industry uses modified starch the most. By incorporating mono- or difunctional chemical blocking groups into the starch polymer backbone, a type of chemically modified starch known as cross-linked starch is produced. Cross-linked starch doesn't just display gel formation, water holding, and making viscous flow streams, and can additionally lower the food's calorie count due to its lower absorbability (Peng & Yao, 2017). Low-dextrose-equivalence

Native and Modified Starches for Food Applications. https://doi.org/10.1016/B978-0-443-27431-2.00004-3

(DE < 20) maltodextrins, produced through enzymatic or chemical hydrolysis of native starch, are found to be the ideal fat replacers (Chavan et al., 2015, pp. 589–595). The function of maltodextrins differs from that of native or other modified starches due to their lack of a globular structure, as they are hydrolyzed products. Regardless of this underlying contrast, maltodextrins show a good water-retention capacity, empowering them to develop gels using water in food. These maltodextrin gels, with molecule sizes going from 1 to 3 μm, closely mimic the size of fat globules, making them viable fat replacers and contributing to a fatlike texture in foods (Peng & Yao, 2017). Notably, the ratio of maltodextrin to water in maltodextrin gel typically only consists of one part maltodextrin to three parts waters, significantly reducing the number of calories per gram in food products from nine to one. Low DE maltodextrins can also improve physicochemical properties like softness, viscosity and lubricity (Grzelak et al., 2017). Several commercial maltodextrin-based fat replacers are available, including Paselli SA2 from potato starch, N-oil from tapioca starch, Maltrin M040 from cornstarch, and Oatrim from oat starch. It is critical to consider a few limitations of maltodextrin-based fat replacers, including their low stability under freeze-thaw conditions and their sensitivity to acid and heat (Chavan et al., 2015, pp. 589–595).

1.3 Importance of healthier food formulations

Cardiovascular illnesses (CVDs) are the main cause of death globally, representing about one-third of all deaths around the world (Tsao et al., 2023). A significant behavioral risk factor that contributes to CVDs is an unhealthy eating regimen, which serves as a primary contributor to obesity. Nearly 30% of the world's population as of 2020 is affected by obesity, which has emerged as a major health crisis (Chen et al., 2020). As worries about serious diseases related to unhealthy diets and insufficient nutritional food intake grow, there is a growing popularity of foods with improved nutritional properties, which can largely be attributed to increased consumer awareness of the links between a nutritious diet, health, and overall well-being. In spite of this pattern, deep-fat fried consumables, which are high in cholesterol, remain among the most preferred food products globally. However, due to the fact that some products may contain fat that accounts for as much as 50 percent of their total weight, the high oil content of fried foods (donuts, chips, fried chicken) remains a significant health concern (Oke et al., 2018). To meet customer expectations, health and sensory qualities must be addressed simultaneously. The advancements in fried food preparation that reduce oil absorption while frying are crucial developments in formulating healthier options. Batter can be described as a type of liquid dough that often includes a leavening agent, causing it to expand upon frying. Typically made from a mixture of flour and water, it also contains additives and seasonings to enhance the flavor of the food. Notably, there are two primary types of batter recognized in culinary practices: adhesion batter and tempura-type batter. Adhesion batter effectively forms an interface layer between the product and its breadcrumb coating (Varela & Fiszman, 2011), while tempura-type batter acts as a leavening agent to create a puffed texture (Taniguchi et al., 2023). The development of batter has advanced into complex systems in which the choice and interplay of ingredients are pivotal to the final quality of the food product. This evolution is especially significant in reducing oil absorption during the frying process. Various functional ingredients like plant and animal proteins, cellulose and its derivatives, hydrocolloids, breadcrumbs, and dietary fiber are used to upgrade the quality of the fried foods (Bhuiyan & Ngadi, 2024). Among them, dietary fiber has drawn huge interest. Unlike other nutrients, dietary fiber is not broken down by digestive enzymes, and it reaches the colon intact, preserving its nutrients and contributing to a healthier diet (Puhlmann & de Vos, 2022). Today, dietary fiber is widely recognized in the food industry as a fat replacer, fat reduction agent during frying, volume enhancer, binder, bulking agent, and as a stabilizer (da Silva et al., 2024). The worldwide market for dietary fiber was valued at USD 4564.3 million in 2020 and is projected to reach USD 9562.2 million by 2026, growing at a compound annual growth rate of 11% (Ching et al., 2021). While there is a developing interest in dietary fiber, research on its applications in food products remains restricted. This is particularly true when it comes to battering for fried foods, where dietary fiber's potential to reduce oil uptake merits more in-depth research. Existing studies are more frequently focused on the usage of proteins and hydrocolloids in minimizing oil absorption during frying (Dehghannya & Ngadi, 2023); however, scientific examinations on dietary fiber are not as extensive.

2. Fundamentals of starch

2.1 Structure and properties of native starch

Native starches are long-chain carbohydrates that exist in their purest starch form and can be obtained from various plant sources, including corn, maize, rice, wheat, potato, cassava, and tapioca. The process of extracting starch from these sources yields a white, tasteless, and odorless substance, available in both powder and liquid forms. Native starches do not dissolve in cold water and exhibit varying degrees of swelling based on their type and the surrounding temperature (Jia et al., 2023). Native starches possess a range of beneficial properties. They can act as thickeners, texturizers, and gelling agents. They are also valuable for moisture retention, extending the shelf life of baked goods, stabilizing mixtures, forming edible films, providing a dusting medium, and enhancing the binding of dough. These versatile functions make them invaluable in various culinary and industrial applications (Cornejo-Ramírez et al., 2018). To gain insights into the functional characteristics of starches in various applications, it is imperative to examine their distinct properties viz., rheological, paste, and gelatinization properties. These properties provide insight into the starch behavior under various conditions and its suitability for specific applications.

2.1.1 Rheological properties

The rheological properties characterize its viscoelastic behavior under applied stress or strain. The storage modulus (G') and loss modulus (G'') values act as an indicator of elastic and viscous behaviors, respectively, of viscoelastic materials. The Tan δ value, a ratio of viscous to elastic modulus, helps determine whether the material behaves more like a liquid or a solid, reflecting the stability and structural integrity of the material (Mo et al., 2024).

2.1.2 Paste properties

Rapid Viscosity Analysis (RVA) assesses how starch viscosity behaves in water under varying temperature conditions. Observations indicate that starches typically begin to thicken at around 60 °C after about 5 min, a common initial pasting temperature for starches. However, variations exist depending on the starch source. Notably, lower pasting temperatures suggest smaller granule sizes, which confer reduced thermal resistance during the swelling process (Choque-Quispe et al., 2024).

2.1.3 Gelatinization properties

Starch, being water-insoluble undergoes gelatinization when heated above 52 °C in water, which is often analyzed using differential scanning calorimetry (Bashir & Aggarwal, 2019). This process occurs in two endothermic phases: (i) initially, starch granules swell as hydrogen bonds in the amorphous areas break, and (ii) subsequently, water acts as a plasticizer, further hydrating and expanding these regions. Therefore, gelatinization leads to a breakdown of starch molecular bonds, increasing water absorption (Donmez et al., 2021). This heating alters the crystalline structure of amylopectin, reduces crystallinity, and results in several changes such as starch solubilization, granule swelling, and loss of birefringence (Bashir & Aggarwal, 2019).

2.2 Modifications and their effects on starch functionality

Native starches pose several challenges that limit their effectiveness in food processing due to their high viscosity, difficulty in handling, opaque gels, etc. In order to overcome these issues, starches are often modified before being used as food ingredients (Compart et al., 2023). This modification enhances a range of properties, leading to improvements in digestibility, emulsifying capabilities, emulsion stability, and encapsulation qualities. Additionally, modifications can increase cold-water swellability, charge the starch molecules, and enhance cooking attributes such as film formation and encapsulation wall materials (Guo, et al., 2021). Modifications enhance the pH, temperature, and shear tolerance, as well as the water-holding ability of starch (Tiozon et al., 2021). Consuming starch in appropriate quantities has been associated with potential health benefits, such as reducing the risks of cancer, diverticulitis, and coronary heart disease (Lockyer & Nugent, 2017). Currently, several methods are employed to produce modified starch, each tailored to meet specific requirements and enhance functionalities in food processing (Fig. 6.1).

2.2.1 Chemical modification of starch

This methodemploys techniques like cross-linking, graft copolymerization, cationization, stabilization/substitution, conversion, and transglycosidation, utilizing chemicals such as phosphoryl chloride, sodium tripolyphosphate, vinyl chloride, mono-sodium phosphate, and epichlorohydrin (Compart et al., 2023). These modifications are nondestructive and involve attaching functional groups such as acetyl-, carboxyl-, and ethyl-groups to the starch's hydroxyl groups, which act as active sites. The effectiveness of these modifications is influenced by several factors, including the source of starch, the amylose-to-amylopectin ratio, granule morphology, reaction conditions, and the type and concentration of the modifying agent. Cross-linking, a prevalent method for altering starch properties, is widely used across industries such as

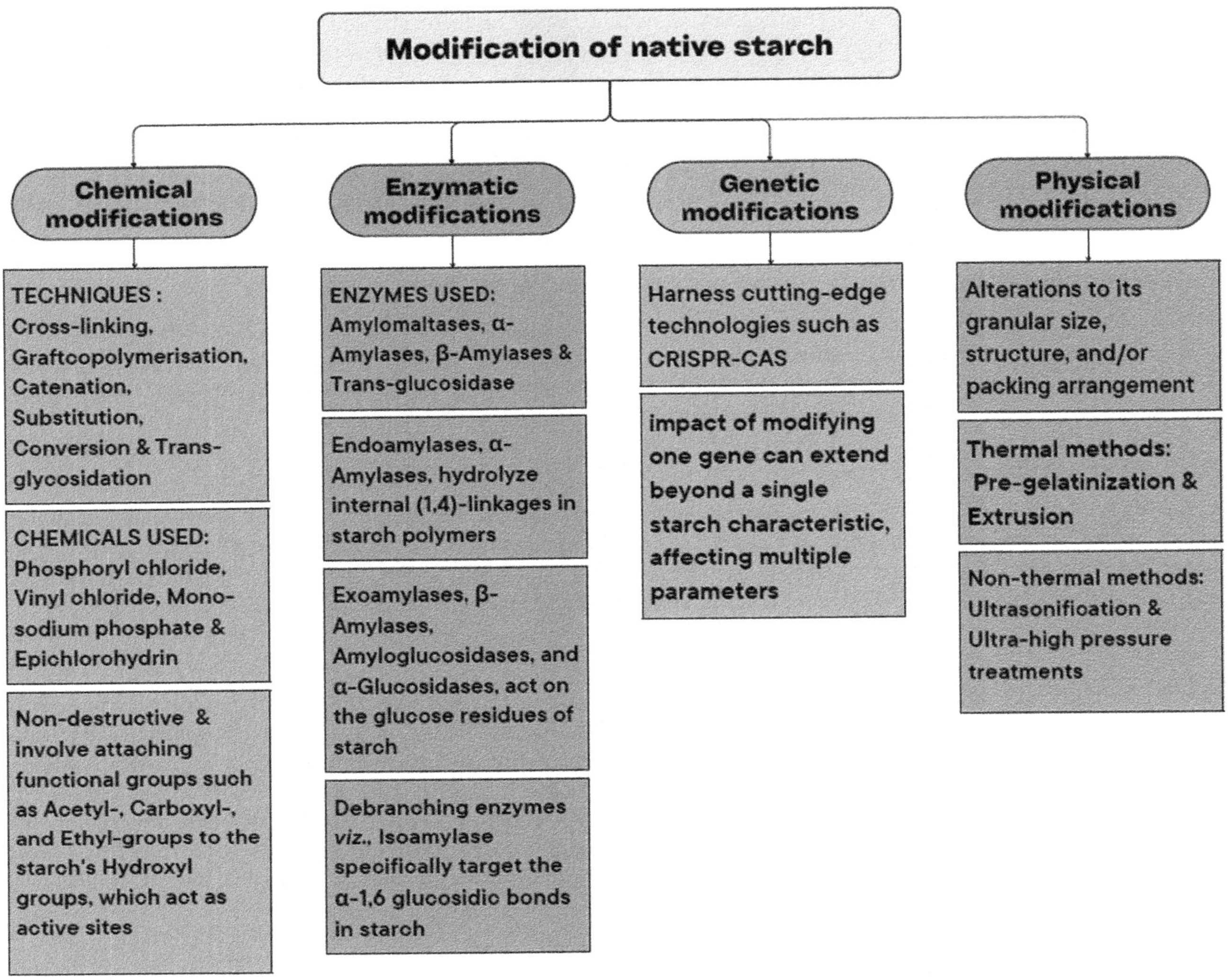

FIGURE 6.1 Comprehensive overview of native starch modifications. Chemical, enzymatic, genetic, and physical approaches.

pharmaceuticals, food production, wastewater treatment, and bioplastics. This process involves creating ether or ester linkages through molecular interactions between the starch's reactive hydroxyl groups and multifunctional reagents. This not only occurs along the linear chains of starch molecules but also with side chains, leading to the formation of cross-links. The effects of cross-linking enhance properties such as solubility in organic solvents, molecular mass, and degree of polymerization. Moreover, cross-linking stabilizes and strengthens starch granules, enhancing hydrogen bonds with additional covalent bonds (Compart et al., 2023).

2.2.2 Enzymatic modification of starch

This process, particularly prevalent in the food industry, involves altering starch characteristics through the strategic use of enzymes. The enzymatic modification of starch can be categorized into four main types: endoamylase, exoamylase, debranching enzymes, and transferases (Fig. 6.2). Endoamylases, such as α-amylases, randomly hydrolyze internal (1,4)-linkages in starch polymers—both amylose and amylopectin—reducing the polymer chain length by cleaving at specific sites. This action can create unique textural modifications on the surface of starch granules depending on the enzyme type and concentration used. Exoamylases, including β-amylases, amyloglucosidases, and α-glucosidases, act on the glucose residues of starch (Dura et al., 2014). They generate β-limit dextrins, maltose, and glucose, respectively. This action of β-amylases, for example, can lead to the formation of multiple cracks on the surface of starch granules, thereby altering their physical properties. Debranching enzymes viz., isoamylase, specifically target the α-1,6 glucosidic bonds in starch.

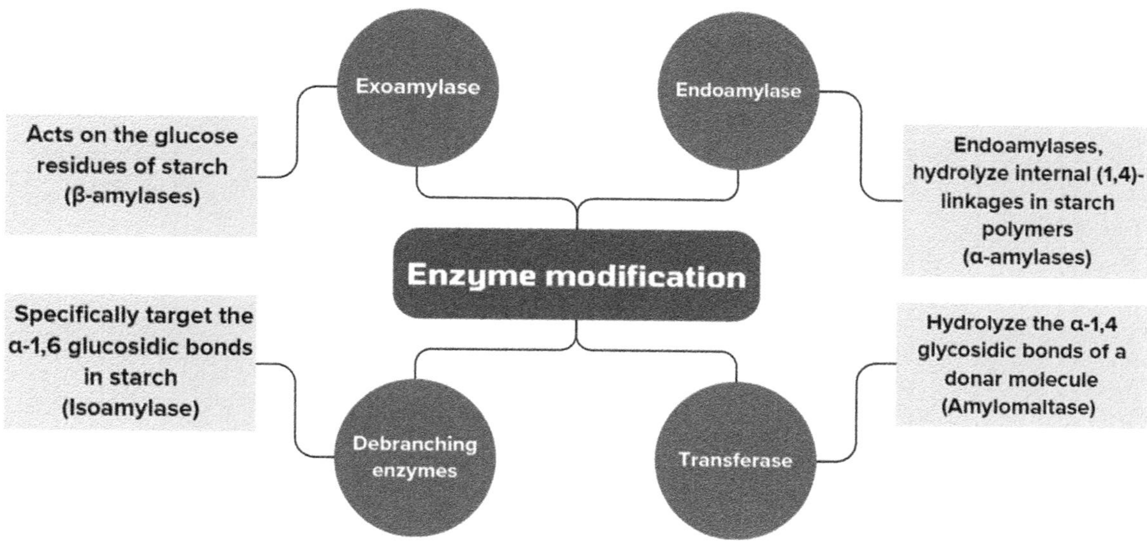

FIGURE 6.2 Types of enzymes involved in starch modification. This diagram categorizes four primary enzyme groups - exoamylases, endoamylases, debranching enzymes, and transferases - detailing their specific catalytic actions on starch structures, such as the hydrolysis of α-1,4 and α-1,6 glycosidic bonds.

By hydrolyzing these bonds, isoamylase promotes the breakdown of amylopectin into linear glucans, fundamentally changing the structure and behavior of the starch (Compart et al., 2023).

2.2.3 Genetic modification of starch

Genetic methods are a transformative approach in the modification of starch, enabling the production of variants such as amylose-free starch, high-amylose starch, and starches having modified amylopectin structures. This technique, known as genetic modification of isolated starch, offers a cleaner alternative to traditional postharvest modifications like chemical, physical, and enzymatic processes, which can be environmentally hazardous. Genetic modifications harness cutting-edge technologies such as CRISPR-CAS, offering significant potential for tailoring starch properties precisely (Adegbaju et al., 2024). However, the application of these genetic technologies is heavily influenced by regulatory environments and is subject to political decisions and opinions (Turnbull et al., 2021). Furthermore, the impact of modifying one gene can extend beyond a single starch characteristic, affecting multiple parameters due to the interconnected nature of genetic pathways. Extensive research is currently being conducted to identify and understand new genes that can regulate starch properties directly within the plants (Compart et al., 2023). This ongoing study underscores the potential of genetic methods to revolutionize starch use in various industries, pending broader regulatory acceptance.

2.2.4 Physical methods of starch modification

Physical methods of modifying the native starches involve alterations to their granular size, structure, and/or packing arrangement (Fonseca et al., 2021). These modifications can be performed without the use of potentially toxic agents, thereby making this process a simpler, cost-effective, and environmentally friendly option. Physical modifications fall into two main categories: (i) thermal and (ii) nonthermal techniques (Fig. 6.3). Thermal methods of modifying starch include techniques such as pregelatinization and extrusion, which are instrumental in the food and packaging industries. Pregelatinization involves boiling starch in water, followed by drying and grinding, a process that significantly reduces hydrogen bonding and results in the fragmentation and molecular disintegration of starch granules (He et al., 2023). Pregelatinized starch dissolves in cold water and is particularly valued in food applications for its quick thickening properties, ideal for instant foods (Liu et al., 2017). It also lacks optical birefringence and features enhanced viscosity and gel stability, making it useful in the microcapsule industry for controlled drug release (Dang et al., 2017). On the other hand, extrusion is a thermomechanical process that forces gelatinized starch under pressure through a shaping aperture. Both methods showcase the versatility of thermal processes in enhancing and diversifying the applications of starch in various industries (Compart et al., 2023). Nonthermal methods for modifying starch, such as ultrasonification and ultrahigh pressure treatments, offer innovative ways to alter the physical properties of starch without using heat

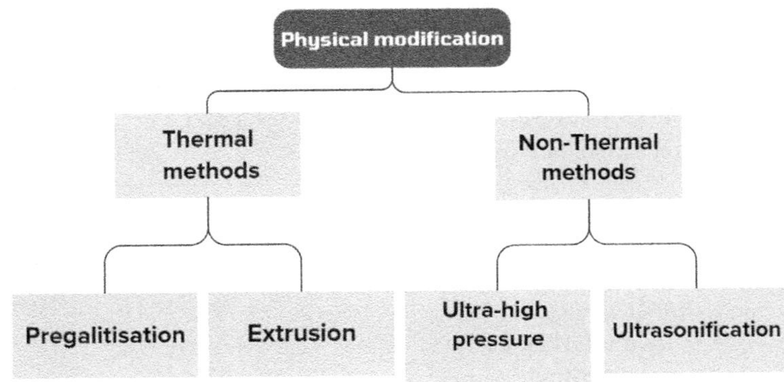

FIGURE 6.3 Methods of physical modification of starch. Thermal and nonthermal methods that can be used to modify starches.

(Raghunathan et al., 2021). Ultrasonification utilizes high-frequency sound waves, while ultrahigh-pressure treatments apply extreme pressure, both effectively modifying characteristics like solubility and texture. These methods do not require the intense energy demands of thermal processes and avoid the use of chemicals, making them more environmentally friendly. Overall, these physical modifications provide versatile and sustainable solutions, enhancing starch properties to meet specific industrial requirements while minimizing environmental impact (Fonseca et al., 2021).

2.3 Comparison of native and modified starch

Native starch and modified starch differ significantly in their sources, processing methods, and applications, reflecting their unique properties and functionalities (Table 6.1).

This table clearly highlights the characteristics, processing differences, and uses of native and modified starches, providing a detailed comparison. Native starch, obtained directly from plants, serves primarily as a thickening agent and stabilizer, maintaining its pure, odorless, and tasteless profile (Cornejo-Ramírez et al., 2018). In contrast, modified starch undergoes additional physical, enzymatic, or chemical treatments, enhancing its functionality to meet specific industrial requirements like improved freeze-thaw stability, acid or alkali resistance, and shear stability (Choque-Quispe et al., 2024). This comparison underscores the significance of starch modification processes in expanding the utility and effectiveness of starch in various culinary and industrial contexts.

3. Mechanisms of fat mimicry by starch

3.1 Physical properties of starch that mimic fat

Starch possesses different structures that influence its characteristics. Specifically, A-type and B-type starches are identified by their orthorhombic and crystalline phases, respectively. Water is crucial in developing these structures. A-type starches contain large granules, while B-type starches have small granules, both contributing to a smooth texture (Werlang et al., 2021). Fig. 6.4 illustrates how starch-based fat replacers simulate the function of fat globules, creating a comparative texture like that of fats. Fat replacers (orange [light gray in print version]) and fat globules (gray) interact with proteins (light blue [very light gray in print version]) to achieve the desired texture. Furthermore, these starches exhibit distinctive physicochemical properties influenced by the molecular structures of amylose and amylopectin, namely, thickening, creating viscous flow, bulking, retaining water, as well as gel-forming abilities, all crucial for fat replacement in various food products. The gelatinization process during industrial processing enhances the suitability of native starches as fat replacers. During gelatinization, the starch granules swell and solubilize, altering their functional properties. This transformation allows the starches to mimic the textural attributes typically provided by fats, such as creaminess and viscosity. Thus, native starch granules, when dispersed among food systems, can simulate the perceptible properties of fat globules, contributing to a desirable mouthfeel (Sun & Fang, 2021, pp. 409–444). Additionally, modified starches enhance these effects by increasing water-holding capacity and minimizing syneresis, making them particularly useful in high-moisture foods such as sauces, baked goods, and sausages (Werlang et al., 2021). These adaptations make starches versatile and valuable ingredients in culinary and food manufacturing settings (Park et al., 2020).

TABLE 6.1 Detailed comparison between native starch and modified starch.

Aspect	Native starch	Modified starch	References
Source and processing	Powder form obtained from starchy plants	Derived from native starches through physical, enzymatic, or chemical processing methods like wet and dry chemical processes, drum drying, and extrusion.	Nsanzabera et al. (2023)
Functionality	Used as a thickening agent and a stabilizer in foods such as custard, desserts, sauces, and instant foods.	Properties like freeze-thaw stability, acid or alkali resistance, or shear stability are enhanced to suit different applications based on the raw materials used.	Compart et al. (2023)
Treatment	Treated hydrothermally, allowing it to swell upon the addition of cold water.	Initially chemically modified, then physically treated to achieve desired properties	Iuga and Mironeasa (2020)
Type of starch	Pregelatinized starches.	Modified pregelatinized starches, which can be customized to impart specific tastes or odor.	Compart et al. (2023)
Purity and sensory profile	The purest form of starch, odorless and tasteless.	Comparatively less pure; can be modified to add odor and taste, enhancing sensory profiles for specific applications.	Dolas et al. (2020)

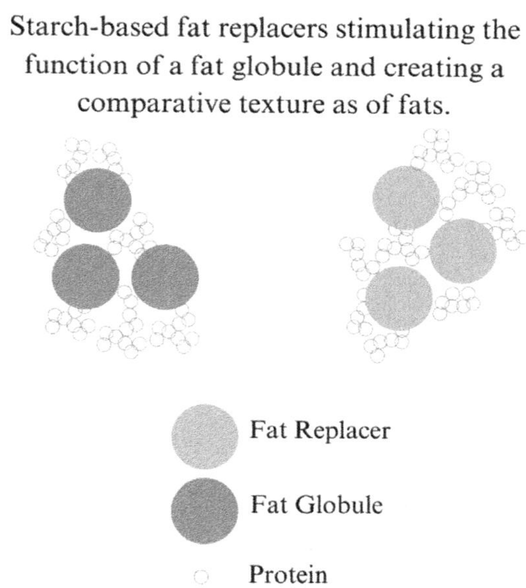

FIGURE 6.4 Starch-based fat replacers mimicking fat globules. The replacers create a similar texture in substituted food products.

3.2 Interaction of starch with other food components

3.2.1 Interaction between proteins

Starch interacts with proteins from various cereals, influencing its swelling, retrogradation process, and granule integrity. On the granule surface, starch combines with proteins, impacting retrogradation—a process where amylose and amylopectin chains in cooked, gelatinized starch realign as they cool. Hydrophilic proteins typically prevent this realignment, helping starch maintain its gel structure. Conversely, proteins with higher surface hydrophobicity tend to expel water from the starch granule, facilitating the realignment and closer association of starch molecules during retrogradation (Scott & Awika, 2023). pH levels also affect protein-starch interactions. At low pH, proteins are positively charged and attach more strongly to starch, reinforcing its structure. At high pH, proteins are negatively charged, hindering their bonding capacity with starch and decreasing stability. Additionally, starch interacts with other proteins, such as gluten and milk proteins (Bello-Perez & Flores-Silva, 2023). These interactions are critical as they significantly modify starch's functionality in food products, impacting its textural qualities.

3.2.2 Interaction with lipids

Amylose, a component of starch, interacts with fatty acids and monoglycerides to form inclusion complexes. These interactions significantly alter the structural and functional characteristics of starch by encapsulating lipid molecules within the amylose helix. While amylose and amylopectin are both components of starch, amylopectin has a weaker binding affinity for fatty acids. Monoglycerides, which act as emulsifiers, play a crucial role in these interactions. They stabilize swollen starch granules and help prevent amylose retrogradation, which is the realignment and solidification of starch molecules as they cool (Krishnan et al., 2021). By slowing down retrogradation, monoglycerides maintain the softness and pliability of starch-based products over time. This property enhances texture and extends the shelf life of starch-rich foods, making emulsifiers indispensable for improving food quality.

3.2.3 Sensory and textural implications

Preserving the quality and texture of starch-based products, especially frozen items, is crucial because low temperatures accelerate starch degradation. Retrogradation causes an opaque, crystallized texture and syneresis, where liquid separates from the gel, resulting in a coarse texture. Effective strategies are needed to mitigate these effects (Xie et al., 2023). Fat substitutions in foods can affect moisture content, leading to increased hardness and firmness. However, certain starches, such as cross-linked cassava starch, enhance the sensory qualities of reduced-fat foods by adding springiness and chewiness, thus improving sensory attributes (Punia Bangar et al., 2024). When used as a fat substitute in whole wheat bread, enzyme-modified maize starch can make the crumb harder and less springy, altering the bread's texture and potentially influencing consumer acceptance (Sheikholeslami et al., 2021). In low-fat cheese substitutes, tapioca starch improves flavor, texture, and overall acceptance. These products have a softer, more tender feel, resembling full-fat cheese, because starch acts as a barrier between the cross-linking proteins (Chen et al., 2020). Maltodextrin is a popular fat substitute across various food industries due to its strong water-holding capacity and fat like texture (Chen et al., 2020).

4. Types of starch used as fat replacers

4.1 Resistant starch

Resistant starch (RS) exists in five main types: RS1, RS2, RS3, RS4, and RS5. These starches are utilized through methods such as gelatinization and retrogradation to enhance food products in various ways. RS1, known as physically embedded starch, is found in natural grains and partially ground beans, with natural starch granules encapsulated within plant components. RS2 is characterized by its raw starch granules, which have a greater density and a unique crystalline structure, making them less sensitive to enzymes. High RS2 content is found in raw potatoes and bananas. When RS2 undergoes gelatinization and retrogradation, it becomes RS3. RS4 and RS5 are modified further with additional functional groups or side chains (Wang et al., 2023). RS2 and RS3 are commonly used in bread formulations to improve softening and weaken the gluten structure. RS2 bread is more marketable due to its texture, while RS3 bread offers higher fiber content. Muffins made with RS2 exhibit superior appearance, texture, and overall acceptance compared to those made without it. Similarly, incorporating resistant starch into cakes results in reduced hardness, with 15% replacement levels being well-received by consumers (Arp et al., 2021). Additionally, baked goods such as muffins and sponge cakes that utilize RS fat replacers, like inulin, benefit from higher moisture content and improved springiness (Yazar & Rosell, 2023). Noodles made with resistant starches from green banana flour and heat moisture-treated (HMT) corn starch show improved dietary fiber content and enhanced chewiness, with around 10% substitution being well-accepted (Li et al., 2022). Pasta incorporating up to 10% RS4 from sources like wheat starch benefits from increased dietary fiber and antioxidant activity, as well as a decreased glycemic index, although it also becomes less firm. In gluten-free pasta, substituting up to 15% with resistant starch increases firmness and reduces stickiness, leading to better overall appearance and acceptance (Zhang & Bao, 2023). Cookies and biscuits made with up to 75% substituted RS2, RS3, and RS4 demonstrate increased dietary fiber content. These products generally show decreased hardness, except in the case of RS3. RS derived from annealed white sorghum starch also increases hardness but results in lower acceptance due to its appearance and taste (Zhang & Bao, 2023). Additionally, unripe banana flour with high RS content can produce highly acceptable gluten-free cookies with up to 70% substitution (Prasajak et al., 2021). Various food types that benefit from the addition of resistant starch are summarized in Table 6.2.

TABLE 6.2 Features improved by adding resistant starch to various food types.

Food type	Features	References
Bread	Better quality, whitening, softening	(Arp et al., 2021)
Cakes and muffins	Better quality, fat replacement, springiness	Yazar and Rosell (2023)
Noodles and pasta	Whitening, more water absorption, reduced loss while cooking	(Li et al., 2022; Zhang & Bao, 2023)
Cookies	Fat replacement, whitening, better quality	Zhang and Bao (2023)

4.2 Modified starch

Modified starches are valuable in the food industry due to their similar morphology to fat globules in terms of size and shape, making them effective fat replacers. As polysaccharides, starches can bind a significant amount of water, imparting a sense of creaminess during consumption and enhancing the quality of low-fat products (Asyrul-Izhar et al., 2023). The following types of modified starches are typically used in formulation for food production.

4.2.1 Cross-linked starches

Cross-linking of starches is a popular method throughout the food science industry. This process involves creating covalent interconnections between starch molecules, resulting in strong bonds that increase resistance to high temperatures, shear forces, and acidity. These starches provide enhanced thickening capabilities, making them ideal for use in sauces, gravies, and desserts. Cross-linked starches are classified as RS4 due to their covalent bonding, which leads to increased resistance to hydrolysis by enzymes (Punia Bangar et al., 2024). The cross-linking process can be achieved using agents such as sodium tripolyphosphate, sodium trimetaphosphate, and monosodium orthophosphate. These agents act as cross-linkers, promoting the formation of ester groups between starch chains (Durmus et al., 2021). The process of forming cross-linked starch is illustrated in Fig. 6.5. Native starch consists of amylose and amylopectin, which undergo cross-linking through the addition of cross-linker agents. The resulting cross-linked starch exhibits enhanced structural integrity, which increases its resistance to high temperatures, shear forces, and enzymatic hydrolysis.

Salad dressing, typically produced under high shear and moderate temperatures, requires modified starches for stability. Waxy modified starches, such as cross-linked starches, are particularly effective additives in this process and are widely used in the food industry. Cross-linked starches are also beneficial in meat products like frankfurters, bologna, and meatloaves. They increase viscosity, absorb water released during cooking, and extend shelf life while maintaining quality. Highly cross-linked starches help prevent low cake volume in baked goods by being resistant to high oven temperatures (Saboonchi et al., 2021). In baked goods, substituting flour and native starches with cross-linked starches

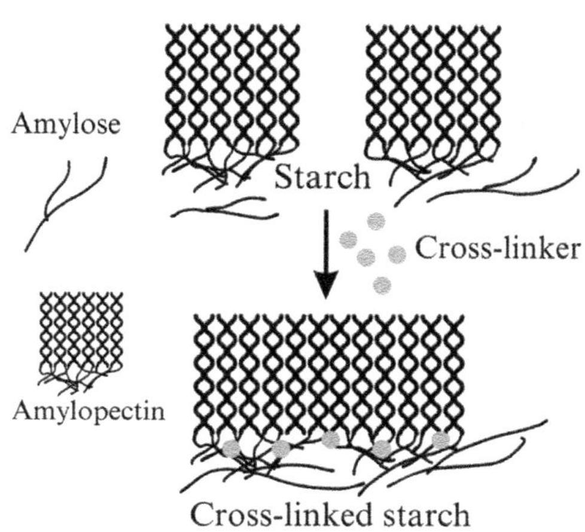

FIGURE 6.5 Formation of cross-linked starch. Native starch, composed of amylose and amylopectin, is chemically modified through the addition of cross-linker agents to form a stabilized network. This structural reinforcement significantly enhances the starch's integrity, resulting in increased resistance to thermal degradation, mechanical shear forces, and enzymatic hydrolysis compared to its native state.

enhances dough development time and mixing tolerance, though it may reduce dough stability and water absorption capacity. This substitution can also increase dietary fiber content while maintaining the desired characteristics of the product. In dairy products such as yoghurt, cross-linked starches improve viscosity, elastic modulus, and other rheological attributes. They also enhance the texture and freeze-thaw stability of low-fat ice creams. In soups and sauces, cross-linked starches provide higher thermal stability, shear resistance, and improved consistency (Punia Bangar et al., 2024).

4.2.2 Pregelatinized starch

Pregelatinized starch (PGS) is a modified starch that undergoes a cooking process of complete gelatinization followed by immediate or subsequent drying. Gelatinization occurs when starch is heated in a moist environment, resulting in the granules losing their orderliness, swelling significantly due to water absorption. This process allows hydrogen bonding sites to bind to more water, irreversibly dissolving the starch granules in water. The process is then halted by cooling, often using coolants like liquid nitrogen, followed by sterile storage methods to prevent moisture sorption (Johnson & Mauer, 2019). PGS can dissolve in cold water without requiring heat treatment to form a paste, making it an excellent substitute for heat-sensitive foods such as low-fat pizza cheese (Akshit et al., 2025). It is also useful in producing softer baked goods due to its improved water absorption capabilities. Additionally, PGS can enhance the crispiness and fluffy structure of snack foods like peanuts and puffed beans (Hong & Liu, 2023). They are also known to improve the phenolic and antioxidant content in the foods to which they are added, although these benefits can diminish over extended storage periods, while maintaining consumer acceptability in terms of color, smell, taste, and overall taste preference when used as a substitute (Supakot et al., 2022). The production of PGS is particularly advantageous in the development of clean-label products, which are increasingly popular as they do not produce chemical by-products (Šárka et al., 2023). PGS is also beneficial in food emulsions with a short shelf life, acting as an effective stabilizer for low-fat oil-in-water emulsions (Hedayati et al., 2020). By incorporating PGS, food manufacturers can enhance the quality, stability, and nutritional profile of their products while catering to the growing demand for clean-label and efficient food production solutions (Hsu et al., 2023).

4.2.3 Oxidized and stabilized starches

Oxidized starches are produced using suitable oxidizing agents such as sodium hypochlorite (NaOCl) (Blumenberg and Emeje, 2020). Factors that affect the oxidation process include pH, reaction time, thermal energy, and the botanical origin of starch. High amylose content may impair the oxidation of amylopectin chains. The degree of polymerization and the presence of carbonyl or carboxyl groups in oxidized starches influence their swelling capacity, solubility, and gelatinization time (Gałkowska et al., 2023). Compared to native starches, oxidized starches offer advantages in fat replacement and texture improvement, such as enhanced hardness, stretch, and meltability, without negatively impacting the sensory profile of foods like cheese slices (Ahsan et al., 2023). Oxidized starches are commonly used as additives in pudding desserts, whipped cream, cake fillings, cake mixes, sauces, soups, frozen dishes, and baby food due to their binding, thickening, and stabilizing properties (Gałkowska et al., 2023).

Stabilized starches are chemically modified to improve their stability, texture, and functional properties. These include starches like monostarch phosphate, octenyl succinic starch, and hydroxypropyl starch. Stabilized starches include acetylated starch, monostarch phosphates, octenyl succinic starches, and hydroxypropyl starches. Acetylated Starch: Produced by treating starch slurry with acetic anhydride or vinyl acetate, acetylated starch results from the substitution of hydroxyl groups in the glucose monomers of starch granules' amorphous domains with hydrophobic acetyl groups. This creates steric hindrance, loosening the starch structure and improving water absorption capabilities. These starches enhance consistency and stability in products like frozen desserts, bakery products, soups, baked goods, curds and dessert concentrates. They are sensitive to high temperatures, acidic environments, and mechanical stresses (Gałkowska et al., 2023). Monostarch Phosphate: These starches produce clear, stable pastes with cohesive structures that remain stable during freeze-thaw processes. Cooking monostarch phosphate with salt can reduce paste viscosity. They have high water-binding and swelling capabilities, improved solubility in cold water, and maintain structural properties even under prolonged mechanical stress. Monostarch phosphates are used in soups, vegetable sauces, dressings, frozen preserves, thermized yogurts, and more (BeMiller, 2019; Gałkowska et al., 2023). Octenyl Succinic Starch: Produced by the esterification of starch, octenyl succinic starch provides resistance to retrogradation and offers amphiphilic and anionic properties. It is cold-water soluble and supports emulsification, making it ideal as a stabilizer in milk products, mayonnaises, and as a smoothing agent in various applications. Primarily used in cold-prepared products, it can also help reduce the glycemic response (Gałkowska et al., 2023). Hydroxypropyl Starch: Formed by reacting starch with propylene oxide in the presence of a strong alkaline catalyst, hydroxypropyl starch introduces hydrophilic ether groups that limit hydrogen bonding and weaken the internal structure of starch granules. This makes them more receptive to water molecules, allowing them to swell and gain viscosity in gel form, with a reduced gelatinization temperature. Hydroxypropyl starches are used as thickeners in water- or milk-based products that need to be frozen and stored at low temperatures (BeMiller, 2019).

4.3 Sources and extraction methods of starch

Typical sources of starch include potatoes, barley, sorghum, corn, wheat, rice, bananas, and cassava. The morphology of starch granules, which are distinct semicrystalline aggregates that vary depending on their source, is one type of carbohydrate. Alteration is therefore essential to managing their characteristics (Lawal, 2019). The main sources of starch are the seeds, pulps, grasses, fruits, and tubers of these plants. The scientific names of the botanical species involved in starch production should be noted because they are widely distributed and have a wide range of qualities (Tagliapietra et al., 2021). Certain plant starches, like those from riceberry rice and bananas, are miscible in liquid forms, which makes them perfect for sauces and salad dressings, and have desirable product-oriented qualities like softness and stability in modified forms (Prasajak et al., 2021; Supakot et al., 2022). These properties make these plant starches ideal for novel industrial applications. The extraction processes for fruits such as bananas, peaches, mangos, jackfruits, and litchis begin with their processing as fruit kernels. The kernels are soaked in sodium metabisulfite at 50 °C for 12 h followed by draining and grinding them. The material is then passed through 100–300 mesh sieves. The supernatant is decanted, and the starch layer left behind is resuspended in distilled water, centrifuged at 8000 RPM for 5 min, washed and centrifuged again until the starch is pure. It is then air-dried for 24 h at 45 °C and pulverized (Tagliapietra et al., 2021).

The techniques differ for sources such as tubers and roots, such as yams, arrowroot, ginger, turmeric, and bamboo. The updated method produces a higher yield and doesn't use any hazardous chemicals. The tuber or root is washed with distilled water and is later processed by cutting and drying. It is then ground, sifted, and filtered. Next, it is precipitated using ethanol in distilled water (ethanol 3:1). The supernatant is collected and concentrated in a rotary evaporator. Again, the extract is filtered, then centrifuged and sonified. Further, the starch is extracted by aqueous methods for 8 h and centrifuged. The precipitate is washed with ethanol and acetone three times. The solvents are evaporated with nitrogen, and the starch extract is collected in a desiccator and dried (Dorantes-Fuertes et al., 2024).

5. Formulation techniques

5.1 Optimizing starch inclusion in different food matrices

5.1.1 Liquid foods

Heating starch solutions above 130 °C can cause molecular dissociation. Starch-water systems can be manipulated to reach a solution state, but starch generally increases viscosity, leading to a semisolid or solid form. To adapt starch for liquid food matrices, methods like pregelatinization, heating, mixing, and high-pressure treatment are used (Fang et al., 2021). High-pressure-treated wheat starch is an effective fat replacer at concentrations of 10%, 15%, and 20% (w/w). It is homogenized with deionized water, mixed with Tween 80 (1%), sunflower oil, and then homogenized again for a stable mixture (Heydari et al., 2021).

5.1.2 Semisolid foods

Mayonnaise, an oil-in-water (O/W) emulsion, can be replaced with high-pressure-treated corn starch powders at varying concentrations. These are mixed with deionized water using a roller mixer, then combined with Tween 80 and sunflower oil (Heydari & Razavi, 2021). Pregelatinization works well for semisolid products like yogurt, where low-fat goat milk with inulin, maltodextrin, and pectin matches the taste and storage qualities of whole milk yogurt (Costa et al., 2022). For pork pâté, boiled meat and fat are mixed with broth, then fat is replaced with modified potato starch (1:9 ratio) and left for 12 h before adding pork liver, sterilizing, and cooling (Rezler et al., 2021). Nanofibril cellulose derivatives are added to low-fat ice cream (0.15%–0.3%) or standard formulations (10%) to enhance texture, reduce melting, and improve fat destabilization (Fang et al., 2021).

5.1.3 Solid foods

RS3 starches, substituted at 10%–15% in cookies, optimize moisture, crumbliness, and spread ratio, while a 5% substitution closely mimics the wheat flour control (Falsafi et al., 2022). Adding 3%–5% modified starch to bread dough reduces the negative effects of dietary fiber, and oxidized starch improves softness and volume. Phenolic compounds in the dough can weaken gluten, enhancing chewiness (Durmus et al., 2021). In low-fat crackers, modified starches, including hydrolyzed and pregelatinized starch, are used at 1%–2% to improve texture and structure (Otero-Guzmán et al., 2023).

5.2 Synergistic use with other hydrocolloids

Sodium alginate and corn starch emulsion gels, combined with 30% oil, show strong gel structure, reduced lipid oxidation, and cooking losses. Calcium ions help form a network that reinforces starch granules (Su et al., 2024). Low-fat ice cream mixes use hydrocolloid stabilizers like guar gum, xanthan gum, and gelatin (4:1 ratio), with carrageenan and soy lecithin (1:1) as emulsifiers. Cross-linked cassava starch replaces 10% fat. The mix is pasteurized, aged at 4–6°C for 24 h, whipped for 16–20 min, and hardened at −25 °C (Marimon-Valverde et al., 2024). Hydrocolloid blends (2% w/w) of tapioca starch, xanthan gum, and pectin, with NaCl, reduce calories in instant soups while keeping creaminess (Ospina-Casas et al., 2022). Starch-based nanogels, made from crosslinked chitosan and carboxymethyl starch, improve bread volume, making them promising for weak wheat flour (Dou et al., 2024). Fat replacers like carboxymethyl cellulose, gum Arabic, carrageenan, and xanthan are used in Labneh, with xanthan offering the best water retention. Hybrid sodium alginate and pectin hydrogels, cross-linked with citric acid, can reduce fat by 50% in chocolate while improving melting resistance (Fang et al., 2021). Some emulsion gels may increase water retention, affecting juiciness, which lipophilic agents can mitigate (Ren et al., 2022).

5.3 Process considerations for maximum effectiveness

Formulation trials for fat-replaced products involve testing methods such as proximate analysis, color, pH, texture profiling, rheology, and sometimes electron microscopy. Cooking losses can be measured using formula (6.1) as follows:

$$\text{Cooking loss } (\%) = \frac{W_1 - W_2}{W_1} \text{ x } 100\% \tag{6.1}$$

where W_1 and W_2 are the product weights before and after cooking (Li et al., 2022). Double emulsions, such as W/O/W (water-in-oil-in-water), are used in meat production to improve storage and reduce oxidation (Ren et al., 2022). When using protein-starch combinations, moisture levels should be tested at different temperatures to ensure stability during processes like freezing, thawing, and drum drying (Bühler et al., 2022). High hydrostatic pressure is promising for minimally processed foods, altering starch structure to reduce retrogradation and improve texture and shelf life compared to thermal treatments (Heydari & Razavi, 2021).

6. Case studies

6.1 Bakery products

Wheat flour is a key ingredient in many foods like bread, cakes, cookies, pastries, sauces, and breading, largely due to its widespread use in Western diets. Bakery goods, popular globally, are excellent carriers of dietary fiber, starch, and bioactive compounds (Lin, 2022). While gluten forms a cohesive, elastic dough when wheat proteins are hydrated and mixed, not all baked goods require gluten. For example, cookies and highly liquid batters often avoid gluten network formation, and scalded doughs lose gluten function when boiled (Gómez, 2022).

There is a rising demand for gluten-free products as around 1% of the population has gluten sensitivity, including celiac disease. Gluten consumption is linked to wheat allergies, nonceliac gluten sensitivity, and celiac disease, driving demand for gluten-free foods (Aleman et al., 2021). Starch is often added to gluten-free recipes to improve texture and volume (Horstmann et al., 2017).

Barley flour, made from pearled barley, is used for various baked goods but doesn't form gluten, as it contains hordeins instead of gliadins (Lukinac & Jukić, 2022). Rice flour, hypoallergenic and easy to digest, is ideal for gluten-free dishes, though its baked products are generally lower in quality than wheat-based items. New varieties like Frontières, a high-protein rice, improve gluten-free baking, particularly cupcakes (Aleman et al., 2021). Gluten-free biscuits using buckwheat, sorghum, and lentil flours are gaining attention, along with efforts to reduce the glycemic index (GI) by replacing sucrose and flour with RS2, maltitol, and inulin (Di Cairano et al., 2022).

6.2 Dairy like products

Yogurt has a lower glycemic index (GI) than many carbohydrate-rich foods, making it a good substitute for reducing a diet's glycemic load. This aligns with recommended dietary patterns that emphasize whole grains, fruits, vegetables, nuts, legumes, fish, vegetable oils, and yogurt (Wolever, 2017). Replacing sugar with inulin in yogurt enhances sweetness and supports gut health (Wan et al., 2021). Adding insoluble fibers like modified okara increases yogurt's hypoglycemic effect, while resistant starch also shows potential for lowering digestion rates (You et al., 2024). Though yogurt is nutritious, it lacks dietary fiber, flavonoids, and iron, prompting recent additions of fiber to improve smoothness and stability (Fan et al., 2023). Adding Lactobacillus paracasei-modified teak peel fiber also enhances yogurt's texture and stability (Tian et al., 2024).

Ice cream is a frozen, aerated oil-in-water emulsion. Starch nanocrystals, which have high crystallinity, reinforce interfacial integrity and can replace fat in ice cream (Lin et al., 2023). Frozen yogurt, lower in fat and containing live bacteria, is seen as a healthier alternative to ice cream. The growing demand for lactose-free ice cream and yogurt faces challenges with lactose crystallization, which can cause a grainy texture during freezing. Using κ-carrageenan and corn starch as stabilizers improves the texture and appearance of lactose-free frozen yogurt (Skryplonek et al., 2019).

6.3 Meat products and analogs

Modified maize starch is a well-known fat substitute due to its globules, which resemble fat in shape and size. It reduces calorie content while offering properties like gel formation, water retention, and viscosity. As a polysaccharide, it binds water, creating a creamy texture, and can deliver lipophilic flavors and nutrients when combined with other ingredients like proteins, enhancing low-fat products (Asyrul-Izhar et al., 2023). Gelatin, a protein-based fat replacer, mimics fat's smoothness and is effective in forming protective coatings, making protein-polysaccharide complexes useful for processed meats (Asyrul-Izhar et al., 2023).

Meat has a short shelf life due to microbial contamination and oxidation, which can be controlled using thermal or nonthermal methods (Guo et al., 2021). Typically, meat lasts about 7 days at 4 °C, but for longer storage and transportation, it needs to be kept at lower temperatures. Studies show that storing meat at −18 °C can extend its shelf life for years. For example, chilled beef packaged in starch film with sea buckthorn pomace extract has shown improved shelf life during superchilled storage (Guo et al., 2021).

7. Challenges and limitations

Although starch-based fat substitutes have demonstrated potential in terms of lowering caloric content and simulating the sensory qualities of fats, several obstacles prevent their widespread use and efficacy. Additionally, research on the application of fat replacers in low-fat food products is limited. One notable issue is that total fat replacement using inulin or maltodextrin (100%) in baked goods significantly decreased consumer acceptance (Colla & Gamlath, 2015). These replacers couldn't fully replicate the mouthfeel, texture, and flavor that fats provide, leading to products that are less appealing to consumers. Although rice starch was only tested at 20% fat replacement, it had no discernible influence on sensory qualities. The physical characteristics of doughs and associated baked goods were significantly impacted by all complex carbohydrate fat replacers. For almost all evaluated items, there were notable increases in density, toughness, breaking strength, wetness, and volume reductions (Lee & Puligundla, 2016). A large number of the evaluated fat replacers have only ever been tested in one kind of meal and once overall (Colla et al., 2018). Additionally, even though recent research in this area has been reviewed, comparing studies' physical and sensory characteristics is challenging. It is difficult to identify the best fat replacer. Moreover, only a few studies have evaluated the usage of fat substitutes in bread, muffins, and croissants, resulting in insufficient information to prescribe the appropriate type of fat replacer for these items. Furthermore, the availability and price of starch-based fat substitutes may restrict how widely they are used. The cost-effectiveness and supply chain management of these alternatives to standard fats may provide difficulties for producers because they may be more expensive or less accessible. The food industry's wider use of starch-based fat replacers may be hampered by this financial obstacle. Furthermore, how consumers perceive low-fat foods is a major factor in their adoption. Foods formulated using starch-based fat replacers are difficult to sell since many consumers mistake low-fat foods for being of worse quality and taste. To change attitudes and boost acceptance of these healthier options, marketing and consumer education initiatives must be implemented effectively. In conclusion, even though starch-based fat substitutes show promise, there are still a lot of application-related obstacles to overcome, necessitating more thorough and varied studies to improve their efficacy and acceptability by consumers.

8. Conclusion and future prospective

Reducing fat in food items is an effective way to lower calorie intake and improve their nutritional value. Nonetheless, fat significantly impacts the flavor, mouthfeel, and texture of food. Carbohydrates can serve as a fat substitute, providing a calorie-reduced alternative without sacrificing flavor. The low-fat food industry relies heavily on starch-based fat replacers, such as native and modified starches. These starch-based fat replacers mimic the sensory and physicochemical properties of fat globules and are commonly used in many food items, including mayonnaise, baked goods, dairy products, and salad dressings. After decades of research, starch-based fat substitutes are widely recognized as safe and effective for replacing fats without compromising human health. Additionally, the market for fat replacements is expanding steadily. However, there is very little evidence from clinical trials that starch-based fat substitutes offer significant health benefits compared to fat. The absence of human intervention trials makes it difficult to determine whether starch-based fat replacers have any additional negative effects while lowering dietary consumption of calories and fat. If such effects exist, their magnitude is also unclear. Thus, while investigating new fat replacers, future research should concentrate on the physiological efficacy and safety of fat replacers based on human intervention studies. Overall, further investigation and advancements in the field of carbohydrate-based fat substitutes may lead to their broader inclusion in food formulations, ultimately helping produce better food options for consumers.

References

Adegbaju, M. S., Ajose, T., Adegbaju, I. E., Omosebi, T., Ajenifujah-Solebo, S. O., Falana, O. Y., Shittu, O. B., Adetunji, C. O., & Akinbo, O. (2024). Genetic engineering and genome editing technologies as catalyst for Africa's food security: The case of plant biotechnology in Nigeria. *Frontiers in Genome Editing, 6*. https://doi.org/10.3389/fgeed.2024.1398813

Ahsan, M., Ali, T. M., & Hasnain, A. (2023). Use of oxidized potato starch as simultaneous fat and casein replacer in analogue mozzarella cheese-II: Impact on functional and sensory properties of cheese. *Food Hydrocolloids, 142*. https://doi.org/10.1016/j.foodhyd.2023.108810

Akshit, F. N. U., Poswal, V., Kaushik, R., Deshwal, G. K., & Huppertz, T. (2025). Influence of starch on the rheological, textural, and microstructural properties of processed and analogue cheeses. *Encyclopedia, 5*(2). https://doi.org/10.3390/encyclopedia5020041

Aleman, R. S., Paz, G., Morris, A., Prinyawiwatkul, W., Moncada, M., & King, J. M. (2021). High protein brown rice flour, tapioca starch & potato starch in the development of gluten-free cupcakes. *Lebensmittel-Wissenschaft & Technologie, 152*. https://doi.org/10.1016/j.lwt.2021.112326

Arp, C. G., Correa, M. J., & Ferrero, C. (2021). Resistant starches: A smart alternative for the development of functional bread and other starch-based foods. *Food Hydrocolloids, 121*, 106949. https://doi.org/10.1016/j.foodhyd.2021.106949

Asyrul-Izhar, A. B., Bakar, J., Sazili, A. Q., Goh, Y. M., & Ismail-Fitry, M. R. (2023). Emulsion gels formed by electrostatic interaction of gelatine and modified corn starch via pH adjustments: Potential fat replacers in meat products. *Gels, 9*(1). https://doi.org/10.3390/gels9010050

Bühler, J. M., Schlangen, M., Möller, A. C., Bruins, M. E., & Jan van der Goot, A. (2022). Starch in plant-based meat replacers: A new approach to using endogenous starch from cereals and legumes. *Starch - Stärke, 74*(1–2). https://doi.org/10.1002/star.202100157

Bashir, K., & Aggarwal, M. (2019). Physicochemical, structural and functional properties of native and irradiated starch: A review. *Journal of Food Science and Technology, 56*(2), 513–523. https://doi.org/10.1007/s13197-018-3530-2

Bello-Perez, L. A., & Flores-Silva, P. C. (2023). Interaction between starch and dietary compounds: New findings and perspectives to produce functional foods. *Food Research International, 172*. https://doi.org/10.1016/j.foodres.2023.113182

BeMiller, J. N. (2019). *Carbohydrate chemistry for food scientists*. Elsevier, 3, Elsevier.

Bhuiyan, M. H. R., & Ngadi, M. (2024). Application of batter coating for modulating oil, texture and structure of fried foods: A review. *Food Chemistry, 453*. https://doi.org/10.1016/j.foodchem.2024.139655

Chavan, R. S., Khedkar, C. D., & Bhatt, S. (2015). *Fat replacer encyclopedia of food and health*. India: Elsevier Inc.. https://doi.org/10.1016/B978-0-12-384947-2.00271-3

Chen, Y., She, Y., Zhang, R., Wang, J., Zhang, X., & Gou, X. (2020). Use of starch-based fat replacers in foods as a strategy to reduce dietary intake of fat and risk of metabolic diseases. *Food Science and Nutrition, 8*(1), 16–22. https://doi.org/10.1002/fsn3.1303

Chen, X. W., & Yang, X. Q. (2023). *Nutritional and functional properties of fat mimetics fat mimetics for food applications*. China: Wiley Blackwell. https://doi.org/10.1002/9781119780045.ch14

Ching, L. W., Zulkipli, N.'. M., Muhamad, I. I., Marsin, A. M., Khair, Z., & Anis, S. N. S. (2021). Dietary management for healthier batter formulations. *Trends in Food Science and Technology, 113*, 411–422. https://doi.org/10.1016/j.tifs.2021.03.054

Choque-Quispe, D., Obregón Gonzales, F. H., Carranza-Oropeza, M. V., Solano-Reynoso, A. M., Ligarda-Samanez, C. A., Palomino-Rínćon, W., Choque-Quispe, K., & Torres-Calla, M. J. (2024). Physicochemical and technofunctional properties of high Andean native potato starch. *Journal of Agriculture and Food Research, 15*. https://doi.org/10.1016/j.jafr.2023.100955

Colla, K., Costanzo, A., & Gamlath, S. (2018). Fat replacers in baked food products. *Foods, 7*(12), 192. https://doi.org/10.3390/foods7120192

Colla, K., & Gamlath, S. (2015). Inulin and maltodextrin can replace fat in baked savoury legume snacks. *International Journal of Food Science and Technology, 50*(10), 2297–2305. https://doi.org/10.1111/ijfs.12892

Compart, J., Singh, A., Fettke, J., Apriyanto, A., & Starch Properties, C. (2023). A review of starch modifications and their applications. *Polymers, 15* (16), 3491. https://doi.org/10.3390/polym15163491

Cornejo-Ramírez, Y. I., Martínez-Cruz, O., Del Toro-Sánchez, C. L., Wong-Corral, F. J., Borboa-Flores, J., & Cinco-Moroyoqui, F. J. (2018). The structural characteristics of starches and their functional properties. *CyTA – Journal of Food, 16*(1), 1003–1017. https://doi.org/10.1080/19476337.2018.1518343

Costa, M. P., Rosario, A. I. L. S., Silva, V. L. M., Vieira, C. P., & Conte, C. A. (2022). Rheological, physical and sensory evaluation of low-fat cupuassu goat milk yogurts supplemented with fat replacer. *Food Science of Animal Resources, 42*(2), 210–224. https://doi.org/10.5851/kosfa.2021.e64

da Silva, V. T., Mateus, N., Freitas, V.de, & Fernandes, A. (2024). Plant-based meat analogues: Exploring proteins, fibers and polyphenolic compounds as functional ingredients for future food solutions. *Foods, 13*(14), 2303. https://doi.org/10.3390/foods13142303

Dang, X., Yang, M., Shan, Z., Mansouri, S., May, B. K., Chen, X., Chen, H., & Woo, M. W. (2017). On spray drying of oxidized corn starch cross-linked gelatin microcapsules for drug release. *Materials Science and Engineering: C, 74*, 493–500. https://doi.org/10.1016/j.msec.2016.12.047

Dehghannya, J., & Ngadi, M. (2023). The application of pretreatments for producing low-fat fried foods: A review. *Trends in Food Science & Technology, 140*, Article 104150. https://doi.org/10.1016/j.tifs.2023.104150

Demiate, I. M., Bet, C. D., & Ito, V. C. (2024). *Commercial starches modified by physical, chemical, enzymatic and combined methods: Processes and applications*. Elsevier BV. https://doi.org/10.1016/b978-0-323-90842-9.00013-3

Di Cairano, M., Condelli, N., Cela, N., Sportiello, L., Caruso, M. C., & Galgano, F. (2022). Formulation of gluten-free biscuits with reduced glycaemic index: Focus on in vitro glucose release, physical and sensory properties. *Lebensmittel-Wissenschaft & Technologie, 154*. https://doi.org/10.1016/j.lwt.2021.112654

Dolas, K. A., Ranveer, R. C., Tapre, A. R., Nandane, A. S., & Sahoo, A. K. (2020). Effect of starch modification on physico-chemical, functional and structural characterization of cassava starch (Manihot esculenta crantz). *Food Research, 4*(4), 1265–1271. https://doi.org/10.26656/fr.2017.4(4).075

Donmez, D., Pinho, L., Patel, B., Desam, P., & Campanella, O. H. (2021). Characterization of starch–water interactions and their effects on two key functional properties: Starch gelatinization and retrogradation. *Current Opinion in Food Science, 39*, 103–109. https://doi.org/10.1016/j.cofs.2020.12.018

Dorantes-Fuertes, M. G., López-Méndez, M. C., Martínez-Castellanos, G., Meléndez-Armenta, R.Á., & Jiménez-Martínez, H. E. (2024). Starch extraction methods in tubers and roots: A systematic review. *Agronomy, 14*(4). https://doi.org/10.3390/agronomy14040865

Dou, X., Yanling, H., Sun, Y., Yang, P., Liu, L., He, Y., Shi, Y., Yang, C., & Chen, F. (2024). A novel baking additive: Preparation, characterization, and application of chitosan hydrochloride/carboxymethyl starch sodium nano-gel for wheat bread. *Food Hydrocolloids, 148*, Article 109459. https://doi.org/10.1016/j.foodhyd.2023.109459

Dura, A., Błaszczak, W., & Rosell, C. M. (2014). Functionality of porous starch obtained by amylase or amyloglucosidase treatments. *Carbohydrate Polymers, 101*(1), 837–845. https://doi.org/10.1016/j.carbpol.2013.10.013

Durmus, Y., Anil, M., & Simsek, S. (2021). Effects of hazelnut skin, cross-linked starch, and oxidized starch on wheat flour and dough quality. *Journal of Food Processing and Preservation, 45*(2). https://doi.org/10.1111/jfpp.14919

Falsafi, S. R., Maghsoudlou, Y., Aalami, M., Jafari, S. M., Raeisi, M., Nishinari, K., & Rostamabadi, H. (2022). Application of multi-criteria decision-making for optimizing the formulation of functional cookies containing different types of resistant starches: A physicochemical, organoleptic, in-vitro and in-vivo study. *Food Chemistry, 393*. https://doi.org/10.1016/j.foodchem.2022.133376

Fan, X., Shi, Z., Xu, J., Li, C., Li, X., Jiang, X., Du, L., Tu, M., Zeng, X., Wu, Z., & Pan, D. (2023). Characterization of the effects of binary probiotics and wolfberry dietary fiber on the quality of yogurt. *Food Chemistry, 406*, Article 135020. https://doi.org/10.1016/j.foodchem.2022.135020

Fang, Y., Zhang, H., & Nishinari, K. (2021). *Food hydrocolloids: Functionalities and applications*. Springer Nature.

Fonseca, L. M., Halal, S. L. M. E., Dias, A. R. G., & Zavareze, E.d. R. (2021). Physical modification of starch by heat-moisture treatment and annealing and their applications: A review. *Carbohydrate Polymers, 274*. https://doi.org/10.1016/j.carbpol.2021.118665

Gómez, M. (2022). Gluten-free bakery products: Ingredients and processes. *Advances in Food and Nutrition Research, 99*, 189–238. https://doi.org/10.1016/bs.afnr.2021.11.005

Gałkowska, D., Kapuśniak, K., & Juszczak, L. (2023). Chemically modified starches as food additives. *Molecules, 28*(22), 7543. https://doi.org/10.3390/molecules28227543

Grzelak, T., Grupińska, J., Pelczyńska, M., Sperling, M., & Czyżewska, K. (2017). The merits of fat replacers in low-calorie food. *Acta Scientiarum Polonorum, Hortorum Cultus, 16*(2), 87–94.

Guo, Z., Ge, X., Gou, Q., Yang, L., Han, M., Han, G., Yu, Q.-li, & Han, L. (2021). Changes in chilled beef packaged in starch film containing sea buckthorn pomace extract and quality changes in the film during super-chilled storage. *Meat Science, 182*, Article 108620. https://doi.org/10.1016/j.meatsci.2021.108620

Guo, Y., Qiao, D., Zhao, S., Zhang, B., & Xie, F. (2021). Starch-based materials encapsulating food ingredients: Recent advances in fabrication methods and applications. *Carbohydrate Polymers, 270*, Article 118358. https://doi.org/10.1016/j.carbpol.2021.118358

He, R., Li, S., Zhao, G., Zhai, L., Qin, P., & Yang, L. (2023). Starch modification with molecular transformation, physicochemical characteristics, and industrial usability: A state-of-the-art review. *Polymers, 15*(13), 2935. https://doi.org/10.3390/polym15132935

Hedayati, S., Shahidi, F., Koocheki, A., Farahnaky, A., & Majzoobi, M. (2020). Influence of pregelatinized and granular cold water swelling maize starches on stability and physicochemical properties of low fat oil-in-water emulsions. *Food Hydrocolloids, 102*, Article 105620. https://doi.org/10.1016/j.foodhyd.2019.105620

Heydari, A., & Razavi, S. M. A. (2021). Evaluating high pressure-treated corn and waxy corn starches as novel fat replacers in model low-fat O/W emulsions: A physical and rheological study. *International Journal of Biological Macromolecules, 184*, 393–404. https://doi.org/10.1016/j.ijbiomac.2021.06.052

Heydari, A., Razavi, S. M. A., & Farahnaky, A. (2021). Effect of high pressure-treated wheat starch as a fat replacer on the physical and rheological properties of reduced-fat O/W emulsions. *Innovative Food Science and Emerging Technologies, 70*. https://doi.org/10.1016/j.ifset.2021.102702

Hong, Y., & Liu, X. (2023). *Physical modifications of starchpre-gelatinized modification of starch*. Singapore: Springer.

Horstmann, S., Lynch, K., & Arendt, E. (2017). Starch characteristics linked to gluten-free products. *Foods, 6*(4), 29. https://doi.org/10.3390/foods6040029

Hsu, J. L., Sung, C. C., & Tseng, J. T. (2023). Willingness-to-pay for ready-to-eat clean label food products at convenient stores. *Future Foods, 7*. https://doi.org/10.1016/j.fufo.2023.100237

Iuga, M., & Mironeasa, S. (2020). A review of the hydrothermal treatments impact on starch based systems properties. *Critical Reviews in Food Science and Nutrition, 60*(22), 3890–3915. https://doi.org/10.1080/10408398.2019.1664978

Jia, R., Congli, C., Qingjie, S., et al. (2023). A review of starch swelling behavior: Its mechanism, determination methods, influencing factors, and influence on food quality. *Carbohydrate Polymers, 321*, Article 121260. https://doi.org/10.1016/j.carbpol.2023.121260

Johnson, K. A., & Mauer, L. J. (2019). Effects of controlled relative humidity storage on moisture sorption and amylopectin retrogradation in gelatinized starch lyophiles. *Journal of Food Science, 84*(3), 507–523. https://doi.org/10.1111/1750-3841.14472

Krishnan, V., Mondal, D., Thomas, B., Singh, A., & Praveen, S. (2021). Starch-lipid interaction alters the molecular structure and ultimate starch bioavailability: A comprehensive review. *International Journal of Biological Macromolecules, 182*, 626–638. https://doi.org/10.1016/j.ijbiomac.2021.04.030

Lawal, M. V. (2019). Modified starches as direct compression excipients – Effect of physical and chemical modifications on tablet properties: A review. *Starch - Stärke, 71*(1–2). https://doi.org/10.1002/star.201800040

Lee, Y. T., & Puligundla, P. (2016). Characteristics of reduced-fat muffins and cookies with native and modified rice starches. *Emirates Journal of Food and Agriculture, 28*(5), 311–316. https://doi.org/10.9755/ejfa.2015-05-227

Li, P. H., Wang, C. W., Lu, W. C., Chan, Y. J., & Wang, C. C. R. (2022). Effect of resistant starch sources on the physical properties of dough and on the eating quality and glycemic index of salted noodles. *Foods, 11*(6). https://doi.org/10.3390/foods11060814

Lin, S. (2022). Dietary fiber in bakery products: Source, processing, and function. *Advances in Food and Nutrition Research, 99*, 37–100. https://doi.org/10.1016/bs.afnr.2021.12.001

Lin, D., Zhao, J., Fan, H., Qin, W., & Wu, Z. (2023). Enhancing starch nanocrystal production and evaluating their efficacy as fat replacers in ice cream: Investigating the influence of high pressure and ultrasonication. *International Journal of Biological Macromolecules, 251*, Article 126385. https://doi.org/10.1016/j.ijbiomac.2023.126385

Liu, Y., Chen, J., Luo, S., Li, C., Ye, J., Liu, C., & Gilbert, R. G. (2017). Physicochemical and structural properties of pregelatinized starch prepared by improved extrusion cooking technology. *Carbohydrate Polymers, 175*, 265–272. https://doi.org/10.1016/j.carbpol.2017.07.084

Lockyer, S., & Nugent, A. P. (2017). Health effects of resistant starch. *Nutrition Bulletin, 42*(1), 10–41. https://doi.org/10.1111/nbu.12244

Lukinac, J., & Jukić, M. (2022). Barley in the production of cereal-based products. *Plants, 11*(24), 3519. https://doi.org/10.3390/plants11243519

Marangoni, A. G., Van Duynhoven, J. P. M., Acevedo, N. C., Nicholson, R. A., & Patel, A. R. (2020). Advances in our understanding of the structure and functionality of edible fats and fat mimetics. *Soft Matter, 16*(2), 289–306. https://doi.org/10.1039/c9sm01704f

Marimon-Valverde, S., Lainez-Ramirez, S., Sepúlveda-Valencia, J. U., Mejia-Villota, A., & Rodriguez-Sandoval, E. (2024). Quality characteristics of low-fat ice cream mixtures as affected by modified cassava starch and hydrocolloids. *International Journal of Food Properties, 27*(1), 123–132. https://doi.org/10.1080/10942912.2023.2293462

Mo, X., Zhu, H., Yi, C., Deng, Y., & Yuan, J. (2024). Rheological properties of indica rice determined by starch structure related enzymatic activities during after-ripening. *International Journal of Biological Macromolecules, 269*(Pt 1), Article 131738. https://doi.org/10.1016/j.ijbiomac.2024.131738

Nsanzabera, F., Manishimwe, A., Mwiseneza, A., & Irakoze, E. (2023). Production of corn (zea mays) starch and cassava (Manihot esculenta) starch and their application as yogurt stabilizer. *Food and Nutrition Sciences, 14*(07), 589–600. https://doi.org/10.4236/fns.2023.147039

Oke, E. K., Idowu, M. A., Sobukola, O. P., Adeyeye, S. A. O., & Akinsola, A. O. (2018). Frying of food: A critical review. *Journal of Culinary Science & Technology, 16*(2), 107–127. https://doi.org/10.1080/15428052.2017.1333936

Ospina-Casas, K.-G., Laguado-Escobar, D.-A., & Narváez-Cuenca, C.-E. (2022). Using a mixture of hydrocolloids to mimic texture and rheological properties of a massive consumption food product. *Journal of Food Processing and Preservation, 46*(4). https://doi.org/10.1111/jfpp.16440

Otero-Guzmán, N., Cadena-Chamorro, E., Ciro-Velásquez, H. J., Mejía-Villota, A., & Rodriguez-Sandoval, E. (2023). Application of modified cassava starch as a fat substitute in cracker production. *Journal of Food Processing and Preservation, 2023*. https://doi.org/10.1155/2023/6888180

Park, J. J., Olawuyi, I. F., & Lee, W. Y. (2020). Characteristics of low-fat mayonnaise using different modified arrowroot starches as fat replacer. *International Journal of Biological Macromolecules, 153*, 215–223. https://doi.org/10.1016/j.ijbiomac.2020.02.331

Peng, X., & Yao, Y. (2017). Carbohydrates as fat replacers. *Annual Review of Food Science and Technology, 8*(1), 331–351. https://doi.org/10.1146/annurev-food-030216-030034

Prasajak, P., Detchewa, P., Sriwichai, W., & Moongngarm, A. (2021). The effects of unripe banana flour on resistant starch content and quality characteristics of gluten-free rice cookies. *Journal of Sustainability Science and Management, 16*(2), 67–78. https://doi.org/10.46754/jssm.2021.02.008

Puhlmann, M. L., & de Vos, W. M. (2022). Intrinsic dietary fibers and the gut microbiome: Rediscovering the benefits of the plant cell matrix for human health. *Frontiers in Immunology, 13*. https://doi.org/10.3389/fimmu.2022.954845

Punia Bangar, S., Sunooj, K. V., Navaf, M., Phimolsiripol, Y., & Whiteside, W. S. (2024). Recent advancements in cross-linked starches for food applications- a review. *International Journal of Food Properties, 27*(1), 411–430. https://doi.org/10.1080/10942912.2024.2318427

Raghunathan, R., Pandiselvam, R., Kothakota, A., & Mousavi Khaneghah, A. (2021). The application of emerging non-thermal technologies for the modification of cereal starches. *Lebensmittel-Wissenschaft & Technologie, 138*. https://doi.org/10.1016/j.lwt.2020.110795

Ren, Y., Huang, L., Zhang, Y., Li, H., Zhao, Di, Cao, J., & Liu, X. (2022). Application of emulsion gels as fat substitutes in meat products. *Foods, 11*(13), 1950. https://doi.org/10.3390/foods11131950

Rezler, R., Krzywdzińska-Bartkowiak, M., & Piątek, M. (2021). The influence of the substitution of fat with modified starch on the quality of pork liver pâtés. *Lebensmittel-Wissenschaft & Technologie, 135*, Article 110264. https://doi.org/10.1016/j.lwt.2020.110264

Šárka, E., Sinica, A., Smrčková, P., & Sluková, M. (2023). Non-traditional starches, their properties, and applications. *Foods, 12*(20), 3794. https://doi.org/10.3390/foods12203794

Saboonchi, S., Mehran, A., Bahramizadeh, P., & Massoud, R. (2021). *Applications of modified starch in food.* Third International Congress on Engineering, Technology and Innovation, 1–7, Available from: https://www.researchgate.net/publication/351194178_Applications_of_modified_starch_in_food.

Scott, G., & Awika, J. M. (2023). Effect of protein–starch interactions on starch retrogradation and implications for food product quality. *Comprehensive Reviews in Food Science and Food Safety, 22*(3), 2081–2111. https://doi.org/10.1111/1541-4337.13141

Sheikholeslami, Z., Mahfouzi, M., Karimi, M., & Ghiafehdavoodi, M. (2021). Modification of dough characteristics and baking quality based on whole wheat flour by enzymes and emulsifiers supplementation. *Lebensmittel-Wissenschaft & Technologie, 139*, Article 110794. https://doi.org/10.1016/j.lwt.2020.110794

Skryplonek, K., Henriques, M., Gomes, D., Viegas, J., Fonseca, C., Pereira, C., Dmytrów, I., & Mituniewicz-Małek, A. (2019). Characteristics of lactose-free frozen yogurt with κ-carrageenan and corn starch as stabilizers. *Journal of Dairy Science, 102*(9), 7838–7848. https://doi.org/10.3168/jds.2019-16556

Su, C.y., Li, D., Wang, L.j., & Wang, Y. (2024). Development of corn starch-sodium alginate emulsion gels as animal fat substitute: Effect of oil concentration. *Food Hydrocolloids, 157*. https://doi.org/10.1016/j.foodhyd.2024.110439

Sun, C., & Fang, Y. (2021). *Replacement of fat or starch food hydrocolloids: Functionalities and applications.* China: Springer Singapore. https://doi.org/10.1007/978-981-16-0320-4_12

Supakot, P., Samappito, J., Rongdon, P., & Yodprom, S. (2022). Effect of fat substitution with pre-gelatinized riceberry rice flour on physicochemical properties, sensory acceptance and shelf life of salad dressing. *Engineering Access, 8*(2), 249–256. https://doi.org/10.14456/mijet.2022.30

Syan, V., Kaur, J., Sharma, K., Patni, M., Rasane, P., Singh, J., & Bhadariya, V. (2024). An overview on the types, applications and health implications of fat replacers. *Journal of Food Science and Technology, 61*(1), 27–38. https://doi.org/10.1007/s13197-022-05642-7

Tagliapietra, B. L., Felisberto, M. H. F., Sanches, E. A., Campelo, P. H., & Clerici, M. T. P. S. (2021). Non-conventional starch sources. *Current Opinion in Food Science, 39*, 93–102. https://doi.org/10.1016/j.cofs.2020.11.011

Taniguchi, A., Isaji, S., Sagawa, T., & Kobayashi, R. (2023). Changes over time in crispiness and moisture content of tempura prepared using barley flour, buckwheat flour, and job's tears flour. *Nihon Reoroji Gakkaishi, 51*(2), 65–72. https://doi.org/10.1678/rheology.51.65

Tian, Y., Sheng, Y., Wu, T., & Wang, C. (2024). Effect of modified okara insoluble dietary fibre on the quality of yoghurt. *Food Chemistry X, 21*, Article 101064. https://doi.org/10.1016/j.fochx.2023.101064

Tiozon, R. J. N., Bonto, A. P., & Sreenivasulu, N. (2021). Enhancing the functional properties of rice starch through biopolymer blending for industrial applications: A review. *International Journal of Biological Macromolecules, 192*, 100–117. https://doi.org/10.1016/j.ijbiomac.2021.09.194

Tsao, C. W., Aday, A. W., Almarzooq, Z. I., Anderson, C. A. M., Arora, P., Avery, C. L., Baker-Smith, C. M., Beaton, A. Z., Boehme, A. K., Buxton, A. E., Commodore-Mensah, Y., Elkind, M. S. V., Evenson, K. R., Eze-Nliam, C., Fugar, S., Generoso, G., Heard, D. G., Hiremath, S., Ho, J. E., … Martin, S. S. (2023). Heart disease and stroke statistics - 2023 update: A report from the American heart association. *Lippincott Williams and Wilkins, undefined Circulation, 147*(8), E93. https://doi.org/10.1161/CIR.0000000000001123

Turnbull, C., Lillemo, M., & Hvoslef-Eide, T. A. K. (2021). Global regulation of genetically modified crops amid the gene edited crop boom – A review. *Frontiers in Plant Science, 12*. https://doi.org/10.3389/fpls.2021.630396

Varela, P., & Fiszman, S. M. (2011). Hydrocolloids in fried foods. A review. *Food Hydrocolloids, 25*(8), 1801–1812. https://doi.org/10.1016/j.foodhyd.2011.01.016

Wan, Z., Khubber, S., Dwivedi, M., & Misra, N. N. (2021). Strategies for lowering the added sugar in yogurts. *Food Chemistry, 344*, Article 128573. https://doi.org/10.1016/j.foodchem.2020.128573

Wang, Z., Wang, S., Xu, Q., Qi, K., Li, F., Lin, L., Xu, Y., & Wei, Y. (2023). Synthesis and functions of resistant starch. *Advances in Nutrition, 14*(5), 1131–1144. https://doi.org/10.1016/j.advnut.2023.06.001

Werlang, S., Bonfante, C., Oro, T., Biduski, B., Bertolin, T. E., & Gutkoski, L. C. (2021). Native and annealed oat starches as a fat replacer in mayonnaise. *Journal of Food Processing and Preservation, 45*(3). https://doi.org/10.1111/jfpp.15211

Wolever, T. M. S. (2017). Yogurt is a low–glycemic index food. *The Journal of Nutrition, 147*(7), 1462S. https://doi.org/10.3945/jn.116.240770

Xie, F., Yuan, C., Zhang, H., Wu, Y., & Ai, L. (2023). Structure-function relationship between galactomannans and their effects on freeze-thaw stability, retrogradation, and texture of corn starch gels during cold storage. *Food Chemistry, 398*, Article 133915. https://doi.org/10.1016/j.foodchem.2022.133915

Yazar, G., & Rosell, C. M. (2023). Fat replacers in baked products: Their impact on rheological properties and final product quality. *Critical Reviews in Food Science and Nutrition, 63*(25), 7653–7676. https://doi.org/10.1080/10408398.2022.2048353

You, M., Peng, Z., Jiang, Y., Yao, C., Yang, B., Ban, Q., & Cheng, J. (2024). The properties of the rice resistant starch processing and its application in skimmed yogurt. *International Journal of Biological Macromolecules, 265*(Pt 2), Article 131087. https://doi.org/10.1016/j.ijbiomac.2024.131087

Zhang, Z., & Bao, J. (2023). Recent advances in modification approaches, health benefits, and food applications of resistant starch. *Starch - Stärke, 75*(9–10). https://doi.org/10.1002/star.202100141

Chapter 7

Resistant starches: A novel approach to lower digestibility, fat intake and glycemic index

Gbeminiyi Olamiti, Shonisani Eugenie Ramashia, Mpho Edward Mashau and Oluwaseun Peter Bamidele
Department of Food Science and Technology, University of Venda, Thohoyandou, South Africa

1. Introduction

Starches reflect their significance in global food consumption as they are major constituents of numerous developing nations' diets (Bojarczuk et al., 2022). Starches are classified as complex carbohydrates due to their multiple glucose units, and are found in numerous food products, including, but not limited to, cereals, legumes, and fruits. Starches may greatly influence metabolism, body weight, and overall health (Eslam et al., 2022). The health impacts of starch consumption are multifaceted. Simple starches are prominent in refined grains and processed carbohydrates. Because of the high glycemic index of these products, blood glucose levels spike, leading to insulin resistance and/or obesity. Conversely, non-digestible (resistant) starches may healthfully impact the body. These starches are fermented in the colon, improving the proliferation of advantageous gut microbiota and the production of short-chain fatty acids (SCFAs), which are important regulators of gut health and metabolism (Chen et al., 2024).

Multiple studies have highlighted the role of starch type in gelling, appetite, and glycemic response (Lai et al., 2022). Consuming dietary sources of resistant starch, along with other high-fiber foods, has been linked to better digestive health, enhanced glycemic control, and a lower risk of chronic diseases, including type 2 diabetes and other cardiovascular diseases. In recent years, as concerns about the effects of the diets one consumes have increased, it has become essential to understand the various roles of starches in enhancing health value and nutritional practices that promote sustainable well-being (Carvalho et al., 2024). Recently, the importance of starch as a dietary component has come into focus with the emergence of resistant starches (RS) in nutritional science. The traditional approach has been that dietary starches are regarded solely as sources of energy, derived from caloric intake, due to their digestibility. However, there appears to be an extensive supply of resistant starches that are not fully digested in the small intestine, fermented in the colon, and can provide more than just calorie-burning energy.

Resistant starches have emerged as a novel class of carbohydrates that share some characteristics with dietary fiber (Bede & Zaixiang, 2021). The emphasis on RS has increased due to research on gut health and its role in overall well-being. Different interactions between resistant starches, the microbiome, and diabetic pathology are now being investigated, and novel mechanisms have been described linking these starches with improved gut health and reduced risk of chronic diseases. Suggestions to integrate resistant starches into dietary guidelines are based on the fact that they can help regulate appetite, glucose metabolism, and the colonization of probiotics in the intestines (Wang et al., 2023). These qualities make RS valuable for the dietary treatment of patients with obesity, type 2 diabetes and gastrointestinal diseases.

In light of the ever-evolving nature of nutrition science, the interest in resistant starches to develop effective nutrition protocols for health protection and promotion is promising. By achieving this robust range of benefits, these starches are beginning to change the conventional way carbohydrates are perceived and offer fascinating potential for future health promotion across a variety of diets. The chapter addresses methods for assimilating RS without digestion, the location of

Native and Modified Starches for Food Applications. https://doi.org/10.1016/B978-0-443-27431-2.00013-4

FIGURE 7.1 **Food sources of resistant starches** Figure showing different sources of resistant starches.

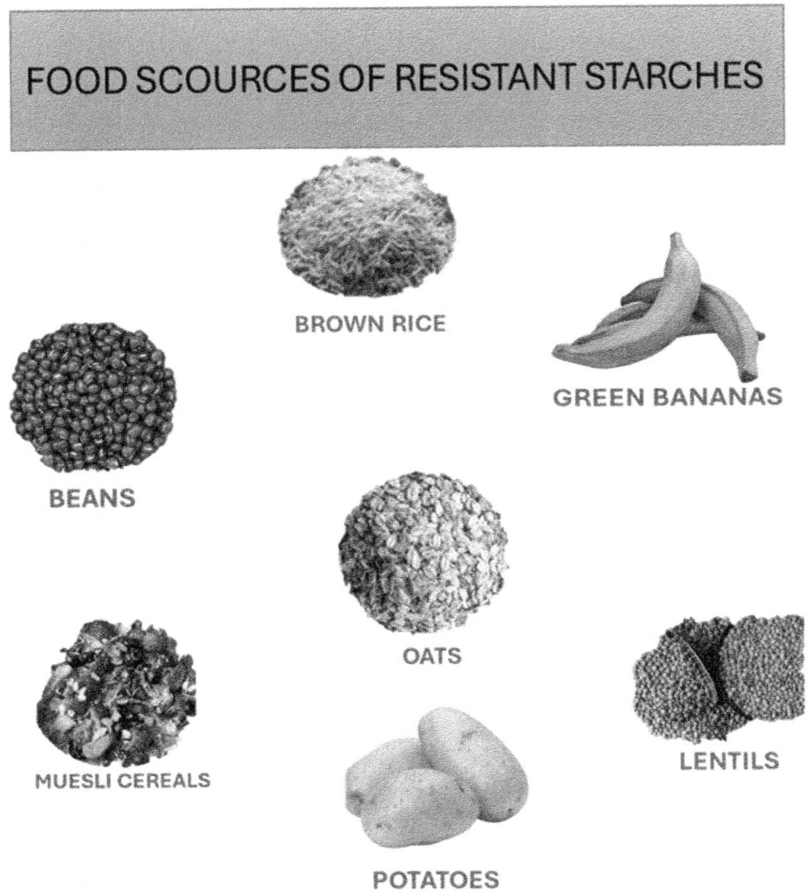

RS in the gut microbiome, treatment methods, including extrusion, to increase RS intake, gelatinization and its effect on the bioavailability of starches, and the nutritional and health value of RS.

2. Understanding resistant starches

2.1 Defining resistant starches: types and sources

According to current understanding, resistant starches are carbohydrates that bypass digestion in the small intestine and reach the colon in almost their original form (Junejo et al., 2022). In contrast to other starches, which are broken down into glucose, resistant starches behave more like fibers, providing a substrate for the gut microbiota and conferring numerous health benefits. Due to their benefits, particularly in improving gut health, these types of starches are now the focus of numerous scholars. Fig. 7.1 shows some food sources of resistant starches.

2.2 Mechanisms of resistance: how resistant starches escape from digestion

Resistant starch (RS) is prevented from entering a digested form by mechanisms that render it enzymatically resistant, as is the case with other carbohydrates in the small intestine. These mechanisms are porous to variability depending on the type of resistant starch. However, Ma and Jiang (2024) argue that these are generally due to their physical and chemical properties and their interactions with other molecules.

2.2.1 Physical inaccessibility (RS1)

RS1 starches are found in whole or lightly milled foodstuffs which have not been intracellularly digested because seed, grain and legume cell walls provide substantial protective fibrous materials that enclose the munching monosaccharides.

RS1 starch granules, fibrous and cellulose materials, inhibit α-amylase digestive enzymes from penetrating and targeting the starch. Because of the lack of availability of certain starches, they do not get digested in the small intestine and end up in the large intestine, where they can be fermented by gut bacteria (Gidley, 2024).

2.2.2 Crystalline granule structure (RS2)

Starch RS2 types exhibit crystalline structures, similar to those found in unripe bananas and raw potato starches, which do not allow enzymatic breakdown due to their tightly ordered, packed structure. The RS2 starch possesses crystalline characteristics that contain large amounts of amylose, which prevent digestive enzymes from attaching and hydrolyzing starch. This is because Amylose is a linear polymer of glucose that prefers to form compact helical structures that are less sensitive to enzyme activity. Cooking or the ripening process will disrupt the ordered/perfect crystalline parts of the structures, making them more resistant to breaking up. However, intact structures can pass through the small intestine without being fully broken down (Junejo et al., 2022).

2.2.3 Retrogradation and molecular reordering (RS3)

RS3-type retrograded starches can be prepared by cooling starchy foods, such as rice, pasta, or potatoes. This mechanism facilitates the reorganization of both amylose and amylopectin polymers into a more resistant structure, one that is less susceptible to enzymatic action due to increased crystallization. Adaptive Forces Starch retrogradation increases the hydrogen bonding between amylose chains, modifying the structure of starch molecules into a nearly compact crystalline network. This network of dense, crystalline structures makes it even more difficult for digestive enzymes to digest. This type of retrograde starch is resistant to digestion when it passes through the small intestine and arrives in the colon, fermented by gut microbiota (Feng et al., 2024).

2.2.4 Chemical modification

RS4-type starches are formed due to chemical alteration procedures such as esterification, cross-linking, or phosphorylation. These interventions inhibit the usual enzymatic breakdown of starch, converting it to a form that cannot be acted upon by amylase and other digestive enzymes. The modified chemical bonds in RS4 starches modulate enzyme binding and hydrolysis. Molecules of esterified starch may also inhibit enzymes from penetrating starch by introducing heavy side chains or cross-linked molecules that are difficult to break down. Chemically induced resistance makes RS4 type of starch evade digestion in the small intestine, but is available for fermentation by colon bacteria (BeMiller, 2020).

2.2.5 Amylose-lipid complexes formation (RS5)

According to the source, the RS5 type of starch originates from the hybridization of amylose-lipid complexes. The unique amylose-lipid complex forms a stable, ordered structure that is resistant to hydrolysis by most enzymes. The lipids located in the amylose matrix impede enzyme access, creating a digestion-resistant hydrophobic complex. Lipids can even fit into the inner helical structure of amylose, creating both a physical and chemical barrier. Defatted amylose-lipid complexes are not hydrolyzed in the small intestine and instead pass into the colon, where they are fermentable by the bacterial microflora (Panesar & Anal, 2022).

According to Chen et al. (2024), resistant starches reach the large intestine mostly intact, having undergone no breakdown in the small intestine. In the large intestine, resistant starches are fermented by the gut microbiota, producing short-chain fatty acids (SCFAs) such as butyrate, acetate, and propionate. SCFAs are utilized by colonic cells as an energy source and have additional benefits, including anti-inflammatory properties and the potential to reduce the risk of colorectal cancer. SCFAs are also absorbed into the blood and support energy expenditure, appetite regulation, and glucose regulation (Gromova et al., 2021). Resistant starches also positively contribute to glycemic control, satiety, and gut health. The slow fermentation and transit of resistant starches positively alter the gut microbiome, providing a broad benefit to gut and systemic health.

2.3 The role of resistant starches in the gut microbiome and the body

Due to their prebiotic properties, resistant starches positively impact the creation and preservation of a gut microbiome. Resistant starches are largely unabsorbed as they pass through the digestive system; they serve as substrates for beneficial gut bacteria, promoting a positive microflora. The fermentation of resistant starches also produces various short-chain fatty acids (SCFAs), including butyrate, as discussed in a previous chapter, which plays a vital role in gut health and

TABLE 7.1 The role of resistant starch in the gut microbiome.

Aspect	Description	Key bacteria involved	Main outcomes/health implications	References
Fermentation in the colon	RS escapes small intestine digestion and undergoes anaerobic fermentation by gut microbiota, producing SCFAs (acetate, propionate, butyrate) and gases.	*Ruminococcus bromii*, *Bifidobacterium*,	Provides energy substrate for colonocytes; lowers gut pH to favor beneficial bacteria and inhibit pathogens.	Venkataraman et al. (2016)
SCFA production (especially butyrate)	The primary metabolite is butyrate, a major energy source for colonocytes with anti-inflammatory and anti-carcinogenic effects.	*Faecalibacterium prausnitzii*, *Roseburia* spp.,	Enhances gut barrier integrity, modulates inflammation, improves insulin sensitivity, and reduces colon cancer risk.	Demartino et al. (2022)
Modulation of microbial composition	Increases beneficial taxa; often raises the Firmicutes/Bacteroidetes ratio; selectively enriches saccharolytic bacteria.	*Faecalibacterium*, *Roseburia*, *Ruminococcus*, *Bifidobacterium*,	Improves microbial diversity in some cases; personalized responses based on baseline microbiome and RS type.	Ma et al. (2025)
Health benefits via microbiome	RS acts as a prebiotic, linking to improvements in metabolic and gut health; variable individual responses highlight the need for personalization.	*Faecalibacteriums*	Supports glucose regulation, weight management, reduced inflammation; potential in obesity, T2DM, IBD, and colorectal cancer prevention.	Demartino et al. (2022); Ma et al. (2025)

overall health (Ashaolu et al., 2021). The roles of some resistant starches in the gut microbiome, along with the microorganisms responsible for these roles, are listed in Table 7.1. The illustration in Fig. 7.2 depicts the functions of resistant starches in the human body.

3. Processing techniques for resistant starches

3.1 Modification of starch by traditional and advanced technology

Food applications rarely use native starches due to their low viscosity, limited processing tolerance, and limited thermal stability. Heat applied during storage can cause thermal deterioration and condensation, making pastes fragile and inconsistent (Gupta et al., 2024). Given these challenges, starch is modified to achieve other desirable traits, such as stability throughout processing and distribution to diverse applications (Zheng et al., 2022). Due to the effects of native starches on quality, effectiveness, and overall functionality, they have been extensively studied (Milanezzi & Silva, 2025). Native starches have limitations in terms of gel strength, viscosity, and solubility, and modifications are likely to help meet industry standards (Gupta et al., 2024). Commercially available modified starch is produced by processes such as acid hydrolysis, etherification, and oxidation, among others. Enzyme treatment is also used to alter starches by creating new molecules within the starch (Ma & Jiang, 2024). Table 7.2 summarizes the advantages and disadvantages of conventional techniques used in starch modification.

The modification of starch using non-thermal methods is an emerging technique to refine its copious, unrelated industrial uses (Milanezzi & Silva, 2025). This emerging technology maintains the functional properties of starch while enhancing others, as it alters starch without the use of harmful chemicals or high temperatures (Maniglia et al., 2021). Nevertheless, consumers' recent interest in functional foods and the absence of chemical additives in foods have prompted scientists and the starch-producing sector to modify starch using physical techniques and non-thermal treatments.

This aligns with the characteristics of the emerging global industrial processing style: clean, using fewer chemicals, and employing almost equally efficient processing techniques (Ma & Jiang, 2024). In non-thermal technology, high-voltage electrodes are applied to starch suspensions to create internal pores within the starch granules and modify their crystalline structure. In addition, this method alters some fundamental characteristics of starch, such as solubility,

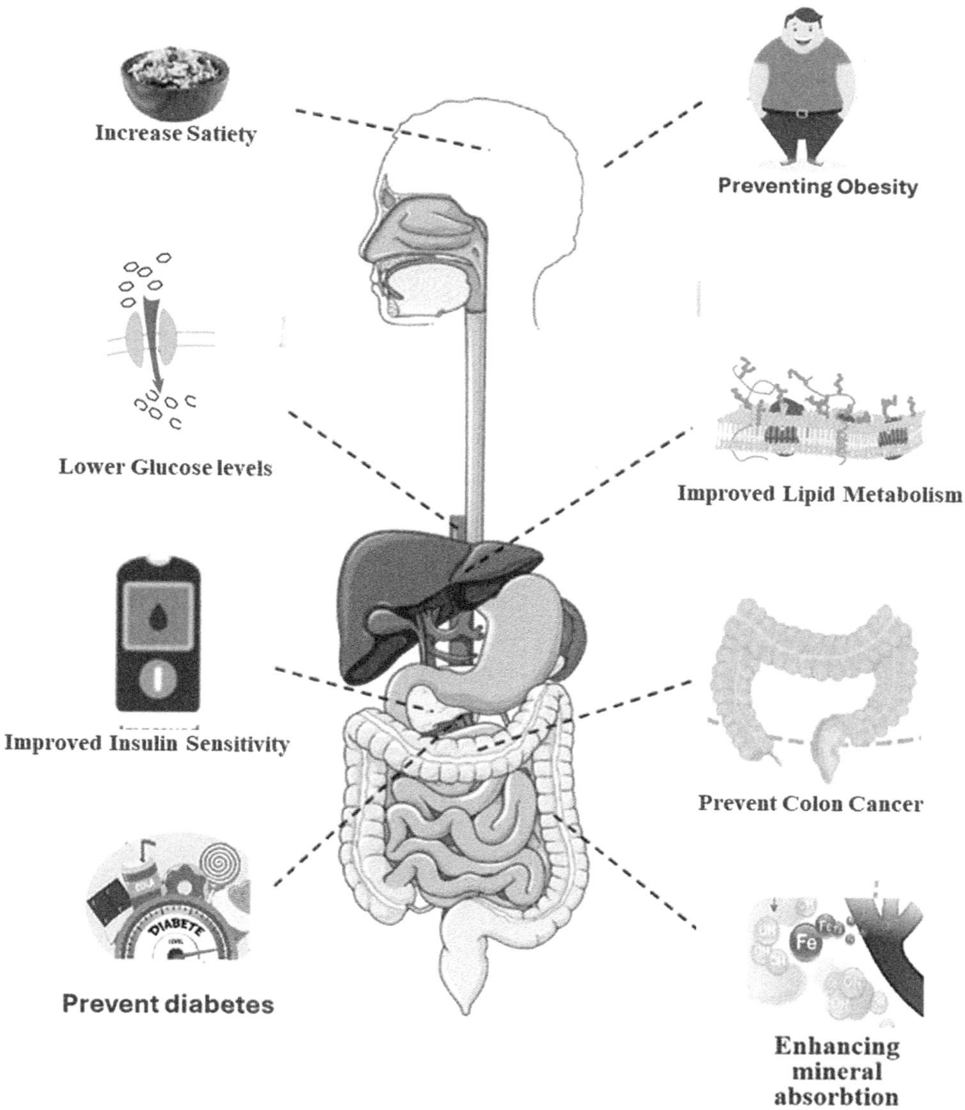

FIGURE 7.2 The roles of resistant starches in the body as a whole The figure shows the different roles played by resistant starches in the body.

swelling power, viscosity, and gelatinization (Almeida et al., 2024; Wu et al., 2022). These modifications are important as they align and adjust starch to serve an array of industrial needs, including those in the food and pharmaceutical industries. Examples of non-thermal technologies include microwaves, pulsed electric fields, high-pressure processing, and cold plasma (Miao et al., 2021; Chen et al., 2021; Li et al., 2018; Wu et al., 2022). The pros and cons of modifying starches via non-thermal methods are shown in Table 7.2.

3.2 The science of extrusion: enhancing starch resistance

Extrusion cooking is a common and reputable method used in the food processing industry. This method aids in cooking by combining high temperature and short cook time, using heat and pressure. Cooking and processing times are short, making food processing more efficient with this method (Xiang et al., 2019). Various changes occur in food, including gelatinization, protein denaturation, and restructuring of extrudates (Wang & Ismail, 2012). This method is used in the production of many foods, including ready-to-eat breakfast cereals and puffed snack foods (Wang & Ismail, 2012). Extruded cereal foods are superior to other industrial food processing technologies. This is due to the low cost, fast rate of food production, high production efficiency, high adaptability, and the unique capacity to save on energy (Dendegh et al., 2022).

TABLE 7.2 Different starch modification techniques.

Starch modification method	Advantages	Disadvantages	Example	References
Physical: Encompasses various processes involving heat, pressure, or mechanical action, both in dry and wet conditions. These methods also include fractionation.	Modifications in the granular structure lead to alterations in its gelatinization behavior, swelling, and solubility characteristics. Induce favorable alterations in starch properties, including modifications in their granular structure and enhancements in solubility levels.Somewhat affordable compared to other methods	With limited efficacy and effectiveness, it falls short compared to the benefits provided by chemical modification. Potential to alter the properties and functionalities of starch, encompassing the characteristics of its hot pastes and gels, as well as their digestibility. Contamination of the starch is a potential concern, particularly when the modified starch is intended for use in food contact or the biomedical industry.	Heat–moisture, annealing, and dry heating	Pandiselvam et al. (2019); Punia (2020)(Garcia et al., 2020; Zhang et al., 2019); (Ojogbo et al., 2021)
Chemical: Involves the treatment of starch in the presence of one or more chemical substances	• Might enhance the resilience of starch granules against mechanical and thermal processing while being pasted and delay retrogradation during storage. • Results in the incorporation of numerous functional groups into the starch polymer, such as acetyl, carboxyl, amide, amine, and hydroxypropyl groups. Cost-effective and user-friendly.	Potential adverse consequences of ingesting chemically modified components. Concerns related to the safety of food, the presence of chemical residues, and the expenses associated with wastewater treatment	Oxidation, esterification, etherification, hydroxypropylation	Ali, Haider, et al. (2024); Garcia et al. (2020); Grgic et al. (2021); Zhang et al. (2019); Gürler et al. (2020)
Enzymatic: Involves subjecting starch to the presence of small quantities of enzymes, resulting in the production of water-soluble or partially hydrolyzed products	• Methods that are regarded as environmentally friendly. • Generates starch with a low absorption rate and glycemic index, making it suitable for diabetics and cardiovascular patients, while also reducing its digestion rate. • Easy procedure, excellent reaction efficiency, and substantial yield.	Enzymes offer a safer and healthier alternative to chemical methods, benefiting both the environment and food consumers.	Dextrin	Pandiselvam et al. (2019); (Garcia et al., 2020); Punia, 2020

Non-thermal: High pressure processing	• Retrogradation of starch is slower compared to the thermal technology • Increases the solubility and swelling of starch • Modification of thermal and pasting properties	Higher pressure levels lead to disruption of the morphology of starch granules Increases starch gelatinization Decreases birefringence of the starch granules		Castro et al. (2020)
Non-thermal: Pulsed electric field	• Modifies the starch with accuracy and reduces processing time • Easy to adjust parameters to suit various sources of starch • Reduces heat-induced degradation while preserving starch's nutritive and sensory characteristics	Some electrical energy input is converted to considerable heat during processing which may affect the quality of the starch Pulsed electric field system that operates at high voltage may raises safety issues, such as the possibility of electrical shocks and possible damage to sensitive constituents of starch The starch may be disrupted during treatment if the electrodes in a chamber become fouled or dirty Protocol needs to be standardized		Aadil et al. (2018). Arshad et al. (2020) Ali, Forsido, et al. (2024)
Non-thermal: Microwave processing	• Changes the structure of starch granules • Improves overall mechanical properties of starch • Low starch retrogradation • Changes the crystalline form of the molecule	Causes a high degree of oxidation in starch paste Decreases the viscosity of starch paste Causes uneven heating and leads to thermal degradation of the material Difficult to maintain proper and uniform temperature		Lewicka et al. (2015)
Non-thermal:Cold plasma	• Has great prospects in 3D printing technology (starch is used as biopolymer) • Can be used to prepare starch gels suitable for 3D printing • All new functional groups and elements introduced to starch by plasma are tested for toxicity to obtain green and clean food condition	Cut-off values that cause dissociation and cross-linking of starch chains are not determined The influence of plasma on the performance of starch-water or starch-other solutions mixed system in various phases is unknown	Microwave discharged, radio frequency plasma, Dielectric barrier discharges, etc.	Ma and Jiang (2024) Rodríguez et al. (2017) Waghmare (2021) Coutinho et al. (2018)

There are numerous studies focused on the impact of ECP on the production of resistant starch (RS). The studies focus on various starches, including, but not limited to, maize, potato, wheat, high-amylose corn, barley, and various cereal grains (Makroo et al., 2021). There are ECP studies on the production of RS in dried bananas and barley, but the yields were reported to be low (Dendegh et al., 2022; Munir et al., 2024). The low-moisture extrusion processing of sorghum flour starch modified in vitro starch digestibility (Bhattarai et al., 2025). Gulzar et al. (2021) reported that resistant starch in extruded brown rice increases with the addition of bean flour and with feed moisture. Neder-Suárez et al. (2020) described how moisture content and temperature during ECP affect RS generation. Since ECP alone produces a low RS yield, it is combined with other processes, such as acid hydrolysis with citric acid. Karma et al. (2022) reported that the combination of extrusion and acid modification increased RS yield from 2.1% to 17.4%, compared with 7.7% for starch modified with acid alone.

In general, RS is usually reduced after ECP. Munir et al. (2024) reported a decrease in RS of extruded banana flour, but it remained higher than those of corn and mango starch (Cabrera-Ramírez et al., 2021). observed the effects of ECP alone and in combination with oleic acid (5% and 10%) on the RS of potato and rice starches. The RS of the isolated rice starch was 15.8 mg/100 mg, but it was reduced to 13.7 mg/100 mg after ECP alone. Furthermore, the RS decreased to 5.56 when ECP was combined with 5% oleic acid and 4.14 mg/100 mg with 10% oleic acid. The RS of native potato starch was 17.51 mg/100 mg, but ECP alone reduced it to 12.41 mg/100 mg. Nevertheless, the addition of oleic acid affected the RS of extruded potato starch, with values ranging from 12.69 to 13.41 mg/100 mg. The decreased RS of extruded rice could be related to the loss of the granular structure noted on the microstructure. On the other hand, the increased RS content of extruded potato starch supplemented with oleic acid may be related to the growth of lamellar crystalline structures arising from the amylose-lipid complex during ECP, resulting in a more thermally stable starch (Neder-Suárez et al., 2020). evaluated the impact of ECP and citric acid on the generation of resistant starch in raw corn starch. The raw corn starch was extruded at constant conditions using different levels (18%–32%) of feed moisture and citric acid levels of 0%–10.3%. The extrudates were stored for 120 h at refrigerated (4 °C) conditions and then at room temperature. There was an increase in resistant starch in extrudates with low moisture content, which could be ascribed to the recrystallization of the liberated starch chains, which favoured retrogradation during low-temperature storage (Karma et al., 2022; Munir et al., 2024). On the other hand, the extrudates had a decreased resistant starch with an increased moisture content. This was likely due to the weak dextrinization of the starch, thereby reducing its retrogradation capacity (Donmez et al., 2021). The added citric acid did not significantly influence the generation of resistant starch formation. This is likely attributed to the greater impact of ECP (temperature and shear stress), which may have resulted in abortive hydrolysis, which failed to break down starch chains.

3.3 Gelatinization and *its impact on starch bioavailability*

The process, known as starch gelatinization, is required to shatter the native starch's crystalline structure and make it hydrolyzable (Liu & Liu, 2020). The granules can absorb water and gelatinize when heated in excess water over the melting point of the crystalline starch structure (>60 °C) (Majzoobi & Farahnaky, 2021). Heat, pressure, microwave heating, and other techniques can all be used to induce starch gelatinization (Liu et al., 2020). Heat-gelatinization can completely disrupt starch granules, resulting in the production of a starch paste and, subsequently, a gel when stored (Wang & Copeland, 2013). Punia (2020) mentioned that the cooking quality of starch is measured by its gelatinization temperature, which is an essential element in food preparation; starches with low gelatinization temperatures have superior cooking quality. Despite its tendency to retrograde, starch is widely used in many food and non-food sectors (Wang et al., 2020).

Starch retrogradation is a recrystallization process that depends on temperature and occurs in three steps: critical nuclei formation (nucleation), crystal growth (propagation), and slow crystal expansion or perfection (maturation) (Ding et al., 2019). Chiravi et al. (2023) noted that retrogradation is most often used to describe the rearrangement of starch's disordered macromolecular components in certain model systems (e.g., amylose or amylopectin) and the subsequent disorganization of the natural order of the starch granule assemblages. Retrogradation is the process by which starch undergoes changes when it is chilled after gelatinization (Zhan et al., 2020). reported that the association of amylose and amylopectin molecules leads to the formation of double helices and the possible formation of crystalline structures.

4. Nutritional profile and health benefits of resistant starches

Unlike most other starches, resistant starches (RS) are similar to dietary fibers because they are not digested in the small intestine. Unlike typical starches, which are generally hydrolyzed into glucose, resistant starch undergoes gut fermentation

TABLE 7.3 Nutritional profile and characteristics of different types of resistant starches.

Resistant starch type	Food sources	Key nutritional components	Fermentation products (SCFAs)	References
RS1 (physically inaccessible starch)	Whole grains (e.g., barley, oats), seeds, legumes	Amylose, cellulose	Butyrate, acetate	Facchin et al. (2024).
RS2 (resistant granular starch)	Raw potatoes, unripe bananas, high-amylose corn	High amylose content, granular structure	Butyrate, propionate, acetate	Zhang et al. (2022).
RS3 (retrograded starch)	Cooked and cooled starchy foods (potatoes, rice, bread)	Retrograded amylose (recrystallized starch)	Butyrate, acetate	Facchin et al. (2024)
RS4 (chemically modified starch)	Processed foods (fiber-enriched baked goods, cereals)	Modified starch (chemically altered structure)	Varies based on modification	Facchin et al. (2024).
RS5 (lipid-complexed starch)	Starch-lipid complexes (processed foods, modified starches)	Complexed starch and lipid molecules	Butyrate, acetate, propionate	Dhital et al. (2017); Zhong et al. (2024).

mostly unaltered. As explained in Table 7.3, this phenomenon occurs due to the presence of beneficial microbiota in the colon. As such, the fermentation of resistant starch is crucial for producing SCFAs such as acetate, propionate, and butyrate, which are key contributors to gut health and metabolism (Facchin et al., 2024).

Improved digestion, better blood glucose control, and prolonged fullness are additional benefits associated with resistant starch fermentation. This makes resistant starch highly important in controlling the diet. Further, the fermentation of resistant starch positively alters gut microbiota composition by increasing Bifidobacteria and Lactobacilli, thereby enhancing immunity (Liu et al., 2020). Examples of foods that contain resistant starch include unripe bananas, legumes, whole grains, certain kinds of corn, and cooked and cooled potatoes (Table 7.3). Processed food products can have their fiber content increased by adding isolated resistant starch (RS), and their taste and texture will not change significantly. RS has positive effects on glycemic control, which is especially beneficial for people with type 2 diabetes and people with insulin resistance (Yu et al., 2020). RS may also help people maintain and manage their weight (Bojarczuk et al., 2022).

Resistant starches are beneficial for metabolic health and can further help mitigate the risk of colorectal cancer. Butyrate, a product of the fermentation of resistant starch, is an important short-chain fatty acid (SCFA) with recognized anti-inflammatory and anticarcinogenic activities in the colon. This ability to confer colon cancer risk reduction and the metabolic disorder benefits is what makes it important to incorporate resistant starch in the diet for individuals with a higher risk of colorectal cancer. The positive effects of resistant starches on health have generated interest in using them as functional ingredients in food products to promote health (Chen et al., 2024).

4.1 Analyzing the nutrient content of resistant starches

Unlike most starches, which are a type of carbohydrate and are considered digestible, resistant starches (RS) function as a carbohydrate and a form of dietary fiber, owing to the fact that they are not digestible in the small intestine and, similarly to fiber, pass through the digestive system unchanged. These starches are made out of polysaccharides, primarily amylose and amylopectin. A distinguishing feature of resistant starches from other types of starch is that the structural composition of RS is linked to the polysaccharide constituent and, more specifically, to its high amylose content. While amylopectin is digestible, amylose is not because of the structure of the glucose unit. Amylose has a linear helical structure made of glucose monomers that encase one another. This structure impedes the action of the digestive enzymes. Conversely, amylopectin, which has a branched structure, is more digestible (Junejo et al., 2022).

Compared with digestible starches, resistant starch has a lower caloric contribution. The digestible starches, on average, produce about 4 kcal/g, while resistant starches produce 2 kcal/g. This lower caloric contribution comes from resistant starch, which is less broken down in the small intestine and more biofermented in the colon (Cronin et al., 2021). A resistant starch contains a polysaccharide constituent, which is the product of a polysaccharide amylose and amylopectin. The polysaccharide content of resistant starch is more closely linked to its resistance to starch-digesting enzymes, such as amylase. The molecular structure and the degree of crystallinity of the starch also influence this resistance (Ding et al., 2019). When resistant starch reaches the colon, it serves as a substrate for gut microbiota, which ferment it, producing short-chain fatty acids (SCFAs) such as acetate, propionate, and butyrate. Butyrate, in particular, is very good for colonic health, as it is a source of energy for colonocytes, improves overall gut health, and decreases inflammation. Butyrate also has several health benefits, including modulation of the immune response, reduced oxidative stress, a lower risk of colorectal cancer, and improved health in inflammatory bowel disease (Liu et al., 2020).

Han et al. (2023) explain how resistant starch can assist with the absorption of specific minerals, particularly magnesium and calcium, which may be because of how the colon's acid, post-fermentative digestion of resistant starch, produces a solution where these minerals may be more soluble and more easily absorbed. Also, Ashaolu et al. (2021) state that resistant starch may help with lipid metabolism and, thus, cardiovascular management, as it may reduce bile acid secretion and blood cholesterol levels. The RS content of food depends on how it is processed and prepared. For example, BeMiller (2020) state that retrogradation, which is a process that increases RS3 in previously processed (gelatinized, cooked, and cooled) starch, is evident in potatoes and rice. Most common foods have RS content that is primarily increased by retrogradation. Also, RS4 is added to many processed foods to increase dietary fiber (Zhang & Bao, 2023).

4.2 Obesity prevention: Strategies for satiety enhancement, fat intake reduction, and gut microbiota influence

The role of resistant starches (RS) in enhancing satiety and reducing subsequent fat absorption helps manage the disease and prevent obesity. In the prevention of obesity, the feeling of fullness after eating (satiety) is important for controlling food intake and preventing subsequent overconsumption. Research indicates that consuming resistant starch (RS) increases the release of the satiety hormones peptide YY (PYY) and glucagon-like peptide-1 (GLP-1), both of which inhibit further food intake (McCarty et al., 2020). These hormones help individuals to consume fewer calories by reducing appetite and increasing the time food remains in the stomach (McCarty et al., 2020).

The steady rise in blood glucose levels and the body's own insulin resistance diminish the possibility of quick blood sugar spikes and falls. This means that energy levels are steady and consistent, and that long-lasting satiety results from the digestion of resistant starch (RS). This effect decreases the chances of snacking between meals (Kumar & Rizvi, 2024). In terms of managing caloric intake, this effect helps control obesity and promote weight loss. Resistance Starch (RS) has an impact on fat metabolism and increases feelings of fullness which is positively correlated with reduced fat storage through the modification of gut microbiota and increasing short-chain fatty acids (SCFAs) levels which include butyrate, acetate, and propionate (Portincasa et al., 2022). More specifically, propionate has been shown to facilitate fat oxidation and inhibit the development of new fat cells (adipogenesis) (Zhan et al., 2020). These alterations to metabolism may be the mechanism to explain the persistent lowering of body fat and why RS may be an important novel treatment for obesity.

In addition, studies have shown that resistant starch decreases the absorption of dietary fats. It has been proposed that this is due to the resistant starch interacting with bile acid in the intestine. Resistant starch is associated with bile acids, which are used for emulsification and absorption of dietary fats, which increases the fecal loss of bile acids. This results in a decline in overall dietary fat absorption. Resistant starch prevents obesity by decreasing fat absorption and increasing fat oxidation, resulting in lower total body fat over time. RS can alter the gut microbiota, which may be linked to changes in body weight. The fermentation of resistant starch by intestinal bacteria yields short-chain fatty acids (SCFAs), which influence fat metabolism and the composition of the gut microbiota (Slavin, 2013). It is documented that resistant starch promotes the proliferation of beneficial strains Bifidobacterium and Lactobacillus, whereas harmful strains linked to obesity and metabolic disorders are diminished. Changes in gut composition have the metabolic effects of improved insulin sensitivity and energy equilibrium, which are protective against obesity.

4.3 Protective effects against colorectal cancer and other metabolic diseases

The recent literature suggests that resistant starches (RS) may also offer protective effects against metabolic disorders and colorectal cancer. The fermentation of resistant starch in the colon results in the formation of short-chain fatty acids (SCFAs) such as butyrate, which have protective effects in colorectal health (Table 7.4). Healthy colon function relies on

TABLE 7.4 Health benefits of resistant starches in disease prevention.

Health benefit	Mechanism of action	Primary types of RS involved	References
Improved glycemic control	Delays carbohydrate digestion and absorption, reducing postprandial blood glucose levels	RS2, RS3	Han et al. (2023); Rashed et al. (2022).
Enhanced insulin sensitivity	Promotes insulin signaling through SCFA production, especially butyrate	RS2, RS3	Liu et al. (2020); Wen et al. (2022).
Obesity prevention (satiety)	Increases PYY and GLP-1 secretion, delaying gastric emptying and reducing appetite	RS3, RS5	Binou et al. (2021); Hughes et al. (2022);
Lipid profile improvement	Lowers LDL cholesterol and triglycerides through SCFA production	RS3, RS5	Wen et al. (2022).
Colorectal cancer protection	Butyrate induces apoptosis in pre-cancerous cells and reduces inflammation in the colon	RS1, RS2	Sánchez-Alcoholado et al. (2020)
Prevention of inflammatory bowel disease (IBD)	SCFAs reduce inflammation and improve gut barrier function	RS2, RS3	Chen et al. (2024).

butyrate, which is the main energy source for colonic enterocyte cells. It helps to nourish and repair the colonic epithelium barrier and enhances gut function (Qu et al., 2023).

In addition to aiding in the prevention of colorectal cancer, resistant starch (RS) helps prevent and manage numerous metabolic disorders, including colorectal cancer, cardiovascular and type 2 diabetes. Resistant starch helps prevent metabolic disorders in many ways, and one of its primary ways is through how it improves lipid metabolism. Resistant starch possesses the ability to positively influence one's lipid profile by decreasing low-density lipoprotein (LDL) cholesterol and minimally affecting (if any) high-density lipoprotein (HDL) cholesterol (Yuan et al., 2018), and a positive way to improve one's lipid profile is by reducing atherosclerosis. Additionally, RS reduces systemic inflammation, which is present in metabolic disorders such as type 2 diabetes and related diseases of the cardiovascular system. RS reduces systemic inflammation (including areas outside the colon) via SCFAs (short-chain fatty acids), which have anti-inflammatory properties and are produced during fermentation of resistant starch(Shahbazi et al., 2021). Since inflammation causes insulin resistance, a major cause of type 2 diabetes, RS helps prevent type 2 diabetes by reducing inflammation and increasing insulin sensitivity, thereby improving blood glucose control.

5. Resistant starches and food quality

Resistant starches (RS) are increasingly used in food products because they support better gut health, increase satiety, and help control blood glucose levels. However, their use in food products also alters sensory characteristics such as texture, flavour, mouthfeel, and other important attributes that affect consumer acceptance (see Table 7.5). Food manufacturers encounter both challenges and opportunities in the distinctive physicochemical characteristics of RS, such as retrogradation, gelatinization, and the formation of resistant granules.

Food quality optimization and an effective product development strategy require an understanding of the impact's food formulation. Resistant starch in food formulations alters product texture, especially in baked goods, dairy products, and extruded foods. Resistant starch modifies food viscosity, hardness, and chewiness, and is primarily responsible for changes in these attributes. Resistant starch enhances the texture of low-fat, high-fiber foods, providing a smooth, creamy mouthfeel that is critical for formulating reduced-calorie versions of foods that are typically high in calories. Resistant starch 2 and Resistant starch 3 have been reported to act as sources of fiber and to contribute to the crispness and firmness of cookies and crackers (Lin, 2022). Conversely, resistant starch in bread can adversely affect dough elasticity, leading to smaller, heavier loaves. This is because RS disrupts the gluten network, which is critical to dough structure and baking performance (Di Cairano et al., 2021). It has been reported that the negative textural effects of resistant starch can be

TABLE 7.5 Influence of resistant starches on texture, flavor, and sensory properties of food products.

Resistant starch type	Food product application	Effect on texture	Effect on flavor	Overall sensory impact	References
Resistant starch 2 (RS2)	Cookies, crackers	Increases crispness, firmness	Neutral	Improves crunch but can reduce moisture retention	Walsh et al. (2022), Rai (2022)
Resistant starch 3 (RS3)	Bread, cakes	Reduces elasticity, increases density	Neutral	Can lead to a drier texture and reduced volume in bread	Wang et al. (2020), Bresciani et al (2021)
RS3/RS5	Low-fat dairy products (yogurt, ice cream)	Mimics fat, providing creaminess and smooth mouthfeel	Neutral	Enhances mouthfeel in reduced-fat products	Zhang and Bao (2023)
RS2/RS3	Beverages, liquid foods	Improves viscosity and body	Neutral	May lead to chalkiness if used in high concentrations	Adjei-Duodu (2015)

alleviated by combining it with other functional components, such as gluten or hydrocolloids, while still retaining the product's other positive attributes (Cervini et al., 2021).

Resistant starch also affects the freeze-thaw stability of foods, an important factor in the texture of frozen foods. Resistant starch has been shown to improve freeze-thaw stability by avoiding the formation of ice crystals and, therefore, helps to maintain the texture of frozen desserts, soups, and sauces after they have been thawed (Joseph et al., 2024, pp. 23–48). Thus, resistant starch aids health and provides functional contributions to the structural integrity and overall quality of a food product. Resistant starch is usually neutral, but the sensory characteristics of foods that contain RS is quite complex, considering RS's concentration and type. RS can add to the sensory quality of the food product or can detract from it. Resistant starch does not add appreciable sweetness; it is a good choice when formulating frozen foods, such as savory snacks or breads that require flavor neutrality. In beverage formulations, resistant starch has been valuable, as it allows an increase in fiber without the associated negative flavors, which is a primary reason it is used in functional beverages and smoothies (Walsh et al., 2022). It is important to ensure that resistant starch is not used in high amounts in order to maintain the mouthfeel of the product, which is critical to the consumers' perceptions.

Products enriched with resistant starch (RS) are in greater demand due to their incorporation into or alignment with specific health trends. As more consumers recognize the importance of gut health and glycemic control, they become more interested in resistant starch products. Customers are more likely to purchase products that highlight being 'high in fiber,' 'low glycemic index,' or 'supporting gut health. In terms of taste, RS is neutral and can be added to almost any food product without altering its flavor. This provides the opportunity to use resistant starch in a variety of products, whether sweet or savory. It is worth noting that excessive resistant starch can lead to a dry mouthfeel, which is especially problematic for baked goods and extruded snacks. In these products, a smooth, crisp texture is highly desired. To mitigate this, resistant starch is often used in combination with gums, fibers, or other starches (Roberfroid, 2007).

5.1 Formulation challenges and solutions in food production

The addition of resistant starches to food formulations is challenging from a technical perspective, especially regarding sensory attributes and structure retention while increasing resistant starch content (Table 7.6). From a formulation perspective, one of the issues is the effect of resistant starch on the dough rheology of bakery products. The addition of RS at higher concentrations negatively affects gluten network formation, thereby influencing dough elasticity and resulting in lower loaf volume in bread. This results in baked products with poor structural quality and density. Food Technologists have created several approaches to overcome these obstacles, including the use of certain enzymes. Amylases, for

TABLE 7.6 Formulation challenges and solutions in food production with resistant starches.

Formulation challenge	Impact on product	Proposed solutions	References
Reduced dough elasticity (bread)	Dense, low-volume bread	Enzyme treatments (amylases), blending with gluten or other starches	Bede & Zaixiang (2021)
Decreased expansion (extruded snacks)	Denser, less crispy snacks	Adjust extrusion parameters (moisture, temperature, screw speed)	Wanikar and Kotwal (2022), Altan and Yağci (2023)
Chalky or dry mouthfeel (beverages)	Unpleasant texture in drinks, especially high resistant starch levels	Microencapsulation, combining resistant starch with gums or fibers	Bojarczuk et al. (2022)
Increased density in baked goods	Heavier, less airy texture	Use of hydrocolloids, optimizing resistant starch concentrations	Wang et al. (2020), Kavya et al. (2024)

example, can partially hydrolyze certain resistant starches to restore dough elasticity, facilitating greater incorporation of resistant starch with little to no sacrifice in bread quality.

Aligning resistant starch-enriched products with current health and wellness trends is vital to consumer acceptance. Consumer interest in functional food products that help achieve health goals, such as managing weight and blood glucose levels, as well as promoting intestinal health, has increased with growing awareness of the health benefits of dietary fiber. The presence of dietary fiber and resistant starch (RS) contributes to a low glycemic index and to health benefits for the digestive system. Resistant starch is a good selection for the desired health outcomes. An increasing number of health-conscious consumers, searching for foods with a functional and nourishing value, have been attracted to products that have high fiber, low glycemic, and digestive-friendly resistant starch (Carlson et al., 2018). Perceived functional food products are preferred and have led to increased innovation in RS applications across different food industry sectors, including dairy, snacks, beverages, and bakery. Resistant starch used: Enriched bread and pasta have become preferred food options for consumer segments that are weight- and blood glucose-conscious and seeking functional foods. They represent a functional food continuum with traditional carbohydrates, with similar mouthfeel and taste. Beyond bakery products, resistant starch has been used in nutrition bars, breakfast cereals, and dairy products.

Increased demand is combined with a build-up of consumer awareness challenges regarding the marketing of resistant starch-enriched products. While consumers in increasing numbers have an appreciation for the diverse roles dietary fiber plays in a balanced diet, an even greater number are unaware of specific health benefits associated with resistant starch, especially as it helps in the reduction of glycemic response to carbohydrates, as well as supporting overall health in terms of improved function of the gut. To build consumer acceptance of resistant starch, educational marketing campaigns should focus on its specific advantages. Labeling resistant starch-enriched products with resistant starch certifications (fiber, glycemic index) allows consumers to make informed choices.

6. Conclusion

Resistant Starch offers an innovative, potentially beneficial dietary solution by modifying products to promote lower digestibility, encourage fat reduction, manage glycemic response, and support health improvement in multiple ways. The effects of resistant starch on lipid metabolism and the lipid profile are crucial in reducing the risk of metabolic disorders, cardiovascular disease and type II diabetes. The fermentation of resistant starch producing short-chain fatty acids works in reducing systemic inflammation and improving insulin sensitivity, reducing the risk of diabetes and cardiovascular disease. This is an important and beneficial effect of resistant starch. Moreover, incorporating resistant starches into food formulations affects positive health outcomes as well as impacts on some positive sensory characteristics, including texture and mouthfeel. Food manufacturers face a primary dilemma: how to balance food quality with the functional benefits of resistant starch in food products. These formulations allow them to provide the market with healthier food products without compromising consumer preferences. To summarize, incorporating resistant starches into dietary routines can be viewed largely as a beneficial, health-promoting practice, which reiterates the significance of nutrient-dense food components in nutritional science and food technology.

References

Aadil, R. M., Zeng, X. A., Han, Z., Sahar, A., Khalil, A. A., Rahman, U. U., Khan, M., & Mehmood, T. (2018). Combined effects of pulsed electric field and ultrasound on bioactive compounds and microbial quality of grapefruit juice. *Journal of Food Processing and Preservation, 42*(2). https://doi.org/10.1111/jfpp.13507

Adjei-Duodu, T. (2015). *Physical, chemical and functional properties of tiger nuts (Cyperus esculentus) selected from Ghana, Cameroon and UK market (Spain)*.

Ali, I. M., Forsido, S. F., Kuyu, C. G., Ahmed, E. H., Andersa, K. N., Chane, K. T., & Regasa, T. K. (2024). Effects of extrusion process conditions on nutritional, anti-nutritional, physical, functional, and sensory properties of extruded snack: A review. *Food Science and Nutrition, 12*(11), 8755–8761. https://doi.org/10.1002/fsn3.4472

Ali, T. M., Haider, S., Shaikh, M., Butt, N. A., & Zehra, N. (2024). Chemical crosslinking, acid hydrolysis, oxidation, esterification, and etherification of starch. In *Advanced research in starch* (pp. 47–94). Singapore: Springer Nature Singapore.

Almeida, R. L. J., Monteiro, S. S., Santos, N. C., Rios, N. S., & dos Santos, E. S. (2024). Starch modification and its application in Pickering emulsion stabilization: A review. *Journal of Food Measurement and Characterization, 18*(6), 4984–5003.

Arshad, R. N., Buntat, Z. B., Dastgheib, A. M., Jusoh, Y. M. M., Munir, A., Aadil, R. M., & Ahmad, M. H. (2020). Continuous flow treatment chamber for liquid food processing through pulsed electric field. *Journal of Computational and Theoretical Nanoscience, 17*(2), 1492–1498. https://doi.org/10.1166/jctn.2020.8829

Ashaolu, T. J., Ashaolu, J. O., & Adeyeye, S. A. (2021). Fermentation of prebiotics by human colonic microbiota in vitro and short-chain fatty acids production: a critical review. *Journal of Applied Microbiology, 130*(3), 677–687.

Altan, A., & Yağci, S. (2023). Physicochemical characteristics and structural changes of fermented faba bean extrudates prepared by twin-screw extrusion. *Food Chemistry, 411, 135502*.

Bede, D., & Zaixiang, L. (2021). Recent developments in resistant starch as a functional food. *Starch - Stärke, 73*(3–4). https://doi.org/10.1002/star.202000139

BeMiller, J. N. (2020). Resistant starch. In *Science and technology of fibers in food systems* (pp. 153–183). Cham: Springer International Publishing.

Bresciani, A., Pagani, M. A., & Marti, A. (2021). Rice: a versatile food at the heart of the mediterranean diet. In *Cereal-based foodstuffs: the backbone of mediterranean cuisine, 12*(6), 193–229.

Bhattarai, R. R., Jayasree Joshi, T., Sruthi, N. U., Rao, P. S., & Johnson, S. (2025). Effects of extrusion cooking on nutritional and health attributes of sorghum and millets: special reference to protein and starch digestibility. *International Journal of Food Science and Technology, 60*(1), vvae093.

Binou, P., Yanni, A. E., Stergiou, A., Karavasilis, K., Konstantopoulos, P., Perrea, D., Tentolouris, N., & Karathanos, V. T. (2021). Enrichment of bread with beta-glucans or resistant starch induces similar glucose, insulin and appetite hormone responses in healthy adults. *European Journal of Nutrition, 60*(1), 455–464. https://doi.org/10.1007/s00394-020-02265-6

Bojarczuk, A., Skąpska, S., Mousavi Khaneghah, A., & Marszałek, K. (2022). Health benefits of resistant starch: A review of the literature. *Journal of Functional Foods, 93*. https://doi.org/10.1016/j.jff.2022.105094

Cabrera-Ramírez, A. H., Cervantes-Ramírez, E., Morales-Sánchez, E., Rodriguez-García, M. E., Reyes-Vega, M.d.l. L., & Gaytán-Martínez, M. (2021). Effect of extrusion on the crystalline structure of starch during RS5 formation. *Polysaccharides, 2*(1), 187–201. https://doi.org/10.3390/polysaccharides2010013

Carlson, J. L., Erickson, J. M., Lloyd, B. B., & Slavin, J. L. (2018). Health effects and sources of prebiotic dietary fiber. *Current Developments in Nutrition, 2*(3). https://doi.org/10.1093/cdn/nzy005

Carvalho, G. P., Costa-Camilo, E., & Duarte, I. (2024). Advancing health and sustainability: A holistic approach to food production and dietary habits. *Foods, 13*(23), 3829.

Castro, L. M. G., Alexandre, E. M. C., Saraiva, J. A., & Pintado, M. (2020). Impact of high pressure on starch properties: A review. *Food Hydrocolloids, 106*, Article 105877. https://doi.org/10.1016/j.foodhyd.2020.105877

Cervini, M., Frustace, A., Garrido, G. D., Rocchetti, G., & Giuberti, G. (2021). Nutritional, physical and sensory characteristics of gluten-free biscuits incorporated with a novel resistant starch ingredient. *Heliyon, 7*(3). https://doi.org/10.1016/j.heliyon.2021.e06562

Chen, L., Dai, Y., Hou, H., Wang, W., Ding, X., Zhang, H., Li, X., & Dong, H. (2021). Effect of high pressure microfluidization on the morphology, structure and rheology of sweet potato starch. *Food Hydrocolloids, 115*, Article 106606. https://doi.org/10.1016/j.foodhyd.2021.106606

Chen, Z., Liang, N., Zhang, H., Li, H., Guo, J., Zhang, Y., Chen, Y., Wang, Y., & Shi, N. (2024). Resistant starch and the gut microbiome: Exploring beneficial interactions and dietary impacts. *Food Chemistry X, 21*, Article 101118. https://doi.org/10.1016/j.fochx.2024.101118

Chiravi, K., Khan, M. A., Wadikar, D. D., Mahesh, C., & Semwal, A. D. (2023). Starch retrogradation: Role of food ingredients and extrinsic factors. *Food Science and Engineering*, 204–215.

Coutinho, N. M., Silveira, M. R., Rocha, R. S., Moraes, J., Ferreira, M. V. S., Pimentel, T. C., Freitas, M. Q., Silva, M. C., Raices, R. S. L., Ranadheera, C. S., Borges, F. O., Mathias, S. P., Fernandes, F. A. N., Rodrigues, S., & Cruz, A. G. (2018). Cold plasma processing of milk and dairy products. *Trends in Food Science and Technology, 74*, 56–68. https://doi.org/10.1016/j.tifs.2018.02.008

Cronin, P., Joyce, S. A., Otoole, P. W., & Oconnor, E. M. (2021). Dietary fibre modulates the gut microbiota. *Nutrients, 13*(5). https://doi.org/10.3390/nu13051655

Demartino, P., Johnston, E. A., Petersen, K. S., Kris-Etherton, P. M., & Cockburn, D. W. (2022). Additional resistant starch from one potato side dish per day alters the gut microbiota but not fecal short-chain fatty acid concentrations. *Nutrients, 14*(3). https://doi.org/10.3390/nu14030721

Dendegh, T. A., Enefola, O. S., Akpapunam, S. O., Yelmi, B. M., & Abdullahi, M. J. B. (2022). Extrusion technology and its application in food processing-an overview. *Tropical Journal of Engineering, Science and Technology, 1*(2), 25–53.

Dhital, S., Warren, F. J., Butterworth, P. J., Ellis, P. R., & Gidley, M. J. (2017). Mechanisms of starch digestion by α-amylase—Structural basis for kinetic properties. *Critical Reviews in Food Science and Nutrition, 57*(5), 875–892. https://doi.org/10.1080/10408398.2014.922043

Di Cairano, M., Caruso, M. C., Galgano, F., Favati, F., Ekere, N., & Tchuenbou-Magaia, F. (2021). Effect of sucrose replacement and resistant starch addition on textural properties of gluten-free doughs and biscuits. *European Food Research and Technology, 247*(3), 707–718. https://doi.org/10.1007/s00217-020-03659-w

Ding, L., Zhang, B., Tan, C. P., Fu, X., & Huang, Q. (2019). Effects of limited moisture content and storing temperature on retrogradation of rice starch. *International Journal of Biological Macromolecules, 137*, 1068–1075. https://doi.org/10.1016/j.ijbiomac.2019.06.226

Donmez, D., Pinho, L., Patel, B., Desam, P., & Campanella, O. H. (2021). Characterization of starch–water interactions and their effects on two key functional properties: Starch gelatinization and retrogradation. *Current Opinion in Food Science, 39*, 103–109.

Eslam, M., El-Serag, H. B., Francque, S., Sarin, S. K., Wei, L., Bugianesi, E., & George, J. (2022). Metabolic (Dysfunction)-associated fatty liver disease in individuals of normal weight. *Nature Reviews Gastroenterology & Hepatology, 19*(10), 638–651. https://doi.org/10.1038/s41575-022-00635-5

Facchin, S., Bertin, L., Bonazzi, E., Lorenzon, G., De Barba, C., Barberio, B., Zingone, F., Maniero, D., Scarpa, M., Ruffolo, C., Angriman, I., & Savarino, E. V. (2024). Short-chain fatty acids and human health: From metabolic pathways to current therapeutic implications. *Life, 14*(5), 559. https://doi.org/10.3390/life14050559

Feng, H., Cheng, B., Lim, J., Li, B., Li, C., & Zhang, X. (2024). Advancements in enhancing resistant starch type 3 (RS3) content in starchy food and its impact on gut microbiota: A review. *Comprehensive Reviews in Food Science and Food Safety, 23*(3). https://doi.org/10.1111/1541-4337.13355

Garcia, M. A. V. T., Garcia, C. F., & Faraco, A. A. G. (2020). Pharmaceutical and biomedical applications of native and modified starch: A review. *Starch-Stärke, 72*(7-8), 1900270.

Gidley, M. J. (2024). Grain factors in food and feed that influence digestion and gut fermentation outcomes. *Cereal Chemistry, 101*(2), 288–298.

Grgic, I., Grec, M., Gryszkin, A., Zieba, T., Kopjar, M., Ackar, D., … Babic, J. (2021). Starches modified by combination of phosphorylation and high-voltage electrical discharge (HVED) treatment. *Polish Journal of Food and Nutrition Sciences, 71*(1).

Gromova, L. V., Fetissov, S. O., & Gruzdkov, A. A. (2021). Mechanisms of glucose absorption in the small intestine in health and metabolic diseases and their role in appetite regulation. *Nutrients, 13*(7), 2474.

Gulzar, B., Hussain, S. Z., Naseer, B., & Naik, H. R. (2021). Enhancement of resistant starch content in modified rice flour using extrusion technology. *Cereal Chemistry, 98*(3), 634–641.

Gupta, R. K., Guha, P., & Srivastav, P. P. (2024). Exploring the potential of taro (*Colocasia esculenta*) starch: Recent developments in modification, health benefits, and food industry applications. *Food Bioengineering, 3*(3), 365–379.

Gürler, N., Paşa, S., Hakkı Alma, M., & Temel, H. (2020). The fabrication of bilayer polylactic acid films from cross-linked starch as eco-friendly biodegradable materials: Synthesis, characterization, mechanical and physical properties. *European Polymer Journal, 127*, Article 109588. https://doi.org/10.1016/j.eurpolymj.2020.109588

Han, J., Wu, J., Liu, X., Shi, J., & Xu, J. (2023). Physiological effects of resistant starch and its applications in food: A review. *Food Production, Processing and Nutrition, 5*(1). https://doi.org/10.1186/s43014-023-00156-x

Hughes, R. L., Horn, W. F., Wen, A., Rust, B., Woodhouse, L. R., Newman, J. W., & Keim, N. L. (2022). Resistant starch wheat increases PYY and decreases GIP but has no effect on self-reported perceptions of satiety. *Appetite, 168*. https://doi.org/10.1016/j.appet.2021.105802

Joseph, T. M., Sathian, A., Joshy, K. S., Kar Mahapatra, D., Haponiuk, J. T., & Thomas, S. (2024). *Modifications of starch and its characterizations handbook of natural polymers, volume 2: Functionalization, surface modification, and properties*. Poland: Elsevier. https://doi.org/10.1016/B978-0-323-99856-7.00024-0

Junejo, S. A., Flanagan, B. M., Zhang, B., & Dhital, S. (2022). Starch structure and nutritional functionality – Past revelations and future prospects. *Carbohydrate Polymers, 277*. https://doi.org/10.1016/j.carbpol.2021.118837

Karma, V., Gupta, A. D., Yadav, D. K., Singh, A. A., Verma, M., & Singh, H. (2022). Recent developments in starch modification by organic acids: A review. *Starch-Stärke, 74*(9-10), 2200025.

Kumar, R., & Rizvi, M. R. (2024). Beyond a meal: The in-depth analysis of breakfast and its impact on obesity prevention- a comprehensive review. *JK Practitioner, 29*(1), 1–12.

Kavya, M., Krishnan, R., Suvachan, A., Sathyan, S., Tozuka, Y., Kadota, K., & Nisha, P. (2024). The art and science of porous starch: understanding the preparation method and structure–function relationship. *Critical Reviews in Food Science and Nutrition*, 1–18.

Lai, J., Li, Q., He, Y., Zou, S., Bai, X., & Rastogi, S. (2022). Glycemic control regimens in the prevention of surgical site infections: a meta-analysis of randomized clinical trials. *Frontiers in Surgery, 9*, 855409.

Lewicka, K., Siemion, P., & Kurcok, P. (2015). Chemical modifications of starch: Microwave effect. *International Journal of Polymer Science, 2015*, 1–10. https://doi.org/10.1155/2015/867697

Li, L., Chen, H., Wang, M., Lv, X., Zhao, Y., & Xia, L. (2018). Development and characterization of irradiated-corn-starch films. *Carbohydrate Polymers, 194*, 395–400. https://doi.org/10.1016/j.carbpol.2018.04.060

Lin, S. (2022). Dietary fiber in bakery products: Source, processing, and function. *Advances in Food and Nutrition Research, 99*, 37–100. https://doi.org/10.1016/bs.afnr.2021.12.001

Liu, K., & Liu, Q. (2020). Enzymatic determination of total starch and degree of starch gelatinization in various products. *Food Hydrocolloids, 103*, Article 105639. https://doi.org/10.1016/j.foodhyd.2019.105639

Liu, H., Zhang, M., Ma, Q., Tian, B., Nie, C., Chen, Z., & Li, J. (2020). Health beneficial effects of resistant starch on diabetes and obesity via regulation of gut microbiota: A review. *Food & Function, 11*(7), 5749–5767.

Ma, S., & Jiang, H. (2024). The effect of cold plasma on starch: Structure and performance. *Carbohydrate Polymers, 340*, Article 122254. https://doi.org/10.1016/j.carbpol.2024.122254

Ma, H. K., Shen, L., Zhao, L. M., & Ji, H. F. (2025). Regulation of resistant starch, non-starch polysaccharides, resistant oligosaccharides and lignin on the gut microbiota and association with their health benefits. *Food & Function, 18*. https://doi.org/10.1039/d5fo02694f

Majzoobi, M., & Farahnaky, A. (2021). Granular cold-water swelling starch; properties, preparation and applications, a review. *Food Hydrocolloids, 111*, Article 106393. https://doi.org/10.1016/j.foodhyd.2020.106393

Makroo, H. A., Naqash, S., Saxena, J., Sharma, S., Majid, D., & Dar, B. N. (2021). Recovery and characteristics of starches from unconventional sources and their potential applications: A review. *Applied Food Research, 1*(1), 100001.

Maniglia, B. C., Pataro, G., Ferrari, G., Augusto, P. E. D., Le-Bail, P., & Le-Bail, A. (2021). Pulsed electric fields (PEF) treatment to enhance starch 3D printing application: Effect on structure, properties, and functionality of wheat and cassava starches. *Innovative Food Science and Emerging Technologies, 68*. https://doi.org/10.1016/j.ifset.2021.102602

McCarty, T. R., Jirapinyo, P., & Thompson, C. C. (2020). Effect of sleeve gastrectomy on ghrelin, GLP-1, PYY, and GIP gut hormones a systematic review and meta-analysis. *Annals of Surgery, 272*(1), 72–80. https://doi.org/10.1097/SLA.0000000000003614

Miao, W.-B., Ning, Y.-Y., Huang, H.-R., Liu, H.-M., Cai, X.-S., & Wang, X.-D. (2021). Effect of dry heat modification and the addition of Chinese quince seed gum on the physicochemical properties and structure of tigernut tuber starch. *Arabian Journal of Chemistry, 14*(11), Article 103407. https://doi.org/10.1016/j.arabjc.2021.103407

Milanezzi, G. C., & Silva, E. K. (2025). Pulsed electric field-induced starch modification for food industry applications: A review of native to modified starches. *Carbohydrate Polymers, 348*. https://doi.org/10.1016/j.carbpol.2024.122793

Munir, H., Alam, H., Nadeem, M. T., Almalki, R. S., Arshad, M. S., & Suleria, H. A. R. (2024). Green banana resistant starch: A promising potential as functional ingredient against certain maladies. *Food Science and Nutrition, 12*(6), 3787–3805.

Neder-Suárez, D., Amaya-Guerra, C. A., Pérez-Carrillo, E., Quintero-Ramos, A., Mendez-Zamora, G., Sánchez-Madrigal, M.Á., Barba-Dávila, B. A., & Lardizábal-Gutiérrez, D. (2020). Optimization of an extrusion cooking process to increase formation of resistant starch from corn starch with addition of citric acid. *Starch/Staerke, 72*(3–4). https://doi.org/10.1002/star.201900150

Ojogbo, E., Jardin, J., & Mekonnen, T. H. (2021). Robust and sustainable starch ester nanocomposite films for packaging applications. *Industrial Crops and Products, 160*, 113153.

Pandiselvam, R., Manikantan, M. R., Divya, V., Ashokkumar, C., Kaavya, R., Kothakota, A., & Ramesh, S. V. (2019). Ozone: An advanced oxidation technology for starch modification. *Ozone Science & Engineering, 41*(6), 491–507. https://doi.org/10.1080/01919512.2019.1577128

Panesar, P. S., & Anal, A. K. (2022). *Starch and its derivatives as potential source of prebiotics probiotics, prebiotics and synbiotics: Technological advancements towards safety and industrial applications*. Wiley. https://doi.org/10.1002/9781119702160.ch16

Portincasa, P., Bonfrate, L., Vacca, M., De Angelis, M., Farella, I., Lanza, E., Khalil, M., Wang, D. Q. H., Sperandio, M., & Di Ciaula, A. (2022). Gut microbiota and short chain fatty acids: Implications in glucose homeostasis. *International Journal of Molecular Sciences, 23*(3). https://doi.org/10.3390/ijms23031105

Punia, S. (2020). Barley starch modifications: Physical, chemical and enzymatic - A review. *International Journal of Biological Macromolecules, 144*, 578–585. https://doi.org/10.1016/j.ijbiomac.2019.12.088

Qu, R., Zhang, Y., Ma, Y., Zhou, X., Sun, L., Jiang, C., Zhang, Z., & Fu, W. (2023). Role of the gut microbiota and its metabolites in tumorigenesis or development of colorectal cancer. *Advanced Science, 10*(23). https://doi.org/10.1002/advs.202205563

Rashed, A. A., Saparuddin, F., Rathi, D. N. G., Nasir, N. N. M., & Lokman, E. F. (2022). Effects of resistant starch interventions on metabolic biomarkers in pre-diabetes and diabetes adults. *Frontiers in Nutrition, 8*. https://doi.org/10.3389/fnut.2021.793414

Rai, V. (2022). *Preparation and shelf-life evaluation of Flaxseed incorporated biscuits* (Doctoral dissertation, Department of Nutrition & Dietetics Central Campus of Technology Institute of Science and Technology Tribhuvan University, Nepal. 2021).

Roberfroid, M. (2007). Prebiotics: The concept revisited1. *The Journal of Nutrition, 137*(3), 830S. https://doi.org/10.1093/jn/137.3.830s

Rodríguez, Ó., Gomes, W. F., Rodrigues, S., & Fernandes, F. A. N. (2017). Effect of indirect cold plasma treatment on cashew Apple juice (Anacardium occidentale L.). *LWT, 84*, 457–463. https://doi.org/10.1016/j.lwt.2017.06.010

Sánchez-Alcoholado, L., Ramos-Molina, B., Otero, A., Laborda-Illanes, A., Ordóñez, R., Medina, J. A., Gómez-Millán, J., & Queipo-Ortuño, M. I. (2020). The role of the gut microbiome in colorectal cancer development and therapy response. *Cancers, 12*(6). https://doi.org/10.3390/cancers12061406

Shahbazi, R., Sharifzad, F., Rana, B., Alsadi, N., Yasavoli-Sharahi, H., & Matar, C. (2021). Anti-inflammatory and immunomodulatory properties of fermented plant foods. *Nutrients, 13*(5), 1516. https://doi.org/10.3390/nu13051516

Slavin, J. (2013). Fiber and prebiotics: Mechanisms and health benefits. *Nutrients, 5*(4), 1417–1435. https://doi.org/10.3390/nu5041417

Venkataraman, A., Sieber, J. R., Schmidt, A. W., Waldron, C., Theis, K. R., & Schmidt, T. M. (2016). Variable responses of human microbiomes to dietary supplementation with resistant starch. *Microbiome, 4*(1). https://doi.org/10.1186/s40168-016-0178-x

Waghmare, R. (2021). Cold plasma technology for fruit based beverages: A review. *Trends in Food Science & Technology, 114*, 60–69. https://doi.org/10.1016/j.tifs.2021.05.018

Walsh, S. K., Lucey, A., Walter, J., Zannini, E., & Arendt, E. K. (2022). Resistant starch—An accessible fiber ingredient acceptable to the Western palate. *Comprehensive Reviews in Food Science and Food Safety, 21*(3), 2930–2955. https://doi.org/10.1111/1541-4337.12955

Wang, S., & Copeland, L. (2013). Molecular disassembly of starch granules during gelatinization and its effect on starch digestibility: A review. *Food & Function, 4*(11), 1564–1580. https://doi.org/10.1039/c3fo60258c

Wang, Q., & Ismail, B. (2012). Effect of Maillard-induced glycosylation on the nutritional quality, solubility, thermal stability and molecular configuration of whey protein. *International Dairy Journal, 25*(2), 112–122.

Wang, Z., Wang, S., Xu, Q., Qi, K., Li, F., Lin, L., Xu, Y., & Wei, Y. (2023). Synthesis and functions of resistant starch. *Advances in Nutrition, 14*(5), 1131–1144. https://doi.org/10.1016/j.advnut.2023.06.001

Wang, H., Yang, Q., Ferdinand, U., Gong, X., Qu, Y., Gao, W., Ivanistau, A., Feng, B., & Liu, M. (2020). Isolation and characterization of starch from light yellow, Orange, and purple sweet potatoes. *International Journal of Biological Macromolecules, 160*, 660–668. https://doi.org/10.1016/j.ijbiomac.2020.05.259

Wanikar, R., & Kotwal, S. (2022). Resistant starch: A promising functional food ingredient. In *Evolution and Recent Advances: Starch*, 113.

Wen, J. J., Li, M. Z., Hu, J. L., Tan, H. Z., & Nie, S. P. (2022). Resistant starches and gut microbiota. *Food Chemistry, 387.* https://doi.org/10.1016/j.foodchem.2022.132895

Wu, J., Xu, S., Huang, Y., Zhang, X., Liu, Y., Wang, H., … Liu, C. (2022). Prevents kudzu starch from agglomeration during rapid pasting with hot water by a non-destructive superheated steam treatment. *Food Chemistry, 386*, 132819.

Xiang, J., Li, W., Ndolo, V. U., & Beta, T. (2019). A comparative study of the phenolic compounds and in vitro antioxidant capacity of finger millets from different growing regions in Malawi. *Journal of Cereal Science, 87*, 143–149. https://doi.org/10.1016/j.jcs.2019.03.016

Yu, Z., Nan, F., Wang, L. Y., Jiang, H., Chen, W., & Jiang, Y. (2020). Effects of high-protein diet on glycemic control, insulin resistance and blood pressure in type 2 diabetes: A systematic review and meta-analysis of randomized controlled trials. *Clinical Nutrition, 39*(6), 1724–1734.

Yuan, H. C., Meng, Y., Bai, H., Shen, D. Q., Wan, B. C., & Chen, L. Y. (2018). Meta-analysis indicates that resistant starch lowers serum total cholesterol and low-density cholesterol. *Nutrition Research, 54*, 1–11. https://doi.org/10.1016/j.nutres.2018.02.008

Zhan, Q., Ye, X., Zhang, Y., Kong, X., Bao, J., Corke, H., & Sui, Z. (2020). Starch granule-associated proteins affect the physicochemical properties of rice starch. *Food Hydrocolloids, 101*, Article 105504. https://doi.org/10.1016/j.foodhyd.2019.105504

Zhang, Z., & Bao, J. (2023). Recent advances in modification approaches, health benefits, and food applications of resistant starch. *Starch - Stärke, 75* (9–10). https://doi.org/10.1002/star.202100141

Zhang, N., Jin, M., Wang, K., Zhang, Z., Shah, N. P., & Wei, H. (2022). Functional oligosaccharide fermentation in the gut: Improving intestinal health and its determinant factors-a review. *Carbohydrate Polymers, 284.* https://doi.org/10.1016/j.carbpol.2021.119043

Zhang, H., Wang, R., Chen, Z., & Zhong, Q. (2019). Enzymatically modified starch with low digestibility produced from amylopectin by sequential amylosucrase and pullulanase treatments. *Food Hydrocolloids, 95*, 195–202.

Zheng, L., Ren, A., Liu, R., Xing, Y., Yu, X., & Jiang, H. (2022). Effect of sodium chloride solution on quality of 3D-printed samples molded using wheat starch gel. *Food Hydrocolloids, 123.* https://doi.org/10.1016/j.foodhyd.2021.107197

Zhong, H., She, Y., Yang, X., Wen, Q., Chen, L., Wang, X. B., & Chen, Z. (2024). Analysis of the mechanism of resistance to enzymatic hydrolysis of RS-5 resistant starch. *Food Chemistry, 452*, Article 139570. https://doi.org/10.1016/j.foodchem.2024.139570

Chapter 8

Modified starches as promising food biopolymers in nutraceuticals applications

R. Rajam[1], P. Subramanian[2], P. Thivya[3] and Pothiyappan Karthik[4]

[1]*Department of Food Technology, Kalasalingam Academy of Research and Education, Virudhunagar, Tamil Nadu, India;* [2]*Riddet Institute, Massey University, Palmerston North, New Zealand;* [3]*Department of Food Process Engineering, National Institute of Technology (NIT), Rourkela, Odisha, India;* [4]*Department of Food Technology, Faculty of Engineering, Karpagam Academy of Higher Education, Coimbatore, Tamil Nadu, India*

1. Introduction

Consumer awareness of human health and nutrition is steadily increasing, driven by the rising prevalence of chronic diseases. As a result, consumers are increasingly favoring foods that not only fulfill nutritional requirements but also offer long-term health benefits. This shift in dietary preference has accelerated the demand for nutraceuticals and functional foods, which are recognized for their potential to prevent, manage, or reduce the risk of chronic diseases. The term, nutraceuticals, was introduced by De Felice in 1995 to describe dietary substances or food-derived bioactive components that confer medicinal or health benefits, including prevention and treatment of various diseases (DeFelice, 1995). Nutraceuticals encompass a wide range of bioactive components including polyphenols, omega-3 fatty acids, carotenoids, and terpenoids, which have been associated with the reduced the risk of chronic diseases such as cardiovascular disorders, cancer, obesity, and neurodegenerative conditions (Das et al., 2011; Shahidi & Ambigaipalan, 2015). Despite their significant health potential, the clinical translation and functional impact of nutraceuticals are often limited due to low solubility in water, limited bioavailability, chemical instability, interaction with food components during processing and storage, low gastrointestinal permeability, and rapid degradation in the digestive tract (Dima et al., 2020; Dissanayake et al., 2022). Thus, effective delivery systems are crucial to enhance the bioavailability, targeted delivery, and controlled release of these compounds within the human body.

Currently, encapsulated delivery systems are extensively explored as a promising strategy to address these challenges with natural biopolymer-based carriers being among the most widely used delivery vehicles. Among the various biopolymers explored for encapsulation and delivery, starch, owing to its abundance, biodegradability, biocompatibility, and food-grade safety, emerges as a highly promising agent (Guo et al., 2021). However, native starch in its unmodified form presents several limitations such as inadequate emulsifying capacity, rapid retrogradation, and susceptibility to pH fluctuations and enzymatic hydrolysis, which restrict its application in complex food systems and nutraceutical delivery (Guo et al., 2021; Ribeiro et al., 2024). Therefore, a significant research effort has focused on modifying starch to improve its physicochemical properties, enabling it to serve as an efficient encapsulation and delivery matrix for bioactive compounds.

Modified starches are chemically, physically, or enzymatically tailored derivatives of native starch with enhanced functionalities such as increased solubility, viscosity, thermal stability, and film-forming capacity (Zia-ud-Din et al., 2017). Based on the modification technique such as acetylation, hydroxypropylation, cross-linking, oxidation, or enzymatic debranching, starches can develop attributes suitable for constructing micro- and nano-encapsulation systems designed to deliver hydrophilic and lipophilic nutraceuticals (Dai et al., 2023; Sedaghat Doost et al., 2020; Zhi et al., 2021). For example, acetylated starch offers improved flexibility and reduced retrogradation, whereas cross-linked starch

Native and Modified Starches for Food Applications. https://doi.org/10.1016/B978-0-443-27431-2.00008-0

enhances resistance to acidic and thermal degradation, making it ideal for controlled delivery applications (Colussi et al., 2017).

With advances in encapsulation technologies such as spray drying, electrospinning, emulsification, coacervation, and nanoprecipitation, modified starches have been effectively employed to encapsulate a wide array of nutraceuticals including curcumin, resveratrol, omega-3 fatty acids, vitamins A and E (Reque & Brandelli, 2021; Viji et al., 2023; Yao & Janaswamy, 2022). These systems not only protect the bioactive ingredients from oxidation, heat, and light but also enable controlled or targeted release at specific gastrointestinal locations, thereby improving their therapeutic efficacy. Starch-based nanocarriers, have demonstrated significant potential in enhancing the oral bioavailability of hydrophobic nutraceuticals by increasing surface area, improving water solubility, and facilitating absorption through intestinal epithelial cells (Sedaghat Doost et al., 2020). Starch nanocrystals and nanoparticles derived from modified starches have been utilized as core-shell carriers and Pickering emulsion stabilizers for efficient nutraceutical delivery. The development of starch-lipid hybrid systems further diversifies the delivery landscape by enabling synergistic release profiles and better mucoadhesive properties (Sun et al., 2025; Zhao et al., 2024).

An additional advantage of modified starches is their compatibility with gastrointestinal enzymes and microbial fermentation pathways, ensuring that nutraceuticals encapsulated within them can be safely metabolized and absorbed. Moreover, the regulatory approval of most starch derivatives as GRAS (Generally Recognized as Safe) substances by food safety authorities worldwide facilitates their translation into commercial food and nutraceutical formulations. In addition to their functional versatility, modified starches meet the growing consumer demand for plant-based, biodegradable delivery systems, making them ideal for incorporation into functional foods, dietary supplements, and fortified beverages (Guo et al., 2021; Montoya-Yepes et al., 2024). Despite the progress, challenges persist in optimizing the molecular architecture and physicochemical interactions between modified starch and the encapsulated nutraceutical to ensure consistent release kinetics and stability. The choice of modification method must align with the specific bioactive's chemical nature, solubility, and target release site. Additionally, understanding the gastrointestinal fate of modified starch carriers, including enzymatic breakdown, mucosal interactions, and microbiota-mediated fermentation, is essential for designing next-generation starch-based delivery vehicles (Montoya-Yepes et al., 2024).

This chapter provides a comprehensive overview of modified starches as promising food-grade biopolymers in the field of nutraceuticals. It explores the major types of starch modifications and their impact on physicochemical and functional properties relevant to delivery systems, advanced encapsulation techniques employing modified starches, encapsulation efficiency, release behavior and bioavailability across different modified starch systems. Also, this chapter highlights the emerging trends, such as smart starch-based delivery systems, synergistic biopolymer combinations, potential applications in functional food design and therapeutic nutrition.

2. Nutraceuticals and bioavailability

2.1 Oral bioavailability

Bioavailability is the key concept in assessing the therapeutic potential of nutraceuticals, which are bioactive compounds sourced from food that provide physiological benefits beyond basic nutrition. Oral bioavailability of nutraceutical is referred as the fraction of an administered compound that reaches systemic circulation in an active form and exerts its intended biological function (Yao et al., 2015). In the development of functional foods and nutraceutical products, ensuring high bioavailability is essential for delivering consistent and effective health benefits. In the context of nutraceuticals, bioavailability should not be limited to just absorption or metabolism; it involves all stages from ingestion to excretion. It includes release from the food matrix, solubilization, absorption in the gastrointestinal tract, metabolism, and tissue distribution. Thus, bio-accessibility, absorption and transformation are the three main stages of bioavailability (Dima et al., 2020). Bioaccessibility represents the initial phase and refers to the proportion of ingested nutraceuticals that is released from the food matrix and becomes available for absorption across the intestinal epithelium (Fig. 8.1). This requires the nutraceutical compound to be released from the food matrix and solubilized in intestinal fluids. Absorption is the second step, mainly occurring in the small intestine through the epithelial membranes by passive or active transport mechanism (Jafari & McClements, 2017). Transformation during digestion and liver metabolism is another factor limiting bioavailability.

2.2 Factors influencing bioavailability

The bioavailability of nutraceuticals is governed by their physicochemical characteristics, the surrounding food matrix and physiological barriers within the human body. Factors such as solubility, stability, molecular size, and interaction with

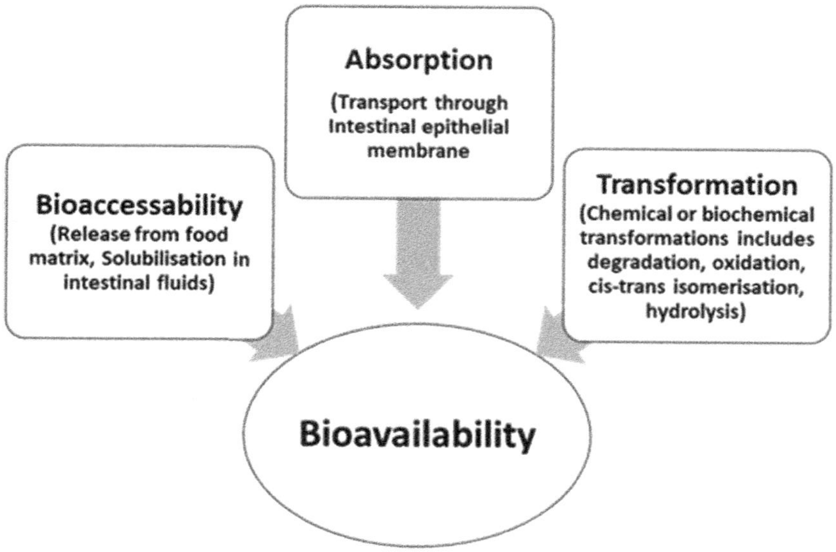

FIGURE 8.1 Main stages influencing the bioavailability of nutraceuticals (Dima et al., 2020).

dietary components significantly influence the extent and rate of absorption. A comprehensive understanding of these determinants is important for the rational design of advanced delivery systems capable of protecting, transporting, and releasing bioactive compounds at specific target sites, thereby enhancing their functional efficacy in food and nutraceutical applications (Dima et al., 2020).

2.2.1 Physicochemical properties of nutraceuticals

Physicochemical attributes such as solubility, lipophilicity, chemical structure, and physical state play a critical role in determining the bioavailability of nutraceuticals. Nutraceuticals, derived from natural plant sources, exist as crystalline or amorphous powders that have high melting points and varying degrees of water solubility. Others, such as conjugated linoleic acids (CLAs) and omega-3/omega-6 polyunsaturated fatty acids (PUFAs), are lipophilic liquids and insoluble in water. Key physicochemical properties, such as molecular weight and chemical structure, substantially impact the bioavailability of these compounds by modulating their solubility, lipophilicity, and cellular permeability. Nutraceuticals are consumed as dietary supplements in the form of tablets, capsules, syrups, and powders or as functional foods enriched with natural ingredients, containing free or encapsulated bioactive components. Achieving high bioavailability requires balancing solubility and lipophilicity. Lipophilic compounds typically have poor solubility in gastrointestinal fluids, leading to low bioaccessibility, while hydrophilic compounds, though more soluble, often face difficulty in permeating through the lipid-rich cell membranes. For example, epigallocatechin-3-gallate and vitamin C are highly soluble in aqueous media yet exhibit poor cellular permeability; curcumin is limited by both low solubility and permeability, while resveratrol, despite its low solubility, demonstrates comparatively high membrane permeability (Dima et al., 2020). Polyphenols commonly present in plants as polymers (e.g., proanthocyanidins) or in glycosylated forms (aglycone attached to a sugar group), generally require enzymatic hydrolysis by intestinal enzymes or gut microbiota for absorption. Furthermore, the molecular structure of triacylglycerols (TAGs), especially the specific positioning of fatty acids at the sn-1, sn-2, and sn-3 positions of the glycerol backbone, significantly influences their enzymatic hydrolysis, intestinal absorption, and the overall bioavailability of lipophilic bioactives. Besides, the chemical reactivity of nutraceuticals plays a crucial role in bioavailability. Components in the food matrix may interact with these compounds, potentially leading to degradation. For instance, PUFAs, CLAs, and retinol, which contain multiple bonds, are susceptible to oxidation when exposed to oxygen, pro-oxidant metals, or enzymes (Badawy et al., 2023). Minerals like calcium and iron may have reduced absorption due to the presence of phytates, oxalates, or phosphates. Furthermore, polyphenols can bind to proteins, influencing biological activities such as creating an astringent taste (Zhang et al., 2024). A deeper understanding of these physicochemical parameters is essential for designing effective delivery systems and optimizing functional food formulations for improved human health outcomes.

2.2.2 Food matrix effect and nutrient interactions

The food matrix is the composite mixture of nutrients and non-nutrients within which nutraceuticals are consumed, plays a crucial role in determining their bioaccessibility and eventual bioavailability. This complex matrix influences solubilization, stability, transport, and cellular uptake of bioactive compounds through physical and chemical interactions. Bioactive compounds are often embedded within food matrices composed of macromolecules like carbohydrates, lipids, and proteins. The chemical interactions between these components and bioactive compounds significantly affect the release and dispersion of nutraceuticals in the gastrointestinal tract (McClements & Xiao, 2014; Shahidi & Pan, 2022). Lipophilic compounds such as carotenoids and curcumin require lipid-based delivery system. Due to their hydrophobic nature, carotenoids require fat in the small intestine to stimulate bile acid release, which emulsifies fats into micelles. This micellization increases the availability of carotenoids for uptake by enterocytes (Corte-Real et al., 2016). Conversely, dietary fibers may bind polyphenols and reduce their availability through complexation or entrapment, whereas certain proteins may act as carriers, improving solubility and stability (Dima et al., 2020; Shahidi & Pan, 2022). Interactions within the matrix can either facilitate or hinder transport across intestinal barriers. Furthermore, processing techniques like homogenization, fermentation, and enzymatic pre-treatment can alter the microstructure of the food matrix and improve nutraceutical delivery (Kasote et al., 2021). Designing functional foods thus requires a systematic approach where the matrix composition and processing methods are tailored to preserve and enhance nutraceutical bioactivity.

2.2.3 Strategies to enhance bioavailability

Low bioavailability of nutraceuticals, primarily due to poor permeability across lipid membranes, high molecular size, and susceptibility to degradation, limits their therapeutic effectiveness. Enhancing their bioavailability is therefore critical for realizing their full therapeutic potential in functional food and nutraceutical formulations (Ting et al., 2014). A variety of strategies have been explored to address these challenges, focusing on improving solubility, protecting bioactives from degradation, and facilitating targeted release and absorption. Nanotechnology-enabled delivery systems offer innovative and effective approaches to address key challenges such as poor solubility, instability, and limited bioavailability of bioactive compounds. Nanoscale delivery systems such as nanoliposomes, nanoemulsions, micelles, and biodegradable polymeric nanoparticles have demonstrated considerable potential in improving the water solubility, physiochemical stability, and site-specific delivery of bioactive nutraceutical compounds (Manocha et al., 2022). These nanostructured systems enable controlled release, improve intestinal absorption, reduce the required dosage, and enhance biological efficacy, thereby advancing the clinical application of nutraceuticals (Manocha et al., 2022; Mishra et al., 2021). Other emerging strategies include food matrix engineering, 3D printing of functional foods, and co-delivery with bioenhancers or excipients that modulate intestinal permeability (Subhasri et al., 2024; Ting et al., 2014). Each technique must be tailored to the specific physicochemical characteristics of the nutraceutical and the intended health outcome. Integrating these strategies within food systems represents a critical step towards delivering bioactive compounds in an effective, safe, and consumer-friendly manner. Modified starch plays a significant role in this context. As a biodegradable, biocompatible, and functional biopolymer, modified starch can serve as an effective encapsulating agent or carrier in delivery systems. It can form nanocomposites or films that stabilize bioactive compounds, facilitate controlled release, and improve their interaction with biological membranes. Additionally, modified starch derivatives can enhance intestinal permeability and resistance to enzymatic breakdown, thereby boosting the overall bioavailability of the encapsulated nutraceuticals.

3. Modified starches in nutraceutical applications

Starch is one of the most abundant sources of stored energy in plants and is widely recognized for its economic viability, renewability, and biodegradability (Forfora et al., 2024). As a naturally occurring polysaccharide, it plays a pivotal role in both the food and pharmaceutical industries. Starch is commonly extracted from various plant sources, including tuberous crops (e.g., potatoes and cassava), cereal grains (e.g., rice, wheat, maize, sorghum, and barley), and unripe fruits (e.g., bananas and plantains) (Nath & Dutta, 2024). Structurally, starch exists in a semi-crystalline granular form, and is primarily composed of two main polysaccharides: amylose and amylopectin. Amylose is mostly a linear polymer of glucose units linked by α-1,4 glycosidic bonds, whereas amylopectin exhibits a highly branched structure containing both α-1,4 and α-1,6 glycosidic linkages (Varghese et al., 2022). Amylose generally comprises 20%–30% of starch, with the remaining 70%–80% represented by amylopectin (Rashwan et al., 2024). The physicochemical characteristics of starch granules, such as size, shape, distribution, and degree of crystallinity, are largely influenced by structural parameters like chain length, branching patterns, and branching density (Li et al., 2023). The crystalline structure of starch granules is categorized into three main polymorphs: A-type, B-type, and C-type, which result from the periodic organization of crystalline and amorphous regions formed by the packing of amylopectin double helices. A-type starch, primarily found in cereals, has tightly packed amylopectin double helices, whereas B-type starch, common in tubers, has a looser packing arrangement. The C-type polymorph, observed in legumes, represents a combination of both A- and B-type structures (Gupta et al., 2021). The degree of polymerization of amylose ranges from 840 to 22,000 glucose units, while amylopectin has an average polymerization degree nearing two million, is considered among the largest naturally occurring macromolecules (Hou & Yan, 2023). Owing to its extensive branching, the molecular weight of amylopectin is estimated to be nearly 1000-fold higher than that of amylose (Gupta et al., 2021).

Starch is a key functional ingredient in the food industry, playing a vital role in enhancing the textural attributes and nutritional quality of food products. Its versatility arises from a range of physiochemical properties, including swelling, thickening, emulsifying, water retention, encapsulation, gelling, gelatinization, and binding capabilities (Singh et al., 2023). Despite its wide applicability, native starch exhibits several limitations, including high hot-paste viscosity, rapid retrogradation, poor stability during cold-storage, and low resistance to acidic conditions, shear forces, and elevated temperatures. These limitations significantly restrict its industrial applications, as they can adversely affect the end-product quality (Bangar et al., 2022). To overcome the inherent limitations of native starch, various modification processes are employed to alter its molecular structure through processes such as chain length adjustment, branch point formation, debranching, and disproportionation. These modifications enhance physicochemical and functional properties of starch, thereby improving its suitability for industrial applications (Bangar et al., 2022; Park & Kim, 2021). Modified starch is extensively used in food processing for its role in thickening, gelling, stabilization, edible film formation, encapsulation, prevention of retrogradation, fat replacement, and enhancement of resistant starch content (Sinhmar et al., 2023).

Starch modification can be accomplished through physical, chemical, enzymatic, or integrated approaches. Physical modifications primarily influence the functional, structural, and physical properties such as solubility, crystallinity, swelling power, viscosity, and thermal stability (He et al., 2023). Chemical modifications involve processes like dextrinization, cross-linking, stabilization, oxidation, and copolymerization, which introduce new functional groups, alter molecular polarity, disrupt the native granular structure, promote cross-link formation, and increase the degree of substitution (Bangar et al., 2022). Traditional chemical modification methods involve the use of acids, acetates, hypochlorites, and phosphates, which generate substantial waste requiring proper recycling to mitigate environmental impact (Compart et al., 2023). Enzymatic modification has recently gained attention at both laboratory and industrial levels due to its ability to impart desired functional characteristics with minimal environmental impact (Olawoye et al., 2023). Physical and enzymatic modification methods are considered as environmentally friendly since they do not involve the use of synthetic additives (Park & Kim, 2021; Zhang & Bao, 2023). Chemically modified starch, through the introduction of functional groups such as amino, carbonyl, carboxyl, and ester groups, enhances starch's ability to adsorb heavy metal ions or dyes, making it useful in industries such as food, textiles, and petrochemicals (Gupta et al., 2021). In food processing, modified starch serves as a cost-effective alternative to conventional hydrocolloid additives, enhancing both the quality and functional properties of a diverse range of processed food products.

Modified starch plays a vital role in formulation of nutraceutical products by improving their stability, bioavailability, and functional performance. It serves as an effective protective barrier for sensitive bioactive ingredients like vitamins, polyphenols, and probiotics, ensuring their controlled and sustained release within the gastrointestinal tract. This protective mechanism helps to prevent the degradation of active compounds before reaching their intended target sites. For example, beta-carotene stabilized in high internal phase emulsions with octenyl succinic anhydride-modified starch (OSA-starch) and tannic acid (TA) significantly enhances its chemical stability against UV light across a range of pH conditions (Liu et al., 2020). This enhanced stability is attributed to the regulation of hydrogen bonding interactions between OSA-starch and TA, as well as the formation of robust complexes within the dual matrix system. In another study, starch extracted from kutki millet (*Panicum sumatrense*) was chemically modified using three amino acids such as lysine (positively charged), aspartic acid (negatively charged), and threonine (neutral) as cross-linking agents, under varying pH conditions. The modified starches were subsequently employed in the development of hydrogels intended for nutraceutical delivery. Among the three amino acids evaluated, lysine demonstrated the highest cross-linking efficiency, resulting in the most effective hydrogel formation. These lysine-modified starch hydrogels exhibited favorable characteristics for encapsulating and delivering bioactive compounds (Mahajan et al., 2021). Similarly, a study on the microencapsulation of vitamin B1 through spray drying using various encapsulating agents found that modified starch achieved an encapsulation efficiency of 66%–100% and released vitamin B1 within 10 min, while chitosan and pectin exhibited a slower release profile (Carlan et al., 2020).

Many nutraceutical compounds exhibit poor water solubility and limited physiochemical stability, which limits their absorption and therapeutic efficacy within the human body. Additionally, these compounds are prone to degradation due to heat, light, and oxidation. Modified starch addresses this challenge by improving solubility and stability, thereby enhancing the bioavailability of these compounds. A study done by (Zhi et al., 2021) demonstrated this by developing self-assembled micelles using dual-modified starch (hydroxypropylation followed by debranching) as a nanocarrier for curcumin encapsulation. This approach significantly improved both the water solubility (approximately 200-fold) and improved its stability by 3–6 times compared to free curcumin. Additionally, modified starch with polyphenols has been shown to slow down starch digestion while enhancing the functional efficacy of incorporated phenolic compounds. Starch modification alters the structural changes in native starch promoting the conversion of rapidly digestible starch into slowly digestible starch and resistant starch. This shift facilitates gradual release of glucose, offering significant advantages for diabetic-friendly and weight management food formulations (Liu et al., 2022). Furthermore, polyphenols contribute to improved immune function due to their diverse biological activities, including antioxidant, anti-inflammatory, and immunomodulatory effects. The modification of maize starch with phenolic extracts from grape pomace and sorghum bran under alkaline conditions (6 and 12 h) resulted in reduced enzymatic hydrolysis and enhanced antioxidant activity (Oladele et al., 2020). As shown in Table 8.1, modified starches are widely utilized in the formulation of various nutraceutical products, primarily through emulsion formation or encapsulation techniques. Typically, following the incorporation of nutraceutical compounds using different biopolymers as wall materials, the formulations are converted into microparticles by spray drying. Recent studies, as summarized in Table 8.1, demonstrate that modified starches improve the water solubility and stability of nutraceutical compounds, while also exhibiting controlled release characteristics. These findings highlight the potential of modified starch as an effective wall material and stabilizer for encapsulating a wide range of bioactive compounds, facilitating their targeted delivery within the body.

4. Encapsulation of nutraceuticals using modified starches

Although nutraceuticals are frequently incorporated into functional foods to enhance bioactive compound consumption, their application remains limited due to unfavorable physicochemical properties, including poor aqueous solubility and thermal instability and degradation under environmental stresses such as light, oxygen, and moisture. To overcome these limitations, the development of encapsulated delivery systems has gained substantial attention. These systems commonly employ natural biocompatible polymers such as proteins and polysaccharides to provide structural protection, reduce environmental degradation, and enhance both the stability and bioavailability of encapsulated nutraceuticals (Sun et al., 2023). Modified starches, derived through chemical, physical, or enzymatic alterations of native starch molecules, offer unique physicochemical properties that make them suitable candidates as encapsulating matrices in both food and pharmaceutical industries (Yang et al., 2024; Zhu, 2017). Starch modification enhances functional attributes such as solubility, gelatinization temperature, retrogradation resistance, and molecular interaction potential, which are critical in designing controlled and targeted delivery systems. By tailoring their structural architecture, modified starches can encapsulate a wide range of bioactive compounds including polyphenols, carotenoids, probiotics, and omega-3 fatty acids, protecting them from degradation and facilitating sustained release during gastrointestinal transit (Guo et al., 2021; Rong et al., 2022; Sun et al., 2023).

4.1 Spray drying technique

Spray drying is recognized as one of the most economical and scalable single-step microencapsulation technique, making it particularly advantageous for applications in the food industry. It is widely employed to protect bioactive compounds and functional ingredients from adverse environmental factors including light, heat, and oxidative degradation. Spray drying transforming the liquid materials into free-flowing dried particles of uniform size in a single processing step (Fig. 8.2). During the process, evaporation occurs as the liquid feed comes into contact with hot air. Owing to the rapid rate of evaporation, the droplet surface temperature remains relatively low, thereby minimizing thermal degradation and preserving product quality.

Spray drying enables the production of high-quality final products characterized by low water activity and reduced weight, which facilitate improved storage stability and ease of transportation. The physicochemical properties of the resulting microencapsulated powders are largely governed by various processing parameters, including feed flow rate, particle size, feed viscosity, drying air inlet and outlet temperatures, atomization pressure, and the type of atomizer employed. In a study, solvent derived oleoresin from *Nigella sativa* L. seeds was emulsified with gum Arabic–maltodextrin (1:1 w/w) solution and subsequently encapsulated into powder form through spray drying. The process resulted in encapsulation efficiencies of $84.2 \pm 1.5\%$ and $96.2 \pm 0.2\%$ indicating a strong potential for application in fortified foods and nutraceutical formulations (Edris et al., 2016; Rehman et al., 2021) developed the emulsions using the

TABLE 8.1 Role of modified starch in nutraceutical products.

Type of nutraceutical products	Modified starch	Nutraceutical compounds	Properties	References
Functional food/Nutraceutical delivery system	Dual-modified starch (acid-ethanol hydrolysis followed by cinnamic acid esterification)	Curcumin	Spherical shaped curcumin loaded starch nano particles (250.7–448.5 nm) exhibited • High encapsulation efficiency (up to 26.7%) • Improved water solubility (18-fold) • Enhanced physical stability (6–8-fold) • Controlled release (korsmeyer–Peppas model) • Excellent biosafety (non-toxic, non-mutagenic)	Du et al. (2023)
Nutraceutical beverages/Nano-lipid carriers (NLCs)	Octenyl modified starch (OMS) and sucrose monostearate (SMS)	Lutein	OMS, and SMS fabricated nano lipid carriers • Improved physical stability (lower creaming velocity $P < 0.05$) of emulsions • Effective encapsulation and stabilization of hydrophobic bioactive compound lutein • Exhibited shear-thinning behavior suitable for beverage formulations	Sedaghat Doost et al. (2020)
Spray-dried nutraceutical powder	Octenyl succinic anhydride (OSA) derived from gros michel banana	Andean blackberry concentrate (rich in total phenolics)	• High encapsulation efficiency (53.01%) • Preserved total phenolic content (≈91 mg GAE/g) and antioxidant capacity (≈89 μg TE/g) • Improved solubility and low hygroscopicity • Small particle size (D10: 22.2 μm; D50: 40.7 μm) • Long shelf life (up to 51.6 months at 15 °C) • High powder yield and suitable for spray-drying applications	Quintero-Castaño et al. (2021)
Excipient emulsion for fruit/vegetable-based nutraceuticals	Octenyl succinic anhydride (OSA) starch	β-carotene from spinach	Emulsion stabilized by OSA resulted • Lower bioaccessibility compared to sodium caseinate but comparable to tween 20 • more stable against droplet aggregation during GIT simulation • Highlighting the importance of emulsifier selection	Yuan et al. (2019)
Pickering emulsion for nutraceutical delivery	Dual-modified starch (acid-ethanol hydrolyzed and cinnamic acid esterified starch – CAES)	Curcumin	• High encapsulation efficiency (up to 96.2%) • Improved photochemical stability of curcumin	Du et al. (2024)

Continued

TABLE 8.1 Role of modified starch in nutraceutical products.—cont'd

Type of nutraceutical products	Modified starch	Nutraceutical compounds	Properties	References
			• Enhanced bioaccessibility and free fatty acid release • Stable emulsion formation through dense interfacial barrier	
Nanocomplex for phenolic delivery	Pea protein isolate–modified starch (PPI–S)	Phenolic extract from pomegranate peel	• High encapsulation efficiency (~73%) • Formation of nanocomplexes with average particle sizes: ~54.5 nm (lyophilized) and ~111.3 nm (in emulsion) • Enhanced antioxidant activity	Guzel et al. (2020)
Lycopene-loaded nanoemulsion	Octenyl succinic anhydride (OSA) starch	Lycopene	• High encapsulation efficiency (77.3%) • Smaller droplet size (125 ± 5.11 nm) compared to protein emulsifiers • Reduced oil–water interfacial tension • greater physical stability based on instability index analysis	Wanyi et al. (2020)
Essential oil-loaded hydrogel microcapsule	Dually modified starch (porous starch + OSA) embedded in alginate hydrogel	Satureja essential oil (SEO)	• High loading efficiency (~70%) • Sustained release profile • Enhanced antibacterial activity • Protection against UV degradation (preserved 60% antioxidant and 90% antibacterial activity after 2 h UV) • Increased crystallinity and stability	Sahraeian et al. (2023)
Microencapsulated β-carotene complex	Octenyl succinic anhydride (OSA)-modified starch (in combination with heated whey protein isolate)	β-carotene	• Significantly enhanced aqueous solubility of β-carotene (264.05 ± 72.53 μg/mL) • Improved chemical stability under low pH conditions • Encapsulation in amorphous form through hydrophobic interactions • good reconstitution after freeze-drying	Lin et al. (2021)
Spray-dried microparticles for food colorants	Modified starches alone and in combination with gum Arabic	Paprika oleoresin (carotenoids)	• High encapsulation efficiency (>90%) • Stable emulsions (turbiscan stability index <3%) and small droplet size • Enhanced stability of carotenoids and color at	Anthero et al. (2021)

TABLE 8.1 Role of modified starch in nutraceutical products.—cont'd

Type of nutraceutical products	Modified starch	Nutraceutical compounds	Properties	References
			25 °C and 35 °C, especially with SN • Suitable for food applications as stable coloring agents	
Spray-dried microparticles for vitamin delivery	Modified starch (compared with other biopolymers like gum Arabic, pectin, chitosan, and sodium alginate)	Folic acid (vitamin B9)	• Lower encapsulation efficiency compared to other biopolymers (~100% in others) • Rapid release profile suitable for immediate delivery applications Lower encapsulation efficiency compared to other biopolymers irregular particle morphology useful in formulations requiring rapid release of water-soluble vitamins.	Estevinho et al. (2020)
Emulsions and alginate gel beads for antioxidant delivery	Dual-modified rice starch	Lycopene	• Enhanced chemical stability of lycopene in alginate beads (35.6%) than in the emulsions (29.5%) • Improved gastrointestinal protection • Higher bioaccessibility in emulsions (20.2%) compared to alginate beads (15.6%) • Suitable for controlled release	Jain et al. (2020)

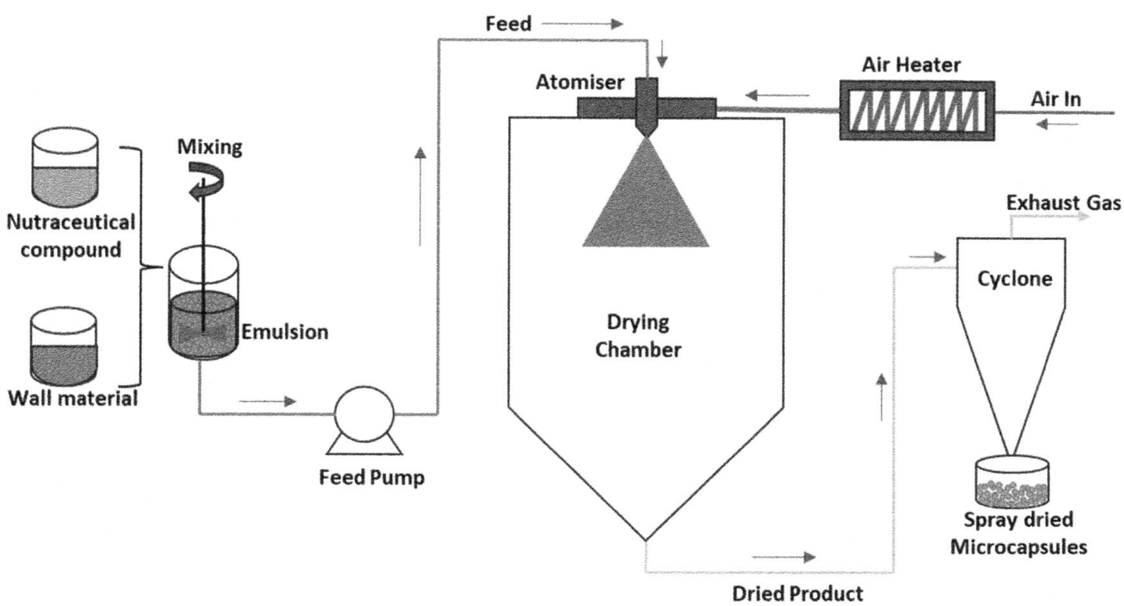

FIGURE 8.2 Typical spray drying technique for encapsulation of nutraceutical substances.

octenyl succinic anhydride modified starches (OSA-starch), specifically Hi-Cap 100 and Purity Gum 2000 as wall materials for co-encapsulation of borage seed oil, resveratrol, and curcumin. These emulsions were subsequently spray-dried into powders that exhibited high encapsulation efficiency (~93.05%). Among them, Hi Cap 100 based microcapsules demonstrated high retention of resveratrol (70.32%), curcumin (81.6%), and γ-linolenic acid (~96%). Although spray drying is widely adopted technique for the encapsulation of nutraceutical compounds due to its scalability and efficiency, there is still a limited number of studies addressing its application for specific bioactives, particularly long-term stability, and controlled release behavior under physiological conditions. Jujube fruit (*Ziziphus jujuba*), rich in antioxidants like flavonoids and polyphenols, is prone to degradation from heat, light, and oxygen. To enhance stability (Parhizkary et al., 2025), explored spray-drying encapsulation using various wall materials: maltodextrin, pectin, Arabic gum, whey protein isolate, and pea protein isolate. Among formulations, maltodextrin/pea protein isolate showed highest total phenolic content (33.95), low moisture (3.65%), low water activity (0.278), and better yield (38.8%). This highlights the importance of wall material choice for developing stable, bioactive-rich powders for food and pharmaceutical use.

4.2 Coacervation technique

Coacervation is the second most widely employed encapsulation technique in the food industry. This method involves phase separation within a colloidal system, resulting in the formation of two immiscible liquid phases: a polymer-rich coacervate phase and a polymer-poor equilibrium phase. The mechanism of coacervation involves the separation of a hydrocolloid from the primary solution, followed by its aggregation into a distinct liquid phase known as the "coacervate." This coacervate serves as a encapsulating matrix, while the residual liquid constitutes the equilibrium phase (Devi et al., 2017). Coacervation is broadly categorized into two main types: simple coacervation and complex coacervation. Simple coacervation occurs through molecular dehydration of a macromolecules, typically induced by altering system parameters such as the addition of micro-ions or non-solvents, changes in temperature, or pH adjustments. These changes reduce the solubility of the polymer, leading to phase separation and the formation of a coacervate phase. Complex coacervates are typically formed through the charge neutralization between two oppositely charged macromolecules such as colloids, proteins, or surfactants in aqueous environments (Devi et al., 2017; Nezamdoost-Sani et al., 2024). The resulting microcapsules are typically recovered using filtration or centrifugation, followed by washing with an appropriate solvent. Subsequent drying is carried out using conventional techniques such as spray drying, freeze drying, or fluidized bed drying, ultimately yielding discrete, free-flowing microcapsules (Fig. 8.3) (Muhoza et al., 2023). The encapsulation of anthocyanins extracted from *Rosa damascena Mill.* L. was investigated using zein, potato starch, and their binary mixtures through both simple and complex coacervation techniques. The highest encapsulation efficiency and antioxidant activity for zein-starch-anthocyanin nanocapsules were observed at pH 8, whereas zein-starch nanocapsules without anthocyanins exhibited maximum antioxidant activity at pH 2 (Farnad & Farhadi, 2023). In another study (Mu et al., 2022), examined the complex coacervation between chitosan and the octenyl succinic anhydride (OSA) modified starch for the microencapsulation of algae oil. Optimal coacervate formation occurred at pH 6.0 with a chitosan to OSA-starch ratio of 1:3 (w/w), yielding encapsulation efficiencies ranging from 42.8 ± 0.8% to 93.1 ± 1.2%, and loading capacities between 30.4 ± 2.7%

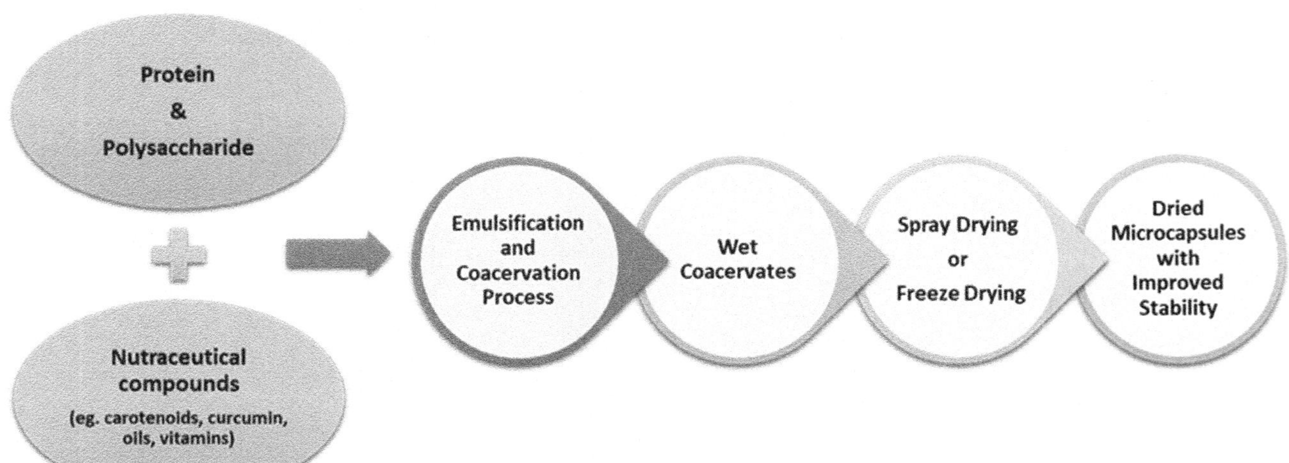

FIGURE 8.3 Spray and freeze-dried microcapsules obtained by complex coacervation.

and 58.3 ± 1.3%. These findings revealed the potential of polysaccharide–protein complex coacervates as a highly effective encapsulation system, offering better protection, and enabling controlled and site-specific delivery of bioactive compounds. More recently, Yang et al., (2023) established a hybrid encapsulation method combining gelatin–sodium hexametaphosphate (SHMP) complex coacervation and freeze-drying, using starch sodium octenyl succinate (SSOS) as a drying aid to produce fish oil microcapsules. The type of gelatin, porcine skin (PSG), bovine skin (BSG), or fish gelatin (FG) are significantly influences the physicochemical properties of the resulting powders. SSOS enhanced the encapsulation efficiency, achieving 98.4% for PSG, 96.4% for FG, and 95.2% for BSG, despite similar oil loading (24.0%–26.2%). In vitro digestion revealed gelatin dependant free fatty acid release, with PSG (55.5%) > FG (34.1%) > BSG (27.0%). This study highlights the feasibility of gelatin–SSOS systems for stable, functional fish oil powders. These results highlight the increasing focus on innovative encapsulation strategies designed to enhance the functionality and stability of nutraceutical components in food systems. Consequently, hot-melt extrusion and complex coacervation have emerged as promising encapsulation techniques in the nutraceutical sector, offering improved efficiency and product stability compared to traditional methods. Hot-melt extrusion technology has been successfully applied in the pharmaceutical industry for the preparation of solid molecular dispersions of active pharmaceutical ingredients. In contrast, complex coacervation involves the phase separation of insoluble coacervates from the continuous liquid phase, driven by electrostatic interactions between oppositely charged biopolymers (Samyor & Haokip, 2023).

4.3 Freeze-drying technique

Freeze-drying is a widely employed encapsulation technique for nutraceutical compounds, primarily focused to improve their physiochemical stability by minimizing degradation during processing and storage. The main advantage of freeze-drying compared to other drying methods is its suitability for processing thermolabile products, as it functions under low temperature and pressure conditions. Conventional freeze-dryers feature a vacuum chamber with trays that hold the product for dehydration (Fig. 8.4).

Freeze drying comprises three sequential stages: freezing, primary drying (sublimation), and secondary drying (desorption). In the first stage, the liquid formulation containing nutraceutical substances rapidly frozen at low temperatures to convert the aqueous phase into a solid ice matrix. As pressure is reduced below the triple point of water and controlled heat is applied through heated shelves, sublimation of ice occurs and this phase is referred to as primary drying. It concludes once all ice crystals have been sublimated, resulting in a porous matrix with residual bound moisture. Secondary drying follows, wherein the shelf temperature is elevated to facilitate the removal of remaining bound moisture through desorption (Massei & Fissore, 2023). Nutraceutical substances entrapped in the resultant porous solid matrix that

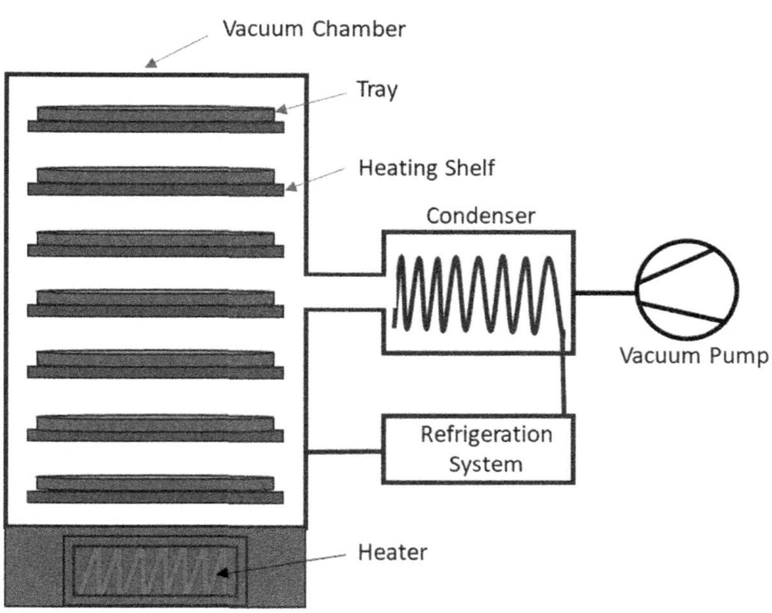

FIGURE 8.4 Freeze drier for encapsulation of nutraceutical substances.

slow down the chemical and physical degradation. This technique is particularly effective for encapsulating heat sensitive nutraceutical compounds, as the vacuum conditions enhance the oxidative stability. However, it is associated with long processing times, substantial energy consumption, and high operational costs.

The effectiveness of freeze drying has been demonstrated in a recent study by Ganje et al., (2025), who employed freeze drying to encapsulate rosemary extract using maltodextrin (MD) and whey protein isolate (WPI) as wall materials at varying MD:WPI ratios (30:70, 50:50, and 70:30) and wall-to-core ratios (1:4, 1:8, and 1:12). Samples C1, F1, and I1, corresponding to a wall-to-core ratio of 1:12 and MD:WPI ratios of 50:50, 30:70, and 70:30 respectively, exhibited the highest viscosities among the tested formulations. The encapsulation efficiencies of these samples were 88.96 ± 1.0% (C1), 88.74 ± 1.0% (F1), and 84.01 ± 0.76% (I1). Among them, the C1 sample showed the highest total phenolic content (TPC) release, while I1 exhibited the lowest. In another study Chlorophyll was microencapsulated using maltodextrin and whey protein isolate as wall materials by freeze-drying and spray-drying. Freeze dried microcapsules resulted smaller particle sizes (1.087–0.165 μm), higher zeta potential (−10.6 to −18.3 mV), greater thermal stability and provided superior protection against pH variations, with retention ranging from 49.67% to 91.28% and light degradation, with a retention of 38.12% (Ledri et al., 2024). Extending this approach to lipid-based compounds, chia seed oil (CSO) was encapsulated using whey protein concentrate (WPC) and modified tapioca starch (MTS) through freeze-drying. Higher WPC concentrations improved the encapsulation efficiency (EE), whereas increased MTS levels enhanced α-linolenic acid (ALA) retention. Fatty acid profiling and stability studies revealed that an MTS:WPC ratio of 25:75 was optimal for effective CSO encapsulation (Anand et al., 2024). Foam mat freeze-drying (FMFD) is an emerging, cost-effective alternative to conventional freeze-drying, which typically involves prolonged dehydration times. FMFD combines the principles of foaming and freeze-drying, where liquid materials are first converted into stable foams using foaming and stabilizing agents, and then subjected to freeze-drying to produce lightweight, porous powders (Hardy & Jideani, 2017). In the development of modified starch-based delivery systems, a recent study examined the influence of wall material type and concentration on the catechin nanoemulsion (CaNE) powders produced through foam mat freeze drying. CaNE containing 0.4 g/100 g green tea catechins was ultrasonically emulsified and foamed using hydroxypropyl methylcellulose followed by encapsulation using maltodextrin (MD), gum arabic (GA), or their mixture (MD + GA) at 5% and 10% concentrations. Among the formulations, 5% MA + GA yielded powders with low moisture, water activity, and hygroscopicity, along with higher encapsulation efficiency, flavonoid recovery, and antioxidant retention (Ruengdech & Siripatrawan, 2022). A sustainable strategy was employed to valorize kinnow peel by microencapsulating its extract by freeze-drying, using MD and GA as wall materials. This process significantly enhanced the extract's physicochemical properties, antioxidant retention, and overall stability, supporting its potential as a functional food ingredient. MD was linked to improved retention of bioactive compounds, while the MD–GA combination contributed to better physical characteristics of the encapsulated powder (Alam et al., 2025).

5. Stability of nutraceuticals encapsulated in modified starches

Encapsulation is a pivotal technique in the food and pharmaceutical industries, employed to protect nutraceutical compounds from environmental factors that may compromise their stability and efficacy. Native starches often require modification to enhance their encapsulation properties, such as solubility, film-forming ability, and resistance to enzymatic degradation. These modifications tailor starch properties to meet specific encapsulation requirements. The stability of encapsulated nutraceuticals is influenced by type of modified starch used and the encapsulation technique employed. For instance, spray-drying is a prevalent method where modified starches serve as wall materials, protecting sensitive compounds like polyunsaturated fatty acids from oxidation and degradation. The structural integrity and barrier properties of the modified starch matrix are crucial in shielding the encapsulated nutraceuticals from environmental factors such as oxygen, light, and moisture (Carlos et al., 2016). A notable example is the encapsulation of curcumin, a hydrophobic polyphenolic compound with potent antioxidant properties, using modified cassava starch nanoparticles. In this approach, cassava starch undergoes heat–moisture treatment to enhance its emulsifying properties, facilitating the formation of stable Pickering emulsions. These emulsions effectively encapsulate curcumin, significantly improving its stability over time and protecting it from degradation (Ramos et al., 2024). Similarly, spray-dried microcapsules containing capsaicin have been developed using whey protein and octenyl succinic anhydride (OSA) modified starch as wall materials. Research demonstrates that varying the protein-to-starch ratio influences the microcapsules' physicochemical characteristics and stability, with certain formulations providing enhanced protection for capsaicin (Zhang et al., 2022). Additionally, starch particle-stabilized Pickering emulsions offer excellent stability and unique properties for delivering bioactive compounds, including polyphenols, carotenoids, fatty acids, and vitamins, thereby preserving their bioactivity and enhancing targeted delivery (Ramos et al., 2025).

The stability of nutraceutical compounds can be further enhanced through encapsulation using modified starch, other biopolymers, and plasticizers. In one study, vitamin B9 (folic acid) was encapsulated using two approaches: one involving zein and the other utilizing modified starch with two plasticizers. The electrospinning technique was employed to improve the stability and antioxidant capacity of vitamin B9. The release profiles of the resulting microstructures were assessed in simulated gastric fluid (37°C), deionized water and ethanol. The combination of modified starch with Tween 80 and glycerol demonstrates potential for enhancing the stability of encapsulated vitamin B9 (Coelho et al., 2022).

Building on this, a study investigated the protective role of gum Arabic and modified corn starch with octenyl succinic anhydride in preserving key bioactive compounds in orange peel during freeze-drying. Findings revealed that both biopolymers enhanced the stability of hydrophilic compounds (hesperidin, naringin, Vitamin C, and other polyphenols); however, they did not protect carotenoid compounds during the freeze-drying process. Additionally, incorporating these biopolymers contributed to improved bioaccessibility of the studied bioactive compounds (García-Martínez et al., 2023). The microencapsulation of anthocyanins from *Echium amoenum* (Iranian borage) petals using a maltodextrin and modified starch blend (1:1 w/w) by spray drying significantly enhanced their stability, with a 43.8% improvement compared to the crude extract. The resulting microcapsules exhibited strong antioxidant stability, and release studies conducted in a simulated gastrointestinal environment revealed that anthocyanins remained largely retained within the microparticles in the gastric phase and were predominantly released in the intestinal phase (Mehran et al., 2020).

A high internal phase emulsion (HIPEs) stabilized with modified starch has been utilized to enhance the stability and bioavailability of nutraceuticals. In one approach, β-carotene was encapsulated within HIPEs using octenyl succinic anhydride modified starch through a simple shear dispersion process. The encapsulation significantly improved the chemical stability of β-carotene under UV exposure, with a consistent degradation rate across different pH levels. Additionally, incorporating β-carotene into HIPEs enhanced its stability and bioaccessibility during in vitro digestion (Yan et al., 2019). Beyond HIPEs, oil-in-water nanoemulsions stabilized by modified starch-based emulsifiers have also been developed using high-pressure homogenization to improve β-carotene stability and bioavailability. Research suggests that a higher molecular density of modified starch in dispersion leads to greater β-carotene retention but lowers bioaccessibility (ranging from 3.1% to 35.6%) within nanoemulsions. This reduction in bioaccessibility is likely due to the formation of a thick and compact interfacial layer surrounding the oil droplets (Liang et al., 2013).

Certain nutraceutical compounds (vitamins, polyphenols, flavonoids, carotenoids, and omega-3 fatty acids) are highly sensitive to oxygen when exposed to air, light, or heat, causing them to degrade or break down. Oxidation leads to a loss of potency, stability, and effectiveness over time. To counter this, microencapsulation has been explored to improve oxidative stability. A study examined the oxidative stability of spray dried microencapsulated *Coffea arabica* (green coffee) oil using spray drying under accelerated storage conditions (32% RH at 25°C). Microfluidization technique adopted to prepare green coffee oil emulsions with reduced particle size, using mesquite gum and octenyl succinic anhydride (OSA) modified starch as wall materials. The results indicated that the green coffee oil microcapsules exhibited lower oxidative stability than that of unencapsulated bulk oil (87.39 meq O_2/kg), with values of 60.83 meq O_2/kg for mesquite gum and 70.67 meq O_2/kg for OSA-starch. Despite the reduction in oxidative stability, the microcapsules demonstrated good potential for application in nutraceutical and cosmetic formulations (Granados-Vallejo et al., 2019).

Likewise, microencapsulation of algae oil with chitosan and octenyl succinic anhydride (OSA) modified starch enhanced the oxidative stability compared to unencapsulated bulk oil. Encapsulation efficiency ranges from 42.8 ± 0.8% to 93.1 ± 1.2%, while the loading capacity varies between 30.4 ± 2.7% and 58.3 ± 1.3% (Mu et al., 2022). To enhance the oxidative stability of borage seed oil (BSO) which is rich in γ-linolenic and linoleic acids, ultrasound-assisted BSO-loaded nanoemulsions were formulated using modified starch and varying concentrations of peppermint oil (PO) as a natural antioxidant. The nanoemulsions were stored at 4, 25, and 40 °C for 30 days to evaluate the stability. Results demonstrated that the incorporation of PO significantly improved the oxidative stability and antioxidant capacity. Among the tested samples, the optimized nanoemulsion (5:5% v/v PO: BSO) exhibited superior physical stability and the highest retained antioxidant capacity (99.42 μg Trolox/mL) when stored at 40 °C, emphasizing the effectiveness of PO in enhancing the stability of BSO in nanoemulsion systems (Rehman et al., 2020). These studies show that modified starch-based emulsion and encapsulation improve the thermal, oxidative, and chemical stability of various nutraceutical compounds during storage.

6. Nanoscale delivery system for nutraceuticals

Conventional emulsions typically have droplet diameters greater than 200 nm, whereas nanoemulsions feature much smaller droplets, ranging from 20 to 200 nm. Both systems are thermodynamically unstable but exhibit metastable kinetic stability, allowing them to remain dispersed for extended periods under certain conditions. A key distinction lies in their surface-to-mass ratios: conventional emulsions range from 0.07 to 70 m^2/g, while nanoemulsions exhibit much higher values between 70 and 330 m^2/g, enhancing interfacial interaction potential. Visually, conventional emulsions appear turbid to opaque, while nanoemulsions are typically transparent to translucent due to their reduced droplet size and uniform distribution (Choi & McClements, 2020). Nanoemulsions, characterized by lipid droplets (10–200 nm) stabilized by surfactants in an aqueous medium, offer great potential to enhance the bioavailability and functionality of bioactive compounds in food applications (Choi & McClements, 2020).

Nanoparticle systems can be developed using two fundamental approaches: the "top-down" and the "bottom-up" methods. In the top-down approach, nanoparticles are generated by mechanically breaking down larger structures into smaller particles through size reduction techniques. In contrast, the bottom-up approach involves the formation of nanoparticles through the self-assembly of smaller molecular units, such as lipids or proteins, by chemical or physicochemical processes (Acosta, 2009). The main advantage of nanocarrier systems is their potential to encapsulate hydrophobic nutraceutical substances and facilitate targeted delivery to specific cells or tissues. The nanoscale droplet size of nano emulsions facilitates the efficient incorporation of lipophilic compounds, and is associated with enhanced bioavailability due to the increased surface area (Ezhilarasi et al., 2012). Achieving smaller droplet sizes typically requires a high emulsifier-to-oil ratio. The selection of emulsifiers is crucial, as it significantly influences the final droplet size, dispersity, and stability of nanoemulsion by reducing interfacial tension through adsorption at the oil–water interface (Pourshamohammad et al., 2024). In the food and pharmaceutical industries, commonly used emulsifiers include synthetic surfactants (e.g., polyoxyethylene sorbitan esters) and natural emulsifiers such as amphiphilic proteins, phospholipids, and polysaccharides. The molecular weight of emulsifiers significantly influences adsorption kinetics during emulsion formation. Low-molecular-weight emulsifiers like Tween, Span, and lecithin are generally more effective at producing smaller droplet sizes compared to high molecular weight protein-based emulsifiers like caseinate and β-lactoglobulin. This is primarily due to their ability to adsorb more rapidly at the oil-water interface during the homogenization process. Recently, natural food biopolymers such as modified starches have gained attention for nanoemulsion preparation due to their favorable emulsifying properties. Driven by consumer preferences for eco-friendly and natural products, there is an increasing interest in substituting synthetic surfactants with naturally derived modified starches in emulsion formulations, due to their perceived safety and sustainability (Flores-Andrade et al., 2021; Sawant et al., 2024). Lutein-loaded nanolipid carriers (NLCs) have been developed using octenyl modified starch (OMS) and sucrose monostearate (SMS) as food-grade emulsifiers to improve dispersibility and stability. Pre-heating SMS had minimal impact on droplet size or physical stability. SMS produced smaller droplets and showed shear-thinning behavior, while OMS-stabilized NLCs exhibited significantly lower creaming velocities ($P < 0.05$), indicating better stability. SMS-based NLCs were more stress-sensitive due to reduced surface charge. These findings emphasize the importance of emulsifier selection in formulating stable nanocarriers for hydrophobic nutraceuticals in functional beverages (Sedaghat Doost et al., 2020).

7. Conclusion

The incorporation of nutraceuticals into functional foods offers significant potential for improving public health; however, their efficacy is often limited by poor solubility, low bioavailability, and instability under gastrointestinal conditions. Modified starches have emerged as versatile, safe, and effective delivery vehicles capable of overcoming these challenges. Through various chemical, physical, and enzymatic modifications, starches can be tailored to enhance encapsulation efficiency, control release kinetics, and improve the stability of bioactive compounds. Advances in nanotechnology and encapsulation techniques have further expanded their utility in delivering both hydrophilic and lipophilic nutraceuticals. Despite considerable progress, challenges remain in optimizing molecular interactions, achieving consistent release profiles, and scaling up production. Future research should focus on designing smart, multifunctional starch-based carriers that respond to physiological cues, ensuring precise and targeted delivery. Overall, modified starches represent a sustainable and promising platform for next-generation nutraceutical delivery systems in functional food and therapeutic applications.

References

Acosta, E. (2009). Bioavailability of nanoparticles in nutrient and nutraceutical delivery. *Current Opinion in Colloid & Interface Science, 14*(1), 3–15. https://doi.org/10.1016/j.cocis.2008.01.002

Alam, M., Sid, S., Giri, S., Das, R., Kishore, A., & Kumar, N. (2025). Encapsulated kinnow peel powder using freeze drying: Effect of maltodextrin and gum Arabic concentrations on physiochemical, functional and thermal properties. *Food and Humanity, 4*, Article 100546. https://doi.org/10.1016/j.foohum.2025.100546

Anand, V., Ksh, V., Kar, A., Varghese, E., Vasudev, S., & Kaur, C. (2024). Encapsulation efficiency and fatty acid analysis of chia seed oil microencapsulated by freeze-drying using combinations of wall material. *Food Chemistry, 430*, Article 136960. https://doi.org/10.1016/j.foodchem.2023.136960

Anthero, A. G.d. S., Bezerra, E. O., Comunian, T. A., Procópio, F. R., & Hubinger, M. D. (2021). Effect of modified starches and gum Arabic on the stability of carotenoids in paprika oleoresin microparticles. *Drying Technology, 39*(12), 1927–1940. https://doi.org/10.1080/07373937.2020.1844227

Badawy, S., Liu, Y., Guo, M., Liu, Z., Xie, C., Marawan, M. A., Ares, I., Lopez-Torres, B., Martínez, M., Maximiliano, J. E., Martínez-Larrañaga, M. R., Wang, X., Anadón, A., & Martínez, M. A. (2023). Conjugated linoleic acid (CLA) as a functional food: Is it beneficial or not? *Food Research International, 172*. https://doi.org/10.1016/j.foodres.2023.113158

Bangar, S. P., Ashogbon, A. O., Singh, A., Chaudhary, V., & Whiteside, W. S. (2022). Enzymatic modification of starch: A green approach for starch applications. *Carbohydrate Polymers, 287*, Article 119265. https://doi.org/10.1016/j.carbpol.2022.119265

Carlan, I. C., Estevinho, B. N., & Rocha, F. (2020). Production of vitamin B1 microparticles by a spray drying process using different biopolymers as wall materials. *Canadian Journal of Chemical Engineering, 98*(8), 1682–1695. https://doi.org/10.1002/cjce.23735

Carlos, J., Ruiz, R., Rubi, M., & Campos, S. (2016). New polymers for encapsulation of nutraceutical compounds. *New Polymers for Encapsulation of Nutraceutical Compounds, 62*, 706–707. https://doi.org/10.1002/9781119227625

Choi, S. J., & McClements, D. J. (2020). Nanoemulsions as delivery systems for lipophilic nutraceuticals: Strategies for improving their formulation, stability, functionality and bioavailability. *Food Science and Biotechnology, 29*(2), 149–168. https://doi.org/10.1007/s10068-019-00731-4

Coelho, S. C., Rocha, F., & Estevinho, B. N. (2022). Electrospinning of microstructures incorporated with vitamin B9 for food application: Characteristics and bioactivities. *Polymers, 14*(20), 4337. https://doi.org/10.3390/polym14204337

Colussi, R., Pinto, V. Z., El Halal, S. L. M., Biduski, B., Prietto, L., Castilhos, D. D., Zavareze, E.d. R., & Dias, A. R. G. (2017). Acetylated rice starches films with different levels of amylose: Mechanical, water vapor barrier, thermal, and biodegradability properties. *Food Chemistry, 221*, 1614–1620. https://doi.org/10.1016/j.foodchem.2016.10.129

Compart, J., Singh, A., Fettke, J., Apriyanto, A., & Starch Properties, C. (2023). A review of starch modifications and their applications. *Polymers, 15*(16), 3491. https://doi.org/10.3390/polym15163491

Corte-Real, J., Iddir, M., Soukoulis, C., Richling, E., Hoffmann, L., & Bohn, T. (2016). Effect of divalent minerals on the bioaccessibility of pure carotenoids and on physical properties of gastro-intestinal fluids. *Food Chemistry, 197*(Pt A), 546–553. https://doi.org/10.1016/j.foodchem.2015.10.075

Dai, Z., Zhang, Y., Meng, Y., Li, S., Suonan, Z., Sun, Y., Ji, J., Shen, Q., Zheng, H., & Xue, Y. (2023). Targeted delivery of nutraceuticals derived from food for the treatment of obesity and its related complications. *Food Chemistry, 418*, Article 135980. https://doi.org/10.1016/j.foodchem.2023.135980

Das, L., Bhaumik, E., Raychaudhuri, U., & Chakraborty, R. (2011). Role of nutraceuticals in human health. *Journal of Food Science and Technology, 49*(2), 173. https://doi.org/10.1007/S13197-011-0269-4

DeFelice, S. L. (1995). The nutraceutical revolution: Its impact on food industry R&D. *Trends in Food Science and Technology, 6*(2), 59–61. https://doi.org/10.1016/S0924-2244(00)88944-X

Devi, N., Sarmah, M., Khatun, B., & Maji, T. K. (2017). Encapsulation of active ingredients in polysaccharide–protein complex coacervates. *Advances in Colloid and Interface Science, 239*, 136–145. https://doi.org/10.1016/j.cis.2016.05.009

Dima, C., Assadpour, E., Dima, S., & Jafari, S. M. (2020). Bioavailability of nutraceuticals: Role of the food matrix, processing conditions, the gastrointestinal tract, and nanodelivery systems. *Comprehensive Reviews in Food Science and Food Safety, 19*(3), 954–994. https://doi.org/10.1111/1541-4337.12547

Dissanayake, T., Sun, X., Abbey, L., & Bandara, N. (2022). Recent advances in lipid-protein conjugate-based delivery systems in nutraceutical, drug, and gene delivery. *Food Hydrocolloids for Health, 2*, Article 100054. https://doi.org/10.1016/j.fhfh.2022.100054

Du, Y., Chu, J., Wang, R., Zhang, C., Zhang, J., & Zhi, K. (2023). Efficient encapsulation of fat-soluble food-derived biofunctional substances (curcumin as an example) in dual-modified starch-based nanoparticles containing large conjugated systems. *International Journal of Biological Macromolecules, 242*(Pt 3), Article 125078. https://doi.org/10.1016/j.ijbiomac.2023.125078

Du, Y., Niu, L., Song, X., Niu, J., Zhang, C., & Zhi, K. (2024). Dual-modified starch as particulate emulsifier for pickering emulsion: Structure, safety properties, and application for encapsulating curcumin. *International Journal of Biological Macromolecules, 266*(Pt 2), Article 131206. https://doi.org/10.1016/j.ijbiomac.2024.131206

Edris, A. E., Kalemba, D., Adamiec, J., & Piaotkowski, M. (2016). Microencapsulation of Nigella sativa oleoresin by spray drying for food and nutraceutical applications. *Food Chemistry, 204*, 326–333. https://doi.org/10.1016/j.foodchem.2016.02.143

Estevinho, B. N., Lazar, R., Blaga, A., & Rocha, F. (2020). Preliminary evaluation and studies on the preparation, characterization and in vitro release studies of different biopolymer microparticles for controlled release of folic acid. *Powder Technology, 369*, 279–288. https://doi.org/10.1016/j.powtec.2020.05.048

Ezhilarasi, P. N., Karthik, P., Chhanwal, N., & Anandharamakrishnan, C. (2012). Nanoencapsulation techniques for food bioactive components: A review. *Food and Bioprocess Technology, 6*(3), 628–647. https://doi.org/10.1007/s11947-012-0944-0

Farnad, N., & Farhadi, K. (2023). Simple and complex coacervation methods for the nanoencapsulation of Rosa damascena mill L. anthocyanin in zein/potato starch: A new approach to enhance antioxidant and thermal properties. *Journal of Food Science, 88*(3), 1019–1032. https://doi.org/10.1111/1750-3841.16463

Flores-Andrade, E., Allende-Baltazar, Z., Sandoval-González, P. E., Jiménez-Fernández, M., Beristain, C. I., & Pascual-Pineda, L. A. (2021). Carotenoid nanoemulsions stabilized by natural emulsifiers: Whey protein, gum Arabic, and soy lecithin. *Journal of Food Engineering, 290*. https://doi.org/10.1016/j.jfoodeng.2020.110208

Forfora, N., Azuaje, I., Kanipe, T., Gonzalez, J. A., Lendewig, M., Urdaneta, I., Venditti, R., Gonzalez, R., & Argyropoulos, D. (2024). Are starch-based materials more eco-friendly than fossil-based? A critical assessment. *Cleaner Environmental Systems, 13*, Article 100177. https://doi.org/10.1016/j.cesys.2024.100177

Ganje, M., Sekhavatizadeh, S. S., Mozafarian, M. R., & Hashemi, S. S. (2025). Physicochemical properties, microstructure, and encapsulation efficiency of rosemary (rosmarinus officinalis. L) extract microencapsulated by freeze-drying using whey protein isolate and maltodextrin. *Carbohydrate Polymer Technologies and Applications, 10*. https://doi.org/10.1016/j.carpta.2025.100861

García-Martínez, E., Camacho, M. del M., & Martínez-Navarrete, N. (2023). In vitro bioaccessibility of bioactive compounds of freeze-dried Orange juice co-product formulated with gum Arabic and modified starch. *Molecules, 28*(2), 810. https://doi.org/10.3390/molecules28020810

Granados-Vallejo, M., Espinosa-Andrews, H., Guatemala-Morales, G. M., Esquivel-Solis, H., & Arriola-Guevara, E. (2019). Oxidative stability of green coffee oil (coffea arabica) microencapsulated by spray drying. *Processes, 7*(10). https://doi.org/10.3390/pr7100734

Guo, Y., Qiao, D., Zhao, S., Zhang, B., & Xie, F. (2021). Starch-based materials encapsulating food ingredients: Recent advances in fabrication methods and applications. *Carbohydrate Polymers, 270*, Article 118358. https://doi.org/10.1016/j.carbpol.2021.118358

Gupta, A. D., Rawat, K. P., Bhadauria, V., & Singh, H. (2021). Recent trends in the application of modified starch in the adsorption of heavy metals from water: A review. *Carbohydrate Polymers, 269*, Article 117763. https://doi.org/10.1016/j.carbpol.2021.117763

Guzel, N., Kahraman, O., & Feng, H. (2020). Solid–liquid extraction by manothermosonication: Recapturing the value of pomegranate peels and nanocomplexation of extracts with pea protein. *ACS Sustainable Chemistry & Engineering, 8*(44), 16671–16679. https://doi.org/10.1021/acssuschemeng.0c06316

Hardy, Z., & Jideani, V. A. (2017). Foam-mat drying technology: A review. *Critical Reviews in Food Science and Nutrition, 57*(12), 2560–2572. https://doi.org/10.1080/10408398.2015.1020359

He, R., Li, S., Zhao, G., Zhai, L., Qin, P., & Yang, L. (2023). Starch modification with molecular transformation, physicochemical characteristics, and industrial usability: A state-of-the-art review. *Polymers, 15*(13), 2935. https://doi.org/10.3390/polym15132935

Hou, C., & Yan, W. (2023). Preparation and application of acrylic polymer modified magnetic starch adsorption materials. *Research Square, 46*. https://doi.org/10.21203/rs.3.rs-3492836/v1

Jafari, S. M., & McClements, D. J. (2017). Nanotechnology approaches for increasing nutrient bioavailability. *Advances in Food and Nutrition Research, 81*, 1–30. https://doi.org/10.1016/bs.afnr.2016.12.008

Jain, S., Winuprasith, T., & Suphantharika, M. (2020). Encapsulation of lycopene in emulsions and hydrogel beads using dual modified rice starch: Characterization, stability analysis and release behaviour during in-vitro digestion. *Food Hydrocolloids, 104*, Article 105730. https://doi.org/10.1016/J.FOODHYD.2020.105730

Kasote, D., Tiozon, R. N., Sartagoda, K. J. D., Itagi, H., Roy, P., Kohli, A., Regina, A., & Sreenivasulu, N. (2021). Food processing technologies to develop functional foods with enriched bioactive phenolic compounds in cereals. *Frontiers in Plant Science, 12*. https://doi.org/10.3389/fpls.2021.771276

Ledri, S. A., Milani, J. M., Shahidi, S. A., & Golkar, A. (2024). Comparative analysis of freeze drying and spray drying methods for encapsulation of chlorophyll with maltodextrin and whey protein isolate. *Food Chemistry X, 21*. https://doi.org/10.1016/j.fochx.2024.101156

Li, M., Daygon, V. D., Solah, V., & Dhital, S. (2023). Starch granule size: Does it matter? *Critical Reviews in Food Science and Nutrition, 63*(19), 3683–3703. https://doi.org/10.1080/10408398.2021.1992607

Liang, R., Shoemaker, C. F., Yang, X., Zhong, F., & Huang, Q. (2013). Stability and bioaccessibility of β-carotene in nanoemulsions stabilized by modified starches. *Journal of Agricultural and Food Chemistry, 61*(6), 1249–1257. https://doi.org/10.1021/jf303967f

Lin, Q., Wu, D., Singh, H., & Ye, A. (2021). Improving solubility and stability of β-carotene by microencapsulation in soluble complexes formed with whey protein and OSA-Modified starch. *Food Chemistry, 352*, Article 129267. https://doi.org/10.1016/j.foodchem.2021.129267

Liu, Y., Yan, C., Chen, J., Wang, Y., Liang, R., Zou, L., McClements, D. J., & Liu, W. (2020). Enhancement of beta-carotene stability by encapsulation in high internal phase emulsions stabilized by modified starch and tannic acid. *Food Hydrocolloids, 109*, Article 106083. https://doi.org/10.1016/j.foodhyd.2020.106083

Liu, M., Yang, Q., Wu, Y., & Ouyang, J. (2022). Effects of endogenous polyphenols in acorn (Quercus wutaishanica blume) kernels on the physicochemical properties of starch. *Starch - Stärke, 74*(5–6). https://doi.org/10.1002/star.202200005

Mahajan, P., Bera, M. B., & Panesar, P. S. (2021). Modification of kutki millet (Panicum sumatrense) starch properties by the addition of amino acids for the preparation of hydrogels and its characterization. *International Journal of Biological Macromolecules, 191*, 9–18. https://doi.org/10.1016/j.ijbiomac.2021.09.072

Manocha, S., Dhiman, S., Singh Grewal, A., & Guarve, K. (2022). Nanotechnology: An approach to overcome bioavailability challenges of nutraceuticals. *Journal of Drug Delivery Science and Technology, 72*, Article 103418. https://doi.org/10.1016/j.jddst.2022.103418

Massei, A., & Fissore, D. (2023). A new model-based approach for the development of freeze-drying cycles using a small-scale freeze-dryer. *Journal of Pharmaceutical Sciences, 112*(8), 2176–2189. https://doi.org/10.1016/j.xphs.2023.05.007

McClements, D. J., & Xiao, H. (2014). Excipient foods: Designing food matrices that improve the oral bioavailability of pharmaceuticals and nutraceuticals. *Food & Function, 5*(7), 1320–1333. https://doi.org/10.1039/c4fo00100a

Mehran, M., Masoum, S., & Memarzadeh, M. (2020). Improvement of thermal stability and antioxidant activity of anthocyanins of Echium amoenum petal using maltodextrin/modified starch combination as wall material. *International Journal of Biological Macromolecules, 148*, 768–776. https://doi.org/10.1016/j.ijbiomac.2020.01.197

Mishra, A., Pradhan, D., Biswasroy, P., Kar, B., Ghosh, G., & Rath, G. (2021). Recent advances in colloidal technology for the improved bioavailability of the nutraceuticals. *Journal of Drug Delivery Science and Technology, 65*, Article 102693. https://doi.org/10.1016/j.jddst.2021.102693

Montoya-Yepes, D. F., Jiménez-Rodríguez, A. A., Aldana-Porras, A. E., Velásquez-Holguin, L. F., Méndez-Arteaga, J. J., & Murillo-Arango, W. (2024). Starches in the encapsulation of plant active ingredients: State of the art and research trends. *Polymer Bulletin, 81*(1), 135–163. https://doi.org/10.1007/s00289-023-04724-6

Mu, H., Song, Z., Wang, X., Wang, D., Zheng, X., & Li, X. (2022). Microencapsulation of algae oil by complex coacervation of chitosan and modified starch: Characterization and oxidative stability. *International Journal of Biological Macromolecules, 194*, 66–73. https://doi.org/10.1016/j.ijbiomac.2021.11.168

Muhoza, B., Yuyang, H., Uriho, A., Harindintwali, J. D., Liu, Q., & Yang, L. (2023). Spray-and freeze-drying of microcapsules prepared by complex coacervation method: A review. *Food Hydrocolloids, 140*, Article 108650. https://doi.org/10.1016/j.foodhyd.2023.108650

Nath, R., & Dutta, K. (2024). *Sources of starch and its morphology: A review*. Informa UK Limited. https://doi.org/10.1201/9781003430469-47

Nezamdoost-Sani, N., Amiri, S., & Mousavi Khaneghah, A. (2024). The application of the coacervation technique for microencapsulation bioactive ingredients: A critical review. *Journal of Agriculture and Food Research, 18*, Article 101431. https://doi.org/10.1016/J.JAFR.2024.101431

Oladele, A. K., Duodu, K. G., & Emmambux, N. M. (2020). Hydrolysis and antioxidant activity of starch modified with phenolic extracts from grape pomace and sorghum bran under alkaline conditions. *Carbohydrate Polymers, 240*, Article 116291. https://doi.org/10.1016/j.carbpol.2020.116291

Olawoye, B., Jolayemi, O. S., Akinyemi, T. Y., Nwaogu, M., Oluwajuyitan, T. D., Popoola-Akinola, O. O., Fagbohun, O. F., & Akanbi, C. T. (2023). *Modification of starch starch: Advances in modifications, technologies and applications*. Nigeria: Springer International Publishing. https://doi.org/10.1007/978-3-031-35843-2_2

Parhizkary, M., Hasanpour, R., Assadpour, E., & Jafari, S. M. (2025). Spray-drying encapsulation of jujube extract: Enhancing bioactivity and stability via maltodextrin-based carriers with polysaccharides and proteins. *Carbohydrate Polymer Technologies and Applications, 10*, Article 100869. https://doi.org/10.1016/j.carpta.2025.100869

Park, S., & Kim, Y. R. (2021). Clean label starch: Production, physicochemical characteristics, and industrial applications. *Food Science and Biotechnology, 30*(1). https://doi.org/10.1007/s10068-020-00834-3

Pourshamohammad, S., Amrani, M., Sarabi-Aghdam, V., Abedi, A.-S., & Mahmoudzadeh, M. (2024). Investigating the surfactant-to-oil ratio, type of ripening inhibitor, and modified starch on the stability of Carum copticum essential oil nanoemulsions: Application in yoghurt drink. *Journal of Agriculture and Food Research, 16*, Article 101207. https://doi.org/10.1016/j.jafr.2024.101207

Quintero-Castaño, V. D., Vasco-Leal, J. F., Cuellar-Nuñez, L., Luzardo-Ocampo, I., Castellanos-Galeano, F., Álvarez-Barreto, C., Bello-Pérez, L. A., & Cortés-Rodriguez, M. (2021). Novel OSA-modified starch from gros michel banana for encapsulation of andean blackberry concentrate: Production and storage stability. *Starch Staerke, 73*(3–4), Article 2000180. https://doi.org/10.1002/star.202000180

Ramos, G. V. C., Pinho, S. C.d., Gomes, A., Dacanal, G. C., Sobral, P. J.d. A., & Moraes, I. C. F. (2024). Designing pickering emulsions stabilized by modified cassava starch nanoparticles: Effect of curcumin encapsulation. *Processes, 12*(7). https://doi.org/10.3390/pr12071348

Ramos, G. V. C., Ramírez-López, S., Pinho, S. C.d., Ditchfield, C., & Moraes, I. C. F. (2025). Starch-based pickering emulsions for bioactive compound encapsulation: Production, properties, and applications. *Processes, 13*(2). https://doi.org/10.3390/pr13020342

Rashwan, A. K., Younis, H. A., Abdelshafy, A. M., Osman, A. I., Eletmany, M. R., Hafouda, M. A., & Chen, W. (2024). Plant starch extraction, modification, and green applications: A review. *Environmental Chemistry Letters, 22*(5), 2483–2530. https://doi.org/10.1007/s10311-024-01753-z

Rehman, A., Jafari, S. M., Tong, Q., Karim, A., Mahdi, A. A., Iqbal, M. W., Aadil, R. M., Ali, A., & Manzoor, M. F. (2020). Role of peppermint oil in improving the oxidative stability and antioxidant capacity of borage seed oil-loaded nanoemulsions fabricated by modified starch. *International Journal of Biological Macromolecules, 153*, 697–707. https://doi.org/10.1016/j.ijbiomac.2020.02.292

Rehman, A., Tong, Q., Jafari, S. M., Korma, S. A., Khan, I. M., Mohsin, A., Manzoor, M. F., Ashraf, W., Mushtaq, B. S., Zainab, S., & Cacciotti, I. (2021). Spray dried nanoemulsions loaded with curcumin, resveratrol, and borage seed oil: The role of two different modified starches as encapsulating materials. *International Journal of Biological Macromolecules, 186*, 820–828. https://doi.org/10.1016/j.ijbiomac.2021.07.076

Reque, P. M., & Brandelli, A. (2021). Encapsulation of probiotics and nutraceuticals: Applications in functional food industry. *Trends in Food Science and Technology, 114*, 1–10. https://doi.org/10.1016/j.tifs.2021.05.022

Ribeiro, A. M., Gonçalves, A., Rocha, F., & Estevinho, B. N. (2024). Statistical simplex centroid experimental design for evaluation of pectin, modified chitosan and modified starch as encapsulating agents on the development of vitamin E-loaded microparticles by spray-drying. *International Journal of Biological Macromolecules, 269*. https://doi.org/10.1016/j.ijbiomac.2024.131792

Rong, L., Liu, W., Shen, M., Xiao, W., Chen, X., Yang, J., & Xie, J. (2022). The effects of mesona chinensis benth gum on the pasting, rheological, and microstructure properties of different types of starches. *Current Research in Food Science, 5*, 2287–2293. https://doi.org/10.1016/j.crfs.2022.11.012

Ruengdech, A., & Siripatrawan, U. (2022). Improving encapsulating efficiency, stability, and antioxidant activity of catechin nanoemulsion using foam mat freeze-drying: The effect of wall material types and concentrations. *Lebensmittel-Wissenschaft & Technologie, 162*, Article 113478. https://doi.org/10.1016/j.lwt.2022.113478

Sahraeian, S., Niakousari, M., Fazaeli, M., & Hosseini, S. M. H. (2023). Fabrication and study on dually modified starch embedded in alginate hydrogel as an encapsulation system for satureja essential oil. *Carbohydrate Polymers, 322*. https://doi.org/10.1016/j.carbpol.2023.121331

Samyor, D., & Haokip, N. (2023). *Hot melt extrusion and complex coacervation system for delivering nutraceuticals and nanonutraceuticals*. Springer Science and Business Media LLC. https://doi.org/10.1007/978-3-030-69677-1_43-1

Sawant, R. B., Nikam, S. P., Roy, A., Kumar, A., Mohammed, O. A., Sharma, K., Rai, A. K., Roy, A., Gaur, A., & Verma, R. (2024). Nanocarriers for nutraceutical delivery: A miniaturized revolution in health. *Nano-Structures & Nano-Objects, 39*. https://doi.org/10.1016/j.nanoso.2024.101321

Sedaghat Doost, A., Afghari, N., Abbasi, H., Nasrabadi, M. N., Dewettinck, K., & Van der Meeren, P. (2020). Nano-lipid carriers stabilized by hydrophobically modified starch or sucrose stearate for the delivery of lutein as a nutraceutical beverage model. *Colloids and Surfaces A: Physicochemical and Engineering Aspects, 605*, Article 125349. https://doi.org/10.1016/j.colsurfa.2020.125349

Shahidi, F., & Ambigaipalan, P. (2015). Phenolics and polyphenolics in foods, beverages and spices: Antioxidant activity and health effects – A review. *Journal of Functional Foods, 18*, 820–897. https://doi.org/10.1016/j.jff.2015.06.018

Shahidi, F., & Pan, Y. (2022). Influence of food matrix and food processing on the chemical interaction and bioaccessibility of dietary phytochemicals: A review. *Critical Reviews in Food Science and Nutrition, 62*(23), 6421–6445. https://doi.org/10.1080/10408398.2021.1901650

Singh, A. K., Kumar, A., Gaikwad, K. K., & Lee, Y. S. (2023). *Enzymatic modification of starch starch: Advances in modifications, technologies and applications*. South Korea: Springer International Publishing. https://doi.org/10.1007/978-3-031-35843-2_17

Sinhmar, A., Kumar Pathera, A., Sharma, S., Nehra, M., Thory, R., & Nain, V. (2023). Impact of various modification methods on physicochemical and functional properties of starch: A review. *Starch - Stärke, 75*(1–2). https://doi.org/10.1002/star.202200117

Subhasri, D., Thirukumaran, R., Maria Leena, M., Moses, J. A., & Anandharamakrishnan, C. (2024). Customizing nutraceutical delivery with 3D food printing. *Food and Humanity, 3*, Article 100430. https://doi.org/10.1016/j.foohum.2024.100430

Sun, C., Du, K., Zhu, Z., Wang, C., He, Z., Hu, Y., Xie, Q., & Du, X. (2025). Starch nanoparticles regulate the steric conformation of soy protein isolate to stabilize high internal phase pickering emulsions for curcumin encapsulation. *International Journal of Biological Macromolecules, 302*, Article 140560. https://doi.org/10.1016/j.ijbiomac.2025.140560

Sun, C., Wei, Z., Xue, C., & Yang, L. (2023). Development, application and future trends of starch-based delivery systems for nutraceuticals: A review. *Carbohydrate Polymers, 308*, Article 120675. https://doi.org/10.1016/j.carbpol.2023.120675

Ting, Y., Jiang, Y., Ho, C. T., & Huang, Q. (2014). Common delivery systems for enhancing in vivo bioavailability and biological efficacy of nutraceuticals. *Journal of Functional Foods, 7*(1), 112–128. https://doi.org/10.1016/j.jff.2013.12.010

Varghese, S., Awana, M., Mondal, D., Rubiya, M. H., Melethil, K., Singh, A., Krishnan, V., & Thomas, B. (2022). *Amylose–amylopectin ratio*. Springer Science and Business Media LLC. https://doi.org/10.1007/978-981-16-6603-2_48-1

Viji, P. C., Chawla, R., Sivakumar, S., Yadav, D. N., Goel, N., & Kumar Anurag, R. (2023). Characterization of ultrasonicated assisted encapsulated omega 3 fatty acids and inulin for food applications. *Carbohydrate Polymer Technologies and Applications, 6*, Article 100336. https://doi.org/10.1016/j.carpta.2023.100336

Wanyi, W., Lu, L., Zehan, H., & Xinan, X. (2020). Comparison of emulsifying characteristics of different macromolecule emulsifiers and their effects on the physical properties of lycopene nanoemulsions. *Journal of Dispersion Science and Technology, 41*(4), 618–627. https://doi.org/10.1080/01932691.2019.1610421

Yan, C., McClements, D. J., Zou, L., & Liu, W. (2019). A stable high internal phase emulsion fabricated with OSA-Modified starch: An improvement in β-carotene stability and bioaccessibility. *Royal Society of Chemistry, China Food and Function, 10*(9), 5446–5460. https://doi.org/10.1039/c9fo00508k

Yang, M., Peng, J., Zi, Y., Shi, C., Kan, G., Wang, X., & Zhong, J. (2023). Encapsulation of fish oil by complex coacervation and freeze drying with modified starch aid. *Food Hydrocolloids, 137*, Article 108371. https://doi.org/10.1016/j.foodhyd.2022.108371

Yang, Z., Xie, F., & Cai, J. (2024). Advances in the delivery of bioactive compounds using starch-based nanogels: Formation mechanism, contemporary challenges, and prospects. *Journal of Agriculture and Food Research, 16*, Article 101142. https://doi.org/10.1016/J.JAFR.2024.101142

Yao, T., & Janaswamy, S. (2022). Ordered hydrocolloids networks as delivery vehicles of nutraceuticals: Optimal encapsulation of curcumin and resveratrol. *Food Hydrocolloids, 126*, Article 107466. https://doi.org/10.1016/j.foodhyd.2021.107466

Yao, M., McClements, D. J., & Xiao, H. (2015). Improving oral bioavailability of nutraceuticals by engineered nanoparticle-based delivery systems. *Current Opinion in Food Science, 2*, 14–19. https://doi.org/10.1016/J.COFS.2014.12.005

Yuan, X., Xiao, J., Liu, X., McClements, D. J., Cao, Y., & Xiao, H. (2019). The gastrointestinal behavior of emulsifiers used to formulate excipient emulsions impact the bioavailability of β-carotene from spinach. *Food Chemistry, 278*, 811–819. https://doi.org/10.1016/j.foodchem.2018.11.135

Zhang, Z., & Bao, J. (2023). Recent advances in modification approaches, health benefits, and food applications of resistant starch. *Starch - Stärke, 75*(9–10). https://doi.org/10.1002/star.202100141

Zhang, K., Huang, J., Wang, D., Wan, X., & Wang, Y. (2024). Covalent polyphenols-proteins interactions in food processing: Formation mechanisms, quantification methods, bioactive effects, and applications. *Frontiers in Nutrition, 11*. https://doi.org/10.3389/fnut.2024.1371401

Zhang, B., Zheng, L., Liang, S., Lu, Y., Zheng, J., Zhang, G., Li, W., & Jiang, H. (2022). Encapsulation of capsaicin in whey protein and OSA-Modified starch using spray-drying: Physicochemical properties and its stability. *Foods, 11*(4), 612. https://doi.org/10.3390/foods11040612

Zhao, X., Mei, T., & Cui, B. (2024). Lipids-modified starch: Advances in structural characteristic, physicochemical property, and application. *Food Research International, 197*, Article 115146. https://doi.org/10.1016/J.FOODRES.2024.115146

Zhi, K., Yang, H., Shan, Z., Huang, K., Zhang, M., & Xia, X. (2021). Dual-modified starch nanospheres encapsulated with curcumin by self-assembly: Structure, physicochemical properties and anti-inflammatory activity. *International Journal of Biological Macromolecules, 191*, 305–314. https://doi.org/10.1016/j.ijbiomac.2021.09.117

Zhu, F. (2017). Encapsulation and delivery of food ingredients using starch based systems. *Food Chemistry, 229*, 542–552. https://doi.org/10.1016/j.foodchem.2017.02.101

Zia-ud-Din, M., Xiong, H., & Fei, P. (2017). Physical and chemical modification of starches: A review. *Critical Reviews in Food Science and Nutrition, 57*(12), 2691–2705. https://doi.org/10.1080/10408398.2015.1087379

Chapter 9

Innovative starch utilization in bakery and confectionery products

Erandi Arvizu-Hernandez[1, 2, 3], Rosa L. González-Díaz[3], Anayansi Escalante-Aburto[1, 2], Jeny A. Rincón-Aguirre[2], Miguel Rebollo-Hernanz[4, 5] and Ivan Luzardo-Ocampo[1, 2, 3]

[1]*Tecnologico de Monterrey, Institute for Obesity Research, Monterrey, Mexico;* [2]*Tecnologico de Monterrey, School of Engineering and Sciences, Monterrey, Mexico;* [3]*Tecnologico de Monterrey, School of Engineering and Sciences, Zapopan, Mexico;* [4]*Department of Agricultural Chemistry and Food Science, Faculty of Science, Universidad Autónoma de Madrid, Madrid, Spain;* [5]*Department of Production and Characterization of Novel Foods, Institute of Food Science Research, CIAL (UAM-CSIC), Madrid, Spain*

1. Introduction

Starch is a versatile ingredient that serves as the most abundant botanical energy source, offering several technological and biological functionalities to the products it is added to, with bakery and confectionery items being prime examples. Typically, starch comes from various sources, and both native and modified starches are used in numerous food products, playing well-known roles; in bakery products, starches are convenient as binding agents, textural modifiers, moisture retainers, or even to reduce staling. In confectionery items, starches can replace fat and gluten, providing new textural features and other benefits (Chakraborty et al., 2022; Chen et al., 2025; Geetha et al., 2024). The bakery and confectionery industries are increasingly adopting modified starches to meet quality and freshness standards while addressing changing consumer demands (Królak et al., 2022; Sabaghi, 2025). Compared to native starches from natural sources, as shown in Fig. 9.1, starches modified through physical, chemical, or enzymatic processes exhibit greater resistance to processing stresses, including heat, shear, pH fluctuations, and freeze-thaw cycles. These modifications also improve water management and deliver more consistent textures, thereby enhancing overall product stability (Paternina-Contreras et al., 2025; Suri & Singh, 2023).

Key techno-functional properties currently sought in bakery and confectionery matrices include extending shelf life, acting as stabilizers and moisture retainers, delaying staling, improving elasticity, and increasing water-holding capacity. In bakery systems, starches serve as binding agents, texture modifiers, humectants, and anti-staling components, while in confectionery products, they can act as fat and gluten replacers, gelling agents, and emulsifiers, enabling the development

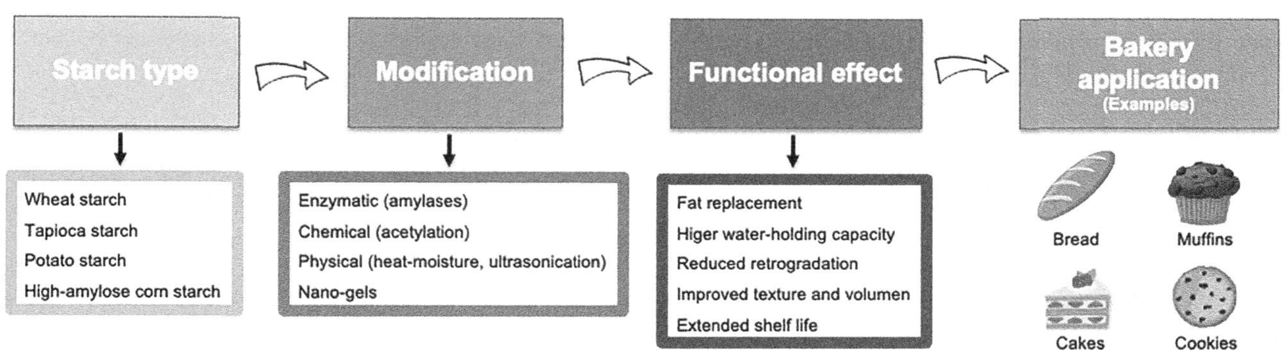

FIGURE 9.1 Technofunctional properties of native and modified starches in bakery and confectionery systems contribute to improved stability, texture, and overall product quality The figure indicates the overall transformation needed from native to modified starches, and how these transformations allow a wide range of applications into bakery and confectionery food products. *Data from the Authors.*

Native and Modified Starches for Food Applications. https://doi.org/10.1016/B978-0-443-27431-2.00019-5

of innovative textures and enhanced sensory qualities. Recent studies have highlighted the diverse applications of modified starches in these sectors (Navaf et al., 2024; Nikolić et al., 2024; Punia et al., 2019; Suárez-Castillo et al., 2024). For example, cross-linked starches have been used to improve the texture and shelf life of gluten-free bakery products, while starches modified with octenyl succinic anhydride (OSA) have been employed as emulsifiers in fat-reduced confectionery items (Balic et al., 2016; Punia et al., 2019). These modifications allow for the creation of products with desirable sensory attributes and extended freshness, appealing to health-conscious consumers seeking low-fat and gluten-free options.

This chapter aims to critically analyze the innovative use of native and modified starches in bakery and confectionery products, emphasizing their technological, nutritional, and functional roles. By incorporating recent scientific advances, it demonstrates how physical, chemical, and enzymatic modifications of starch enhance product quality through improved texture, increased moisture retention, extended shelf life, and reduced fat content, all while meeting consumer demands for healthy, clean-label, and sustainable options. Additionally, this chapter outlines future research directions that position starch not only as a functional ingredient but also as a strategic platform for developing bakery and confectionery products focused on health, innovation, and sustainability.

2. Starch utilization in bakery products

Starch is a key part of bakery systems, affecting texture, structure, and shelf life. The rising demand for healthier and more sustainable baked goods has boosted the use of both native and modified starches as tools for innovation. While native starches mainly provide bulk and gelatinization properties, modified starches—produced through physical, chemical, or enzymatic treatments—show improved process tolerance, higher water-binding capacity, and better interactions with proteins and lipids. These features lead to improved dough rheology, crumb structure, and resistance to staling. The following sections explore how starch functions in bread, cakes, muffins, and cookies, focusing on its molecular mechanisms, techno-functional qualities, and potential to create health-focused and value-added products. This section discusses the use of native and modified starches in bakery items, highlighting their effects on texture, shelf life, nutritional value, and the trend toward functional or health-oriented products.

2.1 Bread

As one of the most widely consumed staple foods, bread quality depends heavily on starch's behavior during dough development and baking. Starch granules absorb water, swell, and gelatinize, forming a viscoelastic network with gluten proteins that stabilizes gas cells and influences crumb texture. Retrogradation of amylopectin during storage causes staling, making starch modification a key strategy to improve freshness (Hu et al., 2023).

2.1.1 Modified starches as fat replacers in bread

Several reports have indicated the feasibility of using modified starches as fat replacers in bread. For instance, Balic et al. (2016) explored the use of modified wheat and tapioca starches as fat replacers in bread formulations. Through their study, they demonstrated that modified starches can effectively mimic the sensory and textural properties of fat, producing bread with reduced fat content while maintaining the desired mouthfeel and softness. The modifications introduced into the starches increased the dough's water-holding capacity, which in turn improved its stability and extended its shelf life. This approach not only addresses health complications associated with high-fat diets but also provides a cost-effective solution for the baking industry.

2.1.2 Revolutionary starch-derived additives for improving bread quality

To introduce novel baking additives, Dou et al. (2024) reported a composite of chitosan hydrochloride and carboxymethyl starch sodium nanogel was reported. The new additive presented an impressive upgrade in the texture, volume, and crumb texture of wheat bread. The creation of nano-gels improved the viscoelastic properties of the dough, resulting in enhanced gas retention during fermentation and baking. Aside from that, the additive manifested antimicrobial behavior, which played a significant role in increasing shelf life by preventing mold growth. This research reveals the prospects of starch-based nano-gels as multifunctional additives in bread-making.

Likewise, Durmus et al. (2023) investigated the novel application of hazelnut skin and modified starches in sourdough bread recipes. The incorporation of these modified starches, along with hazelnut skin, not only enhanced the nutritional value of the bread by increasing its dietary fiber content but also improved its texture. The starch modifications resulted in a softer crumb and retarded staling, while the addition of hazelnut skin imparted a typical flavor and increased the bread's antioxidant level. The study demonstrates the feasibility of utilizing blends of processed starches and food industry waste to produce value-added bakery products.

2.1.3 Enzymatic modification of starch for better bread quality

Enzymatic modification can be applied to several sources of starch. Wang et al. (2025) conducted research on the enzymatic modification of wheat starch by a new maltotetraose-forming amylase from *Atopomonas hussainii*. The enzymatic treatment resulted in starch with decreased retrogradation characteristics, leading to bread with higher softness and extended freshness. The procedure also improved the fermentation quality of the dough, resulting in improved volume and crumb texture. The research highlights the significance of enzymatic starch modification for making high-quality bread with an extended shelf life.

The innovative use of native and modified starches in bread-making has opened new opportunities to improve the quality, nutrition, and shelf life of baked goods. The use of fat substitutes, specialty additives, and enzymatic modifications has helped starch continue to play a key role in advancing the bakery industry. The studies in this section all emphasize different ways starch is used in bread formulation, paving the way for further research and development. Current research shows that starch's role in bread goes beyond simply thickening or gelatinization; it now includes controlling amylopectin retrogradation at the molecular level, managing protein-starch interactions, and optimizing water distribution. The incorporation of enzyme-tailored starches and nano-structured additives creates new possibilities for breads with better quality, longer shelf life, and increased health benefits, leading to a major shift in bakery innovation.

2.2 Cakes

Just as bread, the cake is among the most popular baked goods, largely favored for its soft crumb and sweet flavor. As a key category of baked goods made primarily with starch-rich flours, it is also a trend (Sergiacomo et al., 2025). The sensory quality of baked products such as sponge cake is closely associated with their structural characteristics, particularly crumb tenderness. This attribute emerges from intricate physical and chemical interactions among ingredients during mixing and baking (Guiné, 2022; Olakanmi et al., 2024). Crumb softness largely depends on the batter's capacity to incorporate and retain air bubbles, a process influenced by batter viscosity and foam stability. Central to these interactions is the role of starch, which undergoes gelatinization during baking, contributing to the setting of the crumb structure. Starch gelatinization, alongside protein coagulation facilitated by eggs, fats, and sugar, establishes a matrix that traps air cells, thereby enhancing crumb softness and creating the desirable light and airy texture characteristic of high-quality cakes (Cao et al., 2025; Scott & Awika, 2023).

2.2.1 Starch composition, processing, and functional implications in cake quality

A recent study examined the starch content and its resistant (RS) and digestible (DS) fractions in five different starch sources under model baking conditions that mimicked typical cake preparation, with varying water levels and heat treatment. The findings showed that starch composition and processing significantly influence functional properties related to cake quality (Slade et al., 2021). Native starch granules (RS2) tend to lose their resistance when heated and exposed to moisture, due to gelatinization and structural disruption, thereby affecting their digestibility and texture. Importantly, Hi-Maize corn starch demonstrated thermostability, retaining a high level of RS even after baking, making it a promising ingredient for health-oriented baked products (Gutiérrez-Luna et al., 2024).

The starch modifications directly influence batter viscosity and gelatinization, which are essential factors for air incorporation and crumb structure in cakes (Monnet et al., 2021; Qian et al., 2023). Overall, understanding how starch fractions behave under baking conditions offers valuable insights for optimizing cake formulations to enhance texture, nutritional quality, and storage stability, emphasizing the fundamental role of starch selection and processing in cake manufacturing.

2.2.2 Influence of starch modifications and flour substitutions on batter properties and cake structure

Starch modifications directly impact batter viscosity and gelatinization, which are fundamental for air incorporation and crumb structure. Monnet et al. (2021) investigated how different mixing processes affect the incorporation of air and the consistency of wheat-based cake batter, particularly when substituting part of the wheat flour with pea flour. Among the four tested mixing methods, two were further analyzed because they demonstrated better efficiency in maintaining air bubbles. In this case, wheat-based cakes with dual-modified starch (enzymatic hydrolysis followed by acetylation) exhibited improved moisture retention, enhanced water and oil absorption, and increased resistant starch content, while also reducing paste viscosity. At 5% replacement in wheat flour, it enhanced crumb softness, elasticity, volume, crust color, and structural uniformity through beneficial starch-gluten interactions. Increasing the proportion of pea flour raised the cake density and caused structural changes, including deeper cracks on the surface and altered crumb stiffness. These results highlight its potential as a multifunctional ingredient for producing superior quality, health-oriented wheat-based cakes.

On the other hand, the effect of partially replacing wheat flour with pea flour has been shown to influence the physicochemical properties of steamed oat cakes (Qian et al., 2023). Given the differences in protein composition between pea and wheat flour, the impact of this substitution was assessed across different stages of cake preparation, with particular focus on air incorporation, a key factor influencing cake cell structure and density (Pycarelle & Delcour, 2021; Wang et al., 2024). Four different orders of ingredient incorporation were tested to determine their effects on a standard cake recipe made with 100% wheat flour. Results showed that mixing eggs and sugar first allowed greater air incorporation, but most of it was lost after adding oil and flour. Regardless of the mixing order, the batter density stabilized at approximately 1.1 ± 0.2 g/cm^3; however, the batter consistency and final crumb structure varied. When using the two most energy-efficient mixing sequences with pea flour substitutions, more air was incorporated into the batter (Tiruneh et al., 2025). Interestingly, while the resulting cakes were denser, they were also softer, an effect attributed to changes in cell structure caused by differences in protein composition and air distribution (Maravić et al., 2022).

2.2.3 Applications of modified starches for improved cake quality and health benefits

In the research carried out by Sahnoun et al. (2016), the physicochemical properties of dual-modified starch -produced by enzymatic hydrolysis followed by acetylation- and its effects on wheat-based cake quality were evaluated compared with native and singly modified starches. In wheat-based cakes, dual-modified starch enhanced moisture retention, water absorption, and oil uptake while increasing resistant starch content for potential health benefits. At a 5% replacement level in wheat flour, it improved crumb softness, elasticity, crust color, volume, and uniformity, while reducing hardness and chewiness, with no further gains at 10%. These improvements stemmed from altered starch structure and enhanced starch-gluten interactions, highlighting its value as a multifunctional ingredient for high-quality cake production (Ferron et al., 2022).

Incorporating octenyl succinic anhydride-modified mung bean starch (OSA-MS) as a fat replacer in wheat-based cakes offers significant health benefits by reducing fat content while maintaining desirable texture and volume (Punia et al., 2019). At optimal replacement levels (up to 30%), OSA-MS enhances moisture retention and swelling, contributing to a soft crumb and acceptable sensory qualities. This modification also lowers the risk of retrogradation, potentially improving shelf life and nutritional value. Although higher substitution levels may negatively affect texture, the 30% replacement strikes a balance between healthfulness and product quality. Overall, OSA-MS presents a promising multifunctional ingredient for developing healthier bakery products without compromising consumer acceptance.

The reviewed studies emphasize the crucial role of starch in determining cake quality because it influences batter viscosity, air incorporation, gelatinization, and interactions with other macromolecules. Both native and modified starches, whether added directly or through partial flour substitution, show significant potential to adjust crumb texture, volume, and moisture retention, while also offering nutritional benefits like increased resistant starch or reduced fat content. By optimizing the starch source, modification method, and processing order, targeted improvements in cake structure, sensory qualities, and health benefits can be achieved, supporting the development of high-quality, functionally enhanced bakery products. Starch's function in cakes goes beyond providing structure; it actively influences batter aeration, crumb softness, and nutritional value. Research demonstrates that customizing amylose-amylopectin structure or combining starch with alternative flours allows for cakes that combine indulgence with health benefits, driving the innovation of next-generation bakery items.

2.3 Muffins

Muffins, a popular baked good enjoyed worldwide, have seen notable improvements in their formulation and quality through innovative use of native and modified starches. Starch plays a vital role in muffin production, affecting texture, moisture retention, shelf life, and overall sensory qualities. Recent research has investigated the use of modified starches, starch-based coatings, and alternative starch sources to improve muffin quality, lower fat content, and meet dietary needs such as gluten-free options (Ha et al., 2025; Navaf et al., 2024; Salazar et al., 2021). This section highlights key studies that showcase the various applications of starch in muffin manufacturing.

2.3.1 Modified starches as fat replacers in muffins

A key area of innovation is reducing muffin fat content without compromising texture and sensory properties. Rodriguez-Sandoval et al. (2017) investigated the use of modified cassava starch as a fat replacer in muffin formulations. Their study revealed that modified cassava starch could effectively mimic the functional properties of fat, resulting in muffins with reduced fat content while maintaining desirable softness, moisture levels, and crumb texture. The modified starch also enhanced the viscosity of the batter, resulting in increased air incorporation and uniform texture. This approach offers consumers a healthier option while retaining the richness of conventional muffins. Similarly, Geetanjali et al. (2025) studied the use of ultrasonicated cornstarch as a fat substitute in muffin recipes. Ultrasonication, a new modification process, improved the water-binding ability and gelation properties of starch, leading to muffins with better texture and sensory qualities. The research showed that ultrasonicated corn starch could replace 50% of the fat without harming muffin quality. This emphasizes the potential of modified starches as functional ingredients in low-fat muffin formulations.

2.3.2 Starch-based edible coatings for shelf-life extension

The application of a starch-based edible coating has been developed as a potential method to enhance the shelf life and quality of muffins. Researchers prepared an edible coating using cassava starch and rosemary essential oil, then evaluated its impact on muffin shelf life. The coating significantly reduced moisture loss and delayed staling, helping the muffins stay fresh and soft longer. Additionally, rosemary essential oil showed antimicrobial activity, further extending shelf life by inhibiting mold growth (Hassan et al., 2023). This study demonstrates the effectiveness of starch-based coatings as a natural and eco-friendly way to improve muffin quality and shelf life.

2.3.3 Alternative starch sources for gluten-free muffins

The demand for gluten-free bakery products has led to investigations into alternative starch sources for muffin production. Alvarez et al. (2017) explored using chickpea flour combined with corn starch and egg whites in gluten-free muffin recipes. The addition of corn starch improved the batter's texture and produced a tender crumb, while chickpea flour enhanced the nutritional value by boosting protein and fiber content. The study showed that the combination of egg white and corn starch provided the structure and elasticity typically supplied by gluten, allowing for the production of gluten-free muffins that closely resembled regular muffins in texture and consumer acceptance.

2.3.4 Native and modified starches in reduced-fat muffins

Reducing fat in muffins is challenging, as this bakery product requires a lipid component to display its desirable characteristics, yet several reports have suggested the potential of starch modifications to reduce fat content (Ahsan et al., 2025). Lee and Puligundla (2016) compared the effects of native and modified rice starches in reduced-fat muffin and cookie formulations. In muffins, the modified rice starch exhibited superior water-holding capacity and gelation properties, resulting in improved texture and moisture retention compared to native starch. The study indicated that fat could be successfully replaced with modified rice starch while maintaining muffins' desired sensory qualities. This research highlights the flexibility of modified starches for developing reduced-fat baked products without affecting quality.

In muffins, starch-based innovations demonstrate that process-driven modifications can be just as impactful as molecular changes, expanding the functional uses of starch for both quality preservation and nutritional improvement. The use of native and modified starches in muffin production has opened new opportunities to enhance product quality, nutritional value, and shelf life. From fat substitutes and edible coatings to gluten-free options and alternative starch sources, starch continues to be crucial in advancing muffin formulations (Baixauli et al., 2008; Vidaurre-Ruiz et al., 2022). The studies discussed in this section collectively showcase the varied applications of starch in muffin making, laying a foundation for further research and development in this area. These advancements not only address consumers' preferences for healthier and more sustainable baked goods but also equip the industry with innovative tools to improve product quality and functionality.

2.4 Cookies

Cookies, a popular baked good, have greatly benefited from innovative uses of native and modified starches. Starch is a vital ingredient in cookie recipes because it affects many important traits, such as texture, spreadability, crispiness, and shelf life. Current research has concentrated on modified starches, alternative starch sources, and physical modifications to enhance cookie quality, lower fat content, and meet nutritional goals. In this section, we highlight key studies that showcase the various ways starch is used in cookie manufacturing.

2.4.1 Modified starches in cookie formulations

Modified starches have gained significant attention for their use in cookie products because of their ability to improve functionality and texture. Nadir et al. (2015) studied the modification of potato starch using physical methods, such as heat-moisture treatment and annealing. These modifications increased the starch's water-binding capacity and gelation properties, leading to cookies with better texture, spreadability, and crispness. This study showed that starch from a modified potato can effectively replace part of the wheat flour in cookie recipes, providing a cost-effective and functional alternative.

Similarly, Lee and Puligundla (2016) explored the use of native and modified rice starches in reduced-fat cookie formulations. Modified rice starch exhibited superior water-holding capacity and gelation properties, resulting in cookies with enhanced texture and improved moisture retention compared to those made with native starch. This study highlighted that modified rice starch could effectively replace fat while maintaining the desirable sensory attributes of cookies, underscoring the versatility of modified starches in creating reduced-fat baked goods without compromising quality.

2.4.2 Alternative starch sources for cookie production

The exploration of new starch sources has uncovered new opportunities in cookie production, particularly in meeting nutritional needs and enhancing nutritional value. Mainardes et al. (2024) extracted and physically modified mango Palmer kernel starch for use in cookie formulations. The modified starch improved dough handling and resulted in a softer, more uniform texture in the finished product. Additionally, the mango Palmer kernel starch boosted the nutritional profile of cookies by increasing dietary fiber content. This study highlights the potential of underutilized starch sources to create innovative and nutritious baked goods. On the other hand, Kaur et al. (2016) investigated the extraction and utilization of cassava starch and flour in bakery products, including cookies. Cassava starch, known for its high purity and neutral flavor, was found to improve the texture and crispness of cookies. The study also highlighted the economic benefits of using cassava starch as a partial replacement for wheat flour, particularly in regions where cassava is highly available. This research underscores the potential of cassava starch as a functional and cost-effective ingredient in cookie production.

2.4.3 Starch modifications for improved digestibility and quality

Starch modifications have also been explored to improve the digestibility and quality of cookies. Van Hung et al. (2021) studied the effects of acid and heat-moisture treatment on sweet potato starch and its application in cookie production. The modified starch exhibited reduced digestibility, resulting in cookies with a lower glycemic index. This makes the cookies more suitable for consumers with special dietary requirements, such as those managing diabetes. Additionally, the modified starch improved the cookies' texture and shelf life by enhancing moisture retention and slowing down staling processes. This study highlights the potential of using modified starches to create healthier and more functional baked goods.

Starch modifications in cookies demonstrate dual functionality, improving texture while lowering the glycemic response. This positions cookies as a platform for health-oriented reformulation, aligning indulgence with dietary needs. Furthermore, the application of native and modified starches in cookie making has transformed the baking sector, enhancing product quality, nutrition, and functionality. From modified starches and alternative starch sources to physical modifications and enhanced digestibility, starch remains a pivotal component in advancing cookie formulations. The studies discussed in this section collectively highlight the diverse applications of starch in cookie production, paving the way for future research and development in this field. These developments not only address consumer demand for healthier, more sustainable bakery items but also provide the industry with new means to improve product quality and performance.

Across breads, cakes, muffins, and cookies, starch functions not just as a filler but as a multifunctional biopolymer whose molecular structure and modifications control water behavior, gelatinization, retrogradation, and interactions with proteins and lipids. Recent developments, from nano-gels to enzymatic tweaks, show how starch can be engineered to improve product quality, extend shelf life, and provide nutritional benefits. Unlike earlier studies that mostly viewed starch as a bulk thickener or gelatinizer, this overview emphasizes the molecular mechanisms through which starch influences texture and staling. It also highlights the connection between food engineering and health sciences, illustrating how starch modification allows for the creation of high-quality, shelf-stable products that meet consumer demands for healthier, functional bakery items.

Table 9.1 offers a recent comprehensive overview of starch types and their modifications used in bread, cakes, muffins, and cookies. This compilation summarizes data from the literature, emphasizing the diverse processing methods—such as enzymatic, chemical, and physical treatments, fermentation parameters, and baking protocols—and how they influence the functional properties of the final products. It systematically documents key effects on texture (e.g., hardness, springiness and chewiness), volume, moisture retention, staling behavior, and sensory quality. This detailed characterization highlights the critical role of starch selection and modification in improving product quality and shelf life in the bakery industry.

AACC: American Association of Cereal Chemists; AhMFA: Maltotetraose-forming amylase from *Atopomonas hussainii*; CHC: Chitosan hydrochloride; CMS-Na: Carboxymethyl starch sodium; HMT: Heat-moisture treatment; HPMC: Hydroxyl propyl methyl cellulose; LAB: Lactic acid bacteria; RH: relative humidity; RS: resistant starch; STMP: Sodium trimetaphosphate; STPP: Sodium tripolyphosphate.

2.5 Critical analysis and recommendations

The varied uses of native and modified starches in bakery products emphasize their essential role in influencing texture, structural strength, moisture retention, and overall sensory appeal (Navaf et al., 2024). Starches from traditional sources, such as wheat, tapioca, potato, and corn, as well as alternative sources such as mango kernel, sweet potato, or cassava, can be subjected to physical, enzymatic, or chemical modifications to improve their technological performance.

These modifications improve water retention, reduce retrogradation, and control viscosity, thus influencing dough performance and product quality in various bakery categories. In bread, modified starches serve as fat replacers and stability enhancers, leading to improved softness, volume, and shelf life (Arp et al., 2021). In cakes, starch modification helps incorporate air, improve crumb tenderness, and enhance nutrition through the enrichment with resistant starch (Psimouli & Oreopoulou, 2013). Muffins benefit from starch-based coatings and fat replacement, which help maintain freshness and support healthier formulations. In cookies, starch modifications increase crispness, lower glycemic index, and offer functional alternatives to wheat flour (Colla et al., 2018). Overall, these examples show how starch can act as both a structural and functional ingredient in modern bakery science. Despite these advances, several aspects warrant further exploration to strengthen the scientific basis and industrial relevance of starch applications, particularly in the context of cookie production. Most studies have focused on empirical outcomes such as hardness, moisture retention, or spread ratio, yet have provided limited mechanistic insight into how these properties are linked to starch gelatinization, retrogradation, or molecular interactions with proteins and lipids in the dough matrix. Incorporating mechanistic analyses would not only deepen scientific understanding but also allow more predictive control over product quality (de la Rosa-Millan, 2025; Zhang et al., 2021). Furthermore, inconsistencies in reporting processing parameters (including mixing regimes, resting times, and baking conditions) limit reproducibility, even though these factors critically modulate starch functionality (Arp et al., 2021). Standardization or more detailed documentation of these conditions is therefore essential. From a nutritional perspective, while some studies evaluate fat reduction or glycemic response, systematic analyses of fiber content, resistant starch fractions, and potential prebiotic effects remain scarce, representing a missed opportunity to align starch research with health-driven consumer demands (Tekin & Dincer, 2023).

Equally important are long-term storage studies, as stability, moisture migration, and sensory changes directly affect the commercial viability of starch-enriched cookies and related products. Additionally, the valorization of unconventional starch sources, such as mango kernels or sweet potatoes, provides a dual benefit: enhancing product functionality while promoting sustainability through the reuse of agro-industrial by-products. Expanding this research area would support global priorities for resource efficiency and eco-friendly food production. Therefore, while current evidence highlights the multifunctionality of starches in bakery applications, progressing toward mechanistic understanding, standardized methods, and sustainability-driven innovation will be crucial for establishing starch as a key element in functional bakery science.

TABLE 9.1 A recent description of starch types and modifications in baked products, detailing processing conditions and key functional effects on texture, volume, moisture retention, staling, and sensory quality.

Product type	Starch source	Modification/process conditions	Functional properties observed in formulations	References
Bread				
Wheat bread	Chitosan-HCl/Carboxymethyl starch (nano-gel)	Cross-linking CHC + CMS-Na using EDC·HCl; fermentation: 75 min at 37°C, 75% RH; baking: 160°C. Other ingredients: wheat flour, water, milk powder, eggs, salt, sugar, instant yeast	Increase in specific volume, porosity, and springiness. Decrease in hardness parameter. Sensory score: 85.33/100	Dou et al. (2024)
Wheat-based sourdough bread	Cross-linked and NaClO-oxidized corn starch (3%–5%)	3 %–5% starch replacing part of flour + hazelnut skin; fermentation (LAB), proofing: 150 min at 30°C; baking: 210°C for 25 min; includes flour, sourdough, salt, sugar, yeast, shortening, water	Enhanced hardness, chewiness, and crumb appearance.	Durmus et al. (2023)
Wheat-based bread	Wheat and tapioca starches (4%)	Following AACC 10-09.01 (with fungal α-amylase, instant yeast, ammonium phosphate); fermentation: 2 h, one punch; baking per AACC. Other: patent flour, water.	Improved firmness, starch gel, and dough extension resistance. Decreased stickiness.	Balic et al. (2016)
Wheat-based bread	Wheat starch modified by AhMFA enzyme (0–5 ppm)	Dough with varying AhMFA doses; two 10 min rests; fermentation: 60 min at 38°C, 80% RH; baking: top 180°C, bottom 200°C for 20 min; cooled 2 h at 25°C.	Improved specific volume and decreased bread hardness, contributing to delayed staling.	Wang et al. (2025)
Wheat-based bread	60% RS high-amylose corn starch (10%–20%)	Wheat flour: RS blends (100:0, 90:10, 85:15, 80:20); proofing: 1 h at 32°C; baking: 180°C for 15 min (industrial). Ingredients: Sugar, yeast, oil, salt, water.	Decreased specific volume and retrogradation rate, contributing to increased firmness and crumb moisture.	Barros et al. (2018)
Wheat-based breadsticks	70% RS corn starch (40%–70% flour replacement)	Replacement levels 40%–70%; dough shaped into strips (0.5 × 15 cm), baked: 180°C for 20–25 min (gas oven).	Decreased moisture and hardness. Increased lightness of the resulting product.	Petchoo et al. (2021)
Wheat-based white bread	Spray-dried RS (35% amylose-lipid, retrograded)	RS at 20%, 30%, or 50% (w/w); fermentation: 2 h at 30°C; baking: 200°C for 20 min. Ingredients: shortening, yeast, salt, water.	Decreased porosity and overall acceptability, but lower hardness than control.	Yun et al. (2021)
Rice flour-based bread	Rice flour + tapioca/corn/potato starch (10%–30%)	Mixtures at 10%–30% substitution; fermentation: 45 min at 32°C, 80%–85% RH; baking: top 195°C, bottom 200°C for 30 min. Other: water, sugar, yeast, HPMC, milk, soy protein isolate	Increased crust lightness and acceptability, and decreased crumb hardness, being the 20 %-added potato starch as the most acceptable.	Kim et al. (2015)

Wheat-based steamed buns	Cross-linked rice starch (15% replacement)	Starch cross-linked (12% STMP/STPP) at 45°C, pH 11.5 for 3 h; 15% substitution; proofing until doubled, then steaming 12–15 min.	Increased dietary fiber content. Decreased hardness, yellowness, and resulting products did not differ in sensory acceptability compared to control.	Abdul et al. (2017)
Wheat/composite flour-based white bread	Citric-acid, heat-moisture treated sweet potato starch	Modified starch; fermentation: 90 min at 30°C, 85% RH with punching; blanching 15 min; proofing at 38°C for 33 min; baking: 180°C for 20 min.	Decreased specific volume, increased hardness and gumminess. No differences in color, appearance, texture, odor	Tien et al. (2018)
Wheat-based sponge cake	Corn starch (0%–100%)	Sponge dough formulations; thermal profile via calorimetry: 25 → 100°C at 10°C/min. Other: milk, eggs, water, sugar, baking powder.	Reduced hardness, adhesiveness, cohesiveness, springiness, and chewiness	Guadarrama-Lezama et al. (2016)
Cakes				
Vegan cakes	Hi Maize, potato starch, tapioca starch, green banana flour and potato fiber	Simulated baking: 30%–50% water addition, 180°C for 35 min; 50% wheat flour replacement in cakes; RS measured by official and rapid enzymatic methods	Lower resistant starch at all samples, except for Hi-Maize-added ones.	Gutiérrez-Luna et al. (2024)
Wheat-based cake with pea flour substitution	Wheat flour + pea flour (20%–40% replacement)	Replacement of wheat flour with 0%, 20%, or 40% pea flour; four multi-stage mixing orders tested (industry-type and sponge-type sequences with varied egg, oil, and flour addition steps); planetary mixer; optimized speed and time to control air incorporation; baked at 180°C for 16 min.	Pea flour increased the incorporated air in batter, its density, and softness.	Monnet et al. (2021)
Steamed oat cakes	Naked oat flour (raw and stir-fried)	Oat kernels tempered (16 h) to moisture levels of 25.0%, 27.5%, 32.5%, or 35.0%, then stir-fried at 160°C for 25 min to achieve gelatinization degrees of 35%, 50%, 75%, or 90%. Milled to pass 80-mesh sieve. Batter prepared with varied water addition for constant consistency; fermented with 1% yeast +15% sugar at 37°C for 2 h, then steamed for 25 min.	A higher gelatinization and softness due to increased water uptake was shown. However, resulting products showed lower structure and volume than control.	Qian et al. (2023)
Wheat-based cakes	Dual-modified corn starch (5%–10%)	Enzymatic hydrolysis + acetylation; added at 5%–10% to cake mix; baking: 180°C for 28 min. Ingredients: Sugar, milk, margarine, eggs, yeast, salt.	Decreased hardness, cohesion, adhesion, and chewiness. Increased elasticity, and volume height. Effects were more prominent in the 10 %-added starch formulation.	Sahnoun et al. (2016)
Wheat-based cakes	Heat-treated OSA mungbean starch (10%–40% fat replacer)	Starch pretreated at 120°C for up to 3 h, fat replaced at 10%–40%; baking: 195°C for 25 min.	Increased specific volume, hardness, springiness, cohesiveness, and chewiness. Decreaded overall acceptability.	Punia et al. (2019)
Muffins				
Refined wheat flour-based muffins	Ultrasound corn starch (UCS) (4%–25% fat replacer)	UCS at various levels; batter mixed, baked: 180°C for 20 min. Ingredients: wheat flour, butter, sugar, milk, baking powder, soda, salt.	Enhanced batter viscosity resulted in increased hardness and springiness of the resulting products.	Geetanjali et al. (2025)

Continued

TABLE 9.1 A recent description of starch types and modifications in baked products, detailing processing conditions and key functional effects on texture, volume, moisture retention, staling, and sensory quality.—cont'd

Product type	Starch source	Modification/process conditions	Functional properties observed in formulations	References
Edible coating film for wheat flour-based muffins	Cassava starch film (25%–100%)	Edible film from cassava starch + gelatin, chitosan, glycerol, rosemary oil; applied to muffins and dried.	Films increased shelf life and decreased weight loss and microbial growth. A higher acceptability post-storage was also shown.	Hassan et al. (2023)
Wheat flour-based muffins	Cross-linked cassava starch (CCS) (4.48%–8.96%)	Fat replaced by CCS (8%–16%); baked: 215°C top, 205°C bottom for 19 min. Ingredients: standard muffin base.	Decreaded volume and moisture, resulting in augmented firmness, hardness, springiness, and chewiness.	Rodriguez-Sandoval et al. (2017)
Chickpea flour-based muffins	Corn starch (CS) (25%–75% corn): chickpea flour (CF) blends	CF:CS ratios (75:25, 50:50, 25:75); baked: 185°C for 25 min; cooling: 1 h at 20°C. Ingredients: eggs, milk, oil, sugar, baking powder.	Higher springiness, cohesiveness, and chewiness in the formulations. CS acceptability was 40% higher compared to CF-added products.	Alvarez et al. (2017)
Flour-based muffins	Native, acetylated, hydroxypropylated rice starch (20% fat replaced)	Baked at 204°C for 24 min. Ingredients: flour, sugar, butter, eggs, milk powder, salt, baking powder, water.	Increased weight, crust/crumb brightness, and products displayed a similar acceptability (6 to 6–86 in a 9-point scale).	Lee and Puligundla (2016)
Cookies				
Corn-based cookies	Mango kernel starch (varying substitution)	Mango starch via ultrasound extraction, replacing corn starch; dried at 45°C for 24 h; baked as cookie dough.	The 50% substitution led to softer texture and lighter color than controls.	Mainardes et al. (2013)
Wheat-based crackers	Enzymatically modified cassava starch	Enzyme mix, 6 h hydrolysis; drying at 38°C; crackers via sponge method; baking: 180°C for 4 min then 160°C for 6 min with standard ingredients.	Decreased lipid content, reduced bake weight loss, and enhanced water holding capacity and fracture strength in the formulations.	Otero-Guzmán et al. (2023)
Composite flours-based cookies	Heat-moisture ± citric treated sweet potato starch	20% substitution hot- moisture treated ± citric; baking: 150°C for 10 min then 180°C for 10 min. Ingredients: standard cookie base.	Enhanced water absorption decreased swelling and glycemic index.	Van Hung et al. (2021)
Wheat/soy/cassava-based cookies	Wheat-soy-cassava flour (85:5:10)	Dough blending; baking: 204°C engagement via reel oven. Ingredients: sugar, salt, soda, shortening, dextrose, water.	Lowe width and thickness, while also obtaining the best sensory scores (8.58/9).	Kaur et al. (2016)
Potato starch-based cookies	Modified potato starch through HMT, microwave, ultrasonication, and syneresis.	AACC method; modifications applied; standard cookie preparation.	Decreased oil content, swelling, and solubility. HMT and microwaved samples showed enhanced water absorption, and lower syneresis.	Nadir et al. (2015)
Commercial cookie flour-based cookies	Acetylated, hydroxypropylated rice starch (0%–30%)	Partial fat replacement (0%–30%) per AACC 10–52; baking: 205°C for 11 min. Ingredients: sugar, milk, salts, water.	Higher thickness and lower spread ratio. Products showed a similar acceptability (6.29–7.07 in a 9-point scale)	Lee and Puligundla (2016)

Data from the authors and based on the indicated references.

3. Starch utilization in confectionery

Starch is a plant-derived polysaccharide widely used in the food industry due to its versatile functionality. In confectionery, it plays a crucial role in the formulation of candies, gummies, marshmallows, and other products, acting as a thickener, gelling agent, stabilizer, and texture modifier (Hadnađev et al., 2018; Navaf et al., 2024). Its functionality varies depending on its botanical origin (corn, potato, rice, wheat, tapioca, among others) and whether it is in its native or modified form. Native starches possess gelling and thickening properties, while modified starches have been developed to enhance thermal stability, resistance to retrogradation, and water retention capacity (Hartel et al., 2018).

In confectionery manufacturing, starch plays a crucial role in controlling the viscosity of sugar syrup, forming gels, and retaining moisture. For example, in soft candies and gummies, starch helps create a flexible and elastic structure, while in baked confectionery, it contributes to shape retention and final texture (Wang & Hartel, 2022). Advancements in starch modification have led to the development of variants with greater resistance to cooking and prolonged storage. This has led to the widespread use of oxidized, acetylated, pregelatinized, and phosphorylated starches in the confectionery industry (Hadnađev et al., 2018). This section aims to review the diverse applications of starch in confectionery products, emphasizing its physicochemical roles, modifications, processing conditions, and impact on product quality, texture, and shelf life. We also discuss recent advances in starch functionality and their implications for health-oriented confectionery formulations.

3.1 Hard candies, gummies, and nougat

Starch plays a fundamental role in defining the texture and stability of hard candies, gummies, and nougat. Native corn starch is predominantly used as a thickening agent during dough or syrup preparation, where it enhances viscosity and inhibits sugar crystallization. This function is crucial in preventing recrystallization, which could otherwise compromise the product's smoothness and shelf life (Hartel et al., 2018). Through gelatinization, starch granules swell and form a gel network, contributing to the characteristic chewy yet resilient texture of these confections.

For structurally demanding products like molded candies and nougat, starches derived from potato, wheat, and corn are frequently modified chemically or physically to improve their heat resistance and mechanical stability. These modifications alter the molecular structure of starch, enabling it to better withstand thermal and mechanical stresses during processing and storage, thus maintaining texture uniformity and extending shelf life (Hadnađev et al., 2018). Optimizing processing parameters such as temperature control, mixing shear, and moisture content is essential to preserve these functional properties. Nevertheless, detailed information on these critical variables is often lacking in the literature, highlighting an area for further research to improve reproducibility and product quality.

3.1.1 Physical mechanisms and their processes

Starch is critical in defining the texture and structural stability of hard candies, gummies, and nougat. Native corn starch is primarily employed as a thickening agent during dough or syrup preparation, where it enhances viscosity and suppresses sugar crystallization (O'Connell & Hartel, 2022; Shoukat et al., 2025). This function is essential to prevent recrystallization, which could otherwise compromise product smoothness and shelf life. During gelatinization, starch granules absorb water, swell, and form a gel network, contributing to the characteristic chewy yet resilient texture of these confections (Nath et al., 2023).

For products with demanding structural requirements, such as molded candies and nougat, starches derived from potatoes, wheat, and corn are frequently modified—either chemically or physically—to enhance their thermal stability and mechanical robustness. These modifications alter the molecular structure of starch, allowing it to better withstand thermal and mechanical stresses during processing and storage, thereby preserving texture uniformity and extending shelf life (Hadnađev et al., 2018). Optimizing processing parameters, including temperature control, shear during mixing, and moisture content, is critical to maintaining these functional properties. However, comprehensive data on these key variables remain limited in the literature, underscoring the need for further research to improve reproducibility and overall product quality.

3.1.2 Soft confections and fillings

In soft confections, such as chewing gum, marshmallows, caramel fillings, and chewy candies, starch is essential for texture and sensory quality. Native waxy corn starch serves as an effective gelling agent, imparting elasticity and chewiness—attributes central to consumer acceptance (Hartel et al., 2018). Similarly, native potato starch enhances the creaminess and mouthfeel of caramel fillings and chewy candies by stabilizing emulsions and preventing phase separation, thereby ensuring a smooth and uniform product (Navaf et al., 2024).

Modified starches from corn and potato are vital in shelf-stable gummies and chewy candies due to their ability to inhibit retrogradation during cooking and storage. This preservation of flexibility and moisture significantly extends product shelf life (Hadnađev et al., 2018). Moreover, starches from corn, potato, and wheat increase viscosity and stability during the mixing and cooking of fruit chews and caramel creams, facilitating improved shaping and moldability, which are crucial for consistent product quality (Hadnađev et al., 2018).

Native tapioca starch, recognized for its inherent clarity and elastic texture, is used during gelatinization in gelled candies and fruit snacks, resulting in a smooth, non-cloudy appearance favored by consumers (Navaf et al., 2024). Additionally, native rice starch is incorporated during emulsion preparation for coatings, dessert sauces, and fillings to stabilize fat-containing systems by preventing phase separation, thereby maintaining uniformity and enhancing mouthfeel (Hartel et al., 2018).

3.1.3 *Impact of starches addition on textural parameters*

In soft confections, such as chewing gum, marshmallows, caramel fillings, and chewy candies, starch plays a crucial role in both texture and sensory quality. Native waxy corn starch serves as an effective gelling agent, imparting elasticity and chewiness—key attributes for consumer acceptance (Tarahi et al., 2024; Tireki et al., 2021). Likewise, native potato starch contributes to the creaminess and mouthfeel of caramel fillings and chewy candies by stabilizing emulsions and preventing phase separation, thereby ensuring a smooth and uniform product (Navaf et al., 2024).

Modified starches from corn and potato are particularly important in shelf-stable gummies and chewy candies due to their capacity to inhibit retrogradation during cooking and storage. This preservation of flexibility and moisture substantially extends product shelf life (Suárez-Castillo et al., 2024). Furthermore, starches derived from corn, potato, and wheat enhance viscosity and stability during the mixing and cooking of fruit chews and caramel creams, facilitating improved shaping and moldability, which are critical for consistent product quality (Hadnađev et al., 2018). Native tapioca starch, valued for its inherent clarity and elastic texture, is used during gelatinization in gelled candies and fruit snacks, producing a smooth, non-cloudy appearance that is preferred by consumers (Navaf et al., 2024). Additionally, native rice starch is incorporated during the preparation of emulsions for coatings, dessert sauces, and fillings, where it stabilizes fat-containing systems by preventing phase separation, thereby maintaining uniformity and enhancing the mouthfeel.

3.2 Soft candies, edible films, and baked goods

Modified starches from corn, potato, and tapioca significantly contribute to the softness of candies such as caramels, fudge, and toffees by preventing syneresis. Their inclusion stabilizes texture and preserves chewiness throughout prolonged storage, improving shelf life and consumer satisfaction (Hadnađev et al., 2018). When formulated into edible films or dispersed within emulsions, these starches act as protective barriers, maintaining textural integrity and stabilizing fillings, thereby enhancing product durability and sensory attributes (Navaf et al., 2024). In baked confectionery items, such as cookies and biscuits, native wheat starch is incorporated into dough formulations to retain moisture, reduce dryness, and delay staling. These effects directly translate into prolonged freshness and enhanced consumer acceptance. Nonetheless, optimizing baking parameters, such as temperature and moisture, remains critical to fully realizing these functional benefits, a subject that warrants further comprehensive investigation.

3.2.1 *Role of modified starches in soft candies' texture*

Modified starches from corn, potato, and tapioca play a crucial role in soft candies, including caramels, fudge, and toffees, by preventing syneresis (water migration) (Akdowa et al., 2025). Their incorporation stabilizes texture and preserves chewiness during prolonged storage, thereby enhancing shelf life and consumer satisfaction (Hadnađev et al., 2018). When formulated into edible films or dispersed within emulsions, these starches function as protective barriers that maintain textural integrity and stabilize fillings, improving both product durability and sensory attributes (Navaf et al., 2024). In baked confections, such as cookies and biscuits, native wheat starch is incorporated into dough formulations to retain moisture, reduce dryness, and delay staling. These effects directly contribute to extended freshness and greater consumer acceptance (Paternina-Contreras et al., 2025; Yang et al., 2020). Nevertheless, optimization of baking parameters, including temperature and moisture content, is essential to fully realize these functional properties, highlighting an area in need of further research.

3.3 Snacks, chocolate products, and syrup bases

Modified starches derived from corn and tapioca are indispensable in chocolate-coated snacks and pralines, where they function as emulsifiers and fat stabilizers. Their ability to ensure uniform fat distribution enhances mouthfeel and is crucial for sensory acceptance and consistent product quality (Hadnađev et al., 2018). In high-viscosity syrups and soft candy bases, modified potato and corn starches serve as stabilizers that increase viscosity and prevent phase separation, thereby improving texture and stability in sugar-rich confectionery systems. The careful control of processing variables, such as shear rates during mixing and cooking temperatures, is necessary to maintain starch functionality and product consistency. However, detailed documentation of these parameters remains scarce, indicating a need for further research to optimize industrial practices.

3.3.1 Physical mechanisms and involved processes

Modified starches derived from corn and tapioca are essential in chocolate-coated snacks and pralines, where they serve as emulsifiers and fat stabilizers. Their ability to ensure uniform fat distribution improves mouthfeel and is vital for sensory acceptance and consistent product quality (Hosseini & Ansari, 2019; Lapčíková et al., 2024). In high-viscosity syrups and soft candy bases, modified potato and corn starches serve as stabilizers, increasing viscosity and preventing phase separation, thereby enhancing texture and stability in sugar-rich confectionery systems (Hadnađev et al., 2018). Precise control of processing variables, including shear rates during mixing and cooking temperatures, is necessary to retain starch functionality and maintain product consistency. However, comprehensive documentation of these parameters is limited, emphasizing the need for further research to optimize industrial practices.

Table 9.2 shows key examples of starch sources, their native or modified forms, processing conditions, and the functional properties observed in various confectionery and baked products. It highlights the diverse roles starch plays in product formulation, such as thickening, texture improvement, moisture retention, and stability enhancement. By categorizing starch uses based on product type and processing methods, this overview provides valuable insights into how starch modifications and choices improve the quality and shelf life of candies, gummies, baked goods, and related items. The cited studies support these findings, demonstrating the important role of starch functionality in food manufacturing.

3.4 Limitations and challenges in starch utilization

Despite its broad applicability, starch use in confectionery faces several limitations. Native starches often exhibit variability in composition and performance influenced by botanical source and seasonal conditions, leading to inconsistencies between batches. Although modified starches offer enhanced functionalities, they introduce regulatory challenges due to chemical modifications and potential allergenicity concerns. Furthermore, starch interacts intricately with other ingredients such as sugars, fats, and proteins, requiring meticulous formulation to avoid adverse effects on texture or stability. Long-term effects of storage on starch behavior, including retrogradation and moisture migration, are still insufficiently understood. Additionally, the performance of novel starch modifications designed for health benefits under industrial processing conditions remains underexplored. Addressing these challenges will require multidisciplinary studies integrating formulation science and process engineering.

4. Health benefits and emerging trends

Recent research emphasizes the nutritional advantages of incorporating resistant starches in confectionery products. Resistant starch acts as a dietary fiber, modulating glycemic response and conferring benefits such as improved gut health and reduced metabolic disease risk (Chen et al., 2024; Niu et al., 2025). Concurrently, consumer demand for clean-label, non-genetically modified organisms (GMO), and sustainably sourced starches is driving innovation in starch technology.

These trends are shaping the development of starch modifications that balance functional performance with health promotion and environmental sustainability. This evolving landscape is expected to influence the future of starch utilization in confectionery, facilitating products that harmonize sensory quality with nutritional and ecological benefits.

TABLE 9.2 Summary of starch varieties and their modifications applied in confectionery products, including processing conditions and their main impacts on texture, volume, moisture management, shelf life, and sensory attributes.

Product type	Starch source	Native or modified	Process conditions	Functional properties observed in the formulations	References
Hard candies					
Gummies, nougat	Corn	Native	Added as a thickening agent during dough or syrup preparation	Increased thickening, viscosity, and decreased crystallization.	Hartel et al. (2018)
Molded candies, gummies, and nougats	Potato, wheat, corn	Modified	Incorporated during dough or candy mass formulation	Higher heat resistance, structural stability, and texture enhancement, resulting in lower deformation in formulations, compared to control candies.	Hadnađev et al. (2018)
Gummy candies					
Chewing gum and marshmallows	Waxy corn	Native	Added during the confectionery formulation	Increased gelling agent function, elasticity, and chewiness.	Hartel et al. (2018)
Caramel fillings, chewy candies	Potato	Native	Incorporated into the candy matrix during mixing	Improved textural attributes by increased mouthfeel, creaminess, and emulsion stabilization	Navaf et al. (2024)
Shelf-stable gummies, chewy candies	Corn, potato	Modified	Integrated during formulation and cooking stage	Increased shelf life by preventing retrogradation; augmented flexibility and moisture, compared to control samples.	Hadnađev et al. (2018)
Gummy candies, fruit chews, caramel creams	Corn, potato, wheat	Modified	Added during mixing and cooking stages.	Increased viscosity and stability, enabling improved shaping and moldability in fruit-based and chewy confections	Hadnađev et al. (2018)
Gelled candies, gummy bears, fruit snacks	Tapioca	Native	Incorporated during the gelatinization stage.	Higher clarity and elasticity in formulations compared to control candies, producing smooth, non-cloudy appearance and soft texture	Navaf et al. (2024)
Filled and creamy candies					
Coatings, dessert sauces, fillings	Rice	Native	Added during emulsion preparation.	Increased emulsion stability by preventing phase separation in fat-containing confections	Hartel et al. (2018)
Soft caramels, fudge, toffees	Corn, potato, tapioca	Modified	Incorporated into candy matrix during cooking.	Higher stabilization in soft candies by preventing syneresis, resulting in chewiness maintenance during storage.	Hadnađev et al. (2018)
Edible films, emulsified fillings	Corn, potato, tapioca	Modified	Formulated into edible film matrices or dispersed in emulsions.	Preservation of texture integrity and stabilization of emulsified fillings	Navaf et al. (2024)
Baked goods and cookies					
Baked confectionery, cookies, biscuits	Wheat	Native	Integrated into dough formulations.	Increased moisture retention to prevent dryness, resulting in extended shelf life and staling reduction	Navaf et al. (2024)
Snacks, chocolate products and soft candy bases					
Chocolate-covered snacks, pralines	Corn, tapioca	Modified	Incorporated into chocolate matrix.	Improved emulsification that stabilizes fats, ensuring its uniform distribution, and enhancing mouthfeel	Hadnađev et al. (2018)
High-viscosity syrups, soft candy bases	Potato, corn	Modified	Incorporated as syrup/stabilizer.	Increased viscosity and stability, reduced phase separation, and improvement of texture in high-sugar confections	Hadnađev et al. (2018)

Data from the authors and based on the indicated references.

5. Conclusions and future perspectives

Starch remains a key ingredient in the bakery and confectionery industries because of its multifunctional roles in creating texture, controlling moisture, extending shelf life, and enhancing nutrition. The evidence presented in this chapter shows how native and modified starches improve the technological and sensory qualities of various products, including bread, cakes, gummies, nougat, and chocolate-based confections. Starch modifications have opened new possibilities for replacing fats, preventing retrogradation, strengthening structure, and maintaining stability during industrial processing. Despite these advances, significant gaps still exist. Most studies have concentrated on empirical quality parameters like texture, hardness, or spreadability, but the fundamental molecular mechanisms—such as starch, protein, and lipid interactions, gelatinization kinetics, and retrogradation pathways—are still not fully understood. Gaining mechanistic insights will be essential for predictive control of product performance. Furthermore, inconsistencies in reporting processing conditions hamper reproducibility, highlighting the urgent need for standardized methodologies in starch application research.

Looking forward, three strategic directions should guide future investigations. First, integrating multi-scale characterization tools (rheology, microscopy, molecular modeling) will help link starch structure to functionality and consumer perception. Second, advancing nutrition-oriented innovations, such as resistant starch enrichment, prebiotic potential, and glycemic index reduction, will align starch utilization with health-driven consumer demands. Finally, prioritizing sustainability and circular economy strategies through the valorization of underused starch sources can position starch research at the crossroads of food innovation and environmental responsibility. By connecting mechanistic understanding, nutritional value, and sustainable innovation, starch use in bakery and confectionery products can evolve from merely a technological enhancement to a comprehensive tool for food design, supporting the development of products that balance consumer acceptance, health benefits, and ecological impact.

References

Abdul, F. S., Abdul Refai, S., Shukri, R., Muhammad, K., Mustapha, N. A., Wan Ibadullah, W. Z., & Ramli, N. S. (2017). Dough rheology and physicochemical properties of steamed buns fortified with cross-linked rice starch. *Bioactive Carbohydrates and Dietary Fibre, 12*, 1–6. https://doi.org/10.1016/j.bcdf.2017.10.002

Ahsan, M., Moin, A., Ashraf, H., Khan, A., & Giuffrè, A. M. (2025). Formulation and characterization of reduced fat muffins using a plant based fat replacer. *Journal of Food Science and Technology, 62*(3), 551–561. https://doi.org/10.1007/s13197-024-06045-6

Akdowa, E. P., Didi, A., & Clergé, T. (2025). Comparative study of the impact of stearin-modified starches from maize, ginger, and sweet potato on the physicochemical properties of low-fat mayonnaise. *Proceedings, 118*(12), 1–12. https://doi.org/10.3390/proceedings2025118012

Alvarez, M. D., Herranz, B., Jiménez, M. J., & Canet, W. (2017). End-product quality characteristics and consumer response of chickpea flour-based gluten-free muffins containing corn starch and egg white. *Journal of Texture Studies, 48*(6), 550–561. https://doi.org/10.1111/jtxs.12263

Arp, C. G., Correa, M. J., & Ferrero, C. (2021). Resistant starches: A smart alternative for the development of functional bread and other starch-based foods. *Food Hydrocolloids, 121*, Article 106949. https://doi.org/10.1016/j.foodhyd.2021.106949

Baixauli, R., Salvador, A., & Fiszman, S. M. (2008). Textural and colour changes during storage and sensory shelf life of muffins containing resistant starch. *European Food Research and Technology, 226*(3), 523–530. https://doi.org/10.1007/s00217-007-0565-4

Balic, R., Miljkovic, T., Ozsisli, B., & Simsek, S. (2016). Utilization of modified wheat and tapioca starches as fat replacements in bread formulation. *Journal of Food Processing and Preservation, 41*(1), 1–12. https://doi.org/10.1111/jfpp.12888

Barros, J. H. T., Telis, V. R. N., Taboga, S., & Franco, C. M. L. (2018). Resistant starch: Effect on rheology, quality, and staling rate of white wheat bread. *Journal of Food Science and Technology, 55*(11), 4578–4588. https://doi.org/10.1007/s13197-018-3393-6

Cao, H., Wang, X., Wu, W., Grimi, N., Lu, J., & Guan, X. (2025). Modulation of textural properties in microwave treated gluten-free quinoa sponge cake by alkaline amino acids. *Food Hydrocolloids, 160*, Article 110727. https://doi.org/10.1016/j.foodhyd.2024.110727

Chakraborty, I., Mal, S. S., Paul, U. C., Rahman, M. H., & Mazumder, N. (2022). An insight into the gelatinization properties influencing the modified starches used in food industry: A review. *Food and Bioprocess Technology, 15*(6), 1195–1223. https://doi.org/10.1007/s11947-022-02761-z

Chen, Z., Liang, N., Zhang, H., Li, H., Guo, J., Zhang, Y., Chen, Y., Wang, Y., & Shi, N. (2024). Resistant starch and the gut microbiome: Exploring beneficial interactions and dietary impacts. *Food Chemistry: X, 21*, Article 101118. https://doi.org/10.1016/j.fochx.2024.101118

Chen, H., Yang, Y., & Xu, X. (2025). Citric acid esterification of rice starches from different sources for stabilizing pickering emulsions. *LWT, 230* (118225), 1–13. https://doi.org/10.1016/j.lwt.2025.118225

Colla, K., Costanzo, A., & Gamlath, S. (2018). Fat replacers in baked food products. *Foods, 7*(192), 1–12. https://doi.org/10.3390/foods7120192

Dou, X., Hao, Y., Sun, Y., Yang, P., Liu, L., He, Y., Shi, Y., Yang, C., & Chen, F. (2024). A novel baking additive: Preparation, characterization, and application of chitosan hydrochloride/carboxymethyl starch sodium nano-gel for wheat bread. *Food Hydrocolloids, 148*(109459), 1–10. https://doi.org/10.1016/j.foodhyd.2023.109459

Durmus, Y., Anil, M., & Simsek, S. (2023). Innovative use of hazelnut skin and starch modifications in sourdough bread formulation. *Journal of Food Process Engineering, 47*(1), 1–12. https://doi.org/10.1111/jfpe.14517

Ferron, L., Milanese, C., Colombo, R., Pugliese, R., & Papetti, A. (2022). A new polysaccharide carrier isolated from camelina cake: Structural characterization, rheological behavior, and its influence on purple corn cob extract's bioaccessibility. *Foods, 11*(12), 1736. https://doi.org/10.3390/foods11121736

Geetanjali, G. K., Singh, A., & Khatkar, S. (2025). Effect of ultrasonicated corn starch as a fat replacer on muffin quality and sensory characteristics. *Journal of Food Measurement and Characterization, 19*(1), 1–11. https://doi.org/10.1007/s11694-024-02913-z

Geetha, K., Ilamaran, M., Ramasamy, D. P., Santhi, V. P., Arul Raj, & Ayyanar, M. (2024). Effect of temperature and physical modification on nutritional composition and functional properties of cassava, banana and corn starches. *Journal of Food Measurement and Characterization, 18*(10), 8876–8887. https://doi.org/10.1007/s11694-024-02851-w

Guadarrama-Lezama, A. Y., Carrillo-Navas, H., Pérez-Alonso, C., Vernon-Carter, E. J., & Alvarez-Ramirez, J. (2016). Thermal and rheological properties of sponge cake batters and texture and microstructural characteristics of sponge cake made with native corn starch in partial or total replacement of wheat flour. *LWT, 70*, 46–54. https://doi.org/10.1016/j.lwt.2016.02.031

Guiné, R. P. F. (2022). Textural properties of bakery products: A review of instrumental and sensory evaluation studies. *Applied Sciences (Switzerland), 12*(17). https://doi.org/10.3390/app12178628

Gutiérrez-Luna, K., Ansorena, D., & Astiasaran, I. (2024). Effect of baking conditions on resistant starch: Model systems and cake formulations. *Food Chemistry, 449*(139174), 1–8. https://doi.org/10.1016/j.foodchem.2024.139174

Ha, M., Firdhausa, A. S., & Chung, H.-J. (2025). Microwave treatment modifies the physicochemical properties of starch-protein composite for improved gluten-free muffin quality. *Food Chemistry, 490*, Article 145046. https://doi.org/10.1016/j.foodchem.2025.145046

Hadnađev, M., Dapčević-Hadnađev, T., & Dokić, L. (2018). Functionality of starch derivatives in bakery and confectionery products. In *Biopolymers for food design* (pp. 279–311). Elsevier. https://doi.org/10.1016/B978-0-12-811449-0.00009-8

Hartel, R. W., von Elbe, J. H., & Hofberger, R. (2018). *Confectionery science and technology*. Springer International Publishing. https://doi.org/10.1007/978-3-319-61742-8

Hassan, S. S., El-Hdide, S., & Aboul Anean, H. E. D. (2023). Preparation of edible coating from Cassava Starch and Rosemary essential oil and its effect on shelf life and quality of Muffin. *Food Technology Research Journal, 2*(2), 82–97.

Hosseini, F., & Ansari, S. (2019). Effect of modified tapioca starch on the physicochemical and sensory properties of liquid kashk. *Journal of Food Science and Technology, 56*(12), 5374–5385. https://doi.org/10.1007/s13197-019-04008-w

Hu, X., Cheng, L., Hong, Y., Li, Z., Li, C., & Gu, Z. (2023). An extensive review: How starch and gluten impact dough machinability and resultant bread qualities. *Critical Reviews in Food Science and Nutrition, 63*(13), 1930–1941. https://doi.org/10.1080/10408398.2021.1969535

Kaur, K., Ahluwalia, P., & Singh, H. (2016). Cassava: Extraction of starch and utilization of flour in bakery products. *International Journal of Food and Fermentation Technology, 6*(2), 351. https://doi.org/10.5958/2277-9396.2016.00059.3

Kim, M., Yun, Y., & Jeong, Y. (2015). Effects of corn, potato, and tapioca starches on the quality of gluten-free rice bread. *Food Science and Biotechnology, 24*(3), 913–919. https://doi.org/10.1007/s10068-015-0118-8

Królak, M., Górska-Warsewicz, H., Madra-Sawicka, M., Rejman, K., Żakowska-Biemans, S., Szlachciuk, J., Czeczotko, M., Kwiatkowski, B., Zaremba, R., & Wojtaszek, M. (2022). Towards sustainable innovation in the bakery sector—An example of fibre-enriched bread. *Sustainability (Switzerland), 14*(2743), 1–19. https://doi.org/10.3390/su14052743

Lapčíková, B., Lapčík, L., Valenta, T., & Neuwirth, V. (2024). Chocolate ganaches: Formulation, processing and stability in view of the new production trends. *Foods, 13*(2543), 1–20. https://doi.org/10.3390/foods13162543

Lee, Y. T., & Puligundla, P. (2016). Characteristics of reduced-fat muffins and cookies with native and modified rice starches. *Emirates Journal of Food and Agriculture, 28*(5), 311–316. https://doi.org/10.9755/ejfa.2015-05-227

Mainardes, B., Moreira, L., Delinski Bet, C., Tolentino Marinho, M., Bassetto Bisinella, R. Z., Schiavo Romko, S., & Schnitzler, E. (2013). Extraction and physical modification of Mango palmer kernel starch and its application in cookies. *Ukrainian Food Journal, 2*(1), 1–148.

Maravić, N., Škrobot, D., Dapčević-Hadnađev, T., Pajin, B., Tomić, J., & Hadnađev, M. (2022). Effect of sourdough and whey protein addition on the technological and nutritive characteristics of sponge cake. *Foods, 11*(14). https://doi.org/10.3390/foods11141992

Monnet, A. F., Jeuffroy, M. H., Villemejane, C., & Michon, C. (2021). Effect of the order of incorporation of cake ingredients on the formation of batter and the final properties: Contribution of the addition of pea flour. *Journal of Food Science and Technology, 58*(11), 4252–4262. https://doi.org/10.1007/s13197-020-04899-0

Nadir, A. S., Helmy, I. M. F., Nahed, M., Wafaa, A., Abozeid, M. M., & Ramadan, M. T. (2015). Modification of potato starch by some different physical methods and utilization in cookies production. *International Journal of Current Microbiology and Applied Sciences, 4*(10), 556–569.

Nath, P. C., Debnath, S., Sridhar, K., Inbaraj, B. S., Nayak, P. K., & Sharma, M. (2023). A comprehensive review of food hydrogels: Principles, formation mechanisms, microstructure, and its applications. *Gels, 9*(1). https://doi.org/10.3390/gels9010001

Navaf, M., Sunooj, K. V., Tvisha, M., Bangar, S. P., & George, J. (2024). Native and modified starches for bakery and confectionery products. *Current Food Science and Technology Reports, 2*(3), 333–345. https://doi.org/10.1007/s43555-024-00037-5

Nikolić, I., Šoronja-Simović, D., Zahorec, J., Dokić, L., Lončarević, I., Stožinić, M., & Petrović, J. (2024). Polysaccharide-based fat replacers in the functional food products. *Processes, 12*(12), 2701. https://doi.org/10.3390/pr12122701

Niu, Y., Wang, L., Gong, H., Jia, S., Guan, Q., Li, L., & Cheng, H. (2025). Nutrition and gut health: Preparation and efficacy of resistant starch. *Foods, 14*(3), 471. https://doi.org/10.3390/foods14030471

O'Connell, K., & Hartel, R. W. (2022). The effects of corn syrup, water content and sucrose replacers on sucrose crystallization in starch jellies. *Journal of Food Processing and Preservation, 46*(4). https://doi.org/10.1111/jfpp.16456

Olakanmi, S. J., Bharathi, V. S. K., Jayas, D. S., & Paliwal, J. (2024). Innovations in nondestructive assessment of baked products: Current trends and future prospects. *Comprehensive Reviews in Food Science and Food Safety, 23*(4). https://doi.org/10.1111/1541-4337.13385

Otero-Guzmán, N., Cadena-Chamorro, E., Ciro-Velásquez, H. J., Mejía-Villota, A., Rodriguez-Sandoval, E., & Arshad, M. S. (2023). Application of modified cassava starch as a fat substitute in cracker production. *Journal of Food Processing and Preservation, 2023*, 1–10. https://doi.org/10.1155/2023/6888180

Paternina-Contreras, A. L., Andrade-Pizarro, R. D., & Figueroa-Flórez, J. A. (2025). Physical modification of starch in plant-based flours: Structural, physicochemical, and pasting property changes and potential applications in baked and extruded products. *Comprehensive Reviews in Food Science and Food Safety, 24*(3). https://doi.org/10.1111/1541-4337.70184

Petchoo, J., Jittinandana, S., Tuntipopipat, S., Ngampeerapong, C., & Tangsuphoom, N. (2021). Effect of partial substitution of wheat flour with resistant starch on physicochemical, sensorial and nutritional properties of breadsticks. *International Journal of Food Science and Technology, 56*(4), 1750–1758. https://doi.org/10.1111/ijfs.14800

Psimouli, V., & Oreopoulou, V. (2013). The effect of fat replacers on batter and cake properties. *Journal of Food Science, 78*(10). https://doi.org/10.1111/1750-3841.12235

Punia, S., Siroha, A. K., Sandhu, K. S., & Kaur, M. (2019). Rheological and pasting behavior of OSA modified mungbean starches and its utilization in cake formulation as fat replacer. *International Journal of Biological Macromolecules, 128*, 230–236. https://doi.org/10.1016/j.ijbiomac.2019.01.107

Pycarelle, S. C., & Delcour, J. A. (2021). The role and impact on quality of exogenous and endogenous lipids during sponge cake making. *Trends in Food Science and Technology, 114*, 158–166. https://doi.org/10.1016/j.tifs.2021.05.015

Qian, X., Sun, B., Gu, Y., Ma, S., & Wang, X. (2023). Enhancing the quality of steamed oat cake by partially gelatinized starch in oat flour and its molecular mechanism. *International Journal of Biological Macromolecules, 238*, Article 124139. https://doi.org/10.1016/j.ijbiomac.2023.124139

Rodriguez-Sandoval, E., Prasca-Sierra, I., & Hernandez, V. (2017). Effect of modified cassava starch as a fat replacer on the texture and quality characteristics of muffins. *Journal of Food Measurement and Characterization, 11*(4), 1630–1639. https://doi.org/10.1007/s11694-017-9543-0

Rosa-Millan, J.de la (2025). The role of molecular and structural characteristics of starch, hydrocolloids, and gluten in bread in vitro digestibility. *Polysaccharides, 6*(2), 46. https://doi.org/10.3390/polysaccharides6020046

Sabaghi, M. (2025). Active edible food packaging materials for confectionery products: Applications, challenges, and future directions. In *Polymer Bulletin, 82(7)*. 10.1007/s00289-024-05604-3.

Sahnoun, M., Ismail, N., & Kammoun, R. (2016). Enzymatically hydrolysed, acetylated and dually modified corn starch: Physico-chemical, rheological and nutritional properties and effects on cake quality. *Journal of Food Science and Technology, 53*(1), 481–490. https://doi.org/10.1007/s13197-015-1984-z

Salazar, D., Arancibia, M., Silva, D. R., López-Caballero, M. E., & Montero, M. P. (2021). Exploring the potential of andean crops for the production of gluten-free muffins. *Agronomy, 11*(8), 1642. https://doi.org/10.3390/agronomy11081642

Scott, G., & Awika, J. M. (2023). Effect of protein–starch interactions on starch retrogradation and implications for food product quality. *Comprehensive Reviews in Food Science and Food Safety, 22*(3), 2081–2111. https://doi.org/10.1111/1541-4337.13141

Sergiacomo, A., Bresciani, A., & Marti, A. (2025). Properties of sponge cake enriched in sprouted oats (Avena sativa L.). *LWT, 224*(117791), 1–11. https://doi.org/10.1016/j.lwt.2025.117791

Shoukat, R., Cappai, M., Pilia, L., Pia, G., & Starch Chemistry, R. (2025). Functional properties, and industrial applications: A review. *Polymers, 17*(1), 110. https://doi.org/10.3390/polym17010110

Slade, L., Kweon, M., & Levine, H. (2021). Exploration of the functionality of sugars in cake-baking, and effects on cake quality. *Critical Reviews in Food Science and Nutrition, 61*(2), 283–311. https://doi.org/10.1080/10408398.2020.1729694

Suárez-Castillo, G. M., Salcedo-Guadalupe, J. G., Contreras-Lozano, K. P., Rangel-Pérez, M. G., Cervera-Ricardo, M. A., & Figueroa-Flórez, J. A. (2024). Increase in the degree of substitution of cassava starches by dual modification processes. *Revista Mexicana de Ingeniera Quimica, 23*(3). https://doi.org/10.24275/rmiq/Poly24303

Suri, S., & Singh, A. (2023). Modification of starch by novel and traditional ways: influence on the structure and functional properties. *Sustainable Food Technology, 1*(3), 348–362. https://doi.org/10.1039/d2fb00043a

Tarahi, M., Tahmouzi, S., Kianiani, M. R., Ezzati, S., Hedayati, S., & Niakousari, M. (2024). Current innovations in the development of functional gummy candies. *Foods, 13*(1). https://doi.org/10.3390/foods13010076

Tekin, T., & Dincer, E. (2023). Effect of resistant starch types as a prebiotic. *Applied Microbiology and Biotechnology, 107*(2–3), 491–515. https://doi.org/10.1007/s00253-022-12325-y

Tien, N. N. T., Anh, T. N. Q., Phi, N. T. L., & Hung, P. V. (2018). In vitro and in vivo starch digestibility and quality of bread substituted with acid and heat-moisture treated sweet potato starch. *Starch/Staerke, 70*(9–10). https://doi.org/10.1002/star.201800069

Tireki, S., Sumnu, G., & Sahin, S. (2021). Correlation between physical and sensorial properties of gummy confections with different formulations during storage. *Journal of Food Science and Technology, 58*(9), 3397–3408. https://doi.org/10.1007/s13197-020-04923-3

Tiruneh, A., Ptaszek, P., Żmudziński, D., & Tarko, T. (2025). Peas (Pisum sativum subsp. arvense Asch) and beans (Vicia faba var. minor) as source of quality plant proteins. *Molecules, 30*(9), 2009. https://doi.org/10.3390/molecules30092009

Van Hung, P., Duyen, T. T. M., Van Thanh, H., Widiastuti, D., & An, N. T. H. (2021). Starch digestibility and quality of cookies made from acid and heat-moisture treated sweet potato starch and wheat flour composites. *Journal of Food Measurement and Characterization, 15*(4), 3045–3051. https://doi.org/10.1007/s11694-021-00892-z

Vidaurre-Ruiz, J., Vargas, R. J. Y., Alcázar-Alay, S., Encina-Zelada, C. R., Cabezas, D. M., Correa, M. J., & Repo-Carrasco-Valencia, R. (2022). Andean crops: Kañiwa and tarwi flours used for the development of vegan gluten-free muffins. *Journal of the Science of Food and Agriculture, 102* (15), 7282–7292. https://doi.org/10.1002/jsfa.12094

Wang, L., Guo, Z., Wang, H., Zou, L., Li, Z., Fang, B., & Qiu, J. (2024). Effect of buckwheat hulls at cell-scale on physiochemical and textural properties of steamed rice cake. *LWT, 209*, Article 116818. https://doi.org/10.1016/j.lwt.2024.116818

Wang, R., & Hartel, R. W. (2022). Confectionery gels: Gelling behavior and gel properties of gelatin in concentrated sugar solutions. *Food Hydrocolloids, 124*(107132). https://doi.org/10.1016/j.foodhyd.2021.107132

Wang, Y.c., Ning, H.w., Yan, Q.j., Liu, H.j., Li, Y.x., & Jiang, Z.q. (2025). Enzymatic modification of wheat starch by a novel maltotetraose-forming amylase from Atopomonas hussainii to retard retrogradation and improve bread quality. *Carbohydrate Polymers, 348*. https://doi.org/10.1016/j.carbpol.2024.122909

Yang, B., Guo, M., & Zhao, Z. (2020). Incorporation of wheat malt into a cookie recipe and its effect on the physicochemical properties of the corresponding dough and cookies. *LWT, 117*(108651), 1–8. https://doi.org/10.1016/j.lwt.2019.108651

Yun, P., Devahastin, S., & Chiewchan, N. (2021). In vitro glycemic index, physicochemical properties and sensory characteristics of white bread incorporated with resistant starch powder prepared by a novel spray-drying based method. *Journal of Food Engineering, 294*(110438), 1–8. https://doi.org/10.1016/j.jfoodeng.2020.110438

Zhang, B., Qiao, D., Zhao, S., Lin, Q., Wang, J., & Xie, F. (2021). Starch-based food matrices containing protein: Recent understanding of morphology, structure, and properties. *Trends in Food Science & Technology, 114*, 212–231. https://doi.org/10.1016/j.tifs.2021.05.033

Chapter 10

Contemporary starch applications for dairy and frozen product development

Sony George[1], Girilal Morkattu[2] and Anjaly Paul[3]
[1]*TKM Institute of Technology, Department of Food Technology, Kollam, Kerala, India;* [2]*Saintgits College of Engineering, Department of Food Technology, Kottayam, Kerala, India;* [3]*McGill University, Department of Bioresource Engineering, Sainte-Anne-de-Bellevue, QC, Canada*

Highlights

- Starch as good source of thickening, gelling, and stabilizing agent in dairy and frozen food formulations to enhance the texture, viscosity, and prevent syneresis and ice-crystal formulations.
- Modified starches, including cross-linked, acetylated, and hydroxypropylated types, enhances freeze–thaw stability, firmness, and shelf life in both dairy and frozen systems.
- The recent research and the emergence of clean-label, non GMO, and high-amylose starches have led to natural, sustainable, and functional ingredients with enhanced performance.
- The chapter addresses the challenges, such as retrogradation, production cost, and texture consistency, for developing sustainable and consumer-friendly starch innovations.

1. Introduction

Starch, which is a polysaccharide glucose monomers, is widely used in the food industry due to its versatile functional properties, including thickening, gelling, and stabilizing (Singh et al., 2007). Starches play a vital role in the formulation of dairy and frozen products by offering texture improvement, stability, and functional benefits. They are essential for applications such as thickening, stabilizing, and enhancing mouthfeel. The choice of starch can significantly impact the overall quality and performance of the product. Starch aids in air incorporation in ice cream, increasing volume and enhancing mouthfeel without the need for added fat. Additionally, starch stabilizes formulations by preventing separation and curdling, while also acting as an emulsifier to balance water and fat, ensuring consistent product quality. In frozen products, starch improves viscosity, creaminess, and melting behavior, while inhibiting ice recrystallization. This section discusses the various sources of starches used in dairy and frozen product development.

In dairy products, starch plays a crucial role in improving texture, stability, and sensory attributes. The use of starch in yogurt, cheese, and milk-based beverages enhances product quality and consumer acceptance. Starch modification techniques, such as pregelatinization and cross-linking, allow for better performance in various dairy applications. This chapter explores the various applications of starch in dairy products and frozen products, focusing on its role as a thickening agent, stabilizer, meltability controller, texture enhancer, and viscosity regulator. Fig. 10.1 shows the expanded classification chart outlining the varieties of starches used in food applications.

1.1 Starch applications in dairy products

In dairy products, starch is frequently utilized as a functional ingredient to improve mouthfeel, texture, and stability. A thickening, gelling, and stabilizing chemical, it is used in cheese, ice cream, puddings, and yogurts. In order to ensure a smooth and uniform texture, modified starches increase viscosity and inhibit syneresis. Furthermore, starch helps low-fat dairy formulations replace fat while preserving product quality and providing an appealing sensory experience. Its adaptability makes it a crucial part of dairy processing, maximizing shelf life and quality.

Native and Modified Starches for Food Applications. https://doi.org/10.1016/B978-0-443-27431-2.00024-9

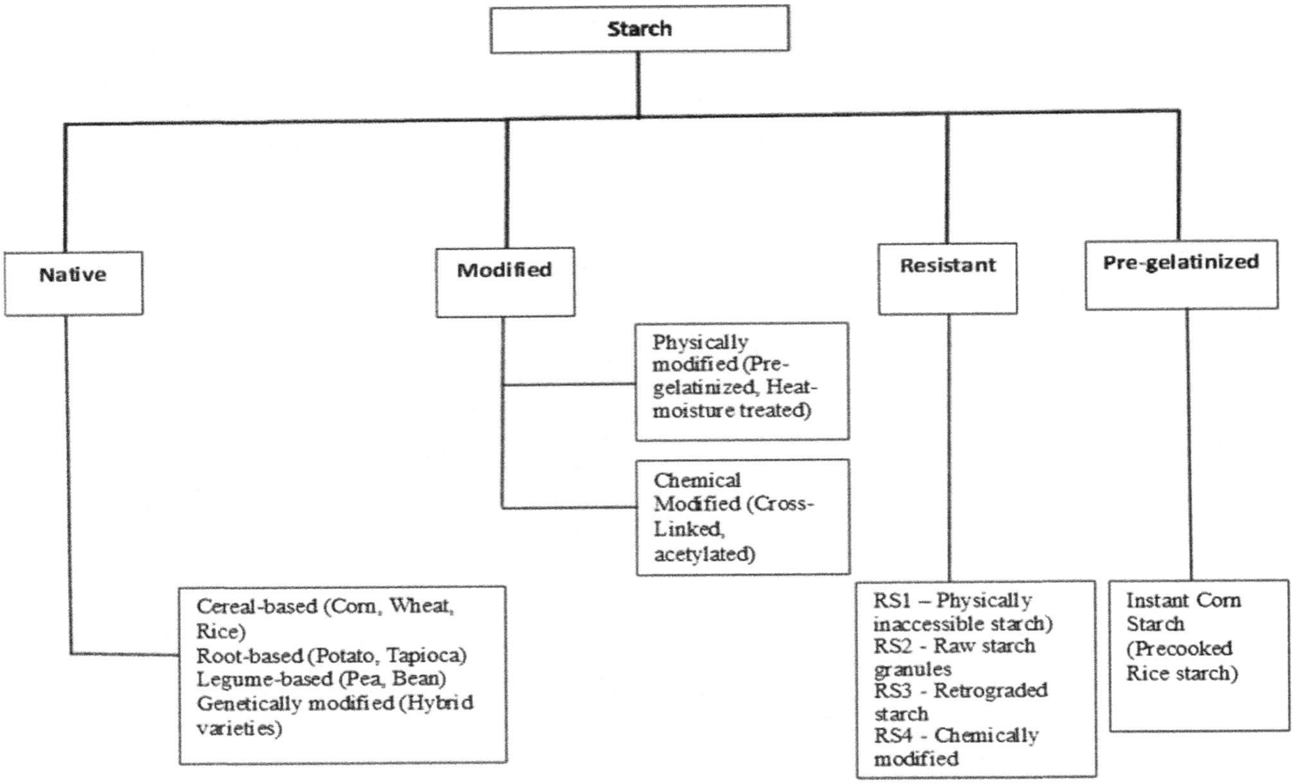

FIGURE 10.1 Variety of starches in the food industry. Starches are categorized as native, modified, resistant, and pregelatinized types. Native starches are derived directly from natural sources such as cereals, tubers, and legumes. Modified starches are modified with physical or chemical treatments to increase functionality. Resistant starches (RS1–RS4) resist digestion and act as dietary fiber. Pregelatinized starches are precooked for instant thickening applications.

1.1.1 Applications of starch in yogurt

Starches from sources like corn, tapioca, and potato are commonly used as thickening agents in yogurt, where they swell in the presence of water and heat, increasing viscosity and enhancing texture and mouthfeel. The thickening effect depends on the starch type, degree of gelatinization, and processing conditions, with modified starches such as cross-linked and pregelatinized starches offering superior performance in maintaining texture over storage (Singh et al., 2007). When starch undergoes gelatinization during heat treatment, its granules absorb water, swell, and rupture, leading to an increase in viscosity, which is crucial in defining the final texture of set and stirred yogurt varieties. The thickening efficiency of starch depends on its source and structural composition; for instance, corn starch, with its high amylose content, forms a firm gel and dense texture, while tapioca starch, which has a higher amylopectin content, provides a smoother and more elastic texture suitable for creamier yogurt. Potato starch, with its large granule size, swells rapidly and exhibits high water-holding capacity, enhancing viscosity and creaminess (BeMiller & Whistler, 2009). The degree of gelatinization is a key factor in starch thickening efficiency, with partial gelatinization through preprocessing methods like precooking or extrusion improving water absorption and stabilizing the gel network in yogurt . Additionally, processing conditions such as heat treatment, shear force, and pH influence starch behavior, where high-temperature processing enhances starch hydration and dispersion, while excessive shear force may degrade starch granules and reduce thickening ability (Singh et al., 2007). Modified starches play a crucial role in yogurt stability, as cross-linked starches have strengthened molecular structures that resist heat, shear, and acidic conditions, preserving yogurt texture over extended storage (Arbuckle, 2013). Pregelatinized starches, which are precooked and dried, disperse easily in cold or warm yogurt formulations, providing instant thickening properties without requiring further cooking (BeMiller & Whistler, 2009). These modifications help prevent undesirable textural changes such as excessive syneresis or phase separation during storage, ensuring consistency in the final product. Starch also enhances sensory properties by forming a stable gel network that contributes to a rich and creamy mouthfeel, increasing consumer acceptance. Manufacturers can leverage different starch types and modifications to fine-tune yogurt texture according to market preferences, whether for thick, spoonable yogurts or lighter, drinkable varieties (Tamime, 2007). The starches from corn, tapioca, and potato are highly effective thickening agents in yogurt, significantly contributing to viscosity, stability, and mouthfeel. The efficiency of starch thickening is influenced by its source, degree of gelatinization, and processing conditions, with modified starches providing superior texture retention and environmental resistance, making them essential in commercial yogurt production. By optimizing starch selection and processing parameters, manufacturers can achieve the desired textural and sensory attributes to enhance consumer satisfaction.

Stabilizers, such as pectin, gelatin, starch, guar gum, locust bean gum, carrageenan and xanthan gum, are essential in yogurt production to prevent whey separation (syneresis) and to maintain a homogeneous texture. Starch stabilizers, particularly hydrocolloid-starch blends, interact with milk proteins, reinforcing the gel network and enhancing product stability. This leads to a smoother mouthfeel and improved shelf life. The use of resistant starch in yogurt has also been studied for its prebiotic potential, contributing to gut health benefits.

1.1.2 Starch applications in cheese

Cheese varieties exhibit diverse textural and melting properties, which can be modified through the incorporation of starch. Starches are used to improve cheese quality, particularly in processed and low-fat cheeses. Starch influences the meltability of cheese by modifying its water-binding capacity and interactions with casein proteins. In processed cheese, starch can limit excessive melting and phase separation, ensuring a consistent texture during heating. Modified starches, acetylated and phosphorylated starches, increase the heat stability of cheese, making them suitable for applications like pizza toppings and ready-to-eat meals (Awasthi, 2025). Controlling meltability is especially important in fast-food applications where cheese consistency must be maintained. Starch enhances the texture of cheese by improving elasticity, firmness, and sliceability. This is particularly important in low-fat cheese formulations, where starch can compensate for the loss of fat. The addition of starch helps retain moisture, preventing a dry and crumbly texture while maintaining a desirable mouthfeel (Singh et al., 2007). In processed cheese, pregelatinized starch is used to improve spreadability and maintain a uniform consistency.

1.1.3 Starch applications in milk beverages

Milk-based beverages often require starch to modify viscosity, improve suspension stability, and enhance sensory properties. Starch plays a vital role in preventing the sedimentation of insoluble components in milk beverages. This is particularly important in flavored milk, cocoa-based drinks, and protein-enriched formulations. Starches with high water-binding capacity help maintain uniform distribution of suspended particles, preventing phase separation (Kawaljit Singh & Sandhu, 2006). The ability of resistant starch to act as a dietary fiber in functional milk beverages has also been explored, offering additional health benefits. The viscosity of milk beverages can be adjusted using starch, ensuring a desirable consistency and drinking experience. Starches with low gelatinization temperatures are preferred for cold-processing applications, while heat-resistant starches are used in UHT-treated beverages. The selection of starch type and concentration depends on the desired rheological properties and product stability (Awasthi, 2025). Modified starches allow for better control of viscosity in ready-to-drink dairy formulations, preventing undesirable phase separation over time. Starch is a versatile ingredient that significantly enhances the quality and functionality of dairy products. Its applications in yogurt, cheese, and milk beverages contribute to improved texture, stability, and consumer appeal. The selection of appropriate starch types and modification techniques is crucial for achieving optimal results in dairy formulations. The growing interest in resistant starch and its health benefits further expands the potential applications of starch in dairy products (Singh et al., 2007).

1.2 Importance of frozen products in the modern food industry

Frozen products have become a cornerstone of the modern food industry, offering convenience, extended shelf life, and nutritional preservation. These products, which include fruits, vegetables, meats, seafood, and ready-to-eat meals, address the demands of busy lifestyles and global food distribution. Freezing technology helps retain the nutritional value of food by preserving it at peak freshness, ensuring that essential vitamins and minerals are not lost. Additionally, frozen foods reduce food waste by extending storage times for both consumers and businesses, while facilitating the global availability of seasonal and exotic foods year-round (Pirsa, 2023). These benefits, coupled with technological advancements like individual quick freezing (IQF), have transformed frozen products into indispensable elements of the modern diet.

Despite their growing popularity, the frozen food industry faces challenges such as high energy consumption, the need for robust cold chain logistics, and lingering consumer perceptions about quality. Maintaining an unbroken cold chain from production to consumption is critical for safety and quality, but requires significant infrastructure investments. Meanwhile, the industry has responded to changing consumer preferences by offering healthier options, including organic, low-calorie, and plant-based frozen foods (Apriyanto et al., 2022). Ready-to-eat frozen meals have seen a surge in demand, driven by urban lifestyles and the desire for convenience. Moreover, sustainability has become a focal point, with brands investing in eco-friendly packaging and energy-efficient freezing technologies to reduce their environmental footprint.

The frozen food industry is expected to continue its growth in accordance with advancements in technology and changing consumer preferences. Smart packaging and Internet of Things (IoT) innovations promise enhanced traceability and quality monitoring, further boosting consumer confidence. The expansion of plant-based frozen products and sustainable practices underscores the industry's commitment to addressing modern dietary trends and environmental concerns. As frozen foods continue to meet the needs for convenience, nutrition, and sustainability, they will remain integral to the global food supply chain, shaping the way we produce, distribute, and consume food in the future. Starch plays a pivotal role in frozen products because it enhances stability, texture, and shelf life, as mentioned in Table 10.1. It is indispensable in frozen products due to its multifunctional properties. It ensures stability by minimizing phase separation and ice crystal growth, enhances texture by contributing to smoothness and elasticity, and extends shelf life by maintaining product integrity over time. Whether in frozen desserts, bakery goods, or ready-to-eat meals, starch serves as a cornerstone ingredient for delivering consistent quality and satisfying consumer expectations (Abbas et al., 2010; Miyazaki et al., 2006). Starch is a critical stabilizer in frozen products, addressing challenges such as ice crystal formation, phase separation, and structural breakdown. Its hydrophilic nature enables it to absorb water and lock it within its molecular structure, reducing free water availability and preventing the formation of large ice crystals, thereby maintaining product uniformity. In multi-phase systems like frozen soups, sauces, or dairy-based products, starch prevents phase separation by creating a gell like matrix that holds ingredients in place, ensuring consistent quality during storage and thawing. Additionally, modified starches designed for freeze-thaw applications inhibit crystal growth, keeping ice crystals small and evenly distributed to preserve texture and visual appeal. Starch also significantly influences the sensory

TABLE 10.1 Importance of starch in frozen food and its mode of action.

Property	Aspect	Mechanism	References
Stability	Water-binding capacity	Absorbs water and locks it within its structure, reducing free water and preventing ice crystal formation, thus maintaining uniformity in frozen products.	Baixauli et al. (2008)
	Prevention of phase separation	Prevents water and oil separation in multi-phase systems by creating a gel like matrix that holds ingredients in place during storage and thawing.	Baixauli et al. (2008)
	Control of ice crystal formation	Inhibits large ice crystal growth during freezing, keeping ice crystals small and evenly distributed, ensuring visual and textural stability.	Abbas et al. (2010)
Texture	Smooth texture	Helps create a smooth, creamy texture by preventing gritty or icy textures caused by large ice crystals.	Gunaratne and Harold (2007)
	Firmness and elasticity	Reinforces gluten in frozen bread dough, cakes, and pastries, ensuring they maintain shape, elasticity, and texture after freezing and reheating.	Miyazaki et al. (2005)
	Creamy consistency	Thickens liquid-based frozen products and helps them maintain a rich, velvety texture after thawing.	Miyazaki et al. (2005)
Shelf life	Freeze-thaw stability	Maintain texture and prevent separation during multiple freeze-thaw cycles, ensuring the product remains stable over time.	Baixauli et al. (2008)
	Prevention of ice recrystallization	Minimizes ice crystal growth during temperature fluctuations, reducing recrystallization and ensuring consistent texture.	Miyazaki et al. (2005)
	Protection of flavor and aroma	Control water migration, starch helps preserve the distribution of flavor compounds, maintaining the intended taste profile during storage.	Gunaratne & Harold (2007)

attributes of frozen products, such as creaminess, smoothness, and firmness, with its type and concentration determining the final texture. In frozen desserts like ice cream and sorbet, starch works alongside other stabilizers to create a smooth, creamy texture by preventing gritty or icy textures caused by large ice crystals. In bakery items such as frozen bread dough, cakes, and pastries, starch reinforces gluten structures, ensuring shape retention, elasticity, and desirable crumb texture after freezing and baking. For liquid-based frozen products like gravies, custards, and soup bases, starch thickens the formulation, providing a rich, velvety consistency even after thawing, with heat-resistant modified starches proving particularly effective in retaining this texture. Beyond texture, starch plays a vital role in extending the shelf life of frozen products by maintaining quality over time. Its freeze-thaw stability prevents syneresis (water separation) and ensures texture and appearance remain intact across multiple freeze-thaw cycles. Furthermore, starch minimizes ice recrystallization by binding water, preventing crystal growth caused by temperature fluctuations, and preserving a consistent texture. Additionally, by controlling water migration, starch protects flavor and aroma, ensuring that flavor compounds remain evenly distributed without dilution or concentration in specific areas, allowing frozen products to retain their intended taste profile throughout storage.

1.2.1 Applications of starch in frozen products

1.2.1.1 Textural improvements

Starch is a versatile carbohydrate with unique properties such as gelatinization, retrogradation, and water-holding capacity, making it a vital ingredient in enhancing the texture and stability of frozen food products (Miyazaki et al., 2006). Its applications span a variety of categories, including ice creams, frozen desserts, ready-to-eat meals, and frozen doughs. Starch plays a significant role as a textural enhancer in frozen food products due to its ability to gelatinize when heated in water and undergo retrogradation upon cooling or freezing, forming a gel like network that stabilizes water and other components. This property helps prevent ice crystal formation, a major cause of texture degradation in frozen foods, by binding free water and limiting its migration, thereby minimizing ice crystal growth. Modified starches, such as pregelatinized or cross-linked varieties, are specifically designed to withstand the freezing and thawing processes, enhancing their effectiveness in frozen applications. In ice creams and frozen desserts, starch acts as a stabilizer, reducing the formation of large ice crystals that contribute to a coarse texture while enhancing creaminess and smoothness by stabilizing emulsions and preventing water separation during storage. Additionally, it improves mouthfeel and supports air incorporation, ensuring a consistent structure (Baixauli et al., 2008). In frozen dough products, starch preserves elasticity and structure throughout freezing and thawing cycles, preventing ice crystal damage that can weaken the gluten network and impair dough rise during baking. Modified starches help retain moisture, resulting in baked goods that remain soft and fresh after freezing. For ready-to-eat frozen meals, starch maintains viscosity and prevents phase separation during freezing and reheating, ensuring consistency in texture and appearance while stabilizing emulsions and suspensions. Starch-based thickeners create smooth and creamy textures in soups, casseroles, and lasagna. In frozen fruits and vegetables, starch-based coatings help reduce ice crystal formation on the surface, preserving texture and appearance while acting as moisture barriers that maintain quality and freshness throughout storage.

1.2.1.2 Freeze-thaw stability

Freeze-thaw stability refers to a product's ability to maintain its physical, textural, and structural integrity during cycles of freezing and thawing, which is crucial for food systems subjected to multiple freeze-thaw cycles during storage, transportation, and consumption. Challenges such as ice crystal formation, moisture migration, and structural breakdown can compromise frozen food quality, but starch plays a vital role in stabilizing water and minimizing phase separation. During freezing, starch forms a viscoelastic gel network that binds water, limiting ice crystal growth and migration, while upon thawing, retrograded starch prevents syneresis by retaining water within its structure. Modified starches, such as cross-linked, acetylated, or hydroxypropylated variants, offer superior freeze-thaw stability by resisting structural breakdown and maintaining water-binding capacity, making them highly effective in frozen food formulations (Abbas et al., 2010). The use of starch in frozen products provides several advantages, including minimizing ice crystal growth to prevent texture damage, reducing syneresis to maintain product consistency, enhancing texture and mouthfeel for smoothness and creaminess, and serving as a cost-effective stabilizer without significantly increasing production costs (BeMiller & Whistler, 2009). Starch is widely applied in dairy-based frozen products, where it prevents large ice crystal formation and stabilizes emulsions to ensure a creamy texture. In thickened sauces and soups, starch forms a stable gel matrix that prevents water separation during freezing and thawing, with modified starches ensuring consistent viscosity and texture retention. For frozen doughs and baked goods, starch helps retain moisture, stabilizes the gluten network, and prevents ice crystal damage, ensuring elasticity and softness after thawing. In ready-to-eat frozen meals, starch stabilizes sauces and fillings, preventing phase separation and maintaining the visual appeal and cohesiveness of complex food systems after reheating (Kawaljit Singh & Sandhu, 2006). Additionally, coating frozen fruits and vegetables with starch-based solutions reduces ice crystal damage on the surface while acting as a barrier to moisture migration, preserving their texture and appearance even after thawing.

1.2.1.3 Thickening and stabilizing

Starch is a key ingredient in the food industry, highly regarded for its thickening and stabilizing properties. These functions are especially critical in frozen products to maintain consistency, texture, and stability throughout freezing, storage, and thawing processes. With unique functional attributes like gelatinization, retrogradation, and water-binding capacity, starch serves as a versatile and efficient solution for enhancing the quality of frozen foods. This essay examines the role of starch in thickening and stabilizing frozen products, its applications across different food categories, and the advantages it brings to frozen food formulation. Starch functions as both a thickener and stabilizer by gelatinizing and forming a gel matrix when heated in water (Waterschoot et al., 2015). During gelatinization, starch granules absorb water, swell, and release amylose and amylopectin, creating a viscous solution that thickens the food system, which is essential for providing structure and viscosity in products like sauces, soups, and desserts. Upon cooling or freezing, starch usually undergoes retrogradation, where the amylose and amylopectin chains realign to form a gel like structure that stabilizes the product by trapping water and other components, reducing phase separation, ice crystal formation, and syneresis. Modified starches, such as cross-linked or pregelatinized varieties, are particularly beneficial in frozen products due to their ability to withstand the stresses of freezing and thawing cycles, preserving their thickening and stabilizing properties in extreme conditions (Jiang et al., 2024).

1.2.1.4 Shelf-life extension

The extension of shelf life is a critical concern in the frozen food industry, as it ensures that products maintain their quality, taste, and nutritional value throughout storage, distribution, and consumption. Starch plays a vital role in enhancing the shelf life of frozen products by stabilizing their structure, preventing moisture loss, and minimizing the degradation caused by freezing and thawing cycles (Ettelaie et al., 2016). Through its unique properties, starch contributes significantly to the preservation of frozen foods, addressing common challenges such as ice crystal formation, syneresis, and texture breakdown. Starch acts as a stabilizing agent in frozen food systems by binding water, forming gel networks, and resisting freeze-thaw damage. Water is a key factor in the deterioration of frozen foods, as large ice crystals that form during freezing can damage the food's cellular structure, leading to texture and quality loss. Starch helps prevent large ice crystal formation by binding to water molecules and creating a gel like matrix that stabilizes the water content, reducing its mobility and minimizing ice crystal growth, thereby preserving the smooth texture and structural integrity of frozen foods and extending their shelf life (Ettelaie et al., 2016). Additionally, starch reduces syneresis, or the separation of water from the food matrix during thawing, preventing flavor, texture, and appearance loss. Its water-retaining ability, especially in modified forms, ensures the product remains intact after thawing. Furthermore, starch's retrogradation properties, where starch molecules realign to form a more rigid structure as they cool, enhance the stability of frozen foods by forming a solid, water-retaining network.

1.2.2 Benefits of starch in frozen food products

Starch provides several benefits for extending the shelf life of frozen products, including improved texture by preventing ice crystal formation, which helps maintain the structural integrity and smooth mouthfeel of products after freezing and thawing. Its water-binding properties also reduce moisture migration, preventing dehydration and freezer burn, which are common causes of quality degradation in frozen foods. Additionally, starch enhances stability by stabilizing emulsions and suspensions, reducing phase separation and syneresis, and ensuring consistent texture and appearance over time. Modified starches, such as cross-linked or pregelatinized varieties, are particularly effective in withstanding repeated freeze-thaw cycles, making them ideal for extended storage periods. Moreover, starch is a cost-effective ingredient that delivers significant improvements in texture, stability, and shelf life, offering manufacturers a valuable solution for maintaining product quality.

Starch-based coatings are widely applied to frozen fruits and vegetables to enhance their stability and shelf life. These coatings form a protective layer that minimizes moisture loss, reduces ice crystal formation, and preserves the texture and appearance of the produce. By preventing dehydration and freezer burn, starch helps extend the shelf life of frozen fruits and vegetables, keeping them fresh and visually appealing.

In frozen doughs and baked goods, starch contributes to the maintenance of dough elasticity and moisture retention during freezing and thawing. The starch helps the dough withstand the stresses of freezing and freezing-thawing cycles, ensuring that it retains its texture and structure. This results in baked goods that remain soft and fresh even after freezing, with minimal loss of quality (Skryplonek et al., 2019). Additionally, starch-based thickeners in fillings and batters help prevent separation and maintain consistency during storage.

In frozen soups and sauces, starch serves as a thickener and stabilizer, helping to maintain the viscosity of the product over time. During freezing and thawing, the starch binds water, preventing the formation of ice crystals and maintaining the smooth texture of the soup or sauce. By stabilizing emulsions and suspensions, starch prevents the separation of ingredients and helps preserve the overall quality of the product, extending its shelf life.

2. Types of starches and applications in frozen and dairy products

Starches play a vital role in the formulation of dairy and frozen products by offering texture improvement, stability, and functional benefits. They are essential for applications such as thickening, stabilizing, and enhancing mouthfeel. The choice of starch can significantly impact the overall quality and performance of the product. As discussed in previous sections, starch aids in air incorporation in ice cream, increasing volume and enhancing mouthfeel without the need for added fat. Additionally, starch stabilizes formulations by preventing separation and curdling, while also acting as an emulsifier to balance water and fat, ensuring consistent product quality. In frozen products, starch improves viscosity, creaminess, and melting behavior, while inhibiting ice recrystallization. This section discusses the various sources of starches used in dairy and frozen product development.

2.1 Native starch

Native starches occur naturally in granular form, with inherent characteristics according to their botanical origins. These starches can act as a fat replacer and improve the texture of dairy and frozen products by retaining water. However, their industrial use is limited due to factors like low shear and thermal resistance, as well as a high tendency for retrogradation. Native starches are typically sourced from plants like corn, potato, tapioca, wheat, and rice, each offering unique properties. As a result, these starches are often modified to suit specific applications.

2.1.1 Sources and applications

Corn starch: Dry and unmodified corn starch is a white powder with a pale-yellow cast, which often undergoes bleaching for food and dairy applications. It is the most widely produced starch, contributing to 85% of global starch production. In dairy applications, corn starch acts as a thickening and gelling agent, enhancing texture and mouthfeel in products like low-fat yogurts. In frozen desserts, it prevents water migration between ice crystals by forming weak gels, ensuring a smooth and creamy texture even in frozen storage. Its low cost, high availability, and functional properties make it a valuable ingredient in food product development and research. Eckhoff and Watson (2009, pp. 373–439) demonstrated the benefits of incorporating corn starch in ice cream-like frozen yogurt, maintaining pH and lactic acid stability over 6 months of storage. However, a decline in the viability of yogurt bacteria during storage was observed. These findings highlight the potential of corn starch as an economical and functional ingredient to improve the quality of dairy and frozen products.

Potato starch: Potato starch is a valuable ingredient in the food industry, known for its neutral flavour and starch paste clarity. It has considerably high starch solubility and paste viscosity with larger starch granules (5–100 μm) compared to other native starch sources (Chartrand et al., 1997). In dairy and frozen product development, it enhances the consistency and smoothness of products like custards, creams, and puddings by providing a stable, creamy texture. Potato starch also serves as an excellent stabilizer, preventing separation and improving the overall viscosity of dairy products. In frozen products such as ice cream, frozen yogurt, and sorbets, it helps to prevent the formation of large ice crystals, contributing to a smoother and velvety mouthfeel. Breuninger et al. (2009) reported that adding potato starch to yogurt could inhibit the growth of lactic acid bacteria by reducing the amount of available water, thereby preventing lactic acid production and acidity rise. It also significantly reduced syneresis in yogurt during 15 days of refrigerated storage.

Tapioca starch: Tapioca starch is unique compared to other starches due to its low levels of residual materials like fat, protein, and ash, lower amylose content, and higher molecular weights of both amylose and amylopectin (Dhivya et al., 2015). Its ability to maintain a stable texture in cold conditions helps prevent the formation of gritty or coarse textures, making it particularly suitable for frozen products. In a study (Mitchell, 2009, pp. 731–770), incorporating tapioca starch into low-fat ice cream increased overrun and improved texture when compared to potato starch. However, the study recommended limiting starch inclusion to 25%, as higher amounts negatively impacted the sensory qualities of the ice cream. Additionally, tapioca starch is a natural, allergen-friendly option, making it an appealing choice for manufacturers aiming for clean-label products while improving quality and stability.

Wheat starch: Wheat starch is a bright white powder free from pigments. It has a higher gel strength than corn starch at concentrations above 6% and forms a clearer paste. As an effective thickening agent, wheat starch provides a smooth, creamy texture, helps stabilize products, and prevents separation. It also enhances the mouthfeel of dairy desserts and can aid in stabilizing emulsions in frozen treats, improving overall product consistency. Its versatility in both cold and warm environments makes it especially useful for frozen foods, offering manufacturers a cost-effective, natural ingredient for enhanced texture and performance. However, the use of wheat starch is limited as it may trigger a reaction in people with the risk of gluten allergy and celiac disease.

Rice starch: Rice starch is particularly beneficial in lactose-free and gluten-free formulations, making it a popular choice for consumers with dietary restrictions. Its superior digestibility, compared to other cereal grains, makes it especially suitable for infant and geriatric foods. The neutral flavour, whiteness, and small granule size of rice starch contribute to its effectiveness in creating products like puddings, which offer excellent mouthfeel and flavor profile (Autio & Eliasson, 2009, pp. 589–599). Traditionally, it has been used in the production of Japanese sweet gels, such as mochi and mochi ice cream, where it is reported to provide better cohesiveness and a softer gel than other native starches (Wurzburg, 2006). Rice starch's mild flavor and hypoallergenic properties make it an ideal alternative to other starches, providing manufacturers with a clean-label option that improves the quality and consistency of dairy and frozen products. Innovative wholegrain rice products like RiceLife have enabled the creation of frozen desserts and even wholegrain nondairy yogurts. However, the production of rice starch is limited due to its high demand as a staple food and whole grain in many cultures (Autio & Eliasson, 2009, pp. 589–599).

Oat starch: Oat starch is particularly advantageous in dairy-free and gluten-free formulations, offering a clean-label, plant-based solution that meets consumer demand for natural and allergen-friendly ingredients. Its mild flavor and excellent water-binding properties make it an ideal choice for enhancing the texture and consistency of both dairy and frozen food products while maintaining a high-quality sensory experience. Oat starch has a higher lipid-binding ability, greater relative crystallinity, and more coiled amylose and amylopectin, though it has shorter amylose chains and smaller granules compared to other cereal starches. Despite its appealing properties for industrial use, the application of oat starch has been limited due to challenges in separating the grain into its starch, fiber, and protein components (Kaur, 2007).

2.2 Modified starch

Modified starches are starches that have been altered through chemical, physical, or enzymatic processes to improve their properties, making them more suitable for specific applications in food products. The use of native starches in food products is often limited due to their inherent physical and chemical properties. For instance, native starch gels lose their water-binding properties over time, leading to syneresis. Additionally, some native starches, like potato starch, are known to have poor digestibility. Modified starches are specifically designed to address these limitations, overcoming the constraints of native starches in food applications. The modifications enhance the starch's functionality, such as its thickening ability, stability under varying conditions, and texture performance, all of which are crucial for maintaining product consistency and quality. In dairy and frozen product development, modified starches are essential for improving mouthfeel, preventing ice crystal formation, enhancing texture, and stabilizing emulsions. The various types of modified starches and their applications provide a wide range of benefits, allowing manufacturers to create high-quality, stable dairy and frozen products.

2.2.1 Chemical modifications

Chemical modifications of starch involve altering the starch molecule through processes such as cross-linking, dextrinization, esterification, etherification, oxidation or acid hydrolysis (Abbas et al., 2010; Zia-ud-Din et al., 2017).

Crosslinked starch: One widely used type of chemically modified starch in dairy and frozen products is cross-linked starch, where chemical bonds are formed between starch molecules to enhance their resistance to heat, shear, and acidic conditions. Cross-linking agents, such as phosphates or epichlorohydrin, create these bonds, making the starch highly effective in applications like ice cream and frozen yogurt. In these products, cross-linked starch prevents ice crystal formation during freezing, contributing to a smoother texture, reduced wheying-off, and improved stability. With enhanced resistance to granule swelling, high temperatures, shear forces, and low pH, cross-linked starches function as reliable thickeners and texturizers in soups, sauces, gravies, bakery goods, and dairy products (BeMiller, 2017, pp. 223–253).

Dextrinized starch: Dextrinized starch is a starch that has been chemically modified through a process called dextrinization, where starch is partially hydrolyzed by heat, acid, or enzymes into dextrins. This process reduces the starch's molecular weight, increases its solubility, and changes its properties, such as improving gel stability. Dextrinized starch is often used as a fat replacer in various food products, such as bakery and dairy items.

Esterified starch: This modification involves the reaction of starch with an esterifying agent, such as acetic or phosphoric acid. It enhances the starch's resistance to retrogradation and increases its stability under varying temperatures and pH levels. Acetate starch (starch esterified with acetic acid) is an example often used in frozen desserts to enhance viscosity and improve the product's resistance to temperature changes.

Etherified starch: Etherification replaces the hydroxyl (-OH) group on the starch molecule with hydroxypropyl, carboxymethyl, or other modified groups, forming an ether chain. Starch reacts with etherifying agents, like acrylonitrile, ethylene oxide or propylene oxide, to form starch ethers. This modification improves solubility, water retention, and texture (Zia-ud-Din et al., 2017).

Oxidized starch: Oxidized or bleached starches are created through controlled oxidation using oxidizing agents like sodium hypochlorite or hydrogen peroxide. These agents break down the starch molecules, reducing their molecular weight. This modification increases the starch's solubility and viscosity, making it ideal for use in dairy-based dressings and sauces. The enhanced water-binding capacity of oxidized starch improves the texture and prevents separation, ensuring a smooth and stable product. Oxidized starches are also valued for their ability to perform well under acidic conditions, further boosting their functionality in various food applications (Zia-ud-Din et al., 2017).

Acid hydrolysis: In this process, starch is treated with a diluted acid to break down the starch granules, producing smaller molecules with increased solubility and modified functional properties. It is used for producing modified starches with specific characteristics like higher clarity or lower viscosity (Zia-ud-Din et al., 2017).

2.2.2 Physical modifications

Physical modifications, such as pregelatinization, heat-moisture treatment, annealing, high-pressure, microwave, or ultrasound, modify the starch's structure without changing its chemical composition (Fonseca et al., 2021; Hong, 2018).

Pregelatinized starch: It is a widely used modified starch in dairy and frozen product development. In this process, starch is cooked and then dried, allowing it to easily dissolve in cold water and thicken without additional heat. This is especially useful in instant dairy desserts, where a quick thickening agent is required. Pregelatinization is applied to both native and chemically modified starches using drum or roll drying, although most starches are sold in granular form. Extrusion can also produce pregelatinized starches, but it is unsuitable for high-viscosity products due to severe processing conditions like high shear and temperature. In contrast, roll drying is gentler, preserving the starch's viscosity. Pregelatinized starch improves convenience and functionality in food applications.

Heat-moisture treated starch: Starch is treated under high temperatures (84–140°C) with low moisture content (<35%) for 1–16 h. This treatment alters the crystalline structure of starch, improving thermal stability and resistance to enzymatic digestion. It also reduces swelling power and increases gelatinization temperature, making it ideal for applications requiring high heat or stability. This modification is beneficial in dairy-based beverages, where stability under varying cooking conditions and during storage is essential (Castro et al., 2020).

Annealed starch: Annealing involves exposing starch to water and heat at subgelatinization temperatures. The annealing temperature of starch varies based on its type, typically ranging between 45 and 55°C. This process enhances the granule's structural stability, increases the gelatinization temperature, and reduces swelling. Annealed starch is used in applications where controlled texture and stability are critical, such as in dairy products, candies and processed foods (Castro et al., 2020).

High-pressure treated starch: High-pressure treatment applies extreme hydrostatic pressure to starch, altering its gelatinization properties and granule structure without heat. The high-pressure treatment alters starch structure by breaking or modifying noncovalent linkages between molecules. This nonthermal method improves texture, stability, and water-holding capacity, making it suitable for dairy gels and frozen products. It enhances starch swelling and solubility, with effects depending on temperature. The treatment also impacts starch's thermal and pasting properties, and it can influence the proportion of resistant, fast, and slow-digestible starch in the products. Starch modified with high pressure can be used as a thickener and fat replacer in dairy products. It is used in high-protein dairy snacks, processed cheese, and frozen desserts to enhance water retention and maintain smooth textures. Its ability to resist granule damage under pressure makes it suitable for advanced processing techniques.

Microwave-modified starch: It is produced by exposing starch to microwave radiation, which induces rapid heating and causes structural changes in the starch granules. This method improves properties such as gelatinization and water absorption, enhancing the starch's functionality in food applications like thickening agents and texture stabilizers. The process is energy-efficient, nontoxic, and allows for precise control over modifications (Abbas et al., 2010; Bravo-Núñez et al., 2019).

Ultrasound-modified starch: It involves the use of high-frequency sound waves to alter the starch's physical and chemical properties. This treatment reduces starch granule size, enhances its solubility, and improves its ability to form gels. Ultrasound modification is a nonthermal method which preserves the starch's functional properties without the use of chemicals while offering faster processing times compared to conventional techniques (Bravo-Núñez et al., 2019).

2.2.3 Enhanced properties and applications in dairy and frozen products

The enhanced properties of modified starches offer a range of benefits to dairy and frozen food formulations, as mentioned in Table 10.2. One of the key advantages is improved stability, which is essential for maintaining product quality throughout its shelf life. In frozen desserts such as ice cream and sorbets, modified starches help minimize the formation of ice crystals, leading to a smoother, creamier texture, even after extended storage periods. Additionally, they contribute to controlling overrun, enhancing texture by making the product lighter and fluffier. Modified starches also improve moisture retention, preventing syneresis in products like yogurt and dairy-based sauces, which not only improves the appearance but also extends shelf life. Moreover, these starches enhance mouthfeel, providing a richer, creamier sensation in dairy products. Furthermore, they help stabilize emulsions in items like salad dressings, preventing the separation of oil and water.

The acetylated starches cross-linked with adipate were found to be more effective in reducing syneresis and improving viscosity and firmness in low-fat yogurt gels compared to those cross-linked with phosphate. Hydroxy propylated starches exhibited the greatest reduction in syneresis, while starch esterified with sodium trimetaphosphate performed poorly. Pregelatinized and chemically modified starches are preferred when increased viscosity or firmness is required in yogurt products. Additionally, starch phosphate and hydroxypropyl starch are reported to improve paste clarity, freeze-thaw stability, and water retention, making them ideal for frozen foods, salad dressings, ice cream, and various dairy applications (Zia-ud-Din et al., 2017).

Enzymatic modification of starch, such as with α-amylase, has shown promising results in improving the stability of ice cream. For instance, taro starch modified with α-amylase at 0.5% levels enhances foam stability, viscosity, and overrun, offering better performance than ice cream stabilized with guar gum. Similarly, potato starch modified with amylosubtilin and amylase can maintain the texture and viscosity of low-fat yogurt. Enzyme-modified and debranched starches contain linear short chains that act like fats, enabling them to replace fats in food products and reduce calorie content. For example, enzyme-modified sweet potato starch can be successfully used up to 20% in low-calorie ice cream

TABLE 10.2 Comparison of native and modified starches in dairy and frozen product development.

Property	Native starch	Modified starch	References
Gelatinization temperature	Lower gelatinization temperature causes premature thickening during milk heating, leading to lumping in custards and puddings.	Higher and controlled gelatinization allows stability during pasteurization, sterilization, and UHT processing.	Singh et al. (2007)
Freeze-thaw stability	Poor freeze-thaw stability; retrogradation and recrystallization lead to syneresis in ice cream and frozen yoghurt.	Improved freeze-thaw stability, preventing ice crystal growth and water separation during frozen storage.	Abbas et al. (2010)
Shear and heat stability	Granule disruption occurs under homogenization and thermal processing, leading to thinning of dairy sauces and desserts.	Stronger shear and heat stability maintain viscosity during pasteurization homogenization, and retort processing.	Zia-ud-Din et al. (2017)
Viscosity control	Inconsistent viscosity depending on starch source; unstable in flavored milk and custard formulations.	Provides consistent viscosity, ensuring smooth flow properties in yoghurt drinks, dairy deserts, and frozen mixes.	Mitchell (2009)
Water holding capacity	Lower water retention leads to whey separation in yoghurt and syneresis in frozen dessert.	Enhanced water holding capacity improves creaminess and prevents whey-off and shrinkage in ice cream.	Zhu (2015)
Texture and mouthfeel	Grainy, pasty, or unstable in dairy matrices.	Provides a smoother, creamier texture in low-fat yoghurt, processed cheese, and ice cream with better sensory appeal.	Skryplonek et al. (2019)

formulations as a fat replacer (Raigond et al., 2015). In summary, the chemical, physical, and enzymatic modifications of starch not only enhance the texture and consistency of dairy and frozen products but also provide significant functional benefits that improve product stability, consumer appeal, and overall quality. However, it is worth noting that most of the modified starches used commercially were developed before the 1980s, with limited recent research and development due to the stringent regulations and high costs associated with safety studies.

2.3 Resistant starch

Resistant starch is a type of starch that resists digestion in the small intestine and reaches the colon, where it acts like dietary fiber. It has gained significant attention in research due to its various health benefits. Resistant starch promotes gut health by fostering the growth of beneficial bacteria, acting as a prebiotic. It also helps control blood sugar by slowing carbohydrate digestion, which is especially helpful for people with diabetes or those seeking stable energy. Additionally, resistant starch has properties such as fine particle size, pale color, bland flavor, good extrusion and film-forming abilities, high gelatinization temperature, low caloric content, and low water binding capacity, making it a valuable functional ingredient. It is also linked to increased satiety, which can aid in weight management. As more consumers seek functional foods that support health, the inclusion of resistant starch in dairy and frozen products offers a way to meet the demand for healthier, more nutritious options. There are five types of resistant starch (RS) of which the first two are naturally available and the remaining three are obtained from the modification of starch.

1. RS1 is the physically inaccessible starch found in partially or whole-milled grains, seeds, and legumes.
2. RS2 is a native granular starch with a highly crystalline structure found in uncooked or partially cooked green bananas, raw potatoes, high-amylose corn, and some legumes.
3. RS3 is the retrograded or recrystallized starch that is found in cooked and cooled rice, potatoes, bread, cornflakes, and other food products that have been cooked with water or moist heat.
4. RS4 is a chemically modified starch that is found in foods that contain modified starches, such as breads and cakes.
5. RS5 is the amylose-lipid complex starch modified by the addition of lipid.

In dairy and frozen product development, resistant starch is increasingly used as a source of dietary fiber. By adding resistant starch to products like yogurt, ice cream, cheese, and other dairy-based desserts, manufacturers can boost the fiber content of these foods, offering consumers a functional product that supports digestive health. This addition is particularly appealing to health-conscious consumers looking to increase their fiber intake, especially in products traditionally low in fiber, such as ice cream or milk-based beverages. In a study, the inclusion of 1% resistant starch acted as a prebiotic for microencapsulated nonfermented ice cream and extended the shelf life of probiotic cultures.

In addition to its health benefits, resistant starch significantly improves the texture of dairy and frozen products. In frozen treats like ice cream, it helps reduce ice crystal formation, contributing to a smoother, creamier mouthfeel (Homayouni, 2013). This texture enhancement is especially valuable in reduced-fat or dairy-free frozen desserts, where achieving a rich, indulgent texture can be difficult. Rezaei et al. (2015) optimized the use of 2% resistant starch in low-fat ice cream, balancing viscosity, overrun, acidity, and reducing the melting rate. Resistant starch is also known to increase viscosity, foam stability, fat destabilization, and hardness while reducing overrun and the meltdown rate in frozen soy yogurt. Furthermore, resistant starch improves the overall structure and stability of dairy products, providing a consistent, desirable texture that enhances the sensory experience (Fig. 10.2).

3. Recent developments

The recent research on starches has made many breakthroughs in the development of new starch forms with unique properties that will enhance the properties of the food enormously. The innovation of blending two starches, either with native and modified or with two or more modified starches, also added colors to enhance the quality of food. Some of the most recent advancements in the Strach are depicted below.

3.1 Clean-label starches

Clean-label starches are becoming increasingly popular in food and dairy product development, as consumers are becoming more conscious about the ingredients in their food. The definition for clean label devised by National Starch in the US, is as follows: "Free from chemical additives, simple ingredient listing (without ingredients that sound chemical or artificial); minimally processed using traditional techniques that are understood by consumers and not perceived as artificial" (Blennow et al., 2020).

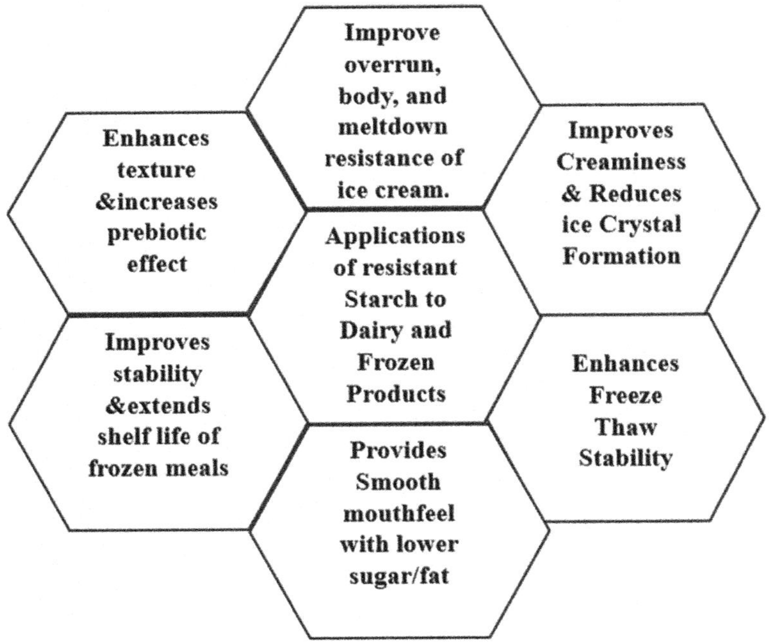

FIGURE 10.2 The outline of the benefits of resistant starch. Resistant starches enhance the texture, stability, and sensory qualities of dairy and frozen goods. It lowers fat and sugar levels while improving mouthfeel, body, and creaminess. It reduces the development of ice crystals and enhances overrun and meltdown resistance in ice cream. Additionally, it prolongs the shelf life of frozen meals, improves freeze-thaw stability, and offers prebiotic advantages that promote intestinal health.

These starches, derived from nonGMO (genetically modified organism) and organic sources, cater to the growing demand for products that are perceived as natural and free from synthetic additives or genetic modification. However, clean-label foods with simple and minimalistic ingredient lists are often expensive to produce and/or may possess less desirable sensory qualities in some cases (Krunic, 2017). For specific applications and improved properties, clean-label starches are manufactured by starch blending, physical, and enzymatic modifications. In dairy and frozen products, clean-label starches are often used as thickening agents, providing a smooth texture without relying on artificial additives. These starches meet the preferences of consumers who seek healthier, more transparent food options while still delivering the functional benefits required for texture, viscosity, and product stability of dairy and frozen products.

3.2 NonGMO starches

NonGMO starches are sourced from plants that have not been genetically modified and have gained popularity due to the growing consumer demand for food products that align with natural and sustainable farming practices. While genetically modified crops use RNA interference technology involving the introduction of foreign genes into crops, alternative methods have been developed to create crops with new starch qualities, such as high amylose content, due to public concerns. One such method is cisgenesis, which uses genes from the same species or closely related plants. Another approach is random mutagenesis, which is considered safe and nonGMO and involves techniques like TILLING to produce high-amylose starch. Recently, precise genome editing, like CRISPR/Cas9, has been used to target specific genes and create nonGM crops. In dairy and frozen products, nonGMO starches are often derived from sources like corn, potato, or tapioca. These starches offer many of the same functional properties as their GMO counterparts, such as thickening, gelling, and stabilizing, while meeting consumer preferences for nonGMO ingredients (Li et al., 2019; Zhong et al., 2023). The use of nonGMO starches helps manufacturers meet clean label trends while ensuring the quality and functionality of their products.

3.3 Organic starches

Organic starches, which are produced without the use of any synthetic fertilizers, pesticides or genetically modified organisms, make them an attractive option for products targeting health-conscious consumers. These starches, often sourced from organically grown native sources like corn, potato, or rice, serve as excellent alternatives to conventional starches. Organic crops have a composition similar to conventional crops, with a modestly higher content of functional components. Organic starches help maintain the desired texture and consistency similar to native starches, while meeting strict certification requirements for organic labeling. As the demand for organic products continues to rise, the use of organic starches in dairy and frozen food products offers both environmental sustainability and marketing advantages, allowing manufacturers to cater to a growing segment of the consumer market that prioritizes natural, eco-friendly ingredients.

3.4 High amylose starches

High amylose starches (HAS) contain over 50% amylose, unlike regular starches, which are predominantly composed of amylopectin (75%). Classified as RS type 2, HAS is recognized as a dietary fiber by food regulatory agencies. Although HAS retains native starch characteristics, such as its white color, fine particle size, and neutral flavor, it demonstrates distinct thermal and pasting behaviors that can influence the sensory qualities of food. The increased amylose content typically leads to higher gelatinization temperatures and a faster rate of retrogradation (Obiro et al., 2012). High amylose starches are produced by genetic engineering methods like gene editing and gene suppression (Jobling, 2004). In addition to their functional benefits in texture and stability, HAS offer other advantages in dairy and frozen product development. The microencapsulation of probiotics using high amylose corn starch (Hylon VII) with chitosan and poly-lysine coatings into ice cream enhanced the survival of probiotics at low temperatures during the storage of ice cream, without negatively affecting its sensory properties. Similarly, Obiro et al. (2012) reported that Hylon VII exhibited the highest resistant starch content and viscosity in milk gel systems, highlighting the potential of HAS in developing healthier yogurt products. The unique properties in controlling moisture retention and providing a desirable mouthfeel further enhance the application of HAS in low-fat or reduced-calorie dairy and frozen products. The enhanced properties of the new and modified starches are to ensure the quality of the food includes the retention of the structure during freezing and thawing, and to prevent syneresis. The new and modified starches help to enhance the freeze-thaw stability and gelling properties.

3.4.1 Freeze-thaw stability

Freeze-thaw stability of starch refers to its ability to maintain its colloidal structure after being frozen and thawed. During freezing, starch molecules form hydrogen bonds, reducing their water-holding capacity and resistance to freezing. Repeated freeze-thaw cycles can impair starch functionality, resulting in texture changes, large ice crystal formation, and water separation. These issues are further exacerbated by thermal fluctuations during transport, which cause water to melt and recrystallize, ultimately lowering product quality. As starch cools, it undergoes recrystallization, leading to syneresis, or the release of water from a gel. Syneresis occurs when starch is stored at low temperatures (4°C) or subjected to freeze-thaw cycles (Fonseca et al., 2021). Modified starches, with improved functional properties, help address these challenges by enhancing freeze-thaw stability and minimizing syneresis. Techniques such as genetic engineering, including down-regulating starch synthase genes, have resulted in the development of cold-resistant starches that offer superior stability, making them ideal for frozen food applications. The starch's botanical source and structural characteristics greatly influence syneresis and stability. High amylose maize starch (HAMS) exhibits lower freeze-thaw stability compared to waxy and native corn starch due to significant retrogradation during frozen storage, as amylose impairs stability. Syneresis in HAMS gels is positively correlated with amylose content, while starches with higher amylopectin content and shorter chains, like waxy corn starch, resist syneresis and have better freeze-thaw stability. Waxy corn starch's highly branched amylopectin structure makes it more stable than HAMS and regular corn starch in frozen applications. HAMS's susceptibility to syneresis limits its use in frozen foods. To address this, enzymatically modified corn starch using cyclodextrin glucotransferase and branching enzymes, creating an amylopectin cluster with enhanced freeze-thaw stability for frozen dough. Additionally, polysaccharide gums such as xanthan gum, guar gum, and sodium alginate can improve the freeze-thaw stability of starch gels (Liu & Lelievre, 1992; Stute & Kern, 1994).

3.4.2 Unique gelling properties

High amylose starches form strong, resilient gels when heated and cooled, making them ideal for applications requiring a firmer, more stable texture. Starch gel strength is influenced by factors like starch composition, molecular interactions with water, and gel formation. The amylose-to-amylopectin ratio plays a key role, with higher amylose content leading to stronger gels. Amylose molecules form hydrogen bonds more easily, resulting in firmer gels, particularly during cooling and short-term storage (Lai, 2001; Waterschoot et al., 2015). High amylose maize starch was reported to exhibit stronger gels compared to waxy corn starch. In dairy-based products like puddings, custards, and yogurts, HAS are used as gelling agents to provide a smooth, consistent texture and prevent syneresis. The high amylose content allows these starches to create gels that are more stable and less prone to degradation, making them highly effective in formulations requiring long shelf life and consistent quality.

3.5 Starch blends

Starch blends combine two or more starches (e.g., native starch with modified starch or resistant starch) to create a product with superior texture, stability, and cost-effectiveness. It has studied the additive melting behavior in wheat and rice starch blends at low concentrations (<30%), but high concentrations caused nonadditive behavior due to competition for water (Jiang et al., 2024). Wheat starch, with a lower gelatinization temperature, melted first under these conditions. By combining starches from different sources, manufacturers can tailor the characteristics of the final product. For instance, combining potato starch, which provides smoothness, with corn starch, which offers stability, may result in the ideal consistency for ice cream or cream-based dairy products. It's observed that the blends of unmodified pea and corn starches reduce syneresis in pudding preparation (Stute & Kern, 1994).

Starch blends, which combine different types of starches such as corn, potato, tapioca, and modified starches, are commonly used in dairy and frozen product development to achieve specific functional benefits. The combination of starches with varying gelatinization temperatures and textural properties ensures that the final product retains its quality and texture under different processing and storage conditions, meeting consumer expectations for consistency and sensory appeal. In addition to improving texture, starch blends are particularly valuable for freeze-thaw stability in dairy and frozen products (Waterschoot et al., 2015).

Blending starch with nonstarch hydrocolloids like xanthan gum, guar gum, Arabic gum, pectin, and gellan can enhance starch properties for specific applications. These high molecular weight polysaccharides interact with starch to modify rheological properties like viscosity and texture, influenced by gum and starch concentration, preparation conditions, and interactions. Hydrocolloids thicken solutions, improve sensory properties, and protect starch granules against shear during processing, reducing syneresis and enhancing moisture retention. Studies report reduced water solubility and absorption capacity in blends with wheat or potato starch, but no impact on swelling power. Gelatinization enthalpy and temperature range increased, while pasting temperature remained unchanged, indicating controlled structural modifications without altering key starch processing traits.

Starch blends also offer flexibility in clean label formulations, providing a natural and effective alternative to synthetic thickeners, stabilizers, and emulsifiers, and meeting consumer demand for more transparent ingredient lists in dairy and frozen foods. This versatility makes starch blends an essential tool for improving the overall quality and performance of a wide range of dairy and frozen products.

4. Challenges in starch applications

Starches play a fundamental role in the dairy and frozen product industries due to their versatility in enhancing texture, stability, and overall product quality. Advances in starch technology, such as clean-label and high-amylose starches, continue to unlock new opportunities for creating high-quality formulations. Modified starches offer functional benefits like improved viscosity, freeze-thaw stability, and fat-mimicking properties, making them indispensable for low-fat and calorie-reduced products. However, achieving consistency in texture and stability remains a challenge due to the variability in starch performance depending on the source, type, and processing methods. Starch's hydrophilic nature also limits its water resistance and mechanical properties in wet environments, posing difficulties in certain applications. Additionally, the rising demand for nonGMO and organic starches often comes with higher costs and supply chain constraints, making it harder for manufacturers to meet consumer expectations while maintaining affordability.

Consumer preferences for clean-label products further complicate starch use, as native or organic starches may lack the functionality of modified starches, such as enhanced texture and stability. Freeze-thaw stability remains a critical challenge, particularly in frozen products, as repeated cycles can lead to syneresis or textural degradation. Health considerations also present opportunities and obstacles; while resistant starches provide dietary benefits, their high incorporation levels may impact product taste or texture. Additionally, sensitivities or allergies to certain modified starches limit their widespread adoption. Despite these challenges, ongoing innovations in starch technology offer promising solutions, including better-performing clean-label starches, improved freeze-thaw resistance, and more sustainable options, ensuring that starch continues to be a key ingredient in dairy and frozen food development.

5. Conclusion

The use of starch is inevitable in the dairy and frozen food products for enhancing texture, stability and shelf life. The major functions which make the starch a key factor include the role of a thickening agent, stabilizing agent, and freeze thaw stabilizer. Various sources of starch, including native, modified and resistant starch, provide intact functionalities that suit specific product requirements. Advances in starch technology, including clean-label choices and modified starches with enhanced freeze-thaw stability, have expanded the potential uses of starch in food compositions. Research and innovation are still influenced by issues such as cost considerations, market preferences for natural and minimally processed products, and retrogradation. Sustainable and consumer-friendly starch solutions will continue to be crucial as the food industry develops, guaranteeing that starch will continue to play a significant role in the creation of dairy and frozen products.

References

Abbas, K. A., Khalil, S. K., Shobirin, A., & Hussin, M. (2010). Modified starches and their usages in selected food products: A review study. *Journal of Agricultural Science, 2*(2), 90–99.

Apriyanto, A., Compart, J., & Fettke, J. (2022). A review of starch, a unique biopolymer – Structure, metabolism and *in planta* modifications. *Plant Science, 318*, Article 111223.

Arbuckle, W. S. (2013). *Ice cream.*

Autio, K., & Eliasson, A. C. (2009). *Oat starch starch.* Finland: Elsevier Inc.. https://doi.org/10.1016/B978-0-12-746275-2.00015-X

Awasthi, R. (2025). Status, development and potential of starch-based analogue cheese: A review. *International Journal of Biological Macromolecules, 310.* https://doi.org/10.1016/j.ijbiomac.2025.143190

Baixauli, R., Sanz, T., Salvador, A., & Fiszman, S. M. (2008). Muffins with resistant starch: Baking performance in relation to the rheological properties of the batter. *Journal of Cereal Science, 47*(3), 502–509. https://doi.org/10.1016/j.jcs.2007.06.015

BeMiller, J. N. (2017). Physical modification of starch. In *Starch in food: Structure, function and applications*. United States: Elsevier. https://doi.org/10.1016/B978-0-08-100868-3.00005-6

BeMiller, J. N., & Whistler, R. L. (2009). *Starch: Chemistry and technology*.

Blennow, A., Skryhan, K., Tanackovic, V., Krunic, S. L., Shaik, S. S., Andersen, M. S., Kirk, H. G., & Nielsen, K. L. (2020). Non-GMO potato lines, synthesizing increased amylose and resistant starch, are mainly deficient in isoamylase debranching enzyme. *Plant Biotechnology Journal, 18*(10), 2096–2108. https://doi.org/10.1111/pbi.13367

Bravo-Núñez, Á., Pando, V., & Gómez, M. (2019). Physically and chemically modified starches as texturisers of low-fat milk gels. *International Dairy Journal, 92*, 21–27.

Breuninger, W. F., Piyachomkwan, K., & Sriroth, K. (2009). Tapioca/cassava starch: Production and use. In *Starch* (pp. 541–568). Elsevier.

Castro, L. M. G., Alexandre, E. M. C., Saraiva, J. A., & Pintado, M. (2020). Impact of high pressure on starch properties: A review. *Food Hydrocolloids, 106.* https://doi.org/10.1016/j.foodhyd.2020.105877

Chartrand, L. J., Russo, P. A., Duhaime, A. G., & Seidman, E. G. (1997). Wheat starch intolerance in patients with celiac disease. *Journal of the American Dietetic Association, 97*(6), 612–618. https://doi.org/10.1016/S0002-8223(97)00156-9

Dhivya, N., Muthu Kumar, V., & Seetha Lakshmi, M. (2015). Role of potato and tapioca in low fat ice cream. *Indian Journal of Dairy Science, 68*(6).

Eckhoff, S. R., & Watson, S. A. (2009). *Corn and sorghum starches: Production starch.* United States: Elsevier Inc.. https://doi.org/10.1016/B978-0-12-746275-2.00009-4

Ettelaie, R., Holmes, M., Chen, J., & Farshchi, A. (2016). Steric stabilising properties of hydrophobically modified starch: Amylose vs. amylopectin. *Food Hydrocolloids, 58*, 364–377. https://doi.org/10.1016/j.foodhyd.2016.03.004

Fonseca, L. M., Halal, S. L. M. E., Dias, A. R. G., & Zavareze, E.d. R. (2021). Physical modification of starch by heat-moisture treatment and annealing and their applications: A review. *Carbohydrate Polymers, 274.* https://doi.org/10.1016/j.carbpol.2021.118665

Gunaratne, A., & Harold, C. (2007). Functional properties of hydroxypropylated, cross-linked, and hydroxypropylated cross-linked tuber and root starches. *Cereal Chemistry, 84*(1). https://doi.org/10.1094/CCHEM-84-1-0030

Homayouni, A. (2013). Resistant starch in food industry: A changing outlook for consumer and producer. *Starch, 66*(1-2). https://doi.org/10.1002/star.201300110

Hong, Y. (2018). Pre-gelatinized modification of starch. In *Physical Modifications of Starch*. https://doi.org/10.1007/978-981-13-0725-6_4

Jiang, X., Liang, Q., & Shi, W. (2024). Influence of starch on freeze-thaw stability of *Hypophthalmichthys molitrix* surimi gel observed via ice crystal distribution and gel properties. *Food Chemistry: X, 24*.

Jobling, S. (2004). Improving starch for food and industrial applications. *Current Opinion in Plant Biology, 7*(2), 210–218.

Kaur, B. (2007). Some properties of corn starches II: Physicochemical, gelatinization, retrogradation, pasting and gel textural properties. *Food Chemistry*. https://doi.org/10.1016/j.foodchem.2006.01.060.

Kawaljit Singh, & Sandhu. (2006). Some properties of corn starches II: Physicochemical, gelatinization, retrogradation, pasting and gel textural properties. *Food Chemistry, 101.* https://doi.org/10.1016/j.foodchem.2006.01.060

Krunic, S. L. (2017). Non-GMO potato lines with an altered starch biosynthesis pathway confer increased-amylose and resistant starch properties. *Starch, 70.* https://doi.org/10.1002/star.201600310

Lai, H. M. (2001). Effects of hydrothermal treatment on the physicochemical properties of pregelatinized rice flour. *Food Chemistry, 72*(4), 455–463.

Li, P., Zhang, B., & Dhital, S. (2019). Starch digestion in intact pulse cells depends on the processing induced permeability of cell walls. *Carbohydrate Polymers, 225*, 115204. https://doi.org/10.1016/j.carbpol.2019.115204

Liu, H., & Lelievre, J. (1993). A model of starch gelatinization linking differential scanning calorimetry and birefringence measurements. Carbohydrate Polymers, 20(1), 1–5. https://doi.org/10.1016/0144-8617(93)90026-z (Accessed 23 March 2020).

Mitchell, J. R. (2009). Starch rheology and texture Starch: Chemistry and technology.

Miyazaki, M. R., Hung, P. V., Maeda, T., & Morita, N. (2005). Recent advances in application of modified starches for breadmaking. *Food Science and Technology Research, 11*(3), 219–226.

Miyazaki, M., Van Hung, P., Maeda, T., & Morita, N. (2006). Recent advances in application of modified starches for breadmaking. *Trends in Food Science and Technology, 17*(11), 591–599. https://doi.org/10.1016/j.tifs.2006.05.002

Obiro, W. C., Sinha Ray, S., & Emmambux, M. N. (2012). V-amylose structural characteristics, methods of preparation, significance, and applications. *Food Reviews International, 28*(4), 412–438.

Pirsa, S. (2023). Hydrocolloids: Structure, preparation method, and application in food industry. *Food Chem.* https://doi.org/10.1016/j.foodchem.2022.133967

Singh, J., Kaur, L., & McCarthy, O. J. (2007). Factors influencing the physico-chemical, morphological, thermal and rheological properties of some chemically modified starches for food applications: A review. *Food Hydrocolloids, 21*(1), 1–22.

Skryplonek, K., Piotrowska, A., Czyżowska, A., & Tarko, T. (2019). Starch as an ingredient in dairy products: Effect on product quality. *Applied Sciences, 9*(24), 5431.

Stute, R., & Kern, H. (1994). U.S. Patent No. 5,324,532. Alexandria, VA: U.S. Patent and Trademark Office.

Tamime, A. Y. (2007). Yoghurt-science and technology. *International Journal of dairy Technology*. https://doi.org/10.1111/j.1471-0307.1999.tb02857.x

Waterschoot, J., Gomand, S. V., Fierens, E., & Delcour, J. A. (2015). Production, structure, physicochemical and functional properties of maize, cassava, wheat, potato and rice starches. *Starch – Stärke, 67*(1–2), 14–29.

Wurzburg, O. B. (2006). Modified starches: Properties and uses.

Zia-ud-Din, Xiong, H., & Fei, P. (2017). Physical and chemical modification of starches: A review. *Critical Reviews in Food Science and Nutrition, 57*(12), 2691–2705.

Zhong, Y., Tai, L., Blennow, A., Ding, L., Herburger, K., Qu, J., & Liu, X. (2023). High-amylose starch: Structure, functionality and applications. *Critical Reviews in Food Science and Nutrition, 63*(27), 8568–8590. https://doi.org/10.1080/10408398.2022.2056871

Zhu, F. (2015). Composition, structure, physicochemical properties, and modifications of cassava starch. *Carbohydrate Polymers, 122*, 456–480. https://doi.org/10.1016/j.carbpol.2014.10.063

Chapter 11

Transforming meat and seafood products with native and modified starch

Olivia Wright[1], Damanjeet Kaur[2], Jincy Abraham[3] and Parnasree Sadhukhan[4]
[1]*School of Human Movement and Nutrition Sciences, Queensland University, Brisbane, QLD, Australia;* [2]*School of Health, Deakin University, Melbourne, VIC, Australia;* [3]*Department of Food Processing Technology, PSG College of Arts & Science, Coimbatore, Tamil Nadu, India;* [4]*Department of Food Science & Nutrition Management, J.D. Birla Institute, Kolkata, West Bengal, India*

1. Introduction to starch in food processing

Starch is the most widely available natural polymer used in a wide range of food applications globally. It is an economical additive that is an easy-to-handle, renewable, and abundantly available popular compound. In the food industry, starch is popular due to its characteristic nature of being adaptable, affordable and renewable. However, starch available in its natural form has several limitations, like low solubility in cold water, high viscosity and paste instability. These properties thus limit its role and functional properties in the food sector. A solution to overcome these limitations is the process of modifying native starch to form modified starches. Researchers have developed different ways to alter the starch structure by various processing techniques like physical, chemical and enzymatic methods (Braşoveanu & Nemţanu, 2018). This chapter will explore the properties of different starches and their application in the meat and seafood industries.

Some of the different sources of native starch are sorghum, rice, maize, and potato, which are influenced by climate and geographical location that impact the composition and functional characteristics. Starch is made up of amylose and amylopectin, wherein amylose promotes gelling and amylopectin contributes to flexibility. Starch, when exposed to moist heat, undergoes an irreversible swelling of the granules, causing loss of granular structure, decline in birefringence and increased solubilization of starch. The change in the structure of gelatinized starch is affected by various factors such as moisture content, temperature, particle size, pH, sugar and certain other factors as depicted in Fig. 11.1 (BeMiller & Whistler, 2009).

Native starches are usually employed as thickeners or moisture regulators; for example, the physicochemical characteristics of Mysore banana starch and Chachafruto native starch have been promising in the meat food industry. Native starches are prone to retrogradation, in which amylose reassociates after gelatinization, resulting in food products being opaque and cloudy. Foods like gravies benefit from this feature of native starch, while products like jellies and fruit-based fillings require greater clarity in texture. Native starches, when compared with modified starches, have a shorter shelf life. Thus, due to the limitations of native starches, modified starches were developed with altered structure and functional qualities. These alterations can be carried out by chemical, physical and enzymatic methods, leading to the formation of novel, varied food products with greater convenience and diversity. Some of the applications include fat substitutes, thickeners, binders, stabilizers, emulsifiers, and texturizing agents.

Physical modifications of native starch are done to change the structure of starch granules using pressure, heat, and mechanical shear. Starch is also modified with the use of spray dryers, microwaves, extruders and drum dryers. Introduction of different chemical groups like acetyl and carboxyl groups, as well as creation of cross-linkages and addition of glycosidic bonds, includes utilization of agents like phosphorus oxychloride, diluted acids, sodium hypochlorite, acetic anhydride, and propylene oxide, leading to significant structural changes. Enzymatic modification involves breaking a number of original starch bonds in order to hydrolyze or rearrange the starch molecule. Different meat and seafood products and their varieties, like meat mixtures, frankfurters, fish cakes, edible films, and frozen products, were developed based on modified starches that have transformed the food sector. These products have a higher texture and quality with

Native and Modified Starches for Food Applications. https://doi.org/10.1016/B978-0-443-27431-2.00026-2

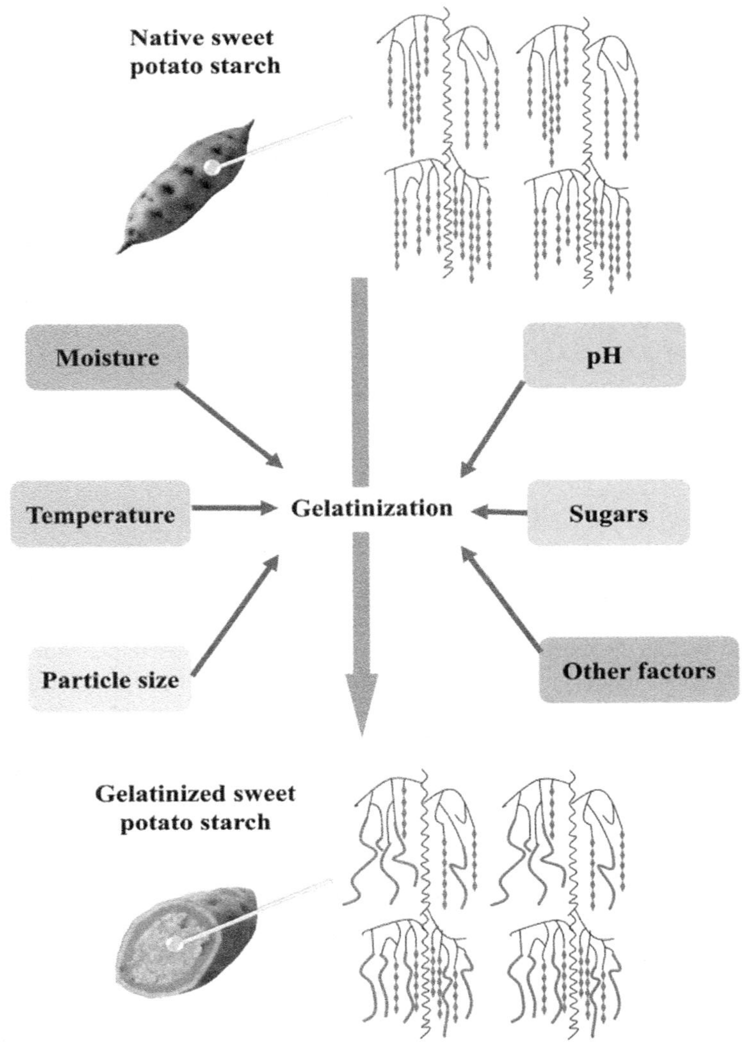

FIGURE 11.1 Gelatinization of a starch molecule (Yang et al., 2025).

better taste. These starches are also ideal in the preparation of foods with low-calorie content, such as spreads and low-fat butters, since they resemble fat (BeMiller & Whistler, 2009). Along with multiple functions, starch in the meat and seafood products also has financial benefits. This is a popular high-density polymer, especially the modified version, which has remarkable effects on the texture and sensory attributes. It imparts the meat and seafood products with the wonderful texture, water-retaining capability and flavor. It increases the shelf life of meat and seafood products substantially since these are highly perishable as the raw products.

The consumption of starch in meat and fish products is strictly regulated and controlled to ensure conformity and compliance with health. New legislation, such as regulation (EC) 1333/2008 and EU Commission Regulation No 231 are significant laws that specify the acceptable reagents, limits of residual and usage requirements of modified starches. In the United States, starch usage is regulated by the CFR Part 318.7 of the Food Safety and Inspection Service of the USDA (United States Department of Agriculture), in which special emphasis is put on texture improvement and water restraint within the products, such as sausages and surimi. Other types of groups, like the AOAC (Association of Official Analytical Chemists) and AACC (American Association of Cereal Chemists) provide standard methods of determining the quantity of starch in food all over the world. Labeling standards dictate the use of the mandatory declaration of modified starches, such as food starch-modified and its functional purpose. Unlike Australia and New Zealand, which use the INS codes, the EU uses E-numbers. Table 11.1 gives a discussion of some of the commercially available chemically

TABLE 11.1 Chemically modified starches as food additives with their E-numbers.

S.No.	Chemically modified starch	E– numbers
	Oxidized starch	E1404
	Monostarch phosphate	E1410
	Distarch phosphate	E 1412
	Phosphated distarch phosphate	E1413
	Acetylated distarch phosphate	E1414
	Acetylated starch	E1420
	Acetylated distarch adipate	E1422
	Hydroxypropyl starch	E1440

modified starches and their E-number. These standards strike a compromise between consumer safety and technological requirements by guaranteeing that starch derivatives enhance the texture, nutritional profiles, and shelf-life of seafood and meat products while stabilizing them without sacrificing compliance. The purpose of the chapter is to discuss the functional characteristics of native and modified starches and their technological application in enhancing the texture, stability, water retention, and sensory quality of meat and seafood products. It also aims to highlight the processing techniques, application approaches, and regulatory considerations that govern the safe and effective integration of starches in value-added meat and marine food systems.

1.1 Overview of native & modified starches

The addition of starch in its native form and modified form has been of great interest in the meat and seafood industry because of its multifunctional characteristics of enhancing the quality of products and lowering the production costs. They are primarily used as water binders, texture enhancers, yield improvement, improved sliceability and increased succulence in the meat and meat-based products. Starches are also employed as extenders as well as fillers in sausages, besides acting as binders. These functional characteristics are attributed to starch gelatinization in a liquid system, which results in a better sensory character by the use of thermal processing methods. Both types of these starches are allowed to be used in standardized and nonstandardized meat products, and regulatory bodies outline the permissible limit of usage.

Chachafruto native starch (CHS) is a unique starch that has a high level of purity and a high extraction yield. Moreover, 3D prints demonstrate excellent textural characteristics, high uniformity and great industrial capabilities as an alternative source of starch (Daza et al., 2023). Mysore banana starch has shown good physicochemical and functional characteristics like increased swelling power, indicating its potential application in foods that require the retention of water, such as in meat products (Fontes et al., 2017). The lowering of the percentage of animal fat in processed meat is a priority in accordance with the increasing interest in healthier food options. Even though animal fat is essential in making meat products soft and pliable, it has also been linked to many health hazards due to its high level of content. Replacement or even partial replacement of fat by starch has been promising in gel formation, freeze-thaw stability and emulsion properties without affecting the product quality (Zhang et al., 2024). Modified starches like hydroxypropyl are frequently used in fish-based surimi gels. Research studies suggest that the incorporation of these starches in surimi enhances its quality by improving their gelling properties (Xiaohuang et al., 2022).

1.2 Properties of starch relevant to meat & seafood products

Starches are beneficial additives in the meat industry that improve the binding of fat in meat mixtures and the binding capacity of meat and meat products, which influences the texture, consistency, and stability of the final product. In meat products, starches are utilized as water-binding agents to boost yields, enhance succulence, improve structure and sliceability, lower cooking losses, and prolong shelf life. For heat-treated beef products to be safe for human consumption, they must be heated for at least 10 min to a core temperature of 70°C. Potato starches are recommended as they gelatinize at temperatures between 50°C and 68°C. While gelatinization of starch in maize occurs at temperatures between 62°C and 72°C, it may result in inadequate gelatinization, especially near the core of meat products. Meat products can contain native or modified forms of several types of starch, including tapioca, wheat, maize, and potatoes.

The functional, rheological, and technological characteristics of the final meat product are all affected differently by native and modified starches, and their applications therefore vary. Simply breaking down natural starches—for example, with the aid of enzymes or high temperatures—produces native starches. Unlike modified starch, they do not have an E number; hence, they are not regarded as food additives. Modified starches made from starches using more complex methods (such as the action of acids) are subject to different conditions than those made by simple cracking. Due to their classification as food additives, these starches must be identified on food packaging by their full name or by their E-number (E−1404–1451). Monosubstituted starches combined with low cross-linked starches are used for smoked meat products because they are required to gelatinize beneath a curing temperature not exceeding 70°C.

When starch reacts with phosphoric acid or its salts, monostarch phosphates are created. Compared to native starch, these starches are more soluble in cold water and have a higher capacity for water binding and swelling. As the modified starch essentially shows no propensity to retrograde, they create transparent pastes that are extremely firm throughout thawing and freezing. Monostarch phosphates are reactive to low pH conditions and retain their ability to form structures even after prolonged contact with mechanical forces. They can stabilize emulsion systems to a certain degree. Monostarch phosphates are used to thicken food, particularly in meat manufacturing and cold-gelling desserts.

Even at low pH, high temperatures, and under strong mechanical pressures, distarch phosphates with a large number of cross-links create gels with excellent stability in rheological properties. In the technological process, this modified starch is utilized to produce goods that undergo heat treatment, specifically pasteurization or sterilization. When meat and vegetable products are thermally processed, it has a unique capacity to stop thermal leakage. In the process of producing meat products, distarch phosphates help to retain acceptable sensory qualities while increasing yield and improving the cohesion and consistency of the final product. Additionally, distarch phosphates increase the stability of defrosted goods, enabling the preservation of the intended consistency of prepared foods. They are also utilized in products as thickeners, binders, stabilizers, or carriers. Acetylated distarch phosphate has good texturizing and stabilizing qualities and is a versatile thickener and binder. Pastes and gels of acetylated distarch adipate and acetylated distarch phosphate are both transparent and impervious to shear forces, acidic conditions, and high temperatures. They are used to make mayonnaises, dressings, ketchups and vegetable and vegetable meat sauces, and they have the advantage of not going through retrogradation processes when stored for an extended time.

Compared to native starch, monostarch phosphate starches exhibit higher levels of water binding and swelling and an increased solubility in low-temperature water. Since the modified starch essentially has no tendency to retrograde, it produces transparent pastes that are extremely firm throughout thawing and freezing. Even after extended exposure to mechanical pressures, they maintain their ability to create structures. They can stabilize emulsion systems to a limited degree and are sensitive to low pH environments. Especially in cold-gelling desserts and the making of vegetable sauces, sauces for canned meat and vegetables, soups, and ketchup, monostarch phosphates are employed as food thickeners. Meat products can be made with modified starches instead of fat.

An enhanced starch composition for processed meat products that combines two cross-linked starches—the first derived from tapioca or potato starch and the second from maize starch—was described in a patent application submitted in the United States. The initial and secondary cross-linked starches' ratio is between 1:3 and 3:1 by weight. The patent implies that using this combination of starches improves the functionality, texture, and appearance of processed meats compared to using either cross-linked or uncross-linked corn or potato starches alone. The improved starch composition is injected into whole muscle tissue, which allows better integration and performance during the cooking process and significantly enhances the texture and visual appeal of processed meats.

Seafood analogue products, such as boiled gel items, are made from fish protein found in surimi. Since it may partially substitute fish proteins while retaining the required gel qualities at a lower cost, starch is the second most important ingredient used in the production of surimi seafood. In most surimi fish formulations, 4%–12% starch is included. The capacity of modified starches to regulate the wetness, stickiness, and thermal stability of surimi shellfish at varying storage and serving temperatures was thoroughly investigated (Hunt et al., 2009). At −20°C, the physicochemical characteristics of fish cakes, called "Eomuk" in Korea, were examined based on their modified starch [physical treatment, acetate (SA-1, SA-2, and SA-3), hydroxypropyl (HS), and acetylated distarch adipate] content. The moisture content dropped during frozen storage when wheat flour (native starch) was employed, which led to a larger expressible drip and a poorer water-holding capacity. Adding modified starch had less of an impact on the moisture content during frozen storage than native starch did. Depending on the type of modified starch used, there were differences in water-holding capacity and expressible drip. HS-modified starch and SA-3 were whiter than native starch, and hydroxypropyl starch had the lowest expressible drip. Consequently, SA-1 and HS are appropriate additives to raise the caliber of frozen fish cakes (Jeong et al., 2020). Mince gels like surimi are vulnerable to the textural alterations brought on by reductions in their ability to retain water when stored in a frozen state. Consequently, a rubbery texture develops, which reduces acceptance. Such a change might be minimized and the quality of gel products preserved throughout the prolonged frozen storage by adding modified starch, especially hydroxy-propylated distarch phosphate, at the ideal amount (Tuankriangkrai & Benjakul, 2010).

Native maize and potato starches were mixed 2:1:1 with two modified starches: acetylated phosphated tapioca di-starch [TAPDS] and acetylated waxy maize di-starch [WADS]. The resulting mixture was then compared to CP at 1:1% and 0% starch (CON). When employed with surimi seafood, TAPDS (4% and 6%) and WADS (2%) performed similarly to CON ($P > .05$) or better than CON ($P < .05$) in preserving textural qualities at 5°C and 55°C. The study demonstrates that employing optimal amounts of modified starch can reduce the surimi content from 75% to 51% while maintaining textural and other physical qualities. This is followed by a rise in the water and starch contents, respectively. Modified starch was found to play a substantial role in preserving the physical characteristics of cooked starch surimi fish protein gels, even though it would be more expensive than native starch. The potential for success as a starch-blend in surimi-based seafood products is suggested by the capacity of WADS (2%) and TAPDS (4% and 6%) starch mixtures to maintain textural characteristics at both cold and hot temperatures and the stability of frozen and refrigerated storage (Hunt et al., 2009).

1.2.1 *Native starch applications in meat and seafood*

Starch, being one of the most easily available polymers in nature, can be widely utilized in multiple applications. In its native form, starch has a semicrystalline structure and is found in the chloroplasts of leaves and amyloplasts of grains, tubers, and seeds (Magallanes-Cruz et al., 2023). The starch directly extracted from these plant sources and utilized in food matrices is termed native starch. The functionality of native starch can also be altered to improve the physico-chemical properties. This is done by physical, chemical, and enzymatic treatments, which result in modified starch (Chavan et al., 2021). The application scale is quite widespread in both its native and modified forms. Being calorigenic in nature, it can be used as a base for various high-calorie recipes in its native form or can be modified for specific purposes in various processed foods.

1.3 Application: functional properties of native starch in meat and seafood products

The application of starches in various foods depends on their physicochemical properties and their consequent functional behavior in the matrix. In its native form, it is primarily used to enhance the properties of water-holding capacity, viscosity, and gel strength, as well as serving as a bulking agent, binding agent, and fat replacer in various meat and seafood products, as shown in Fig. 11.1. Starches are widely used in the manufacturing of seafood and meat products. Surimi is one such processed food where starch is the second most important ingredient. Recent years have seen a gradual adoption of carbohydrates such as oligosaccharides and polysaccharides in the production of surimi and its value-added products. This is primarily due to consumer demands for nutritional value and flavor, as well as producer demands for longer shelf life. These demands include rich taste, low-calorie intake, improved antioxidant properties, and dietary fiber enrichment. Along with antifreezing and improved gelling capabilities, new features like fat replacement, flavor enhancement, anti-oxidant and antibacterial properties, and enhanced three-dimensional printability have been investigated (Sijin et al., 2024).

Surimi, a gel-based processed fish product, uses starch as a water-binder and fish-protein replacer. It is prepared by deboning, cleaning and dehydrating the fish flesh. It results in a stable concentration of myofibrillar protein that can be further processed into a range of other surimi-based products such as kamaboko, sausage and fish balls. These products are mainly consumed by people of Thailand, Japan, Korea and other Asian countries (Mi et al., 2021). These products contain 4%–12% starch, which helps in maintaining the gel like texture at a much lower cost. Studies have shown that increased amylopectin content, along with the large granule size of the starch, also helps in improving the toughness of surimi starch gels (Yang & Park, 1998). The choice of the source of the starch is vital since it determines the texture of the end products. The effect of different starch sources has been studied in the preparation of surimi to see the gel performance. It has been observed that among sweet potato, potato, wheat and corn, sweet potato was most effective in improving the gel strength. It also showed the least expressible moisture content and an intact microstructure (Liu et al., 2021). If the gel has high hardness and low deformability, then it leads to an increased mouthfeel of brittleness, whereas a gel of low hardness and higher deformability gives a chewy mouthfeel (Roy et al., 2000).

A limitation of native starches in the industry is that they readily hydrate and swell, leading to faster rupturing of the starch granules and consequent thinning of the starch paste (Chavan et al., 2021). To combat this drawback, surimi was prepared with starch paired with gums. The gel qualities, texture, chemical forces, secondary structure of the proteins, and organoleptic qualities of silver carp surimi were all significantly impacted by the inclusion of starch and gums, either separately or in combination. The gel's strength, chewiness, hardness, water-holding capacity, cohesiveness, gumminess, and whiteness were enhanced by the addition of starch and gums. Significant changes were observed in the chemical interactions, including hydrogen bonds, disulfide bonds and hydrophobic interactions (Mi et al., 2021).

Traditionally, starches and polysaccharides from various plants have been used in different meat product compositions to improve textural qualities, increase viscosity, and bind water. Meat emulsions benefit from the functional use of starches because of the heat-induced expansion of the starch granules in the protein gel matrix. Myofibrillar protein's continuous gel network is filled with expanded and gelatinized starch granules as the starch-myofibrillar protein complex is heated. Water is bound because of this process, improving the structural qualities of meat products. These changes must take place at temperatures encountered during meat processing, in tandem with thermal alterations and the release of moisture by meat proteins, to form a stable structure. The animal protein and starch, both water-binding biopolymers with radically differing thermal characteristics, have the capacity to form structures that are influenced by temperature and processing time. Synergistic or cooperative functional benefits, when combined into a single product, would depend on how well these two sets of biopolymers work together, with processing parameters regulated to promote interactions between them (Pietrasik & Soladoye, 2021).

Starches in their native form are significantly involved in the production of sausages and processed meat as binders. Its main function is to bind the meat into a noncrumbly mass and to avoid rapid dehydration of the mass during cooking. Here, mainly rice, potato, and maize starches are used. Certain types of canned meat also contain an added number of starches. The starch is added as a bulking agent to prevent shrinkage of the can contents during heat treatment. Generally, the native starch from potatoes is used in these cases (Dickson, 2020). A study was conducted to observe the effect of various native starches on the rheological properties of myofibrillar protein-starch composite. The various sources of native starches used included rice, corn, potato and tapioca starch. Individual composite gels were prepared using $20\ g\ kg^{-1}$ myofibrillar protein in combination with these native starches at pH 6.2. The study reflected that the water holding capacity and gel strength of the myofibrillar protein-starch composite prepared with tapioca starch showed notable improvement at 80°C. Thus, the different native starches reflected a wide range of impact on the microstructure and the rheological properties of the myofibrillar protein-starch composite (Wu et al., 2020).

1.3.1 Modified starch application in meat and seafood

Modified starch finds a wide range of applications in meat and seafood for improving texture, water retention capacity, stability and functionality (Zia-ud-Din & Fei, 2017). The modification is conducted by three techniques—physical, chemical, and enzymatic, each of which imparts a distinct physicochemical property.

1.4 Physical modification

The physical modification of starch includes treatments like pregelatinization, heat-moisture treatment (HMT) and annealing, which alter the starch granules through thermal and mechanical processes without causing any chemical alteration. These treatments help to improve processing functionalities; for example, pregelatinized starch produced using drum or spray drying enables cold water solubility and rapid viscosity development essential for instant food systems (Zia-ud-Din & Fei, 2017). Similarly, HMT, conducted at subgelatinization temperature and controlled humidity level of 15%–30%, helps in reducing retrogradation besides enhancing freeze-thaw stability and resistant starch content in processed meat. This property finds application in products that require stability during processing and storage, like baked goods, retorted foods and frozen foods. The process of annealing, conducted at 40–60°C in excess water, leads to restructuring to a better crystalline order with an elevation in gelatinization temperature of the starch. This helps in improving the texture of noodles and rice-based seafood products. Extrusion is a high temperature, high pressure processing of starch that leads to its gelatinization and fragmentation. This helps to produce dextrin and expanded products like puffed snacks. Application of microwave treatment provides quick and even heating, alters gelatinization kinetics and improves the functional properties of starch used in instant foods and surimi-based foods.

1.5 Chemical modification

Functional groups are added through chemical modification using covalent bonds, which enhance process tolerance and customized application. The cross-linking is done with the use of agents such as phosphorus oxychloride ($POCl_3$) or sodium trimetaphosphate (STMP) that increase the integrity of the granules, enhance the acid, heat and shear resistance of retorted and canned seafood products. In acid hydrolysis, starch is produced in low-viscosity, gel-strengthening forms to be used in confectionery, as well as in seafood gels as a thickening agent (Haq et al., 2019). Oxidation leads to the incorporation of carboxyls and carbonyls, which enhances starch paste clarity and stability. It also stabilizes emulsions and batters in coated meats. Esterified (acetylated) and etherified (hydroxypropylated) starches have enhanced freeze-thaw stability and water-binding properties, which are crucial in frozen seafood texture and juiciness (Haq et al., 2019). Cationic starches increase adhesion and are used in batters when using fried seafood.

1.6 Enzymatic modification

Enzymatic modification is applied using specific enzymes like α-amylase, β-amylase, glucoamylase, and debranching enzymes, in order to rearrange starch chains by the process of hydrolyzation (Choton et al., 2024; (Hutabarat & Stevensen, 2023)). These enzyme-treated starch molecules exhibit altered granule morphology, improved crystallinity and pasting properties. As a result of this, there can be precise adjustment of water retention, viscosity and gel-network formation that is critical for high-yield processed meats and seafood emulsions (Choton et al., 2024). Besides this, enzymatic modification also increases the concentration of resistant starch and modulates digestibility, with applications in health-oriented seafood formulations.

2. Applications in meat and seafood

Modified starches offer excellent water-binding properties, gel strength, and texture, with purge and freeze-thaw in processed meat and seafood products being reduced (Zia-ud-Din & Fei, 2017; Zhang et al., 2024). Their addition enhances sensory, structural integrity, and high-moisture, low-fat formulae in the new seafood and modern meat products (Zia-ud-Din & Fei, 2017).

2.1 Functional properties of starch in meat products

Starch is a polysaccharide that is universal in plants, and it has more than just the functions of thickening and binding in meat products. The presence of starch-water, starch-protein and starch-lipid interaction and phase changes makes it undergo a multifaceted interaction between textural and sensory properties. It remarkably contributes to a property such as textural, water-binding property, and general palatability of processed meat. It is active owing to its special capacity to be gelatinized, retrograded and interact with other meat components. The functions of starch in meat products are as follows.

2.1.1 Gelatinization

Starch granules, when heated in the presence of water, undergo irreversible transformation between an ordered crystalline structure and a disordered amorphous structure. This is the so-called gelatinization process and means the breaking of intermolecular hydrogen bonds within the starch granule, which leads to the swelling and subsequent solubilization of amylose, the linear component of starch. Gelatinized starch is used in meat products in order to enhance viscosity, binding, and texture. Gelatinization degree depends on such factors as heating temperature, presence of other ingredients such as salts and sugars, concentration and type of starch (Zhang et al., 2024).

Although temperature and water availability are the major factors, other factors also have a significant effect on this process. Salt, which is often employed in the preparation of meat, may also affect the temperature and the degree of gelatinization. In most cases, salts elevate the gelatinization temperature, which is presumably because of competition with water molecules. The exact effect, however, differs based on the type and concentration of salt. Certain meat products contain sugars, which also may have a similar impact on gelatinization, tending to raise the gelatinization temperature and starch granules' swelling behavior. Gelatinization can also be hindered by the presence of fats, which coat starch granules and prevent the entry of water (Zhang et al., 2024).

2.1.1.1 Water-holding capacity

Gelatinized starch is also important in the improvement of the water-holding capacity (WHC) of meat products. Swollen gelatinized starch grain particles confine water in their swollen structure, which is referred to as occluded water. Moreover, leached amylose can create a three-dimensional network that further increases the water retention, which is termed as 'bound water'. The starch granules are swollen, trapping the water molecules within the matrix, which lowers the water loss during processing and storage. It is very crucial in the comminuted meat products such as sausages, where water binding is crucial in attaining the required texture and juiciness. The starch amylose (initially lost during the process of gelatinization) can be used as a three-dimensional network, which additionally helps retain water. Starches that are modified, e.g., cross-linked or substituted, may have even higher WHC than the native starches, and so have a specific application in meat products. Special purposes of modified starches include cross-linked starches whose modified properties include greater resistance to shear and heat, enabling them to hold more water throughout processing. Water binding of starch is a process that occurs due to the interactions of the hydroxyl groups of the compound and water molecules. The proportion of amylose to amylopectin in starch and the presence of other components influence the extent to which this interaction takes place.

2.1.1.2 Texture modification

The effect of starch on meat products involves a significant change in meat products texture after the interaction with the meat proteins. Gelatinized starch causes the formation of a good, firm and elastic texture. Further modification of the textural properties may be made in relation to the interaction of starch and meat protein during processing. An example is that starch may inhibit the undesirable protein-protein interactions, which may cause toughness of cooked meats. During heating, gelatinized starch can form a matrix that encapsulates meat proteins, preventing the excessive protein-protein interactions contributing to firmness and cohesiveness. The interactions of the starch and proteins are complicated and require being dependent on factors such as the type of starch, the type of proteins, the temperature of processing of the product, and its pH. Studies indicate that the hydrophobic interactions and hydrogen bonding have a significant role in the interactions.

Also, the retrogradation of starch and the reassociation of starch molecules following the gelatinization process may affect the long-term product texture. Although a certain amount of retrogradation may be a positive attribute toward a firmer texture, too much retrogradation may result in an unpleasant staling and a gritty mouthfeel. The other functional areas of meat product quality can also be enhanced by starch. It may serve as a fat mimetic in low-fat products without compromising the creamy texture. In addition, it is capable of enhancing freeze-thaw stability by minimizing the growth of ice crystals, which may ruin the texture of the product.

2.2 Functional properties of starch in seafood products

There are various uses of starch in seafood products. In addition to its conventional applications as a source of carbohydrates, starch also makes a significant contribution to the texture, water-binding properties and overall sensory qualities of diverse types of seafood preparations. It is functional because it passes through gelatinization, retrogradation, and reactions with other food ingredients, especially proteins and water. Due to the presence of biochemical components and microbial activities, seafood products are highly perishable as the quality of the food deteriorates easily. Hence, it is quite challenging to keep seafood fresh while extending its shelf life. Edible coatings or films made out of modified starches have found application in addressing these challenges (Kim et al., 2019).

2.2.1 Gelatinization

Gelatinized starch in seafood products helps to raise viscosity and binding, as well as better texture. This is especially significant in products made of surimi (as gels), where starch is a contributing factor to the production of a pleasant, elastic, and binding texture. The type of starch, the quantity, and the heating temperature of starch, and other ingredients such as sugars and salts often used during the processing of seafood, have an influence on the extent of gelatinization (BeMiller & Whistler, 2009; Kim et al., 2019).

2.2.2 Water-holding capacity

Gelatinized starch in seafood products helps to increase the water-holding capacity (WHC) to a significant extent. The starch granules are swollen so that they entrap the water inside the matrix, thereby minimizing moisture loss in processing and storage. It is especially essential to retain the tenderness and taste of cooked fish. Amylose in the starch, which is leached during gelatinization, has the potential of creating a three-dimensional meshwork that also helps to retain water. Structurally altered starches, including cross-linked and substituted ones, may have even greater WHC than the native starches and are found to be of particular use in seafood packaging where moisture control is important (BeMiller & Whistler, 2009; Kim et al., 2019).

2.2.2.1 Texture modification

The texture of seafood products is significantly affected by starch. Gelatinized starch helps in the enhancement of a desirable, firm and elastic texture. The increased interaction between starch and protein caused by the processing of seafood can also alter the textural properties. As an example, in surimi, the starch forces the myofibrillar proteins to form a gel network, which adds to the typical texture of such products as crab sticks and fish balls. In addition, retrograding and reassociation of starch molecules after gelatinization can also affect the long term texture of the product. Whereas a certain level of retrogradation may be preferable in the pursuit of a firmer texture, too much retrogradation may result in objectionable staling and a gritty mouthfeel (Bemiller & Huber, 2015; BeMiller & Whistler, 2009; Kim et al., 2019).

3. Application of edible starch films in the meat and seafood industry

There are multiple challenges that the meat and seafood industries face due to their low shelf-stability. Modification of starches and their incorporation in edible films have been considered as an innovative path that can enhance the quality and shelf-life of these perishable products. The application of modified starches in edible films in the meat and seafood industries, along with their functional characteristics are discussed below.

3.1 Functional properties of modified starches in edible films for meat and seafood

Modified starches provide an assortment of desirable functionality in seafood products when incorporated in edible films. The advantages include improvement of barrier properties, better preservation, carriers of additives and enhancement of shelf life. The moisture, oxygen and carbon dioxide transfer is controlled by the use of starch-based films, which provide a barrier between the food and its surrounding atmosphere. They have good oxygen and carbon dioxide barrier properties, but can have less good moisture barrier properties since they are hydrophilic. Edible films keep seafood products safe against physical, chemical, microbiological, and mechanical degradation. Modified starch films can be added with food additives such as flavors, colourants, antimicrobials, and antibrowning agents. By providing a protective barrier and delivering functional additives, modified starch films can prolong the shelf life of seafood products.

The thermoplastic properties of starch find extensive application in edible films. The edible films themselves, being odorless, colorless, and tasteless, do not alter the flavor, taste or appearance of food. Starch-based films with additives of chitosan and lauric acid showed a significant inhibition of *Bacillus subtilis* and *Escherichia coli*, which proved the strong antibacterial effect of the film. Modified starches can be used to produce edible coatings and films which enhance seafood products and extend their shelf-life. Their capabilities as barriers, sources of functional additives, and general preservation enhance their significance to the meat and seafood processing sector.

4. Modified starches in meat and seafood products

Natural starch supplies can be transformed into modified starches by enzymatic, chemical, or physical means as discussed in the previous section. These changes improve their functional qualities and qualify them for a range of food sector applications.

4.1 Functional properties of modified starches

Modified starches can be used as effective thickeners and gelling agents in meat and seafood items. They enhance mouthfeel and texture, besides adding stability to fillings, sauces, and gravies. The ability of modified starches to retain moisture is a primary benefit to them. The latter property is especially significant in meat products, where the loss of water may cause drying and loss of quality in processing and storage. Refined starches are also binding agents in processed meat products like sausages and meatballs. They assist in the preservation of the integrity of the product by eliminating ingredient separation (BeMiller & Whistler, 2009; Kim et al., 2019). In low-fat meat products, modified starches are incorporated to replace fat without any adverse effect on the texture and mouthfeel. Pregelatinized starch is a modified starch that is utilized as a bulking agent. They are incorporated into meat products like sausages and burgers to enhance their texture, moisture retention, juiciness, and yield at a lower production cost. The applications are summarized in Fig. 11.2.

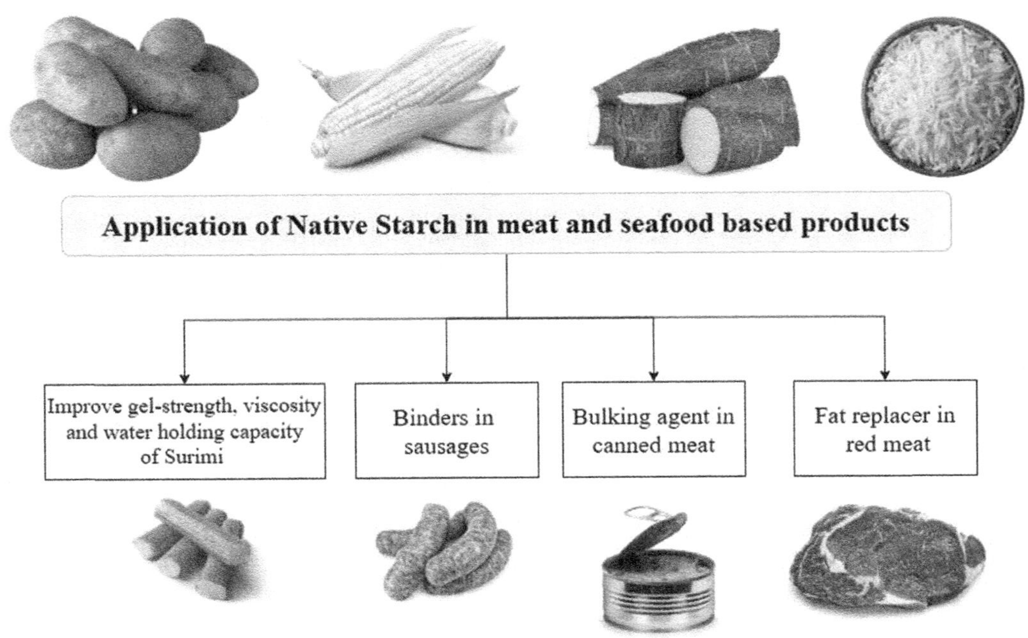

FIGURE 11.2 Application of native starches in meat and seafood products.

5. Edible films in meat and seafood preservation

Edible films made from polysaccharides, proteins, or lipids have emerged as effective tools for preserving the quality of meat and seafood products. These films can be applied directly to food surfaces or used as packaging materials.

Benefits of Edible Films

1. Barrier Properties: It functions as a barrier against moisture, oxygen and other gases and thus prolongs shelf life and prevents spoiling. They can efficiently regulate the movement of gases and moisture between food ingredients or between food and its surroundings.
2. Preservation of Quality: The edible films help maintain the sensory attributes of meat and seafood products by providing a protective shield. It helps in the prevention of oxidation, microbial contamination, and moisture loss.
3. Incorporation of Functional Additives: Various antimicrobial agents and antioxidants, incorporated in edible films, can effectively enhance their protective properties. For example, incorporating natural extracts like essential oils into edible films can improve their antimicrobial efficacy.

6. Applications in meat products

The use of modified starches in meat products has demonstrated good results in the improvement of the quality attributes. Starch plays a vital role in meat processing and improves moisture retention, succulence, and firmness, regulates purge, and minimizes shrinkage. Refined corn starch is harder to bite, and very stable; slightly crosslinked tapioca starches or waxy maize are more effective in holding moisture. It is important to note that meat with added starches has high steam table performances in comparison to other texture modifiers. As much as 3.5% of modified food starch can be used in bologna and frankfurters, which enables them to hold more water and reduces purging. The preference for cook-up starches is due to the maintenance of batter viscosity and the ability to withstand shear as batter is rapidly mixed, and also to undergo swelling in a smokehouse. Also, with the introduction of gelatinization after the smokehouse, modified food starch reduces purging in "Ham Water Added" products up to 2.0%.

Also, shake-on coating materials include dextrins, octenyl succinylated starches, and pregelatinized materials, and chicken coating materials have employed derivatized pregelatinized waxy maize starch in order to impart the fried appearance on baking. In sausages, it is common to use modified starches in order to improve binding properties, texture and moisture retention. They assist in maintaining the juiciness while lowering fat content. In meatball recipes, modified starches are used to provide the required texture, as well as to avoid the separation of ingredients during cooking. Edible films may be applied to fresh or processed meat as a packaging material. These films provide a barrier against spoilage while allowing for extended shelf life without compromising quality.

7. Applications in seafood products

The seafood industry also benefits significantly from the use of modified starches. The basis for imitation crab legs and other seafood goods is surimi, a fish paste made from deboned fish. The paste is blended with cryoprotectants such as sorbitol, phosphates, and sucrose, and then frozen to guarantee preservation. To make the final product, the frozen paste is thawed, mixed with starch, and extruded as a film onto a conveyor. After that, the film is moved by this belt into an oven, where the starch is cooked, and the fish protein is heat-denatured. The film is then cut to the appropriate shape, molded, dyed, and rolled to produce striations. The product is either frozen or refrigerated based on distribution needs. Modified starches are used as binders in surimi-based products, like imitation shrimp, crab sticks, or fish cakes, to increase texture and water-holding ability (Hunt et al., 2009).

Since they can create a cohesive and elastic matrix that matches the required seafood texture, potato and tapioca starches have been historically utilized in surimi products. The usage of highly stabilized, somewhat cross-linked tapioca starch has become more common since the introduction of frozen distribution, frequently in conjunction with native tapioca starch. To improve cuttability, both unmodified corn starch and modified waxy maize products are used. The viscosity of starch paste and its gel-strengthening capacity are correlated, underscoring the significance of starch selection in obtaining the ideal surimi texture (Hunt et al., 2009; Kong et al., 2016). Edible coatings made from modified starches can be applied to fish fillets to enhance moisture retention during storage. This helps maintain texture and flavor while extending shelf life (Kim et al., 2019). As shown in Fig. 11.3, edible films can be used to coat frozen seafood products to prevent freezer burn by reducing moisture loss during storage (Hunt et al., 2009; Kong et al., 2016).

Modified starches, achieved through genetic, enzymatic, physical, and chemical processes, dramatically expand the range of applications for native starch. Physical modifications offer cleaner label solutions, while chemical modifications enable precise control over starch properties. Enzymatic methods offer specificity in the alterations of the starch molecule, and genetic modification allows for tailored starch production within the plant itself. These diverse approaches are vital in industries ranging from food to textiles, and their development will continue to advance with technological innovation and evolving consumer demands.

Modified starches are highly capable of improving product quality and increasing shelf life in the meat and seafood industries. By enhancing moisture retention, texture and overall preservation properties, these innovative materials contribute to better consumer satisfaction while addressing challenges related to food safety. As research continues to explore new formulations and applications for modified starches and edible films in these industries, their role is likely to expand further. Future research into the interactions between starch and meat and seafood protein under different processing conditions will continue to optimize the use of starch in seafood product formulations.

7.1 Impact of starch on nutritional profile

Starch functions as a primary source of energy in most diets across the world. It is the supplier of 50%–60% of calories from the diet consumed by most of the world's population. It is metabolized into sugars, which are readily absorbed by the

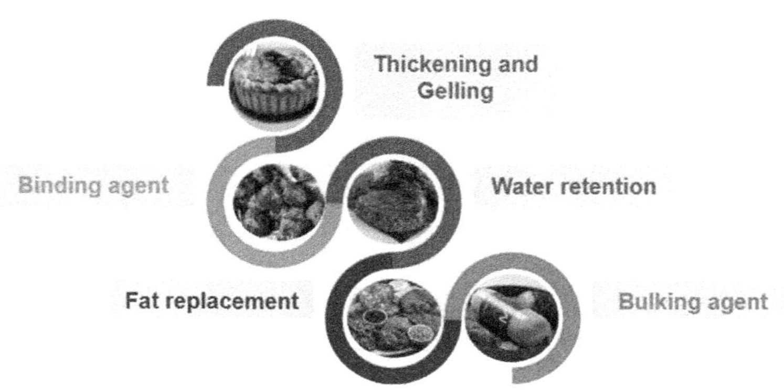

FIGURE 11.3 Application of various functional properties of modified starch.

blood (Calvin, 2016). Although native starch, especially in potatoes, is resistant to digestion, it can be easily metabolized in cooked starch due to its structural modification. Because of this, most processed starch from potato products has a high glycemic index (GI), making them unsuitable for those with type 2 diabetes and obesity (Dupuis & Liu, 2019).

7.2 Sensory aspects of modified starch

The properties of native starches, such as low stability in drastic conditions like heat, acid, and shear, prevent them from being used for various industrial purposes in food processing. It also displays retrogradation and syneresis, which can progress to unwanted textural modification. Modified starch, on the other hand, can be used as a polymer in a variety of food matrices. Several physicochemical properties of the homopolymers—amylose and amylopectin—are altered on modifications, which helps in eliminating the undesirable properties. Food businesses use modified starches to preserve gel strength, create edible food coatings and clarity, texturize and thicken meals, delay retrogradation, stabilize emulsions, enhance freeze-thaw stability, replace fat, encapsulate fragrance compounds, and raise the resistant starch content (Mohammadi et al., 2022).

Complex systems consisting of water, salts, phosphates, muscle fibers, solubilized muscle proteins, broken myofibrils, fat cells, and fat droplets make up meat batters. Various forms of starch have been extensively used as binders, extenders, and fillers in pulverized meat products to decrease the amount of meat mass lost owing to shrinkage and to improve emulsion stability, nutritional value, emulsifying capacity, slicing properties, and water-binding potential. Low-fat meat products have long employed tapioca starch, namely "Tapiocaline", a modified tapioca starch product, to increase succulence and juiciness. However, since low ionic strength reduces the solubility of proteins, it may also be advantageous when utilized in the production of low-salt goods. In a study, tapioca was used primarily to produce low-salt sausage and improve its textural properties. Tapioca starch and potassium chloride were found to partially substitute sodium chloride, improving the water binding and holding capacities while lowering the mass of total expressible fluid (Cruz-Romero et al., 2022).

Since modified starch affects the dough's rheological characteristics, texture, degree of starch gelatinization, water absorption, and bread staling, it may be used as a substitute for native wheat starch. The granular form of starch is preserved when heated over 100°C with water present in the treatment of starch with heat and moisture. Starch treatment with moisture and heat increases its hydrophobic characteristics while reducing its hydrophilic characteristics, thus making the starch more stable against mechanical agitation. Modified starch is increasingly replacing native starch due to improved functional properties while boosting the rheological properties and scope of application in various processed foods. Table 11.1 shows distinct types of native and modified starches and their application.

7.3 Regulatory & safety considerations

On a global scale, many organizations govern the modification of starch and its residual components. Depending on the country of preparation, the chemical or process, and the area of usage, different rules apply. Before utilizing a starch product for export in a food chain, it is vital to understand all international regulatory issues, constraints, and classifications.

7.4 Regulations governing the use of starch in meat and seafood

Like other food additives, chemically modified starches must adhere to certain legal standards outlined in Commission Regulation (EU) No 231 for nations that are members of the European Union. The reagents that can be used, as well as the amount of added functional groups and/or residue from the reagents that can be utilized, are all specified in this regulation. The use of modified starches in food is governed by European legislation, which has been combined into a single European Parliament and Council Regulation (EC) 1333/2008 on food additives, and Directive 4/2008, which describes the types and conditions of use of additives and extraction solvents in the production of foodstuffs, as amended by Decrees 130/2010 and 122/2011.

Around the world, starch has been approved for widespread usage in a variety of meat and analogue products. Currently, several commercially produced beef products must meet extremely stringent identity rules. Starch and associated ingredients are authorized for usage in standardized and nonstandard meat products to enhance processing and storage. In the US, the Food Safety and Inspection Service of the US Department of Agriculture regulates the use of starch, whether it is natural or modified, in meat products. The Code of Federal Regulations, Part 318.7, contains the specifications. Before starch and related products were approved, excessive salt concentrations were used to control water

TABLE 11.2 Application of modified starch on rheological properties of meat and seafood.

S. No.	Source of native starch	Mode of modification	Modified starch	E-number	Functional and rheological properties studied	References
1.	Potato	Chemical modification	Oxidized starch	E 1404	Texturization, 3D and 4D printing, and electrospinning for fish analogs	Nowacka et al. (2023)
2.	Maize	Chemical modification	Distarch phosphate	E 1412	Water holding capacity, fat holding capacity and texture properties in processed meat products	Petracci et al. (2013)
3.	Potato starch, cereal starch,	Chemical modification	Acetylated distarch adipate	E 1422	Gel strength, breaking force, water-holding capacity, and the three-dimensional network structure of the shrimp myofibrillar protein gel	Mi et al. (2022)
4.	Tapiocacorn, waxy maize	Chemical modification	Hydroxy-propyl distarch phosphate	E 1442	Shrimp myofibrillar protein gel's breaking force, gel strength, water-holding capacity and breaking distance.	Mi et al. (2022)
5.	Tapioca, corn	Chemical modification	Acetylated distarch phosphate	E 1414	The in vitro digestibility and gel characteristics of kung-wan, a Chinese pork product that has been emulsified	Wei et al. (2023)
6.	-Include the starch name	Chemical modification	Hydroxypropyl starch	E 1440	Tilapia (*Oreochromis mossambicus*) fish surimi's gel characteristics, oil retention, water loss rate, color changes, protein content, and gas composition	Xiaohuang et al. (2022)
7.	-Include the starch name	Chemical modification	Octenyl-succinylated starch	E 1450	Ground beef patties' physical-chemical characteristics, microstructure image, firmness, color, cooking loss, and sensory assessment	Eshag Osman et al. (2021)
8.	-Include the starch name	Chemical modification	Sodium octenyl succinate	E 1450	Emulsified surimi gels' mechanical, microstructural, and taste characteristics	Yu et al. (2022)
9.	Cassava, potato, corn	Chemical modification	Acetylated distarch phosphate, starch acetate esterified with acetic anhydride, hydroxypropyl distarch phosphate	E 1414, E 1420, E 1442	Water holding capacity, gel-strength, microstructure, lipid oxidation of ground beef	Zhang et al. (2024)
10.	Include the starch name.	Chemical modification (etherification) and enzymatic modification (hydrolysis)	Enzyme-modified carboxymethyl starch	-	Sausage's ability to retain water, emulsion stability, and sensory qualities (juiciness, oiliness, texture, flavor, and general appeal)	Luo and Xu (2011)

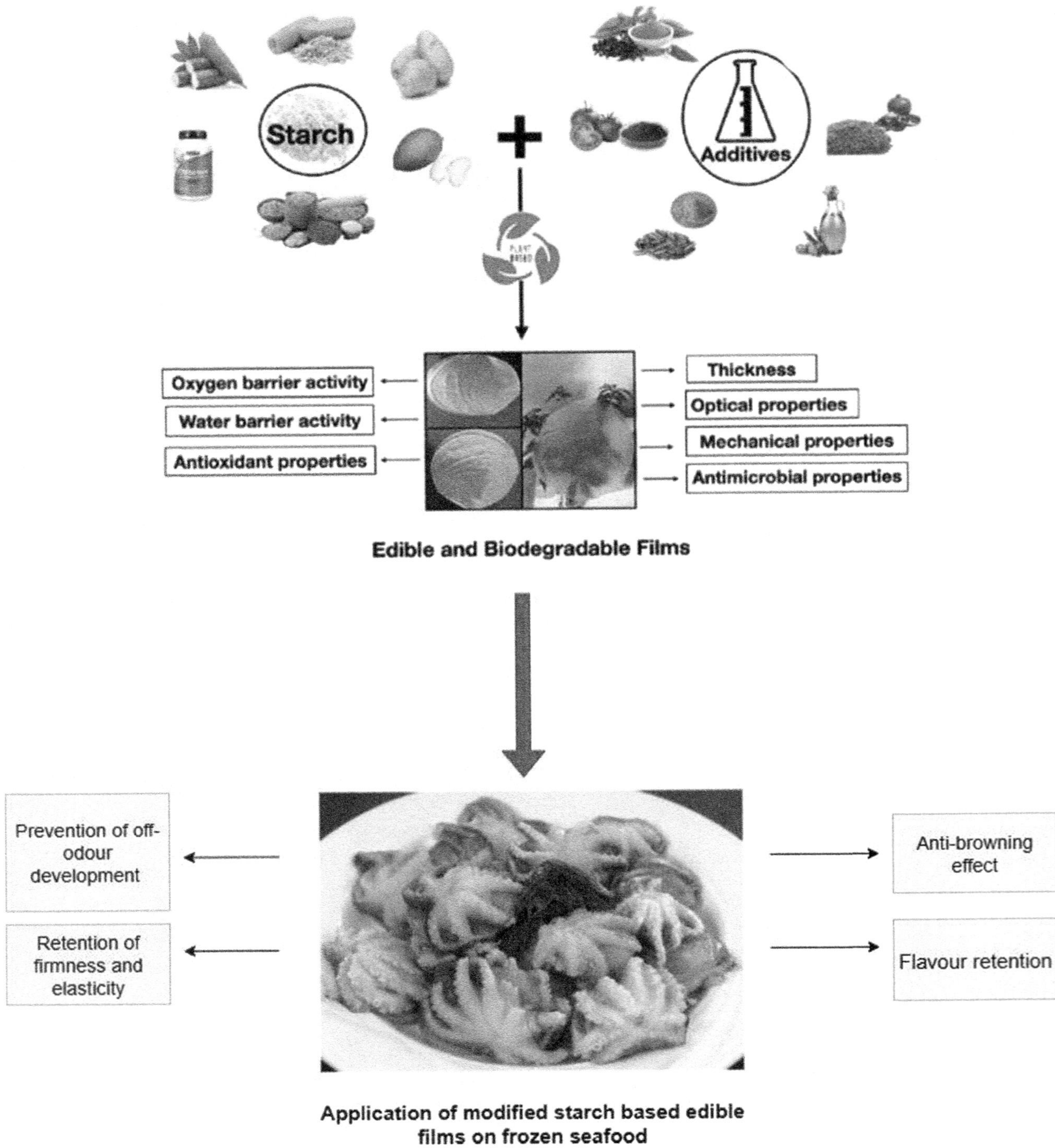

FIGURE 11.4 Application of modified starch-based edible film in seafood (Singh et al., 2022).

management in many meat products. Starch helps reduce and manage purge (free water/brine) in meats like frankfurters, bologna, and luncheon bread (emulsified or diced). Over the years, flour and other grain-based materials have been used instead of modified starches, but none of them has improved the capacity to absorb and regulate water.

Starch does more than just help with water management in seafoods like surimi (seafood analogs). Adding starches has improved texture, improved the process, and provided economic benefits. It has been found that considerable amounts of native (common/unmodified) starches can also be utilized in addition to modified starches. To create the functional qualities required in a meat substitute, unmodified tapioca, potato, and wheat are frequently combined with one or more modified starches. Tapioca, potatoes, and modified waxy maize are popular starches for meat analogs. Furthermore, the final product is processed, and the required texture determines the nature of starches that are employed. Such a meat analogue is a complex formulation that necessitates understanding the usefulness of starches before their application. Finding the best combination of starches and associated ingredients to create a commercial meat analogue may need multiple assessments.

Several approved procedures have been adopted by the Association of Approved Analytical Chemists (AOAC) to determine the amount of starch present in most foods. A specific method is suggested based on the kind of food to be examined, some of which differ slightly from the original process. However, standard techniques for determining the amount of starch in cereals and derivatives were also recognized by the American Association of Cereal Chemists International. They should be used for almost all goods that include starch and have recently been updated and modified. The International Association for Cereal Science and Technology has also taken similar steps.

7.4.1 Labeling requirements and health claims

Worldwide, chemically modified starches are regulated by a number of organizations; consequently, limitations vary according to the country of origin, function in food products, and level of chemical modification. Starch that is chemically modified is classified as a food additive, and its name must be labeled on the box as "food starch-modified," along with the food additive's purpose. The majority of food product labels in Europe have E numbers, which do not specify a maximum numerical level, except for "cream and cream powder with fat content less than 20%" and "cereal-based foods and baby foods for infants and young children" that have a fixed amount of addition (The European Parliament and Council on Food Additives, 2008). New Zealand and Australia utilize the International Numbering System (INS) number, which is defined by the Food and Agriculture Organisation of the United Nations and the World Health Organisation (WHO). According to FAO/WHO Standards (2017), most food additives have to be identified by their class name, which is followed by the food additive number or code. It is possible to modify any starch source to give a dietary fiber function by using high quantities of cross-linking that are yet within prescribed limits. The addition of these modified starches adds caloric value and also reduces the total calories, while providing colon health properties. These are known as resistant starches (RSs) and include RS-I, RS-II, RS-III, RS-IV, and RS-V (Table 11.2).

8. Conclusion

Technological advances in the area of food technology have made modified starches possible, helping in an array of processed food products by contributing to the rheological characteristics and organoleptic properties while adding to a shelf-stable product. The modification in the functional properties of starches contributes to the development of innovative value-added products. The meat and seafood industry has particularly benefited from modified starches, as they help with textural challenges faced when using native starches traditionally. Consumers also benefit by having a range of options of innovative products to choose from. Modified starches have thus contributed significantly to the food industry (Fig. 11.4).

References

Bemiller, J. N., & Huber, K. C. (2015). Physical modification of food starch functionalities. *Annual Review of Food Science and Technology, 6*, 19–69. https://doi.org/10.1146/annurev-food-022814-015552

BeMiller, J., & Whistler, R. (2009). Starch: Chemistry and technology. *Food Science and Technology.*

Braşoveanu, M., & Nemţanu, M. R. (2018). Aspects on starches modified by ionizing radiation processing. In *Applications of modified starches. InTech.* https://doi.org/10.5772/intechopen.71626

Calvin, O. (2016). Starch and modified starch in bread making: A review. *African Journal of Food Science, 10*(12), 344–351. https://doi.org/10.5897/AJFS2016.1481

Chavan, P., Sinhmar, A., Nehra, M., Thory, R., Pathera, A. K., Sundarraj, A. A., & Nain, V. (2021). Impact on various properties of native starch after synthesis of starch nanoparticles: A review. *Food Chemistry, 364*. https://doi.org/10.1016/j.foodchem.2021.130416

Choton, S., Bandral, J. D., Singh, J., Bhat, A., Sood, M., Gupta, N., Reshi, M., & Kaur, D. (2024). Enzymatic modification of starch: A review. *Saudi Journal of Medical and Pharmaceutical Sciences, 10*(01), 1–8. https://doi.org/10.36348/sjmps.2024.v10i01.001

Cruz-Romero, M. C., O'flynn, C. C., Troy, D., Mullen, A. M., & Kerry, J. P. (2022). The use of potassium chloride and tapioca starch to enhance the flavour and texture of phosphate-and sodium-reduced low fat breakfast sausages manufactured using high pressure-treated meat. *Foods, 11*(1). https://doi.org/10.3390/foods11010017

Daza, L. D., Montealegre, M.Á., Reche, C., Sandoval-Aldana, A., Eim, V. S., & Váquiro, H. A. (2023). Chachafruto starch: Physicochemical characterization, film-forming properties, and 3D printability. *International Journal of Biological Macromolecules, 247*. https://doi.org/10.1016/j.ijbiomac.2023.125795

Dickson, J. K. (2020). *Food biochemistry* (1st ed.). CBS Publisher.

Dupuis, J. H., & Liu, Q. (2019). Potato starch: A review of physicochemical, functional and nutritional properties. *American Journal of Potato Research, 96*(2), 127–138. https://doi.org/10.1007/s12230-018-09696-2

Eshag Osman, M. F., Mohamed, A. A., Alamri, M. S., Ali, I., Hussain, S., Ibraheem, M. I., & Qasem, A. A. (2021). Quality characteristics of beef patties prepared with octenyl-succinylated (osan) starch. *Foods, 10*(6). https://doi.org/10.3390/foods10061157

Fontes, S.d. M., Cavalcanti, M. T., Candeia, R. A., & Almeida, E. L. (2017). Characterization and study of functional properties of banana starch green variety of Mysore (musa AAB - Mysore). *Food Science and Technology, 37*(2), 224–231. https://doi.org/10.1590/1678-457X.18916

Haq, F., Yu, H., Wang, L., Teng, L., Haroon, M., Khan, R. U., Mehmood, S., Ul-Amin, B., Ullah, R. S., Khan, A., & Nazir, A. (2019). Advances in chemical modifications of starches and their applications. *Carbohydrate Research, 476*, 12–35. https://doi.org/10.1016/j.carres.2019.02.007

Hunt, A., Getty, K. J. K., & Park, J. W. (2009). Roles of starch in surimi seafood: A review. *Food Reviews International, 25*(4), 299–312. https://doi.org/10.1080/87559120903155834

Hutabarat, D. J. C., & Stevensen, J. (2023). Physicochemical properties of enzymatically modified starch: A review. *IOP Conference Series: Earth and Environmental Science, 1169*(1). https://doi.org/10.1088/1755-1315/1169/1/012093

Jeong, Y. G., Park, J., Jeong, Y., Chun, B., Lee, H., et al. (2020). Effect of modified starch on the frozen storage stability of fish cake eomuk. *Korean Journal of Fisheries and Aquatic Sciences, 53*(3), 290–296. https://doi.org/10.5657/KFAS.2020.0290

Kim, S., Choi, G., Kim, J., Lee, J., Kim, S. B., Kim, Y. M., Kim, J. S., Lee, J. S., & Cho, S. (2019). Physicochemical characteristics of fried-fish snacks with different types of starch. *Korean Journal of Fish Aquatic Science, 52*(6), 580–586. https://doi.org/10.5657/KFAS.2019.0580

Kong, W., Zhang, T., Feng, D., Xue, Y., Wang, Y., Li, Z., Yang, W., & Xue, C. (2016). Effects of modified starches on the gel properties of Alaska Pollock surimi subjected to different temperature treatments. *Food Hydrocolloids, 56*, 20–28. https://doi.org/10.1016/j.foodhyd.2015.11.023

Liu, Z., Yuan, Y., Qin, Y., Feng, A., He, Y., Zhou, D., Dong, X., Shen, X., Cao, J., & Li, C. (2021). Sweet potato starch addition together with partial substitution of tilapia flesh effectively improved the golden pompano surimi quality. *Journal of Texture Studies, 52*(2), 197–206. https://doi.org/10.1111/jtxs.12574

Luo, Z., & Xu, Z. (2011). Characteristics and application of enzyme-modified carboxymethyl starch in sausages. *LWT - Food Science and Technology, 44*(10), 1993–1998. https://doi.org/10.1016/j.lwt.2011.07.016

Magallanes-Cruz, P. A., Duque-Buitrago, L. F., & del Rocío Martínez-Ruiz, N. (2023). Native and modified starches from underutilized seeds: Characteristics, functional properties and potential applications. *Food Research International, 169*, 112875. https://doi.org/10.1016/J.FOODRES.2023.112875

Mi, H., Yi, L., Wang, C., Yi, S., Li, X., & Li, J. (2021). The interaction of starch-gums and their effect on gel properties and protein conformation of silver carp surimi. *Food Hydrocolloids, 112*, Article 106290. https://doi.org/10.1016/j.foodhyd.2020.106290

Mi, J., Zhao, X., Huang, P., Hong, J., Jia, R., Deng, S., Yu, X., Wei, H., & Yang, W. (2022). Effect of hydroxypropyl distarch phosphate on the physicochemical characteristics and structure of shrimp myofibrillar protein. *Food Hydrocolloids, 125*, Article 107417. https://doi.org/10.1016/j.foodhyd.2021.107417

Mohammadi, M., Khorshidian, N., Yousefi, M., & Khaneghah, A. M. (2022). Physicochemical, rheological, and sensory properties of gluten-free cookie produced by flour of chestnut, date seed, and modified starch. *Journal of Food Quality, 2022*. https://doi.org/10.1155/2022/5159084

Nowacka, M., Trusinska, M., Chraniuk, P., Piatkowska, J., Pakulska, A., Wisniewska, K., Wierzbicka, A., Rybak, K., & Pobiega, K. (2023). Plant-based fish Analogs—A review. *Applied Sciences, 13*(7). https://doi.org/10.3390/app13074509

Petracci, M., Bianchi, M., Mudalal, S., & Cavani, C. (2013). Functional ingredients for poultry meat products. *Trends in Food Science & Technology, 33*(1), 27–39. https://doi.org/10.1016/j.tifs.2013.06.004

Pietrasik, Z., & Soladoye, O. P. (2021). Use of native pea starches as an alternative to modified corn starch in low-fat Bologna. *Meat Science, 171*, Article 108283. https://doi.org/10.1016/j.meatsci.2020.108283

Roy, E. M., Emily, P. C., & George, J. F. (2000). *Marine and freshwater products handbook*. CRC Press.

Sijin, Z., Zhang, L., Yin, T., You, J., Liu, R., Wang, L., Huang, Q., Wang, W., & Ma, H. (2024). Exploring the versatility of carbohydrates in surimi and surimi products: Novel applications and future perspectives. *Journal of the Science of Food and Agriculture, 104*(4), 1874–1883. https://doi.org/10.1002/jsfa.13081

Singh, G. P., Bangar, S. P., Yang, T., Trif, M., Kumar, V., & Kumar, D. (2022). Effect on the properties of edible starch-based films by the incorporation of additives: A review. *Polymers, 14*(10). https://doi.org/10.3390/polym14101987

Tuankriangkrai, S., & Benjakul, S. (2010). Effect of modified tapioca starch on the stability of fish mince gels subjected to multiple freeze-thawing. *Journal of Muscle Foods, 21*.

Wei, S., Liang, X., Xu, Y., Kong, B., Li, X., Zhang, H., Liu, Q., & Wang, H. (2023). In-depth insight into the effects of tapioca or corn acetylated distarch phosphate on the gel properties and in vitro digestibility of kung-wan. *International Journal of Biological Macromolecules, 253*(Pt 3), Article 126997. https://doi.org/10.1016/j.ijbiomac.2023.126997

Wu, M., Wang, J., Hu, J., Li, Z., Liu, R., Liu, Y., Cao, Y., Ge, Q., & Yu, H. (2020). Effect of typical starch on the rheological properties and NMR characterization of myofibrillar protein gel. *Journal of the Science of Food and Agriculture, 100*(1), 258–267. https://doi.org/10.1002/jsfa.10033

Xiaohuang, C., Azam, M. S., & Islam, M. N. (2022). Effect of chitosan and hydroxypropyl starch complex polysaccharide on the physico-chemical properties of Tilapia fish (*Oreochromis mossambicus*) gel. *Food Research, 6*(5), 340–348. https://doi.org/10.26656/fr.2017.6(5).663

Yang, S., Hu, W., Qiao, S., Song, W., & Tan, W. (2025). Advances in processing techniques and determinants of sweet potato starch gelatinization. *Foods, 14*(4), 545. https://doi.org/10.3390/foods14040545

Yang, H., & Park, J. W. (1998). Effects of starch properties and thermal-processing conditions on surimi–starch gels. *LWT - Food Science and Technology, 31*(4), 344–353. https://doi.org/10.1006/fstl.1997.0366

Yu, J., Xiao, H., Xue, Y., & Xue, C. (2022). Effects of soybean phospholipids, ovalbumin, and starch sodium octenyl succinate on the mechanical, microstructural, and flavor properties of emulsified surimi gels. *Lebensmittel-Wissenschaft & Technologie, 161*, Article 113260. https://doi.org/10.1016/j.lwt.2022.113260

Zhang, S., Zhang, L., Yin, T., You, J., Liu, R., Wang, L., Huang, Q., Wang, W., & Ma, H. (2024). Exploring the versatility of carbohydrates in surimi and surimi products: Novel applications and future perspectives. *Journal of the Science of Food and Agriculture, 104*(4), 1874–1883. https://doi.org/10.1002/jsfa.13081

Zia-ud-Din, X. H., & Fei, P. (2017). Physical and chemical modification of starches: A review. *Critical Reviews in Food Science and Nutrition, 57*(12), 2691–2705. https://doi.org/10.1080/10408398.2015.1087379

Chapter 12

Starches in the extruded food system

Oluwaseun Peter Bamidele
Food Science and Technology, University of Venda, Thohoyandou, South Africa

1. Introduction

For thousands of years, human beings have depended on starches as they synthesize the predominant forms of energy utilized by the body, as well as vital ingredients needed to tailor countless food innovations (Ali et al., 2024, pp. 241–260). Perhaps the most significant example of this versatility and importance is in modern extruded food products where starch serves as the primary 'filler' serving as the supportive matrix for the development of various attributes relating to structure, rheological characteristics, expansion and mouthfeel, nutrition and stability. There have been significant advancements in how starches are utilized in food. This is attributed to the thermal and mechanical energy input in extrusion cooking. This process involves the use of moist, starchy food, whereby, under extreme mechanical and thermal energy (90–200°C), extrusion produces rapid swelling, gelatinization, and melting, as well as degradation within the starch of amylose and amylopectin (Leonard et al., 2020).

These reactions are rapid, occurring in mere seconds rather than in hours as is in traditional cooking, and allow for the mass production of various food products, including but not limited to puffed breakfast cereals, extruded snacks, texturized vegetable proteins, fortified pasta, and confectionery products and pet food (Huang et al., 2022). The ability of extrusion to customize starch functionality is unparalleled. Manufacturing can adjust starch behavior based on moisture levels (12%–35%), temperature profiles, screw configurations, and residence times. Products that can be created include those that are highly expanded, crispy, and low-density, as well as those that are dense, chewy, and high-moisture meat analogs (Zhao et al., 2024). The versatility is the result of extrusion's ability to crystallize regions, reduce molecular weight, and promote phenomena (Wang et al., 2022). The starch transformation cascade includes the formation of amylose–lipid complexes and the cooling of retrograded multicomplexes. The ability of extrusion to rearrange these molecules directly affects the properties, such as solubility and water absorption indices, pasting properties, expansion ratio, and consumer acceptance of the product.

Nutrition is another area where extrusion alters the digestibility of starch. The processes of gelatinization and retrogradation, as well as the interaction between starch and protein or fiber, can significantly increase the amount of resistant starch (RS) (Pandey et al., 2025). RS3 and RS5 can also be created and contribute to the reduced available energy. The duality of these qualities in extruded starch-based foods makes them ideal for addressing worldwide health issues, such as diabetes management, obesity prevention, and gut health. Plant-based foods are clearly fulfilling consumer demand for 'better for you' foods with clean labels. As major sources of native starch, maize, wheat, and rice are gradually being replaced with new alternatives, such as high amylose maize, waxy starches, legumes (peas, lentils, chickpeas), tubers (tapioca, potato), and pseudo-cereals (quinoa, amaranth), for specific nutrition and texture (Li et al., 2025). Starch-containing products that are gluten-free, high in dietary fiber, and enriched with protein can be produced using extrusion and a staged treatment process for crystallinity and structural modification, which includes heat and moisture, amylose and amylopectin pre-treatment, and enzymes (Castro-Campos et al., 2024).

For the unique structure, behavior, and properties of starches, they are capable of yielding high-quality, low-cost, and functional foods. The objective of this chapter is to describe the scientific basis for the diversity of starches and the interactions that they can have with thermomechanical stress, fiber, protein, and lipids. Ultimately, we aim to shed light on the fundamentals of starch, texture, and molecular rearrangement, guiding product developers and food scientists in the creation of advanced extruded products for the global health market.

Native and Modified Starches for Food Applications. https://doi.org/10.1016/B978-0-443-27431-2.00010-9

2. Extrusion and the function of starch

Extruded foods, such as snacks, cereals, pasta, and even pet foods, largely depend on starch, which not only adds value but also offers texture and expansion (Choton et al., 2020). Starch consists of two components, amylose and amylopectin. During the extruding process (which takes place at ultra-high temperatures and under significant Vertical pressure), starch molecules are gelatinized, melted, and broken apart (Qiu et al., 2024). Starch also acts as the matrix of expansion and texture in extruded products, becoming gelatinized at temperatures ranging from 100 to 160°C with a moisture content of 15%–30% (Qiu et al., 2024). High-moisture extrusion, a common process in protein-starch blends, alters the flow of starch. In high-moisture food extrusion, starch-based raw materials expand because of the evaporation of moisture. If the product is over-gelatinized, its density is lower. Moisture content in starchy materials is utilized in extrusion cooking technology, which then transforms the starchy materials into products with high expansion and a porous, cross-polymerized structure. For millets, the extrusion process enhances the functionality of high-value nutrient products by gelatinizing starch and enriching extruded products (Gulati et al., 2020, pp. 415–443).

2.1 Effects of extrusion on starch structure and properties

Molecular degradation, disorder, and order transitions, as well as digestibility changes, alter starch structures. Disruption of starch granules results in contact swelling and melting, fragmentation, and gelatinization with varying starch temperatures and moisture levels (Nawaz et al., 2020). Cutting-edge techniques in starch extrusion systems and processes alter the physical and chemical properties of extruded food, thereby changing its texture and shelf life. Antecedent processes and extrusion techniques enhance starch's complexation, resulting in the formation of resistant starch (RS) as RS3 and retrograde RS during cooling (Ma et al., 2020). In the extrusion of blends of maize and cowpea, the extruded starches and flours enhance the physicochemical properties. Increased inclusion of sugar, however, limits the extent of starch transformation. Cooking extrudates increases the nutritional quality by enhancing the digestibility of starch, or, in the case of extrusion cooking, by reducing the formation of potentially harmful chemicals, such as acrylamide. Resistant starch is also generated during the extrusion of waxy maize, and its RS enhances starch and protein digestion in the small intestine (Nandane et al., 2025).

2.2 Interactions of starch with other macromolecules during extrusion

Starch protein, lipid and phenolic interactions during extrusion are important because they affect digestibility and texture of the end products. In blends of starch and protein, the changes in rheological and pasting properties of the systems are attributed to protein-starch interactions (Kumar et al., 2017). Starch-lipid digestion-resistant complexes formed in model systems and in real food aid in the retention of more beneficial nutrients. In the extrusion of sorghum-based products, more digestible macromolecules, such as fatty acids, Fig. 12.1).

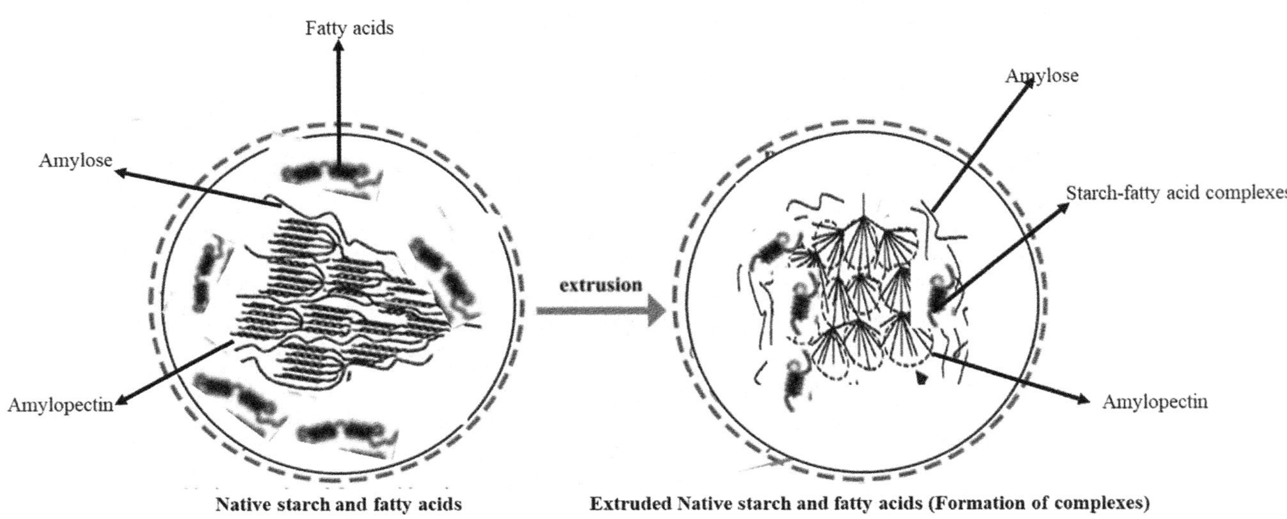

FIGURE 12.1 Extrusion of native starch with fatty acid and formation of complexes after extrusion. Formation of amylose-lipid complexes.

Proteins and phenolics, bind to starch and extruded products (Ortiz-Cruz et al., 2020). Starch in extrudates of rice bran also assimilates with added fibers and oil to enhance the nutritional quality of the extrudates. High moisture extrusion of plant protein modifies the structure of starch by thermally denaturing.

3. The extrusion cooking process

Extrusion cooking is one of the most diversified and applied technologies in food processing. Raw materials are transformed into finished products by combining mechanical and thermal forces to achieve desired textures, shapes, and functionalities (Choton et al., 2020). It has applications in producing snacks, breakfast cereals, pasta, pet foods, and even meat analogs. The general configuration of a system extruder includes a number of important components that help in the successful transformation of raw ingredients into value-added products (Fig. 12.2). The components are Feeder System, Extruder Barrel, and Screw(s). The screw may be a single-screw or twin-screw extruder, a Die, a Cutting System, and cooling and drying units.

4. Phases of the extrusion cooking process

Extrusion parameters are critical in altering the degree of starch gelatinization, which in turn influences starch digestibility. Increased barrel temperatures drive starch gelatinization, which in turn increases the amounts of rapidly digestible starch and the rate of digestion within the crystalline structures of starch, thereby making it more accessible to enzymatic action (Huang et al., 2022). This concept can, for example, be observed in maize-based extrudates, where increasing the temperature from 80°C to 140°C increases the rate of starch hydrolysis, as the elevated temperature causes the granules to swell and amylose to leach out (Yadav et al., 2022).

The amount of moisture content during the extrusion of normally extruded material is typically 16%–23%. This is another important parameter; loss of moisture is inversely correlated to the degree of shear stress, which increases the amount of starch fragmentation savagely digested while still retaining some unbound moisture, thereby preserving the level of resistant starch formation by repressing the degree of gelatinization (Wang et al., 2024). Screw speeds, which vary from 250 to 600 rpm, also relate to the amount of mechanical energy supplied to the extruder. Higher values increase the dominance of friction and shear strains, which are hypothesized to increase the rate constant of starch digestion by more than 10 times that attained from non-extruded grains (Wang et al., 2024). Below are the phases of the extrusion cooking process.

4.1 Formulation of ingredients and feeding

The first step includes mixing all these ingredients into a uniform mass with adequate moisture content, usually achieved by adding water, oil, or steam. Most formulas contain many starch-containing ingredients for gelatinization, texturation or

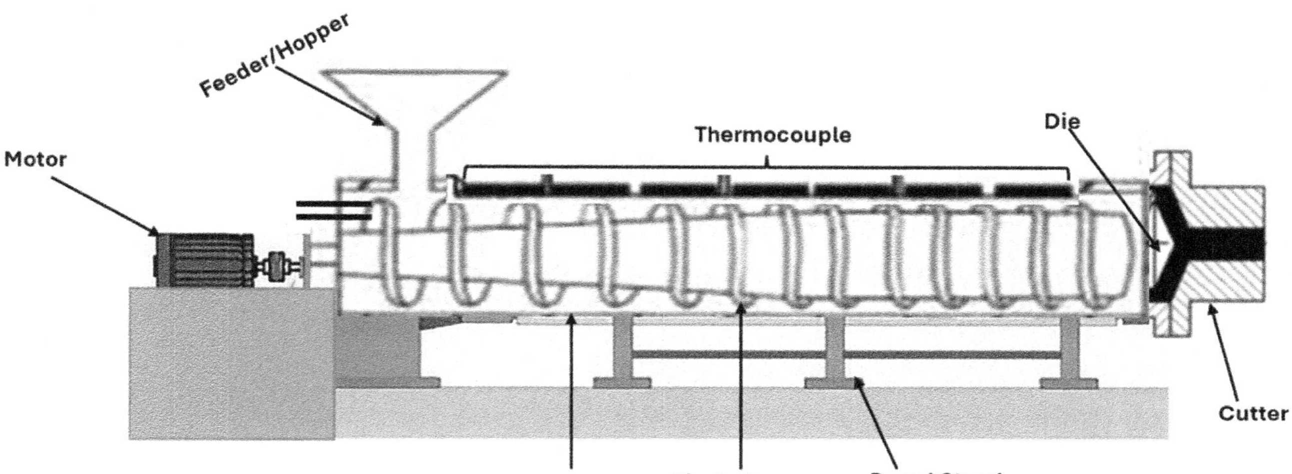

FIGURE 12.2 Single screw extruder. Single screw extruder used in food companies.

functional use as binding agents with the other ingredients during processing (Navale et al., 2015). Discharging the prepared feedstock mixture into the extruder under the control of a feeder, which regulates the feeding rate. Improper action generally causes defects in the final product (Navale et al., 2015).

4.2 Mixing, thermal transfer and shear

Inside the extruder, the screw effect transports and mixes the material. The geometry of the screw is of significant importance in setting flow and mixing characteristics, such as pitch and flight depth (Navale et al., 2015). On its journey down the barrel, it picks up heat energy supplied by heated barrel segments, which interact with mechanical energy due to screw rotation, applying intense shear forces. This simultaneous heating up and shearing enables the starch to gelatinize and the proteins to denature, which is particularly important in texturization and product formation (Navale et al., 2015).

4.3 Other operations

The screw compresses the matter, hence increasing pressure and diminishing voids within it. This is why it effectively melts starch and brings all ingredients into a homogeneous mixture (Navale et al., 2015). As soon as it comes out of the die, the sudden release of pressure causes moisture to evaporate, causing the material to puff. The degree of puff depends on factors like moisture content, temperature, and starch composition.The extrudate is cut into desired shapes and sizes. It is then cooled and dried to remove residual moisture and ensure structural stability (Navale et al., 2015).

5. Basic principles of extrusion cooking

Heat transfer is one of the basic principles of extrusion cooking. It supplies heat directly or indirectly to the food matrix using heated barrels, mechanical energy from screw rotation, or steam injection (Abilmazhinov et al., 2023). Rapid heat transfer of the extruder facilitates starch gelatinization, protein denaturation and more such transformations. After heat transfer, extremely high shear forces occur due to the interaction between the rotating screw(s) and the barrel's inner walls. These forces cause the mixing and breakdown of particle size and the ordering of molecules, which is crucial to texture formation and high-pressure generation (Pan et al., 2025).

There is pressure built up in the barrel by the compressive action of the screw. The melting happens due to a sufficient combination of thermal energy transfer provided by pressure; thus, a straining action pushes the material ahead. When the material leaves the die, there is an instantaneous pressure drop, which causes expansion through water vaporization (Patil et al., 2016). Raw water content is important for starch gelatinization, protein solubilization and product expansion. Moisture levels vary between 10%–40% depending on the product, which can be influenced by drying the product after the extrusion cooking. Residence time is fundamental to extrusion cooking. The residence time, typically a few seconds to a few minutes, determines the extent of heat and shear applied to the material. Short residence times reduce nutrient loss while achieving the desired structural transformations (Kristiawan et al., 2020, pp. 141–167).

6. Starch modification during extrusion

Extrusion is a widely used food processing method that utilizes elevated temperatures and specialized grinding machinery to transform basic materials into a diverse range of products. During this process, starch undergoes several modifications that dramatically alter its activity. Improves texture, digestibility and overall finished product. Understanding these modifications is critical to developing endogenously extracted foods that meet consumer demands for nutrition, taste, and texture.

6.1 Gelatinization

Gelatinization occurs when starch granules absorb water and swell when heated. This causes a loss of crystalline structure (Fig. 12.3). This is particularly important in outdoor air, as high temperatures and humidity facilitate these changes. Gelatinization creates a viscous gel-like matrix that contributes to the desired stability of the extract. This is especially important for snacks and breakfasts. The gelatinization degree affects the starch's solubility and water retention capacity, as well as the shelf life of the product (Donmez et al., 2021).

6.2 Dextrinization

The breakdown of starch into shorter glucose chains (dextrin) occurs through heat, shear, and moisture (Li et al., 2023). We call this process dextrinization. During extrusion, people sometimes do this to get specific product features. When

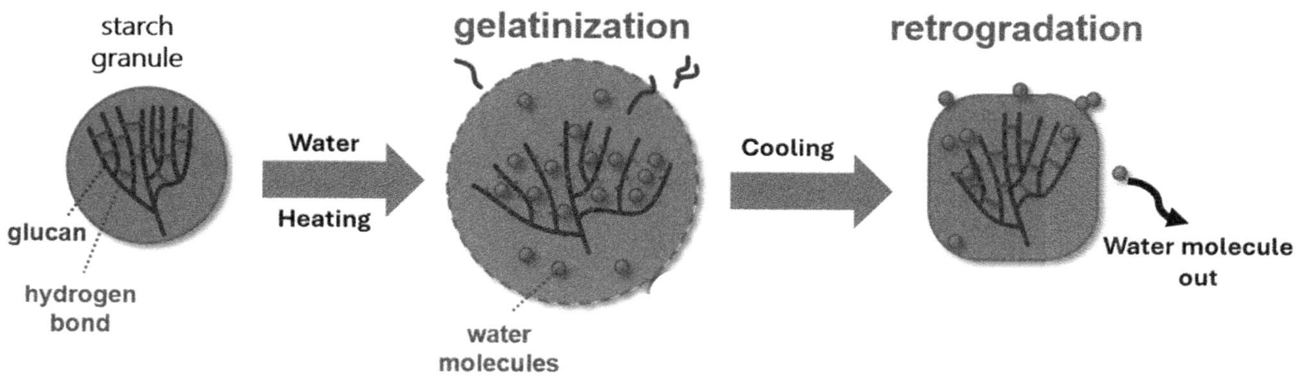

FIGURE 12.3 Gelatinization and retrogradation of starch. Gelatinization and strach regrodation during extrusion.

starch breaks down during extrusion, it can improve the taste of extruded products. This happens because it helps the Maillard reaction between amino acids and reducing sugars. When dextrin forms, it can change color. Often, this leads to a golden-brown color in many snack foods (Shukla et al., 2022).

6.3 Retrogradation

Retrogradation occurs when gelatinized starch molecules reassociate as they cool, forming a junction zone(Zhang et al., 2024). This process can create a gel or a crystal-like structure. Retrogradation affects how firm and textured starch-based products feel. It can be beneficial for certain products, such as pancakes, but excessive use can make extruded pasta rigid or not bendable enough (Hirao et al., 2021). Starch can form structures that hold onto water during this process. This might help products last longer on the shelf by keeping moisture from escaping.

6.4 Synthesis of amylose-lipid complexes

One of the processes during the extrusion of starches, which is crucial in the functionality and properties of extruded products, is the formation of amylose-lipid complexes (Qiu et al., 2024). During the extrusion process, the starch granules are subjected to high temperature and mechanical shear and, thus, get gelatinized, and their granular structure is disrupted. So, as starch gelatinizes, one of its fractions, amylose, solubilizes and oozes out of the starch matrix surrounding, complexing with lipids and forming amylose-lipid complexes (Resistant starch III/V).

This is a type of complex that is usually formulated through physical entrapment or hydrophobic interactions and, in turn, modifies the solubility, digestibility, and retrogradation of the starch. The presence of such complexes improves the texture and stability of extruded products, improving the mouthfeel while increasing the resistance to moisture absorption (Lazou, 2024). Also, lipids have been shown to modify the thermal characteristics of the starch matrix, which is a key variable in both the extrusion process and the final product characteristics (Huang et al., 2022).

7. Nutritional aspects of extruded starch

Extruded starch ginseng has a complex set of attributes concerning the energy value of the food, its digestion rate, glycemic response and health aspects. Extrusion has been known to affect the structural properties of starch, which modulates its digestion. Extrusion cooking and its mechanical shear create a sufficient force to gelatinize the starch granules, which enhances its susceptibility to digestive enzymes (Qiu et al., 2024). This could result in a faster infusion of glucose level in the body, elevating the GI of the extruded food products (Table 12.1). Nevertheless, the amount of extruded starch granule gelatinization and the retrogradation that may occur when the food product cools still leaves some portions of starch undigested when it reaches the small intestine (Nadia et al., 2021). This inversely correlates with the GI and may be in a position to improve blood glucose levels. Evidence-based research substantiates that resistant starch, which is formed during extrusion and other processing (Pasting, Spray drying, Heat Moisture treatment, etc), ameliorates the gut as a prebiotic that feeds colonies of beneficial bacteria (Kim & Mills, 2024). Resistant starch obtained through extrusion may increase satiety, reduce food cravings and assist in effective weight control.

TABLE 12.1 Nutritional aspects of extruded starch.

Effects of extrusion	Description	Impact on nutrition	References
Digestibility	Extrusion cooking breaks down starch molecules, increasing gelatinization.	Enhances digestibility, leading to faster glucose release and a higher glycemic index.	Wang et al. (2023)
Glycemic index (GI)	Extruded starches often have a high GI due to rapid digestion and absorption.	May cause quicker blood sugar spikes, less suitable for sustained energy release.	Huang et al. (2022)
Resistant starch content	Extrusion can reduce resistant starch, though specific conditions may retain it.	Lower resistant starch reduces prebiotic benefits and negatively impacts gut health.	Liu et al. (2020)
Nutrient retention	High temperatures in extrusion may degrade heat-sensitive nutrients like vitamins.	Potential loss of vitamins (e.g., B vitamins) and minerals, reducing nutritional value.	Nandane et al. (2025)
Protein interaction	Starch-protein interactions during extrusion can form complexes.	May reduce protein bioavailability, affecting the nutritional quality of extruded products.	Mejía Terán and Blanco-Lizarazo (2021)
Fiber content	Extrusion may disrupt dietary fiber structure, depending on processing conditions.	Reduced fiber functionality can lower satiety and digestive health benefits.	Nikinmaa et al. (2023)
Caloric density	Extruded starches are often energy-dense due to high carbohydrate content.	High caloric density may contribute to weight gain if consumed in excess.	Shah et al. (2019)

As resistant starch passes through the gut, bacteria may be able to ferment it into short-chain fatty acids, which are known to have positive effects on gut and metabolic health (Chen et al., 2024). Additionally, they may help lower cholesterol and increase insulin sensitivity, which can be advantageous in preventing certain chronic conditions like type 2 diabetes or heart disease (Bojarczuk et al., 2022). Furthermore, there is the potential for extruded starch to be modified and improved in nutritional value as well. For example, adding dietary fibers, proteins, or micronutrients during the extrusion cooking process can improve the overall value of the final product. This strategy is also valuable for preparing functional foods to meet public health concerns, such as high-fiber foods that enhance good health or food products fortified with vitamins and minerals.

This demonstrates that extruding starch can alter food texture, which positively influences taste and acceptance by consumers (Qiu et al., 2024). A better texture might also lead to an increased intake of selected extruded products, enhancing diet variability and nutritional quality. The other ingredients used in the formulation determine the nutritional composition of the extruded starch products. For example, whole grains, legumes, or seeds in the extruder can boost protein and polyunsaturated fatty acids, vitamins, and minerals, consequently improving the product's nutritional composition (Gulati et al., 2020, pp. 415–443).

8. Functionality and application of modified starch

Modified starches aim to improve a particular property, unlike the original starches, which are native (Garcia et al., 2020). They differ from native starches because they have enhanced water retention, emulsification, and film-forming capabilities, enabling their use in various applications. Their applications range from food products such as soups, sauces and baked goods to thickeners, texture enhancers, stabilizers and even drug delivery systems in place of excipients and as tablet binding agents (Bangar et al., 2023). The applications that make viscosity control and ameliorating properties beneficial include textiles, biodegradable packaging, and paper production. Their advancement allows them to penetrate reinforcing mechanisms throughout other industries (Adewale et al., 2022).

8.1 Recent advances in starch utilization in extruded foods

Extracting starch from food to create extruded products has undergone robust growth over time due to the quest for novel means of processing food more deeply (Lazou, 2024). The recent focus of starch utilization in extruded foods pivots to improving nutrition, sustainability and functionality through unique processing, modification, and source techniques to facilitate the diversification of starch. Extrusion cooking is a high-shear, high-temperature processing method that modifies starch to achieve texturization and expansion in the production of snacks and breakfast cereals. Existing techniques, however, are suboptimal due to the loss of resistant starch (RS) and the generation of waste. As of 2025, a shift in focus to optimizing resistant starch in waste minimization aligns with global trends in demand for healthier, low-glycemic-index (GI) food, which is the primary goal.

A major innovation in the creation of RS-enriched extrudates stems from the implementation of physical-level alterations, such as conjugate annealing and ultrasound-assisted extrusion, which yield a 71% increase in RS levels in extrudates of corn and bean starches (Zailani et al., 2023). This leads to an even lower glycemic response due to the decrease in digestibility, while the enhancement is retained, as evidenced by kidney bean extrudates, which show 16.3% RS (compared to 2.5% in corn). The incorporation of α-amylase in reactive extrusion results in improved protein digestion in quinoa flour, while preserving the retention of phenolic compounds alongside RS3 (retrograded starch), which is ideal for producing functional snack foods (Pismag et al., 2023).

One of the primary factors driving the development of extruded biodegradable packaging is its contribution to environmental sustainability. Starches modified to include blends with polyvinyl alcohol or nanoclay have been shown to form films with enhanced barrier properties, while also reducing water vapor permeability to approximately 50% and exhibiting some antimicrobial properties (Karnwal et al., 2025). Some films made from RS extrudates containing starch from jackfruit seeds also incorporate essential oils into the matrix to enhance the level of active packaging and thereby improve the shelf life of the final product (Ye et al., 2025). The newly defined starches, potentially derived from neglected sources such as Tartary buckwheat, possess unique and even desirable properties and improved digestibility for gluten-free foods after extrusion (Zhang et al., 2023). The use of hydrothermal processing of cassava and yam starches has also been shown to enhance the RS of the extruded flours (Acevedo-Viloria et al., 2025). RS retention with extrusion will continue to present challenges in isolating the particular desired texture and may require more focused extrusion with AI-optimized processes (Paternina-Contreras et al., 2025). Sustainable, health-focused foods will now be significantly aided by the further development of extruding starches.

8.2 Functionalized starch: enhancing texture and expansion

Traditionally, starches have been poorly modified by cross-linking, acetylation, and hydroxypropylation; however, more modern techniques have recently emerged that have profoundly repaired their functional extrusion properties (Kaur et al., 2024, pp. 97–117). On the other hand, starch modifications enhance the puffing expansion and the structural integrity of puffy snacks by minimizing the stickiness (Table 12.2). The modifications (physical, chemical and enzymatic) make altering starch's gelation and retrogradation characteristics easier, which helps control the texture (Bangar et al., 2023). This provides a more appealing snack and cereal mouthfeel by increasing the crispiness without compromising structural integrity during processing. Meat analogs, high-protein snacks or other plant-based foods have been developed by optimizing starch-protein blends during extrusion to enhance their textures (Guyony et al., 2023).

TABLE 12.2 Extruded products, starch type and the role of starch in the food products.

Extruded product	Starch type used	Role of starch	References
Breakfast cereals	High-amylose maize starch, potato starch	Provides expansion, crispness, and retrogradation potential	Beech et al. (2022)
Puffed snacks (e.g., chips, curls)	Modified starches, high-amylose starch	Improves expansion, texture, and fiber enrichment	Li et al. (2023)
Pasta/noodles (extruded)	Native wheat starch, high-amylose maize starch	Structure and texture, with retrogradation during cooling	Lubowa et al. (2018)
Snack bars /extruded fiber-enriched products	RS4 starches (cross-linked, esterified, etc.)	Adds fiber, improves texture and binding	BeMiller (2020)
Gluten-free extruded products	RS4, pregelatinized starches	Provides structure, replaces gluten functionality	Roman and Martinez (2019)
Infant cereals/instant mixes	Pre-gelatinized starch, high-amylose starch	Provides digestibility control, viscosity	Bresciani et al. (2023)

Additionally, these functionalized starches can enhance food formulations in terms of viscosity and stability. This is particularly useful in processes like extrusion, as high temperatures and shear conditions can significantly alter the material's properties. Using modified starches greatly assists food scientists in formulating new products, as they can tailor the texture and expand the characteristics to meet consumer needs and expectations (Adewale et al., 2022).

9. Resistant starch in functional foods

Resistant starch (RS) refers to a special type of dietary starch that, unlike normal starch, does not get digested but passes through to the colon (Bojarczuk et al., 2022). This makes it distinct from regular starch, which is highly significant in the development of functional foods. There are four types of RS (RS1, RS2, RS3 and RS4) based on the formation's source and conditions (Raigond et al., 2015). Resistant starch in modifying food formulation can enhance texture and mouthfeel. It can also function as a bulking agent that may help increase satiety in weight management products (Walsh et al., 2022). In addition, resistant starch is beneficial for functional foods since it improves moisture retention and increases fiber content. As awareness of health and well-being rises, more consumers seek functional foods containing resistant starch (Subhasri et al., 2022, pp. 17–39). This trend is increasing the demand for food products modified with resistant starch, catering to health-conscious consumers who seek digestive and metabolic enhancements (Yan et al., 2021).

10. Starch in gluten-free and protein enriched products

Starch is important in gluten-free and high-protein products, as it is a basic functional ingredient that adds value to these foods. Native and modified starches, particularly those derived from rice, tapioca, and corn, have successfully developed gluten-free extruded products with good texture and structural integrity (Obadi & Xu, 2021). Using starches and plant proteins like peas, lentils, and soy enriches the nutritional value of extruded snacks and cereals to meet dietary protein requirements (Tas & Shah, 2021).

As in gluten-free products, starches are important because they replace the gluten-connective material ordinarily present in baked goods in a glutinous manner (Witczak et al., 2016). A fundamental characteristic of gluten-free flours is that they may vary significantly in composition, resulting in the absence of one of the key attributes of gluten and elasticity. To enhance the texture, moisture retention and coating, starches such as tapioca, corn, or potato starch are included to produce better-tasting and more acceptable food products (Ali et al., 2024, pp. 241–260). The texture characteristics, crumb structure and final product quality depend more highly on the type and ratio of starch and high-protein products (Zhang et al., 2021).

For high-protein goods, such as bars and additives, high-protein snack consumers seek starches that assist in retaining the product's structure and simultaneously act as a source of energy (Ananthan et al., 2021, pp. 309–332). Furthermore, in the cooking or formulation processes, it is common to use modified starches that aid in maintaining the mouthfeel and moisture content. The protein content of these products is not enhanced in any way because starches help create flavors for protein-based bars and other snacks by supplementing taste without altering the protein content of such products (Ananthan et al., 2021, pp. 309–332). These products can significantly benefit from the addition of starches. The effective use of starches in gluten-free and protein-rich assortments makes it easier to enhance the intended physical characteristics. Improving and optimizing the eating experience and catering to the current market for better living, is looking for tasty yet healthy options.

10.1 Starches with a dual modification for better processing

The food sector has been reporting an increase in the industry's overall focus on incorporating trend- and market-oriented starches into the functional composition of food products (Almanza-Oliveros et al., 2024). To address this challenge, dual–modified starches have emerged as a viable option due to their optimal development and application in food products (He et al., 2023). These starches are also utilized when specific needs, such as fortified and extruded foods with enhanced stability, texture, and nutritional profile, are required. Utilizing two modification techniques can produce additive effects, resulting in dual-modified starches with superior characteristics than what a single modification could yield. Dual-modified starches with both physical and chemical modifications, as in the case of heat-moisture treatment and cross-linking, exhibit better thermal stability than non-modified starches, making them ideal in applications involving high temperatures such as extrusion processes (Lin et al., 2024).

These starches provide perfect flow and consistency in extruded foods that are dense or have a rigid structure, such as pasta and breakfast cereals. Dual-modified starches help achieve the expected properties in various food products, such as snacks and sauces, by refining the ratios of amylose and amylopectin and altering their interaction properties (Lin et al., 2024). The modifications help reduce the chances of retrogradation and syneresis, thereby increasing the shelf life of products while maintaining their quality over time. On the other hand, starch modifications can lead to animal and human health food products that are more nutritious and have a lower glycemic response. In extrusion cooking, the contribution of dual-modified starches becomes far more important. In extrusion processes, starches that meet the necessary requirements regarding shear and temperature must be used instead of traditional types to achieve the desirable attributes. Depending on the required outcome, dual-modified starches can help increase the expansion of extrudates, enhance ingredient binding, and provide the desired crunchy and chewy texture (He et al., 2023).

10.2 A study on the role of starch in low-fat and reduced-calorie products

Starch contributes well to the design of low-fat and reduced-calorie foods by leveraging the functional attributes of fats while simultaneously reducing calories (Thondre & Clegg, 2019, pp. 7–64). The process by which starches replace fat is through gel formation, which provides a silky and creamy feel, compensating for the absence of fat and retaining moisture; furthermore, it enhances the juiciness and moisture of the product. Certain types of starches, for example, the cross-linked alteration or the hydroxy propylated types, are effective substitutes for fats because they replicate the creamy feel of fats in the mouth (Sun & Fang, 2021, pp. 409–444). Starches are included in puffed and extruded snack products to reduce the amount of fat without compromising the texture of the products, which remain non-greasy and crunchy. Calories are decreased, and the fiber content of foods is increased with the use of resistant starch and dextrin (Witczak et al., 2016).

Fat- and oil-free formulations take advantage of starch's capacity to provide thickness in the mouth and a satisfying feel without extra calories (Wang et al., 2023). Starches, such as resistant starch, not only reduce calories but also offer functional advantages, making them suitable for health-conscious consumers. They promote good gut health and reduce glucose secretion after eating. Starch-based fat mimics are very affordable and easily accessible options for developing low-calorie products without compromising appealing taste and texture (Sun & Fang, 2021, pp. 409–444).

10.3 Starch texturing a meat substitute

Starch is gaining popularity as a primary ingredient in the construction of meat substitutes, aligning with the growing interest in plant-based protein food sources. Applying specific modifications to starch enables the products to approximate the texture and mouthfeel of meat. Thus, they offer appealing options for healthy or eco-friendly food (Gallego et al., 2022). Starch is processed with protein isolates through high-moisture extrusion to produce fibrous, meat-like structures in plant-based proteins (Sun et al., 2022). Starch has been found to improve water retention and binding capabilities, which are crucial for the juiciness and physical structure of meat substitutes.

In meat substitute production, extraction and processing are important steps in producing texturized starch. Starch slurries are extruded and subjected to high temperatures and shear forces, increasing their molecular temperatures and transforming them into fibrous, meat-like structures (Farooq & Yu, 2024). The produced product mimics the chewiness of real meat, aiding in moisture retention and overall mouthfeel. The source of starch, whether pea, potato or corn starch and its ratio of amylose to amylopectin determine the meat analog's texture and other physical and functional characteristics (Bühler et al., 2022). Furthermore, plant proteins, flavorings and other components can augment the final product's health and taste properties. (Caporgno et al., 2020) reported extruded meat analogs based on yellow heterotrophically cultivated (*Auxenochlorella protothecoides*) microalgae. They revealed that microalgae combined with soy concentrations helped to create fibrillar-textured extrudates (meat analogs). In the same vein (Wittek et al., 2021), reported on the morphology development and flow characteristics during the high-moisture extrusion of a plant-based meat analog. Additionally, they suggested that the plant-based meat analog formed was influenced by the extruder die, which may be specifically designed to suit each type of plant-based meat analog.

10.4 Enhanced sensory and shelf-life properties

Increased sensory properties in extruded foods refer to improving the food's texture, taste, aroma, making eating and more appealing to eat (Alam et al., 2016). During the extrusion process, starches can be modified to produce various textures, including 'crunchy' and 'chewy' (Offiah et al., 2019). The textural variation has a significant impact on consumer acceptance and satisfaction levels. Furthermore, the procedure can foster the formation of desirable flavor and aroma through the Maillard reaction and other high-temperature chemical transformations. Extruded snacks are crispy and retain their texture even in high humidity, thanks to modified starches. Ready-to-eat cereals and pasta have a longer shelf life due to cross-linked starches, which minimize syneresis and retrogradation (Bangar et al., 2023). Similarly, extrusion plays a role in extending the shelf life of food products. When raw food is extruded, moisture levels drop, and other harmful microbes are killed in the process, which ultimately helps keep the products fresh for a longer period (Alam et al., 2016). As a bonus, extrusion changes the starch, strengthening other parts of the product and securing its quality and freshness. The food products are protected from spoiling and remain in good condition for a long time, satisfying the modern man's needs for resistant and convenient food items. To summarize, the modern consumer's diverse wants and needs for quality and durability are met by enhanced extruded properties.

11. Case studies

One of the case studies is that of Tapioca starch as a functional component in gluten-free extruded snacks. Among food ingredients, tapioca starch lends an unusual quality to the manufacturing of gluten-free extruded snacks (Milde et al., 2012). Its excellent gelatinization and expansion capabilities make it crucial for achieving desirable textural and sensory qualities (Pongjaruvat et al., 2014). Additionally, finding a suitable combination with other ingredients is equally important in mitigating drawbacks such as low nutrient density. Sensory properties, along with structural stability, are desirable in products recommended for gluten-free diets. Being tasteless, high-amylopectin tapioca starch is well-suited for the extrusion processing of gluten-free snacks (Milde et al., 2012). They were extruded at a screw speed of 400 rpm and a barrel temperature ranging from 150 to 180°C, with varying ratios of pea protein, rice flour, and tapioca starch, and a 12% feed moisture content. When extruded, the tapioca starch swells and binds air to the food, making the end products crispier and crunchier while compensating for the loss of gluten with its binding properties.

The products with Tapioca Starch in the 40%–50% range showed good expansion and texture. The addition of starch (Tapioca) enhanced the water absorption and solubility indices of the extrudate, resulting in a product with a better taste and chewability (Patel et al., 2016). There are some drawbacks, including high dependence on tapioca starch, which can result in a more brittle structure with low moisture content and nutrient density in starch-rich formulated products. In addition, it could be helpful to consider supplementing the products with resistant starch or fiber for a more favorable nutritional content. It is also proposed that starch's enzymatic or chemical modification, as discussed previously (subsection 7.4), be investigated to achieve better material properties during extrusion. Another case study is utilizing corn starch in snack food production through extrusion.

Due to its high amylopectin content (approximately 72%), cornstarch is frequently used in the extruded food industry, as it possesses thickening and gelling capabilities. This case study uses corn starch for extruded snack foods, particularly puffed corn snacks (Pęksa et al., 2016). This case study focuses on analyzing the importance of corn starch during the extrusion process while considering its influence on the texture, flavors and quality of puffed snacks. The puffed snacks were prepared using corn flour and corn starch as the main ingredients. The formulation had varying ratios of corn starch to achieve different degrees of texture profiles. Extrusion conditions were also varied in terms of heating temperatures (130°C, 150°C, and 180°C), screw speeds (100 rpm, 150 rpm, and 200 rpm), and moisture contents (12%, 14%, and 16%). This approach facilitated the investigation of the implications of the different conditions tested during the extrusion process.

The color quality of the extruded products was evaluated using a colourimeter, and the texture's crispness and hardness were also determined. During the study, taste panellists were asked to score the extruded products based on their flavor, aroma and overall acceptability. The results showed that the extrusion temperature and moisture content significantly influenced the textural properties of the snacks. High temperatures (180°C) resulted in a lighter and crispier product due to increased gelatinization and more significant starch expansion. At intermediate temperatures of 150°C, more flavor is obtained due to a good degree of the Maillard reaction being activated without excessive burning. When reading the colorimeters, it was found that the golden-brown color, which is quite popular for snack foods, was produced from high extruding temperatures. On the other hand, lower temperatures yielded a faint color. As part of the sensory evaluations conducted during the study, corn starch and snacks made from 30% corn starch, which were either extruded or thermally tempered at 150°C, were rated above the rest in terms of taste and overall acceptability. A balanced mixture of texture and flavor enabled a superior feeding response.

The case studies demonstrate the significant role of starch in the extrusion of snack foods. The findings align with the fundamental principles of starch behavior under varying thermal and mechanical conditions, illustrating how extrusion conditions can be optimized to tailor products to specific consumer preferences. Moreover, the results emphasize the potential for product differentiation in the competitive snack market. As consumers increasingly seek healthier and more flavorful options, the effective utilization of starch in extrusion presents an opportunity for innovation in product formulations.

12. Health and nutritional consequences of starches in extruded food systems

Starches constitute a vital part of extruded food systems, as they are the major source of energy and the scaffold in extruded snacks, cereals, pasta, and ready-to-eat meals (Mandliya et al., 2024, pp. 35–51). During the extrusion process, which involves cooking temperatures, shearing, and pressurizing the mixture, changes in the starch, such as gelatinization, melting, and fragmentation, occur, altering the nutritional profile and influencing the health impacts of the food product (Qiu et al., 2024). Starch digestion changes in a way that is RS, or resistant. The alteration of the glucose response and gut is the result of health changes in the gut microbiome.

12.1 Nutritional implications

The process of extrusion cooking tends to enhance the nutritive value (Table 12.3) of starch-centered foods by improving the digestibility of starches, thus improving the energy and nutrients available to the body (Dendegh et al., 2022). The extrusion process causes starch to gelatinize, thus making it more easily digestible through enzymatic hydrolysis (Ye et al., 2018). For example, the extrusion of sorghum-based extrudates increases protein and starch digestion while decreasing the anti-nutritional factors associated with tannins and other compounds, thereby improving the bioavailability of minerals and vitamins. The process can also create resistant starch (particularly RS3) through retrogradation, thereby promoting the dietary fiber and protein quality of the extrudate and decreasing the overall caloric density (Feng et al., 2024).

The retention of nutrients varies depending on the extrusion conditions. For instance, severe conditions will degrade the heat-sensitive compounds of millet extrudates, while gentler conditions preserve the vitamins and antioxidants. The extrusion of millet extrudates, for example, enhances functional properties, such as the bioactive compounds, and starch serves as the carrier (Dey et al., 2024). This blend of extrudates, such as maize with cowpea, indicates enhanced nutrition, where extrusion also lowers phytic acid while improving the bioavailability of zinc and iron. In summary, extruded starches enhance the value of supplemented foods, but excessive processing can cause substantial nutrient loss; thus, a need for proper conditions to ensure nutrient retention.

The analysis of the health implications of starches in extruded foods is complex, and concerns glycemic control, gut health, and certain associated risks (Himat et al., 2021). Starch extrudates are promptly digestible and associated with a high glycemic index and are claimed to worsen insulin resistance and increase the risk of diabetes (Huang et al., 2022). The health benefits of the gut are related to the colon fermentation of starch, wherein butyric acid is formed. Butyric acid is one of the SCFAs that promotes gut health and encourages the healing of the gut epithelial barrier, while reducing gut inflammation. In addition, extrusion in certain cases helps reduce the anti-nutritional factors present in legumes and cereal grains, which are used for food in nutritional health (Ozolina et al., 2023).

13. Starch in extruded food products: future perspectives

The ordinate primary direction could be the intentional design of extruded foods to maximize the content of Resistant Starch (RS), particularly RS3 (retrograded starch) and RS4 (chemically altered starch), to counter the rising public health concerns of obesity and diabetes. Controllable extrusion conditions, such as temperature, moisture, and screw speed, can be adjusted to encourage retrogradation during cooling, thereby enhancing these prebiotic effects. For animal studies,

TABLE 12.3 Nutritional and health implications of starches in extruded food systems.

Aspect	Nutritional implications	Health implications	Food samples	References
Digestibility enhancement	Increases RDS, improving energy availability; reduces anti-nutritional factors.	Rapid glucose release may raise GI; balanced with RS for controlled blood sugar.	Maize-cowpea extrudates show higher digestibility.	Himat et al. (2021)
Resistant starch formation	Boosts fiber-like content (RS3/RS4), lowering caloric density.	Prebiotic effects, SCFAs production for gut health, and reducing diabetes risk.	Waxy maize extrudates increase RS.	Liu et al. (2020)
Nutrient retention	Preserves bioactive under mild conditions; enhances mineral bioavailability.	Supports metabolic health; hypocholesterolemic benefits.	Sorghum extrudates retain antioxidants.	Bhattarai et al. (2025)

high-amylose extruded starch snacks are predicted to lower the glycemic index by 20%–30% and thus promote metabolic health (Gulzar et al., 2021). Due to the stable chemical structure of resistant starch, its use will be expanded for controlling slow digestibility, as in the case of certain maize-based extrudates, where modification improves fiber-like properties without disrupting the texture (George & Ravishankar, 2022). By 2030, AI-optimized extrusion is expected to facilitate the production of starch-enriched gut microbiome modulators, which can reduce the risk of colon cancer associated with gut-derived short-chain fatty acids. Chief among these is the retarded formation of resistant starch with certain desirable sensory properties. Retrogradation (staling) can be excessive, but not in fusion processes that marry extrusion and annealing, which are especially encouraging.

13.1 The optimized integration to smart and sustainable extrusion

The optimization of starch performance in smart extrusion technologies and its real-time monitoring via AI and smart sensors is now within reach. Tomorrow's systems will dynamically and in real-time adjust parameters to control energy and waste in the gelatinization and starch complex formation. Starch incorporated into high-moisture extruded meat analogs serves as a binder, enhancing texture and driving the sustainable development of meat analogs from plant sources. Twin-screw extruders with variable zones will create multi-functional products such as starch-enriched probiotic cereals. Replacing starch with bioplastics in extruded foods to lighten the "plastic-like" environmental burden will enhance sustainability. Approaches to the circular economy will be a cost-effective means of decreasing the carbon footprint from extrusion by recycling extruded by-products as starch feed by 2030.

13.2 Personalized and functional food applications

Shifting the focus for extruded systems toward starch nutrition personalization is the future of these technologies, particularly with individualized goals, such as tailoring low-glycemic index diets for diabetics. Custom 3D printed extrudates, such as variable starch blend extrudates, will enhance the versatility of snacks. Starch-based functional extrudates with resistant starches as prebiotic fiber will have targeted applications in nutrition for infants and the elderly to support gut health. Ethically sourced and sustainable starch blends, such as quinoa, as well as global breadfruit, will diversify the diet for the increasingly large population with allergies. Starch innovations will address the health risk of acrylamide mitigation.

14. Conclusion

In conclusion, starches play a crucial role in the extrusion of food systems. They serve as important nutritional sources and greatly determine the physical and functional properties of the extruded goods. This chapter highlights the limitations and insights that extrusion, a cooking method that combines high temperature and short duration, significantly alters the structures and functionalities of starches through gelatinization and other processes. These changes fundamentally alter the texture, digestibility, and nutritional value of these foods, enabling the creation of better products suitable for the current market, as demonstrated using Tapioca starch. The tapioca starch swells during extrusion, binding air to the food and making the end products crispier and crunchier, while compensating for the loss of gluten with its binding properties. In this same vein, corn starch was extruded, and extrusion improves the color of the final products. The shift toward a more sustainable and healthier food system will likely increase the importance of starches in extruded food systems. This emphasizes that starches should be further studied and integrated into food systems as a multi-functional component. As a result, functional and versatile food constructs are continually evolving, with starches at the core of innovation in food technology, underscoring the significance they hold as ingredients.

References

Abilmazhinov, Y., Bekeshova, G., Nesterenko, A., Dibrova, Z., Ermolaev, V., Ponomarev, E., & Vlasova, V. (2023). A review on the improvement of extruded food processing equipment: Extrusion cooking in food processing. *Food Science and Technology, 43.* https://doi.org/10.5327/fst.80621

Acevedo-Viloria, N., Figueroa-Flórez, J., Salcedo-Mendoza, J., Hernández-Ruydiaz, J., & Ortega-Quintana, F. (2025). Effect of hydrothermal processing on the native starches of cassava (Manihot esculenta) and yam (Dioscorea alata). *Revista Facultad Nacional de Agronomía Medellín, 78*(1), 10977–10987. https://doi.org/10.15446/rfnam.v78n1.112802

Adewale, P., Yancheshmeh, M. S., & Lam, E. (2022). Starch modification for non-food, industrial applications: Market intelligence and critical review. *Carbohydrate Polymers, 291.* https://doi.org/10.1016/j.carbpol.2022.119590

Alam, M. S., Kaur, J., Khaira, H., & Gupta, K. (2016). Extrusion and extruded products: Changes in quality attributes as affected by extrusion process parameters: A review. *Critical Reviews in Food Science and Nutrition, 56*(3), 445–473. https://doi.org/10.1080/10408398.2013.779568

Ali, A., Tahir, A., Khalid, W., Manzoor, M. F., Arshad, M. S., Khan, A., & Raza, A. (2024). *Gluten-free pasta nutritional and bioactive profile development of gluten-free pasta*. China: Elsevier. https://doi.org/10.1016/B978-0-443-13238-4.00008-3

Almanza-Oliveros, A., Bautista-Hernández, I., Castro-López, C., Aguilar-Zárate, P., Meza-Carranco, Z., Rojas, R., Michel, M. R., & Martínez-Ávila, G. C. G. (2024). Grape Pomace—Advances in its bioactivity, health benefits, and food applications. *Foods, 13*(4). https://doi.org/10.3390/foods13040580

Ananthan, P., Kumar Sharma, G., & Dutt Semwal, A. (2021). *Energy bars: A perfect choice of nutrition to all*. UK Limited: Informa. https://doi.org/10.1201/9781003245513-12

Bangar, S. P., Ali, N. A., Olagunju, A. I., Pastor, K., Ashogbon, A. O., Dash, K. K., Lorenzo, J. M., & Ozogul, F. (2023). Starch-based noodles: Current technologies, properties, and challenges. *Journal of Texture Studies, 54*(1), 21–53. https://doi.org/10.1111/jtxs.12730

Beech, D., Beech, J., Gould, J., & Hill, S. (2022). Effect of amylose/amylopectin ratio and extent of processing on the physical properties of expanded maize starches. *International Journal of Food Science and Technology, 57*(4), 2298–2309. https://doi.org/10.1111/ijfs.15581

BeMiller, J. N. (2020). *Resistant starch food engineering series* (pp. 153–183). United States: Springer. https://doi.org/10.1007/978-3-030-38654-2_7

Bhattarai, R. R., Jayasree Joshi, T., Sruthi, N. U., Rao, P. S., & Johnson, S. (2025). Effects of extrusion cooking on nutritional and health attributes of sorghum and millets: Special reference to protein and starch digestibility. *International Journal of Food Science and Technology, 60*(1). https://doi.org/10.1093/ijfood/vvae093

Bojarczuk, A., Skąpska, S., Mousavi Khaneghah, A., & Marszałek, K. (2022). Health benefits of resistant starch: A review of the literature. *Journal of Functional Foods, 93*. https://doi.org/10.1016/j.jff.2022.105094

Bresciani, A., Giuberti, G., Cervini, M., & Marti, A. (2023). Using pre-gelatinised red lentils in pasta production: Connecting starch features and pasta cooking behaviour. *International Journal of Food Science and Technology, 58*(4), 1803–1809. https://doi.org/10.1111/ijfs.16312

Bühler, J. M., Schlangen, M., Möller, A. C., Bruins, M. E., & Jan van der Goot, A. (2022). Starch in plant-based meat replacers: A new approach to using endogenous starch from cereals and legumes. *Starch - Stärke, 74*(1–2). https://doi.org/10.1002/star.202100157

Caporgno, M. P., Böcker, L., Müssner, C., Stirnemann, E., Haberkorn, I., Adelmann, H., ... Mathys, A. (2020). Extruded meat analogues based on yellow, heterotrophically cultivated auxenochlorella protothecoides microalgae. *Innovative Food Science and Emerging Technologies, 59*, 102275.

Castro-Campos, F. G., Esquivel-Fajardo, E. A., Morales-Sánchez, E., Rodríguez-García, M. E., Barron-Garcia, O. Y., Ramirez-Gutierrez, C. F., Loarca-Piña, G., & Gaytán-Martínez, M. (2024). Resistant starch type 5 formation by high amylopectin starch–lipid interaction. *Foods, 13*(23). https://doi.org/10.3390/foods13233888

Chen, Z., Liang, N., Zhang, H., Li, H., Guo, J., Zhang, Y., Chen, Y., Wang, Y., & Shi, N. (2024). Resistant starch and the gut microbiome: Exploring beneficial interactions and dietary impacts. *Food Chemistry X, 21*, Article 101118. https://doi.org/10.1016/j.fochx.2024.101118

Choton, S., Gupta, N., Bandral, J. D., Anjum, N., & Choudary, A. (2020). Extrusion technology and its application in food processing: A review. *The Pharma Innovation, 9*(2), 162–168. https://doi.org/10.22271/tpi.2020.v9.i2d.4367

Dendegh, T. A., Enefola, O. S., Akpapunam, S. O., Yelmi, B. M., & Abdullahi, M. J. B. (2022). Extrusion technology and its application in food processing-an overview. *Tropical Journal of Engineering, Science and Technology, 1*(2), 25–53.

Dey, D., Singh, G., Ishwarya, P., & Mateen, A. (2024). Functionality and extrusion processing of millets - A review. *Food Reviews International, 40*(10), 3605–3633. https://doi.org/10.1080/87559129.2024.2367643

Donmez, D., Pinho, L., Patel, B., Desam, P., & Campanella, O. H. (2021). Characterization of starch–water interactions and their effects on two key functional properties: Starch gelatinization and retrogradation. *Current Opinion in Food Science, 39*, 103–109. https://doi.org/10.1016/j.cofs.2020.12.018

Farooq, M. A., & Yu, J. (2024). Recent advances in physical processing techniques to enhance the resistant starch content in foods: A review. *Foods, 13*(17). https://doi.org/10.3390/foods13172770

Feng, H., Cheng, B., Lim, J., Li, B., Li, C., & Zhang, X. (2024). Advancements in enhancing resistant starch type 3 (RS3) content in starchy food and its impact on gut microbiota: A review. *Comprehensive Reviews in Food Science and Food Safety, 23*(3). https://doi.org/10.1111/1541-4337.13355

Gallego, M., Barat, J. M., Grau, R., & Talens, P. (2022). Compositional, structural design and nutritional aspects of texture-modified foods for the elderly. *Trends in Food Science and Technology, 119*, 152–163. https://doi.org/10.1016/j.tifs.2021.12.008

Garcia, M. A. V. T., Garcia, C. F., & Faraco, A. A. G. (2020). Pharmaceutical and biomedical applications of native and modified starch: A review. *Starch/Staerke, 72*(7–8). https://doi.org/10.1002/star.201900270

George, C., & Ravishankar, C. (2022). Effect of extrusion processing parameters on physical and nutritional properties of fish based Snack—A review. *Fishery Technology, 59*, 225–237.

Gulati, P., Brahma, S., & Rose, D. J. (2020). *Impacts of extrusion processing on nutritional components in cereals and legumes: Carbohydrates, proteins, lipids, vitamins, and minerals*. Elsevier BV. https://doi.org/10.1016/b978-0-12-815360-4.00013-4

Gulzar, B., Hussain, S. Z., Naseer, B., Bashir Shikari, A., Nazir, N., & Gani, G. (2021). Investigation of process and product parameters on physical attributes, resistant starch, and in vitro starch digestibility of modified rice flour-based extruded snacks. *Journal of Food Processing and Preservation, 45*(11). https://doi.org/10.1111/jfpp.15953

Guyony, V., Fayolle, F., & Jury, V. (2023). High moisture extrusion of vegetable proteins for making fibrous meat analogs: A review. *Food Reviews International, 39*(7), 4262–4287. https://doi.org/10.1080/87559129.2021.2023816

He, R., Li, S., Zhao, G., Zhai, L., Qin, P., & Yang, L. (2023). Starch modification with molecular transformation, physicochemical characteristics, and industrial usability: A state-of-the-art review. *Polymers, 15*(13), 2935. https://doi.org/10.3390/polym15132935

Himat, A. S., Gautam, S., Chavez Garcia, J. P., Vidrio-Sahagún, A. X., Liu, Z., Bressler, D., & Vasanthan, T. (2021). Starch-based novel ingredients for low glycemic food formulation. *Bioactive Carbohydrates and Dietary Fibre, 26*. https://doi.org/10.1016/j.bcdf.2021.100275

Hirao, K., Kondo, T., Kainuma, K., & Takahashi, S. (2021). Starch gel foods in cookery science: Application of native starch and modified starches. *Journal of Biorheology, 35*(1), 29–41. https://doi.org/10.17106/jbr.35.29

Huang, X., Liu, H., Ma, Y., Mai, S., & Li, C. (2022). Effects of extrusion on starch molecular degradation, order–disorder structural transition and Digestibility—A review. *Foods, 11*(16), 2538. https://doi.org/10.3390/foods11162538

Karnwal, A., Rauf, A., Jassim, A. Y., Selvaraj, M., Al-Tawaha, A. R. M. S., Kashyap, P., Kumar, D., & Malik, T. (2025). Advanced starch-based films for food packaging: Innovations in sustainability and functional properties. *Food Chemistry X, 29*. https://doi.org/10.1016/j.fochx.2025.102662

Kaur, L., Kaur, R., & Singh, J. (2024). *Chemical modification of starch*. Elsevier BV. https://doi.org/10.1016/b978-0-323-96102-8.00015-2

Kim, Y. T., & Mills, D. A. (2024). Exploring the gut microbiome: Probiotics, prebiotics, synbiotics, and postbiotics as key players in human health and disease improvement. *Food Science and Biotechnology, 33*(9), 2065–2080. https://doi.org/10.1007/s10068-024-01620-1

Kristiawan, M., Chaunier, L., Sandoval, A. J., & Della Valle, G. (2020). *Extrusion-cooking and expansion breakfast cereals and how they are made: Raw materials, processing, and production*. France: Elsevier Inc.. https://doi.org/10.1016/B978-0-12-812043-9.00007-2

Kumar, L., Brennan, M. A., Mason, S. L., Zheng, H., & Brennan, C. S. (2017). Rheological, pasting and microstructural studies of dairy protein–starch interactions and their application in extrusion-based products: A review. *Starch/Staerke, 69*(1–2). https://doi.org/10.1002/star.201600273

Lazou, A. E. (2024). Food extrusion: An advanced process for innovation and novel product development. *Critical Reviews in Food Science and Nutrition, 64*(14), 4532–4560. https://doi.org/10.1080/10408398.2022.2143474

Leonard, W., Zhang, P., Ying, D., & Fang, Z. (2020). Application of extrusion technology in plant food processing byproducts: An overview. *Comprehensive Reviews in Food Science and Food Safety, 19*(1), 218–246. https://doi.org/10.1111/1541-4337.12514

Li, F., Muhmood, A., Akhter, M., Gao, X., Sun, J., Du, Z., Wei, Y., Zhang, T., & Wei, Y. (2023). Characterization, health benefits, and food applications of enzymatic digestion- resistant dextrin: A review. *International Journal of Biological Macromolecules, 253*(Pt 4), Article 126970. https://doi.org/10.1016/j.ijbiomac.2023.126970

Li, H., Zhu, F., & Li, G. (2025). Beverages developed from pseudocereals (Quinoa, buckwheat, and amaranth): Nutritional and functional properties. *Comprehensive Reviews in Food Science and Food Safety, 24*(1). https://doi.org/10.1111/1541-4337.70081

Lin, Z., Cheng, H., He, K., McClements, D. J., Jin, Z., Xu, Z., Meng, M., Peng, X., & Chen, L. (2024). Recent progress in the hydrophobic modification of starch-based films. *Food Hydrocolloids, 151*, Article 109860. https://doi.org/10.1016/j.foodhyd.2024.109860

Liu, H., Zhang, M., Ma, Q., Tian, B., Nie, C., Chen, Z., & Li, J. (2020). Health beneficial effects of resistant starch on diabetes and obesity via regulation of gut microbiota: A review. *Food & Function, 11*(7), 5749–5767. https://doi.org/10.1039/d0fo00855a

Lubowa, M., Yeoh, S. Y., & Easa, A. M. (2018). Textural and physical properties of retort processed rice noodles: Influence of chilling and partial substitution of rice flour with pregelatinized high-amylose maize starch. *Food Science and Technology International, 24*(6), 476–486. https://doi.org/10.1177/1082013218766984

Ma, Z., Hu, X., & Boye, J. I. (2020). Research advances on the formation mechanism of resistant starch type III: A review. *Critical Reviews in Food Science and Nutrition, 60*(2), 276–297. https://doi.org/10.1080/10408398.2018.1523785

Mandliya, S., Vishwakarma, S., Bora, M., Bag, S. K., Dalbhagat, C. G., & Mishra, H. N. (2024). *Ready-to-eat foods development through composite extrusion technological development recent advances in ready-to-eat food technology*. India: CRC Press. https://doi.org/10.1201/9781032622408-3

Mejía Terán, A., & Blanco-Lizarazo, C. M. (2021). Considerations for functional food design based on starch-protein interactions: A systematic review. *Rica International Journal of Food Sciences and Nutrition, 72*(8), 1009–1018. https://doi.org/10.1080/09637486.2021.1905784

Milde, L. B., Ramallo, L. A., & Puppo, M. C. (2012). Gluten-free bread based on tapioca starch: Texture and sensory studies. *Food and Bioprocess Technology, 5*(3), 888–896. https://doi.org/10.1007/s11947-010-0381-x

Nadia, J., Bronlund, J., Singh, R. P., Singh, H., & Bornhorst, G. M. (2021). Structural breakdown of starch-based foods during gastric digestion and its link to glycemic response: In vivo and in vitro considerations. *Comprehensive Reviews in Food Science and Food Safety, 20*(3), 2660–2698. https://doi.org/10.1111/1541-4337.12749

Nandane, A. S., Ganorkar, P. M., Ranveer, R. C., Patil, H., Al-Asmari, F., Sangsawad, P., Nirmal, N., & Ozogul, F. (2025). Impact of extrusion process on the macro- and micro-nutrient in extruded food products: Challenges and future trends. *Food and Bioprocess Technology, 18*(8), 6939–6963. https://doi.org/10.1007/s11947-025-03886-7

Navale, S. A., Swami, S. B., & Thakor, N. J. (2015). Extrusion cooking technology for foods: A review. *Journal of Ready to Eat Food., 2*(3), 66–80.

Nawaz, H., Waheed, R., Nawaz, M., & Shahwar, D. (2020). *Physical and chemical modifications in starch structure and reactivity*. IntechOpen. https://doi.org/10.5772/intechopen.88870

Nikinmaa, M., Zehnder-Wyss, O., Nyström, L., & Sozer, N. (2023). Effect of extrusion processing parameters on structure, texture and dietary fibre composition of directly expanded wholegrain oat-based matrices. *Lebensmittel-Wissenschaft & Technologie, 184*, Article 114972. https://doi.org/10.1016/j.lwt.2023.114972

Obadi, M., & Xu, B. (2021). Review on the physicochemical properties, modifications, and applications of starches and its common modified forms used in noodle products. *Food Hydrocolloids, 112*, Article 106286. https://doi.org/10.1016/j.foodhyd.2020.106286

Offiah, V., Kontogiorgos, V., & Falade, K. O. (2019). Extrusion processing of raw food materials and by-products: A review. *Critical Reviews in Food Science and Nutrition, 59*(18), 2979–2998. https://doi.org/10.1080/10408398.2018.1480007

Ortiz-Cruz, R. A., Ramírez-Wong, B., Ledesma-Osuna, A. I., Torres-Chávez, P. I., Sánchez-Machado, D. I., Montaño-Leyva, B., López-Cervantes, J., & Gutiérrez-Dorado, R. (2020). Effect of extrusion processing conditions on the phenolic compound content and antioxidant capacity of sorghum (Sorghum bicolor (L.) moench) bran. *Plant Foods for Human Nutrition, 75*(2), 252–257. https://doi.org/10.1007/s11130-020-00810-6

Ozolina, K., Sarenkova, I., & Muizniece-Brasava, S. (2023). Latvia university of life sciences and technologies Latvia the anti-nutritional factors of legumes and their treatment possibilities: A review. *Research for Rural Development, 38*, 68–76. https://doi.org/10.22616/RRD.29.2023.010

Pan, W., Wang, S., Huang, T., Jiang, X., & Jian, R. (2025). Effect of screw design on the mixing and heat efficiencies in an extrusion system by tailoring ductile forming with field synergy principle. *Applied Thermal Engineering, 262*, Article 125291. https://doi.org/10.1016/j.applthermaleng.2024.125291

Pandey, V. K., Ashoka, S., Nath, P. C., Singh, K., Dar, A. H., & Rustagi, S. (2025). Exploring the significance of resistant starch in ready-to-eat meals for enhanced functioning of gut microbiota: A comprehensive review. *Food Science and Biotechnology, 26*, 1–20. https://doi.org/10.1007/s10068-025-01978-w

Patel, J. R., Patel, A. A., & Singh, A. K. (2016). Production of a protein-rich extruded snack base using tapioca starch, sorghum flour and casein. *Journal of Food Science and Technology, 53*(1), 71–87. https://doi.org/10.1007/s13197-015-2012-z

Paternina-Contreras, A. L., Andrade-Pizarro, R. D., & Figueroa-Flórez, J. A. (2025). Physical modification of starch in plant-based flours: Structural, physicochemical, and pasting property changes and potential applications in baked and extruded products. *Comprehensive Reviews in Food Science and Food Safety, 24*(3). https://doi.org/10.1111/1541-4337.70184

Patil, H., Tiwari, R. V., & Repka, M. A. (2016). Hot-melt extrusion: From theory to application in pharmaceutical formulation. *AAPS PharmSciTech, 17* (1), 20–42. https://doi.org/10.1208/s12249-015-0360-7

Pęksa, A., Kita, A., Carbonell-Barrachina, A. A., Miedzianka, J., Kolniak-Ostek, J., Tajner-Czopek, A., Rytel, E., Siwek, A., Miarka, D., & Drożdż, W. (2016). Sensory attributes and physicochemical features of corn snacks as affected by different flour types and extrusion conditions. *LWT - Food Science and Technology, 72*, 26–36. https://doi.org/10.1016/j.lwt.2016.04.034

Pismag, R., Pico, J., Fernández, A., Hoyos, J. L., & Martinez, M. M. (2023). α-Amylase reactive extrusion enhances the protein digestibility of saponin-free quinoa flour while preserving its total phenolic content. *Innovative Food Science and Emerging Technologies, 88*. https://doi.org/10.1016/j.ifset.2023.103448

Pongjaruvat, W., Methacanon, P., Seetapan, N., Fuongfuchat, A., & Gamonpilas, C. (2014). Influence of pregelatinised tapioca starch and transglutaminase on dough rheology and quality of gluten-free jasmine rice breads. *Food Hydrocolloids, 36*, 143–150. https://doi.org/10.1016/j.foodhyd.2013.09.004

Qiu, C., Hu, H., Chen, B., Lin, Q., Ji, H., & Jin, Z. (2024). Research progress on the physicochemical properties of starch-based foods by extrusion processing. *Foods, 13*(22), 3677. https://doi.org/10.3390/foods13223677

Raigond, P., Ezekiel, R., & Raigond, B. (2015). Resistant starch in food: A review. *Journal of the Science of Food and Agriculture, 95*(10), 1968–1978. https://doi.org/10.1002/jsfa.6966

Roman, L., & Martinez, M. M. (2019). Structural basis of resistant starch (RS) in bread: Natural and commercial alternatives. *Foods, 8*(7). https://doi.org/10.3390/foods8070267

Shah, F.u. H., Sharif, M. K., Bashir, S., & Ahsan, F. (2019). Role of healthy extruded snacks to mitigate malnutrition. *Food Reviews International, 35* (4), 299–323. https://doi.org/10.1080/87559129.2018.1542534

Shukla, A. D., Hossain, S. A., Kumari, A., Rao, K. J., & Bharti, B. K. (2022). Comparisons of baking and frying effects on food. *Indian Food Indus Mag., 4*, 64–75.

Subhasri, D., Moses, J. A., & Anandharamakrishnan, C. (2022). *New food structures and their influence on nutrition, health and well-being food structure engineering and design for improved nutrition, health and well-being*. India: Elsevier. https://doi.org/10.1016/B978-0-323-85513-6.00014-1

Sun, C., & Fang, Y. (2021). *Replacement of fat or starch food hydrocolloids: Functionalities and applications*. China: Springer Singapore. https://doi.org/10.1007/978-981-16-0320-4_12

Sun, C., Fu, J., Chang, Y., Li, S., & Fang, Y. (2022). Structure design for improving the characteristic attributes of extruded plant-based meat analogues. *Food Biophysics, 17*(2), 137–149. https://doi.org/10.1007/s11483-021-09692-w

Tas, A. A., & Shah, A. U. (2021). The replacement of cereals by legumes in extruded snack foods: Science, technology and challenges. *Trends in Food Science & Technology, 116*, 701–711. https://doi.org/10.1016/j.tifs.2021.08.016

Thondre, P. S., & Clegg, M. E. (2019). *Reformulation of foods for weight loss: A focus on carbohydrates and fats reformulation as a strategy for developing healthier food products: Challenges, recent developments and future prospects*. United Kingdom: Springer International Publishing. https://doi.org/10.1007/978-3-030-23621-2_2

Walsh, S. K., Lucey, A., Walter, J., Zannini, E., & Arendt, E. K. (2022). Resistant starch—An accessible fiber ingredient acceptable to the Western palate. *Comprehensive Reviews in Food Science and Food Safety, 21*(3), 2930–2955. https://doi.org/10.1111/1541-4337.12955

Wang, B., Dong, Y., Fang, Y., Gao, W., Kang, X., Liu, P., Yan, S., Cui, B., & Abd El-Aty, A. M. (2022). Effects of different moisture contents on the structure and properties of corn starch during extrusion. *Food Chemistry, 368*, Article 130804. https://doi.org/10.1016/j.foodchem.2021.130804

Wang, K., Ma, J., Wang, L., Yue, X., Ma, X., Huo, J., Duan, Y., Wang, P., Yu, X., & Xiao, Z. (2024). Insight into the relationship between the starch crystalline structure and textural quality and physicochemical properties of reconstituted rice: Influence of feed moisture content. *International Journal of Biological Macromolecules, 280*(Pt 1), Article 135758. https://doi.org/10.1016/j.ijbiomac.2024.135758

Wang, J., Shang, M., Li, X., Sang, S., Julian McClements, D., Chen, L., Long, J., Jiao, A., Ji, H., Jin, Z., & Qiu, C. (2023). Polysaccharide-based colloids as fat replacers in reduced-fat foods. *Trends in Food Science & Technology, 141*, Article 104195. https://doi.org/10.1016/j.tifs.2023.104195

Witczak, M., Ziobro, R., Juszczak, L., & Korus, J. (2016). Starch and starch derivatives in gluten-free systems – A review. *Journal of Cereal Science, 67*, 46–57. https://doi.org/10.1016/j.jcs.2015.07.007

Wittek, P., Ellwanger, F., Karbstein, H. P., & Emin, M. A. (2021). Morphology development and flow characteristics during high moisture extrusion of a plant-based meat analogue. *Foods, 10*(8). https://doi.org/10.3390/foods10081753

Yadav, G. P., Dalbhagat, C. G., & Mishra, H. N. (2022). Effects of extrusion process parameters on cooking characteristics and physicochemical, textural, thermal, pasting, microstructure, and nutritional properties of millet-based extruded products: A review. *Journal of Food Process Engineering, 45*(9). https://doi.org/10.1111/jfpe.14106

Yan, R., Andrew, L., Marlow, E., Kunaratnam, K., Devine, A., Dunican, I., & Christophersen, C. (2021). Dietary fibre intervention for gut microbiota, sleep and mental health in adults with irritable bowel syndrome: A scoping review. *Nutrients, 13*(7). https://doi.org/10.3390/nu13072159

Ye, S., Kulkarni, R. A. P., Nawaz, M. A., Ali, I., Yadav, D., Moore, C., & Suleria, H. A. R. (2025). Enhancing jackfruit shelf-life through intelligent packaging: A review. *Food Reviews International, 42*, 1–27. https://doi.org/10.1080/87559129.2025.2502432

Ye, J., Liu, C., Luo, S., Hu, X., & McClements, D. J. (2018). Modification of the digestibility of extruded rice starch by enzyme treatment (β-amylolysis): An in vitro study. *Food Research International, 111*, 590–596. https://doi.org/10.1016/j.foodres.2018.06.002

Zailani, M. A., Kamilah, H., Husaini, A., Awang Seruji, A. Z. R., & Sarbini, S. R. (2023). Starch modifications via physical treatments and the potential in improving resistant starch content. *Starch/Staerke, 75*(1–2). https://doi.org/10.1002/star.202200146

Zhang, B., Qiao, D., Zhao, S., Lin, Q., Wang, J., & Xie, F. (2021). Starch-based food matrices containing protein: Recent understanding of morphology, structure, and properties. *Trends in Food Science & Technology, 114*, 212–231. https://doi.org/10.1016/j.tifs.2021.05.033

Zhang, L., Zhao, J., Li, F., Jiao, X., Zhang, Y., Yang, B., & Li, Q. (2024). Insight to starch retrogradation through fine structure models: A review. *International Journal of Biological Macromolecules, 273*(Pt 1), Article 132765. https://doi.org/10.1016/j.ijbiomac.2024.132765

Zhang, Z., Zhu, M., Xing, B., Liang, Y., Zou, L., Li, M., Fan, X., Ren, G., Zhang, L., & Qin, P. (2023). Effects of extrusion on structural properties, physicochemical properties and in vitro starch digestibility of tartary buckwheat flour. *Food Hydrocolloids, 135*, Article 108197. https://doi.org/10.1016/j.foodhyd.2022.108197

Zhao, Y., Li, K., Zhang, X., Zhang, T., Zhao, J., Jiang, L., & Sui, X. (2024). Protein blend extrusion: Crafting meat analogues with varied textural structures and characteristics. *Food Chemistry, 460*(Pt 3), Article 140709. https://doi.org/10.1016/j.foodchem.2024.140709

Chapter 13

Enhancing texture and mouthfeel: the role of starches in soups, sauces, and salad dressing

Rakesh Raghunathan[1], Rohan M. Shah[2], Snehal R. Jadhav[3] and Rajaraman Eri[4]

[1]International Rice Research Institute, ISARC, Centre of Excellence in Rice Value Addition (CERVA), Varanasi, Uttar Pradesh, India; [2]Department of Chemistry and Biotechnology, School of Science, Computing and Emerging Technologies, Iverson Health Innovation Research Institute, Swinburne University of Technology, Melbourne, VIC, Australia; [3]Consumer Analytical Safety Sensory (CASS) Food Research Centre, School of Exercise and Nutrition Sciences, Deakin University, Melbourne, VIC, Australia; [4]School of Science, STEM College, RMIT University, Melbourne, VIC, Australia

1. Introduction

Humans consume food not only for energy and nutrition but also for pleasurable sensory experiences. The appreciation for food is governed primarily by texture and mouthfeel. While *texture* is generally used for solid and semisolid foods, *mouthfeel* comprises all tactile properties perceived from the time the food enters the mouth until swallowing. These sensations depend on the properties of the food during early consumption but extend until food breakdown and subsequent changes during oral processing (Chen, 2009; Guinard & Mazzucchelli, 1996; Janssen et al., 2007). Therefore, both texture and other sensory properties (appearance, aroma, and taste) are key factors that drive the consumers' preference and acceptance. Because oral processing is dynamic, early-stage textural perceptions are governed by bulk-phase rheology, whereas latter-stage perceptions depend on product-saliva interactions. Hence, sensory aspects of food cannot be explained by rheological properties alone, and tribological data are necessary to understand lubrication and friction mechanisms (Xu et al., 2022).

Food manufacturers and researchers strive to gain a comprehensive grasp of the role that food structure plays in oral processing and how these features correlate with textural properties. This understandings enable the design of products with desired functionalities (Koç et al., 2013). For instance, when utilizing starch in sauces and soups, the primary objective is to preserve the granular structure of starch for effective thickening. In powdered sauce/soup products, only minimum processing occurs to achieve the desired viscosity; thus, it often does not need modification. However, simmered soup requires some modification because it needs initial heating followed by a holding time (Imeson, 2011). In this case, starch modification can help keep the granules intact while preventing retrogradation (the holding period might lead to skin formation on the soup surface).

Starch has several key functions, such as gelling, thickening, emulsifying, and enhancing mouthfeel. Optimizing starch usage is essential for desired final product properties. In soups and sauces, it acts as a thickening agent, with potato, tapioca, and waxy maize starches being commonly preferred sources for their stability in texture changes. However, for puddings and cheese, dent corn starch is preferred for gelling. In products such as salad dressings, a combination of gelling and non-gelling starches can provide slight cuttability. Starch is also used in canned foods to stabilize, thicken, and enhance the mouthfeel of soups, gravies, and sauces. Spoonable salad dressings are prepared by cooking starch with salt, vinegar, sugar, and water to form a paste which is then cooled. Oil and an emulsifier (e.g., egg yolk) are added to form a stable emulsion. Apart from thickening and stabilizing, starch can act as a fat replacer enhancing mouthfeel and improving flowability in salad dressings. Modified waxy maize starch is commonly used due to its acid and shear tolerance, whereas tapioca and potato starches typically do not tolerate such conditions. Hence, the starch selection and modification methods

Native and Modified Starches for Food Applications. https://doi.org/10.1016/B978-0-443-27431-2.00005-5

for salad dressing are critical because they involve high temperature, shear and low pH. Limited water solubility of native starch means upon heating, granules swell, fragment, and solubilize, depending on heat and shear. The balance between swollen/fragmented granules and soluble polymers is crucial for the appearance, stability and texture of food products. Starch modification endows the desired appearance and texture despite stress conditions encountered in processing. Modified starches can help maintain smoothness, clarity and pleasant mouthfeel for longer durations. Additionally, modification aids film formation and emulsification (Biliaderis, 1991). Starch is used in foods as a thickener, binder, stabilizer and for providing structure and shape in baked foods and so the study of starch modification in improving the textural and sensory attributes of final product becomes important. This chapter focuses primarily on the advantages of modified starches, types of modification, and proposals for future research.

2. Physical modification

As native starch has inherent limitations, it is modified physically or chemically to produce more desirable properties. Although chemically modified starches find industrial applications, the pharmaceutical and food sectors often favor reduced chemical residues. Therefore, physical modification methods provide a safe choice because they leave no chemical residues.

2.1 Hydrothermal treatments

One of the commonly used physical modifications is a hydrothermal treatment, which includes heat moisture treatment (HMT) and annealing (ANN). Both processes modify starch properties without damaging its granular structure. Key controlling parameters include starch: water ratio, holding time and temperature. The two processes differ in their water levels and temperature. ANN typically uses 50%–60% water at temperatures below gelatinization, whereas HMT occurs with 10%–30% water at high temperatures (90–120°C). HMT increases the gelatinization transition temperatures, decreases swelling, and reduces amylose leaching, also improving thermal stability across botanical origins. Fig. 13.1 below depicts SEM images of palmyra (*Borassus flabellifer*) starch granules including, NTPS (native palmyra starch granules) and HMT modified granules with different moisture levels (15%, 20% and 25%). In this particular instance, the appearance of the native starch was not altered after the HMT-15 treatment; however, HMT 20 and 25 granules exhibited some disruption to the granular structure.

Fig. 13.1 A study focused on HMT in rice starches (high-, medium- and low-amylose varieties) equilibrating them at 15%, 20% and 25% moisture and heating for 1 h at 110°C. Effects of HMT were strongest in the high-amylose starch, showing lower breakdown and setback values. Hence, HMT rice starch is suitable for products demanding low swelling, low viscosity, and good thermal stability (Zavareze et al., 2010). A key advantage of hydrothermal treatments is the resultant waxy appearance, which enhances stability by limiting retrogradation during storage. For example, unfermented cassava starch was annealed at 50°C (24–192 h, with a starch: water ratio of 1:5), leading to lower peak viscosity, implying less amylose leaching. Strengthening of amylose-amylose (AM-AM) interactions reduced solubility and swelling. Such lower swelling, lower peak viscosity and reduced setback are useful in sauces, soups and frozen foods (Gomes et al., 2004).

African yam bean, cultivated in Africa, remains underutilized. Its seeds (~50% carbohydrate, predominantly starch) were subjected to HMT at 18%, 21%, 24% and 27% moisture, plus an ANN condition, significantly decreasing breakdown and setback values. Structural reorganization in modified starches improves shear resistance. HMT also restricted swelling and decreased solubility. In sauces and gravies, higher water absorption is favorable and the lower retrogradation suits sauces and soups by preventing viscosity loss or precipitation (Adebowale et al., 2009). HMT results in structural rearrangements of starch chains that modify physicochemical properties. The increased shear and thermal stability are useful in products such as salad dressings. For instance, HMT (25% moisture, 3 h and 110°C) applied to pearl millet starch improved thermostability and reduced retrogradation, suiting products needing higher solid content with limited thickening (Sandhu et al., 2020).

As opposed to packaged soups and sauces, canned food products demand heat-stable thickeners, as they are sterilized at high temperatures. HMT-modified potato starch has been shown to provide improved thickening in tomato sauce at 121°C for 1 h, producing greater stability and sensory properties. In a study conducted by Subroto et al. (2020), an optimal level of 1.5% HMT potato starch offered both stability and good organoleptic attributes. The results indicated a reduction in viscosity upon addition of the modified potato starch. In an independent study, Peruvian carrot and cassava starches subjected to HMT (90°C, 8 h, 30%–35% moisture) showed enhanced thermal stability and partial gelatinization, also suitable for canned soups (Moraes et al., 2014).

FIGURE 13.1 SEM images of native and HMT treated palmyra granules. *Adapted from Medha et al. (2025).*

Sweet potato is a traditional crop grown in Sri Lanka and is rich in starch, vitamins and minerals. Applying HMT (30% moisture, 6 h, 85°C) on different sweet potato varieties (Swp 1 (Wariyapola Red), Swp 3 (Wariyapola white), Swp 4 (Pallepola sample), Swp 5 (Malaysian sample) and Swp 7 (CARI 273)) increased viscosity in most samples (except Swp 3). The authors hypothesized that the high temperature treatment would have modified the crystalline arrangement of starch, thus strengthening its hydrogen bonding with water. Sensory analysis indicated that some varieties (Swp 4 and Swp 5) offered superior appearance, aroma, taste, and overall acceptability. Hence, physically modified sweet potato starch serves as an excellent thickener in vegetable soup mixture, sometimes surpassing corn starch (Senanayake et al., 2014). Tartary buckwheat starch (TBS) also benefits from HMT (20%–35% moisture), showing a decrease in swelling power and solubility, while improving thermal stability and lowering hardness, helpful for the production of soups (Liu et al., 2015).

Very often, glutinous rice is the preferred choice for thickening soups and desserts because of its opacity and stickiness. To reduce damage during freezing (20°C for 6 months), in a research study by Shi et al. (2021), glutinous rice flour used in dumplings was annealed under three conditions (ANN1 = 55% moisture, 4 h, 50°C; ANN2 = 55% moisture, 6 h, 50°C; ANN3 = 50% moisture, 6 h, 55°C). ANN1 increased crystallinity, improved soup transmittance, and ANN2 had the most desirable properties but the thixotropic property was found to decrease in ANN3.

2.2 Pregelatinization

Pregelatinization is frequently employed to increase the absorption and swelling capacity of starch, leading to higher viscosity. The three common methods for pregelatinized starches are spray drying, drum drying, and extrusion. Spray drying rapidly converts starch suspension into (semi)amorphous particles, preserving granular integrity more effectively than other methods. Spray drying often simplifies post-processing by eliminating extra screening and crushing procedures (Ma et al., 2022). On the other hand, extrusion employs high temperature, pressure, and shear, leading to granule disruption, covalent bond breakage, and partial depolymerization. Drum drying distributes starch slurry on a heated drum, followed by steam injection. The evaporation of water leads to the formation of a starch film that is scraped off into granules or flakes.

Pregelatinization partially or fully gelatinizes native starch, forming a gel at ambient temperature. During gelatinization, inter- and intramolecular bonds rearrange between starch and water, disrupting crystallites. Pregelatinized starches are used as thickeners in low-heat or instant foods (soups and desserts). Their preparation is simple, requiring only hot water or brief high shear mixing. For instance, sweet potato starch can be heated between 57 and 69°C using spray drying. The optimal results in improving the starch functionality were obtained by preheating at ~67°C, suitable for products requiring greater thickness (Dos Santos et al., 2020). Pregelatinized starches exhibit rapid water solubility and high viscosity, which is a key to instant products. However, powdered starches may clump in cold water. Agglomeration can alleviate this issue by increasing particle size. For instance, fluidized bed agglomeration with 0%–40% maltodextrin on pregelatinized potato starch improved its flowability and porosity, thus increasing its solubility (Lee & Yoo, 2023).

The influence of pregelatinized potato starch concentration (PSC, 0–5 wt%) on the rheology of salad dressings stabilized by dried egg yolk (DEY) or sodium caseinate (SC) was studied (Bortnowska et al., 2014). Both systems showed similar rheological trends. Increasing PSC from 0% to 5% produced a pronounced increase in shear resistance, reflecting gel formation via intermolecular amylose-amylopectin interactions and starch-lipid-protein complexes. Thus, 5% potato starch stabilized low-fat salad dressings with sufficient creaming and fat-holding capacities (Bortnowska et al., 2014).

A recent study prepared pregelatinized rice starch (PRS) co-spray dried with various hydrocolloids (sodium alginate, xanthan gum, pectin and carrageenan). The cold paste viscosity of these PRS-hydrocolloid mixtures was much higher than that of PRS alone, presumably due to starch-hydrocolloid interactions. Xanthan gum and carrageenan produced the greatest thickening. Thus, PRS-hydrocolloid blends can be applied to baby foods, puddings, and pastry dressings (He et al., 2020). Drum drying can also be used in the case of rice starch pregelatinization. Three rice starches with varying amylose percentages (glutinous: 1.4%, Jasmine: 15%, and Chiang: 20.2%) were pregelatinized at 110, 117, and 123°C using a double drum dryer. Glutinous and Jasmine starch formed an amorphous structure, creating cold pastes, whereas high-amylose Chiang rice showed hot-paste viscosity and amylose-lipid complexation. Thus, additional crosslinking may be required for high-amylose starches (Nakorn et al., 2009). Wheat starch is widely used globally, and one research study used a twin drum dryer to produce pregelatinized wheat starch (10% w/w slurry) (Majzoobi et al., 2011). The final product was milled and sieved (150–250 μm). SEM images showed structural disruption, and the resulting pregelatinized starch exhibited cold water viscosity at 25°C, improved water absorption, and swelling (Majzoobi et al., 2011).

Palmyrah tuber is an underutilized starch source derived from the palmyrah palm in Sri Lanka. Acid modification, pregelatinization and dextrinization were used to modify palmyrah palm in a recent study (Sobini et al., 2022). In comparison with pregelatinized starch, Palmyrah dextrinized starch (PDS) had minimal swelling but excessive syneresis, making it unsuitable for products requiring low-temperature storage. Thus, pregelatinized starch was preferred as a soup thickener (Sobini et al., 2022). Similarly, a study evaluating pregelatinized starch (PGS, 2%–8% w/w) in glutinous rice flour dumplings found that PGS formed an elastic gel network, increasing soup transparency (70%–92%) and thickening (Wang, Ren, et al., 2019). For the elderly, an instant soup from riceberry flour was formulated. Drum drying at 110°C, 120°C, and 130°C showed that higher temperatures reduced total phenol content and antioxidant capacity (due to degradation of anthocyanins). Hence, 110°C was recommended to maintain phenolics. High breakdown viscosity indicates reduced tolerance to heat and shear (Wiriyawattana et al., 2018). The functional properties of pregelatinized starches are beneficial in quality improvement of the products and so the manufacturers must exercise caution while selecting the optimal method for starch modification.

Developing a gluten-free (GF) diet for celiac patients often involves legumes, which provide key nutritional and functional benefits. Mung bean, for instance, is rich in dietary fiber, protein, bioactive compounds, vitamins and minerals. Partial replacement of rice flour with mung bean flour in a drum-dried GF tarhana (instant soup) improved radical scavenging, total phenolics, water absorption, and dietary levels (Taşkın & Savlak, 2022). The functional aspects of pregelatinized starch can be modified by other components in the food matrix. One study investigated the impact of glucose and sucrose (10%–40% of starch, dry weight basis) on pregelatinized maize starch. Sugars bind water, limiting starch chain mobility and increasing viscosity. Sugars (with more equatorial hydroxyl groups) exert a stronger effect than glucose, improving freeze-thaw stability and reducing syneresis. Sucrose, being a disaccharide can interact more with water molecules than glucose which is a monosaccharide. These features of sucrose (large molecular weight and more equatorial hydroxyl groups) allow food manufacturers to formulate pregelatinized starch in sauces and instant desserts (Hedayati et al., 2016).

2.3 Granular cold-water swelling

Granular cold-water swelling (GCWS) starch is a physically modified starch designed to retain its granular structure while exhibiting enhanced ability to swell and hydrate in cold or ambient water. This unique property allows GCWS starch to mimic the functional characteristics of gelatinized starch without requiring extensive heating. The controlled physical modifications imparted to GCWS starch make it ideal for a variety of food applications, particularly in instant and convenience products where rapid hydration and specific textural properties are desirable. GCWS starch is produced through advanced physical modification techniques that retain the native chemical composition of starch while partially disrupting its granular architecture (Qian et al., 2019). Key methods include spray drying, alcoholic-alkaline treatment, alcoholic treatment under elevated pressure, ball milling, freeze-drying and ultrasonication.

Spray drying is one of the most used methods to produce GCWS starch. In this process, a starch slurry (20%–40% solids) is atomized into a chamber where it is exposed to steam at a high temperature (~160°C) (Pitchon et al., 1981). The starch granules are rapidly gelatinized at a moisture content of around 60% and subsequently dried as they move through the chamber. By adjusting the droplet size or slurry concentration, different granular sizes can be achieved. This method is efficient and scalable, making it suitable for industrial applications. Developed as early as the 1930s, aqueous monohydric alcohol treatment utilizes a mixture of starch, water, and monohydric alcohols such as ethanol, propanol, or isopropanol. The slurry is heated at 140–180°C under high pressure (35–52 atm) for 1–30 min (Eastman & Moore, 1984). During this process, the double-helical structure of amylopectin is converted to single helices, which form V-complexes with alcohol molecules. The evaporation of alcohol leaves behind granules with micro-channels, enhancing water solubility.

Alcoholic-alkaline treatment is widely used due to its simplicity, cost-effectiveness, and efficiency. Starch is dispersed in an alcohol-water mixture (e.g., ethanol or propanol) with an alkali such as sodium hydroxide or potassium hydroxide. The process is conducted at ambient pressure and a temperature of 20–40°C for 20–60 min (Chen & Jane, 1994). The alkali induces swelling by creating repulsive forces between negatively charged starch molecules, while the alcohol stabilizes the dissociated chains and prevents excessive swelling. This results in deformed granules with high solubility and cold-water swelling properties. Ball milling involves mechanical forces such as shear, abrasion, and friction to disrupt the crystalline structure of starch granules (Huang et al., 2008). This environmentally friendly method enhances the interactions between water molecules and free hydroxyl groups, thereby increasing cold-water solubility and swelling power. Studies have shown that corn and tapioca starches treated with ball milling achieve water solubility rates of up to 95% (Huang et al., 2007). Surface modification involves grafting starch granules with water-soluble polymers such as polyethylene glycol (Wang et al., 2011). This technique preserves the granules' native features, including size and shape, while significantly enhancing cold-water solubility to nearly 90%. This method is particularly useful for applications requiring high solubility without altering granular morphology.

In the sodium hydroxide-urea method, starch is stirred in a solution of 14% NaOH and 12% urea for 10 min at room temperature (Chen et al., 2019a, 2019b). The strong alkaline and urea combination breaks hydrogen bonds within the granules, enhancing water interactions and cold-water solubility. This technique is particularly effective for high-amylose and waxy starches. Recent advancements have focused on combining traditional methods with modern techniques to improve GCWS starch properties. Fig. 13.2 below depicts an example workflow used by Liu et al. (2023) to modify cassava starch. The combinatorial approach improved starch solubility to 92.45%.

GCWS starch is first produced using the alcoholic-alkaline method, followed by ultrasonic treatment alone or a combination of it (Liu et al., 2023; Zhu et al., 2017). Ultrasonic treatment significantly enhances water solubility and cold-water swelling properties, particularly for corn and rice starches. The ultrasonic waves disrupt the granular structure further, improving hydration and reducing processing time. Whereas microwave treatment can alter the molecular structure of starch thus affecting its particle size, crystalline properties, thermal and swelling properties, its overall solubility, viscosity, and digestibility. Considering that technology providers can now provide a combination of both treatments; it has been used widely for starch medication.

Alternatively, after producing GCWS starch through alcoholic-alkaline treatment, a vacuum freezer is used to recover the starch. The vacuum freezing process, conducted at 20–100 Pa for 6–18 h, enhances the structural integrity and water solubility of the starch. This method is effective for corn, potato, and wheat starches, yielding a product with superior freeze-thaw stability and reduced retrogradation (Zhu et al., 2017). In another variation, the method begins with enzymatic treatment using a combination of α-amylase and amyloglucosidase (3:1) at 50°C for 12 h, pH 10. The resulting porous starch is then modified using the alcoholic-alkaline method. This dual treatment enhances the water solubility, encapsulation properties, and overall functionality of the starch, making it particularly suitable for waxy corn starch (He et al., 2020).

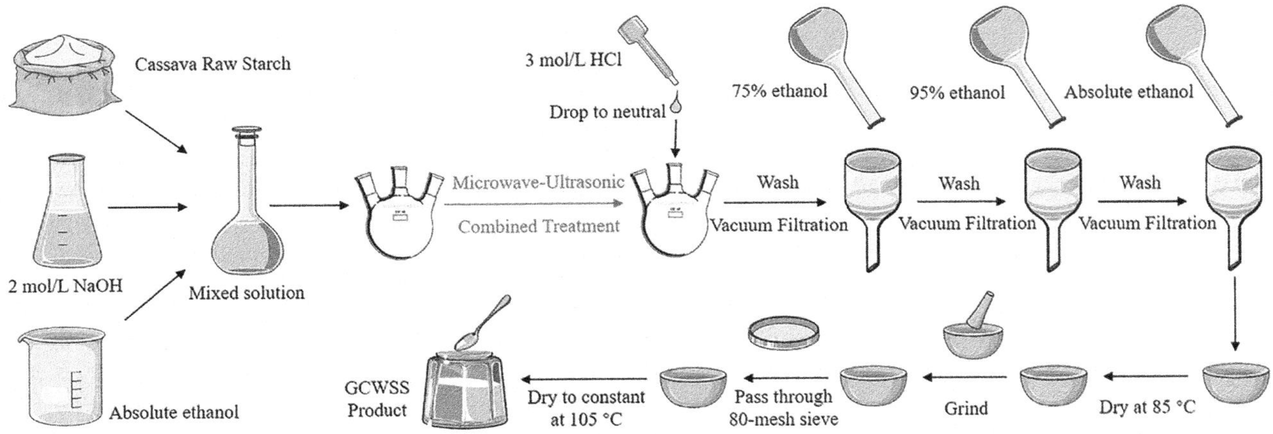

FIGURE 13.2 Cassava starch modification using a combination of alcohol alkali treatment along with combined microwave and ultrasonic treatment. *Adapted from Liu, Y., Zhu, L., Yu, X., Xu, Y. and Zhong, T. (2023). A Combined Ultrasound–Microwave Assisted Alcoholic–Alkaline Method for Preparation of Granulated Cold-Water-Soluble Starch. Starch-Stärke, 75(11–12).*

In yet another approach, starch (100 g) is first treated with citrates or lactates (20%–40% in water) at 25°C for 12 h. This pre-treatment improves the starch's structural flexibility and interaction with modifying agents. The starch is then subjected to alcoholic-alkaline treatment, yielding a product with enhanced functional properties such as improved gelation and cold-water solubility. This method has been effectively applied to rice starch (Butt et al., 2019). While combination treatments offer promise, further research is needed to optimize these methods for tailored properties. GCWS starch is versatile and finds applications in various industries, particularly in food manufacturing. Its ability to hydrate without heating makes GCWS starch ideal for instant foods such as soups, sauces, baby foods, and cold desserts. It also functions as a thickener, stabilizer, and viscosifier in salad dressings and pudding mixes. GCWS starch is used to encapsulate bioactive compounds, nutrients, and gases (e.g., ethylene for controlled fruit ripening). Its encapsulation efficiency and stability make it suitable for functional food applications. Enhanced freeze-thaw stability allows GCWS starch to maintain texture and moisture in frozen products. Its ability to form rigid gels with glucose or sucrose syrups is exploited in gum candies, pie fillings, and other confectionery products.

3. Chemical modification

Chemical modifications address the limitations of native starch and typically involve the introduction of new chemical groups or the formation of covalent linkages. These changes can alter the granule structure, intrinsic polymer chain mobility, and intermolecular bonding. Even though chemically modified starches are widely used across industries, stringent regulations in food and pharmaceutical sectors limit or govern the level of chemical reagents/residues allowed in food applications.

3.1 Acetylation

Acetylation introduces acetyl groups (-$COCH_3$) into starch, disrupting inter- and intramolecular hydrogen bonding, which weakens the granule structure and improves chain mobility in the amorphous region (Fig. 13.3).

Consequently, the modified starch shows higher swelling power, solubility, and freeze-thaw stability, while reducing retrogradation. Acetylation depends on pH, temperature, time, and reactant concentrations (e.g., acetic anhydride) (Han et al., 2012). Acetylated starches are categorized based on their degree of substitution (DS). Low-DS (<0.2) acetylated starches are utilized in foods (salad dressings, gravies, cakes, sauces, bakery products, soups), medium-DS (<1) can substitute cellulose acetate, and high-DS (2–3) find use in nonfood sectors (biodegradable plastics development, coatings, pharmaceuticals, adhesives) (Trela et al., 2020). In practice, low-degree acetylation (i.e., DS < 0.2) is permissible in foods, whereas higher DS leads to less frequent food use (but more industrial applications) (Han et al., 2012).

A research study by Arocas et al. (2009) compared the freeze/thaw versus thermal stabilities of white sauces made from native versus modified corn starches (acetylated distarch adipate and hydroxypropyl distarch phosphate). The modified starch maintained viscoelastic properties under freeze/thaw cycles or heating to 80°C, whereas native starches

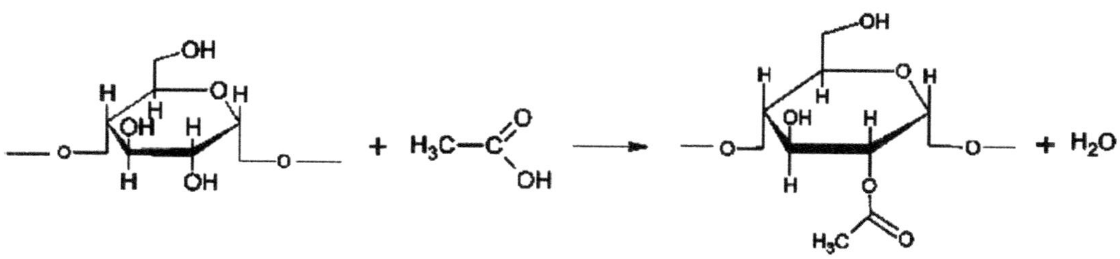

FIGURE 13.3 Acetylation of starch. *Adapted from Morán et al. (2011).*

showed syneresis. Similarly, Betancur et al. (1997) acetylated *Canavalia ensiformis* starch, decreasing gelatinization temperature, preventing retrogradation, and increasing viscosity, useful for sauces, ice creams and frozen products. Likewise (Cisse et al., 2023), acetylated purple maize starches, improving swelling, clarity and solubility, suitable for sauces, ice creams and bakery items.

Acetylation of rice starch at 5, 10 and 20 g acetic anhydride/100 g of starch increased peak viscosity, final viscosity, breakdown, and pasting temperature, but reduced gel hardness. The Food and Drug Administration (FDA) requires acetyl levels <2.5 g/100 g in food applications. Because acetylated starch provides higher final viscosity than native starch, there is a clear advantage for thickening (soups, sauces and broths) after heating (Colussi et al., 2015). Acetylated distarch phosphate (ADSP) and acetylated distarch adipate (ADA) yield crosslinked, stabilized starches resistant to high temperature, shear and low pH, ideal for sauces and ketchups (Gałkowska et al., 2013). In addition to this, pectin is also incorporated as it is a good gelling and thickening agent. The interactions occurring from pectin and modified starch have a crucial role to play in texture stabilization and flavor development of food products. The modified starch-pectin-sucrose gels exhibited greater peak and final viscosities and thus were found to also positively influence the rheological properties.

Treatment of canna (*Canna edulis*) starches with various concentrations (5%, seven% and 9% w/w) of acetic anhydride was performed was performed in a study (Saartrat et al., 2005). The Rapid Visco Analyzer data (RVA) data of native and acetylated starches (6%) were compared and it was observed that the pasting temperature decreased after acetylation and this could be explained by the incorporation of acetyl groups in the amorphous region. The presence of acetyl groups prevented the hydrogen bonding formation and so the extent of starch swelling was faster than native starches. This property is useful for products requiring a gel formation at low temperatures. Another interesting aspect is the lower setback viscosities (15%–46%) of acetylated starches when compared with the native ones. The reaggregation occurring between the amylose chains was significantly reduced after acetylation. The acetyl groups hindered the realignment of starch molecules after cooling. Visual observation indicated a soft texture and good clarity of acetylated pastes/gels and therefore canna starch acetate could be used in sauces, jellies, ice creams etc. for stabilization and thickening.

Avocado seeds possess high levels of starch but they remain unutilized and so a study directed toward characterizing the properties of native and acetylated avocado starch. The data indicated a significant improvement in the functional properties of acetylated starch when compared to the native ones. The breakdown value exhibited by acetylated starch (116 cP) was lower when compared to native starch (121 cP) indicating its better stability. Also, the setback values of the acetylated starch (303.58 cP) were lower than the native starch (304.08 cP). This reiterates the fact that acetylation improves the viscosity and breakdown stability. The retrogradation rates are reduced after acetylation. In addition to that, acetylated starches showed a high solubility index and greater oil absorption when compared to the native starch. These factors contribute to the inclusion of acetylated avocado seed starch in desserts, puddings and frozen items (Silva et al., 2017).

As per the literature, modified starch, when combined with natural gums ameliorate the standard of salad dressings. One such study was carried out by (Wang et al., 2016) on the effect of modified starch-gum in yellow mustard sauce. The research involved a combination of 4% (w/w) modified starch (MS) with xanthan gum (0.1%–0.5% w/w) or guar gum (GG) (0.1%–0.5% w/w). It was observed that MS/XG blends had a better effect on texture than MS/GG. Three combinations – 0.4% MS/0.4% GG, 0.4% MS/0.3% XG, 0.4% MS/0.4% XG were found to be as excellent thickeners for yellow mustard sauces. The choice of two hydrocolloids – XG and GG relies on their high molecular weights and viscosities. Increasing XG content led to the formation of a more rigid, solid-like texture. Increasing the gum concentration resulted in better interaction and a denser network. A research study that focused on including red sweet pepper as an ingredient in white sauces using native and modified starches observed that crosslinking with acetylated distarch adipate resulted in a more creamy and consistent texture. Also, the presence of antioxidants in red pepper improved the nutritional aspects in addition to good rheology and reduced syneresis leading to wider acceptability (Hernández-Carrión et al.,

2015). Another example of a high value-added product is the utilization of oyster in sauce products (oyster sauce). Oysters have a high myofibrillar protein level and are incorporated in a variety of products. A research study tested the effect of ADSP with concentrations ranging from 0% to 5% on oyster sauce. The data revealed that the presence of ADSP increases the firmness, cohesiveness, consistency and water-holding ability of oyster sauce. An improvement in shelf life and stability was also observed in the modified waxy corn starch (Li et al., 2024).

To sum it up, acetylation improves the swelling power and volume, oil/water absorption and offers retrogradation stability. The thermal stability can be further enhanced by subjecting the starch to dual modification – acetylation + -crosslinking or acetylation + hydrothermal treatments (Subroto et al., 2023).

3.2 Crosslinking

Crosslinking is a commonly used type of chemical modification to overcome the limitations of native starch (low heat, shear, pH stability and thermal decomposition). Crosslinking establishes new covalent bonds that offer stability and strength to the starch granules. Typically, the hydroxyl groups in starches react with the crosslinking reagents resulting in the formation of covalent bonds within chains and between chains. Crosslinking enables the formation of ether or ester linkages between the hydroxyl groups of starch (Fig. 13.4).

As a result, the modified starch is viscous and is ideal for thickening and stabilizing an array of food products (sauces, soups and puddings, etc.) Some commonly used crosslinking agents are sodium tripolyphosphate (STPP), sodium trimetaphosphate (STMP), phosphoryl chloride ($POCl_3$), a combination of acetic anhydride and adipic acid. One study utilized crosslinked tapioca starch (CLTS) by suspending the native starch in an alkaline medium (pH 11, 41.67% w/w) with a combination (99:1 w/w) of STMP and STPP with concentrations ranging from 0.25% to 6% for 3 h at 45°C. The CLTS developed with 1% STMP/STPP exhibited the formation of the strongest gel, produced high shear resistance and possessed the greatest thermal stability. Therefore, 1% CLTS was found to enhance the textural and sensory attributes of soups (Wongsagonsup et al., 2014). Crosslinking stabilizes the starch granule and restricts swelling and breakdown. The effect of crosslinking in waxy starches is more advantageous as congelation does not take place in the absence of amylose. Another study reported the pasting properties of *Amaranthus paniculatus* (Rajgeera) and the influence of crosslinking the starch with phosphorous oxychloride ($POCl_3$). It was found that 5 mL $POCl_3$/100 g starch enhanced the stability of canned foods and so was found to be suitable in salad dressings that demanded pH and shear stability. The incorporation of this crosslinked starch in tomato soup improved the thickening capacity but was found unsuitable for frozen foods because of poor freeze-thaw stability (Singhal & Kulkarni, 1991). An independent study focused on the modification of tapioca and sweet potato starches by crosslinking with STMP. It was observed that the pasting and freeze-thaw stability of these modified starches matched with the commercially available modified corn starch used for sauces. Also, the phosphate and phosphorous residual levels were much lower than the permissible levels set by FAO. The modified sweet potato and

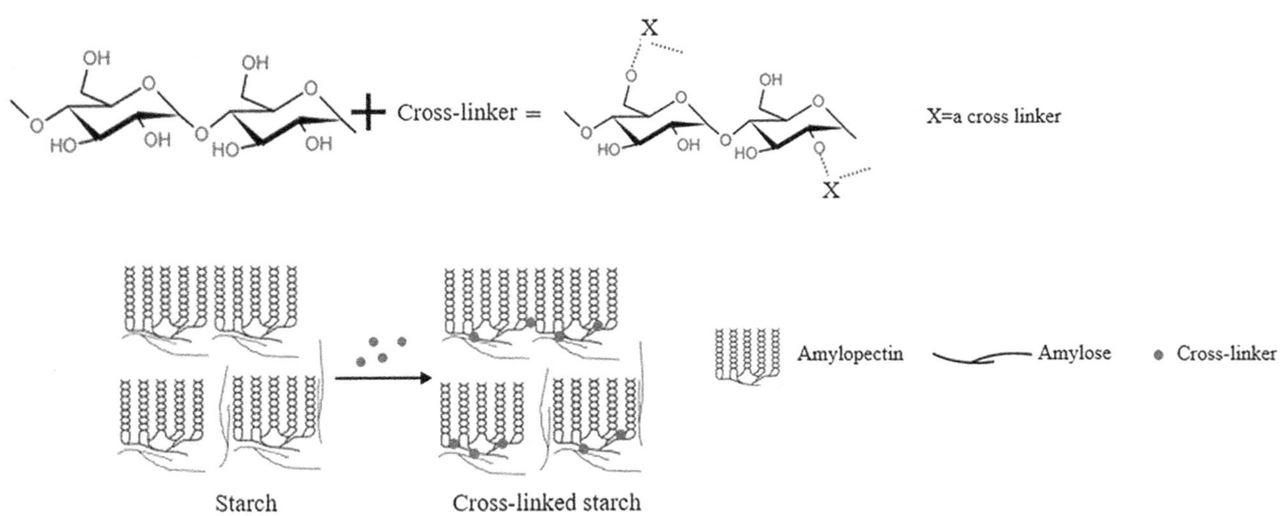

FIGURE 13.4 Schematic representation of crosslinked starch (sourced from Punia Bangar et al., 2024). *Adapted from Punia Bangar, S., Sunooj, K.V., Navaf, M., Phimolsiripol, Y. and Whiteside, W.S. (2024). Recent advancements in cross-linked starches for food applications-a review. International Journal of Food Properties, 27(1), 411–430. https://doi.org/10.1080/10942912.2024.2318427.*

modified tapioca starches were recommended for use in foods with pH $\geq$ 3.7 and pH $\geq$ 3.5 (Khatijah & Patimah, 1997). The thermal and rheological properties of crosslinked sweet potato starches were observed by a research group. Crosslinking was found to increase the gelatinization temperature (0.8–1.1°C) and this is explained by the insertion of phosphate groups that led to the development of extra covalent bonds and tightening of the granular structure. This restricted the molecular movement. The enthalpy (ΔH) also increased with a rise in the concentration of $POCl_3$ as the crosslinked starches require more energy to unwind and disrupt the structure of starch. The crosslinked starches were more elastic as compared to native starch and so the desirable aspects for industrial applications can be achieved using crosslinking (Kim & Yoo, 2010). The consumption of avocado has been on the rise, but the seeds have been discarded and remain underutilized in the food industry. The seeds are rich in starch (30%) and so can be modified to expand the functional properties. Crosslinking using sodium tripolyphosphate (STPP) (2%, 4% and 6%) for 1, 2 and 3 h was performed by Cornelia and Christianti (2018) on avocado starch. The results indicated that 6% STPP for 1 h yielded the best results for cream soup and the viscosity stability of the modified starch was higher than the native one. The presence of the phosphate group has resulted in the strengthening of starch to reduce the breakdown viscosity. To reduce retrogradation and produce starches tolerant of freeze-thaw and thermal operations, crosslinking and stabilization are commonly employed. Crosslinking covalently bonded starch polymers enhances structural rigidity to resist acid, temperature and shear. Stabilization adds chemical substituents, hindering retrogradation upon cooling.

The literature is replete with crosslinking studies on corn, tapioca and wheat starch. Potato starch, which possesses several functional aspects (gelling, thickening, adhesion, encapsulation, etc.) has not been explored much to our knowledge. One study (Heo et al., 2017) determined the properties of potato starch after crosslinking with STMP/STPP at different concentrations ranging from 0.0125% to 10%. Scanning electron microscopy (SEM) images showed a stark difference between the native and crosslinked starches. Light microscopy revealed that the native starches were totally ruptured but the crosslinked ones at different concentrations were hardly disintegrated after 95°C heat treatment for 30 min. The breakdown viscosities of crosslinked starches were also much lower than the native ones which reiterates the fact that crosslinking strengthens the swollen starch granules and offers stability against high temperature and shear. Therefore, crosslinked potato starches at 0.0125%, 0.025% and 0.05% were recommended for use as thickeners, binders and stabilizers in food products such as canned foods, salad dressings, etc.

The rice milling factories churn out byproducts such as flour and broken grains that are not properly utilized apart from serving as a feedstock. As rice contains a high percentage of starch, these byproducts could be chemically modified to improve functional properties. One research group used phosphorylation to modify the physicochemical properties. Starch phosphates have applications in sauces, soups, gravies, etc. Different ratios ranging from 0.018 to 0.1 g of phosphorous oxychloride/100 g flour were used for modification. The phosphorylated rice flour (PRF) exhibited a low viscosity, but the stability was significantly improved after modification and so it is useful for utilizing it as an ingredient in salad dressing and cream soups (Nabeshima et al., 2010). Phosphate starches are one of the most commonly available, widely used modified starches and have very important uses in frozen and canned foods. One study delved into the properties of tuber starches (arrowroot, cassava and sweet potato) after STMP/STPP modification with the reaction period ranging from 7.5 to 120 min. With an increase in the time of reaction, swelling power, peak viscosity and solubility were enhanced irrespective of the starch source. The setback of arrowroot and cassava starches declined when the reaction was extended for a longer time. The modified cassava starch exhibited a more pronounced change with less reaction time, whereas sweet potato starch required a much longer time to achieve the desired attributes (Leonel et al., 2021).

3.3 Hydroxypropylation

Hydroxypropylation is one of the commonly used methods of chemical modification in starch. Hydroxypropylated starches (HPS) are formed by subjecting the native starch to etherification with propylene oxide. An alkaline catalyst is also used in this process. The hydroxypropyl groups incorporated into the starch chains are hydrophilic in nature and this results in the weakening of bonds existing between the granules, thereby modifying the properties. Hydroxypropylation lowers the gelatinization temperature, enhances the swelling power, viscosity and improves solubility and freeze-thaw stability of starch. These properties make it suitable for the utilization of HPS in frozen foods and refrigerated items such as gravies, sauces, puddings and salad dressings. The rheological properties of the modified (hydroxypropylated) rice starch at 5% w/w with various molar substitution (MS) from 0.03 – 0.142 were observed. The swelling power of the modified starch was found to be much higher (35.5–52.8 g/g) as compared to native rice starch (26.6 g/g) and the solubility increased (8.19%–10.7% for modified starch and 7.78% for native rice starch) (Chun & Yoo, 2007). Hydroxypropylation confers several advantages such as more solubility in cold water, retarding retrogradation, easy gelatinization, etc. There is a growing interest in formulating healthier sauces that are gluten-free with low fat content. A study attempted to utilize two Pakistan rice varieties – Basmati and Irri for hydroxypropylation. The treatment of the starches isolated from broken rice grains with 10% propylene oxide enhanced the functional properties. The modified starches contain 1.06% and 1.16% of hydroxypropyl groups in Basmati and Irri which are much lower than the limits set by FDA (Moin et al., 2024).

Hydroxypropylation has been tested on various starches with different MS. One such study was performed on sweet potato starch (SPS) which is an important constituent in sauces, noodles, soups and snacks. To expand the applications in the food industry, SPS was hydroxypropylated with MS ranging from 0.042 to 0.153 and their properties were examined. The hydroxypropylated sweet potato starch (HSPS) exhibited low gelatinization temperatures, good paste clarity and greater swelling power. HSPS gels had lower rates of retrogradation and syneresis as compared to the native gel. It was thus reported that a low degree of hydroxypropylation produces desirable aspects in SPS for various industrial uses (Lee & Yoo, 2011). Hydroxypropylation decreases the number of hydroxyl groups, and this results in the weakening of hydrogen interactions between molecules. This change influences clarity and freeze-thaw stability. One study that carried out hydroxypropylation using propylene oxide (concentrations of 2.5%, 5%, 7.5% and 10% w/v) on apios starches observed an increase in light transmittance. This was due to the loosening of particles occurring after modification. The starch clarity was enhanced after gelatinization which is an important requirement in the manufacture of sauces, jellies etc (Park & Kim, 2020). Water yam, a plant that yields tubers having a starch content of 70%–80% is underutilized and so a research study involved in extracting starch from water yam, subjected it to hydroxypropylation and observed the properties of modified starch. It was evident from the data that hydrxypropylation was effective in retarding the retrogradation tendency of starch gel, improving the water holding ability and freeze thaw stability. Gels formed from 8% – 12% HPS showed lower syneresis as compared to 4%–6% HPS. This establishes the fact that a higher concentration of propylene oxide (8%–12% HPS) provides more stability to freeze-thawing and, therefore more suitable for incorporation in frozen food products (Arueya & Ojesanmi, 2019). Edible canna, native to South America is a traditional food that is used for starch production in factories. It has been reported in the literature that canna starch gave a clear paste and exhibited a higher viscosity than cassava starch at a similar concentration. This high viscosity makes it suitable as a thickening agent but is limited by high retrogradation rates. Though acetylation reduces the retrogradation tendency, it also reduces the peak viscosity by 13%–20% as compared to the native starch. Hydroxypropylation was however, found to limit starch retrogradation and this is due to the steric effects caused by the presence of bulky hydroxypropyl groups. These groups hinder the proper starch chain alignment during aggregation and crystallization. As a result of this, a decrease in setback by 16%–59% was observed. In addition to this, a lowering of pasting temperature and a rise in peak viscosity were also noted. This can be attributed to the weakening of bonding forces between starch granules and hydrophilicity. The freeze-thaw stability data also reinforce the fact that the retrogradation in canna starch could be limited by hydroxypropylation. As the permissible level for the incorporation of hydroxyl groups is set to 7% by the FDA, hydroxypropylated canna starch can be used for stabilization and thickening in sauces, fruit jellies, ice creams, etc., with a low degree of substitution (Chuenkamol et al., 2007). A recent study focused on incorporating the native (NAS), hydroxypropylated (HYPP) and heat-moisture treated (HMT) anchote starches in tomato sauce at different concentrations (0.5%, 1% and 1.5%). A reduction in breakdown viscosity was observed in tomato sauce that was subjected to modification. This indicates that the resulting product is stable against shear and heat. Therefore, the modified anchote starch is a suitable choice for products that require high viscosity (e.g., tomato sauce) (Hailu & Admassu, 2024). Insights into the discarded seeds of jackfruit and starch isolation revealed the problems of low solubility and swelling power, high rates of syneresis and low freeze-thaw stability. To circumvent this, a research group used hydroxypropylation to modify the starch with propylene oxide (5–50 mL). It was observed that hydroxypropyl starches exhibited a higher paste clarity when compared with native starches. The hydroxypropyl groups restrict the reformation of the inter and intramolecular bonds present in starch molecules. Paste clarity improves consumer acceptability and is an important factor in the extension of shelf life. Another important outcome is the stability of the gel under low temperature conditions. As a result of limited water release of hydroxypropyl starches during refrigeration conditions, they are preferred for storage (Naknaen, 2014).

Etherification enables the incorporation of lipophilic groups that present starch with higher stability. This feature permits their utilization in water-oil emulsions. One study compared the properties of native corn, native cassava and hydroxypropylated cassava starch (HPS). Hydroxypropylation has been shown to enhance the water retention ability and this is brought about by the establishment of the intermolecular bond between water and starch. Hydroxypropyl starch and dihydroxypropyl starch (DHPS) exerted positive effects on swelling power. When tested for syneresis, native corn starch had the highest water loss but DHPA had the least water loss during freeze-thaw cycles and so is an ideal choice for application in frozen foods (Schmitz et al., 2006). An important research study was focused on the impact of salt and sugar on the gelatinization and retrogradation of native and modified starches. Native tapioca starch and hydroxypropylated crosslinked tapioca starches were used with MS of 0.05 and 0.08. The influence of sodium chloride (NaCl) and sucrose on the properties was observed. Hydroxypropylated crosslinked tapioca starch hindered retrogradation to a greater extent than the native tapioca starch and it also exhibited a low gelatinization temperature and enthalpy. There was no significant effect of sucrose on starch retrogradation; however, NaCl lowered retrogradation rates in both the starches (Thirathumthavorn & Trisuth, 2008).

3.4 Succinylation

Succinylation is another method that is used to present a hydrophilic nature to starch by adding succinyl groups (4% w/w succinic anhydride) that are negatively charged, and the reaction is performed in alkaline conditions (Fig. 13.5).

Succinylated starches do not produce hard gels on storage and are excellent thickening agents for soups. Succinylated starches have superior thickening ability, lesser gelatinization temperatures and enhanced cold storage stability. These features make it useful for addition to food products like sauces, puddings etc. Pearl millet is a cereal grain cultivated for food and fodder uses and sustains severe temperature, drought and saline conditions. The starch isolated from pearl millet is subjected to different chemical modifications (acetylation, succinylation, oxidation and octenyl succinylation). The results indicate an increase in swelling power and solubility upon succinylation, improved paste clarity with octenylsuccinylation, and a reduction in the reassociation of starch components after acetylation. The important application of OSA starches is in the formation of strong films at the oil-water interface. This is due to the binding of the hydrophilic glucose portion with water and the hydrophobic octenyl portion with oil. Succinylated and OSA starches are thus very useful for thickening purposes in ketchup, sauces, etc (Shaik, Ali, & Hasnain, 2017).

White sorghum is a cereal crop that is grown in Asia, America and Africa and starch is the major carbohydrate storage form. Sorghum starch retrogrades rapidly, and this is due to the higher amount of long amylopectin B-chains. Modification of sorghum starch would facilitate more applications. Succinylation on amaranth and corn starches was studied by a research group (Bhandari & Singhal, 2002) with DS from 0.05 to 0.20. Succinylation imparts hydrophilicity and weakens the internal bonding responsible for holding the starch granules together. Starch succinates enhance the cold-water solubility. Paste clarity is an important property in the food system and varies with product type. A transparent paste is preferable in pie-fillings, whereas opacity is required in salad dressings. The introduction of succinate groups hinders the ordered arrangement of starch chains and thus prolongs retrogradation. Succinylation was found to enhance the swelling power. The decrease in retrogradation is associated with the steric effect caused by the insertion of a bulky group (succinyl) that hinders the starch chain alignment.

Esterification of Jackbean starch (*Canavalia ensiformis*) with succinic anhydride (3% and 4% concentrations), pH (8–8.5, 9–9.5), temperatures (20°C and 30°C) and reaction periods (1 and 3 h) were carried out and the properties of modified starch were observed. The conditions – 4% concentration, 8–8.5 pH, 1 h, 30°C of reaction produced 1.58% succinylation. A decrease in the gelatinization temperature range (67–73°C from 77% – 83%) and prevention of retrogradation were also noticed. Therefore, succinylated Jack bean starches are an ideal option for thickening and stabilizing sauces, ice creams and frozen products (Betancur-Ancona et al., 2002).

Lima bean is a legume that originates from Central America and is rich in carbohydrates (50%–60%). The starch derived from lima bean is not widely used in the food industry owing to its high gelatinization temperature and so a study was directed toward the use of octenyl succinic anhydride (OSA) in starch modification (40% w/w starch, 3% OSA, pH 7, 30 min). This produced a starch with improved gel clarity (37.3% from 23.8%), enhanced viscosity (1000 BU from 700 BU), decreased enthalpy (9.73 J/g from 10.72 J/g) and gelatinization temperature (64.6°C from 75.3°C). These changes in

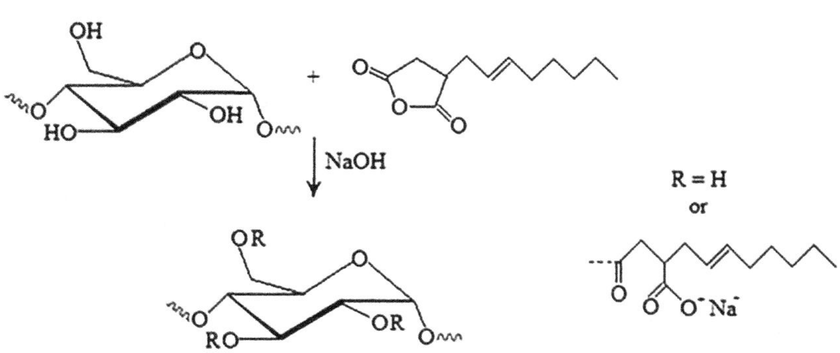

FIGURE 13.5 Succinylation of starch with octenyl succinic anhydride. *Adapted from Ačkar et al. (2015).*

properties make OSA starches suitable for thickening purposes in sauces, soups and salad dressings. These starches however, cannot be useful for refrigerated or frozen food products because of their syneresis and less firmness (Segura-Campos et al., 2010). OSA-starch has its applications in emulsion stability and also is useful in improving paste clarity and forming strong films in the oil-water interface. These properties are desirable in salad dressings and creams. One study explored the pre-treatment of potato starch with calcium chloride ($CaCl_2$) and its impact on succinylation. Different concentrations of $CaCl_2$ ranging from 0.05 to 1M were tested and it was observed that the pretreatment with 1M $CaCl_2$ produced the highest reaction efficiency (RE) and degree of substitution (DS). The replacement of hydroxyl group with the succinyl group also restricts the rearrangement of starch chains, hinders retrogradation and produces a better fluid paste. Thus, 1.0 M OSA-modified starch produced the highest emulsification activity of 0.74 (Wang, Ren, et al., 2019). An interesting study on the effect of OSA in granular and debranched forms of two starches (waxy potato and waxy rice) was performed. Starch debranching was achieved using pullulanase (45 U/g of starch). OSA was used at two concentrations (3% and 15%). It was found that the reaction efficiency (RE) of debranched starches used at 3% OSA was significantly greater (83.6%–87.4%) than that of granular starches (51.2%–56.1%). The same pattern was observed with 15% OSA modification but with less RE. The molecular weights were higher for the modified debranched starches due to the presence of substituted groups as well as starch chain association. It was reported that the substitution of 90% soybean oil with debranched and OS debranched starches produced salad dressings equivalent to those of the commercially available dressings (Klaochanpong et al., 2017).

OSA can also be used as an alternative to emulsifiers. Emulsions contain two immiscible phases: dispersed and continuous. Emulsions are common in a variety of food products including milk and beverages (liquid), sauces and salad dressings (viscous) and in cheese and mayonnaise (gel). As emulsions are not thermodynamically stable, emulsifiers are added to impart stability because they contain both hydrophobic and hydrophilic parts. Some examples of common emulsifiers are soy protein, gum Arabic, gelatin, carrageenan, lecithin etc. However, each one has a limitation associated with it. For instance, gum arabic is overpriced and scarce. Protein-based emulsions employed in beverages are prone to droplet aggregation when the reaction conditions (ionic strength, temperature, pH) change during processing. It is therefore necessary to find alternative emulsifiers and a research group (Cheng et al., 2021) selected starch because of its abundant supply and renewability. However, starch in its native form does not possess the amphiphilic properties required for stabilizing the emulsions. Therefore, modification using OSA (1%, 3% and 5% w/w) was performed in three starches with varying amylose contents: pea (41.5%), corn (32.9%) and waxy corn (1.7%). Increasing DS led to the lowering of gelatinized temperatures and retrogradation in OS-starches. OS-pea starch displayed a poor ability to stabilize emulsions, and this could be because of the higher amylose content leading to stronger reassociation and higher retrogradation at the interface contributing to less steric repulsion. It was thus suggested that higher DS contributes to better emulsion stabilizing ability. These observations would pave the way for further studies on OS-pea starch to enhance the emulsifying properties and expand the industrial applications.

4. Dual treatments

Dual modification is preferred to achieve better results than single methods. For example, crosslinking (chemical) combined with ANN (physical) delays retrogradation, elevates gelatinization temperature, and broadens applications in sauces, soups and gravies. Crosslinking wheat starch (0%–0.4% w/w $POCl_3$) followed by ANN (40°C, 72 h) lowered setback and final viscosities, hence reducing retrogradation and forming stronger gels, suitable for use in both frozen and canned products (Majzoobi et al., 2012). Another study combined hydrothermal pretreatment (35% w/w water, 48–62°C, 3 h) with 3% octenylsuccinic anhydride (OSA) esterification (chemical modification) in corn starches, finding that 60°C optimized peak viscosity. Such H-OS-starch is useful as a thickening agent in salad oil (Chen et al., 2014). Because of the rising concern over high calorie processed foods, developing fat replacers is increasingly critical. Starches in native or modified forms are being used as fat mimetics or replacers. Rice starch, with its bland taste, smoothness, creaminess and glossy texture, is ideal for many products. Dual-modified rice starches often show higher crystallinity and thermal stability. They can partially replace fats in whipped cream, while also offering a controlled release of glucose. Further research is necessary to clarify the mechanisms (Iftikhar & Dutta, 2020).

A dual treatment was performed by Mehfooz et al. (2020) in barley starch to improve the properties and applications. Hydroxypropylation leads to the insertion of hydroxypropyl groups in starch. The hydrophilicity of these groups weakens the internal bonding in starch and traps water molecules and binds them in the starch gel offering freeze-thaw stability. This modification lowers the gelatinization temperature and enhances paste clarity. This study crosslinked the hydroxypropylated (6%) barley starches at 1%, 1.5% and 2% mixture of STPP and STMP. The dual modification (hydroxypropylation and crosslinking) produces synergistic effects on limiting the retrogradation rates and so is beneficial as a thickening agent in soups. A similar study was performed on barley starches with propylene oxide (8% and 12%) and crosslinking with 1% STMP and STPP. The hydroxypropylated distarch phosphates were added at 5% and 8% in soups (8 HPCL and 12 HPCL). The dual modification significantly improved the solubility, swelling capacity and water retention. The paste clarity of 12 HPCL was improved five times as compared with the native starch. A higher extent of hydroxypropylation also reduced the rate and extent of retrogradation. Also, 8HPCL at 8% addition in soups enhanced the thickening ability in the food system as compared to other combinations (Mehfooz et al., 2021). Another study on potato starch by a research group employed a single and dual modification with 3% w/v calcium chloride ($CaCl_2$) solution (5 h, 25°C) and dry heat (DH) method for 1, 2 and 4 h at 130°C. The large size and high swelling nature of potato starch granules contribute to high viscosity but also high fragility. Treatment with $CaCl_2$ was only found to increase the shear viscosity and thermal transition values and decrease the breakdown and peak viscosity. This is attributed to the crosslinking reaction that establishes electrostatic interactions with the phosphate and hydroxyl groups in the potato starch. However, the final viscosity was found to be similar to control (untreated potato starch) and which led to high setback. DH only resulted in partial disintegration of the crystalline structure and double helices unwinding in the amorphous domain followed by subsequent rearrangement of the disrupted crystallites. This treatment decreased the paste viscosity, thermal transition values and shear viscosity but prolonged heating is required for further reduction. The synergy of DH and $CaCl_2$ treatments led to the complete dissociation of starch granules, producing low swelling and shear viscosity. These unique properties obtained after dual modification enabled the utilization of potato starch in salad dressings and sauces (Zhang et al., 2021). A unique study was performed by a research team on barley starch. The initial step was to perform a single modification with oxidation, succinylation and crosslinking. This was followed by a dual modification of the resulting succinylated and oxidized starches with crosslinking. It was observed that the crosslinked barley starch exhibited resistance toward high temperature and swelling. This characteristic feature makes it suitable for utilization in soups and sauces. The oxidized barley starch enhanced the paste clarity and so could be used in confectionery. Succinylated starches had low gelatinization temperatures and retrogradation in addition to a lower tendency for gel formation. This property makes it useful for incorporation in frozen items and refrigerated products. Dual modification methods aided in delaying the retrogradation during the storage period and demonstrated high thermal stability. These advantages are required for canned, frozen and refrigerated foods, puddings, salad dressings. It was thus concluded that single and dual modifications are required to overcome the limitations of native barley starch (Mehfooz et al., 2019).

A study on sago starch adopted a dual modification method – first hydroxypropylation using propylene oxide (6%–12%) followed by crosslinking using three different agents (a combination of sodium trimetaphosphate (STMP) and sodium tripolyphosphate (STPP), epichlorohydrin and phosphorous oxychloride). The native sago starch is prone to retrogradation forming long cohesive gels that are unsuitable for utilization in canned foods. It is therefore necessary to overcome the limitation by the dual modification approach. It was found that 2% STMP +5% STPP combination yielded a better crosslinking effect than epichlorohydrin and phosphorous oxychloride. Therefore, hydroxypropylation (10%–12% propylene oxide) with crosslinking (2% STMP + 5% STPP) modified the sago starch with desirable properties such as high acid resistance, freeze-thaw stability and superior gel texture (Wattanachant et al., 2003). The addition of starch is preferred in a variety of sauces, particularly in white sauce due to its thickening ability. Corn starch is by far the most widely used commodity in food products. However, a recent study proposed an alternative starch source owing to the rising demand for the utilization of starch in food sectors. Pearl millet is cost effective, easy to obtain and possesses great tolerance toward climatic conditions, salt and drought. In order to enhance the functional properties, a research group performed two kinds of starch modification on corn and pearl millet starches such as succinylation and hydroxypropylation. Propylene oxide (20 g in 100 g starch, db) and succinic anhydride (2 g in 100 g starch, db) were treated with native corn and pearl millet starches. The data revealed that there was no syneresis in hydroxypropylated pearl millet (HPWS) and hydroxypropylated corn (HCWS) starch after 14 days storage period. Also, sauces having native pearl millet (PWS) and succinylated pearl millet starches (SPWS) exhibited a much lower syneresis than native corn (CWS) and succinylated corn starch (SCWS). The absence of the retrogradation peak was observed in HCWS and HPWS. To conclude, chemically modified pearl millet starches offer excellent texture, storage and sensory properties in white sauce (Shaik, Ali & Hasnain, 2020).

One study preferred to test dual modifications with acid-thinning and succinylation in addition to single treatments. Acid hydrolysis has been used to form thin boiled starch. This method involves hydrolyzing the starch with a low concentration of mineral acid and a lower temperature than that of the gelatinization temperature. The amorphous portion of the granule is disrupted leading to a lower molecular weight and an increase in relative crystallinity. In this study, hydrochloric acid with concentrations ranging from 0.1 to 1 M was used. It was evident from the results that the succinylated lintnerized starches (Su-$AT_{1.0M}$) can enhance the textural properties of dressings. This is attributed to the incorporation of succinyl groups which improves the water binding ability and helps in emulsification and prevention of phase separation in salad dressings (Mehboob et al., 2015). Another dual modification study was performed on white sorghum starch with varying concentrations of propylene oxide (10%, 20% and 30% w/v) and succinic anhydride (1%, 2% and 3% dry basis) to enhance the functional aspects. It was observed that the water retention ability and swelling power increased with the addition of hydroxypropyl and succinyl groups whereas sucinylation alone did not lead to improvement in starch solubility. The modified starches exhibited better clarity than the native ones. The retrogradation percentage (%R) was found to decrease upon succinic anhydride addition (3% and 4%) and in hydroxypropylated starches. The major finding was that the propylene oxide used at 30% concentration produced the desired properties and the lowest rates of retrogradation. Thus, hydroxypropylation and succinylation used in tandem can serve as a suitable option for stabilizing and thickening sauces (Mehboob et al., 2021).

A unique study was performed on two Pakistan rice varieties – Basmati and Irri. A dual modification (succinylation and crosslinking) was performed and the resulting alterations in structure and properties were reported. The crosslinking process alone does not overcome the problems associated with increasing retrogradation rates and improving freeze-thaw stability. The combined effect of succinylation and crosslinking will enable us to enhance the functional properties. Succinylation was performed with 4% succinic anhydride and the modified starches were then treated with 99:1 STMP/STPP (4% crosslinking). This treatment resulted in low values of breakdown which denotes high stability and so the resulting succinylated-crosslinked starches are suitable options for utilization in soups. The lower setback values also indicate enhanced cold storage stability and so can be incorporated in frozen foods (Moin et al., 2019). Though succinylation offers several advantages (solubility, thickening, paste clarity and freeze-thaw stability), it suffers from instability during shear at high temperatures. This limitation can be overcome by including another type of modification, annealing (ANN). ANN is a preferred choice as it helps to increase the shear and thermal stability of starch. The effect of succinylation-annealing method was studied by Ariyantoro et al. (2018). It was observed that the dual treatment in corn starches enhanced several parameters such as swelling power, paste clarity, water binding ability, and gelatinization temperature but no noticeable difference was observed in granule shape and solubility. The outcome of this research would be useful in the manufacture of food products like sauce as it offers resistance to heat and shear while processing. This is evident from the low breakdown values after dual modification.

5. Nonthermal methods

Of late, nonthermal methods such as ultrasonication (US), pulsed electric field (PEF), cold plasma and high hydrostatic pressure (HHP) are gaining momentum due to their rapid modification and the effects they confer on the granular structure of starch. For instance, the modified starch after US treatment has its application as a gelling and thickening agent in soups and sauces. The texture and stability are significantly improved as a result of the starch modification. In addition to that, US corn starches offer an increased apparent viscosity, a feature that is desirable in the food industry (Bangar et al., 2024). In starch, HHP brings about cold gelatinization, contrary to heat-induced gelatinization. In the case of low-fat salad dressings, mouthfeel can be enhanced while using starch-based fat mimetics to replace vegetable oil which in turn modifies the gelling and rheological aspects. Wheat starch, after HHP treatment served as an excellent fat replacer in oil-in-water emulsions, providing good stability and homogeneity. The effect was more pronounced when higher concentrations (15%–20%) were used (Heydari et al., 2021). A similar finding was noticed in HHP-treated corn and waxy corn starches (Heydari & Razavi, 2021a), thereby limiting the calories in low-fat emulsions. Table 13.1 below indicates the various non-thermal methods employed in enhancing the texture and mouthfeel of sauces, soups and salad dressings.

6. Future directions

Investigating nonthermal and low-chemical processes (e.g., ultrasound, high-pressure treatments, and cold plasma) may reduce reliance on harsh reagents, aligning with clean-label trends and sustainability goals. Examining how these emerging techniques can synergize with conventional methods (e.g., annealing, and crosslinking) will likely yield valuable insights. Hybridizing GCWS with chemical modifications (e.g., acetylation + GCWS or crosslinking + GCWS) may produce unique textural or emulsifying properties. Future studies can evaluate synergistic effects on rheology, freeze–thaw stability, and shear resistance in complex food matrices including soups/salad dressings. Developing new starch sources through breeding or biotechnology could yield tailor-made starches with predetermined amylose/amylopectin ratios or specific granule morphologies. These novel starches could minimize the need for extensive downstream modifications. Understanding interactions between modified starches and other biopolymers (e.g., xanthan gum, pectin, and plant proteins) under different pH, ionic strength, and thermal conditions remains crucial. This knowledge can guide the rational design of food emulsions and gels with tailored texture and nutritional profiles. Low-glycemic or resistant

TABLE 13.1 The various non-thermal methods employed in enhancing the texture and mouthfeel of sauces, soups and salad dressings.

Type of modification	Starch type	Treatment conditions	Observations	References
Gamma irradiation	Rice starch	OSA treatment was initially performed on rice starch followed by gamma irradiation (18°C, 5 h, 10 kGy/h).	OS starch is used as an emulsifier in salad dressings as it produces a strong film at the oil-water interface. The OS starch exhibited higher emulsion capacity and better stability and this is attributed to its amphipathic nature. The irradiation treatment enhanced the emulsification property and 10 kGy was found to be the best dose.	Han and Lim (2012)
Cold plasma	Cassava starch	Cassava starch (CS) solution was prepared to 30% (w/v) concentration and then subjected to atmospheric pressure plasma jet (APPJ) for 1, 3, 5, 10 and 15 min and then modified with 3% OSA.	APPJ treatment modified the morphology and crystallinity of CS. APPJ-OSA-CS produced good emulsion stability and activity. It was thus concluded that the modified starch could be utilized as a food grade emulsifier.	Ji et al. (2022)
UV irradiation	Maize starch	The prepared OSA samples were subjected to thermal treatment (30 min, 210°C) and then irradiated (30 min, 2.88 J/cm^2)	The modified OSA possessed functional properties that are useful in food industry. The viscosity reduction of these OSA starches was pronounced and so could be used in beverage emulsions and salad dressings. UV irradiation enabled better water binding ability at higher temperatures.	Kurdziel et al. (2020)
Ultrasound and microwave	Corn starch	The ultrasound conditions used were 20 kHz, 25°C, 600 W and 4 min and microwave at 2450 MHz, 1 min, 700 W.	Ultrasonic + microwave treatment was found to be useful in preserving the flavonoid, polyphenol, capsaicin and antioxidant activity of chilli sauce. It is also effective in retaining the nutrition and flavor of chilli sauce because of its ability to decrease the loss of volatile compounds.	Tang et al. (2023)
Ultrasonication	Rice starch	Native rice starch (40 g) dispersed in distilled water (100 mL) was treated at various temperatures (15–75°C) for 20 min by an ultrasonic processor (20 kHz frequency, 500 W power).	The textural aspects such as springiness, stickiness etc. were found to enhance upon ultrasonication in addition to increasing the thermal and freeze-thaw stabilities. This enables the utilization of ultrasound-treated starches in sauces.	Okonkwo et al. (2021)
Gamma irradiation	Potato starch	Gamma irradiation was performed in a cobalt-60 irradiator with doses of 0, 5, 10 and 15 and 0.997 kGy/h dose rate.	Ionizing radiation generates free radicals that have the ability to bring about molecular changes and starch fragmentation. The scission of glycosidic bonds in starch and the subsequent degradation occurs during gamma irradiation. A reduction in viscosity was noticed.	Teixeira et al. (2015)
Ultrasound	Sweet potato starch	Annealing (PA) was performed separately and then combined with enzyme (EA), α-amylase (0–10 U/g) and ultrasound (360 W power, 25°C, 30 min) (UEA).	EA was found to be efficient than PA for starch modification and ultrasound enhanced the EA treatment. UEA exhibited the highest effect on starch. This produced lower swelling, greater thermal stability and strong shear which make it suitable for incorporation in canned foods as well as refrigerated foods.	Wang et al. (2020)

starch variants, particularly from whole-grain or leguminous sources, can be modified to enhance digestibility and reduce postprandial glucose spikes. Research linking modified starch structure to human gut microbiome or glycemic response will be increasingly relevant.

Through innovative research on greener processes, advanced modification strategies, and molecular tailoring, modified starches will continue to facilitate the development of high-performance, health-conscious, and environmentally responsible soups, sauces, and salad dressings—ultimately meeting the evolving needs of consumers and the food industry alike.

7. Conclusions

Starch remains central to soup, sauce, and salad dressing formulations due to its versatile thickening, stabilizing, and emulsifying capacities. Native starch, while abundant and inexpensive, often exhibits limited resistance to heat, shear, and acid, prompting widespread adoption of physical (e.g., heat moisture treatment, annealing, pregelatinization, and GCWS) and chemical (e.g., acetylation, crosslinking, hydroxypropylation, and succinylation) modifications. Physical modifications generally lack chemical residues, increasing consumer acceptance, while chemical routes allow fine-tuning of properties to enhance freeze–thaw stability, emulsion formation, and acid tolerance. Dual or multi-step modifications further expand functional possibilities, enabling manufacturers to offer consistent, high-quality products with extended shelf life and pleasant mouthfeel. Choosing the optimal modification, individually or in combination, can significantly improve textural properties and organoleptic attributes in soups, sauces, and salad dressings, thereby enhancing consumer acceptance and market competitiveness.

References

Ačkar, Đ., Babić, J., Jozinović, A., Miličević, B., Jokić, S., Miličević, R., & Šubarić, D. (2015). Starch modification by organic acids and their derivatives: A review. *Molecules, 20*(10), 19554–19570.

Adebowale, K. O., Henle, T., Schwarzenbolz, U., & Doert, T. (2009). Modification and properties of African yam bean (Sphenostylis stenocarpa hochst. Ex A. rich.) harms starch I: Heat moisture treatments and annealing. *Food Hydrocolloids, 23*(7), 1947–1957. https://doi.org/10.1016/j.foodhyd.2009.01.002

Ariyantoro, A. R., Katsuno, N., & Nishizu, T. (2018). Effects of dual modification with succinylation and annealing on physicochemical, thermal and morphological properties of corn starch. *Foods, 7*(9), 133.

Arocas, A., Sanz, T., & Fiszman, S. M. (2009). Influence of corn starch type in the rheological properties of a white sauce after heating and freezing. *Food Hydrocolloids, 23*(3), 901–907.

Arueya, G. L., & Ojesanmi, A. A. (2019). Evaluation of effects of increasing molar substitution of hydroxypropylene on physicochemical, functional and morphological properties of starch from water yam (Dioscorea alata). *Journal of Food Research, 8*(4), 58. https://doi.org/10.5539/jfr.v8n4p58

Bangar, S. P., Sunooj, K. V., & Siroha, A. K. (2024). *Starch: Structure, properties, and modifications for food applications*. CRC Press.

Betancur, A. D., Chel, G. L., & Cañizares, H. E. (1997). Acetylation and characterization of *Canavalia ensiformis* starch. *Journal of Agricultural and Food Chemistry, 45*(2), 378–382.

Betancur-Ancona, D., García-Cervera, E., Cañizares-Hernández, E., & Chel-Guerrero, L. (2002). Chemical modification of jack bean (*Canavalia ensiformis*) starch by succinylation. *Starch Staerke, 54*(11), 540–546.

Bhandari, P. N., & Singhal, R. S. (2002). Effect of succinylation on the corn and amaranth starch pastes. *Carbohydrate Polymers, 48*(3), 233–240. https://doi.org/10.1016/S0144-8617(01)00310-1

Biliaderis, C. G. (1991). Canada the structure and interactions of starch with food constituents. *Canadian Journal of Physiology and Pharmacology, 69*(1), 60–78. https://doi.org/10.1139/y91-01100084212

Bortnowska, G., Balejko, J., Schube, V., Tokarczyk, G., Krzemińska, N., & Mojka, K. (2014). Stability and physicochemical properties of model salad dressings prepared with pregelatinized potato starch. *Carbohydrate Polymers, 111*, 624–632. https://doi.org/10.1016/j.carbpol.2014.05.015

Butt, N. A., Ali, T. M., & Hasnain, A. (2019). Rice starch citrates and lactates: A comparative study on hot water and cold water swelling starches. *International Journal of Biological Macromolecules, 127*, 107–117.

Chen, J. (2009). Food oral processing-A review. *Food Hydrocolloids, 23*(1), 1–25. https://doi.org/10.1016/j.foodhyd.2007.11.013

Chen, H., Chen, L., Dang, X., Shan, Z., Dai, R., & Wang, Y. (2019). The structure and properties of granular cold-water-soluble starch by a NaOH/urea aqueous solution. *International Journal of Biological Macromolecules, 141*, 732–737.

Chen, H., Chen, L., Dang, X., Shan, Z., Dai, R., & Wang, Y. (2019). The structure and properties of granular cold-water-soluble starch by a NaOH/urea aqueous solution. *International Journal of Biological Macromolecules, 141*, 732–737. https://doi.org/10.1016/j.ijbiomac.2019.08.226

Chen, X., He, X., & Huang, Q. (2014). Effects of hydrothermal pretreatment on subsequent octenylsuccinic anhydride (OSA) modification of cornstarch. *Carbohydrate Polymers, 101*(1), 493–498. https://doi.org/10.1016/j.carbpol.2013.09.079

Chen, J., & Jane, J. J. C. C. (1994). Properties of granular cold-water-soluble starches prepared by alcoholic-alkaline treatments. *Cereal Chemistry, 71*(6), 623–626.

Cheng, F., Ai, Y., & Ghosh, S. (2021). Utilization of octenyl succinic anhydride-modified pea and corn starches for stabilizing oil-in-water emulsions. *Food Hydrocolloids, 118*, Article 106773. https://doi.org/10.1016/j.foodhyd.2021.106773

Chuenkamol, B., Puttanlek, C., Rungsardthong, V., & Uttapap, D. (2007). Characterization of low-substituted hydroxypropylated canna starch. *Food Hydrocolloids, 21*(7), 1123–1132. https://doi.org/10.1016/j.foodhyd.2006.08.013

Chun, S.-Y., & Yoo, B. (2007). Effect of molar substitution on rheological properties of hydroxypropylated rice starch pastes. *Starch - Stärke, 59*(7), 334–341. https://doi.org/10.1002/star.200700603

Cisse, M., Michel Koffi, D., & Aya Akaffou, F. (2023). Functional properties of acetylation-modified starches of three purple maize cultivars from Côte d'ivoire. *European Journal of Nutrition & Food Safety, 15*(10), 63–72. https://doi.org/10.9734/ejnfs/2023/v15i101346

Colussi, R., El Halal, S. L. M., Pinto, V. Z., Bartz, J., Gutkoski, L. C., Zavareze, E.da R., & Dias, A. R. G. (2015). Acetylation of rice starch in an aqueous medium for use in food. *LWT - Food Science and Technology, 62*(2), 1076–1082. https://doi.org/10.1016/j.lwt.2015.01.053

Cornelia, M., & Christianti, A. (2018). Utilization of modified starch from avocado (Persea americana mill.) seed in cream soup production. In *IOP conference series: Earth and environmental science*. IOP Publishing, Article 012074. Vol. 102.

Dos Santos, T. P. R., Franco, C. M. L., & Leonel, M. (2020). Gelatinized sweet potato starches obtained at different preheating temperatures in a spray dryer. *International Journal of Biological Macromolecules*, 1339–1346.

Eastman, J. E., & Moore, C. O. (1984). *U.S. patent no. 4,465,702*. Washington, DC: U.S. Patent and Trademark Office.

Gałkowska, D., Długosz, M., & Juszczak, L. (2013). Effect of high methoxy pectin and sucrose on pasting, rheological, and textural properties of modified starch systems. *Starch-Stärke, 65*(5-6), 499-508.

Gomes, A. M. M., Mendes da Silva, C. E., Ricardo, N. M. P. S., & Sasaki, J. M. (2004). Rogério germani, impact of annealing on the physicochemical properties of unfermented cassava starch (" polvilho doce "). *Starch - Stärke, 56*(9), 419–423. https://doi.org/10.1002/star.200300271

Guinard, J. X., & Mazzucchelli, R. (1996). The sensory perception of texture and mouthfeel. *Trends in Food Science and Technology, 7*(7), 213–219. https://doi.org/10.1016/0924-2244(96)10025-X

Hailu, G. G., & Admassu, H. (2024). Potential stabilization effect of native and modified anchote (Coccinia abyssinica) starches in tomato sauce. *International Journal of Food Properties, 27*(1), 12–33. https://doi.org/10.1080/10942912.2023.2293658

Han, J. A., & Lim, S. T. (2012). Effect of γ-irradiation on pasting and emulsification properties of octenyl succinylated rice starches. *Carbohydrate Polymers, 90*(4), 1480–1485.

Han, F., Liu, M., Gong, H., Lu, S., Ni, B., & Zhang, B. (2012). Synthesis, characterization and functional properties of low substituted acetylated corn starch. *International Journal of Biological Macromolecules, 50*(4), 1026–1034. https://doi.org/10.1016/j.ijbiomac.2012.02.030

He, X.h., Xia, W., Chen, R.y., Dai, T.t., Luo, S.j., Chen, J., & Liu, C.m. (2020). A new pre-gelatinized starch preparing by gelatinization and spray drying of rice starch with hydrocolloids. *Carbohydrate Polymers, 229*. https://doi.org/10.1016/j.carbpol.2019.115485

Hedayati, S., Shahidi, F., Koocheki, A., Farahnaky, A., & Majzoobi, M. (2016). Comparing the effects of sucrose and glucose on functional properties of pregelatinized maize starch. *International Journal of Biological Macromolecules, 88*, 499–504. https://doi.org/10.1016/j.ijbiomac.2016.04.026

Heo, H., Lee, Y. K., & Chang, Y. H. (2017). Rheological, pasting, and structural properties of potato starch by cross-linking. *International Journal of Food Properties, 20*, 2138–2150. https://doi.org/10.1080/10942912.2017.1368549

Hernández-Carrión, M., Sanz, T., Hernando, I., Llorca, E., Fiszman, S. M., & Quiles, A. (2015). New formulations of functional white sauces enriched with red sweet pepper: A rheological, microstructural and sensory study. *European Food Research and Technology, 240*(6), 1187–1202. https://doi.org/10.1007/s00217-015-2422-1

Heydari, A., & Razavi, S. M. A. (2021a). Evaluating high pressure-treated corn and waxy corn starches as novel fat replacers in model low-fat O/W emulsions: A physical and rheological study. *International Journal of Biological Macromolecules, 184*, 393–404.

Heydari, A., Razavi, S. M. A., & Farahnaky, A. (2021). Effect of high pressure-treated wheat starch as a fat replacer on the physical and rheological properties of reduced-fat O/W emulsions. *Innovative Food Science and Emerging Technologies, 70*, Article 102702.

Huang, Z. Q., Lu, J. P., Li, X. H., & Tong, Z. F. (2007). Effect of mechanical activation on physico-chemical properties and structure of cassava starch. *Carbohydrate Polymers, 68*(1), 128–135.

Huang, Z. Q., Xie, X.l., Chen, Y., Lu, J.p., & Tong, Z. F. (2008). Ball-milling treatment effect on physicochemical properties and features for cassava and maize starches. *Comptes Rendus Chimie, 11*(1–2), 73–79. https://doi.org/10.1016/j.crci.2007.04.008

Iftikhar, S. A., & Dutta, H. (2020). Use of raw and physically modified rice starches as fat replacer in whipping cream. *Current Research in Nutrition and Food Science, 8*(1), 122–130. https://doi.org/10.12944/CRNFSJ.8.1.11

Imeson, A. (2011). *Food stabilisers, thickeners and gelling agents*. John Wiley and Sons.

Janssen, A. M., Terpstra, M. E. J., De Wijk, R. A., & Prinz, J. F. (2007). Relations between rheological properties, saliva-induced structure breakdown and sensory texture attributes of custards. *Journal of Texture Studies, 38*(1), 42–69. https://doi.org/10.1111/j.1745-4603.2007.00085.x

Ji, S., Xu, T., Huang, W., Gao, S., Zhong, Y., Yang, X., & Lu, B. (2022). Atmospheric pressure plasma jet pretreatment to facilitate cassava starch modification with octenyl succinic anhydride. *Food Chemistry, 370*, Article 130922.

Khatijah, I., & Patimah, H. (1997). Physico-chemical properties of cross-linked tuber starches. *Journal of Tropical Agriculture and Food Science, 25*, 179–196.

Kim, B.-Y., & Yoo, B. (2010). Effects of cross-linking on the rheological and thermal properties of sweet potato starch. *Starch - Stärke, 62*(11), 577–583. https://doi.org/10.1002/star.201000035

Klaochanpong, N., Puncha-arnon, S., Uttapap, D., Puttanlek, C., & Rungsardthong, V. (2017). Octenyl succinylation of granular and debranched waxy starches and their application in low-fat salad dressing. *Food Hydrocolloids, 66*, 296–306. https://doi.org/10.1016/j.foodhyd.2016.11.039

Koç, H., Vinyard, C. J., Essick, G. K., & Foegeding, E. A. (2013). Food oral processing: Conversion of food structure to textural perception. *Annual Review of Food Science and Technology, 4*(1), 237–266. https://doi.org/10.1146/annurev-food-030212-182637

Kurdziel, M., Królikowska, K., Łabanowska, M., Pietrzyk, S., & Michalec, M. (2020). The effect of thermal and irradiation treatments on structural and physicochemical properties of octenyl succinate maize starches. *Food Chemistry, 330*, Article 127242.

Lee, H. L., & Yoo, B. (2011). Effect of hydroxypropylation on physical and rheological properties of sweet potato starch. *LWT - Food Science and Technology, 44*(3), 765–770. https://doi.org/10.1016/j.lwt.2010.09.012

Lee, H., & Yoo, B. (2023). Particle agglomeration and properties of pregelatinized potato starch powder. *Gels, 9*(2), 93. https://doi.org/10.3390/gels9020093

Leonel, M., Del Bem, M. S., dos Santos, T. P. R., & Franco, C. M. L. (2021). Preparation and properties of phosphate starches from tuberous roots. *International Journal of Biological Macromolecules, 183*, 898–907. https://doi.org/10.1016/j.ijbiomac.2021.05.045

Li, X., Zou, F., Kang, X., Gao, W., Cui, B., & Sui, J. (2024). Effects of acetylated distarch phosphate on the physicochemical characteristics and stability of the oyster sauce system. *Frontiers in Nutrition, 11*. https://doi.org/10.3389/fnut.2024.1412314

Liu, H., Lv, M., Peng, Q., Shan, F., & Wang, M. (2015). Physicochemical and textural properties of tartary buckwheat starch after heat–moisture treatment at different moisture levels. *Starch - Stärke, 67*(3–4), 276–284. https://doi.org/10.1002/star.201400143

Liu, Y., Zhu, L., Yu, X., Xu, Y., & Zhong, T. (2023). A combined ultrasound–microwave assisted alcoholic–alkaline method for preparation of granulated cold-water-soluble starch. *Starch Staerke, 75*(11–12), Article 2200257.

Ma, H., Mei, L., Liang, Y., Zheng, X., Sun, Le, Dang, W., Li, J., Li, L., & Liu, C. (2022). Research progress on properties of pre-gelatinized starch and its application in wheat flour products. *Grain & Oil Science and Technology, 5*(2), 87–97. https://doi.org/10.1016/j.gaost.2022.01.001

Majzoobi, M., Radi, M., Farahnaky, A., Jamalian, J., Tongdang, T., & Mesbahi, G. (2011). Physicochemical properties of pre-gelatinized wheat starch produced by a twin drum drier. *Journal of Agricultural Science and Technology A, 13*(2), 193–202.

Majzoobi, M., Sabery, B., Farahnaky, A., & Tongdang Karrila, T. (2012). Physicochemical properties of cross-linked-annealed wheat starch. *Iranian Polymer Journal (English Edition), 21*(8), 513–522. https://doi.org/10.1007/s13726-012-0056-8

Medha, T., Prakruthi, L. Y., Kumari, T., Alavilli, H., Surekha, C., Xu, B., & Reddy, C. K. (2025). Effect of heat-moisture treatment on structural, morphological, thermal, and physicochemical properties of non-conventional palmyra (*Borassus flabellifer*) starch. *Biomass Conversion and Biorefinery, 15*, 1–9.

Mehboob, S., Ali, T. M., Alam, F., & Hasnain, A. (2015). Dual modification of native white sorghum (Sorghum bicolor) starch via acid hydrolysis and succinylation. *LWT - Food Science and Technology, 64*(1), 459–467. https://doi.org/10.1016/j.lwt.2015.05.012

Mehboob, S., Ali, T. M., Shaikh, M., & Hasnain, A. (2021). Effects of varying levels of succinylation and hydroxypropylation on functional, thermal and textural characteristics of white sorghum starch. *Cereal Chemistry, 98*(3), 624–633. https://doi.org/10.1002/cche.10406

Mehfooz, T., Ali, T. M., Ahsan, M., Abdullah, S., & Hasnain, A. (2021). Morphological, functional and thermal characteristics of hydroxypropylated-crosslinked barley starches. *Journal of Food Measurement and Characterization, 15*(1), 237–246. https://doi.org/10.1007/s11694-020-00624-9

Mehfooz, T., Ali, T. M., & Hasnain, A. (2019). Effect of cross-linking on characteristics of succinylated and oxidized barley starch. *Journal of Food Measurement and Characterization, 13*(2), 1058–1069. https://doi.org/10.1007/s11694-018-00021-3

Mehfooz, T., Ali, T. M., Shaikh, M., & Hasnain, A. (2020). Characterization of hydroxypropylated-distarch phosphate barley starch and its impact on rheological and sensory properties of soup. *International Journal of Biological Macromolecules, 144*, 410–418.

Moin, A., Ali, T. M., & Hasnain, A. (2019). Thermal, morphological, and physicochemical characteristics of succinylated–crosslinked rice starches. *Cereal Chemistry, 96*(5), 885–894. https://doi.org/10.1002/cche.10191

Moin, A., Ali, T. M., & Hasnain, A. (2024). Enhancing functional properties of rice starches through hydroxypropylation for development of reduced-fat white sauces. *Food Chemistry, 446*. https://doi.org/10.1016/j.foodchem.2024.138860

Morán, J. I., Cyras, V. P., Giudicessi, S. L., Erra-Balsells, R., & Vazquez, A. (2011). Influence of the glycerol content and temperature on the rheology of native and acetylated starches during and after gelatinization. *Journal of Applied Polymer Science, 120*(6), 3410–3420.

Moraes, J., Branzani, R. S., & Franco, C. M. L. (2014). Behavior of Peruvian carrot (Arracacia xanthorrhiza) and cassava (Manihot esculenta) starches subjected to heat-moisture treatment. *Starch - Stärke, 66*(7–8), 645–654. https://doi.org/10.1002/star.201300207

Nabeshima, E. H., Bustos, F. M., Hashimoto, J. M., & El Dash, A. A. (2010). Improving functional properties of rice flours through phosphorylation. *International Journal of Food Properties, 13*(4), 921–930. https://doi.org/10.1080/10942910902927094

Naknaen, P. (2014). Physicochemical, thermal, pasting and microstructure properties of hydroxypropylated jackfruit seed starch prepared by etherification with propylene oxide. *Food Biophysics, 9*(3), 249–259. https://doi.org/10.1007/s11483-014-9347-2

Nakorn, K. N., Tongdang, T., & Sirivongpaisal, P. (2009). Crystallinity and rheological properties of pregelatinized rice starches differing in amylose content. *Starch - Stärke, 61*(2), 101–108. https://doi.org/10.1002/star.200800008

Okonkwo, V. C., Kwofie, E. M., Mba, O. I., & Ngadi, M. O. (2021). Impact of thermo-sonication on quality indices of starch-based sauces. *Ultrasonics Sonochemistry, 73*, Article 105473.

Park, M. H., & Kim, M. (2020). Physicochemical properties of hydroxypropylated apios starches. *Preventive Nutrition and Food Science, 25*(3), 286–292. https://doi.org/10.3746/pnf.2020.25.3.286

Pitchon, E., O'rourke, J. D., & Joseph, T. H. (1981). *U.S. patent no. 4,280,851*. Washington, DC: U.S. Patent and Trademark Office.

Punia Bangar, S., Sunooj, K. V., Navaf, M., Phimolsiripol, Y., & Whiteside, W. S. (2024). Recent advancements in cross-linked starches for food applications- a review. *International Journal of Food Properties, 27*(1), 411–430.

Qian, S. Y., Tang, M. Q., Gao, Q., Wang, X. W., Zhang, J. W., Tanokura, M., & Xue, Y. L. (2019). Effects of different modification methods on the physicochemical and rheological properties of Chinese yam (Dioscorea opposita thunb.) starch. *Lebensmittel-Wissenschaft & Technologie, 116*. https://doi.org/10.1016/j.lwt.2019.108513

Saartrat, S., Puttanlek, C., Rungsardthong, V., & Uttapap, D. (2005). Paste and gel properties of low-substituted acetylated canna starches. *Carbohydrate Polymers, 61*(2), 211–221. https://doi.org/10.1016/j.carbpol.2005.05.024

Sandhu, K. S., Siroha, A. K., Punia, S., & Nehra, M. (2020). Effect of heat moisture treatment on rheological and in vitro digestibility properties of pearl millet starches. *Carbohydrate Polymer Technologies and Applications, 1*. https://doi.org/10.1016/j.carpta.2020.100002

Schmitz, C. S., De Simas, K. N., Santos, K., João, J. J., De Mello Castanho Amboni, R. D., & Amante, E. R. (2006). Cassava starch functional properties by etherification - Hydroxypropylation. *International Journal of Food Science and Technology, 41*(6), 681–687. https://doi.org/10.1111/j.1365-2621.2005.01136.x

Segura-Campos, M., Chel-Guerrero, L., & Betancur-Ancona, D. (2010). Effect of octenylsuccinylation on functional properties of Lima bean (*Phaseolus lunatus*) starch. *Journal of Food Process Engineering, 33*(4), 712–727.

S.A. Senanayake, K.K.D.S Ranaweera, A. Gunaratne, A. Bamunuarachchi, Formulation of vegetable soup mixture using physically modified sweet potato starch as a thickener. (2014),

Shaikh, M., Ali, T. M., & Hasnain, A. (2017). Effects of different modification reagents on functional properties of pearl millet starches. *Starch Staerke, 69*(3–4), Article 1600142.

Shaikh, M., Ali, T. M., & Hasnain, A. (2020). Comparative study on the application of chemically modified corn and pearl millet starches in white sauce. *Journal of Food Processing and Preservation, 44*(4), Article e14393.

Shi, C., Zhu, S., Ding, G., Du, C., Huang, D., & Li, Y. (2021). Modulating structure and properties of glutinous rice flour and its dumpling products by annealing. *Processes, 9*(12), 2248. https://doi.org/10.3390/pr9122248

Silva, I. R. A., Magnani, M., de Albuquerque, F. S. M., Batista, K. S., Aquino, J.d. S., & Queiroga-Neto, V. (2017). Characterization of the chemical and structural properties of native and acetylated starches from avocado (Persea americana mill.) seeds. *International Journal of Food Properties, 20*, Article S279. https://doi.org/10.1080/10942912.2017.1295259

Singhal, R. S., & Kulkarni, P. R. (1991). Studies on cross-linked A. Paniculatas (rajgeera) starch. *Starch - Stärke, 43*(1), 15–18. https://doi.org/10.1002/star.19910430107

Sobini, N., Darsiga, S., Kananke, T. C., & Srivijeindran, S. (2022). Characterization of modified palmyrah tuber starch by pre-gelatinization, acid and dextrinization processes and its applicability. *Food Chemistry Advances, 1*, Article 100143. https://doi.org/10.1016/j.focha.2022.100143

Subroto, E., Cahyana, Y., Indiarto, R., & Aray Rahmah, T. (2023). Modification of starches and flours by acetylation and its dual modifications: A review of impact on physicochemical properties and their applications. *Polymers, 15*(14), 2990. https://doi.org/10.3390/polym15142990

Subroto, E., Indiarto, R., Marta, H., & Shalihah, S. (2020). Application of heat-moisture treatment potato starch to improve the heat stability of tomato sauce. In *IOP conference series: Earth and environmental science*. IOP Publishing, Article 012076. Vol. 443, No. 1.

Taşkın, B., & Savlak, N. (2022). Functional, chemical, and sensorial properties of gluten-free fermented instant soup powders developed by use of mung bean and drum drying process. *Food Bioscience, 47*, Article 101677. https://doi.org/10.1016/j.fbio.2022.101677

Tang, T., Zhang, M., Law, C. L., & Liu, Y. (2023). Effects of ultrasound combined technology on quality and volatile compound properties of chili sauce. *Food Bioscience, 53*, Article 102771.

Teixeira, B. S., Inamura, P. Y., & Mastro, N. L. D. (2015). The influence of gamma irradiation on texture, color and viscosity properties of potato starch (International Nuclear Atlantic Conference - INAC 2015).

Thirathumthavorn, D., & Trisuth, T. (2008). Gelatinization and retrogradation properties of native and hydroxypropylated crosslinked tapioca starches with added sucrose and sodium chloride. *International Journal of Food Properties, 11*(4), 858–864. https://doi.org/10.1080/10942910701659567

Trela, V. D., Ramallo, A. L., & Albani, O. A. (2020). Synthesis and characterization of acetylated cassava starch with different degrees of substitution. *Brazilian Archives of Biology and Technology, 63*. https://doi.org/10.1590/1678-4324-2020180292

Wang, X., Hu, A., Zheng, J., Li, L., Li, L., & Li, Y. (2020). Physicochemical properties and structure of annealed sweet potato starch: Effects of enzyme and ultrasound. *Starch Staerke, 72*(11–12), Article 1900247.

Wang, J., Ren, F., Huang, H., Wang, Y., Copeland, L., Wang, S., & Wang, S. (2019). Effect of CaCl2 pre-treatment on the succinylation of potato starch. *Food Chemistry, 288*, 291–296.

Wang, J., Zhai, W., & Zheng, W. (2011). Preparation of granular cold-water-soluble corn starch by surface modification with poly (ethylene glycol). *Starch Staerke, 63*(10), 625–631.

Wang, T., Zhang, M., Fang, Z., Liu, Y., & Gao, Z. (2016). Rheological, textural and flavour properties of yellow mustard sauce as affected by modified starch, xanthan and guar gum. *Food and Bioprocess Technology, 9*(5), 849–858. https://doi.org/10.1007/s11947-016-1673-6

Wattanachant, S., Muhammad, K., Mat Hashim, D., & Rahman, R. A. (2003). Effect of crosslinking reagents and hydroxypropylation levels on dual-modified sago starch properties. *Food Chemistry, 80*(4), 463–471. https://doi.org/10.1016/S0308-8146(02)00314-X

Wiriyawattana, P., Suwonsichon, S., & Suwonsichon, T. (2018). Effects of drum drying on physical and antioxidant properties of riceberry flour. *Agriculture and Natural Resources, 52*(5), 445–450. https://doi.org/10.1016/j.anres.2018.11.008

Wongsagonsup, R., Pujchakarn, T., Jitrakbumrung, S., Chaiwat, W., Fuongfuchat, A., Varavinit, S., Dangtip, S., & Suphantharika, M. (2014). Effect of cross-linking on physicochemical properties of tapioca starch and its application in soup product. *Carbohydrate Polymers, 101*(1), 656–665. https://doi.org/10.1016/j.carbpol.2013.09.100

Xu, W., Yu, S., & Zhong, M. (2022). A review on food oral tribology. *Friction, 10*(12), 1927–1966. https://doi.org/10.1007/s40544-022-0594-9

Zavareze, E.d. R., Storck, C. R., de Castro, L. A. S., Schirmer, M. A., & Dias, A. R. G. (2010). Effect of heat-moisture treatment on rice starch of varying amylose content. *Food Chemistry, 121*(2), 358–365. https://doi.org/10.1016/j.foodchem.2009.12.036

Zhang, H., He, F., Wang, T., & Chen, G. (2021). Thermal, pasting, and rheological properties of potato starch dual-treated with CaCl2 and dry heat. *Lebensmittel-Wissenschaft & Technologie, 146*, Article 111467. https://doi.org/10.1016/j.lwt.2021.111467

Zhu, B., Liu, J., & Gao, W. (2017). Process optimization of ultrasound-assisted alcoholic-alkaline treatment for granular cold water swelling starches. *Ultrasonics Sonochemistry, 38*, 579–584.

Chapter 14

Starches as emulsifiers and stabilizers in liquid food and beverages

Swetha Vasudevan[1], G.C. Jeevitha[1], Aanchal Mittal[2], Faraz Ahmad[3], Shafiul Haque[4] and Esra Capanoglu[5]
[1]*Vellore Institute of Technology, Department of Bio-Sciences, Vellore, Tamil Nadu, India;* [2]*Vellore Institute of Technology, Department of Biotechnology, Vellore, Tamil Nadu, India;* [3]*Department of Integrative Biology, School of Bio Sciences and Technology, Vellore Institute of Technology, Vellore, Tamil Nadu, India;* [4]*Jazan University, Research and Scientific Studies Unit, College of Nursing and Health Sciences, Jazan, Saudi Arabia;* [5]*Istanbul Technical University, Department of Food Engineering, Faculty of Chemical and Metallurgical Engineering, Maslak, Turkey*

1. Introduction

Emulsion systems are often employed structures that transport bioactive compounds efficiently, protecting them from unfavorable environmental conditions (Niu et al., 2020). An emulsion is a mixture of incompatible liquid phases that acquire the form of tiny droplets. An emulsion requires three necessary components, namely an oil phase, a water phase, and a suitable emulsifier. These can be a protein, lipid, surfactant, or a peptide (Sneha & Kumar, 2022). Several studies reported that starch, protein, cellulose, and other substances are obtained as natural emulsifiers, which could be used for preparing natural food emulsions (Xu et al., 2020). Emulsifiers are incorporated into products to improve their stability and to maintain an adequate shelf life (Kumar, 2014). By modulating fat crystallization, changing the functional characteristics of other food ingredients, and lowering the surface tension at the interface of oil and water, food emulsifiers play a critical role in maintaining and creating emulsions (Kumar, 2014). Food industries manufacture several products, including sauces, spreads, dips, beverages, dressings, and creams, where the emulsion technology is extensively employed (Tan & McClements, 2021).

During storage, the instability of the food emulsions causes them to separate into liquid and oil phases. Evenly dispersed emulsions can be prepared using proteins, polymers, and emulsifiers, thereby adhering to the oil droplets and reducing interfacial tension (Wilde, 2019). Additionally, they offer kinetic stability over the long term by inhibiting Ostwald ripening, flocculation, and coalescence (Wilde, 2019). Although chemically produced surfactants are frequently used to stabilize emulsions, their alleged health hazards cause controversy in the pharmaceutical and food industries (Aryanti et al., 2024). Therefore, choosing the appropriate emulsifying agent is difficult because there are few safety and toxicity standards. For emulsion stabilization, food-derived particles are suggested as a novel substitute for synthetic surfactants (Aryanti et al., 2024). Natural substances like proteins, gums, starches, and fibers are being studied due to the need to reduce artificial additives in food (Wilde, 2019). The world's food supply uses more than hundreds of emulsifying agents as food additives. Frequently used food additive emulsifiers include xanthan gum (E415), mono- and diglycerides of fatty acids (E471), carboxymethylcellulose (E460-E469), guar gum (E412), lecithin (E322), and carrageenan (E407) (Cox et al., 2021). These substances help to stabilize emulsions, preventing them from separating into their component phases.

Plant photosynthesis produces starch, a naturally occurring biological polymer that is the second-largest biomass in the environment and an essential matrix for plant energy storage. It is crucial to living things and the main dietary source of carbohydrates for humans. The molecular, crystalline, and particle size contribute to the complex structure of starch (Zhou et al., 2023). Starch is utilized extensively in various industries since it is a completely biodegradable, renewable, and beneficial biopolymer. The versatility of starch is especially valued in the food sector, where it is utilized both in its native

Native and Modified Starches for Food Applications. https://doi.org/10.1016/B978-0-443-27431-2.00014-6

state and with various kinds of alterations that are intended to enhance functional qualities (Gałkowska et al., 2023). Food emulsifiers made with starch are widely employed due to their low cost, easy availability as a raw material, and their suitability for combining with various food additives (Fang et al., 2014). Starch grains, which are produced through physical and chemical alterations and whose size is below 1000 nm, are known as starch nanoparticles (Campelo et al., 2020). Incorporating various physical modification techniques, such as radiation, pressure, or heat treatment, changes the structure and characteristics of starch. New functional groups like phosphates or carboxyl are produced through chemical alterations, where the hydroxyl groups of the starch molecules react. Therefore, these alterations enhance the starch properties, and they are used in various applications. The drawbacks of native starch granules include size and hydrophobicity, which make them less potent when compared with the chemically treated nanosized starch in stabilizing emulsions (Li, Li, et al., 2014). Chemically treated hydrophobic starch granules can efficiently stabilize the pickering-type emulsions (Timgren et al., 2013). The starch-based nanoparticle stabilized emulsions are effective in stabilizing the beverage emulsion when compared to stabilized surfactant and biopolymer emulsions (Chaudhari et al., 2015). Starch is often regarded as a food-grade substance.

The stability of the emulsion is determined by the size and water affinity of the starch particles (Song et al., 2015). The COO-groups found in octenyl succinic chains, a charged group of polyelectrolytes, are referred to as octenyl succinic anhydride (OSA) starches (Shih & Daigle, 2003). Various industries, such as food, pharmaceuticals, cosmetics, and other industries, utilize the OSA starches as emulsifiers and stabilizing agents (Dokic et al., 2008). The emulsion stability can be enhanced by preventing the separation of phases and decreasing the size of the droplet. In addition, the application of starch includes stabilizing the liquid food by thickening and preventing syneresis. This chapter delves into the starch modification techniques and their function as an emulsifier, stabilizer in a wide range of liquid food and beverages.

2. Emulsion

Tiny droplets of one inseparable liquid are mixed with another to form a heterogeneous suspension known as an emulsion (Cox et al., 2021). The rapid agitation or stirring of the oil and water creates an emulsion. One liquid acts as the continuous phase, while the other is the dispersion phase. Once created, emulsions are defined by their spatial distribution. The emulsion is forced through extremely tiny openings by high-pressure homogenizers at pressures ranging from 10 to 500 MPa, and the typical droplet diameter of food emulsions is 0.1–10 mm (Wilde, 2019).

2.1 Different types of emulsion

Emulsions are of four forms, such as water-in-oil, oil-in-water, oil-water-oil, and water-oil-water (Fig. 14.1). The dispersion phase is water, and the continuous phase is fat, as in vinaigrettes and butter, forming a water-in-oil emulsion. In contrast, an oil-in-water emulsion is formed when fat droplets, like milk or mayonnaise, are distributed in a water phase. Water-oil-water and Oil-water-oil emulsions are the two forms of multilayer emulsions. Water-oil-water emulsions have water as a continuous phase and oil as a dispersed phase, whereas oil-water-oil emulsions contain water as a dispersed phase and oil as a continuous phase (Cox et al., 2021).

Emulsions are naturally unstable during formation and tend to return to their distinct phases. The instability is due to the creaming (the rise of lighter droplets), sedimentation (the settling of heavier droplets), flocculation (the aggregation of droplets without merging), coalescence (the merging of droplets into larger ones), and finally total separation into distinct oil and water phases (Cox et al., 2021). Mayonnaise separation is a typical example, with vinegar sinking to the bottom and oil floating to the top. In such cases, emulsifiers must be added to prevent the separation of the emulsion. Nanoemulsions are systems with droplet sizes usually less than 100 nm that are kinetically stable but thermodynamically unstable. They need low ratios of surfactant to oil and are produced by high-energy homogenization techniques. Droplet sizes in microemulsions range from 100–500 nm, making them thermodynamically stable systems. High surfactant-to-oil ratios are necessary for their formation, which involves low-energy processes. Pickering emulsions are kinetically inert but thermally unstable systems that are stabilized by solid particles (such as cellulose or proteins) rather than surfactants, while their droplet size ranges from 10 μm to 10 nm (Wang, Ai, et al., 2023).

3. Starch as an emulsifier in food products

Emulsifiers are molecules that naturally adsorb onto the interface between the water and oil phases of an emulsion (Wilde, 2019). During food preparation, they are often employed to improve the taste, appearance, and consistency and create a satisfying mouthfeel (Cox et al., 2021). In the food sector, the quality of the product and shelf life are extensively

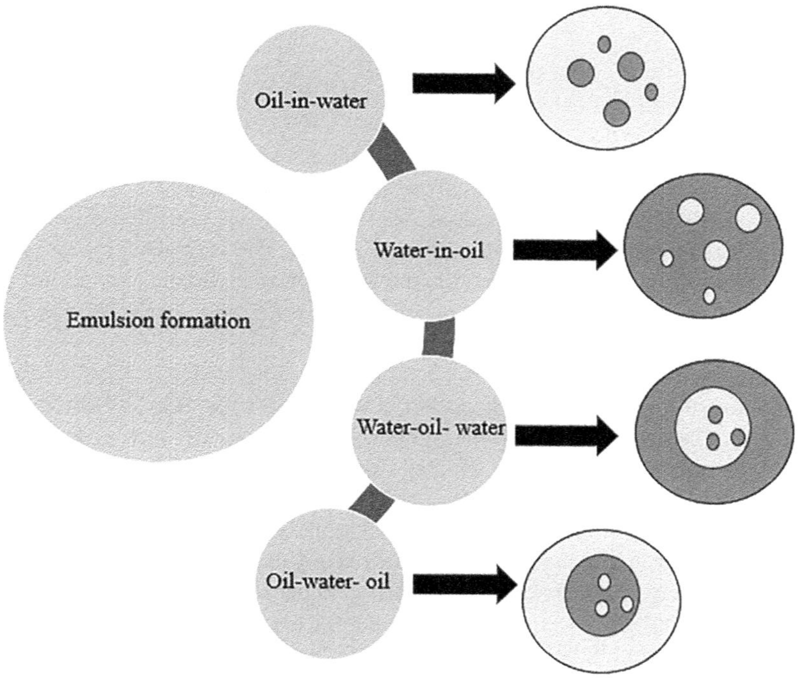

FIGURE 14.1 Schematic representation of emulsion formation. Emulsion formation.

enhanced with the application of emulsifiers, thereby playing a vital role. In the bread manufacturing process, diacetyl tartaric acid aids in the production of carbon dioxide bubbles, which fortify the dough, thereby increasing the texture and volume (Garzon et al., 2018).

Emulsifiers facilitate the inhibition of the components' separation in low- and no-fat applications (Hasenhuettl & Hartel, 2019). They also provide lubrication, reduction of the fat globule size, fat dispersion, perception, and complexation with starch and proteins (Hasenhuettl & Hartel, 2019). Another vital role of emulsifiers in the preparation of confectionery products includes the stabilization of fat droplets by serving as plasticizers, lubricants, and hydration agents. In addition to this, during the compound coating, they influence the fat crystallization, thereby enhancing the chewiness and texture of the gum, and help to regulate the viscosity and flow properties of the chocolates (Hasenhuettl & Hartel, 2019). The addition of emulsifiers to the shortenings has several advantages, such as reducing the mixing time and improving the longevity, water absorption, texture, volume, rate of hydration, and release of flavor (Hasenhuettl & Hartel, 2019).

An emulsion was prepared using modified waxy maize starch (2% w/w) along with whey protein isolate (1% w/w) as an emulsifier. The mixture was stirred for 2 h and hydrated overnight at 4°C. Then, orange oil (0.033% w/w) was added along with sunflower oil (20% w/w) and homogenized for 2 min. To prevent microbial growth, sodium benzoate (0.1%) was incorporated into it. The resultant emulsion was passed through a microfluidizer (Hu et al., 2019). The maize starch emulsion underwent 5 cycles at 300 bar, and the whey protein isolate emulsion underwent 4 cycles at 375 bar. Through both in vitro and in vivo studies, the stability and flavor release of emulsions were investigated. The findings showed that the oral behavior of a protein-stabilized emulsion differs significantly from that of a starch-stabilized emulsion. The modified starch acted as an emulsifier, and this innovative approach could be used to manipulate food emulsions' oral behavior (Hu et al., 2019).

Octenyl succinic anhydride is currently the only substance that is legally approved to hydrophobically modify starch for usage as a food emulsifier. Even yet, the FDA in the US limits this to less than 3% by weight (Mu et al., 2021). Modified starch was used as an emulsifier in combination with whey protein isolate, along with two different oils, namely orange oil and sunflower oil, to prepare an oil-in-water emulsion through the microfluidization technique, where it was used for the controlled flavor release when mixed with saliva (Karthik et al., 2019). Waxy maize starch underwent pullulanase debranching followed by octenyl succinic anhydride esterification modification to make it a superior emulsifier (Wang, Zhu, et al., 2023). The α-amylase-resistant modified starch stabilized the oil-in-water emulsions and remained stable for more than 3 weeks at 4 and 25°C, but it had a lesser emulsifying property than the altered natural starch (Wang, Zhu, et al., 2023).

4. Starch as a stabilizer in food products

Food stabilizers help to improve the critical aspect of the overall quality by enhancing the consistency, texture, and shelf life of products. The sensory properties of various food items, including drinks, dressings, sauces, and ice creams, can be improved by the addition of stabilizing agents. Among several sensory qualities, stabilizing agents are often utilized to inhibit separation and to enhance the viscosity and satisfactory mouthfeel (Samal et al., 2023). They are also employed to preserve the quality of food items during storage and transportation. The modified starches have stabilizing properties in various dairy products where they play a vital role in enhancing their sensory attributes, texture, and consistency (Saleh et al., 2020).

The study by Zhao et al. (2017) developed a formulation for preparing a stabilized oil-in-water emulsion using gelatinized kudzu starch. Commercially available kudzu root starch was dispersed in water and gelatinized by stirring at 90°C for 20 min; this served as the continuous phase. The dispersed phase was composed of various oils, such as medium-chain triglyceride, limonene, and soybean oil. These emulsions were homogenized using a high-pressure homogenizer at 10,000 rpm for 5 min. Various oil and starch concentrations were prepared, and the kudzu starch emulsion was compared with other starches such as corn, rice, and wheat starch. The results showed that 3% gelatinized starch gave better stability when compared to other formulations. It has also been proven that the gelatinized kudzu root starch could be employed in stabilizing the oil-in-water emulsion without the use of any commercial emulsifier. It also demonstrated that rotor-stator homogenization followed by high-pressure homogenization was a well-suited technique for the preparation of gelatinized oil-in-water starch emulsion. This stabilization technique could be employed to yield a high-quality food product at a low cost. To enhance the consistency and firmness of a variety of food products, including ice cream, salad dressings, and sauces, the food sector employs the application of hydrocolloids that acquire a gel-like structure when hydrated. A few examples of hydrocolloids obtained from plant exudates, microorganisms, and seaweeds include pectin, guar gum, carrageenan, xanthan gum, and agar (Samal et al., 2023).

The starch nanocrystals stabilized through pickering emulsions are regarded as green or eco-friendly, as the starch is non-allergenic, edible, and Generally Recognized as Safe (GRAS) (Zhu, 2019). In addition, in the food industry, the starch-based pickering emulsions provide increased stability to food items, maintain prolonged shelf life and enhance sensory properties (Zhu et al., 2024). Fig. 14.2 shows the native, modified, and nano-starch stabilized pickering emulsions and their application in food products (Almeida et al., 2024). The starch or modified starch plays a vital role as a thickener and stabilizer, improving the texture of colloidal emulsions. This provides stability of the product and thereby increases consumer appeal (Almeida et al., 2024).

Xian et al. (2020) reported that the (food-grade) Pickering emulsion can be stabilized using modified potato starches. To produce surfactant-free particle-stabilized emulsion, various starches such as potato, sweet potato, and tapioca starches were selected. These starches were modified using four different treatments, such as dry-heat treatment, hydrothermal treatment, OSA modification, and hydroxypropylation. The result revealed that the OSA modification of starch was superior in offering hydrophobic properties to the tuberous starches. The OSA-modified starches had a great viscosity, and they resulted in stable emulsion formation when compared to other treatments. Native potato starch had superior emulsifying ability and was capable of stabilizing the emulsion. In contrast, the hydrothermally treated starch could not stabilize the O/W emulsion.

The use of proteins obtained from plant or animal sources is shown to enhance the stability and the texture of several food products manufactured in the food sector. Casein, whey protein, gelation, and soy protein are a few examples of proteins that function as foam stabilizers, emulsifiers, and gelling agents (Samal et al., 2023). Whey protein improves the texture of cheese and the shelf-life of oil-in-water emulsions (Samal et al., 2023).

5. Starch

Many plants (rice, wheat and corn), roots, and tubers contain starch, which is a complex carbohydrate (Rahaman et al., 2021). Due to its wide availability and versatility, it is a renewable, low-cost natural polymer extensively used in various industries, including food, chemicals, textiles, and medicine (Gilet et al., 2018). However, its industrial applications are frequently limited by its inherent form. Many physical (heating and irradiation) and chemical (etherification, esterification and oxidation) changes are employed to overcome these limitations. These alterations seek to improve its technical properties by disrupting molecular order, degrading the polymer, rearranging molecular structures, and introducing cross-

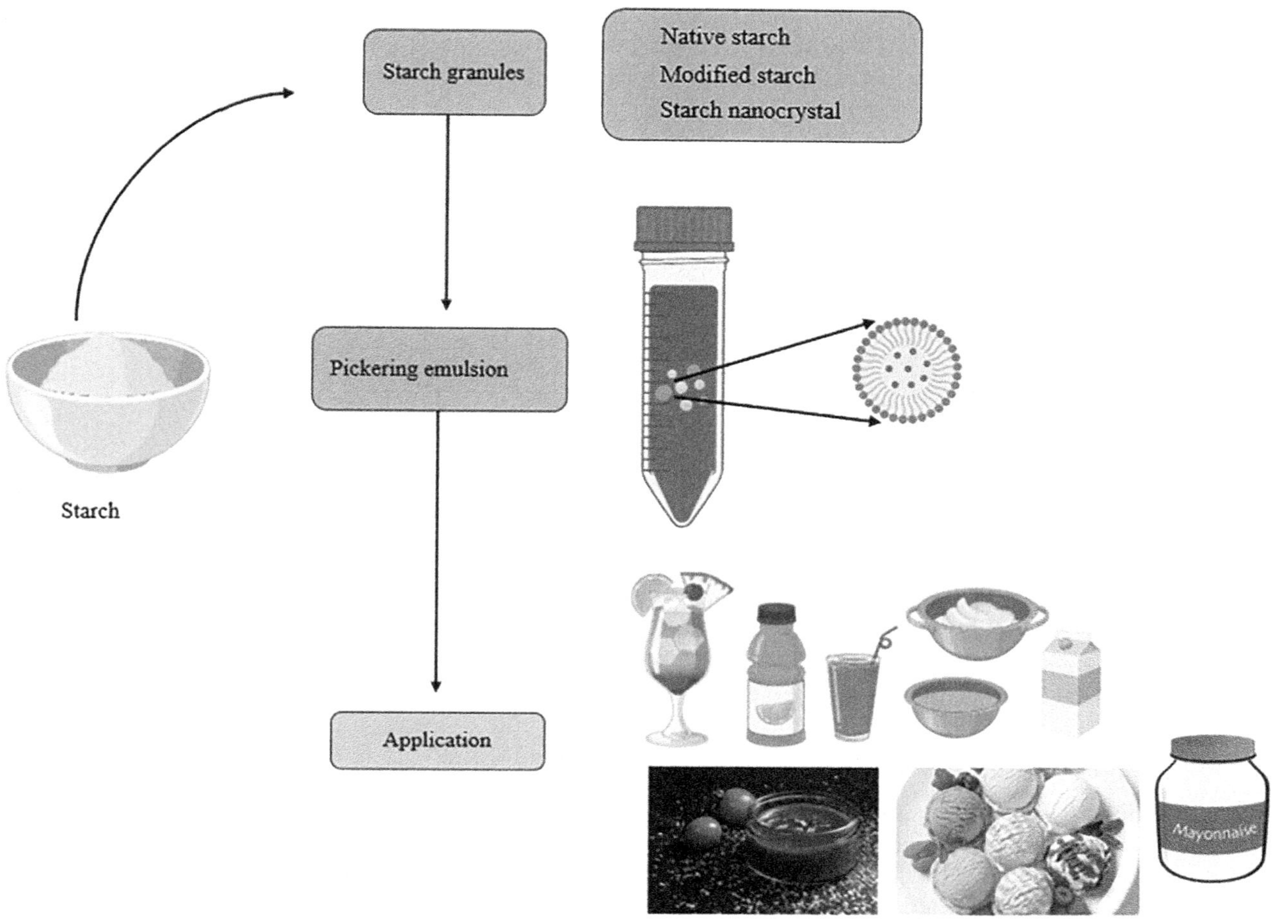

FIGURE 14.2 **The native, modified, and nano-starch stabilized pickering emulsions and their application in food products.** Applications of pickering emulsion.

links or chemical groups (Rahaman et al., 2021). Eventually, these changes improve solubility, boost thermal stability, improve shear stress resistance, and reduce challenges like syneresis and retrogradation (Rahaman et al., 2021).

5.1 Structure and composition

The two forms of glucose polymers that comprise starch are highly branched amylopectin and linear unbranched amylose. Granules are formed by the complex arrangement of these elements within a semi-crystalline structure (Yılmaz & Tugrul, 2023). Two chains of polymers, especially the sugar amylose (alpha 1–4 links) and amylopectin (both alpha 1–4 linkages and alpha 1–6 linkages), combine to form starch (Rahaman et al., 2021). Amylose (10%–30%) and amylopectin (70%–90%) make up the majority of starch; however, different types have variable proportions. The only factors affecting the structural and functional characteristics of starch are the structure and quantity of amylose and amylopectin. The two types of α-glucans, amylose and amylopectin, constitute around 98%–99% of the dry mass of starch granules. The amount of amylose varies: waxy starches have the least (<15%), normal starches have moderate (20%–35%), and high-amylose starches have the most (>40%) (Tester et al., 2004). Cereals and some roots and tubers have a moisture content of 10%–12% and 14%–18%, respectively, for air-equilibrated starches (Tester et al., 2004).

5.2 Functional properties of starch

The functional characteristics of the starch are affected by the presence of fiber, proteins, and lipids that are present after extraction. The swelling index, water solubility, retrogradation, gelling ability, and gelatinization temperature are modified when they form compounds with lipids and amylose (Bashir & Aggarwal, 2019). The starch is insoluble in cold water, and it alters into a plasticizer when it is heated above 52°C; this process is known as gelatinization. This process is affected by various factors such as water, starch concentration, pH, and salts (Bashir & Aggarwal, 2019). Due to the presence of excess water, solubilization and maximum swelling take place in foods such as sauces, gravies, and puddings. Extremely high viscosities could be attained in extruded or baked products as they have lower moisture content with incomplete starch hydration (Hasenhuettl & Hartel, 2019). Considering the pasting properties of starch is vital as it has a crucial effect on the viscosity in various sectors, including the food industry. The significant function of the amylose is to increase the final viscosity and the propensity of starch to retrograde (Bashir & Aggarwal, 2019). In general, the increase in the amylose content results in a decrease in the peak viscosity and temperature of gelatinization (Bashir & Aggarwal, 2019). Regardless of their size and shape, native starch granules have some undesirable properties that restrict their industrial use, including poor solubility, syneresis, retrogradation, thermal breakdown, and excessive viscosity upon gelatinization (Compart et al., 2023). Therefore, there is a need for modification of starch for its wide range of applications in the food industry, especially for liquid foods and beverages.

6. Modification of starch

The various processing, operational, and environmental factors that are used to make any changes to the starch's molecular structure are known as modifications. The modifications may result in either advantageous or disadvantageous effects on the structure and characteristics of starch molecules (Nawaz et al., 2020). The native starches obtained from different plants will vary in their structures and functions. The structurally modified starches have enhanced functional and structural properties that help them acquire beneficial results in various formulations (Nawaz et al., 2020). The functional quality of starch can be enhanced through mechanical, chemical, or physical alterations (Korma, 2016; Nabeshima & Grossmann, 2001; Singh & Ali, 2000). The processing and storage conditions of products also influence the functional properties of starch (Villareal et al., 1976).

6.1 Physical modification

Physical modifications are generally more secure and highly acceptable than chemical modifications because of the chemical structure modification, which in turn restricts their application for several food and nutraceutical formulations (Nawaz et al., 2020). Physical modifications refer to alterations in the morphology and structure of starch caused by several factors, including pH, temperature, moisture, pressure, and irradiation. These changes can affect the starch's suitability for different applications (Nawaz et al., 2020). Some of the frequently used physical alteration techniques for starch are instantaneous controlled pressure treatment, microwave treatment, superheating, gamma and ultraviolet irradiation, freeze-thaw treatment, thermal inhibition, osmotic and high pressure, pulsed electric field treatment, micronization in a vacuum ball mill, annealing, and mechanical activation by ball mill. Physical modifications can be divided into two types such as thermal and non-thermal modifications.

6.1.1 Thermal modification

Thermal modifications comprise heat treatment, hydrothermal treatment, heat-moisture, and radiation treatment. Different heat treatments include simple oven heating, superheating, extrusion heating, and fluidized bed heating. Radiation treatment includes exposure to microwave, ultraviolet radiation, and gamma irradiation (Nawaz et al., 2020). Mild heating of starch at (45–65°C) modifies the amylopectin-to-amylose ratio and slightly affects its structure. It has no discernible impact on the physico-mechanical characteristics of starch (Metta et al., 2022). Superheating of starch to an adequately high temperature (180–220°C) causes it to solidify into spreadable gel particles that cool to a creamy texture. It enhances the gelatinization and pasting quality of starch (Majzoobi et al., 2011). In hydrothermal treatment, starch is heated in a liquid medium, where the granules of starch undergo a physical rearrangement as a result which facilitates easier amylase digestion by enhancing the stability, motility, and size of the granules. Additionally, it enhances the ability of starch to gelatinize (Loisel et al., 2006; Priya et al., 2014; Tester & Debon, 2000). Microwave treatment at different temperature ranges and moisture influences the starch's dielectric constant and enhances its morphological and granular crystallinity properties. It further enhances the paste viscosity and pasting temperature, lowers peak viscosity and gelatinization, and promotes the solubility, gelling, and swelling ability, emulsifying activity, and oil and water holding capacity (Nawaz et al., 2020).

The study by Peng et al. (2024) applied dry heat treatment at 190°C to alter corn starch, and it thereby increasing hot water soluble amylopectin. Through the Pickering mechanism, they were able to form nanoaggregates approximately

33 nm that played a role as a particle stabilizer in an oil-in-water emulsion. This emulsion was found to be stable during the storage period of 180 days. Therefore, the dry heat-modified corn starch that was hot water soluble could be utilized as a natural emulsifier in a wide range of applications.

The heat moisture treatment of starch from different sources, namely trifoliate yam, white yam, and sweet potato, was shown to have enhanced water absorption capacity compared to the native and chemically modified starch through acetylation (Okereke et al., 2022). This can be used as stabilizers and thickening agents in gravies, sauces, and beverages (Okereke et al., 2022). In another study, the quinoa starch was treated with hydrochloric acid followed by microwave dry heating. This modified starch enhanced the emulsifying capacity of starch. Therefore, the nano starch could be utilized for stabilizing Pickering emulsion (Han et al., 2025).

6.1.2 Non-thermal modification

The non-thermal modification comprises mechanical treatment, annealing, thawing, pH variation, pressure treatment, sonication, deep freezing, pulsed electric field treatment and cold plasma treatment (Nawaz et al., 2020). The radius of gyration and molar size of starch particles decrease in response to a high pH, indicating partial disintegration. Starch undergoes hydrolysis at low pH levels, particularly in the amorphous portion of granules, which decreases the starch's molecular weight. The swelling, solubility, and compression attributes are also improved by an increase in pH. Treatment with a low pH increases the starch's capacity to gel (Lee et al., 2009; Singh & Ali, 2000; Wang & Wang, 2001; Metta et al., 2022). The effect of moisture on starch films can be both plasticizing and anti-plasticizing. Tensile strength is one indicator of an anti-plasticizing effect, whereas tensile modulus, water vapor permeability, and linear expansion, are indicators of a plasticizing effect (Chang et al., 2000). Simple grinding reduced the amorphous or crystalline ratio and viscosity of starch. It also increased the water-holding capacity and susceptibility to both chemical and physical characteristics of starch (Grant, 1998; Liu et al., 2011).

The crystal structure of starch is broken down into amorphous particles when mechanical force is applied in a stirring ball mill. This process lowers the temperature at which gelatinization occurs, and increases its solubility in cold water, shear-thinning, and viscosity (Huang et al., 2007). Hydrostatic pressure treatment, which involves employing 400–900 MPa of high pressure, breaks down and retrogrades starch granules. It also prevents swelling, lowers viscosity, and maintains the nutrient content and taste of the starch (Stute et al., 1996). Granules of starch are distorted by ultrasonication, which also raises the temperature of gelatinization, and enhances solubility, viscosity, and swelling properties while decreasing pasting efficiency and digestibility (Jambrak et al., 2010; Nadir et al., 2015; Yu et al., 2013). In high temperatures, thermal inhibition leads to the dehydration of starch, thereby stabilizing viscosity, reducing granular size, increasing cohesive texture, and producing an anhydrous condition (Chiu et al., 1998; Lim et al., 2002). Starch treated with low-temperature plasma resulted in altered physical and functional characteristics, like elevated pasting and viscosity but decreased tendency for retrogradation, free radical-induced cross-linking, increased amylose leaching, and decreased crystallinity (Sarangapani et al., 2015; Thirumdas et al., 2017; Zou et al., 2004).

Electric field-modified starches stabilize oil-in-water emulsions in foods like drinks, mayonnaise, and salad dressings by demonstrating an improved emulsifying capacity (Zhu, 2018). The pulsed electric field-based esterification leads to greater hydrophobicity and alters the crystallinity, thereby decreasing droplet size and helping to stabilize emulsions. It also aids in maintaining consistency by offering gelling, stabilizing, and thickening attributes in products like soups, sauces, and desserts (Zhu, 2018). Starch extracted from sweet potato, trifoliate yam, and white yam was modified chemically by the acetylation process, which shows they are well-suited to be used as fillers, stabilizers, and thickeners in beverages, dairy products, gravies, and sauces (Okereke et al., 2022). Another study by Villamonte et al. (2016) utilized high-pressure treatment at 600 MPa for 6 min at 20°C to modify the corn starch, and it was shown to stabilize the emulsion through homogenization. The study also demonstrated the potential for producing emulsions from gelatinized corn starch without the need for surfactants or heat.

6.2 Chemical modification

In the chemical modification of starch, new functional groups are added without changing the size or shape of the molecule. Important areas for modification are the three reactive hydroxyl groups present in the glucose units of amylose and amylopectin. Stabilizing intramolecular and intermolecular bonding changes the physical characteristics of starch, like salting, retrogradation, and gelatinization. Chemical modifications of starch that are frequently employed include oxidation using different oxidizing agents, cationization, etherification, cross-linking, esterification, and graft-polymerization of starch with synthesized polymers (Cizova et al., 2008; Greenwood et al., 1977; Hernandez, 2018; Jyothi et al., 2006; Korma, 2016; Liu et al., 2015; Paleos et al., 2017; Páramo-Calderón et al., 2016; Waliszewski et al., 2003; Zanella Pinto et al., 2019).

6.2.1 Methods of the chemical modification of starch

The oxidation process provides native starch with additional carboxyl and carbonyl groups, which depolymerize and inhibit recrystallization. It improves the stability, clarity, and binding qualities but decreases the dispersion viscosity (Boukhalfa et al., 2018; Kuakpetoon & Wang, 2001; Rutenberg & Solarek, 1984). The addition of a hydroxyethyl group (hydroxyethylation) alters the granular structure of the starch, and enhances the binding of anticancer and pharmacological compounds (Paleos et al., 2017). Interaction between starch and fatty acids causes the development of amylose-fatty acyl complexes and modifies the starch's thermal properties and optical function (Bulpin et al., 1982). Monophosphate or diphosphate starch is produced when a phosphate group is introduced to the hydroxyl groups in starch. The resulting starch has resistance to high temperature, a relatively low pH and high shear. It also has stability during freezing and thawing, improved texture, viscosity and paste transparency. In addition to this, it also avoids the chain linearity (Nabeshima & Grossmann, 2001; Waliszewski et al., 2003). The treatment with cationic molecules like amino, imino, ammonium, sulfonium, or phosphonium has been shown to enhance the starch's clarity, dispersibility, solubility, and stability (Korma, 2016).

The application of an aqueous alkaline slurry containing a combination of sodium tripolyphosphate and sodium trimetaphosphate etherifies and esterifies the starch granules by cross-linking. This procedure minimizes water solubility, improves internal structure and stability, and fortifies the granules. Moreover, it increases melting enthalpy, glass transition, stability at high temperatures, and gelatinization (Boukhalfa et al., 2018; Chung et al., 2004; Jyothi et al., 2006; Klein et al., 2010; Nabeshima & Grossmann, 2001; Páramo-Calderón et al., 2016).

Recently, it was suggested that OSA-modified starches with tiny granule sizes are better alternatives for emulsion stabilization (Rayner et al., 2012; Timgren et al., 2013). The combination of OSA-modified maltodextrins and waxy starch may be used as an emulsion stabilizer, but this necessitates further research (Agama-Acevedo & Bello-Perez, 2017). An oil-in-water emulsion was prepared using 10% waxy corn octenyl succinic anhydride starch modified through enzymatic hydrolysis, which resulted in reduced molar mass, thereby enhancing emulsification properties (Li, Hong, et al., 2014). A mayonnaise-like emulsion (o/w) was prepared using 10% OSA starch, which formed a Pickering emulsion. During the prolonged storage, the emulsion stability of the starch was found to be lower when compared to Arabic gum and soy soluble polysaccharides (Chivero et al., 2016). The incorporation of xanthan gum along with 10% OSA starch as an emulsifier improves its stability by altering the continuous phase viscosity (Krstonošić et al., 2015). The inclusion of starch nanoparticles produced by mechanical processing or acid hydrolysis, for example, treatment by milling or ultrasound, could serve as an alternative to stabilize the emulsion after OSA modification (Agama-Acevedo & Bello-Perez, 2017).

7. Application of starch as an emulsifier in liquid food and beverages

A solution made with added sugar and carbonated water is known as a beverage emulsion. In drinks like orange soda, the oil-in-water emulsion adds both flavor and opacity. To maintain the stability of these emulsions and avoid phase separation, precipitation, and flavor loss, modified starch can be utilized (Saboonchi et al., 2021). The most widely utilized polysaccharide in the preparation of emulsions is starch granules because of their inexpensiveness and biodegradable properties. For stable systems, native starch must be modified as it is not amphiphilic (Xie et al., 2020). An emulsifier is added to decrease the surface tension between the two phases, which in turn helps with the formation of a stable emulsion. The starch components' strong hydrophilicity is anticipated to result in less stability of the emulsion (Aryanti et al., 2024). Heat induced starch gelatinization alters the structures, which facilitates the formation of a highly stable emulsion when compared to proteins and lipids (Liu et al., 2018). Starch nanoparticles (SNPs) are defined as the granule size of starch created through different physical, chemical or biologically modified starches, whose size is less than 600–1000 nm. To produce a starch nanoparticle, it is shown that a "bottom-up" approach, like the anti-solvent method, is more suitable. The lesser particle size with high reproducibility, minimal equipment requirement and simple process shows the significant development potential of SNPs. Since native starch has minimal emulsifying ability, in order to produce a stabilized emulsion, the starch can be combined with another bio-based emulsifier (Lee et al., 2021).

An emulsifier composed of a 1:2 ratio of cyclodextrin was developed for the study, and after 6 months of storage, the emulsification index remained high and starch emulsification was greatly improved (Li et al., 2022). Raw starch has a highly hydrophilic surface and contains a higher number of hydroxyl groups, whereas SNP can potentially be an emulsifier for food (Zhou et al., 2023). Modified starch is a potent emulsifier that reduces creaming and oil ring formation while absorbing water in the continuous phase to produce a uniform and stable emulsion (Nasrolahi et al., 2022). They are also used in the preparation of an orange terpene-based model citrus beverage emulsion (Reiner et al., 2010). Acetylated starch acts as an emulsion stabilizer, while physically modified starch is known to function as an emulsifier or fat substitute in emulsions in addition to stabilizing them (Małyszek et al., 2021).

The gum acacia altered with OSA starch can be utilized as an additive for unstandardized drinks, and it has been approved by the government of Canada (Vilela et al., 2018). Due to the enhanced solubility and their capability to produce less viscous dispersions, the hydrolyzed OSA starch is an excellent emulsifier and stabilizer (Prochaska et al., 2007). The OSA-modified pineapple starch has better emulsifying qualities than unmodified starch, and it can be used in medicinal applications to stabilize the emulsions(Latt et al., 2019). Compared to various starches like corn, rice, and cassava, the pineapple stem starch is found to have superior properties. It has the lowest paste viscosity and the most amylose content, retrogradation, pasting, syneresis rates, and gelatinization enthalpy. It also has an abundant amount of resistant starch and slowly digested starch. The emulsifying ability of modified starch is superior to cassava starch and is suitable for the preparation of low-moisture foods, including cereal, bread, pasta, and cookies (Sriprablom et al., 2023). Fruit drinks, soup, milk, and liquid eggs are some of the liquid foods with low electrical conductivity and low viscosity that can be sterilized using pulsed electric field technology. Recently, it was discovered that starches treated with a pulsed electric field produced resistance, quickly digested, and slowly digested starches (Mhaske et al., 2023).

Various starch-based food emulsifiers, such as beta-cyclodextrin, sodium starch phosphate, octenyl succinic anhydride starch, carboxymethylated starch, hydroxypropyl starch, and acetylated starch, are used in rice and flour products, frozen foods, and beverages due to emulsifying properties such as amphipathy, stability, dissolubility, and other physicochemical characteristics. The use of starches can reduce syneresis and increase firmness during storage, but sensory defects like foreign flavors may also occur. Starches can increase firmness and decrease syneresis during storage, but they can also cause sensory flaws like foreign flavors (Imbachi-Narvaez et al., 2018).

In food industries, starch is mainly used as a food additive to control the stability, uniformity, and texture of soups and sauces, helping in the resistance of gel breaks down during the process. Because of the high adaptive physicochemical characteristics, maize starch is one such type that is highly valued and used as a thickener, stabilizer, colloidal gelling and water retention agent, and as an adhesive (Egharevba, 2020). For low-fat versions of traditionally fatty foods, such as industrial milk-based desserts like yogurt (Lin et al., 2024) or, reduced-fat hard salami, a suitably modified starch is used as a fat substitute, having roughly one-third the typical fat content. It is a substitute for the product Olestra for the latter kind of applications (He et al., 2019).

Instant desserts are thickened with pre-gelatinized starch, which allows the food to thicken when cold water or milk is added. The same is true for cheese sauce granules, like those in lasagna, macaroni and cheese, or gravy granules, which can be thickened with boiling water without becoming lumpy. Commercial pizza toppings that contain modified starch will become runny when cooled after thickening in the oven and remaining on top of the pizza (Zhang et al., 2024).

7.1 Different types of starches used as emulsifiers

7.1.1 Corn starch

Food-grade emulsions could be made with starch instead of surfactants. Corn starch treated with high-pressure stabilized emulsions. Creaming stability increased with the treated starch concentration (Villamonte et al., 2016). Pre-gelatinized corn starch functions better than gelatinized starch to produce stable emulsions, suggesting that starch-based emulsifiers could replace traditional surfactants. The greater stable emulsion was achieved with Tween 80 (3%) and pre-gelatinized corn starch (5%) (Aryanti et al., 2024).

7.1.2 Modified tapioca starches

Polysaccharide gums are used as stabilizers in ice cream to improve viscosity in the ice cream mixture and in the unfrozen phase of the ice cream after freezing. The application of modified tapioca starch as a stabilizer has been shown to effectively prevent freeze-thawing, improve texture, and lower the melting point. As a result, frozen food products frequently contain these modified starches, such as octenyl succinic anhydride starch (OSA), acetylated starch (AS), and hydroxypropylated starch (HPS). The OSA, a starch derivative, is made by a typical esterification process that involves mixing starch and cyclic dicarboxylic acid anhydride in water under alkaline conditions. The dual functional hydrophilic and hydrophobic groups of OSA make it an effective emulsifier (Thaiudom et al., 2008).

7.1.3 OSA modified starch

An OSA-modified starch is utilized in emulsion systems and the delivery of bioactive compounds. Homogenization makes the starch amphiphilic and carries oil to the water interface (Cheng et al., 2021). The starch polymers become surface active and can adsorb at the oil-water interface due to the presence of OSA groups. Adsorbed starch provides effective steric stabilization of emulsions because of the large size of the starch polymers (Nilsson & Zielke, 2024). Low-fat mayonnaise was made by replacing 75% of the fat with OSA-modified starches. The emulsifying qualities of mayonnaise were improved by OSA-modified starches (Bajaj et al., 2019). The OSA-based starch emulsion is a good substitute for protein because its stability is unaffected by temperature, pH, or ionic strength (Wang, Ai, et al., 2023).

7.1.3.1 Amphiphilic starch

Modified starches, like amphiphilic starches, carry both hydrophilic and lipophilic moieties, so they can help stabilize oil-in-water emulsions. These amphiphilic fat-loving gum Arabic are mainly used to stabilize beverage emulsions like orange sodas, lemonades, etc. In native form, starch is hydrophilic in nature, having no affinity toward the hydrophobic substances like fats and oils (Kasprzak et al., 2018). The uses of different polysaccharides and emulsifiers are used to improve the stability of different vegetable protein beverages (Liu et al., 2024).

8. Application of starch as a stabilizer in liquid food and beverages

Native starches lack emulsifying qualities due to the lack of hydrophobicity and excess larger particle sizes, which makes it difficult to create an outer coating surrounding the oil globules in the emulsion. Their potential to raise the viscosity of the continuous phase makes them applicable as emulsion stabilizers (Shao et al., 2020). Reduced starch particle size forms emulsions with tiny droplets and enhances the emulsion storage duration (Zhu, 2019). A less stable emulsion might be due to the greater hydrophilicity of starch (Aryanti et al., 2024). Compared with stabilized surfactant and biopolymer emulsions, starch-based nanoparticle-stabilized emulsions can further enhance beverage emulsion stability (Chaudhari et al., 2015). Medium-milled SNPs (700 nm) can be utilized to prepare a Pickering emulsion with good emulsion stability (Lu et al., 2018). Citrate-esterified cassava starch contains increased resistant starch content and acts as a good dietary fiber source. The incorporation of citrate-esterified cassava starch into a dairy beverage has not modified the sensory and rheological properties (Oltramari et al., 2017). The modified cassava starch in sweet whey fermented dairy beverage controls syneresis over product storage without affecting the soluble solids, acidity, pH, or consistency index, indicating its effectiveness as an emulsifier (Imbachi-Narvaez et al., 2018). The modified cassava starch in fermented soybean beverages improved the structural and rheological properties compared to the native starch (Rodriguez-Ruiz et al., 2023). Fermented dairy drinks made from pullulanase-modified cassava starch as a stabilizer exhibited increased viscosity, decreased syneresis, and increased overall acceptance (Lopez-Ochoa et al., 2022).

The stability of yoghurt can be increased by the addition of corn starch without affecting the nutritional content (Nsanzabera et al., 2023). The different types of starch used in various beverages are shown in Table 14.1. The quality of soup production was found to be improved with the use of tapioca starch (Li et al., 2013). In fruit fillings, the tapioca-based natural starch can be employed as a new thickener (Oladunmoye et al., 2014). The addition of hydrocolloids like sodium carboxymethylcellulose (CMC) and hydroxypropyl methylcellulose (HPMC) to rice starch improves the gelling ability as well as stabilizes the emulsions (Shrivastava et al., 2018; Sun et al., 2017).

TABLE 14.1 Types of starch in various beverages.

S. no.	Types of beverages	Source of starch	References
1	Soy-based beverage	Cassava starch	Drunkler et al. (2012)
2	Dairy beverage	Citrate-esterified cassava starch	Oltramari et al. (2017)
3	Sweet whey fermented dairy beverage	Modified cassava starch	Imbachi-Narvaez et al. (2018)
4	Yogurt	Corn starch, cassava starch	Nsanzabera et al. (2023)
5	Fermented soybean beverage	Modified cassava starch	Rodriguez-Ruiz et al. (2023)
6	Fermented dairy beverage	Enzymatically modified cassava starch	Lopez-Ochoa et al. (2022)
7	Sweet potato beverage	Cross-linked starch	Jia-yan et al. (2020)
8	Soursop beverage	Modified starch	Cheong et al. (2015)
9	Yoghurt-like beverages	Potato starch	Saeedyzadeh et al. (2017)

TABLE 14.2 Modified starch products and their applications in beverages.

S. no.	Modified starch products	Application in beverages	References
1	Starch acetate	Used as a stabilizer, thickener, and flavor encapsulating agent in soups, sauces, creamers, and beverages.	Egharevba (2020)
2	Starch succinate	Used as a thickener in soups and custard creams.	Ackar et al. (2015)
3	OSA starch	Emulsion stabilizer in beverages	Egharevba (2020)
4	Distarch phosphate	Used as stabilizers and thickeners in soups, sauces.	Neelam et al. (2012)
5	Etherified starches	Used as stabilizers for gravies, sauces and dips, beverages.	Korma (2016)
6	Crosslinked starches	Used as thickeners in gravies, soups and sauces.	Egharevba (2020)
7	Pre-gelatinized starch	Used in sauces and instant soups.	Lee and Yoo (2023)

In liquid foods and beverages, the application of modified starches like esterified, crosslinked, and etherified starches is widely employed. Modified starch products and their applications in beverages are shown in Table 14.2. In the manufacturing of a sweet whey fermented drink, the OSA-based altered cassava starch is used (Imbachi-Narvaez et al., 2018). The incorporation of modified cassava starches has been shown to improve the quality of soy bean drink, where the starch functions as a stabilizer and/or thickener (Rodriguez-Ruiz et al., 2023). In liquid foods like cereal porridges, the gamma irradiation-based treatment of starch has been shown to decrease the viscosity, and has yielded a higher cereal content (Mhaske et al., 2023). Dairy sweets made with starch are popular throughout Europe; examples include "Natillas" in Spain, "Vanilla Vla" in the Netherlands, and "Crème dessert" in France. Milk, thickeners (modified starch), sugar, fragrance, and colorants are essentially used in its formulation (Abbas et al., 2010). Using the amylomaltase-treated starch (ATS) domain improves creaminess four times more effectively than maltdextrin, which is now a common fat substitute in set-style yoghurt (Abbas et al., 2010).

8.1 Different types of starches used as stabilizers

8.1.1 Annealed Bambara starch

Bambara groundnut (Vigna subterranea (L.) verdc) is a neglected legume of African origin that is gradually gaining attention because of its relatively high nutritional content and drought resistance. There are potential uses for Bambara grain's 18%–45% starch in both food and medicine. Annealing is the process of heating a mixture of starch and water, typically in ratios of one to two, for varied periods of time. According to reports, the resultant annealed starch exhibits enhanced thermal stability and a lower rate of retrogradation. Annealed Bambara starch may be used alone or in mixtures to stabilize ice cream, especially when it is produced at 45°C (Nwaogazie et al., 2024).

8.1.2 Quinoa and rice starch

Rice starch is the most effective emulsifier among the native starches. Rice starch works better than other natural starches like wheat, waxy corn, and potatoes. When the starch content in the water phase was greater than 3%, emulsions stabilized with rice starch remained stable against droplet coalescence for several months (Li et al., 2013).

8.1.3 Potato starch

In the dairy products, the stabilizers play a vital role as they can increase the ratio of total solids in these products, enhance the viscosity and sensory qualities, and inhibit or lower the whey separation during storage. The addition of a suitable quantity of potato starch of about 0.75% and 1% has been shown to enhance the pH, overall acidity, syneresis, and sensory properties. Additionally, utilizing varying concentrations of extracted starch during storage significantly decreased the number of bacteria in yogurt (Altemimi, 2018).

8.1.4 Cornstarch and cassava starch

Cornstarch and cassava starch were used as stabilizers to prepare a yogurt beverage. The yogurt without the incorporation of starch was found to have higher water-holding capacity and syneresis. Whereas, the cornstarch incorporated yoghurt was effective in improving water-holding capacity and preventing syneresis when compared with cassava starch. The sensorial attributes of yogurt incorporated with cornstarch were more acceptable to customers (Nsanzabera et al., 2023). In soy-based beverages, cassava starch could be employed to increase viscosity without compromising sensory appeal. Even a lesser quantity of cassava starch (0.4%–0.6%) significantly increased the viscosity, suggesting cassava starch is a promising natural stabilizer for soy-based beverages (Drunkler et al., 2012). Another study by Rahayu et al. (2024) stated that the cornstarch along with guar gum acted as a stabilizer, and it improved the quality of the yogurt drink. The yogurt beverage quality was maintained for 6 weeks. In addition to this, the combination of cornstarch and guar gum as a stabilizer has been shown to produce high viscosity.

8.1.5 Etherified and esterified starch

Etherification of starch has properties such as enhanced starch paste clarity, freeze-thaw stability, higher viscosity, and lower syneresis. Etherified starches can be utilized in various food applications, including sauces, dips, gravies, and fruit pie fillings Korma (2016) (Fig. 14.3). An example of etherified starch is hydroxypropyl starch (Singh et al., 2007). The hydroxypropyl starch can withstand high temperatures and remain stable, particularly at pH 6, when compared to starch acetate. It has applications in the dairy industry and frozen foods (Egharevba, 2020). Esterification of starch has properties such as greater paste clarity, retrogradation, decreased gelatinization temperature, and lesser capacity to gel. The applications of esterified starches include emulsion stabilizers and the encapsulation of beverage clouds Korma (2016). Examples of esterified starch include starch acetate, starch sodium octenylsuccinate, and acetylated distarch adipate (Singh et al., 2007). Starch acetate is used as a stabilizer, thickener, and flavor encapsulating agent in soups, sauces, creamers, and beverages (Egharevba, 2020).

8.1.6 Cross-linked starch

Cross-linked starches are found to be stable at high temperatures, shear, and acidic environments (Korma, 2016). Due to these properties, they are used as viscosifiers and texturizers in the preparation of soups, gravies, and sauces Korma (2016). Cross-linked starch was used to develop the sweet potato turbid juice, and it was shown to have improved stability and reduced sedimentation (Jia-yan et al., 2020).

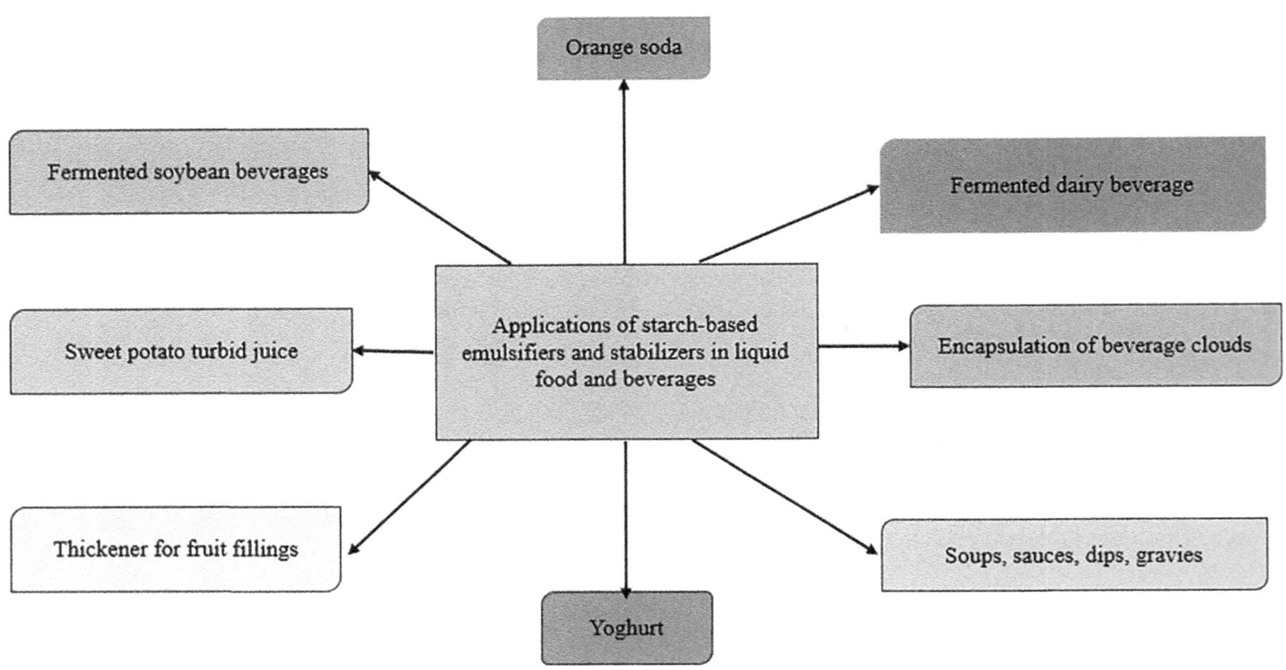

FIGURE 14.3 Applications of starch-based emulsifiers and stabilizers in liquid food and beverages. Starch emulsifiers in foods.

8.1.7 Sweet potato starch

Sweet potato starch that has undergone hydrothermal modification provides a higher viscosity than commercial corn starch, making it a useful alternative thickener for soup preparation (Senanayake et al., 2014). So, this could be used as an additive for the enhancement of viscosity (Senanayake et al., 2014). Gelatinized sweet potato starch provides the creamy texture and helps to maintain and enhance the stability of food items during thermal and mechanical processing stages. They have a wide application in various food products such as puddings, dairy alternatives, frozen desserts, and plant-derived yogurts (Li et al., 2019; Paixao e Silva et al., 2021). A study by Torres-Álvarez et al. (2025) stated that sweet potato starch could be employed as a novel additive in emulsion-based foods like dressings.

8.1.8 Pregelatinized wheat starch

The common techniques used for preparing pregelatinized starch include extrusion, spray drying, and drum drying (Ma et al., 2022). Pregelatinized wheat starch that has been drum-dried shows low crystallinity, high solubility in water, and cold-watery viscosity, which implies it is a great gelling or thickening agent in cold desserts, salad dressings, and baby foods (Majzoobi et al., 2011). Low starch fragmentation, high water absorbing capacity, and swelling at room temperature are a few of the properties of drum-dried pregelatinized starch. In low-fat mayonnaise-type emulsions, this starch acts as a clean-label oil replacer by enhancing the emulsion stability, increasing the viscosity, and lowering the droplet size of oil (Roman et al., 2022). In addition to their application in beverages, the pregelatinized wheat starch was also used in the preparation of fermented and non-fermented flour products such as bread and wheat-based noodles and tartary buckwheat noodles, respectively (Ma et al., 2022).

9. Conclusion

In the food sector, the starch-based stabilizers and emulsifiers play a vital role, mainly in various liquid foods and beverages, namely gravies, creamers, soups, sauces, yogurt, and dressings. The drawbacks of native starches, such as size and hydrophobicity, poor solubility, syneresis, retrogradation, and thermal breakdown, were overcome by the various types of physically or chemically modified starches, which offer them better stability and emulsification properties, thereby expanding their application in the development of novel food products. Further research is required to discover new starch sources, optimize alteration methods, and emulsification effects in various food products.

References

Abbas, K. A., Khalil, S. K., & Hussin, A. S. M. (2010). Modified starches and their usages in selected food products: A review study. *Journal of Agricultural Science, 2*(2), 90.

Ackar, D., Babic, J., Jozinovic, A., Milicevic, B., Jokic, S., Milicevic, R., & Subaric, D. (2015). Starch modification by organic acids and their derivatives: A review. *Molecules, 20*(10), 19554–19570.

Agama-Acevedo, E., & Bello-Perez, L. A. (2017). Starch as an emulsions stability: The case of octenyl succinic anhydride (OSA) starch. *Current Opinion in Food Science, 13*(1), 78–83.

Almeida, R. L. J., Santos, N. C., Monteiro, S. S., Rios, N. S., & dos Santos, E. S. (2024). Exploring the potential of native and modified starch and starch nanocrystals in Pickering emulsions: Current advances, future perspectives, and challenges. *Food Bioscience, 61*(1), Article 104675.

Altemimi, A. B. (2018). Extraction and optimization of potato starch and its application as a stabilizer in yogurt manufacturing. *Foods, 7*(2), 14.

Aryanti, N., Wardhani, D. H., Nafiunisa, A., Putri, N. A. Y., Firmansyah, R. A., Cahyani, N. D., & Nabila, S. (2024). Application of modified corn starch in stabilization of ultrasonic-assisted pickering oil in water emulsion. *Food Research, 8*(1), 78–89.

Bajaj, R., Singh, N., & Kaur, A. (2019). Properties of octenyl succinic anhydride (OSA) modified starches and their application in low fat mayonnaise. *International Journal of Biological Macromolecules, 131*, 147–157.

Bashir, K., & Aggarwal, M. (2019). Physicochemical, structural and functional properties of native and irradiated starch: A review. *Journal of Food Science and Technology, 56*(2), 513–523.

Boukhalfa, F., Kadri, N., Bouchemel, S., Ait Cheikh, S., Chebout, I., & Madani, K. (2018). Chemically modified starch and utilization in food stuffs. *Mediterranean Journal of Nutrition and Metabolism, 11*(1), 37–50.

Bulpin, P. V., Welsh, E. J., & Morris, E. R. (1982). Physical characterization of amylose-fatty acid complexes in starch granules and in solution. *Starch-Stärke, 34*(10), 335–339.

Campelo, P. H., Sant'Ana, A. S., & Clerici, M. T. P. S. (2020). Starch nanoparticles: Production methods, structure, and properties for food applications. *Current Opinion in Food Science, 33*(1), 136–140.

Chang, Y. P., Cheah, P. B., & Seow, C. C. (2000). Plasticizing—Antiplasticizing effects of water on physical properties of tapioca starch films in the glassy state. *Journal of Food Science, 65*(3), 445–451.

Chaudhari, A., Pan, Y., & Nitin, N. (2015). Beverage emulsions: Comparison among nanoparticle stabilized emulsion with starch and surfactant stabilized emulsions. *Food Research International, 69*(1), 156–163.

Cheng, F., Ai, Y., & Ghosh, S. (2021). Utilization of octenyl succinic anhydride-modified pea and corn starches for stabilizing oil-in-water emulsions. *Food Hydrocolloids, 118*(1), Article 106773.

Cheong, K. W., Mirhosseini, H., Leong, W. F., Hamid, N. S. A., Osman, A., Basri, M., & Tan, C. P. (2015). Rheological properties of modified starch—Whey protein isolate stabilized soursop beverage emulsion systems. *Food and Bioprocess Technology, 8*(6), 1281–1294.

Chiu, C. W., Schiermeyer, E., Thomas, D. J., & Shah, M. B. (1998). *Thermally inhibited starches and flours and process for their production.* Washington, DC: U.S. Patent and Trademark Office. U.S. Patent No. 5,725,676.

Chivero, P., Gohtani, S., Yoshii, H., & Nakamura, A. (2016). Assessment of soy soluble polysaccharide, gum arabic and OSA-starch as emulsifiers for mayonnaise-like emulsions. *LWT-Food Science and Technology, 69*(1), 59–66.

Chung, H. J., Woo, K. S., & Lim, S. T. (2004). Glass transition and enthalpy relaxation of cross-linked corn starches. *Carbohydrate Polymers, 55*(1), 9–15.

Cizova, A., Srokova, I., Sasinkova, V., Malovikova, A., & Ebringerova, A. (2008). Carboxymethyl starch octenylsuccinate: Microwave-and ultrasound-assisted synthesis and properties. *Starch-Stärke, 60*(8), 389–397.

Compart, J., Singh, A., Fettke, J., & Apriyanto, A. (2023). Customizing starch properties: A review of starch modifications and their applications. *Polymers, 15*(16), 3491.

Cox, S., Sandall, A., Smith, L., Rossi, M., & Whelan, K. (2021). Food additive emulsifiers: A review of their role in foods, legislation and classifications, presence in food supply, dietary exposure, and safety assessment. *Nutrition Reviews, 79*(6), 726–741.

Dokic, P., Dokic, L., Dapcevic, T., & Krstonosic, V. (2008). Colloid characteristics and emulsifying properties of OSA starches. In *Colloids for nano- and biotechnology* (pp. 48–56). Berlin, Heidelberg: Springer Berlin Heidelberg.

Drunkler, N. L., Leite, R. S., Mandarino, J. M. G., Ida, E. I., & Demiate, I. M. (2012). Cassava starch as a stabilizer of soy-based beverages. *Food Science and Technology International, 18*(5), 489–499.

Egharevba, H. O. (2020). Chemical properties of starch and its application in the food. *Chemical Properties of Starch, 63*.

Fang, F., Gu, Z., Hong, Y., Li, Z., & Cheng, L. (2014). Starch-based food emulsifier and its application. *Journal of the Chinese Cereals and Oils Association, 29*, 110–114.

Gałkowska, D., Kapuśniak, K., & Juszczak, L. (2023). Chemically modified starches as food additives. *Molecules, 28*(22), 7543.

Garzon, R., Hernando, I., Llorca, E., & Rosell, C. M. (2018). Understanding the effect of emulsifiers on bread aeration during breadmaking. *Journal of the Science of Food and Agriculture, 98*(14), 5494–5502.

Gilet, A., Quettier, C., Wiatz, V., Bricout, H., Ferreira, M., Rousseau, C., & Tilloy, S. (2018). Unconventional media and technologies for starch etherification and esterification. *Green Chemistry, 20*(6), 1152–1168.

Grant, L. A. (1998). Effects of starch isolation, drying, and grinding techniques on its gelatinization and retrogradation properties. *Cereal Chemistry, 75* (5), 590–594.

Greenwood, C. T., Muir, D. D., & Whitcher, H. W. (1977). Hydroxyethyl starch as a cryoprotective agent for human red blood cells the relation between the molecular properties and the cryoprotective effect. *Starch-Stärke, 29*(10), 343–347.

Han, L., Fu, W., Wang, Y., Yang, X., Li, J., Yang, X., & Wang, Y. (2025). Effect of microwave irradiation assisted with pre-acidification on the molecular structure and emulsification properties of quinoa starch. *Journal of Food Science, 90*(11), Article e70699.

Hasenhuettl, G. L., & Hartel, R. W. (Eds.). (2019). *Food emulsifiers and their applications* (Vol 19, pp. 11–37). New York: Springer.

He, J., Han, Y., Liu, M., Wang, Y., Yang, Y., & Yang, X. (2019). Effect of 2 types of resistant starches on the quality of yogurt. *Journal of Dairy Science, 102*(5), 3956–3964.

Hernandez, A. R. (2018). Chemical modification of starch with synthetic. *Applications of Modified Starches, 2*, 3–22.

Hu, X., Karthik, P., & Chen, J. (2019). Manipulating oral behaviour of food emulsions using different emulsifiers. *International Journal of Food Science and Technology, 54*(7), 2408–2415.

Huang, Z. Q., Lu, J. P., Li, X. H., & Tong, Z. F. (2007). Effect of mechanical activation on physico-chemical properties and structure of cassava starch. *Carbohydrate Polymers, 68*(1), 128–135.

Imbachi-Narvaez, P. C., Sepulveda-Valencia, J. U., & Rodriguez-Sandoval, E. (2018). Effect of modified cassava starch on the rheological and quality properties of a dairy beverage prepared with sweet whey. *Food Science and Technology, 39*(1), 134–142.

Jambrak, A. R., Herceg, Z., Šubarić, D., Babić, J., Brnčić, M., Brnčić, S. R., & Gelo, J. (2010). Ultrasound effect on physical properties of corn starch. *Carbohydrate Polymers, 79*(1), 91–100.

Jia-yan, Z. H. A. N. G., Jian-wen, X. I. O. N. G., Na, C., & Zheng-pei, C. (2020). Effect of cross-linked starch on the stability of sweet potato turbid juice beverage. *Food and Machinery, 36*(2), 187–192.

Jyothi, A. N., Moorthy, S. N., & Rajasekharan, K. N. (2006). Effect of cross-linking with epichlorohydrin on the properties of cassava (Manihot esculenta Crantz) starch. *Starch-Stärke, 58*(6), 292–299.

Karthik, P., Ettelaie, R., & Chen, J. (2019). Oral behaviour of emulsions stabilized by mixed monolayer. *Food Research International, 125*(1), Article 108603.

Kasprzak, M. M., Macnaughtan, W., Harding, S., Wilde, P., & Wolf, B. (2018). Stabilisation of oil-in-water emulsions with non-chemical modified gelatinised starch. *Food Hydrocolloids, 81*(1), 409–418.

Klein, M. I., DeBaz, L., Agidi, S., Lee, H., Xie, G., Lin, A. H. M., & Koo, H. (2010). Dynamics of Streptococcus mutans transcriptome in response to starch and sucrose during biofilm development. *PLoS One, 5*(10), Article e13478.

Korma, S. A. (2016). Chemically modified starch and utilization in food stuffs. *International Journal of Nutrition and Food Sciences, 5*(4), 264.

Krstonošić, V., Dokić, L., Nikolić, I., & Milanović, M. (2015). Influence of xanthan gum on oil-in-water emulsion characteristics stabilized by OSA starch. *Food Hydrocolloids, 45*(1), 9–17.

Kuakpetoon, D., & Wang, Y. J. (2001). Characterization of different starches oxidized by hypochlorite. *Starch-Stärke, 53*(5), 211–218.

Kumar, P. (2014). Emulsion and its applications in food processing: A review. *International Journal of Engineering Research and Applications, 4*, 241–248.

Latt, S. S., Patomchaiviwat, V., Sriamornsak, P., & Piriyaprasarth, S. (2019). Modification of pineapple starch from stem and rhizome using multiple desired response optimization and its characterization. *Pharmaceutical Sciences Asia, 46*(4).

Lee, J. H., Han, J. A., & Lim, S. T. (2009). Effect of pH on aqueous structure of maize starches analyzed by HPSEC-MALLS-RI system. *Food Hydrocolloids, 23*(7), 1935–1939.

Lee, Y. S., Tarté, R., & Acevedo, N. C. (2021). Synergistic effects of starch nanoparticles and chitin nanofibers on the stability of oil-in-water Pickering emulsions. *Food Chemistry, 363*(1), Article 130301.

Lee, H., & Yoo, B. (2023). Particle agglomeration and properties of pregelatinized potato starch powder. *Gels, 9*(2), 93.

Li, Z., Hong, Y., Gu, Z., Tian, Y., Li, Z., & Cheng, L. (2014). Emulsification properties of enzymatically treated octenyl-succinic anhydride starch. *Starch-Stärke, 66*(11–12), 1089–1095.

Li, Q., Huang, Y., Du, Y., Chen, Y., Wu, Y., Zhong, K., Huang, Y., & Gao, H. (2022). Food-grade olive oil Pickering emulsions stabilized by starch/β-cyclodextrin complex nanoparticles: Improved storage stability and regulatory effects on gut microbiota. *LWT, 155*, 112950.

Li, C., Li, Y., Sun, P., & Yang, C. (2013). Pickering emulsions stabilized by native starch granules. *Colloids and Surfaces A: Physicochemical and Engineering Aspects, 431*, 142–149.

Li, C., Li, Y., Sun, P., & Yang, C. (2014). Starch nanocrystals as particle stabilisers of oil-in-water emulsions. *Journal of the Science of Food and Agriculture, 94*(9), 1802–1807.

Li, S., Ye, F., Zhou, Y., Lei, L., & Zhao, G. (2019). Rheological and textural insights into the blending of sweet potato and cassava starches: In hot and cooled pastes as well as in fresh and dried gels. *Food Hydrocolloids, 89*(1), 901–911.

Lim, S. T., Han, J. A., Lim, H. S., & BeMiller, J. N. (2002). Modification of starch by dry heating with ionic gums. *Cereal Chemistry, 79*(5), 601–606.

Lin, Q., Jiang, L., Li, X., Sang, S., Ji, H., Jin, Z., & Qiu, C. (2024). Starch based fat replacers in food system: Modification, structured design, and application. *Food Bioscience, 59*(1), Article 104149.

Liu, W., Li, Y., Chen, M., Xu, F., & Zhong, F. (2018). Stabilizing oil-in-water emulsion with amorphous and granular octenyl succinic anhydride modified starches. *Journal of Agricultural and Food Chemistry, 66*(35), 9301–9308.

Liu, T. Y., Ma, Y., Yu, S. F., Shi, J., & Xue, S. (2011). The effect of ball milling treatment on structure and porosity of maize starch granule. *Innovative Food Science & Emerging Technologies, 12*(4), 586–593.

Liu, Y., Wu, Q., Zhang, J., & Mao, X. (2024). Effect of different amphiphilic emulsifiers complexed with xanthan gum on the stability of walnut milk and structural characterization of their complexes. *Food Chemistry, 455*(1), Article 139873.

Liu, J., Yang, R., & Yang, F. (2015). Effect of the starch source on the performance of cationic starches having similar degree of substitution for papermaking using deinked pulp. *Bioresources, 10*(1), 922–931.

Loisel, C., Maache-Rezzoug, Z., Esneault, C., & Doublier, J. L. (2006). Effect of hydrothermal treatment on the physical and rheological properties of maize starches. *Journal of Food Engineering, 73*(1), 45–54.

Lopez-Ochoa, J. D., Cadena-Chamorro, E., Ciro-Velasquez, H., & Rodríguez-Sandoval, E. (2022). Enzymatically modified cassava starch as a stabilizer for fermented dairy beverages. *Starch-Stärke, 74*(7–8), Article 2100242.

Lu, X., Xiao, J., & Huang, Q. (2018). Pickering emulsions stabilized by media-milled starch particles. *Food Research International, 105*(1), 140–149.

Małyszek, Z., Lewandowicz, J., Le Thanh-Blicharz, J., Walkowiak, K., Kowalczewski, P.Ł., & Baranowska, H. M. (2021). Water behavior of emulsions stabilized by modified potato starch. *Polymers, 13*(13), 2200.

Ma, H., Liu, M., Liang, Y., Zheng, X., Sun, L., Dang, W., & Liu, C. (2022). Research progress on properties of pre-gelatinized starch and its application in wheat flour products. *Grain & Oil Science and Technology, 5*(2), 87–97.

Majzoobi, M., Radi, M., Farahnaky, A., Jamalian, J., Tongtang, T., & Mesbahi, G. (2011). Physicochemical properties of pre-gelatinized wheat starch produced by a twin drum drier. *Journal of Agricultural Science and Technology A, 13*(2), 193–202.

Metta, S., Reddy, M. N., Fatima, N., Manasa, N., Bala, N. A., & Maneesha, S. (2022). A review on: Disintegrants and super disintegrants with prominence role of starch as disintegrant and its methods of modification. *World Journal of Pharmaceutical Research, 11*(13), 468–500.

Mhaske, P., Majzoobi, M., & Farahnaky, A. (2023). Starch, modified starch, and extruded foods. In *Food texturology: Measurement and perception of food textural properties* (pp. 393–420). Cham: Springer International Publishing.

Mu, M., Karthik, P., Chen, J., Holmes, M., & Ettelaie, R. (2021). Effect of amylose and amylopectin content on the colloidal behaviour of emulsions stabilised by OSA-modified starch. *Food Hydrocolloids, 111*(1), Article 106363.

Nabeshima, E. H., & Grossmann, M. V. E. (2001). Functional properties of pregelatinized and cross-linked cassava starch obtained by extrusion with sodium trimetaphosphate. *Carbohydrate Polymers, 45*(4), 347–353.

Nadir, A. S., Helmy, I. M. F., Nahed, M. A., Wafaa, M. M. A., & Ramadan, M. T. (2015). Modification of potato starch by some different physical methods and utilization in cookies production. *International Journal of Current Microbiology and Applied Sciences, 4*(10), 556–569.

Nasrolahi, S., Sadeghizadeh-Yazdi, J., Ehrampoush, M. H., Madadizadeh, F., & Khalili, E. (2022). Evaluation of rheological and optical properties plus stability of beverage cloud emulsions prepared with corn oil, gum rosin, and modified starch. *Food Science and Nutrition, 11*(2), 806–816.

Nawaz, H., Waheed, R., Nawaz, M., & Shahwar, D. (2020). Physical and chemical modifications in starch structure and reactivity. In *Chemical properties of starch*. IntechOpen.

Neelam, K., Vijay, S., & Lalit, S. (2012). Various techniques for the modification of starch and the applications of its derivatives. *International Research Journal of Pharmacy, 3*(5), 25–31.

Nilsson, L., & Zielke, C. (2024). Starch-based food emulsifiers. In *Starch in food* (pp. 525–543). Woodhead Publishing.

Niu, B., Shao, P., & Sun, P. (2020). Ultrasound-assisted emulsion electrosprayed particles for the stabilization of β-carotene and its nutritional supplement potential. *Food Hydrocolloids, 102*(1), Article 105634.

Nsanzabera, F., Manishimwe, A., Mwiseneza, A., & Irakoze, E. (2023). Production of corn (Zea mays) starch and cassava (Manihot esculenta) starch and their application as yogurt stabilizer. *Food and Nutrition Sciences, 14*(7), 589–600.

Nwaogazie, F. O., Akinwande, B. A., Adebo, O. A., & Oyeyinka, S. A. (2024). Application of annealed bambara starch as a stabilizer in ice cream production. *Macromol, 4*(3), 533–543.

Okereke, G. O., Igbabul, B. D., Ikya, J. K., & Araka, O. (2022). Physicochemical and functional properties of modified starches of white yam, trifoliate yam and sweet potato. *Journal of Food Chemistry and Nanotechnology, 8*(2), 50–60.

Oladunmoye, O. O., Aworh, O. C., Maziya-Dixon, B., Erukainure, O. L., & Elemo, G. N. (2014). Chemical and functional properties of cassava starch, durum wheat semolina flour, and their blends. *Food Science and Nutrition, 2*(2), 132–138.

Oltramari, K., Madrona, G. S., Neto, A. M., de Morais, G. R., Baesso, M. L., de Cássia Bergamasco, R., & de Moraes, F. F. (2017). Citrate esterified cassava starch: Preparation, physicochemical characterisation, and application in dairy beverages. *Starch-Stärke, 69*(11–12), Article 1700044.

Paixao e Silva, G.D.L., Bento, J.A.C., Ribeiro, G.O., Liao, L.M., Soares Junior, M.S., & Caliari, M. (2021). Application potential and technological properties of colored sweet potato starches. *Starch-Stärke, 73*(1–2), Article 2000100.

Páramo-Calderón, D. E., Carrillo-Ahumada, J., Juárez-Arellano, E. A., Bello-Pérez, L. A., Aparicio-Saguilán, A., & Alvarez-Ramirez, J. (2016). Effect of cross-linking on the physicochemical, functional and digestibility properties of starch from Macho (Musa paradisiaca L.) and Roatan (Musa sapientum L.) banana varieties. *Starch-Stärke, 68*(7–8), 584–592.

Paleos, C. M., Sideratou, Z., & Tsiourvas, D. (2017). Drug delivery systems based on hydroxyethyl starch. *Bioconjugate Chemistry, 28*(6), 1611–1624.

Peng, Z., Wu, M., Liao, Q., Zhu, N., Li, Y., Huang, Y., & Wu, J. (2024). Hot-water soluble fraction of starch as particle-stabilizers of oil-in-water emulsions: Effect of dry heat modification. *Carbohydrate Polymers, 336*, Article 122130.

Priya, B., Gupta, V. K., Pathania, D., & Singha, A. S. (2014). Synthesis, characterization and antibacterial activity of biodegradable starch/PVA composite films reinforced with cellulosic fibre. *Carbohydrate Polymers, 109*, 171–179.

Prochaska, K., Kędziora, P., Le Thanh, J., & Lewandowicz, G. (2007). Surface properties of enzymatic hydrolysis products of octenylsuccinate starch derivatives. *Food Hydrocolloids, 21*(4), 654–659.

Rahaman, A., Kumari, A., Zeng, X. A., Farooq, M. A., Siddique, R., Khalifa, I., & Manzoor, M. F. (2021). Ultrasound based modification and structural-functional analysis of corn and cassava starch. *Ultrasonics Sonochemistry, 80*, Article 105795.

Rahayu, P. P., Unur, D., Fathimah, F., Nursafitri, E., Asadel, B., Yunita, C. N., Purwadi, P., Tohari, I., & Jaya, F. (2024). The quality of yogurt drink with guar gum and cornstarch as stabilizer. *Livestock and Animal Research, 22*(3), 156–166.

Rayner, M., Sjöö, M., Timgren, A., & Dejmek, P. (2012). Quinoa starch granules as stabilizing particles for production of Pickering emulsions. *Faraday Discussions, 158*(1), 139–155.

Reiner, S. J., Reineccius, G. A., & Peppard, T. L. (2010). A comparison of the stability of beverage cloud emulsions formulated with different gum Acacia-and starch-based emulsifiers. *Journal of Food Science, 75*(5), E236–E246.

Rodriguez-Ruiz, J. D., Rodriguez-Sandoval, E., & Hernández, M. S. (2023). Influence of modified cassava starch on the physicochemical properties of a fermented soybean beverage. *Agronomía Colombiana, 41*(1), 1.

Roman, L., Walker, M. R., Detlor, N., Best, J., & Martinez, M. M. (2022). Pregelatinized drum-dried wheat starch of different swelling behavior as clean label oil replacer in oil-in-water emulsions. *Foods, 11*(14), 2044.

Rutenberg, M. W., & Solarek, D. (1984). Starch derivatives: Production and uses. In *Starch: Chemistry and technology* (pp. 311–388). Academic Press.

Saboonchi, S., Mehran, A., Bahramizadeh, P., & Massoud, R. (July 2021). Applications of modified starch in food. In *3rd International congress on Engineering, Technology & Innovation* (pp. 1–8).

Saeedyzadeh, N., Zamindar, N., Pezeshkzadeh, M., & Tahmourespour, A. (2017). Evaluation of yogurt-like beverages made of potato starch waste and grape must. *Journal of Food Measurement and Characterization, 11*(4), 1787–1794.

Saleh, A., Mohamed, A. A., Alamri, M. S., Hussain, S., Qasem, A. A., & Ibraheem, M. A. (2020). Effect of different starches on the rheological, sensory and storage attributes of non-fat set yogurt. *Foods, 9*(1), 61.

Samal, D., Malviya, S., & Meesala Krishna Murthy, P. (2023). Use of stabilizers in food industry and their biosynthesis pathways with health impact. *Chelonian Research Foundation, 18*, 502–514.

Sarangapani, C., Devi, Y., Thirundas, R., Annapure, U. S., & Deshmukh, R. R. (2015). Effect of low-pressure plasma on physico-chemical properties of parboiled rice. *LWT-Food Science and Technology, 63*(1), 452–460.

Senanayake, S. A., Ranaweera, K. K. D. S., Gunaratne, A., & Bamunuarachchi, A. (2014). Application of hydrothermally modified sweet potato starch as a substitute additive for soup mixture. *Journal of Food Processing, 1*(1), Article 904125.

Shao, P., Feng, J., Sun, P., Xiang, N., Lu, B., & Qiu, D. (2020). Recent advances in improving stability of food emulsion by plant polysaccharides. *Food Research International, 137*(1), Article 109376.

Shih, F. F., & Daigle, K. W. (2003). Gelatinization and pasting properties of rice starch modified with 2-octen–1-ylsuccinic anhydride. *Food, 47*(1), 64–67.

Shrivastava, M., Yadav, R. B., Yadav, B. S., & Dangi, N. (2018). Effect of incorporation of hydrocolloids on the physicochemical, pasting and rheological properties of colocasia starch. *Journal of Food Measurement and Characterization, 12*(2), 1177–1185.

Singh, V., & Ali, S. Z. (2000). Acid degradation of starch. The effect of acid and starch type. *Carbohydrate Polymers, 41*(2), 191–195.

Singh, J., Kaur, L., & McCarthy, O. J. (2007). Factors influencing the physico-chemical, morphological, thermal and rheological properties of some chemically modified starches for food applications—A review. *Food Hydrocolloids, 21*(1), 1–22.

Sneha, K., & Kumar, A. (2022). Nanoemulsions: Techniques for the preparation and the recent advances in their food applications. *Innovative Food Science & Emerging Technologies, 76*(1), Article 102914.

Song, X., Pei, Y., Qiao, M., Ma, F., Ren, H., & Zhao, Q. (2015). Preparation and characterizations of Pickering emulsions stabilized by hydrophobic starch particles. *Food Hydrocolloids, 45*(1), 256–263.

Sriprablom, J., Suphantharika, M., Smith, S. M., Amornsakchai, T., Pinyo, J., & Wongsagonsup, R. (2023). Physicochemical, rheological, in-vitro digestibility, and emulsifying properties of starch extracted from pineapple stem agricultural waste. *Foods, 12*(10), 2028.

Stute, R., Heilbronn, Klingler, R. W., Boguslawski, S., Eshtiaghi, M. N., & Knorr, D. (1996). Effects of high pressures treatment on starches. *Starch-Stärke, 48*(11-12), 399–408.

Sun, J., Zuo, X. B., Fang, S., Xu, H. N., Chen, J., Meng, Y. C., & Chen, T. (2017). Effects of cellulose derivative hydrocolloids on pasting, viscoelastic, and morphological characteristics of rice starch gel. *Journal of Texture Studies, 48*(3), 241–248.

Tan, C., & McClements, D. J. (2021). Application of advanced emulsion technology in the food industry: A review and critical evaluation. *Foods, 10*(4), 812.

Tester, R. F., & Debon, S. J. (2000). Annealing of starch—A review. *International Journal of Biological Macromolecules, 27*(1), 1–12.

Tester, R. F., Karkalas, J., & Qi, X. (2004). Starch—Composition, fine structure and architecture. *Journal of Cereal Science, 39*(2), 151–165.

Thaiudom, S., Singchan, K., & Saeli, T. (2008). Comparison of commercial stabilizers with modified tapioca starches on foam stability and overrun of ice cream. *Asian Journal of Food Agro Industry, 1*(01), 51–61.

Thirumdas, R., Trimukhe, A., Deshmukh, R. R., & Annapure, U. S. (2017). Functional and rheological properties of cold plasma treated rice starch. *Carbohydrate Polymers, 157*(1), 1723–1731.

Timgren, A., Rayner, M., Dejmek, P., Marku, D., & Sjöö, M. (2013). Emulsion stabilizing capacity of intact starch granules modified by heat treatment or octenyl succinic anhydride. *Food Science and Nutrition, 1*(2), 157–171.

Torres-Álvarez, C., García-Alanís, K. G., Amaya-Guerra, C. A., Cabello-Ruiz, E. D., Chávez-Montes, A., Castillo-Hernández, S. L., & Bautista-Villarreal, M. (2025). Effect of sweet potato starch on rheological properties and emulsion stability of salad dressings. *Polysaccharides, 6*(2), 51.

Vilela, A., Cosme, F., & Pinto, T. (2018). Emulsions, foams, and suspensions: The microscience of the beverage industry. *Beverages, 4*(2), 25.

Villamonte, G., Jury, V., & de Lamballerie, M. (2016). Stabilizing emulsions using high-pressure-treated corn starch. *Food Hydrocolloids, 52*(1), 581–589.

Villareal, R. M., Resurreccion, A. P., Suzuki, L. B., & Juliano, B. O. (1976). Changes in physicochemical properties of rice during storage. *Starch-Stärke, 28*(3), 88–94.

Waliszewski, K. N., Aparicio, M. A., Bello, L. A., & Monroy, J. A. (2003). Changes of banana starch by chemical and physical modification. *Carbohydrate Polymers, 52*(3), 237–242.

Wang, Y., Ai, C., Wang, H., Chen, C., Teng, H., Xiao, J., & Chen, L. (2023). Emulsion and its application in the food field: An update review. *EFood, 4*(4), Article e102.

Wang, L., & Wang, Y. J. (2001). Structures and physicochemical properties of acid-thinned corn, potato and rice starches. *Starch-stärke, 53*(11), 570–576.

Wang, L., Zhu, S., Chen, Y., Karthik, P., & Chen, J. (2023). Fabrication and characterization of O/W emulsion stabilized by Octenyl Succinic Anhydride (OSA) modified resistant starch. *Food Hydrocolloids, 141*(1), Article 108750.

Wilde, P. J. (2019). Improving emulsion stability through selection of emulsifiers and stabilizers.

Xian, L. X., Shariffa, N. Y., Azwan, M. L., & Uthumporn, U. T. R. A. (2020). Modified tuber starches as potential stabilizer for food-grade pickering emulsions. *Food Research, 4*(3), 753–763.

Xie, Y., Liu, H., Li, Y., Tian, J., Qin, X., Shabani, K. I., Liao, C., & Liu, X. (2020). Characterization of Pickering emulsions stabilized by OSA-modified sweet potato residue cellulose: Effect of degree of substitute and concentration. *Food Hydrocolloids, 108*(1), Article 105915.

Xu, B., Liu, C., Sun, H., Wang, X., & Huang, F. (2020). Oil-in-water Pickering emulsions using a protein nano-ring as high-grade emulsifiers. *Colloids and Surfaces B: Biointerfaces, 187*(1), Article 110646.

Yılmaz, A., & Tugrul, N. (2023). Effect of ultrasound-microwave and microwave-ultrasound treatment on physicochemical properties of corn starch. *Ultrasonics Sonochemistry, 98*(1), Article 106516.

Yu, S., Zhang, Y., Ge, Y., Zhang, Y., Sun, T., Jiao, Y., & Zheng, X. Q. (2013). Effects of ultrasound processing on the thermal and retrogradation properties of nonwaxy rice starch. *Journal of Food Process Engineering, 36*(6), 793–802.

Zanella Pinto, V., Goncalves Deon, V., Moomand, K., Levien Vanier, N., Pilatti-Riccio, D., da Rosa Zavareze, E., da, R., Lim, L.-T., & Dias, A. R. G. (2019). Characteristics of modified carioca bean starch upon single and dual annealing, heat-moisture-treatment, and sonication. *Starch-Stärke, 71*(3–4), Article 1800173.

Zhang, H., Fan, H., Xu, X., & Xu, D. (2024). Deterioration mechanisms and quality improvement methods in frozen dough: An updated review. *Trends in Food Science & Technology, 143*(1), Article 104251.

Zhao, Y., Khalid, N., Shu, G., Neves, M. A., Kobayashi, I., & Nakajima, M. (2017). Formulation and characterization of oil-in-water emulsions stabilized by gelatinized kudzu starch. *International Journal of Food Properties, 20*(Suppl. 2), 1329–1341.

Zhou, J., Guo, M., Qin, Y., Wang, W., Lv, R., Xu, E., & Wu, Z. (2023). Advances in starch nanoparticle for emulsion stabilization. *Foods, 12*(12), 2425.
Zhu, F. (2018). Modifications of starch by electric field based techniques. *Trends in Food Science & Technology, 75*(1), 158–169.
Zhu, F. (2019). Starch based Pickering emulsions: Fabrication, properties, and applications. *Trends in Food Science & Technology, 85*(1), 129–137.
Zhu, Y., Du, C., Jiang, F., Hu, W., Yu, X., & Du, S. K. (2024). Pickering emulsions stabilized by starch nanocrystals prepared from various crystalline starches by ultrasonic assisted acetic acid: Stability and delivery of curcumin. *International Journal of Biological Macromolecules, 267*(1), Article 131217.
Zou, J. J., Liu, C. J., & Eliasson, B. (2004). Modification of starch by glow discharge plasma. *Carbohydrate Polymers, 55*(1), 23–26.

Chapter 15

Novel applications of gluten-free starches in food systems

Ravi Sanjana[1], Kalaiselvi Krishnamoorthy[3], Khalid Bashir[4] and Pothiyappan Karthik[1, 2]

[1]Department of Food Technology, Faculty of Engineering, Karpagam Academy of Higher Education, Coimbatore, Tamil Nadu, India; [2]Centre for Food Nanotechnology, Karpagam Academy of Higher Education, Coimbatore, Tamil Nadu, India; [3]Centre for Herbal Pharmacology and Environmental Sustainability, Chettinad Hospital and Research Institute, Chettinad Academy of Research and Education, Kelambakkam, Tamil Nadu, India; [4]Department of Food Technology, Jamia Hamdard, New Delhi, India

1. Introduction

The Gluten-free (GF) range of food products has become a paramount need to deal with gluten intolerance and insensitivity. For people with celiac disease, consumption of even a small quantity of gluten may lead to bowel-related irritability and inflammation. A diet consisting of gluten-free food is the sole solution to address this condition (Šmídová & Rysová, 2022). The Business Research Company (2024) reports the worldwide annual average growth and market of gluten-free products. Celiac illness is an autoimmune condition that affects the lining of the small intestine in the digestive system, inhibiting the normal absorption and digestion levels that lead to nutrient deficiencies. Other manifestations of gluten-related disorders are "dermatitis herpetiformis" and "gluten ataxia" (Al-Toma et al., 2019). Dermatitis herpetiformis is characterized by skin rashes with bumps and blisters as an allergic reaction due to the immunological responses (IgA) caused by gluten; while gluten ataxia is an autoimmune disease that gradually harms the brain, leading to unstable coordination and poor motor skills. The demand for gluten-free alternatives that simulate similar functional and sensory attributes to gluten products in the food sector is expanding immensely, leading to further investigation and advancements. Gluten-free starches are mainly used in the formulation of gluten-free food products. Removal or elimination of gluten from the starch sources will have a significant effect on the structure and texture of the gluten-free foods.

Gluten is a mixture of different protein structures, mainly made of gliadin (prolamin) and glutenin in wheat (Biesiekierski, 2017). This is the storage protein in most cereals and grains. Gluten protein is further categorized into "alpha, beta, gamma, and omega gliadins" according to their structure and interactive forces. Gluten (90% of the wheat protein) is the major contributor to the structural, chemical, and physical characteristics of grain. Gluten acts as a binding agent, affecting the flavor and textural profile of the food products. According to Wieser (2006), gluten mixture (ratio of glutenin and gliadin) is attributed to the dough's rheological and other functional attributes, such as viscoelastic properties, moisture content, and strength. According to the Codex Alimentarius Commission (1979), gluten-free foods are those that:

1. "Consist or made of only one or more ingredients which don't have *Triticum* species, barley, rye, oats and their other cross-bred species, and the level of gluten of these products should not surpass 20 mg/kg".
2. "Consist or made of one or more ingredients of *Triticum* species, barley, rye, oats and their cross-bred species, which are processed to eliminate gluten so that their level of gluten does not surpass 20 mg/kg".
3. "Consist or made of one or more ingredients from *Triticum* species, barley, rye, oats and their cross-bred species which are processed to reduce gluten so that their level of gluten does not surpass 100 mg/kg".

Gluten can be extracted from the dough mix by washing it off with water or NaCl, which separates the protein bulk (gluten) and starch and other soluble particles (Bathula, 2018). For wheat flour, a batter consistency is made, which is then forced into a series of screens to separate the bulk particles, followed by a decantation (or hydro-cyclones) procedure to

Native and Modified Starches for Food Applications. https://doi.org/10.1016/B978-0-443-27431-2.00028-6

separate the gluten. A small-scale gluten separation system was fabricated by Weegels et al. (1988). This design works on the principle of gluten coagulation followed by the separation of gluten and hemicellulose. The desired component can be freeze-dried to obtain the product. This technique is widely used in the production of gluten-free starch and starch purification techniques. Some other commonly practiced methods of gluten extraction, including ethanol extraction and cocktail extraction, were discussed by Segura et al. (2021). The Aqueous ethanol extraction system is deemed inefficient as the protein components, such as glutelins and gliadins, cannot be completely removed. Hence, a reducing agent, an acid or an enzyme is usually applied to break the disulfide bonds. Whereas, the cocktail extraction method contains some toxic components such as "guanidine hydrochloride" and "2-mercaptoethanol". Considering these points, a novel method of "UGES (Universal Gluten Extraction Solution)" is being used widely to remove gluten from other starch components. Segura et al. (2021) verified the efficiency of the UGES method in various commercially available food products using the ELISA technique. Recent advancements in gluten removal to produce gluten-free starches were investigated with biological and enzymatic techniques. Endogenous enzymes extracted from bacteria, fungi and plant sources are used to hydrolyze the gluten protein, which can be further removed (Bradauskiene et al., 2021).

Many researchers are focused on finding alternative novel gluten-free starch sources in place of gluten-rich cereals, such as pseudocereals like quinoa and millet and processing technologies such as starch modification and the usage of hydrocolloids to combat the gluten sensitivity issue by developing gluten-free foods under the above-mentioned standards (Azeem et al., 2024; Cao et al., 2024; Zhu et al., 2024). Tapioca, corn, potato and rice are some widely studied gluten-free starches which are easily digested and commonly available for the manufacturing of gluten-free breads, noodles, pastas, crackers, etc. Various starch modification approaches using physical, chemical, and enzymatic methods help in better interaction of the molecules and enhance the attributes of gluten-free foods (Elgeti et al., 2013). It is well established that severe nutritional deficiency is observed in people consuming gluten-free diets. Macro-nutrients, including dietary fibers, proteins, and micronutrients, including vitamin B, zinc, calcium, and iron, are deficient in the majority of the gluten-free foods commercially available in the market. Nutrient enrichment through fortification is highly necessary to enhance the well-being of people on a Gluten-Free Diet (GFD). This chapter discusses the various sources of gluten-free starch in detail and their properties. Besides, gluten-free starch applications in the food industry, along with the nutritional value fortification and consumer acceptability are examined.

2. Properties of gluten-free starch

Starch is the primary reservoir of carbohydrates found in plants, which is broadly distributed in different plant parts, like seeds, stems, leaves, flowers, fruits, and roots, acting as an important energy source (Tester & Debon, 2000; Yadav et al., 2024). Starch is made of two primary polymers, viz., "amylopectin and amylose". Amylopectin is the dominant constituent and highly branched polymer made of linear chains composed of α-1,4 glycosidic bonds and α-1,6 linkages. Amylopectin forms soft gels and weak films in water dispersion and is more stable. In contrast, amylose, a linear polymer with α-D-(1–4) glycosidic bond linkage, constitutes approximately 20%–30% starch (Bashir et al., 2017). Amylose affects starch physicochemical properties, including water absorption, swelling, and gelation during food processing of gluten-free foods (Singh et al., 2003). These parameters determine the elasticity and structure of the gluten-free foods. Starch is nutritionally classified into "Slowly Digestible Starch (SDS), Rapidly Digestible Starch (RDS), and Resistant Starch (RS)". RDS is found in foods like cooked potatoes, a gluten-free starch source, and is quickly converted to glucose within 20 min of consumption. Starch quality is frequently evaluated using particle size/distribution or starch degradation with their water-binding ability. Nevertheless, minimal information is known about the interactions between gluten and starch and how they affect the creation of networks. By acting as a physical and mechanical barrier for delaying the binding of starch and enzymes, glutenin elasticity serves as the functional foundation of the gluten network. Furthermore, the barrier function of the gluten network blocks the water absorption of starch granules, reducing starch hydration and subsequent gelatinization. Due to consumer choices and related ailments, gluten-free products are in considerable demand. However, there is a strong link between type 1 diabetes and celiac disorder since foods made of gluten-free starch are higher in carbohydrates and lower in protein, fiber, minerals, and vitamins. As a result, there is an increased significance in modifying starch digestion, particularly in gluten-free starch-based diets, which will influence food quality features, nutritional properties, and consumer acceptability.

3. Gluten-free starches and their classifications

An essential carbohydrate source is starch, which is extensively applied in the food sector for the manufacturing process because of specific characteristics such as binding, thickening, gelling, and texture (Zhang et al., 2021). Gluten-free starches are categorized into two types: native and modified. Native starches are unprocessed and utilized in their natural form. In contrast, modified starches are chemically or physically treated to improve qualities such as heat stability, thickening capacity, and freeze-thaw resistance. Modified starches include pregelatinized corn starch and cross-linked starches, which are designed for specific industrial and culinary applications. Overall, gluten-free starches are a beneficial and vital component in both food preparation and medicinal applications, notably in the manufacturing of gluten-free food products.

Gluten is a useful component in many food product development, recognized for its viscoelasticity derived from the Latin word "Gluten". Specifically, gluten-free starches are derived from various types of grains, legumes, etc. The texture, shelf life, and general quality of GF foods are greatly affected by the type of starch used. Because of their distinct gelatinization characteristics, swelling patterns, and resistance to retrogradation, different starches are appropriate for use (Chang et al., 2021). Starch is a vital energy source in human diets (Yadav et al., 2024). It is found in a variety of plant-based sources, which may be divided into common and rare categories depending on worldwide use and distribution. An exhaustive list of common and rare gluten-free starches used in the food industry is provided below.

In food manufacturing, gluten-free starch plays a vital part in achieving the user's requirements. These starches are synthesized from various sources of plant types, which have potent food quality as well as shelf life. Understanding the unique characteristics of various starches is important for improving food compositions. Future research should concentrate on enhancing the functionality of gluten-free starches through changes and novel processing to increase the sensory and consumer acceptability, along with the nutritional aspect of gluten-free meals.

Fig. 15.1 Categories of gluten starch sources into common and uncommon types based on their prevalence in human diets.

4. Techniques for enhancing properties of gluten-free starch

Gluten-free starch form corn and rice were treated to improve the physical characteristics. The gluten-free breads made after hydrothermally treating the starch showed increased bread volume with better height/width ratios. Additionally, the bread was less chewy and softer when compared to non-treated breads made from gluten-free starch (Bourekoua et al., 2016). Fasakin et al. (2024) investigated improvements in gluten-free starch from faba bean when introduced to dry treatments. A decrease in the amylose content was noticed along with an enhancement in the crystallinity of the GF starch. Other physical modification techniques, such as annealing, hydro-thermal treatments and high pressure, also impact the attributes of gluten-free starch. It is used in a variety of gluten-free sources such as millets, tapioca, sago, and potato starch that improve the solubility, swelling capacity and water binding capability (Hu et al., 2020; Zailani et al., 2020).

Chemical modification techniques such as oxidation, esterification, acetylation, cross-linking, and acid hydrolysis aid in enhancing the properties of the gluten-free starch by adding or removing the functional groups. The overall retrogradation, swelling ratio, gelatinization and the pasting attributes of the GF starch are shown to be enhanced. Oxidation has a vital role in increasing the quality and performance of food products produced from gluten-free starch, particularly in imitating the structural features of gluten in typical baked goods. Gluten proteins in doughs create strong, elastic networks that trap gas during fermentation, resulting in good volume and texture. Esterification is mostly used in the manufacturing of gluten-free products to change starches and lipids, hence improving the functional qualities of dough made from gluten-free starch and baked items. Gluten-free starches frequently lack gluten's structural and moisture-retaining capabilities, resulting in issues such as poor texture, low volume, and dryness in the end product (Singh et al., 2007).

Enzymatic modification is a technique used in the production of food items from gluten-free starch to improve texture, structure, and quality. Without it, gluten-free dishes made from rice, corn, millet, or other grains are dry, heavy, and crumbly. To compensate for these inadequacies and produce a superior product, enzymes are added to change gluten-free dough ingredients and increase dough strength and elasticity. Some enzymes, such as transglutaminase, facilitate the production of a gluten like protein network by protein cross-linking, hence increasing dough strength and elasticity (Altındağ et al., 2014).

5. Innovations in gluten-free starch blends

Gluten, a protein found in cereals like wheat, rye, and barley, plays a crucial role in providing rheological and viscoelastic properties to food products. The gluten formation improves the gas retention capacity, which is essential in baked goods. Once the gluten is removed, these properties are relied upon by the starch. Thus, the study of starch in all forms, including modified and resistant, is important (Horstmann et al., 2017). Starch provides the gelling property in addition to adhesion

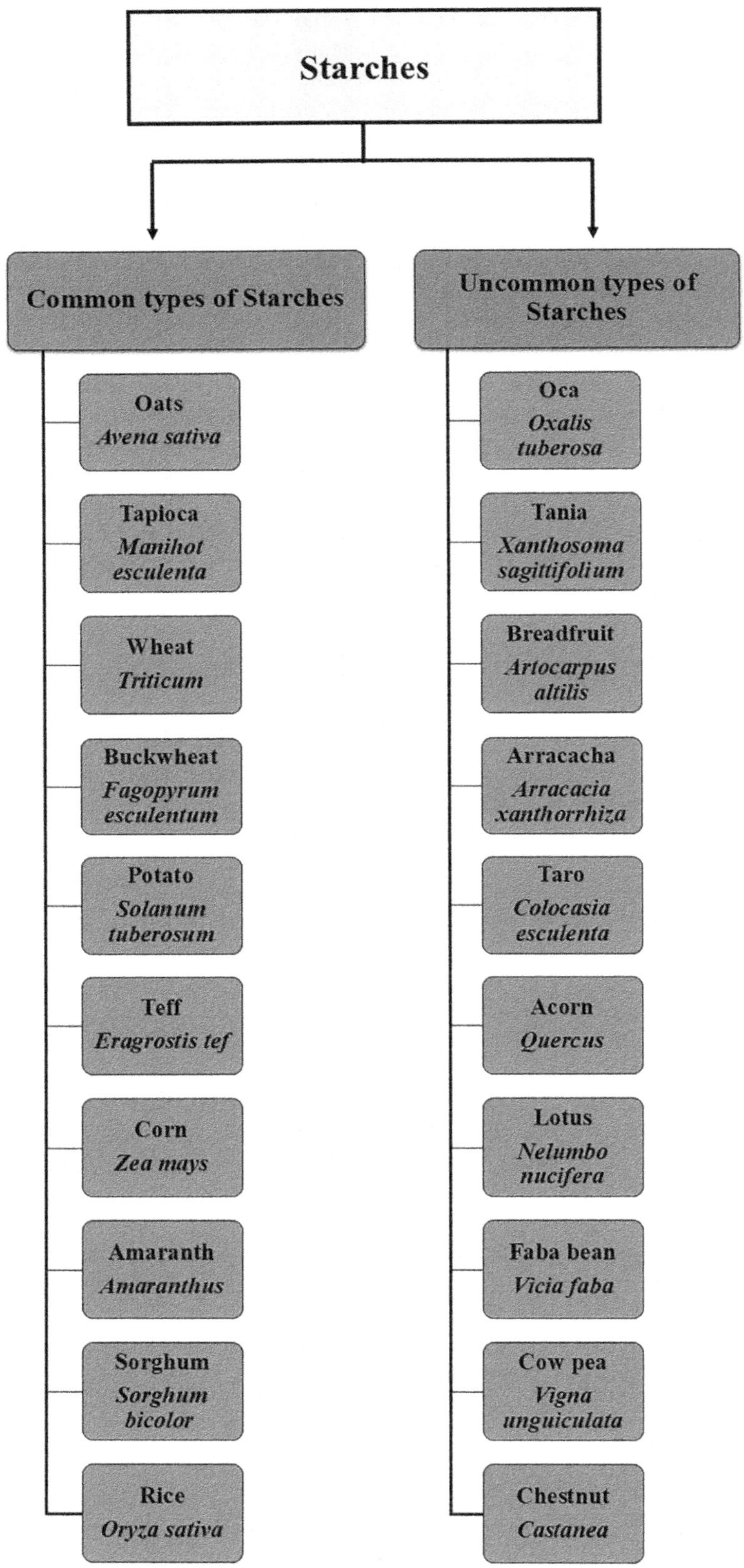

FIGURE 15.1 Categories of gluten starch sources into common and uncommon types based on their prevalence in human diets.

and thickening properties. Besides, they aid in retaining the moisture of the product and act as a stabilizing agent in the foods. The incorporation of emulsifiers such as hydroxypropyl methylcellulose, xanthan gum, guar gums, and hydrocolloids improves the dough strength (Khoury et al., 2018). Proteins and fiber are incorporated to enhance the nutritional profile, but more importantly, to enhance the textural properties. Dietary fiber improves the gelling property through water-binding capacity. Protein sources added to the gluten-free products are eggs, dairy, and legumes. These are also involved in improving the rheological properties, along with the sensory aspect of the products. Some widely used naturally available sources of gluten-free starch from cereals are corn, rice, potato, tapioca, and millet. Other novel gluten-free starch sources are legumes such as chickpeas, faba beans, soy, peas, etc. (Imam et al., 2024). Gluten-free starch extracted from oats is a controversial topic. Oats in their natural form are gluten-free, but it is exposed to other gluten-containing cereals during harvest and processing, making them unfit to be declared as completely gluten-free (Khoury et al., 2018). Gluten-free starch extracted from fruits, vegetables, seeds and nuts is also used in the manufacturing of GF food products and is discussed below in detail.

5.1 Sources of gluten-free starch

5.1.1 Gluten-free starch from cereals

Rice, a staple cereal, is a common source of gluten-free starch, consumed in many regions, mainly in Asian cuisines. They are colorless and lack any prominent flavor. Although they are minimal in protein and fibers, gluten-free starch extracted from rice is used because of their rich source of carbohydrate as an energy source, which makes them a good base for producing gluten-free pasta, bread, noodles, etc. It is to be noted that the prolamin content being low in rice, addition of emulsifier or enzymes is required to enhance the required rheological attributes (Sivaramakrishnan et al., 2003).

Similarly, gluten-free starch from millet is a nutritious substitute used in the GF food products, including breads, cookies, pasta, noodles and could also be used as a dairy substitute. It can be used in any form, like soaked, roasted and germinated (Griffith et al., 1998). In addition to gluten-free starch sources, the mineral and amino acid profile of millets such as foxtail and pearl millet makes them the best gluten-free source for infants. Teff is another source of gluten-free starch, similar to millet and can be used to produce GF foods rich in fiber, protein, and minerals. The retrogradation rate of teff is very low, which delays the bread spoilage condition (Alaunyte et al., 2011).

5.1.2 Gluten-free starch from pseudo cereals

Pseudocereals are a group of crops similar to cereals but that contain lesser prolamin, an important gluten storage protein, making them suitable for the extraction of gluten-free starch. Amaranth is highly nutritious in proteins, fibers and minerals than most other vegetables. The baked goods, like bread, produced from gluten-free starch of amaranth, have increased volume and uniform crumb (De La Barca et al., 2010). Quinoa, another gluten-free source of starch, is rich in vitamins and minerals like iron, calcium and phosphorus. Their high fat content makes them suitable for the manufacturing of extruded GF food products with high shear rate and increased volume of the product (Doğan & Karwe, 2003). GF starch extracted from buckwheat is a recommended source of gluten-free starch for the celiac population with diabetes, as they have low glycemic index. This directly helps maintain blood pressure and cholesterol. Additionally, utilizing buckwheat starch in gluten-free bread and crackers was studied to improve the sensory attributes and delay the spoilage parameters (Sedej et al., 2010; Wronkowska et al., 2012).

5.1.3 Tubers and legume-based gluten-free starch

Gluten-free starch extracted from potato and tapioca is used in various gluten-free food products (Table 15.1). Tapioca starch has the ability to retain moisture and provide the gelling property along with thickening attributes in gluten-free baked goods (Ali et al., 2014). Contrasting to this, potato starch did not react as expected with gelatinization (Kim et al., 2017). It is also to be noted that the substitution ratios should be monitored as potatoes are considered high in glycemic index. Soya, beans, peas and chickpeas are highly considered as legume-based sources for gluten-free starch. Additionally, they are rich in nutrients and have commendable physicochemical attributes (Hosseini et al., 2018). Gularte et al. (2012) evaluated the rheological properties of GF cakes made with gluten-free starch of legumes and rice in the ratio of 50:50. Lentil, as the crucial source of protein, reduced the viscosity of batter; whereas it improved the volume of the cakes. Other properties, such as crumb texture, were found to be similar to those of the gluten-containing cakes. Another novel source of protein-rich legume-based gluten-free starch is obtained from carob, a member of the Fabaceae family. Tsatsaragkou et al. (2014) showed that using carob instead of rice as a gluten-free source of starch produced results that were better in color and crumb hardness.

TABLE 15.1 Studies on various gluten-free food products and their characteristics.

Product category	Sources of gluten-free substitute	Additive used/processing technology	Findings	References
Bread	Rice, potato, tapioca, corn and gluten-free wheat	–	Increased particle size and the baking properties.	Horstmann et al. (2016)
Bread	Rice	HPMC	The rheological properties of rice bread with 3% HPMC were similar to those of control wheat bread.	Sivaramakrishnan et al. (2003)
Bread	Rice	Incorporation of gums (CMC and k-Carrageenan)	Improved shelf-life	Mirzaei and Movahed (2013)
Bread	Rice	Skim milk powder	Formation of the protein phase contributes to the structure with improved volume and structure of the crumb.	Moore et al. (2004)
Cakes	Quinoa	Addition of basic amino acid with microwave treatment	Regulated protein aggregation and enabled a porous structure for the sponge cake	Cao et al. (2024)
Cakes	Millet	Addition of chitosan oligosaccharide	Improved shelf life	Zhu et al. (2024)
Pasta	Rice	Egg albumin	Reduced cooking loss and increased protein content	Phongthai et al. (2017)
Noodles	Rice	Annealing at 55°C for 24 h	Improvement in textural properties, such as gelling and swelling characteristics	Hormdok and Noomhorm (2007)
Noodles	Rice	Hydrothermal treatment at 110°C for 90 min	Improvement in textural properties, such as gelling and swelling characteristics	Hormdok and Noomhorm (2007)
Noodles	Rice	Fermentation with *Lactobacillus* spp.	Improvement in pasting property	Yang and Tao (2007)
Noodles	Sweet potato	Particle size reduction	Improved pasting property, reduced hardness, and enhanced amino acid content.	Azeem et al. (2024)
Cookies	Konjac	Milk protein as an additive	Improved the spread ratio and volume	Akesowan (2016)
Cookies	Buckwheat, maize and rice	Enzymatic treatment with transglutaminase	Improved the spread ratio and tractability	Altındağ et al. (2014)
Beer	Rice and Buckwheat	Malting	Increase the sugar and amino acids for beverage production	Agu et al. (2012)
Nonalcoholic beverage	Sorghum	Souring method	Increased the bioavailability of protein	Adeyanju et al. (2018)

5.1.4 Gluten-free fruits and vegetables-based starch

Fruit and vegetable-based starches are also used as a gluten-free base for the production of cakes and breads. Gluten-free starch from jackfruit seed, flesh of soursop and banana fruit has been used to produce bread and cake (Noor, 2014; Yazid et al., 2018). The jackfruit seed starch is known for its gelatinizing property, while the banana starch is known for its emulsifying and pasting properties, apart from producing flavor. Bread produced with the starch of acorn, when substituted in 20% ratio, showed a positive influence on the volume, while anything more than 20% reduced the proportion of gluten system responsible for volume and decreased the bread volume (Korus et al., 2015). The addition of fruit- and vegetable-based gluten-free starch provides nutritional benefits apart from textural properties. Martín-Diana et al. (2025) studied the valorizing of peach peel as a source of gluten-free starch in the GF flour mix. They exhibited high levels of fiber content, making them fit to be claimed as "source of fiber" in the labeling claim practices. The total phenolic content also increased along with the antioxidant attributes. Gluten-free starch from peach also enhanced the gelling property suitable for a baked product.

5.1.5 Other novel gluten-free starch sources

Novel starches used in gluten-free food formulation are from chia seed, flax seed, chestnut and sorghum. Both chestnut and chia seeds are found more in European and American regions. GF starch processed from chestnut at low temperatures is similar to corn starch in terms of functional properties (Spada et al., 2017). Protein content of these sources is on the higher side, with 4%–7% in chestnut and 18%–25% in chia seeds. The dietary fibers of chia seeds span from 20% to 37%, while those of chestnuts are 4%–10%. Other nutrients such as minerals and vitamins, including magnesium, phosphorus, potassium, and vitamins B and E, are found in chestnut; while chia seeds are rich in omega-3-fatty acids and α-linoleinc acid (Hosseini et al., 2018). Substituting the conventional gluten-free starches like rice and potato with starch from chia seed in varied ratios led to minimal differences in the crumb texture, such as moisture, firmness, specific volume, baking loss and sensory scores. It was also reported that using 100% of chia-based gluten-free starch did not yield suitable results in terms of texture and sensory attributes (Da Mota Huerta et al., 2016; Sandri et al., 2017). Alternatively, other studies suggest that the incorporation of 2.5%–7.5% of chia seed GF starch in chestnut-based gluten-free bread showed improved rheological properties like elasticity and viscosity. In place of rice starch, 15% of chia seed starch improved the hardness of the bread, which decreased the volume, whereas the consumer acceptance identified through sensory analysis was unaltered (Moreira et al., 2012; Steffolani et al., 2014). Besides, sorghum based starch are used in gluten-free products like biscuits, chips, weaning foods, etc. While formulating sorghum-based breads, a softer batter consistency is preferred over a hard dough. This gives better viscoelastic properties and proofing of the bread. However, when correlated to breads made of wheat, sorghum-based gluten-free counterparts produced a decreased volume. Additionally, incorporation of hydrocolloids improved the quality of sorghum-based gluten-free products (Taylor et al., 2006). As discussed above, many gluten-free starch substitutes are available from various sources such as cereals, legumes, vegetables and fruits for the formulation of gluten-free food products. Such micro-nutrient-rich sources should be carefully chosen to compensate for the nutritional loss during the processing stages.

6. Application of gluten-free starch in food products

Gluten-free foods are being consumed by an extensive population due to various reasons. One of the main logical grounds is that a gluten-free diet (GFD) is the best way to medically manage celiac diseases and related disorders, while many are consuming GFD as a personal and lifestyle choice. This trend has shown an immense surge in the global demand for gluten-free products and is expected to expand manyfold, as they are nutritionally and texturally fit to compete with their gluten-containing counterparts. American and European countries are forwarding their research and product development and focusing on improving the scope for GFD, as the recent awareness of diagnosing and dietary restrictions in consumers has increased. A broad range of products, including but not limited to baked goods such as cakes, breads, and cookies, starch foods such as pasta, noodles, and pizza, and extruded foods like chips and snack range, can be made as gluten-free products. However, the biggest challenges seen in this industry are nutritional deficiency and unsatisfactory consumer response. A comprehensive review of the study on GF food products in the food industry is provided in this section and briefed in Table 15.2.

6.1 Baked products

Gluten-free baked products like breads and cakes are universally preferred as a staple food by the CD community. The incorporation of emulsifiers and or hydrocolloids in various combinations will improve the dough properties and act as gluten substitutes. They can also be added as a humectant to retain moisture and as a thickening agent (Houben et al., 2012). The type of hydrocolloid used and the quantity added, along with its modification techniques, including the processing characters will influence its effect on the baked goods, such as volume, crumb softness and color, dough strength and hardness, etc. (Gallagher et al., 2002). Bread made from rice starch has similar properties to those of wheat bread, such as crust color and texture, although the specific volume was found to be lower, which could be because the gluten, which is responsible for gas retention, is missing in rice starch. Thus, the product becomes less cohesive with reduced visco-elastic properties, leading to decreased volume and structures (Clerici et al., 2008). In this context, the

TABLE 15.2 Classification of various plant-based starch sources based on starch type, notable benefits, and uses.

Plant sources	Starch type	Notable benefits and uses	References
Oats	Resistant	Heart-healthy lowers cholesterol and gluten-free cereals	Qian et al. (2025)
Corn	Digestible	High in fiber, antioxidants, and supports digestion, biofuel and thickening agent	Ukwatta et al. (2025)
Wheat	Digestible	Used in gluten-free baking and pasta production	Huang et al. (2022)
Rice	Digestible	Easily digestible and good for sensitive stomachs, cosmetics, and gluten-free baking	Rhowell et al. (2021)
Potato	Resistant and digestible	Gluten-free baking, thickening, rich in resistant starch and supports gut health.	Tong et al. (2023)
Teff	Resistant	High in protein, fiber, iron, calcium, and resistant starch. Used in injera (Ethiopian bread), porridge, gluten-free baking, and as a thickener.	Adepoju et al. (2024)
Tapioca	Digestible	Used in food, textiles, and adhesives, low in protein but provides energy	Javadian et al. (2021)
Amaranth	Resistant and digestible	Rich in protein, lysine (an essential amino acid), fiber, antioxidants, and minerals like magnesium and phosphorus. Used in gluten-free flour, breakfast cereals, energy bars, and soups	Sindhu et al. (2021)
Sorghum	Resistant and digestible	Bioethanol, gluten-free flour, high in antioxidants, may help to lower the inflammation.	Zahid et al. (2025)
Buckwheat	Resistant	High protein, fiber-rich, good for diabetics and supports heart health	Gao et al. (2023)
Acorn	Resistant and digestible	High fiber, rich in tannins, has antiinflammatory properties	Castro et al. (2023)
Lotus	Digestible	Asian cuisine, medicinal uses, supports digestion, and high in fiber	Chen et al. (2021)
Faba bean	Resistant and digestible	High protein, fiber-rich, gluten-free flour alternative, high in protein and fiber and supports muscle growth.	Nilsson et al. (2023)
Cowpea	Resistant	Rich in protein, dietary fiber, and antidiabetic properties.	Oyeyinka et al. (2021)
Chestnut	Digestible	Unique nut, gluten-free desserts, high in vitamin C. Low in fat, high in vitamin C and fiber, it has immune-boosting properties.	Liu et al. (2024)
Oca	Resistant and digestible	Andean root, prebiotic benefits and tangy-sweet flavor.	Dini et al. (2024)
Tania	Resistant	Gut health benefits, resistant starch source, helps manage blood sugar and improve digestion.	Espinosa-Solis et al. (2021)
Breadfruit	Digestible	Sustainable tropical crop, high in fiber and vitamin C	Otemuyiwa and Aina (2021)
Arracacha	Digestible	Easy to digest, ideal for baby food easily digestible and rich in vitamin A	Sánchez Riaño et al. (2022)
Taro	Resistant and digestible	High fiber, rich in antioxidants, staple food and prebiotics for gut health	Nagar et al. (2021)

usage of hydroxypropyl methylcellulose (HPMC) was examined in many studies. Thus, these molecules create a gel-like network that enhances the dough viscosity and retention of gas, which will increase the bread volume (Lazaridou & Biliaderis, 2009). The farinograph results show that the dough strength of the rice-based GF bread with HPMC were like the conventional wheat product. Besides, the flowability results of the rice dough with 3% HPMC from the creep graph showed similar characteristics to those of the wheat dough, which establishes the significance of starch and protein in GF food products. It was also observed that the HPMC, when used more than 3%, showed only additive properties rather than those of an emulsifier or a stabilizer (Sivaramakrishnan et al., 2003; Kumar et al., 2024).

In another investigation, the rheological attributes of the rice-based bread were investigated with different combinations of gums, hydrocolloids, and emulsifiers and compared with the control (wheat bread). The viscosity of the bread increased when xanthan gum and guar gum were added, while it decreased when other gums like pectin and HMPC were added. The increased viscosity improves the overall quality of bread by providing a denser structure and shape. Additionally, a thick dough enables a greater fermentation rate (Demirkesen et al., 2009). The usage of Diacetyl Tartaric Acid Esters of Mono- and Diglycerides (DATEM) enhanced the gas retention and increased the volume and fermentation rate in gluten-free breads. The addition of xanthan gum also enhanced the shelf life (Lee & Lee, 2006). Other gums, such as carboxymethyl cellulose gum and k-Carrageenan, when incorporated at 1%, inhibited the staling rate and improved the shelf life of the gluten-free rice breads (Mirzaei & Movahed, 2013).

Proteins can also be employed as gluten alternatives in bread manufacturing due to their chemical nature. Egg, soya, and milk are the most used protein sources that mimic the gluten properties by swelling and forming a network (Arendt et al., 2006). Fig. 15.2 depicts the positive effect of milk proteins on the bread structure. The dough morphology and rheology of the gluten-free bread with milk protein are like the control wheat bread. The Confocal laser-scanning microscopy (CLSM) images discovered that the structure of the crumb of GF bread made from dairy milk was very similar to that of the control bread made of wheat, due to the gluten protein network (Moore et al., 2004).

High-pressure treatment enabled protein denaturation, but application of pressure over 350 MPa decreased the complete quality of the oats bread. The covalent bonds gelatinized the starch molecule and increased the dough's hardness and elastic properties (Hüttner et al., 2010). Apart from the nutritional benefits, the addition of milk proteins improved the crust texture and color, volume, water-binding capacity, and shelf life. But care should be taken for lactose-intolerant patients allergic to milk lactose for whom other sources of protein, such as soya, egg protein, surimi, legumes, etc. (Houben et al., 2012). Fig. 15.3 illustrates the bread models from numerous starch sources such as potato, corn, rice, tapioca, and gluten-free wheat. The SEM images represent the morphological properties (Horstmann et al., 2016). As the image depicts a proper shape, dough structure and rheology were obtained in bread made of gluten-free wheat starch and potato starch. As the particle size increases, better baking properties are observed. This result correlated with the finding of Sadullayev et al. (2024), in which the authors optimized the starch particle size to be 46 μm to have the best baking character for the volume and softness of the crumb. It was also reported that as the particle size reduced, the bread spoilage and staling quickened. The granule size was greater in potato starch, followed by wheat, tapioca, corn, and rice. As depicted in the SEM images of Fig. 15.3, agglomeration of the starch molecules reduced the baking performance as

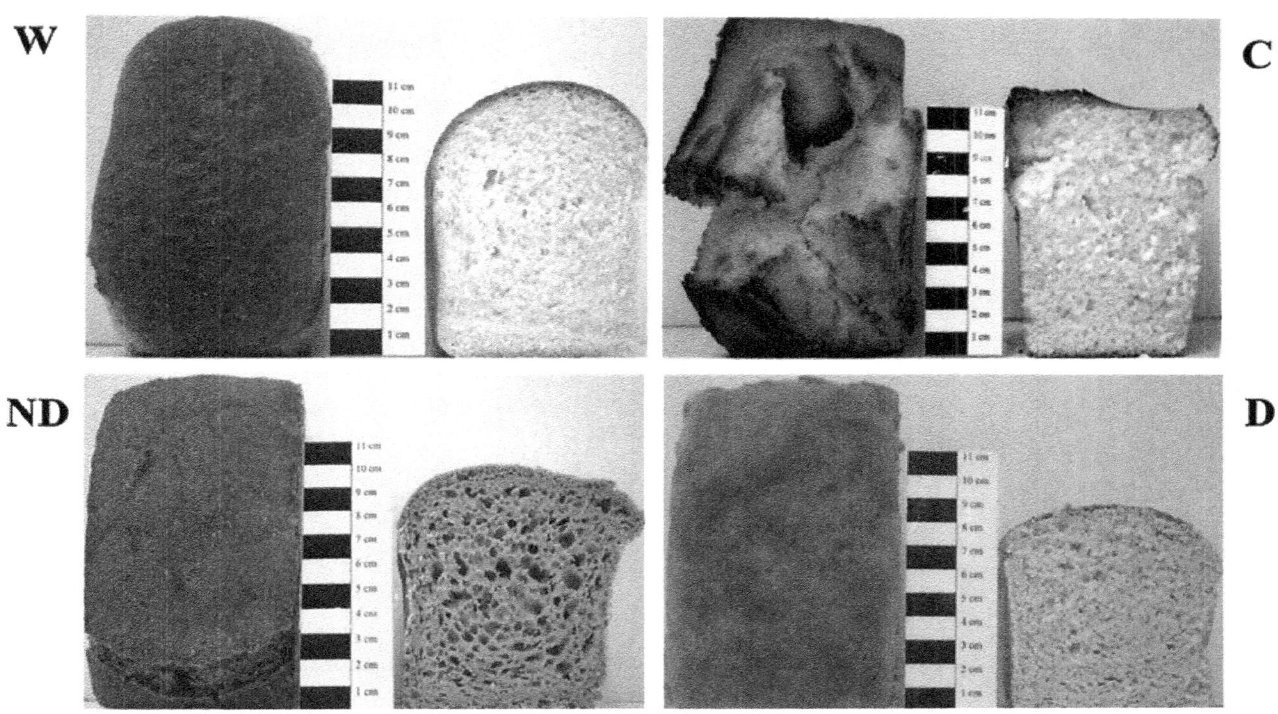

FIGURE 15.2 Digital pictures of wheat bread (W), commercial gluten-freebread (C), non dairy gluten-free bread (ND), and dairy gluten-freebread (D) (Moore et al., 2004).

FIGURE 15.3 SEM morphology of bread models (a) potato starch, (b) tapioca starch, (c) corn starch, (d) rice starch, (e) wheat starch (Horstmann et al., 2016).

seen in rice starch. Apart from the particle size, the lipid-amylose content plays a crucial part in the bread morphology. This has a direct correlation to the crumb hardness and shelf-life of these bread models (Copeland et al., 2008).

Cao et al. (2024) investigated the addition of alkaline amino acids like arginine and lysine into gluten-free quinoa-based sponge cakes to improve their functional properties. Since quinoa is gluten-free, its protein content (gluten), contributing to structure development, will be minimal. Heat processing techniques such as extrusion and cooking at high pressures have been used in the production of many quinoa products, which may be sustainably inefficient in terms of energy usage and pollution (Itusaca-Maldonado et al., 2024). Cao et al. (2023) studied the utilization of microwave irradiation as an alternate treatment to improve the textural properties. The author also investigated amino acid fortification to improve the surface area. The results revealed that the incorporation of 0.6% Lys/Arg in the sponge cake improved the protein interaction and enhanced the volume of the cake by improving gas retention. Also, it produced a porous structure ideal for a sponge cake due to increased cell density (Cao et al., 2024). Apart from physical treatments to improve the properties of GF products, the inclusion of additives can also perform the same task. The addition of chitosan oligosaccharide in the production of millet cakes was experimented with by Zhu et al. (2024) to improve the shelf-life by 6 days and its antimicrobial properties. It also enhanced the protein interaction to build an integral structure of the cake through the hydrogen bonding of COS with millets (Zhu et al., 2024).

6.2 Extruded snacks

Extruded foods such as pasta, noodles, crackers, and other cereal products are also made from gluten-free starch. Maize is a cereal crop utilized in the making of extruded gluten-free food products such as puff balls, chips, and other snack forms such as granules, shreds, etc. (Arendt & Bello, 2008). Noodles and pasta are usually characterized by elasticity, chewiness properties, and less loss during cooking (Sozer, 2008). Potential gluten alternatives are added as additives and/or processing techniques such as enzymatic treatment, fermentation, and heat treatments are applied to GF extruded products (Hormdok & Noomhorm, 2007; Yang & Tao, 2007). Rice starch, as a gluten-free alternative used in the making of noodles and pasta, was studied by many researchers. The inclusion of soy isolate, egg albumin components, and legumes was specifically investigated to enhance the protein content and to minimize the cooking loss (Phongthai et al., 2017). Advanced Processing technologies, such as extrusion and hot-air drying, improved the characteristic features of the pasta, like firmness, and reduced the cooking loss. Other heat treatments, such as annealing at 60°C and hydrothermal procedure at 110°C, improved the textural attributes of the rice-based noodles (Hormdok & Noomhorm, 2007).

Another approach to reducing the particle size to improve noodle quality was experimented by Azeem et al. (2024). Starch from sweet potato used as a gluten-free substitute in noodles tends to become harder during the cooking stage. Fine granule size made the noodles softer and elastic, and improved the chewability and springiness of the noodles, exhibiting similar properties to wheat noodles. An overall improvement in the texture due to better gelling properties and high viscosity was observed (Wang, Ding, et al., 2021). Although the amino acid profile and the sugar content increased with decreased size, the carbohydrate and protein values decreased due to starch retrogradation (Warren et al., 2015).

The spread ratio of the dough is a vital element in determining the cooking quality of the product. Spread proportions, where the higher the ratio, the better quality is interpreted due to the addition of "Partially hydrolyzed guar gum" PHGG as an additive (Mudgil et al., 2017). GF cookies prepared out of rice starch and elevated levels of Konjac glucomannan (KGM) as additives increased this ratio, along with the volume and the hardness of the cookies (Akesowan, 2016). Similarly, enzymatic treatment with transglutaminase on buckwheat (BW) based cookies in different combinations of rice starch (R) and maize (M) starch was used to investigate the influence of the spread ratio on the hardness of the cookie. An equal proportion of BW and M starch had a greater spread ratio and lesser hardness, contributing to a softer cookie texture (Altındağ et al., 2014).

6.3 Gluten-free beverages

Gluten-free beverages have become well-known in the past decade due to their novelty and innovation, although the fermented beverages, their precursor, were primarily produced only for food preservation. GF beverages can be categorized as alcoholic (beer) and nonalcoholic types (health drinks and probiotic drinks) (Khairuddin & Lasekan, 2021). Beer is usually produced from rice, sorghum, sake, maize, and other millets. Malting the rice releases enzymes such as α and β-amylases and enables the release of sugar and amino acids required for beer production (Agu et al., 2012). Health drink mix (nonalcoholic beverage) sourced from millets, quinoa, etc., offers a variety of health benefits against hypertension and diabetes and promotes antioxidant properties (Kaur & Tanwar, 2016). Xiong et al. (2019) formulated a tea drink from the black sorghum variety that yielded better phenolic components. Probiotic drinks are another vital category of nonalcoholic gluten-free beverages that enhance the nutritional value of the products. Beverages made from sprouted oats along with *Lactobacillus plantarum* species improved the protein, vitamin and antioxidant properties (Aparicio-García et al., 2021). The bioavailability of minerals like zinc and iron multiplied exponentially due to the addition of LAB in the fermentation of millet (sorghum) drinks (Adeyanju et al., 2018). The lowering of pH due to fermentation also enhanced the digestibility of protein (Ogodo et al., 2019).

Although many gluten-free products are conveniently accessible in the market, there is a huge extent of development in their processing techniques. Thermal and nonthermal treatments can be explored further to optimize their protein interaction and to develop a structurally integral product. The future of gluten-free food products relies on further developments using additives such as gums, hydrocolloids, enzyme technology, fermentation, and 3D food printing, as discussed above in the section. Nutritional fortification with various types of gluten-free flours should be explored to produce novel gluten-free products. Sensory analysis done for market available gluten-free products shows the high interest and growing likeness towards this sector in the food industry. With the immense increase in consumer awareness of the GF diet, customer acceptance should be focused on proper labeling and promoting the adoption of sustainable choices with gluten-free diets.

7. Conclusion and future perspective

Awareness of gluten intolerance and related disorders has grown over the past few decades. Patients with celiac-related disorders understand the importance of dietary restriction by eliminating gluten to avoid further complications. Starch-based foods like bread and pasta are the familiar staple foods in most of the communities around the world. Thus, finding a nutritionally filling and compelling diet made of gluten-free starch, without compromising the organoleptic and textural properties, is the need of the hour. Substituting gluten-rich cereals with gluten-free starch-based alternatives, including millets, legume and vegetable sources such as tapioca and potato starch, improves the nutritional aspect of the product. Selection of raw gluten-free starch sources should be done carefully to avoid any other potential allergenic source as a substitute for wheat and gluten starch. Although GF foods are currently present in retail, most of them lack essential nutrients. Thus, focus should be shifted to produce nutrition-fortified gluten-free foods with alternate sources like quinoa, almond flour, etc. and other underutilized sources. In order to achieve gluten-free products, textural modification using enzymatic, physical, and chemical techniques is used. The addition of HPMC and xanthan gum as hydrocolloids to the gluten-free starch has a positive textural impact and presents encouraging opportunities for further improvement. Optimizing the ratios of different mixtures of these hydrocolloids has a great impact on the rheological and textural properties. Processing technologies to reduce or eliminate gluten should be mentioned on the labels of the products to promote transparency to the customers. Inclusion of policymakers and collaboration of industries with researchers are highly essential to innovate more gluten-free starch products. Economic strategies to promote affordable production and processing techniques could attract a wide range of demographics. By investigating and addressing the above-said areas, the application of gluten-free starch in the food system can grow and continue to innovate, then multiply the quality and acceptance of gluten-free products by fulfilling the dietary needs of consumers globally.

Acknowledgments

The author (PK) wishes to thank the Faculty of Engineering, Karpagam Academy of Higher Education, for their support and help.

References

Adepoju, M., Verheecke-Vaessen, C., Pillai, L. R., Phillips, H., & Cervini, C. (2024). Unlocking the potential of teff for sustainable, gluten-free diets and unravelling its production challenges to address global food and nutrition security: A review. *Foods, 13*(21), 3394.

Adeyanju, A. A., Kruger, J., Taylor, J. R. N., & Duodu, K. G. (2018). Effects of different souring methods on the protein quality and iron and zinc bioaccessibilities of non-alcoholic beverages from sorghum and amaranth. *International Journal of Food Science and Technology, 54*(3), 798–809.

Agu, R. C., Chiba, Y., Goodfellow, V., MacKinlay, J., Brosnan, J. M., Bringhurst, T. A., Jack, F. R., Harrison, B., Pearson, S. Y., & Bryce, J. H. (2012). Effect of germination temperatures on proteolysis of the gluten-free grains rice and buckwheat during malting and mashing. *Journal of Agricultural and Food Chemistry, 60*(40), 10147–10154.

Akesowan, A. (2016). Influence of konjac flour on foaming properties of milk protein concentrate and quality characteristics of gluten-free cookie. *International Journal of Food Science and Technology, 51*(7), 1560–1569.

Al-Toma, A., Volta, U., Auricchio, R., Castillejo, G., Sanders, D. S., Cellier, C., Mulder, C. J., & Lundin, K. E. A. (2019). European Society for the Study of Coeliac Disease (ESsCD) guideline for coeliac disease and other gluten-related disorders. *United European Gastroenterology Journal, 7* (5), 583–613.

Alaunyte, I., Stojceska, V., Plunkett, A., Ainsworth, P., & Derbyshire, E. (2011). Improving the quality of nutrient-rich Teff (Eragrostis tef) breads by combination of enzymes in straight dough and sourdough breadmaking. *Journal of Cereal Science, 55*(1), 22–30.

Ali, A., Wani, T. A., Wani, I. A., & Masoodi, F. A. (2014). Comparative study of the physico-chemical properties of rice and corn starches grown in Indian temperate climate. *Journal of the Saudi Society of Agricultural Sciences, 15*(1), 75–82.

Altındağ, G., Certel, M., Erem, F., & Konak, Ü.İ. (2014). Quality characteristics of gluten-free cookies made of buckwheat, corn, and rice flour with/without transglutaminase. *Food Science and Technology International, 21*(3), 213–220.

Aparicio-García, N., Martínez-Villaluenga, C., Frias, J., & Peñas, E. (2021). Production and characterization of a novel Gluten-Free fermented beverage based on sprouted oat flour. *Foods, 10*(1), 139.

Arendt, E. K., & Bello, F. D. (2008). *Gluten-free cereal products and beverages*. In Elsevier eBooks.

Arendt, E. K., Moore, M., Hui, Y. H., Corke, H., De Leyn, I., Nip, W.-K., & Cross, N. (2006). Gluten-free cereal-based products. In *Bakery products: Science and technology* (1st ed.).

Azeem, M., Wadood, S. A., Qureshi, T. M., Rehman, M. a. U., Ahmed, I. a. M., Aljobair, M. O., & Ali, A. A. (2024). FTIR characteristics of sweet flour as affected by particle size distribution and in vitro digestibility of gluten-free sweet potato noodles. *International Journal of Food Properties, 28*(1), 16–31.

Bashir, K., Jan, K., & Aggarwal, M. (2017). Thermo-rheological and functional properties of gamma-irradiated wholewheat flour. *International Journal of Food Science and Technology, 52*(4), 927–935.

Bathula, S. R. (2018). Extraction of gluten from food material. *MOJ Proteomics & Bioinformatics, 7*(3).

Biesiekierski, J. R. (2017). What is gluten? *Journal of Gastroenterology and Hepatology, 32*(S1), 78–81.

Bourekoua, H., Benatallah, L., Zidoune, M. N., & Rosell, C. M. (2016). Developing gluten free bakery improvers by hydrothermal treatment of rice and corn flours. *LWT, 73*, 342–350.

Bradauskiene, V., Vaiciulyte-Funk, L., Shah, B., Cernauskas, D., & Tita, M. (2021). Recent advances in biotechnological methods for wheat gluten immunotoxicity abolishment – A review. *Polish Journal of Food and Nutrition Sciences*, 5–20.

Cao, H., Wang, X., Ruan, Y., Huang, K., Zhang, Y., Song, H., Li, S., & Guan, X. (2023). Alkaline amino acids present unique aggregation behavior for quinoa protein gelation under microwave irradiation. *Food Hydrocolloids, 144*, Article 108951.

Cao, H., Wang, X., Wu, W., Grimi, N., Lu, J., & Guan, X. (2024). Modulation of textural properties in microwave treated Gluten-Free quinoa sponge cake by alkaline amino acids. *Food Hydrocolloids, 160*, Article 110727.

Castro, L. M., Caço, A. I., Pereira, C. F., Sousa, S. C., Brassesco, M. E., Machado, M., & Pintado, M. (2023). Modification of acorn starch structure and properties by high hydrostatic pressure. *Gels, 9*(9), 757.

Chang, Q., Zheng, B., Zhang, Y., & Zeng, H. (2021). A comprehensive review of the factors influencing the formation of retrograded starch. *International Journal of Biological Macromolecules, 186*, 163–173.

Chen, C., Li, G., & Zhu, F. (2021). A novel starch from lotus (Nelumbo nucifera) seeds: Composition, structure, properties and modifications. *Food Hydrocolloids, 120*, Article 106899.

Clerici, M. T. P. S., Airoldi, C., & El-Dash, A. A. (2008). Production of acidic extruded rice flour and its influence on the qualities of gluten-free bread. *LWT, 42*(2), 618–623.

Codex Alimentarius Commission. (1979). *Standard for foods for special dietary use for persons intolerant to gluten*. In CXS 118-1979.

Copeland, L., Blazek, J., Salman, H., & Tang, M. C. (2008). Form and functionality of starch. *Food Hydrocolloids, 23*(6), 1527–1534.

Da Mota Huerta, K., Alves, J. D. S., Da Silva, A. F. C., Kubota, E. H., & Da Rosa, C. S. (2016). Sensory response and physical characteristics of gluten-free and gum-free bread with chia flour. *Food Science and Technology, 36*(Suppl. 1), 15–18.

De La Barca, A. M. C., Rojas-Martínez, M. E., Islas-Rubio, A. R., & Cabrera-Chávez, F. (2010). Gluten-free breads and cookies of raw and popped amaranth flours with attractive technological and nutritional qualities. *Plant Foods for Human Nutrition, 65*(3), 241–246.

Demirkesen, I., Mert, B., Sumnu, G., & Sahin, S. (2009). Rheological properties of gluten-free bread formulations. *Journal of Food Engineering, 96*(2), 295–303.

Dini, C., Garzón, R., & Rosell, C. M. (2024). Breadmaking potential of andean roots and tuber starches from ahipa (Pachyrhizus ahipa), oca (Oxalis tuberosa), and arracacha (Arracacia xanthorrhiza). *Starch-Stärke*, Article 2400085.

Doğan, H., & Karwe, M. V. (2003). Physicochemical properties of quinoa extrudates. *Food Science and Technology International, 9*(2), 101–114.

Elgeti, D., Nordlohne, S. D., Föste, M., Besl, M., Linden, M. H., Heinz, V., Jekle, M., & Becker, T. (2013). Volume and texture improvement of gluten-free bread using quinoa white flour. *Journal of Cereal Science, 59*(1), 41–47.

Espinosa-Solis, V., Zamudio-Flores, P. B., Espino-Díaz, M., Vela-Gutiérrez, G., Rendón-Villalobos, J. R., Hernández-González, M., & Ortega-Ortega, A. (2021). Physicochemical characterization of resistant starch type-III (RS3) obtained by autoclaving malanga (Xanthosoma sagittifolium) flour and corn starch. *Molecules, 26*(13), 4006.

Fasakin, O. B., Uchenna, O. F., Ajayi, O. M., Onarinde, B. A., Konar, S., Seung, D., & Oyeyinka, S. A. (2024). Optimisation of dry heat treatment conditions for modification of faba bean (*Vicia faba* L.) starch. *Heliyon, 10*(16), Article e35817.

Gallagher, E., Gormley, T., & Arendt, E. (2002). Crust and crumb characteristics of gluten free breads. *Journal of Food Engineering, 56*(2–3), 153–161.

Gao, L., Van Bockstaele, F., Lewille, B., Haesaert, G., & Eeckhout, M. (2023). Characterization and comparative study on structural and physicochemical properties of buckwheat starch from 12 varieties. *Food Hydrocolloids, 137*, Article 108320.

Griffith, L. D., Castell-Perez, M. E., & Griffith, M. E. (1998). Effects of blend and processing method on the nutritional quality of weaning foods made from select cereals and legumes. *Cereal Chemistry, 75*(1), 105–112.
Gularte, M. A., De La Hera, E., Gómez, M., & Rosell, C. M. (2012). Effect of different fibers on batter and gluten-free layer cake properties. *LWT, 48*(2), 209–214.
Hüttner, E. K., Bello, F. D., & Arendt, E. K. (2010). Fundamental study on the effect of hydrostatic pressure treatment on the bread-making performance of oat flour. *European Food Research and Technology, 230*(6), 827–835.
Hormdok, R., & Noomhorm, A. (2007). Hydrothermal treatments of rice starch for improvement of rice noodle quality. *LWT, 40*(10), 1723–1731.
Horstmann, S., Belz, M., Heitmann, M., Zannini, E., & Arendt, E. (2016). Fundamental study on the impact of gluten-free starches on the quality of gluten-free model breads. *Foods, 5*(2), 30.
Horstmann, S., Lynch, K., & Arendt, E. (2017). Starch characteristics linked to gluten-free products. *Foods, 6*(4), 29.
Hosseini, S. M., Soltanizadeh, N., Mirmoghtadaee, P., Banavand, P., Mirmoghtadaie, L., & Shojaee-Aliabadi, S. (2018). Gluten-free products in celiac disease: Nutritional and technological challenges and solutions. *Journal of Research in Medical Sciences, 23*(1), 109.
Houben, A., Höchstötter, A., & Becker, T. (2012). Possibilities to increase the quality in gluten-free bread production: An overview. *European Food Research and Technology, 235*(2), 195–208.
Hu, A., Wang, X., Li, L., Xu, T., & Zheng, J. (2020). Effects of annealing time on structure and properties of sweet potato starch. *Cereal Chemistry, 97*(3), 573–580.
Huang, J., Wang, Z., Fan, L., & Ma, S. (2022). A review of wheat starch analyses: Methods, techniques, structure and function. *International Journal of Biological Macromolecules, 203*, 130–142.
Imam, Y. T., Irondi, E. A., Awoyale, W., Ajani, E. O., & Alamu, E. O. (2024). Application of legumes in the formulation of gluten-free foods: Functional, nutritional and nutraceutical importance. *Frontiers in Sustainable Food Systems, 8*.
Itusaca-Maldonado, Y. M., Apaza-Humerez, C. R., Pumacahua-Ramos, A., & Pinto, E. M. (2024). Technological and textural properties of gluten-free quinoa-based pasta (Chenopodium quinoa Wild). *Heliyon, 10*(7), Article e28363.
Javadian, N., Mohammadi Nafchi, A., & Bolandi, M. (2021). The effects of dual modification on functional, microstructural, and thermal properties of tapioca starch. *Food Science and Nutrition, 9*(10), 5467–5476.
Kaur, I., & Tanwar, B. (2016). Quinoa beverages: Formulation, processing and potential health benefits. *Romanian Journal of Diabetes Nutrition and Metabolic Diseases, 23*(2), 215–225.
Khairuddin, M. a. N., & Lasekan, O. (2021). Gluten-free cereal products and beverages: A review of their health benefits in the last five years. *Foods, 10* (11), 2523.
Khoury, D. E., Balfour-Ducharme, S., & Joye, I. J. (2018). A review on the gluten-free diet: Technological and nutritional challenges. *Nutrients, 10*(10), 1410.
Kim, H., Oh, S., Bae, J., Yeom, J., Kim, B., Kim, H., & Baik, M. (2017). Preparation and characterization of amorphous granular potato starches (AGPS) and cross-linked amorphous granular potato starches (CLAGPS). *Carbohydrate Polymers, 178*, 41–47.
Korus, J., Witczak, M., Ziobro, R., & Juszczak, L. (2015). The influence of acorn flour on rheological properties of gluten-free dough and physical characteristics of the bread. *European Food Research and Technology, 240*(6), 1135–1143.
Kumar, A., Swarnalatha, A. P., Shwetha, J., Saravanan, P., Hatamleh, A. A., Al-Dosary, M. A., … Ravindran, B. (2024). Pyrolysis behaviour and synergistic effect in co-pyrolysis of wheat straw and polyethylene terephthalate: A study on product distribution and oil characterization. *Heliyon, 10*(17).
Lazaridou, A., & Biliaderis, C. G. (2009). Gluten-free doughs: Rheological properties, testing procedures – Methods and potential problems. In *Gluten-Free food science and technology* (pp. 52–82).
Lee, & Lee. (2006). Bread-making properties of rice flours produced by dry, wet and semi-wet milling. *Journal of the Korean Society of Food Science and Nutrition, 35*(7), 886–890.
Liu, C., Liu, S., Li, R., Zhang, X., & Chang, X. (2024). A mechanistic study of chestnut starch retrogradation and its effects on in vitro starch digestion. *International Journal of Biological Macromolecules, 276*, Article 133803.
Martín-Diana, A. B., Jiménez-Pulido, I. J., Aguiló-Aguayo, I., Abadías, M., Pérez-Jiménez, J., & Rico, D. (2025). Peach peel extrusion for the development of sustainable gluten-free plant-based flours. *Molecules, 30*(3), 573.
Mirzaei, M., & Movahed, S. (2013). Evaluation of staling rate and quality of gluten-free toast breads on rice flour basis. *Research Journal of Applied Sciences, Engineering and Technology, 5*(1), 224–227.
Moore, M. M., Schober, T. J., Dockery, P., & Arendt, E. K. (2004). Textural comparisons of gluten-free and wheat-based doughs, batters, and breads. *Cereal Chemistry, 81*(5), 567–575.
Moreira, R., Chenlo, F., & Torres, M. (2012). Effect of chia (Sativa hispanica L.) and hydrocolloids on the rheology of gluten-free doughs based on chestnut flour. *LWT, 50*(1), 160–166.
Mudgil, D., Barak, S., & Khatkar, B. (2017). Cookie texture, spread ratio and sensory acceptability of cookies as a function of soluble dietary fiber, baking time and different water levels. *LWT, 80*, 537–542.
Nagar, C. K., Dash, S. K., Rayaguru, K., Pal, U. S., & Nedunchezhiyan, M. (2021). Isolation, characterization, modification and uses of taro starch: A review. *International Journal of Biological Macromolecules, 192*, 574–589.
Nilsson, K., Johansson, M., Sandström, C., Röhnisch, H. E., Hedenqvist, M. S., & Langton, M. (2023). Pasting and gelation of faba bean starch-protein mixtures. *Food Hydrocolloids, 138*, Article 108494.
Noor, F. (2014). Physicochemical properties of flour and extraction of starch from jackfruit seed. *International Journal of Nutrition and Food Sciences, 3*(4), 347.

Ogodo, A. C., Ugbogu, O. C., Onyeagba, R. A., & Okereke, H. C. (2019). Microbiological quality, proximate composition and in vitro starch/protein digestibility of Sorghum bicolor flour fermented with lactic acid bacteria consortia. *Chemical and Biological Technologies in Agriculture, 6*(1).

Otemuyiwa, I. O., & Aina, A. F. (2021). Physicochemical properties and in-vitro digestibility studies of microwave assisted chemically modified breadfruit (Artocarpus altilis) starch. *International Journal of Food Properties, 24*(1), 140–151.

Oyeyinka, S. A., Kayitesi, E., Adebo, O. A., Oyedeji, A. B., Ogundele, O. M., Obilana, A. O., & Njobeh, P. B. (2021). A review on the physicochemical properties and potential food applications of cowpea (Vigna unguiculata) starch. *International Journal of Food Science and Technology, 56*(1), 52–60.

Phongthai, S., D'Amico, S., Schoenlechner, R., Homthawornchoo, W., & Rawdkuen, S. (2017). Effects of protein enrichment on the properties of rice flour based gluten-free pasta. *LWT, 80*, 378–385.

Qian, X., Sun, B., Zhou, M., Ma, S., & Wang, X. (2025). Understanding the dual impact of oat protein on the structure and digestion of oat starch at pre- and post-retrogradation stages. *International Journal of Biological Macromolecules, 288*, Article 138762.

Rhowell Jr, N. T., Bonto, A. P., & Sreenivasulu, N. (2021). Enhancing the functional properties of rice starch through biopolymer blending for industrial applications: A review. *International Journal of Biological Macromolecules, 192*, 100–117.

Sánchez Riaño, A. M., Alegría Vivas, E. M., Bermeo Andrade, H. P., Bohórquez Pérez, Y., Villada Castillo, H. S., Daza, L. D., & Valenzuela Real, C. P. (2022). Physicochemical, structural, and thermal characterization of biodegradable film prepared using arracacha thermoplastic starch and polylactic acid. *Journal of Food Measurement and Characterization, 16*(5), 3597–3606.

Šmídová, Z., & Rysová, J. (2022). Gluten-free bread and bakery products technology. *Foods, 11*(3), 480.

Sadullayev, S., Ravshanov, S., Mirzayev, J., Ibragimov, A., Baxromova, L., & Yuldashova, R. (2024). *Impact of flour particle size and starch damage on baking properties of wheat flour grown in dry climates: A Uzbekistan case study* (Vol 12, p. 47).

Sandri, L. T. B., Santos, F. G., Fratelli, C., & Capriles, V. D. (2017). Development of gluten-free bread formulations containing whole chia flour with acceptable sensory properties. *Food Science and Nutrition, 5*(5), 1021–1028.

Sedej, I., Sakač, M., Mandić, A., Mišan, A., Pestorić, M., Šimurina, O., & Čanadanović-Brunet, J. (2010). Quality assessment of gluten-free crackers based on buckwheat flour. *LWT, 44*(3), 694–699.

Segura, V., Díaz, J., Ruiz-Carnicer, Á., Muñoz-Suano, A., Carrillo-Carrión, C., Sousa, C., Cebolla, Á., & Comino, I. (2021). Rapid, effective, and versatile extraction of gluten in food with application on different immunological methods. *Foods, 10*(3), 652.

Sindhu, R., Devi, A., & Khatkar, B. S. (2021). Morphology, structure and functionality of acetylated, oxidized and heat moisture treated amaranth starches. *Food Hydrocolloids, 118*, Article 106800.

Singh, J., Kaur, L., & McCarthy, O. J. (2007). Factors influencing the physico-chemical, morphological, thermal and rheological properties of some chemically modified starches for food applications—A review. *Food Hydrocolloids, 21*(1), 1–22.

Singh, N., Singh, J., Kaur, L., Sodhi, N. S., & Gill, B. S. (2003). Morphological, thermal and rheological properties of starches from different botanical sources. *Food Chemistry, 81*(2), 219–231.

Sivaramakrishnan, H. P., Senge, B., & Chattopadhyay, P. (2003). Rheological properties of rice dough for making rice bread. *Journal of Food Engineering, 62*(1), 37–45.

Sozer, N. (2008). Rheological properties of rice pasta dough supplemented with proteins and gums. *Food Hydrocolloids, 23*(3), 849–855.

Spada, F. P., Zerbeto, L. M., Ragazi, G. B. C., Gutierrez, É. M. R., Souza, M. C., Parker, J. K., & Canniatti-Brazaca, S. G. (2017). Optimization of postharvest conditions to produce chocolate aroma from jackfruit seeds. *Journal of Agricultural and Food Chemistry, 65*(6), 1196–1208.

Steffolani, E., De La Hera, E., Pérez, G., & Gómez, M. (2014). Effect of chia (Salvia hispanica L) addition on the quality of gluten-free bread. *Journal of Food Quality, 37*(5), 309–317.

Taylor, J. R., Schober, T. J., & Bean, S. R. (2006). Novel food and non-food uses for sorghum and millets. *Journal of Cereal Science, 44*(3), 252–271.

Tester, R. F., & Debon, S. J. (2000). Annealing of starch—A review. *International Journal of Biological Macromolecules, 27*(1), 1–12.

The Business Research Company. (2024). *Gluten free food global market report 2024*. The Business Research Company. https://www.thebusinessresearchcompany.com/report/gluten-free-food-global-market-report.

Tong, C., Ma, Z., Chen, H., & Gao, H. (2023). Toward an understanding of potato starch structure, function, biosynthesis, and applications. *Food Frontiers, 4*(3), 980–1000.

Tsatsaragkou, K., Gounaropoulos, G., & Mandala, I. (2014). Development of gluten free bread containing carob flour and resistant starch. *LWT, 58*(1), 124–129.

Ukwatta, R. H., Yuan, R., Ma, Y., Xiong, X., Hu, Y., Li, C., & Xue, F. (2025). Effect of lipid addition on the physiochemical, structural, and photoactive antibacterial properties of cornstarch-chlorophyllin composite film. *Food Research International, 202*, Article 115699.

Wang, H., Ding, F., Ma, L., & Zhang, Y. (2021). Edible films from chitosan-gelatin: Physical properties and food packaging application. *Food Bioscience, 40*, Article 100871.

Warren, F. J., Gidley, M. J., & Flanagan, B. M. (2015). Infrared spectroscopy as a tool to characterise starch ordered structure—A joint FTIR–ATR, NMR, XRD and DSC study. *Carbohydrate Polymers, 139*, 35–42.

Weegels, P. L., Marseille, J. P., & Hamer, H. R. J. (1988). Small scale separation of wheat flour in starch and gluten. *Starch - Stärke, 40*(9), 342–346.

Wieser, H. (2006). Chemistry of gluten proteins. *Food Microbiology, 24*(2), 115–119.

Wronkowska, M., Haros, M., & Soral-Śmietana, M. (2012). Effect of starch substitution by buckwheat flour on gluten-free bread quality. *Food and Bioprocess Technology, 6*(7), 1820–1827.

Xiong, Y., Zhang, P., Johnson, S., Luo, J., & Fang, Z. (2019). Comparison of the phenolic contents, antioxidant activity and volatile compounds of different sorghum varieties during tea processing. *Journal of the Science of Food and Agriculture, 100*(3), 978–985.

Yadav, P., Prabahar, A. E., & Verma, A. (2024). Chemical modification and pharmaceutical evaluation of starch obtained from edible natural sources. *South Eastern European Journal of Public Health*, 303–310. https://doi.org/10.70135/seejph.vi.2664

Yang, Y., & Tao, W. (2007). Effects of lactic acid fermentation on FT-IR and pasting properties of rice flour. *Food Research International, 41*(9), 937–940.

Yazid, N. S. M., Abdullah, N., Muhammad, N., & Matias-Peralta, H. M. (2018). Application of starch and starch-based products in food industry. *Journal of Science and Technology, 10*(2).

Zahid, M. K., Ahmad, D., Amin, R., & Bao, J. (2025). Sorghum starch: Composition, structure, functionality, and strategies for its improvement. *Comprehensive Reviews in Food Science and Food Safety, 24*(1), Article e70101.

Zailani, M. A., Kamilah, H., Husaini, A., Seruji, A. Z. R. A., & Sarbini, S. R. (2022). Functional and digestibility properties of sago (*Metroxylon sagu*) starch modified by microwave heat treatment. *Food Hydrocolloids, 122*, Article 107042.

Zhang, B., Qiao, D., Zhao, S., Lin, Q., Wang, J., & Xie, F. (2021). Starch-based food matrices containing protein: Recent understanding of morphology, structure, and properties. *Trends in Food Science & Technology, 114*, 212–231.

Zhu, Y., Sang, L., Sharafeldin, S., Zhao, L., Chen, R., Wang, C., Xue, Y., & Shen, Q. (2024). The effects of chitosan oligosaccharides on the structure and shelf-life of whole-millet cakes. *Food Chemistry, 466*, Article 142267.

Chapter 16

Starches in baby, adult, and elderly foods: Nutritional benefits and formulation considerations

Sakthi Priyanka Palanivel[1], Shubasree Manickavel[1], Ravi Sai Nandhini[1], A. Saravana Raj[2] and Vaishnavi Mohanadas Nair[3]

[1]*Department of Biotechnology, Vel Tech Rangarajan Dr. Sagunthala R&D Institute of Science and Technology, Chennai, Tamil Nadu, India;*
[2]*Vel Tech High Tech Dr.Rangarajan Dr.Sakunthala Engineering College, Department of Chemical Engineering, Chennai, Tamil Nadu, India;*
[3]*Technical University of Munich, Department of Chemical Engineering, Bavaria, Germany*

1. Introduction

Starch serves as one of the most extensive reserves of carbohydrates in plants, making up a significant portion of human meals across all age groups. Its significance in food systems has been established by its adaptability as a renewable, multipurpose polysaccharide, acting as a vital energy source and a crucial functional component in a variety of formulations (Saboonchi et al., 2021). The starch content, granule size, and physical features differ across grains, tubers, roots, and fruits. Notwithstanding their widespread availability, native starches have drawbacks such as low stability and solubility, and modifications to improve functional properties. Starch is composed of two types of polymers: amylose (linear-chain polymer) and amylopectin (branched-chain polymer), which depend on the proportions for certain functional properties, including gelatinization temperature, retrogradation rate, and digestibility (Saboonchi et al., 2021).

Other health benefits of resistant starches, a form of dietary fiber, include improved blood sugar control, cholesterol levels, and gut microbiota. Research on the development of starch derivatives with enhanced nutritional and functional attributes continues in an effort to meet a range of dietary needs and consumers with health issues. Food for the elderly should include modified starches to address their unique dietary, sensory, and health needs. Food formulations that ensure safety, ease of consumption, and adequate nutrition are required due to physiological alterations such as reduced saliva secretion, restricted chewing and swallowing abilities, and altered digestion (Makran et al., 2022). The production of texture-modified foods (TMF) that are soft, cohesive, and easy to swallow is facilitated by modified starches, particularly those used for thickness and gel formation. This significantly reduces the risk of choking or inhalation of air, which could lead to severe health complications. Cross-linked and pregelatinized starches are extensively utilized because they can provide consistent textures in various situations, which makes them ideal for hot soups, pureed foods, and thickened drinks that older adults commonly consume. Apart from modifying texture, modified starches enhance nutrient bioavailability, which increases the nutritional value of food for the elderly (Araiza-Calahorra et al., 2023). Most gastrointestinal problems in the elderly are addressed by enzymatically modified starches, particularly resistant starches, which function as prebiotics and promote gut health by cultivating a healthy microbiota (Liu et al., 2022).

Modified starches also act as transporters for bioactive substances, safeguarding delicate nutrients and regulating their release throughout digestion to maximize absorption. By producing smooth textures and improving flavor release, they also enhance sensory experiences, making up for the diminished sensory awareness that comes with ageing and making common staples like pasta, bread, and pureed meals more palatable. The chemical alterations ensure their suitability and reliability for ingestion by simulating biological functions. A variety of senior food items, such as pureed meals, thickened beverages, fortified foods, and nutritional foods, contain modified starches (Methacanon et al., 2021). They avoid syneresis in soft-textured food items, give patients with dysphagia balance in liquids, and supply vital nutrients to correct

Native and Modified Starches for Food Applications. https://doi.org/10.1016/B978-0-443-27431-2.00016-X

common deficits in older persons. Notwithstanding their adaptability, there are still difficulties in satisfying customer demands for clean-label products while combining textural, nutritious, and sensory features. In order to make sure that foods are safe, nourishing, and aesthetically pleasing, advancements in starch formulation concentrate on improving functional qualities that are suited to the needs of the elderly, such as better nutrient encapsulation and sophisticated texture evaluation techniques (Alfa, Strang, Tappia, Olson, et al., 2018).

From infant formulae and complementary foods to processed foods for adults and TMFs for the elderly, the importance of starch across all stages of life emphasizes its significant role in food systems. Technological advances in starch modification have made it possible to tailor its use to meet the evolving functional and nutritional demands of different ages. To maintain starch's ability to support well-being and health throughout life, additional research is required to tackle the specific challenges of digestion, safety, and palatability to taste in speciality formulations (Araiza-Calahorra et al., 2023; Liu et al., 2022; Makran et al., 2022; Methacanon et al., 2021).

Starch Digestion and Metabolic Pathway Overview: Starch digestion is a process where there is a sequence of enzyme actions. The process starts with salivary amylase in the mouth. This enzyme breaks down starch to maltose and carbohydrate compounds. Pancreatic amylase then continues this process in the stomach. Brush border enzymes finish this process by producing glucose for blood uptake by the liver.

The rate of glucose appearance plays a determining role in the glycemic response:

- Rapidly Digestible Starch (RDS): Glucose absorbed after 20 min → high glucose and insulin peaks
- Slowly Digestible Starch (SDS): Slow release of glucose in 20–120 min → moderate response
- Resistant Starch (RS): Resistant to enzymatic hydrolysis → fermented in the large intestine forming short-chain fatty acids

The chapter will discuss the role of native and modified starch in three different stages of human life: infancy, adulthood, and old age, and will underscore how native and modified starch chemistry is a response to the different physiological requirements in different stages. The chapter will strongly focus on the digestibility, glycemic index, texture, and health-related properties in different stages.

2. The role of starch in baby foods

Infants are known to pose distinct challenges with regard to nutrition and physiology compared to older children and adults. The physiological system of infants produces a lower amount of pancreatic α-amylase before 6 months of age, with underdeveloped swallowing reflexes. The above-mentioned physiological limitations simulated the need to create specialized starches with properties of easier digestion, texture safety, and nutritional adequacy. This section will discuss how native and modified starches are required to meet infant needs. According to Galen, starches have been an essential part of baby nutrition since the second century AD. Starches were used in early recipes, such as panada and diluted milk with barley water to thicken, improve meal consistency, and decrease vomiting (Saboonchi et al., 2021). In order to improve digestibility and replicate breast milk, starch was introduced to infant formulae in the 19th century. To lower infant mortality, companies such as Liebig and Nestlé employed starch derived from wheat (Kathuria et al., 2018). Large casein curds are avoided, intestinal comfort is increased, viscosity is improved, and vital energy is provided by the starches in baby formula. Cereals such as rice, corn, oats, wheat and barley were utilized in the past; barley was prized for its ability to shield the lining of the stomach. Maltodextrin and other processed starches are now included in modern formulae to increase digestibility and calorie density. Starch is gelatinized by cooking, which turns complicated carbohydrates into glucose that can be absorbed (Chen et al., 2020). With components including sucrose, glucose, and dextrin, modern infant formulas meet a variety of nutritional needs and lactose intolerance while promoting a baby's healthy growth, digestion, and energy (Gałkowska et al., 2023).

2.1 Starch types applied to baby foods

As a vital form of carbohydrates and an additive for thickening, starch is crucial to the formulation of baby food. Starch is derived from a variety of cereals and improves nutritional content, texture, and digestion (Gutiérrez, 2020). Different kinds of starch found in infant meals are provided below in Fig. 16.1.

Native starches in infant foods: Natural starches, which are made solely from grains, are essential ingredients in infant food because of their high amount of carbohydrates, quick digestion, and beneficial qualities. Native starches have varying levels of amylose -amylopectin content, with waxy cereals such as rice and maize having lower amylose content and being helpful for infants (Saboonchi et al., 2021). Baby meals contain natural starches derived from a variety of cereals, each

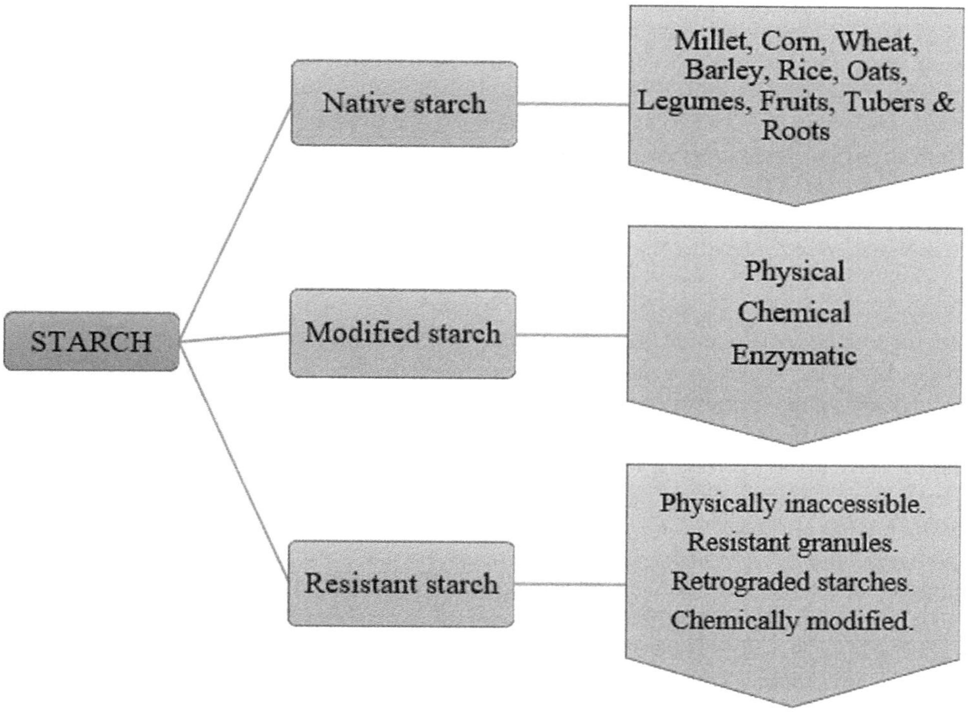

FIGURE 16.1 Types of starch.

with unique qualities (Gutiérrez, 2020; Quann and Carvalho., 2018). Rice has become one of the most often utilized grains in starch extraction, especially in infant meals. It has tiny, easily digestible starch granules that are about 5 μm in diameter. Because of this, rice is perfect for babies whose digestive systems are still developing. Its neutral flavor further increases its adaptability for creating baby food, such as puddings and simple cereals (Pascari et al., 2019; Gutiérrez, 2020).

Wheat contains medium-sized granules ranging from 25 to 40 μm, making it a considerable source of starch. A lot of baby cereals, baked goods, and porridges made for babies contain wheat. Additionally, its composition includes insoluble fiber and resistant starch, which support good gut flora and increase stool volume, therefore improving gut health. Whole wheat is still abundant in nutrients for the majority of babies, providing energy and vital nutrients, even though the level of gluten may be problematic (Pascari et al., 2019; Gutiérrez, 2020; Quann and Carvalho., 2018). Corn (maize) starch is a cost-effective and versatile ingredient in baby food compositions. It is frequently used to make baby purees, soups and liquid meals. From a nutritional point of view, this gives an economical, high-energy food, and the resistant starch profile serves as dietary fiber to ensure intestinal well-being. Corn is a foundation in infant food manufacture because it is readily available (Gutiérrez, 2020; Quann and Carvalho., 2018). Modified starches in infant food: The increased functional properties of modified starches, for example, solubility, consistency, and stability under a range of manufacturing conditions, have contributed to their popularity in infant food preparation. The limitations of native starches, i.e., poor solubility, susceptibility to retrogradation, and resistance to enzymatic hydrolysis, are improved upon by these modifications. Physical or chemical modifications can be carried out on starches to enhance their use as infant food, leading to the development of products that are safe, nutritious, and easy to digest (Saboonchi et al., 2021; Kathuria et al., 2018).

Physical changes techniques, including pregelatinization, high-pressure processing, and ultrasound treatment, that change the size and shape of starch granules. The process of being cooked and getting dried starch results in pregelatinized starch, which is perfect for thickened soups and instant baby cereals since it increases its solubility and dissolves in cold water (Saboonchi et al., 2021). Although nonthermal techniques, such as high-pressure treatment, enhance compound texture and viscosity without sacrificing nutrients, ultrasonic alters solubility and viscosity through depolymerizing granules, improving digestion and preserving the taste. These methods are easy to use, safe, and free of synthetic chemicals, which makes them safe for human consumption, especially for young children (Chen et al., 2020; Gałkowska et al., 2023). However, in order to enhance characteristics like stability, solubility, and durability during industrial processing, chemical changes modify the functional groups of the starch molecules. By forming a structure with three dimensions, cross-linked starch improves resistance to heat, strain, and acids, guaranteeing stability throughout handling and storage. By introducing acetyl groups, acetylated starch enhances solubility, thickening capacity, and digestibility while preventing retrogradation, maintaining the freshness of infant foods. To maintain the appropriate thickness and texture in baby food formulations, these modifications allow starches to serve as thickeners, stabilizers, and texturizers. As far as the application is concerned, the use of starches is the result of dual physical and chemical modifications. Both physical modification (pregelatinization and high-pressure treatment) and chemical modifications (cross-linking and acetylation) enable higher digestibility through distinct mechanisms: the former through the partially gelatinized granules in the body, and the latter through the nonretrograde molecular structure. This convergence of different modification processes targets the same functional outcomes, indicating the prime significance of texture and stability in infant nutrition (Kathuria et al., 2018).

Resistant starches: A unique type of carbohydrate called Resistant Starches (RS) resists degradation in the small intestine and is used as a substrate for fermentation in the large intestine, resulting in a range of beneficial byproducts such as short-chain fatty acids (SCFAs), including butyrate, acetate, and propionate. Due to these characteristics, RS is an important functional ingredient of baby food, supporting improved metabolic outcomes, gastrointestinal health, and overall well-being. By specifically favoring the growth of beneficial gut bacteria such as Lactobacilli and Bifidobacteria, which are important for maintaining a healthy microbiome, RS functions as a prebiotic. This prebiotic effect enhances a healthy gut ecology and allows babies to establish a robust immune system (Garg et al., 2017). RS produces SCFAs such as butyrate through large intestine microbial fermentation, which supplies colonocytes (intestinal cells) with energy. Butyrate maintains a healthy gut, protects against colonic disease, and maintains the normal integrity of the intestinal barrier. In addition, SCFAs synthesis reduces fecal pH, which suppresses the growth of pathogenic bacteria and enhances digestive comfort (Garg et al., 2017; Rodríguez et al., 2022).

Comparative context: Unlike the mature digestive system of adults, infants have limited pancreatic α-amylase before 6 months; therefore, cereal starch selection should focus on small granule size and pregelatinization to ensure digestibility. With fully developed enzymatic systems, adults efficiently digest a wide range of starches, including resistant starches. This fundamental difference in enzyme production establishes distinct starch formulation approaches with respect to age groups.

2.2 Functions of modified starches in baby formulas

Modified starches, particularly gelatinized starches, are critical components in infant formulae, especially Anti-Regurgitation Formulae (AR-F), which treat regurgitation, feeding intolerance, and gastro-esophageal reflux disease (GER). They thicken the formula, making it viscous, helping to reduce regurgitation by preventing the formula from flowing back to the esophagus. Gelatinized starches are easier to digest, and resistant starches manage digestion by slowing down the emptying of the stomach. Resistant starches enhance digestion by making the formula simpler for infants to digest. Modified starches in AR-F minimize frequent feeding complications, enhance baby comfort, and enhance healthy growth through enhanced viscosity and digestive tolerance (Chakraborty et al., 2022; Moin et al., 2019). The Progress of the Modified Starch Inclusion in the Infant Formulas: Modified starch inclusion in the infant formulas started in the early 1900s after the advancement in industrial technology and nutrition. Some of the major inventions include the engineering of Dextri Maltose in 1912, which was invented as a solution for a person with lactose intolerance, and the multiple strains of Maize Starch engineered during World War 1. By the end of the 1920s, these starches made it easier to improve the thickness of the infant formulas as well as assist in increasing the nutrient concentration, which aided the growth of the infant (Pascari et al., 2019; Nguyen, 2017).

Starch-Induced Thickness: Rice starch, locust bean gum, maize starch and other starches were used to compound infant formulas, making them thicker in viscosity, which greatly reduced GER symptoms as well as increased the time required for them to empty from the stomach. Slow digestion was key. These starches were important as they were able to increase the acidity of the stomach and continue thickening the compounds, which resulted in improving digestion (Gałkowska et al., 2023; Himat et al., 2021). Digestibility and absorption: Breast milk can be difficult for the immature enzymes of infants to digest breast milk thoroughly. It is easier for the body to break down precooked or gelatinized starches in terms of food absorption. Various types of starches differ in the amount of amylose and amylopectin present in them and help in the difference in the rate of absorption among infants (Salvatore et al., 2018; Nguyen, 2017).

GER treatment: Modified starches, such as rice starch, maize starch, and carob bean gum, have been used in infant formulas to treat GER, a common condition characterized by recurrent regurgitation in babies. These starches increase the viscosity of the infant formula, keeping it inside the stomach for longer, reducing the chance of reflux into the esophagus. Studies have established that when these starches are included in the AR-F fed to babies, they experience a great reduction in episodes of regurgitation. The frequency of regurgitation episodes fell from 5.43 per day to as low as 2.5, or sometimes even as few as 0.6 to 1.8 episodes per day. These reductions in frequency of regurgitation suggest that AR-F thickened with starch do indeed offer clinically important symptom reduction without compromising infant growth or nutrient absorption. This is particularly vital, especially for babies who have difficulty eating and digesting, as it enhances their overall wellness and standard of living (Salvatore et al., 2018).

Infant formulas containing starch modification with locust bean gum, rice starch, and corn starch alter gastric emptying and retard the process. The higher quantity of thickeners delays the gastric emptying and reduces vomiting because it delays the formula's stay in the stomach (Ménard et al., 2018; Salvatore et al., 2018). Casein-based formulas have a delayed emptying than whey-based formulas as proteins coagulate differently. However, thicker feeds provide greater digestive comfort by enhancing nutrient retention and digestion without exacerbating GER symptoms (Salvatore et al., 2018).

2.3 Safety concerns and health effects

Essential to infant nutrition, starches and modified starches have become fundamental ingredients in special formulae for GER management and in supplemental feeding. Because of better digestibility, the absorption of vitamins, and support to gut health, they should be formulated, though, appropriately and within stricter standards. Digestion, energy, gut health, and feeding considerations of starches in infant nutrition. Benefits of modified starches in infant foods: Modified starches have been used extensively for infant foods in the improvement of digestibility and control of GER of such babies and babies receiving long-life energy, particularly in antiregurgitation (AR) formulae and supplemental feeding products. The different types of modified starches include pregelatinized, cross-linked, and oxidized starches, which are specially processed to meet the needs of the newborn but have digestive organs under constant growth. Even though these changes have numerous advantages, there are a number of difficulties and safety issues that need to be resolved to protect the health and welfare of the baby.

Enhanced digestion: Modified starches are treated with specific alterations through either physical or chemical methods to enhance their digestibility. For example, pregelatinized or gelatinized starches are more easily digested by the digestive system. These activities help to increase the availability of carbohydrates for newborns whose digestive enzymes, especially pancreatic α-amylase, are immature. This is particularly critical since the digestive systems of newborns cannot break down complex starch molecules in their native form. By making such modifications, the discomfort from undigested carbohydrates is avoided, and even neonates with minimal activity of enzymes can split and absorb the starches efficiently (Chakraborty et al., 2022; Moin et al., 2019).

Management of GER by Thickening: Modified starches, especially in AR preparations, play a crucial role in minimizing the occurrence of reflux. This is because they thicken the formula and reduce reflex episodes by prolonging its stay in the stomach. In infants, this comes in handy for common conditions during infancy. Modified starches help keep down the contents of the stomach by thickening the formula, which prevents them from running back up into the oesophagus and reduces reflux symptoms. Infant comfort and feeding tolerance are improved (Salvatore et al., 2018). Sustained Release of Energy: Modified starches do give a long-term, progressive glucose supply within the bloodstream, unlike simple sugars, which are fast absorbed or digested. In infants, high and sudden jumps of energy have potential risks of harm; these are prevented owing to delayed digestive breakdown. The slow degradation of the starch derivatives ensures a steady supply of glucose, which is necessary for promoting growth and development, especially at the initial stages when the energy requirement is the highest.

2.4 Challenges and safety considerations of modified starches in infant foods

Digestive intolerance: Excessive intake of resistant or incompletely digested starches might cause digestive disturbances in infants, even if resistant starch exerts favorable prebiotic effects. Unabsorbed carbohydrates may be fermented in the colon, releasing gases and causing gastrointestinal discomfort. Infants suffering from overfermentation of carbohydrates through gas, bloating, and cramping would experience difficulties while feeding or fussing. Moreover, osmotic diarrhea, where water is drawn into the intestines and causes loose stool, could be due to overfermentation. This can lead to dehydration and poor feeding acceptance. Therefore, the level of resistant starch within the baby food or complementary diet should be at safe levels to eliminate these adverse effects (Saboonchi et al., 2021; Chen et al., 2020).

Limitations of Concentration: To ensure safety, the amount of modified starch used in infant food is very limited. Research has shown that formulas containing over 2 g of starch per 100 mL lead to digestive disorders, especially among younger babies. The reasoning behind this limitation is that there is a great possibility that high concentrations of starch may be overburdened for the guts of a newborn, causing incomplete digestion or colon fermentation. It can further lead to stomach upset. As such, it is upon these manufacturers to show respect for the concentration limitations above. To balance the advantages of modified starches against the possible hazards of overuse, this needs to be carefully formulated (Gałkowska et al., 2023; Moin et al., 2019).

There are several sources of modified starches, such as rice, maize, and wheat. Wheat starches, which include gluten, can be associated with low allergic hazards, especially in toddlers who are genetically disposed to wheat allergies or celiac disease. However, to ensure the products are allergy-free, a lot of focus has to be placed on ingredients' origin despite the fact that gluten-containing starches are traditionally excluded from baby formulas. On safety grounds, nongluten starches as those found in rice, are applied more often when it comes to baby food. Testing the infant's tolerance to certain formulations is crucial because some babies may also become sensitive to additional sources of modified starch (Nguyen, 2017).

3. The role of starch in adult nutrition

The improvement of digestive capacity and variation in dietary habits occur as infants grow into adulthood. The key nutritional use of starch changes from dealing with enzymatic immaturity and safe feeding to maintaining energy functions and the management of blood glucose reactions. The following section discusses how normally functioning individuals with a properly working digestive system can tap into different kinds of starch, ranging from easily digestible starch to resistant starch.

Starch is a fundamental carbohydrate and an excellent source of sustained energy when consumed from high-fiber, minimally processed sources like whole grains, legumes, and vegetables. These starches are composed of two significant chemicals, i.e., amylose and amylopectin, which have differing degrees of digestibility in the stomach. It is a form of energy, and thus, starch is converted into glucose for use in metabolism processes. Starch is not only a source of energy, but it also contains other components such as fibers, calcium, and iron. Furthermore, some metabolic benefits of starch improve glycemic control, increase satiety and support gut health.

3.1 Starch types applied to adult foods

In the context of adult nutrition, starch plays a critical role not only as an energy source but also as a functional ingredient contributing to health benefits. Based on the extent of digestibility provided, starches can be classified into rapidly digestible starch, slowly digestible starch and resistant starch. Cereal grains, root vegetables, legumes, and processed starches are some of the sources of similar starches. RDS breaks down easily into glucose, leading to a quick increase in blood glucose concentrations. RDS is usually present in sports nutrition supplements, energy drinks, and in nutrition formulations meant for people who need quick access to energy, as well as in people with impaired ability to digest their food. RDS ought not to be used by people with metabolic syndrome, such as diabetes, as they would require quick glucose and insulin resistance, which would ideally need slowly digestible starch (SDS) or resistant starch (RS) (Liu et al., 2015).

The role of resistant starch in adult nutrition and health is an area of great concern to researchers. As a type of starch, RS acts more like a dietary fiber in that it is also not digestible within the small intestine. Instead of being absorbed in the form of glucose in the large intestine, gut microbes ferment RS to short-chain fatty acids (SCFAs) such as acetate, propionate and butyrate. SCFAs consumption has been shown to enhance gut health by stimulating the growth of healthy gut bacteria that have a broad range of specific health benefits. RS is used in many prebiotic-enriched foods targeting to optimize the balance of the gut microbiota and thereby achieve metabolic homeostasis (Wang et al., 2019; Wei et al., 2022). Some functional snacks, such as crackers and cookies, contain RS to provide fiber and control blood glucose levels. Various studies have reported that the consumption of RS is related to the suppression of insulin resistance, enhanced lipid profiles, and lower levels of inflammation and oxidative stress (Du et al., 2020; Gourineni et al., 2020). In summary, the current evidence confirms resistant starch as a diet component that is useful in preventing lifestyle-related diseases such as obesity, hyperlipidemia, and type 2 diabetes.

Slowly digestible starch (SDS) is widely used in many food products as it works efficiently to provide energy while regulating blood sugar levels. Also, the suitability of the ingredient in bread, cereal, and pasta dough stems from the glycemic index it has. Gastric emptying will be delayed when taking the ingredient, which leads to satiety. Additionally, the energy thus created is sustained in the case of the drink component of the ingredient, which is meant to be a meal replacement. This will prevent fluctuations in blood glucose levels, which may affect the mind (Goux et al., 2020).

3.2 Starch metabolism in adults

For adults with fully developed digestive abilities, the starch breakdown will follow the enzyme pathway discussed in the Overview of starch digestion and metabolic pathways, as illustrated in Fig. 16.2. The efficacy of this will depend on aspects such as physical exercise levels, health conditions, and the type of starch ingested (RDS, SDS, or RS). Physically

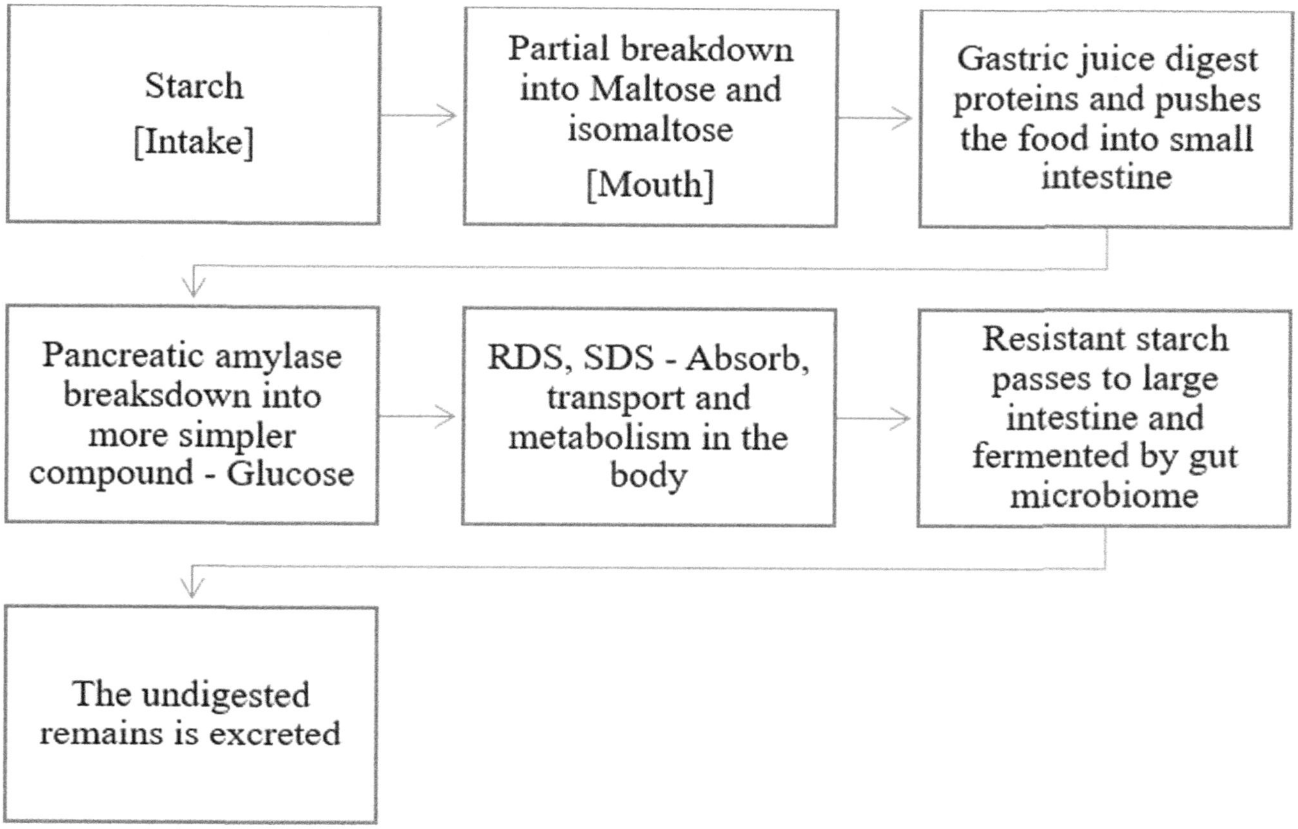

FIGURE 16.2 Human starch digestion and metabolic pathways. This figure presents the stepwise enzymatic hydrolysis of starch from the mouth via small intestine to glucose absorption. This integrated pathway is common to the three age groups covered in this textbook, but with modifications depending on life stage, e.g., infants and elderly have lower levels of some enzyme productions.

active adults have the most effective starch energy conversion, while gym or less-active adults will store the starch in the form of glycogen in the fat tissue.

The efficiency of the starch metabolism is affected by factors such as physical activity, medical conditions, types of starch consumed, etc. The adults who engage in exercise and other physical activity tend to utilize the starch efficiently, whereas the more dormant people tend to accumulate starch in the form of glycogen, contributing to the adipose tissue. Peculiar medical conditions, such as type 2 diabetes individuals have varying effects on the intake and storage of glucose. It's interesting to note that RS improves fat oxidation, increases insulin sensitivity, and lowers inflammation in adults, all of which support metabolic health (Rashed et al., 2022; Apriyanto et al., 2022; Ai & Jane, 2024). These benefits highlight the importance of incorporating diverse starch types in a balanced diet.

The biosynthesis of starch begins with glucose-1-phosphate, which is converted to ADP-glucose that undergoes modification by starch and branching enzymes. These enzymes form α linkages between amylose and amylopectin. These enzymatic actions determine starch granule size, composition and structure.

3.3 Nutritional benefits and impacts

Starch is the primary staple food across the globe; its nutritional benefits are predominantly linked to metabolism, gut health, energy release and disease management in adults. Dietary sources include resistant starch from legumes, green bananas, rice, and pasta, potatoes, and slowly digestible starch from whole grains and minimally processed cereals (Zhang & Bao, 2023). Starch in adults has various health benefits, including resistant starch (RS), slowly digestible starch (SDS), high-amylose starch, and starch for specific populations. Resistant starch (RS) reduces postprandial blood glucose levels, improves insulin sensitivity, gut health, and weight management. Slowly digestible starch (SDS) provides sustained energy, reduced diabetes risk, and improved satiety (Zhang & Hamaker, 2013). High-amylose starch selectively stimulates the growth of beneficial gut bacteria, potentially reducing inflammatory diseases and cancer prevention. RS and SDS can help combat age-related insulin resistance, support gut health, improve endurance and recovery, and help individuals with chronic diseases. Advances in starch modification techniques enhance starch properties for specific health benefits, such as increasing its resistant fraction. Starch metabolism and structural diversity underpin its health benefits, making it a versatile component in nutrition and therapy. Dietary sources of beneficial starch include resistant starch from green bananas, legumes, potatoes, rice, and pasta, and slowly digestible starch from whole grains and minimally processed cereals. Resistant starch in long-term consumption showed significant maintenance of insulin secretion and sensitivity. They enhance the gut barrier and reduce inflammation, which provides extended satiety by stimulating GLP-1 and PYY hormone secretion (Keskin et al., 2022; Li et al., 2023).

4. The role of starch in elderly foods

The digestive and sensory system shows large degeneration changes in an aging population. Although adults have ideal enzyme secretion and swallowing abilities, seniors display reduced enzyme secretion, improper gastrointestinal movement, an imbalance of intestinal bacteria, and even swallowing difficulties. In view of such changes, a novel formulation requirement emerges due to competing needs of easy swallowing, sufficient nutritional delivery, prolonged energy release, and support of intestinal health. To meet nutritional needs and physiological obstacles, starch and modified starches are essential components of diets for the aged. They maintain energy levels by releasing glucose steadily, which is crucial for people who consume fewer calories or have poor digestion. While readily digestible starches (such as boiling or gelatinized) are required because of age-related decreases in pancreatic α-amylase activity, resistant starch supports gut health by cultivating good bacteria. For people who have trouble chewing or swallowing, modified starches made by enzymatic, chemical, or physical processes improve texture, increase digestibility, and facilitate oral processing. Cross-linked starches stay consistent while under strain, but too much resistant starch might upset your stomach, so you need to watch how much you eat. Advances in tribology help optimize starch formulations for improved oral comfort and swallowing, and sensory acceptability is still crucial (Alfa, Strang, Tappia, Olson, et al., 2018).

4.1 Starch digestibility challenges in aging

Nutritional concerns for elderly people result from the physiological and dietary problems that come with aging, which have a major impact on starch digestibility. Although starch, the main source of carbohydrates, is essential for energy production, aging-associated modifications in the gastrointestinal system impact how it is metabolized, which affects glucose management and general health. Physiological changes and their effect on starch digestibility: Being able to digest and absorb starches is greatly impacted by a number of physiological changes that take place in the digestive tract as people age. These include structural modifications in the small intestine, gut microbiota composition, gastrointestinal transfer, and enzymatic activity. When taken together, such physiological changes as lowered enzyme secretion, motility disorders, and dysbiosis pose synergistic obstacles to starch digestibility, especially among seniors. Each of these may be overcome by proper modification of starches:

- Reduce enzyme production: Cross-linked and pregelatinized starches have prehydrolyzed glucose content, thereby lowering the demand for enzyme secretion due to reduced pancreatic secretion of amylase (Alfa, Strang, Tappia, Olson, et al., 2018).
- Altered motility: Resistant starch preparations enhance the production of SCFAs by colon bacteria, which positively contributes to colon health in spite of the slower transit time (Araiza-Calahorra et al., 2023; Makran et al., 2022).
- Dysbiosis: Resistant starch has prebiotic properties, supporting the growth of beneficial bacteria and overcoming the problem of age-related dysbiosis (Liu et al., 2022).

Reduced Production of Enzymes: Important enzymes like pancreatic and salivary amylase are produced less often as humans age. The loss of salivary and pancreatic amylase with age contributes to lowered starch digestibility in elderly populations. The small intestine's ability to convert carbohydrates into glucose is also impaired by reduced amylase output, a result of poor pancreatic function. Because of this, unabsorbed starches enter the colon and are digested by gut flora, which may lead to energy deficits (Araiza-Calahorra et al., 2023; Makran et al., 2022).

Modified Transit of the Gastrointestinal System: Food passage through the digestive tract is impacted by gastrointestinal motility, which declines with age. Reduced intestinal motility and slower stomach emptying are the results of weak GI muscles. This delayed transit can postpone starch's arrival in the small intestine, where the majority of digestion takes place, but it also gives more time for enzymatic activity. Long-term stomach retention can cause undigested starch to ferment in the colon, which can result in bloating, gas, and pain. This can lead to lower glucose availability and inadequate starch digestion (Alfa, Strang, Tappia, Olson, et al., 2018; Araiza-Calahorra et al., 2023).

Dysbiosis of the Gut Microbiota: Dysbiosis, or microbial imbalance, arises in the elderly because of the reduced presence of beneficial bacteria (*Bifidobacterium, Faecalibacterium prausnitzii*) with aging. Additionally, the reduced production of butyric acid, an SCFAs, brought on by this dysbiosis, contributes to the weakening of the integrity of colonic epithelial cells, the impaired regulation of inflammation, and the amplified presence of toxins from the fermentation of carbohydrates by some pathogenic bacterial populations. Several studies demonstrate the promotion of age-related inflammatory diseases by dysbiosis, impairing nutritional absorption (Aguilera & Park., 2016; Alfa, Strang, Tappia, Olson, et al., 2018). These observations emphasize the importance of prebiotic-resistant starch preparations, those containing beneficial bacteria, in the nutritional management of the elderly.

4.2 Formulation concerns for elderly foods

Foods for elderly people must be carefully formulated with texture, nutrition, convenience of preparation, and sensory appeal in mind, especially for people who have dysphagia (having trouble swallowing). Because of their adaptability, affordability, and capacity to satisfy rheological and sensory needs, modified starches are essential in resolving these issues. Important criteria for foods that benefit the elderly are:

Management of Dysphagia: Aspiration risk is elevated in older persons with dysphagia, a disorder that impairs swallowing capacity. Certain textures are necessary for foods to be safe to swallow. Modified starches serve to provide controlled viscosity and texture, allowing the creation of soft, gel like food textures that can be easily swallowed. Answering the question of what the resulting boluses will be like, elastic starch gels will produce boluses that can retain their shape during swallowing, slowing aspiration, while Granular Cold-Water Swelling (GCWS) starches are highly appreciated due to their fast-swelling properties, enabling the creation of smooth, moist textures that are essential in the treatment of dysphagia (Liu et al., 2024; Methacanon et al., 2021).

Nutritional Deficiency: Because of decreased appetite, sickness, or difficulty in feeding, many elderly people suffer from nutritional deficits. Modified starches can be combined with bioactive materials such as flaxseed gum and dietary fibers in order to enhance nutritional profiles. These mixes reduce postmeal glucose spikes, increase glucose absorption, and increase the fiber content. Starch-modified blends can also have proteins, vitamins, and minerals incorporated into them in order to meet the nutritional demands, while the texture of the product is not altered, which makes swallowing easier and aids in meeting the dysphagia challenge (Araiza-Calahorra et al., 2023).

Diminished Sensory Perception: With age, some people may experience a decrease in the sensitivity of their taste and smell, which can make the prospect of eating less appealing. To counteract this, the texture and sensory characteristics of the food have to be increased. In food products, the mouthfeel and taste can be improved by adding modified starches, which have creamy, silky, and elastic textures. The mechanical hardness, as measured with stress tests, bears a strong relation to the sensory hardness as reported by older patients. Certain formulations of modified starch that are designed to meet specific sensory requirements are made to ensure that the textures are not only palatable but also easy to swallow. Considering the palate and the capacity of the target consumers to swallow, elderly food formulations can be classified into textural categories such as moist, smooth and thin, or firm and springy (Liu et al., 2022).

Convenience: Elderly people and their caregivers often prefer food products that are easy to prepare and do not require elaborate cooking methods. Modified starches cater to a large part of this need. The Granular Cold-Water Swelling enables the rapid hydration and thickening of starches in cold water. This makes it simple to prepare food purees or thickened drinks. This modified starch-thickened foods remain stable without water separation, or syneresis, guaranteeing uniform textures during storage (Kim et al., 2017), as shown in Fig. 16.3.

Using modified starches in elderly food formulation:

Modification of starches is vital for the formulation of foods for the older population because of the special challenges that are associated with working with gnarled elderly people. They can be chemically or physically modified to meet the nutritional, culinary, and textural requirements of elderly feasts, especially for people dealing with swallowing difficulties (Rodríguez et al., 2022).

Thickening and gelling:

Modified starches are the best for producing gels and other semifluid materials with certain rheological properties. Because of these properties, these materials are especially useful for creating soft, thick, cohesive textures for dysphagia-friendly diets. Granular Cold-Water Swelling (GCWS) starches, which are a type of modified starch, are cold thickeners. The ability to thicken in cold water makes it very useful for the preparation of dysphagia food since it does not need to be heated. Starch-based gels have been gelatinized with higher G' than G'', allowing nonrigid food structures to assist in the sucking and swallowing roles within the body (Abu Zarim et al., 2018; Majzoobi & Farahnaky., 2021; Ren et al., 2024)

Textural Alteration:

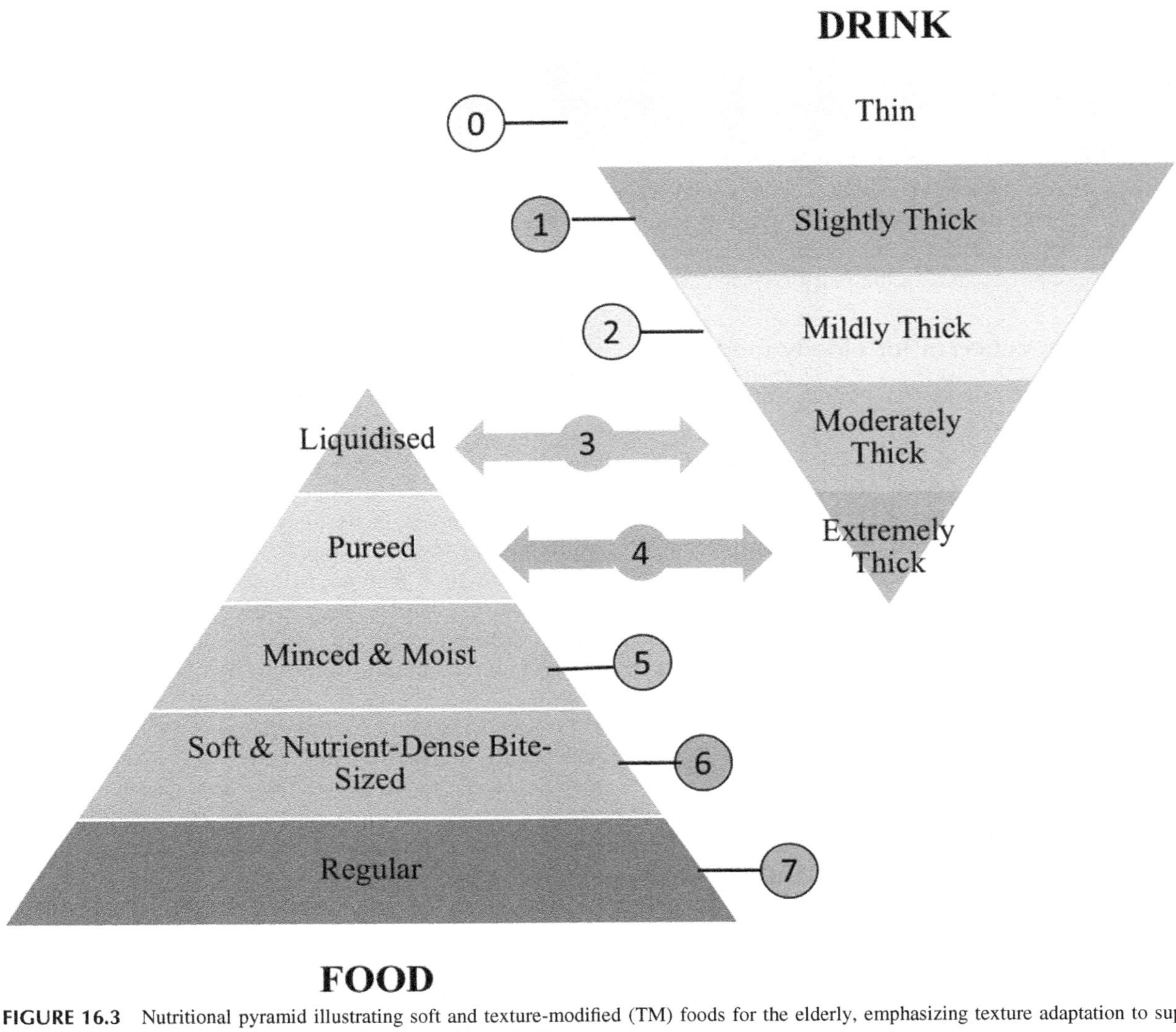

FIGURE 16.3 Nutritional pyramid illustrating soft and texture-modified (TM) foods for the elderly, emphasizing texture adaptation to support safe swallowing, improved digestibility, and balanced nutritional intake.

Modified starches can be utilized to create food textures that are appealing to people with different forms of dysphagia. These textures can range from hard and springy for sufferers with moderate dysphagia, to wet and smooth for those with severe forms of the condition. Modified starches can be used to formulate textures which are "firm" and "springy" (stress values $> 5.0\ N/m^2$) or "moist" and "smooth" (low stress values, $<3.5\ N/m^2$). This flexibility is critical to fulfill the different textural requirements of diets for dysphagia. Because the objective hardness of starch gels is well correlated with sensory perception, textural qualities could be controlled with precision to meet customer expectations (Kim et al., 2017; Liu et al., 2022).

The Role of Nutrition: Along with other components, when combined with modified functional starches, the nutritional intake, both in modified form and in diets for elderly people, is enhanced. These modified starches are able to promote intestinal health and aid glucose management by enhancing the fiber content while incorporating flaxseed. In addition to providing these health benefits, these formulations may also help with appetite control and glucose absorption, which goes beyond altering texture (Rodríguez et al., 2022).

4.3 Safety concerns and health effects

Physiological changes affect elderly people, for instance, a reduction in metabolism, immune suppression, and even a dread of swallowing. The hygiene of the food, as well as its health risks and effects, needs to be put under extra scrutiny for these reasons. Below, there is a description of the health and safety risks and concerns posed to the elderly by some foods, such as modified starches, thickeners, and diets for those with Dysphagia.

Concerns on Choking and Aspiration: Modified starches are important for dysphagia diets because they provide controlled and safe textures. Granular Cold-Water Swelling (GCWS) Starches are particularly useful because they hydrate quickly in both hot and cold water, making them easy to swallow. GCWS Starches produce smooth textures that are ideal for those who have difficulty swallowing. Crosslinking also further enhances the texture by providing increased strength at various pH and temperature levels that prevent disintegration. They assist in maintaining the consistently required strength of semisolid foods and thickened fluids, preventing concerns for uneven strength distribution that could lead to an aspiration obstacle. Moreover, the long-term stability of these starches decreases the risk of aspiration by ensuring that thickened fluids and soft solids remain viscous during storage and use. Thus, CGWS Starches are particularly beneficial for these formulations. Additionally, gelatinous elastic properties of starch-based gels diminish the impact of external forces during swallowing, making them suitable for use by clients with dysphagia (Aguilera & Park, 2016; Methacanon et al., 2021; Warman et al., 2022).

Relationships with Saliva: The enzyme Salivary amylase splits starch to produce glucose, resulting in a decreased thickening ability and possibly dangerous textures. This may compromise the quality and safety of starch-based thickeners used in dysphagia diets. However, Granular Cold-Water Swelling (GCWS) starches are purposely developed to resist enzymatic degradation, so they can thicken even in the presence of saliva. In a similar vein, cross-linked starches are resistant to hydrolysis and maintain their structural integrity. Although substitutes such as scleroglucan or xanthan gum are less vulnerable to salivary amylase, they might not offer the same range of textures or simplicity of preparation as starches. Maintaining texture is important when eating; modified starches are made to resist the interaction of food and saliva, resulting in smooth, uniform textures and lowering the possibility of texture breakdown or viscosity loss during ingestion (Gallego et al., 2022; Kim et al., 2017).

Preventing Aggregates: Foods that have been thickened improperly may congregate or create clumps, which could be choking hazards. Even under freezing preparation conditions, GCWS starches guarantee smooth hydration without clumping. Cross-linked starches create stable gels and uniformly disperse liquids by preventing the creation of sticky clumps (Liu et al., 2024; Warman et al., 2022). The formation of sticky, mucoadhesive textures by specific hydrocolloids or incorrectly modified starches can make swallowing more difficult and raise the risk of aspiration. According to your inputs, scleroglucan avoids mucoadhesive qualities, which makes it a safer option for applications involving dysphagia. When used with other thickeners (such as curdlan or flaxseed gum), modified starches can improve texture while decreasing mucoadhesion (Methacanon et al., 2021).

5. Techniques and applications of modified starches

Starches possess specific restrictions in their native form that limit their use in food systems. These restrictions include a low protuberance, nondecomposition stability and low solubility within various solvents. To tackle the issues that arise due to the limitations, a plethora of modification techniques are available and can be used to improve the performance of the starch within a variety of food matrices (Punia, 2020). Generally, the starch is made by various methods, such as Physical modification (osmotic-pressure treatment, deep-freezing and thrashing), Chemical modification (etherification, esterification, crosslinking, oxidation, cationization, and grafting), Enzymatic modification and Genetic modifications (Neelam et al., 2012) as described in Fig. 16.4. Such modification strategies can lead to some progressive development in reactivity, functionality and application of starch in various sectors like nutritional, pharmacological, biomedical and industrial (Nawaz et al., 2020). It is important to note that the treatments mentioned above are each designed specifically to enhance different characteristics of starch, including resistant starch, digestibility and texturizing effects in food. These techniques allow for starches to be tailored for different nutritional needs across age groups, ensuring improved stability and functional versatility in baby, adult, and elderly foods (Masina et al., 2017; Krithika & Ratnamala, 2019).

5.1 Chemical modifications of starches

Chemical modifications of starch emerged in the past decade as an attractive solution to the significant drawbacks of native starches, including susceptibility to acids, enzymes, shear stress, and thermal fluctuations and low resistance to digestion, weak solubility in organic solvents, and retrogradation. Every anhydroglucose unit has three hydroxyl groups located on C2, C3, and C6 positions, which allows the possibility of chemical modification of starch molecules by reactions with different functional groups. These techniques include esterification, etherification, acid hydrolysis, cross-linking and oxidation. These modifications improve the functional and nutritional characteristics of the starches and

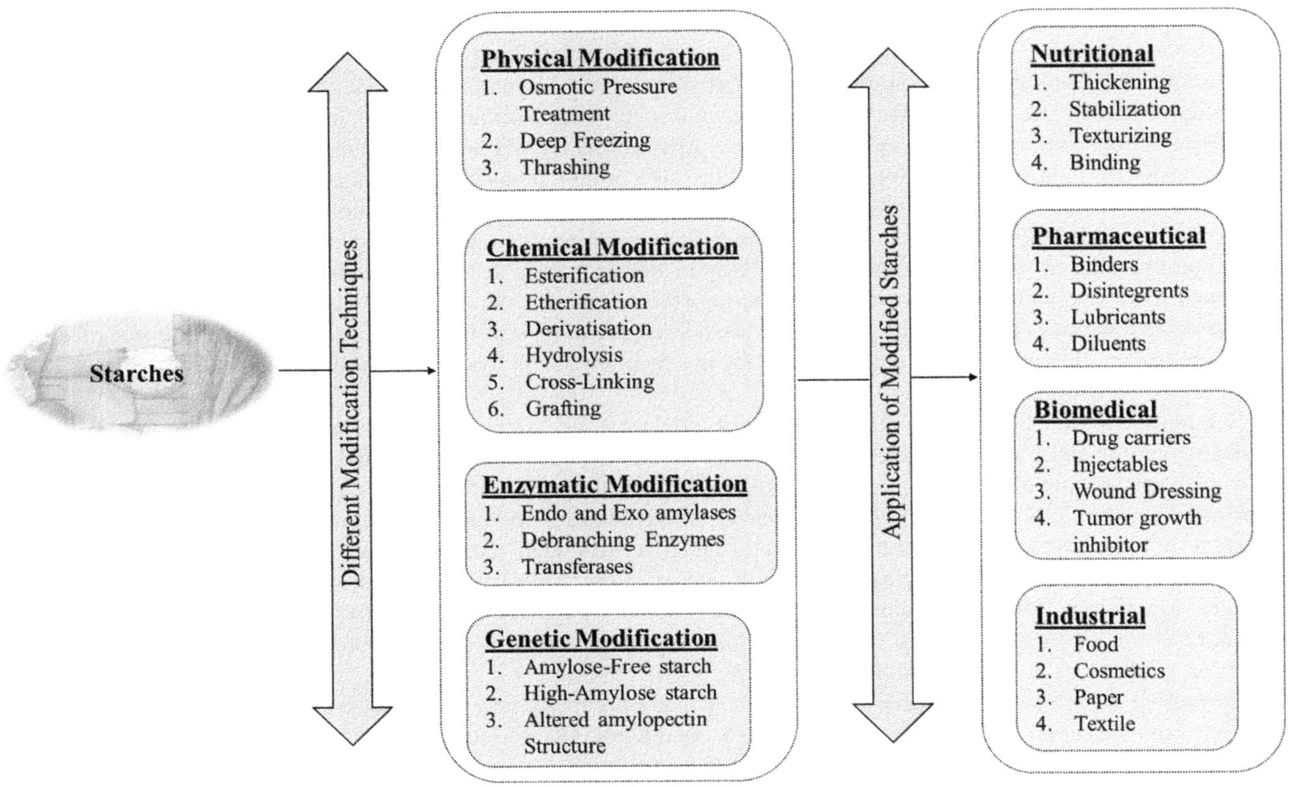

FIGURE 16.4 Starch modification techniques and their applications.

their applications in food formulations by choosing suitable reagents and reaction conditions (Ai & Jane, 2024; Kaur et al., 2024). Table 16.1, summarizing the chemical modification of starches for enhancing their functional properties for various applications, is provided. In all modification techniques, esterification has been commonly applied to various starches. Some specific examples include the oleic acid-treated corn starch, which showed improved hydrophobicity and thermoplastic properties and was therefore useful for advanced material applications. On the other hand, esterification of cassava starch with acetic, propionic, and butyric anhydrides resulted in significantly enhanced hydrophobic properties, whereas saturated fatty acid chlorides-treated waxy maize starch provided enhanced hydrophobicity and properties suitable for water-resistant applications (Mirzaaghaei et al., 2022). Esterification of potato starch with oleic acid, however, results in decreased Thermal stability, thereby limiting its application at high temperatures (Zarski et al., 2016). Conversely, modified tapioca starch with maleic anhydride and DMAP was of lower thermal stability than its native form but can be beneficial for low-temperature applications (Triwises et al., 2016).

Etherification can also enhance starch properties. Due to its structural advantages, Cassava starch that underwent modification by the use of diethyl, dipropyl and dibutyl maleate demonstrated high potential in its application to bioactive packaging as well as encapsulation (Clasen et al., 2018). In the same manner, sago starch modified with benzyl chloride showed good thermal stability and improved flowability and was considered of industrial importance (Misman et al., 2015). Double etherification of corn starch using N-(3-chloro-2-hydroxypropyl) trimethylammonium chloride and 3-chloro-2-hydroxy-1-propanesulfonic acid sodium salt provided warp sizing for textiles with better adhesion properties (Li et al., 2018). Additionally, cassava starch modified with CHPTAC produced a cost-effective cationic starch suitable for widespread industrial use (Rahayu et al., 2024).

Acid hydrolysis is another method employed to modify starch characteristics. Potato starch hydrolyzed with HCl exhibited increased crystallinity, enhanced thermal transitions, and improved enthalpy, ideal for high-performance applications (Yang et al., 2018). Like so, with HCl treatment of corn starch, viscosity stability and bonding properties improve. Therefore, it finds application in wood adhesives (Wang et al., 2017). For potato starch, hydrolysis enhances the solubility and reduces viscosity, swelling power, and crystallinity with significantly improved applicability in the food and pharmaceutical sectors (Li et al., 2017). Cross-linking techniques have been used to enhance starch functionality to serve a specific application. Corn starch cross-linked with sodium trimetaphosphate and sodium tripolyphosphate exhibited wider

TABLE 16.1 Chemical modification of starches and their outcomes.

Substrate	Modification method	Reagent	Outcomes	References
Corn	Esterification	Oleic acid	Enhanced hydrophobicity and thermoplastic properties.	Adak and Banerjee (2016)
Cassava starch	Esterification	Acetic, propionic, and butyric anhydrides	Improved hydrophobic properties.	Jerachaimongkol et al. (2006)
Waxy maize starch	Esterification	Saturated Fatty acid	Increased hydrophobicity with desirable characteristics.	Mirzaaghaei et al. (2022)
Potato starch	Esterification	Oleic acid	Reduced thermal stability of modified starch.	Zarski et al. (2016)
Tapioca starch	Esterification	Maleic anhydride and DMAP	Lower thermal stability compared to native starch.	Triwises et al. (2016)
Cassava starch	Etherification	Diethyl, dipropyl, and dibutyl maleate	Potential applications in bioactive packaging and encapsulation.	Clasen et al. (2018)
Sago starch	Etherification	Benzyl chloride	Exhibited high thermal stability and improved flow properties.	Misman et al. (2015)
Corn starch	Double Etherification	N-(3-chloro-2-hydroxypropyl) trimethylammonium chloride and 3-chloro-2-hydroxy-1-propanesulfonic acid sodium salt	Improved adhesion to cotton and polyester fibers for warp sizing.	Li et al. (2018)
Cassava starch	Etherification	CHPTAC	Produced cost-effective cationic starch.	Rahayu et al. (2024)
Potato starch	Acid hydrolysis	HCl	Increased crystallinity, improved thermal transition, and enhanced enthalpy.	Yang et al. (2018)
Corn starch	Acid hydrolysis	HCl	Improved viscosity stability and bonding for wood adhesives.	Wang et al. (2017)
Potato starch	Acid hydrolysis	HCl	Enhanced solubility, reduced paste viscosity, swelling power, and crystallinity.	Li et al. (2017)
Corn starch	Cross-linking	Sodium trimetaphosphate/sodium tripolyphosphate	Broadened potential applications and improved physiological performance.	Koo et al. (2010)
Banana starch	Cross-linking	Sodium trimetaphosphate/sodium tripolyphosphate and epichlorohydrin	Reduced solubility levels.	Carmona-Garcia et al. (2009)
Sorghum starch	Cross-linking	Epichlorohydrin	Modified starch granules with rough surfaces, cavities, and cracks.	Sandhu et al. (2021)
Potato starch	Oxidation	Sodium hypochlorite	Films with reduced tensile strength, water solubility, and vapor permeability compared to native starch.	Fonseca et al. (2015)

Continued

TABLE 16.1 Chemical modification of starches and their outcomes.—cont'd

Substrate	Modification method	Reagent	Outcomes	References
Potato starch	Oxidation	Sodium hypochlorite	Films exhibited reduced solubility and elongation properties.	Zavareze et al. (2012)
Banana starch	Oxidation	Sodium hypochlorite	Films showed higher permeability compared to native starch films.	Argüello-García et al. (2014)

potential applications and enhanced physiological performance, making it suitable for nutritional products (Koo et al., 2010). Similarly, banana starch treated with the same reagents along with epichlorohydrin showed decreased solubility, ensuring better stability under varying conditions (Carmona-Garcia et al., 2009). Sorghum starch cross-linked with epichlorohydrin possessed rough surface granules with cavities, which provided improved structural characteristics (Sandhu et al., 2021).

In addition, oxidation processes have been applied to starches and are being used to create modified materials with tailored properties. Oxidation of potato starch with sodium hypochlorite provided films with reduced tensile strength, water solubility, and permeability to vapors, which would be appropriate for specific packaging applications (Fonseca et al., 2015; Zavareze et al., 2012). On the other hand, banana starch films oxidized with the same reagent had higher permeability than native films, which may be useful in controlled-release systems (Argüello-García et al., 2014). Such chemical modification methods demonstrate the flexibility of starches and their capability to fulfill a wide range of industrial requirements through the improvement of their physicochemical properties.

5.2 Applications in food processing with safety concerns

Modified starches are of immense significance in the food processing chain, since they impart texture and stability to such a wide diversity of food items. Oxidized, cross-linked, and acetylated are some of the chemically modified starches that are used to enhance food functionality by inhibiting freeze-thaw stability, retrogradation, and the rheological properties of foods. These starches are used for thickening purposes, stabilizing, and gelling agents in such food systems as soups, sauces, baked goods, and baby food (Konował et al., 2024). For instance, oxidized starches are prepared using sodium hypochlorite. This produces a good derivative that has improved solubility, viscosity stability, and a reduced tendency for retrogradation. These properties make it useful for cake fillings, puddings, and whipped creams. Better freeze-thaw stability is associated with oxidized starches that ensure product consistency on storage and transportation. Despite these chemical modifications, regulatory mechanisms have ensured that the content of carboxyl groups is below permissible levels of about 1.1% dry weight basis (Gałkowska et al., 2023).

This leads to improved thermal and mechanical stability due to cross-linking with agents such as sodium trimetaphosphate and sodium tripolyphosphate. These types of improved properties are very useful in high-temperature and high-shear applications such as instant soups, sauces, and confectionery fillings. On the other hand, cross-linking reduces the swelling and solubility of starch granules, which makes them resistant to extreme processing conditions. Compliance by such starches to very minimum residual phosphorus content and very high regulations followed as stated food safety requirements ensures its safety (Punia Bangar et al., 2024). Acetylated starches show improved water-uptake potential, swelling strength, and depressed retrogradation while prepared from a native starch subjected to acetic anhydride or vinyl acetate treatments. These modifications are useful to make frozen desserts, yoghurt and fruit fillers. The regulatory tolerance of acetyl groups up to 2.5% ensured the safe usage of these modified starches for human consumption. The gelled structure of this acetylated starch is remarkably stable, showing utility in the production of canned and thermally treated food products (Subroto et al., 2023).

Apart from the functionality improvements, modified starches are being looked into for some innovative applications, like encapsulation of bioactive compounds and producing biodegradable films. It goes hand in hand with the food industry's focus on sustainability and consumer health. But the safety of modified chemically starch continues to be a consumer concern. Such apprehensions are countered through intense legal and scientific regulation so that the chemical reactions used reflect the metabolic processes occurring in the living world, and the resultant product does not carry harmful residues (Gałkowska et al., 2023; Wang & Zhang, 2021).

6. Types of starches based on their digestibility

Starchy foods are very important for human beings; their digestibility is affected by various determining factors, including endogenous nonstarch components, food chemicals, additives, and storage (Bello-Perez et al., 2020). Starch is divided according to the release and absorption of glucose in the gut. These are Rapidly Digestible Starch (RDS), Slowly Digestible Starch (SDS), and Resistant Starch (RS) (Englyst et al., 1992). A large percentage of the starch in the diet is digested rapidly. RDS refers to starch that is hydrolyzed in the compartment to glucose molecules by the enzymatic action for less than 20 min. When a food contains too much RDS, it can cause a sharp increase in blood glucose and insulin concentration, which can be detrimental to one's health. On the other hand, slowly digestible starch (SDS) is more resistant to digestion for a longer period of time, and complete digestion occurs in the small intestine. SDS is considered a starch that is broken down enzymatically to glucose, but only after 120 min of digestion (Englyst & Cummings et al.,

TABLE 16.2 Resistant Starch (RS) is distinguished by the characteristics it possesses, its occurrence plus how it is digested in the small intestine.

Types	Description	Occurrence	Digestion in the small intestine	References
Type I RS	Physically inaccessible starch	Whole or partly milled grain and seeds, legumes	Slow rate, partial degree, totally digested	Englyst et al. (1992)
Type II RS	Raw granular starch with a B- or C-type polymorph	Raw potato, green banana, some legumes, high amylose corn	Very slow, little degree, tattily digested when freshly cooked	Englyst et al. (1992)
Type III RS	Retrograded starch	Cooked and cooled potatoes, bread, cornflakes, and food products With repeated moist heat treatment	Slow rate, partial degree, reversible digestion, digestibility improved by reheating.	Woo and Seib (2002)
Type IV RS	Chemically modified starch	Food in which modified starches have been used	A result of chemical modification it can resist hydrolysis	Han and BeMiller (2007)
Type V RS	Amylose-lipid complex and resistant maltodextrin	Food with high amylose content	Can resist digestion	Seneviratne and Biliaderis (1991)

1986). Resistant starch (RS), which is starch that was not digested in the small intestine for 120 min or longer, instead passes to the large intestine and serves as a substrate for fermentation by gut microbiota (Bojarczuk et al., 2022). The versatile applications of chemically modified starches in food processing, as well as meeting the stringent standards of safety, make them particularly useful. Its utility in the improvement of quality, shelf life, and sensory properties of foods underlines their significance in modern food technology.

6.1 Resistant starch and its benefits

Resistant starch (RS) introduced by Englyst was defined as "a small fraction of starch that was resistant to hydrolysis by exhaustive treatment with amylase and pullulanase in vitro" (Englyst et al., 1996). Starch combined with some food or starch on its own is referred to as resistant starch, which remains undigested as it moves through the gastrointestinal tract. As we could see, there are four subgroups of RS, such as RS1, RS2, RS3, RS4, and RS5, which are grouped for easier classification, taking note of the various barriers to digestion forming in the human body (Hasjim et al., 2013). In conjunction with its wide variety of applications in both food and nonfood industries, RS has potential beneficial effects toward health and aids in the prevention of noncommunicable diseases.

Resistant starch (RS) is a type of food that resists enzymatic digestion and can be classified into five types, as shown in Table 16.2. Type I is considered as RS that is embedded within the food matrix, making it physically inaccessible to the enzymes because they are trapped, making it difficult to break down. It is stable to heat even after cooking, and only becomes digestible after milling the food. Enzymatic breakdown does affect the native starch granules that make up Type II RS; however, the crystalline structures still remain intact. Unripe bananas, raw potatoes and High-amylose maize are some of the sources of raw starches. Uncooked potatoes and unripe bananas lose most of their resistance when exposed to heat or other forms of thermal processing. The long-chain double-helical crystallites that make up HAM maize starch remain intact even while being cooked or prepped for food (Birt et al., 2013; Li et al., 2018). Type III RS occurs through the gelatinization of starch after it has been boiled in water during the process of retrogradation. It is emphasized that it is possible to manufacture Type III RS by using linear amylose because it is heat-stable and will not be damaged during most of the food processing operations (Ma et al., 2020). Chemically modified starches, which are created via esterification, etherification or cross-linking, also fall under the Suite IV RS category. These modifications enhance the resistance to enzymatic digestion by changing the starch structure. such as OSA, which adds hydrophobic groups onto the starch to reduce the enzymatic activity (Coulon et al., 2020; Remya et al., 2017). These RS Type V amylose lipid complexes, which are superresistant and nondigestible maltodextrin, improve functional properties that can be exploited in a variety of food applications. Unlike OSA, other types of resistant starches have some nonneglecting benefits and features (Li et al., 2018).

It is possible to choose from a wide variety of resistant starches, with each having its own merits and positive impacts on the body's health, making it easy to incorporate into food products and diets aimed especially at improving glycemic as well as digestive health.

Benefits of RS: It is able to inhibit the growth of colorectal cancer and regulate cholesterol levels and lipid metabolism. It is a type of starch which has significant potential to improve certain factors, especially glycemic and insulin responses. Such factors greatly vary with age. A good, proper diet is vital to them at different ages.

Glycemic and insulin responses: The consumption of Resistant Starch (RS) has been shown to lower postprandial glucose levels by modifying the insulin and glycemic response (Bodinham et al., 2010). In healthy subjects, this RS consumption also results in decreased postbreakfast insulin concentrations. Studies show that long-term consumption of RS, that is, RS2 in particular, results in the regulation of blood glucose concentrations, where a decrease was seen in the calorie-restrictive overweight subjects' fasting blood glucose after only 4 weeks of consuming Resistant starch II (Bodinham et al., 2012). There is also some evidence of increased insulin sensitivity associated with longer periods of RS consumption, in particular, Resistant starch II (Bodinham et al., 2010). In addition to this, RS fermentation by gut bacteria generates short-chain fatty acids (SCFAs), which act as an avenue for the intestinal gluconeogenesis, lowering the gluconeogenesis in the liver, thereby providing better control of glycemic levels (De Vadder et al., 2014). SCFAs absorption in the colon also has the potential to minimize the concentration of free fatty acids and glycerol released during fat breakdown, improving insulin sensitivity (Mathew et al., 2010). Also, the effects of gut bacteria on inflammation and permeability of the colon also affect insulin sensitivity, which further strengthens the effects of RS on metabolic health.

Regulating lipid metabolism: It has been established that resistant starch (RS) plays a crucial role in lipid metabolism through cholesterol intervention and protection against cardiovascular diseases. RS is fermented by gut microbiota, which enhances bacterial numbers and creates short-chain fatty acids, especially propionate, that inhibit cholesterol biosynthesis (DeMartino & Cockburn, 2020; Zhou et al., 2015). Furthermore, RS elevates digesta thickness, which interferes with sugars' intermolecular interaction with fatty acids, hence lowering endogenous cholesterol synthesis and absorption. It is also capable of bonding or absorbing bile acids, making it impossible for those acids to be reabsorbed, which increases their excretion in the faeces, thus decreasing the total amount of cholesterol in the body (Bindels et al., 2015). In contrast with traditional bile acid sequestering agents, the use of RS does not elevate triglyceride levels in serum, indicating that it may have a novel mechanism of ameliorating the lipid profile (Mudgil, 2017).

Gut health: There is a growing interest among researchers regarding the relationship between gut health and microbiomes and the role prebiotics have in combating obesity and chronic diseases. RS has been shown to increase the growth of probiotic microbes occupying the small intestine. and enhances the makeup of gut microbiota in obese mice. The microbiome, however, does decrease obesity related microbiota (Yuganthra & Ravi, 2026). and blocks the reproduction of some bacteria that are pH sensitive. Compounds produced from RS fermentation, such as SCFAs, which are derived from microbial metabolites, are known for enhancing gut health through certain processes like hormone secretion, regulation of glucose levels and reduction of inflammation and oxidative stress.

6.2 Slowly digestible starch and its benefits

SDS is a slow, complete starch that is digested in the human small intestine (Dona et al., 2010) and is a desirable form of starch. It is digested after the RDS but in no longer than 120 min under standard conditions (Englyst et al., 1992). SDS can assist in regulating diabetes, especially in the case of these patients, with an improvement in the overall metabolism with regard to carbohydrates. The digestion of starch takes place in three phases, which include: Intraluminal phase, the brush border phase and the glucose absorption phase. Starch from raw foods is converted to oligomers, and thereafter, pancreatic enzymes degrade them so as to facilitate easy digestion (Miao et al., 2015). The purported characteristic of SDS of being slow-digestible can be attributed to its structure and chemical composition, which lessens enzyme-SDS interaction (Huang et al., 2018). SDS enables a more enhanced glucose metabolism, better diabetes control, improves mental performance and provides a sense of fullness (Lehmann & Robin et al., 2007).

Stable glucose metabolism: One of the promising strategies of carbohydrate metabolism involves the Glycemic index (GI), which depicts the extent to which carbohydrates are absorbed postmeal. Particular attention is channeled to any shifts in carbohydrate metabolism. Low GI diets help in reducing the risk of diabetes, cardiovascular diseases, as well as colon and breast cancers. It is evident that slowly swallowing starch consumes a moderate amount of carbohydrates, which ultimately lowers the GI index of food. More fats are released in the blood after the consumption of rapidly digestible starch (RDS), there are also rapid elevations of blood glucose and insulin, as well as an increase in NEFA concentration (Goux et al., 2020). Likewise, R almond milk lowers postprandial insulin in obese, insulin-resistant subjects, where it was found that Slowly Available Glucose improves a metabolic profile in such patients with lower circulating triacylglycerols and apolipoproteins (Wang et al., 2022). There is also evidence for differential effects of slowly digestible starches with reference to the stimulation of the secretion of gut incretin hormones.

Diabetes management: In the management of diabetes, associated hyperglycaemia during a meal can be reduced by lowering the postmeal glucose peak, raising insulin sensitivity, maintaining normal lipid response, lowering episodes of hypoglycaemia, and reaching lower concentrations of glycosylated hemoglobin and fructosamine levels. SDS has a positive effect on the previously mentioned conditions, thus serving as a recommended measure for both prevention and management of diabetes. Soups containing SDS for breakfast, for example, significantly enhance carbohydrate metabolism and decrease the insulin requirement in type 2 diabetes patients who are under insulin treatment (Goux et al., 2020). Since there are no appropriate sources available, raw corn flour is recommended for use in diabetic patients under insulin therapy to enhance glycemic response and reduce the risk of evening hypoglycemia in these patients (Goux et al., 2020).

Satiety: The glucostatic theory indicates that carbohydrate intake is relatively the stronger determinant of satiety, while a low blood glucose concentration stimulates feeding, and it has been found that animal models and humans have similar behavior patterns when food is in short supply (Wang et al., 2022). In this instance, a low blood sugar level is said to cause hunger while high levels lead to satiety. It has been demonstrated that fasting and the level of blood glucose within the body do, indeed, interact. For instance, it is pointed out that a slight fall in glucose concentrations in the bloodstream will indeed lead to an increased desire for food, while avoiding a sequence of taking meals leaves an insulin response that is constant (Vinoy et al., 2017). Additionally, there are intermediaries such as postmeal concentrations of glucose and insulin, as well as the response of metabolism to carbohydrates of slow-release type (Wang et al., 2022), Satiety is not however, only generated by such other mechanisms alongside gastric dyskinesia, for instance, contact with the small intestine, characteristics of absorption, hormones and even composition of the meal. There are only scanty proofs, but they are there to support the hypothesis that satiety is positively affected by slow digesting starch. Here, slowly digestible starch (SDS) dialysis removes more glucose while simultaneously controlling blood glucose levels and thereby improving diabetes treatment in the long run.

7. Future trends and developments

There is a need to develop appropriate in vitro analyses for measuring resistant starch (RS) content, as current in vivo methods can be costly and invasive. In the near future, research could orient toward validating in vitro methods against in vivo results to ensure accuracy and reliability in determining RS levels in foods. Further research on the dependence of diverse processing conditions on the yield and functional properties of various kinds of resistant starches is necessary. This involves understanding the way variables like temperature and starch-to-water ratio influence the preparation of RS-based food products (Chung et al., 2011). Research should relate to innovative foods that meet the specific nutrition requirements for older adults, considering the specific challenges they experience, such as taste decline, dysphagia, xerostomia, and anorexia, which explain the "anorexia of aging. Research on how to make food more acceptable among seniors would be important since these specific issues influence their dietary habits and status of health (Lutz et al. (2019). Future research will consider creating food products friendly for age, ensuring that the products are nutritious and palatable for consumption to the oral physiology changes that an older person experiences. This includes exploring the interactions among oral processing, food structure and sensory perception to enhance the eating experience for seniors. Also, an opportunity for the future revolves around the protein requirements of the elderly population, particularly in combating sarcopenia through soft cereal food fortification (Assad-Bustillos et al., 2022).

The sugar concentration and energy density of baby biscuits are to be given attention to sustain the naturalness of the current market offerings, nutritional value and quality of both children's and adult biscuits (Klerks et al., 2023). Physiological effects of various dietary fibers and their specific properties can be effectively utilized in food formulations and processing of high-fiber food products to ensure that the nutritional benefits of fiber are preserved throughout production. Different food matrices affect fiber fortification and the overall palatability and processability of these products, aiming to improve consumer acceptance and public health outcomes (Figs. 16.1 and 16.2, Bordenave et al., 2020). As a specific case, tests are to be conducted from consumers for the developed RS-3 rice starch beverage formulation to evaluate its market viability and gather feedback from the elderly population regarding taste, texture, and overall satisfaction randomized controlled trials to be implement to assess the health benefits of the RS-3 rice starch beverage, particularly in relation to blood sugar control and the management of dysphagia (Tandhanskul et al., 2021). A better understanding of the molecular structure of starch from various crops is crucial when digestibility and functionality in complementary feeding for young children are concerned. Through which, effective infant food formulations shall be developed with optimized digestion properties. Investigating starch and other food component interactions while cooking, processing and storage could provide insights into improving the nutritional quality of complementary foods. It may also help to enhance the health benefits of indigestible starch for colonic health (Li et al., 2018).

Novel ingredients and techniques need to be developed to enhance the textural characteristics of texture-modified foods (TMF). Ensuring the softness, safety and nutritional quality of TMF will be crucial in designing products that effectively meet the dietary needs of the elderly population. The long-term effects of incorporating the developed geriatric foods into the daily diets of elderly individuals, particularly examining how sustained consumption impacts muscle mass, fat mass, and overall health outcomes (Gallego et al., 2022).

8. Conclusion

Starches play a vital role in human nutrition across all stages of life, serving as an essential source of energy and offering unique functional benefits in food formulations. In baby foods, starches contribute to the digestibility, texture, and stability of infant formulas, with modified variants offering enhanced nutritional outcomes. Adult nutrition benefits significantly from resistant and slowly digestible starches, which support metabolic health, weight management, and sustained energy release. For elderly populations, starch-based solutions address problems like reduced enzyme production, dysphagia, and nutrient deficiencies, enhancing overall quality of life. Advancements in techniques to modify starch, including enzymatic and physical processes, have further expanded its use while decreasing chemical treatment dependence. However, safety, regulatory compliance, and other constraints are crucial factors, especially for sensitive populations, such as infants and the elderly. Clean-label and sustainable food trends push the development of natural and minimally processed starches toward more environmentally friendly and consumer-acceptable solutions.

References

Abu Zarim, N., Zainul Abidin, S., & Ariffin, F. (2018). Rheological studies on the effect of different thickeners in texture-modified chicken rendang for individuals with dysphagia. *Journal of Food Science and Technology, 55*(11), 4522–4529. https://doi.org/10.1007/s13197-018-3386-5

Adak, S., & Banerjee, R. (2016). A green approach for starch modification: Esterification by lipase and novel imidazolium surfactant. *Carbohydrate Polymers, 150*, 359–368.

Aguilera, J. M., & Park, D. J. (2016). Texture-modified foods for the elderly: Status, technology and opportunities. *Trends in Food Science and Technology, 57*, 156–164. https://doi.org/10.1016/j.tifs.2016.10.001

Ai, Y., & Jane, J. (2024). Understanding starch structure and functionality. In *Starch in food* (pp. 55–77). Elsevier. https://doi.org/10.1016/B978-0-323-96102-8.00018-8

Alfa, M. J., Strang, D., Tappia, P. S., Graham, M., Van Domselaar, G., Forbes, J. D., Laminman, V., Olson, N., DeGagne, P., Bray, D., Murray, B. L., Dufault, B., & Lix, L. M. (2018). A randomized trial to determine the impact of a digestion resistant starch composition on the gut microbiome in older and mid-age adults. *Clinical Nutrition, 37*(3), 797–807. https://doi.org/10.1016/j.clnu.2017.03.025

Alfa, M. J., Strang, D., Tappia, P. S., Olson, N., DeGagne, P., Bray, D., Murray, B.-L., & Hiebert, B. (2018). A randomized placebo controlled clinical trial to determine the impact of digestion resistant starch MSPrebiotic® on glucose, insulin, and insulin resistance in elderly and mid-age adults. *Frontiers of Medicine, 4*, 260. https://doi.org/10.3389/fmed.2017.00260

Apriyanto, A., Compart, J., & Fettke, J. (2022). A review of starch, a unique biopolymer – Structure, metabolism and in planta modifications. *Plant Science, 318*, Article 111223. https://doi.org/10.1016/j.plantsci.2022.111223

Araiza-Calahorra, A., Mackie, A. R., Feron, G., & Sarkar, A. (2023). Can tribology be a tool to help tailor food for elderly population? *Current Opinion in Food Science, 49*, Article 100968. https://doi.org/10.1016/j.cofs.2022.100968

Argüello-García, E., Solorza-Feria, J., Rendón-Villalobos, J. R., Rodríguez-González, F., Jiménez-Pérez, A., & Flores-Huicochea, E. (2014). Properties of edible films based on oxidized starch and zein. *International Journal of Polymer Science, 2014*. https://doi.org/10.1155/2014/292404

Assad-Bustillos, M., Feron, G., & Della Valle, G. (2022). Guiding the formulation of soft cereal foods for the elderly population through food oral processing: Challenges and opportunities. In *Advances in food and nutrition research* (pp. 137–188). Elsevier. https://doi.org/10.1016/bs.afnr.2021.11.003. Vol. 99.

Bello-Perez, L. A., Flores-Silva, P. C., Agama-Acevedo, E., & Tovar, J. (2020). Starch digestibility: Past, present, and future. *Journal of the Science of Food and Agriculture, 100*(14), 5009–5016. https://doi.org/10.1002/jsfa.8955

Bindels, L. B., Walter, J., & Ramer-Tait, A. E. (2015). Resistant starches for the management of metabolic diseases. *Current Opinion in Clinical Nutrition and Metabolic Care, 18*(6), 559–565. https://doi.org/10.1097/MCO.0000000000000223

Birt, D. F., Boylston, T., Hendrich, S., Jane, J. L., Hollis, J., Li, L., McClelland, J., Moore, S., Phillips, G. J., Rowling, M., Schalinske, K., Paul Scott, M., & Whitley, E. M. (2013). Resistant starch: Promise for improving human health. *Advances in Nutrition, 4*(6), 587–601. https://doi.org/10.3945/an.113.004325

Bodinham, C. L., Frost, G. S., & Robertson, M. D. (2010). Acute ingestion of resistant starch reduces food intake in healthy adults. *British Journal of Nutrition, 103*(6), 917–922. https://doi.org/10.1017/S0007114509992534

Bodinham, C. L., Smith, L., Wright, J., Frost, G. S., & Robertson, M. D. (2012). Dietary fibre improves first-phase insulin secretion in overweight individuals. *PLoS One, 7*(7). https://doi.org/10.1371/journal.pone.0040834

Bojarczuk, A., Skąpska, S., Mousavi Khaneghah, A., & Marszałek, K. (2022). Health benefits of resistant starch: A review of the literature. *Journal of Functional Foods, 93*, Article 105094. https://doi.org/10.1016/j.jff.2022.105094

Bordenave, N., Lamothe, L. M., & Kale, M. S. (2020). Dietary fibers in foods – Formulating and processing for nutritional benefits. In J. Welti-Chanes, S. O. Serna-Saldívar, O. Campanella, & V. Tejada-Ortigoza (Eds.), *Science and technology of fibers in food systems* (pp. 437–457). Springer International Publishing. https://doi.org/10.1007/978-3-030-38654-2_19

Carmona-Garcia, R., Sanchez-Rivera, M. M., Méndez-Montealvo, G., Garza-Montoya, B., & Bello-Pérez, L. A. (2009). Effect of the cross-linked reagent type on some morphological, physicochemical and functional characteristics of banana starch (Musa paradisiaca). *Carbohydrate Polymers, 76*(1), 117–122. https://doi.org/10.1016/j.carbpol.2008.09.029

Chakraborty, I., Mal, S. S., Paul, U. C., Rahman, M. H., & Mazumder, N. (2022). An insight into the gelatinization properties influencing the modified starches used in food industry: A review. *Food and Bioprocess Technology, 15*(6), 1195–1223.

Chen, Y., She, Y., Zhang, R., Wang, J., Zhang, X., & Gou, X. (2020). Use of starch-based fat replacers in foods as a strategy to reduce dietary intake of fat and risk of metabolic diseases. *Food Science and Nutrition, 8*(1), 16–22. https://doi.org/10.1002/fsn3.1303

Chung, H.-J., Donner, E., & Liu, Q. (2011). Resistant starches in foods. In *Comprehensive biotechnology* (pp. 527–534). Elsevier. https://doi.org/10.1016/B978-0-08-088504-9.00309-3

Clasen, S. H., Müller, C. M. O., Parize, A. L., & Pires, A. T. N. (2018). Synthesis and characterization of cassava starch with maleic acid derivatives by etherification reaction. *Carbohydrate Polymers, 180*, 348–353. https://doi.org/10.1016/j.carbpol.2017.10.016

Coulon, D. B, Ryan, P., Anne, M. R., Justin, G., Brain, M., Vishnupriya, G., Maria, L. S., & Michael, J. K. (2020). Novel resistant starch type 4 products of different starch origins, production methods, and amounts are not equally fermented when fed to sprague-dawley rats. *Molecular Nutrition & Food Research, 64*(2), 1900901. https://doi.org/10.1002/mnfr.201900901

De Vadder, F., Kovatcheva-Datchary, P., Goncalves, D., Vinera, J., Zitoun, C., Duchampt, A., Bäckhed, F., & Mithieux, G. (2014). Microbiota-generated metabolites promote metabolic benefits via gut-brain neural circuits. *Cell, 156*(1–2), 84–96. https://doi.org/10.1016/j.cell.2013.12.016

DeMartino, P., & Cockburn, D. W. (2020). Resistant starch: Impact on the gut microbiome and health. *Current Opinion in Biotechnology, 61*, 66–71. https://doi.org/10.1016/j.copbio.2019.10.008

Dona, A. C., Pages, G., Gilbert, R. G., & Kuchel, P. W. (2010). Digestion of starch: In vivo and in vitro kinetic models used to characterise oligosaccharide or glucose release. *Carbohydrate Polymers, 80*(3), 599–617. https://doi.org/10.1016/j.carbpol.2010.01.002

Du, Y., Wu, Y., Xiao, D., Guzman, G., Stewart, M. L., Gourineni, V., Burton-Freeman, B., & Edirisinghe, I. (2020). Food prototype containing resistant starch type 4 on postprandial glycemic response in healthy adults. *Food & Function, 11*(3), 2231–2237. https://doi.org/10.1039/c9fo02674f

Englyst, H., & Cummings, J. (1986). Digestion of the carbohydrates of banana (Musa paradisiaca sapientum) in the human small intestine. *The American Journal of Clinical Nutrition, 44*(1), 42–50. https://doi.org/10.1093/ajcn/44.1.42

Englyst, H. N., Kingman, S. M., & Cummings, J. H. (1992). Classification and measurement of nutritionally important starch fractions. *European Journal of Clinical Nutrition, 46*(Suppl. 2), S33–S50.

Englyst, H. N., Kingman, S. M., Hudson, G. J., & Cummings, J. H. (1996). Measurement of resistant starch in vitro and in vivo. *British Journal of Nutrition, 75*(5), 749–755. https://doi.org/10.1079/BJN19960178

Fonseca, L. M., Gonçalves, J. R., El Halal, S. L. M., Pinto, V. Z., Dias, A. R. G., Jacques, A. C., & Zavareze, E.d. R. (2015). Oxidation of potato starch with different sodium hypochlorite concentrations and its effect on biodegradable films. *Lebensmittel-Wissenschaft & Technologie, 60*(2), 714–720. https://doi.org/10.1016/j.lwt.2014.10.052

Gałkowska, D., Kapuśniak, K., & Juszczak, L. (2023). Chemically modified starches as food additives. *Molecules, 28*(22), 7543. https://doi.org/10.3390/molecules28227543

Gallego, M., Barat, J. M., Grau, R., & Talens, P. (2022). Compositional, structural design and nutritional aspects of texture-modified foods for the elderly. *Trends in Food Science and Technology, 119*, 152–163. https://doi.org/10.1016/j.tifs.2021.12.008

Garg, N. K., Singh, A., & Chaudhary, D. P. (2017). Resistant starch: A potential impact on human health. *International Journal of Current Microbiology and Applied Sciences, 6*(5), 2046–2057. https://doi.org/10.20546/ijcmas.2017.605.228

Gourineni, V., Stewart, M. L., Wilcox, M. L., & Maki, K. C. (2020). Nutritional bar with potato-based resistant starch attenuated post-prandial glucose and insulin response in healthy adults. *Foods, 9*(11). https://doi.org/10.3390/foods9111679

Goux, A., Breyton, A. E., Meynier, A., Lambert-Porcheron, S., Sothier, M., Van Den Berghe, L., Brack, O., Normand, S., Disse, E., Laville, M., Nazare, J. A., & Vinoy, S. (2020). Design and validation of a diet rich in slowly digestible starch for type 2 diabetic patients for significant improvement in glycemic profile. *Nutrients, 12*(8), 1–13. https://doi.org/10.3390/nu12082404

Gutiérrez, T. J. (2020). *Food science, technology and nutrition for babies and children*. Springer Nature.

Han, J.-A., & BeMiller, J. N. (2007). Preparation and physical characteristics of slowly digesting modified food starches. *Carbohydrate Polymers, 67*(3), 366–374. https://doi.org/10.1016/j.carbpol.2006.06.011

Hasjim, J., Ai, Y., & Jane, J. (2013). Novel applications of amylose-lipid complex as resistant starch type 5. In Y. Shi, & C. C. Maningat (Eds.), *Resistant starch* (1st ed., pp. 79–94). Wiley. https://doi.org/10.1002/9781118528723.ch4

Himat, A. S., Gautam, S., Chavez Garcia, J. P., Vidrio-Sahagún, A. X., Liu, Z., Bressler, D., & Vasanthan, T. (2021). Starch-based novel ingredients for low glycemic food formulation. *Bioactive Carbohydrates and Dietary Fibre, 26*. https://doi.org/10.1016/j.bcdf.2021.100275

Huang, J., Yang, Q., & Pu, H. (2018). Slowly digestible starch. In Z. Jin (Ed.), *Functional starch and applications in food* (pp. 27–61). Singapore: Springer. https://doi.org/10.1007/978-981-13-1077-5_2

Jerachaimongkol, S., Chonhenchob, V., Naivikul, O., & Poovarodom, N. (2006). Modification of cassava starch by esterification and properties of cassava starch ester films. *Agriculture and Natural Resources, 40*(5), 148–151.

Kathuria, D., Gautam, S., & Sharma, K. D. (2018). Utilization and health benefits of modified starch. *The Review, 9.*

Kaur, L., Kaur, R., & Singh, J. (2024). Chemical modification of starch. In *Starch in food* (pp. 97–117). Elsevier. https://doi.org/10.1016/B978-0-323-96102-8.00015-2

Keskin, S. O., Ali, T. M., Ahmed, J., Shaikh, M., Siddiq, M., & Uebersax, M. A. (2022). Physico-chemical and functional properties of legume protein, starch, and dietary fiber—A review. *Legume Science, 4*(1). https://doi.org/10.1002/leg3.117

Kim, Y., Kim, H. J., Cho, W., Ko, S., Park, S. K., & Lee, S. (2017). Classification of starch gel texture for the elderly diets based on instrumental and sensory methodology. *Journal of Texture Studies, 48*(5), 357–361. https://doi.org/10.1111/jtxs.12248

Klerks, M., Román, S., & Sánchez-Siles, L. (2023). Baby, children, and adult biscuits. Differences in nutritional quality and naturalness. *Food Science and Nutrition, 11*(12), 7946–7956. https://doi.org/10.1002/fsn3.3711

Konował, E., Sulej-Chojnacka, J., & Prochaska, K. (2024). New derivatives of modified starch for food technology. *Molecules, 29*(14), 3292. https://doi.org/10.3390/molecules29143292

Koo, S. H., Lee, K. Y., & Lee, H. G. (2010). Effect of cross-linking on the physicochemical and physiological properties of corn starch. *Food Hydrocolloids, 24*(6–7), 619–625. https://doi.org/10.1016/j.foodhyd.2010.02.009

Krithika, P., & Ratnamala, D. K. V. (2019). Modification of starch: A review of various techniques. *International Journal of Research and Analytical Reviews, 6*(1), 32–45.

Lehmann, U., & Robin, F. (2007). Slowly digestible starch – Its structure and health implications: A review. *Trends in Food Science & Technology, 18* (7), 346–355. https://doi.org/10.1016/j.tifs.2007.02.009

Li, W., Xu, Z., Wang, Z., Liu, X., Li, C., & Ruan, F. (2018). Double etherification of corn starch to improve its adhesion to cotton and polyester fibers. *International Journal of Adhesion and Adhesives, 84*, 101–107. https://doi.org/10.1016/j.ijadhadh.2018.03.001

Li, D., Yang, N., Jin, Y., Guo, L., Zhou, Y., Xie, Z., Jin, Z., & Xu, X. (2017). Continuous-flow electro-assisted acid hydrolysis of granular potato starch via inductive methodology. *Food Chemistry, 229*, 57–65. https://doi.org/10.1016/j.foodchem.2017.02.047

Li, H.-T., Zhang, W., Zhu, H., Chao, C., & Guo, Q. (2023). Unlocking the potential of high-amylose starch for gut health: Not all function the same. *Fermentation, 9*(2), 134. https://doi.org/10.3390/fermentation9020134

Liu, R., Li, J., Wu, T., Li, Q., Meng, Y., & Zhang, M. (2015). Effects of ultrafine grinding and cellulase hydrolysis treatment on physicochemical and rheological properties of oat (Avena nuda L.) β-glucans. *Journal of Cereal Science, 65*, 125–131. https://doi.org/10.1016/j.jcs.2015.07.002

Liu, F., Yin, J., Wang, J., & Xu, X. (2022). Food for the elderly based on sensory perception: A review. *Current Research in Food Science, 5*, 1550–1558. https://doi.org/10.1016/j.crfs.2022.09.014

Liu, T., Zheng, J., Du, J., & He, G. (2024). Food processing and nutrition strategies for improving the health of elderly people with dysphagia: A review of recent developments. *Foods, 13*(2), 215. https://doi.org/10.3390/foods13020215

Lutz, M., Petzold, G., & Albala, C. (2019). Considerations for the development of innovative foods to improve nutrition in older adults. *Nutrients, 11*(6), 1275. https://doi.org/10.3390/nu11061275

Ménard, O., Bourlieu, C., De Oliveira, S. C., Dellarosa, N., Laghi, L., Carrière, F., Capozzi, F., Dupont, D., & Deglaire, A. (2018). A first step towards a consensus static in vitro model for simulating full-term infant digestion. *Food Chemistry, 240*, 338–345. https://doi.org/10.1016/j.foodchem.2017.07.145

Ma, Z., Hu, X., & Boye, J. I. (2020). Research advances on the formation mechanism of resistant starch type III: A review. *Critical Reviews in Food Science and Nutrition, 60*(2), 276–297. https://doi.org/10.1080/10408398.2018.1523785

Majzoobi, M., & Farahnaky, A. (2021). Granular cold-water swelling starch; properties, preparation and applications, a review. *Food Hydrocolloids, 111*, Article 106393. https://doi.org/10.1016/j.foodhyd.2020.106393

Makran, M., Miedes, D., Cilla, A., Barberá, R., Garcia-Llatas, G., & Alegría, A. (2022). Understanding the influence of simulated elderly gastrointestinal conditions on nutrient digestibility and functional properties. *Trends in Food Science & Technology, 129*, 283–295. https://doi.org/10.1016/j.tifs.2022.10.005

Masina, N., Choonara, Y. E., Kumar, P., du Toit, I. C., Govender, M., Indermun, S., & Pillay, V. (2017). A review of the chemical modification techniques of starch. *Carbohydrate Polymers, 157*, 1226–1236. https://doi.org/10.1016/j.carbpol.2016.09.094

Mathew, M., Tay, E., & Cusi, K. (2010). Elevated plasma free fatty acids increase cardiovascular risk by inducing plasma biomarkers of endothelial activation, myeloperoxidase and PAI-1 in healthy subjects. *Cardiovascular Diabetology, 9*(1), 9. https://doi.org/10.1186/1475-2840-9-9

Methacanon, P., Gamonpilas, C., Kongjaroen, A., & Buathongjan, C. (2021). Food polysaccharides and roles of rheology and tribology in rational design of thickened liquids for oropharyngeal dysphagia: A review. *Comprehensive Reviews in Food Science and Food Safety, 20*(4), 4101–4119. https://doi.org/10.1111/1541-4337.12791

Miao, M., Jiang, B., Cui, S. W., Zhang, T., & Jin, Z. (2015). Slowly digestible Starch—A review. *Critical Reviews in Food Science and Nutrition, 55* (12), 1642–1657. https://doi.org/10.1080/10408398.2012.704434

Mirzaaghaei, M., Nasirpour, A., Keramat, J., Goli, S. A. H., Dinari, M., Desobry, S., & Durand, A. (2022). Chemical modification of waxy maize starch by esterification with saturated fatty acid chlorides: Synthesis, physicochemical and emulsifying properties. *Food Chemistry, 393*. https://doi.org/10.1016/j.foodchem.2022.133293

Misman, M. A., Azura, A. R., & Hamid, Z. A. A. (2015). Physico-chemical properties of solvent based etherification of sago starch. *Industrial Crops and Products, 65*, 397–405. https://doi.org/10.1016/j.indcrop.2014.11.009

Moin, A., Ali, T. M., & Hasnain, A. (2019). Thermal, morphological, and physicochemical characteristics of succinylated–crosslinked rice starches. *Cereal Chemistry, 96*(5), 885–894. https://doi.org/10.1002/cche.10191

Mudgil, D. (2017). The interaction between insoluble and soluble fiber. In *Dietary fiber for the prevention of cardiovascular disease* (pp. 35–59). Elsevier. https://doi.org/10.1016/B978-0-12-805130-6.00003-3

Nawaz, H., Waheed, R., Nawaz, M., & Shahwar, D. (2020). *Physical and chemical modifications in starch structure and reactivity*. IntechOpen. https://doi.org/10.5772/intechopen.88870

Neelam, K., Vijay, S., & Lalit, S. (2012). Various techniques for the modification of starch and the applications of its derivatives. *International Research Journal of Pharmacy, 3*(5), 14–20.

Nguyen,, T. P. T. (2017). Digestibility and structural changes of ingredients in infant formulae during the gastrointestinal digestion. *The University of Queensland*.

Pascari, X., Marín, S., Ramos, A. J., Molino, F., & Sanchis, V. (2019). Deoxynivalenol in cereal-based baby food production process. A review. *Food Control, 99*, 11–20. https://doi.org/10.1016/j.foodcont.2018.12.014

Punia, S. (2020). Barley starch modifications: Physical, chemical and enzymatic - A review. *International Journal of Biological Macromolecules, 144*, 578–585. https://doi.org/10.1016/j.ijbiomac.2019.12.088

Punia Bangar, S., Sunooj, K. V., Navaf, M., Phimolsiripol, Y., & Whiteside, W. S. (2024). Recent advancements in cross-linked starches for food applications- a review. *International Journal of Food Properties, 27*(1), 411–430. https://doi.org/10.1080/10942912.2024.2318427

Quann, E., & Carvalho, R. (2018). Starch consumption patterns in infants and young children. *Journal of Pediatric Gastroenterology and Nutrition, 66* (S3). https://doi.org/10.1097/MPG.0000000000001971

Rahayu, S., Wirjosentono, B., Oktavia, E., Zuhra, C. F., Pasaribu, K. M., Rahman Piliang, A. F., Sihotang, N. S., Tarigan, J. B., Siow, J., Goei, R., Tok, A. I. Y., & Gea, S. (2024). Temperature-dependent etherification of cassava starch with CHPTAC for cationic starch production. *Case Studies in Chemical and Environmental Engineering, 10*. https://doi.org/10.1016/j.cscee.2024.100779

Rashed, A. A., Saparuddin, F., Rathi, D. N. G, Nasir, N. N. M., & Lokman, E. F. (2022). Effects of resistant starch interventions on metabolic biomarkers in pre-diabetes and diabetes adults. *Frontiers in Nutrition, 8*, 793414.

Remya, R., Jyothi, A. N., & Sreekumar, J. (2017). Comparative study of RS4 type resistant starches derived from cassava and potato starches via octenyl succinylation. *Starch - Stärke, 69*(7–8), Article 1600264. https://doi.org/10.1002/star.201600264

Ren, Y., Jia, F., & Li, D. (2024). Ingredients, structure and reconstitution properties of instant powder foods and the potential for healthy product development: A comprehensive review. *Food & Function, 15*(1), 37–61. https://doi.org/10.1039/D3FO04216B

Rodríguez, M. D., León, A. E., & Bustos, M. C. (2022). Starch digestion in infants: An update of available in vitro Methods—A mini review. *Plant Foods for Human Nutrition, 77*(3), 345–352. https://doi.org/10.1007/s11130-022-01001-1

Saboonchi, S., Mehran, A., Bahramizadeh, P., & Massoud, R. (2021). Applications of modified starch in food. *In 3rd International Congress on Engineering, Technology & Innovation*, pp.1–8.

Salvatore, S., Savino, F., Singendonk, M., Tabbers, M., Benninga, M. A., Staiano, A., & Vandenplas, Y. (2018). Thickened infant formula: What to know. *Nutrition, 49*, 51–56. https://doi.org/10.1016/j.nut.2017.10.010

Sandhu, K. S., Siroha, A. K., Punia, S., Sangwan, L., Nehra, M., & Purewal, S. S. (2021). Effect of degree of cross linking on physicochemical, rheological and morphological properties of sorghum starch. *Carbohydrate Polymer Technologies and Applications, 2*. https://doi.org/10.1016/j.carpta.2021.100073

Seneviratne, H. D., & Biliaderis, C. G. (1991). Action of α-amylases on amylose-lipid complex superstructures. *Journal of Cereal Science, 13*(2), 129–143. https://doi.org/10.1016/S0733-5210(09)80030-1

Subroto, E., Cahyana, Y., Indiarto, R., & Aray Rahmah, T. (2023). Modification of starches and flours by acetylation and its dual modifications: A review of impact on physicochemical properties and their applications. *Polymers, 15*(14), 2990. https://doi.org/10.3390/polym15142990

Tandhanskul, A., Yasurin, P., Chavanon, P., Watanakijcharoenman, P., Sriariyanun, M., Rattanakom, S., Lindayani, I., Sriariyanun, M., Cheng, Y. S., Rodiahwati, W., & Gundupalli, M. P. (2021). Utilization of RS-3 rice starch into development of food for elderly: A ready-to-mix beverage case. *E3S Web of Conferences, 302*, Article 02003. https://doi.org/10.1051/e3sconf/202130202003

Triwises, P., Ruksakulpiwat, Y., & Ruksakulpiwat, C. (2016). The modification of tapioca starch by esterification technique. *Suranaree Journal of Science & Technology, 23*(2).

Vinoy, S., Meynier, A., Goux, A., Jourdan-Salloum, N., Normand, S., Rabasa-Lhoret, R., Brack, O., Nazare, J.-A., Péronnet, F., & Laville, M. (2017). The effect of a breakfast rich in slowly digestible starch on glucose metabolism: A statistical meta-analysis of randomized controlled trials. *Nutrients, 9*(4), 318. https://doi.org/10.3390/nu9040318

Wang, Y., Chen, J., Song, Y. H., Zhao, R., Xia, L., Chen, Y., Cui, Y. P., Rao, Z. Y., Zhou, Y., Zhuang, W., & Wu, X. T. (2019). Effects of the resistant starch on glucose, insulin, insulin resistance, and lipid parameters in overweight or Obese adults: A systematic review and meta-analysis. *Nutrition & Diabetes, 9*(1). https://doi.org/10.1038/s41387-019-0086-9

Wang, Y., Xiong, H., Wang, Z., ud-Din, Z.-, & Chen, L. (2017). Effects of different durations of acid hydrolysis on the properties of starch-based wood adhesive. *International Journal of Biological Macromolecules, 103*, 819–828. https://doi.org/10.1016/j.ijbiomac.2017.05.102

Wang, Y., & Zhang, G. (2021). The preparation of modified nano-starch and its application in food industry. *Food Research International, 140*, Article 110009. https://doi.org/10.1016/j.foodres.2020.110009

Wang, Y., Zhou, X., Xiang, X., & Miao, M. (2022). Association of slowly digestible starch intake with reduction of postprandial glycemic response: An update meta-analysis. *Foods, 12*(1), 89. https://doi.org/10.3390/foods12010089

Warman, D. J., Jia, H., & Kato, H. (2022). The potential roles of probiotics, resistant starch, and resistant proteins in ameliorating inflammation during aging (inflammaging). *Nutrients, 14*(4), 747. https://doi.org/10.3390/nu14040747
Wei, Y., Zhang, X., Meng, Y., Wang, Q., Xu, H., & Chen, L. (2022). The effects of resistant starch on biomarkers of inflammation and oxidative stress: A systematic review and meta-analysis. *Nutrition and Cancer, 74*(7), 2337–2350. https://doi.org/10.1080/01635581.2021.2019284
Woo, K. S., & Seib, P. A. (2002). Cross-linked resistant starch: Preparation and properties. *Cereal Chemistry, 79*(6), 819–825. https://doi.org/10.1094/CCHEM.2002.79.6.819
Yang, L., Xia, Y., Tao, Y., Geng, H., Ding, Y., & Zhou, Y. (2018). Multi-scale structural changes in lintnerized starches from three coloured potatoes. *Carbohydrate Polymers, 188*, 228–235. https://doi.org/10.1016/j.carbpol.2018.02.017
Yuganthra, B., & Ravi, S. N. (2026). Exploring microbial probiotics for functional foods enhancing health benefits and industrial processing trends. *Food and Humanity, 10096*. https://doi.org/10.1016/j.foohum.2025.100960
Zarski, A., Ptak, S., Siemion, P., & Kapusniak, J. (2016). Esterification of potato starch by a biocatalysed reaction in an ionic liquid. *Carbohydrate Polymers, 137*, 657–663. https://doi.org/10.1016/j.carbpol.2015.11.029
Zavareze, E. D. R., Pinto, V. Z., Klein, B., El Halal, S. L. M., Elias, M. C., Prentice-Hernández, C., & Dias, A. R. G. (2012). Development of oxidised and heat-moisture treated potato starch film. *Food Chemistry, 132*(1), 344–350. https://doi.org/10.1016/j.foodchem.2011.10.090
Zhang, Z., & Bao, J. (2023). Recent advances in modification approaches, health benefits, and food applications of resistant starch. *Starch - Stärke, 75* (9–10), Article 2100141. https://doi.org/10.1002/star.202100141
Zhang, G., & Hamaker, B. R. (2013). Slowly digestible starch and health benefits. In Y. Shi, & C. C. Maningat (Eds.), *Resistant starch* (1st ed., pp. 111–130). Wiley. https://doi.org/10.1002/9781118528723.ch06
Zhou, Z. K., Wang, F., Ren, X. C., Wang, Y., & Blanchard, C. (2015). Resistant starch manipulated hyperglycemia/hyperlipidemia and related genes expression in diabetic rats. *International Journal of Biological Macromolecules, 75*, 316–321. https://doi.org/10.1016/j.ijbiomac.2015.01.052

Chapter 17

Starch formulation strategies for canned and instant foods

Gangaraju Divyashri[1], Harini Hutti[1], Likitha V.[2], Lalitha Prasanna Madduri Venkata[1], E.D. Mahima[2], Shreya Mechri[2] and Murat Karamese[3]

[1]*Iom Bioworks Pvt. Ltd., Centre for Cellular and Molecular Platforms (C-CAMP), GKVK Campus, Bengaluru, Karnataka, India;* [2]*Ramaiah Institute of Technology, Bengaluru, Karnataka, India;* [3]*Kafkas University, Faculty of Medicine, Department of Medical Microbiology, Kars, Turkey*

1. Introduction to starch and its role in processed foods

Starch is a vital carbohydrate widely used in the food industry due to its diverse functional properties. Derived from various plant sources, it serves as a texturizer, stabilizer, thickener, gelling agent, and in numerous food applications (Emeje, 2019). After cellulose, starch is the second-largest natural biopolymer and the main substance used by most higher plants to store carbohydrates (Patricio et al., 2024). It is a fundamental component of the human diet, acting as the main source of energy required for sustaining life (Yazid et al., 2018). Starch is primarily composed of two molecular constituents: amylose and amylopectin, which differ in structure and properties. Amylose, a linear polysaccharide, typically constitutes 20%–30% of starch, whereas amylopectin, a highly branched molecule, makes up the remaining portion. The proportion of amylose to amylopectin significantly influences the functional properties of starch, including gelatinization, pasting, and retrogradation behaviors in food applications (Varghese et al., 2022). To meet specific industrial requirements, starch can be modified through chemical, physical, or enzymatic processes, thereby expanding its functionality beyond its native form (Nawaz et al., 2020). In processed foods, starch plays a crucial role in ensuring product stability, texture, and consistency. In canned foods, it prevents ingredient separation, maintains texture, and enhances shelf stability. Similarly, in instant food products such as instant noodles and mashed potatoes, starch facilitates rapid hydration and helps achieve the desired consistency when mixed with hot water. Due to its versatility, affordability, and ability to adapt to various processing conditions, starch remains a cornerstone ingredient in the processed food industry (Rashwan et al., 2024).

1.1 Overview of starch types

Starches, based on their source and structural modifications, exhibit unique characteristics that determine their use in food processing. Table 17.1 outlines various starch types, their sources, characteristics, and primary applications. Common sources include corn, potato, rice, and tapioca, each imparting distinct qualities such as gelatinization temperature, paste viscosity, and resistance to retrogradation (Yazid et al., 2018). Understanding these characteristics is essential for selecting the appropriate starch type for specific food processing application (Fig. 17.1).

1.1.1 Native starches

Native starches are unmodified starches extracted directly from plant sources such as rice, corn, wheat, and potatoes, without any chemical or physical processing. Each type of native starch has distinct functional properties based on factors such as granule size, gelatinization temperature, and amylose-to-amylopectin ratio (Cornejo-Ramírez et al., 2018). For example, potato starch is ideal for thickening soups and sauces due to its large granules and high amylose content, which contribute to excellent gelling properties and high viscosity.

Native and Modified Starches for Food Applications. https://doi.org/10.1016/B978-0-443-27431-2.00022-5

TABLE 17.1 Starch types and their applications in food processing.

Starch type	Source	Characteristics	Primary applications in food processing	References
Native starch	• Corn • Potato • Wheat • Tapioca	Natural form, unmodified, variable properties	• Thickening agents in soups, sauces, and gravies • Texture modification in baked goods	Alcázar-Alay and Meireles (2015)
Modified starch	Any native starch source	Chemically, physically, or enzymatically altered for improved functionality	• Stabilizers in dressings and sauces • Thickening in canned products • Freeze–thaw stability in frozen foods	Zia-Ud-Din et al. (2015);Bangar et al. (2023)
Pregelatinized starch	• Corn • Potato • Tapioca	Precooked, instant thickening, cold swelling	• Instant soups, desserts, and sauces • Bakery fillings for rapid preparation	Ma et al. (2022)
Cross-linked starch	• Corn • Wheat	Increased resistance to heat, acid, and shear	• Canned foods • Acidic sauces • Products subjected to intense processing	Bangar et al. (2024)
Resistant starch	• Maize • Legumes • Bananas	Resistant to digestion, acts as dietary fiber	• Functional foods for gut health • Low-glycemic index products • High-fiber snacks	Scott & Awika (2023)
Waxy starch	• Waxy maize • Waxy rice	High amylopectin content, low amylose, high viscosity	• Thickening in dairy products • Pie fillings • Salad dressings for a creamy texture	Hsieh et al. (2019)

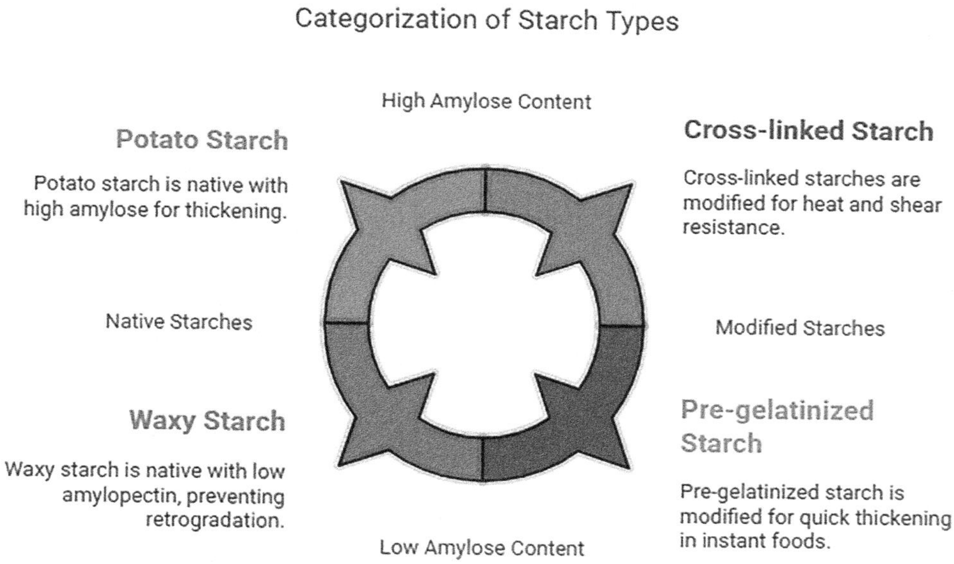

FIGURE 17.1 Classification of starch types based on amylose content and modification. Native and modified starch.

Corn starch: Extracted from the endosperm of corn kernels, corn starch typically consists of 75% amylopectin and 25% amylose. It is widely used in sauces, gravies, and puddings for its reliable thickening abilities. The gelatinization temperature range for corn starch is between 62 °C and 72 °C (Yu & Moon, 2021)

Potato starch: Known for its large granules (ranging from 10 to 100 mm) and high amylopectin content (80%–90%), potato starch imparts a glossy appearance and is highly viscous. Due to its superior binding properties, it is commonly used in processed foods such as soups, sauces, and instant mashed potatoes.

Wheat starch: Derived as a by-product from gluten manufacturing, wheat starch retains some residual protein that imparts a flavor, aroma, and appearance similar to flour. Its rheological properties resemble those of corn starch, although its viscosity and gel stability are generally lower. Wheat starch is primarily used in baking to replace part of the wheat flour (Kim & Kim, 2021).

1.1.2 Modified starches

Modified starches are created through chemical, physical, or enzymatic modifications to enhance specific functional properties such as heat stability, shear resistance, and gel clarity. These modifications make starches suitable for a broader range of food processing applications, including those requiring high-temperature processing or extended shelf life (Wang et al., 2020). Examples include the following:

Pregelatinized starch: Produced by heating starch in the presence of moisture, pregelatinized starch is commonly used in instant foods for quick thickening.

Cross-linked starch: Chemically modified to resist mechanical shear, heat, and acidic conditions, cross-linked starches are particularly useful in canned products (Bangar et al., 2024).

1.1.3 Resistant starches

Resistant starches are a type of dietary fiber that is difficult for the small intestine to break down. They are classified into five types (RS1-RS5) based on their origin and digestion resistance mechanisms. Resistant starches offer several health benefits, including enhancing satiety, helping regulate blood sugar levels, and promoting gut health by supporting the growth of beneficial gut bacteria (Zaman & Sarbini, 2015).

1.1.4 Waxy starch

Waxy starches are derived from waxy maize and waxy rice, which are distinguished by their extremely high amylopectin content and minimal amylose content. These starches exhibit high viscosity and create a smooth, creamy texture without gelling when cooled, in contrast to starches rich in amylose. The absence of amylose prevents retrogradation. Waxy starches are primarily used as thickening agents, bread enhancers, or as a base for Asian cereal products (Šárka & Dvořáček, 2017).

1.2 Functional properties of starch in food products

Starch exhibits several important functional properties in food processing, including gelatinization, retrogradation, thickening, film formation, and syneresis. Gelatinization occurs when starch granules absorb water and expand during heating, breaking down their crystalline structure. This process, typically occurring between 60 °C and 90 °C, is essential for thickening gravies, soups, and sauces, contributing to the desired texture and mouthfeel by forming a viscous paste (Li, 2022). Retrogradation, on the other hand, refers to the regrouping of gelatinized amylose molecules as they cool, forming a gel-like structure. While this can improve the texture in gelled desserts, it can also lead to syneresis, the separation of water, in stored products. The extent of retrogradation depends on the amylose content, with higher levels typically increasing the likelihood of syneresis (Wang et al., 2015). Starch also has excellent water-binding capacity, allowing it to retain moisture, which is particularly beneficial for products that require longer shelf life, such as processed meats and baked goods (Egharevba, 2019). This characteristic also contributes to the juiciness and tenderness of foods, especially those that are frozen and reheated (Zhang et al., 2024). In addition, starch's thickening ability plays a significant role in providing body and texture to liquid-based food products. The viscosity is influenced by both gelatinization and the ratio of amylose to amylopectin, with higher amylose levels leading to thicker textures. Starch is also used to create films for biodegradable packaging and edible coatings, a sustainable solution that has gained popularity in the food industry. These starch-based films help retain moisture, prevent microbial growth, and reduce oxidation, extending shelf life and maintaining product quality (Bangar et al., 2021; Thakur et al., 2019). Finally, syneresis, the separation of water from a starch gel, is an important factor in food products, especially in items such as yogurt, custards, and sauces, where stable gel structures are desired (Kashi & Razavi, 2024a, 2024b). Starches that contain higher levels of amylose are more prone to syneresis due to increased retrogradation, while those with higher amylopectin content are less likely to experience this issue.

1.3 Importance of starch in canned and instant food industries

Starches play a crucial role in the canned and instant food industries, contributing to the desired texture, viscosity, and overall product stability. In canned soups and sauces, starches are essential for achieving the right thickness and viscosity, which are critical for maintaining the consistency and overall standard of the product throughout the canning process. The canning process involves exposure to high temperatures, which can affect the texture and stability of food mixtures. Cross-linked starches are particularly advantageous in these cases due to their improved stability when subjected to heat and mechanical shear. This enhanced stability prevents syneresis, or the separation of liquid, ensuring the product maintains its quality and consistency throughout its shelf life (Bangar et al., 2024). Additionally, the thickening properties of starches improve consumer acceptability by enhancing mouthfeel and palatability. Starches are also integral in a wide variety of packaged foods including desserts such as puddings, fillings for pies, gravies, sauces, and soup. They serve to thicken and stabilize these products, ensuring the proper viscosity and texture during the filling process, which also allows for accurate measurement of suspended particles and minimizes splashing (Himashree et al., 2022). In canned desserts, starches contribute to the gel-like consistency found in fruit fillings, custards, and puddings. In the instant food industry, starches are fundamental for providing convenience and facilitating quick preparation. They serve as a primary source of energy and influence the cooking properties of products such as instant noodles and rice. Pregelatinized starches are essential in products such as instant noodles and soup, as they facilitate rapid rehydration when hot water is added. These starches are precooked and dried, allowing them to absorb water quickly and expand, resulting in an appealing texture without the need for extended cooking time (Ma et al., 2022). The inclusion of pregelatinized starches ensures that instant products maintain quality, achieving the right consistency and flavor profile expected by consumers. In instant noodles, starch affects the dough's flexibility, hardness, and chewiness. The amount of starch in noodle dough directly influences its water absorption during cooking, and starches with higher amylose content help maintain the noodles' shape and texture after rehydration (Wang et al., 2024).

2. Key properties of starch relevant to canned and instant foods

Starch is a versatile carbohydrate polymer and a cornerstone ingredient in food science. It is known for its unique properties that profoundly influence food texture, consistency, and stability, particularly in processed, canned, and instant foods. The physical and chemical properties of starch significantly influence its functionality and applications in various industries (Fig. 17.2).

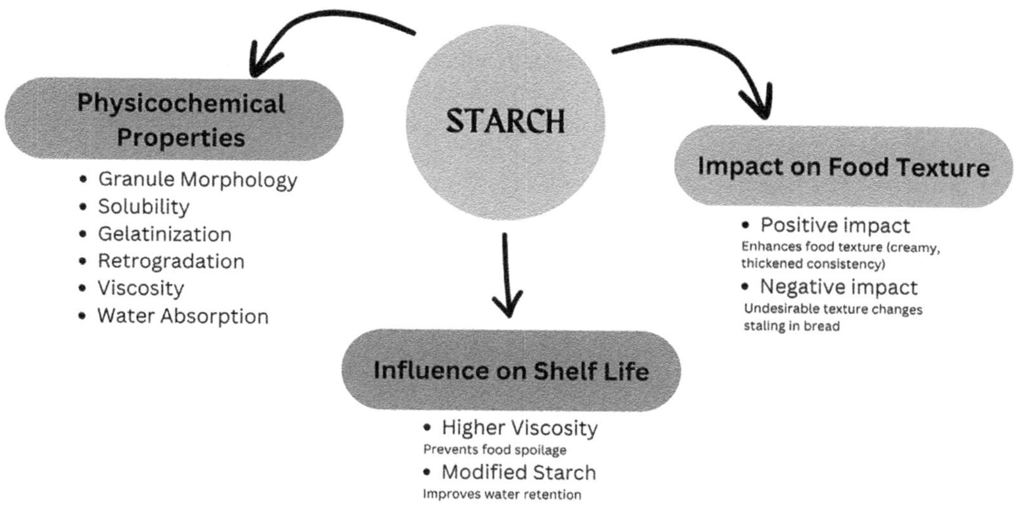

FIGURE 17.2 Key properties and effects of starch in foods.

2.1 Physiochemical properties of starch

Physicochemical properties refer to the physical and chemical characteristics of starch, which influence its behavior in various food applications. Physical properties are observed without altering the chemical structure of starch and include granule morphology, solubility, gelatinization, retrogradation, viscosity, swelling, and water absorption (Emeje, 2019). Chemical properties involve changes in starch due to chemical reactions, which require bond formation or breakage. Key chemical properties include polymeric composition, hydrolysis, oxidation, esterification, and etherification.

2.1.1 Granule morphology

Starch exists as granules that vary in size and shape depending on their botanical source. These granules consist of two glucan polymers: linear amylose and branched amylopectin, which are packed into starch granules with distinct sizes and forms. The granule morphology significantly influences the starch's physical properties. For example, smaller starch granules have a larger surface area, often containing surface pores and channels that facilitate better water absorption (Cornejo-Ramírez et al., 2018).

2.1.2 Solubility

The ability of starch to dissolve in water or other solvents under particular circumstances is referred to as solubility. The closely packed granular nature of starch usually makes it insoluble in cold water. The amylose-to-amylopectin ratio, granule size (Cornejo-Ramírez et al., 2018), and botanical source are some of the variables that affect starch solubility. Through enzymatic, chemical, and physical alterations, starch solubility can be altered. It is intimately related to other functional characteristics like viscosity, gelatinization, and swelling power.

2.1.3 Gelatinization

When natural starch is heated to between 50 °C and 100 °C in the presence of excess water, its semicrystalline structure is disrupted, causing gelatinization (Wertz & Goffin, 2020). Hydrogen bonds are broken during this process, changing the starch granules' structure from ordered crystalline to disordered amorphous. Nuclear magnetic resonance (NMR) spectroscopy, polarized light microscopy, differential scanning calorimetry (DSC), and thermomechanical analysis are among methods that can be used to examine the change from granular starch to a paste or gel. The gelatinization temperature and enthalpy change (ΔH), which represents the energy required to dissociate the crystalline structure, are frequently measured using DSC. Starch should be combined with at least twice its weight in water for precise results. Starch composition, water content, temperature, pH, agitation, additives, and starch source are some of the variables that affect gelatinization (Ai & Jane, 2015).

2.1.4 Retrogradation

The process by which the molecules of gelatinized starch realign into an ordered structure is known as retrogradation. Amylose and amylopectin are released into the solution when starch granules swell and rupture during the gelatinization process. The molecules of amylose and amylopectin reassociate to create hydrogen bonds when the mixture cools, leading to recrystallization, which mostly involves amylose. As a result, crystalline structures are formed, giving the starch gel a harder, opaque consistency. DSC can be used to study retrogradation, much like it can be used to study gelatinization. The amount of amylose and amylopectin, the starch type, the moisture content, the temperature, and storage time are the main factors influencing retrogradation (Chang et al., 2021).

2.1.5 Viscosity

Viscosity refers to starch's resistance to flow, which is a crucial characteristic for food and industrial uses that affects the texture and thickening of foods made from starch. Temperature plays a significant role, with lower final temperatures resulting in more viscous gels (Jiang et al., 2020). Stirring is important when analyzing starch in aqueous suspensions (Chiravi et al., 2023). Pasting profiles are influenced by the starch–water ratio; larger water concentrations facilitate the transition from powder to paste (Yang et al., 2022).

2.1.6 Water absorption

Water absorption, also known as water absorption capacity (WAC), is the term used to describe starch's capacity to absorb and hold onto water (Zhang et al., 2024). This property is critical for starch functionality in food applications, influencing thickening, stabilizing, binding, and enameling properties (Institute of Food Resources of NAAS et al., 2021). Water retention capacity (WRC), swelling measurement, and sophisticated methods such as Raman spectroscopy to examine the short-range order of starch are ways to measure WAC (Li et al., 2024). WAC is influenced by crystalline structures, hydrophilic groups, granule surface microstructure, and starch content (Zhang et al., 2024). Understanding water absorption is essential for optimizing starch performance, especially in food products.

2.2 Impact gelatinization and retrogradation on food texture and stability

Among the various physical properties of starch, gelatinization and retrogradation are considered the most significant in the context of food products. The impact of gelatinization and retrogradation can be either positive or adverse to food quality, and the following section explores their consequences for texture and stability.

2.2.1 Positive impacts of gelatinization and retrogradation on food quality

Foods such as sauces, soups, gravies, and custards benefit from gelatinization because it increases viscosity, which enhances mouthfeel. Starch granules soften foods containing starch for easier absorption by absorbing water, swelling, and producing a thicker consistency that adds to a creamy texture (Chiravi et al., 2023). Additionally, starch gelatinization improves starch digestibility by making it more susceptible to enzymatic breakdown, which is important for products such as instant meals and canned foods, where quicker digestion is beneficial (Pellegrini et al., 2020). Additionally, gelatinization is essential for enhancing the shelf life of food items such as prepared meals. At the macro- and microlevels, it causes irreversible structural alterations that could improve stability. This is particularly significant for canned and instant foods, where extended shelf life is a key factor in maintaining product quality (Chakraborty et al., 2022). To preserve the texture and nutritional content of canned and instant foods, gelatinized starch, for instance, retains substantially more water than native starch (22.35% at 100 °C against 4.3%) (Donmez et al., 2021). Conversely, retrogradation plays a key role in the production of resistant starch, especially type III (RS3), which has several health advantages. Because it is less digestible and functions similarly to dietary fiber, retrograded starch helps control blood sugar, particularly in foods that are suitable for diabetics. This process also affects the texture of food; for example, retrograded starch improves the texture of cooked rice and potatoes by increasing hardness and cohesiveness while decreasing adhesiveness (Jiang et al., 2024). Retrogradation, for example, improves the cooking and rehydration qualities of rice products by lowering adhesiveness by 41.6% and raising hardness by up to 37.3%.

2.2.2 Adverse effects of gelatinization and retrogradation

Although gelatinization is usually advantageous, there are some circumstances in which it may be disadvantageous. Excessive retrogradation and gelatinization in frozen foods can result in a coarse texture upon thawing, which reduces the products appeal for consumption. Food processing may also be complicated by the energy needed for gelatinization and the requirement for exact temperature and moisture control. Unwanted textural changes may result from the process of retrogradation, in which starch molecules reassociate and form a more organized structure. Retrogradation, for instance, causes starch-based products to become stiffer and harder, which is undesirable for products such as bread since it causes the softness to be lost and the hardness to grow (Yamaguchi et al., 2019). Freshness and softness are eventually impacted by this process, which also leads to moisture loss, staleness, and altered mouthfeel. Retrogradation shortens the shelf life of baked goods and reduces their suitability for long-term storage by speeding up the staling process. Although retrogradation may have certain advantages, such as improving resistant starch formation for health benefits, it is typically viewed negatively in food science due to its effects on texture, moisture retention, and digestibility. Therefore, managing retrogradation is essential in ensuring the quality and stability of food products.

2.3 Impact of viscosity and water absorption on the shelf life of canned and instant foods

The shelf life of canned and instant foods is significantly influenced by the viscosity and water absorption capacity of starch. By decreasing water migration and maintaining the consistency of the meal, higher starch viscosity in food products, such as sauces or soups, can assist in improving texture and stability and extend shelf life (Chinachoti et al., 2018; Santamaria et al., 2022). Higher viscosity, for instance, slows down enzymatic activities in instant soups, reducing staling and spoiling and extending shelf life. Food preservation is also impacted by the WAC, particularly for dried and canned foods. By inhibiting microbial growth and preserving quality, modified starches such as acetylated or octenyl succinylated starches improve the stability of low-moisture foods and increase their shelf life (Królikowska et al., 2019). However, the inherent WAC of native starch can lead to moisture loss and degradation in perishable canned or instant foods, such as fruits or vegetables, thereby shortening their storage stability (Bangar et al., 2023). Using the right modification techniques to control starch viscosity and water absorption is crucial for maximizing food preservation and product longevity.

3. Starch formulation strategies for instant foods

The development of starch formulations for instant foods is a key factor in enhancing both the functionality and quality of these products. One primary strategy involves modifying starches to optimize their viscosity, stability, and texture, which is crucial for products that experience freeze–thaw cycles or extended storage (Wang, Li, and Zheng 2020, pp. 131–149). By boosting their ability to retain water and avoiding unwanted changes such as syneresis or separation during rehydration, modified starches such as acetylated, oxidized, or cross-linked starches can greatly enhance the texture and stability of instant food products (Obadi & Xu, 2020). By offering advantages such as better digestion and a lower glycemic index, resistant starch is another strategy that raises the nutritional content of instant foods. This is especially crucial for formulations aimed at consumers with particular dietary requirements or those who are health-conscious. Furthermore, manufacturers can fine-tune product attributes including gel formation, mouthfeel, and rehydration capabilities by combining various starches, such as maize, rice, or potato starch, in starch blends (Tiozon et al., 2021). Starch-based films and coatings are also being investigated in instant meal compositions to further enhance the product consistency and shelf life. By acting as moisture barriers, these films can lessen the absorption of water and assist in maintaining the quality of food while it is being stored and transported (Chen et al., 2021). When combined, these tactics guarantee that, even under difficult storage situations, instant foods retain the texture, stability, and nutritional content that consumers expect.

Agglomerated instant starch is one such strategy where the formulation includes agglomerated instant starch, which is designed to provide immediate dispersibility in food products. This is important for instant food as it allows consumers to reduce preparation time, omitting prolonged mixing or cooking. A combination of pregelatinized starches is a very flexible strategy, which involves one or more pregelatinized starches. The starches can be modified or native leaving scope for chemical alterations or usage in natural states to achieve desired properties. This strategy allows customization based on specific food applications. The usage of water soluble coating agents is one such strategy where the aspect of the formulation is the inclusion of water-soluble food ingredients that have surface-active functionality. The coating agent can be derived from various sources including Gum Arabic, which is a natural gum derived from the hardened sap belonging to the Acacia tree. I i's widely used in the food and beverage industry as a stabilizer, emulsifier, and thickening agent due to its excellent solubility and ability to improve texture in various products. The other sources can be treated as starches or maltodextrins. The presence of these agents helps improve the dispersibility and stability of the starch in food products. The high-percentage formulation of pregelatinized starch is designed to contain more than 50% of this ingredient, ensuring the thickener provides the necessary viscosity and texture in food products. This concentration is critical for achieving the desired thickening effect without compromising food quality. A unique feature of this formulation is its ability to achieve "Instant Viscosity Development," enabling it to develop high viscosity almost immediately upon mixing with water (Reyniers et al., 2019). This property is particularly advantageous for instant foods, allowing quick preparation while ensuring a smooth, lump-free texture. Additionally, the formulation exhibits excellent storage stability, a crucial attribute for commercial products requiring long shelf lives. This stability ensures the thickener retains its functional properties over time, making it reliable and well-suited for instant food production. Several studies have explored the formulation of instant gelling starches, particularly using tapioca or potato starch. Aqueous slurry preparation begins with native tapioca or potato starch (Tian et al., 2023), with the pH adjusted to a critical range of 5–12 to facilitate subsequent steps. The slurry is then drum-dried, a process that requires precise pH control to ensure ingredient stability and effective interactions, ultimately achieving the desired quality and functionality in the final product. Following drum drying, the starch undergoes heat treatment to reduce its viscosity. This process, conducted in a conventional heating device such as an electric or gas oven, typically lasts 1.5–24 h at controlled temperatures between 125 °C and 180 °C. Heat treatment is essential for developing the starch's gel strength and viscosity, ensuring it meets the consistency and performance standards required in various applications. However, maintaining a pH above 6.5, along with selecting appropriate time and temperature conditions during heat treatment, is crucial to achieving the desired gelling properties. These strategies have been effectively implemented in instant foods, as instant gelling starches provide rapid thickening and desirable textures. Such starches find wide application in products such as pie fillings and jellies, where they contribute to quick setting and an appealing taste. Additionally, they are well-suited for puddings, offering a creamy and thick consistency upon preparation.

Starch plays an essential role in the formulation of instant foods, with its adaptable properties making it invaluable across diverse applications. The strategies employed to optimize starch in instant foods are wide-ranging and meticulously designed to enhance both functionality and consumer appeal. One prominent approach involves the use of cold-water-soluble starches, which dissolve effectively in cold liquids, making them particularly suited for instant food applications such as puddings and microwaveable meals. The ability of these starches to hydrate rapidly not only facilitates quick preparation but also aligns with the growing demand for convenience in the fast-paced food market, where time-saving products are increasingly sought after by consumers (Ma et al., 2022). Another significant strategy involves the utilization of small-crystallite starches, which serve as effective fat substitutes in various food products. By modifying the crystalline structure of the starch, manufacturers can enhance the creamy texture and mouthfeel of instant foods while simultaneously reducing the overall fat content. This caters to health-conscious consumers who are looking for lower-fat options without sacrificing sensory quality or indulgence. Moreover, a variety of physical modification methods have been developed to create starches with distinct characteristics tailored for specific applications. Methods such as heat treatment, mechanical processing, and the application of certain solvents allow food technologists to alter the properties of starch, resulting in enhanced thickening, gelling, or stabilizing capabilities (López-Pedrouso et al., 2022). Each of these techniques modifies starch to meet precise functional requirements, enabling manufacturers to achieve desired textures and functionalities in their products. Additionally, a comprehensive understanding of the granular structure of starch is vital for optimizing its application in food products. By studying how different modifications affect the starch granule structure, food scientists can better predict how these starches will behave under various cooking and processing conditions, which is essential for achieving consistent quality and texture in the final product.

The versatility of modified starches significantly contributes to their broad application spectrum, as they can be adapted for use in an extensive range of instant food products, from soups and sauces to desserts and snacks. This inherent flexibility not only supports innovation within the food industry but also enables manufacturers to respond to diverse consumer preferences and dietary needs by developing instant foods with specific textural and sensory profiles. Ultimately, the effective formulation of starches in instant foods relies on a strategic combination of specialized cold-water-soluble and small-crystallite starches, the employment of advanced modification techniques, and a thorough examination of granule structure. Together, these strategies not only enhance the functional role of starch in instant foods but also promote the creation of healthier, high-quality products that meet both nutritional standards and sensory expectations, thereby enriching the overall consumer experience in the realm of convenience foods.

Starch plays a fundamental role in the formulation of rice-based instant foods, significantly impacting both texture and flavor. A comprehensive understanding of starch structure is essential, as its multiscale characteristics—including particle morphology, crystalline arrangement, and molecular configuration—greatly influence how starch interacts with other ingredients and contributes to the overall flavor profile of the food. Several mechanisms by which starch facilitates flavor development are notable. For instance, during processing, reducing sugars generated from starch can engage in the Maillard reaction (Li, 2022), a critical pathway that fosters the synthesis of complex flavor compounds through the interaction of sugars with amino acids. Additionally, when starch is subjected to heat, caramelization can occur, producing a diverse array of aromatic compounds that enhance flavor complexity. Furthermore, the thermal degradation of starch under high temperatures may yield new flavor compounds, enriching the sensory attributes of the instant food product.

The formulation of starch also affects flavor release during consumption, which is crucial for the overall eating experience. For example, starch granules have the capacity to adsorb flavor compounds, which can be released upon cooking or eating, thereby augmenting the flavor experience. Moreover, retrograded V-type crystals formed during starch processing can encapsulate these flavors, controlling their release and contributing to a sustained flavor profile.

The formation of starch gels during cooking is another critical factor, as it facilitates the regulated release of flavors, enhancing the sensory experience. The paper illustrates these strategies with specific examples of rice-based instant foods, such as instant rice, where the optimization of starch properties ensures quick cooking while preserving flavor; fermented rice cake, where the fermentation process interacts with starch to influence both texture and flavor; and rice noodles, fried rice, and rice dumplings, each utilizing starch in unique ways to achieve desired flavor and texture profiles. Collectively, these examples underscore the versatility of starch in food formulation. In conclusion, implementing effective starch formulation strategies is vital for enhancing the flavor and quality of rice-based instant foods. A thorough understanding of the processes through which starch contributes to flavor formation and release can pave the way for improved processing methods and the progression of higher-quality products. The most popular strategies for starch formulation in instant food production are summarized in Table 17.2.

3.1 Precooked and instantized starches for quick rehydration

Precooked and instantized starches are engineered to enhance rehydration properties, making them highly convenient for food applications. Research on instant starch derived from Taro (*Colocasia esculenta*) highlights efficient preparation methods and their versatility in food formulations (Gupta et al., 2024). The process begins with wet milling extraction to isolate starch, followed by pregelatinization via autoclaving for varying durations. Among the samples, 3-h autoclaved starch (IS-3) demonstrated the best functional properties, including higher resistant starch (RS) content (30.14%) and superior water absorption capacity (1.72 g/g). Resistant starch promotes gut health, acting as a prebiotic, while improved swelling power and solubility enhance the rapid rehydration required for instant food products. Incorporating IS-3 into cake formulations improved volume, texture, and sensory qualities, establishing its utility in quick-preparation foods.

Precooked and instantized starches are widely used in instant soups, sauces, and snacks due to their rapid water absorption and desirable textural properties. Their production typically involves gelatinization, where starch granules swell in water before drying. Techniques such as spray drying or drum drying further optimize solubility and functionality. For example, substituting instant corn bakwan flour containing pregelatinized sweet corn flour enhanced water absorption and solubility, contributing to better consistency and nutritional value, including increased protein and β-carotene levels. The swelling behavior of pregelatinized starches is critical for rehydration and other applications like oil replacement in emulsions. Their ability to replace up to 60% of oil in low-fat formulations, while maintaining texture and stability, underscores their value (Yusuf et al. 2023). Variations in swelling properties significantly influence consistency and rehydration speed, making these starches ideal for quick-cooking applications. The use of pregelatinized red lentil flours in pasta production demonstrates their potential in food formulations. With lower gelatinization temperatures (~57 °C) and enhanced water absorption (1.23 g/g), these flours improve rehydration without compromising cooking quality (Bresciani et al., 2023). Although pregelatinization reduces resistant starch content and increases digestibility, it aligns with consumer demand for quick meal solutions.

In summary, precooked and instantized starches provide significant advantages in convenience, functionality, and nutritional potential. Their rapid rehydration capability and versatility make them indispensable in modern food applications, addressing the growing demand for quick and easy meal preparation.

3.2 Formulating with starch for stability, consistency, and mouthfeel

Starch plays a crucial role in food formulation by enhancing stability, consistency, and mouthfeel across various products. It acts as a nutritive stabilizer, influencing viscosity and texture, particularly in items such as sauces, soups, and desserts, where a creamy or thick texture is essential. The functional properties of starch vary depending on its form. Gelatinized starches typically provide a smoother mouthfeel, facilitating flavor incorporation, while uncooked granular starches contribute distinct textures beneficial for specific applications (Gałkowska et al., 2023). Starch also functions as a processing aid, preventing sticking and improving manufacturing efficiency. For example, powdered corn starch is often used to dust surfaces, streamlining operations during food production. The commercialization of specialty starches, which can

TABLE 17.2 Strategies for starch formulation in instant food production.

Formulation strategy	Description	Application	References
Pregelatinization	Prepares starch by partially cooking it, enabling it to dissolve quickly in hot or cold liquids, which results in rapid thickening.	Suitable for instant soups, sauces, and gravies that require quick preparation.	Ma et al. (2022)
Cross-linking	Chemically reinforces starch granules, enhancing their durability against heat, stirring, and acidic environments, which boosts stability.	Ideal for products with intense processing or longer shelf-life needs.	Bangar et al. (2024)
Hydroxypropylation and acetylation	Introduces hydroxypropyl or acetyl groups to starch, which improves its ability to retain moisture and resist separation during freezing and thawing.	Often used in frozen or refrigerated instant foods to retain consistency over time.	Gałkowska et al. (2023)
Combination blends	Combines different starch types, like tapioca and corn, to achieve a balance of texture, stability, and cost-effectiveness.	Used in products where specific textures, such as creaminess or thickness, are desired.	Tiozon et al. (2021)
Enzyme-modified starch	Alters starch with enzymes to fine-tune its viscosity and texture, allowing for precise control over thickening properties.	Commonly applied in dairy-based instant products, dressings, and foods requiring consistent texture.	Iuga and Mironeasa (2020)

be modified to improve gelling properties, highlights starch's versatility in various food applications. High-amylose starch, optimizes jelly candy production by significantly reducing drying times and improving efficiency, enabling manufacturers to increase output without additional equipment (Hu et al., 2024). This illustrates starch's dual role in enhancing sensory attributes and improving production efficiency, crucial for meeting the demands of the growing food industry. The adaptability of starch solidifies its importance in food formulation, offering both product quality and operational benefits. As shown in Fig. 17.3, the process of formulating with starch is categorized into three key components: stability, consistency, and mouthfeel. These elements highlight the integral role of starch in ensuring product quality and enhancing the sensory experience for consumers.

4. Modified starches for enhanced performance

The diverse structure and function of native starch, obtained from various plants, can be enhanced to achieve better results in formulations in the food industry using various environmental, operational, and processing factors. This is termed as starch modification, which may exert an overall positive or negative effect on the functional qualities of native starch (Nawaz et al., 2020). Native starch can be modified to overcome its disadvantages such as insolubility in cold water, loss of viscosity and thickening power after cooking, low shear resistance, thermal resistance, and high tendency toward retrogradation (Korma et al., 2016). Starch modification relies on changing the natural starch to make it suitable for different applications in the field of packaged foods, paper, and textile industries. Starch modification normally affects the phenomena of starch gelatinization—breaking of the intermolecular bonds in the starch structure when starch is mixed with water (Chakraborty et al., 2022).

4.1 Types of modifications and their benefits

To prepare modified starches with desirable properties for applications in the food industry, there are numerous physical and chemical methods of modification. Each method works on varying the properties of native starch, which could be its solubility, stickiness, texture, and gelatinization temperatures (Fig. 17.4).

4.2 Chemical modification of starch

Chemical modification of starch involves introducing additional chemical groups to the native starch structure, which results in a change of composition, gelatinization temperature, pasting properties, or retrogradation of the native starch. The commonly observed chemical modification processes are acid treatment, cross-linking, oxidation, and substitution,

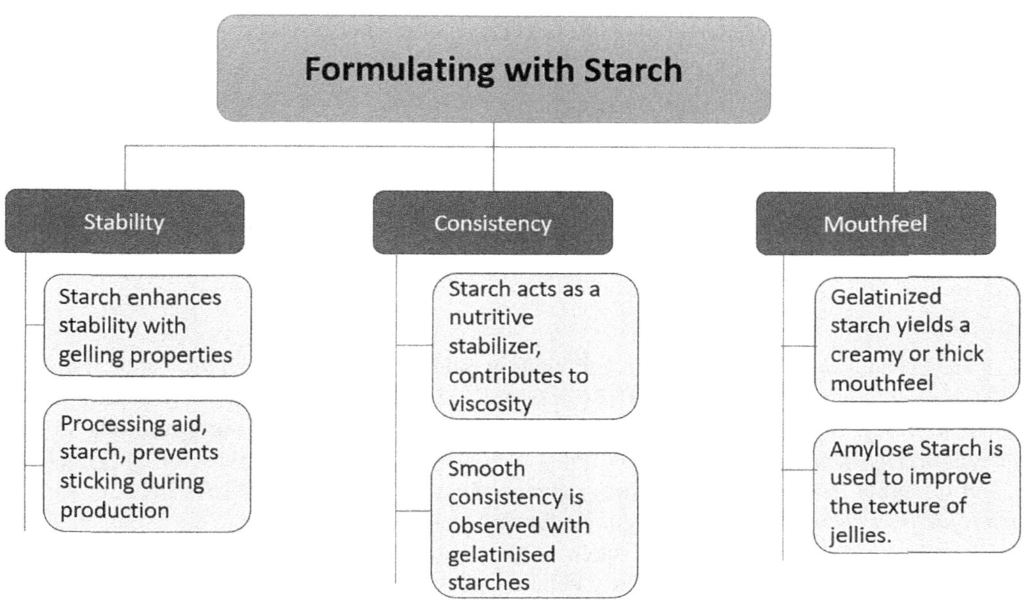

FIGURE 17.3 Key components of formulating with starch. Stability, consistency, and mouthfeel.

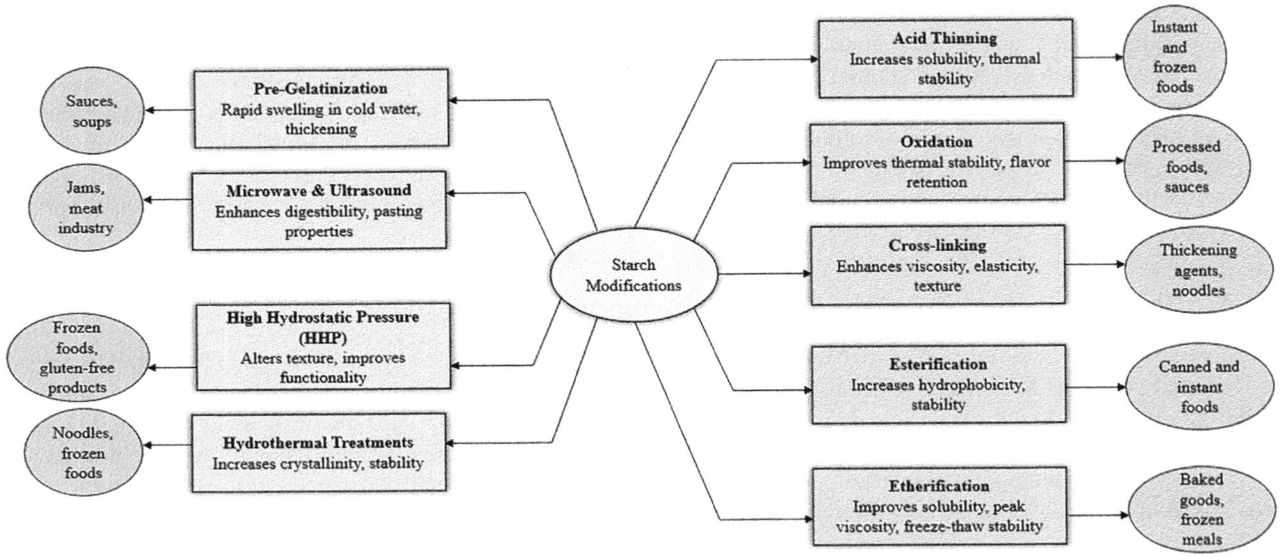

FIGURE 17.4 Overview of chemical and physical starch modifications. Benefits and applications in the food industry.

including esterification and etherification (Korma et al., 2016). Stabilization is a vital step in frozen food production because cold temperatures accelerate starch retrogradation, leading to an opaque texture and, eventually, liquid separation (syneresis) from the gel. Preventing retrogradation helps extend the shelf life of the food product by improving its texture and preventing syneresis (Almanza-Benitez et al., 2015). Stabilization involves the addition of special chemical groups to the native starch resulting in a rearrangement of its molecular structure, achieving the objective of inhibiting starch retrogradation and enhancing the shelf life of the final product (Korma et al., 2016). Starch stabilization includes the mechanisms of acetylation, phosphorylation, hydroxypropylation, cross-linking, etc.

4.2.1 Acid thinning of starch

The depolymerization of the native starch structure caused on reacting it with acids such as hydrochloric acid or sulfuric acid is termed acid-thinning of starch (Chakraborty et al., 2022). Oderinde et al. in 2020 performed the chemical modification of the African breadfruit (*Treculia africana*) kernel starch by acid-thinning mechanism using 0.15 M HCl, which resulted in enhanced solubility and thermal stability of starch, thereby playing an important role in instant and frozen foods, where it can be used to prepare gels without releasing too much water on thawing it (syneresis) (Oderinde et al., 2020).

4.2.2 Oxidation

Oxidized starches are formed when native starch is made to react with suitable oxidizing agents (usually NaOCl, in food applications) under controlled temperature and pH. In 2020, Oderinde et al. chemically modified the African breadfruit (*T. africana*) kernel starch using sodium hypochlorite (NaOCl), 4% w/w active chlorine, and concluded that the modification enhanced the thermal stability of native starch and can function as flavor retention agent (Oderinde et al., 2020).

4.2.3 Cross-linking

Cross-linking of starch relies on introducing chemical bonds at random locations within the starch molecule, resulting in a stabilized and strong structure (Qiu L. et al., 2013). Ali et al. (2020) subjected native lotus seed starch to citric acid treatment, which improved its overall functionality, increasing its applications in the processing of food products as thickening and stabilizing agents. The cross-linked lotus seed starch showed an increase in its crystalline structure, gel adhesiveness, and solubility but a decline in swelling power and viscosity (Ali et al., 2020). In 2018, Alimi et al. extracted starch from *acha* (*Digitaria exilis*) and *iburu* (*Digitaria iburua*), cereal crops native to West Africa, and subjected to citric acid treatment and observed that it resulted in starch polymerization and reduced viscosities, useful for products like candies. The modified starches increased lightness indices and whiteness indices of the starch, useful in the case of products such as ice creams and juices, where color clarity is crucial (Alimi & Workneh, 2018).

Research has demonstrated various practical applications of cross-linked starch. For instance (Wandee et al., 2015), it was shown that native canna starch modified by cross-linking enhances the tensile strength and elasticity of rice noodles. Additionally, Chao-Feng (Hsieh et al., 2020) found that incorporating 10%–40% cross-linked phosphorylated RS4 wheat starch into wheat flour maintained the noodles' texture. Furthermore, when 2%–8% vital wheat gluten was added to a mix containing 40% RS4, the noodles became firmer and more flexible, retaining their cohesiveness and springiness.

4.2.4 Esterification

The mechanism of starch modification involving the replacement of hydroxyl groups in native starch with ester groups, thereby resulting in hydrophobicity due to the substitution of free hydroxyl groups, is termed esterification (Korma et al., 2016). Acetylation, succinylation, phosphorylation, etc. come under the broader category of esterification of starch. Rincón-Aguirre et al. (2018), studied the characteristics of taro starch chemically modified by acetylation (using acetic anhydride), phosphorylation (using sodium tripolyphosphate), and succinylation (using 2-octen-1-ylsuccinic anhydride). The acetylated taro starch, irrespective of the degree of substitution, caused an overall decrease in viscosity and damage in the structure, which could be beneficial (enhances the mixing capability of starch during processing) or detrimental (unstable structure affecting the texture and appearance) in canned and instant foods (Rincón-Aguirre et al., 2018).

Oh et al. (2019) described that acetylation of chestnut starch demonstrated an enhanced solubility and peak viscosity and showed a decline in pasting temperature when compared to the native chestnut starch. In addition, acetylated chestnut starch was more tolerant toward damage from repeated freeze–thaw cycles and possessed the properties of inhibiting retrogradation (Oh et al., 2019). Low-DS (degree of substitution) acetylated starches serve as thickening, stabilizing, and binding agents, making them highly versatile for applications in the field of instant and canned foods such as baked goods, sauces, and frozen meals (Obadi and Xu, 2020).

The effects of succinylated (treating native starch with succinic anhydride under alkaline conditions) and octenyl succinylated (treating native starch with OSA (octenyl succinic anhydride)) pearl millet (*bajra*) starch on improving the overall swelling power and viscosity were discussed in the study conducted by Shaikh et al., resulting in these starches being used as thickening agents in sauce and ketchups (Shaikh et al., 2017). This study also discussed the acetylated pearl millet starch decreasing the gelatinization enthalpy and inhibiting retrogradation, making it suitable for applications in frozen foods (Shaikh et al., 2017).

4.2.5 Etherification

Hydroxypropylated starches are produced by the etherification of native starch molecules with propylene oxide (in the presence of an alkaline catalyst). The hydroxypropyl groups introduced during this process damage the hydrogen bonds within the starch molecule, thereby weakening the structure, increasing the rotational freedom of starch chains (Korma et al., 2016). Oh et al. (2019) investigated the properties of hydroxypropylated chestnut starch and stated that it showed a decline in pasting temperature and an improved peak viscosity when compared with the native chestnut starch. In addition, the modified chestnut starch was more tolerant toward damage from repeated freeze–thaw cycles and possessed the properties of inhibiting retrogradation (Oh et al., 2019).

4.3 Physical modification of starch

Physical modification of starch involves processes such as extrusion cooking, radiation, sonication, and pressure treatments. These methods are widely accepted due to their ability to modify starch without the use of chemical reagents, making it simple and safe for human consumption. The impact of physical modifications on molecular integrity varies depending on the method used.

4.3.1 Pregelatinized starch

Pregelatinization is a physical starch modification method that results in starch rapidly swelling in cold water (gelatinization and then dried) and forming a thick paste with high viscosity. This is one of the most widely used techniques in food manufacturing. A study conducted by Okunlola and Adewusi in 2019 uses pregelatinized breadfruit starch (*Artocarpus altilis*) in the controlled release of the drug theophylline. Some insights gained from the study such as the increase in viscosity and flow of pregelatinized starch make it useful for applications in foods such as sauces and soups, where it can serve the role of a thickener and stabilizer (Okunlola & Abidemi Adewusi, 2019).

4.3.2 Microwave-treated and ultrasound-treated

Microwaves are a type of electromagnetic radiations with frequencies in the range of 300 MHz to 300 GHz, which produce heat within the starch granule due to its alternating electromagnetic fields. The advantage of microwave-heating is that the waves heat starch from the inside out (faster heating), unlike traditional methods of heating. Zhang et al., in 2020 studied the effects of microwave-treatment on native cassava starch-sugar mixtures and concluded that when the samples were mixed before microwaving, they showed higher peak viscosity, and pasting temperature as compared to the samples which were mixed after microwave irradiation of starch (Zhang et al., 2020). Overall, the study provided insights into the uses of microwave irradiation to modify native starch for applications in the food industry.

The usage of ultrasound to modify starch relies on high-frequency sound waves (~20 kHz) causing a disruption in the starch structure through the process of cavitation—tiny bubbles growing and collapsing rapidly due to pressure changes, producing shock waves. Li et al. (2019) subjected millet starch to microwave and ultrasound modifications and resolved that both modifications enhanced the enzymatic digestibility of starch resulting in improved starch properties for use in jam fillers and the meat industry (Li et al., 2019).

4.3.3 High hydrostatic pressure treatment

High hydrostatic pressure (HHP) treatment relies on exposing native starch slurry to extreme pressures (100–600 MPa) without heating, significantly affecting the shape and properties of starch granule. A study conducted by Zeng et al. in 2018 stated that HHP treatment (400 MPa) of waxy rice starch played a role in enhancing its functionality when compared with its native counterpart (Zeng, Chen, et al., 2018). For most native starches to fully gelatinize, it requires a pressure level of 600 MPa in HHP treatment. Another study conducted by Du et al. (2019) provided insightful information on how the addition of tea polyphenols (10%–30%) resulted in rice starch gelatinizing at lower pressure levels (400 MPa), thereby increasing the efficiency of HHP treatment.

4.3.4 Hydrothermal treatments

The most commonly used hydrothermal treatments for cereals and pseudocereals starches and flours are heat moisture treatment (HMT) and annealing (Iuga and Mironeasa, 2020). HMT relies on heating at a high temperature (hot air oven) under controlled moisture conditions resulting in a change in the structure and functionality of the native starch. Ali et al. (2020) studied the properties of HMT-modified lotus seed starch, an eco-friendly and consumer-health friendly option, and concluded that there was an increase in its crystalline structure, gel adhesiveness, and solubility but a decline in swelling power and viscosity. In addition, thermal properties also significantly increased when compared to native lotus seed starch (Ali et al., 2020).

Alimi & Workneh (2018) subjected starch from *acha* (*D. exilis*) and *iburu* (*D. iburua*), cereal crops native to West Africa, to HMT and stated that it resulted in starch polymerization and decreased viscosities, useful for products like candies. The modified starches increased lightness and whiteness indices of the starch, useful in the case of products such as ice creams and juices, where color clarity is crucial (Alimi & Workneh, 2018).

HMT was stated to be an impactful strategy to improve the quality of texture in sweet potato starch vermicelli, owing to its increased crystallinity and elasticity when compared with native sweet potato starch (Liao et al., 2019).

Hydrothermal modification techniques are designed to achieve specific temperature and moisture levels in native starch molecules over a set period (Obadi & Xu, 2020). Pasting characteristics have shown that HMT-modified starches are stable under continuous heating, which is particularly beneficial for buckwheat food products like noodles (Goel et al., 2020; Obadi & Xu, 2020). On the contrary, annealing involves heating the native starch at low temperature (lower than gelatinization temperature) under high moisture levels for a long period of time. A study conducted in 2018 by Xu et al., where the authors modified native potato starch using repeated annealing treatments (starch heated at 55°C for 12 h repeated for eight cycles) and continuous annealing treatment (55°C for 24, 48, 72, and 96 h), concluded that the repeatedly annealed starch showed potential usage in food applications owing to its properties of increased thermal stability, peak viscosity, and reduced rate of starch digestion (Xu et al., 2018). In addition, Choi and Koh (2017) found that annealing effectively improved the texture of rice flour noodles (Choi & Koh, 2017; Obadi and Xu, 2020). The applications of annealed starches in the food industry are mainly as emulsifying and bulking agents (Chakraborty et al., 2022; Schmiele et al., 2018).

Incorporating modified starch into cereal starch or flour during noodle-making can enhance the noodles' taste and texture. While mixing native starches or cereal flours is common in noodle production, predicting the final noodle quality is challenging due to the complex interactions taking place between starch and flour. As consumers seek safer, longer-lasting options, modified starches are increasingly used in noodles and other instant foods. These starches improve texture and quality, and we are likely to see more of them in noodle products in the future. The most popular modification for enhanced performance is summarized in Table 17.3.

5. Challenges and innovations in starch-based formulations

5.1 Challenges in starch-based formulations

Starch is a fundamental component of the human balanced diet and is a chief ingredient used when formulating instant and canned foodstuffs. Despite its benefits, starch-based formulations present several challenges that impact product stability, processing, and overall quality. Some of the key challenges are discussed in the following.

5.1.1 Retrogradation

Retrogradation is a critical factor in determining the texture as well as the stability of food items depending on starch. While it can be beneficial in some cases, it often leads to undesirable changes in others. Retrogradation is the process by which the gelatinized starch molecules undergo a transformation into an ordered state from a disordered state (Zeng, Chen, et al., 2018). Fully gelatinized starch is in a disordered state. As gelatinized starch cools down, the highly disordered, high-energy amylose and amylopectin chains slowly recombine to form crystals. This ordered, stable system now achieved is different when compared with the starch before gelatinization (Chang et al., 2021). Starch retrogradation is affected by processing conditions as well as many food ingredients such as water, salt ions, proteins, and lipids. The occurrence of both starch gelatinization and retrogradation during the production and storage of food products has an influence on processes affecting the taste, texture, digestion, and other nutritional properties of the food product (Zhu, 2017). Depending on the type of food, either moderate retrogradation or suppression of retrogradation is required. In foods such as breakfast cereals, mashed potato flakes, and rice noodles, retrogradation is introduced intentionally to enhance crunchiness, rehydration, and reduce stickiness (Wang et al., 2015). However, there are many food items, such as bread and cakes that undergo unfavorable changes due to retrogradation. In meat products, such as meat batters, starch is frequently used to enhance viscosity, product yield, texture, and water-holding capacity (Genccelep et al., 2015). However, a high amount of starch retrogradation has been shown to decrease the quality of the meat by increasing syneresis and disrupting the local protein network. The collecting of amylose during retrogradation may be the primary reason for the disarrangement of the protein matrix (Pietrasik and Soladoye, 2020). Similarly, in meat imitation products, such as burgers, sausages, and chicken nuggets, starch increases viscosity, water-holding capacity, as well as the stability of the matrix (Bühler et al., 2022). Uncontrolled starch retrogradation in plant-based meat alternatives negatively affects the texture and stability of the product. This is due to the formation of coarse-textured proteins with a reduced degree of fibrosity (the ratio of lengthwise to crosswise strength) (Chen et al., 2022).

5.1.2 Syneresis

Syneresis is the spontaneous expulsion of liquid, usually water, from a gel or colloidal solution due to contraction (Wani et al., 2016). During gelatinization, a swollen amorphous starch structure is formed when starch is subjected to high temperatures in the presence of water. However, during retrogradation, this unstable starch structure loses its swelling pressure capacity, creating ideal conditions for syneresis to occur. At low temperatures, the degree of syneresis functions as a measure of the extent of starch retrogradation and phase separation. The capacity of starch to withstand unwelcome physical changes during freezing and thawing is known as freeze–thaw stability. Freeze–thaw stability tests are particularly essential in frozen food products, where phase separation occurs due to ice crystal formation. If the melted ice crystals are not reabsorbed into the matrix after thawing, syneresis takes place. In the food sector, syneresis is generally undesirable when formulating starch-based products, as it affects texture, consistency, and overall quality. The degree of syneresis is dependent on the concentration of starch in the gel. A higher starch concentration typically results in less liquid expulsion. Differences in amylose composition and intermolecular interactions lead to varying degrees of syneresis. Increased interaction between amylose chains tends to enhance syneresis (Wani et al., 2016). Syneresis can be controlled in starch gels by interfering with the crystallization process, which stabilizes the amorphous structure. In food products where starch is an added ingredient, regular starch can be replaced with a modified version to improve stability. In formulations where starch is a natural component, surfactants that form complexes with amylose may be used. Additionally, research by Mohamed et al. (2024) has shown that the addition of gum Cordia and gum Ziziphus to wheat, potato, and chickpea starches reduces syneresis, offering a potential solution for improving the quality of starch-based food products.

TABLE 17.3 Modified starches for enhanced performance.

Classification of starch modification mechanisms	Method of modification	Source	Benefits of the modified starch	References
Chemical modification	Acid-thinning of starch	Native starch source: African breadfruit Acid-thinned starch produced using 0.15 M HCl	• Prevents syneresis • Enhanced solubility • Improved thermal stability	Oderinde et al. (2020)
	Oxidized starch	Native starch source: African breadfruit Oxidized starch prepared using NaOCl	• Enhanced thermal stability • Can function as flavor retention agent	Oderinde et al. (2020)
	Cross-linking	Native starch source: Lotus seed Modified by citric acid treatment	• Improved overall functionality • Increased crystalline structure, gel adhesiveness and solubility • Thickening and stabilizing agent	Ali et al. (2020)
		Native starch source: *Acha* and *Iburu* Modified by citric acid treatment	• Reduced viscosity • Increased lightness and whiteness indices • Useful in ice creams, candies and juices	Alimi and Workneh (2018)
		Native starch source: Canna Modified by cross-linking	• Improved tensile strength • Enhanced elasticity of rice noodles	Wandee et al. (2015)
	Esterification	Native starch: Taro Modified by acetylation, phosphorylation, succinylation	• Acetylated taro starch showed a decrease in viscosity	Rincón-Aguirre et al. (2018)
		Native starch: Chestnut Modified by acetylation	• Improved solubility and peak viscosity • Decline in pasting temperature • Retrogradation inhibition	Oh et al. (2019)
		Native starch: Pearl millet (bajra) Modified by acetylation, succinylation, and octenyl succinylation	• Improved overall swelling power and viscosity • Thickening agents in sauce and ketchups • Frozen foods (acetylated showed retrogradation inhibition)	Shaikh et al. (2017)
	Etherification	Native starch: Chestnut Modified by hydroxypropylation	• Improved peak viscosity • Retrogradation inhibition	Oh et al. (2019)
Physical modification	Pregelatinized starch	Native starch: Breadfruit starch Starch fully gelatinized and then dried	• Increases viscosity and flow • Useful in sauces and soups, as thickener and stabilizer	Okunlola and Abidemi Adewusi (2019)

	Microwave-treated	Native cassava starch–sugar mixture Modified by microwave irradiation	• Higher peak viscosity and pasting temperature (mixed before microwave)	Zhang et al. (2020)
	Ultrasound-treated	Native millet starch Modified by ultrasound	• Enhanced enzymatic digestibility of starch • Jam fillers and meat industry	Li et al. (2019)
	High hydrostatic pressure (HHP) treatment	Native starch: Waxy rice starch Modified by HHP treatment (400 MPa)	• Enhanced overall functionality	Zeng, Chen, et al. (2018)
	Hydrothermal treatment—heat moisture treatment (HMT)	Native starch: Lotus seed HMT-modified lotus seed starch	• Increase in crystallinity • Increase in gel adhesiveness and solubility • Improved thermal properties	Ali et al. (2020)
		Native starch source: *Acha* and *Iburu* Modified by HMT	• Decreased viscosity (useful for candies) • Increased lightness and whiteness indices (useful in ice creams and juices)	Alimi and Workneh (2018)
		HMT-modified starches	• Stable under continuous heating • Beneficial for instant foods like noodles	Goel et al. (2020); Obadi & Xu (2020)
	Hydrothermal treatment—annealing	Native starch: Rice Modified by annealing	• Improved texture of rice flour noodles	(Choi & Koh, 2017; Obadi and Xu (2020)).
		Native: Potato starch Modified by annealing	• Increased thermal stability • Reduced rate of starch digestibility	Xu et al. (2018)

5.1.3 Shelf stability

An optimal end product in a food formulation should have a long shelf life while maintaining key properties such as flavor, texture, viscosity, and water-holding capacity. This requires minimizing retrogradation during storage, as excessive retrogradation negatively impacts product quality. However, reducing retrogradation also affects the gelatinization process, requiring careful formulation adjustments (Ziobro et al., 2012). During storage, bread staling occurs due to moisture loss, starch retrogradation, and gluten network changes (network plasticization), leading to increased hardness over time (Ziobro et al., 2012). The quality changes associated with starch retrogradation include increased firmness over time and variations in the enthalpy required to melt retrograded amylopectin (Ziobro et al., 2012). Wheat proteins, such as gluten, have an inhibitory effect on bread staling, but this effect depends on the constitution of the wheat proteins. Stronger flours exhibit better antistaling properties compared with weaker flours. Gluten-free bread is more prone to staling than gluten-containing bread due to its heterogenous composition. The addition of proteins in gluten-free formulations can either accelerate or delay starch retrogradation, depending on aspects such as the type of protein and starch, protein-to-starch ratio, and batter hydration. Proteins that stimulate phase separation can accelerate the retrogradation of starch, while proteins that cause the immobilization of water within the starch matrix delay it, thereby reducing staling (Scott & Awika, 2023) Staling in gluten-free bread can be suppressed by adding highly water-soluble proteins, which minimize phase separation. Additionally, matrix stabilization using proteins with strong abilities to form gels, for instance, egg albumin, in combination with xanthan gum, can reduce water mobility while enhancing interactions between starch and proteins. However, while this approach decreases staling, it can also lead to increased crumb hardness (Masure et al., 2019).

Starch is frequently used as a fat replacer, bulking agent, and stabilizer in milk-based products such as cheese, yogurt, and ice cream. However, its addition can negatively impact product quality by causing phase separation, which shortens shelf life. Interactions between starch and dairy proteins contribute to starch retrogradation, affecting the rheological properties and protein-gel structures of dairy products. In whey-based products, starch retrogradation can lead to syneresis and reduced product stability over time. Products with higher whey content are particularly susceptible to these issues (Scott & Awika, 2023). In plant-based meat alternatives, amylose self-aggregation during phase separation results in increased starch retrogradation, negatively affecting texture during storage. To mitigate this issue, starches with high amylopectin content are preferred, as they contribute to improved texture and stability in plant-based meat analogs (Scott & Awika, 2023).

5.2 Innovations to improve product performance

Several strategies have been developed to reduce retrogradation and enhance product performance. These include the use of additives, modified starches, optimized processing conditions, and alternative starch sources. In recent years, significant innovations have emerged to improve the quality and stability of starch-based formulations. Some of these advancements are discussed in the following.

5.2.1 Clean label starches

In recent years, growing consumer awareness of health and sustainability has led to an increased demand for all-natural, clean-label food products. Clean-label formulations emphasize the use of natural, familiar ingredients while avoiding chemically modified or synthetic additives (Herrero et al., 2023). Native starches, despite their susceptibility to thermal and shear instability and their proclivity to retrograde during cooling, have gained popularity due to consumer preferences for clean-label foods. This trend has driven the food industry to incorporate native starches in various formulations, despite their challenges. Plant-based alternatives align with the consumer demand for healthier and more sustainable food products. Starch plays a crucial role in improving the nutritional profile and sensory properties of these alternatives by mimicking the mouthfeel and texture of meat and dairy products. Additionally, dual modification of starch can improve its functionality and alter its digestibility to meet specific formulation needs (Scott & Awika, 2023). A study by Dufour et al. (2022) highlighted the excellent bread-making potential of amylose-free waxy cassava starches. The research revealed that the number of linear chains in starch determines its bread-making performance, with increased branching improving functionality. Both native and oven-dried forms of this starch were found to be cost-effective and advantageous over more processed variants.

Starch is also widely used as a thickener in sauces, soups, and other processed foods. A comparison of the appearance and viscoelastic properties of sauces made with waxy corn, rice starch, potato, and corn starch demonstrated that while freeze–thaw cycles slightly affected the properties of waxy corn and rice starch sauces, reheating after thawing nearly eliminated syneresis, although it slightly reduced viscoelasticity. These findings suggest that the use of native starches can lead to high-quality clean-label industrial sauces, depending on the starch source. Furthermore, research by Li et al. (2021) demonstrated the potential of granular cold-water-swelling starch as an affordable, clean-label structurant for oil, enabling the reduction of solid fats in diets. The study found that the rheological behavior of starch-structured oil closely resembled that of butter spreads, with texture modulation possible through ingredient composition adjustments.

6. Future perspectives and market strategies

The ever-increasing consumer demand for sustainable and healthier foods has resulted in various innovations in starch-based formulations for canned and instant foods. Recent research provides insights into using diverse or alternative sources of starch and different modification techniques to improve the texture, shelf-life, and overall performance of the product, thereby shaping future market strategies. Starch from diverse sources such as sweet potato for instant rice noodles (Zhang et al., 2025), *kudzu* (He et al., 2024), and sago palm (Mohamed et al., 2024) provides unique properties that enhance texture and digestibility while also providing a clean-label appeal to consumers. Future research should focus on improving the processing and methods of extraction of starch from these sources in an effective and efficient manner. Along with alternative starches, studies on cassava starch (Lu et al., 2025) and nonstarch polysaccharides (Li et al., 2024) help improve viscosity and shelf-life, thereby reducing food waste. Therefore, hydrocolloid–starch interactions must be further explored to enhance the overall performance of the product. Integrating seaweed-based starches (Cullano et al., 2025) have proved that sustainability plays a role in framing starch formulation strategies, by reducing the use of chemically modified ingredients and supporting eco-friendly food production. Additionally, it helps in targeting health-conscious consumers, leading to opportunities for functional food development. Dual modification (He et al., 2024) and enzyme-assisted processes have proved to be breakthroughs in the field of starch modification, and by highlighting and optimizing these processes and innovations, the food industry can move toward more sustainable, healthier, and consumer-friendly starch-based products.

7. Conclusion

Starch is a crucial component of the food industry known for its various functional properties of thickening, gelling, texturizing, and stabilizing. It has a wide range of applications in baked goods, dairy products, sauces, soups, and processed foods. Both native starch and modified starch have been extensively studied and the developments, as discussed in the previous sections, demonstrate the role of starch in product quality and consumer appeal. Chemical and physical modifications of starch such as acid thinning, oxidation, cross-linking, esterification, pregelatinization, microwave, ultrasound, and high hydrostatic pressure help in overcoming challenges such as retrogradation, syneresis, and shelf stability, improve overall performance of starch under different processing conditions, and provide viscosity control and freeze–thaw stability. Alternative starch sources, such as sweet potato, *kudzu*, sago palm, cassava, and sea-weed based starches have been driving innovation due to the emergence of clean-labeling and sustainable market strategies. Overall, the future of starch-based formulations is dependent on the combination of advanced modification techniques with suitable sustainable practices, thereby ensuring that the food industry continues to meet evolving consumer demands for quality, functionality, and eco-friendly solutions.

References

Ai, Y., & Jane, J.-lin (2015). Gelatinization and rheological properties of starch. *Starch - Stärke, 67*(3–4), 213–224. https://doi.org/10.1002/star.201400201

Alcázar-Alay, S. C., & Meireles, M. A. A. (2015). Physicochemical properties, modifications and applications of starches from different botanical sources. *Food Science and Technology (Campinas), 35*(2), 215–236. https://doi.org/10.1590/1678-457X.6749

Ali, N. A., Dash, K. K., & Routray, W. (2020). Physicochemical characterization of modified lotus seed starch obtained through acid and heat moisture treatment. *Food Chemistry, 319*, Article 126513. https://doi.org/10.1016/j.foodchem.2020.126513

Alimi, B. A., & Workneh, T. S. (2018). Structural and physicochemical properties of heat moisture treated and citric acid modified acha and iburu starches. *Food Hydrocolloids, 81*, 449–455. https://doi.org/10.1016/j.foodhyd.2018.02.001

Almanza-Benitez, S., Osorio-Díaz, P., Méndez-Montealvo, G., Islas-Hernández, J. J., & Bello-Perez, L. A. (2015). Addition of acid-treated unripe plantain flour modified the starch digestibility, indigestible carbohydrate content and antioxidant capacity of semolina spaghetti. *LWT - Food Science and Technology, 62*(2), 1127–1133. https://doi.org/10.1016/j.lwt.2015.02.031

Bühler, J. M., Schlangen, M., Möller, A. C., Bruins, M. E., & van der Goot, A. J. (2022). Starch in plant-based meat replacers: A new approach to using endogenous starch from cereals and legumes. *Starch-Stärke, 74*(1–2), Article 2100157. https://doi.org/10.1002/star.202100157

Bangar, S. P., Scott Whiteside, W., Suri, S., Barua, S., & Phimolsiripol, Y. (2023). Native and modified biodegradable starch-based packaging for shelf-life extension and safety of fruits/vegetables. *International Journal of Food Science and Technology, 58*(2), 862–870. https://doi.org/10.1111/ijfs.16219

Bangar, S. P., Singh Purewal, S., Trif, M., Maqsood, S., Kumar, M., Manjunatha, V., & Rusu, A. V. (2021). Functionality and applicability of starch-based films: An eco-friendly approach. *Foods, 10*(9), 2181. https://doi.org/10.3390/foods10092181

Bangar, S. P., Sunooj, K. V., Navaf, M., Phimolsiripol, Y., & Whiteside, W. S. (2024). Recent advancements in cross-linked starches for food applications: A review. *International Journal of Food Properties, 27*(1), 411–430. https://doi.org/10.1080/10942912.2024.2318427

Bresciani, A., Giuberti, G., Cervini, M., & Marti, A. (2023). Using pre-gelatinised red lentils in pasta production: Connecting starch features and pasta cooking behaviour. *International Journal of Food Science and Technology, 58*(4), 1803–1809. https://doi.org/10.1111/ijfs.16312

Chakraborty, I., Pooja, N., Sib, S. M., Paul, U. C., Hafizur Rahman, Md, & Mazumder, N. (2022). An insight into the gelatinization properties influencing the modified starches used in food industry: A review. *Food and Bioprocess Technology, 15*(6), 1195–1223. https://doi.org/10.1007/s11947-022-02761-z

Chang, Q., Zheng, B., Zhang, Y., & Zeng, H. (2021). A comprehensive review of the factors influencing the formation of retrograded starch. *International Journal of Biological Macromolecules, 186*, 163–173. https://doi.org/10.1016/j.ijbiomac.2021.06.089

Chen, J., Zeng, X., & Chen, L. (2021). Regulation nature of water-choline amino acid ionic liquid mixtures on the disaggregation behavior of starch. *Carbohydrate Polymers, 272*, 118474.

Chen, Q., Zhang, J., Zhang, Y., Kaplan, D. L., & Wang, Q. (2022). Protein-amylose/amylopectin molecular interactions during high-moisture extruded texturization toward plant-based meat substitutes applications. *Food Hydrocolloids, 127*, Article 107559. https://doi.org/10.1016/j.foodhyd.2022.107559

Chinachoti, P., Alberto, S., & Dimitrios, F. (2018). *Mechanism of staling: An overview* (1st ed., pp. 1–18). CRC Press. https://doi.org/10.1201/9781351070348-1. In , edited by Chinachoti Pavinee and Vodovotz Yael.

Chiravi, K., Khan, M. A., Wadikar, D. D., Mahesh, C., & Semwal, A. D. (2023). Starch retrogradation: Role of food ingredients and extrinsic factors. *Food Science and Engineering, 18*, 204–215. https://doi.org/10.37256/fse.4220232939

Choi, S. Y., & Koh, B. K. (2017). Effects of annealing and removal of the water-soluble fraction of dry-milled rice flour on the texture of cooked rice noodles. *Food and Nutrition Sciences, 8*(10), 889–900. https://doi.org/10.4236/fns.2017.810064

Cornejo-Ramírez, Y. I., Martínez-Cruz, O., Toro-Sánchez, C. L. Del, Wong-Corral, F. J., Borboa-Flores, J., & Javier Cinco-Moroyoqui, F. (2018). The structural characteristics of starches and their functional properties. *CyTA - Journal of Food, 16*(1), 1003–1017. https://doi.org/10.1080/19476337.2018.1518343

Cullano, R. A., Tiu, A. M., Evangelista, S. S., & Ocampo, L. (2025). Informing starch-based food product designs with seaweeds using an analytical kano-quality function deployment model. *Journal of Texture Studies, 56*(2). https://doi.org/10.1111/jtxs.70007

Donmez, D., Pinho, L., Patel, B., Desam, P., & Campanella, O. H. (2021). Characterization of starch–water interactions and their effects on two key functional properties: Starch gelatinization and retrogradation. *Current Opinion in Food Science, 39*, 103–109. https://doi.org/10.1016/j.cofs.2020.12.018

Du, J., Yang, Z., Xu, X., Wang, X., & Du, X. (2019). Effects of tea polyphenols on the structural and physicochemical properties of high-hydrostatic-pressure-gelatinized rice starch. *Food Hydrocolloids, 91*, 256–262. https://doi.org/10.1016/j.foodhyd.2019.01.020

Dufour, D., Rolland-Sabaté, A., Cordoba, H. A. M., Luis Luna Melendez, J., Larry Moreno Alzate, J., Pizzaro, M., Dubois, S. G., et al. (2022). Native and fermented waxy cassava starch as a novel gluten-free and clean label ingredient for baking and expanded product development. *Food & Function, 13*(18), 9254–9267.

Egharevba, H. O. (2019). *Chemical properties of starch and its application in the food industry*. IntechOpen eBooks. https://doi.org/10.5772/intechopen.87777

Emeje, M. (2019). Chemical properties of starch. In *IntechOpen eBooks*. https://doi.org/10.5772/intechopen.78119

Gałkowska, D., Kapuśniak, K., & Juszcz, L. (2023). Chemically modified starches as food additives. *Molecules, 28*(22), 7543. https://doi.org/10.3390/molecules28227543

Genccelep, H., Turker Saricaoglu, F., Anil, M., Agar, B., & Turhan, S. (2015). The effect of starch modification and concentration on steady-state and dynamic rheology of meat emulsions. *Food Hydrocolloids, 48*, 135–148. https://doi.org/10.1016/j.foodhyd.2015.02.002

Goel, C., Semwal, A. D., Khan, A., Kumar, S., & Sharma, G. K. (2020). Physical modification of starch: Changes in glycemic index, starch fractions, physicochemical and functional properties of heat-moisture treated buckwheat starch. *Journal of Food Science and Technology, 57*(8), 2941–2948. https://doi.org/10.1007/s13197-020-04326-4

Gupta, R. K., Guha, P., & Srivastav, P. P. (2024). Physical action of nonthermal cold plasma technology for starch modification. *Food Physics, 1*, Article 100011. https://doi.org/10.1016/j.foodp.2024.100011

He, R., Du, C., Li, S., Guo, L., Wang, K., & Yang, L. (2024). Structural-functional analysis of modified kudzu starch as a novel instant powder: Role of modified technology. *Food Chemistry X, 24*, Article 101912. https://doi.org/10.1016/j.fochx.2024.101912

Herrero, M., Hugas, M., Lele, U., Wirakartakusumah, A., & Torero, M. (2023). A shift to healthy and sustainable consumption patterns. *Science and innovations for food systems transformation, 59*.

Herrero, M., Hugas, M., Lele, U., Wirakartakusumah, A., & Torero, M. (2023). *A shift to healthy and sustainable consumption patterns science and innovations for food systems transformation* (pp. 59–85). United States: Springer International Publishing. https://doi.org/10.1007/978-3-031-15703-5

Himashree, P., Sengar, A. S., & Sunil, C. K. (2022). Food thickening agents: Sources, chemistry, properties and applications-A review. *International Journal of Gastronomy and Food Science, 27*, 100468.

Hsieh, C.-F., Liu, W., Whaley, J. K., & Shi, Y.-C. (2019). Structure and functional properties of waxy starches. *Food Hydrocolloids, 94*, 238–254. https://doi.org/10.1016/j.foodhyd.2019.03.026

Hsieh, C.-F., Wang, L.-K., Xu, B., Seib, P. A., & Shi, Y.-C. (2020). Preparation and textural properties of white salted noodles made with hard red winter wheat flour partially replaced by different levels of cross-linked phosphorylated RS4 wheat starch. *Journal of the Science of Food and Agriculture, 100*(15), 5334–5343. https://doi.org/10.1002/jsfa.10581

Hu, A., Liu, Y., & Wu, S. (2024). A review on polysaccharide-based jelly: Gell food. *Food Chemistry: X, 23*, 101562.

Iuga, M., & Mironeasa, S. (2020). A review of the hydrothermal treatments impact on starch based systems properties. *Critical Reviews in Food Science and Nutrition, 60*(22), 3890–3915. https://doi.org/10.1080/10408398.2019.1633095

Jiang, H., Julian McClements, D., Dai, L., Qin, Y., Ji, N., Xiong, L., Qiu, C., & Sun, Q. (2024). Effects of moisture content and retrogradation on structure and properties of indica rice flour and starch gels. *Food Hydrocolloids, 150*, Article 109657. https://doi.org/10.1016/j.foodhyd.2023.109657

Jiang, J., Zeng, J., Gao, H., Zhang, L., Wang, F., Su, T., Xiang, F., & Li, G. (2020). Effect of low temperature on the aging characteristics of a potato starch gel. *International Journal of Biological Macromolecules, 150*, 519–527. https://doi.org/10.1016/j.ijbiomac.2020.02.077

Kashi, R., & Razavi, S. M. A. (2024a). Development of a state-of-art custard formulation: Investigating the rheological, physicochemical, sensory, and in-vitro digestibility properties. *Starch - Stärke, 77*(1). https://doi.org/10.1002/star.202400107

Kashi, R., & Razavi, S.M.A. (2024b). A new dietary low-fat custard formulation made by waxy corn starch and xylitol: Rheological, sensory and in vitro digestibility investigation. *LWT, 203*, Article 116296. https://doi.org/10.1016/j.lwt.2024.116296.

Kim, K.-H., & Kim, J.-Y. (2021). Understanding wheat starch metabolism in properties, environmental stress condition, and molecular approaches for value-added utilization. *Plants, 10*(11), 2282. https://doi.org/10.3390/plants10112282

Korma, S. A. (2016). Chemically modified starch and utilization in food stuffs. *International Journal of Nutrition and Food Sciences, 5*(4), 264. https://doi.org/10.11648/j.ijnfs.20160504.15

Królikowska, K., Červenka, L., Pietrzyk, S., Fortuna, T., & Khachatryan, G. (2019). Relationship between sorption characteristic and selected functional properties of chemically modified waxy maize starches. *Journal of Food Processing and Preservation, 43*(7). https://doi.org/10.1111/jfpp.13948

Institute of Food Resources of NAAS, Kyiv, U., Khomichak, L., Kuznietsova, I., Zajchuk, L., & Yarmolyuk, M. (2021). Study of water absorption capability of swellable starch. *Food Resources, 9*(16), 205–211. https://doi.org/10.31073/foodresources2021-16-19

López-Pedrouso, M., Lorenzo, J. M., Moreira, R., & Franco, D. (2022). Potential applications of pickering emulsions and high internal-phase emulsions (HIPEs) stabilized by starch particles. *Current Opinion in Food Science, 46*, Article 100866. https://doi.org/10.1016/j.cofs.2022.100866

Li, P., Kierulf, A., & Abbaspourrad, A. (2021). Application of granular cold-water-swelling starch as a clean-label oil structurant. *Food Hydrocolloids, 112*, Article 106311. https://doi.org/10.1016/j.foodhyd.2020.106311.

Li, C. (2022). Recent progress in understanding starch gelatinization - An important property determining food quality. *Carbohydrate Polymers, 293*, Article 119735. https://doi.org/10.1016/j.carbpol.2022.119735

Li, Y., Hu, A., Zheng, J., & Wang, X. (2019). Comparative studies on structure and physiochemical changes of millet starch under microwave and ultrasound at the same power. *International Journal of Biological Macromolecules, 141*, 76–84. https://doi.org/10.1016/j.ijbiomac.2019.08.218

Li, S., Zhang, M., Ren, X., Guan, L., Ye, G., & Li, Y. (2024). Improvement in storage stability of fresh instant rice using non-starch polysaccharides. *International Journal of Food Science and Technology, 59*(5), 2961–2970. https://doi.org/10.1111/ijfs.17021

Liao, L., Liu, H., Gan, Z., & Wu, W. (2019). Structural properties of sweet potato starch and its vermicelli quality as affected by heat-moisture treatment. *International Journal of Food Properties, 22*(1), 1122–1133. https://doi.org/10.1080/10942912.2019.1626418

Lu, W.-C., Chan, Y.-J., Li, Z.-E., & Li, P.-H. (2025). Influence of different hydrocolloids on the pasting, rheological, and morphological characteristics of heat gelatinized cassava starch. *Food Chemistry, 463*(Pt 4), Article 140933. https://doi.org/10.1016/j.foodchem.2024.140933

Ma, H., Liu, M., Liang, Y., Zheng, X., Sun, L., Dang, W., Li, J., Li, L., & Li, C. (2022). Research progress on properties of pre-gelatinized starch and its application in wheat flour products. *Grain & Oil Science and Technology, 5*(2), 87–97. https://doi.org/10.1016/j.gaost.2022.01.001

Masure, H. G., Wouters, A. G., Fierens, E., & Delcour, J. A. (2019). Impact of egg white and soy proteins on structure formation and crumb firming in gluten-free breads. *Food Hydrocolloids, 95*, 406–417. https://doi.org/10.1016/j.foodhyd.2019.04.062

Mohamed, A. M. D., Rashid, R. S. A., Nair, S., Alkarkhi, A. F. M., & Shaharuddin, S. (2024). American institute of physics Malaysia sago palm (Metroxylon sagu Rottb.) as a new industrial source for value-added product. *AIP Conference Proceedings, 2923*(1). https://doi.org/10.1063/5.0195491

Nawaz, H., Waheed, R., Nawaz, M., & Shahwar, D. (2020). *Physical and chemical modifications in starch structure and reactivity*. IntechOpen. https://doi.org/10.5772/intechopen.88870

Obadi, M., & Xu, B. (2020). Review on the physicochemical properties, modifications, and applications of starches and its common modified forms used in noodle products. *Food Hydrocolloids, 112*, Article 106286. https://doi.org/10.1016/j.foodhyd.2020.106286

Oderinde, A. A., Ahmed Ibikunle, A., Bakre, L. G., & Adesola Babarinde, N. A. (2020). Modification of African breadfruit (Treculia africana, Decne) kernel starch: Physicochemical, morphological, pasting, and thermal properties. *International Journal of Biological Macromolecules, 153*, 79–87. https://doi.org/10.1016/j.ijbiomac.2020.02.293

Oh, S.-M., Kim, H.-yun, Bae, J.-E., Ye, S.-J., Kim, B.-Y., Choi, H. D., Choi, H.-W., & Baik, M.-Y. (2019). Physicochemical and retrogradation properties of modified chestnut starches. *Food Science and Biotechnology, 28*(6), 1723–1731. https://doi.org/10.1007/s10068-019-00622-8

Okunlola, A., & Abidemi Adewusi, S. (2019). Development of theophylline microbeads using pregelatinizedbreadfruit starch (Artocarpus altilis) as a novel co-polymer for controlled release. *Advanced Pharmaceutical Bulletin, 9*(1), 93–101. https://doi.org/10.15171/apb.2019.012

Patricio, P. S.de O., de Souza, P. P., Couto, M. L. S. D., & Rocha Coura, Í. (2024). *The polymeric carbohydrate: Starch – Extraction and modifications* (pp. 775–807). Springer Science and Business Media LLC. https://doi.org/10.1007/978-981-99-6727-8_31

Pellegrini, N., Vittadini, E., & Fogliano, V. (2020). Designing food structure to slow down digestion in starch-rich products. *Current Opinion in Food Science, 32*, 50–57. https://doi.org/10.1016/j.cofs.2020.01.010

Pietrasik, Z., & Soladoye, O. P. (2021). Use of native pea starches as an alternative to modified corn starch in low-fat bologna. *Meat Science, 171*, Article 108283. https://doi.org/10.1016/j.meatsci.2020.108283

Qiu, L., Hu, F., & Peng, Y. (2013). Structural and mechanical characteristics of film using modified corn starch by the same two chemical processes used in different sequences. *Carbohydrate Polymers, 91*(2), 590–596. https://doi.org/10.1016/j.carbpol.2012.08.060

Rashwan, A. K., Younis, H. A., Abdelshafy, A. M., Osman, A. I., Eletmany, M. R., Hafouda, M. A., & Chen, W. (2024). Plant starch extraction, modification, and green applications: A review. *Environmental Chemistry Letters, 22*(5), 2483–2530. https://doi.org/10.1007/s10311-024-01753-z

Reyniers, S., De Brier, N., Matthijs, S., Brijs, K., & Delcour, J. A. (2019). Impact of mineral ions on the release of starch and gel forming capacity of potato flakes in relation to water dynamics and oil uptake during the production of snacks made thereof. *Food Research International, 122*(March), 419–431. https://doi.org/10.1016/j.foodres.2019.03.06

Rincón-Aguirre, A., Bello Pérez, L. A., Mendoza, S., del Real, A., & Rodriguez Garcia, M. E. (2018). Physicochemical studies of taro starch chemically modified by acetylation, phosphorylation, and succinylation. *Starch-Stärke, 70*(3-4), 1700066.

Šárka, E., & Dvořáček, V. (2017). Waxy starch as a perspective raw material (a review). *Food Hydrocolloids, 69*, 402–409. https://doi.org/10.1016/j.foodhyd.2017.03.001

Santamaria, M., Montes, L., Garzon, R., Moreira, R., & Rosell, C. M. (2022). Unraveling the impact of viscosity and starch type on the in vitro starch digestibility of different gels. *Food & Function, 13*(14), 7582–7590. https://doi.org/10.1039/d2fo00697a

Schmiele, M., Sampaio, U. M., Gomes, P. T. G., & Pedrosa Silva Clerici, M. T. (2018). *Physical modifications of starch starches for food application: Chemical, technological and health properties*. Brazil: Elsevier. https://doi.org/10.1016/B978-0-12-809440-2.00006-X

Scott, G., Awika, J. M., et al. (2023). Effect of protein–starch interactions on starch retrogradation and implications for food product quality. *Comprehensive Reviews in Food Science and Food Safety*. https://doi.org/10.1111/1541-4337.13141

Shaikh, M., Ali, T. M., & Hasnain, A. (2017). Effects of different modification reagents on functional properties of pearl millet starches. *Starch-Stärke, 69*(3–4), Article 1600142. https://doi.org/10.1002/star.201600142

Thakur, R., Pristijono, P., Scarlett, C. J., Bowyer, M., Singh, S., & Vuong, Q. V. (2019). Starch-based films: Major factors affecting their properties. *International Journal of Biological Macromolecules, 132*, 1079–1089. https://doi.org/10.1016/j.ijbiomac.2019.03.190

Tian, Y., Liu, X., Kirkensgaard, J. J. K., Khakimov, B., Enemark-Rasmussen, K., Hebelstrup, K. H., Blennow, A., & Zhong, Y. (2023). Characterization of different high amylose starch granules. Part I: Multi-scale structures and relationships to thermal properties. *Food Hydrocolloids, 146*, Article 109286. https://doi.org/10.1016/j.foodhyd.2023.109286

Tiozon, R. N., Bonto, A. P., & Sreenivasulu, N. (2021). Enhancing the functional properties of rice starch through biopolymer blending for industrial applications: A review. *International Journal of Biological Macromolecules, 192*, 100–117. https://doi.org/10.1016/j.ijbiomac.2021.09.194

Varghese, S., Awana, M., Mondal, D., Rubiya, M. H., Melethil, K., Singh, A., Krishnan, V., & Thomas, B. (2022). *Amylose–amylopectin ratio* (pp. 1–30). Springer Science and Business Media LLC. https://doi.org/10.1007/978-981-16-6603-2_48-1

Wandee, Y., Uttapap, D., Puncha-arnon, S., Puttanlek, C., Rungsardthong, V., & Wetprasit, N. (2015). Quality assessment of noodles made from blends of rice flour and canna starch. *Food Chemistry, 179*, 85–93. https://doi.org/10.1016/j.foodchem.2015.01.119

Wang, S., Li, C., Copeland, L., Niu, Q., & Wang, S. (2015). Starch retrogradation: A comprehensive review. *Comprehensive Reviews in Food Science and Food Safety, 14*(5), 568–585. https://doi.org/10.1111/1541-4337.12088

Wang, J., Li, Y., Guo, X., Zhu, K., & Wu, Z. (2024). A review of the impact of starch on the quality of wheat-based noodles and pasta: From the view of starch structural and functional properties and interaction with gluten. *Foods, 13*(10), 1507. https://doi.org/10.3390/foods13101507

Wang, S., Wang, J., Liu, Y., & Liu, X. (2020). *Starch modification and application*. Springer Science and Business Media LLC. https://doi.org/10.1007/978-981-15-0622-2_8

Wani, I. A., Singh Sogi, D., Hamdani, A. M., Gani, A., Ahmad Bhat, N., & Shah, A. (2016). Isolation, composition, and physicochemical properties of starch from legumes: A review. *Starch - Stärke, 68*(9–10), 834–845. https://doi.org/10.1002/star.201600007

Wertz, J., & Goffin, B. (2020). Starch in the bioeconomy. https://doi.org/10.1201/9781003111986.

Wertz, J.-L., & Goffin, B. (2020). *Starch in the bioeconomy*. CRC Press. https://doi.org/10.1201/9781003111986

Xu, M., Saleh, A. S. M., Gong, B., Li, B., Jing, L., Gou, M., Jiang, H., & Li, W. (2018). The effect of repeated versus continuous annealing on structural, physicochemical, and digestive properties of potato starch. *Food Research International, 111*, 324–333. https://doi.org/10.1016/j.foodres.2018.05.052

Yamaguchi, Y., Okawa, Y., Ninomiya, K., Kumagai, H., & Kumagai, H. (2019). Evaluation and suppression of retrogradation of gelatinized rice starch. *Journal of Nutritional Science and Vitaminology, 65*(Suppl. ment), S134–S138. https://doi.org/10.3177/jnsv.65.S134

Yang, S., Dhital, S., Zhang, M. N., Wang, J., & Chen, Z. G. (2022). Structural, gelatinization, and rheological properties of heat-moisture treated potato starch with added salt and its application in potato starch noodles. *Food Hydrocolloids, 131*, 107802.

Yazid, N. S. M., Abdullah, N., Muhammad, N., & Matias-Peralta, H. M. (2018). Application of starch and starch-based products in food industry. *Journal of Science and Technology, 10*(2). https://doi.org/10.30880/jst.2018.10.02.023

Yu, J., & Moon, Y. (2021). Corn starch: Quality and quantity improvement for industrial uses. *Plants, 11*(1), 92. https://doi.org/10.3390/plants11010092

Yusuf, F., Kubo, A., Abdulrashid, F., Madu, S., & Muazu, J. (2023). Studies on the physicochemical properties of coprocessed starch obtained from *Ipomoea batatas* Nigerian. *Journal of Basic & Applied Sciences, 30*(2), 35–41. https://doi.org/10.4314/njbas.v30i2.5

Zaman, S. A., & Sarbini, S. R. (2015). The potential of resistant starch as a prebiotic. *Critical Reviews in Biotechnology, 35*(1), 1–7. https://doi.org/10.3109/07388551.2014.993590

Zeng, H., Chen, P., Chen, C., Huang, C., Lin, S., Zheng, B., & Zhang, Y. (2018). Structural properties and prebiotic activities of fractionated lotus seed resistant starches. *Food Chemistry, 251*, 33–40. https://doi.org/10.1016/j.foodchem.2018.01.050

Zeng, F., Li, T., Gao, Q., Liu, B., & Yu, S. (2018). Physicochemical properties and in vitro digestibility of high hydrostatic pressure treated waxy rice starch. *International Journal of Biological Macromolecules, 120*, 1030–1038. https://doi.org/10.1016/j.ijbiomac.2018.08.076

Zhang, X., Ma, Q., Liu, X., Zhang, D., Ma, L., Luo, D.-lin, & Liu, X. (2020). Effect of microwave irradiation on the pasting, thermal, and rheological properties of cassava starch–sugar mixtures. *Journal of Food Process Engineering, 43*(8). https://doi.org/10.1111/jfpe.13431

Zhang, R., Wang, Z., He, X., Xu, Z., Li, M., Liu, X., Xiong, L., & Sun, Q. (2025). Effects of sweet potato starch on the physicochemical properties and edible qualities of instant fresh rice noodles. *International Journal of Biological Macromolecules, 286*, Article 138553. https://doi.org/10.1016/j.ijbiomac.2024.138553

Zhang, S., Wang, L., Wang, Q., Wang, Y., Wang, L., & Du, R. (2024). Effects of different types of starch on physicochemical properties and microstructure of beef during cold storage. *Foods, 13*(17), 2767. https://doi.org/10.3390/foods13172767

Zhu, F. (2017). Structures, properties, and applications of lotus starches. *Food Hydrocolloids, 63*, 332–348. https://doi.org/10.1016/j.foodhyd.2016.10.015

Zia-Ud-Din, N., Xiong, H., & Fei, P. (2015). Physical and chemical modification of starches: A review. *Critical Reviews in Food Science and Nutrition, 57*(12), 2691–2705. https://doi.org/10.1080/10408398.2015.1087379

Ziobro, R., Korus, J., Witczak, M., & Juszczak, L. (2012). Influence of modified starches on properties of gluten-free dough and bread. Part II: Quality and staling of gluten-free bread. *Food Hydrocolloids, 29*(1), 68–74. https://doi.org/10.1016/j.foodhyd.2012.02.009

Chapter 18

Utilization of starches for bioactive delivery systems and flavor processing

Sohel Rana Sheakh, Ulaganathan Gnana Moorthy Eswaran and M. Maria Leena

School of Biosciences and Technology, Faculty of Engineering and Technology, SRM Institute of Science and Technology, Tiruchirappalli, Tamil Nadu, India

1. Introduction

There is a growing interest in the production and utilization of food containing bioactive compounds, as natural functional ingredients are used to enhance health-promoting attributes (Alongi & Anese, 2021; Venkatakrishnan et al., 2019). A diverse array of pharmaceuticals, dietary supplements, and functional food components is increasingly significant in the prevention of prevalent chronic diseases such as diabetes and hypertension (Li et al., 2021; Selvasekaran & Chidambaram, 2021). As health consciousness increases, the trend of creating meals with functional ingredients has become more prevalent in the food industry (Beikzadeh et al., 2020; Xiong et al., 2020). Natural bioactive food components, while beneficial to health, are highly susceptible to environmental factors such as oxygen, humidity, temperature, and pH, which can result in degradation or loss of efficacy, thereby constraining their use in the food industry (Charoux et al., 2019; Comunian et al., 2021; Xiong et al., 2020).

Starch is a naturally occurring, renewable polymer that is prevalent in nature and has numerous applications in food industry (Jobling, 2004; Tiozon et al., 2021a). The aforementioned packaging material offers robust protection for several substances, including herbicides, proteins, bacteria, and probiotics (Kumari et al., 2020; Miao et al., 2021). The widespread usage of starch, attributed to its omnipresence, affordability, and functional characteristics—such as water retention and viscosity—underpins its application in encapsulation (Leyva-López et al., 2019). Starch has been explored as encapsulating agent for various bioactive and drug compounds as an effective delivery system Fig. 18.1 (Montoya-Yepes et al., 2024). Starch-derived coating agents, including cyclodextrin, amylose, and octenyl succinic anhydride (OSA)-modified starch, are utilized to impart a diverse array of advantageous components to food products. Furthermore, precise modification point adjustments facilitate the development of novel starch-based wall materials, hence expanding their applicability and versatility. In bioactive delivery, starch functions to safeguard active compounds, regulate their release, and mitigate undesirable sensory attributes such as bitterness and acidity (Tian et al., 2022). The inherent low solubility of starch and its incapacity to retain water restrict its practical applications; however, contemporary research demonstrates that structural modifications such as altering aggregation and chain conformations, augmenting the amylose weight fraction, or incorporating lipophilic groups can enhance its solubility and aqueous affinity, thereby rendering starch a more viable encapsulation material (Rehman et al., 2021; Xiao et al., 2019). This chapter discuss recent developments in starch-based encapsulation systems for effective delivery of bioactives and flavor compounds, emphasizing their classifications, encapsulation methods, and practical uses in the food industry.

2. Starch sources and characteristics

Starch is a naturally occurring polysaccharide that is abundant, economical, biodegradable, non-toxic, and renewable. It serves as an energy reservoir in plants and a crucial element in the cosmetics, pharmaceutical, and notably the food sectors (Guo et al., 2021). Starch comprises two components: amylose (20%–30%) and amylopectin (70%–80%) Fig. 18.2. Amylose is a linear molecule composed of D-glucopyranose units linked by α-1, 4-glycosidic bonds, with some α-1, 6-glycosidic connections present. Amylopectin, conversely, is an extensively branched polymer containing around 5%–6% α-1, 6-glycosidic bonds. It can be classified as A, B, or C chains based on the number of branches(Hsieh et al., 2019). The

Native and Modified Starches for Food Applications. https://doi.org/10.1016/B978-0-443-27431-2.00018-3

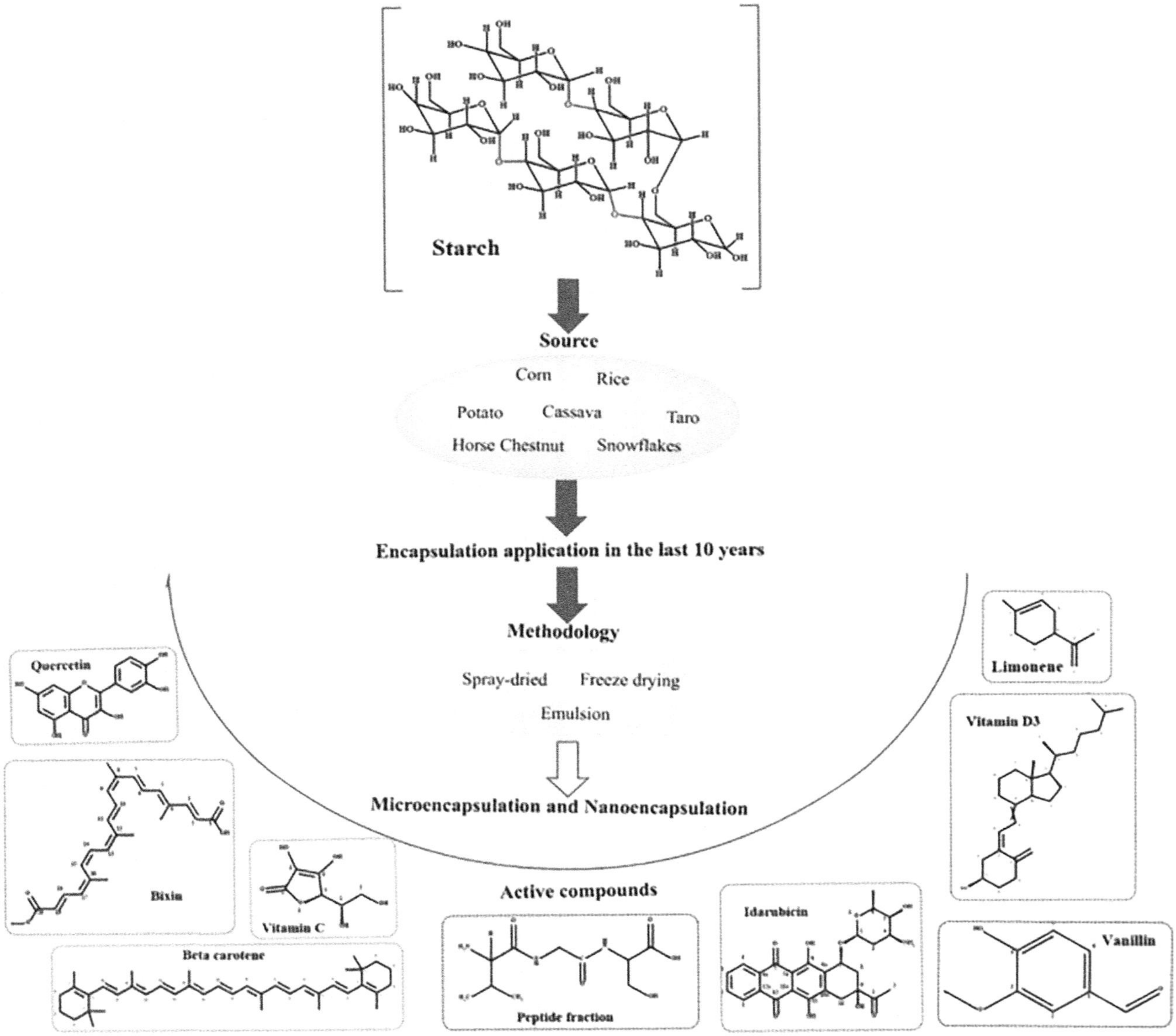

FIGURE 18.1 Schematic representation of starch encapsulation (Montoya-Yepes et al., 2024). *Adapted from Montoya-Yepes, D. F., A. A. Jiménez-Rodríguez, A. E. Aldana-Porras, L. F. Velásquez-Holguin, J. J. Méndez-Arteaga and W. Murillo-Arango (2024). "Starches in the encapsulation of plant active ingredients: state of the art and research trends." Polymer Bulletin 81(1): 135-163.*

size and form of starch granules vary depending on their plant source, influencing their possible use. Starch granules range in diameter from 1 to 100 μm (Koksel et al., 2023; Wang & Guo, 2020). The amalgamation of amylose and amylopectin molecules produces water-insoluble starch granules characterized by layered amorphous and semi-crystalline areas, measuring 100–400 nm in thickness (Li et al., 2021; Wen et al., 2020). The extent of amorphousness and crystallinity differs among sources, affecting digestion via gelatinization and retrogradation. The relative crystallinity of purple yam starch is 45.5%, compared to 33.1% for wheat starch (Lu & Wang, 2022).

3. Type of starch used as encapsulating wall material

3.1 Native starch

Native starch, a widely utilized natural polymer in encapsulation, offers high entrapment efficiencies, efficiently facilitates the intelligent encapsulation of volatile substances, serves as a lipid alternative, and stabilizes emulsions. Consequently, it is

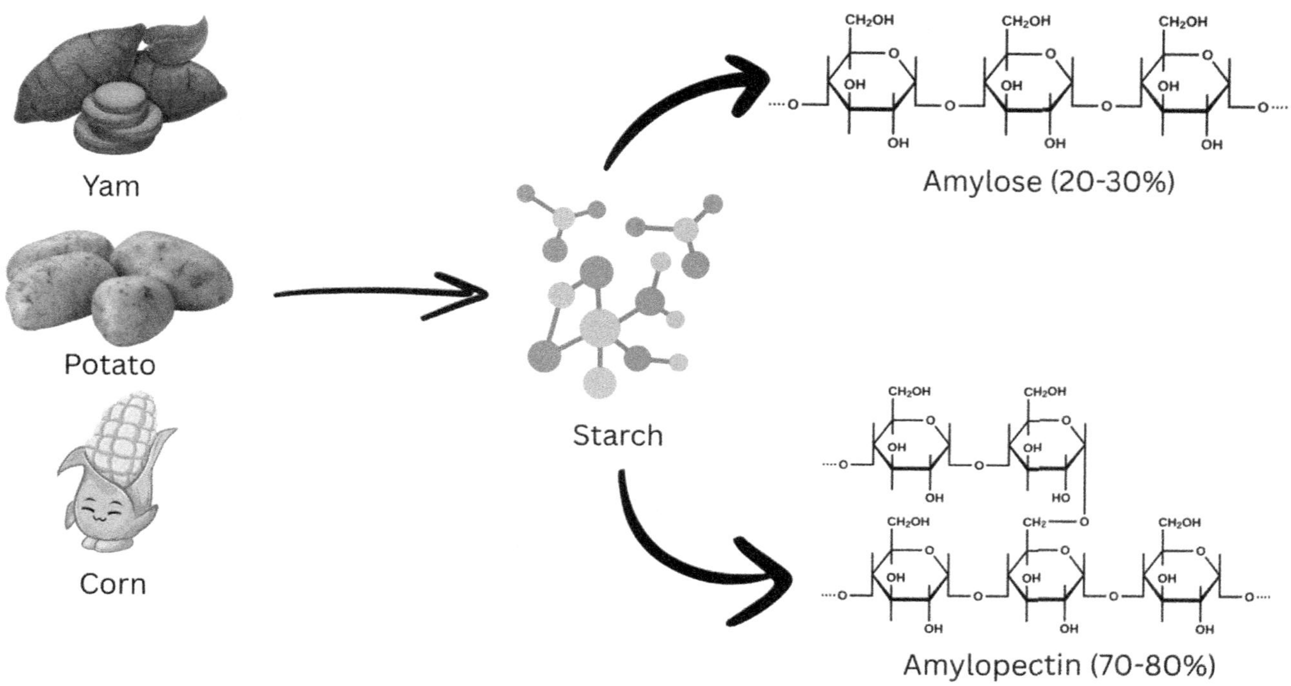

FIGURE 18.2 Composition of starch.

particularly relevant in sectors necessitating biodegradability, food safety, and pharmaceutical-grade standards (Montoya-Yepes et al., 2024). Microencapsulation is the process of enclosing solid or liquid substances within an exterior covering, thereby forming a protective shell that protects the interior from the external environment. In this configuration, the active components constitute the core, encased by a chemically inert, nonreactive barrier referred to as a microcapsule, often measuring less than 1000 μm in diameter (Augustin et al., 2014; Gangurde et al., 2017). The efficacy of encapsulation is contingent upon various parameters, including the encapsulation method, the physicochemical properties of the encapsulation medium—typically a natural polymer—and the attributes of the encapsulated bioactive compound. These systems enhance the bioavailability of functional elements by addressing issues that impede their bioavailability, such as restricted intestinal absorption, metabolite-induced clearance, and fast bioavailability. Numerous microencapsulation techniques exist, primarily distinguished by the interaction between the wall material and core, encompassing chemical ways (e.g., suspension and emulsion polymerization) and physical methods (e.g., solvent evaporation and spray drying). Although starches play crucial roles in encapsulation, native starches exhibit considerable drawbacks, including inadequate solubility in aqueous and organic solvents, retrogradation, elevated viscosity, and restricted thermal stability. Consequently, structural modifications necessitate investigation to address these shortcomings (Montoya-Yepes et al., 2024).

3.2 Modified starches as encapsulating materials

Modified starches are described as altered starches that have undergone chemical or physical transformations to improve their functional properties for specific applications. Starch modification is carried out by physical, chemical, or enzymatic means. The majority of physical modification involves high temperature or high pressure processes. Chemical changes, such as acetylation and hydroxypropylation, markedly enhance the swelling capacity and light transmission of starches, hence improving their efficacy across various applications. Conversely, cross-linking typically reduces swelling and solubility, with the magnitude of these effects contingent upon the cross-linking reagent and the degree of alteration. The incorporation of acetyl groups into the starch polymer disrupts the crystalline structure due to steric hindrance, leading to molecular repulsions that facilitate water penetration into amorphous regions, thereby augmenting swelling (Mayilvahanan et al., 2019; Rincón Aguirre et al., 2018). Both physical and chemical alteration methodologies encounter significant obstacles in pharmaceutical and cosmetic applications due to exposure to severe alkaline or acidic environments and hazardous substances. These interactions frequently yield enantiomeric by-products that may be teratogenic, compromise biocompatibility, and induce adverse consequences, including toxicity, hypersensitivity, cytotoxicity, and bioaccumulation (Adak & Banerjee, 2016; Alissandratos & Halling, 2012). The alteration of polysaccharide chains via acetylation and acylation with organic acids or their respective molecules represents one of the most versatile modifications (Zarski et al., 2016). These alterations affect the water affinity of the polymer, resulting in significant modifications to its mechanical strength and thermal properties (Montoya-Yepes et al., 2024).

3.3 Polymer combinations for improved wall structures

Agri-food industrial application of starch is high due to its hypoallergenic properties, however its poor solubility, lower resistant starch, reduced retrogradation limits in functionality for its application. Physico-chemical properties of starch can be improvized with endo/exogenous mixing of biopolymers, synthetic polymers and plasticizers, leading to functional blends or composites in developing starch with enhanced functional attributes (Tiozon et al., 2021b). The encapsulation efficiency and optimization of regulated distribution of active substances in starch based encapsulates can be enhanced when combined with various biopolymers, such as gums, proteins, and carbohydrates (Montoya-Yepes et al., 2024). Hydrolyzed starch is primarily utilized as a construction material or wall structure due to its appealing attributes, including affordability, lack of flavor, low viscosity, and capacity to safeguard food contents from oxidation during microencapsulation. Nonetheless, its primary deficiency lies in its inadequate emulsifying properties. Hence, hydrolyzed starch is typically utilized alongside proteins such as sodium caseinate and whey protein concentrate, modified starches, and surfactant biopolymers for the microencapsulation of fish oils (Hasani et al., 2015). Cassava starch when combined with gum Arabic and whey protein demonstrated the capacity to encapsulate limonene at concentrations above 40% through spray drying. This approach resulted in the creation of uniform, crack-free capsules, indicating that the encapsulation system and the matrices employed provide sufficient protection for the active ingredient. The inclusion of starch is essential for preserving the crystalline structure of gum Arabic, stabilizing whey protein concentrate, and improving the retention of the encapsulated component. While the gum Arabic mixture exhibited superior encapsulation capacity compared to the wheat protein alone, it demonstrated reduced stability under accelerated storage conditions relative to the cassava starch-whey protein concentrate blend, which offered enhanced shelf-life extension potential (Ordoñez & Herrera, 2014). The encapsulation of proteinaceous compounds, such as oyster peptides, utilizing a blend of alginate, chitosan, and starch represents an innovative method via an emulsion-gelation mechanism, conducted either externally or internally. The incorporation of starch effectively enhances the swelling capacity of microcapsules, leading to increased water absorption. This is closely linked to peptide release rates, which illustrate an effective release control mechanism applicable in various contexts (Zhang et al., 2009). The advancement of biopolymer-based delivery systems currently enables the co-encapsulation of several natural chemicals, significantly improving the efficacy of conventional encapsulation methods. The dual delivery of zinc and an antioxidant compound (yerba mate extract) was achieved by first incorporating zinc in a starch carrier followed by its entrapment into the calcium alginate beads with yerba mate extract. The beads exhibited a strong protection of antioxidants and free radial scavenging activity, making it promising approach when interaction between compounds should be avoided in functional food formulation (López-Córdoba et al., 2014).

4. Different encapsulation techniques

Starch based encapsulation can be executed using several techniques like spray drying, freeze drying, spray freeze drying, electrospinning/spraying, coating, coacervation, extrusion as shown in Fig. 18.3, with the source of core and wall materials affecting encapsulation efficacy.

4.1 Physical method

4.1.1 Microwave

Microwave-assisted encapsulation using starch-based materials is an innovative and eco-friendly technology designed to protect sensitive bioactive compounds, improve shelf life, and provide controlled release. This method is prioritized in modern food science due to its ability to process materials in an "enormously short time" with low energy requirements compared to conventional techniques (Aldana Heredia et al., 2024; Mohd Nawi, Muhamad and Mohd Marsin, 2015). For microwave-assisted processes, several forms are commonly utilized: Maltodextrin is a starch derivative valued for its low viscosity at high solid ratios and its ability to act as an effective binder and stabilizer. It helps minimize "stickiness" during processing, which is a common issue in microwave-assisted drying (Mohd Nawi, Muhamad and Mohd Marsin, 2015). Specialized starches, such as Octenyl Succinic Anhydride (OSA) modified faba bean starch, can be produced rapidly using microwave energy in seconds rather than the hours required by conventional methods. These starches exhibit high

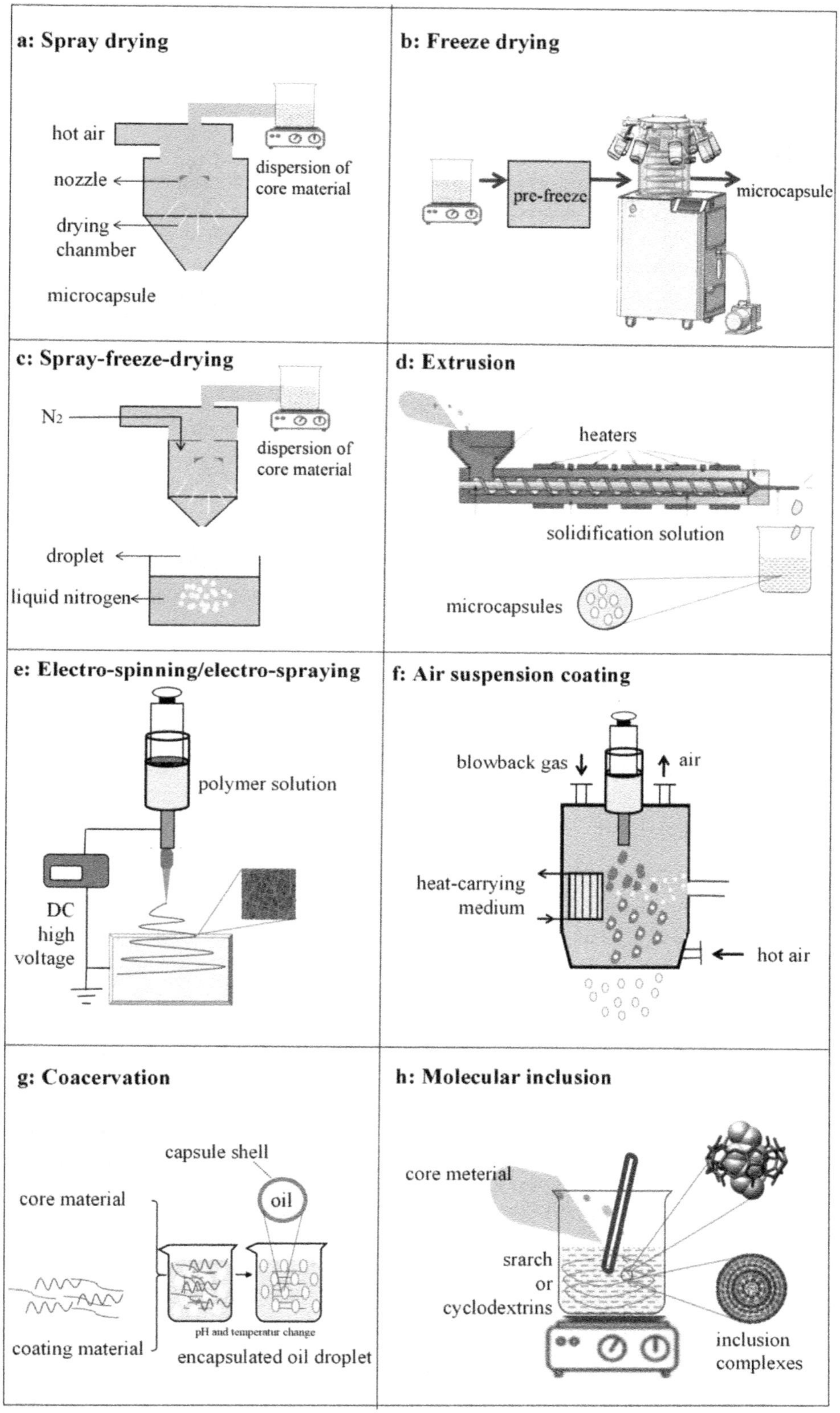

FIGURE 18.3 Schematic diagram of some common encapsulation methods (Shan et al., 2025). *Adapted from Shan, Y., Li, J., Nie, M., Li, D., Zhang, Y., Li, Y., Wang, L., Liu, L., Wang, F. & Tong, L.-T. (2025). A comprehensive review of starch-based technology for encapsulation of flavor: From methods, materials, and release mechanism to applications. Carbohydrate Polymers, 348. https://doi.org/10.1016/j.carbpol.2024.122816*

solubility and low viscosity, making them excellent protective matrices for probiotics like *Lactobacillus rhamnosus GG* (González Mendoza et al., 2025). Natural starches (potato, tapioca, waxy corn) can serve as microcapsules when short microwave heating causes the granules to swell and form an amorphous matrix that traps guest molecules without chemically reacting with them (Kapuśniak & Tomasik, 2006). β-Cyclodextrin is capable of molecular encapsulation, this derivative is highly suitable for entrapping temperature-sensitive food bioactives due to its ability to form inclusion complexes. The technique relies on the interaction between materials and an electromagnetic field, governed by their dielectric properties. The process often exploits differences in the dielectric constants of the core and shell materials, which should ideally differ by a factor of nearly 10. Microwave radiation selectively heats the core material with the higher dielectric constant, causing it to indirectly fuse or dry the surrounding shell material (Abbasi & Rahimi, 2008; Mohd Nawi, Muhamad and Mohd Marsin, 2015). For starch to be an effective wall material in this system, its loss tangent ($\tan\delta$) should generally be higher than 0.1 at 2.45 GHz to ensure proper energy dissipation and heat generation. The quality of microwave-encapsulated starch powders is evaluated through several metrics: Effective combinations, such as maltodextrin mixed with gum arabic, produce powders with low moisture content and water activity, which are crucial for inhibiting microbial growth (Mohd Nawi, Muhamad and Mohd Marsin, 2015). Starch-based microcapsules produced by microwave drying often exhibit a smooth surface and flake-like structure (Mohd Nawi, Muhamad and Mohd Marsin, 2015). In other instances, such as with OSA-modified faba bean starch, the resulting microcapsules are shriveled and spherical, measuring between 5 and 20 μm (González Mendoza et al., 2025). The technology successfully shields thermolabile compounds—such as anthocyanins and probiotics—from environmental stressors like oxygen and light (Aldana Heredia et al., 2024; Mohd Nawi, Muhamad and Mohd Marsin, 2015). Microwave technology has been successfully applied to encapsulate a variety of substances such as fatty acids in granular starches can range from 11% to 94% (Kapuśniak & Tomasik, 2006), citric acid crystals using hydrocolloids to prevent premature reactions with other food ingredients (Abbasi & Rahimi, 2008) and enhancing the survival rates of probiotics during stressful processing conditions, with encapsulation efficiencies reaching up to 90% (González Mendoza et al., 2025).

4.1.2 Spray drying

Starch-based encapsulation by spray drying is a widely utilized industrial technique that accounts for approximately 80%–90% of encapsulated products in the flavor industry (Chen et al., 2025). Starch is a primary candidate for this process because it is low-cost, biodegradable, non-toxic, and "Generally Recognized as Safe" (GRAS) (Lopes et al., 2025a). The process involves converting a liquid slurry or emulsion into a powder by atomizing the feed into a hot drying medium, which rapidly evaporates the moisture to form solid microparticles (Mohamad et al., 2018). Schematic of spray drying process is shown in Fig. 18.4.

Native starches are naturally hydrophilic and often possess a high viscosity that limits their effectiveness as carriers for hydrophobic flavor oils (Verdalet-Guzmán et al., 2013). The efficiency of the encapsulation process is heavily dependent on optimizing drying conditions and emulsion characteristics. Ordinary spray drying (OSD) typically uses high inlet temperatures (150–250 °C) that may degrade heat-sensitive compounds. Electrostatic spray drying (ESD) is an advanced alternative that uses electrostatic forces to facilitate drying at much lower temperatures (e.g., 85 °C), leading to superior aroma retention and smaller particle sizes (Chen et al., 2025). The concentration of total solids in the feed emulsion—usually between 20% and 40%—must be carefully balanced; excessive wall material can hinder solubility, while insufficient solids may lead to poor core protection. Starch is frequently complexed with other materials like gum arabic, maltodextrin, or inulin to improve film-forming capabilities, oxidative stability, and encapsulation efficiency (Ahmadi et al., 2025; Chen et al., 2025).

Ideal starch microcapsules are nearly spherical with smooth, dense surfaces, though rapid water evaporation can sometimes cause surface indentations or wrinkles. Internally, these structures often appear as a core-shell or multicapsule matrix where the starch forms a protective barrier around the core. Optimized starch-based systems can achieve efficiency levels exceeding 80%–90%, successfully protecting core materials from environmental stresses such as UV light, high salinity, and extreme pH (Lopes et al., 2025). Starch encapsulation significantly extends the shelf-life of volatile oils and sensitive bioactives, such as thymoquinone, by preventing degradation during storage (Abedi et al., 2021). These delivery systems are critical in the food, pharmaceutical, and cosmetic industries for producing stable powders from vegetable and marine oils that would otherwise be prone to oxidation and unpleasant tastes.

4.1.2.1 High pressure homogenization

High-pressure homogenization produces significant shear, impact, and cavitation forces by propelling liquid through constricted openings in the homogenizer valve. This procedure breaks starch granules, solubilizes amylose, and facilitates the

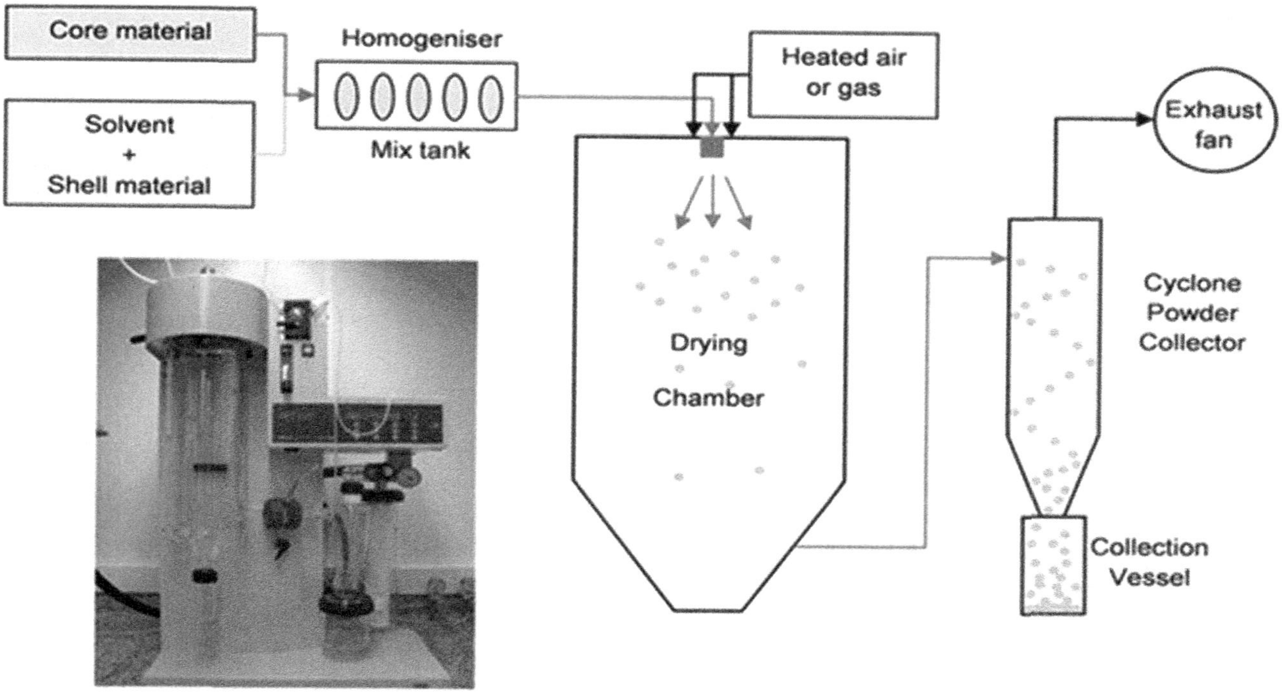

FIGURE 18.4 Schematic of spray drying process as encapsulation technique adopted from (Chávarri, Marañón and Villarán, 2012). *Adapted from Chávarri, M., Marañón, I. & Villarán, M. C. (2012). Chávarri, M., I. Marañón and M. C. Villarán (2012). Encapsulation technology to protect probiotic bacteria. Probiotics, IntechOpen.*

formation of inclusion complexes with guest molecules (Zhao et al., 2019). To fabricate starch-based nano-microcapsules, peanut starch and lysolecithin were subjected to high-pressure homogenization. The technique resulted in a V-type crystalline structure, which improved the starch's ability to retain water by enthalpy and reduced its viscosity. Moreover, homogenization markedly improved the capacity of lysophosphatidylcholine to engage with peanut starch, hence diminishing the occurrence of syneresis in the resultant nano-microcapsules. These characteristics enhance texture and stability, rendering homogenized starch nano-microcapsules applicable in frozen and dessert goods (Oyeyinka et al., 2016).

4.1.3 Ultrasonic waves

Ultrasound technology has emerged as a high-yield, rapid physical modification tool that enhances starch-based encapsulation without the need for complex chemical purification (Zhu, 2017). Operating at frequencies above the human hearing threshold (>16 kHz), the technology relies on acoustic cavitation. This phenomenon involves the formation and violent collapse of microbubbles, generating localized high pressure, high temperature, and intense shear forces (Remanan & Zhu, 2021). In starch systems, these forces break macromolecular C–C bonds to form free radicals and eliminate air within granules, facilitating the entry of bioactive molecules into internal channels (Li et al., 2022). Ultrasound is used to produce starch nanoparticles (SNPs) through two primary methodologies - Top-down methods involve the mechanical reduction of large starch granules into nanoscale particles, a process that can take several hours. Bottom-up approaches utilize ultrasonic irradiation to drive the autonomous self-assembly of molecular interactions (Liu et al., 2017). These ultrasound-generated SNPs serve as edible delivery vehicles for hydrophobic compounds such as rutin, curcumin, and vitamin D3 (Rehman et al., 2020). Rutin-loaded SNPs prepared from quinoa and maize starch via ultrasound showed high encapsulation efficiency (63.1%–67.4%) and improved bioavailability due to the formation of hydrogen bonds between the polyphenol and starch chains. Ultrasonic emulsification of cheese aroma in modified starch resulted in significantly higher retention of volatile fatty acids (94.3%) compared to mechanical methods (83.3%) (Mongenot et al., 2000). Starch-based Pickering emulsions are stabilized by solid starch particles at the oil-water interface (Ramos et al., 2025). Ultrasound-assisted sonoprecipitation can produce modified SNPs that stabilize high internal phase emulsions (HIPEs) for carrying bioactives like ferulic acid (Rehman et al., 2020). Research on bixin (annatto extract) shows that while low-power ultrasound (150 W) can achieve an encapsulation efficiency of 92.6% in a high-amylose corn starch matrix, excessive power can actually decrease efficiency due to re-coalescence.

A novel "green" strategy involves the synergy between ultrasound and enzymatic treatments to construct porous starch (PS) carriers. The cavitation effect of ultrasound roughens the starch surface and creates depressions, which promote directed enzymatic action by α-amylase and glucoamylase. This synergistic approach can shorten reaction times by 50% and reduce enzyme dosage by 20%. Carriers such as dual-modified corn starch have successfully encapsulated gentiopicroside, achieving an encapsulation efficiency of 77.8 9% (Yang et al., 2025). Encapsulation via ultrasound technology significantly enhances the bio-efficacy and physicochemical stability of food ingredients (Remanan & Zhu, 2021). Ultrasound-assisted encapsulation provides a physical barrier that protects sensitive compounds like betalains and gentiopicroside from degradation during heating or UV exposure (Benhabiles et al., 2025; Yang et al., 2020). These systems often demonstrate a biphasic release pattern, resisting the acidic environment of the stomach but allowing for accelerated release in the small intestine due to enzymatic degradation of the starch wall (Remanan & Zhu, 2021; Yang et al., 2025). Sonicated microparticles have been shown to exhibit significantly higher radical scavenging activity than unsonicated controls, confirming that ultrasound promotes stronger core-shell interactions (Benhabiles et al., 2025).

4.1.3.1 Ultra-high pressure (UHP)

UHP treatment is a non-thermal processing technique that utilizes water to apply pressure. This approach subjects sample materials to either ambient or low temperature settings while applying pressures that surpass 100 MPa. The elevated pressure compresses the food matrix macromolecules, so altering molecular interactions, including hydrophobic forces, ionic bonds, and hydrogen bonds, resulting in structural and functional transformations (Błaszczak et al., 2015). In a separate publication, SFPLotus seed starch and fatty acid nano-microcapsules were synthesized using the UHP method; a solution of starch and fatty acid in 10% ethanol was exposed to 600 MPa at 25°C. The resultant particles formed the intended nano-microcapsules, characterized by a V6-type crystalline structure, with fatty acids encapsulated within amylose helices and distributed in amorphous regions of the starch (Jia et al., 2018).

4.1.4 High speed shearing

High-speed shearing techniques reduce starch particles through grinding, collision, shear force, and pulverization, which is significantly influenced by slurry concentration, velocity, and the characteristics of the grinding beads employed. The resultant starch-based nano-microcapsules exhibit irregular shapes and poor crystallinity, which improve dispersibility, accelerate dissolving, and facilitate gelatinization at ambient temperature. This method is reagent-free, efficient, and user-friendly; yet, it has not achieved widespread use due to its significant energy consumption. Potato starch nanocapsules measuring 120 nm were effectively produced using a high-energy ball mill with ZrO_2 as the grinding medium (Lin et al., 2016).

4.1.5 Freeze drying

Encapsulation is a technique used to entrap bioactive core materials, such as sensitive oils or pigments, within a protective wall material to enhance stability and control their release (Eun et al., 2020). Freeze drying, or lyophilization, is an attractive dehydration method for starch-based encapsulation because it operates at low temperatures and under a vacuum, which minimizes the risk of thermal degradation and oxidation for heat-sensitive ingredients like fish oils and carotenoids (Charles et al., 2021). Starch serves as an effective wall material due to its low cost, high capacity to form a matrix, and its ability to act as a significant oxygen barrier (Charles et al., 2021). Modified starches, particularly those esterified with octenyl succinic anhydride (OSA), are frequently utilized because they possess hydrophobic groups that improve emulsifying properties and stabilize the oil-water interface (Perrechil et al., 2021). In some formulations, native starches such as arrowroot starch are used for their ability to function as cryoprotectants, helping to stabilize the wall material and the core against the stresses of the freezing process (Charles et al., 2021). A common approach involves creating an oil-in-water (O/W) emulsion or a water-in-oil-in-water (W/O/W) double emulsion using starch as the primary stabilizer or as a Pickering emulsifier (Marefati et al., 2015). In Pickering emulsions, OSA-modified starch granules adsorb irreversibly at the interface, forming a rigid mechanical barrier that prevents droplet coalescence during the freezing and drying stages (Marefati et al., 2015).

The stability of starch-encapsulated products during freeze drying is influenced by several factors such as applying in situ heat treatment to starch-stabilized emulsions causes partial gelatinization of the granules at the interface, creating a more robust coating that protects the core during ice crystal formation. The melting point of the encapsulated core significantly affects stability; for instance, solid shea oil demonstrates higher physical stability during freeze-thaw cycles and freeze drying compared to liquid oils, which are more susceptible to destabilization as the external water phase crystallizes. The typical freeze-drying process involves freezing the emulsion (often at $-18°C$ to $-80°C$) followed by sublimation at low pressures (e.g., 0.5 mmHg) for durations ranging from 24 to 72 h (Perrechil et al., 2021). Unlike the spherical particles produced by spray drying, freeze-dried starch microcapsules often exhibit irregular, porous, and flake-like structures with limited shrinkage. This high porosity facilitates rapid reconstitution and solvent uptake but may require careful storage to prevent oxygen diffusion (Rezvankhah et al., 2020). Encapsulation efficiency (EE) for starch-based freeze-dried powders is generally high, often ranging between 80% and 91% for various oils and bioactive compounds, and can reach over 97% in stabilized Pickering emulsions. These powders typically show low moisture content (less than 5.5%) and low water activity, which are critical for inhibiting microbial growth and chemical degradation. Starch-based matrices provide superior oxidative stability compared to spray-dried versions because they avoid the high air-inlet temperatures that can trigger lipid oxidation during the drying process (Rezvankhah et al., 2020).

4.2 Chemical methods

Chemical modification of starch encompass oxidation, crosslinking, esterification, etherification, and biotechnological processes, as illustrated in Fig. 18.5 (Masina et al., 2017).

4.2.1 *Esterified starch*

The chemical modification of natural starch yields esterified starch, typically through an esterification reaction involving starch and octenyl succinic anhydride (OSA). Under alkaline conditions, OSA covalently interacts with specific functional groups on starch, resulting in OSA starch that displays markedly distinct features compared to its normal form. The modified starch exhibits enhanced hydrophilicity, superior emulsion-stabilizing capabilities, and improved gel-forming properties. Consequently, OSA starch is regarded as a modern, safe, and economical thickening and emulsifying agent. Its aqueous dispersions exhibit a notable strength, achieving elevated solid content while maintaining relatively low viscosity (Tian et al., 2022). Nutmeg oleoresin was encapsulated using a mixture of carrier materials, specifically gum arabic and

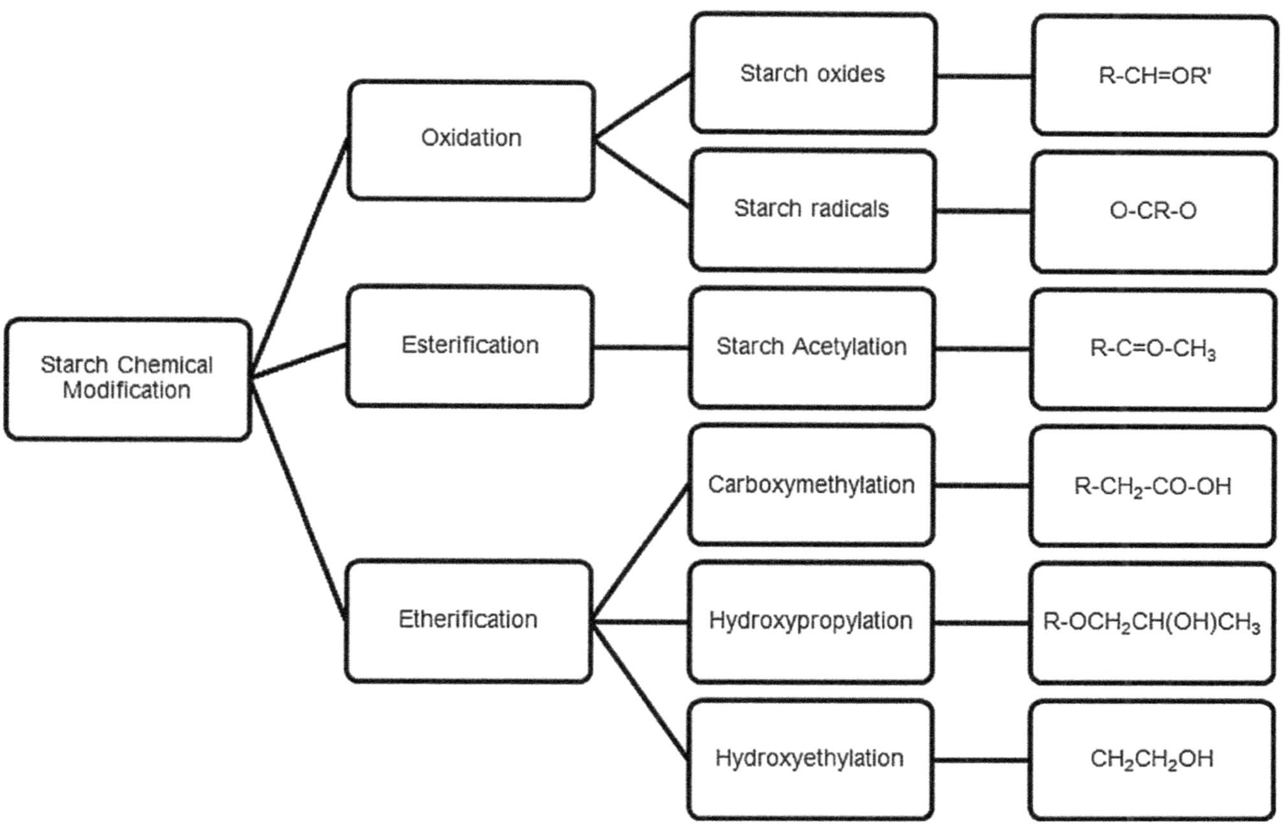

FIGURE 18.5 Schematic summarizing the classical chemical methods for starch modification (Masina et al., 2017). *Adapted from Masina, Nonhlanhla, Choonara, Yahya E., Kumar, Pradeep, du Toit, Lisa C., Govender, Mershen, Indermun, Sunaina & Pillay, Viness. (2017). A review of the chemical modification techniques of starch. Carbohydrate Polymers, 157, 1226–1236. https://doi.org/10.1016/j.carbpol.2016.09.094.*

OSA-modified sorghum starch. The starch underwent chemical acetylation, was subsequently dissolved with the oleoresin, and then underwent spray drying to produce nano-to micro-sized capsules. The capacity of capsules formulated with OSA starch, in contrast to unmodified starch, to preserve bioactive compounds and uphold functional attributes led to sustained antioxidant activity over extended storage durations of up to 60 days, including the retention of elevated amounts of phenols and flavonoids. The capsules showed antibacterial efficacy against *Escherichia coli* and *Bacillus cereus* (Arshad et al., 2018).

4.2.2 Oxidized starch

Oxidation is a crucial modification process in starch, wherein hydroxyl groups at the C2, C3, and C6 positions of glucose units are biologically transformed into C2, C3, and C6 carboxyl and carbonyl functional groups. This oxidative reaction may disrupt the glycosidic linkages in the starch chain, hence altering its molecular structure and potentially leading to substantial modifications in its physical and chemical properties (Tian et al., 2022). Oxidation often modifies the properties of starch regarding enthalpy, crystallinity, gelatinization temperature, and molecular size, resulting in improved heat stability and decreased retrogradation (Hoyos-Leyva et al., 2018; Wang et al., 2015). The creation of nano-microcapsules with a spray-drying process from a solution of gelatin and oxidized corn starch (G/OCS) served as carriers for functional food components. Particle shape was assessed using scanning electron microscopy (SEM), while heat resistance was evaluated by thermogravimetric analysis (TGA). The swelling capacity and vitamin C efflux were also evaluated. The utilization of gelatin and oxidized starch led to the formation of a stable structure that exhibited resistance to deterioration at 326°C. The enhanced capsule swelling at alkaline pH and decreased swelling in high-salt settings can be attributed to the chemical interactions between starch and gelatin, together with the activation of ionizable groups. The release of vitamin C was regulated, with the highest release observed within the initial 3 hours of immersion in water (Chen et al., 2017).

4.2.3 Acid hydrolyzed starch

It is a form of starch that is produced by subjecting native starch to mineral acids at low temperatures below the gelatinization point (e.g., 35–50°C) and changing its structure and functionalities. The resulting starch form has lesser viscosity and a lower thickness; this has made it suitable for the making of thick pastes or gels which are highly dense(Shabana et al., 2019; Soler et al., 2020). In a recent study hydrolyzed starch with different dextrose equivalent (DE) values (DE-8, DE-18, and DE-38) was used wall materials wall material to *Bacillus subtilis*, as biofertilizers for sustainable agricultural practices. This study reveled that degree hydrolysis, and dextrose equivalent has direct impact on spray dried microcapsules morphology producing smaller, homogenous and smoother microparticles. Microencapsulates effectively protected cells against acidic pH stresses and high salinity, enhanced thermal stability, UV protection and storage stability (Lopes et al., 2025). Phosphorylation of starch makes it significantly resistant to both acid and enzymatic hydrolysis process. In another technique, the phosphorylated starch was used as a carrier to encapsulate anthocyanin. Phosphorylation decreased the crystalline structure and viscosity of the starch, which altered its functional properties. Findings showed that an increase in the concentration of starch phosphorylation provided more protection to anthocyanin substances, showing a direct relationship between the extent of starch modification and encapsulation efficiency (García-Tejeda et al., 2016).

4.2.4 Cross linked starch

Chemical crosslinking markedly enhances the structural integrity of starch granules and modifies critical physical properties, such as gelatinization temperature, drying resistance, and mechanical stress resistance (e.g., shear force). The abundance of hydroxyl groups on the starch surfaces facilitates the formation of chemical bonds, such as diesters or diacid bonds. These bonds link numerous starch molecules of differing molecular sizes, forming a continuous and interconnected network. Consequently, the starch develops a cross-linked configuration with a three-dimensional spatial arrangement, so augmenting its mechanical stability and functional efficacy in diverse industrial and food applications (Tian et al., 2022). Recent studies indicate that starch nano particles can be synthesized via the emulsion-crosslinking method, as depicted in Fig. 18.6 (Morán et al., 2021).

4.3 Enzymatic method

4.3.1 Amylolytic starch

Starch-degrading enzymes such as α-amylase, β-amylase, and glucosidase are extensively utilized in the production of nano- and microcapsules. Recently, researchers have utilized these enzymes to partially degrade natural starch, yielding

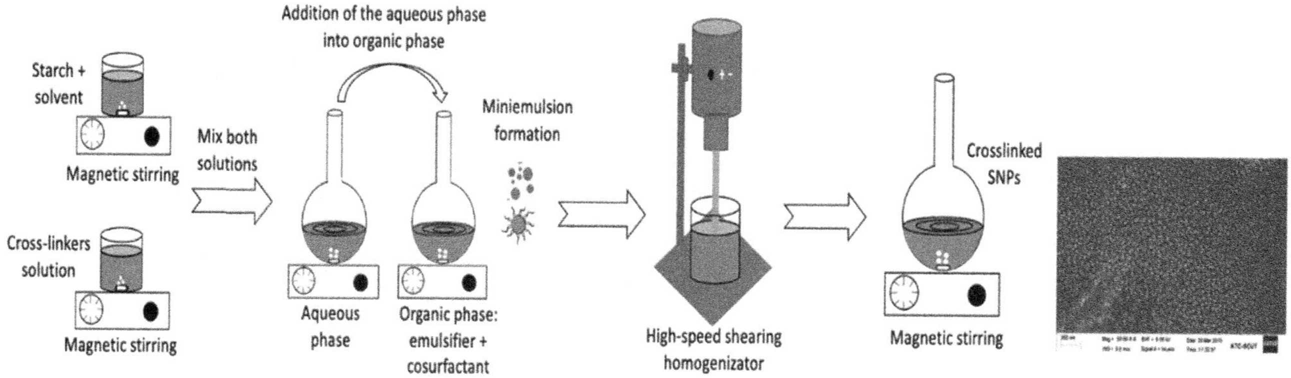

FIGURE 18.6 Starch nanoparticles synthesis using the emulsion cross-linking technique. Adopted from (Morán et al., 2021). *Adapted from Morán, Diana, Gutiérrez, Gemma, Blanco-López, María Carmen, Marefati, Ali, Rayner, Marilyn & Matos, María. (2021). Synthesis of Starch Nanoparticles and Their Applications for Bioactive Compound Encapsulation. Applied Sciences, 11(10). https://doi.org/10.3390/app11104547.*

products such as porous starch, which demonstrates significant potential as a wall material for nano-microencapsulation. Porous starch exhibits superior adsorption capacity compared to other carriers, due to its many pores extending from the surface to the particle core, which increases the specific surface area and establishes a robust internal structure. Due to these characteristics, it has been widely employed in the food industry for encapsulating beneficial compounds such as resveratrol, anthocyanin, probiotics, olive oil, and vitamins (Tian et al., 2022). A study was conducted in which microcapsules were created by covering particles composed of α-amylase, resveratrol, and extruded corn starch with chitosan. This approach employed low-temperature extrusion followed by α-amylase treatment to mitigate common limitations associated with extruded starch, including low solubility, reduced gelatinization temperature, and diminished water absorption. The use of chitosan enhanced the structural integrity of the capsules while also providing additional biological functionality. Thus, the incorporation of enzymatic modification with chitosan coating significantly improved the release efficiency of resveratrol from the encapsulated system (Song et al., 2021).

4.3.2 Debranching starch

Debranched starch is typically produced by treating conventional starch with specialized enzymes that dismantle its branching architecture. During this process, the starch transitions from a convoluted structure to a more organized single-helix configuration. The helical configuration of starch creates a cavity capable of accommodating tiny molecules such as fatty acids, alcohols, pharmaceuticals, iodine, and fragrances. The hydrophobic inside of the helix can encapsulate and shield non-polar molecules in this manner. Enzymatic debranching of starch increases the proportion of linear chains and promotes the formation of nano- and microcapsules. Isoamylase, also known as debranching enzyme, is frequently employed to cleave α-1,6-glycosidic bonds that connect side chains to the principal starch backbone, hence releasing both long and short linear glucan chains (Bhatt et al., 2025). In a study, researchers examined the impact of several durations of pullulanase exposure (0, 4, 8, and 16 h) on the ability of high-amylose maize starch (HAMS) to form complexes with ascorbate palmitate (AP), a fat-soluble derivative of vitamin C. An 8-hour debranching duration enhanced the encapsulation efficiency to 8.63%, indicating a greater incorporation of active pharmaceutical ingredients into the starch matrix. The treatment also enhanced starch heat absorption and crystallinity, suggesting the formation of more organized structures. In contrast, smaller treatments (0 and 4 h) mitigated starch retrogradation, resulting in undesired recrystallization. Extending the enzyme treatment to 16 h yielded an adverse effect, perhaps due to the excessive formation of short chains, which compromised the structure (Zhou et al., 2020).

4.3.3 Cyclodextrin

Cyclodextrins (CDs) are cyclic oligosaccharides produced through the enzymatic modification of starch, primarily via the degradation and cyclization of α-1,4-glucans by the enzyme cyclodextrin glucanotransferase. The most common naturally occurring forms are α-, β-, and γ-cyclodextrins, which consist of six, seven, and eight glucose units, respectively. Structurally, these molecules resemble a toroid truncated cone, featuring a relatively hydrophobic inner cavity and a hydrophilic outer surface. This unique amphipathic nature allows CDs to serve as "ready encapsulating wall materials" capable of trapping hydrophobic guest molecules within their cavities (dos Santos et al., 2017; Marques, 2010;

Muñoz-Shugulí et al., 2021; Jeong et al., 2025). The primary mechanism for encapsulation is molecular inclusion, a dynamic process where a guest molecule is enclosed within the host cavity without forming covalent. The leading driving force for complex formation is the displacement of high-enthalpy water molecules from the hydrophobic cavity, which significantly reduces the overall energy of the system. Stability is further maintained through van der Waals forces, hydrophobic interactions, and hydrogen bonding (dos Santos et al., 2017; Marques, 2010). Recent research has highlighted V-type starch-cyclodextrin composites as superior carrier systems. Unlike single carriers, these composites utilize the flexible folding of V-type starch and the rigid structure of β-CD to create a dual high-efficiency encapsulation system. The introduction of β-CD during starch recrystallization promotes the formation of single helices and improves both long- and short-range ordered structures, which enhances the loading capacity for molecules such as p-anisaldehyde (Gao et al., 2025). The effectiveness of encapsulation often depends on the preparation method employed. Common techniques include enzymatic encapsulation utilizing cyclodextrin glucanotransferase to produce large cyclic maltooligosaccharides known as cycloamylose (CA) directly from starch or dextrin. This method is inexpensive, food-grade safe, and allows for the high-yield production of complexes with compounds like isoflavones (Jeong et al., 2025). Encapsulation in starch-derived cyclodextrins offers critical functional benefits for food and pharmaceutical applications. CDs dramatically enhance the water solubility of hydrophobic compounds—for instance, increasing the solubility of daidzein by up to 89,000 times (Jeong et al., 2025). This directly improves the bioaccessibility and bioavailability of bioactive compounds like curcumin and spice oleoresins (Kulal et al., 2025; Xu et al., 2024). The host cavity protects sensitive molecules from environmental stressors such as light, heat, oxidation, and enzymatic degradation (dos Santos et al., 2017; Marques, 2010; Kulal et al., 2025). CDs are widely used to mask unpleasant odors and bitter tastes associated with plant extracts and essential oils (dos Santos et al., 2017; Marques, 2010). By altering the interaction between the host and bioactive compound, starch-based systems can achieve sustained or site-specific release of active ingredients (Gao et al., 2025; Marques, 2010; Xu et al., 2024). In conclusion, the versatility of starch-based encapsulation using cyclodextrins makes it a promising strategy for developing green, biodegradable, and high-load carriers to enhance the industrial utilization of functional food materials (Gao et al., 2025; Jeong et al., 2025).

5. Encapsulation of different food components

5.1 Fatty acids

Oils abundant in omega fatty acids, particularly those sourced from plants and fish, are significantly susceptible to oxidative damage and breakdown. Microencapsulation utilizing starch-derived materials is extensively utilized to safeguard delicate lipids and mitigate the development of off-flavors due to oxidation (Li et al., 2022; Yildiz et al., 2018). The efficacy of these encapsulation systems is evaluated based on entrapment capacity and oxidative stress resistance (Guo et al., 2021). Flawless, unblemished capsule surfaces signify enhanced structural integrity, providing superior protection for the encapsulated oils (Wang et al., 2020). OSA-modified starch, exemplified by Capsul, is recognized for its robust emulsifying and coating properties, rendering it appropriate for the spray-drying of oil-based constituents (Guo et al., 2021). The dispersibility of the substance in aquatic environments may be constrained by its hydrophobic characteristics; however, this limitation can be mitigated by combining it with water-soluble polysaccharides such as chitosan, maltodextrin, or inulin (Wang et al., 2020). utilized this method to encapsulate algal oil within microgel particles with diameters ranging from 223.70 to 644.46 nm. The combination of modified starch derived from oligostearic acid, chitosan, and inulin (OSA/CS/IN) demonstrates the most significant capacity for oxidative stability, extending the induction period nearly threefold compared to free oil. This formulation resulted in particles with a finer surface shape, attributed to flexible molecular connections and electrostatic interactions among the constituent polymers (Wang et al., 2020). Conjugated linoleic acid (CLA) is a nutritionally advantageous fatty acid that is predominantly absorbed in the small intestine, despite its low water solubility and susceptibility to oxidation, undergoing breakdown in the stomach prior to absorption (Yang et al., 2020). reported that CLA was effectively delivered to the gastrointestinal system via an emulsion of OSA-modified waxy maize starch and xanthan gum (He et al., 2016). optimized this delivery mechanism through spray-drying, enhancing its protective properties against non-absorptive regions of the digestive tract. In contrast, products made with native starch exhibited diminished flowability and surface wrinkling, thereby impairing their antioxidant protective properties and overall efficacy (Arshad et al., 2018; Wang & Guo, 2020). On the other study done by (Yildiz et al., 2018) demonstrated that the composite system (1:1 w/w) achieved the lowest peroxide levels and the highest fatty acid release rates in simulated gastric (37.9 w) and intestinal (91.3 w) fluids, outperforming previously utilized formulations containing Tween 20 or SDS. This encapsulant mixture shows robust antioxidant properties and is applicable in health-oriented food products (Yildiz et al., 2018).

5.2 Antioxidants

5.2.1 Phenolic compounds

Phenolic compounds are plant secondary metabolites categorized by their structural diversity, containing at least one aromatic ring with one or more hydroxyl groups. As noted by (Grgić et al., 2020), these molecules offer significant health benefits, including antioxidant, anti-inflammatory, and antimicrobial activities. However, their application in functional foods is limited by high sensitivity to light, heat, oxygen, and the harsh conditions of the gastrointestinal tract (Schlindweinn et al., 2022). highlight that starch is an ideal wall material for protecting these bioactive compounds because it is abundant, low-cost, biodegradable, and biologically compatible. Encapsulation within starch-based systems improves water solubility and ensures the targeted delivery and controlled release of phenolics in specific regions of the digestive system. According to Shi et al., (2021) treatments like acetylation and succinylation (using Octenyl Succinic Anhydride) convert hydrophilic native starch into hydrophobic oramphiphilic derivatives, which are more effective for trapping hydrophobic polyphenols (García-Tejeda et al., 2016). demonstrated that phosphorylated sorghum starch provides superior thermal stability and higher encapsulation efficiency for red sorghum phenolics compared to other esterified starches. Enzymatically acylated rice starch can form hydrophobic conglomerates that induce resistance to enzymatic degradation, making them suitable for the long-term administration of active compounds. Etherification techniques such as carboxymethylation enhance water absorption and stability, making them ideal for encapsulating hydrophilic pigments like anthocyanins (Lian et al., 2022). found that the development of starch nanoparticles through high-speed jet or anti-solvent precipitation allows for direct absorption by intestinal epithelial cells, significantly increasing uptake and bioaccessibility (Wang et al., 2023).explain that V-type Inclusion Complexes form when a hydrophobic ring of a phenolic ligand enters the cavity of the amylose single helix. The cavity size—classified as V6, V7, or V8—is determined by the structure of both the starch and the phenolic compound. Non-inclusion Complexes are primarily driven by intermolecular hydrogen bonds, CH–π interactions, or van der Waals forces. In these systems, phenolic compounds are often excluded from the hydrophobic helix and instead associate with the amorphous regions or the outer chains of amylopectin. According to (Shi et al., 2021) starch-based encapsulation facilitates pH-responsive release. At acidic gastric pH (2–4), the protonation of carboxylic groups forms a compact outer layer that prevents the medium from penetrating the capsule, shielding the active ingredients. In the basic environment of the small intestine, outer groups are deprotonated and hydrated, causing the capsule to swell and release its payload through erosion or disintegration. Encapsulated phenolics that reach the lower digestive tract act as prebiotics that they can alter the abundance of gut microbiota, such as by reducing the Firmicutes/Bacteroidetes (F/B) ratio and increasing beneficial populations like *Bifidobacterium.*

5.2.2 Carotenoids

Starch is a widely utilized wall material in the food industry for the encapsulation of carotenoids due to its abundant availability, low cost, and effective emulsifying properties during drying processes (Arshad et al., 2018; Eun et al., 2020). As a carbohydrate, it is categorized as a generally recognized as safe (GRAS) material, which is ideal for protecting sensitive bioactive compounds like carotenoids from environmental stressors (Rehman et al., 2020) Starch is composed of two primary fractions: the linear amylose and the branched amylopectin (Eun et al., 2020; Kong et al., 2018). There are two primary ways starch is utilized to encapsulate carotenoids: the formation of inclusion complexes and its use as a shell material in microencapsulation (Eun et al., 2020). Amylose has the unique ability to form left-handed single helices in the presence of hydrophobic ligands, featuring a hydrophilic outer surface and a hydrophobic inner channel that can accommodate guest molecules (Kong et al., 2018). These helices arrange to form a crystalline structure known as the V-type (Eun et al., 2020; Kong et al., 2018). Because carotenoids like β-carotene have extremely low solubility, research has explored ternary systems using amphiphilic materials, such as surfactants or ascorbyl palmitate (AscP), to improve their accessibility to the amylose helical core (Kong et al., 2018). In these systems, carotenoids are suggested to be homogeneously immobilized within the polycrystalline inclusion complexes, which act as a protective barrier against oxidative stress and acidic environments (Kong et al., 2018). Modified starches, particularly those esterified with octenyl succinic anhydride (OSA), are frequently utilized as economic, non-allergenic substitutes for gum arabic (Anthero et al., 2021). These starches are amphiphilic, colorless, and tasteless, providing excellent protection against oxidation (Anthero et al., 2021; Eun et al., 2020). For instance, pregelatinized waxy corn starch-OSA has demonstrated superior protection of paprika oleoresin carotenoids compared to other wall materials during storage at temperatures up to 35°C (Anthero et al., 2021). Encapsulation in starch-based matrices significantly enhances the storage stability of carotenoids by acting as a physical barrier against oxygen, light, and heat (Coelho et al., 2022; Eun et al., 2020; Kong et al., 2018). For example, β-carotene encapsulated in starch-AscP inclusion complexes remains more stable over a 6-week period compared to physical mixtures (Kong et al., 2018). Regarding bioavailability, the dense gel matrix of some starch systems can reduce immediate bioaccessibility by slowing the release of the compound. However, microencapsulation using starches like those from amaranth grain has been shown to result in higher bioaccessibility than the free, crystalline form of the carotenoid after simulated digestion (Coelho et al., 2022).

5.3 Flavors and essential oils

Natural taste molecules frequently degrade during food processing as a result of contact with heat and moisture (Belingheri et al., 2015). The incorporation of flavor ingredients often diminishes sensory qualities; however, liquid flavors are especially susceptible to degradation by oxidation, evaporation, or interaction with other components (Guo et al., 2021). Encapsulation effectively safeguards sensitive chemicals from environmental stressors and facilitates controlled release (Pellicer et al., 2019). Starch is extensively utilized for this purpose owing to its capacity to create diverse structures, such as amylose helices (Guo et al., 2021), microporous networks (Belingheri et al., 2015), and porous ordinary particles (Zhu et al., 2018), rendering it an exceptional carrier for volatile flavor compounds. Starch-based systems have effectively encapsulated several flavor profiles, including tomato, strawberry, vanilla oil, and limonene (Belingheri et al., 2015; Pellicer et al., 2019; Yeo et al., 2016; Zhu et al., 2018) characterized liquid tomato flavor using a straightforward plating technique with porous starch and evaluated its impact on the sensory attributes of tomato sauce after 3 and 6 months of storage. Their findings indicated that retention was significantly influenced by the solvent employed to convey flavor. Nonpolar solvents containing medium-chain triglycerides (MCT) exhibited higher quantities of the nonpolar compound p-cymene (0.974) compared to polar solvents such as propylene glycol (0.051) and triacetin (0.360). These results underscore the necessity of aligning solvent polarity with that of the flavor ingredient for effective retention in porous starch systems (Zhu et al., 2018). and Zhu (unpublished) examined the volatility and instability of vanilla oil, a prevalent food preservative and additive, by encapsulating it with jackfruit seed starch, β-cyclodextrin, and chitosan by solution mixing, sonication, and freeze-drying techniques. The starch-based normal particles exhibited a slower and more regulated flavor release, but β-cyclodextrin and chitosan facilitated a more rapid fragrance release during the initial storage phase. The starch from jackfruit seeds exhibited superior shelf life (250 days) and encapsulation efficiency (79.33%), surpassing those of β-cyclodextrin (160 days, 77.92%) and chitosan (160 days, 76.64%). This is associated with the low crystallinity of starch, superior performance, robust gel strength, excellent flexibility, and the ability to produce films.

5.4 Probiotics

Starch and its modified derivatives have emerged as prominent wall materials for probiotic encapsulation due to their high biocompatibility, non-toxicity, low cost, and low allergenicity. These biopolymers act as an effective physical barrier, shielding sensitive microorganisms like *Lactobacillus* spp. from harsh environmental stressors, including high processing temperatures, oxidative conditions, and the acidic environment of the gastrointestinal tract (Ashwar et al., 2018). Particularly resistant starch, produced via chemical modifications like phosphorylation or acetylation, is highly effective for targeted delivery to the colon because it survives enzymatic digestion in the small intestine (Ashwar et al., 2018). Resistant starch microcapsules have demonstrated exceptional thermal stability, protecting probiotics during high-heat processes such as extrusion for functional rice flour products (Cao et al., 2025). Porous Starch fabricated through enzymatic hydrolysis (e.g., using α-amylase or glucoamylase), porous starch features a network of micro- and nano-scale pores that facilitate the physical adsorption and entrapment of probiotic cells (Firdaus et al., 2025). Botanical sources like arrowroot (*Maranta arundinacea*) are preferred for porous starch production due to a high amylose content (~40%), which supports the formation of uniform pores (Firdaus et al., 2025). Starch nanoparticles synthesized through green methods like ultrasonication, from sources such as rice, maize, or water chestnut offer a high surface area and improved cell permeability. These nanoparticles significantly prolong viability under hostile conditions and do not negatively affect the sensory attributes of foods (Firdaus et al., 2025; Noman et al., 2023). Porous starch carriers protect probiotics (e.g., *L. plantarum*) from the melting temperatures (40–50 °C) and shearing forces inherent in 3D-printing chocolate or producing rice extrudates. Freeze-drying and spray drying methods are used to create stable probiotic powders. Studies on jackfruit seed starch found it superior to conventional materials like maltodextrin, achieving encapsulation efficiencies of up to 96% and maintaining viability levels of 7–8 log CFU/g during 28-day storage (Singh et al., 2025). Using resistant starch as a wall material in emulsification systems has been shown to keep probiotic counts high (>7 log CFU/g) for up to 2 months when stored at refrigerated temperatures. Starch-based encapsulation not only enhances the shelf life and stability of functional foods but also ensures that an adequate number of viable cells (typically >106 CFU/g) reach the large intestine to confer health benefits (Ashwar et al., 2018).

5.5 Other food ingredients

Iron deficiency is a leading cause of anemia, prompting extensive efforts to improve iron supplementation and absorption in humans. Although widely used, food fortification can lead to unwanted changes in sensory properties and chemical reactions with other food components (Gupta et al., 2015). Additionally, the lack of nutrients such as essential minerals, vitamins, and coenzymes can trigger a number of health conditions (Guo, Qiao et al., 2021). In order to address these shortcomings, researchers have come up with starch based encapsulation systems that can incorporate the sensitive micronutrients without compromising the food quality (Bryszewska et al., 2019). also used thermally modified starch (TMS) to entrap ferrous sulfate and ferrous lactate in an attempt to form iron-loaded particles which in turn were added to loaves fermented by sourdough and conventional yeast. The TMS encapsulant has the capability of effectively sequestratingiron and food-based ligands or chelators and preserving nutrient integrity. The bioavailability of encapsulated iron was assessed simulated digestion system using Caco2 cell models to analyze the transport mechanism and efficiency of the encapsulated iron. The transport of iron was higher from yeast-bread (6.09%) compared to sourdough (1.79%) which explains the effect of processing on the uptake of nutrients. The encapsulation of food can also preserve shelf life by using starch. In another work, fresh tiger nut milk was encapsulated with a mixture of inulin and OSA-modified tiger nut starch, which produced lyophilized microspheres with a mean diameter of 1.01 μm (da Costa Neto et al., 2019; Gomes et al., 2019). These encapsulated milk stored at a stable pH (6.88–6.99) during 60 days old and could not be found after 10 days old with the milk pH reducing to 4.03 as a result of the proliferation of fungi. The inulin-OSA starch blend was used to create the smooth, crack-free particles of low permeability, which can effectively shield the core material and control the amount of the active compounds to be released.

6. Application of starch based encapsulates

Starch based micro and nano encapsulates are significant contributors to the development of functional foods. These nano-microcapsules exist in various forms, with different features that are determined by their composition and design. Their applications can be classified into four main domains depending on the bioactive ingredients embedded. They include antioxidant delivery, flavor retention, nutritional enhancement and food preservation (Ahmad et al., 2017; García-Tejeda et al., 2016). Starch based encapsulation offers wide range of advantages including bioactivity and thermal stability enhancement, flavor protection, masking unpleasant flavor, enhance anti fungal activity, controlled and targeted delivery to enhance bioavailability Fig. 18.7 (Guo et al., 2021). Different starch based encapsulates of active ingredients in the functional foods application are listed in Table 18.1.

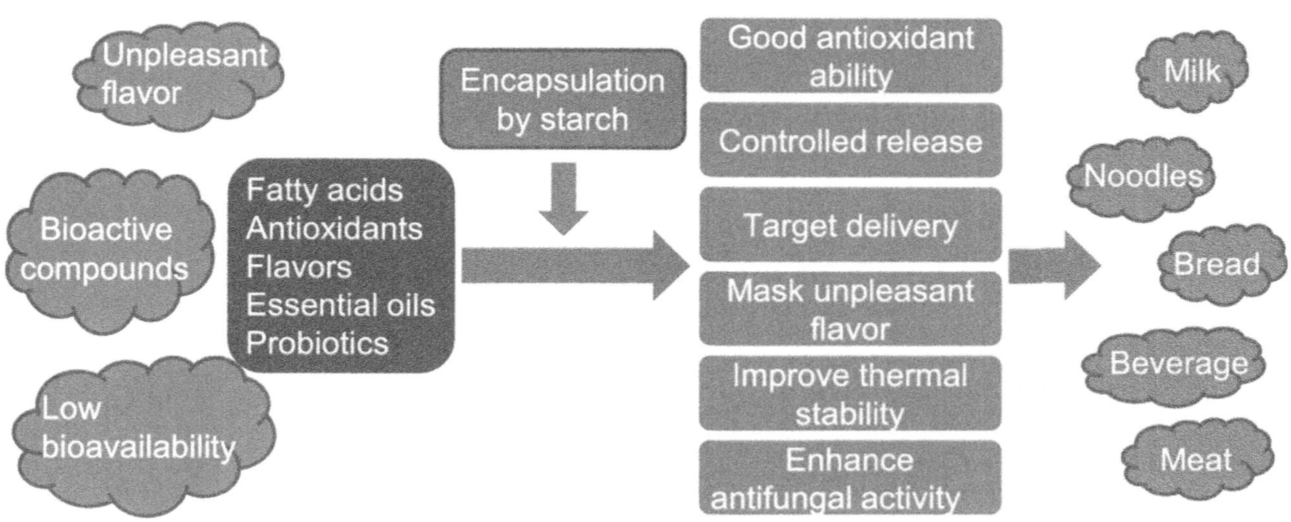

FIGURE 18.7 Role of starch based encapsulation in designing functional food ingredients (Guo et al., 2021). *Adapted from Guo, Y., Qiao, D., Zhao, S., Zhang, B. & Xie, F. (2021). Starch-based materials encapsulating food ingredients: Recent advances in fabrication methods and applications. Carbohydrate Polymers, 270. https://doi.org/10.1016/j.carbpol.2021.118358.*

TABLE 18.1 Starch based encapsulation of active ingredients and its applications.

Wall materials	Core materials	Applications	References
Highland barley starch	Cinnamon essential oil	Food antioxidant and stabilizer	Li et al. (2020)
OSA-esterified taro starch	Avocado oil	Fat-soluble bioactives	Romero-Hernandez et al. (2021)
Arrowroot starch compared to maltodextrin	Tuna fish oil	Oxidation-prone components	Charles et al. (2021)
Potato starch	Thyme oil	Chilled meat	Yuan et al. (2021)
Horse chestnut–based starch in combination with β-cyclodextrin	Folic acid	Antioxidant	Ahmad et al. (2017)
Resistant starch	Three probiotic strains (lactobacillus casei, lactobacillus brevis and lactobacillus plantarum)	Probiotic oral delivery	Ashwar et al. (2018)

From Tian, S., Xue, X., Wang, X. & Chen, Z. (2022). Preparation of starch-based functional food nano-microcapsule delivery system and its controlled release characteristics. Frontiers in Nutrition, 9. https://doi.org/10.3389/fnut.2022.982370

6.1 Antioxidants encapsulation

Phenols, flavonoids, anthocyanins, vitamins, and conjugated linoleic acid (CLA) are good examples of antioxidant-active compounds in the development of functional foods, but their use is often limited by a low stability to environmental stresses including heat, light, among others. The micro and nano encapsulation technology based on starch has been effective in improving the stability of these compounds and expanding their application with food systems (Guo et al., 2021; Tian et al., 2022). Folic acid is highly sensitive to pH fluctuations, making it prone to rapid degradation during digestion; however, when encapsulated in a starch-based nanocarrier, it demonstrates enhanced stability at elevated temperatures, evidenced by higher transition and melting temperatures compared to free folic acid. Simulated gastrointestinal studies also found that antioxidant and folic acid levels were large in intestinal fluid compared to gastric fluid and confirmed the selective and controlled release of bioactive substances in the intestine (García-Tejeda et al., 2016).

6.2 Flavor encapsulation

Flavors are critical to consumer satisfaction but are highly sensitive to environmental factors such as temperature, oxygen, and light (Aslam et al., 2022; Madene et al., 2006). Encapsulation in starch-based materials provides an effective strategy to protect these volatile compounds from degradation, mask unpleasant off-flavors, and extend a product's shelf life Fig. 18.8. Starch is regarded as an ideal carrier because it is inexpensive, widely available, biodegradable, and easily modified to achieve specific delivery properties. The starch-based encapsulation has been classified through both the materials used and the methods of preparation as follows - Typical wall materials include natural starch, cyclodextrins (CDs) (which have size-selective lipophilic cavities), maltodextrins (used for their film-forming abilities), OSA starches (modified with octenyl succinic anhydride for superior emulsification), and porous starch (which offers high surface area for physical adsorption). Common techniques include spray drying (the industrial standard due to low cost), freeze-drying (preferred for heat-sensitive volatiles), extrusion, and emerging technologies like electrospinning and spray-freeze-drying, which produce high-efficiency nanofibers or microporous particles. Starch-based encapsulation techniques offer innovative solutions for incorporating flavors into food products, particularly those with strong odors, by mitigating direct mouth irritation and enhancing flavor retention. Various encapsulation methods yield different forms, including powders, pastes, and capsules, with powdered forms being predominant for applications in bakery items, instant beverages, and snacks (Aslam et al., 2022; Madene et al., 2006). Notable studies highlight the use of β-cyclodextrin (β-CD) for flavor retention during cooking processes, such as in snack foods and confections (Aytac et al., 2017). In terms of food preservation, natural antimicrobial agents like allicin and essential oils have shown effectiveness against harmful bacteria. Encapsulated essential oils have demonstrated enhanced antimicrobial activity and flavor retention in food products, such as sausages and various meats, even after heat treatment. The interactions between released flavor compounds and

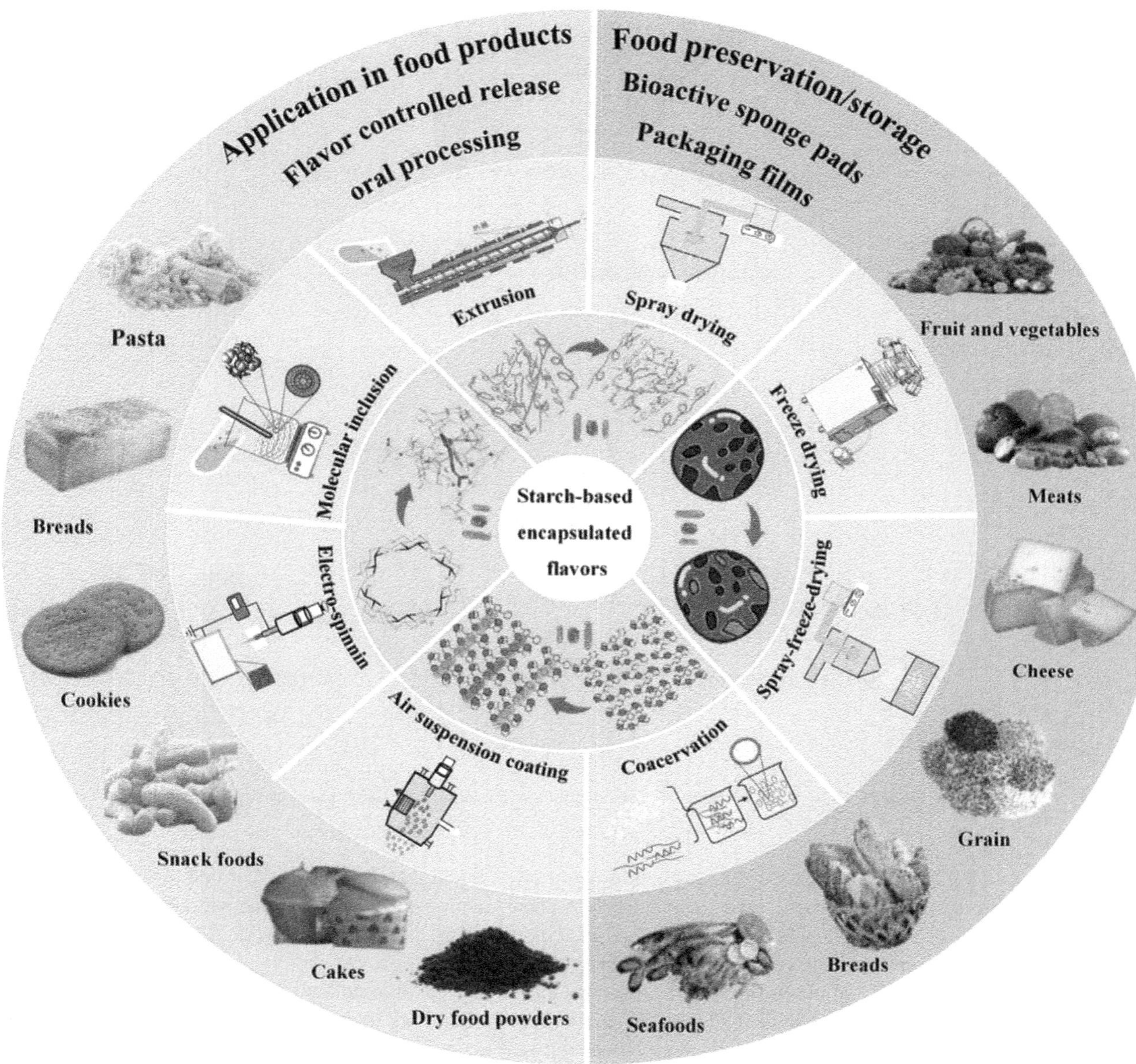

FIGURE 18.8 Schematic overview of starch-based flavor encapsulation technologies, illustrating processing methods, encapsulation structures, and their applications in food products and food preservation systems (Shan et al., 2025). *Adapted from Shan, Y., Li, J., Nie, M., Li, D., Zhang, Y., Li, Y., Wang, L., Liu, L., Wang, F. & Tong, L.-T. (2025). A comprehensive review of starch-based technology for encapsulation of flavor: From methods, materials, and release mechanism to applications. Carbohydrate Polymers, 348. https://doi.org/10.1016/j.carbpol.2024.122816*

microbial cell membranes are complex, involving potential disruptions to cell integrity and enhanced oil concentration due to microcapsule properties (Qiu et al., 2022). However, the structural limitations of starch microcapsules may restrict their broader application in food preservation, particularly compared to synthetic alternatives. Overall, the integration of starch-based encapsulation in food technology presents promising avenues for flavor enhancement and preservation.

6.3 Nutrient encapsulation

Functional food substances including micronutrients and bioactive compounds undergo digestion and affected by acid and enzymes in gastrointestinal tract which limits its therapeutic benefits (Fayed et al., 2019; Samakradhamrongthai et al., 2016). As a result, developing delivery mechanisms capable of protecting the biological activity of these ingredients and releasing them at controlled rate at targeted sites in the digestive tract is one of the challenge to enhance its bioavailability,

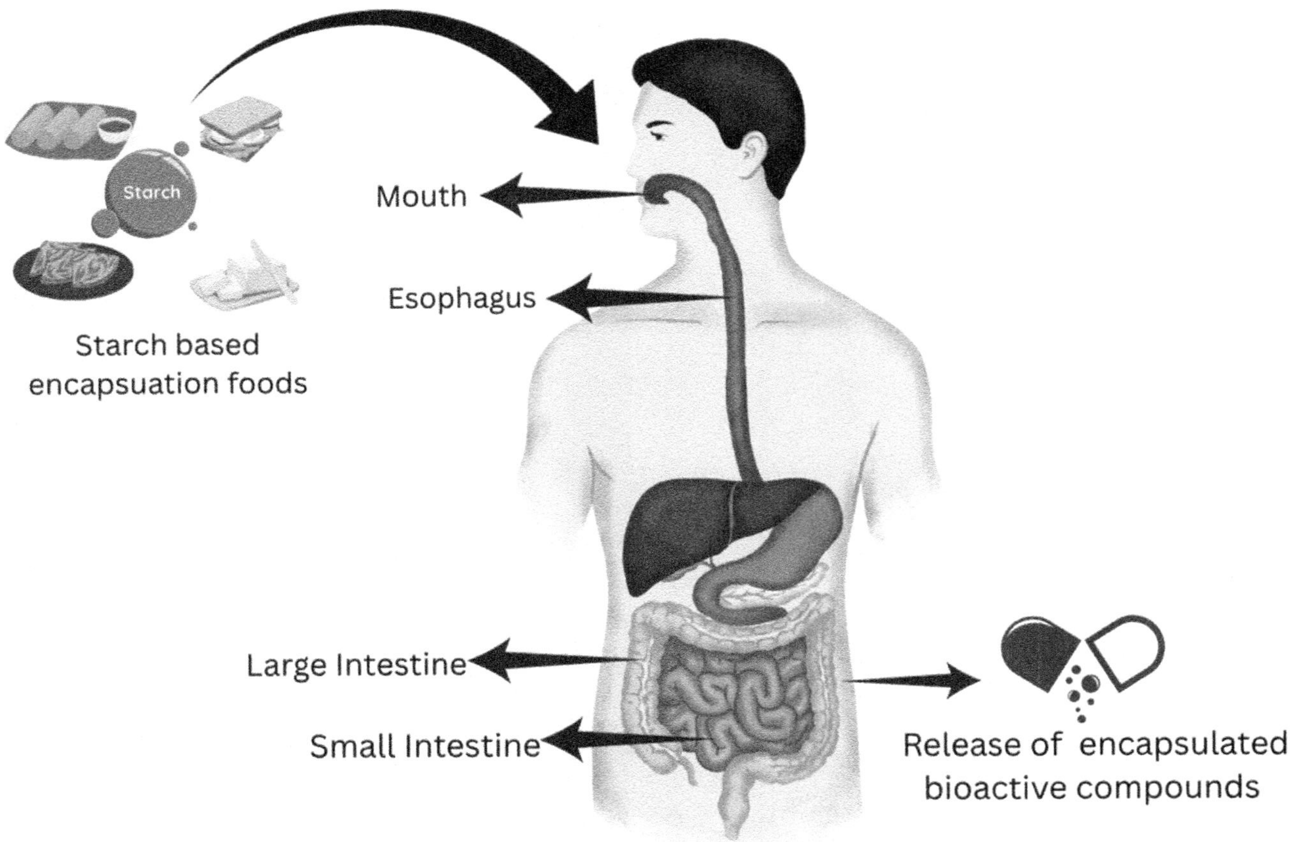

FIGURE 18.9 An illustration of controlled bioactive release from starch based carriers.

as shown in Fig. 18.9 (Tian et al., 2022). Encapsulation using crosslinked starch has demonstrated potential in altering bioactives digestibility to enable it to serve as an effective encapsulation agent to facilitate targeted delivery. As an example, three probiotic strains namely *Lactobacillus casei*, *Lactobacillus brevis* and *Lactobacillus plantarum* has been encapsulated using resistant starch as a microencapsulated system showing efficacy as a delivery system and protectant of probiotics administered orally. Microencapsulated probiotics survived well in simulated gastrointestinal conditions and adverse heat conditions, proving that resistant starch is the potential new delivery carrier for oral administration of probiotics (Ashwar et al., 2018). In another study, In recent study, folic acid stability under gastrointestinal digestion and thermal stability during food processing was achieved via microencapsulation with melt-coating of glycerol monostearate-porous starch (GMS-PS) composite. In vitro digestion proved that, starch based microencapsulation reduced gastric loss of folic acid to 12.19% (compared to 89.36% for free folic acid), while enabling 77.14% release in the intestine demonstrating a gastric-protective effect of starch based encapsulates (Tang et al., 2025).

6.4 Food preservation

Bacterial growth has a profound impact on the shelf life and quality of products in the process of food storage (Huang et al., 2021). However, traditional sterilization is typically practiced through high-temperature processing and ultraviolet lamps, which require expensive and energy consuming equipment (Li et al., 2020; Romero-Hernandez et al., 2021). In place of this, natural antibacterial substances like curcumin, cinnamaldehyde, allicin, and clove oil are commonly added to food additives, since they are effective against food pathogens like E. coli, Salmonella enteritidis and Staphylococcus aureus. But they have low solubility in water and low flavor thresholds which limit their antimicrobial efficacy and sensory acceptability. To counter such issues, the concept of nano-microencapsulation has been considered as a method of preservation, allowing such active compounds to be delivered effectively (Tian et al., 2022). Under one such method, allicin encapsulation was carried out in porous starch (1:7 ratio) coated by β -cyclodextrin to serve as the wall material.

The preparation of nano-microcapsules included the preparation of allicin 20% solution, emulsifying the solution, then adding the solution to the carrier matrix. The efficacy of these capsules as antimicrobials was evaluated against common food samples like bread, pork, chicken and tofu. The results revealed that the nano-microcapsules retained high sterilizing potency after high temperatures were exposed to with a proportion of mold spore reductions (Ro) ranging between 14.5% and 26.3% thus highlighting a promising future in food preservation (Charles et al., 2021).

7. Conclusion

Encapsulation technologies based on starch have proven to be ultra-promising tools in producing functional food products. These systems use active substances, like probiotics, vitamins, and functional lipids, to add stability, absorption capacity, and controlled release to modified starch systems. Several wall materials with desired qualities can be prepared through modification of a variety of plant-based starches using oxidation, acetylation, hydrolysis and enzymatic treatment. Starches have been modified hydrophobically, which is regarded as a surfactant and thus encapsulation rate is enhanced hugely. These regimes are commonly manufactured through technologies which include spray drying, high-pressure homogenization, ultrasound, microwave and ultra-high pressure. Compared to the traditional encapsulating ingredients such as gelatin, acacia gum, and xanthan gum, starch has a more benefits, which encompass higher antioxidant preservation, probiotic viability, superior processing and storage stability, and high masking of undesired flavors. Recent advancements in food delivery systems have concentrated on employing microgels, which are assemblies of starch granules and composite particles, to target bioactive compounds such as carotenoids, phenolics, fatty acids, iron, essential oils, coenzymes, and caffeine for incorporation into diverse food systems, including meat products, dairy, baked goods, and beverages. These encapsulation methods provide better sensory properties and nutrition as well as extending the shelf-life and targeted release of the products. Starch, being widespread, cheap, and familiar to the food industry, has its unique benefits. Resistant starch may be unfermentable in the small intestine, rendering it particularly useful where direct delivery is needed e.g., probiotics, whereas gelatinized starch has great film-forming properties and processability. Collectively, starch-based encapsulation emerges as a versatile and economically feasible platform with significant potential for advancing functional food innovation and commercialization.

References

Abbasi, S., & Rahimi, S. (2008). Microwave-assisted encapsulation of citric acid using hydrocolloids. *International Journal of Food Science and Technology, 43*(7), 1226–1232. https://doi.org/10.1111/j.1365-2621.2007.01595.x

Abedi, A.-S., Rismanchi, M., Moosavi, M. H., Mousavi Khaneghah, A., Mohammadi, A., & Mahmoudzadeh, M. (2021). A mixture of modified starch and maltodextrin for spray drying encapsulation of Nigella sativa seeds oil containing thymoquinone. *Starch - Stärke, 73*(3–4). https://doi.org/10.1002/star.201900255

Adak, S., & Banerjee, R. (2016). A green approach for starch modification: Esterification by lipase and novel imidazolium surfactant. *Carbohydrate Polymers, 150*, 359–368. https://doi.org/10.1016/j.carbpol.2016.05.038

Ahmad, M., Qureshi, S., Maqsood, S., Gani, A., & Masoodi, F. A. (2017). Micro-encapsulation of folic acid using horse chestnut starch and β-cyclodextrin: Microcapsule characterization, release behavior & antioxidant potential during GI tract conditions. *Food Hydrocolloids, 66*, 154–160. https://doi.org/10.1016/j.foodhyd.2016.11.012

Ahmadi, M., Khajeh, F., Sohrabi, S., Shahbahrami, R., Mirmoghtadaie, L., Shahraz, F., Bahmanyar, F., Hashempour-Baltork, F., & Hosseini, S. M. (2025). Spray-dried probiotic microcapsules with calcium cross-linked oxidized starch and inulin. *Carbohydrate Polymer Technologies and Applications, 10*, Article 100760. https://doi.org/10.1016/j.carpta.2025.100760

Aldana Heredia, J. F., Hernández-Carrión, M., & Gómez-Franco, J. D. (2024). Carlos-eduardo Narváez-Cuenca, andrea del Pilar sánchez-camargo, microwave-assisted extraction, encapsulation, and bioaccessibility of carotenoids from organic tomato industry by-product. *Innovative Food Science & Emerging Technologies, 95*, Article 103706. https://doi.org/10.1016/j.ifset.2024.103706

Alissandratos, A., & Halling, P. J. (2012). Enzymatic acylation of starch. *Bioresource Technology, 115*, 41–47. https://doi.org/10.1016/j.biortech.2011.11.030

Alongi, M., & Anese, M. (2021). Re-thinking functional food development through a holistic approach. *Journal of Functional Foods, 81*, Article 104466. https://doi.org/10.1016/j.jff.2021.104466

Anthero, A. G.da S., Oliveira Bezerra, E., Aline Comunian, T., Ramalho Procópio, F., & Dupas Hubinger, M. (2021). Effect of modified starches and gum Arabic on the stability of carotenoids in paprika oleoresin microparticles. *Drying Technology, 39*(12), 1927–1940. https://doi.org/10.1080/07373937.2020.1844227

Arshad, H., Ali, T. M., Abbas, T., & Hasnain, A. (2018). Effect of microencapsulation on antimicrobial and antioxidant activity of nutmeg oleoresin using mixtures of gum arabic, OSA, and native sorghum starch. *Starch - Stärke, 70*(7–8). https://doi.org/10.1002/star.201700320

Ashwar, B. A., Gani, A., Gani, A., Shah, A., & Ahmad Masoodi, F. (2018). Production of RS4 from rice starch and its utilization as an encapsulating agent for targeted delivery of probiotics. *Food Chemistry, 239*, 287–294. https://doi.org/10.1016/j.foodchem.2017.06.110

Aslam, S., Akhtar, A., Nirmal, N., Khalid, N., & Maqsood, S. (2022). Recent developments in starch-based delivery systems of bioactive compounds: Formulations and applications. *Food Engineering Reviews, 14*(2), 271–291. https://doi.org/10.1007/s12393-022-09311-5

Augustin, M. A., Sanguansri, L., Rusli, J. K., Shen, Z., Cheng, L. J., Keogh, J., & Clifton, P. (2014). Digestion of microencapsulated oil powders: In vitro lipolysis and in vivo absorption from a food matrix. *Food & Function, 5*(11), 2905–2912. https://doi.org/10.1039/c4fo00743c

Aytac, Z., Ipek, S., Durgun, E., Tekinay, T., & Uyar, T. (2017). Antibacterial electrospun zein nanofibrous web encapsulating thymol/cyclodextrin-inclusion complex for food packaging. *Food Chemistry, 233*, 117–124. https://doi.org/10.1016/j.foodchem.2017.04.095

Beikzadeh, S., Khezerlou, A., Jafari, S. M., Pilevar, Z., & Mortazavian, A. M. (2020). Seed mucilages as the functional ingredients for biodegradable films and edible coatings in the food industry. *Advances in Colloid and Interface Science, 280*. https://doi.org/10.1016/j.cis.2020.102164

Belingheri, C., Ferrillo, A., & Vittadini, E. (2015). Porous starch for flavor delivery in a tomato-based food application. *LWT - Food Science and Technology, 60*(1), 593–597. https://doi.org/10.1016/j.lwt.2014.09.047

Benhabiles, N., Boudries, N., Mokrane, H., & Chaput, F. (2025). The performance of combined ultrasound and nonsolvent precipitation for betalain encapsulation in different types of starches: Thermal stability and antioxidant activity. *Journal of Food Measurement and Characterization, 19*(1), 276–292. https://doi.org/10.1007/s11694-024-02968-y

Bhatt, P., Kumar, V., Singh, S., Garg, S., Kumar, M., Wong, L. S., Kumarasamy, V., Pahwa, S., & Subramaniyan, V. (2025). Enzymatic debranching of starch: Techniques for improving drug delivery and industrial applications. *Starch/Staerke, 77*(4). https://doi.org/10.1002/star.202400224

Błaszczak, W., Buciński, A., & Górecki, A. R. (2015). In vitro release of theophylline from starch-based matrices prepared via high hydrostatic pressure treatment and autoclaving. *Carbohydrate Polymers, 117*, 25–33. https://doi.org/10.1016/j.carbpol.2014.09.031

Bryszewska, M. A., Tomás-Cobos, L., Gallego, E., Villalba, M. P., Rivera, D., Taneyo Saa, D. L., & Gianotti, A. (2019). In vitro bioaccessibility and bioavailability of iron from breads fortified with microencapsulated iron. *Lebensmittel-Wissenschaft & Technologie, 99*, 431–437. https://doi.org/10.1016/j.lwt.2018.09.071

Cao, H., Wen, M., Zhu, Q., Yao, S., Feng, J., Li, D., Chen, J., Zhou, J., Liu, D., & Xu, E. (2025). Porous starch-probiotics encapsulation for 3D-printed chocolate. *Journal of Food Engineering, 387*, Article 112334. https://doi.org/10.1016/j.jfoodeng.2024.112334

Charles, A. L., Ahadi Abdillah, A., Rizky Saraswati, Y., Sridhar, K., Balderamos, C., Dewi Masithah, E., & Amin Alamsjah, M. (2021). Characterization of freeze-dried microencapsulation tuna fish oil with arrowroot starch and maltodextrin. *Food Hydrocolloids, 112*, Article 106281. https://doi.org/10.1016/j.foodhyd.2020.106281

Charoux, C. M. G., O'Donnell, C. P., & Tiwari, B. K. (2019). Effect of airborne ultrasonic technology on microbial inactivation and quality of dried food ingredients. *Ultrasonics Sonochemistry, 56*, 313–317. https://doi.org/10.1016/j.ultsonch.2019.03.025

Chávarri, M., Marañón, I., Villarán, M. C., Chávarri, M., Marañón, I., & Villarán, M. C. (2012). *Encapsulation technology to protect probiotic bacteria*. USA: Probiotics, IntechOpen.

Chen, H., Yang, M., Shan, Z., Mansouri, S., May, B. K., Chen, X., Chen, H., & Woo, M. W. (2017). On spray drying of oxidized corn starch cross-linked gelatin microcapsules for drug release. *Materials Science and Engineering: C, 74*, 493–500. https://doi.org/10.1016/j.msec.2016.12.047

Chen, Q., Zhuang, H., Yao, L., Wang, H., Sun, M., Song, S., Kang, W., & Feng, T. (2025). Microencapsulation of cornmint oil with octenyl succinate anhydride debranched starch by electrostatic spray drying: Characterization and release kinetics of the aroma components. *International Journal of Biological Macromolecules, 307*(Pt 2), Article 142093. https://doi.org/10.1016/j.ijbiomac.2025.142093

Coelho, L. M., Gonçalves, I., Ferreira, P., Pinheiro, A. C., Vicente, A. A., & Martins, J. T. (2022). Exploring the performance of amaranth grain starch and protein microcapsules as β-carotene carrier systems for food applications. *Food Structure, 33*, Article 100287. https://doi.org/10.1016/j.foostr.2022.100287

Comunian, T. A., Silva, M. P., & Souza, C. J. F. (2021). The use of food by-products as a novel for functional foods: Their use as ingredients and for the encapsulation process. *Trends in Food Science and Technology, 108*, 269–280. https://doi.org/10.1016/j.tifs.2021.01.003

da Costa Neto, J. J. G., Gomes, T. L. M., Justo, T. F., Pereira, K. S., Amaral, P. F. F., Rocha Leão, M. H. M., & Fontes Sant'Ana, G. C. (2019). Microencapsulation of tiger nut milk by lyophilization: Morphological characteristics, shelf life and microbiological stability. *Food Chemistry, 284*, 133–139. https://doi.org/10.1016/j.foodchem.2019.01.110

dos Santos, C., Buera, P., & Mazzobre, F. (2017). Novel trends in cyclodextrins encapsulation. Applications in food science. *Current Opinion in Food Science, 16*, 106–113. https://doi.org/10.1016/j.cofs.2017.09.002

Eun, J.-B., Ahmed, M., Das, P. R., & Nam, S.-H. (2020). A review of encapsulation of carotenoids using spray drying and freeze drying. *Critical Reviews in Food Science and Nutrition, 60*(21), 3547–3572. https://doi.org/10.1080/10408398.2019.1698511

Fayed, B., El-Sayed, H. S., Abood, A., Hashem, A. M., & Mehanna, N. S. H. (2019). The application of multi-particulate microcapsule containing probiotic bacteria and inulin nanoparticles in enhancing the probiotic survivability in yoghurt. *Biocatalysis and Agricultural Biotechnology, 22*. https://doi.org/10.1016/j.bcab.2019.101391

Firdaus, T. R., Pranoto, Y., Manuel Brueck, W., & Witasari, L. D. (2025). Probiotic encapsulation using porous starch from arrowroot (Maranta arundinacea). *Food Chemistry Advances, 8*, Article 101077. https://doi.org/10.1016/j.focha.2025.101077

Gangurde, A. B., Ali, M. T., Pawar, J. N., & Amin, P. D. (2017). Encapsulation of vitamin E acetate to convert oil to powder microcapsule using different starch derivatives. *Journal of Pharmaceutical Investigation, 47*(6), 559–574. https://doi.org/10.1007/s40005-016-0287-3

Gao, Q., Ou, Z., Jia, X., Ahmed Junejo, S., Zhang, B., & Huang, Q. (2025). β-Cyclodextrin synergistically enhanced solid encapsulation of p-anisaldehyde into V-type starch for sweet cherry preservation. *Carbohydrate Polymers, 358*, Article 123513. https://doi.org/10.1016/j.carbpol.2025.123513

García-Tejeda, Y. V., Salinas-Moreno, Y., Hernández-Martínez, Á. R., & Martínez-Bustos, F. (2016). Encapsulation of purple maize anthocyanins in phosphorylated starch by spray drying. *Cereal Chemistry, 93*(2), 130–137. https://doi.org/10.1094/CCHEM-04-15-0072-R

Gomes, S., Finotelli, P. V., Sardela, V. F., Pereira, H. M. G., Santelli, R. E., Freire, A. S., & Torres, A. G. (2019). Microencapsulated Brazil nut (Bertholletia excelsa) cake extract powder as an added-value functional food ingredient. *Lebensmittel-Wissenschaft & Technologie, 116*. https://doi.org/10.1016/j.lwt.2019.108495

González Mendoza, M. E., Martínez-Bustos Fernando Castaño-Tostado Eduardo Cortez-Trejo María, D., & Amaya-Llano Silvia, L. (2025). Microwave energy OSA starches faba bean probiotic microencapsulation spray drying 8181 microwave–assisted OSA–Faba bean starch production for probiotic microencapsulation. *Polysaccharides, 6*(3), 2673–4176. https://doi.org/10.3390/polysaccharides6030081

Grgić, J., Šelo, G., Planinić, M., Tišma, M., & Bucić-Kojić, A. (2020). Role of the encapsulation in bioavailability of phenolic compounds. *Antioxidants, 9*(10), 923. https://doi.org/10.3390/antiox9100923

Guo, Y., Qiao, D., Zhao, S., Zhang, B., & Xie, F. (2021). Starch-based materials encapsulating food ingredients: Recent advances in fabrication methods and applications. *Carbohydrate Polymers, 270*. https://doi.org/10.1016/j.carbpol.2021.118358

Gupta, C., Chawla, P., Arora, S., Tomar, S. K., & Singh, A. K. (2015). Iron microencapsulation with blend of gum Arabic, maltodextrin and modified starch using modified solvent evaporation method - Milk fortification. *Food Hydrocolloids, 43*, 622–628. https://doi.org/10.1016/j.foodhyd.2014.07.021

Hasani, M., Rad, A. H. E., Hosseini, M. M., & Noghabi, M. S. (2015). Physicochemical characteristic of microencapsulated fish oil by freeze-drying using different combinations of wall materials. *Biosciences Biotechnology Research Asia, 12*, 45–51. https://doi.org/10.13005/bbra/2171

He, H., Hong, Y., Gu, Z., Liu, G., Cheng, L., & Li, Z. (2016). Improved stability and controlled release of CLA with spray-dried microcapsules of OSA-Modified starch and xanthan gum. *Carbohydrate Polymers, 147*, 243–250. https://doi.org/10.1016/j.carbpol.2016.03.078

Hoyos-Leyva, J. D., Bello-Pérez, L. A., Alvarez-Ramirez, J., & Garcia, H. S. (2018). Microencapsulation using starch as wall material: A review. *Food Reviews International, 34*(2), 148–161. https://doi.org/10.1080/87559129.2016.1261298

Hsieh, C. F., Liu, W., Whaley, J. K., & Shi, Y. C. (2019). Structure, properties, and potential applications of waxy tapioca starches – A review. *Trends in Food Science and Technology, 83*, 225–234. https://doi.org/10.1016/j.tifs.2018.11.022

Huang, R. M., Feng, K., Li, S. F., Zong, M. H., Wu, H., & Han, S. Y. (2021). Enhanced survival of probiotics in the electrosprayed microcapsule by addition of fish oil. *Journal of Food Engineering, 307*. https://doi.org/10.1016/j.jfoodeng.2021.110650

Jeong, H.-M., Kang, H.-N., Kim, M.-Y., Cho, H.-S., Baek, J.-W., Yu, Y.-J., & Shim, J.-H. (2025). Enhanced physicochemical properties of isoflavone derivatives through enzymatic encapsulation using cyclodextrin glucanotransferase. *Food Hydrocolloids, 161*, Article 110874. https://doi.org/10.1016/j.foodhyd.2024.110874

Jia, X., Sun, S., Chen, B., Zheng, B., & Guo, Z. (2018). Understanding the crystal structure of lotus seed amylose–long-chain fatty acid complexes prepared by high hydrostatic pressure. *Food Research International, 111*, 334–341. https://doi.org/10.1016/j.foodres.2018.05.053

Jobling, S. (2004). Improving starch for food and industrial applications. *Current Opinion in Plant Biology, 7*(2), 210–218. https://doi.org/10.1016/j.pbi.2003.12.001

Kapuśniak, J., & Tomasik, P. (2006). Lipid microencapsulation in starch. *Journal of Microencapsulation, 23*(3), 341–348. https://doi.org/10.1080/02652040600687571

Koksel, H., Muti, S. O., & Kahraman, K. (2023). *Starch ICC handbook of 21st century cereal science and technology*. Turkey: Elsevier. https://doi.org/10.1016/B978-0-323-95295-8.00026-5

Kong, L., Bhosale, R., & Ziegler, G. R. (2018). Encapsulation and stabilization of β-carotene by amylose inclusion complexes. *Food Research International, 105*, 446–452. https://doi.org/10.1016/j.foodres.2017.11.058

Kulal, R., Bhaskarrao Borse, B., & Serva Peddha, M. (2025). Nano-encapsulation of spice oleoresins using hydroxypropyl β-cyclodextrin (HPBCD) and its bioavailability studies. *Measurement: Food, 18*, Article 100219. https://doi.org/10.1016/j.meafoo.2025.100219

Kumari, S., Yadav, B. S., & Yadav, R. B. (2020). Synthesis and modification approaches for starch nanoparticles for their emerging food industrial applications: A review. *Food Research International, 128*. https://doi.org/10.1016/j.foodres.2019.108765

Leyva-López, R., Palma-Rodríguez, H. M., López-Torres, A., Capataz-Tafur, J., Bello-Pérez, L. A., & Vargas-Torres, A. (2019). Use of enzymatically modified starch in the microencapsulation of ascorbic acid: Microcapsule characterization, release behavior and in vitro digestion. *Food Hydrocolloids, 96*, 259–266. https://doi.org/10.1016/j.foodhyd.2019.04.056

Li, S., Li, Q., Zhu, F., Song, H., Wang, C., & Guan, X. (2022). Effect of vacuum combined ultrasound treatment on the fine structure and physiochemical properties of rice starch. *Food Hydrocolloids, 124*. https://doi.org/10.1016/j.foodhyd.2021.107198

Li, H., Zhai, F., Li, J., Zhu, X., Guo, Y., Zhao, B., & Xu, B. (2021). Physicochemical properties and structure of modified potato starch granules and their complex with tea polyphenols. *International Journal of Biological Macromolecules, 166*, 521–528. https://doi.org/10.1016/j.ijbiomac.2020.10.209

Li, L., Zhang, W., Peng, J., Xue, B., Liu, Z., Luo, Z., Lu, D., & Zhao, X. (2020). A novel shell Material—Highland barley starch for microencapsulation of cinnamon essential oil with different preparation methods. *Materials, 13*(5), 1192. https://doi.org/10.3390/ma13051192

Lian, F., Gong, E., Liang, H., Lin, Y., Chen, J., He, Y., Hebelstrup, K. H., & Wen, X. (2022). Nano-encapsulation of polyphenols in starch nanoparticles: Fabrication, characterization and evaluation. *Food & Function, 13*(14), 7762–7771. https://doi.org/10.1039/d1fo04197e

Lin, H., Qin, L., Hong, H., & Li, Q. (2016). Preparation of starch nanoparticles via high-energy ball milling. *Journal of Nano Research, 40*, 174–179. https://doi.org/10.4028/www.scientific.net/JNanoR.40.174

Liu, C., Li, M., Ji, N., Liu, J., Xiong, L., & Sun, Q. (2017). Morphology and characteristics of starch nanoparticles self-assembled via a rapid ultrasonication method for peppermint oil encapsulation. *Journal of Agricultural and Food Chemistry, 65*(38), 8363–8373. https://doi.org/10.1021/acs.jafc.7b02938

Lopes, M. M., Sanchez Farinas, C., Bueno, M. M., García-Moreno, P. J., & Guadix, E. M. (2025a). Microencapsulation of Bacillus subtilis by spray-drying using starch hydrolysates with different dextrose equivalent values. *International Journal of Biological Macromolecules, 320*(Pt 3), Article 146082. https://doi.org/10.1016/j.ijbiomac.2025.146082

Lopes, M. M., Sanchez Farinas, C., Bueno, M. M., García-Moreno, P. J., & Guadix, E. M. (2025b). Microencapsulation of Bacillus subtilis by spray-drying using starch hydrolysates with different dextrose equivalent values. *International Journal of Biological Macromolecules, 320*. https://doi.org/10.1016/j.ijbiomac.2025.146082

López-Córdoba, A., Deladino, L., & Martino, M. (2014). Corn starch-calcium alginate matrices for the simultaneous carrying of zinc and yerba mate antioxidants. *Lebensmittel-Wissenschaft & Technologie, 59*(2P1), 641–648. https://doi.org/10.1016/j.lwt.2014.06.021

Lu, K., & Wang, J. (2022). Physicochemical properties, structural and in vitro digestibility of starch from five different sources. *Science of Advanced Materials, 14*(4), 695–702. https://doi.org/10.1166/sam.2022.4258

Madene, A., Jacquot, M., Scher, J., & Desobry, S. (2006). Flavour encapsulation and controlled release – A review. *International Journal of Food Science and Technology, 41*(1), 1–21. https://doi.org/10.1111/j.1365-2621.2005.00980.x

Marefati, A., Sjöö, M., Timgren, A., Dejmek, P., & Rayner, M. (2015). Fabrication of encapsulated oil powders from starch granule stabilized W/O/W pickering emulsions by freeze-drying. *Food Hydrocolloids, 51*, 261–271. https://doi.org/10.1016/j.foodhyd.2015.04.022

Marques, H. M. C. (2010). A review on cyclodextrin encapsulation of essential oils and volatiles. *Flavour and Fragrance Journal, 25*(5), 313–326. https://doi.org/10.1002/ffj.2019

Masina, N., Choonara, Y. E., Kumar, P., du Toit, L. C., Govender, M., Indermun, S., & Pillay, V. (2017). A review of the chemical modification techniques of starch. *Carbohydrate Polymers, 157*, 1226–1236. https://doi.org/10.1016/j.carbpol.2016.09.094

Mayilvahanan, A., Ramchary, A., Niraikulam, A., Marichetti Kuppuswami, G., & Numbi Ramudu, K. (2019). A green process for starch oleate synthesis by cryptococcus sp. MTCC 5455 lipase and its potential as an emulsifying agent. *Starch - Stärke, 71*(1–2). https://doi.org/10.1002/star.201700325

Miao, Z., Zhang, Y., & Lu, P. (2021). Novel active starch films incorporating tea polyphenols-loaded porous starch as food packaging materials. *International Journal of Biological Macromolecules, 192*, 1123–1133. https://doi.org/10.1016/j.ijbiomac.2021.09.214

Mohamad, S. N. H., Idayu Muhamad, I., Jusoh, Y. M. M., & Khairuddin, N. (2018). Dielectric properties for selected wall material in the development of microwave-encapsulation-drying. *Journal of Food Science and Technology, 55*(12), 5161–5165. https://doi.org/10.1007/s13197-018-3327-3

Mohd Nawi, N., Idayu Muhamad, I., & Marsin, A. M. (2015). The physicochemical properties of microwave-assisted encapsulated anthocyanins from Ipomoea batatas as affected by different wall materials. *Food Science and Nutrition, 3*(2), 91–99. https://doi.org/10.1002/fsn3.132

Mongenot, N., Charrier, S., & Chalier, P. (2000). Effect of ultrasound emulsification on cheese aroma encapsulation by carbohydrates. *Journal of Agricultural and Food Chemistry, 48*(3), 861–867. https://doi.org/10.1021/jf990494n

Montoya-Yepes, D. F., Jiménez-Rodríguez, A. A., Aldana-Porras, A. E., Velásquez-Holguin, L. F., Méndez-Arteaga, J. J., & Murillo-Arango, W. (2024). Starches in the encapsulation of plant active ingredients: State of the art and research trends. *Polymer Bulletin, 81*(1), 135–163. https://doi.org/10.1007/s00289-023-04724-6

Morán, D., Gutiérrez, G., Carmen Blanco-López, M., Marefati, A., Rayner, M., & Matos, M. (2021). Synthesis of starch nanoparticles and their applications for bioactive compound encapsulation. *Applied Sciences, 11*(10), 4547. https://doi.org/10.3390/app11104547

Muñoz-Shugulí, C., Vidal, C. P., Cantero-López, P., & Lopez-Polo, J. (2021). Encapsulation of plant extract compounds using cyclodextrin inclusion complexes, liposomes, electrospinning and their combinations for food purposes. *Trends in Food Science & Technology, 108*, 177–186. https://doi.org/10.1016/j.tifs.2020.12.020

Noman, M., Afzaal, M., Saeed, F., Ahmad, A., Ali, I., Akram, N., Asghar, A., Shah, Y. A., Ateeq, H., Khan, M. R., Lorenzo, J. M., & Asres, D. T. (2023). Effect of starch-based nanoparticles on the viability and stability of probiotics under adverse conditions. *International Journal of Food Properties, 26*(1), 1841–1854. https://doi.org/10.1080/10942912.2023.2236316

Ordoñez, M., & Herrera, A. (2014). Morphologic and stability cassava starch matrices for encapsulating limonene by spray drying. *Powder Technology, 253*, 89–97. https://doi.org/10.1016/j.powtec.2013.11.005

Oyeyinka, S. A., Singh, S., Ma, Y., & Amonsou, E. O. (2016). Influence of high-pressure homogenization on the physicochemical properties of Bambara starch complexed with lysophosphatidylcholine. *Lebensmittel-Wissenschaft & Technologie, 74*, 120–127. https://doi.org/10.1016/j.lwt.2016.07.035

Pellicer, J. A., Fortea, M. I., Trabal, J., Rodríguez-López, M. I., Gabaldón, J. A., & Núñez-Delicado, E. (2019). Stability of microencapsulated strawberry flavour by spray drying, freeze drying and fluid bed. *Powder Technology, 347*, 179–185. https://doi.org/10.1016/j.powtec.2019.03.010

Perrechil, F., Louzi, V. C., Paiva, L. A.da S., Natal, G. S. V., & Braga, M. B. (2021). Evaluation of modified starch and rice protein concentrate as wall materials on the microencapsulation of flaxseed oil by freeze-drying. *Lebensmittel-Wissenschaft & Technologie, 140*, Article 110760. https://doi.org/10.1016/j.lwt.2020.110760

Qiu, L., Ma, H., Luo, Q., Bai, C., Xiong, G., Jin, S., Wang, J., Zu, X., Li, H., & Liao, T. (2022). Preparation, characterization, and application of modified starch/chitosan/sweet Orange oil microcapsules. *Foods, 11*(15), 2306. https://doi.org/10.3390/foods11152306

Ramos, G. V. C., Ramírez-López, S., de Pinho, S. C., Ditchfield, C., & Moraes, I. C. F. (2025). Starch-based pickering emulsions for bioactive compound encapsulation: Production, properties, and applications. *Processes, 13*(2), 342. https://doi.org/10.3390/pr13020342

Rehman, A., Tong, Q., Jafari, S. M., Assadpour, E., Shehzad, Q., Aadil, R. M., Iqbal, M. W., Rashed, M. M. A., Mushtaq, B. S., & Ashraf, W. (2020). Carotenoid-loaded nanocarriers: A comprehensive review. *Advances in Colloid and Interface Science, 275*. https://doi.org/10.1016/j.cis.2019.102048

Rehman, A., Tong, Q., Jafari, S. M., Korma, S. A., Khan, I. M., Mohsin, A., Manzoor, M. F., Ashraf, W., Mushtaq, B. S., Zainab, S., & Cacciotti, I. (2021). Spray dried nanoemulsions loaded with curcumin, resveratrol, and borage seed oil: The role of two different modified starches as encapsulating materials. *International Journal of Biological Macromolecules, 186*, 820–828. https://doi.org/10.1016/j.ijbiomac.2021.07.076

Remanan, M. K., & Zhu, F. (2021). Encapsulation of rutin using quinoa and maize starch nanoparticles. *Food Chemistry, 353*, Article 128534. https://doi.org/10.1016/j.foodchem.2020.128534

Rezvankhah, A., Emam-Djomeh, Z., & Askari, G. (2020). Encapsulation and delivery of bioactive compounds using spray and freeze-drying techniques: A review. *Drying Technology, 38*(1–2), 235–258. https://doi.org/10.1080/07373937.2019.1653906

Rincón Aguirre, A., Pérez, L. A. B., Mendoza, S., Real, A.del, & Enrique Rodríguez García, M. (2018). Physicochemical studies of taro starch chemically modified by acetylation, phosphorylation, and succinylation. *Starch - Stärke, 70*(3–4). https://doi.org/10.1002/star.201700066

Romero-Hernandez, H. A., Sánchez-Rivera, M. M., Alvarez-Ramirez, J., Yee-Madeira, H., Yañez-Fernandez, J., & Bello-Pérez, L. A. (2021). Avocado oil encapsulation with OSA-Esterified taro starch as wall material: Physicochemical and morphology characteristics. *Lebensmittel-Wissenschaft & Technologie, 138*. https://doi.org/10.1016/j.lwt.2020.110629

Samakradhamrongthai, R., Thakeow, P., Kopermsub, P., & Utama-ang, N. (2016). Microencapsulation of white champaca (Michelia Alba D.C.) extract using octenyl succinic anhydride (OSA) starch for controlled release aroma. *Journal of Microencapsulation, 33*(8), 773–784. https://doi.org/10.1080/02652048.2016.1264493

Schlindweinn, E. B., Chacon, W. D. C., Luiza Koop, B., Fonseca, J.de M., Monteiro, A. R., & Valencia, G. A. (2022). Starch-based materials encapsulating anthocyanins: A review. *Journal of Polymers and the Environment, 30*(9), 3547–3565. https://doi.org/10.1007/s10924-022-02474-6

Selvasekaran, P., & Chidambaram, R. (2021). Food-grade aerogels obtained from polysaccharides, proteins, and seed mucilages: Role as a carrier matrix of functional food ingredients. *Trends in Food Science and Technology, 112*, 455–470. https://doi.org/10.1016/j.tifs.2021.04.021

Shabana, S., Prasansha, R., Kalinina, I., Potoroko, I., Bagale, U., & Shirish, S. H. (2019). Ultrasound assisted acid hydrolyzed structure modification and loading of antioxidants on potato starch nanoparticles. *Ultrasonics Sonochemistry, 51*, 444–450. https://doi.org/10.1016/j.ultsonch.2018.07.023

Shan, Y., Li, J., Nie, M., Li, D., Zhang, Y., Li, Y., Wang, L., Liu, L., Wang, F., & Tong, L.-T. (2025). A comprehensive review of starch-based technology for encapsulation of flavor: From methods, materials, and release mechanism to applications. *Carbohydrate Polymers, 348*. https://doi.org/10.1016/j.carbpol.2024.122816

Shi, Y., Zhou, S., Fan, S., Ma, Y., Li, D., Tao, Y., & Han, Y. (2021). Encapsulation of bioactive polyphenols by starch and their impacts on gut microbiota. *Current Opinion in Food Science, 38*, 102–111. https://doi.org/10.1016/j.cofs.2020.11.001

Singh, A., Swer, T. L., Kumar, L., Rana, A., Taneja, N. K., & Jha, P. (2025). Exploring the performance of jackfruit seed starch and protein microcapsules as probiotic carrier systems: Enhancing viability, shelf life, and gastrointestinal resilience. *Journal of Food Processing and Preservation, 2025*(1). https://doi.org/10.1155/jfpp/9970101

Soler, A., Valenzuela-Díaz, E. D., Velazquez, G., Huerta-Ruelas, J. A., Morales-Sanchez, E., Hernandez-Gama, R., & Mendez-Montealvo, G. (2020). Double helical order and functional properties of acid-hydrolyzed maize starches with different amylose content. *Carbohydrate Research, 490*. https://doi.org/10.1016/j.carres.2020.107956

Song, J., Zong, J., Ma, C., Chen, S., Li, H., & Zhang, D. (2021). Microparticle prepared by chitosan coating on the extruded mixture of corn starch, resveratrol, and α-amylase controlled the resveratrol release. *International Journal of Biological Macromolecules, 185*, 773–781. https://doi.org/10.1016/j.ijbiomac.2021.06.154

Tang, R., Xu, Y., Zhang, Z., Liu, J., Pu, Y., Zhao, R., Liu, Z., & Jiang, H. (2025). Encapsulation of folic acid-loaded porous starch by glycerol monostearate improves its thermal stability and bioavailability in soda crackers. *International Journal of Biological Macromolecules, 333*. https://doi.org/10.1016/j.ijbiomac.2025.148766

Tian, S., Xue, X., Wang, X., & Chen, Z. (2022). Preparation of starch-based functional food nano-microcapsule delivery system and its controlled release characteristics . Frontiers media S.A. *China Frontiers in Nutrition, 9*. https://doi.org/10.3389/fnut.2022.982370

Tiozon, R. J. N., Bonto, A. P., & Sreenivasulu, N. (2021a). Enhancing the functional properties of rice starch through biopolymer blending for industrial applications: A review. *International Journal of Biological Macromolecules, 192*, 100–117. https://doi.org/10.1016/j.ijbiomac.2021.09.194

Tiozon, R. J. N., Bonto, A. P., & Sreenivasulu, N. (2021b). Enhancing the functional properties of rice starch through biopolymer blending for industrial applications: A review. *International Journal of Biological Macromolecules, 192*, 100–117. https://doi.org/10.1016/j.ijbiomac.2021.09.194

Venkatakrishnan, K., Chiu, H. F., & Wang, C. K. (2019). Popular functional foods and herbs for the management of type-2-diabetes mellitus: A comprehensive review with special reference to clinical trials and its proposed mechanism. *Journal of Functional Foods, 57*, 425–438. https://doi.org/10.1016/j.jff.2019.04.039

Verdalet-Guzmán, I., Martínez-Ortiz, L., & Martínez-Bustos, F. (2013). Characterization of new sources of derivative starches as wall materials of essential oil by spray drying. *Food Science and Technology, 33*(4), 757–764. https://doi.org/10.1590/S0101-20612013000400023

Wang, S., Chen, Y., Liang, H., Chen, Y., Shi, M., Wu, J., Liu, X., Li, Z., Liu, B., Yuan, Q., & Li, Y. (2015). Intestine-specific delivery of hydrophobic bioactives from oxidized starch microspheres with an enhanced stability. *Journal of Agricultural and Food Chemistry, 63*(39), 8669–8675. https://doi.org/10.1021/acs.jafc.5b03575

Wang, S., & Guo, P. (2020). *Botanical sources of starch starch structure, functionality and application in foods*. China: Springer Singapore. https://doi.org/10.1007/978-981-15-0622-2_2

Wang, R., Li, M., Brennan, M. A., Dhital, S., Kulasiri, D., Brennan, C. S., & Guo, B. (2023). Complexation of starch and phenolic compounds during food processing and impacts on the release of phenolic compounds. *Comprehensive Reviews in Food Science and Food Safety, 22*(4), 3185–3211. https://doi.org/10.1111/1541-4337.13180

Wang, Y., Zheng, Z., Wang, K., Tang, C., Liu, Y., & Li, J. (2020). Prebiotic carbohydrates: Effect on physicochemical stability and solubility of algal oil nanoparticles. *Carbohydrate Polymers, 228*. https://doi.org/10.1016/j.carbpol.2019.115372

Wen, Y., Yao, T., Xu, Y., Corke, H., & Sui, Z. (2020). Pasting, thermal and rheological properties of octenylsuccinylate modified starches from diverse small granule starches differing in amylose content. *Journal of Cereal Science, 95*. https://doi.org/10.1016/j.jcs.2020.103030

Xiao, Z., Kang, Y., Hou, W., Niu, Y., & Kou, X. (2019). Microcapsules based on octenyl succinic anhydride (OSA)-Modified starch and maltodextrins changing the composition and release property of rose essential oil. *International Journal of Biological Macromolecules, 137*, 132–138. https://doi.org/10.1016/j.ijbiomac.2019.06.178

Xiong, K., Zhou, L., Wang, J., Ma, A., Fang, D., Xiong, L., & Sun, Q. (2020). Construction of food-grade pH-sensitive nanoparticles for delivering functional food ingredients. *Trends in Food Science and Technology, 96*, 102–113. https://doi.org/10.1016/j.tifs.2019.12.019

Xu, H., Ma, Q., Qiu, C., Wang, J., Jin, Z., & Hu, Y. (2024). Encapsulation and controlled delivery of curcumin by self-assembled cyclodextrin succinate/chitosan nanoparticles. *Food Hydrocolloids, 157*, Article 110465. https://doi.org/10.1016/j.foodhyd.2024.110465

Yang, J., He, H., Gu, Z., Cheng, L., Li, C., Li, Z., & Hong, Y. (2020). Conjugated linoleic acid loaded starch-based emulsion nanoparticles: In vivo gastrointestinal controlled release. *Food Hydrocolloids, 101*. https://doi.org/10.1016/j.foodhyd.2019.105477

Yang, Y., Yang, X., Yang, L., Tian, H., Li, J., & Zhao, X. (2025). Ultrasound-enzyme synergistic and octenyl succinic anhydride dual-modified acorn starch: A green carrier for enhanced stability and sustained release of gentiopicroside. *ACS Sustainable Chemistry & Engineering, 13*(42), 18116–18128. https://doi.org/10.1021/acssuschemeng.5c07588

Yeo, L., Thompson, D. B., & Peterson, D. G. (2016). Inclusion complexation of flavour compounds by dispersed high-amylose maize starch (HAMS) in an aqueous model system. *Food Chemistry, 199*, 393–400. https://doi.org/10.1016/j.foodchem.2015.12.054

Yildiz, G., Ding, J., Gaur, S., Andrade, J., Engeseth, N. E., & Feng, H. (2018). Microencapsulation of docosahexaenoic acid (DHA) with four wall materials including pea protein-modified starch complex. *International Journal of Biological Macromolecules, 114*, 935–941. https://doi.org/10.1016/j.ijbiomac.2018.03.175

Yuan, L., Feng, W., Zhang, Z., Peng, Y., Xiao, Y., & Chen, J. (2021). Effect of potato starch-based antibacterial composite films with thyme oil microemulsion or microcapsule on shelf life of chilled meat. *Lebensmittel-Wissenschaft & Technologie, 139*. https://doi.org/10.1016/j.lwt.2020.110462

Zarski, A., Ptak, S., Siemion, P., & Kapusniak, J. (2016). Esterification of potato starch by a biocatalysed reaction in an ionic liquid. *Carbohydrate Polymers, 137*, 657–663. https://doi.org/10.1016/j.carbpol.2015.11.029

Zhang, L., Liu, Y., Wu, Z., & Chen, H. (2009). Preparation and characterization of coacervate microcapsules for the delivery of antimicrobial oyster peptides. *Drug Development and Industrial Pharmacy, 35*(3), 369–378. https://doi.org/10.1080/03639040802369255

Zhao, B., Wang, B., Zheng, B., Chen, L., & Guo, Z. (2019). Effects and mechanism of high-pressure homogenization on the characterization and digestion behavior of lotus seed starch–green tea polyphenol complexes. *Journal of Functional Foods, 57*, 173–181. https://doi.org/10.1016/j.jff.2019.04.016

Zhou, D., Liu, S., Song, H., Liu, X., & Tang, X. (2020). Effect of pullulanase debranching on complexation, structure, digestibility, and release of starch-ascorbyl palmitate inclusion complexes. *Journal of Food Processing and Preservation, 44*(11). https://doi.org/10.1111/jfpp.14878

Zhu, F. (2017). Encapsulation and delivery of food ingredients using starch based systems. *Food Chemistry, 229*, 542–552. https://doi.org/10.1016/j.foodchem.2017.02.101

Zhu, H., Zhang, Y., Tian, J., & Chu, Z. (2018). Effect of a new shell material—Jackfruit seed starch on novel flavor microcapsules containing vanilla oil. *Industrial Crops and Products, 112*, 47–52. https://doi.org/10.1016/j.indcrop.2017.10.060

Chapter 19

Recent advancements in thermoplastic starches for food packaging

P. Poojapushkarni[1], Abhishek Biswal[1], Naveen Krishna[2], Ashwina Shankar[2], Vinoth Kumar Vaidyanathan[2], Periyar Selvam Sellamuthu[1] and Emmanuel Rotimi Sadiku[3]

[1]*Department of Food Process Engineering, Postharvest Technology Research Lab, School of Bioengineering, SRM Institute of Science and Technology, Chennai, Tamil Nadu, India;* [2]*Integrated Bioprocessing Laboratory, Department of Biotechnology, School of Bioengineering, SRM Institute of Science and Technology, Chennai, Tamil Nadu, India;* [3]*Institute of NanoEngineering Research (INER) and Department of Chemical, Metallurgical and Materials Engineering, Tshwane University of Technology, Pretoria West Campus, Pretoria, Republic of South Africa*

1. Introduction

The pervasive accumulation of petroleum-based plastics in landfills and marine environments poses a significant threat to the ecological systems, wildlife, and human health. According to the U.S. Environmental Protection Agency (US EPA), marine organisms frequently ingest plastic debris, resulting in injuries, malnutrition, and fatalities to marine life (Koelmans et al., 2019). The pervasive issue of plastic pollution extends beyond visible debris, with microplastics posing a concerning hazard to the environment. These minute particles, fragmented from larger plastic items, tend to readily infiltrate the ecosystems. This contamination allows microplastics to enter the food chain, creating a pathway for human exposure. The consumption of contaminated seafood, produce grown in affected soil and even drinking water exposes individuals to these particles. Consequently, microplastics present a significant risk to human health, raising concerns about potential long-term impacts (Wang et al., 2021).

The major contributors of microplastics are petroleum-based single-use plastics (Mutmainna et al., 2024). This plastic packaging is naturally nonbiodegradable and releases harmful chemicals during the degradation process (Idris et al., 2023). Substances such as bisphenol A (BPA) and phthalates are leached into soils and waterways, functioning as an endocrine disruptor that poses severe health risks, including reproductive and developmental issues in both wildlife and humans (Gündoğdu et al., 2024). In accordance with the increasing global population and the current trends in waste management, plastic production is estimated to triple by 2050 (Kibria et al., 2023; Subramaniyan et al., 2025). The escalating threat of plastic pollution has instigated researchers and industries to explore sustainable alternatives, so as to reduce environmental impact (Idris et al., 2023).

As naturally derived materials, proteins and polysaccharides are the future in the production of biodegradable polymers (Baranwal et al., 2022). Plant-based polysaccharides are used widely for the synthesis of films and coatings in the food sector (Yao et al., 2024). Due to its abundance and affordability, starch is preferred over other traditional materials (Schutz et al., 2024). The mechanical characteristics and barrier properties of starch-based films can be manipulated to enhance the shelf life of food (Surendren et al., 2022). Incorporating plasticizers into starch-based films improves their flexibility and processability, making them more suitable for use as advanced biodegradable packaging solutions (Benitez et al., 2024). Adding essential oils (EOs) to TPS formulations can help to achieve antimicrobial properties. EOs also help in extending the shelf life of products by restricting the growth of spoilage organisms and maintaining the freshness. Food packaging materials processed, including the biodegradability of TPS with active functionalities, would be a viable, environmentally sustainable solution in the future (Bangar et al., 2021a, 2021b; Todhanakasem et al., 2022). This chapter focuses on the limitations of native starch (NSs), thermoplastic starch (TPS) and the process of conversion of NSs or modified starch into TPS. Besides, it examines the possibilities of adding bioactive compounds and nanoparticles to improve its properties and make it a feasible and sustainable option for food packaging materials.

Native and Modified Starches for Food Applications. https://doi.org/10.1016/B978-0-443-27431-2.00025-0

2. Basic concepts of thermoplastic starches

As a major polysaccharide, starch acts as the main energy reserve in plants. It comprises amylose and amylopectin, two glucose polymers. The physicochemical properties and their behavior during processing, especially during the conversion of starch into TPS for food packaging applications, are significantly affected by the ratio of amylose to amylopectin and their arrangement (Rashwan et al., 2024). The semicrystalline form of starch consists of alternating amorphous and crystalline regions. The hydrogen bonds between the hydroxyl groups of the glucose units are responsible for the stability of the structure (Li et al., 2023). During thermal processing, starch granules are subjected to alterations by water diffusion, granule expansion, gelatinization, decomposition, melting, and crystallization. Other processes are influenced by gelatinization. Starch is not suitable for conventional plastic processing without undergoing gelatinization as it decomposes at temperatures lower than its melting point (Jiang et al., 2020). The crystalline structure of starch (which is observed due to the amylose and amylopectin structures) is changed into an amorphous form before molding. Plasticization is done with the aim of reducing the intermolecular forces within the starch structure, as it leads to flexibility (Surendren et al., 2022).

2.1 Starch plasticization

Plasticization converts native starch into a malleable thermoplastic material (Huneault & Li, 2010). This is done by disrupting the semicrystalline structure of starch granules and adding plasticizers, which leads to a reduction of intermolecular forces and thermoplastic behavior (Guarás et al., 2020). The process of starch plasticization is characterized by the gelatinization of starch granules, the subsequent creation of hydrogen bonds, and the resulting decrease in the glass transition temperature (Tg).

The process of plasticization begins by disrupting the starch granules under heat and mechanical forces (Guarás et al., 2020). The semicrystalline structure of native starch comes from the arrangement of amylose and amylopectin (Varghese et al., 2023). Extrusion causes breakdown of these granules, converting their semicrystalline structure into an amorphous phase. This enhances the mobility of the polymer chain and, in turn, leads to better plasticization (Donmez et al., 2021; Sejidov et al., 2005).

Plasticizers are essential to modify the structure and function of starch (Yang et al., 2024). Low-molecular-weight plasticizers such as water, glycerol, sorbitol, and ionic liquids (ILs) penetrate the starch matrix and disrupt the intermolecular hydrogen bonds between starch molecules. This results in the increase of chain mobility, reduces stiffness, and enhances thermoplastic properties (Da Róz et al., 2006; Vieira et al., 2011). Gelatinization and melting occur as a result of heat and plasticizer interaction, breaking the crystalline regions and allowing the starch to form a flexible, processable matrix. This transformation typically occurs between 60 and 70°C, depending on starch type and plasticizer concentration. The hydrogen bond formation between plasticizers and starch hydroxyl (-OH) groups is a key factor in stabilizing TPS. Intermolecular forces weaken, and the energy density reduces upon adding plasticizers. Hydrogen bonds break between starch molecules, and new hydrogen bonds form between the starch and plasticizer molecules. This increases flexibility and reduces rigidity (Nafchi et al., 2013; Shanks & Kong, 2012). The glass transition temperature in native starch is mostly above its degradation temperature. This makes it unsuitable for processing. However, plasticizers reduce the glass transition temperature to a range of 50–100°C, depending on the formulation, enabling the processing of TPS in a suitable temperature range (Zdanowicz & Sałasińska, 2023). This plasticized starch is comparatively more ductile and less brittle than native starch (Chaléat et al., 2014). By adjusting the type and amount of plasticizer, these properties can be manipulated (Csiszár et al., 2021).

2.2 Types of plasticizers used in thermoplastic starch

Plasticizers are compounds included in polymers to enhance their flexibility, durability, and processability. They achieve this by reducing the polymer's glass transition temperature (Tg). Table 19.1 explain about the types of plasticizers and their influence on thermoplastic starch properties. These additives are often low molecular weight liquids or resins that engage with the polymer chains via secondary bonds; therefore, augmenting the interstitial space (Eslami et al., 2023). The reduction of polymer-polymer chain interactions enhances the mobility of macromolecules, leading to a softer and more malleable material. The effectiveness of a plasticizer depends on factors such as its molecular weight, polarity, and compatibility with starch (Bangar et al., 2021a, 2021b).

2.3 Limitations of pure TPS in food packaging

TPS has natural advantages like availability, affordability, and biodegradability. However, pure TPS has many limits that impede its extensive application in food packaging, requiring alterations and integration with other materials (Chen et al., 2020).

TABLE 19.1 Types of plasticizers and their influence on thermoplastic starch (TPS).

Plasticizer type	Examples	Key characteristics	Influence on TPS properties	References
Polyols	Glycerol, sorbitol, xylitol, mannitol, and PEG	Comprise many hydroxyl (-OH) groups, facilitating hydrogen bonding with starch molecules. Frequently utilized due to its accessibility and superior plasticizing efficacy.	Decrease Tg: improve processability and flexibility. Enhance ductility and elongation properties. Enhance water sensitivity based on hygroscopic properties.	Eslami et al. (2023), Yang et al. (2024)
Sugars	Glucose, sucrose, and fructose	Exhibits lower efficacy as plasticizers compared to polyols. Commonly utilized in conjunction with alternative plastic	Enhance flexibility and elongation; however, their efficiency as sole plasticizers is limited.	Ozeren et al. (2020)
Water	-	Low molecular weight highly volatile , and readily available	Increases Water Vapor Permeability [WVP] due to its inherent nature. Often used in conjunction with other plasticizers. It regulates the temperature during the hot compounding of thermoplastics.	Gao et al. (2017a, 2017b)
Organic acids and esters	Citric acid, lactic acid, and esters	Nontoxic food safe nonvolatile	Enhances plasticization and improves compatibility with other polymers. Prevents starch retrogradation and modifies molar mass. Reduces crystallinity, improves solubility, and increases flexibility.	Da Róz et al. (2011)
Ionic liquids (ILs)	-	Recyclable Low vapor pressure Nonflammable cation and anion structures can be altered to dissolve diverse solutes and adjust miscibility with solvents.	Forms hydrogen bonds with starch hydroxyl groups, disrupting inter and intramolecular interactions, improving solubility, and modifying crystallinity and thermal behavior.	Mahmood et al. (2017)

2.3.1 Poor mechanical property

Pure TPS is fundamentally fragile and has poor mechanical characteristics relative to traditional plastics. It demonstrates minimal tensile strength, elongation at break, and tear resistance. These constraints hinder its capacity to endure the forces experienced during packing, shipping, and handling. The poor mechanical qualities arise from the alteration of the original starch granule architecture during thermoplastic processing, leading to a less cohesive and more amorphous configuration. The robust intermolecular forces in natural starch are reduced during gelatinization and melting, resulting in a substance with less mechanical integrity. These TPS films are brittle and therefore are more likely to crack or rip under stress, harming the sterility of the food, possibly causing deterioration (Si & Zhang, 2024; Thongsomboon et al., 2023).

2.3.2 High sensitivity to moisture

A defining feature of starch's molecular structure is its large number of hydroxyl (-OH) groups. This abundance of polar groups is responsible for the polymer's inherently hydrophilic nature and its strong affinity for absorbing water (Ren et al., 2024). These groups readily form hydrogen bonds with water molecules, allowing the material to soak up a lot of moisture. The hydrophilic nature of TPS, or the biopolymer's seeming preoccupation with taking on water, is particularly problematic in terms of food packaging applications. This becomes far more of an issue among moist or foods that need to be protected from humidity, where water picked up by the package can saturate the food quality inside (Amaraweera et al., 2021).

The water absorption adversely impacts the barrier properties of flake TPS, which becomes more permeable to water vapor. The consequence of this is that as a film becomes more permeable, more moisture can be exchanged between the food and its surrounding environments, significantly reducing the quality and longevity of the food. This may lead to issues such as texture degradation, enhanced microbial growth, and overall reduction in freshness (de Freitas et al., 2021; Ribba et al., 2017). This increased moisture content now provides a favorable environment that can enhance microbial growth, hence increasing the risk of food spoilage (Carvalho et al., 2024).

2.3.3 Barrier property limitations

In addition to water vapor permeability, pure TPS has inadequate barrier characteristics against gases, including oxygen and carbon dioxide. This restricts its use for packaging oxygen-sensitive goods, or those that necessitate modified atmosphere packaging. The comparatively open and amorphous configuration of TPS facilitates the flow of gas molecules, undermining the protective capability of the packaging. This inadequate gas barrier may result in fat oxidation, loss of aromatic chemicals and other adverse alterations in food quality (de Freitas et al., 2021).

2.3.4 Processing challenges

Processing pure TPS is tricky. It has a high melt viscosity and usually degrades at high temperatures. High viscosity especially prevents the use of typical industrial operations such as extrusion and injection molding, but when the polymer degrades due to heat exposure, it can discolor, change the molecular weight distribution, and lose mechanical properties. Therefore, process parameters must be controlled to include temperature, shear rate, and time of heat exposure in order for the final product to meet specifications (Nafchi et al., 2013).

2.3.5 Thermal instability

The low thermal stability of TPS is a major drawback, limiting both its processability and performance for use in food packaging. Starch starts to degrade at relatively low temperatures (around 200–250 °C), which limits the use of common thermal processing methods (Nafchi et al., 2013). In addition, even a fully manufactured TPS product may exhibit dimensional instability when exposed to temperature fluctuations during storage (Ocieczek et al., 2022).

2.3.6 Retrogradation

One of the main issues around TPS during storage is what is called retrogradation. It is simply the way that starch molecules seek to reorder and crystallize, but this process occurs much more quickly in a cooler environment (Zhang & Rempel, 2012). There is a major drawback in the newly obtained crystal structure — it turns out that crystallization makes TPS packaging more brittle and thus more prone to cracking or breaking (Lai et al., 2022). Besides, this structural phenomenon can also reduce the performance of the material in the form of a water vapor barrier, in which reorganization of the crystalline starch matrix changes networks for access of water molecules (Bangar et al., 2021a, 2021b; Fatima et al., 2024).

2.3.7 Economic and industrial challenges

The commercial viability and cost-effectiveness of pure TPS for food packaging still face significant hurdles. When you modify TPS to improve its properties—usually through blending or chemical changes—it drives up manufacturing costs. As a result, producers using TPS-based materials remain at a fair disadvantage toward producers using lower-cost synthetic polymer products. The industrial infrastructure for TPS-based products remains underdeveloped in comparison with conventional plastics, limiting their availability and increasing their cost (Garavito et al., 2024).

3. Recent innovations in TPS formulations

Blending with other polymers is mainly done to overcome the limitations of TPS. This study aims to improve its mechanical, thermal, and barrier properties without compromising its biodegradability. Fig. 19.1 explains the recent advancements in the development of thermoplastic starch materials. For instance, combining thermoplastic starch (TPS) with polymers like polylactic acid (PLA) and polycaprolactone (PCL) improves the strength, water resistance, and thermal stability of the material. The resulting composite effectively possesses the beneficial features of all its constituents (Averous et al., 2000). The PLA, being rigid and hydrophobic, proves to be beneficial with respect to improving the flexibility and biodegradability of TPS, making it excellent for food packaging applications (Baniasadi et al., 2024). Plasticizers like glycerol and sorbitol reduce the brittleness and improve the flexibility of TPS. By breaking hydrogen linkages, the plasticizer functions as a mobile agent that enhances the mobility of starch molecules and reduces the glass

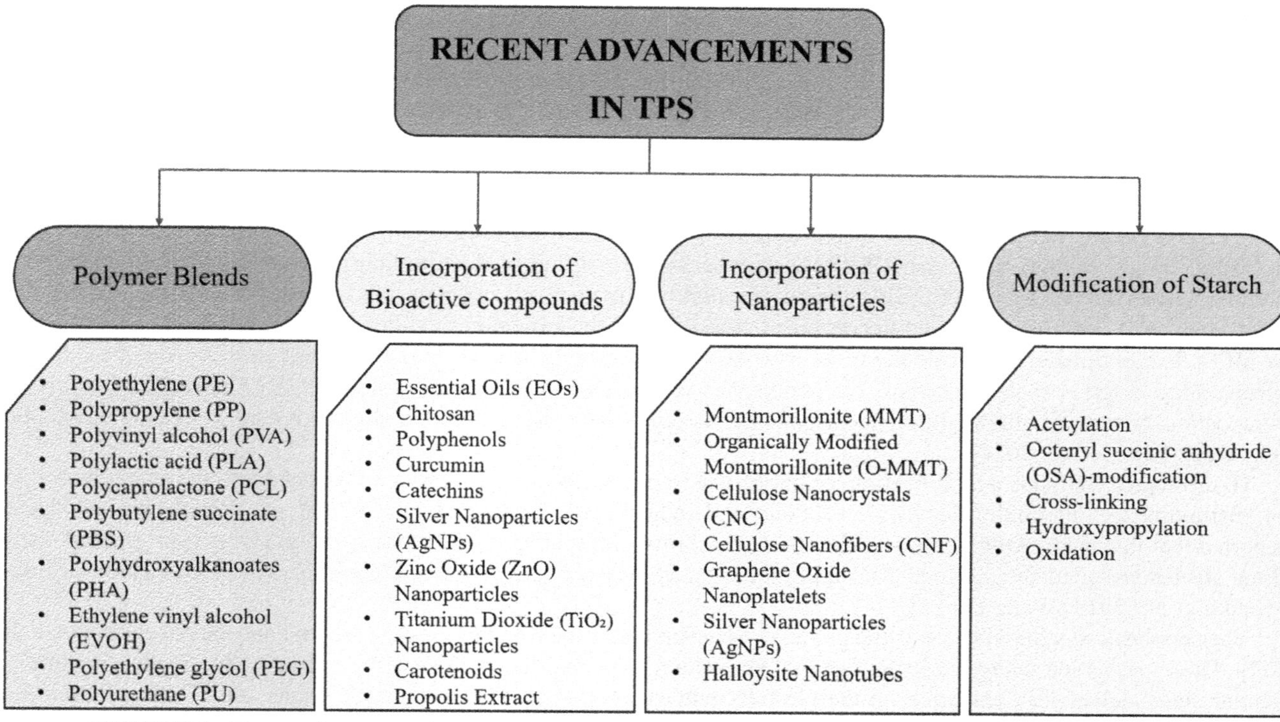

FIGURE 19.1 Recent advancement in thermoplastic starch. Different types of blending material are used to construct TPS.

transition temperature (Tg) (Tarique et al., 2021). Although a high plasticizer percentage enhances flexibility, it increases water uptake in TPS as plasticizers are hydrophilic in nature.

A carefully tailored formulation is therefore key to optimizing the overall combination of material characteristics (Eslami et al., 2023). Due to these limitations, a lot of research has focused on modifying TPS to make it a better food packaging material. Acetylation and esterification have simultaneously improved TPS's polymer compatibility and barrier properties (Diah et al., 2024). Enhancements in the mechanical, thermal, and barrier properties of TPS have been noticed upon the introduction of nanofillers such as clay, cellulose nanocrystals (CNCs), and graphene oxide. The incorporation of hydrophobic or functional coatings onto TPS films significantly reduced water affinity and improved overall barrier properties. The functional enhancements in TPS can be achieved using polymer blends, bio and synthetic plasticizers, chemical modification, nanocomposites, and protective surface coatings. Ongoing research focuses on advancing and altering existing methods to develop TPS-based materials that can act as an environmentally sustainable alternative to petroleum-based plastics, while meeting the standards of diverse food packaging systems (Diyana et al., 2021; Liu et al., 2022; Ghozali et al., 2022).

3.1 Modification of starch

Modifying starch involves changing its physical, chemical, or functional properties in order to remediate some of the issues, such as brittleness, crystallinity, and sensitivity to water, to make it suitable for industrial applications (Bangar et al., 2021a, 2021b), including food packaging. Starch modification is necessary to improve polymer compatibility with other polymers, mechanical properties, and resistance to water. There are many different methods that have been developed to expand upon the applications of TPS. Acetylation refers to a modification of starch where the native hydroxyl (-OH) groups of the starch are replaced with acetyl groups. This modification makes starch inherently more resistant to both heat and moisture, desirable attributes for developing biodegradable films and coatings. Furthermore, the acetylation of starch produces a starch that is more flexible and resistant to moisture absorption, thus making it easier to blend with other biodegradable polymers (Fringant et al., 1998).

Bulky, hydrophobic octenyl succinyl groups are introduced to starch. These are commonly used as emulsifiers to mix emulsions (Li et al., 2020). An edible film developed from native and octenyl succinylated potato starches incorporating various honeybee extracts (HBEs) and exhibited improved pasting and rheological behavior (Fringant et al., 1998).

Similarly, Siroha et al. (2022) studied the effects of OSA modification on pearl millet starch properties and reported that increased pasting viscosities were observed in modified starches, while a reduction in storage and loss moduli was observed during heating. Rheological analysis revealed shear-thinning behavior in all pastes. Modification with OSA, increased starch-resistant content and decreased slow-digesting starch. An increase in solubility was seen after treatment with OSA, suggesting film biodegradability and a lesser environmental impact.

A chemical bond is formed due to cross-linking between the starch molecules, which makes it more heat and shear-resistant. Agents like epichlorohydrin, sodium trimetaphosphate, citric acid, adipic acid, borax, and boric acid can be used to make cross-links (Yin et al., 2020). Starch-based bio-composites reinforced with micro fibrillated cellulose (MFC) use polyethylene glycol diglycidyl ether (PEGDE) to enhance the stability of the film matrix, as it is soluble in water and a nontoxic cross-linking agent. PEDGE helps to crosslink the matrix and maintain the biodegradable nature. The FTIR, NMR and XRD analyses confirmed the chemical crosslink and the structural modification achieved by the incorporation of MFC. The optimized films containing 12 wt% of MFC showed enhanced mechanical and barrier properties, and the biodegradation test revealed that most bio-composites achieved over 80% degradation within 15 days under controlled composting conditions, highlighting their potential as environmentally friendly alternatives to conventional plastics (González-Pérez et al., 2024).

Hydroxypropylation is a modification technique where hydroxypropyl groups are introduced into the starch molecules by disrupting the intermolecular hydrogen bond and reducing the gelatinization temperature. While previous studies reported that this modification enhances the film's flexibility, transparency, and processability. Hydroxypropylated starch films are better suited for the food packaging industry as they are less brittle and show improved elongation at break (Shaikh et al., 2019; Wang et al., 2024).

Thermoplastic starch oxidation is the process of conversion of hydroxyl groups to carbonyl and carboxyl groups in the TPS. This is achieved using oxidizing agents like sodium hypochlorite, potassium periodate, or hydrogen peroxide. A similar study delves into oxidation on the physicochemical properties of thermoplastic canna starch using sodium hypochlorite (NaClO) as an oxidizing agent. In this study, oxidation increased the carbonyl and carboxyl content and enhanced the interaction between oxidized starch and plasticizers. The resulting thermoplastics had improved flexibility due to their better elongation at break. Therefore, it was concluded that oxidation improves TPS performance and makes it more suitable for biodegradable film and packaging applications in the food industry (Pham et al., 2024).

3.2 Polymer blending

Thermoplastic starch and biodegradable polymer have distinct advantages over conventional petroleum-derived plastics as they create less of an impact on the environment. Polymer blends decrease the production costs and enhance other characteristics. Considerable progress has been achieved in integrating hydrophilic natural polymers with hydrophobic synthetic biodegradable polymers (Bulatović et al., 2021). There are various studies have shown time and time again the combinations of TPS with different biodegradable polymers, including PLA, PCL, PHB, PBAT, and PBS, pinpointing the viability of this method to create economical and sustainable products.

Palai et al. (2019) analyzed biodegradable blown films based on PLA/TPS blends, including in situ compatibilization using GMA grafting. A successful interaction between these phases was recorded. The results demonstrated interaction between PLA and TPS phases. Upon the addition of high levels of TPS (30%), there was an increase in film elongation. The tensile strength and modulus were maintained during this period. Notably, the 30 wt% cassava starch blends exhibited the highest elongation (144.4%) in the transverse direction. Furthermore, the study revealed improved oxygen permeability for both cassava and maize starch blends, with a 41.7% and 38.1% reduction, respectively. However, water vapor permeability increased twofold for both starch blends at 30 wt%. DSC analysis revealed an enhanced crystallization rate of PLA in the presence of TPS. These findings demonstrate the potential of in situ compatibilization to improve the processability and certain mechanical properties of PLA/TPS blends, while also influencing their barrier properties.

Another approach, stated by Parlak et al. (2024) examined how TPS coatings affected PLA film characteristics. It was suggested that upon increasing the TPS levels, the oxygen barrier property of PLA showed a significant increase. This resulted in a decrease of 2.5%–29.6% of oxygen permeability. A rise in water vapor permeability up to 59.2%–94.4% was observed. Other properties, like moisture absorption and UV barrier characteristics were also enhanced with TPS coating. According to research by FTIR, no major chemical interactions between PLA and TPS were seen. Modifications observed in the properties are due to the hydrophilic characteristics of TPS and their role in governing film morphology and crystallinity. The inherent hydrophilicity of thermoplastic starch (TPS), which increases water absorption, can alter the shape of the film by increasing the polarity or disrupting the crystallinity, and hence altering its properties. Although TPS films improve the oxygen barrier, further research is required to optimize the barrier properties.

Similarly, Esmaeili et al. (2019) studied how the combination of sorbitol and glycerol affects the rheological and mechanical characteristics of PLA/TPS blends. Melt blending leads to improved melt processability, reduced viscosity, complex, and storage modulus. Plasticizer migration from TPS to the PLA phase in melt mixing results in improved PLA matrix processability. Decreased tensile strength and increased ductility were noted in this when compared with blends achieved through chemical compatibilization. This approach was economical as well as industrially efficient and exhibited better mechanical properties. Bai et al. (2021) examined reactive epoxy compatibilizer and extrusion techniques on PBAT/TPS mixtures. According to the results, improved the compatibility of the blend, giving less phase separation and better distribution of TPS/PBAT. The extrusion process significantly improves the morphology of the blend. The resulting mixture had better phase structure and adhesion. Due to this, better mechanical properties were seen. This points toward the importance of compatibilizer selection and optimal processing techniques in PBAT/TPS blends.

Kodsangma et al. (2023) reported a better biodegradable TPS with improved mechanical and water-resistant properties. The blending of zinc oxide (ZnO) with carboxymethyl cellulose (CMC) enhanced tensile strength, water resistance, and miscibility. Ionic interactions and effective ZnO dispersion amplified the TPS matrix, with a strength of 14 MPa. TPS or polyvinyl alcohol (PVOH) can be used in food packaging as it is economically feasible in comparison with ethylene vinyl alcohol (EVOH) (Ge et al., 2021). Films containing 25%–75% TPS in the central layer showcased better oxygen barrier capacities. Specifically, those having 25 wt% TPS show high tensile strength and impact resistance.

3.3 Incorporating bioactive materials

The incorporation of bioactive compounds into TPS is a new and optimistic approach to creating a multifunctional material useful across various fields. They can be altered to have characteristics like antimicrobial activity and antioxidant properties (Almeida et al., 2023; Andrew & Dhakal, 2022). Grape cane extract, abundant in stilbene compounds, was combined with TPS matrices to produce new bioactive materials (Díaz-Galindo et al., 2020). Results showed a minimal impact on thermal stability upon adding 15 wt% grape cane extract. A reduction in mechanical strength was also noted. Some notable observations were the antifungal activity against *Botrytis cinerea* and antibacterial activity against *Staphylococcus aureus* in the TPS/WE15 combinations. This integration of grape cane extract into TPS showed higher antibacterial properties, pointing toward use in food packaging.

Improved stiffness of around 3.8 GPa and tensile strength of 38.6 MPa were observed in the starch-ethylene vinyl alcohol copolymer (SEVA-C) blend (Leonor et al., 2002). Bioactivity at minimal filler concentration was significant with 10 wt% of 45S5 Bioglass, as evidenced by the production of hydroxylapatite (HA) layers, analyzed using SEM, TF-XRD, and ICP spectroscopy. While the bioactivity was maintained for a period of 30 days, and the composites were found to be biodegradable, highlighting the potential of SEVA-C/Bioglass composites for bioactive and temporary medical or environmental applications.

Oxygen scavenging bio-based thermoplastic starch blended with gallic acid (GA) and linear-low-density polyethylene (LLDPE) for controlling oxygen levels in food packaging (Promsorn & Harnkarnsujarit, 2022). Where GA facilitated starch melting and plasticized in the TPS phase, while the interaction with LLDPE occurred via C–H bonded aromatic rings. Higher temperatures and gallic acid levels boost oxygen scavenging speed, while minimum residual oxygen contents were achieved by 5 days at 50°C. Grape seed oil, alpha-tocopherol, and rice husk as antioxidant additives incorporated into TPS and poly (butylene adipate-co-terephthalate (PBAT) films properties were investigated (Tekin & Ersus, 2024). The results showed films containing grape seed oil exhibited the best in balancing the mechanical strength and antioxidant properties.

Gui et al. (2024) developed a sustained-release active packaging film by cross-linked porous starch (ScPS) adsorbed with clove essential oil (CEO) into a PLA/PBAT/TPS matrix using industrial melt extrusion. Where it increased the controlled release of the CEO by 100%, leading to an increase in the antimicrobial and antioxidant properties. The film effectively extended the shelf life of salmon by 100% at 4°C by reducing the proteolysis, lipid oxidation, and microbial growth.

3.4 Nanocomposite-based thermoplastic starch

Nanofillers are compounds that are widely used in enhancing the mechanical and barrier properties of the film (Table 19.2). Recently, these fillers have been extensively researched due to their wide array of benefits, including enhanced stability, water resistance, strength, and biodegradability (Rivadeneira-Velasco et al., 2021).

The FeO and ZnO nanoparticle were found to be a potential active ingredient for the development of a double-layer nano-bio thermoplastic packaging film (Chandrasekar et al., 2024). The amorphous property of the starch was masked by

TABLE 19.2 Nanofillers used in thermoplastic starch base packaging materials.

Nanofillers	Properties enhanced	Mechanism of enhancement	Applications	References
Montmorillonite (MMT)	Increased tensile strength, elastic modulus, and barrier properties	Layered structure creates a tortuous path for gas and moisture diffusion, and reinforcing the matrix	Food packaging, and biodegradable films	Aouada et al. (2011), Garusinghe et al. (2018)
Organically modified montmorillonite (O-MMT)	Improved mechanical and thermal stability	Surface modification enhances dispersion and interfacial adhesion with TPS	High-performance biodegradable packaging	Tan et al. (2020)
Cellulose nanocrystals (CNC)	High mechanical strength, barrier properties, and transparency	Hydrogen bonding with the TPS matrix, reinforcing effect due to high crystallinity	Transparent packaging films, and high-strength bioplastics	Popescu et al. (2020), Vaezi et al. (2020)
Cellulose nanofibers (CNF)	Flexibility, elongation, barrier properties, and compatibility	It acts as a reinforcing phase, effectively transferring stress and improving the overall mechanical properties of the composite	Flexible and biodegradable food packaging	AbdulRasheed-Adeleke et al. (2022)
Graphene oxide nanoplatelets	Conductivity, mechanical reinforcement, and antimicrobial properties	High surface area enhances interaction with TPS, creates a reinforcing network	Smart packaging, and antimicrobial films	Poongodi et al. (2024), Ramezani et al. (2020), Tarawneh et al. (2020)
Silver nanoparticles (AgNPs)	Antimicrobial properties, and barrier enhancement	Interaction with microbial cells disrupts growth and enhances the oxygen barrier.	Antibacterial packaging, and medical applications	González-Ceballos et al. (2022)
Halloysite nanotubes	Increased tensile strength, good biocompatibility and nontoxicity	A hollow structure can be used as a delivery system for active ingredients	Biodegradable structural materials	Schmitt et al. (2015)

the crystalline characteristics of nanoparticles, which helped in modifying their functional behavior in the film. While the nanoparticle diffusion from the packaging system significantly affected the microbial, chemical, and color stability of mutton and chicken meat, extending their shelf life at 4°C.

Similarly, the influence of cellulose nanocrystal on low-density polyethylene and thermoplastic starch composites was studied by Gray et al. (2018). Also, it showed significant improvement in thermal, mechanical and barrier properties of the composites. Therefore, the addition of 1% CNC blend showed enhanced tensile strength and modulus of elasticity (MOE), leading to a superior composite blend performance.

An earlier study conducted by Nasri-Nasrabadi et al. (2014) used a blend of starch with cellulose nanofiber extracted from rice straw and investigated the effect of nanofibers content on the nanocomposite films. The incorporation of 15 wt% CNF increased Young's modulus by 233% compared to pure starch, which helped in the improvement of the dynamic mechanical property of the film. The permeability coefficient ($P \times 10^8$) decreased from 1.31 (pure starch) to 0.62 (TPS + 10 wt% CNF) but increased slightly to 1.06 (TPS + 15 wt% CNF). This indicates a better humidity resistance with optimized CNF content. Additionally, it also reduces the transparency of the film due to the light scattering of the nanofibers.

4. Application of the starch-based thermoplastics in food packaging

Films made of thermoplastic starch have numerous potential applications as food packaging materials (Fig. 19.2). For successful utilization of thermoplastic starch films in food packaging, these materials must exhibit a range of certain desirable properties, which include appropriate mechanical strength, effective barrier properties against oxygen and water vapor, and potentially antimicrobial and antioxidant functionalities (Cheng et al., 2021). Current research is focused on extending the shelf life of food products packaged in starch-based materials by incorporating active components like

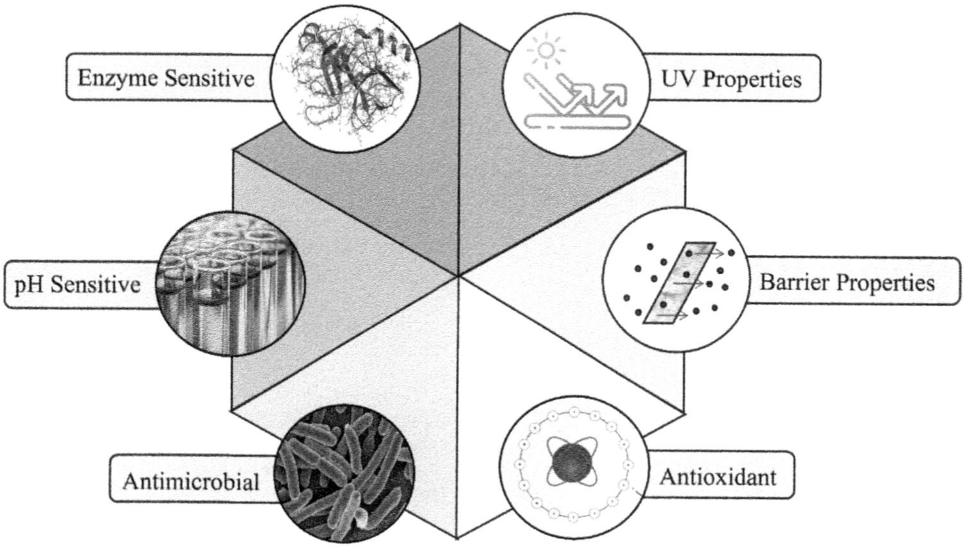

FIGURE 19.2 Functional properties of thermoplastic starch. Different types of applications using TPS.

antioxidants or antibacterial agents directly into the starch matrix for controlled release and enhanced preservation, and enhancing barrier properties against oxygen and water vapor transmission through blending starch with other polymers or additives (Dutta & Sit, 2024). Table 19.3 demonstrates the promising commercial potential of TPS-based packaging by showcasing its diverse applications across various food categories, including fruits, vegetables, baked goods, and meat products.

TABLE 19.3 Applications of thermoplastic starch-based food packaging.

Film components	Method of preparation	Function	Food material	Application	References
TPS-PVA-cellulose nano-composite films	Solvent casting	Extending the shelf-life	Fresh red chili	Storage at ambient temperature maintained red chili quality for 12 days, while chilled storage extended freshness up to 15 days.	Fahma et al. (2020)
Thermoplastic corn starch with chitosan oligomers	Thermo compression molding by hydraulic press	Controlled-release delivery system for the chitosan oligomers	Strawberries, ricotta, and bread	Chitosan oligomers incorporated film sachet provided improved antimicrobial protection compared to spraying technique.	Castillo et al. (2017)
Acetylated cassava TPS and green tea	Blown extrusion process	Antimicrobial activity and inhibiting lipid oxidation	Soybean oil and bacon	The film matrix effectively inhibited microbial growth while minimizing the metmyoglobin-induced browning, which preserved the bacon color and reduced lipid oxidation up to 38% in packed soybean oil.	Panrong et al. (2019)

Continued

TABLE 19.3 Applications of thermoplastic starch-based food packaging.—cont'd

Film components	Method of preparation	Function	Food material	Application	References
TiO_2 nanoparticle incorporated poly (butylene adipate-co-terephthalate) [PBAT] thermoplastic cassava starch (TPS)-blended films	Blown-film extrusion	Extending the shelf-life	Banana	The addition of TiO_2 nanoparticle improved the mechanical strength and effectively reduced oxygen and carbon dioxide permeability, lowering the color change and increasing the shelf life of the fruit.	Phothisarattana et al. (2021)
Starch (TPS) blended polybutylene adipate-co-terephthalate (PBAT) and thermoplastic ZnO (1%–5%) film	Blown extrusion	Active meat packaging	Pork meat	The addition of 3% ZnO nanoparticle effectively increased lipid stability and reduced microbial activity, leading to an extension of shelf life of more than 3 days.	Phothisarattana et al. (2022)
Maltol (MT) and ethyl maltol (EM) infused thermoplastic starch blend with polybutylene adipate terephthalate biodegradable film	Blown extrusion	Extending the shelf-life	Butter cake	The addition of flavoring compounds MT, EM effectively delayed the fungal growth and inhibited the microbial growth.	Promhuad et al. (2025)
Banana peel extracted and reconstituted cellulose-reinforced starch-based bioactive thermoplastic packaging films	Solvent casting	Extending the shelf-life	bread	Films showed excellent antimicrobial properties, mechanical and barrier properties, while demonstrating effective UV blocking capacity and extending the shelf life.	Chandrasekar et al. (2023)

5. Future perspectives

Further studies and research need to focus on improving mechanical and barrier properties. The search to find novel nanofillers, compatible techniques, and bioplasticizers needs to be continued to improve the durability and moisture resistance of TPS. For cost-effective production of TPS-based materials, techniques like melt blending and reactive extrusion need to be practiced. The shelf life and functionality can be improved by incorporating bioactive compounds, antioxidants, and antimicrobial agents into TPS films. There is room for exploration of the use of nontraditional starch sources to generate TPS with lesser environmental impact and economic costs. Despite TPS being biodegradable, further efforts are required to reduce the environmental impacts and the slow deterioration rate. Advancements in electrospinning and 3D printing have introduced novel pathways to develop tailored, biodegradable packaging materials, enabling on-demand production and hence causing minimal waste generation.

Therefore, TPS demonstrates considerable prospects for sustainable packaging applications. Ongoing research is focused on material optimization, processing efficiency, and functional innovations, which will replace conventional options and strengthen its role as a biodegradable and environmentally responsible packaging material. Improvements in this field depend upon multidisciplinary research and regulatory guidance.

6. Conclusion

Although TPS is a biodegradable and sustainable alternative to petroleum-based plastics, its poor mechanical strength, water sensitivity, and retrogradation pose significant barriers to its large-scale industrial applications. Despite that, considerable progress has been made in recent years to address these limitations. Strategies including chemical modification, nanofiller reinforcement, and incorporation of bioactive compounds have been analyzed, leading to substantial improvements in the functional properties of TPS. These have significantly improved the functional properties of TPS. Key improvements involve the use of nanofillers, particularly montmorillonite (MMT), cellulose nanocrystals (CNC), and cellulose nanofibers (CNF). The hydrophobic nature, mechanical and barrier properties are improved because of these changes. This makes TPS a more cutthroat material for packaging applications. Better results have been observed after incorporation with biodegradable polymers like PLA, PHB, or PBAT. These blended polymers are environmentally sustainable, flexible, and durable. More research is required to optimize and further test these techniques to avoid phase separation and improve interfacial adhesion.

References

AbdulRasheed-Adeleke, T., Egwim, E. C., Sadiku, E. R., & Ochigbo, S. S. (2022). Optimization of lignin–cellulose nanofiber-filled thermoplastic starch composite film production for potential application in food packaging. *Molecules, 27*(22), 7708.

Almeida, T., Karamysheva, A., Valente, B. F. A., Silva, J. M., Braz, M., Almeida, A., Silvestre, A. J. D., Vilela, C., & Freire, C. S. R. (2023). Biobased ternary films of thermoplastic starch, bacterial nanocellulose and gallic acid for active food packaging. *Food Hydrocolloids, 144*. https://doi.org/10.1016/j.foodhyd.2023.108934

Amaraweera, S. M., Gunathilake, C., Gunawardene, O. H. P., Fernando, N. M. L., Wanninayaka, D. B., Dassanayake, R. S., Rajapaksha, S. M., Manamperi, A., Fernando, C. A. N., Kulatunga, A. K., & Manipura, A. (2021). Development of starch-based materials using current modification techniques and their applications: A review. *Molecules, 26*(22). https://doi.org/10.3390/molecules26226880

Andrew, J. J., & Dhakal, H. N. (2022). Sustainable biobased composites for advanced applications: Recent trends and future opportunities – A critical review. *Emirates Composites Part C: Open Access, 7*. https://doi.org/10.1016/j.jcomc.2021.100220

Aouada, F. A., Mattoso, L. H. C., & Longo, E. (2011). New strategies in the preparation of exfoliated thermoplastic starch–montmorillonite nanocomposites. *Industrial Crops and Products, 34*(3), 1502–1508.

Averous, L., Moro, L., Dole, P., & Fringant, C. (2000). Properties of thermoplastic blends: Starch-polycaprolactone. *Polymer, 41*(11), 4157–4167. https://doi.org/10.1016/S0032-3861(99)00636-9

Bai, J., Pei, H., Zhou, X., & Xie, X. (2021). Reactive compatibilization and properties of low-cost and high-performance PBAT/thermoplastic starch blends. *European Polymer Journal, 143*, 110198.

Bangar, S. P., Kumar, M., & Whiteside, W. S. (2021a). Mango seed starch: A sustainable and eco-friendly alternative to increasing industrial requirements. *International Journal of Biological Macromolecules, 183*, 1807–1817.

Bangar, S. P., Whiteside, W. S., Ashogbon, A. O., & Kumar, M. (2021b). Recent advances in thermoplastic starches for food packaging: A review. *Food Packaging and Shelf Life, 30*. https://doi.org/10.1016/j.fpsl.2021.100743

Baniasadi, H., Äkräs, L., Madani, Z., Silvenius, F., Fazeli, M., Lipponen, S., Vapaavuori, J., & Seppälä, J. (2024). Development and characterization of polylactic acid/starch biocomposites – From melt blending to preliminary life cycle assessment. *International Journal of Biological Macromolecules, 279*(Pt 1), Article 135173. https://doi.org/10.1016/j.ijbiomac.2024.135173

Baranwal, J., Barse, B., Fais, A., Delogu, G. L., & Kumar, A. (2022). Biopolymer: A sustainable material for food and medical applications. *Polymers, 14*(5). https://doi.org/10.3390/polym14050983

Benitez, J. J., Florido-Moreno, P., Porras-Vázquez, J. M., Tedeschi, G., Athanassiou, A., Heredia-Guerrero, J. A., & Guzman-Puyol, S. (2024). Transparent, plasticized cellulose-glycerol bioplastics for food packaging applications. *International Journal of Biological Macromolecules, 273*. https://doi.org/10.1016/j.ijbiomac.2024.132956

Bulatović, V. O., Mandić, V., Kučić Grgić, D., & Ivančić, A. (2021). Biodegradable polymer blends based on thermoplastic starch. *Journal of Polymers and the Environment, 29*(2), 492–508. https://doi.org/10.1007/s10924-020-01874-w

Carvalho, T. B. de, Silva, B. N., Tomé, E., & Teixeira, P. (2024). Preventing fungal spoilage from raw materials to final product: Innovative preservation techniques for fruit fillings. *Foods, 13*(17), 2669.

Castillo, L. A., Farenzena, S., Pintos, E., Rodríguez, M. S., Villar, M. A., García, M. A., & López, O. V. (2017). Active films based on thermoplastic corn starch and chitosan oligomer for food packaging applications. *Food Packaging and Shelf Life, 14*, 128–136.

Chaléat, C., Halley, P. J., & Truss, R. W. (2014). Mechanical properties of starch-based plastics. In P. J. Halley, & L. Averous (Eds.), *Starch polymers: From genetic engineering to green applications* (pp. 187–209). Elsevier.

Chandrasekar, C. M., Krishnamachari, H., Farris, S., & Romano, D. (2023). Development and characterization of starch-based bioactive thermoplastic packaging films derived from banana peels. *Carbohydrate Polymer Technologies and Applications, 5*, 100328.

Chandrasekar, C. M., Nespoli, L., Bellesia, T., Ghaani, M., Farris, S., & Romano, D. (2024). Fabrication of double layer nanoparticle infused starch-based thermoplastic food packaging system for meat preservation. *International Journal of Biological Macromolecules, 254*(Pt 1), Article 127689. https://doi.org/10.1016/j.ijbiomac.2023.127689

Chen, J., Wang, X., Long, Z., Wang, S., Zhang, J., & Wang, L. (2020). Preparation and performance of thermoplastic starch and microcrystalline cellulose for packaging composites: Extrusion and hot pressing. *International Journal of Biological Macromolecules, 165*(Pt B), 2295–2302. https://doi.org/10.1016/j.ijbiomac.2020.10.117

Cheng, H., Chen, L., McClements, D. J., Yang, T., Zhang, Z., Ren, F., Miao, M., Tian, Y., & Jin, Z. (2021). Starch-based biodegradable packaging materials: A review of their preparation, characterization and diverse applications in the food industry. *Trends in Food Science and Technology, 114*, 70–82. https://doi.org/10.1016/j.tifs.2021.05.017

Csiszár, E., Kun, D., & Fekete, E. (2021). The role of structure and interactions in thermoplastic starch–nanocellulose composites. *Polymers, 13*(18), 3186.

Da Róz, A. L., Carvalho, A. J. F., Gandini, A., & Curvelo, A. A. S. (2006). The effect of plasticizers on thermoplastic starch compositions obtained by melt processing. *Carbohydrate Polymers, 63*(3), 417–424. https://doi.org/10.1016/j.carbpol.2005.09.017

Da Róz, A. L., Zambon, M. D., Curvelo, A. A., & Carvalho, A. J. (2011). Thermoplastic starch modified during melt processing with organic acids: The effect of molar mass on thermal and mechanical properties. *Industrial Crops and Products, 33*(1), 152–157.

de Freitas, A. de S. M., Rodrigues, J. S., Maciel, C. C., Pires, A. A. F., Lemes, A. P., Ferreira, M., & Botaro, V. R. (2021). Improvements in thermal and mechanical properties of composites based on thermoplastic starch and Kraft Lignin. *International Journal of Biological Macromolecules, 184*, 863–873.

Diah, S. B. M., Abdullah, S. H. Y. S., Seok, Y. B., Fatah, I. Y. A., Rahman, N. I. A., Hafizulhaq, F., & Alias, N. N. (2024). Towards sustainable food packaging: A review of thermoplastic starch (TPS) as a promising bioplastic material, its limitations, and improvement strategies with bio-fillers and essential oils. *Journal of Advanced Research in Fluid Mechanics and Thermal Sciences, 119*(1), 80–104. https://doi.org/10.37934/arfmts.119.1.80104

Díaz-Galindo, E. P., Nesic, A., Cabrera-Barjas, G., Mardones, C., Von Baer, D., Bautista-Baños, S., & Garcia, O. D. (2020). Physical-chemical evaluation of active food packaging material based on thermoplastic starch loaded with grape cane extract. *Molecules, 25*(6). https://doi.org/10.3390/molecules25061306

Diyana, Z. N., Jumaidin, R., Selamat, M. Z., Ghazali, I., Julmohammad, N., Huda, N., & Ilyas, R. A. (2021). Physical properties of thermoplastic starch derived from natural resources and its blends: A review. *Polymers, 13*(9), 1396.

Donmez, D., Pinho, L., Patel, B., Desam, P., & Campanella, O. H. (2021). Characterization of starch–water interactions and their effects on two key functional properties: Starch gelatinization and retrogradation. *Current Opinion in Food Science, 39*, 103–109.

Dutta, D., & Sit, N. (2024). Comprehensive review on developments in starch-based films along with active ingredients for sustainable food packaging. *Sustainable Chemistry and Pharmacy, 39*, Article 101534. https://doi.org/10.1016/j.scp.2024.101534

Eslami, Z., Elkoun, S., Robert, M., & Adjallé, K. (2023). A review of the effect of plasticizers on the physical and mechanical properties of alginate-based films. *Molecules, 28*(18), 6637. https://doi.org/10.3390/molecules28186637

Esmaeili, M., Pircheraghi, G., Bagheri, R., & Altstädt, V. (2019). Poly (lactic acid)/coplasticized thermoplastic starch blend: Effect of plasticizer migration on rheological and mechanical properties. *Polymers for Advanced Technologies, 30*(4), 839–851.

Fahma, F., Wening, O. P., Lisdayana, N., Purwoko, & Sugiarto. (2020, March). Thermoplastic starch-PVA-cellulose nanocomposite film for extending the shelf life of red chili. In *IOP Conference Series: Earth and Environmental Science* (Vol. 460, No. 1, p. 012036). IOP Publishing.

Fatima, S., Khan, M. R., Ahmad, I., & Sadiq, M. B. (2024). Recent advances in modified starch based biodegradable food packaging: A review. *Heliyon, 10*(6). https://doi.org/10.1016/j.heliyon.2024.e27453

Fringant, C., Rinaudo, M., Foray, M. F., & Bardet, M. (1998). Preparation of mixed esters of starch or use of an external plasticizer: Two different ways to change the properties of starch acetate films. *Carbohydrate Polymers, 35*(1–2), 97–106. https://doi.org/10.1016/s0144-8617(97)00250-6

Gao, C., Pollet, E., & Avérous, L. (2017a). Innovative plasticized alginate obtained by thermo-mechanical mixing: Effect of different biobased polyols systems. *Carbohydrate Polymers, 157*, 669–676.

Gao, C., Pollet, E., & Avérous, L. (2017b). Properties of glycerol-plasticized alginate films obtained by thermo-mechanical mixing. *Food Hydrocolloids, 63*, 414–420.

Garavito, J., Peña-Venegas, C. P., & Castellanos, D. A. (2024). Production of starch-based flexible food packaging in developing countries: Analysis of the processes, challenges, and requirements. *Foods, 13*(24), 4096.

Garusinghe, U. M., Varanasi, S., Raghuwanshi, V. S., Garnier, G., & Batchelor, W. (2018). Nanocellulose-montmorillonite composites of low water vapour permeability. *Colloids and Surfaces A: Physicochemical and Engineering Aspects, 540*, 233–241.

Ge, C., Lansing, B., & Lewis, C. L. (2021). Thermoplastic starch and poly(vinyl alcohol) blend-centered barrier film for food packaging applications. In *Innovations in sustainable food packaging materials* (pp. 145–165). Elsevier.

Ghozali, M., Meliana, Y., & Chalid, M. (2022). Novel in situ modification for thermoplastic starch preparation based on *Arenga pinnata* palm starch. In *Advances in thermoplastic starch and biopolymer materials* (pp. 221–240). Springer.

González-Ceballos, L., Guirado-moreno, J. C., Guembe-García, M., Rovira, J., Melero, B., Arnaiz, A., Diez, A. M., García, J. M., & Vallejos, S. (2022). Metal-free organic polymer for the preparation of a reusable antimicrobial material with real-life application as an absorbent food pad. *Food Packaging and Shelf Life, 33*, 100910.

González-Pérez, M. M., Lomelí-Ramírez, M. G., Robledo-Ortiz, J. R., Silva-Guzmán, J. A., & Manríquez-González, R. (2024). Biodegradable bio-composite of starch films cross-linked with polyethylene glycol diglycidyl ether and reinforced by microfibrillated cellulose. *Polymers, 16*(9). https://doi.org/10.3390/polym16091290

Gray, N., Hamzeh, Y., Kaboorani, A., & Abdulkhani, A. (2018). Influence of cellulose nanocrystal on strength and properties of low density polyethylene and thermoplastic starch composites. *Industrial Crops and Products, 115*, 298–305. https://doi.org/10.1016/j.indcrop.2018.02.017

Guarás, M. P., Ludueña, L. N., & Alvarez, V. A. (2020). Recent advances in thermoplastic starch biodegradable nanocomposites. In N. H. M. Ekka, S. K. Shukla, & V. K. Singh (Eds.), *Handbook of nanomaterials and nanocomposites for energy and environmental applications* (pp. 1–24). Springer.

Gui, H., Ma, W., Cao, Y., Chao, H., Fan, M., Dong, Q., & Li, L. (2024). Sustained release, antimicrobial, and antioxidant properties of modified porous starch-based biodegradable polylactic acid/polybutylene adipate-co-terephthalate/thermoplastic starch active packaging film. *International Journal of Biological Macromolecules, 267*, 131657.

Gündoğdu, S., Bour, A., Köşker, A. R., Walther, B. A., Napierska, D., Mihai, F. C., Syberg, K., Hansen, S. F., & Walker, T. R. (2024). Review of microplastics and chemical risk posed by plastic packaging on the marine environment to inform the Global Plastics Treaty. *Science of the Total Environment, 946*. https://doi.org/10.1016/j.scitotenv.2024.174000

Huneault, M. A., & Li, H. (June 29, 2010). *Process of producing thermoplastic starch/polymer blends*. U.S. Patent and Trademark Office. U.S. Patent No. 7,745,528.

Idris, S. N., Amelia, T. S. M., Bhubalan, K., Lazim, A. M. M., Zakwan, N. A. M. A., Jamaluddin, M. I., Santhanam, R., Amirul, A. A. A., Vigneswari, S., & Ramakrishna, S. (2023). The degradation of single-use plastics and commercially viable bioplastics in the environment: A review. *Environmental Research, 231*. https://doi.org/10.1016/j.envres.2023.115988

Jiang, T., Duan, Q., Zhu, J., Liu, H., & Yu, L. (2020). Starch-based biodegradable materials: Challenges and opportunities. *Advanced Industrial and Engineering Polymer Research, 3*(1), 8–18. https://doi.org/10.1016/j.aiepr.2019.11.003

Kibria, M. G., Masuk, N. I., Safayet, R., Nguyen, H. Q., & Mourshed, M. (2023). Plastic waste: Challenges and opportunities to mitigate pollution and effective management. *International Journal of Environmental Research, 17*(1). https://doi.org/10.1007/s41742-023-00507-z

Kodsangma, A., Thajai, N., Punyodom, W., Worajittiphon, P., Jantrawut, P., Ruksiriwanich, W., Sommano, S. R., Sringarm, K., Thanakkasaranee, S., Rachtanapun, P., & Jantanasakulwong, K. (2023). Mechanical properties and water resistance improvement of thermoplastic modified starch, carboxymethyl cellulose, and zinc oxide nanometal particles by reactive blending. *International Journal of Biological Macromolecules, 253*, 126783.

Koelmans, A. A., Mohamed Nor, N. H., Hermsen, E., Kooi, M., Mintenig, S. M., & De France, J. (2019). Microplastics in freshwaters and drinking water: Critical review and assessment of data quality. *Water Research, 155*, 410–422. https://doi.org/10.1016/j.watres.2019.02.054

Lai, D. S., Osman, A. F., Adnan, S. A., Ibrahim, I., Salimi, M. N. A., & Alrashdi, A. A. (2022). Effective aging inhibition of the thermoplastic corn starch films through the use of green hybrid filler. *Polymers, 14*(13). https://doi.org/10.3390/polym14132567

Leonor, I. B., Sousa, R. A., Cunha, A. M., Reis, R. L., Zhong, Z. P., & Greenspan, D. (2002). Novel starch thermoplastic/Bioglass® composites: Mechanical properties, degradation behavior and in-vitro bioactivity. *Journal of Materials Science: Materials in Medicine, 13*(10), 939–945. https://doi.org/10.1023/a:1019800411229

Li, H., Ma, Y., Yu, L., Xue, H., Wang, Y., Chen, J., & Zhang, S. (2020). Construction of octenyl succinic anhydride modified porous starch for improving bioaccessibility of β-carotene in emulsions. *RSC Advances, 10*(14), 8480–8489. https://doi.org/10.1039/c9ra10079b

Li, C., Guo, Y., Chen, M., Wang, S., Gong, H., Zuo, J., Zhang, J., & Dai, L. (2023). Recent preparation, modification and application progress of starch nanocrystals: A review. *International Journal of Biological Macromolecules, 250*, Article 126122. https://doi.org/10.1016/j.ijbiomac.2023.126122

Liu, J., Ma, Z., Liu, Y., Zheng, X., Pei, Y., & Tang, K. (2022). Soluble soybean polysaccharide films containing in-situ generated silver nanoparticles for antibacterial food packaging applications. *Food Packaging and Shelf Life, 31*, 100800.

Mahmood, H., Moniruzzaman, M., Yusup, S., & Welton, T. (2017). Ionic liquids assisted processing of renewable resources for the fabrication of biodegradable composite materials. *Green Chemistry, 19*(9), 2051–2075.

Mutmainna, I., Gareso, P. L., Suryani, S., & Tahir, D. (2024). Microplastics from petroleum-based plastics and their effects: A systematic literature review and science mapping of global bioplastics production. *Integrated Environmental Assessment and Management, 20*(6), 1892–1911. https://doi.org/10.1002/ieam.4976

Nafchi, A. M., Moradpour, M., Saeidi, M., & Alias, A. K. (2013). Thermoplastic starches: Properties, challenges, and prospects. *Starch/Staerke, 65*(1–2), 61–72. https://doi.org/10.1002/star.201200201

Nasri-Nasrabadi, B., Behzad, T., & Bagheri, R. (2014). Preparation and characterization of cellulose nanofiber reinforced thermoplastic starch composites. *Fibers and Polymers, 15*(2), 347–354.

Ocieczek, A., Mesinger, D., & Toczek, H. (2022). Hygroscopic properties of three cassava (*Manihot esculenta* Crantz) starch products: Application of BET and GAB models. *Foods, 11*(13), 1966.

Ozeren, H. D., Guivier, M., Olsson, R. T., Nilsson, F., & Hedenqvist, M. S. (2020). Ranking plasticizers for polymers with atomistic simulations: PVT, mechanical properties, and the role of hydrogen bonding in thermoplastic starch. *ACS Applied Polymer Materials, 2*(5), 2016–2026.

Palai, B., Biswal, M., Mohanty, S., & Nayak, S. K. (2019). In situ reactive compatibilization of polylactic acid (PLA) and thermoplastic starch (TPS) blends; synthesis and evaluation of extrusion blown films thereof. *Industrial Crops and Products, 141*, 111748.

Panrong, T., Karbowiak, T., & Harnkarnsujarit, N. (2019). Thermoplastic starch and green tea blends with LLDPE films for active packaging of meat and oil-based products. *Food Packaging and Shelf Life, 21*, Article 100331.

Parlak, M. E., Uzuner, K., Ozdemir, S., Kirac Demirel, F. T., Dundar, A. N., Sahin, O. I., Dagdelen, A. F., Odabas, H. I., & Saricaoglu, F. T. (2024). Characterization of biodegradable bi-layer films from thermoplastic starch and poly-l-lactic acid. *Journal of Applied Polymer Science, 141*(20), e55378.

Pham, T. T. H., Nguyen, T. T., Nguyen, V. K., Nguyen, T. D., Pham, T. T., Nguyen, T. H., & Pham, N. A. (2024). Effect of oxidation on physico-chemical characteristics of thermoplastic canna starch. *Vietnam Journal of Chemistry, 62*, 15–22. https://doi.org/10.1002/vjch.202300260

Phothisarattana, D., Wongphan, P., Promhuad, K., Promsorn, J., & Harnkarnsujarit, N. (2021). Biodegradable poly (butylene adipate-co-terephthalate) and thermoplastic starch-blended TiO2 nanocomposite blown films as functional active packaging of fresh fruit. *Polymers, 13*(23), 4192.

Phothisarattana, D., Wongphan, P., Promhuad, K., Promsorn, J., & Harnkarnsujarit, N. (2022). Blown film extrusion of PBAT/TPS/ZnO nanocomposites for shelf-life extension of meat packaging. *Colloids and Surfaces B: Biointerfaces, 214*, 112472.

Poongodi, T., Paulpandi, M., Kavithaa, K., Kalaiarasi, G., Rajaganesh, R., Dhayalan, S., Ramabhai, V., & Sharmila, C. (2024). Reduced graphene oxide nanoplatelets derived from the Ziziphus mauritiana: Bio-mediated synthesis and antimicrobial and anticancer activities. *Biocatalysis and Agricultural Biotechnology, 56*, 103038.

Popescu, M.-C., Dogaru, B.-I., & Popescu, C.-M. (2020). Effect of cellulose nanocrystals nanofiller on the structure and sorption properties of carboxymethyl cellulose–glycerol–cellulose nanocrystals nanocomposite systems. *Materials, 13*(13), 2900.

Promhuad, K., Ebel, L., & Harnkarnsujarit, N. (2025). Thermoplastic starch/poly(butylene adipate-co-terephthalate) blown film with maltol and ethyl maltol preserving cake quality: Morphology and antimicrobial function. *Food Chemistry, 464*, Article 141646.

Promsorn, J., & Harnkarnsujarit, N. (2022). Oxygen absorbing food packaging made by extrusion compounding of thermoplastic cassava starch with gallic acid. *Food Control, 142*, Article 109273. https://doi.org/10.1016/j.foodcont.2022.109273

Ramezani, H., Behzad, T., & Bagheri, R. (2020). Synergistic effect of graphene oxide nanoplatelets and cellulose nanofibers on mechanical, thermal, and barrier properties of thermoplastic starch. *Polymers for Advanced Technologies, 31*(3), 553–565.

Rashwan, A. K., Younis, H. A., Abdelshafy, A. M., Osman, A. I., Eletmany, M. R., Hafouda, M. A., & Chen, W. (2024). Plant starch extraction, modification, and green applications: A review. *Environmental Chemistry Letters, 22*(5), 2483–2530. https://doi.org/10.1007/s10311-024-01753-z

Ren, J., Cao, Y., Zhang, Y., Xiao, X., Zhang, Q., Yang, Q., & Liu, F. (2024). Dual cross-linked network of the active starch film by incorporating palmitic acid and geraniol: A comprehensive evaluation of the hydrophobic mechanism and antimicrobial activity. *International Journal of Biological Macromolecules, 280*(Pt 2), Article 135830. https://doi.org/10.1016/j.ijbiomac.2024.135830

Ribba, L., Garcia, N. L., D'Accorso, N., & Goyanes, S. (2017). Disadvantages of starch-based materials, feasible alternatives in order to overcome these limitations. In *Starch-based materials in food packaging* (pp. 37–76). Academic Press.

Rivadeneira-Velasco, K. E., Utreras-Silva, C. A., Díaz-Barrios, A., Sommer-Márquez, A. E., Tafur, J. P., & Michell, R. M. (2021). Green nanocomposites based on thermoplastic starch: A review. *Polymers, 13*(19). https://doi.org/10.3390/polym13193227

Schmitt, H., Creton, N., Prashantha, K., Soulestin, J., Lacrampe, M. F., & Krawczak, P. (2015). Melt-blended halloysite nanotubes/wheat starch nanocomposites as drug delivery system. *Polymer Engineering & Science, 55*(3), 573–580.

Schutz, G. F., de Ávila Gonçalves, S., Alves, R. M. V., & Vieira, R. P. (2024). A review of starch-based biocomposites reinforced with plant fibers. *International Journal of Biological Macromolecules, 261*. https://doi.org/10.1016/j.ijbiomac.2024.129916

Sejidov, F. T., Mansoori, Y., & Goodarzi, N. (2005). Esterification reaction using solid heterogeneous acid catalysts under solvent-less condition. *Journal of Molecular Catalysis A: Chemical, 240*(1–2), 186–190.

Shaikh, M., Haider, S., Ali, T. M., & Hasnain, A. (2019). Physical, thermal, mechanical and barrier properties of pearl millet starch films as affected by levels of acetylation and hydroxypropylation. *International Journal of Biological Macromolecules, 124*, 209–219. https://doi.org/10.1016/j.ijbiomac.2018.11.135

Shanks, R., & Kong, I. (2012). Thermoplastic starch. In A. Z. El-Sonbati (Ed.), *Thermoplastic elastomers*. IntechOpen.

Si, W., & Zhang, S. (2024). The green manufacturing of thermoplastic starch for low-carbon and sustainable energy applications: a review on its progress. *Green Chemistry, 26*(3), 1194–1222.

Siroha, A. K., Bangar, S. P., Sandhu, K. S., Lorenzo, J. M., & Trif, M. (2022). Octenyl succinic anhydride modified pearl millet starches: An approach for development of films/coatings. *Polymers, 14*(12), 2478.

Subramaniyan, V., Sellamuthu, P. S., Jarugala, J., & Sadiku, E. R. (2025). Effect of PVA-based films incorporated with postbiotics, flax seed mucilage and guar gum to enhance the postharvest quality of fig fruits. *Food Chemistry, 465*. https://doi.org/10.1016/j.foodchem.2024.142018

Surendren, A., Mohanty, A. K., Liu, Q., & Misra, M. (2022). A review of biodegradable thermoplastic starches, their blends and composites: Recent developments and opportunities for single-use plastic packaging alternatives. *Green Chemistry, 24*(22), 8606–8636.

Tan, L. J., Zhu, W., & Zhou, K. (2020). Development of organically modified montmorillonite/polypropylene composite powders for selective laser sintering. *Powder Technology, 369*, 25–37.

Tarawneh, M. A. A., Saraireh, S. A., Chen, R. S., Ahmad, S. H., Al-Tarawni, M. A., Yu, L. J., Alsobhi, B. O., & Hui, D. (2020). Mechanical reinforcement with enhanced electrical and heat conduction of epoxy resin by polyaniline and graphene nanoplatelets. *Nanotechnology Reviews, 9*(1), 1550–1561.

Tarique, J., Sapuan, S. M., & Khalina, A. (2021). Effect of glycerol plasticizer loading on the physical, mechanical, thermal, and barrier properties of arrowroot (*Maranta arundinacea*) starch biopolymers. *Scientific Reports, 11*, Article 13204.

Tekin, İ., & Ersus, S. (2024). Development of thermoplastic starch/poly(butylene adipate-co-terephthalate) biobased active packaging films and potential usage in the food industry. *Journal of Applied Polymer Science, 141*(33). https://doi.org/10.1002/app.55824

Thongsomboon, W., Srihanam, P., & Baimark, Y. (2023). Preparation of flexible poly(l-lactide)-b-poly(ethylene glycol)-b-poly(l-lactide)/talcum/thermoplastic starch ternary composites for use as heat-resistant and single-use bioplastics. *International Journal of Biological Macromolecules, 230*, Article 123172.

Todhanakasem, T., Jaiprayat, C., Sroysuwan, T., Suksermsakul, S., Suwapanich, R., Maleenont, K. K., Koombhongse, P., & Young, B. M. (2022). Active thermoplastic starch film with watermelon rind extract for future biodegradable food packaging. *Polymers, 14*(16). https://doi.org/10.3390/polym14163232

U.S. Environmental Protection Agency. (n.d.). U.S. Environmental Protection Agency. Retrieved January 29, 2025, from https://www.epa.gov/.

Vaezi, K., Asadpour, G., & Sharifi, S. H. (2020). Bio nanocomposites based on cationic starch reinforced with montmorillonite and cellulose nanocrystals: Fundamental properties and biodegradability study. *International Journal of Biological Macromolecules, 146*, 374–386.

Varghese, S., Awana, M., Mondal, D., Rubiya, M. H., Melethil, K., Singh, A., Krishnan, V., & Thomas, B. (2023). *Amylose–Amylopectin Ratio*. Springer Science and Business Media LLC. https://doi.org/10.1007/978-981-19-0710-4_48

Vieira, M. G. A., Da Silva, M. A., Dos Santos, L. O., & Beppu, M. M. (2011). Natural-based plasticizers and biopolymer films: A review. *European Polymer Journal, 47*(3), 254–263. https://doi.org/10.1016/j.eurpolymj.2010.12.011

Wang, C., Zhao, J., & Xing, B. (2021). Environmental source, fate, and toxicity of microplastics. *Journal of Hazardous Materials, 407*, Article 124357.

Wang, L., Li, D., Ye, L., Zhi, C., Zhang, T., & Miao, M. (2024). Characterizations and film-forming properties of different fractionated high-amylose maize starches subjected to hydroxypropylation. *Food Chemistry, 440*, Article 138177. https://doi.org/10.1016/j.foodchem.2023.138177

Yang, Y., Fu, J., Duan, Q., Xie, H., Dong, X., & Yu, L. (2024). Strategies and methodologies for improving toughness of starch films. *Foods, 13*(24), 4036. https://doi.org/10.3390/foods13244036

Yao, Q. B., Huang, F., Lu, Y. H., Huang, J. M., Ali, M., Jia, X. Z., Zeng, X. A., & Huang, Y. Y. (2024). Polysaccharide-based food packaging and intelligent packaging applications: A comprehensive review. *Trends in Food Science and Technology, 147*. https://doi.org/10.1016/j.tifs.2024.104390

Yin, P., Chen, C., Ma, H., Gan, H., Guo, B., & Li, P. (2020). Surface cross-linked thermoplastic starch with different UV wavelengths: Mechanical, wettability, hygroscopic and degradation properties. *RSC Advances, 10*(73), 44815–44823. https://doi.org/10.1039/d0ra07549c

Zdanowicz, M., & Sałasińska, K. (2023). Characterization of thermoplastic starch plasticized with ternary urea-polyols deep eutectic solvent with two selected fillers: Microcrystalline cellulose and montmorillonite. *Polymers, 15*(4), 972. https://doi.org/10.3390/polym15040972

Zhang, Y., & Rempel, C. (2012). *Retrogradation and antiplasticization of thermoplastic starch*. IntechOpen.

Part III

Future trends

Chapter 20

Plant-based alternatives: Role of native and modified starches in plant-based meat and dairy analogs

R. Saran, M. N. Tejaswini, R. Poorvasruthi, R. Devadharshini and M. Maria Leena
School of Biosciences and Technology, Faculty of Engineering and Technology, SRM Institute of Science and Technology, Tiruchirappalli, Tamil Nadu, India

1. Introduction

The significant growth in the usage of animal-based foods has a direct impact on human health. The principal factor of mortality due to direct intake of animal-derived food is heart attack. This can be notably reduced by switching to a vegan diet. Many 'diseases of nutritional extravagance', which include obesity, high chances of cancer, diabetes, and heart disease, are closely linked to increased intake of animal products consumption, under situations of food abundance, where diets based on plant foods are linked with longevity, the shift toward diets richer in animal products results in unhealthy populations. Investigations have found that even very small amounts of animal-derived diets are associated with an increase in chronic degenerative death rates. This has become clear in the current scenario of China, where diets high in animal products have contributed to a rise in heart disease and cancer rates (Machovina et al., 2015). Recent interest in plant-based alternatives has received substantial acceptance and market share due to the diversification of diets through plant-based alternatives such as starch, which is rich in fat. This can tackle rising lactose intolerance, allergies, elevated levels of saturated fatty acids, and sustainability concerns (Kailaje et al., 2022).

Plants are considered the best source of polysaccharides because of their widespread availability, low cost, and simple extraction technique. Among the other polysaccharides, starch (also known as amylum) is suitable for utilization across various sectors, such as medicine and food, as it has fewer side effects. Starch is a carbohydrate commonly found in potatoes, corn, wheat, and cassava (tapioca) (see Fig. 20.1), with a polymeric structure composed of glucose connected by glycosidic linkages. Starch granules typically comprise two polysaccharides: amylose (20%–30%) and amylopectin (70%–80%). The two starch components have unique features and are inappropriate for identical usage. Overall, starch content and the relative proportions of amylose and amylopectin differ drastically between plant species and organs, depending on organ growth and developmental circumstances (Gerrano et al., 2014). Starch is a necessary energy storage molecule in plants and an important human food source and nutritious supplement (Rashwan et al., 2024).

Starches produced by different plant sources often differ in composition and structural properties, which alter physicochemical characteristics like solubility, hydrolysis, swelling, thermal behavior, degradation, and rheology. These variations are essential for the development and production of starch-based products. Furthermore, the different aspects of the starch structure normally regulate the functional qualities of starch granules, which include hydrolysis, syneresis, swelling power, retrogradation, gelatinization, and rheological behavior. Native starch, derived from plants, is usually unable to endure harsh processing conditions that include high shear rates, extreme temperatures, strong acid-alkali treatments, and freeze-thaw cycles. As a result, its use is restricted and inappropriate in many industrial applications. However, various methods are used to modify the native starch to improve or inhibit its natural properties or to enrich it with particular characteristics to meet the standards of industrial applications (Liu et al., 2017).

Native and Modified Starches for Food Applications. https://doi.org/10.1016/B978-0-443-27431-2.00021-3

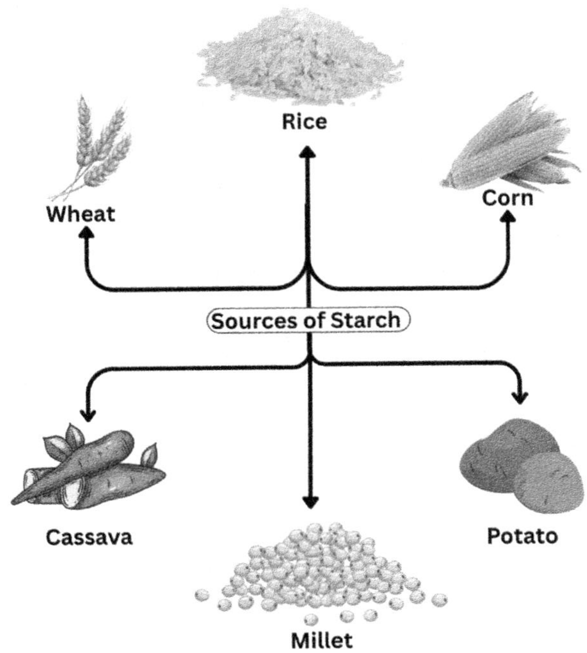

FIGURE 20.1 **Sources of starch.** Plant-based starch sources rice, wheat, corn, cassava, potato, and millet form the biochemical backbone for native and modified starch applications in plant-based meat and dairy analogs.

Starch, used as a nutritional and probiotic carrier, is utilized in dairy analogs, which undergo modifications to mimic the flavor, texture, and appearance of usual dairy foods. This modification frequently ensures that these alternatives closely resemble the nutritional profile of their dairy equivalents by mimicking the smoothness and consistency similar to traditional dairy-based products (milk, yoghurts, and cheese).

However, with more meat products reaching consumers through food service establishments, Potential meat alternative starch is mostly used to help provide more consistent juiciness, tenderness, and texture in products and improve restaurant yields and hold times. As in the past, starch products are used largely in the meat industry for moisture management (Kyriakopoulou et al., 2021). Other than utilizing the protein already present, starches have always been one of the most economical ways to retain water in processed meat products. Most starch products can increase yield if there is the proper amount of time, temperature, and water available for the starch granule to cook and swell. On the other hand, plant-based products are especially challenging to produce due to their compositional and structural intricacy. The terms plant-based meat and dairy analogs are solely utilized to add clarification (Grossmann & McClements, 2021).

2. Types of starches

Starch is a vital carbohydrate polymer commonly seen in both food and non-food industries due to its availability, biodegradability, renewability, non-toxicity, and cost-effectiveness (Wang & Tong, 2024). Starch is produced in plants through a process referred to as condensation polymerization of glucose using starch-synthesizing enzymes. Starch comprises little white granules with dimensions that vary between 1 and 100 μm, seen under a microscope that varies in shape with a differentiated amylose and amylopectin content (Table 20.1). After cellulose, starch is the second most prevalent substance generated by plant cells. Eq. (20.1) depicts the synthesis of glucose in plants.

$$nC_6H_{12}O_6 \underset{\text{enzymes}}{\longrightarrow} (C_6H_{10}O_5)n + nH_2O \quad \text{Glucose} \rightarrow \text{Starch} + \text{Water} \tag{20.1}$$

Starch is an important source of carbohydrates in the human diet and serves as a major source of energy. Depending on the in vivo digestion of starch, it is classified into rapidly digestible starch, slowly digestible starch, and resistant starch as per its rate of glucose release and its absorption in the gastrointestinal tract. The completely digested starch fractions included rapidly digestible starch and slowly digestible starch, while resistant starch is the portion that resists digestion and absorption in the small intestines of healthy individuals and undergoes fermentation in the large bowel by human colonic microflora into short-chain fatty acids, mainly propionate, acetate, and butyrate (Fm, 2023).

TABLE 20.1 Comparative starch composition and granule morphology across major plant sources.

Plant source	Shape of granules	Total starch content (%)	Amylose composition (%)	Amylopectin composition (%)	References
Potato	Semi-crystalline and amorphous structure	10.8–22.1	11.4–36.6	63.4–88.6	Tong et al. (2023)
Common wheat	Disc-like lenticular shape, spherical or oval and uneven or cuboidal.	60–75	18.2–28.8	71.2–81.8	Shevkani et al. (2017)
Rice	Polyhedral, usually angular, irregular, smooth, oval, or irregular in shape.	70–80	37	70–85	Shoukat et al. (2025)
Maize	Polyhedral, irregular, and round.	70–75	25–30	72–73	Luo (2022)
Cassava	polyhedral or truncated-oval facets and a central hilum	80–90	17–24	76–83	Wambua et al. (2023)
Corn	polyhedral, with regular facets and a central or slightly off-center hilum. Rare oval or truncated spheres.	68–74	29.4–24.64	75.36–72	Mohammed et al. (2022)

2.1 Native starches

Native starch is the unaltered starch polysaccharide because it occurs naturally in the plant cells as the primary energy reserve in the higher plants by virtue of its semi-crystalline granular structure that is made up of amylose and amylopectin polymers (Bashir & Aggarwal, 2019). Amylose is a minimally branched linear model bridged by α-1,4 bonds that are packed closely within the granules located in the endosperm of grains in cereals (Harris, 2019). Amylose exhibits a brittle helical shape, which appears to be more difficult to digest than ordinary starch constituents, making it a key attribute of resistant starch. The basic structure of the amylose helix is identified in the amylose molecular groups. As a result, strong molecular interactions are observed due to its simpler structure; amylose molecules can build hydrogen bonds more readily than amylopectin. Whereas amylose contains fewer branches than amylopectin, which makes it more difficult to digest, it crystallizes rapidly in most meals. Amylose is a part of starch, making up roughly between 20% and 30% (Subroto, 2020).

Amylopectin has a nuclear-like unit, indicating a heavily structured chain this amylopectin possesses highly branched links formed via α-1,6 joined to linear glucans units, which form an orderly structure the amylopectin branched links are known as large long (b1) mid-length (b2) and short (b3) and (b4) chains of the branch chain length can alternate depending on the starch source the branches of amylopectin link for each 20 to 25 glucose linear unit; however, the length or degree of polymerization (DP) of the glucan chains differ according to the source of starch and the deed of the enzymes liable for the synthesis of the starch granules due to genetic differences and the environmental and nutritional conditions presenting throughout the growth of the plants and seed formation (Cornejo et al., 2018).

2.2 Modified starches

Native starch is unable to meet certain industrial requirements. Therefore, it may be modified either physically or chemically. Chemical modification is commonly used to enhance starch pasting properties and extend shelf life. Overall, starch processing quality may be improved or modified through variances in their pasting, gelatinization, and retrogradation properties. Commercial starches like corn starch and other cereals and grains may have undergone physical modification or simple chemical modification in order to be used within a specified food or other industrial application. Modifications of starch through physical or chemical changes allow for improved heat stability, heat resistance, stable freeze-thaw characteristics, and complete dissolution with either hot or cold suspensions (Yazid et al., 2018). Modified starches are employed in various processed foods as they have significant functional properties compared to native starches. Annealing (ANN) and heat-moisture treatment (HMT) modify starches without causing gelatinization or damage to the granules. These treatments do not pose any harm to shape, size, and birefringence. Both approaches cause significant changes to the interior granule structure. The osmotic pressure treatment on the starches led to modifications in starch chain interactions, amylase leaching, acid, granule swelling, viscosity, retrogradation, gelatinization temperatures, and enzyme hydrolysis. However, several modifications were applied to maize, potato, and taro starch to enhance their attributes, which can be valuable for their industrial applications (Karmakar et al., 2014).

2.2.1 Physical modification

Physical modification remodels starch granules and their internal structure by imposing selected physical forces, each on a distinct functional axis (Fig. 20.2). Heat processes (superheating, thermal inhibition, microwave treatment, annealing) reorganize crystalline and amorphous regions, altering gelatinization and pasting profiles. Moisture processes (freeze–thaw, heat–moisture treatment, osmotic-pressure protocols) exploit the presence of water to alter swelling power, solubility, and water-holding capacity. Radiation processes (gamma and UV irradiation) impose controlled microfractures in the granule matrix, facilitating greater enzyme access and surface activity. Mechanical processes (high-energy or vacuum ball milling, micronization) mill the particles, enhance surface area, and modify pasting viscosity. Pressure methods, high-hydrostatic–pressure treatment, and instant controlled pressure drop (DIC) form porous, fissured structures that hydrate quickly. Lastly, pulsed-electric-field treatment perforates the network of granules by brief, high-voltage pulses, enhancing dispersibility and gel strength. By grouping each process by its dominant physical driver thermal, moisture, radiation, mechanical, pressure, or electromagnetic scientists can customize solubility indices, swelling behavior, pasting curves, and gelation properties to dairy and meat analogs (Nawaz et al., 2020).

2.2.2 Chemical modification

Chemical modification of starch involves a series of reactions etherification and esterification, cross-linking, acid hydrolysis, oxidation, and dual-modification processes (Fig. 20.3), that modify both granule chemistry and function simultaneously by covalent modification (Chen et al., 2015). In etherification and esterification, native hydroxyls are substituted by hydrophobic groups (acetyl, hydroxypropyl, hydroxyethyl, carboxymethyl, succinyl, citrate), which form cold-water-soluble, retrogradation-resistant products like sodium carboxymethyl starch (sodium starch glycolate) for pharmacy and textiles; inter-chain hydrogen bonding is blocked by acetylation to enhance paste clarity and freeze-thaw stability (Olawoye et al., 2023). Hydroxypropylation (propylene oxide) and hydroxyethylation (ethylene oxide) randomize granular ordering to enhance chain mobility and gel strength, while phosphorylation adds mono- or di-esters (via $POCl_3$ or sodium trimetaphosphate) to sterically hinder crystallization, form covalent cross-bridges, and impart resistance to heat, shear, and acidic pH.

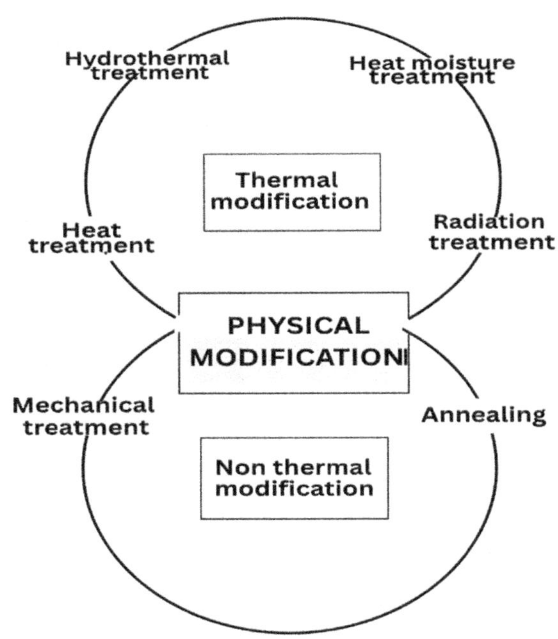

FIGURE 20.2 Types of physical modification. A circular diagram divided into two halves shows physical modification methods such as the upper half lists thermal treatments (hydrothermal treatment, heat-moisture treatment, direct heat treatment, and radiation treatment), and the lower half lists temperature-independent methods (mechanical treatment and annealing) used to enhance functionality in plant-based meat and dairy analogs.

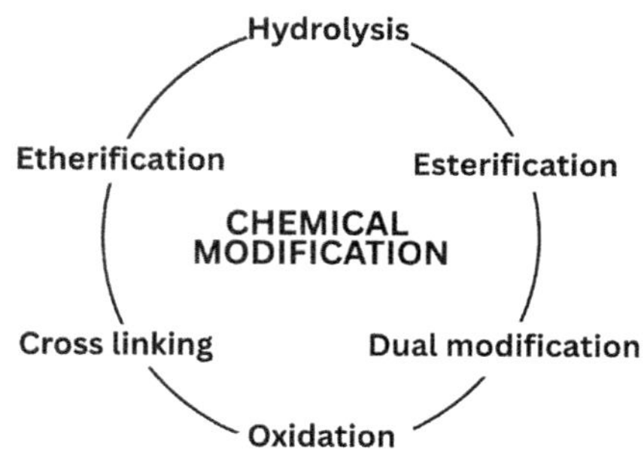

FIGURE 20.3 Types of chemical modification. This is a circular schematic of key chemical starch modification pathways such as hydrolysis, esterification, etherification, cross-linking, oxidation, and dual modifications, each altering starch functional properties for application in plant-based meat and dairy analogs.

Cross-linking with bifunctional reagents (e.g., epichlorohydrin) further densifies molecular networks, increasing gelatinization temperature, decreasing swelling, and stabilizing viscosity, though hyper-ordered matrices anomalously increase syneresis. Acid treatment uses mineral acids (HCl, H_2SO_4, H_3PO_4) to hydrolyze glycosidic bonds from the granule surface to the interior, producing lower-molecular-weight fractions with better alkali fluidity, wider gelatinization endotherms, and bespoke viscosity. Oxidation with oxidants like sodium hypochlorite or hydrogen peroxide de-polymerizes starch chains and adds carbonyl and carboxyl groups, which decrease paste viscosity, enhance clarity, and slow recrystallization. Lastly, dual-modification processes, microwave- or ultrasound-promoted esterification, concurrent cross-linking and phosphorylation, or enzymic hydrolysis followed by etherification, dramatically reduce reaction times, enhance degrees of substitution, and impart starches with bespoke emulsifying, thickening, or thermoplastic performance for food, pharmacy, and industry (Masina et al., 2017).

2.2.3 Enzymatic modification

Enzymatic modification employs biocatalysts to break, modify, or append functional groups to starch polymers in milder, more benign conditions to yield highly selective modifications are mentioned in Fig. 20.4. Amylolytic enzymes (α-amylase, β-amylase and glucoamylase) hydrolyze α-1,4-glycosidic linkages to produce glucose, dextrin, maltose, and reducing molecular weight and viscosity, but enhancing fermentability and digestibility. Pullulanase and isoamylase hydrolyze α-1,6-branches selectively, amylopectin debranching, and reducing molecular entanglement, and permitting crystallization or complex formation vital in resistant starch design. Transferases such as cyclodextrin glycosyltransferase (CGTase) catalyze linear chains into cyclic oligosaccharides (cyclodextrins), which act as an initiator for hydrophobic molecules in food formulations. Branching enzymes introduce α-1,6-linkages to linear chains to provide cold water solubility and paste stability by lowering gel point and retrogradation kinetics. Glucan phosphorylases and maltogenic amylases further extend the polymer architecture by altering chain length distribution or increasing preferred oligomer formation. Enzymic de-polymerization is frequently implied to maintain granular structure while optimizing rheological characteristics such as elasticity, transparency, and freeze–thaw stability. Such processes, performed under aqueous,

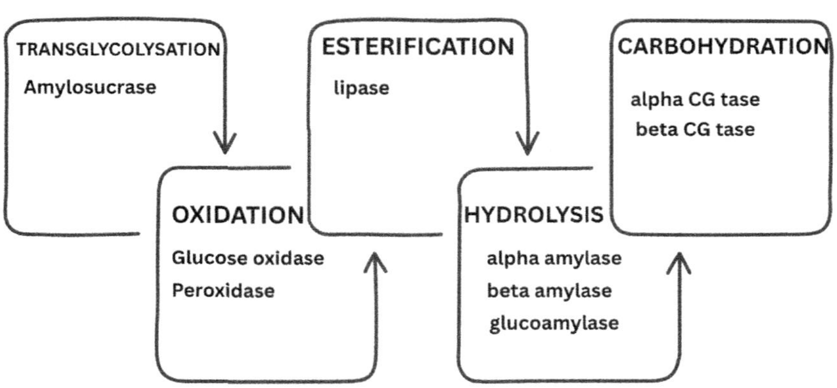

FIGURE 20.4 Types of enzymatic modification. This flowchart represents enzymatic modification pathways of starches in plant based meat and dairy analogs. representing transglycosylation, esterification, carbohydration, oxidation, and hydrolysis reactions for improved functional properties.

temperature, and pH-controlled conditions, offer substrate specificity, regio-selectivity, and minimal by-products, thus enabling the production of clean-label, digestibility-enhanced starches for tailoring meat and dairy analogs (Vitolo, 2020).

3. Role of starch in plant-based meat analogs

Starch is a primary structuring component in plant-based meat analogs, heat or shear gelatinizing to form a continuous hydrocolloid matrix that entraps proteins and lipids and so replicates meat-like juicy texture and bite. Native starch provides early gel strength, while modified starches (cross-linked or phosphorylated, for instance) enhance thermal and shear stability, enhance water-holding capacity, and reduce syneresis upon cooking and storage. A recent study illustrates that starch–protein ratio adjustment in mung bean and pea isolate systems directly regulates moisture retention, hardness and springiness in analog gels.

3.1 Textural enhancement

The fundamental learning of textural attributes is critical for the production of alternatives for meat-based that possess identical characteristics as conventional meat-based products, such as similar cutting ability or mouthfeel features, for example, bite strength, certain valuable aspects were noted in the elongation studies, which provide a framework for the degree of anisotropic structures in meat alternatives for the texture of meat substitutes is affected by the mixture and the starch interaction protein and oil content. Textural properties, such as hardness, chewiness, and springiness, are crucial for meat replacers to replicate the eating experience of meat. The types of modifications tailored to achieve various improvements in meat texture are listed in Table 20.2. Studies investigate the influence of starch on the textural properties of protein extrudates (Schreuders et al., 2021).

3.1.1 Extrusion-mediated fibrousness enhancement

Starch from different sources on the textural properties of extruded soy protein isolate/wheat gluten mixtures report variations in the degree of fibrousness between gated starch sources, connecting the different thermal transition properties of the starch sources with the extrusion response parameters, such as specific mechanical energy (SME) and die pressure. They conclude that the thermal transition properties of the starch affected the textural properties of the extrudates by influencing the extrusion response parameters. Several Extrusion trials found that the addition of 50% starch gave a

TABLE 20.2 Starch modification strategies and their functional impact on plant-based meat analogs.

Type of modification	Purpose of modification	Functional properties	References
Cross-linking of starch	Improves texture stability in extruded meat analogs	Decreases hardness, gumminess, springiness, and cohesiveness	Liu et al. (2025b), Zhang et al. (2025a)
Hydroxypropylation of starch	Enhances rheological behavior in protein extrudates	Modifies viscosity, and improves processability	Wu et al. (2025)
Amylose-enriched starch	Increases firmness and chewiness in meat alternatives	Strengthens gel structures, and enhances bite strength	Wang et al. (2024)
Amylopectin-enriched starch	Softens structure for improved mouthfeel	Reduces hardness, and increases elasticity	Yang et al. (2025)
Pregelatinized starch (PS)	Enhances water retention and appearance in ground meats	Improves water binding, and prevents sedimentation	Lee and Yoo (2023)
Resistant starch incorporation	Modifies rigidity and springiness in pork batters	Increases cohesiveness, chewiness, and gel stability	Xie et al. (2024), Liu et al. (2025a)
Thermal modification	Improves extrusion response and fibrous texture	Alters starch-protein interactions, and regulates hydration	Hao et al. (2024)
Industrial PI compound treatment	Modifies viscosity for protein denaturation	Improves solubility, enhances stickiness, and water-binding	Zhang et al. (2025b)

maximum compression force of the extrudate when using 28% moisture and 194°C in the extruder. The addition of wheat starch could alter the firmness of the product (Huang et al., 2022). This is attributed to the formation of bonds between the starch and the hydrophilic groups of the protein, made possible by the structural modification of both biopolymers caused by the high temperatures in the extruder. These bonds were responsible for the limited hydration capacity of the protein, a higher initial viscosity, and the dense and rigid structures observed after the addition of starch (Zhang et al., 2016).

3.1.2 Tailoring texture through starch modification

A study on soy-protein isolate (SPI)—based meat analogs replacing gluten with five different variety of hydrocolloids including hydroxypropyl starch (HPS) and cross-linked tapioca starch (CLS) increased the viscosity and degree of plasticity of the meat analog compared with the gluten-based control sample. Addition of 1%–4% HPS and CLS decreased the hardness, gumminess, springiness, and cohesiveness of the extruded meat analogs. The changes in textural properties correlated well with the rheological data obtained directly from the extruder. The starches were not found to be suitable replacements for gluten in the recipes (Nanta et al., 2021). Another study found that the addition of amylopectin to pea protein isolate decreased the hardness, springiness, and chewiness of the extrudates, while amylose had the opposite effect, underlining that the effect of starch depends on the composition and origin of the starch. Depending on the desired outcome, it is possible to adjust several textural properties of meat-replacing products by adding the right amount of the right starch. Generalizing the effect of starch on textural properties is difficult, especially since the effect is indirect via the extrusion response parameters. The polymer blending law can indicate the effect of starch on the textural properties, the rheological properties of the starch, and the other ingredients at process conditions. In general, firmness can be increased or decreased by the addition of amylose or amylopectin-rich starches. Modified starches could be able to deliver tailor-made textural properties for meat-replacing products (De Angelis et al., 2020).

3.1.3 Gelation-mediated textural enhancement of pork batter through pig backfat replacement with resistant starch

The texture of meat-based goods is a significant factor that influences consumer acceptance and can be determined by several controlling factors. The textural features of pork batters was altered with various quantities of pig backfat and resistant starch. Incorporating resistant starch substantially improved the rigidity, springiness, cohesiveness, and chewiness of the pork batter this can be due to the gelation of resistant starch at high temperatures which causes starch granules to expand and alters the structural features of the gel-like substances formed (Xie et al., 2024).

3.1.4 Hydration and shear sensitivity in plant-based ingredients

The two independent industrial Protein Isolate (PI) compounds have viscosity characteristics that were a little bit identical. They demonstrated that the full protein denaturation, limited solubility, and elevated Water Binding Capacity (WBC) were all related to their strong initial viscosity, the greater protein quantity and WBC of PI which already forms a very sticky network at 25°C. Hydration and swelling of the PI particles may cause the difference in initial viscosity between the PI and (Textured Protein concentrate (TPC) in this investigation. Additional mechanical shearing appears to break down the created network quite readily at 25°C the TPC scattering was far less susceptible to fluctuations during the first mixing phase. re-association of the existing starch molecules may be the reason for the increased, ultimate viscosity of TPC in the material (Pöri et al., 2023).

3.2 Binding mechanisms

3.2.1 Starch-gelatinization binding mechanism

Starch is a multi-functional ingredient with many different possible functionalities (i.e., binding agent, emulsifying agent, fat replacer, gelling agent, and water-retaining agent) used in the meat industry. In meat emulsions, the heat-induced swelling of starch granules within the protein gel matrix is likely to be advantageous (Xu et al., 2021). Starch granules also absorb water during gelatinization, as well as augment the water-retaining ability of meat products. When heating myofibrillar protein with starch, notice that the starch granules gelatinize, swell, and occupy the entire protein network, binding additional water and developing the texture of the myofibrillar protein gel. Starch has been observed to improve the water-binding capability of meat patties. Several studies have shown that when corn flour is added to pork patties as a water-binding agent, there is an increase in water binding and shear value but a decrease in cooking and thawing loss. Several studies reported that the discrete addition of modified pea and corn starch as binding agents would reduce cooking loss, expressible moisture, and purge loss in low-fat bologna and, interestingly, increase hardness (representing the strength of each bite), chewiness (representing the need to chew), and cohesiveness (representing the toughness of the bite). Studies using crosslinked and acetylated tapioca starch reported improvement in the textural and gelation properties of Kung-wan and improved rheological properties of the finished batter. In these studies, modified starches act as fillers for the gelling matrix in the gel produced from the crosslinked starch (Punia Bangar et al., 2024).

3.2.2 Starch–protein binding for fibrous structure formation

Meat analogs are enriched with a mixture of polysaccharides derived from plants, including dietary fiber, starches, and sugars, when incorporated into the formulation of meat analogs to enhance the interactions between proteins, water, and lipids during extrusion, which contributes to the process of obtaining the unique fibrous structure of meat. When starch is included in meat analog products, it contributes to fibrous structures with a ratio of less than 10%. The main function is to be a binding agent and alter the structure of the extrudate through gelation and breakdown, and the extrusion process. On the contrary, when the wheat starch in a product was increased from 0% to 8%, the extrudes in fibrousness significantly decreased and noticeably, the texture was poorer owing to the reduced hardness and increased chewability. Adding polysaccharides like starch, soy arabinogalactans, or maltodextrins before extrusion into a food mix can increase fiber. During extrusion, starch changes. Initially, water is absorbed by the starch particles, causing them to expand after being melted in the feed of other components in the material. In the melting zone, starch forms a gel with the breakdown of intermolecular hydrogen bonds. The viscosity of the melt is important for fiber structure development, and it is dependent on protein-starch interactions. These interactions inhibit protein molecules from growing in size and/or aggregating. Unlike low-moisture extrusion, which is restricted by the limited water absorption of starch, high-moisture conditions promote disulfide bonding, hydrogen bonding, and hydrophobic interactions that could promote high degrees of fibrous development. High-moisture extrusion of pea protein Isolate and amylose/amylopectin demonstrated that increasing the proportion of amylopectin increased the fibrousness from 1.32, possibly due to its contribution to developing an anisotropic fibrous structure, while the opposite effect was seen when mixing amylose with pea protein. Isolates, decreasing fibrousness from 1.30 to 1.12 and further reduced with increasing amylose (Dinali et al., 2024).

3.2.3 Retrograded resistant starch regulated binding mechanism

Retrogradation is a process that determines the textural attributes and nutritional aspects of the starch. The binding mechanism of retrogradation, particularly when used in the preparation of food products, typically enhances the food's quality by improving its eating quality. Adding them to the meat analogs creates a preferable texture by altering their structure to best suit consumers' taste needs. Retrogradation of the textural attributes and nutritional aspects of the starch. The binding mechanism of retrogradation, particularly when used in the preparation of food products, typically enhances the food's quality by improving its eating quality. Adding them to the meat analogs creates a preferable texture by altering their structure to best suit consumers' taste needs. The process of gelatinization and subsequent retrogradation results in the formation of retrograded resistant starch (RS3), which plays a significant role in protein binding and exhibits heightened intensity within emulsion networks. Through its affinity for oil droplets enveloped in myofibrillar protein, RS3 reduces the interaction of hydrophobic groups, thereby enhancing interactions within protein-protein networks. This contact enhances hydrophobic forces and facilitates the formation of disulfide bonds. Consequently, this enhances the robustness and connectivity of the gel network. The improved network fortifies the gel, enabling it to retain a greater volume of water while preserving its structural integrity, a vital factor in the development of meat analogs that do not compromise protein digestibility (Chen et al., 2024).

3.3 Flavor and moisture retention

3.3.1 Role of starch in the Maillard reaction and flavor retention in extruded meat analogs

Starch is a multi-functional ingredient in high moisture extrusion processes used to produce meat analogs. Besides being the reason behind structural hardness and water retention, starch is also a source of reducing sugars that are responsible for the Maillard reaction. During extrusion, the reducing sugars interact with amino acids and proteins to yield non-volatile Maillard reaction products. For example, it has been established that starch blended with glucose and lysine produces extrudates that lead to typical flavor compounds, which are crucial to attain a desirable meat flavor.

Hence, this study explains the indirect contribution of starch in the flavor-preserving properties, starch use in the manufacture of non-volatile products implies a secondary role as a flavor stabilizer. Starch-derived carbonyls' Maillard reaction products are responsible for having a stable matrix upon formation, and this can limit flavor-active compound loss when processing. Water-binding capacity of starch-treated matrices also retains volatile compounds, thereby providing improved moisture and flavor retention in the final product (Jiang et al., 2024).

3.3.2 Enhancing water binding and freeze-thaw stability with pregelatinized starches

Pregelatinized starches (PS) disintegrate in freezing water. They are not meant to be cooked unlike other types of starches, these substances enhance the appearance of ground meat items and assist in water binding water binding is crucial for the thermal treatment of meat-based items as proteins exhibit water during denaturation the purity of the item is determined by how much and what kind of starch is incorporated pregelatinized starches (PS) are included to the solution of brine to enhance stickiness and avoid sedimentation potato starch is commonly utilized for its high-water holding capacity. In contrast, the waxy form of maize starch is more popular for its capability to endure freezing and thawing conditions (Babic et al., 2013).

4. Role of starch in plant-based dairy analogs

Starch acts as a major emulsifier and texturizer in frozen foods, yoghurts, cheeses, and non-dairy milks, enhancing viscosity, stabilizing oil-in-water dispersions, and creating transparent gels that replicate casein networks. Engineered derivatives like oxidized or acetylated starch deliver freeze-thaw stability, creaminess, and controlled melt behavior, along with syneresis inhibition and spreadability enhancement in analog cheeses and ice-cream analogs.

4.1 Texture and creaminess

4.1.1 Enhancing texture and creaminess with native and modified starches

The Replication of the mouthfeel, flavor, and texture that customers anticipate is the biggest obstacle when it comes to plant-based alternatives for dairy. As plant-based dairy products have less protein content than their dairy counterparts, this is where starch plays an essential for improving their texture and creaminess. The thickening properties of certain starches, such as rice, potatoes, corn, and tapioca, enhance viscosity and structural stability. An example of clean-label native rice starch is Beneo's RemyPure S51, which has a creamy texture similar to dairy and great resistance to high temperatures, high shear, and low pH (Radeloff & Beck, 2016).

The potential use of starch as a fat alternative in fresh cheese and milk was explored by scientists in the context of different types of native starches (normal waxy and high-amylose maize starch) and modified starches (acetylated/adipate maize starch on a regular base and hydroxypropylated/phosphate maize starch on a waxy basis) in combination with calcium caseinate in a casein matrix the modified starch blends with the calcium caseinate exhibited low peak temperature and low enthalpy change with a need for low gelatinization temperatures and high viscosity peak and final it is an indicator of high water retention capacity in cheese they were also endowed with good thermomechanical properties high swelling capability and low retrogradation tendency the rheological characteristics of the starch-caseinate blends such as shear-thinning or viscoelasticity would also endow it with easier processability and simulation of texture (Diamantino et al., 2019).

4.1.2 Enzymatic modification for creaminess enhancement

Among enzymatic starch modifiers, amylomaltase (EC 2.4.1.25) offers unique potential to re-engineer fat-like textures in meat analogs. The enzyme hydrolyzes native starch α-(1 $\rightarrow$ 4) glucan chains and re-anchors them on new acceptors, producing highly branched or cyclic glucans, which are easily soluble and produce thermoreversible gels. Blended (5%–10%) into pea- or soy-protein-based patties or nuggets, these starch gel models mimic animal fat lubrication and mouth-coating, maintaining moisture, juiciness, and clean "creaminess" without the addition of lipids. Their dense network resists high-shear extrusion and cooking, preventing dryness and syneresis in finished analogs (Gong et al., 2024).

4.2 Stability and emulsification

4.2.1 Gelatinization as a stability enhancing mechanism

Starch granules remain undissolved in cold water and begin swelling reversibly when heated in an aqueous medium, maintaining characteristics such as birefringence. Upon reaching the gelatinization temperature, which generally varies by 5–10°C because of the heterogeneity of granules, the swelling becomes irreversible, leading to a disruption of the internal framework. Water passes through the amorphous areas, initiating the release of amylose and causing a significant increase in viscosity. In progress heating further downgrades hydrogen bonds, resulting in complete granule hydration and the disintegration of the micellar network (Chakraborty et al., 2022). This process leads to the creation of a cohesive gel as linear molecules are released and engage with one another. Marchant and Blanshard outlined this transformation in three simultaneous stages: water diffusion, a helix-to-coil transition (melting), and swelling resulting from the disintegration of crystallites. According to Olkku and Rha, gelatinization encompasses granule swelling, the loss of birefringence, enhanced clarity and viscosity of the mixture, diffusion of amylose, and the creation of a consistent gel-like matrix upon cooling, which is essential for the stability and texture of dairy analogs (Donmez et al., 2021).

4.2.2 Role of whey-protein starch interactions in thermal stability

The interaction of whey protein with starch utilizes complex physicochemical mechanisms impacted by the protein type, starch concentration, and thermomechanical processing. Three primary binding mechanisms have been hypothesized in these systems. First, proteins may adsorb onto starch granules through physical insertion, controlled by endogenous starch-bound lipids and proteins, as seen in α-caseinate systems. Second, whey proteins, notably β-lactoglobulin, tend to accumulate within starch matrices, influencing the gel structure and viscosity, which accompanies an increase in the concentration of amylopectin, stimulating clustering and shear-thinning behavior. Third, covalent and non-covalent bonding occurs between the starch molecules and whey proteins, where hydrophilic functional groups enable hydrogen bonding. On the other hand, non-covalent interactions in WPI-wheat starch gels show that whey proteins may function as fillers, making it harder for the network of the starch to develop. Whey proteins can also bind with the exposed branches of amylose and amylopectin chain extensions through non-covalent bonding. This bond prevents starch from recrystallizing during cooling. Additionally, it lowers the final viscosity because whey proteins present a barrier that prevents hydrogen bonds from forming between starch molecules. Studies indicate that these interactions enhance thermal stability; however, they do not necessarily imply the existence of direct molecular networks. Understanding. These binding mechanisms are crucial for tailoring functional properties in food. Systems, mainly in plant-based dairy analogs (Kumar et al., 2017).

4.2.3 Polysaccharide function in emulsion gel stability

Polysaccharides have typically been used as the structural element of an emulsion gel alone or in a mixed gel with protein. In a mixed gel, the emulsion is stabilized with a surfactant before adding a polysaccharide, and gelation occurs. The disperse phase may also function as both an active filler and an inactive filler, depending on the interactions between the particular surfactant/protein selection and the continuous polysaccharide phase. Numerous plant-based polysaccharides (e.g., methylcellulose, inulin, pectin, native and modified starches) have been identified in the literature as suitable additives for emulsion gels that have the ability to deliver stability and mechanical properties. In addition, the polysaccharide sugars added to create an increase in viscosity reported a decrease in coalescence of dispersed oil droplets during processing. Smaller droplet size is positively correlated with stability and gel strength in emulsion gels. At the end, possible uses of plant-based emulsion gels, like dairy alternative applications, were proposed and considered from a sensory and consumer acceptance standpoint (Yiu et al., 2023).

5. Blend techniques for optimizing starch usage

5.1 Synergistic effects

Synergistic effects in starch systems are achieved through starch interaction with other food components, leading to enhanced functional properties greater than the combined effect of individual effects. Such effects have profound impacts on digestibility, texture, and physicochemical properties in food systems. Various Hydrocolloids along with their functional properties are mentioned in Table 20.3. Environmental factors, notably temperature and stress caused by drought, regulate starch gelatinization properties, altering its beginning temperature, final temperature, and enthalpy (ΔH). This work shows that pre-anthesis temperature and heat stress during grain loading induce structural changes in starch, prompting the inclusion of hydrocolloids for stabilization in food formulas. Heat-stressed starch has a heightened transition initial temperature of approximately 62.5°C, while starch from moderate growing circumstances remains below 60.5°C.

Hydrocolloids support the gelatinization behavior, thereby ensuring consistency in plant-based formulations. Elevated final temperature values (~72.4–75.0°C) under stressful conditions indicate that hydrocolloids may facilitate the development of heat-resistant gels, which are beneficial for dairy mimics. The use of hydrocolloids with modified starch enhances stability, texture, and functional attributes in plant-based food applications, mirroring contemporary trends in sustainable food innovation (Franz et al., 2025). Hydrocolloids are said to be non-starch polysaccharides, including a heterogeneous collection of long-chain polymers that yield viscous fragments and/or gels upon dispersion in water. Hydrocolloids encompass pectin, gellan, guar gum, xanthan gum, carrageenan, gum arabic, locust bean gum, and konjac-glucomannan, among others (See Fig. 20.5 Types of Hydrocolloids). Each of the hydrocolloids offers various benefits in the tailoring of meat and dairy analogs, as mentioned in the table (Yang et al., 2021).

TABLE 20.3 List of hydrocolloids with its functional properties.

Type of hydrocolloid	Synergistic effect with starch	References
Pectin	Improves texture and stability, especially in acidic conditions; may prevent syneresis.	Lin et al. (2022)
Carrageenan	Modifies gel texture, prevents retrogradation, increases viscosity, and improves freeze-thaw stability.	Yap et al. (2024)
Gellan gum	Forms strong, clear gels even at low concentrations and is useful in structured desserts and fillings. It has been selected as one of the most common plant-derived thickening agents.	Bagnolo et al. (2023)
Guar gum	Increases viscosity and water-binding capacity; enhances texture and mouthfeel in sauces and soups. The synergistic effects of the gums and modified starch were hypothesized by considering the molecular incompatibility and molecular interactions between the gums and HPSPS.	Zhang et al. (2024)
Gum arabic	Acts as an emulsifier and stabilizer; improves texture and shelf life.	Singh et al. (2017)
Konjac glucomannan	Greatly increases viscosity and can form thermally reversible gels. Has been chosen among the most common thickening agents derived from plants	Kapoor et al. (2024)
Locust bean gum	Forms synergistic gels with xanthan gum and carrageenan, improving gel strength and elasticity.	Barak and Mudgil (2014)

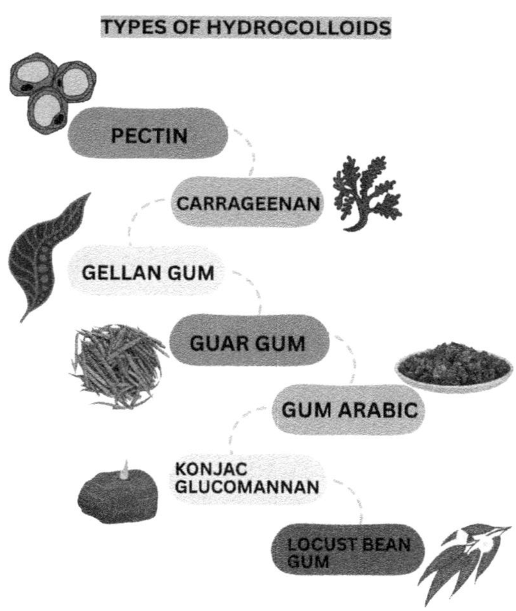

FIGURE 20.5 Types of hydrocolloids used in plant-based meat and dairy analogs. This is a schematic representation of common hydrocolloids such as pectin, carrageenan, gellan gum, guar gum, gum Arabic, konjac glucomannan, and locust bean gum used as gelling, thickening, and stabilizing agents. Also, these hydrocolloids alter moisture, texture, and viscosity to achieve required structural and sensory properties in meat and dairy analogs.

5.1.1 Pectin

Pectins, also known as pectic polysaccharides, build up as a pile of polysaccharides over the cell walls in higher plants, and their structure contains galacturonic acid (GalA) and 16 monosaccharides as co-monomers. Pectin forms into several block-like structures that possess xylogalacturonans (XGA), homogalacturonan (HG), rhamnogalacturonan-I (RG-I), and rhamnogalacturonan-II (RG-II). Pectin obtained from various locations, sources, processing formulas, and growth constraints exhibits a different degree of acetylation (DA) and degree of methyl esterification (DE), resulting in a complex structure (Bai & Gilbert, 2022). Pectin powder, when mixed with water, goes through hydration, dissolution, gelation, and swelling (Seetapan et al., 2024) and is used in dairy formulations by starch-pectin interaction. However, the presence of pectin in starch can impact gelatinization and swelling; when added in higher amounts, the gelatinization is significantly affected. Therefore, making it valuable for plant-based yoghurts, drinkable dairy alternatives, and fermented products.

5.1.2 Carrageenans

Carrageenans are water-soluble, sulfated polysaccharides produced from the red algae (Udo et al., 2023). They comprise a recurring block of disaccharide, which consists of sulfate esters of galactose and 3,6-anhydrogalactose polysaccharides; these hexoses are connected alternatively via α-1,3 and β-1,4 in the copolymer. These organic polysaccharides express themselves in three primary forms: kappa, iota, and lambda. With k- and i-Carrageenan can create thermos-reversible gels in the presence of gel-enhancing cations. This categorization into three varieties (k-, i-, or l-carrageenan) is based on the quantity (one, two, or three) of sulfate groups per disaccharide repeat. The interaction of carrageenan significantly influences the swelling behavior in aqueous systems. The addition of κ-carrageenan or ι-carrageenan to starch pasting makes the starch granules swell further, likely because of contact between the granules. Native waxy maize starch has the lowest swelling ratio of all the samples, while protein-free acetylated distarch adipate has the highest swelling behavior, suggesting that removing protein makes the material more porous and able to hold more moisture (Lascombes et al., 2017). Chemical modifications, such as acetylation and hydroxypropylation, also play an important role in altering the swelling and gelatinization properties. When ionic strength shifts (for example, the presence of NaCl), swelling increases a bit. Swelling occurs due to the osmotic pressure affecting the entropy of the counterion. These results demonstrate the potential of starch-carrageenan synergy in controlling the texture and stability of plant-based dairy products (Rupert et al., 2022).

5.1.3 Gellan gum

Gellan Gum is an organically toxin-free, negatively charged polysaccharide made up of tetrasaccharide repeats of two residues of β-D-glucose, one of β-D-glucuronate, and one of α-L-rhamnose. Gellan gel is created thermally through the synthesis of double helix structures, which is facilitated by cooling. The gelation mechanism occurs through the transition from double helices to cation-mediated aggregation of those helices as a result of the cooling process. However, this process is influenced by the presence of acids or salts. The gel network is enhanced by the presence of monovalent or divalent cations, resulting in increased stability over a broad range of pH and temperature levels (Souza Almeida & Kawazoe Sato, 2019). Gellan gum can be used for texture modification, stabilization, and as a gelling agent. We determine its effect on starch by using water as a solvent, typically involving dairy formulations. Therefore, this study concludes that the addition of gellan gum to cassava starch paste in milk-based systems exhibits maximum viscosity, higher gelatinization, and lower thermal stability when compared to the milk-based systems (Coronato et al., 2012).

5.1.4 Gum Arabic

Gum Acacia or gum arabic is a complex and natural branched-link polysaccharide. The structural backbone of gum Arabic is composed of 1,3-linked β-D-galactopyranosyl units, with 2–5 of the same units on the side chain that connect to the backbone via 1,6-linkages. The presence of abundant calcium, magnesium, and potassium salts on its backbone makes gum Arabic either neutral or slightly acidic. Arabinose protein, galactose, glucuronic acid, and rhamnose are the primary parts of the composition. Due to this, it is believed to be a natural glycoprotein that exhibits excellent emulsifying properties. Incorporating gum Arabic into tapioca starch suspension reduced its Peak viscosity, swelling power, and solubility index, proving its ability to be a stabilizer and a viscosity mediating agent (Kiprop et al., 2021).

5.1.5 Konjac glucomannan

Konjac glucomannan (KGM) is a plant polysaccharide from the Konjac plant. The molecular structure of KGM is composed of β-mannose and β-glucose, and scarce amounts of acetyl groups, and has a single chain structure with a random curl state (Sharma & Wadhwa, 2022). The single–chain structure of KGM is dominated by dihedral angle energy, electrostatic interaction in vacuum, and hydrogen bonding in aqueous solutions. KGM is high in viscosity, solubility, and swelling capacity, has good film-forming ability, and has enhanced gelling properties in aqueous solutions. The water solubility of KGM is due to the existence of 5%–10% acetyl substituent residues at mannose on the glucomannan main block (Sun et al., 2023). Konjac is known to be a low-calorie meal that can act as a fat replacer, providing a rich mouthfeel without adding excess calories. That can decrease appetite by maximizing feelings of fullness. Konjac glucomannan is used in low-fat dairy analogs and in meat analogs to improve firmness and chewiness. Two of its important characteristics are having thermally stable gelling properties and a high consistency-building ability (Hu et al., 2025).

5.1.6 Locust bean gum

Locust Bean Gum is a high-molecular-weight, non-ionic galactomannan polysaccharide isolated from seeds of Ceratonia Siliqua. It consists of β-(1-4)-mannose backbones branched with α-(1-6)-galactose in a random fashion. It is primarily employed as an additive (E410) in the food and beverage industry, most commonly as a thickener, stabilizer, gelling agent, or emulsifier. LBG in water gels only in the presence of other substances, for instance, under high sugar concentration or in the presence of other hydrocolloids like κ-C. The addition of LBG to κ-C has many effects (Russo Spena et al., 2024). In this study, locust bean gum, or when combined with guar gum, is used in frozen dairy-based analogs to tailor the required textural properties, such as improving viscosity and suppressing ice recrystallization. Hence, it is a hydrocolloid with differentiated functional properties, making it a unique stabilizer and thickener.

5.2 Innovative blending methods

Novel blending techniques have recently emerged as an important process to enhance the functional utility of starch in food systems. These novel procedures focus on blending starch from different sources through the use of identified physical and enzymatic treatments to create products with targeted rheological and structural properties, which can successfully enhance the function of a food analog or clean-label product. For example, A recent study employed ultrasonic treatment of rice and potato starch at various ratios (25:75 ratio) to develop a clean-label starch alternative that exhibited greater viscosity and stability than xanthan gum. Their methods used a physical treatment (ultrasound) and a unique blending method and produced a product that was described where viscosity modeling could be carried out with high R^2 predictability ($R^2 = 0.988$) (Park & Kim, 2025).

Recent research focused on the various physical modifications, mainly before blending, including high-pressure processing, cold plasma treatment, and pulsed electric fields. These modifications have shown improvements in the morphology of the granules and overall enhancements in gelation, water absorption, and network formation of the starch granules. Each of the physical treatments improves the functional properties of the individual starches, and when blended, they act together synergistically. A recent review on starch modifications showed that, with the introduction of new blending techniques and some of the physical methods mentioned above, we can create a starch matrix with better performance properties, with improved stability, low retrogradation, and improved texture profiles or characteristics of the starch in food formulations. Likewise, a great deal of the research literature has pursued clean-label starches with minimal processing. The evidence suggests that when chemical additives can be eliminated and the amount of processing is significantly reduced, we can produce a more suitable starch blend for health-conscious consumers (Miao et al., 2025). Such clean-label strategies underscore the value of physical and enzymatic modification methods when integrated into an optimized blending platform, enabling the delivery of starch systems with high emulsification and water-binding capacities. Moreover, recent results on starch nanocomposites demonstrate that new blending approaches can also be utilized to aid in designing hybrid matrices. Namely, by using nanoscale reinforcements in conjunction with starch blends, researchers have tailored the stability, bioactivity, and even sensory aspects of food products (Sun et al., 2025). In line with these possibilities, research on starch modification via both new and traditional approaches suggests that blending strategies, combined with targeted enzyme treatments and controlled hydrothermal processes, can lead to distinct advantages, such as improved gelatinization behavior and network formation (Suri & Singh, 2023). These new blending methodologies, supported by advances in current research, demonstrate that the ability to finely tune starch characteristics through the use of physical pretreatments and clean-label blending methods will optimize starch functionality in food applications. The degree of blending ratios and the conditions of pretreatment will enable the development of starch-based ingredients that can match the sensory and textural parameters of food analogs while also providing additional nutritional and processing characteristics.

5.3 Industrial practices

Industrially produced food relies on industrial processes that transform native and modified starches into functional ingredients in plant-based meat and dairy substitutes. At the industry level, starch functionality is uniformly modified and essentially through industrial means, which include methods such as high-pressure processing, cooking, extrusion cooking, ultrasound treatments, and any other physical modification means available to the industry. A recent study showed that ultrasound-treated starches could create a clean-label product with viscosity and gel stability similar to commercially available hydrologies when combined at a defined ratio of their base starches as a means to satisfy both processability and consumers' need for sensibility. Chemical and enzymatic modifications of starch are sometimes applied, along with physical treatments, to improve water binding, rheological characteristics and textural properties while reducing formulation costs. As said in the literature, including compilations summarizing the science of modified starch usage, they do not report new data themselves, but show that chemical modifications (e.g., esterification, cross-linking or oxidation) create starch derivatives capable of being used at an industrial scale that are able to perform very well. Modified starches were planned in dairy substitutes (i.e., processed cheese) to provide desirable textures and water retention properties while diverting costly protein fractions (Mohamed, 2015). A portion of the protein fraction might be replaced entirely or partially with native or modified starch to derive up to 40% cost reductions whilst retaining favorable sensory qualities in alternative processed cheese. Industrial procedures are also challenged to meet evolving regulations with consumer expectations for clean-label ingredients. Regulatory-based protocols now apply to both native and modified starches. As described from a regulatory perspective, starch in meat products is advantageous because starch provides a predictable ability to control water and enhance textural properties while achieving food safety standards outlined by regulatory bodies such as the U.S. Department of Agriculture (Vilpoux et al., 2023).

Furthermore, food processors are taking advantage of clean-label products, such as through relatively new studies on the production of clean-label starches. Together, food processors can produce starches that meet performance and "natural" ingredient attributes by using both physical pre-treatments, blending, and controlled blending without significant chemical modification. Together, these industry practices, which integrate advanced processing technologies, economical formulation practices, and compliance with regulatory practices, ensure starch-based ingredients have met the optimal manufacturability for plant-based meat and dairy alternatives. The final product form that can result achieves not only the texture and structure of the animal products it seeks to replace but also navigates sustainability, economic, and clean-label demand in a modern food market (Sołowiej et al., 2016).

6. Applications of starch

Native and custom-designed starches provide critical functionality throughout plant-based foods and into other applications. In meat substitutes, starches are used as binders, fats and moisture regulators working in a blend with proteins and hydrocolloids to manage texture, juiciness, and cook-loss profiles. In dairy alternatives from almond-milk yoghurts to plant-based cheeses, custom-designed starches stabilize emulsions, manage viscosity, inhibit syneresis, and manage retrogradation, thereby improving mouthfeel, shelf life, and consumer acceptability. High-end 3D food printing utilizes heat-stable, shear-thinning starch gels for custom-designed form and layered architecture, while edible films and thermoplastic starch composites provide exciting opportunities for sustainable packaging. Beyond food, pharmaceutical tablets depend on starch derivatives for binders, disintegrants, and controlled-release matrices, while industrial uses utilize modified starches as rheology modifiers for paints, adhesives, and wastewater treatment.

6.1 Plant-based dairy products

Since the seventh millennium BC, bovine milk, mainly cow's milk, has supplied high-biological-value proteins, lipids, vitamins, and minerals (notably calcium) essential for human health. However, milk protein allergy (CMPA) and lactose intolerance (LI), which affect over 75% of the global population, alongside environmental and ethical concerns, have driven demand for plant-based milk alternatives as replacements for routine daily use. Plant-based cheese alternatives (PBCAs) aim to reproduce the physicochemical and sensory characteristics of conventional cheeses (e.g., cheddar) by tailoring processing methods and ingredient blends. Polysaccharides from plant starches deliver texture, viscosity and stability; proteins from legumes (peas, soy, lupin), potatoes, nuts, seeds and zein provide structure, elasticity and mouthfeel; and a mix of solid fats (cocoa butter, coconut oil, palm oil) with liquid oils (avocado, canola, sesame, soybean, sunflower) imparts creaminess and melt behavior. These formulations serve vegetarians, vegans, and those with CMPA or LI, and they offer added benefits for rheumatoid arthritis and metabolic and dermatological disorders, and, with appropriate nutritional safeguards, can even be adapted for infant feeding (Plamada et al., 2023).

6.1.1 Tapioca starch in almond milk yoghurt analogs

In this study, the effects of varying tapioca starch content on almond milk yoghurt-like products (AYML) were examined through a comparison of physicochemical, microbial, and sensory traits. Almond milk, a dairy alternative, naturally suffers from textural problems as an oil-in-water emulsion predisposed to phase separation, typically corrected by the addition of stabilizers such as xanthan gum. In this study, tapioca starch was investigated as a substitute for a stabilizer. Almond milk (Farm Fresh Milk Sdn Bhd, Malaysia), fine sugar, and tapioca starch (Cap Kapal ABC, Malaysia) were ingredients in formulations, with commercial dairy yoghurt (Malaysia Milk Sdn Bhd) used as a control, and Streptococcus thermophilus and Lactobacillus bulgaricus starter culture added after pasteurization. Samples were fermented at 40–43°C for 8 h, then cooled and refrigerated at 4°C for 14 days to maintain quality preservation. Analyses included pH, viscosity, color, and texture measurements, and LAB counts on MRS agar plates. A sensory test was also conducted by 50 untrained panelists using a 9-point hedonic scale, and differences were significant at $P < .05$. The main findings were that dairy yoghurt (DY) was of lower pH and higher viscosity due to post-acidification. However, the addition of tapioca starch increased the viscosity of AYML, although not to DY levels, and LAB counts were erratic, with 3% starch samples initially having undetectable bacteria, possibly due to the inhibitory effects of starch. Sensory tests consistently scored DY higher in color, aroma, texture, flavor, and overall acceptability, indicating that texture improvements failed to meet consumer standards for traditional DY in terms of AYML. In summary, this research illustrates the utilization of tapioca starch as a stabilizer in plant-based dairy foods, its potential to provide added texture and viscosity, and the necessity for further optimization of formulation to closely mimic the sensory and microbiological characteristics of conventional dairy yoghurt (Zainuddin et al., 2024).

6.1.2 Starch in processed cheese formulations

Heating starch granules with water produces gelatinization; the starch granule swells and ruptures, and in the presence of amylose, the viscosity increases. Upon cooling, the starch undergoes retrogradation, forming a stable, more ordered structure, with the temperature of gelatinization varying with the type of starch of interest. Starches from sources including rice, potatoes, barley, wheat, corn, maize, oats, and tapioca each have different ratios of amylose to amylopectin with unique functional characteristics concerning processed cheese formulations. The interactions of milk proteins (casein and/or whey) with starch happen through different pathways based on the particular system: adsorption, hydrogen bond formation, and phase separation (Sun et al., 2016). Casein systems with starch and potato granules show interactions through hydrogen bonding and competition for waters of hydration, while when casein is interacting with modified starches (e.g., phosphate starch, hydroxypropyl starch), the interactions happen through electrostatic forces and steric stabilization. Whey protein-starch interactions are predominantly hydrogen bonding and may affect gelation properties and retrogradation. Whey proteins tend to adsorb onto rice and oat starch granules and affect starch's crystalline structure, in turn affecting viscosity (low and high) and gel-forming strength (Liyanaarachchi & Vasiljevic, 2018). Fat level and the addition of lactose, with concentration, are very influential on starch viscosity and gelatinization. Studies indicate that a mixture of wheat starch and milk with a higher concentration of sucrose, glucose, fructose, and milk fat induces a significant viscosity pasting effect, with sugars with axial hydroxyl groups inducing viscosity, while sugars with equatorial hydroxyl groups decrease viscosity; lactose and soluble minerals contribute to higher gelatinization temperature and reduced swelling power of starch and water absorption. An example of a plant-based cheese alternative with a heat treatment simulates consumer preparation conditions and illustrates the role of protein–tapioca starch interactions. Zein imparts viscoelastic stretchability, while tapioca starch contributes to water-binding and structural integrity, together achieving cheese-like meltability and performance, as illustrated in Fig. 20.6 (Grasso et al., 2023).

Processed cheese viscosity is associated with casein, fat, and emulsifying salts, whereas levels of viscosity with starch increased due to improvements in the water-binding capacity. There is strong evidence that the larger size of starch granules results in an increase in viscosity, and high-amylose starches decrease meltability due to their linear chain structure in the melted cheese matrix. The linear chains work against the protein and carbohydrate interactions and create an increase in viscosity. Pre-gelatinized starches, as seen in the use of rice and waxy maize starches, further increase viscosity and impact flow behavior and spreadability found in cheese, of melt and processed cheese varieties. Meltability is an important property when considering end-use applications for pizza, burgers, and toasted sandwiches, and pregelatinized starches decrease meltability by 30%–50%, while starch immobilizes water and decreases protein hydration. It has been shown that potato starch substantially inhibits meltability compared to wheat starch, which loses meltability on a per-fat basis in grams of cheese proportion because of so much emulsified fat per gram of starch; and further, cross-linked and acetylated waxy starches decrease meltability, which also would require adjustments in pH to lessen. In conclusion, starches have an important role in the formulation of processed cheese; they not only offer modifiers for textural properties but are also modifiers for viscosity, meltability, and microstructure. Generally, high-amylose starches make firmer cheese with less meltability, and low-amylose starches create the cheese softer and spreadable. pH changes can be used to modify the performance of starch to obtain specific desired sensory and functional properties, and starch also has the potential to act as a fat replacer for processed cheese formulations, a pathway that can yield exciting new products that deliver on specific consumer needs and anticipated quality characteristics (Akshit et al., 2025).

6.1.3 Role of modified cassava starch in probiotic yoghurt

This research analyzed the function of enzymatically modified cassava starch as a texture modifier in fat-free milk products, specifically for Symbilact probiotic yoghurt. Fat-free milk varies in texture and mouthfeel, and cassava starch,

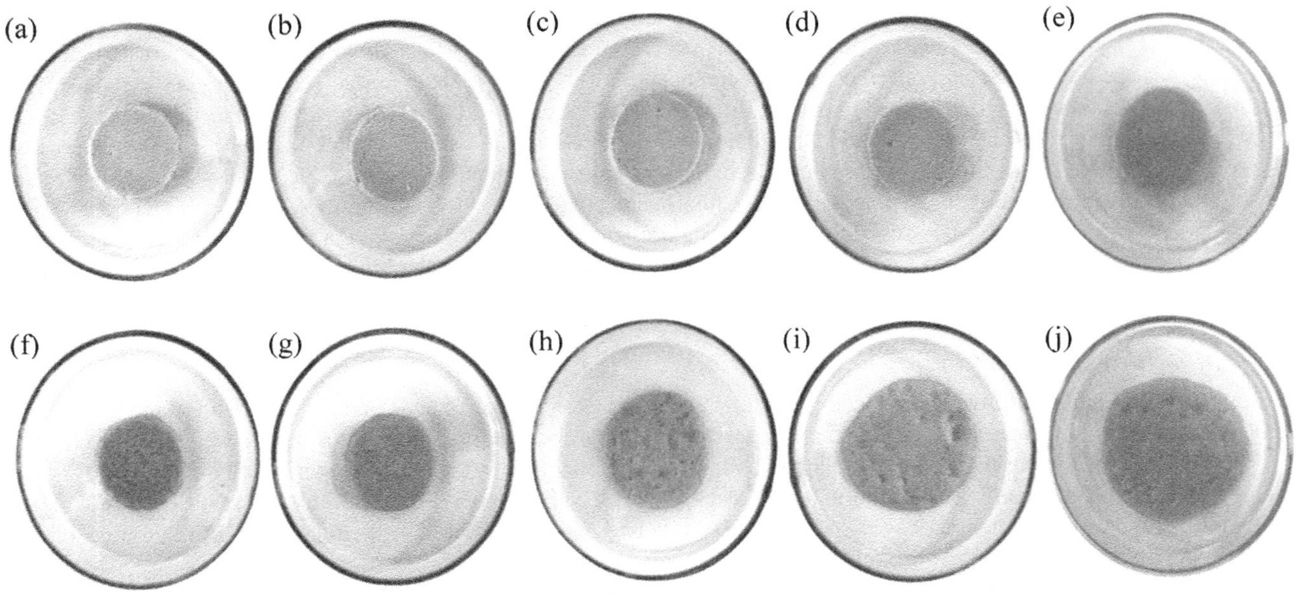

FIGURE 20.6 Visual appearance of plant-based cheese alternatives formulated with different zein and chickpea protein concentrate ratios (0Z-100C to 100Z-0C), demonstrating how protein blend composition affects final product characteristics in starch-based systems. All samples contained identical tapioca starch concentrations (Grasso et al., 2023). Foods, CC BY 4.0. This image represents the formulation of plant-based cheese alternatives with different zein, and chickpea protein concentrate ratios, depicting how differentiated protein blend composition affects final product characteristics in the respective starch-based systems. *Data From Grasso, N., Bot, F., Roos, Y. H., Crowley, S. V., Arendt, E. K., & O'Mahony, J. A. (2023). Plant-based alternatives to cheese formulated using blends of zein and chickpea protein ingredients.* Foods, *12(7). (2023). https://doi.org/10.3390/foods12071492.*

known for its viscosity and gelling capability, was tested for structure and stability enhancement. Native cassava starch, however, can contribute to a starchy aftertaste, lowering consumer acceptability. Hydrolyzed cassava starch with Amylosubtilin and *Bacillus licheniformis* amylases was investigated by the study for enhanced solubility without affecting protein characteristics. Enzymatic hydrolysis at 40°C for 1 h was followed by neutralization, filtration, and drying, with modified starch variants identified by enzyme concentration. Physicochemical analysis analyzed acidity, pH, viscosity, and texture; microbiological tests determined lactic acid bacteria (LAB) viability; sensory testing employed a 9-point hedonic scale with 50 untrained panelists, and antioxidant capacity was quantified by DPPH and FRAP assays.

The result concluded that native cassava starch contributed to an elevated pH (4.72) compared to the modified starch samples, which showed differences in acidification. Titratable acidity increased after 28 days because of LAB post-acidification, but was lower in modified starch samples, indicating higher stability. Native starch exhibited elevated viscosity but was a retrograded neutral starch, resulting in a grainy texture over time. In contrast, AT-0.5 and AT-1 modified starches improved smoothness and consistency. Modified starches also reduced syneresis and improved water-holding capacity (WHC), thereby contributing to product stability. LAB viability remained intact in modified samples, but viability was diminished in native starch samples, possibly because water activity was lowered, impacting microbial metabolism. Sensory appraisal confirmed that modified starches provided superior texture and taste, with the most preferred formulations being AT-0.5 and BT-1. Native starch ranked last due to its starchy aftertaste. Antioxidant properties were notably higher in BT-starch formulations, particularly BT-0.1 and BT-1, while AT-starches improved texture but exhibited lower antioxidant activity (Tsyganov et al., 2022).

6.2 Plant-based meat products

The production of meat substitutes as plant-based food innovations is a new movement in food innovation that aims to mimic the taste, texture, color, and appearance of meat. These products drive changes in food consumption along the ethical, environmental, and health dimensions of a tasty option for plant-based meat alternatives. To formulate products that meet customer expectations, starch plays a crucial role in the formulation of plant-based meat analogs. These include chewiness, holding shape, moisture retention, and enhanced sensory appeal. Starch is fundamental for wholly plant-based formulations because it is important to mimic some of the raw structuring or binding capabilities of animal protein. The introduction of starch considerably improved the quality, textural properties, and organoleptic properties of plant-based meat substitutes (Kyriakopoulou et al., 2019). Microstructural visualization of meat alternatives formulated with mung bean and pea protein isolates via heat-induced gelation shows the fibrous structure of plant-based meat alternatives Fig. 20.7 (Baig et al., 2024).

6.2.1 Starch-based emulsion gels as adipose tissue mimetics

In this study, the authors investigated the production of adipose tissue mimetics (ATMs) for plant-based meat analogs (PBMAs) using emulsified gels made with pea starch and chickpea flour. Most PBMAs use free plant oils because they encompass all of the liquid fat properties while lacking the structural properties of animal fat. Excessive loss of oil and a lack of texture post-cooking ensue. The authors sought to address an imbalance in sustainable and low-cost fat mimetics by developing pea starch emulsified gels that mimic the properties of beef adipose tissue. The authors demonstrated—using X-ray micro-computed tomography—that pea starch gels produced oil entrapment networks similar to adipocytes in animal fat tissue. The authors only incorporated chickpea flour to help reduce oil loss while retaining the texture and elasticity of the PBMAs. The final formulation 6% pea starch, 4% chickpea flour, and 40% oil (7:3 coconut: sunflower oil) exhibited thermal behavior that was most similar to beef adipose tissue and significantly altered the texture of lean ground beef to a medium ground beef quality. Although we later found pea starch to be the most sustainable and available ingredient, it was ideally first because synchrotron tomography indicated that the gels made from pea starch had visual qualities of adipose tissue and formed oil pockets that were pooled within a gelatinous-based starch matrix, demonstrating similar structural habits to where the fat is induced in beef. Pea starch gel showed elasticity across a range of temperatures, ranging from 5 to 85°C, indicating thermal stability more similar to meat adipose tissue. Chickpea flour showed reduced oil loss and retained juiciness while maintaining texture. Overall, we demonstrated that the best formulation (6% pea starch, 4% chickpea flour, 40% oil) had the optimal combination of texture and oil retention in a gel, confirming its suitability for application in PBMA (Czapalay et al., 2025).

6.2.2 Modified starch in Bake-Only Chicken Nuggets

In the study "Effect of Modified Starch on Bake-Only Chicken Nuggets," the researchers described a series of key factors. Native starch was enzymatically hydrolyzed initially with amyloglucosidase to create a porous structure with a limited gelatinization range and lower pasting viscosities, thus creating efficient oil absorption. Oil-plated, enzyme-modified starch was then incorporated into the chicken nugget batter, where it enhanced water retention and adhesion, two of the most important factors in mimicking the sensory attributes typically provided by frying, but in a bake-only system. Instrumental texture analysis and sensory panel testing confirmed that modified starch significantly enhanced crispness and mouth-coating properties to produce a product that closely mimicked the texture and flavor profile of partially fried nuggets. Overall, these findings demonstrate that starch modification targeted to specific properties can play a substantial role in the development of healthier, bake-only products with higher textural and sensory performance, thereby paving the way to greater utility of starch in refashioned meat products (Purcell, 2013).

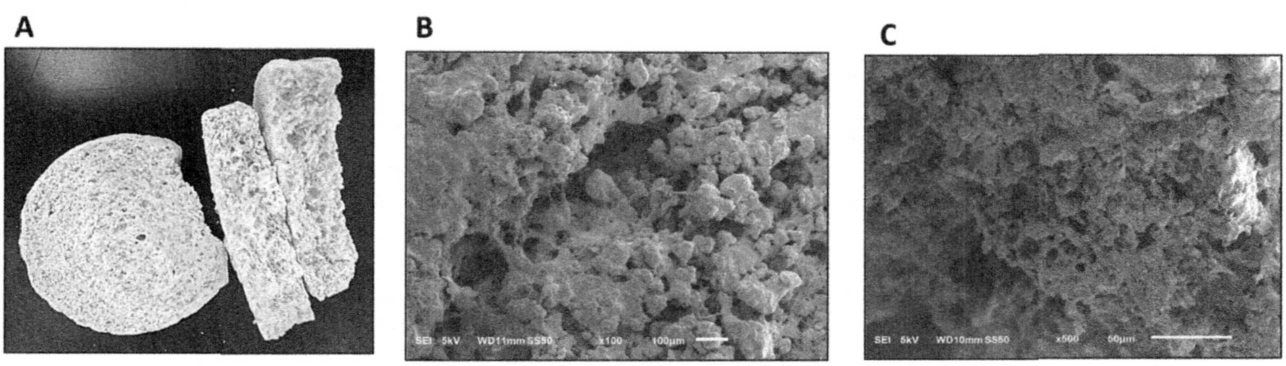

FIGURE 20.7 Microstructural visualization of meat alternatives formulated with mung bean and pea protein isolates via heat-induced gelation (Baig et al., 2024). This image represents the microstructural view of the meat alternative formulated using the mung bean and pea protein isolates through the heat induced gelation process increased the quality, textural properties, and organoleptic properties of plant-based meat substitutes. *Data from Baig, M. A., Mostafa, H., Sivapragasam, N., Aslam, R., Zhou, W., & Maqsood, S. (2024). Investigating the role of starch in the structuring of meat alternatives from mung bean and pea protein isolates via heat-induced gelation.* Frontiers in Sustainable Food Systems, 8*(2024). https://doi.org/10.3389/fsufs.2024.1473663.*

6.2.3 Incorporation of starch in plant-based beef patties

Beef burger simulation foods are a flagship category in the new meat analog market. The product is designed to replicate the visual, sensory, and nutritional profile of the traditional beef burger (De Marchi et al., 2021). Levels of nutrients of reference foods like cooked ground beef (93% lean, 7% fat) and a completely prepared foodservice product typical of the McDonald's beef patty have been used as a reference in comparison testing. The USDA Nutrient Database shows that typically cooked ground beef has approximately 182 kcal/100 g, 25.56 g/100 g total protein, 8.01 g/100 g total fat, 3.29 g/100 g saturated fat, and negligible amounts of carbohydrates and dietary fiber. McDonald's beef patty, in contrast, has higher energy density of approximately 266.67 kcal/100 g, lower protein (23.33 g/100 g), higher total and saturated fat (20.00 g/100 g total fat and 8.33 g/100 g saturated fat), and significantly higher sodium content (approximately 5.5 times that of unseasoned ground beef) (Tang et al., 2024).

The modern meat analog products that were examined i.e., commonly consumed types such as the Beyond Burger and the Impossible Burger nutrient profiles differ from traditional formulations in several respects. Specifically, while these analogs accurately mimic the composition of fat and protein in cooked beef patties, these analogs contain more total carbohydrate and dietary fiber. Increased carbohydrate content primarily results from the addition of plant-derived ingredients, i.e., starches and flours, as fillers due to water- and fat-binding ability through physical entrapment. Starches incorporated into these formulations have a twofold purpose. They enhance textural quality, aid in binding, and aid in moisture retention during preparation and cooking. Second, although the added dietary fiber, which to a great extent is starch-derived carbohydrates, holds some health-directed advantages, there would appear to be no other macronutrient advantage over traditional meat patties. In essence, beyond allowing for favorable product texture and aiding in water retention, the starch-derived carbohydrate portion does not provide added nutritional superiority on a macronutrient basis (Bohrer, 2019).

7. Potential challenges and considerations

7.1 Technical and processing challenges

Starch is mainly used as a functional ingredient in meat and dairy replacement foods because it performs better than native starch. Although used in small amounts, modified starch is an essential ingredient in real meat and commercially available ones as a meat replacer. Achieving proper consistency, flavor, and color is one of the most difficult steps in the manufacturing of these products. The result depends on the manufacturing processes, ingredients employed, and structural content of the products. Apart from being functional, starch also has a significant role in the context of sustainability, especially in plant-based replacement foods. Endogenous starch, occurring naturally as an element after purification, is preserved to some extent. Nevertheless, removal and subsequent addition of starch sequentially may be energy-intensive and cause material loss, pointing toward the inefficiency of such methodology. Reassembly of the individual elements as a unifying ingredient mixture to attain the desired structure and functionality has been a more cost-effective and sustainable strategy (Hamaker, 2021).

Starch is highly appropriate for alternative meat products to modify sensory characteristics such as smoothness and texture. Through its application of rheological performance under various temperatures, consistency characteristics of the food products are regulated in balance with an appropriate quality and quantity of starch. Thermal transition properties of starch have been found to contribute to the extrusion response parameters, thus modifying the textural consistency of the extrudates. Starch is used differently in dairy and meat products, and mildly fractionated materials have been found to have high functionality and sustainability potential. Issues like component interaction, water distribution, and lack of functionality may be overcome by establishing fractionation methods in crops having high protein and starch content, especially meat replacers. These technologies provide benefits to meat substitute quality, lifespan, and cost (Bühler et al., 2021).

The fibrous texture of peanut protein extrudates was adversely affected by including native wheat starch, reducing their toughness and firmness. Wheat starch has been reported to induce protein aggregation by cutting intramolecular disulfide bonds, enhancing hydrophobic interactions, and raising apparent viscosity to fix the assumed shape. Potato starch was found to have substantially decreased the size of the fiber, but the size of the protein aggregate remained unchanged. The movement toward lower processing plant protein content is both undesirable and promising, where new functionalization processes have to deliver a quality product. The difficulty lies in ensuring that starch-protein interactions keep the desired properties at processing.

Starch is a plant carbohydrate with high value due to its low price, favorable characteristics, and total hydrolysis without producing toxic byproducts. Physical starch modifications are simpler and less expensive, and hence are favored for bulk uses. Process modifications like removing the OH group from the starch molecule have succeeded in this context. New green chemistry methods like cold plasma, pulsed electric field, microwave processing, and high hydrostatic processing have been in the spotlight for their green, efficient starch modification processes. These processes also offer the convenience of being carried out under various environmental conditions and applications (Chen et al., 2018).

Starch is a liquid food flavor enhancer, managing consistency, textural transformation, and stability. Starch stops the degradation of the gels during processing and contributes solidity to the product. Thermal filtering of starch in a saturated condition can transform it from an unstructured form to starch crystals, leading to thickening. Ethanoylation, which increases the water-absorbing character of the starch, increases its bulking, gelling, thickening, and textural characteristics, as well as its market acceptability. Additionally, the incorporated functional groups in ethanoylated starches increase their functionality compared to the native starches. Ethanoylation increases the swelling capacity of the starch, while cross-linking processes lead to reductions in apex and breakdown viscosities of faba bean starches (Suri & Singh, 2023).

Problems of starch modification are based on two considerations: The first is that new compound reagents or their derivatives are not allowed, at least to food starches and possibly to some bulk-volume industrial starches. The second is that approved processing levels cannot be changed. Biological modification alone can be implemented to produce commercially modified starches. This is particularly evident in chemically processed and biologically modified starches, that is made into new, higher-value products with allowable reagents and levels of addition. Altering nature to achieve some of these desirable modifications is promising. All starches will need to be modified to become more valuable and useful, but any new derivatives or substitution levels are not likely to be approved. The key issues arise from regulatory and safety concerns. Newly introduced derivatives or substitution levels will require rigorous testing to ensure that they meet the necessary standards and pose no potential threat to health when consumed. Therefore, modifications must be made within the approved limits to meet the existing guidelines, simultaneously enhancing the starch's functionality.

7.2 Consumer acceptance and market trends

The usage of pulse- and legume-based material in meat alternatives is highly used driven by the need to reduce environmental footprint without sacrificing product quality less processed material is subject to physical chemical or enzymatic modification with opportunities for innovative solutions interactions between starch and protein are different interactions are different on a molecular and macroscopic scale depending on their native structure processing step and concentration endogenous starch effects in meat alternatives and protein gels are still under research and there is a demand for additional research pre-gelatinization or chemical modification can enhance endogenous starch functionality thermal pre-treatments can also be promising in functionalizing less processed raw materials offering synergies with protein functionalization.

The modernization of starch usage has the potential to alter consumer perception of starch-based foods. Innovations can change the consumer dynamics from perceiving starches as unsatisfying, fattening, or delivering "empty calories" to describing them as satisfying and possibly beneficial to weight control. Non-digestible starches, especially, have the potential to enhance gut microbiota and overall well-being. Further investigation is required, however, to realize their potential in gut health as well as to create them as attractive, consumer-acceptable food products. Overcoming these challenges could render non-digestible starches a regular component of diets for the maintenance of a healthy gut flora, ultimately adapting to technical advancement with evolving market trends (Hamaker, 2021).

Although the importance of modification has developed continuously, native starch still accounts for more than 60% of the market. In 2000, production shares of the US and EU markets, presented for both product categories, achieved impressive levels, even when compared with starch for starch-based sweeteners, other hydrolyzates, and ethanol. The status of wheat starch and gluten production and uses in Europe changed significantly between 1996 and 2002, while total starch production grew by approximately 40% over these 6 years. The increasing importance of the future can be attributed to the section on 'fermentation, ' where glucose syrups serve as feedstock for lactic, citric, fumaric, propionic, and gluconic acid. Glucose syrups also act as the raw material for L-ascorbic acid in a combined chemical/fermentative process. The brewing industry, which follows British brewing rules, employs adjuncts based on glucose syrups high in maltose (Bergthaller et al., 2007).

8. Future trends and technologies

The applications of Native and modified starch in plant-based food systems are increasing continuously, driven by technological advancements, advanced processing technologies, and emerging applications that extend beyond the traditional meat and dairy analogs. Novel starch processing technique enhances digestibility, stability, and texture while meeting sustainable needs.

8.1 Technological advancements in starch modification

Modern starch modification techniques employ sophisticated physical, chemical, and enzymatic approaches, providing precise control over polymer structure. Unique dual-modification methods, such as ultrasonic treatment and cold plasma processing, have been demonstrated to modify amylopectin chain lengths and molecular weight distributions, which are significant parameters that determine gelation, mouthfeel, and water retention in meat analogs. The researchers demonstrated that such consecutive treatments may modify rheological characteristics for extrusion procedures, with important implications for the fabrication of fibrous, meat-like textures (Gupta et al., 2025). In addition to these physical treatments, enzymatic modifications and chemical modifications have increasingly become prominent. A study (Ashogbon & Akintayo, 2014) examined novel techniques that utilize oxidation and grafting methods to improve the stability and functional performance of starch within protein-rich matrices, with constant improvement for plant-based dairy implications. Furthermore, a study (Negi et al., 2024) offered new insights into enzymatic techniques that produce starch derivatives with enhanced digestion and solubility, thus broadening their utility in low-glycemic index compositions (Punia Bangar et al., 2022).

8.2 Emerging applications of starch in the food industry

The food industry has undergone modifications, leading to the application of modified starch, which extends beyond the production of traditional meat and dairy substitutes. Modification parameter optimization yields improved textural properties in extruded plant-based meat products, resulting in enhanced fibrous structures and mouthfeel characteristics (Himashree et al., 2022). Recently, researchers have enhanced the texture and sensory attributes of plant-based meat by improving the choice and combinations of diverse plant proteins and polysaccharides, while deploying advanced processing technologies such as 3D printing, spinning, shearing, and extrusion (Benković et al., 2023). Consequently, a comprehensive knowledge of these interactions and their impact on the texture of plant-based meat products is crucial for creating plant-based meat products with improved flavor and societal acceptance (Tang et al., 2025). The advent of 3D food printing creates possibilities for the production of customized food supplements. Utilizing modified starch's shear-thinning and gelation properties enables the creation of starch-based printing inks, allowing for the fabrication of complex, meat-like textural structures. This development makes plant-based diets much more personalized (Dzhatdoeva et al., 2016).

9. Conclusion

Briefly, starch is not merely an ingredient but the master texturizing designer of moisture management, sensory attraction, and texture in meat and dairy analogs. Its chemical structure allows for the development of strong networks that can replicate animal fat, allowing for clean-label and sustainable formulation of products. Breakthroughs in pre-treatment technologies via ultrasonic, cold plasma treatments, enzymatic modification, and nanocomposite incorporation enhance its functionality by allowing for improved gelation and emulsification. These breakthroughs guarantee that plant-based alternatives can achieve the rich, creamy mouthfeel and stability of dairy and meat products. As breakthroughs continue to drive these processes, starch will pave the way for the next wave of high-quality, sustainable analog foods. Lately, societal and scientific standpoints have changed to accepting that meat analogs made from plant-based, insect-based, or mycoprotein origins generally have minimized ecological effects compared to traditional meat production products. Hence, they imply that electing meat analogs crafted from alternative sources elevates animal welfare by reducing the demand for animal-based products.

References

Akshit, F. N. U., Poswal, V., Kaushik, R., Deshwal, G. K., & Huppertz, T. (2025). Influence of starch on the rheological, textural, and microstructural properties of processed and analogue cheeses. *Encyclopedia, 5*(2), 2673–8392. https://doi.org/10.3390/encyclopedia5020041

Ashogbon, A., & Akintayo, E. (2014). Recent trend in the physical and chemical modification of starches from different botanical sources: A review. *Starch - Stärke, 66*. https://doi.org/10.1002/star.201300106

Babic, J., Šubarić, D., Ackar, D., Jozinović, A., Miličević, B., Pajin, B., & Aličić, D. (2013). Application of starch based additives in meat industry. *Meat, 15*, 224–229.

Baig, M. A., Mostafa, H., Sivapragasam, N., Aslam, R., Zhou, W., & Maqsood, S. (2024). Investigating the role of starch in the structuring of meat alternatives from mung bean and pea protein isolates via heat-induced gelation. *Frontiers in Sustainable Food Systems, 8*. https://doi.org/10.3389/fsufs.2024.1473663

Bagnolo, L. M., Souza Almeida, F., Silva, K. C. G., & Kawazoe Sato, A. C. (2023). Starch can act differently when combined with alginate or gellan gum to form hydrogels. *Food Research International, 173*, Article 113333. https://doi.org/10.1016/j.foodres.2023.113333

Bai, Y., & Gilbert, R. G. (2022). Mechanistic understanding of the effects of pectin on in vivo starch digestion: A review. *Nutrients, 14*(23), 2072–6643. https://doi.org/10.3390/nu14235107

Barak, S., & Mudgil, D. (2014). Locust bean gum: Processing, properties and food applications—A review. *International Journal of Biological Macromolecules, 66*, 74–80. https://doi.org/10.1016/j.ijbiomac.2014.02.017

Bashir, K., & Aggarwal, M. (2019). Physicochemical, structural and functional properties of native and irradiated starch: A review. *Journal of Food Science and Technology, 56*(2), 513–523. https://doi.org/10.1007/s13197-018-3530-2

Benković, M., Jurinjak Tušek, A., Sokač Cvetnić, T., Jurina, T., Valinger, D., & Gajdoš, K. J. (2023). An overview of ingredients used for plant-based meat analogue production and their influence on structural and textural properties of the final product. *Gels, 9*(12), 2310–2861. https://doi.org/10.3390/gels9120921

Bergthaller, W., Hollmann, J., & Kamerling, H. (2007). *2.18 - Starch*. Oxford: Elsevier. https://doi.org/10.1016/B978-044451967-2/00139-2

Bohrer, B. M. (2019). An investigation of the formulation and nutritional composition of modern meat analogue products. *Food Science and Human Wellness, 8*(4), 320–329. https://doi.org/10.1016/j.fshw.2019.11.006

Bühler, J., Schlangen, M., Möller, A., Bruins, M., & Goot, A. J. (2021). Starch in plant-based meat replacers A new approach to using endogenous starch from cereals and legumes. *Starch - Starke, 74*. https://doi.org/10.1002/star.202100157

Chakraborty, I., Pooja, N., Mal, S. S., Paul, U. C., Rahman, M. H., & Mazumder, N. (2022). An insight into the gelatinization properties influencing the modified starches used in food industry: A review. *Food and Bioprocess Technology, 15*(6), 119t–1223. https://doi.org/10.1007/s11947-02p-02761-z

Chen, J., Hu, F., Guo, J., Zhang, W., & Wu, Z. (2024). Retrograded resistant starch improves emulsion stability and emulsion gel properties stabilized by myofibrillar proteins without degrading in vitro protein digestibility. *Foods, 13*(23), 2304–8158. https://doi.org/10.3390/foods13233739

Chen, Y.-F., Kaur, L., Singh, J., Sjöö, M., & Nilsson, L. (2018). Chapter 7: Chemical modification of starch. In *Woodhead Publishing series in food science, technology and nutrition* (pp. 283–321). Woodhead Publishing. https://doi.org/10.1016/B978-0-08-100868-3.00007-X

Chen, Q., Yu, H., Wang, L., Abdin, Z.ul, Chen, Y., Wang, J., Zhou, W., Yang, X., Khan, R. U., Zhang, H., & Chen, X. (2015). Recent progress in chemical modification of starch and its applications. *RSC Advances, 5*(83), 67459–67474. https://doi.org/10.1039/C5RA10849G

Cornejo, Y., Martínez-Cruz, O., Toro-Sánchez, C., Wong Corral, F., Borboa-Flores, J., & Francisco, C. M. (2018). The structural characteristics of starches and their functional properties. *CyTA - Journal of Food, 16*, 1003–1017. https://doi.org/10.1080/19476337.2018.1518343

Coronato, R., Biasutti, E. A. R., Carvalho, C. W. P., & Grossmann, M. V. E. (2012). Gellan gum/cassava starch mixtures in water systems and in milk systems. *Starch - Stärke, 64*(5), 359–366. https://doi.org/10.1002/star.201100102

Czapalay, E. S., Dobson, S., & Marangoni, A. G. (2025). Legume starch and flour-based emulsion gels as adipose tissue mimetics in plant-based meat products. *Future Foods, 11*, Article 100578. https://doi.org/10.1016/j.fufo.2025.100578

De Angelis, D., Kaleda, A., Pasqualone, A., Vaikma, H., Tamm, M., Tammik, M.-L., Squeo, G., & Summo, C. (2020). Physicochemical and sensorial evaluation of meat analogues produced from dry-fractionated pea and oat proteins. *Foods, 9*(12), 2304–8158. https://doi.org/10.3390/foods9121754

De Marchi, M., Costa, A., Pozza, M., Goi, A., & Manuelian, C. L. (2021). Detailed characterization of plant-based burgers. *Scientific Reports, 11*(1), 2049. https://doi.org/10.1038/s41598-021-81684-9

Diamantino, V. R., Costa, M. S., Taboga, S. R., Vilamaior, P. S. L., Franco, C. M. L., & Penna, A. L. B. (2019). Starch as a potential fat replacer for application in cheese: Behaviour of different starches in casein/starch mixtures and in the casein matrix. *International Dairy Journal, 89*, 129–138. https://doi.org/10.1016/j.idairyj.2018.08.015

Dinali, M., Liyanage, R., Silva, M., Newman, L., Adhikari, B., Wijesekara, I., & Chandrapala, J. (2024). Fibrous structure in plant-based meat: High-moisture extrusion factors and sensory attributes in production and storage. *Food Reviews International, 40*(9), 2940–2968. https://doi.org/10.1080/87559129.2024.2309593

Donmez, D., Pinho, L., Patel, B., Desam, P., & Campanella, O. H. (2021). Characterization of starch–water interactions and their effects on two key functional properties: Starch gelatinization and retrogradation. *Current Opinion in Food Science, 39*, 103–109. https://doi.org/10.1016/j.cofs.2020.12.018

Dzhatdoeva, A. A., Polimova, A. M., Proskurnina, E. V., Proskurnin, M. A., & Vladimirov, Y. A. (2016). Determination of lipids and their oxidation products by IR spectrometry. *Journal of Analytical Chemistry, 71*(6), 542–548. https://doi.org/10.1134/S1061934816060058

Fm, B. (2023). Functional and structural characteristics of rice starch: A review. *International Journal of Pharmacognosy & Chinese Medicine, 7*(2), 1–4. https://doi.org/10.23880/ipcm-16000259

Franz, V., Hör, S., Petermeier, H., Neugrodda, C., Becker, T., & Gastl, M. (2025). The influence of systematic heat and drought applications at defined growth stages on malting barley starch properties. *Journal of the Science of Food and Agriculture, 105*(8), 4493–4503. https://doi.org/10.1002/jsfa.14183

Gerrano, A. S., Labuschagne, M. T., van Biljon, A., & Shargie, N. G. (2014). Genetic variability among sorghum accessions for seed starch and stalk total sugar content. *Scientia Agricola, 71*(6), 472–479. https://doi.org/10.1590/0103-9016-2013-0322

Gong, Y., Xiao, S., Yao, Z., Deng, H., Chen, X., & Yang, T. (2024). Factors and modification techniques enhancing starch gel structure and their applications in foods: A review. *Food Chemistry: X, 24*, Article 102045. https://doi.org/10.1016/j.fochx.2024.102045

Grasso, N., Bot, F., Roos, Y. H., Crowley, S. V., Arendt, E. K., & O'Mahony, J. A. (2023). Plant-based alternatives to cheese formulated using blends of zein and chickpea protein ingredients. *Foods, 12*(7). https://doi.org/10.3390/foods12071492

Grossmann, L., & McClements, D. J. (2021). The science of plant-based foods: Approaches to create nutritious and sustainable plant-based cheese analogs. *Trends in Food Science & Technology, 118*, 207–229. https://doi.org/10.1016/j.tifs.2021.10.004

Gupta, R. K., Gnana Moorthy Eswaran, U., Kumar, S., Pipliya, S., Kaur, B., & Prakash Srivastav, P. (2025). Synergistic modification of starch polysaccharide via sequential treatments with ultrasonic and cold plasma technology. *Journal of Food Processing and Preservation, 2025*(1), Article 9063007. https://doi.org/10.1155/jfpp/9063007

Hamaker, B. R. (2021). Current and future challenges in starch research. *Current Opinion in Food Science, 40*, 46–50. https://doi.org/10.1016/j.cofs.2021.01.003

Hao, S., Zheng, Y., Li, M., Feng, X., & Yang, X. (2024). Effects of heat-moisture extrusion on the structure and functional properties of protein-fortified whole potato flour. *Food Chemistry: X*, (24), Article 102048. https://doi.org/10.1016/j.fochx.2024.102048

Harris, K. F. (2019). An introductory review of resistant starch type 2 from high-amylose cereal grains and its effect on glucose and insulin homeostasis. *Nutrition Reviews, 77*(11), 748–764. https://doi.org/10.1093/nutrit/nuz040

Himashree, P., Sengar, A. S., & Sunil, C. K. (2022). Food thickening agents: Sources, chemistry, properties and applications - A review. *International Journal of Gastronomy and Food Science*, (27), Article 100468. https://doi.org/10.1016/j.ijgfs.2022.100468

Hu, Q., Huang, G., & Huang, H. (2025). Extraction, structure, activity and application of konjac glucomannan. *Ultrasonics Sonochemistry, 116*, Article 107315. https://doi.org/10.1016/j.ultsonch.2025.107315

Huang, X., Liu, H., Ma, Y., Mai, S., & Li, C. (2022). Effects of extrusion on starch molecular degradation, order–disorder structural transition and digestibility—A review. *Foods, 11*(16), 2304–8158. https://doi.org/10.3390/foods11162538

Jiang, W., Yang, X., & Li, L. (2024). Flavor of extruded meat analogs: A review on composition, influencing factors, and analytical techniques. *Current Research in Food Science*, (8), Article 100747. https://doi.org/10.1016/j.crfs.2024.100747

Kailaje, J., Sonawane, S. K., Athul Abraham George, P., Ashish, D., & Bhushette, P. (2022). Development of a dairy analog with combinations of non-dairy milk. *Industrial Biotechnology, 18*(5), 286–292. https://doi.org/10.1089/ind.2022.0018

Kapoor, D. U., Sharma, H., Maheshwari, R., Pareek, A., Gaur, M., Prajapati, B. G., Castro, G. R., Thanawuth, K., Suttiruengwong, S., Sriamornsak, P., & Glucomannan, K. (2024). A comprehensive review of its extraction, health benefits, and pharmaceutical applications. *Carbohydrate Polymers, 339*, Article 122266. https://doi.org/10.1016/j.carbpol.2024.122266

Karmakar, R., Ban, D., & Ghosh, U. (2014). Comparative study of native and modified starches isolated from conventional and nonconventional sources. *International Food Research Journal, 21*, 597–602.

Kiprop, V., Omwamba, M., & Mahungu, S. (2021). Influence of gum Arabic from Acacia Senegal var. kerensis on the modifications of pasting and textural properties of cassava and corn starches. *Food and Nutrition Sciences, 12*, 1098–1115. https://doi.org/10.4236/fns.2021.1211081

Kumar, L., Brennan, M. A., Mason, S. L., Zheng, H., & Brennan, C. S. (2017). Rheological, pasting and microstructural studies of dairy protein–starch interactions and their application in extrusion-based products: A review. *Starch - Stärke, 69*(1–2), Article 1600273. https://doi.org/10.1002/star.201600273

Kyriakopoulou, K., Dekkers, B., Jan van der Goot, A., & Galanakis, C. M. (2019). *Chapter 6: Plant-based meat analogues*. Academic Press. https://doi.org/10.1016/B978-0-12-814874-7.00006-7

Kyriakopoulou, K., Keppler, J. K., & van der Goot, A. J. (2021). Functionality of ingredients and additives in plant-based meat analogues. *Foods, 10*(3), 2304–8158. https://doi.org/10.3390/foods10030600

Lascombes, C., Agoda-Tandjawa, G., Boulenguer, P., Le Garnec, C., Gilles, M., Mauduit, S., Barey, P., & Langendorff, V. (2017). Starch-carrageenan interactions in aqueous media: Role of each polysaccharide chemical and macromolecular characteristics. *Food Hydrocolloids, 66*, 176–189. https://doi.org/10.1016/j.foodhyd.2016.11.025

Lee, H., & Yoo, B. (2023). Particle agglomeration and properties of pregelatinized potato starch powder. *Gels, 9*(2), 2310–2861. https://doi.org/10.3390/gels9020093

Lin, D., Ma, Y., Wen, Q., Loy, D. A., Chen, H., & Zhang, Q. (2022). The structure, properties and potential probiotic properties of starch-pectin blend: A review. *Food Hydrocolloids, 129*, Article 107644. https://doi.org/10.1016/j.foodhyd.2022.107644

Liu, G., Gu, Z., Hong, Y., Cheng, L., & Li, C. (2017). Structure, functionality and applications of debranched starch: A review. *Trends in Food Science & Technology, 63*, 70–79. https://doi.org/10.1016/j.tifs.2017.03.004

Liu, G., Fan, J., Wang, L., Liang, M., Xie, C., & Kang, Z. (2025a). Mitigating quality deterioration of reduced-fat pork sausages during cold storage via resistant starch incorporation: Gel properties and protein conformation study. *Gels, 11*(10), 2310–2861. https://doi.org/10.3390/gels11100763

Liu, J., Wang, Z., Hu, C., Xu, J., & Chang, R. (2025b). Manipulation of the structure of type 4 resistant starch using cross-linking and acylation treatment and its diverse influence on digestion and in vitro fermentation characteristics. *Current Research in Food Science*, (11), Article 101125. https://doi.org/10.1016/j.crfs.2025.101125

Liyanaarachchi, W. S., & Vasiljevic, T. (2018). Caseins and their interactions that modify heat aggregation of whey proteins in commercial dairy mixtures. *International Dairy Journal, 83*, 43–51. https://doi.org/10.1016/j.idairyj.2018.03.006

Luo, L. (2022). Sciencing what are the chemicals in cornstarch?. https://www.sciencing.com/chemicals-cornstarch-15444/.

Machovina, B., Feeley, K. J., & Ripple, W. J. (2015). Biodiversity conservation: The key is reducing meat consumption. *Science of The Total Environment, 536*, 419–431. https://doi.org/10.1016/j.scitotenv.2015.07.022

Masina, N., Choonara, Y. E., Kumar, P., du Toit, L. C., Govender, M., Indermun, S., & Pillay, V. (2017). A review of the chemical modification techniques of starch. *Carbohydrate Polymers, 157*, 1226–1236. https://doi.org/10.1016/j.carbpol.2016.09.094

Miao, M., Chen, L., & BeMiller, J. N. (2025). *Handbook of starch science and technology*. CRC Press. https://doi.org/10.1201/9781003464396

Mohamed, M. E. (2015). Using native and modified starch in processed cheese analogue manufacture to reduce cost. *Journal of Food and Dairy Technology*.

Mohammed, A. A. B. A., Hasan, Z., Omran, A. A. B., Kumar, V. V., Elfaghi, A. M., Ilyas, R. A., & Sapuan, S. M. (2022). Corn: Its structure, polymer, fiber, composite, properties, and applications. *Polymers, 14*(20), 2073–4360. https://doi.org/10.3390/polym14204396

Nanta, P., Skolpap, W., & Kasemwong, K. (2021). Influence of hydrocolloids on the rheological and textural attributes of a gluten-free meat analog based on soy protein isolate. *Journal of Food Processing and Preservation, 45*(3). https://doi.org/10.1111/jfpp.15244

Nawaz, H., Waheed, R., Nawaz, M., & Shahwar, D. (2020). *Physical and chemical modifications in starch structure and reactivity*. London: IntechOpen. https://doi.org/10.5772/intechopen.88870

Negi, A., Barthwal, R., Kathuria, D., & Singh, N. (2024). Enzymatic advances in starch modification: Creating functional derivatives and exploring applications. *Food Bioscience*, (62), Article 105074. https://doi.org/10.1016/j.fbio.2024.105074

Olawoye, B., Jolayemi, O. S., Yemisi Akinyemi, T., Nwaogu, M., Oluwajuyitan, T. D., Popoola-Akinola, O. O., Fisoye Fagbohun, O., & Taiwo Akanbi, C. (2023). *Modification of starch* (pp. 11–54). Cham: Springer International Publishing. https://doi.org/10.1007/978-3-031-35843-2_2

Park, S., & Kim, Y.-R. (2025). Production of clean-label starch using physically treated starch blending and its application as a xanthan gum substitute. *Food Science and Biotechnology, 34*(9), 1949–1956. https://doi.org/10.1007/s10068-025-01825-y

Plamada, D., Teleky, B.-E., Nemes, S. A., Mitrea, L., Szabo, K., Călinoiu, L.-F., Pascuta, M. S., Varvara, R.-A., Ciont, C., Martău, G. A., Simon, E., Barta, G., Dulf, F. V., Vodnar, D. C., & Nitescu, M. (2023). Plant-based dairy alternatives—A future direction to the milky way. *Foods, 12*(9), 2304–8158. https://doi.org/10.3390/foods12091883

Pöri, P., Aisala, H., Liu, J., Lille, M., & Sozer, N. (2023). Structure, texture, and sensory properties of plant-meat hybrids produced by high-moisture extrusion. *LWT, 173*, Article 114345. https://doi.org/10.1016/j.lwt.2022.114345

Punia Bangar, S., Ashogbon, A. O., Singh, A., Chaudhary, V., & Scott Whiteside, W. (2022). Enzymatic modification of starch: A green approach for starch applications. *Carbohydrate Polymers, 287*, Article 119265. https://doi.org/10.1016/j.carbpol.2022.119265

Punia Bangar, S., Sunooj, K. V., Navaf, M., Phimolsiripol, Y., & Scott Whiteside, W. (2024). Recent advancements in cross-linked starches for food applications—A review. *International Journal of Food Properties, 27*(1), 411–430. https://doi.org/10.1080/10942912.2024.2318427

Purcell, S. (2013). *Effect of modified starch on bake-only chicken nuggets*. Fayetteville: University of Arkansas. https://scholarworks.uark.edu/etd/986.

Radeloff, M. A., & Beck, R. H. F. (2016). "Clean label" – Starches and their functional diversity. *Sugar Industry*, 209–215. https://doi.org/10.36961/si17334

Rashwan, A. K., Younis, H. A., Abdelshafy, A. M., Osman, A. I., Eletmany, M. R., Hafouda, M. A., & Chen, W. (2024). Plant starch extraction, modification, and green applications: A review. *Environmental Chemistry Letters, 22*(5), 2483–2530. https://doi.org/10.1007/s10311-024-01753-z

Rupert, R., Rodrigues, K. F., Yee Thien, V., & Yong, W. T. L. (2022). Carrageenan from Kappaphycus alvarezii (Rhodophyta, Solieriaceae): Metabolism, structure, production, and application. *Frontiers in Plant Science, 13*. https://doi.org/10.3389/fpls.2022.859635

Russo Spena, S., Pasquino, R., & Grizzuti, N. (2024). K-Carrageenan/locust bean gum gels for food applications—A critical study on potential alternatives to animal-based gelatin. *Foods, 13*(16), 2304–8158. https://doi.org/10.3390/foods13162575

Schreuders, F. K. G., Schlangen, M., Kyriakopoulou, K., Boom, R. M., & Jan van der Goot, A. (2021). Texture methods for evaluating meat and meat analogue structures: A review. *Food Control, 127*, Article 108103. https://doi.org/10.1016/j.foodcont.2021.108103

Seetapan, N., Limparyoon, N., Raksa, P., Gamonpilas, C., & Methacanon, P. (2024). Influence of cellulose and pectin on anisotropic texture of high-moisture meat analogue from soy protein isolate. *International Journal of Food Science and Technology, 59*(10), 7619–7628. https://doi.org/10.1111/ijfs.17549

Sharma, S., & Wadhwa, N. (2022). Application of glucomannan. *Journal of Pharmacy Research, 21*(1), 1–5. https://doi.org/10.18579/jopcr/v21i1.glucomannan

Shevkani, K., Singh, N., Bajaj, R., & Kaur, A. (2017). Wheat starch production, structure, functionality and applications—A review. *International Journal of Food Science and Technology, 52*(1), 38–58. https://doi.org/10.1111/ijfs.13266

Shoukat, R., Cappai, M., Pilia, L., & Pia, G. (2025). Rice starch chemistry, functional properties, and industrial applications: A review. *Polymers, 17*(1), 2073–4360. https://doi.org/10.3390/polym17010110

Singh, A., Geveke, D. J., & Yadav, M. P. (2017). Improvement of rheological, thermal and functional properties of tapioca starch by using gum arabic. *LWT, 80*, 155–162. https://doi.org/10.1016/j.lwt.2016.07.059

Sołowiej, B., Dylewska, A., Kowalczyk, D., Sujka, M., Tomczyńska-Mleko, M., & Mleko, S. (2016). The effect of pH and modified maize starches on texture, rheological properties and meltability of acid casein processed cheese analogues. *European Food Research and Technology, 242*(9), 1577–1585. https://doi.org/10.1007/s00217-016-2658-4

Souza Almeida, F., & Kawazoe Sato, A. C. (2019). Structure of gellan gum–hydrolyzed collagen particles: Effect of starch addition and coating layer. *Food Research International, 121*, 394–403. https://doi.org/10.1016/j.foodres.2019.03.057

Subroto, E. (2020). Review on the analysis methods of starch, amylose, amylopectin in food and agricultural products. *International Journal of Emerging Trends in Engineering Research, 8*(7), 3519–3524. https://doi.org/10.30534/ijeter/2020/103872020

Sun, N.-xia, Liang, Y., Yu, B., Tan, C.-ping, & Cui, B. (2016). Interaction of starch and casein. *Food Hydrocolloids, 60*, 572–579. https://doi.org/10.1016/j.foodhyd.2016.04.029

Sun, Y., Wang, Z., Ye, J., Li, Y., Wang, L., & Cao, R. (2025). Advances in starch-based nanocomposites for functional food systems: Harnessing AI and nuclear magnetic resonance technologies for tailored stability and bioactivity. *Foods, 14*(5), 2304–8158. https://doi.org/10.3390/foods14050773

Sun, Y., Xu, X., Zhang, Q., Zhang, Di, Xie, X., Zhou, H., Wu, Z., Liu, R., & Pang, J. (2023). Review of konjac glucomannan structure, properties, gelation mechanism, and application in medical biology. *Polymers, 15*(8), 2073–4360. https://doi.org/10.3390/polym15081852

Suri, S., & Singh, A. (2023). Modification of starch by novel and traditional ways: Influence on the structure and functional properties. *Sustainable Food Technology, 1*(3), 348–362. https://doi.org/10.1039/D2FB00043A

Tang, M., Miri, T., Soltani, F., Onyeaka, H., & Al-Sharify, Z. T. (2024). Life cycle assessment of plant-based vs. beef burgers: A case study in the UK. *Sustainability, 16*(11). https://doi.org/10.3390/su16114417, 2071-1050.

Tang, W., Pan, Q., He, J., & Liu, J. (2025). Plant-based meat: The influence on texture by protein-polysaccharide interactions and processing techniques. *Food Research International, 202*, Article 115673. https://doi.org/10.1016/j.foodres.2025.115673

Tong, C., Ma, Z., Chen, H., & Gao, H. (2023). Toward an understanding of potato starch structure, function, biosynthesis, and applications. *Food Frontiers, 4*(3), 980–1000. https://doi.org/10.1002/fft2.223

Tsyganov, M. S., Ezhkova, G. O., Kharitonova, M. A., & Nikitina, E. V. (2022). Cassava starch as an effective texture corrector of fat-free dairy products based on symbiotic starter culture. *International Journal of Food Science, 2022*(1), Article 1087043. https://doi.org/10.1155/2022/1087043

Udo, T., Mummaleti, G., Mohan, A., Singh, R. K., & Kong, F. (2023). Current and emerging applications of carrageenan in the food industry. *Food Research International, 173*, Article 113369. https://doi.org/10.1016/j.foodres.2023.113369

Vilpoux, O. F., Pascoli Cereda, M., & François Vilpoux, O. (2023). *Chapter 6: International legislations for the use of modified starches in food and standards for native starches*. Academic Press. https://doi.org/10.1016/B978-0-323-90058-4.00005-0

Vitolo, M. (2020). Enzymatic modification of starch. *World Journal of Pharmacy and Pharmaceutical Sciences, 9*, 1341. https://doi.org/10.20959/wjpps20204-15958

Wambua, M., Mulwa, R., & Nduko, J. M. (2023). Nutritional composition, antinutritive compounds, and starch properties of flour from cassava varieties grown in Nakuru County, Kenya. *Research Square*. https://doi.org/10.21203/rs.3.rs-3606733/v1

Wang, Y., Ou, X., Ali Al-Maqtari, Q., He, H.-J., & Othman, N. (2024). Evaluation of amylose content: Structural and functional properties, analytical techniques, and future prospects. *Food Chemistry: X*, (24), Article 101830. https://doi.org/10.1016/j.fochx.2024.101830

Wang, L., & Tong, L. (2024). Production and properties of starch: Current research. *Molecules, 29*(3), 1420–3049. https://doi.org/10.3390/molecules29030646

Wu, Y., Xiao, S., Jia, Z., Zhao, K., Hou, L., Ding, W., & Wang, X. (2025). Effects of modified starches on the dough rheological properties of wheat flour and frozen storage stability of frozen raw noodles. *Food Chemistry X*, (27), Article 102392. https://doi.org/10.1016/j.fochx.2025.102392

Xie, C., Liu, G.-H., Liang, M.-H., Li, S.-H., & Kang, Z.-L. (2024). Applying resistant starch to improve the gel and water retention of reduced-fat pork batter. *Gels, 10*(5), 2310–2861. https://doi.org/10.3390/gels10050347

Xu, F., Zhang, L., Liu, W., Liu, Q., Wang, F., Zhang, H., Hu, H., & Blecker, C. (2021). Physicochemical and structural characterization of potato starch with different degrees of gelatinization. *Foods, 10*(5), 2304–8158. https://doi.org/10.3390/foods10051104

Yang, S., Hu, W., Qiao, S., Song, W., & Tan, W. (2025). Advances in processing techniques and determinants of sweet potato starch gelatinization. *Foods, 14*(4), 2304–8158. https://doi.org/10.3390/foods14040545

Yang, T., Wang, P., Zhou, Q., Wang, X., Cai, J., Huang, M., & Jiang, D. (2021). Investigation on the molecular and physicochemical changes of protein and starch of wheat flour during heating. *Foods, 10*(6), 2304–8158. https://doi.org/10.3390/foods10061419

Yap, X. Y., Khalid, M., Raju, G., Ti Gew, L., & Yow, Y.-Y. (2024). Synergistic effects of starch and carrageenan from Kappaphycus alvarezii in composite film formation: Physicochemical and degradable properties. *International Journal of Biological Macromolecules*, (278), Article 135205. https://doi.org/10.1016/j.ijbiomac.2024.135205

Yazid, N. S., Abdullah, N., Muhammad, N., & Monica Peralta, H. (2018). Application of starch and starch-based products in food industry. *Journal of Science and Technology, 10*. https://doi.org/10.30880/jst.2018.10.02.023

Yiu, C. C., Liang, S. W., Mukhtar, K., Kim, W., Wang, Y., & Selomulya, C. (2023). Food emulsion gels from plant-based ingredients: Formulation, processing, and potential applications. *Gels, 9*(5), 2310–2861. https://doi.org/10.3390/gels9050366

Zainuddin, N., Khalid, N. I., Mohsin, A. Z., Azman, E. M., Mahmud, N. K., Rashid, A., & Abd Rahim, M. H. (2024). Physicochemical, microbiological, and sensory properties of almond milk yogurt-like products with varied concentrations of tapioca starch as stabilizer. *Journal of Biochemistry, Microbiology and Biotechnology, 12*(SP1), 90–92. https://doi.org/10.54987/jobimb.v12iSP1.951

Zhang, G., Chen, Y., Pan, J., Qiu, Y., Cui, P., Lyu, F., & Zhang, J. (2024). Effect of guar gum on the physicochemical properties and in vitro digestive characteristics of extruded starches. *International Journal of Biological Macromolecules, 275*, Article 133176. https://doi.org/10.1016/j.ijbiomac.2024.133176

Zhang, W., Li, S., Zhang, B., Drago, S. R., & Zhang, J. (2016). Relationships between the gelatinization of starches and the textural properties of extruded texturized soybean protein-starch systems. *Journal of Food Engineering, 174*, 29–36. https://doi.org/10.1016/j.jfoodeng.2015.11.011

Zhang, X., Liang, Y., Bai, H., Huang, Q., Liu, D., Ma, G., & Liu, X. (2025a). In situ cross-linked porous starch microencapsulation enhances the colonization of lactobacillus in vivo. *Foods, 14*(12), 2304–8158. https://doi.org/10.3390/foods14122031

Zhang, J., Liu, Y., Wang, P., Zhao, Y., Zhu, Y., & Xiao, X. (2025b). The effect of protein–starch interaction on the structure and properties of starch, and its application in flour products. *Foods, 14*(5), 2304–8158. https://doi.org/10.3390/foods14050778

Chapter 21

Nanotechnology innovations: Advancing nanoscale modifications of starch for enhanced functionality in food

Sahar Akhavan Mahdavi
Department of Food Materials and Process Design Engineering, Gorgan University of Agricultural Science and Natural Resources, Gorgan, Iran

1. Introduction

Nanotechnology is recognized as a transformative agent in food science, offering innovative solutions to improve the quality, safety, and performance of food. Among the numerous applications of this technology, nanoscale modification of starch has attracted much attention due to its ability to overcome the inherent limitations of natural starches. Starch, a widely used carbohydrate polymer in food processing, plays an important role in improving texture, thickening, stabilization, and encapsulation of bioactive compounds (Wei et al., 2018). However, its natural form faces challenges such as poor solubility, susceptibility to retrogradation, and limited thermal and mechanical stability. However, by using nanoscale engineering, starch can be modified to exhibit superior functional properties and make it more suitable for various applications in the food industry. Advances in nanotechnology have made it possible to manipulate starch at the molecular level and improve its solubility, stability, and performance in various processes, in a way that traditional modification techniques have not been able to achieve this level of improvement (Hassan et al., 2022).

The application of nanotechnology in food science has evolved significantly, with the initial focus being on improving food texture, increasing nutrient bioavailability, and improving shelf life. As research in this area has progressed, it has become clear that nanoscale interventions offer significant advantages over traditional modification methods. Nanostructured foods have been designed to improve emulsification, control release mechanisms, and enhance sensory properties (Yu et al., 2021). In particular, nanoscale starch modifications show the potential to create foods with improved properties such as higher thickening ability, greater stability to freeze-thaw processes, and better resistance to enzymatic digestion. These advances will result in food products that remain consistent in quality over time, meet consumer demand for healthier formulations, and help develop novel textures and structures in foods (Bhavya & Raman, 2024).

Starch is of great importance in food systems due to its high versatility, but its inherent properties often limit its full potential in industrial applications. Starch, extracted from plant sources such as corn, potato, rice, and wheat, is mainly composed of amylose and amylopectin, which determine its physicochemical behavior (Campelo et al., 2020). While native starch acts as an excellent thickening and gelling agent, under severe processing conditions, it may undergo undesirable changes such as retrogradation, syneresis, and degradation. These limitations necessitate the modification of starch to improve its performance in food applications. Traditional starch modification methods include physical, chemical, and enzymatic methods that are widely used to improve starch properties. However, these techniques usually involve harsh processing conditions, chemical treatments, or partial degradation of starch, which can affect its natural function or pose safety concerns (Bajer, 2023; Kumari et al., 2020).

Nanotechnology provides a more precise and effective approach to starch modification by changing the structure of starch at the nanoscale. Nanoscale starch particles are suitable for applications requiring high stability and controlled performance due to their properties such as improved water dispersion, larger surface area, and greater interaction with other food components (Tian et al., 2022). Various nanoscale techniques, including the production of starch nanoparticles, nanocrystalline starch, and starch nanocomposites, have been developed to tailor starch properties for specific food

Native and Modified Starches for Food Applications. https://doi.org/10.1016/B978-0-443-27431-2.00006-7

applications. These nanosized starch structures improve mechanical and thermal stability, increase the encapsulation capacity of bioactive compounds, and improve emulsification properties. These advances provide new possibilities for the use of starch in food processing, packaging, and the development of functional foods (Adeyeye et al., 2023; Wang & Zhang, 2021).

Given the growing interest in starch modification using nanotechnology, this chapter is dedicated to reviewing the latest developments in this field. In this chapter, various nanoscale techniques used for starch modification, their effects on starch performance, and their applications in food formulations will be reviewed. In addition, potential regulatory and safety concerns associated with the use of nanotechnology in food will be addressed.

2. Types of nanoscale starch modifications

Nanoscale starch modification involves a variety of structural changes that transform native starch granules into materials with improved properties and broader applications. These modifications, which include nanoparticles, nanocrystalline starch, starch nanocomposites, and surface-modified starch nanoparticles, have revolutionized food applications by improving stability, texture, and bioactive delivery (Torres & De-la-Torre, 2022). Nanosized starch granules, also known as starch nanoparticles, are obtained by reducing the size of native starch granules to the nanometer scale, typically below 100 nm. This size reduction results in a significant increase in surface area, changes in crystallinity, and improved functional properties (Kim et al., 2015). Nano-sized starch granules exhibit improved dispersibility, high solubility, and more effective interactions with other food ingredients, making them suitable for applications such as emulsions, texture enhancement, and encapsulation systems. Due to the high surface reactivity of these particles, they also play a key role in the fabrication of biodegradable films and controlled release systems in food formulations (Wei et al., 2018). These nanoparticles effectively stabilize emulsions, improve the texture of food products, and act as efficient carriers for bioactive compounds. Their ability to form nanostructured films offers broad potential in active food packaging, extending shelf life, and improving food safety (Ahmad et al., 2023).

Nanocrystalline starch (NCS) is composed of highly ordered crystalline regions extracted from native starch granules by selective removal of amorphous regions. This process produces nanocrystals with high strength, reduced solubility, and greater resistance to enzymatic degradation (Sujka, 2017). Due to its excellent mechanical and barrier properties, NCS is widely used in food packaging, making it an ideal component for the production of biodegradable films and edible coatings that prevent moisture loss and improve product shelf life. In addition, NCS is used in controlled release systems, where its high thermal stability ensures the sustained release of bioactive compounds. Its crystalline nature makes it heat-resistant, making it very useful for heat-resistant food formulations where stability at high temperatures is essential (Xu et al., 2023). Starch nanocomposites are hybrid materials in which nanoscale reinforcements, such as nanoclays, nanocellulose, or graphene oxide, are incorporated into the starch matrix. These nanomaterials significantly enhance the mechanical, thermal, and barrier properties of starch-based materials by creating a synergistic effect between starch and the reinforcement agents (Olayil et al., 2022). Starch nanocomposites have become a key component in biodegradable food packaging, edible coatings, and active packaging systems with antimicrobial properties. Their ability to improve moisture resistance and gas permeability makes them suitable for smart food packaging applications, where these materials protect products from environmental factors while maintaining their quality and freshness (Pal et al., 2021).

Surface-modified starch nanoparticles are modified through chemical, physical, or enzymatic processes to improve their compatibility, dispersion, and interaction with other food ingredients. These modifications, which include changes in the surface charge, hydrophobicity, and reactivity of the nanoparticles, lead to improved properties such as emulsification, gelation, and stability in food formulations. Surface-modified nanoparticles act as stabilizers in emulsions, thickeners in sauces and condiments, and controlled-release carriers for nutrients and bioactive compounds. These nanoparticles also find application in food packaging, where surface modifications enhance their adhesion to film matrices and improve their barrier properties, which in turn increases the shelf life of food products (Chavan et al., 2021; Oh et al., 2020).

In addition to the main categories, other nanoscale starch modifications include cross-linked starch nanoparticles, hydrogel-based nanostarch, and enzyme-modified starch nanoparticles. Cross-linked starch nanoparticles are engineered through chemical modification to increase their stability, shear strength, and thermal stability, making them suitable for food processing at high temperatures (Hussain et al., 2024). Hydrogel-based nanostarch formulations create nanostructured hydrogels that are used for encapsulation, water retention, and controlled release applications in functional foods. Enzyme-modified starch nanoparticles, functionalized through enzymatic processes, increase solubility, improve digestibility, and provide prebiotic benefits that contribute to gut health and the development of functional foods (Choton et al., 2024). These nanoscale starch modifications generally offer significant advances in food processing by increasing performance, stability, and durability. They overcome the limitations of native starch and open up new possibilities for the development of high-performance food and packaging materials. As summarized in Table 21.1, a wide range of nanoscale starch modifications offer viable solutions for improving food texture, increasing shelf life, and effectively delivering bioactive ingredients (Torres & De-la-Torre, 2022).

Incorporation of zinc oxide nanoparticles (ZnO NPs) into starch matrices significantly enhances mechanical strength, thermal stability, UV-blocking capacity, and antibacterial activity against both Gram-positive and Gram-negative bacteria, making them ideal for active food packaging (Dhiman et al., 2025).

3. Modified nanostarch (NS) and their preparation methods

Nanostarch (NS) has attracted a lot of attention due to its potential applications in various industries including food, pharmaceutical and cosmetics. However, the inherent hydrophilic nature of NS limits its dispersibility in non-polar solvents and its compatibility with hydrophobic polymers. To overcome these limitations and improve its performance, various chemical and physical modification methods are used, which are designed to change the hydroxyl groups on the NS surface. These modifications improve properties such as hydrophobicity, dispersibility and stability of NS, which ultimately expand its application range (Hassan et al., 2022). Enzymatic modification using α-amylase, glucoamylase, and pullulanase under mild conditions (pH 4–7, 30–65 °C) offers a precise, eco-friendly alternative to chemical methods, enabling tailored digestibility, improved freeze–thaw stability, and enhanced solubility while reducing environmental impact by up to 70% (Bhati et al., 2025).

3.1 Chemical modification

3.1.1 Esterification and acylation

Esterification is one of the widely used methods in which an organic acid reacts with an alcohol to form an ester. This method modifies the starch structure by replacing hydroxyl groups (C-2, C-3 and/or C-6) with ester groups. Acetylation, which is a form of esterification using acetic acid or its derivatives (such as acetic anhydride), is particularly common for

TABLE 21.1 Types of nanoscale starch modifications and their applications.

Type of nanoscale starch modification	Description	Applications in food industry	References
Nanoparticles and nano-sized starch granules	Starch particles reduced to nanometer size (<100 nm), with increased surface area and altered crystallinity.	Used as stabilizers in emulsions, thickeners in food formulations, and carriers for bioactive compounds in encapsulation systems.	Wei et al. (2018)
Nanocrystalline starch (NCS)	Crystalline regions of starch isolated by removing amorphous components, leading to rigid and thermally stable structures.	Applied in food packaging for improved barrier properties, edible coatings, and controlled-release bioactive systems.	(Wang, Hong, et al., 2020)
Starch nanocomposites	Starch-based materials reinforced with nanoparticles such as nanoclays, nanocellulose, or graphene oxide.	Used in biodegradable and smart food packaging, improving moisture resistance and mechanical strength.	Ortega et al. (2022)
Surface-modified starch nanoparticles	Chemically, physically, or enzymatically treated starch nanoparticles with enhanced functional properties.	Employed as emulsifiers, texture enhancers, and controlled-release carriers in food formulations.	Rashid et al. (2019)
Cross-linked starch nanoparticles	Starch nanoparticles chemically cross-linked to enhance thermal stability and shear resistance.	Used in high-temperature food processing, thickening agents, and texturizers in sauces and soups.	Lu et al. (2019)
Hydrogel-based nanostarch	Starch-based hydrogels with nanostructures capable of retaining water and controlling release of active compounds.	Applied in moisture-retentive edible films, encapsulation of flavors, and functional foods.	Wei et al. (2022)
Enzyme-modified starch nanoparticles	Starch modified by enzymatic treatments to improve digestibility and bioavailability.	Used in prebiotic formulations, gut health products, and low-glycemic index food systems.	Choton et al. (2024)

the modification of nanostarch (NS) (Fig. 21.1). Acetylation of nanostarch (SNPs) gives rise to additional absorption bands in the FTIR spectrum at 1750, 1435, 1375 and 1024 cm-1, which are related to carbonyl bending, CH3 group bending and C-O stretching vibrations. These changes lead to a decrease in starch crystallinity, which is visible through X-ray diffraction patterns, and is useful for applications such as bioactive or drug delivery where substances are mainly released from amorphous regions. In addition, the zeta potential of acetylated SNPs increases, which increases their stability and prevents particle aggregation (zeta potential > -18 mV is desirable) (Saartrat et al., 2005; Wang & Zhang, 2021).

Other esterifications such as octenyl succinic anhydride (OSA) also show a similar acetylation process, decreasing crystallinity and increasing zeta potential. Esterification of SNPs with OSA increases dispersion and hydrophobicity, which is particularly important for the stabilization of Pickering emulsions (Ye et al., 2017). In contrast, oxidation modifies starch by introducing aldehyde or carboxyl groups, usually via peroxides such as hydrogen peroxide, leading to depolymerization and particle size reduction. This modification significantly reduces the crystalline phase of starch, which is very important for specific applications such as drug encapsulation (Abidin et al., 2018).

3.1.2 Hydrolysis and fatty acid modification

Hydrolysis of starch with long-chain fatty acids such as stearic acid results in smaller particle sizes (from 2.4 to 10 nm depending on the starch source) and hydrophobic properties. Increased hydrophobicity is useful in emulsion stabilization, where hydrolyzed SNPs show a higher affinity for nonpolar solvents such as chloroform, in contrast to unmodified SNPs that prefer the polar aqueous phase (Namazi & Dadkhah, 2010) (Table 21.2).

3.2 Grafting modification

Grafting is another modification method that involves the covalent attachment of monomers onto the NS backbone. Grafting can be achieved through two approaches: "grafting onto," where polymer chains are coupled to the surface, and "grafting from," where new chains are polymerized from the surface using initiators. This technique has been employed to modify NS with various copolymer types. PLA-grafted SNPs, synthesized through ring-opening graft-polymerization, retained the crystalline nature of starch while improving thermal stability and solubility in organic solvents (Wang, Hong, et al., 2020). Other examples include acrylonitrile-grafted SNPs, where microwave irradiation was used as an initiator without the need for toxic chemicals, resulting in high graft yields (Haroon et al., 2016). Another significant example is the grafting of 2-(dimethylamine) ethyl methacrylate (DMAEMA) onto carboxymethylated SNPs. The grafting led to a reduction in the relative crystallinity from 27.8% to 1.6%, improving the Zeta potential and shifting the SNPs from hydrophilic to hydrophobic. These grafted SNPs are promising for applications in Pickering emulsions, providing stability and dispersibility in hydrophobic environments (Xiao et al., 2020) (Fig. 21.2).

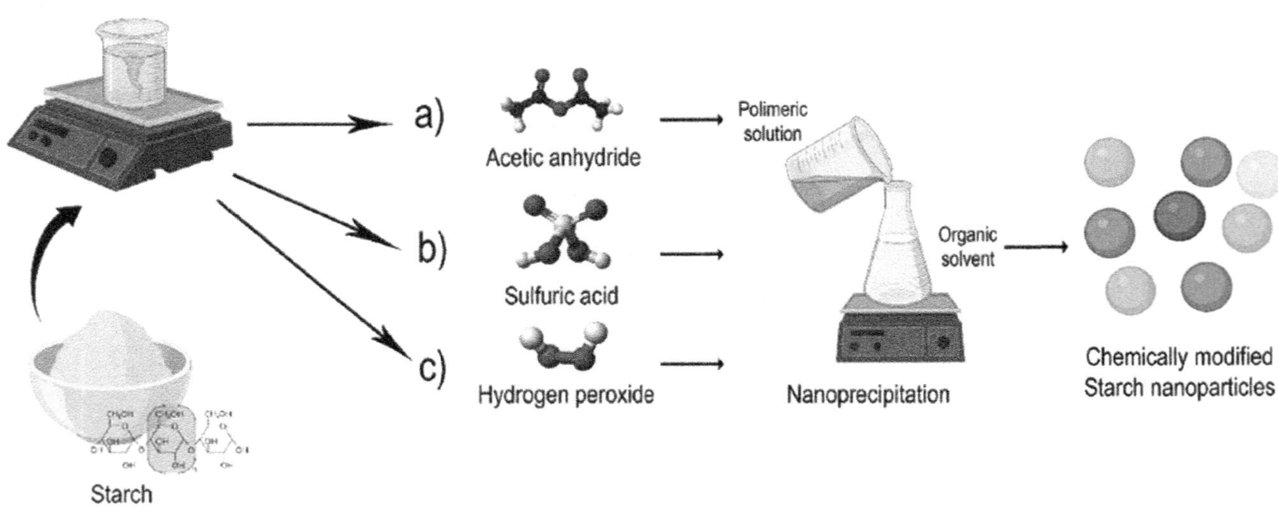

FIGURE 21.1 Schematic overview illustrating the processes of acetylation (a), hydrolysis (b), and oxidation (c) (Torres & De-la-Torre, 2022).

TABLE 21.2 Comparison of modification purposes and outcomes for starch and nanostarch (NS).

Modification technique	Purpose of modification	Outcome for starch	Outcome for nanostarch (NS)	References
Acylation	To improve water resistance and create modified starch derivatives	Limited solubility in water, improved hydrophobicity	Enhanced surface properties, improved solubility, better compatibility with hydrophobic compounds	Yao et al. (2023)
Esterification	To increase hydrophobicity and thermal stability	Reduced water solubility and increased thermal stability	Increased water resistance, enhanced emulsifying properties, and improved dispersion	Yao et al. (2023)
Crosslinking	To increase the mechanical strength and stability of starch	Increased viscosity, and limited gel stability under stress	Enhanced mechanical properties, higher shear strength, more stable nanoparticle dispersion	Lu and Tian (2021)
Oxidation	To modify surface charge and introduce functional groups	Increased oxidation resistance and slight structural change	Higher surface reactivity, better dispersion, and introduction of functional groups (e.g., carboxyl groups)	Li et al. (2023)
Hydrolysis	To produce smaller molecules (e.g., oligosaccharides)	Production of simple sugars or oligosaccharides	Formation of smaller nanostructures with controlled size and shape, better bioavailability	Dukare et al. (2021)
Complexation	To form complexes with other bioactive substances for delivery	Formation of complexes with limited bioactive components	Enhanced bioactive compound encapsulation and controlled release properties	Marta et al. (2023)
Amylose/amylopectin ratio adjustment	To alter the ratio for modifying starch's gelling properties	Altered gelling, swelling, and texture properties	Enhanced control over particle aggregation, controlled gelation at nanoscale	Zhang et al. (2019)
Nanoparticle formation	To reduce particle size to nano range for specific applications	No nanoparticles formed in native starch	Formation of nanoparticles (<250 nm), enhanced surface area, reactivity, and specific interactions	Chavan et al. (2021)
Grafting	To attach functional groups to starch for specific applications	Limited functionalization, and surface interaction with some molecules	Improved surface functionalization, higher stability, and increased specific interactions with target molecules	Cazotti et al. (2020)

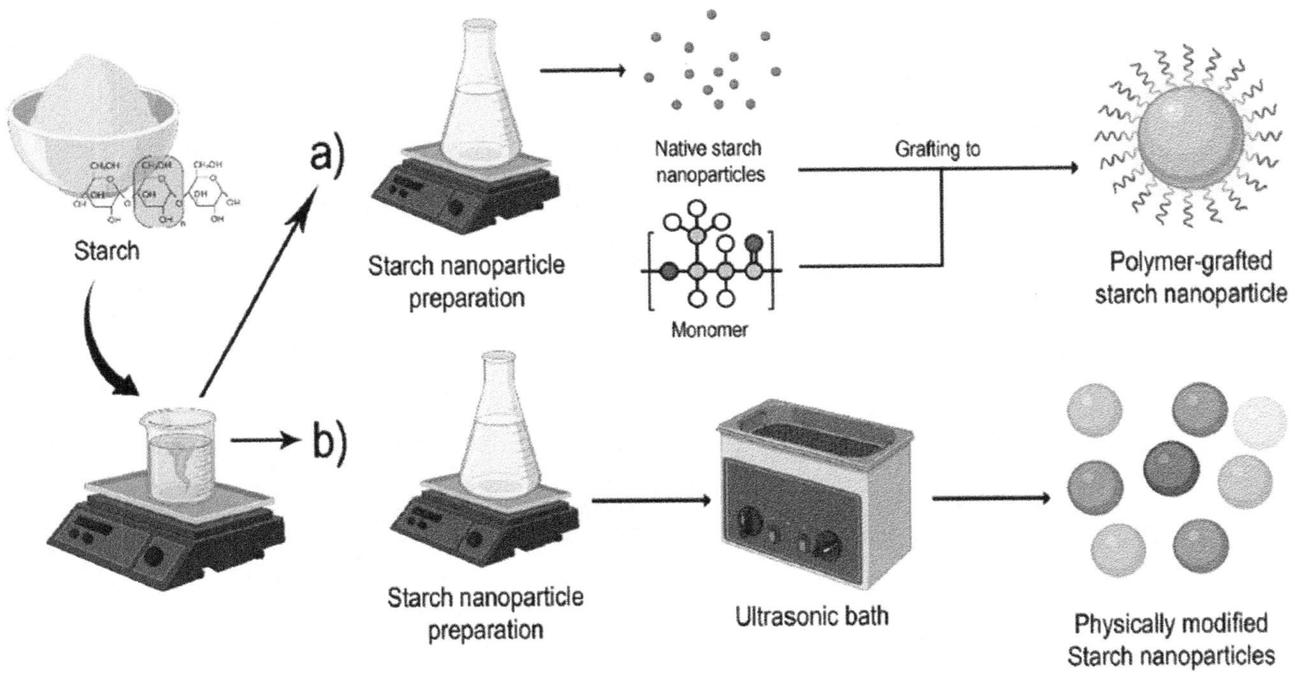

FIGURE 21.2 Schematic illustration of (a) surface polymer grafting and (b) ultrasonication (Torres & De-la-Torre, 2022).

3.3 Physical modifications

3.3.1 Cold plasma treatment

Cold plasma (CP) treatment is a non-thermal modification technique that modifies the surface of NS to enhance its hydrophobicity and functional properties without the use of toxic chemicals. CP processing, particularly under oxygen, leads to oxidation and the introduction of additional functional groups (carbonyl and carboxyl groups), significantly increasing the Zeta potential from −9.1 to −21.6 mV after 21 min of treatment. This modification is advantageous for increasing the stability of nanoparticle suspensions (Table 21.3) (Chang et al., 2019).

3.3.2 Ultrasound treatment

Ultrasonic treatment is another environmentally friendly physical modification method that reduces particle size and alters the surface structure of starch granules. The combination of ultrasound-assisted modification with acid hydrolysis has been shown to reduce particle size by about 50%, while maintaining the crystalline structure. Furthermore, ultrasound-assisted enzymatic hydrolysis of lotus seed SNPs has enhanced the crystallinity to 74.9%, a significant increase compared to non-ultrasonic methods. This method can improve the stability of NS and provide better control over particle dispersion and structure (Li et al., 2022; Shabana et al., 2019; Sujka, 2017). Recent advances in green synthesis of nanostarch have emphasized eco-friendly physical techniques including ultrasonication, cold plasma treatment, high-pressure homogenization, ball milling, nanoprecipitation, and enzymatic hydrolysis, which produce highly biocompatible and biodegradable nanoparticles suitable for food applications without toxic solvents (Sreejit et al., 2025).

4. Characterization techniques for nanoscale starches

Starch nanoparticles (SNPs) have emerged as versatile materials in various applications, including food, pharmaceuticals, and biodegradable packaging, due to their biocompatibility, recyclability, and tunable properties. However, the performance and efficiency of SNPs are largely influenced by their structural and physicochemical properties. Therefore, understanding the fundamental properties of SNPs requires the use of advanced characterization techniques. Characterization of SNPs includes techniques such as scanning electron microscopy (SEM) and transmission electron microscopy (TEM) to investigate the surface morphology and particle size, while X-ray diffraction (XRD) reveals the crystal structure and degree of crystallinity. Thermal behavior is usually analyzed using differential scanning calorimetry

TABLE 21.3 Summary of particle sizes, dispersibility, and stability of modified nanostarch (ns) prepared using various methods.

Modification method	Particle size (nm)	Dispersibility	Stability	References
Acylation (e.g., acetylation)	<250	Moderate to high	Improved hydrophobicity, reduced aggregation	Yao et al. (2023)
Esterification (e.g., fatty acid chloride)	360–500	Moderate	Enhanced thermal stability, hydrophobicity	Namazi et al. (2011)
Crosslinking (e.g., sodium trimetaphosphate)	50–250	High	Increased mechanical strength, stable under stress	Shi et al. (2011)
Oxidation (e.g., TEMPO-mediated oxidation)	20–60	High	Enhanced surface reactivity, good stability under various conditions	Guo et al. (2024)
Hydrolysis (e.g., acid hydrolysis)	50–100	High	Reduced stability over time, increased solubility	Jiang et al. (2016)
Nanoparticle formation (e.g., direct nanoparticle synthesis)	<250	Excellent	Stable colloidal dispersion, good encapsulation potential	Escobar-Puentes et al. (2020)
Grafting (e.g., surface modification with functional groups)	100–200	Moderate to high	Improved dispersibility in different solvents, stable under varied conditions	Pei et al. (2019)
Oxidation (e.g., sodium hypochlorite oxidation)	183–225	Excellent	Stronger adsorption capacity for ions, stable in water	Liu et al. (2018)
Complexation (e.g., starch maleate complexation)	87	Moderate	Stable under specific conditions, improved bioactive compound encapsulation	Rutkaite et al. (2012)

(DSC) and thermogravimetric analysis (TGA), which provide insight into gelatinization, decomposition, and thermal stability (Hebeish et al., 2014). In addition, dynamic light scattering (DLS) helps to assess particle size distribution and zeta potential, which indicate colloidal stability. Rheological properties are evaluated to understand the flow behavior and viscosity of SNPs, which affect their application in various systems. In addition, functional and digestibility properties of SNPs, such as water retention, solubility, and enzymatic resistance, play an important role in determining their potential applications in the food and pharmaceutical industries. The amylose content of SNPs, which significantly affects their crystallinity, thermal properties, and digestibility, is also a fundamental parameter in starch characterization (Szymońska et al., 2009). Table 21.4 summarizes the key description techniques, along with their analytical purposes and the types of information they provide.

4.1 Morphology and particle size analysis of starch nanoparticles

The morphology and particle size of starch nanoparticles (SNPs) are key parameters that directly affect their performance and stability in various applications, especially in the food, pharmaceutical and biomedical industries. Physical modification techniques such as ultrasonication, gamma irradiation and high-pressure homogenization (HPH) are commonly used to reduce the particle size and change the morphology of SNPs, which consequently improves their behavior in different matrices. For example, ultrasonication of starches extracted from sources such as cassava, quinoa and maize produces diverse particle size distributions depending on the sonication parameters. This method breaks down large starch granules into smaller particles and increases their dispersion and stability in different formulations (Hassan et al., 2022).

Gamma irradiation, especially at doses of about 20 kGy, alters the structure of cassava SNPs and creates lamellar aggregates with particle sizes between 30 and 50 nm (Puspita et al., 2019). In addition to reducing the particle size, this process adds hydroxyl groups to the surface of SNPs, strengthening hydrogen bonds and improving the stability of nanoparticles. High-pressure homogenization (HPH) also significantly reduces the size of starch granules by applying high pressure and multiple cycles (Apostolidis & Mandala, 2020). Studies on sago starch have shown that HPH effectively converts large granules into nanoparticles, improving their application in food systems. In addition to physical methods,

TABLE 21.4 Main characterization techniques for nanoscale starches.

Characterization technique	Key properties analyzed	Application in nanoscale starch analysis	References
SEM (scanning electron microscopy)	Morphology, and surface structure	Evaluates particle size, aggregation, surface roughness, and morphology of starch nanoparticles	Dukare et al. (2021)
TEM (transmission electron microscopy)	Internal nanostructure	Assesses dispersion, crystallinity, particle integrity, and nanoparticle aggregation	Qin et al. (2016)
XRD (X-ray diffraction)	Crystallinity patterns	Determines starch type (A-, B-, C-type), degree of crystallinity, and structural modification	Marimuthu et al. (2013)
FTIR (fourier-transform infrared spectroscopy)	Functional group changes	Identifies chemical interactions, bonding patterns, and modifications in starch nanoparticles	Pumacahua-Ramos et al. (2024)
DSC (differential scanning calorimetry)	Gelatinization temperature, and enthalpy	Measures thermal properties, including gelatinization, retrogradation, and phase transitions	Fuentes et al. (2019)
Rheological analysis	Flow and viscosity characteristics	Analyzes shear thinning, gel strength, viscoelasticity, and flow behavior of starch nanoparticles in various applications	Kumari et al. (2022)
DLS (dynamic light scattering)	Particle size, and zeta potential	Determines nanoparticle size distribution, colloidal stability, and dispersion	Hassan et al. (2022)
AFM (atomic force microscopy)	Surface roughness, and molecular interactions	Provides high-resolution imaging and nanoscale characterization of starch surfaces	Chichti et al. (2013)
TGA (thermogravimetric analysis)	Thermal degradation profile	Assesses thermal stability, weight loss, and decomposition kinetics of starch nanoparticles	Pumacahua-Ramos et al. (2024)
Zeta potential analysis	Surface charge, and colloidal stability	Determines nanoparticle surface charge and colloidal stability in aqueous suspensions	Masoudipour et al. (2017)
NMR (nuclear magnetic resonance spectroscopy)	Molecular mobility, and chemical structure	Analyzes amylose/amylopectin ratio, molecular mobility, and conformational changes at the nanoscale	El Nokab et al. (2022)
UV-vis spectroscopy	Optical properties, and absorbance	Monitors nanoparticle stability, bioactive interaction, and light absorption properties	Bajer (2023)
Functional and digestibility tests	Water solubility, and enzymatic degradation	Evaluates water absorption, enzymatic hydrolysis, and digestibility properties of starch nanoparticles	Tian et al. (2022)
Amylose content measurement	Starch composition analysis	Determines amylose/amylopectin ratio, affecting physicochemical properties of starch nanoparticles	Qin et al. (2016)

chemical and enzymatic hydrolysis also play an important role in shaping the morphology of SNPs. Acid hydrolysis gradually reduces the size of the granules while maintaining the structural integrity of the starch, resulting in smaller and more uniform nanoparticles, ideal for applications such as encapsulation and controlled release (Perez Herrera et al., 2016). On the other hand, enzymatic hydrolysis using amylase causes particle surface disruption, formation of cracks, and increased porosity (Chorfa et al., 2022). However, excessive enzymatic hydrolysis can lead to particle aggregation, which increases the size and alters the final properties of the nanoparticles.

Several advanced techniques are used to accurately characterize the morphology and size of starch nanoparticles (SNPs). Atomic force microscopy (AFM) allows high-resolution 3D imaging of SNPs and provides valuable information on surface features and particle size, without the need for complex sample preparation (Tomoaia-Cotisel et al., 2010). This method is particularly useful for analyzing very small structures. Transmission electron microscopy (TEM) allows for precise analysis of the thickness, grain composition, and size of nanoparticles using a focused electron beam and is ideal for investigating the internal structure of SNPs and their interactions with other compounds (Kim et al., 2015). In contrast, scanning electron microscopy (SEM) evaluates the surface morphology and distribution of SNPs and provides detailed images of particle shape and size. SEM studies have shown that SNPs prepared by nanoprecipitation and hydrolysis have different shapes and sizes ranging from 30 to over 250 nm, depending on the starch source and preparation method (Gutiérrez et al., 2020).

Dynamic light scattering (DLS) is a technique that determines the size distribution of nanoparticles in suspension by measuring fluctuations in light intensity due to Brownian motion. These fluctuations are related to the particle diffusion coefficient and can be used to calculate the hydrodynamic diameter of SNPs, which is typically between 50 and 250 nm. DLS is non-invasive, rapid, and requires minimal sample preparation, making it suitable for application in food systems. However, there are limitations such as aggregation of nanoparticles, which can affect the results, and the assumption of particle sphericity, which is not always accurate. These challenges can be overcome by proper sample preparation and employing complementary techniques such as TEM or AFM to validate the results (Qiu et al., 2016).

Du et al. (2023) evaluated the morphology of native starches and starch nanoparticles (SNPs) prepared via nanoprecipitation and ultrasonication from sago (C-type), corn (A-type), and potato (B-type) starches. SEM showed native sago and potato starches as spherical or oval (34.47 and 44.65 μm), while corn starch was polygonal and irregular (14.19 μm). The SNPs were uniformly spherical, with sizes of 59.4 nm (sago), 68.1 nm (corn), and 55.5 nm (potato), indicating starch type minimally affected shape but influenced size distribution. Sago and potato SNPs had homogeneous distributions, while corn SNPs showed higher polydispersity due to aggregation, likely because A-type starch is less susceptible to ultrasonication than B- and C-type starches. Smaller SNP sizes in sago and potato starches were linked to higher amylose content, aligning with findings that higher amylose correlates with smaller nanoparticles (Du et al., 2023) (Fig. 21.3).

Qin et al. (2016) examined the morphologies of native starch granules and SNPs prepared by nanoprecipitation from seven starch types with varying amylose content (0.8%–69.0%): high amylose corn starch (HAS), pea starch (PES), potato starch (PS), normal corn starch (CS), sweet potato starch (SPS), tapioca starch (TS), and waxy corn starch (WX). SEM images of the native starches (Fig. 21.4, A–G) revealed mostly spherical, oval, and irregular granular shapes with smooth surfaces, with mean diameters ranging from 10.5 μm (HAS) to 49.9 μm (PS). In contrast, TEM images of the corresponding SNPs (Fig. 21.4, a–g) showed predominantly spherical and elliptical shapes with particle sizes ranging from 20 to 200 nm, though some aggregation was observed. The mean particle sizes of the SNPs, determined by DLS, were 30 nm (HASNP), 35 nm (CSNP, WXNP), 45 nm (TSNP), 60 nm (PESNP), 65 nm (SPSNP), and 75 nm (PSNP), indicating that smaller native starch granules generally produced smaller SNPs. This trend aligns with the blocklet model of starch granules, where blocklets (20–500 nm) are gelatinized and precipitated into SNPs, with their size consistent with the blocklet dimensions in native starches.

Preparation of starch nanoparticles using deep eutectic solvents (choline chloride-oxalic acid) from different botanical origins (wheat, rice, cassava, potato) results in reduced relative crystallinity, crystallization transition (e.g., B to C type in potato SNPs), and altered thermal stability, making them promising for Pickering emulsion stabilization and biodegradable packaging reinforcements (Wang et al., 2025).

The particle size and morphology of SNPs directly influence their properties and applications. Smaller SNPs, typically in the range of 50–250 nm, often exhibit enhanced solubility, encapsulation efficiency, and stability, making them suitable for a wide range of applications such as nutrient delivery, emulsion stabilization, and controlled release systems. However, the tendency of SNPs to aggregate in aqueous suspensions, as revealed by DLS, highlights the need for strategies to improve their stability, such as surface modification or the use of stabilizers. By employing a combination of characterization techniques like AFM, TEM, SEM, and DLS, researchers can better optimize SNPs for specific applications, ensuring consistent performance and stability in food formulations and beyond (Chavan et al., 2021; Dong et al., 2021).

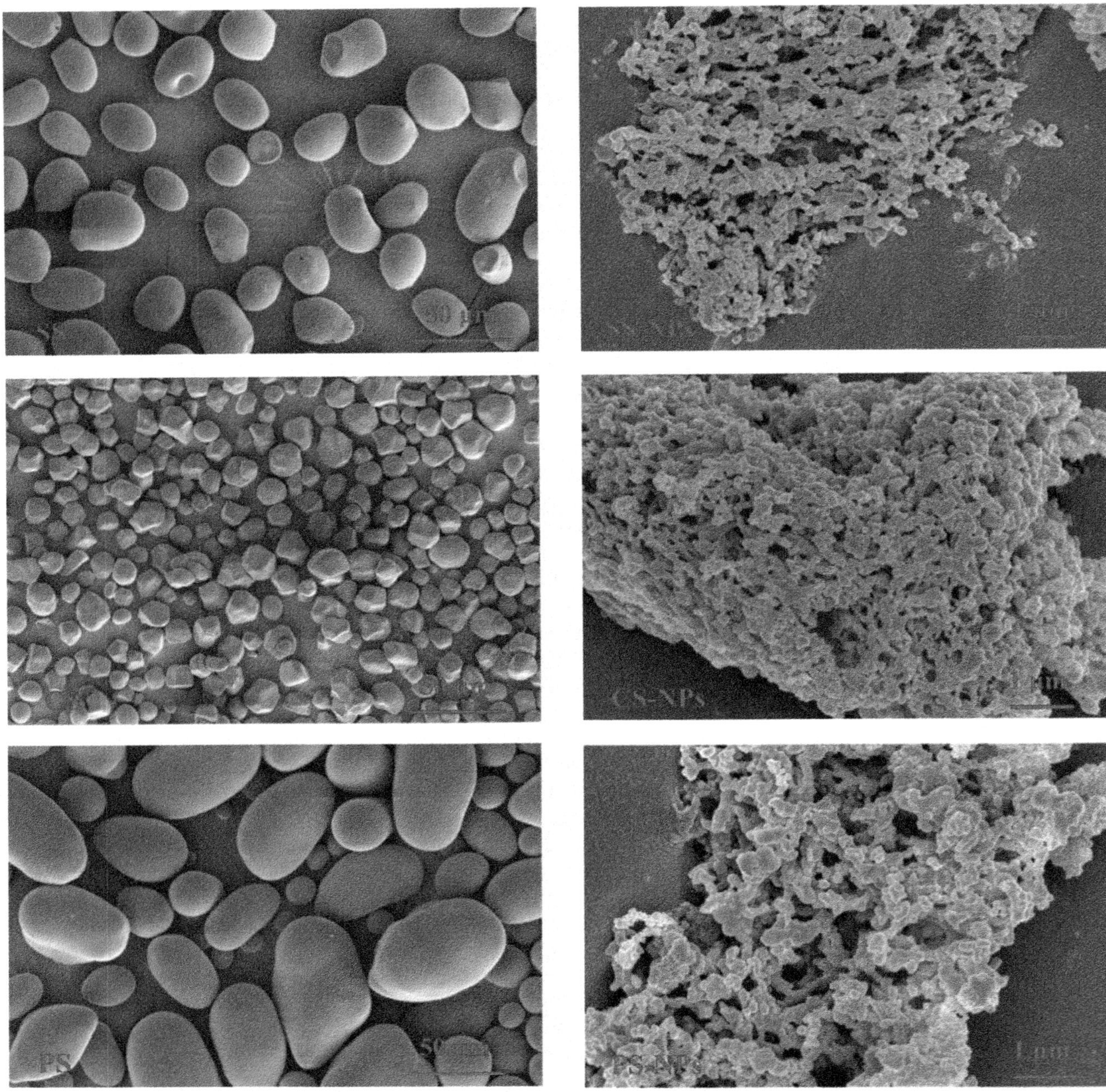

FIGURE 21.3 SEM of native starch (×1500) and starch nanoparticles (×50,000) (Du et al., 2023) (A), (B), and (C) show native sago, corn, and potato starches at 1500× magnification, respectively, images (D), (E), and (F) depict the corresponding sago, corn, and potato starch nanoparticles at 500,00× magnification (Du et al., 2023).

4.2 Starch crystallinity and structural analysis

The crystallinity of starch nanoparticles (SNPs) significantly influences their physicochemical properties, impacting their use in encapsulation, film formation, and texture enhancement. Crystallinity, primarily driven by amylopectin, is the ratio of crystalline to amorphous regions and is analyzed using X-ray diffraction (XRD), which classifies starch into A, B, C, and V types. A-type starches (e.g., waxy maize) show tightly packed double helices with XRD peaks at 2θ of 15°, 17°, 18°, and 23°, while B-type starches (e.g., high-amylose maize) have a loosely packed hexagonal structure with peaks at 5.6°, 15°, 20°, 22°, and 24°. C-type starches (e.g., mung beans) combine A and B patterns, and V-type arises from amylose-lipid or amylose-ethanol complexes during nanoprecipitation (Marta et al., 2023). Crystallinity depends on

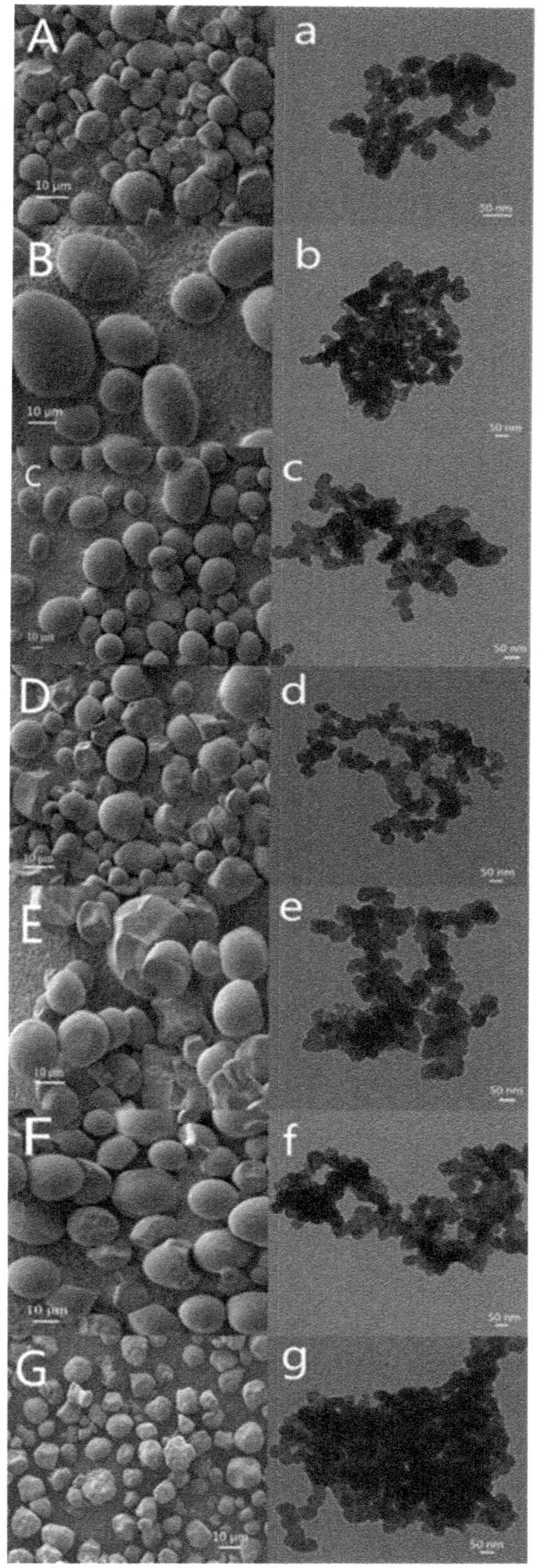

FIGURE 21.4 Morphologies of native starch granules and starch nanoparticles prepared by nanoprecipitation. Scanning electron microscopy images of native starches: (A) High amylose corn starch, (B) pea starch, (C) potato starch, (D) normal corn starch, (E) sweet potato starch, (F) tapioca starch, and (G) waxy corn starch. Transmission electron microscopy images of corresponding starch nanoparticles: (a) High amylose corn starch nanoparticles, (b) pea starch nanoparticles, (c) potato starch nanoparticles, (d) normal corn starch nanoparticles, (e) sweet potato starch nanoparticles, (f) tapioca starch nanoparticles, and (g) waxy corn starch nanoparticles (Qin et al., 2016).

amylopectin content, chain length, crystal size, and double helix orientation, with higher amylopectin leading to greater crystallinity due to fewer amorphous regions. Preparation methods like gamma irradiation, ultrasonication, and high-pressure homogenization reduce crystallinity, creating more amorphous structures, while heat moisture treatment increases crystallinity, shifting B-type to A-type with rising temperature (Du et al., 2023).

Chemical and enzymatic methods significantly affect the crystallinity of starch nanoparticles (SNPs). Acid hydrolysis increases crystallinity by selectively degrading amorphous regions, as seen in starch nanocrystals (SNCs) from amadumbe corn starch, where crystallinity rose but remained below 50% due to residual amorphous material. Nanoprecipitation induces V-type crystallinity through amylose-ethanol complexes, independent of the native starch type. Enzymatic hydrolysis, often with pullulanase, enhances crystallinity by breaking down amorphous regions, typically yielding a B + V-type pattern. Temperature during enzymatic hydrolysis influences the outcome, with high temperatures promoting A-type crystallinity and lower temperatures favoring B-type patterns (Chavan et al., 2021; Duyen & Van Hung, 2021). Hedayati et al. (2020) studied the crystallinity of tapioca starch SNPs prepared by nanoprecipitation and nanoprecipitation-sonication, with XRD patterns shown in Fig. 21.5. Native tapioca starch displayed an A-type pattern (peaks at $2\theta = 15.05°, 17.10°, 18.05°, 23.6°$), while SNPs exhibited a V-type pattern, indicating a crystalline transformation. The relative crystallinity decreased, with the lowest crystallinity in SNPs from 1 g starch, becoming amorphous at lower concentrations (Hedayati et al., 2020).

4.3 Thermal properties of starch nanoparticles

The thermal properties of starch nanoparticles (SNPs) are vital for their application in food systems, such as nanocomposite production, polymeric blend reinforcement, and encapsulation, and are influenced by starch source, preparation method, and modifications. Native starch and SNPs show distinct thermograms due to differences in crystalline and molecular structures, with factors like sulfate ester groups from acid hydrolysis reducing degradation temperature by destabilizing crystallinity. High amylose starches, like high amylose corn starch, exhibit higher gelatinization enthalpy and thermal stability due to dense crystalline structures, while amylose content and amylopectin chain length affect endothermic enthalpy, reflecting double helical stability (Shi et al., 2013). Physical methods such as gamma irradiation, ultrasonication, and high-pressure homogenization (HPH) generally lower thermal stability; for instance, gamma irradiation causes a drop in degradation temperature due to surface hydroxyl group degradation, and ultrasonication of cassava, quinoa, and waxy maize starches reduces gelatinization temperatures and enthalpy by weakening hydrogen bonds. Conversely, HPH slightly increases the degradation temperature of sago starch, enhancing thermal stability through structural reorganization. Chemical methods like acid hydrolysis increase peak (Tp) and conclusion (Tc) temperatures while decreasing onset temperature (To), as seen in waxy maize, normal maize, and unripe plantain SNPs, though prolonged hydrolysis can reduce enthalpy (Xu et al., 2014). Nanoprecipitation lowers gelatinization enthalpy due to the single helical structure's susceptibility to disintegration, except in high amylose SNPs, which retain higher stability. Enzymatic hydrolysis with recrystallization, as in proso millet and waxy maize SNPs, increases gelatinization temperatures but reduces thermal stability, with degradation temperatures dropping to 260.81–292.73 °C from 303.63 °C in native starch (Campelo et al., 2020). Thermal analysis techniques like Differential Scanning Calorimetry (DSC),

FIGURE 21.5 X-ray diffraction pattern of native tapioca starch and SNPs prepared by nanoprecipitation and nanoprecipitation-sonication treatments (Hedayati et al.,2020).

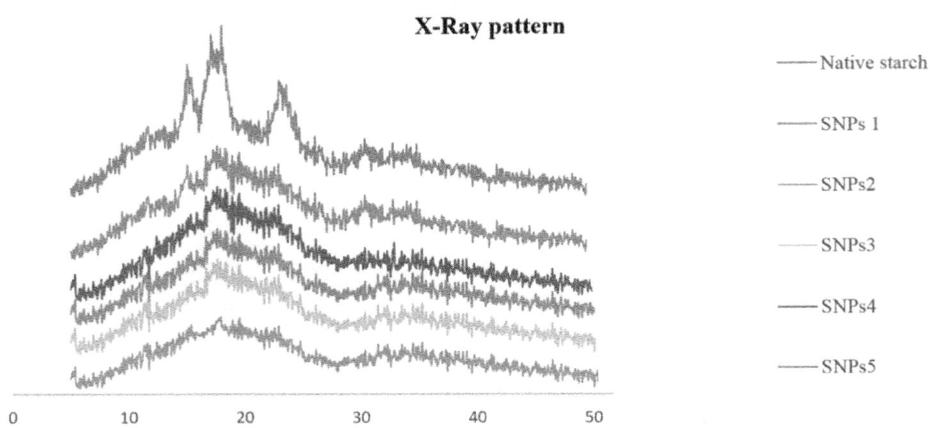

Thermogravimetric Analysis (TGA), and Differential Thermal Analysis (DTA) provide insights into these properties, with DSC showing higher gelatinization temperatures in SNPs (90.38–117.49°C vs. 65.59–79.57 °C for native starch) and TGA revealing three-stage degradation patterns, aiding in optimizing SNP processing for applications like food packaging and drug delivery (Ji et al., 2015; Mahdavi et al., 2021).

Dukare et al. (2021) used DSC to compare the thermal properties of native starches and nanostarch prepared by enzymatic hydrolysis from maize, potato, and cassava, with thermograms. The gelatinization peaks for native maize, potato, and cassava starches were observed at 84°C, 78°C, and 82°C, respectively, but shifted to below 50°C for maize nanostarch, and to 52°C and 56°C for potato and cassava nanostarch, respectively. The increase in melting enthalpy after enzymatic hydrolysis suggested preferential erosion of the amorphous regions, making water more accessible to crystalline regions, although the overall gelatinization enthalpy of nanostarch decreased compared to native starch, except in high amylose maize starch (Dukare et al., 2021).

4.4 Functional and digestibility properties of starch nanoparticles

The functional and digestibility properties of starch nanoparticles (SNPs) significantly influence their role in food systems, affecting texture, encapsulation, and nutrition. Functionally, SNPs show enhanced swelling volume (5.28–7.92 g/g in arrowroot SNPs) and solubility (9.43%–16.89%) compared to native starch, as seen in nanoprecipitation and mechanical treatments, making them ideal for sauces and beverages (Marta et al., 2023; Winarti et al., 2014). However, acid-hydrolyzed waxy rice SNPs exhibit reduced water binding capacity due to increased crystallinity (Jeong & Shin, 2018), and they achieve high encapsulation efficiency (91.63% for vitamin E, over 97% for conjugated linoleic acid), enhancing bioactive delivery (Ahmad & Gani, 2021). Digestibility varies: pullulanase-treated SNPs show increased digestibility (41.29%–43.24%) for rapid energy release (Suriya et al., 2018), while methods like butanol precipitation reduce digestibility, forming resistant starch in arrowroot and maize SNPs, aiding glycemic control (Oh et al., 2020; Winarti et al., 2014). SNPs also inhibit α-amylase, reducing glucose absorption, which is beneficial for diabetic-friendly foods (Oh et al., 2020). These properties enable SNPs to improve texture, stabilize bioactives, and meet diverse nutritional needs in food applications.

4.5 Rheological properties of starch nanoparticles

The rheological properties of starch nanoparticle (SNP) suspensions are crucial for determining their suitability in various food and industrial applications, including flexible packaging materials, coatings, and filler matrices. These properties such as viscosity, shear behavior, and gelation are influenced by several factors, including particle size, concentration, preparation methods, and the presence of ionic additives. Understanding the rheological behavior of SNP suspensions allows for the optimization of their performance in specific applications.

Smaller SNPs (≤600 nm) exhibit shear-thinning behavior, meaning their viscosity decreases as the shear rate increases. This characteristic is beneficial for applications that require easy flow under high shear, such as coating and spraying processes. Shear-thinning behavior enhances the application and spreadability of SNP-based materials, making them ideal for use in food formulations and coatings. Conversely, larger SNPs (>1000 nm) exhibit more Newtonian-like behavior, where viscosity remains constant regardless of changes in shear rate. This behavior makes them more suitable for applications that require a stable, unchanging viscosity under varying shear conditions. The concentration of SNPs also plays a significant role in the rheological profile of the suspensions. Higher concentrations of SNPs lead to increased pasting viscosity, which is important for achieving the desired texture and consistency in food products or industrial formulations. For example, corn SNPs prepared via acid hydrolysis exhibit reduced pasting viscosity compared to native starch. However, a higher SNP concentration can compensate for this reduction and increase the viscosity. These properties make SNP suspensions highly useful in industries like paper and paint production, where a high solid content with manageable viscosity is necessary for efficient processing. Preparation methods also influence the rheological behavior of SNP suspensions. SNPs prepared using vacuum freeze-drying exhibit higher viscosity, stronger shear-thinning behavior, and reduced thixotropic behavior when compared to those prepared using spray-drying. The vacuum freeze-drying process helps preserve the particle structure, minimizing aggregation and allowing for a more robust suspension. In contrast, spray-drying tends to cause more aggregation, leading to a suspension with lower viscosity and less shear-thinning behavior. The rheological behavior of SNP suspensions also varies depending on the method of preparation. Acid-hydrolyzed starch nanocrystals (SNCs) generally form suspensions with lower viscosity due to the disruption of the starch's amorphous regions during the hydrolysis process. In contrast, self-assembled SNPs tend to exhibit more complex rheological behavior due to molecular reorganization and interactions. This variation emphasizes the importance of tailoring the preparation method to achieve the desired rheological properties for specific applications. For example, certain food formulations may require suspensions with specific viscosities and shear behaviors, while industrial applications may prioritize other properties like flowability or thermal resistance.

In food applications, the shear-thinning behavior of SNP suspensions plays a vital role in enhancing the texture and stability of products like sauces, dressings, and emulsions. Shear-thinning suspensions ensure smooth application and optimal mouthfeel in food products, improving the sensory experience for consumers. Moreover, SNP suspensions with high solid content but lower viscosity are particularly useful in food manufacturing, where achieving a high solid content with manageable viscosity is crucial for efficient processing and desired product texture. Dong et al. (2021) investigated the rheological properties of starch nanoparticle (SNP) suspensions from pulse (faba bean, field pea) and cereal (corn, wheat) starches prepared by rapid antisolvent nanoprecipitation. At 1% (w/v), SNP suspensions exhibited Newtonian behavior, shifting to shear-thinning at 3% for pulse SNPs and 5% for all SNPs, with pulse SNPs showing higher viscosity and elasticity due to smaller particle sizes and stronger inter-particulate hydrogen bonding. These findings highlight the influence of starch source and concentration on SNP rheology, with pulse SNPs offering potential for applications requiring enhanced elasticity and stability, such as emulsion stabilization (Dong et al., 2021). Jiang et al. (2022) further explored the rheology of 4% (w/v) SNP suspensions from coarse cereal starches (quinoa, waxy proso millet, non-waxy proso millet, maize) prepared by rapid nanoprecipitation, with and without ultrasonic pretreatment. All suspensions exhibited shear-thinning behavior, with quinoa SNPs (QSNPs) displaying the highest viscosity and viscoelasticity (higher G' and G''), attributed to a stable network structure that immobilized more water. Temperature-dependent tests revealed that QSNPs maintained viscous-dominated stability across all temperatures, while waxy proso millet and maize SNPs transitioned at around 60 °C, highlighting QSNPs' potential as emulsifiers for Pickering emulsions. These findings underscore the influence of starch source, particle size, and processing conditions on SNP rheology, with pulse and quinoa SNPs offering enhanced elasticity and stability for applications like emulsion stabilization and texture modification (Jiang et al., 2022).

4.6 Amylose content and its impact on starch nanoparticles

Amylose content significantly influences the physicochemical properties of starch nanoparticles (SNPs), affecting gel formation, adhesion, and structural organization, which are crucial for food applications like encapsulation and film formation. High amylose content, as in high amylose maize (up to 69%–70%), enhances gel strength, adhesion, and relative crystallinity ($R^2 = 0.95$ correlation with amylose), ideal for edible films, while low amylose content, as in waxy maize (~0.8%–3%), results in softer gels suited for beverages (Bajer, 2023; Qin et al., 2016). Preparation methods alter amylose content: acid hydrolysis reduces it by up to 80% in Andean potato SNPs by targeting amorphous regions, increasing crystallinity and dispersibility (Bajer, 2023); reported that corn starch (22.5% amylose) and waxy corn starch (3% amylose) subjected to acid hydrolysis, ultrasonication, or both showed reduced amylose content, with corn starch exhibiting greater amylose loss due to its unbranched chains' higher susceptibility to HCl and ultrasound, while waxy corn starch's amylopectin-rich structure was more resistant, supporting its use in applications requiring stability. Nanoprecipitation with butanol increases amylose in arrowroot SNPs, improving gel-forming ability (Qin et al., 2016; Winarti et al., 2014) found that nanoprecipitation of starches (0.8%–69.0% amylose) produced SNPs with V-type crystallinity independent of native starch type, with higher amylose content linked to increased crystallinity but lower gelatinization enthalpy (except in high amylose corn starch), enhancing suitability as nanocarriers for bioactive compounds. High-pressure homogenization increases amylose in corn SNPs, enhancing adhesion for coatings (Apostolidis & Mandala, 2020). These variations allow tailoring of SNP properties for specific food applications, from high-dispersibility systems to strong gel networks.

Molecular structures of amylose and amylopectin significantly influence the emulsification performance of starch nanoparticles; rice-derived SNPs exhibit near-neutral wettability (contact angle ~90°) and 100% emulsifying stability index, with amylopectin size positively correlated to emulsion stability for clean-label food applications (Li et al., 2025).

5. Enhancing functional properties of starch nanostructures in food systems

Starch, as a key biopolymer in food systems, is widely used due to its outstanding functional properties such as viscosity generation, gelation, and emulsification, all of which are essential for achieving desired texture, stability, and sensory properties in food processing processes. However, native starch often faces limitations such as low solubility, thermal instability, and limited functionality in some food matrices (Sothornvit, 2019). Nanotechnology has fundamentally transformed starch applications by producing nanostructures such as starch nanoparticles (SNPs) and starch nanocrystals (SNCs), which offer improved properties thanks to their nanoscale dimensions and increased specific surface area (Table 21.5).

Nanoscale starch modification significantly enhances the performance of starch-based materials in food systems by enhancing their physicochemical and functional properties, and contributes to improved texture, stability, and better delivery of bioactive compounds. One of the most important advantages of nanoscale starch is its significant enhancement of thickening and gelling properties. Due to their higher specific surface area and modified structure, nanostarches interact more effectively with hydrocolloids, leading to increased viscosity and stronger gel formation (Sharma et al., 2019). These interactions provide greater stability under variable temperature and shear conditions, making nanostarches highly suitable for use in products such as soups, sauces, broths, dairy products, and jelly desserts, where textural stability is critical.

Starch nanostructures also show excellent performance in encapsulating bioactive compounds and act as carriers of vitamins, antioxidants and probiotics; this allows for targeted and controlled release in functional foods (Cornejo-Ramírez et al., 2018). Qin et al. (2016) reported that SNPs with higher amylose content show lower enthalpy of gelatinization and are therefore suitable for nanocarriers of hydrophobic bioactive compounds; Bajer (2023) also showed that starch nanoparticles can encapsulate probiotics and increase their bioavailability under simulated digestion conditions. This ability to protect bioactive substances from degradation improves the effectiveness of nutrients, ensures higher bioavailability and controlled absorption in the gastrointestinal tract; These properties are particularly valuable in health-oriented food formulation and fortification strategies. Smart oral nano-delivery systems based on starch nanoparticles can overcome multiple gastrointestinal barriers (mucus layer, tight junctions, and P-glycoprotein efflux) through tailored particle size, surface charge, wettability, and ligand functionalization, significantly improving bioavailability of encapsulated bioactive compounds (Yu et al., 2025).

From an emulsion stabilization perspective, SNPs act as Pickering stabilizers, forming stable emulsions without the need for synthetic surfactants; this is very useful for clean-label food products. Jiang et al. (2022) observed that quinoa SNPs formed a stable network in the aqueous phase and effectively stabilized Pickering emulsions, making them ideal for natural sauces and condiments. The rheological properties of SNP suspensions also help modify texture and provide a desirable mouthfeel or consistency. Dong et al. (2021) reported that SNPs derived from legumes (e.g., fava beans and peas) exhibited higher viscosity and elasticity compared to cereal SNPs; This difference is attributed to the smaller particle

TABLE 21.5 Functional improvements of nanoscale starch modifications in food systems.

Functional improvement	Description	Applications in food systems	References
Enhanced thickening and gelling properties	Nano-starches interact effectively with hydrocolloids, improving viscosity and gel formation.	Sauces, soups, gravies, dairy products, and gel-based desserts	Campelo et al. (2020)
Controlled release and encapsulation	SNPs act as carriers for bioactives (vitamins, antioxidants, probiotics), ensuring sustained release.	Nutraceuticals, functional foods, and fortified beverages	Rostamabadi et al. (2019)
Improved water retention	Superior hydration and swelling behavior enhance water retention and reduce moisture loss.	Bakery products, ready-to-eat meals, and meat analogs	Marta et al. (2023)
Reduced retrogradation and freeze-thaw stability	Nano-starches reduce retrogradation and syneresis, improving stability in frozen products.	Frozen foods, refrigerated products, and convenience foods	Bajer (2023)
Emulsion stability	Quinoa SNPs form a stable network, ideal for clean-label products.	Natural sauces, and dressings	Jiang et al. (2022)
Texture modification	Pulse SNPs exhibit higher viscosity and elasticity, suitable for texture enhancement.	Soups, sauces, and salad dressings	Dong et al. (2021)
Packaging strength	Improves film strength, flexibility, and barrier properties.	Sustainable food packaging	Bajer (2023)
Digestibility control	Up to 45.2% resistant starch in waxy rice SNPs, beneficial for glycemic control.	Diabetic-friendly foods	Jeong and Shin (2018)

size and stronger interparticle hydrogen bonds, making them suitable for use in soups, sauces, and salad dressings. Starch-based nanomaterials also enhance the mechanical and barrier properties of biopolymer films, making them ideal for sustainable food packaging. In addition, nanostarches play an important role in water retention and moisture management. Their high specific surface area and improved hydration and swelling behaviors increase water retention capacity; this directly has a positive impact on food texture and increased shelf life in processed products such as breads, meat products, and ready-to-eat foods, where moisture control is of paramount importance.

The nanostructure of starch can affect its digestibility and provide health benefits by controlling the glycemic index. Acid-hydrolyzed SNPs, as shown by Jeong & Shin, 2018; Torres & De-la-Torre, 2022, increase crystallinity, reduce digestibility, and increase resistant starch content. In addition, nanoscale modifications help reduce retrogradation and increase freeze-thaw stability. Retrogradation, where starch molecules rearrange and cause undesirable textural changes, is a major challenge in frozen and refrigerated foods. Nanostarches prevent syneresis (water loss) in such products, improve freeze-thaw stability, and increase usability; this property is very beneficial for frozen foods, refrigerated products, and ready-to-eat foods where texture and quality maintenance after freezing are essential (Sun, Cong, et al., 2025). Encapsulation of quercetin in ginkgo-derived starch nanoparticles significantly enhances solubility, bioavailability, and targeted delivery, promoting anti-inflammatory effects and cellular homeostasis in functional food applications with demonstrated in vivo biocompatibility (Sun, Cong, et al., 2025, Sun, Wang, et al., 2025).

Starch nanoparticles loaded with green propolis extracts (via pressurized liquid extraction) demonstrate enhanced polyphenol stability under thermal and pH stress, with low degradation rates and extended shelf-life predictions, highlighting their potential as robust delivery systems in functional foods (Chacon et al., 2025). Plant polysaccharide-capped nanoparticles, including starch-based ones, offer a sustainable strategy to modulate gut microbiota by promoting beneficial bacteria and short-chain fatty acid production, enhancing nutrient absorption and supporting next-generation functional foods (Bamigbade et al., 2025).

6. Applications in food product development

Nanoscale starch has emerged as a versatile and efficient material that improves quality, performance, and sensory properties in a wide range of food systems. Its wide applications in various food product groups effectively address challenges such as gluten replacement, fat reduction, moisture retention, and stability enhancement (Liu et al., 2017). Nanoscale starch significantly improves food formulations by improving texture, increasing shelf life, and enhancing product stability in a variety of applications (Fig. 21.6).

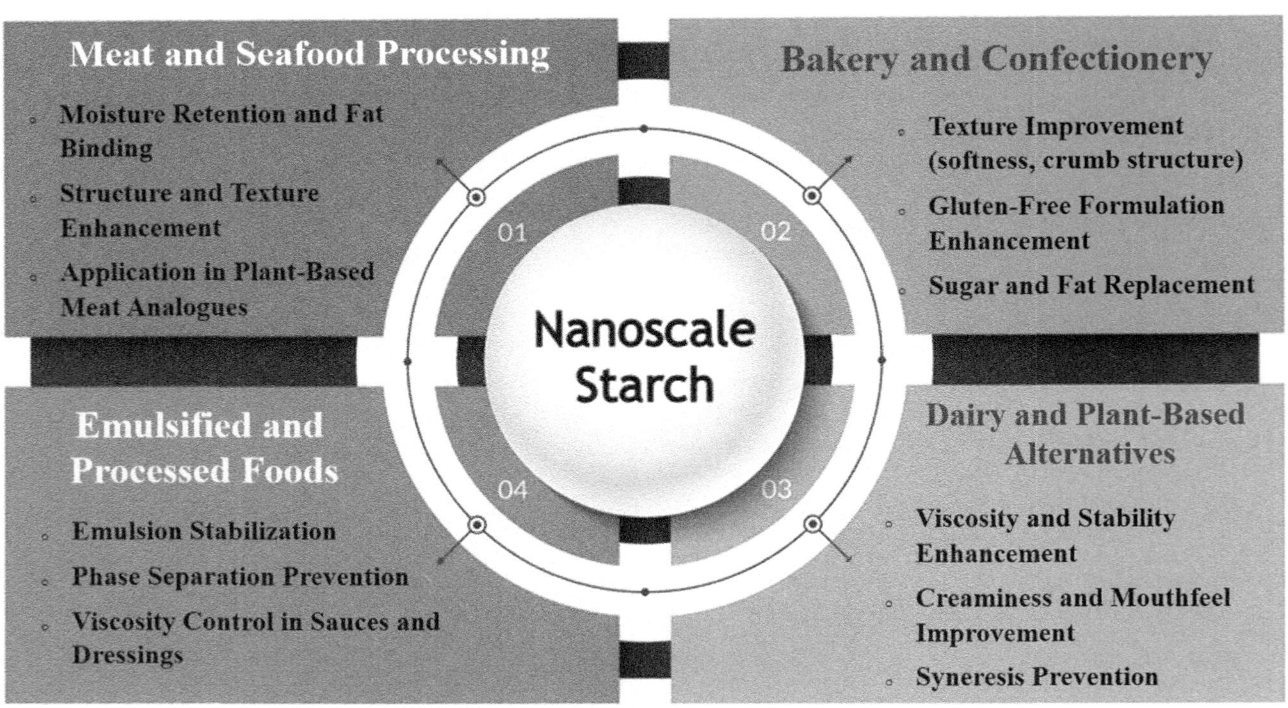

FIGURE 21.6 Applications of nanoscale starch in food product development.

Nanoscale starch plays a prominent role in improving the quality of gluten-free bakery products, which often face problems such as poor texture, low volume, and undesirable sensory properties. Starch nanocrystals and nanocomposites significantly improve the viscoelastic properties of gluten-free doughs by improving water retention capacity, increasing elasticity, and enhancing film-forming ability. These changes lead to better crumb structure, longer shelf life, and increased softness in products such as gluten-free bread, cookies, and cakes. In addition, nanoscale starch acts as an efficient replacement for fat and sugar in bakery and confectionery products; its high specific surface area and strong gelling ability allow it to mimic the texture and mouthfeel of fats, enabling calorie reduction without compromising sensory quality (Joye, 2020).

The use of nanostarch in dairy products and plant-based alternatives significantly improves texture, stability, and sensory properties. In plant-based beverages such as almond, soy, and oat milk, nano-starch acts as a thickener and stabilizer, minimizing phase separation and sedimentation. It improves mouthfeel by increasing viscosity and creating a smooth, creamy texture that closely resembles traditional dairy formulations (Mehany et al., 2024). In plant-based yogurts and cream substitutes, nano-starch also increases viscosity, prevents runoff, stabilizes emulsions and helps create a more uniform structure. These improvements are particularly valuable in fermented products where stability and texture are critical for consumer acceptance (Soni et al., 2022). Nano-starch significantly improves the quality of processed meats and seafood by enhancing moisture retention and fat binding. Its high water-binding capacity prevents water loss, increases product hydration, and maintains moisture during cooking and storage (Hashemi et al., 2023). This property is especially important in heated and frozen products, where moisture loss is a major concern. Furthermore, in plant-based meat analogs, nanostarch helps to better bind plant proteins and fats, improving the texture and firmness of the final product. By mimicking the fibrous structure of meat, retaining moisture, and increasing hydration, it has become a key component in the formulation of high-quality meat substitutes (Li et al., 2023). Nanoscale starch effectively enhances the stability, texture, and consistency of emulsions and processed food systems. It stabilizes emulsions in products such as salad dressings, sauces, and gravies by forming a protective layer around oil droplets, preventing droplet aggregation, and reducing phase separation. The result is a uniform consistency and a desirable mouthfeel with long-term shelf life. In addition, nano-sized starch improves the viscosity and gel strength of processed foods, helping to enhance sensory properties and consumer satisfaction (Aliabbasi & Emam-Djomeh, 2024).

In addition to improving food formulations, nano-sized starch also has a wide range of applications in food packaging and encapsulation systems. Starch nanocomposites are used in the production of biodegradable packaging films and provide superior barrier properties, extending the shelf life of packaged products. Valorization of agro-industrial by-products such as litchi seed starch into nanoparticles enables the development of fully biodegradable edible films and coatings that extend tomato shelf life up to 23 days under ambient conditions while improving mechanical properties, water barrier performance, and microbial stability (Chavan et al., 2025). Recent advances in starch-based sustainable packaging include polymer modifications, bio-nanocomposite incorporation, and innovative processing to overcome moisture sensitivity and mechanical fragility, with advanced characterization (FTIR, DSC, NMR) enabling better monitoring of retrogradation and structure-performance relationships (Arruda et al., 2025). Starch-based nanofilms with antimicrobial properties inhibit the growth of microorganisms and further enhance shelf life. Starch nanoparticles also act as efficient carriers for bioactive compounds such as vitamins and antioxidants, improving the stability and bioavailability of encapsulated nutrients. These innovative applications contribute significantly to the development of sustainable diets and functional foods (Remanan & Zhu, 2021).

Nanoscale starch plays a pivotal role in improving the texture and rheological properties of sauces, soups, and dairy products; it ensures a desirable mouthfeel and consistency by increasing viscosity, gel strength, and stability. In addition, nanoscale starches are widely used in the development of functional and fortified foods, including low-calorie, gluten-free, and digestion-resistant formulations. These materials also enable the delivery of probiotics and prebiotics, adding value to functional food applications. The versatility of nanoscale starch in improving texture and performance in different food groups is well-known (Hassan et al., 2022).

7. Challenges and regulatory considerations

Although the use of nanoscale starch offers innovative opportunities to improve the texture, stability, and shelf life of food products, the technology faces significant barriers in terms of safety, regulatory requirements, and general consumer acceptance. Ensuring that nanoscale starch complies with international safety standards and strict regulatory compliance is a prerequisite for expanding its application in the food industry (Jafari et al., 2017).

A key concern in this regard is the potential toxicity of nanomaterials in food systems. Due to their extremely small size and high surface-to-volume ratio, nanoscale starch exhibits different physicochemical properties than its macroscopic form, which can alter the way it interacts with biological systems (Jafari, 2021). Although it is derived from natural sources, the nanofabrication process may affect cellular responses and cause issues such as cytotoxicity, genetic damage, or bioaccumulation. Therefore, comprehensive toxicological assessments are necessary to determine the safety of nanoscale starch (Dave et al., 2021). These assessments generally include laboratory (in vitro) and animal (in vivo) studies that examine parameters such as cell viability, genotoxicity, bioavailability, and long-term effects of exposure. The results of existing research indicate that nanoscale starch exhibits negligible adverse effects at the permitted consumption levels; however, there is a need for continued research to improve assessment models and ensure its long-term safety in various food applications (Singh & Kumar, 2023). Global regulatory agencies have developed specific frameworks for managing the application of nanotechnology in food to ensure the safety and transparency of nanomaterials. Agencies such as the US Food and Drug Administration (FDA), the European Food Safety Authority (EFSA), and Food Standards Australia and New Zealand (FSANZ) require rigorous pre-market safety assessments before granting commercialization authorization (Singh & Kumar, 2023). These frameworks include guidance on risk assessment of nanomaterials, post-market surveillance, and mandatory labeling requirements to inform consumers of the presence of nanomaterials in food products (Jafari, 2021). Despite these regulatory measures, consumer attitudes and perceptions of nanotechnology remain a major barrier. Public skepticism is often driven by a lack of awareness about the technology and potential health concerns. Research shows that consumers are more likely to reject products containing nanomaterials if there is a lack of transparency or inadequate labeling. To overcome this challenge, the food industry needs to build trust by providing transparent and accessible information about the safety and benefits of nano-sized starch. Public education programs can also play a key role in increasing awareness and building trust, empowering consumers to make more informed choices about nano-sized products (Dave et al., 2021).

Consumer acceptance is a key factor in the successful commercialization of nano-sized starch in the food industry. While it offers several functional benefits, including improved texture, moisture retention, and increased nutrient transfer, ambiguities and misconceptions about the safety of nanotechnology can limit its adoption (Jafari et al., 2017). To build public trust, industry players should prioritize transparency and act through accurate and clear labeling of products containing nano-sized starch. Additionally, implementing targeted communication strategies can greatly help educate consumers about the benefits and safety of this technology in food applications (Dave et al., 2021).

8. Conclusion

This chapter is dedicated to exploring the transformative role of nanotechnology in enhancing the performance of nanoscale starches for food applications through the application of various modification methods. Chemical methods such as esterification, acylation, hydrolysis, and modification with fatty acids, along with crosslinking techniques and physical methods like cold plasma and ultrasound, have been widely used to modify starch nanoparticles (SNPs) and starch nanocrystals (SNCs) to improve their physicochemical properties. Results from descriptive techniques including morphology, particle size distribution, degree of crystallinity, thermal behavior, functional properties, digestibility, rheological properties, and amylose content show the extent to which these modifications can significantly enhance properties such as gel-forming ability, adhesion, and stability. This chapter highlights that nanoscale modified starches significantly enhance the performance of food systems by providing capabilities i.e., effective thickening, gelation, encapsulation of active ingredients, stabilization of emulsions, water retention, improved freeze-thaw stability, and digestion control. These features enable the development of clean-label products, sustainable packaging, and functional foods with health benefits, such as improved glycemic control. Despite these significant advances, challenges remain, including industrial scalability, process cost-effectiveness, and regulatory issues related to safety and labeling requirements for nanomaterials in food products, necessitating further research to optimize production methods and ensure consumer safety. Future directions will include exploring novel refinement techniques, improving production scalability, and expanding the scope of applications in areas like plant-based foods and personalized nutrition. By overcoming these challenges and taking advantage of the improved properties of nanoscale starches, nanotechnology has the potential to fundamentally transform the food industry, meeting consumers' growing demand for natural, sustainable, and health-focused food solutions, while helping to create a more innovative and environmentally friendly food industry.

References

Abidin, M. N. Z., Goh, P. S., Ismail, A. F., Said, N., Othman, M. H. D., Hasbullah, H., Abdullah, M. S., Ng, B. C., Kadir, S. H. S. A., & Kamal, F. (2018). Highly adsorptive oxidized starch nanoparticles for efficient urea removal. *Carbohydrate Polymers, 201*, 257–263. https://doi.org/10.1016/j.carbpol.2018.08.069

Adeyeye, S. A. O., Surendra Babu, A., Guruprasath, N., Guruprasath, N., & Sankar Ganesh, P. (2023). Starch nanocrystal and its food packaging applications. *Current Research in Nutrition and Food Science Journal, 11*(1), 01–21. https://doi.org/10.12944/crnfsj.11.1.1

Ahmad, M., & Gani, A. (2021). Ultrasonicated resveratrol loaded starch nanocapsules: Characterization, bioactivity and release behaviour under in-vitro digestion. *Carbohydrate Polymers, 251*, Article 117111. https://doi.org/10.1016/j.carbpol.2020.117111

Ahmad, M., Hassan, I., Shah, M. A., Gani, A., & Muthukumarappan, K. (2023). Co-encapsulation of multivitamins in micro & nano-sized starch, target release, capsule characterization and interaction studies. *International Journal of Biological Macromolecules, 240*, Article 124367. https://doi.org/10.1016/j.ijbiomac.2023.124367

Aliabbasi, N., & Emam-Djomeh, Z. (2024). Application of nanotechnology in dairy desserts and ice cream formulation with the emphasize on textural, rheological, antimicrobial, and sensory properties. *eFood, 5*(4), Article e170.

Apostolidis, E., & Mandala, I. (2020). Modification of resistant starch nanoparticles using high-pressure homogenization treatment. *Food Hydrocolloids, 103*, Article 105677. https://doi.org/10.1016/j.foodhyd.2020.105677

Arruda, T. R., Machado, G.d. O., Marques, C. S., Souza, A. L.d., Pelissari, F. M., Oliveira, T. V.d., & Silva, R. R. A. (2025). An overview of starch-based materials for sustainable food packaging: Recent advances, limitations, and perspectives. *Multidisciplinary Digital Publishing Institute (MDPI), Germany Macromol, 5*(2). https://doi.org/10.3390/macromol5020019

Bajer, D. (2023). Nano-starch for food applications obtained by hydrolysis and ultrasonication methods. *Food Chemistry, 402*, Article 134489. https://doi.org/10.1016/j.foodchem.2022.134489

Bajer, D. (2023). Nano-starch for food applications obtained by hydrolysis and ultrasonication methods. *Food Chemistry, 402*, 134489.

Bamigbade, G. B., Abdin, M., Subhash, A., Arachchi, M. P., Ullah, N., Gan, R.-Y., Ali, A., Kamal-Eldin, A., & Ayyash, M. (2025). Plant polysaccharide-capped nanoparticles: A sustainable approach to modulate gut microbiota and advance functional food applications. *Comprehensive Reviews in Food Science and Food Safety, 24*(2). https://doi.org/10.1111/1541-4337.70156

Bhati, N., Bhatt, P., Kumar, V., Tomar, S., Arora Sethi, V., Mittal, A., Maguluri, L. P., Goel, R., & Tyagi, L. K. (2025). Eco-friendly enzymatic starch modification: Precision strategies for tailored industrial and biomedical functionalities. *Starch - Stärke, 77*(9). https://doi.org/10.1002/star.70087

Bhavya, E. P., & Raman, M. (2024). Comprehensive review on synthesis of nano starch and its applications in food packaging industry. *Journal of Packaging Technology and Research, 8*(3), 153–166. https://doi.org/10.1007/s41783-024-00173-2

Campelo, P. H., Sant'Ana, A. S., & Clerici, M. T. P. S. (2020). Starch nanoparticles: Production methods, structure, and properties for food applications. *Current Opinion in Food Science, 33*, 136–140.

Cazotti, J. C., Fritz, A. T., Garcia-Valdez, O., Smeets, N. M. B., Dubé, M. A., & Cunningham, M. F. (2020). Graft modification of starch nanoparticles using nitroxide-mediated polymerization and the "grafting to" approach. *Biomacromolecules, 21*(11), 4492–4501. https://doi.org/10.1021/acs.biomac.0c00462

Chacon, W. D. C., Sorita, G. D., Monteiro, A. R., Verruck, S., Montero, L., Cifuentes, A., Álvarez-Rivera, G., & Valencia, G. A. (2025). Brazilian green propolis-loaded starch nanoparticles: Influence of extraction conditions on bioactivity and kinetic stability under thermal and pH stress. *International Journal of Biological Macromolecules, 322*. https://doi.org/10.1016/j.ijbiomac.2025.146861

Chang, R., Ji, N., Li, M., Qiu, L., Sun, C., Bian, X., Qiu, H., Xiong, L., & Sun, Q. (2019). Green preparation and characterization of starch nanoparticles using a vacuum cold plasma process combined with ultrasonication treatment. *Ultrasonics Sonochemistry, 58*, Article 104660. https://doi.org/10.1016/j.ultsonch.2019.104660

Chavan, P., Sawant, P., Singh, G. P., Aayush, K., Sharma, S., Thory, R., & Sinhmar, A. (2025). Sustainable polymer innovation: Developing edible films from Indian deshi litchi seed starch nanoparticles for extending fresh produce shelf life. *Journal of Food Process Engineering, 48*(4). https://doi.org/10.1111/jfpe.70082

Chavan, P., Sinhmar, A., Nehra, M., Thory, R., Pathera, A. K., Sundarraj, A. A., & Nain, V. (2021). Impact on various properties of native starch after synthesis of starch nanoparticles: A review. *Food Chemistry, 364*. https://doi.org/10.1016/j.foodchem.2021.130416

Chichti, E., George, M., Delenne, J. Y., Radjai, F., & Lullien-Pellerin, V. (2013). Nano-mechanical properties of starch and gluten biopolymers from atomic force microscopy. *European Polymer Journal, 49*(12), 3788–3795. https://doi.org/10.1016/j.eurpolymj.2013.08.024

Chorfa, N., Nlandu, H., Belkacemi, K., & Hamoudi, S. (2022). Physical and enzymatic hydrolysis modifications of potato starch granules. *Polymers, 14*(10), 2027. https://doi.org/10.3390/polym14102027

Choton, S., Bandral, J. D., Singh, J., Bhat, A., Sood, M., Gupta, N., Reshi, M., & Kaur, D. (2024). Enzymatic modification of starch: A review. *Saudi Journal of Medical and Pharmaceutical Sciences, 10*(01), 1–8. https://doi.org/10.36348/sjmps.2024.v10i01.001

Cornejo-Ramírez, Y. I., Martínez-Cruz, O., Del Toro-Sánchez, C. L., Wong-Corral, F. J., Borboa-Flores, J., & Cinco-Moroyoqui, F. J. (2018). The structural characteristics of starches and their functional properties. *CyTA – Journal of Food, 16*(1), 1003–1017. https://doi.org/10.1080/19476337.2018.1518343

Dave, V., Sur, S., & Gupta, N. (2021). Current framework, ethical consideration and future challenges of regulatory approach for nano-based products. *Nanopharmaceutical advanced delivery systems, 12*, 447–472.

Dhiman, S., Kumari, A., Kumari, S., & Sharma, R. (2025). Advanced biodegradable starch-based nanocomposite films incorporating zinc oxide nanoparticles: Synthesis, characterization, and efficacy in antibacterial food packaging applications. *Journal of Environmental Chemical Engineering, 13*(3), Article 116296. https://doi.org/10.1016/j.jece.2025.116296

Dong, H., Zhang, Q., Gao, J., Chen, L., & Vasanthan, T. (2021). Comparison of morphology and rheology of starch nanoparticles prepared from pulse and cereal starches by rapid antisolvent nanoprecipitation. *Food Hydrocolloids, 119*, Article 106828. https://doi.org/10.1016/j.foodhyd.2021.106828

Du, C., Jiang, F., Hu, W., Ge, W., Yu, X., & Du, S.-kui (2023). Comparison of properties and application of starch nanoparticles optimized prepared from different crystalline starches. *International Journal of Biological Macromolecules, 235*, Article 123735. https://doi.org/10.1016/j.ijbiomac.2023.123735

Dukare, A. S., Arputharaj, A., Bharimalla, A. K., Saxena, S., & Vigneshwaran, N. (2021). Nanostarch production by enzymatic hydrolysis of cereal and tuber starches. *Carbohydrate Polymer Technologies and Applications, 2*, Article 100121. https://doi.org/10.1016/j.carpta.2021.100121

Duyen, T. T. M., & Van Hung, P. (2021). Morphology, crystalline structure and digestibility of debranched starch nanoparticles varying in average degree of polymerization and fabrication methods. *Nam Carbohydrate Polymers, 256*. https://doi.org/10.1016/j.carbpol.2020.117424

El Nokab, M. E. H., Alassmy, Y. A., Abduljawad, M. M., Al-shamrani, K. M., Alnafisah, M. S., Asgar Pour, Z., Tucker, C. L., & Sebakhy, K. O. (2022). Solid-state NMR spectroscopy: Towards structural insights into starch-based materials in the food industry. *Polymers, 14*(21). https://doi.org/10.3390/polym14214686

Escobar-Puentes, A. A., García-Gurrola, A., Rincón, S., Zepeda, A., & Martínez-Bustos, F. (2020). Effect of amylose/amylopectin content and succinylation on properties of corn starch nanoparticles as encapsulants of anthocyanins. *Carbohydrate Polymers, 250*, Article 116972. https://doi.org/10.1016/j.carbpol.2020.116972

Fuentes, C., In Kang, J. L., Song, D., Sjöö, M., Choi, J., Lee, S., & Nilsson, L. (2019). Fractionation and characterization of starch granules using field-flow fractionation (FFF) and differential scanning calorimetry (DSC). *Analytical and Bioanalytical Chemistry, 411*(16), 3665–3674. https://doi.org/10.1007/s00216-019-01852-9

Guo, T., Xie, F., & Chen, L. (2024). Oxidation-induced starch molecular degradation: A comprehensive kinetic investigation using NaClO/NaBr/TEMPO system. *International Journal of Biological Macromolecules, 277*, Article 134283.

Gutiérrez, G., Morán, D., Marefati, A., Purhagen, J., Rayner, M., & Matos, M. (2020). Synthesis of controlled size starch nanoparticles (SNPs). *Carbohydrate Polymers, 250*, Article 116938. https://doi.org/10.1016/j.carbpol.2020.116938

Haroon, M., Wang, L., Yu, H., Abbasi, N. M., Ul-Abdin, Z.-, Saleem, M., Khan, R. U., Ullah, R. S., Chen, Q., & Wu, J. (2016). Chemical modification of starch and its application as an adsorbent material. *RSC Advances, 6*(82), 78264–78285. https://doi.org/10.1039/c6ra16795k

Hashemi, S. M. B., Kaveh, S., Abedi, E., & Phimolsiripol, Y. (2023). Polysaccharide-based edible films/coatings for the preservation of meat and fish products: Emphasis on incorporation of lipid-based nanosystems loaded with bioactive compounds. *Foods, 12*(17). https://doi.org/10.3390/foods12173268

Hassan, N. A., Darwesh, O. M., Smuda, S. S., Altemimi, A. B., Hu, A., Cacciola, F., Haoujar, I., & Abedelmaksoud, T. G. (2022). Recent trends in the preparation of nano-starch particles. *Molecules, 27*(17). https://doi.org/10.3390/molecules27175497

Hebeish, A., El-Rafie, M. H., EL-Sheikh, M. A., & El-Naggar, M. E. (2014). Ultra-fine characteristics of starch nanoparticles prepared using native starch with and without surfactant. *Journal of Inorganic and Organometallic Polymers and Materials, 24*(3), 515–524. https://doi.org/10.1007/s10904-013-0004-x

Hedayati, S., Niakousari, M., & Mohsenpour, Z. (2020). Production of tapioca starch nanoparticles by nanoprecipitation-sonication treatment. *International Journal of Biological Macromolecules, 143*, 136–142.

Hussain, S., Akhter, R., & Maktedar, S. S. (2024). Advancements in sustainable food packaging: From eco-friendly materials to innovative technologies. *Sustainable Food Technology, 2*(5), 1297–1364.

Jafari, S. M. (2021). *Safety and regulatory issues of nanoencapsulated food ingredients.* (Vol. 7). Academic Press.

Jafari, S. M., Katouzian, I., & Akhavan, S. (2017). 15 - Safety and regulatory issues of nanocapsules. In S. M. Jafari (Ed.), *Nanoencapsulation technologies for the food and nutraceutical industries* (pp. 545–590). Academic Press.

Jeong, O., & Shin, M. (2018). Preparation and stability of resistant starch nanoparticles, using acid hydrolysis and cross-linking of waxy rice starch. *Food Chemistry, 256*, 77–84. https://doi.org/10.1016/j.foodchem.2018.02.098

Ji, N., Li, X., Qiu, C., Li, G., Sun, Q., & Xiong, L. (2015). Effects of heat moisture treatment on the physicochemical properties of starch nanoparticles. *Carbohydrate Polymers, 117*, 605–609.

Jiang, S., Dai, L., Qin, Y., Xiong, L., Sun, Q., & Choi, J. (2016). Preparation and characterization of octenyl succinic anhydride modified taro starch nanoparticles. *PLoS One, 11*(2), Article e0150043. https://doi.org/10.1371/journal.pone.0150043

Jiang, F., Feng, J., Hu, W., Ren, Y., Ma, C., Chang, L., Liu, Y., Lv, X., Yu, X., & Du, S.-kui (2022). Effects of morphology and rheology of starch nanoparticles prepared from various coarse cereals on emulsifying ability. *Carbohydrate Polymers, 298*, Article 120137. https://doi.org/10.1016/j.carbpol.2022.120137

Joye, I. J. (2020). *Application of nano/microencapsulated ingredients in cereal flours and bakery products application of nano/microencapsulated ingredients in food products.* Canada: Elsevier. https://doi.org/10.1016/B978-0-12-815726-8.00006-4

Kim, H.-Y., Park, S. S., & Lim, S.-T. (2015). Preparation, characterization and utilization of starch nanoparticles. *Colloids and Surfaces B: Biointerfaces, 126*, 607–620.

Kumari, S., Yadav, B. S., & Yadav, R. B. (2020). Synthesis and modification approaches for starch nanoparticles for their emerging food industrial applications: A review. *Food Research International, 128*, Article 108765.

Kumari, S., Yadav, B. S., & Yadav, R. (2022). Characterization of acid hydrolysis based nano-converted mung bean (Vigna radiata L.) starch for morphological, rheological and thermal properties. *International Journal of Biological Macromolecules, 211*, 450–459.

Li, X., Fan, M., Huang, Q., Zhao, S., Xiong, S., Yin, T., & Zhang, B. (2022). Effect of micro-and nano-starch on the gel properties, microstructure and water mobility of myofibrillar protein from grass carp. *Food Chemistry, 366*, 130579.

Li, S., Wang, Z., Pan, Y., Sun, C., Li, E., & Gilbert, R. G. (2025). Effects of amylose and amylopectin molecular structures on the emulsification performance of starch nanoparticles. *International Journal of Biological Macromolecules, 308*(Pt 4), Article 142717. https://doi.org/10.1016/j.ijbiomac.2025.142717

Li, G., Zhan, J., Hu, Z., Huang, J., Yao, Q., Yuan, C., Chen, J., & Hu, Y. (2023). Effects of nano starch-lutein on 3D printing properties of functional surimi: Enhancement mechanism of printing effects and anti-oxidation. *Journal of Food Engineering, 346*, Article 111431. https://doi.org/10.1016/j.jfoodeng.2023.111431

Liu, Q., Fang, L., Lu, H., Man, L., Liu, J., Zhang, S., Sun, Q., & Xiong, L. (2018). Enhanced dispersion stability and heavy metal ion adsorption capability of oxidized starch nanoparticles. *Food Chemistry, 242*, 256–263. https://doi.org/10.1016/j.foodchem.2017.09.071

Liu, G., Gu, Z., Hong, Y., Cheng, L., & Li, C. (2017). Structure, functionality and applications of debranched starch: A review. *Trends in Food Science & Technology, 63*, 70–79. https://doi.org/10.1016/j.tifs.2017.03.004

Lu, H., Ji, N., Li, M., Wang, Y., Xiong, L., Zhou, L., Qiu, L., Bian, X., Sun, C., & Sun, Q. (2019). Preparation of borax cross-linked starch nanoparticles for improvement of mechanical properties of maize starch films. *Journal of Agricultural and Food Chemistry, 67*(10), 2916–2925. https://doi.org/10.1021/acs.jafc.8b06479

Lu, H., & Tian, Y. (2021). Nanostarch: Preparation, modification, and application in pickering emulsions. *Journal of Agricultural and Food Chemistry, 69*(25), 6929–6942. https://doi.org/10.1021/acs.jafc.1c01244

Mahdavi, S., Aalami, M., Ziaiifar, A. M., & Maghsoudlou, Y. (2021). Effect of heat-moisture treatment of brown rice flour and natural additives on the properties of gluten-free frozen cake batter. *Food Processing and Preservation Journal, 13*(3), 115–131.

Marimuthu, M., Sundaram, U., & Gurumoorthi, P. (2013). X-ray diffraction and starch analysis of nano sized seed powder of velvet bean (Mucuna pruriens). *Indo American Journal of Pharmaceutical Research, 3*(2), 4319–4329.

Marta, H., Rizki, D. I., Mardawati, E., Djali, M., Mohammad, M., Cahyana, Y., & Nanoparticles, S. (2023). Preparation, properties and applications. *Polymers, 15*(5), 1167. https://doi.org/10.3390/polym15051167

Masoudipour, E., Kashanian, S., Azandaryani, A. H., Omidfar, K., & Bazyar, E. (2017). Surfactant effects on the particle size, zeta potential, and stability of starch nanoparticles and their use in a pH-responsive manner. *Cellulose, 24*(10), 4217–4234. https://doi.org/10.1007/s10570-017-1426-3

Mehany, T., Siddiqui, S. A., Olawoye, B., Olabisi Popoola, O., Hassoun, A., Manzoor, M. F., & Punia Bangar, S. (2024). Recent innovations and emerging technological advances used to improve quality and process of plant-based milk analogs. *Critical Reviews in Food Science and Nutrition, 64*(20), 7237–7267. https://doi.org/10.1080/10408398.2023.2183381

Namazi, H., & Dadkhah, A. (2010). Convenient method for preparation of hydrophobically modified starch nanocrystals with using fatty acids. *Carbohydrate Polymers, 79*(3), 731–737. https://doi.org/10.1016/j.carbpol.2009.09.033

Namazi, H., Fathi, F., & Dadkhah, A. (2011). Hydrophobically modified starch using long-chain fatty acids for preparation of nanosized starch particles. *Scientia Iranica, 18*(3), 439–445.

Oh, S. M., Lee, B. H., Seo, D. H., Choi, H. W., Kim, B. Y., & Baik, M. Y. (2020). Starch nanoparticles prepared by enzymatic hydrolysis and self-assembly of short-chain glucans. *Food Science and Biotechnology, 29*(5), 585–598. https://doi.org/10.1007/s10068-020-00768-w

Olayil, R., Arumuga Prabu, V., DayaPrasad, S., Naresh, K., & Rama Sreekanth, P. S. (2022). A review on the application of bio-nanocomposites for food packaging. *Materials Today: Proceedings, 56*, 1302–1306. https://doi.org/10.1016/j.matpr.2021.11.315

Ortega, F., Sobral, P., Jios, J. L., Arce, V. B., & García, M. A. (2022). Starch nanocomposite films: Migration studies of nanoparticles to food simulants and bio-disintegration in soil. *Polymers, 14*(9). https://doi.org/10.3390/polym14091636

Pal, K., Sarkar, P., Anis, A., Wiszumirska, K., & Jarzębski, M. (2021). Polysaccharide-based nanocomposites for food packaging applications. *Materials, 14*(19), 5549. https://doi.org/10.3390/ma14195549

Pei, X., Zhai, K., Wang, C., Deng, Y., Tan, Y., Zhang, B., Bai, Y., Xu, K., & Wang, P. (2019). Polymer brush graft-modified starch-based nanoparticles as pickering emulsifiers. *Langmuir, 35*(22), 7222–7230. https://doi.org/10.1021/acs.langmuir.9b00413

Perez Herrera, M., Vasanthan, T., & Hoover, R. (2016). Characterization of maize starch nanoparticles prepared by acid hydrolysis. *Cereal Chemistry, 93*(3), 323–330.

Pumacahua-Ramos, A., Demiate, I. M., Travalini, A. P., Granza, A. G., Farias, F. O., Schnitzler, E., & Lopes-Filho, J. F. (2024). Morphological, thermal, and physicochemical characteristics of nano starch from cañihua (Chenopodium pallidicaule aellen). *Starch Staerke, 76*(11–12), Article 2300095.

Puspita, I., Winarti, C., Maddu, A., & Kurniati, M. (2019). Synthesis of cassava starch based nano-hydrogels using gamma irradiation. *IOP Conference Series: Earth and Environmental Science*.

Qin, Y., Liu, C., Jiang, S., Xiong, L., & Sun, Q. (2016). Characterization of starch nanoparticles prepared by nanoprecipitation: Influence of amylose content and starch type. *Industrial Crops and Products, 87*, 182–190. https://doi.org/10.1016/j.indcrop.2016.04.038

Qiu, C., Yang, J., Ge, S., Chang, R., Xiong, L., & Sun, Q. (2016). Preparation and characterization of size-controlled starch nanoparticles based on short linear chains from debranched waxy corn starch. *Lebensmittel-Wissenschaft & Technologie, 74*, 303–310. https://doi.org/10.1016/j.lwt.2016.07.062

Rashid, U. S., Simsek, S., Kanel, S. R., & Bezbaruah, A. N. (2019). Modified tapioca starch for iron nanoparticle dispersion in aqueous media: Potential uses for environmental remediation. *SN Applied Sciences, 1*(11). https://doi.org/10.1007/s42452-019-1364-9

Remanan, M. K., & Zhu, F. (2021). Encapsulation of rutin using quinoa and maize starch nanoparticles. *Zealand Food Chemistry, 353*. https://doi.org/10.1016/j.foodchem.2020.128534

Rostamabadi, H., Falsafi, S. R., & Jafari, S. M. (2019). Nanostructures of starch for encapsulation of food ingredients. In *Biopolymer nanostructures for food encapsulation purposes* (pp. 419–462). Elsevier.

Rutkaite, R., Bendoraitiene, J., Klimaviciute, R., & Zemaitaitis, A. (2012). Cationic starch nanoparticles based on polyelectrolyte complexes. *International Journal of Biological Macromolecules, 50*(3), 687–693. https://doi.org/10.1016/j.ijbiomac.2012.01.037

Saartrat, S., Puttanlek, C., Rungsardthong, V., & Uttapap, D. (2005). Paste and gel properties of low-substituted acetylated canna starches. *Carbohydrate Polymers, 61*(2), 211–221. https://doi.org/10.1016/j.carbpol.2005.05.024

Shabana, S., Prasansha, R., Kalinina, I., Potoroko, I., Bagale, U., & Shirish, S. (2019). Ultrasound assisted acid hydrolyzed structure modification and loading of antioxidants on potato starch nanoparticles. *Ultrasonics Sonochemistry, 51*, 444–450.

Sharma, S., Jaiswal, S., Duffy, B., & Jaiswal, A. K. (2019). Nanostructured materials for food applications: Spectroscopy, microscopy and physical properties. *Bioengineering, 6*(1). https://doi.org/10.3390/bioengineering6010026

Shi, A. M., Li, D., Wang, L. J., Li, B. Z., & Adhikari, B. (2011). Preparation of starch-based nanoparticles through high-pressure homogenization and miniemulsion cross-linking: Influence of various process parameters on particle size and stability. *Carbohydrate Polymers, 83*(4), 1604–1610. https://doi.org/10.1016/j.carbpol.2010.10.011

Shi, A. M., Wang, L. J., Li, D., & Adhikari, B. (2013). Characterization of starch films containing starch nanoparticles part 1: Physical and mechanical properties. *Carbohydrate Polymers, 96*(2), 593–601. https://doi.org/10.1016/j.carbpol.2012.12.042

Singh, R., & Kumar, S. (2023). Regulatory and safety concerns regarding the use of active nanomaterials in food industry. In *Nanotechnology advancement in agro-food industry* (pp. 269–306). Springer.

Soni, M., Maurya, A., Das, S., Prasad, J., Yadav, A., Singh, V. K., Singh, B. K., Dubey, N. K., & Dwivedy, A. K. (2022). Nanoencapsulation strategies for improving nutritional functionality, safety and delivery of plant-based foods: Recent updates and future opportunities. *Plant Nano Biology, 1*. https://doi.org/10.1016/j.plana.2022.100004

Sothornvit, R. (2019). Nanostructured materials for food packaging systems: New functional properties. *Current Opinion in Food Science, 25*, 82–87. https://doi.org/10.1016/j.cofs.2019.03.001

Sreejit, V., Preetha, R., Mubeena, S. A., & Dhananjay, S. (2025). Green synthesized nano starch: Preparation and emerging avenues in food industry. *Journal of Nanoparticle Research, 27*(4). https://doi.org/10.1007/s11051-025-06285-3

Sujka, M. (2017). Ultrasonic modification of starch–impact on granules porosity. *Ultrasonics Sonochemistry, 37*, 424–429.

Sun, Y., Cong, K., Wang, T., Li, X., Li, T., Fan, G., Zhou, D., & Wu, C. (2025). Quercetin-loaded ginkgo starch nanoparticles: A promising strategy to improve bioactive delivery and cellular homeostasis in functional foods. *Foods, 14*(11), 1890.

Sun, Y., Wang, Z., Ye, J., Li, Y., Wang, L., & Cao, R. (2025). Advances in starch-based nanocomposites for functional food systems: Harnessing AI and nuclear magnetic resonance technologies for tailored stability and bioactivity. *Foods, 14*(5), 773.

Suriya, M., Reddy, C. K., Haripriya, S., & Harsha, N. (2018). Influence of debranching and retrogradation time on behavior changes of Amorphophallus paeoniifolius nanostarch. *International Journal of Biological Macromolecules, 120*, 230–236. https://doi.org/10.1016/j.ijbiomac.2018.08.059

Szymońska, J., Targosz-Korecka, M., & Krok, F. (2009). Characterization of starch nanoparticles. *Journal of Physics: Conference Series, 146*.

Tian, S., Xue, X., Wang, X., & Chen, Z. (2022). Preparation of starch-based functional food nano-microcapsule delivery system and its controlled release characteristics. *Frontiers in Nutrition, 9*. https://doi.org/10.3389/fnut.2022.982370

Tomoaia-Cotisel, M., Cioica, N., Cota, C., Racz, C., Petean, I., Bobos, L. D., Mocanu, A., & Horovitz, O. (2010). Structure of starch granules revealed by atomic force microscopy. *Studia Universitatis Babes-Bolyai, Chemia, 2*(2), 313–324.

Torres, F. G., & De-la-Torre, G. E. (2022). Synthesis, characteristics, and applications of modified starch nanoparticles: A review. *International Journal of Biological Macromolecules, 194*, 289–305. https://doi.org/10.1016/j.ijbiomac.2021.11.187

Wang, K., Hong, Y., Gu, Z., Cheng, L., Li, Z., & Li, C. (2020). Stabilization of pickering emulsions using starch nanocrystals treated with alkaline solution. *International Journal of Biological Macromolecules, 155*, 273–285.

Wang, X., Huang, L., Zhang, C., Deng, Y., Xie, P., Liu, L., & Cheng, J. (2020). Research advances in chemical modifications of starch for hydrophobicity and its applications: A review. *Carbohydrate Polymers, 240*, Article 116292.

Wang, J., Xiao, Q., & Zhao, Z. (2025). Physicochemical properties of starch nanoparticles from different botanical origins starch prepared by the deep eutectic solvent method: Wang et al. *Journal of Food Measurement and Characterization, 19*, 1–9.

Wang, Y., & Zhang, G. (2021). The preparation of modified nano-starch and its application in food industry. *Food Research International, 140*, Article 110009. https://doi.org/10.1016/j.foodres.2020.110009

Wei, B., Cai, C., & Tian, Y. (2018). Nano-sized starch: Preparations and applications. *Functional Starch and Applications in Food, 21*, 147–176.

Wei, H., Li, S., Liu, Z., Chen, H., Liu, Y., Li, W., & Wang, G. (2022). Preparation and characterization of starch-cellulose interpenetrating network hydrogels based on sequential diels-alder click reaction and photopolymerization. *International Journal of Biological Macromolecules, 194*, 962–973. https://doi.org/10.1016/j.ijbiomac.2021.11.154

Winarti, C., Sunarti, T. C., Mangunwidjaja, D., & Richana, N. (2014). Preparation of arrowroot starch nanoparticles by butanol-complex precipitation, and its application as bioactive encapsulation matrix. *Indonesia International Food Research Journal, 21*(6), 2207–2213.

Xiao, Z., Wang, L., Lv, C., Guo, S., Lu, X., Tao, L., Duan, Q., Yang, Q., & Luo, Z. (2020). Preparation and characterization of pH-responsive pickering emulsion stabilized by grafted carboxymethyl starch nanoparticles. *International Journal of Biological Macromolecules, 143*, 401–412. https://doi.org/10.1016/j.ijbiomac.2019.10.261

Xu, Y., Sismour, E. N., Grizzard, C., Thomas, M., Pestov, D., Huba, Z., Wang, T., & Bhardwaj, H. L. (2014). Morphological, structural, and thermal properties of starch nanocrystals affected by different botanic origins. *Cereal Chemistry, 91*(4), 383–388. https://doi.org/10.1094/CCHEM-10-13-0222-R

Xu, H., Zongwei, H., Zhang, J., Liu, H., Deng, C., Yu, Z., Zheng, M., Liu, Y., Zhou, Y., & Xiao, Y. (2023). Influence pathways of nanocrystalline cellulose on the digestibility of corn starch: Gelatinization, structural properties, and α-amylase activity perspective. *Carbohydrate Polymers, 314*, Article 120940. https://doi.org/10.1016/j.carbpol.2023.120940

Yao, X., Lin, R., Liang, Y., Jiao, S., & Zhong, L. (2023). Characterization of acetylated starch nanoparticles for potential use as an emulsion stabilizer. *Food Chemistry, 400*, Article 133873. https://doi.org/10.1016/j.foodchem.2022.133873

Ye, F., Miao, M., Lu, K., Jiang, B., Li, X., & Cui, S. W. (2017). Structure and physicochemical properties for modified starch-based nanoparticle from different maize varieties. *Food Hydrocolloids, 67*, 37–44. https://doi.org/10.1016/j.foodhyd.2016.12.041

Yu, M., Chi, C., Chen, L., & Li, X. (2025). Tailoring smart oral nano/micro delivery system to address multifaceted gastrointestinal barriers for improved delivery of food bioactive compounds. *Comprehensive Reviews in Food Science and Food Safety, 24*(5). https://doi.org/10.1111/1541-4337.70275

Yu, M., Ji, N., Wang, Y., Dai, L., Xiong, L., & Sun, Q. (2021). Starch-based nanoparticles: Stimuli responsiveness, toxicity, and interactions with food components. *Comprehensive Reviews in Food Science and Food Safety, 20*(1), 1075–1100. https://doi.org/10.1111/1541-4337.12677

Zhang, S., Zhu, J., Liu, Y., Zou, S. Y., & Li, L. (2019). Hierarchical structure and thermal property of starch-based nanocomposites with different amylose/amylopectin ratio. *Polymers, 11*(2). https://doi.org/10.3390/polym11020342

Chapter 22

Nanostructured packaging: Innovations in starch-based nanocomposite films and coatings for extended food shelf life

Rachna Gupta[1], Kanchan Goswami[2], Aashish Pratap Singh[3] and Vivek Kumar Gupta[4]

[1]Department of Food Technology, School of Biosciences and Technology, Galgotias University, Greater Noida, Uttar Pradesh, India; [2]Human Nutrition, Krishi Vigyan Kendra, Jaipur-I, Home Science, Jaipur, Rajasthan, India; [3]Master of Business Admistration, Narsee Monjee Institute of Management Studies (NMIMS), Mumbai, India; [4]Department of Biological Engineering, Biohybrid Systems Research Centre (BSRC), Inha University, Incheon, Republic of South Korea

1. Introduction

Starch is a common naturally occurring biomacromolecule with a wide range of industrial applications. In plants, it is a durable, biodegradable carbohydrate. Molecularly, starch is expressed as $(C_6H_{10}O_5)_n$. It is composed of two polymeric components: amylopectin and amylose. Amylose is a long-chain, linear glucose polysaccharide with a relatively low molecular weight, whereas amylopectin is a shorter, highly branched polymer connected by glycosidic linkages (Marinopoulou et al., 2019). The ratio between these two components is not fixed; as highlighted by Aldao et al. (2018), the botanical origin of the starch influences its typical composition of 20%–30% amylose and 70%–80% amylopectin. Inside the granule, these molecules arrange into tightly coiled chains, or double helices, which stack to form a repeating pattern of crystalline and amorphous layers.

Plant genotype, cultivation methods, and environmental interactions largely determine the size (0.5–175 μm), shape, and structure of starch granules. While wheat, sorghum, barley, and other cereals account for a smaller portion of global starch production, the primary industrial sources are maize, cassava, potato, millet, and sweet potato. Additional sources include rice, arrowroot, mung beans, and sago (Miyazaki et al., 2006). The specific properties of corn starch, for instance, make it especially valuable as a thickener in dairy products and beverages.

The Global Starch Market Research Report (2020–2028) estimates that the starch market will be worth 97.85 billion USD in 2020 and is expected to grow by 7% between 2020 and 2028. Of this, 75% comes from corn starch, the remaining 14% from potatoes, 7% from rice, and 4% from potatoes (Yazid et al., 2018). In light of this information, there is interest in researching new starch products for industrial use like fat replacers and additives in baked goods, snacks, dairy products, desserts, soups, noodles, mayonnaise, and other products (Santana & Meireles, 2014). The physicochemical, functional and thermal properties of starch affect its many uses. However, since the use of starch in its natural form is very little, it has been modified for wider use (Yazid et al., 2018). Create nanoparticles or nanocrystals. Starch-based nanomaterials have recently attracted attention due to their unique properties and biodegradable properties (Punia et al., 2019). According to Le Corre and Angellier-Coussy (2014), starch-based nanomaterials can be produced using several physical, chemical and enzymatic methods. They can be utilized as supports in biodegradable or non-biodegradable polymer matrices and as food quality sensors. According to crystallinity, starch-based nanosystems are divided into two groups: starch nanoparticles and starch nanocrystals. When starch granules are heated below the gelatinization temperature, they transform into semi-crystalline starch nanocrystals. Starch nanocrystals are usually created by acid hydrolysis and are widely used as emulsion stabilizers in foods (Haaj et al., 2014) and nanofillers to improve nanocomposites (Chen, Zhou, et al., 2019).

Native and Modified Starches for Food Applications. https://doi.org/10.1016/B978-0-443-27431-2.00023-7

However, starch nanoparticles function as incredibly small, non-crystalline particles, possessing a crystallinity level of less than 10%. Due to their tiny size and properties, they act as effective delivery vehicles for other compounds. Researchers have developed several methods to produce these nanoparticles, including high-pressure homogenization, nanoprecipitation, and ultrasonic treatment (Gardouh et al., 2021). Some advantages of starch-based nanosystems include increased surface area, reduced viscosity, and improved chemical performance (Nieto-Suaza et al., 2019). Many plant sources including potato starch (Shabana et al., 2019), cassava starch (Hedayati et al., 2020), banana starch (Acevedo-Guevara et al., 2018), corn starch (Minakawa et al., 2019), rice flour (Momenpoor et al., 2019), peanut flour (Ahmad, Gani, et al., 2020), and lotus seed starch (Guo et al., 2019) have been fully utilized to construct starch-based nanosystems. The integration of nanoscale (1–100 nm) organic or inorganic particles into starch-based thermoplastics has led to the development of a novel and promising category of "nano-enhanced packaging materials" (de Moraes et al., 2012). These materials are recyclable, biodegradable, and possess a low molecular weight. Their breaking strength is comparable to that of conventional low-density polyethylene (LDPE). Research has demonstrated that such nano-packaging systems also provide effective protection against moisture, oxygen, and microbial contamination (de Moraes et al., 2012).

2. Starch as a base material for food packaging

2.1 Chemical components and structure of starch

Two polysaccharides, amylose (amorphous) and amylopectin (crystalline), make up starch. Amylopectin and amylose are made of glucose units. About 20%–30% of all starch consists of amylose, a linear polymer connected by α-D-(1–4) glycosidic bonds, and amylopectin, a linear polymer with α-D-(1–6) branched chains and α-D-(1–4) glycosidic bonds in linear chains (Saunders, 2010; Singh, Isono, et al., 2008; Singh, Nakaura, et al., 2008); Starch is rich in amylopectin and is also found in some plant species (e.g., waxy maize and potato). A low-amylose starch has the advantage of being paste-like and less likely to retrograde; a high-amylose starch can be used to replace starch. Since the amylose: amylopectin ratio varies from plant to plant, a wide variety of properties occur in starches obtained from these plants. The rigidity of the amylopectin structure determines the size of starch granules. During development, the endosperm produces starch granules of various sizes and shapes. The ratio of long to short chains determines the form of amylopectin, which in turn affects the size and density of starch granules. The functional attributes of starch are largely influenced by the phospholipid content, phosphate monoesters, particle structure, distribution of branched chains in amylopectin, and lipids. The shapes of starch granules are polygonal, spherical, lenticular and elliptical, but their diameters vary between 1 and 100 nm (Singh, Isono, et al., 2008, 2010).

2.2 Characteristics of starch

Starch granules can be round, flat, angular, lenticular, elliptical or rectangular depending on the plant. Their dimensions on the nanometer scale range from 0.1 to 200 μm (Singh et al., 2003). They can have unimodal, bimodal or multimodal size distributions. Starch granules can be located singly or in clusters within the amyloplast (Pérez & Bertoft, 2010). Two types of starch are found in foods such as wheat, barley and rye: (i) type A, which is large and lens-shaped, and (ii) type B, which is spherical and small (Vamadevan & Bertoft, 2015). Usually, starch is isolated and then examined under a microscope. The separation method is very important as it will affect the size of the starch particles (Gao et al., 2014). The hilum, often called the "Maltese Cross", is the central point of the nuclear morphology. One or more Maltese crosses can be found in each starch granule, which reduces the birefringence potential of the granule (Jiang et al., 2010). The complexity is based on the differences in structure, amorphous and crystalline regions, and their composition (proteins, lipids, phosphorylation, water, and α-glucan). The higher the percentage of amylopectin in the starch granule, the higher the crystallinity, because amylose chains attached to the large branches of amylopectin molecules form the amorphous region of the granule, while amylopectin molecules with short branches form crystalline regions (Cheetham & Tao, 1998). According to (Singh et al., 2003), there are three different types of starch: type A found in cereal starches, type B found in tubers and type C found in legumes. The double helical structure formed by amylopectin molecules forms the basis of crystal formation. Branched amylopectin in the type A structure is connected by α-1,6 bonds and has a short degree of polymerization of 6–15. Pullulan derivatives have type A. The more magnetic pullulan in type B can form starch or act as a base in type A. B1, B2, B3 and B4 are branches of the type B chain. The degree of polymerization of B1 and B2 chains is between 15 and 25, while B3 and B4 have the highest degree of polymerization, usually between 40 and 50 (Cheetham & Tao, 1998).

2.3 Properties of starch

There is a variation in the ratio of amylose to amylopectin within and between species of starch. Starch still contains some protein, fiber, and lipid fractions after extraction. The presence of these elements can affect the functionality of the starch. When lipids and amylose combine to form complexes, they can change properties such as swelling index, retrogradation, gel strength, decreased water solubility and increased gelatinization temperature (Dar et al., 2018; Kumar et al., 2017; Singh et al., 2007). Some important properties of starch are.

2.3.1 Gelatinization

Although starch does not dissolve in cold water, it can undergo a process called gelatinization when heated above 52 °C. As a result of gelatinization, starch granules swell in two stages as shown in Fig. 22.1. During the first stage, amorphous starch loosens its hydrogen bonds. Second, water acts as a plasticizer, hydrating and swelling the amorphous regions. As a result of gelatinization, the bonds between starch molecules are broken down and the starch becomes more water-absorbent. The crystal structure of amylopectin changes when heated, resulting in decreased crystallinity and increased water absorption. This causes changes such as starch dissolution, swelling of the starch granules, and loss of birefringence. Water penetrates into the starch granules, causing them to become larger and more numerous. When amylose finally separates from the broken starch granules, starch paste is formed, which increases the viscosity of the solution. Factors such as starch, moisture, pH value, and salt will affect the temperature of starch. During the gelatinization process, water serves as a plasticizer (Chung & Liu, 2009; Mahto & Das, 2014; Singh et al. 2011).

2.3.2 Pasting properties

A rheometer or Rapid Visco Analyzer (RVA) can be used to analyze the gelatinization (rheological) properties of starch. When starch is heated under continuous rotation, the size of the starch increases and eventually breaks down, allowing amylose to escape from the matrix and form a viscous paste (Bashir, Jan, & Aggarwal, 2017). The gelatinization ability of starch is key in the food and other industries and is indicated by viscosity (final viscosity, final shear viscosity, loss viscosity, etc.). The gelatinization temperature is the temperature at which viscosity increases significantly. Peak viscosity is the highest viscosity reached. Peak viscosity can affect the ability of starch to bind water and the final attributes of products made with starch. The lowest viscosity value reached in a constant temperature period (i.e., at the highest temperature) is called the minimum viscosity. The viscosity measured at the end of the cooling phase is referred to as the final or cold dough viscosity. This value indicates how effectively the starch will set into a gel or thicken after it has been cooked and then cooled. The difference between the peak viscosity and the final viscosity is known as the breakdown,

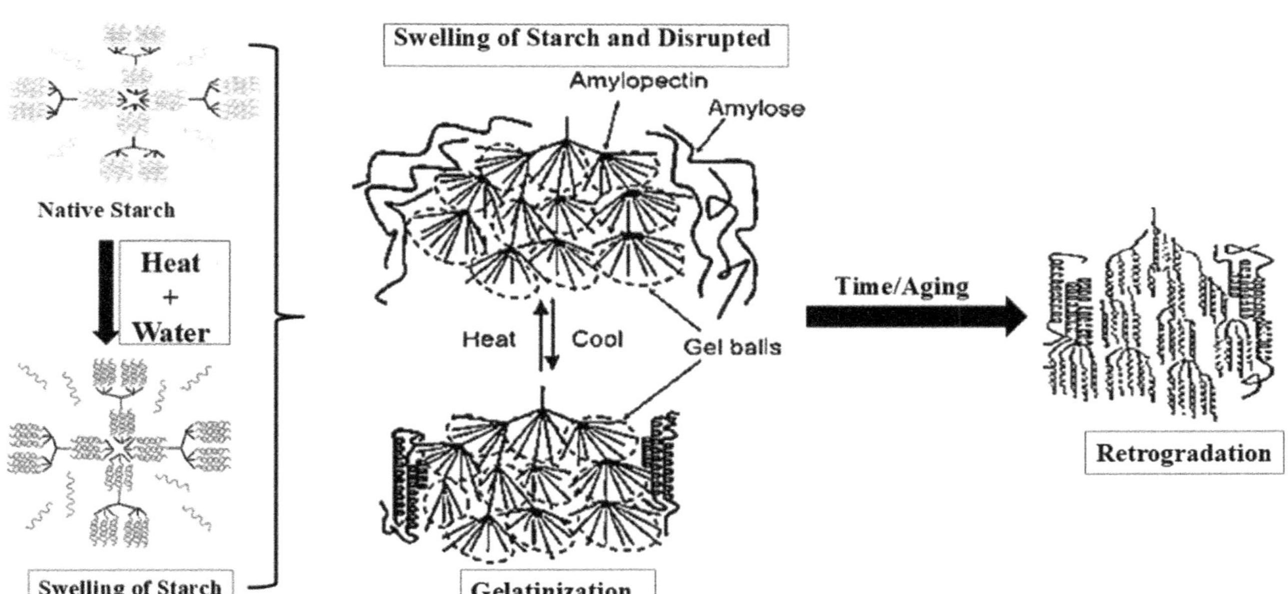

FIGURE 22.1 Gelatinization of starch.

while the setback or total viscosity change is calculated by adding the final viscosity and the minimum viscosity together. Measurement of viscosity loss during storage or thawing can be used as an indicator of recovery or syneresis. The main component contributing to the final viscosity and the high retrogradation rate is amylose. Peak viscosity, gelatinization temperature, and amylose concentration are generally negatively correlated. Disulfide bonds between proteins and the proteins themselves are another factor affecting gel texture. Gel turbidity, Fourier transform infrared (FTIR) spectroscopy or scattered light have been used to monitor retrograde motion (Bashir, Swer, et al., 2017; Sofi et al., 2013).

2.4 Advantages of starch-based packaging

Starch has numerous benefits in food packaging. Since it is a natural polymer derived from plants and foods, it can be utilized in food contact applications. As it is abundant in nature, it is quite cheap in comparison to other synthetic and bio-based polymers on the market. It also does not add any flavor that will affect the quality of the product. It is a widely used polymer with a simplified structure consisting of interconnected α-d-glucose monomers that can be modified chemically, physically, enzymatically or in combination for numerous uses (such as smart and active packaging). Owing to its biodegradable and compostable nature, this material serves as an excellent choice for food packaging and offers a practical way to address existing waste issues. The hydroxyl groups in starch make starch-based polymers very sensitive to moisture, which helps break down hydrogen bonds which cause the material to break down into smaller particles. This form makes starchy products an easy food for bacteria, which in turn creates favorable conditions for bacteria, such as high humidity (55%–60% RH), adequate oxygen, and temperatures that favor the biodegradation of the material (Castro-Aguirre et al., 2016; Grima et al., 2000; Shah et al., 2008).

2.5 Disadvantages of starch-based packaging

In addition to these advantages, starch also has many disadvantages when it comes to food packaging. The low water content of starch is due to its hydrophilicity property, which originates from the hydroxyl groups in its chemical structure (Glenn et al., 2014). When the relative humidity reaches medium to high levels, starchy substances tend to absorb water from their surroundings. Therefore, the polymer matrix first swells and then the hydrogen bonds break, which causes the chains to move and freely fill the gaps in the matrix. Therefore, in addition to the decrease in moisture barrier strength, the oil barrier strength is also affected. For this reason, air-sensitive products and dry foods cannot be packaged with starch products. In a humid environment, the physical integrity of starch-based packaging will gradually deteriorate. In addition, starch polymers have poor properties that limit their application in food packaging and preservation. In addition, public and international organizations such as the Food and Agriculture Organization are concerned about the use of starch in packaging because it can cause food shortages during use (Redazione, 2016). In addition, even if starch is cheap, the price of the final product can be high due to processing and refining. Therefore, starch packaging products have advantages and disadvantages like other products used today. There are many issues to consider when designing starch packaging materials for food packaging.

3. Nanoparticles in starch-based nanocomposites

Starch-based nanosystems are often divided into two main groups: starch nanoparticles (SNp) and starch nanocrystals (SNc). SNc is produced by hydrolyzing the amorphous phase of starch granules and removing most of the amylose until nanosized particles are obtained. In contrast, SNp consists mostly of amorphous particles formed by carefully tuned nanoprecipitation of gelatinized starch. It has been reported that SNp particles are generally larger than SNc particles and the synthesis process may affect their size and morphology. However, the properties of the non-SNc or SNp regions are significantly affected by the origin (Le Corre , 2012a; Le Corre et al., 2012b).

Also, according to Le Corre et al. (2010), it is stated that the plant origin affects the shape and properties of the final nanoparticles. They are used more than natural starches due to their nano-size; Therefore, starches are highly preferred due to their high relative size and high nutritional value (Dularia et al., 2019). Many research works have indicated that SNPs offer a promising method to enhance emulsion stability, increase bioactive ingredient availability in products, and develop effective packaging for manufacturing food-grade films with strong resistance to mechanical stress (Campelo et al., 2020).

3.1 Types of nanoparticles

3.1.1 Nanoclays

Nanoclays, a category of layered silicate nanofillers, offer a practical solution to many limitations of starch-based materials. Valued for their low cost, biocompatibility, and versatility, they are commonly integrated into synthetic and biodegradable films and coatings to significantly improve thermal stability and barrier performance (Nath et al., 2022).

This application is at the forefront of polymer nanocomposite technology for food packaging (Huang et al., 2015). Even minimal additions of nanoclay can dramatically enhance a film's barrier qualities; studies note reductions in gas permeability by 50–500 times (Choudalakis & Gotsis, 2009). This improvement arises because dispersed nanoclay plates force gases and water vapor to follow a longer, more tortuous path through the material, thereby slowing their transfer (Tang et al., 2008). The effectiveness of these nanocomposites is further boosted by using nanoclays with a high aspect ratio and at optimal loading levels (Choudalakis & Gotsis, 2009). Beyond simple barriers, nanoclay composites can be engineered for advanced functionality. By blending them with active compounds, researchers create films with antimicrobial properties (Girdthep et al., 2015). The unique structure of the nanoclays can also help control the release of antibacterial agents from the polymer matrix, potentially extending a package's protective effect.

3.1.2 Cellulose nanocrystals

Cellulose, the most abundant natural polysaccharide, is composed of β-1,4 linked D-glucose chains that form a strong crystalline structure through hydrogen bonding (Klemm et al., 2005). It is found in wood, plants, agricultural waste, and some microorganisms (Shanks, 2014). Cellulose forms as fibrils assemble into microfibrils, with crystalline regions providing strength and amorphous areas imparting flexibility (Habibi et al., 2010). Nanoscale forms like bacterial nanocellulose (BNC), cellulose nanofibrils (CNF), and cellulose nanocrystals (CNC) are now widely studied for their unique properties (Gregory et al., 2021). CNCs, also called nanowhiskers (Eichhorn, 2011), are characterized by high aspect ratios, hydrophilicity, and large surface areas. They retain beneficial properties of cellulose, such as biocompatibility and recyclability, though these are influenced by nanoparticle size and source (Liu et al., 2023). Similar to starch, cellulose is commonly converted to CNCs via a two-step acid hydrolysis process. Introducing a small percentage of CNC wood (1.5%–2.5%) into extruded potato starch has been shown to reinforce the hydrogen bonding between hydroxyl groups. According to Nessi et al. (2019), it reduces swelling and enzyme degradation while improving overall properties. However, when CNC concentration exceeds 5 wt%, nanocrystals aggregate and eventually cause physical damage (Nessi et al., 2019).

According to Azeredo et al. (2017), it was revealed that CNC aggregation is mainly due to hydrogen bonding intermolecular interactions between hydroxyl (OH) groups on the surface of nanocrystals. According to Trinh et al. (2022), it was found that beeswax CNC/cornstarch coating of canned fruits improved their tensile strength and modulus, as well as air and water barrier capacity. Although, their hydrophilicity and self-assembly tendency hinder this process. To address these issues and provide additional functionality, several chemical modifications and copolymerization strategies have been described (Chen, Hao, et al., 2019), such as: surface modification of CNC using vinyl silane grafting (Dhali et al., 2022); combination of CNC with citric acid and amidation of chitosan result in improved nanocomposites with higher tensile strength, lower water sorption and water vapor permeability and resistance to Escherichia coli and Staphylococcus aureus (Azeredo et al., 2017).

3.1.3 Metal nanoparticles

3.1.3.1 Silver nanoparticles (Ag NPs)

Various polymeric nanocomposites have been formulated with metal nanoparticles, including gold, copper, silver, palladium, and cobalt, for various applications when it comes to metallic nano-reinforcements. Silver is by far the most commonly used metal nano-reinforcement (Peponi et al., 2014). Silver nanoparticles are an important additive to starch as they can be produced using green technology (Rafique et al., 2017). Ahmed et al. (2016) showed that this synthesis method also has the advantages of being low cost, not requiring chemicals, high pressure, energy or heat, and being a simple one-step method that can be easily expanded for large-scale synthesis. Silver is a potent broad-spectrum antimicrobial, effective against over 650 bacterial species including both Gram-positive and Gram-negative types as well as viruses and fungi. Silver nanoparticles are more effective as compared to essential oils and nano-emulsions in upgrading the antibacterial attributes of thermoplastic starch used as the main component in food packaging. Several edible films containing either silver nanoparticles or essential oils were made using solvent casting and starch as a matrix. Silver nanoparticle-containing edible films showed an inhibitory zone width that was almost three times wider, according to measurements of these films antibacterial activity (Hassan et al., 2011).

Integrating silver nanoparticles into starch-based films enhances their mechanical strength and functional properties. Research shows these nanocomposites exhibit improved physical resistance, confirming silver's role in reinforcing material structure (Mohseni et al., 2020). Recent eco-friendly synthesis methods, such as using *S. asoca* leaf extract, produce silver nanoparticles that, when added to starch films, lower water solubility and vapor permeability while imparting antibacterial activity (Khurshid et al., 2022). Similarly, incorporating silver into mushroom starch/agar composites boosts their antibacterial effect, water vapor barrier, and hydrophobicity (Roy & Rhim, 2022).

3.1.3.2 Zinc oxide (ZnO) nanoparticles

Of the various metal oxide nanoparticles studied for food packaging, zinc oxide (ZnO) has emerged as a leading candidate. This is due to its non-toxic nature, biocompatibility, stability under heat and UV light, and its official designation as "Generally Recognized as Safe" (GRAS) by the U.S. Food and Drug Administration (Yadav et al., 2022). When integrated into starch-based films, ZnO nanoparticles impart significant functional improvements. For instance, incorporating ZnO into buckwheat starch films enhanced their physical strength, water resistance, and antimicrobial activity. Films with 3% ZnO effectively inhibited *L. monocytogenes*, helping preserve the freshness of fresh-cut mushrooms (Kim & Song, 2018). Similarly, ZnO has been used to extend bread shelf life and improve the tensile strength of starch films (Anugrahwidya et al., 2022). Research shows these composites can be fine-tuned. While ZnO alone may degrade slowly in certain environments, adding materials like chitosan can accelerate biodegradation while further improving mechanical properties. The combination of ZnO nanoparticles with fennel essential oil (FEO) in potato starch films, for example, created a material with superior tensile strength, enhanced barrier properties against water vapor and oxygen, and strong antibacterial and antifungal activity (Babapour et al., 2021). These innovations extend beyond flexible films. A cornstarch coating containing 5% ZnO nanoparticles created hydrophobic and antimicrobial paper for food packaging, with zinc migration remaining within legal limits (Ni et al., 2018). Direct coating of mangoes with a ZnO-starch blend successfully prolonged shelf life by suppressing fungal decay and maintaining fruit quality (Rosman et al., 2022). Together, these applications highlight ZnO's versatile role in developing smarter, more active biodegradable packaging.

3.1.3.3 Titanium dioxide (TiO_2) nanoparticles

Titanium dioxide (TiO_2) is widely employed as a white pigment in paints and coatings due to its high refractive index and low light absorption, which together produce strong opacity and thermal stability . Beyond its visual role, TiO_2 nanoparticles are increasingly used in packaging to enhance the properties of biodegradable films, including those made from starch (Vallejo-Montesinos et al., 2020).

A key advantage of TiO_2 in food packaging is its light-activated antimicrobial effect. When exposed to sufficient light, TiO_2 generates reactive oxygen species (ROS) that effectively inactivate bacteria, viruses, and fungi. Studies have demonstrated its efficacy against pathogens such as *E. coli*, *S. aureus*, *P. putida*, and *L. innocua* (Ni et al., 2018). For example, adding just 0.01% w/w TiO_2 to starch films helped extend the shelf life of fresh fruit more effectively than petroleum-based films, with the added benefit that the composite fully biodegraded within 2 weeks, leaving no nanoparticle residue in the soil (Tunma, 2018). However, higher nanoparticle concentrations can make the packaging brittle. The combination of TiO_2 with natural extracts can further improve functionality. Arezoo et al. (2020) developed a sago starch film containing both TiO_2 nanoparticles and cinnamon essential oil (CEO), which showed strong antibacterial action against *S. aureus*, *S. typhimurium*, and *E. coli*. This blend of antioxidant and antimicrobial properties helped better preserve fresh pistachios and extend their shelf life.

3.1.3.4 Copper oxide (CuO) nanoparticles

They serve as more than just the polymeric matrix's reinforcement. Nanoparticles possess significant antibacterial capabilities because of their small size and extensive surface area, enabling them to interact closely with cell membranes (Ramyadevi et al., 2012). According to Garcia et al. (2018) copper ions cause significant damage to the structural components of microbes due to their rapid electron-acceptance and electron-donation properties.

Lots of researches demonstrated that edible films containing Copper oxide nanoparticles possess higher antioxidant attributes compared to other films. For example, a case study in medicine was presented by Rehana et al. (2017) to prevent the development of cancerous cells. In a different scenario, nanoparticles were incorporated into a chitosan edible film as an enhancement. In contrast to clean chitosan films, reinforced films exhibited stronger antibacterial effects against both gram-positive and gram-negative bacteria, as well as enhanced antioxidant attributes and increased thermal stability (Revathi & Thambidurai, 2019). When incorporated as reinforcements or additions in thermoplastic starch matrices, they improve mechanical and antioxidant properties, while also generating a superior antibacterial response (Bumbudsanpharoke et al., 2023).

3.2 Properties of starch nanoparticles

3.2.1 Amylose content

The functional properties of starch, including its ability to form gels and act as an adhesive, are directly influenced by its botanical source (Chavan et al., 2021). Processing starch into nanoparticles can significantly alter its composition, particularly its amylose content. For instance, research on Andean potato starch demonstrated that acid hydrolysis reduced the amylose content in the resulting nanoparticles by as much as 80% compared to the native starch (Torres et al., 2019). A further combination of hydrolysis and ultrasonication led to an even greater decline. This is attributed to the hydrolysis preferentially degrading the amorphous regions formed by amylose molecules, while the crystalline areas of the granule remain more resistant (Kim et al., 2015). Ultimately, the characteristics of starch nanoparticles are governed by the amylose-to-amylopectin ratio, the specific organization of the polymer chains, and their resulting crystalline structure.

3.2.2 Particle morphology and size distribution

The shape and size of starch nanoparticles (SNPs) vary based on their plant source and how they are prepared. They can appear round, oval, irregular, or porous, and are typically analyzed using microscopy techniques such as SEM and TEM (Alcázar-Alay & Meireles, 2015; Sun, 2018). Generally, smaller starch granules yield smaller SNPs, which often possess distinct functional properties suited for different applications (Kim et al., 2015; Torres et al., 2019). Additionally, higher amylose content tends to produce larger nanoparticles (Kim et al., 2015).

3.2.3 Starch crystallinity

The crystallinity of starch nanoparticles (SNPs) reflects the proportion of ordered crystalline regions relative to the total mass, primarily influenced by amylopectin content (Dufresne, 2014; Kumari et al., 2020). X-ray diffraction (XRD) is used to analyze this crystalline structure, classifying starch into types A, B, C, and V—based on diffraction patterns. Typically, about 70% of starch exists in an amorphous state (Liu et al., 2015). These crystal types vary by botanical origin: Type A, common in cereals, has a densely packed structure; Type B, found in tubers and high-amylose starches, forms a more open hexagonal arrangement; Type C is a mix of A and B; and Type V appears under certain saturation conditions (Dufresne, 2014). During SNP production, physical treatments such as gamma irradiation, ultrasound, and high-pressure homogenization often reduce crystallinity, leading to a more amorphous structure.

3.2.4 Thermal properties

The thermal properties of starch nanoparticles (SNPs) are commonly studied using differential scanning calorimetry (DSC) and thermogravimetric analysis (TGA). These analyses help define suitable processing conditions and predict the durability of final products (Tagliapietra et al., 2021).Thermal stability is strongly influenced by how SNPs are produced. Treatments like gamma irradiation and ultrasonication generally lower degradation temperatures, often due to the breakdown of starch granules and weakened hydrogen bonds (Ahmad, Lim, et al., 2020; Da Silva et al., 2017; Lamanna et al., 2013).On the other hand, methods such as high-pressure homogenization can improve thermal resistance (Ahmad, Lim, et al., 2020), while enzymatic hydrolysis combined with recrystallization tends to reduce stability (Sun, 2018).

3.2.5 Functional properties of starch nanoparticles

The characteristics and abilities of SNPs have been studied using different tests and evaluations. Winarti et al. (2014) produced arrowroot starch nanoparticles (SNPs) using butanol-assisted nanoprecipitation and evaluated their properties. Their analysis showed an increase in both swelling volume, which ranged from 5.28 to 7.92 g/g, and solubility, between 9.43% and 16.89%, compared to the native starch nanoparticles. The swelling volume was notably enhanced in different tests, including those on mechanically treated potato and cassava SNPs (Szymonska and Targosz-Korecka, 2009). As noted by Jeong and Shin (2018), SNPs from waxy rice created through the acid hydrolysis method exhibit reduced capacity for water binding. Surprisingly, there was no significant variation in the duration of hydrolysis.

3.2.6 Digestibility properties

In order to digest starch granules, enzymes must diffuse into the substrate, the substrate's porosity must be altered, enzymes must be absorbed into the starch material, and hydrolysis must take place. The in vitro digestibility of SNPs was improved compared to conventional starch, which was mainly due to the increase in the surface area of nano starch (Li et al., 2014). Research indicates that treating starch with pullulanase can increase the digestibility of starch nanoparticles

(SNPs) to 41%–43% within 12–24 h, due to high digestibility and short retrogradation (Suriya et al., 2018). Generally, starch digestion depends on factors like content, granule size, and amylose ratio(Ding et al., 2016). However, treatments such as linearization can reduce starch content, as seen in arrowroot SNPs (Winarti et al., 2014). Similarly, maize SNPs show low hydrolysis, likely because recrystallization increases short-chain glucans that resist digestion (Liu et al., 2016).

4. Enhancement of starch-based films and coatings

Indeed, starch-based films and coatings are increasingly recognized as a sustainable substitute for traditional petroleum-derived plastics. Their inherent biodegradability and minimal ecological footprint position them as promising materials for broader implementation, particularly in areas like food packaging. To fully realize their potential, however, key properties such as mechanical strength, controlled degradability, and antimicrobial activity must be enhanced. Incorporating nanoparticles offers a promising route to improve these characteristics through synergistic effects. This review examines recent advances in nanoparticle-modified starch films, focusing on how they improve mechanical, barrier, and antibacterial performance.

4.1 Mechanical properties

To determine whether a starch-based film is practical for real-world use, researchers evaluate several key mechanical properties. These include tensile strength (resistance to pulling), elongation at break (how much it can stretch), Young's modulus (stiffness), and dynamic mechanical parameters like tan δ (which relates to energy dissipation). Together, these measurements reveal the material's durability, flexibility, and overall structural integrity—qualities essential for surviving the physical demands of handling, shipping, and storage (Salmieri et al., 2014; Santana et al., 2019). TS and EAB are particularly important indicators: tensile strength reflects resistance to breaking, while elongation at break describes flexibility before failure. Instruments such as texture analyzers, universal testing machines, and dynamic mechanical analyzers (DMA) are commonly used to measure these and other properties like water vapor permeability (WVP) and solubility. Research shows that incorporating nanomaterials can significantly alter these characteristics. For example, while nanocellulose can enhance tensile strength, starch nanoparticles may reduce elongation at break, as observed by Sanyang et al. (2015). Understanding these relationships helps tailor starch films to meet practical packaging needs and consumer expectations.

4.2 Barrier properties

Effective barrier properties are crucial in food packaging, as they safeguard against microbial contamination and regulate moisture exchange, key factors in preserving quality and extending shelf life . However, many films and coatings made from proteins and carbohydrates have a naturally hydrophilic nature, which can compromise their moisture barrier effectiveness (Ribeiro et al., 2007). To overcome this limitation, hydrophobic components such as lipids are often incorporated to improve water resistance. Additionally, surfactants can be used to enhance a film's moisture barrier and adhesion while reducing surface tension, leading to more uniform and functional coatings (Andreuccetti et al., 2010).

Effective food packaging relies heavily on barrier properties, particularly resistance to water vapor and oxygen, to protect contents and extend shelf life. Starch-based films, however, face a fundamental challenge: their natural hydrophilicity often results in high water vapor permeability (WVP), which can limit their protective function. Research on mung bean starch (MBS) films, for example, shows their WVP is typically higher than conventional plastics like polyethylene (Majzoobi et al., 2015). This is largely due to the hydrophilic nature of starch and, in composite films, the presence of components like lipids and proteins that can create a porous structure (Sandhu et al., 2020). Therefore, a primary goal in developing starch packaging is to reduce WVP.

Several modification strategies show promise.

- Chemical modification: Cross-linking starch with agents like sodium trimetaphosphate can successfully lower WVP (Colussi et al., 2017).
- Nanocomposites: Incorporating nanoparticles like montmorillonite nanoclay or using cross-linkers like citric acid can reduce hygroscopicity and improve the water barrier (Rompothi et al., 2017; Suriyatem et al., 2019).
- Plasticizers and Additives: While glycerol is a common plasticizer that increases flexibility, it can also raise WVP. Alternatives like sorbitol may offer better barrier properties (Thirathumthavorn et al., 2010). Adding hydrophobic substances such as lipids, fats, or certain essential oils creates a more tortuous path for water vapor, significantly improving moisture resistance (Basiak et al., 2015; Do Evangelho et al., 2019).

The oxygen barrier is equally critical. Native starch films can have favorable oxygen vapor permeability (OVP), but this property is also influenced by formulation. The addition of lipids or oils can dramatically reduce oxygen permeability by introducing hydrophobic barriers within the film matrix (Basiak et al., 2015). In summary, while native starch films have inherent permeability limitations, their barrier performance can be effectively engineered through cross-linking, the addition of nanofillers, and the strategic use of hydrophobic components and plasticizers.

4.3 Optical materials

The visual clarity of starch-based films is an important functional property, especially for edible coatings and transparent packaging where product appearance matters. In general, plasticized starch can form simple, digestible films with good transparency, but their optical properties can be deliberately adjusted through formulation and processing (Gontard, 1991). Opacity and transparency are primarily influenced by the film's composition. The addition of nanoparticles, such as calcium carbonate, typically increases opacity and reduces light transmittance (Mbey et al., 2012; Shi et al., 2013). Conversely, certain additives like chitosan can enhance a film's transparency (Oleyaei et al., 2016). Plasticizers also play a key role: while glycerol is essential for flexibility, increasing its concentration usually decreases transparency, as demonstrated in studies measuring visible light transmission (Geleta et al., 2020).

Other ingredients introduce specific visual characteristics. Lipids and oils, such as sunflower oil, tend to reduce transparency due to light scattering (Ahmad Shapi'i & Othman, 2016). Proteins can sometimes impart a yellow hue, potentially through Maillard reactions (Cheftel et al., 1985). The botanical source of the starch and the drying conditions further affect the final film's clarity and uniformity (Prabhu et al., 2021). Ultimately, by carefully selecting and balancing components—such as starch, gums, plasticizers, and nanoparticles films can be engineered to achieve the desired optical properties, whether high transparency for visibility or increased opacity to protect contents from light-induced degradation.

5. Fabrication techniques for starch-based nanocomposites

Diverse biopolymer-based films and coatings have been created using a variety of manufacturing processes, each of which affects the material's mechanical barrier and antibacterial qualities. Solvent casting, electrospinning, extrusion, and layer-by-layer assembly are notable instances of these methods (Fig. 22.2). As developments in this area continue to improve the performance of bio-based polymer films, these techniques offer clear benefits dependent on the intended use (Table 22.1). The fabrication methods covered in this part are restricted to those that are pertinent to the creation of films and edible coatings designed for use in food packaging. Fabrication techniques with film properties.

5.1 Solvent casting

The solvent casting method (SCM) is a traditional, popular, laboratory-scale method and technology used to produce biocomposite films. Although SCM is a very old technology, multilayer and electrospinning methods are widely used in

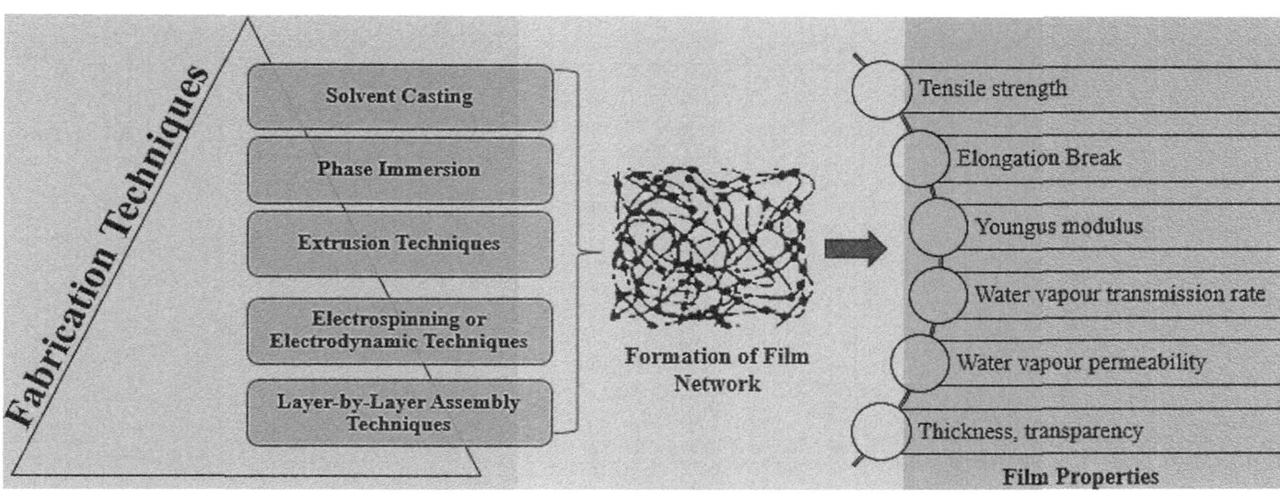

FIGURE 22.2 Fabrication techniques with film properties.

TABLE 22.1 Comparison of various fabrication techniques based on description, advantages, disadvantages, and applications.

Fabrication techniques	Description	Advantages	Disadvantages	Application	References
Solvent casting	Casting the starch into a mold after dissolving it in an appropriate solvent and letting the solvent evaporate.	Easy, inexpensive process that creates consistent coatings and films.	Takes a long time and requires the use of organic solvents; the drying conditions are crucial for quality.	Biodegradable materials, biomedical uses, and packaging films.	(Lizundia et al., 2020; Tan et al., 2022)
Phase immersion casting	To create a gel-like structure and promote phase separation, a starch solution is submerged in a non-solvent.	Makes it possible to create porous structures, which are useful for applications requiring controlled release.	Restricted scalability; circumstances may have an impact on the phase procedure.	Scaffolding in tissue engineering, and drug delivery methods.	(Gao et al., 2023; Kim et al., 2023)
Extrusion	To create continuous forms, starch is melted, combined with additives and nanofillers, and then forced through a die.	Good dispersion of nanofillers, scalable and continuous manufacturing, and product form diversity.	Higher shear, which calls for a lot of equipment and energy, can break down starch.	Consumer items, car parts, and food packaging.	(Chachlioutaki et al., 2020; Li et al., 2020)
Electrospinning	By applying a voltage to a polymer solution, electrostatic forces are used to form nanofibers.	Generates nanofibers with a large surface area that are appropriate for tissue engineering and fine filtration applications.	Fibrous structures alone; complicated machinery and low manufacturing rate.	Filtration materials, sensors, and tissue engineering.	(Fei et al., 2017; Li et al., 2021)
Layer-by-layer assembly	Building up layers by alternating the adsorption of positively and negatively charged molecules, usually using compounds based on starch.	Enables fine-grained control over the content and thickness of layers; ideal for intricate architectures.	Time-consuming; interlayer adhesion problems may arise; assembly conditions must be carefully controlled.	Coatings, medication delivery methods, and sensor applications.	(Lipton et al., 2020; Zhang et al., 2019)

the production of polysaccharide films today. Solvent casting is a more elegant process to produce thin biopolymer-based films with excellent structure. The basic principle of this synthetic process is the crystallization of biopolymers in immiscible solvents, which should produce a low-content, viscous liquid after homogenization and form a film that can be easily removed with the casting support (Lizundia et al., 2020).

The advantages of the solvent casting method over the extrusion method are: homogeneous thickness distribution, high optical speed and very good. However, this technique has more disadvantages than advantages; long drying time, shrinkage and brittleness of the film after drying, uneven distribution of the drug, and many other disadvantages of the film. One of the disadvantages of this technology is the "rippling effect", which causes the shear stress in the film to become problematic and the consistency to be lost. The appearance of bubbles, commonly called "nucleate boiling", can be caused by excessive drying (Lizundia et al., 2020). These problems can be solved by adding plasticizers to the polymer solution. This will reduce the cohesive intermolecular forces in the polymer network, thereby reducing the tensile strength of the film and increasing its flexibility (Tan et al., 2022).

5.2 Phase immersion casting technique

Immersion phase processes generally transform liquid (polymer and solvent) solutions into solids due to molecular exchange between the solvent and the insoluble matter. After precipitation, the films were washed and dried (Gao et al., 2023). "Liquid-liquid stratification" refers to a phenomenon that occurs throughout the entire process. The separation of the liquid-liquid mixture does not have a significant effect on the quality of the resulting film; for example, this method can control the porosity of protective films (Kim et al., 2023). The advantages of this technology include low energy cost, shorter film production time, and ease of use. It also has great potential for future applications involving measurement.

5.3 Extrusion technology

According to Scarfato et al. (2017), the co-extrusion process was born to produce multilayer films using more than two extruders and feeding them through a die. The number of different components that make up the co-extruded film determines the number of extruders required. Only two extruders are needed to produce two layers, and three-layer co-extrusion includes a barrier core (Dziadowiec et al., 2023). The adapter is attached to the film extrusion die. A good mold ensures that the sheep is completely flowable, clear and smooth. Most film extrusion dies are subject to heat and temperature control (Morris, 2022). In air extrusion, the melt enters a circular die from the bottom. The sheep is extruded from a circular die and passes through a spiral groove around the mandrel and through the die. Increasing the number of spiral grooves can increase the breakage of the sheep (Ghosh et al., 2016). Most polyethylene-blown film molds are oriented vertically to force the tube upward and outward. According to Li et al. (2020), the distance between the ring mold and the mandrel is between 0.5 and 3 mm. In some molds, the mandrel can be moved along the length of the mold to change the opening process. The mandrel in the mold can be moved along the length to change the mold mouth. Most molds require mandrel adjustment by changing the mold gap. Production can be easily increased by using a larger ring, but special instructions and tee lines will be more difficult due to irregularity. There is also a challenge in supporting films with a thickness of less than 0.5 mils (Chachlioutaki et al., 2020).

5.4 Electrospinning or electrodynamic method

Electrospinning technology has gained popularity in recent years. Temperature, effect of electric current, effect of fluid flow, polymer concentration, solution viscosity, distance between the needle and the substrate, and needle diameter are some of the changes that affect this process (Yadav et al., 2019). This method requires a syringe that also acts as a pump, a ground or second electrode fitting, and a high-voltage power source with a good terminal. The beginning of the electrospinning stage is the injection of electric charges into the polymer solution using a syringe. As electrons are generated, the charges form a conical shape called a Taylor cone (Li et al., 2021). On the other hand, ultrafine nanofibers emerge from the tip of the needle, forming a cone, where they are collected by the spinning material. The electric charge destabilizes the liquid medium. The surface tension is simultaneously balanced by mutual repulsion forces, causing the liquid to move in the direction of the electric field. Due to the very high surface-to-volume ratio and the maximum number of intrafiber/interfiber pores, the droplet material is transformed into a spherical plane, making this process more efficient than casting and dipping (Yadav et al., 2019). Electrospinning is a labor-intensive process; only because the production process takes more than 24 hours to convert dry material into film. This method has significant advantages over the traditional casting method. Films produced by electrospinning have superior properties compared to films produced by solvent spinning, including smooth surfaces, increased flexibility, viscoelastic properties, and moderate strength (Fei et al., 2017).

5.5 Layer-by-layer assembly techniques

A widely used technique for creating thin films is layer-by-layer assembly, which is straightforward, favored, and recognized in commercial applications. It depends on various factors including electrostatic interactions, hydrogen-hydrogen bonds, covalent bonds, hydrophobic attractions, or molecular entanglements (Zhang et al., 2019). The type of chemicals that assist in creating the film dictates how the polymers engage with one another. Producing thin films or layers typically involves a sequence of several steps for forming films on the surface. This method is among the most commonly employed techniques for creating thin films through the electrostatic attraction between positive and negative electrode layers (Tian et al., 2019). In this approach, the substrate is soaked in a negatively charged solution, leading to the formation of positively charged compounds that are drawn toward the substrate's surface. Once the surplus material is eliminated, the positively charged substrate creates a negatively charged entity in a different solution, aiding in the development of new structures. Through the continual repetition of this method, film layers can be created (Zhang et al., 2019). In comparison to alternative production techniques, like creating thin films through the selection of various types and quantities of film-forming substances and altering their physicochemical and structural characteristics (thickness, strength), the layer-by-layer method offers numerous benefits. This approach does not need specialized tools and is straightforward and affordable (Lipton et al., 2020).

6. Applications of biodegradable starch-based nanocomposites

Starch-based biodegradable nanocomposites find applications in numerous sectors such as packaging, biomedical, agriculture, and environmental industries (Fig. 22.3).

6.1 Agricultural

Scientists have made significant advancements in agricultural technology by creating biodegradable films that meet various farming needs. These films can be used for mulching, covering greenhouses, and applying crop-boosting substances such as fertilizers and pesticides, ultimately supporting agricultural output. Control soil temperature and restrict plant growth (Gamage et al., 2022). Various important elements need to be considered concerning mulches. For example, their relationship with water is essential, considering factors such as water vapor permeability, contact angle, and ability to absorb or resist water. Their ability to withstand environmental changes such as temperature fluctuations and expansion/contraction also plays a role. Mulches must have low water vapor permeability to reduce evaporation and help retain soil moisture. Improving thermal stability is essential for these films since they are often affected by changes in outside temperatures (Merino et al., 2019).

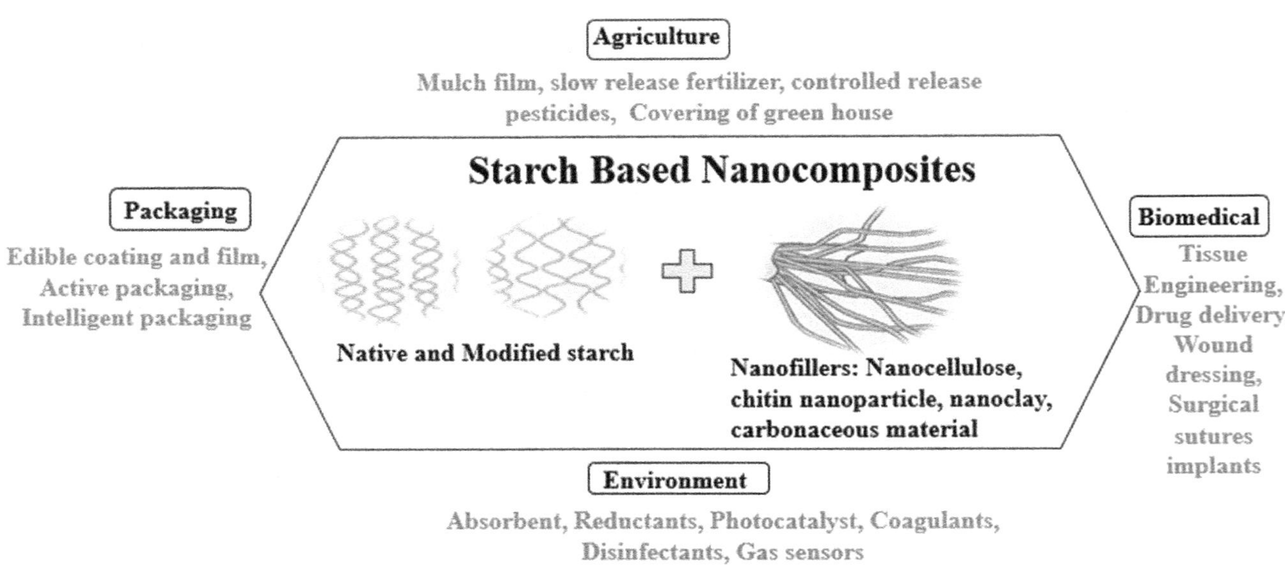

FIGURE 22.3 Uses of starch based on nanocomposites with application.

Nonetheless, these substances frequently encounter issues in reaching their intended target because of processes like degradation, erosion, and leaching that diminish their efficacy. Excessive loss may impact plant development. To address these issues, researchers have suggested a novel approach that merges nanomaterials with biopolymers. Enhance water safety. These slow-release systems may diminish losses and enhance the utilization of vital nutrients and pesticides by embedding nanomaterials within the biopolymer matrix. The ecological impacts appear encouraging for farming. Examine how light and moisture interact with coatings created from cornstarch. It was discovered that incorporating chitosan/bentonite nanofillers into standard and oxidized thermoplastic cornstarch enhanced antibacterial, radiation, and water resistance while not altering the healing properties or water vapor permeability. The inclusion of bentonite/chitosan in two distinct matrices (native and oxidized thermoplastic corn starch) enhances the crystallinity and also modestly increases thermal stability in comparison to either matrix on its own. High absorbency hydrogels are commonly utilized in slow-release fertilizers for dual purposes because of their ability to retain water and nutrients (Merino et al., 2018). Pimsen et al. (2021) found that a nanocomposite containing chitosan (CS), sago starch (ST), and nano zeolite (NZ) improved water retention. The nanocomposite performed better than the direct application of urea and showed good growth in Philodendron spp., proving its effectiveness as a slow-release fertilizer.

6.2 Packaging

Food packaging plays an important role in preserving and improving food quality throughout production and storage using technologies such as gas and fire detectors, and lightning detectors. It also protects food from moisture, temperature changes and other environmental factors (Galstyan et al., 2018). Starch is often used as food packaging due to its strong, translucent properties and lack of taste or odor (Yousefi et al., 2019). Nonetheless, its use is constrained by certain limitations arising from its fragility and inadequate resistance to moisture. To address these restrictions, nanoparticles have emerged as vital materials in creating films and coatings, along with active and intelligent solutions. These nanoparticles enhance mechanical strength, hydrophobic properties, moisture resistance, oxygen barrier capabilities, UV protection, and thermal stability (Pirsa et al., 2018).

Inorganic nanofillers comprise different metals like silver (Ag), gold (Au), and copper (Cu), in addition to zinc oxide (ZnO), titanium dioxide (TiO2), silver oxide (Ag2O), magnesium oxide (MgO), vanadium oxide (VOx), copper oxide (CuO), and tin oxide (SnO2). Conversely, organic nanofillers comprise nanoclays (like montmorillonite and halloysite), natural biopolymers including chitosan and cellulose, and antimicrobial substances such as nisin, which are widely investigated for food packaging purposes (Galstyan et al., 2018).

Films intended for packaging uses are typically assessed considering their water vapor and oxygen barrier characteristics along with their heat sealability (Patil et al., 2022). The film is both flexible and mechanically resilient, shielding against physical harm like impacts. Research indicates that incorporating titanium dioxide nanoparticles (TiO2 NPs) into potato starch-based composite films enhances their tensile strength and UV resistance, while simultaneously decreasing their water solubility, hygroscopicity, and water vapor permeability, applicable for food packaging purposes. Likewise, the combination of starch with sodium montmorillonite (Na-MMT) and TiO2 NPs provides enhanced artificial characteristics and boosts barriers against water vapor and UV light, while decreasing hydrophilicity, making it suitable for food packaging (Yousefi et al., 2019). This serves as a beneficial UV barrier packaging at elevated temperatures (Babaei-Ghazvini et al., 2018). The addition of flaxseed polyol improved the contact properties, water absorption capacity, thermal stability, and biodegradability of PVA/corn starch. Ag NPs also boost their antibacterial effectiveness against different pathogens such as Proteus mirabilis and Klebsiella pneumoniae (Sharmin et al., 2020).

Polyethyl methacrylate-co-starch nanocomposite films refine mechanical strength, flexibility, and significant resistance to water vapor and oxygen, which helps reduce the oxidation of oil during storage (Mohanty & Swain, 2019). This corn starch can store usable oil for more than 3 months at room temperature (27 ± 3 °C and $65 \pm 5\%$ RH) without affecting the oil quality such as color, viscosity, or deterioration (Patil et al., 2022). Starch/CNC nanocomposites are used as food packaging materials since cellulose nanocrystals (CNCs) made from sugarcane bagasse and mixed with starch improve the material properties, water resistance and reduce water affinity (Slavutsky & Bertuzzi, 2014). A combination of Cassava starch/glycerol, cinnamon essential oil, sodium bentonite nanocomposites showed antimicrobial activity against *Salmonella typhimurium*. It also reduced the bacterial growth in pork stored in cold environments (Iamareerat et al., 2018).

6.3 Biomedical

Biodegradable polymers, especially starch bio-nanocomposites, are increasingly used in tissue engineering, medicine and other industries due to excellent properties, biocompatibility and biodegradability. These include dispensing, dressings, surgical sutures and implants. These materials are ideal for medical applications because they can produce non-toxic, biodegradable by-products (Taherimehr et al., 2021).

One of the key benefits of biopolymers is the ability to optimize the healing process and eliminate the need for surgical implant removal. These plant materials and their biodegradable components should be biocompatible and noncytotoxic (Taherimehr et al., 2021). For example, when 10% bioactive β-tricalcium phosphate (β TCP) nanoparticles were added to thermoplastic starch (TPS), mechanical attributes were significantly improved, exquisite biocompatibility was achieved, and no cytotoxic effects were observed (Taherimehr et al., 2021). Additionally, Waghmare et al. (2018) reported the success of starch nanofiber scaffolds for wound healing applications using electrospinning technology was reported. However, its brittleness and limited flexibility currently pose challenges. One way to reduce brittleness is to combine hydroxyapatite with starch, as the polarity of starch improves the adhesion between the two materials. Sadjadi et al. (2010) developed a nanocomposite containing starch and nanohydroxyapatite that mimics natural bone's mechanical and biological attributes. According to Abdel-Halim and Al-Deyab (2014) Nanocomposite hydrogel produced from silver nanoparticles (Ag NP), starch and polyacrylamide is effective against bacteria such as Staphylococcus aureus and Escherichia coli and pathogens such as Aflatoxin and Candida albicans. Furthermore, a mixture of polyvinyl alcohol (PVA) and starch supported with green synthesized Ag NPs (using palm fruit extract) has been shown to have potential as a bandage as it can increase swelling and moisture retention, reduce water vapor permeability and promote wound healing. It also helps prevent wound desiccation and has antibacterial properties against Escherichia coli and Staphylococcus aureus (Batool et al., 2019). Mallakpour (2018) prepared glucose-modified starch/multiwalled carbon nanotube (MWCNT) nanocomposites (MWCNT-G).

6.4 Environment

Agricultural and industrial activities result in the formation of pollutants, including heavy and trace metals such as Cr6+, Hg2+, Zn2+, Pb2+, Co2+, Cd2+, and Cu2+. Other pollutants include organic chemicals such as dyes, pesticides, fertilizers, fragrances, aliphatic hydrocarbons, volatile organic compounds (VOCs) and oil spills, and pathogens and pollutants (Guerra et al., 2018). These materials include metals such as silver, gold, and palladium nanoparticles, bimetallic compounds such as silver-gold, metal oxides such as titanium dioxide, zinc oxide, iron oxide, and manganese oxide, zero-valent metals at the nanometer scale, and carbon-based materials such as carbon materials. Single-walled and multi-walled carbon nanotubes (CNTs), graphene and graphene oxide. Also, nano clays (such as montmorillonite and halloysite) and biopolymers (such as chitin and cellulose nanowhiskers) are important nanofillers. These nanocomposites are versatile materials that can be recycled and used in applications such as filtration, adsorption, catalysis, coagulation and flocculation, disinfection, and oxygen sensing to monitor and treat various bacteria such as dyes, heavy metal ions (such as arsenic, and lead, chromium, copper, cadmium, mercury, nickel, and cobalt), various fragrances, fertilizers such as urea, and other solvents play an important role in preventing organic pollution (Adeola & Nomngongo, 2022). A polymer derived from hydroxyethyl starch and poly (acrylamide-acrylic acid) has proven effective in the eco-friendly creation of silver-gold bimetallic nanocomposites. These composites exhibit strong catalytic performance, successfully converting 4-nitrophenol into 4-aminophenol and aiding in the synthesis of azo dyes like Congo red, Sudan-1, and methyl orange by breaking azo bonds. This highlights its promising role in water treatment technologies (Tripathy et al., 2017). In a related area of research, Gomes et al. (2015) developed a hydrogel composite from starch and cellulose nanowhiskers, which demonstrated a high capacity for adsorbing methylene blue dye, further underscoring the value of starch-based materials in addressing water pollution.

The reusable adsorbents prepared by the present invention can effectively remove malachite green (MG) and similar cationic dyes from water using starch-grafted polyacrylamide (PAM), graphene oxide (GO), and hydroxyapatite nanoparticles (nHAp). Integrating nano-hydroxyapatite (nHAp) into PAM/GO composites enhanced their biocompatibility, though it also reduced other key properties as the concentration increased namely biodegradability, porosity, water content, and water sorption capacity.

7. Conclusion

Edible starch-based films offer a biodegradable and renewable alternative to traditional plastics for food packaging. Their transparency, flexibility, and ability to act as a barrier make them suitable for specific uses like wraps, pouches, and coatings for fresh produce. However, their wider application is limited by sensitivity to moisture, relatively low mechanical strength, and concerns about stability. Research is actively addressing these limitations. Incorporating nanomaterials such as nanocellulose or nanoclays can improve strength and reduce water sensitivity, while adding natural oils can introduce antibacterial properties. Innovations like integrating pH-sensitive indicators are also creating "smart" packaging that can monitor food freshness. For these materials to move beyond the lab, challenges in large-scale production, cost, and compliance with food safety regulations must be resolved. Future progress hinges on scaling up manufacturing, enhancing performance for diverse foods, and aligning with circular economy principles using waste-derived starch. Therefore, continued development positions starch-based packaging as a practical component in achieving more sustainable food systems.

References

Abdel-Halim, E. S., & Al-Deyab, S. S. (2014). Antimicrobial activity of silver/starch/polyacrylamide nanocomposite. *International Journal of Biological Macromolecules, 68*, 33–38.

Acevedo-Guevara, L., Nieto-Suaza, L., Sanchez, L. T., Pinzon, M. I., & Villa, C. C. (2018). Development of native and modified banana starch nanoparticles as vehicles for curcumin. *International Journal of Biological Macromolecules, 111*, 498–504. https://doi.org/10.1016/j.ijbiomac.2018.01.063

Adeola, A. O., & Nomngongo, P. N. (2022). Advanced polymeric nanocomposites for water treatment applications: A holistic perspective. *Polymers, 14* (12), 2462.

Ahmad, M., Gani, A., Hassan, I., Huang, Q., & Shabbir, H. (2020). Production and characterization of starch nanoparticles by mild alkali hydrolysis and ultra-sonication process. *Scientific Reports, 10*(1), 1–11.

Ahmad, A. N., Lim, S. A., Navaranjan, N., Hsu, Y. I., & Uyama, H. (2020). Green sago starch nanoparticles as reinforcing material for green composites. *Polymer, 202*, Article 122646.

Ahmad Shapi'i, R., & Othman, S. H. (2016). Effect of concentration of chitosan on the mechanical, morphological, and optical properties of tapioca starch film. *International Food Research Journal, 23*(Suppl. 1), S187–S193.

Ahmed, S., Ahmad, M., Swami, B. L., & Ikram, S. (2016). A review on plants extract mediated synthesis of silver nanoparticles for antimicrobial applications: A green expertise. *Journal of Advanced Research, 7*(1), 17–28.

Alcázar-Alay, S. C., & Meireles, M. A. A. (2015). Physicochemical properties, modifications and applications of starches from different botanical sources. *Food Science and Technology, 35*, 215–236.

Aldao, D. C., Šárka, E., Ulbrich, P., & Menšíková, E. (2018). Starch nanoparticles – Two ways of their preparation. *Czech Journal of Food Sciences, 36* (2), 133–138.

Andreuccetti, C., Carvalho, R. A., & Grosso, C. R. (2010). Effect of hydrophobic plasticizers on functional properties of gelatin-based films. *Food Research International, 42*(8), 1113–1121.

Anugrahwidya, R., Armynah, B., & Tahir, D. (2022). Composites bioplastic film for various concentrations of zinc oxide (ZnO) nanocrystals towards physical properties for high biodegradability in soil and seawater. *Journal of Polymers and the Environment, 30*, 2589–2601.

Arezoo, E., Mohammadreza, E., Maryam, M., & Abdorreza, M. N. (2020). The synergistic effects of cinnamon essential oil and nano TiO2 on antimicrobial and functional properties of sago starch films. *International Journal of Biological Macromolecules, 157*, 743–751.

Azeredo, H. M. C., Rosa, M. F., & Mattoso, L. H. C. (2017). Nanocellulose in bio-based food packaging applications. *Industrial Crops and Products, 97*, 664–671.

Babaei-Ghazvini, A., Shahabi-Ghahfarrokhi, I., & Goudarzi, V. (2018). Preparation of UV-protective Starch/Kefiran/znO nanocomposite as a packaging film: Characterization. *Food Packaging and Shelf Life, 16*, 103–111.

Babapour, H., Jalali, H., & Mohammadi Nafchi, A. (2021). The synergistic effects of zinc oxide nanoparticles and fennel essential oil on physicochemical, mechanical, and antibacterial properties of potato starch films. *Food Science and Nutrition, 9*, 3893–3905.

Bashir, K., Jan, K., & Aggarwal, M. (2017). Thermo-rheological and functional properties of gamma-irradiated wholewheat flour. *International Journal of Food Science and Technology, 52*(4), 927–935.

Bashir, K., Swer, T. L., Prakash, K. S., & Aggarwal, M. (2017). Physico-chemical and functional properties of gamma irradiated whole wheat flour and starch. *LWT-Food Science & Technology, 76*, 131–139.

Basiak, E., Debeaufort, F., & Lenart, A. (2015). Effect of oil lamination between plasticized starch layers on film properties. *Food Chemistry, 195*, 56–63.

Batool, S., Hussain, Z., Niazi, M. B. K., Liaqat, U., & Afzal, M. (2019). Biogenic synthesis of silver nanoparticles and evaluation of physical and antimicrobial properties of Ag/PVA/starch nanocomposites hydrogel membranes for wound dressing application. *Journal of Drug Delivery Science and Technology, 52*, 403–414.

Bumbudsanpharoke, N., Harnkarnsujarit, N., Chongcharoenyanon, B., Kwon, S., & Ko, S. (2023). Enhanced properties of PBAT/TPS biopolymer blend with CuO nanoparticles for promising active packaging. *Food Packaging and Shelf Life, 37*, Article 101072.

Campelo, P. H., Sant'Ana, A. S., & Pedrosa Silva Clerici, M. T. (2020). Starch nanoparticles: Production methods, structure, and properties for food applications. *Current Opinion in Food Science, 33*, 136–140.

Castro-Aguirre, E., Iniguez-Franco, F., Samsudin, H. E. A., Fang, X., & Auras, R. (2016). Poly (lactic acid)—Mass production, processing, industrial applications, and end of life. *Advanced Drug Delivery Reviews, 107*, 333–366.

Chachlioutaki, K., Tzimtzimis, E. K., Tzetzis, D., Chang, M. W., Ahmad, Z., Karavasili, C., & Fatouros, D. G. (2020). Electrospun orodispersible films of isoniazid for pediatric tuberculosis treatment. *Pharmaceutics, 12*(5), 470.

Chavan, P., Sinhmar, A., Nehra, M., Thory, R., Pathera, A. K., Sundarraj, A. A., & Nain, V. (2021). Impact on various properties of native starch after synthesis of starch nanoparticles: A review. *Food Chemistry, 364*, Article 130416.

Cheetham, N. W. H., & Tao, L. (1998). Variation in crystalline type with amylose content in maize starch granules: An X-ray powder diffraction study. *Carbohydrate Polymers, 36*(4), 277–284. https://doi.org/10.1016/S0144-8617(98)00007-1

Cheftel, J. C., Cuq, J. L., & Lorient, D. (1985). Amino acids, peptides, and proteins. In O. R. Fennema (Ed.), *Food chemistry* (pp. 245–369). Marcel Dekker.

Chen, Y., Hao, Y., Ting, K., Li, Q., & Gao, Q. (2019). Preparation and emulsification properties of dialdehyde starch nanoparticles. *Food Chemistry, 286*, 467–474.

Chen, Q., Zhou, L., Zou, J., & Gao, X. (2019). The preparation and characterization of nanocomposite film reinforced by modified cellulose nanocrystals. *International Journal of Biological Macromolecules, 132*, 1155–1162.

Choudalakis, G., & Gotsis, A. D. (2009). Permeability of polymer/clay nanocomposites: A review. *European Polymer Journal, 45*(4), 967–984. https://doi.org/10.1016/j.eurpolymj.2009.01.027

Chung, H. J., & Liu, Q. (2009). Effect of gamma irradiation on molecular structure and physicochemical properties of corn starch. *Journal of Food Science, 74*(5), C353–C361.

Colussi, R., Pinto, V. Z., El Halal, S. L. M., Biduski, B., Prietto, L., Castilhos, D. D., & Dias, A. R. G. (2017). Acetylated rice starches films with different levels of amylose: Mechanical, water vapor barrier, thermal, and biodegradability properties. *Food Chemistry, 221*, 1614–1620.

da Silva, N. M. C., Correia, P. R. C., Druzian, J. I., Fakhouri, F. M., Fialho, R. L. L., & de Albuquerque, E. C. M. C. (2017). PBAT/TPS composite films reinforced with starch nanoparticles produced by ultrasound. *International Journal of Polymer Science, 2017*(1), Article 4308261.

Dar, M. Z., Deepika, K., Jan, K., Swer, T. L., Kumar, P., Verma, R., Prakash, K. S., Jan, S., & Bashir, K. (2018). Modification of structure and physicochemical properties of buckwheat and oat starch by γ-irradiation. *International Journal of Biological Macromolecules, 108*, 1348–1356.

de Moraes, J. O., Müller, C. M. O., & Laurindo, J. B. (2012). Influence of the simultaneous addition of bentonite and cellulose fibers on the mechanical and barrier properties of starch composite films. *Journal of Food Science and Technology, 18*(1), 35–45.

Dhali, K., Daver, F., Cass, P., & Adhikari, B. (2022). Surface modification of the cellulose nanocrystals through vinyl silane grafting. *International Journal of Biological Macromolecules, 200*, 397–408.

Ding, Y., Zheng, J., Zhang, F., & Kan, J. (2016). Synthesis and characterization of retrograded starch nanoparticles through homogenization and miniemulsion cross-linking. *Carbohydrate Polymers, 151*, 656–665.

do Evangelho, J. A., da Silva Dannenberg, G., Biduski, B., El Halal, S. L. M., Kringel, D. H., Gularte, M. A., Fiorentini, A. M., & da Rosa Zavareze, E. (2019). Antibacterial activity, optical, mechanical, and barrier properties of corn starch films containing Orange essential oil. *Carbohydrate Polymers, 222*, Article 114981.

Dufresne, A. (2014). Crystalline starch based nanoparticles. *Current Opinion in Colloid & Interface Science, 19*(5), 397–408. https://doi.org/10.1016/j.cocis.2014.06.001

Dularia, C., Sinhmar, A., Thory, R., Pathera, A. K., & Nain, V. (2019). Development of starch nanoparticles based composite films from non-conventional source - Water chestnut (Trapa bispinosa). *International Journal of Biological Macromolecules, 136*, 1161–1168. https://doi.org/10.1016/j.ijbiomac.2019.06.169

Dziadowiec, D., Matykiewicz, D., Szostak, M., & Andrzejewski, J. (2023). Overview of the cast polyolefin film extrusion technology for multi-layer packaging applications. *Materials, 16*(3), 1071.

Eichhorn, S. J. (2011). Cellulose nanowhiskers: Promising materials for advanced applications. *Soft Matter, 7*(2), 303–315.

Fei, L., Yoo, S. H., Villamayor, R. A. R., Williams, B. P., Gong, S. Y., Park, S., & Joo, Y. L. (2017). Graphene oxide involved air-controlled electrospray for uniform, fast, instantly dry, and binder-free electrode fabrication. *ACS Applied Materials & Interfaces, 9*(11), 9738–9746.

Galstyan, V., Bhandari, M. P., Sberveglieri, V., Sberveglieri, G., & Comini, E. (2018). Metal oxide nanostructures in food applications: Quality control and packaging. *Chemosensors, 6*(2), 16.

Gamage, A., Liyanapathiranage, A., Manamperi, A., Gunathilake, C., Mani, S., Merah, O., & Madhujith, T. (2022). Applications of starch biopolymers for a sustainable modern agriculture. *Sustainability, 14*(10), 6085.

Gao, H., Cai, J., Han, W., Huai, H., Chen, Y., & Wei, C. (2014). Comparison of starches isolated from three different trapa species. *Food Hydrocolloids, 37*, 174–181. https://doi.org/10.1016/j.foodhyd.2013.11.001

Gao, C., Guo, X., Nie, L., Wu, X., Peng, L., & Chen, J. (2023). A novel Pt loading method for WO_3 gasochromic film and the morphology control of the film by two-step solvothermal synthesis. *International Journal of Hydrogen Energy, 48*(7), 2849–2860.

Garcia, C. V., Shin, G. H., & Kim, J. T. (2018). Metal oxide-based nanocomposites in food packaging: Applications, migration, and regulations. *Trends in Food Science & Technology, 82*, 21–31.

Gardouh, A., El-Din, A. S., Moustafa, Y., & Gad, S. (2021). Formulation factors of starch-based nanosystems preparation and their pharmaceutical application. *Records of Pharmaceutical and Biomedical Sciences, 5*(1), 28–39. https://doi.org/10.21608/rpbs.2020.51097.1080

Geleta, T. T., Habtegebreil, S. A., & Tolesa, G. N. (2020). Physical, mechanical, and optical properties of enset starch from bulla films influenced by different glycerol concentrations and temperatures. *Journal of Food Processing and Preservation, 44*(8), Article e14586.

Ghosh, A. K. (2016). Extrusion of multicomponent product. In *Multicomponent polymeric materials* (pp. 109–131). Elsevier.

Girdthep, S., Worajittiphon, P., Molloy, R., Leejarkpai, T., & Punyodom, W. (2015). Formulation and characterization of compatibilized poly (lactic acid)-based blends and their nanocomposites with silver-loaded kaolinite. *Polymer International, 64*(2), 203–211.

Glenn, G. M., Orts, W., Imam, S., Chiou, B. S., & Wood, D. F. (2014). *Starch plastic packaging and agriculture applications starch polymers: From genetic engineering to green applications*. United States: Elsevier B.V.. https://doi.org/10.1016/B978-0-444-53730-0.00032-4

Gomes, R. F., de Azevedo, A. C. N., Pereira, A. G., Muniz, E. C., Fajardo, A. R., & Rodrigues, F. H. (2015). Fast dye removal from water by starch-based nanocomposites. *Journal of Colloid and Interface Science, 454*, 200–209.

Gontard, N. (1991). *Films et enrobages comestibles: Étude et amélioration des propriétés filmogènes du gluten [Edible films and coatings: Study and improvement of gluten film-forming properties]*. Université Montpellier II [Doctoral dissertation.

Gregory, D. A., Tripathi, L., Fricker, A. T., Asare, E., Orlando, I., Raghavendran, V., & Roy, I. (2021). Bacterial cellulose: A smart biomaterial with diverse applications. *Materials Science and Engineering: R: Reports, 145*, Article 100623.

Grima, S., Bellon-Maurel, V., Feuilloley, P., & Silvestre, F. (2000). Aerobic biodegradation of polymers in solid-state conditions: A review of environmental and physicochemical parameter settings in laboratory simulations. *Journal of Polymers and the Environment, 8*(4), 183–195.

Guerra, F. D., Attia, M. F., Whitehead, D. C., & Alexis, F. (2018). Nanotechnology for environmental remediation: Materials and applications. *Molecules, 23*(7), 1760.

Guo, Z., Zhao, B., Chen, L., & Zheng, B. (2019). Physicochemical properties and digestion of lotus seed starch under high-pressure homogenization. *Nutrients, 11*(2), 371. https://doi.org/10.3390/nu11020371

Haaj, S. B., Thielemans, W., Magnin, A., & Boufi, S. (2014). Starch nanocrystal stabilized pickering emulsion polymerization for nanocomposites with improved performance. *ACS Applied Materials & Interfaces, 6*(11), 8263–8273.

Habibi, Y., Lucia, L. A., & Rojas, O. J. (2010). Cellulose nanocrystals: Chemistry, self-assembly, and applications. *Chemical Reviews, 110*, 3479–3500.

Hassan, A. H. A., Omnia, A., Ahmed, A. M., Khalafalla, F. A., & Ali, F. H. M. (2011). Correspondence antimicrobial activity of starch-based biodegradable antimicrobial films incorporated with biosynthesized silver nanoparticles against multiple drug-resistant Staphylococcus aureus food isolates. *Journal of Advanced Veterinary Research, 13*, 41–46.

Hedayati, S., Niakousari, M., & Mohsenpour, Z. (2020). Production of tapioca starch nanoparticles by nanoprecipitation-sonication treatment. *International Journal of Biological Macromolecules, 143*, 136–142.

Huang, J. Y., Li, X., & Zhou, W. (2015). Safety assessment of nanocomposite for food packaging application. *Trends in Food Science & Technology, 45*(2), 187–199.

Iamareerat, B., Singh, M., Sadiq, M. B., & Anal, A. K. (2018). Reinforced cassava starch-based edible film incorporated with essential oil and sodium bentonite nanoclay as food packaging material. *Journal of Food Science and Technology, 55*, 1953–1959.

Jeong, O., & Shin, M. (2018). Preparation and stability of resistant starch nanoparticles, using acid hydrolysis and cross-linking of waxy rice starch. *Food Chemistry, 256*, 77–84.

Jiang, H., Jane, J. L., Acevedo, D., Green, A., Shinn, G., Schrenker, D., Srichuwong, S., Campbell, M., & Wu, Y. (2010). Variations in starch physicochemical properties from a generation-means analysis study using amylomaize v and VII parents. *Journal of Agricultural and Food Chemistry, 58*(9), 5633–5639. https://doi.org/10.1021/jf904531d

Khurshid, S., Arif, S., Ali, T. M., Iqbal, H. M., Shaikh, M., Khurshid, H., Akber, Q. U. A., & Yousaf, S. (2022). Effect of silver nanoparticles prepared from Saraca asoca leaf extract on morphological, functional, mechanical, and antibacterial properties of rice starch films. *Starch Staerke, 74*(5–6), Article 2100228.

Kim, H. Y., Park, S. S., & Lim, S. T. (2015). Preparation, characterization, and utilization of starch nanoparticles. *Colloids and Surfaces B: Biointerfaces, 126*, 607–620.

Kim, H., Park, S., Song, Y., Jang, W., Choi, K., Lee, K. G., & Im, S. G. (2023). Vapor-phase synthesis of a robust polysulfide film for transparent, biocompatible, and long-term stable anti-biofilm coating. *Korean Journal of Chemical Engineering, 40*(2), 412–418.

Kim, S., & Song, K. B. (2018). Antimicrobial activity of buckwheat starch films containing zinc oxide nanoparticles against Listeria monocytogenes on mushrooms. *International Journal of Food Science and Technology, 53*(6), 1549–1557. https://doi.org/10.1111/ijfs.13737

Klemm, D., Heublein, B., Fink, H. P., & Bohn, A. (2005). Cellulose: Fascinating biopolymer and sustainable raw material. *Angewandte Chemie International Edition, 44*(22), 3358–3393.

Kumar, P., Prakash, K. S., Jan, K., Swer, T. L., Jan, S., Verma, R., Deepika, K. M., Dar, M. Z., Verma, K., & Bashir, K. (2017). Effects of gamma irradiation on starch granule structure and physicochemical properties of brown rice starch. *Journal of Cereal Science, 77*, 194–200.

Kumari, S., Yadav, B. S., & Yadav, R. B. (2020). Synthesis and modification approaches for starch nanoparticles for their emerging food industrial applications: A review. *Food Research International, 128*, Article 108765.

Lamanna, M., Morales, N. J., García, N. L., & Goyanes, S. (2013). Development and characterization of starch nanoparticles by gamma radiation: Potential application as starch matrix filler. *Carbohydrate Polymers, 97*(1), 90–97.

Le Corre, D., & Angellier-Coussy, H. (2014). Preparation and application of starch nanoparticles for nanocomposites: A review. *Reactive and Functional Polymers, 85*, 97–120.

Le Corre, D., Bras, J., & Dufresne, A. (2010). Starch nanoparticles: A review. *Biomacromolecules, 11*, 1139–1153.

Le Corre, D., Vahanian, E., Dufresne, A., & Bras, J. (2012b). Enzymatic pretreatment for preparing starch nanocrystals. *Biomacromolecules, 13*(1), 132–137.

LeCorre, D. S., Bras, J., & Dufresne, A. (2012a). Influence of the botanic origin of starch nanocrystals on the morphological and mechanical properties of natural rubber nanocomposites. *Macromolecular Materials and Engineering, 297*(10), 969–978.

Li, B. M., Yildiz, O., Mills, A. C., Flewwellin, T. J., Bradford, P. D., & Jur, J. S. (2020). Iron-on carbon nanotube (CNT) thin films for biosensing E-Textile applications. *Carbon, 168*, 673–683.

Li, Y., Zhu, J., Cheng, H., Li, G., Cho, H., Jiang, M., Gao, Q., & Zhang, X. (2021). Developments of advanced electrospinning techniques: A critical review. *Advanced Materials Technologies, 6*(11), Article 2100410.

Li, F., Zhu, A., Song, X., & Ji, L. (2014). Novel surfactant for preparation of poly (L-lactic acid) nanoparticles with controllable release profile and cytocompatibility for drug delivery. *Colloids and Surfaces B: Biointerfaces, 115*, 377–383.

Lipton, J., Weng, G. M., Röhr, J. A., Wang, H., & Taylor, A. D. (2020). Layer-by-layer assembly of two-dimensional materials: Meticulous control on the nanoscale. *Matter, 2*(5), 1148–1165.

Liu, C., Jiang, S., Han, Z., Xiong, L., & Sun, Q. (2016). In vitro digestion of nanoscale starch particles and evolution of thermal, morphological, and structural characteristics. *Food Hydrocolloids, 61*, 344–350. https://doi.org/10.1016/j.foodhyd.2016.06.018

Liu, A., Wu, H., Naeem, A., Du, Q., Ni, B., Liu, H., Li, Z., & Ming, L. (2023). Cellulose nanocrystalline from biomass wastes: An overview of extraction, functionalization and applications in drug delivery. *International Journal of Biological Macromolecules, 241*, Article 124557.

Liu, Y., Xu, Y., Yan, Y., Hu, D., Yang, L., & Shen, R. (2015). Application of raman spectroscopy in structure analysis and crystallinity calculation of corn starch. *Starch Staerke, 67*(7–8), 612–619.

Lizundia, E., Costa, C. M., Alves, R., & Lanceros-Mendez, S. (2020). Cellulose and its derivatives for lithium ion battery separators: A review on the processing methods and properties. *Carbohydrate Polymer Technologies and Applications, 1*, Article 100001.

Mahto, R., & Das, M. (2014). Effect of gamma irradiation on the physico-mechanical and chemical properties of potato (Solanum tuberosum L.), Cv.'Kufri sindhuri', in non-refrigerated storage conditions. *Postharvest Biology and Technology, 92*, 37–45.

Majzoobi, M., Pesaran, Y., Mesbahi, G., Golmakani, M. T., & Farahnaky, A. (2015). Physical properties of biodegradable films from heat-moisture-treated rice flour and rice starch. *Starch Staerke, 67*(11–12), 1053–1060.

Mallakpour, S. (2018). Ultrasonic-assisted fabrication of starch/MWCNT-glucose nanocomposites for drug delivery. *Ultrasonics Sonochemistry, 40*, 402–409.

Marinopoulou, A., Papastergiadis, E., & Raphaelides, S. N. (2019). Inclusion complexes of non-granular maize starch with fatty acids and ibuprofen. A comparative study of their morphology and structure. *Starch Staerke, 71*(1–2), Article 1800100.

Mbey, J. A., Hoppe, S., & Thomas, F. (2012). Cassava starch–kaolinite composite film: Effect of clay content and clay modification on film properties. *Carbohydrate Polymers, 88*(1), 213–222.

Merino, D., Gutiérrez, T. J., & Alvarez, V. A. (2019). Structural and thermal properties of agricultural mulch films based on native and oxidized corn starch nanocomposites. *Starch Staerke, 71*(7–8), Article 1800341.

Merino, D., Gutiérrez, T. J., Mansilla, A. Y., Casalongue, C. A., & Alvarez, V. A. (2018). Critical evaluation of starch-based antibacterial nanocomposites as agricultural mulch films: Study on their interactions with water and light. *ACS Sustainable Chemistry & Engineering, 6*(11), 15662–15672.

Minakawa, A. F., Faria-Tischer, P. C., & Mali, S. (2019). Simple ultrasound method to obtain starch micro-and nanoparticles from cassava, corn and yam starches. *Food Chemistry, 283*, 11–18.

Miyazaki, M., Hung, P. Van, Maeda, T., & Morita, N. (2006). Recent advances in application of modified starches for breadmaking. *Trends in Food Science & Technology, 17*(11), 591–599. https://doi.org/10.1016/j.tifs.2006.05.002

Mohanty, F., & Swain, S. K. (2019). Nano silver embedded starch hybrid graphene oxide sandwiched poly (ethylmethacrylate) for packaging application. *Nano-Structures & Nano-Objects, 18*, Article 100300.

Mohseni, M. S., Khalilzadeh, M. A., Mohseni, M., Hargalani, F. Z., Getso, M. I., Raissi, V., & Raiesi, O. (2020). Green synthesis of Ag nanoparticles from pomegranate seeds extract and synthesis of Ag-Starch nanocomposite and characterization of mechanical properties of the films. *Biocatalysis and Agricultural Biotechnology, 25*, Article 101569.

Momenpoor, B., Danafar, F., & Bakhtiari, F. (2019). Size controlled preparation of starch nanoparticles from wheat through precipitation at low temperature. *Journal of Nano Research, 56*, 131–141.

Nath, D., Santhosh, R., Pal, K., & Sarkar, P. (2022). Nanoclay-based active food packaging systems: A review. *Food Packaging and Shelf Life, 31*, Article 100803. https://doi.org/10.1016/j.fpsl.2021.100803

Nessi, V., Falourd, X., Maigret, J. E., Cahier, K., D'orlando, A., Descamps, N., Gaucher, V., Chevigny, C., & Lourdin, D. (2019). Cellulose nanocrystals-starch nanocomposites produced by extrusion: Structure and behavior in physiological conditions. *Carbohydrate Polymers, 225*, Article 115123.

Ni, S., Zhang, H., Dai, H., & Xiao, H. (2018). Starch-based flexible coating for food packaging paper with exceptional hydrophobicity and antimicrobial activity. *Polymers, 10*(11), 1260.

Nieto-Suaza, L., Acevedo-Guevara, L., Sánchez, L. T., Pinzón, M. I., & Villa, C. C. (2019). Characterization of aloe vera-banana starch composite films reinforced with curcumin-loaded starch nanoparticles. *Food Structure, 22*. https://doi.org/10.1016/j.foostr.2019.100131

Oleyaei, S. A., Zahedi, Y., Ghanbarzadeh, B., & Moayedi, A. A. (2016). Modification of physicochemical and thermal properties of starch films by incorporation of TiO_2 nanoparticles. *International Journal of Biological Macromolecules, 89*, 256–264.

Pérez, S., & Bertoft, E. (2010). The molecular structures of starch components and their contribution to the architecture of starch granules: A comprehensive review. *Starch Staerke, 62*(8), 389–420.

Patil, S., Bharimalla, A. K., Nadanathangam, V., Dhakane-Lad, J., Mahapatra, A., Jagajanantha, P., & Saxena, S. (2022). Nanocellulose reinforced corn starch-based biocomposite films: Composite optimization, characterization, and storage studies. *Food Packaging and Shelf Life, 33*, Article 100860.

Peponi, L., Puglia, D., Torre, L., Valentini, L., & Kenny, J. M. (2014). Processing of nanostructured polymers and advanced polymeric based nanocomposites. *Materials Science and Engineering: R: Reports, 85*, 1–46.

Pimsen, R., Porrawatkul, P., Nuengmatcha, P., Ramasoot, S., & Chanthai, S. (2021). Efficiency enhancement of slow release of fertilizer using nanozeolite–chitosan/sago starch-based biopolymer composite. *Journal of Coatings Technology and Research, 18*, 1321–1332.

Pirsa, S., Karimi Sani, I., & Khodayvandi, S. (2018). Design and fabrication of starch-nanoclay composite films loaded with methyl Orange and bromocresol green for determination of spoilage in milk package. *Polymers for Advanced Technologies, 29*(11), 2750–2758.

Prabhu, S. V., Hemalatha, P., Tizazu, B. Z., & Gosu, A. (February 2021). Development of Teff-starch based edible film: Mechanical and optical properties. In *IOP conference series: Materials science and engineering*. IOP Publishing, Article 012050.

Punia, S., Sandhu, K. S., Dhull, S. B., & Kaur, M. (2019). Dynamic, shear and pasting behaviour of native and octenyl succinic anhydride (OSA) modified wheat starch and their utilization in preparation of edible films. *International Journal of Biological Macromolecules, 133*, 110–116.

Rafique, M., Sadaf, I., Rafique, M. S., & Tahir, M. B. (2017). A review on green synthesis of silver nanoparticles and their applications. *Artificial Cells, Nanomedicine and Biotechnology, 45*(7), 1272–1291.

Ramyadevi, J., Jeyasubramanian, K., Marikani, A., Rajakumar, G., & Rahuman, A. A. (2012). Synthesis and antimicrobial activity of copper nanoparticles. *Materials Letters, 71*, 114–116.

Redazione. (2016). Bioplastics for food packaging: Production, applications, and limits. *Italian Food Materials and Machinery, 32*, 128–141. http://italianfoodmaterialsandmachinery.com/bioplastics-for-food-packaging-production-applications-and-limits/.

Rehana, D., Mahendiran, D., Kumar, R. S., & Rahiman, A. K. (2017). Evaluation of antioxidant and anticancer activity of copper oxide nanoparticles synthesized using medicinally important plant extracts. *Biomedicine & Pharmacotherapy, 89*, 1067–1077.

Revathi, T., & Thambidurai, S. (2019). Cytotoxic, antioxidant and antibacterial activities of copper oxide incorporated chitosan-neem seed biocomposites. *International Journal of Biological Macromolecules, 139*, 867–878. https://doi.org/10.1016/j.ijbiomac.2019.07.214

Ribeiro, C., Vicente, A. A., Teixeira, J. A., & Miranda, C. (2007). Optimization of edible coating composition to retard strawberry fruit senescence. *Postharvest Biology and Technology, 44*(1), 63–70.

Rompothi, O., Pradipasena, P., Tananuwong, K., Somwangthanaroj, A., & Janjarasskul, T. (2017). Development of non-water soluble, ductile mung bean starch-based edible film with oxygen barrier and heat sealability. *Carbohydrate Polymers, 157*, 748–756.

Rosman, N. F., Nurfazianawatie, M. Z., Omar, H., Malek, N. S. A., Hajar, N., Buniyamin, I., Abdullah, S., Rusop, M., & Asli, N. A. (2022). Fungal growth physicochemical properties inhibition by novel zinc oxide/glutinous tapioca starch composite. *J. Adv. Res. Appl. Sci. Eng. Technol, 29*(1), 76–89.

Roy, S., & Rhim, J.-W. (2022). Starch/Agar-Based functional films integrated with enoki mushroom-mediated silver nanoparticles for active packaging applications. *Food Bioscience, 49*, Article 101867. https://doi.org/10.1016/j.fbio.2022.101867

Sadjadi, M. S., Meskinfam, M., & Jazdarreh, H. (2010). Hydroxyapatite - Starch nano biocomposites synthesis and characterization. *International Journal of Nano Dimension, 1*, 57–63.

Salmieri, S., Islam, F., Khan, R. A., Hossain, F. M. M., Miao, C., Hamad, W. Y., & Lacroix, M. (2014). Antimicrobial nanocomposite films made of poly(lactic acid)–cellulose nanocrystals (PLA–CNC) in food applications. Part B: Effect of oregano essential oil release on the inactivation of Listeria monocytogenes in mixed vegetables. *Cellulose, 21*(6), 4271–4285.

Sandhu, K. S., Sharma, L., Kaur, M., & Kaur, R. (2020). Physical, structural, and thermal properties of composite edible films prepared from pearl millet starch and carrageenan gum: Process optimization using response surface methodology. *International Journal of Biological Macromolecules, 143*, 704–713.

Santana, J. S., de Carvalho Costa, É. K., Rodrigues, P. R., Correia, P. R. C., Cruz, R. S., & Druzian, J. I. (2019). Morphological, barrier, and mechanical properties of cassava starch films reinforced with cellulose and starch nanoparticles. *Journal of Applied Polymer Science, 136*(4), Article 47001.

Santana, Á. L., & Meireles, M. A. A. (2014). New starches are the trend for industry applications: A review. *Food and Public Health, 4*(5), 229–241.

Sanyang, M. L., Sapuan, S. M., Jawaid, M., Ishak, M. R., & Sahari, J. (2015). Effect of plasticizer type and concentration on tensile, thermal, and barrier properties of biodegradable films based on sugar palm (Arenga pinnata) starch. *Polymers, 7*(6), 1106–1124.

Saunders, J. (2010). *Physicochemical properties of wheat starches and their relationship to liquefaction and fermentative bioethanol performance* (MSc Thesis). Winnipeg, Manitoba, Canada: University of Manitoba.

Scarfato, P., Di Maio, L., Milana, M. R., Giamberardini, S., Denaro, M., & Incarnato, L. (2017). Performance properties, lactic acid specific migration and swelling by simulant of biodegradable poly (lactic acid)/nanoclay multilayer films for food packaging. *Food Additives & Contaminants: Part A, 34*(10), 1730–1742.

Shabana, S., Prasansha, R., Kalinina, I., Potoroko, I., Bagale, U., & Shirish, S. H. (2019). Ultrasound assisted acid hydrolyzed structure modification and loading of antioxidants on potato starch nanoparticles. *Ultrasonics Sonochemistry, 51*, 444–450. https://doi.org/10.1016/j.ultsonch.2018.07.023

Shah, A. A., Hasan, F., Hameed, A., & Ahmed, S. (2008). Biological degradation of plastics: A comprehensive review. *Biotechnology Advances, 26*, 246–265.

Shanks, R. A. (2014). *Chemistry and structure of cellulosic fibres as reinforcements in natural fibre composites*. Elsevier BV. https://doi.org/10.1533/9780857099228.1.66

Sharmin, E., Kafyah, M. T., Alzaydi, A. A., Fatani, A. A., Hazazzi, F. A., Babgi, S. K., & Mojally, M. (2020). Synthesis and characterization of polyvinyl alcohol/corn starch/linseed polyol-based hydrogel loaded with biosynthesized silver nanoparticles. *International Journal of Biological Macromolecules, 163*, 2236–2247.

Shi, A. M., Wang, L. J., Li, D., & Adhikari, B. (2013). Characterization of starch films containing starch nanoparticles: Part 1: Physical and mechanical properties. *Carbohydrate Polymers, 96*(2), 593–601.

Singh, N., Isono, N., Srichuwong, S., Noda, T., & Nishinari, K. (2008). Structural, thermal and viscoelastic properties of potato starches. *Food Hydrocolloids, 22*(6), 979–988.

Singh, N., Nakaura, Y., Inouchi, N., & Nishinari, K. (2007). Fine structure, thermal and viscoelastic properties of starches separated from indica rice cultivars. *Starch Staerke, 59*(1), 10–20.

Singh, N., Nakaura, Y., Inouchi, N., & Nishinari, K. (2008). Structure and viscoelastic properties of starches separated from different legumes. *Starch Staerke, 60*(7), 349–357.
Singh, S., Singh, N., Ezekiel, R., & Kaur, A. (2011). Effects of gamma-irradiation on the morphological, structural, thermal and rheological properties of potato starches. *Carbohydrate Polymers, 83*(4), 1521–1528.
Singh, S., Singh, N., Isono, N., & Noda, T. (2010). Relationship of granule size distribution and amylopectin structure with pasting, thermal, and retrogradation properties in wheat starch. *Journal of Agricultural and Food Chemistry, 58*(2), 1180–1188.
Singh, N., Singh, J., Kaur, L., Sodhi, N. S., & Gill, B. S. (2003). Morphological, thermal and rheological properties of starches from different botanical sources. *Food Chemistry, 81*(2), 219–231. https://doi.org/10.1016/S0308-8146(02)00416-8
Slavutsky, A. M., & Bertuzzi, M. A. (2014). Water barrier properties of starch films reinforced with cellulose nanocrystals obtained from sugarcane bagasse. *Carbohydrate Polymers, 110*, 53–61.
Sofi, B. A., Wani, I. A., Masoodi, F. A., Saba, I., & Muzaffar, S. (2013). Effect of gamma irradiation on physicochemical properties of broad bean (Vicia faba L.) starch. *LWT-Food Science & Technology, 54*(1), 63–72.
Sun, Q. (2018). Starch nanoparticles. In *Starch in food* (pp. 691–745). Woodhead Publishing. https://doi.org/10.1016/B978-0-08-100868-3.00017-1
Suriya, M., Reddy, C. K., Haripriya, S., & Harsha, N. (2018). Influence of debranching and retrogradation time on behavior changes of Amorphophallus paeoniifolius nanostarch. *International Journal of Biological Macromolecules, 120*, 230–236.
Suriyatem, R., Auras, R. A., & Rachtanapun, P. (2019). Utilization of carboxymethyl cellulose from durian rind agricultural waste to improve physical properties and stability of rice starch-based film. *Journal of Polymers and the Environment, 27*, 286–298.
Tagliapietra, B. L., de Melo, B. G., Sanches, E. A., Plata-Oviedo, M., Campelo, P. H., & Clerici, M. T. P. S. (2021). From micro to nanoscale: A critical review on the concept, production, characterization, and application of starch nanostructure. *Starch Staerke, 73*(11–12), Article 2100079.
Taherimehr, M., Bagheri, R., & Taherimehr, M. (2021). In vitro evaluation of thermoplastic starch/beta-tricalcium phosphate nano-biocomposite in bone tissue engineering. *Ceramics International, 47*(11), 15458–15463.
Tan, S. X., Andriyana, A., Ong, H. C., Lim, S., Pang, Y. L., & Ngoh, G. C. (2022). A comprehensive review on the emerging roles of nanofillers and plasticizers towards sustainable starch-based bioplastic fabrication. *Polymers, 14*(4), 664.
Tang, X., Alavi, S., & Herald, T. J. (2008). Barrier and mechanical properties of starch-clay nanocomposite films. *Cereal Chemistry, 85*(3), 433–439.
Thirathumthavorn, D., Charoenrein, S., & Krochta, J. M. (2010). Influence of glass transition on oxygen permeability of starch-based edible films. *Water Properties in Food, Health, Pharmaceutical and Biological Systems: ISOPOW, 10*, 641–646.
Tian, W., VahidMohammadi, A., Wang, Z., Ouyang, L., Beidaghi, M., & Hamedi, M. M. (2019). Layer-by-layer self-assembly of pillared two-dimensional multilayers. *Nature Communications, 10*(1), 2558.
Torres, F. G., Arroyo, J., Tineo, C., & Troncoso, O. (2019). Tailoring the properties of native andean potato starch nanoparticles using acid and alkaline treatments. *Starch Staerke, 71*(3–4), Article 1800234.
Trinh, B. M., Smith, M., & Mekonnen, T. H. (2022). A nanomaterial-stabilized starch-beeswax pickering emulsion coating to extend produce shelf-life. *Chemical Engineering Journal, 431*, Article 133905.
Tripathy, T., Kolya, H., Jana, S., & Senapati, M. (2017). Green synthesis of Ag-Au bimetallic nanocomposites using a biodegradable synthetic graft copolymer; hydroxyethyl starch-g-poly (acrylamide-co-acrylic acid) and evaluation of their catalytic activities. *European Polymer Journal, 87*, 113–123.
Tunma, S. (2018). Starch-based nanocomposites in active packaging for extended shelf life of fresh fruits. *Walailak Journal of Science and Technology, 15*, 273–281.
Vallejo-Montesinos, J., Gámez-Cordero, J., Zarraga, R., Pérez Pérez, M. C., & Gonzalez-Calderon, J. A. (2020). Influence of the surface modification of titanium dioxide nanoparticles TiO2 under efficiency of silver nanodots deposition and its effect under the properties of starch–chitosan (SC) films. *Polymer Bulletin, 77*(1), 107–133.
Vamadevan, V., & Bertoft, E. (2015). Structure-function relationships of starch components. *Starch/Staerke, 67*(1–2), 55–68. https://doi.org/10.1002/star.201400188
Waghmare, V. S., Wadke, P. R., Dyawanapelly, S., Deshpande, A., Jain, R., & Dandekar, P. (2018). Starch-based nanofibrous scaffolds for wound healing applications. *Bioactive Materials, 3*(3), 255–266.
Winarti, C., Sunarti, T. C., Mangunwidjaja, D., & Richana, N. (2014). Preparation of arrowroot starch nanoparticles by butanol-complex precipitation, and its application as bioactive encapsulation matrix. *International Food Research Journal, 21*(6).
Yadav, N., Singh, S., Saini, O., & Srivastava, S. (2022). Technological advancement in the remediation of heavy metals employing engineered nanoparticles: A step towards cleaner water process. *Environmental Nanotechnology, Monitoring & Management, 18*, Article 100757.
Yadav, T. C., Srivastava, A. K., Mishra, P., Singh, D., Raghuwanshi, N., Singh, N. K., & Pruthi, V. (2019). Electrospinning: An efficient biopolymer-based micro-and nanofibers fabrication technique. In *Next generation biomanufacturing technologies* (pp. 209–241). American Chemical Society.
Yazid, N. S. M., Abdullah, N., Muhammad, N., & Matias-Peralta, H. M. (2018). Application of starch and starch-based products in the food industry. *Journal of Science and Technology, 10*(2), 144–174.
Yousefi, A. R., Savadkoohi, B., Zahedi, Y., Hatami, M., & Ako, K. (2019). Fabrication and characterization of hybrid sodium montmorillonite/TiO2 reinforced cross-linked wheat starch-based nanocomposites. *International Journal of Biological Macromolecules, 131*, 253–263.
Zhang, X., Xu, Y., Zhang, X., Wu, H., Shen, J., Chen, R., & Guo, S. (2019). Progress on the layer-by-layer assembly of multilayered polymer composites: Strategy, structural control and applications. *Progress in Polymer Science, 89*, 76–107.

Chapter 23

Starch based nanocomposite materials for enhanced quality and safety of food products during storage

Abhishek Biswal[1], P. Poojapushkarni[1], Aravinth Sithivinayagam[2], V. Sneha[2], Vinoth Kumar Vaidyanathan[2], Periyar Selvam Sellamuthu[1] and Emmanuel Rotimi Sadiku[3]

[1]*Department of Food Process Engineering, Postharvest Technology Research Lab, School of Bioengineering, SRM Institute of Science and Technology, Chennai, Tamil Nadu, India;* [2]*Integrated Bioprocessing Laboratory, Department of Biotechnology, School of Bioengineering, SRM Institute of Science and Technology, Chennai, Tamil Nadu, India;* [3]*Tshwane University of Technology, Department of Chemical, Metallurgical and Materials Engineering, Pretoria, South Africa*

1. Introduction

Food grade packaging serves a pivotal part in prolonging the preservation period by acting as a protective barrier against the physical and chemical elements that can cause contamination. It serves as a crucial component of the modern food supply chain by maintaining the product integrity, enhancing safety, and supporting the long term sustainability of food products (Jafarzadeh et al., 2021). As material science has advanced, traditional packaging techniques have changed, giving rise to intelligent and active packaging solutions that improve food preservation (Firouz et al., 2021). The researchers developed active packaging incorporated with biomaterials that actively modify the internal environment of packaging, thereby extending the shelf life. In addition, intelligent packaging includes sensors and indicators that monitor food freshness by reducing food waste and improving supply chain efficiency (Taherimehr et al., 2021).

Biodegradable and nanocomposite based packaging solutions are gaining increasing attention due to environmental concerns with traditional/regular plastic packaging materials. These novel materials were constructed using the integration of biopolymers, essential oils, cellulose nanocrystals, and nanocomposites. Natural polymers are mainly derived from polysaccharides, proteins, and lipids that offer ecofriendly solutions by decomposing into nontoxic by products under microbial degradation (Satdive et al., 2022). The incorporation of nanocomposite materials such as nanoclay, metallic nanoparticles, and biofibers into the polymeric matrices significantly enhances mechanical and barrier properties, imparts antimicrobial functionality, and improves tensile strength and thermal stability by providing resistance to oxygen and water vapor transmission (Chi et al., 2022). Improves the physical and barrier attributes, imparts antimicrobial active functionality, and elevates tensile strength and thermal stability by providing resistance to oxygen and moisture transmission.

Starch based packaging materials have emerged as an eco friendly solution to conventional plastic packaging owing to their biodegradability, cost effectiveness, and renewable nature. Starch is a polysaccharide that possesses excellent film forming properties, especially those obtained from agricultural sources like corn, potato, and cassava, positioning them as a sustainable food packaging material. The primary drawbacks of the starch lie in its high hydrophobicity, which restricts its application in humid conditions, leading to a reduction in both mechanical and barrier performance. To overcome these limitations, starch based films are blended with polymer matrices such as poly lactic acid (PLA), polyvinyl alcohol (PVA) and polycaprolactone (PCL). Plasticization can also be done using glycerol and sorbitol used as plasticizers that enhance the flexibility, making them suitable for food packaging. Additionally, cross linking agents like boric acid and citric acid improve the structural integrity of films. However, large scale industrial adoption still faces challenges related to cost, performance under humid conditions and regulatory approvals. Current studies mainly focus on optimizing formulations

Native and Modified Starches for Food Applications. https://doi.org/10.1016/B978-0-443-27431-2.00020-1

and processing techniques with the aim of enhancing mechanical and functional attributes of the starch derived biocomposites. Emerging trends for the development of starch based intelligent packaging, incorporating freshness indicators and sensors, are expected to revolutionize food packaging by improving the quality control and reducing food waste. The ongoing progress in material science and nanotechnology is likely to lead to the use of starch based biodegradable packaging for sustainable food packaging in the future (Park et al., 2022; Yong, 2022).

In this chapter, we will explore the advancements focusing on starch based composite film as a sustainable approach for prolonging. The shelf life of the food products from 2020 to 2025. The chapter then delves into the integration of nanomaterials in the different matrices and also highlights active and smart packing technologies like integrating essential oils, bioactive agents, and nanosensors, which contribute to food safety and quality monitoring. Additionally, future trends in nanostructured emphasize sustainability and smart functionalities that are explored to provide insights into the next generation of eco friendly food preservation technologies. This chapter also serves as a comprehensive model for researchers and industry professionals interested in leveraging nanotechnology for enhanced food packaging solutions.

2. Fabrication method for starch derived nanocomposite films and coatings

Starch derived coatings are extensively applied in food packaging and pharmaceutical industries owing to their isotropic nature, odorless and tasteless properties, as well as being colorless, nontoxic, and biodegradable. Starch derived coatings are extensively employed in food packaging and pharmacological sectors owing to their isotropic nature, odorless and tasteless attributes, as well as being colorless, nontoxic, and environmental degradability. Biodegradable films and coatings can be developed using both native and modified starches. Notably, starch film exhibits low oxygen permeability and the coating solutions are safe, nutritious and cost effective, making them valuable for food storage and marketing. Their physical characteristics, chemical resistance and mechanical properties closely resemble those of plastic films. A range of different films and coating methods are applied for the sustainable packaging materials (Dai et al., 2024).

2.1 Film formation through the solution casting technique

Among the available film forming techniques, the solution casting method is the most frequently applied one as it is simple and produces uniform films effectively (Fig. 23.1). For the solution casting technique, the starch is suspended in water and heated to form a gelatinized solution and to avoid brittleness, a plasticizer like glycerol or sorbitol is added to improve flexibility. Nanomaterials such as clay, nanocrystals, or metallic oxide compounds are dispersed into starch materials to enhance mechanical barrier and antimicrobial properties (Table 23.1). The prepared solution is poured on the glass plate and set for drying at room temperature (Wang, Zhang, Li, Deng, et al., 2023; Wang, Zhang, Li, Li, et al., 2023).

2.2 Extrusion blow molding

The processing of hydrolyzed corn starch pellets was carried out with a single screw extrusion system. To prevent agglomeration and to improve the tensile strength, cellulose was added. The extrusion process was maintained at a temperature profile of 35–55°C with the scree rotation speed of 30 rpm, which results in the production of extruded material. Using the extrusion blow molding, the infused nanomaterials are well distributed during the process. The molten starch composite is extruded into a tube-like shape. The compressed air is blown into parison inside a mold expanding into the desired shape (Fig. 23.2). Table 23.2 explains the various starch based biocomposite films using the extrusion blow molding method.

2.3 Electrospinning

Production of nanofibrous starch films using a sophisticated fabrication method electrospinning (Fig. 23.3). In this setup, the biomaterial is placed in a syringe with a fine needle, and a high voltage electric field is established between the needle and the grounded collector. The electrostatic force overcomes surface tension, drawing the solution into ultrafine nanofibers. The solvent evaporates as the fibers travel toward the collector, forming a film. The main limitation of electrospinning is that it controls the fiber uniformity, and the mechanical properties can be challenging (Cao et al., 2022). Table 23.3 explains the various starch based bionanocomposite produced using the electrospun technique.

2.4 Layer by layer assembly

A state of the art method for multilayer fruit coatings that provides exact control over the structure and functionality of the coating is layer by layer (LBL) assembly (Fig. 23.4). This technique produces highly ordered and useful coatings by

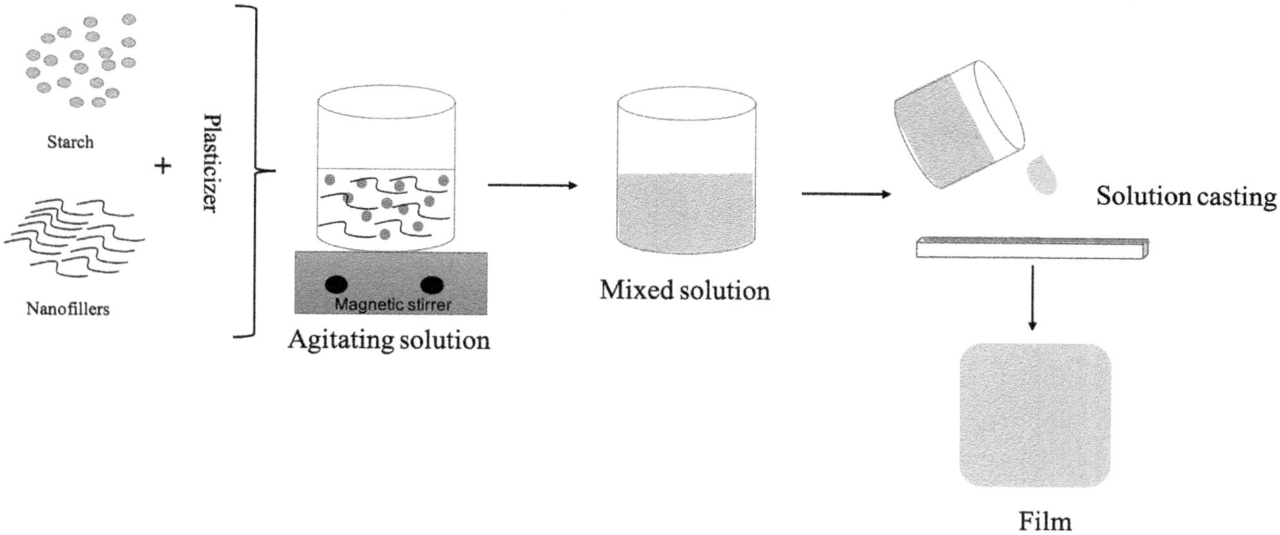

FIGURE 23.1 Schematic presentation of the solution casting method. Technique of casting the solvent on the glass plate.

TABLE 23.1 Starch based biocomposite using the solution casting method.

Formulations	Analysis	Output	References
Starch/Beeswax	Phase morphology, surface wettability and water vapor permeability	Starch/BW serves as a foundation for the development of novel hydrophobic formulas.	Luchesi et al. (2024)
Corn starch/Chlorogenic acid	Rheological properties, SEM analysis, FTIR, encapsulation efficiency and thermal stability	Starch derived film system enhances both the stability and encapsulation efficiency.	Zhang et al. (2024)
AgO/CuO/PVA/Starch hydrogel	Dynamic light scattering, SEM, XRD, FTIR and mechanical properties	AgO/CuO/PVA/starch hydrogel offers an effective solution for wound dressing.	Uzair et al. (2024)
β cyclodextrin/curry leaf essential oil/chitosan microcapsules	Encapsulation efficiency, SEM and thermal stability	Multi functional film was developed using controlled and gradual release of CLEO for active fruit packaging applications.	Bansal et al. (2025)
Starch/gly/CMC films	Tensile properties and soluble properties	Develop sustainable alternatives to plastic using Plackett Burman Design.	Magnaghi et al. (2024)
Starch/SEBS-*g*-MA/Graphene oxide	FTIR, XRD and DSC	Enhanced mechanical strength exhibited by the cassava starch based bio-nanocomposite films due to a hydrogen bonding network.	Bibi et al. (2024)
Oil palm trunk starch modified with citric acid	Diffraction peaks, thermal stability and degradation time	Optimal conditions of the biocomposite film were achieved with 30% glycerol.	Hernando et al. (2024)
CMC/Starch/Selenium oxide nanocomposite films	HR TEM, XRD, FTIR and FE SEM	The conduction mechanism improves the CMC/St/SeO_2 composites.	El Sayed et al. (2024)

alternating the deposition of oppositely charged elements, such as polyelectrolytes, biopolymers, and nanoparticles. The LbL method primarily uses hydrophobic forces, hydrogen bonds, or electrostatic interactions to deposit thin layers of materials. The fruit is sprayed or submerged in a coating solution to apply each layer. By regulating the interchange of carbon dioxide and oxygen, LbL coatings reduce moisture loss and improve barrier qualities. The LbL coverings for

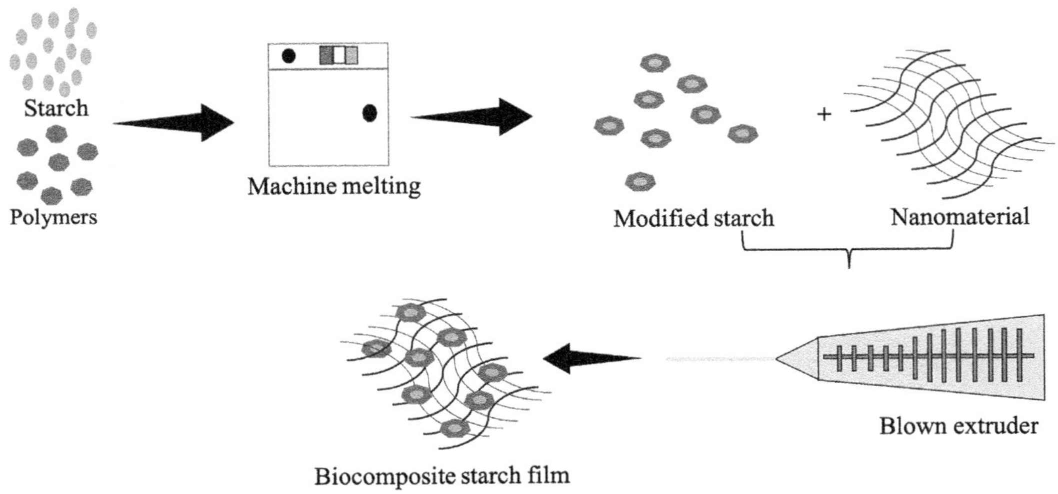

FIGURE 23.2 **Extrusion blown method using starch and polymers.** Fabrication of biodegradable film using extrusion blown technique.

TABLE 23.2 Starch based biocomposite film using extrusion blow molding.

Formulations	Analysis	Output	References
High performance composite materials based on PBAT and corn starch	SEM, thermal stability and water absorption	Enhanced compatibility of the composite material upon the addition of polycaprolactone based polyurethane compatibilizer.	Ren et al. (2024)
Modifying starch and gelatin	Mechanical and barrier properties	Optimal conditions for gelatin content and extrusion temperature producing edible films through the extrusion casting process.	Aguilar et al. (2025)
.Hydroxypropyl-β-cyclodextrin/ethyl lauroyl arginate/cinnamon essential oil incorporated starch/PBAT film.	Encapsulation efficiency, thermal and antibacterial properties	SP films with HPCD/LAE/CEO microcapsules exhibit potential use in the degradable bioactive food packaging materials.HPCD/LAE/CEO microcapsules with SP films exhibit feasible use in the degradable bioactive food packaging materials.	Hu et al. (2024)
Maltol and ethyl maltol loaded starch/PBAT blown film.	SEM, FTIR, XRD and antibacterial activity	Integrating volatile compounds like MT/EM into PBAT/TPS films has significant importance in creating active bakery packaging.	Promhuad et al. (2025)
Biofiller of starch and gelatin using eggshell powder	FTIR, SEM of film matrix, water vapor permeability and oxygen transmission rate	Ideal physicochemical properties and optimal biodegradability obtained using thermo processed PS/G/ESP composite films.	Li et al. (2023)

strawberries that used chitosan and alginate successfully reduced weight loss and postponed ripening (Ganjeh et al., 2024). The details of starch biocomposite coating on fruits for enhancing the shelf life are shown in Table 23.4.

2.5 Spray coating

Coatings on fruits and vegetables are widely used; they are cost efficient and adaptable to various types of coatings, including waxes, biopolymers and emulsions (Fig. 23.5). The details of coating of fruits using the spraying method are given in Table 23.5. Starch is dissolved in water or ethanol with plasticizers. Functional additives such as nanoparticles,

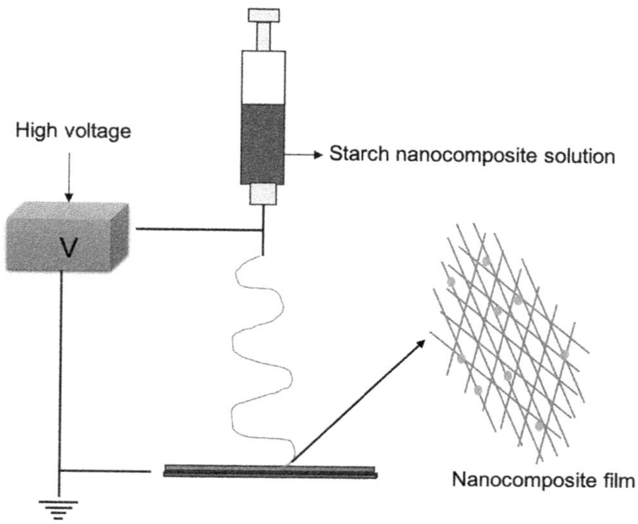

FIGURE 23.3 **Electrospinning of nanocomposite solution.** Advanced technology for fabricating the film with different carrier molecules.

TABLE 23.3 Starch based bionanocomposite using electrospun technique.

Formulations	Analysis	Output	References
Starch water optimization	RSM and SEM studies	Nanofibers can serve as an effective supporting structure for biologically active compounds in the food and pharmaceutical industries.	Balderas-Guerrero et al. (2025)
Starch catechin based green nanofiber mats	DPPH, antimicrobial activity and thermal stability	Mechanical attributes, and hydrophobicity of the nanofiber mats were enhanced via the vapor phase crosslinking method.	Lv et al. (2024)
Taro starch/formic acid and DMSO	Surface analysis, fiber morphology and size	Effective fabrication of taro starch nanofibers and optimization of process parameters.	Sawitri et al. (2025)
Cinnamaldehyde essential oil/octenyl succinylated starch/pullulan	Viscosity, increased conductivity and surface tension	Electrospun starch based nanofiber mats incorporated with essential oils can have promising applications.	Huang et al. (2024)
Geranium essential oil was successfully extracted and incorporated into ultrafine bean starch fibers	GCMS, Antibacterial activity	GEO [40%] showed a significant reduction in tested bacteria	Dos Santos et al. (2024)
Polyurethane (PU), propolis extract (PE), hyaluronic acid (HA)	SEM, TEM and contact angle of the films	The electrospun structure, comprising a PU shell and ST/PE/HA core, exhibits promise as a versatilemulti purpose platform.	Hajipour et al. (2025)

essential oils or proteins can be incorporated for enhanced properties. For spray coating, the solutions are loaded into a spray gun or nozzle system using compressed air or ultrasonic energy that breaks the liquid into droplets, ensuring even distribution. The coatings should be sprayed under controlled temperature and humidity to form continuous coatings (Nain et al., 2021).

3. Incorporation techniques for nanostructures

Embedding nanostructure into the biopolymer matrix has shown significant enhancement in the performance of the biodegradable material. Various incorporation techniques are utilized to ensure uniform dispersion, stability and interaction between nanostructures and polymer matrix. The selection of the incorporation method depends on factors such as

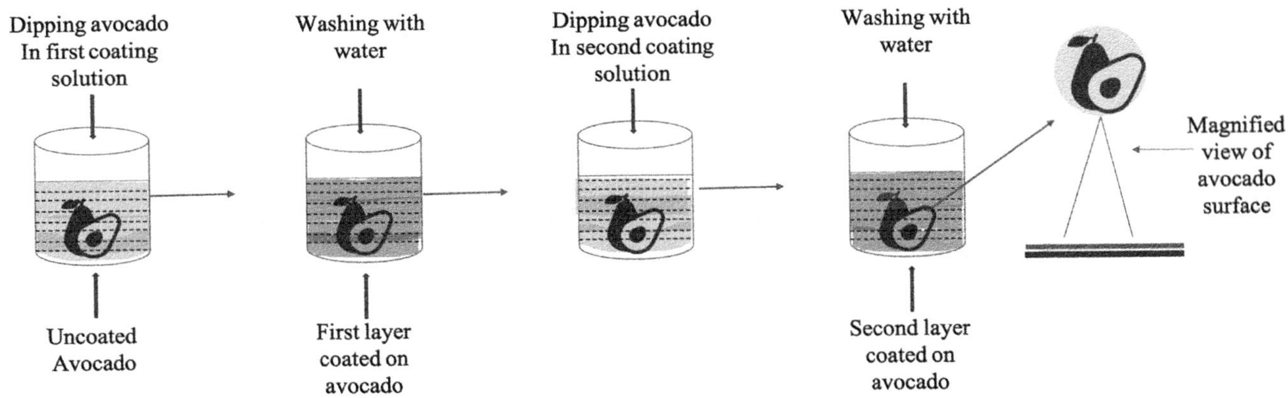

FIGURE 23.4 Schematic illustration of the sequential coating process on avocado fruits using an emulsion-based solution to enhance preservation and extend shelf life. Coating of fruits using the layer-by-layer technology.

TABLE 23.4 Starch biocomposite coating on fruits for enhancing the shelf life.

Formulations	Analysis	Output	References
Chitosan/CMC	Chlorophyll, color, gas exchange and fruit parameters	Mangoes coated with CH- and CMC based formulations exhibited higher textural integrity and reduced activity of enzymes such as cellulase, pectin methylesterase, polygalacturonase, and β-galactosidase.	Ali et al. (2024)
SA/CMC/Graphene oxide	Mechanical and barrier properties	Prolonging the preservation period of ready to eat apples.	Wang et al. (2024)
Chitosan/*Opuntia stenopetala* polysaccharides/Alginate	Confocal microscope, FTIR, WVP, opacity, color, and SEM	Extract from *F. microphylla* showed inhibitory action against *F. oxysporum.*	Carrillo-Lomelí et al. (2024)
Chitosan/Sodium alginate multilayer/Anthocyanin	Mechanical and barrier properties	Development of fully biodegradable, multifunctional films for sustainable food packaging	He et al. (2024)

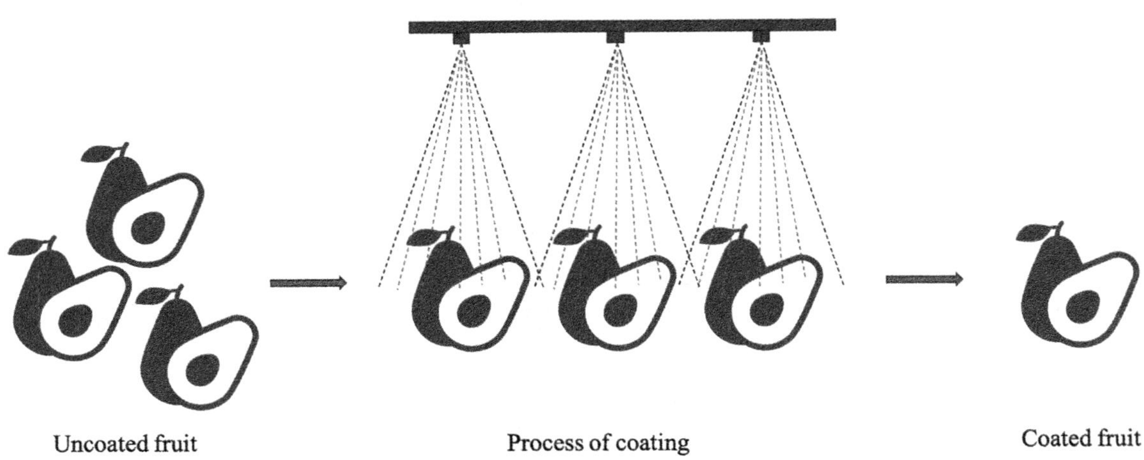

FIGURE 23.5 Spray coating on fruits for enhancing their shelf life. Coating of avocado fruits using the spraying technology for even distribution.

TABLE 23.5 Coating of fruits using spraying method.

Formulations	Analysis	Output	References
Cellulose nanofiber/citric acid/starch	Morphological, WVP, OTR, thermal stability and transparency	Characteristics of the composite film and storage stability of tomatoes can be improved by using CNFs as fillers for cross linked starch.	Wardak et al. (2024)
Soyabean starch	Physiochemical parameters of emulsion	The preservation period of the sapota was increased by 2 weeks in comparison with the control.	Dai et al. (2024)
Mung bean starch	Particle size, zeta potential, FTIR, thermal stability and morphological analysis	Papaya fruits are being preserved by coating them in mung bean starch.	Sharma et al. (2024)
Konjac glucomannan nanocomposite spray coating	Antibacterial, mechanical properties, hydrophobicity and UV shielding properties	An increased preservation period of banana can be achieved by coating it in the K NPs 15 composite.	Deng et al. (2024)
Alginate/chitosan/pectin	Physicochemical, bioactive, microbiological, and textural properties	Edible coating method for increasing the shelf life of prickly pears.	Rodrigues et al. (2025)
Corn starch/Mango kernel/Lithi starch/Clove EO	physicochemical, rheological, and microbial properties	Significant enhancement of antimicrobial and some key functional properties of edible coating using Clove EO and ultrasonic application.	Kumar et al. (2024)

compatibility of nanostructures with the polymer matrix, processing conditions and intended application in food packaging, pharmaceutical and biomedical coatings. Nonencapsulated clove essential oil incorporated in to nanonetwork porous starch based material offers dual function as a natural food preservative and improves antimicrobial activities. These findings emphasize the potential of starch derived nanomaterials as carriers of bioactive compounds, offering improved stability, slow release and long lasting antimicrobial activities. The study contributes to the development of sustainable, biodegradable and high performance biopolymer films for use in food preservation and medical applications (Fang et al., 2020). From the study carried out, it was observed that the starch grafted polyacrylic acid doped BaO nanostructures, synthesized via coprecipitation, effectively reduced the toxicity of methylene blue, with microbial activity also being assessed (Ikram et al., 2023).

The influence of the nanozinc oxide morphology on the functional properties and antimicrobial characteristics of the topioca starch film was evaluated by Tamimi and coworkers. They were able to determine that the nanosphere forms of ZnO have the strongest influence on the functional and antimicrobial characteristics (Tamimi et al., 2021). Alinaqi developed a biocomposite film by electro spraying the zein nanoparticles containing clove essential and then it was incorporated in a starch matrix at different concentrations (Alinaqi et al., 2021). Subsequently, the work done by Fallah reported the development of a rey starch nanorod incorporated with ZnO, which exhibited improved barrier properties and strong light protective capacity, rendering the film suitable for food packaging applications (Fallah et al., 2023). The pH indicators were developed based on cassava starch and anthocyanins from jambolana fruit by the thermo compression method. Films that are manufactured with laponite and jambolana fruit are more colorful, which indicates laponite can be used to preserve purple color (Gaviria et al., 2021). In this study, a bilayer of bioactive film with antioxidant and antimicrobial properties was formulated by utilizing environmentally degradable biopolymers. In the bilayer composite film, the outer layer consists of thermoplastic starch reinforced with ZnO nanorods obtained through flat die extrusion, and the inner layer contains an electrospun polyvinyl alcohol matrix loaded with rosemary polyphenols (Estevez-Areco et al., 2020).

Through the extrusion process, modified starch based biopolymer composites were produced that enhanced the properties of the film, where the starch was gelatinized with glycerol and polybutylene succinate to achieve plasticization and compatibilization (Estevez-Areco et al., 2020). To achieve active packing with modified starch, native and acetylated cassava starch, which contains potassium sorbate, was produced via the ultrasound method. Incorporation of ferulic and cinnamic acids in the starch and poly lactic acid to promote sustained release of active biomolecule (Zhang, Muiruri, et al., 2023; Zhang, Zou, et al., 2023).

4. Functional enhancements of starch based nanocomposites

4.1 Mechanical properties

A considerable enhancement in the mechanical characteristics of the starch based nanocomposite film is achieved through the introduction of various nanoparticles and plasticizers. The durability and flexibility of the films are mainly determined by their tensile strength (TS) and elongation at break (EAB%). The addition of zein nanoparticles (ZNP) enhances TS by forming hydrogen bonds and electrostatic interactions with the starch matrix while also restricting starch chain movement at higher concentrations, leading to reduced elongation. Olive oil, incorporated in the matrix, reduces TS due to the heterogeneity, whereas it can increase film flexibility at lower concentrations (Babaee et al., 2022; Farajpour et al., 2020; García-Ramón et al., 2021). Chitosan and cellulose nanoparticles significantly improve both TS and EAB%, with cellulose reinforced films showing a remarkable increase in TS from 7.68 to 47.05 MPa, though higher concentrations result in brittle films. Silver nanoparticles (AgNPs) and starch nanoparticles (StNPs) further contribute to mechanical reinforcement, with optimized formulations achieving superior stress resistance and tensile properties (Mohseni et al., 2020; Pirsa et al., 2020). Adding tragacanth and clay nanoparticle improve the film properties by increasing the thickness and tensile strength while reducing elongation. ZnO, Fe_3O_4 incorporating hybrid bionanocomposites form a dense and compact structure, increasing mechanical integrity. Similarly, nanocellulose (NCC) at 5 wt.% in sorbitol plasticized starch films significantly enhances TS (4.69 MPa) and elongation (56%). Overall, the integration of nanoparticles and plasticizers effectively strengthens starch based nanocomposite films, making them viable for applications in biodegradable packaging and sustainable materials (Arifin et al., 2022; Filip et al., 2023).

4.2 Barrier properties: Resistance to water vapor, oxygen, and other gases

Starch derived nanocomposite films with the inclusion of many reinforcing agents have shown significant improvements in their resistance to water vapor and oxygen. By the incorporation of chitosan nanoparticles (CNPs), the water vapor permeability (WVP) and oxygen transmission rate (OTR) of the topioca starch films were observed to be reduced by 4 times and 0.51 times, respectively (Shapi et al., 2022). Similarly, kaolin integration into cassava starch blend films lowered WVP and enhanced structural integrity in water. The blending of starch with polyvinyl alcohol (PVA) resulted in a 52% and 57% reduction in WVP and OP, respectively, demonstrating effective moisture and gas barrier capabilities (Rammak et al., 2021). Carnauba wax nanoparticles (CW70) enhanced starch crystallinity, leading to the lowest relative WVP (0.86) and OP (0.49). Cellulose nanofibers (CNF) and thymol incorporation in corn starch films reduced WVP by approximately 60%, while sugarcane cellulose improved the hydrophobicity of sweet potato starch films (Liu et al., 2022). Retrograded starch treated with energetic neutral argon (EAr) atoms enhanced crystallinity, contributing to better barrier properties (Sudheesh et al., 2023). The addition of silica nanoparticles (SiO_2) in cassava starch/PVA films significantly increased hydrophobicity, with a water contact angle reaching 113°. Sugar palm nanofibrillated cellulose (SPNFC) reinforcement reduced water absorption and moisture uptake. Moreover, multilayer thermoplastic starch (TPS)/poly (butylene adipate co terephthalate) [PBAT] exhibited enhanced moisture barrier properties by 86.8% and oxygen barrier properties by 74.3%. These results show that incorporating nanoparticles with polymer blending enhances the barrier performance of starch-based nanocomposite films, making them suitable for sustainable food packaging.

4.3 Thermal and UV resistance improvements

The thermal and UV resistance of starch based nanocomposite films has been greatly improved through various modifications by incorporating lignin, ZnO nanoparticles, cellulose nanofibers (CNF) and specialized chemical treatments. Lignin, when integrated as core shell nanoparticles or within starch/PBAT and starch/PVA films, provided superior UV shielding and enhanced hydrophobicity, with water contact angles exceeding 120° (Chang et al., 2021; Lin et al., 2021). The ZnONPs synthesized from *Centella asiatica* improved UV barrier properties while also increasing thermal stability (Homthawornchoo et al., 2022). Dual-modified cassava starch coatings with ferulic acid achieved excellent transparency (87.06% at 600 nm) while blocking over 92% of UV radiation at 280 nm. Similarly, starch carbamate films demonstrated enhanced UV resistance and thermal stability with increasing degrees of substitution. Moreover, CNF reinforced starch films, without requiring complex chemical modifications, exhibited improved mechanical strength and UV blocking due to hydrogen bonding between CNF, lignin, and the starch matrix (Lin et al., 2021). Coating thermoplastic starch (TPS) films with biodegradable polyesters enhances water resistance while maintaining UV barrier properties, offering a sustainable alternative to conventional petroleum plastics (Chang et al., 2021). This leads to the development of high-performance, biodegradable packaging materials with better durability and protection.

4.4 Antimicrobial and antioxidant functionalities

Starch based nanocomposite films showed antimicrobial and antioxidant properties when combined with bioactive compounds, essential oils, or nanoparticles. Addition of rice straw phenolic extracts, containing ferulic, p coumaric, and protocatechuic acids, improved antioxidant activity and oxygen barriers but made the films slightly more brittle (Menzel et al., 2020). Molecular docking investigations revealed that lemongrass essential oil incorporated into chitosan Dioscorea hispida starch films exhibited antimicrobial potential against *E. coli*, *S. typhi*, *S. aureus*, and *S. epidermidis*, primarily through predicted interactions with the bacterial FtsA protein (Istiqomah et al., 2022). Tea polyphenols in electrospun starch nanofibrous films improved antioxidant capabilities while maintaining optimal mechanical properties and water resistance (Zhang et al., 2021). Propolis Extract and anthocyanin in intelligent food packaging films provided both antibacterial protection and pH-sensitive color response, effectively detecting food spoilage in pasteurized milk (Mustafa et al., 2021). Starch films of oregano essential oil in *Dioscorea zingiberensis* exhibited improved shelf life of fresh chicken due to enhanced UV shielding and antioxidant activity (Shen et al., 2022).

Incorporating olive extract (OE) into corn starch films imparted strong antioxidant activity and raised thermal stability by 50°C, whereas the addition of extra virgin olive oil to starch chitosan blend films exhibited remarkable increases in tensile strength (158.1%) and elongation (224.6%), accompanied by improved antioxidant and antimicrobial activities (García et al., 2020). Cinnamaldehyde loaded chitosan nanoparticles (CCNPs) in PVA/fish gelatin matrices provided a sustained release mechanism, prolonging the storage life of rainbow trout fillets up to 12 days (Hosseini et al., 2022). Caraway essential oil in carrageenan-starch edible films exhibited strong antibacterial effects, particularly against Grampositive bacteria like *B. cereus* and *S. aureus*.) Integrating *Baccharis dracunculifolia* extract in starch PVA films provides antimicrobial activity against *E. coli* and *Salmonella enterica* (Casagrande et al., 2021). Additionally, tannic acid (TA) in chitosan films improved antioxidant and antibacterial properties, reducing enzymatic browning and moisture loss in banana storage applications. Incorporating Ag ZrP nanoparticles into starch/PVA electrospun films enhanced surface hydrophobicity and showed strong antibacterial activity against both Grampositive and negative strains, highlighting their potential for food packaging. These findings highlight the effectiveness of incorporating natural bioactives and nanomaterials into starch based films, offering biodegradable, antimicrobial, and antioxidant-rich alternatives for sustainable food packaging solutions.

4.5 Intelligent and active packaging features

Starch based nanocomposite films are increasingly recognized as intelligent and active packaging materials, incorporating pH sensitive indicators, antimicrobial agents, and antioxidant functionalities to improve food safety and extend shelf life (Sivapragasam et al., 2024). Starch/PVA films enriched with anthocyanins from *Lycium ruthenicum*, purple sweet potato, red cabbage, and *Aronia melanocarpa* enabled smart packaging systems, offering visual colorimetric monitoring of food quality and freshness (Qin et al., 2020; Huang et al., 2025). By responding to changes in pH and volatile nitrogen compounds released during spoilage, these films provide a reliable approach for assessing the freshness of shrimp, beef, milk, and bass fillets (Huang et al., 2025). Dual-modified starch matrices, such as acetylated distarch phosphate, cross-linked oxidized starch, and hydroxypropyl distarch phosphate, improved the mechanical stability and color retention of anthocyanin-based intelligent films (Huang et al., 2025). Slow-release active packaging has been achieved by encapsulating bioactive compounds like tea polyphenols in porous starch/maltodextrin microcapsules (PSTP@MD), cinnamon essential oil in β cyclodextrin, and thymol in bacterial cellulose, ensuring extended antibacterial activity and sustained antioxidant release over prolonged storage periods (Sivapragasam et al., 2024). Furthermore, propolis extract, nisin, and curcumin zein quercetin chondroitin sulfate nanoparticles were incorporated into starch based films to provide dual antibacterial and antioxidant properties, effectively inhibiting *E. coli*, *S. aureus*, and *Salmonella* (Sivapragasam et al., 2024). These films also exhibited improved UV shielding, water vapor barrier properties, and mechanical strength, ensuring enhanced food preservation. Additionally, steam induced crosslinking with glutaraldehyde (GTA) in starch/PVA films improved hydrophobicity, while copigmentation with ferulic acid (FA) in blueberry anthocyanin loaded films enhanced color stability and pH sensitivity, increasing their practical application value. Integrating bacteriostatic and freshness indicators makes these films sustainable, biodegradable options for real-time food monitoring and extended shelf life (Sivapragasam et al., 2024).

5. Applications in food shelf life extension

Starch based nanocomposites help preserve fresh produce by extending shelf life, maintaining quality, and reducing microbial spoilage (Bangar et al., 2023). These films act as moisture barriers, antimicrobial coatings, and modified atmosphere materials, protecting fruits and vegetables from texture loss, dehydration, metabolic changes, and decay (Bangar et al., 2023). Mechanical integrity, mechanical adaptability, and other antimicrobial characteristics of starch film increased with the incorporation of the biopolymers, plasticizer, essential oils, and nanomaterials (Zhou et al., 2021). Chitosan nanoparticles and curcumin integrated into the zein/potato starch film extended the preservation of *Schizothorax prenati* fillets for 15 days, at the same time freshness of the blueberries and raspberries was sustained up to 7 days using polyurethane-encapsulated essential oils in starch films (Wang, Zhang, Li, Deng, et al., 2023; Wang, Zhang, Li, Li, et al., 2023; Xin et al., 2020). Arrowroot starch based coatings effectively minimized weight loss and respiration rates in plums stored at 5°C for 35 days, while tapioca starch xanthan gum coatings were shown to enhance the shelf life of fresh produce (Bangar et al., 2023).

Yam starch films incorporated with glycerol plasticizer prolonged the storage period of strawberries from 14 to 21 days, and banana starch chitosan films exhibited antimicrobial activity by suppressing *Staphylococcus aureus* growth on fresh cut vegetables such as asparagus, baby corn, and Chinese cabbage (Bangar et al., 2023). Cassava starch coatings effectively preserved pineapple slices for 12 days, reduced respiration rates in mangoes, and maintained water retention in apple slices (Bangar et al., 2023). Additionally, pregelatinized potato based starch coatings with antifungal agents protected mandarins from blue/green mold and sour rot, while cassava starch alginate coatings reduced fruit weight loss and delayed browning (Bangar et al., 2023). These results emphasize the promise of starch based biodegradable films as eco friendly alternatives to conventional plastic packaging, providing improved moisture regulation, antimicrobial defense, and prolonged shelf life for both fresh-cut and whole produce (Bangar et al., 2023; Wang, Zhang, Li, Deng, et al., 2023; Wang, Zhang, Li, Li, et al., 2023; Zhou et al., 2021).

5.1 Application in dairy products

Starch based nanocomposite films show considerable promise in the dairy application by improving microbial safety, delaying lipid oxidation, and extending shelf life. This is achieved through the incorporation of bioactive agents, essential oils, and smart indicators (Francis et al., 2024). Aloe vera based edible films effectively inhibited *E. coli* and improved the storage stability of frozen dairy products such as kulfi, reducing microbial counts, lipid oxidation, and free fatty acids over 6 months of frozen storage (Mahajan et al., 2021). Similarly, clove oil and peppermint oil in whey based coatings for soft cheese, and ginger oil in starch-based coatings for butter, significantly improved antioxidant activity, reduced microbial spoilage, and maintained sensory attributes during storage (Francis et al., 2024). Anthocyanin-immobilized starch films, particularly those using black carrot extracts, were developed as intelligent freshness indicators for milk, showing visible color changes correlating with pH, acidity, and microbial growth, providing real time spoilage detection (Francis et al., 2024).

Sodium alginate based edible films enriched with essential oils effectively inhibited *Salmonella*, *Staphylococci* and molds, while also reducing weight loss and hardness in cheese (Kaur et al., 2024). Sago starch films with *Bauhinia kockiana* anthocyanins successfully monitored coconut milk freshness, undergoing visible color changes as pH declined (Francis et al., 2024). Additionally, natamycin loaded cellulose/PVA nanocomposite films exhibited antifungal activity against yeast and mold, prolonging the shelf life of Minas cheese by inhibiting *S. cerevisiae* growth. The use of polyurethane encapsulated essential oil microcapsules in starch films enabled sustained antimicrobial release, effectively preserving fresh blueberries, raspberries, and soft cheese for extended periods (Maria et al., 2024). The incorporation of bioactive agents and smart indicators in the starch derived nanocomposite films presents an eco friendly and effective solution for extending the shelf life, ensuring food safety, and reducing spoilage in dairy products, positioning them as viable solutions for active and intelligent food packaging (Francis et al., 2024; Kaur et al., 2024; Maria et al., 2024).

5.2 Use in meat and seafood packaging

Starch based nanocomposite films hold strong potential in the storage and packaging of meat and seafood, where they contribute to extended shelf life, improved quality retention, and enhanced food safety (Umaraw et al., 2020). These films act as biodegradable, antimicrobial, and antioxidant active packaging materials, offering protection against moisture loss, lipid oxidation, microbial spoilage, and structural deterioration (Prakoso et al., 2023). The integration of essential oils, bioactive agents, and nanomaterials has greatly enhanced the films' resistance to deterioration by improving their mechanical strength, barrier performance, and functional attributes (Sreekanth et al., 2024). Cassava starch fish gelatin films incorporated with cinnamon essential oil (CEO) nanoemulsion prolonged the shelf life of fish fillets up to 10 days. Similarly, chitosan/starch based films enriched with CEO and cellulose nanofibers displayed strong antibacterial effects against *Staphylococcus aureus* and *Escherichia coli*, significantly reducing the microbial load in stored beef (Arul Raj et al., 2024; Sreekanth et al., 2024). Curcumin modified rice starch films functioned as intelligent indicators for hypoxanthine detection in chicken and fish, changing color from yellow to reddish brown as spoilage progressed, enabling real time freshness monitoring (Erna et al., 2022). Edible coatings infused with proteins, polysaccharides, essential oils, and plant extracts effectively inhibited microbial spoilage in fresh and processed meat, while also preventing lipid, protein, and pigment oxidation (Umaraw et al., 2020).

Adding seafood derived biopolymers like collagen, gelatin, and chitosan to starch films improves antimicrobial and antioxidant effects, reducing spoilage and extending seafood shelf life (Gulzar et al., 2023). This shows starch nanocomposites as sustainable alternatives to synthetic packaging, providing better preservation, freshness monitoring, and food safety for meat and seafood (Umaraw et al., 2020). Starch based nanocomposites help maintain quality and extend shelf life in beef, poultry, seafood, and processed meats (Prakoso et al., 2023). These films act as biodegradable, antimicrobial, and intelligent packaging solutions, offering protection against moisture loss, lipid oxidation, microbial contamination, and spoilage indicators (Umaraw et al., 2020).

The addition of natural extracts, essential oils, and nanoparticles has further reinforced the mechanical strength, barrier efficiency, and functional attributes of starch based nanocomposite films (Sreekanth et al., 2024). Starch agar films incorporated with shikonin demonstrated pH responsive color changes, effectively signaling shrimp spoilage. Similarly, chitosan starch films enriched with pomegranate peel extract and thymus oil enhanced the shelf life of beef by reducing lipid oxidation and suppressing microbial growth (Sreekanth et al., 2024). Similarly, sea buckthorn pomace extract starch films provided superior preservation for superchilled beef and beef jerky, reducing bacterial spoilage and oxidative degradation (Arul Raj et al., 2024; Prakoso et al., 2023; Sreekant et al., 2024; Umaraw et al., 2020; Zainal Arifin et al., 2023). For seafood applications, microporous starch based films loaded with oregano essential oil successfully maintained the freshness of sea bass, slowing lipid oxidation and microbial growth.

Cassava starch and carboxymethyl cellulose sodium films incorporated with *Bacillus* spp. extended the shelf life of salmon by 3 days, while grape pomace extract starch films provided real time color based monitoring of fish freshness (Erna et al., 2022; Gulzar et al., 2023). Thermoplastic starch films incorporated with FeO and ZnO nanoparticles, which have a high level of oxygen scavenging and antimicrobial properties, showed a reduction in the microbial contamination of meat products such as mutton and chicken (Gulzar et al., 2023). Additionally, semirefined carrageenan/cassava starch films with SiO_2 ZnO nanoparticles successfully preserved minced chicken meat for 6 days by improving UV screening and antimicrobial effects (Gulzar et al., 2023). Furthermore, taro starch duck bone gelatin films provided an environmentally friendly packaging solution for beef meatballs, effectively predicting shelf life using the Arrhenius kinetic model and extending freshness under various storage temperatures (Gulzar et al., 2023). Taro starch sodium alginate films incorporated with nisin were optimized to inhibit *Listeria monocytogenes*, extending the shelf life of refrigerated chicken by 15 days. These findings highlight that starch-based nanocomposite films offer eco friendly, intelligent, and functional packaging solutions capable of prolonging the shelf life of meat and seafood products, while reducing food waste and dependence on conventional plastics (Erna et al., 2022; Gulzar et al., 2023).

6. Sustainability of starch based nanocomposites

Starch based nanocomposites are gaining attention for sustainable applications due to their biodegradability, reusability, and low ecological impact (Gamage et al., 2022). As substitutes for petroleum polymers, they are used in packaging, agriculture, biomedicine, electronics, and bioremediation (Dong et al., 2024). Current development focuses on strengthening starch nanocomposites with nanocellulose, chitin nanoparticles, nanoclay, and carbon-based materials (Khoo et al., 2023). All these enhancements increased their application by keeping the ecocompactible features, facilitating the transition toward eco friendly material options. The incorporation of Mxene nanoplates into starch-derived materials is emerging as a promising solution in the electronic sector to develop biodegradable materials and enhance the structural strength of films for wearable sensors, such as biosensors for glucose or ECG monitoring, and transient electronic devices (Dong et al., 2024). By incorporating MXene at varying concentrations, these films achieved tunable properties, including increased tensile strength (up to 11.2 MPa) and electrical conductivity ($3.98\ S\ m^{-1}$). These films decompose within a month in soil, highlighting their potential in reducing electronic waste (Dong et al., 2024).

Additionally, MXene starch nanocomposites enhanced packaging applications by significantly improving barrier properties, with reductions of water vapor permeability [92.9%] and oxygen permeability [74.0%], while still maintaining biodegradability with decomposition occurring within 6 weeks (Dong et al., 2024). Environmental remediation has also benefited from starch based nanocomposites, particularly in wastewater treatment. Functionalized starch derivatives have demonstrated exceptional adsorption capabilities, removing over 90% of contaminants such as heavy metals, organic solvents, pesticides, dyes, and pharmaceuticals (Khoo et al., 2023). With adsorption capacities reaching up to 25,000 mg/g for dyes and 2000 mg/g for heavy metals, starch based adsorbents present a viable alternative for large scale water purification, though further research is needed for commercial application (Khoo et al., 2023).

Elevated physical, antimicrobial, and barrier properties are observed in the gluten based bionanocomposite films integrated with starch nanocrystals and chitosan (Sharma et al., 2023). These films exhibited improved material characteristics, including high tensile strength (40.82 MPa), notable elongation at break (31.09%), and reduced water vapor transmission rates, supporting their use as sustainable food packaging materials (Sharma et al., 2023). Their demonstrated antibacterial activity against *S. aureus* and *E. coli* further reinforces their potential for commercial food packaging applications (Sharma et al., 2023). Remarkably, with a degradation rate exceeding 95% in 55 days under soil conditions, these films demonstrated their potential as a sustainable substitute for plastic films (Gamage et al., 2022). Altogether, starch based nanocomposites provide an eco friendly, biodegradable solution to the conventional synthetic materials (Gamage et al., 2022). Starch based nanocomposites have a broad spectrum of applications, ranging from biodegradable food packaging materials and transient electronics to ecological rehabilitation, illustrating their relevance and contributions in minimizing the environmental pollution rate and reliance on fossil based plastics (Dong et al., 2024; Khoo et al., 2023). By exploring and diving deep into this area and refining it, the starch based nanocomposites can play a vital role in fostering global sustainability efforts (Gamage et al., 2022).

Starch based nanocomposites have attracted growing attention for their sustainable attributes as they can provide more biodegradable, renewable, and eco friendly substitutes in place of petroleum based plastics (Gamage et al., 2022). Leveraging its widespread availability and environmental degradability, starch extracted from corn, potatoes, rice, cassava, and tapioca has been extensively used in many applications such as food packaging, agro industry, biomedical sector, and environmental areas (Dong et al., 2024). Despite their broad applications, poor structural strength, high water sensitivity, and limited barrier properties are some of the drawbacks of this starch based material (Khoo et al., 2023). These limitations can be overcome via the incorporation of supporting materials like cellulose fibers, nanoclays, chitin nanoparticles, carbon based materials, and fatty agents (Khoo et al., 2023). Through the integration of these strengthening materials, barrier properties, thermal stability, and mechanical strength of starch-based nanocomposites can be enhanced, increasing their suitability for industrial applications (Dong et al., 2024).

The addition of cellulose fibers, natural fibers (hemp, jute, sisal, cotton, bamboo, and bagasse), and agricultural by products (such as potato and cassava peels) has been shown to significantly improve the rigidity and mechanical strength of starch derived films (Gamage et al., 2022). By adding clays and nanocomposites, the gas permeability resistance and thermal resistance can be enhanced, whereas the moisture barriers can be improved by adding fatty agents like waxes and paraffins (Dong et al., 2024). Sourcing materials locally in developing nations can not only cut the production costs but also reduce import reliance and foster environmental sustainability. This approach aligns with the principles of a zero waste economy by transforming agricultural waste into valuable resources. One of the major aspects of starch based nanocomposites is their promising application in food packaging (Dong et al., 2024).

Starch derived films are biodegradable and effective in elevating the preservation period of the high moisture content or short lived food commodities like fruits, vegetables, meat, and dairy products compared to conventional synthetic plastics (Gamage et al., 2022). The integration of natural antimicrobial agents, essential oils, and chitosan into the starch-based films provides advantages like antimicrobial and antioxidant activities (Sharma et al., 2023). For instance, composite films containing polyurethane encapsulated essential oils demonstrated prolonged antibacterial activity while maintaining the sensory attributes of blueberries and raspberries (Sharma et al., 2023). Similarly, aloe vera-incorporated edible films contributed to enhanced microbial control and reduced lipid oxidation in frozen dairy desserts like kulfi (Gamage et al., 2022). Advanced packaging methods have been executed in the meat industry using starch based nanocomposites. It has been observed that the preservation period of the beef has been enhanced by starch based films infused with pomegranate peel extract, chitosan, and thymus essential oils, as they prevent microbial growth and lipid oxidation (Dong et al., 2024). Likewise, progression in the color stability, lipid oxidation resistance, and microbial control has been seen in starch films composed of sea buckthorn pomace extract, used for deep chilled storage of beef (Khoo et al., 2023).

The integration of microporous starch incorporated with oregano essential oils in PVA films has exhibited efficacy in storing sea bass by lowering microbial contamination and oxidative damage (Sharma et al., 2023). Apart from food packaging, starch based nanocomposites have proved their potential in the areas of environmental cleanup strategies like wastewater treatment (Khoo et al., 2023). As a result, the nontoxicity character and biodegradability potential of starch based derivatives have been examined for their adsorption ability in removing pollutants like heavy metals, aromatic compounds, and dyes from water sources (Khoo et al., 2023). Introducing functional groups like amide, amino, and carboxyl into the starch molecules increases their adsorption capacity, enabling them as a potential alternative in sustainable water purification methods (Khoo et al., 2023). Along with this, starch based nanocomposites are progressing as relevant and promising materials for electronic systems.

In electronics, starch films with conductive Mxene nanoplates serve as biodegradable components for motion tracking, tactile sensing, and electronic skin (Dong et al., 2024). These films offer adjustable conductivity, strength, and full biodegradability, decomposing within a month (Dong et al., 2024). However, large scale production faces challenges from inconsistent feedstock, a lack of standardization, and competition with traditional plastics (Gamage et al., 2022). Their biodegradability also depends on factors like temperature, moisture, microbes, and nondegradable additives. Researchers are investigating ways to enhance the industrial applicability of the starch based materials by optimizing reactive extrusion, crosslinking, and integration of performance maximizing additives. Overall, starch nanocomposites hold promise in packaging, environmental, and electronic applications. These biopolymers can play major roles in lowering the level of plastic waste, reducing carbon footprints, and fostering sustainable futures by applying the use of renewable sources, improving formulations, and addressing the challenges while scaling up (Dong et al., 2024; Sharma et al., 2023).

7. Conclusion

Progress in the starch derived nanocomposites and coatings provides a dual benefit by mitigating the environmental impacts of plastic packaging while improving sustainable food packaging. These biodegradable materials gain improved strength, barrier performance and functionality when reinforced with nanomaterials like cellulose nanofibers, metal nanoparticles, essential oils and bioactive compounds. Through the integration of active and intelligent packaging attributes, these films contributed to prolonging the shelf life of the food products and also enhanced the food safety by exerting antimicrobial and antioxidant properties. The application of starch based nanocomposite films has shown promising results across various food categories, including fresh produce, dairy, meat, and seafood. The integration of bioactive additives such as aloe vera, essential oils, and plant-derived extracts has enhanced microbial inhibition and lipid oxidative stability, providing effective preservation solutions. Additionally, the use of intelligent indicators, such as anthocyanins and pH responsive pigments, enables real time freshness monitoring, further improving consumer confidence in food quality. Beyond food packaging, starch based nanocomposites have demonstrated potential in environmental sustainability through applications in water remediation, biodegradable electronics, and transient packaging solutions. Their ability to replace petroleum-based plastics makes them a viable alternative for reducing plastic pollution and supporting circular economy initiatives. Despite these advancements, challenges remain in large scale production, cost effectiveness, and performance optimization under humid conditions. Future research should focus on improving the structural integrity of starch-based films, exploring novel reinforcement techniques, and integrating emerging technologies such as nanostructured sensors and IoT enabled packaging. Sustained innovation and collaboration are required to address commercialization challenges and enable the successful transition of strain based nanocomposites from laboratory research to industry level packaging solutions. With ongoing advancements, these materials have the potential to redefine food preservation strategies while minimizing environmental impact, making them a cornerstone of the next generation of eco friendly packaging solutions.

AI disclosure

During the preparation of this work, the author(s) used ChatGPT in order to improve the grammar and continuity. After using this tool/service, the author(s) reviewed and edited the content as needed and take(s) full responsibility for the content of the publication.

References

Aguilar-Palazuelos, E., Fitch-Vargas, P. R., Delgado-Nieblas, C. I., López-Díaz, A., Gastélum-Ávila, A., Sánchez-Chilero, M. A., & Calderón-Castro, A. (2025). Edible films based on corn starch and gelatin obtained by the combination of extrusion and casting process: Characterization and applications. *Food Biophysics, 20*(1), 2.

Ali, S., Ishtiaq, S., Nawaz, A., Naz, S., Ejaz, S., Haider, M. W., Shah, A. A., Ali, M. M., & Javad, S. (2024). Layer by layer application of chitosan and carboxymethyl cellulose coatings delays ripening of mango fruit by suppressing cell wall polysaccharides disassembly. *International Journal of Biological Macromolecules, 256*. https://doi.org/10.1016/j.ijbiomac.2023.128429

Alinaqi, Z., Khezri, A., & Rezaeinia, H. (2021). Sustained release modeling of clove essential oil from the structure of starch-based bio-nanocomposite film reinforced by electrosprayed zein nanoparticles. *International Journal of Biological Macromolecules, 173*, 193–202.

Arifin, H. R., Djali, M., Nurhadi, B., Hasim, S. A., Hilmi, A., & Puspitasari, A. V. (2022). Improved properties of corn starch-based bio-nanocomposite film with different types of plasticizers reinforced by nanocrystalline cellulose. *International Journal of Food Properties, 25*(1), 509–521. https://doi.org/10.1080/10942912.2022.2052085

Arul Raj, J. S., Kasi, M., Karuppiah, P., & Hirad, A. H. (2024). Green packaging solutions for extending the shelf life of fish fillet: Development and evaluation of cinnamon essential oil-infused cassava starch and fish gelatin edible films. *ACS Omega, 9*(46), 45898–45910. https://doi.org/10.1021/acsomega.4c05249

Babaee, M., Garavand, F., Rehman, A., Jafarazadeh, S., Amini, E., & Cacciotti, I. (2022). Biodegradability, physical, mechanical and antimicrobial attributes of starch nanocomposites containing chitosan nanoparticles. *International Journal of Biological Macromolecules, 195*, 49–58. https://doi.org/10.1016/j.ijbiomac.2021.11.162

Balderas-Guerrero, M. V., Ramos-Ramírez, E. G., González-Martínez, R., San Martín-Martínez, E., & Salazar-Montoya, J. A. (2025). Application of the electrospinning process to obtain nanofibers from triticale (Triticosecale Wittmack) starch obtained using the response surface methodology and morphological evaluation. *Industrial Crops and Products, 223*. https://doi.org/10.1016/j.indcrop.2024.120133

Bangar, S. P., Scott Whiteside, W., Suri, S., Barua, S., & Phimolsiripol, Y. (2023). Native and modified biodegradable starch-based packaging for shelf-life extension and safety of fruits/vegetables. *International Journal of Food Science and Technology, 58*(2), 862–870. https://doi.org/10.1111/ijfs.16219

Bansal, H., Sharma, A., Singh, S., & Mehta, S. K. (2025). Fabrication and application of buckwheat starch-based sustained-release composite films infused with curry leaf essential oil/β-cyclodextrin micro-capsules for preservation of green grapes. *Food Hydrocolloids, 159*, Article 110666. https://doi.org/10.1016/j.foodhyd.2024.110666

Bibi, M., Cochran, E. W., Forrester, M., Mushtaq, I., Fatima, K., Sarwar, M. I., & Zulfiqar, S. (2024). Unravelling the potential of innovative eco-friendly thermoplastic starch/SEBS-g-MA/graphene oxide bionanocomposites. *Starch-Stärke, 76*(1–2), Article 2300066.

Cao, P., Wu, G., Yao, Z., Wang, Z., Li, E., Yu, S., Liu, Q., Gilbert, R. G., & Li, S. (2022). Effects of amylose and amylopectin molecular structures on starch electrospinning. *Carbohydrate Polymers, 296*, Article 119959. https://doi.org/10.1016/j.carbpol.2022.119959

Carrillo-Lomelí, D. A., Cerqueira, M. A., Moo-Huchin, V., Bourbon, A. I., Souza, V. G., Lestido-Cardama, A., & de Rodríguez, D. J. (2024). Influence of edible multilayer coatings with *Opuntia stenopetala* polysaccharides and *Flourensia microphylla* extract on the shelf-life of cherry tomato (*Solanum lycopersicum* L.). *Scientia Horticulturae, 332*, Article 113224.

Casagrande, M., Zanela, J., Wagner Júnior, A., Yamashita, F., Busso, C., Wouk, J., Radaelli, J. C., & Malfatti, C. R. M. (2021). Optical, mechanical, antioxidant and antimicrobial properties of starch/polyvinyl alcohol biodegradable film incorporated with Baccharis dracunculifolia lyophilized extract. *Waste and Biomass Valorization, 12*(7), 3829–3848. https://doi.org/10.1007/s12649-020-01269-x

Chang, C. C., Trinh, B. M., & Mekonnen, T. H. (2021). Robust multiphase and multilayer starch/polymer (TPS/PBAT) film with simultaneous oxygen/moisture barrier properties. *Journal of Colloid and Interface Science, 593*, 290–303.

Chi, Z., Ju, S., Liu, X., Sun, F., & Zhu, Y. (2022). Graphene oxide supported sulfidated nano zero-valent iron (S-nZVI@ GO) for antimony removal: The role of active oxygen species and reaction mechanism. *Chemosphere, 308*, 136253.

Dai, L., Wang, X., Mao, X., He, L., Li, C., Zhang, J., & Chen, Y. (2024). Recent advances in starch-based coatings for the postharvest preservation of fruits and vegetables. *Carbohydrate Polymers, 328*, Article 121736. https://doi.org/10.1016/j.carbpol.2023.121736

Deng, P., Wang, Z., Bu, J., Fan, Y., Kuang, Y., & Jiang, F. (2024). Konjac glucomannan-based nanocomposite spray coating with antimicrobial, gas barrier, UV blocking, and antioxidation for bananas preservation. *International Journal of Biological Macromolecules, 265*(Pt 1), Article 130895. https://doi.org/10.1016/j.ijbiomac.2024.130895

Dong, M., Mastroianni, G., Bilotti, E., Zhang, H., & Papageorgiou, D. G. (2024). Biodegradable starch-based nanocomposite films with exceptional water and oxygen barrier properties. *ACS Sustainable Chemistry & Engineering, 12*(29), 11056–11066. https://doi.org/10.1021/acssuschemeng.4c04198

Dos Santos, F. N., Fonseca, L. M., Jansen-Alves, C., Crizel, R. L., Pires, J. B., Kroning, I. S., & Rosa Zavareze, E. (2024). Antimicrobial activity of geranium (*Pelargonium graveolens*) essential oil and its encapsulation in carioca bean starch ultrafine fibers by electrospinning. *International Journal of Biological Macromolecules, 265*, Article 130953.

El Sayed, & Alanazi, T. I. (2024). Improving the structural, optical, and electrical properties of carboxymethyl cellulose/starch/selenium oxide nanocomposites for flexible electronic devices. *Scientific Reports, 14*(1), 3398.

Erna, K. H., Felicia, W. X. L., Rovina, K., Vonnie, J. M., & Huda, N. (2022). Development of curcumin/rice starch films for sensitive detection of hypoxanthine in chicken and fish meat. *Carbohydrate Polymer Technologies and Applications, 3*. https://doi.org/10.1016/j.carpta.2022.100189

Estevez-Areco, S., Guz, L., Candal, R., & Goyanes, S. (2020). Active bilayer films based on cassava starch incorporating ZnO nanorods and PVA electrospun mats containing rosemary extract. *Food Hydrocolloids, 108*, Article 106054. https://doi.org/10.1016/j.foodhyd.2020.106054

Fallah, N., Marvizadeh, M. M., Jahangiri, R., Zeinalzadeh, A., & Nafchi, A. M. (2023). High-barrier and light-protective bionanocomposite film based on rye starch/nanorod-ZnO for food packaging applications. *Journal of Chemical Health Risks, 13*(2).

Fang, Y., Fu, J., Liu, P., & Cu, B. (2020). Morphology and characteristics of 3D nanonetwork porous starch-based nanomaterial via a simple sacrifice template approach for clove essential oil encapsulation. *Industrial Crops and Products, 143*, Article 111939. https://doi.org/10.1016/j.indcrop.2019.111939

Farajpour, R., Emam Djomeh, Z., Moeini, S., Tavahkolipour, H., & Safayan, S. (2020). Structural and physico-mechanical properties of potato starch-olive oil edible films reinforced with zein nanoparticles. *International Journal of Biological Macromolecules, 149*, 941–950. https://doi.org/10.1016/j.ijbiomac.2020.01.175

Filip, D., Macocinschi, D., Luminita Nica, S., Asandulesa, M., Condurache, B., Stoleru, E., Mihaela Rata, D., Bargan, A., & Zaltariov, M.-F. (2023). Hybrid green bionanocomposites based on chitosan/starch/gelatin and metallic nanoparticles for biological applications. *International Journal of Biological Macromolecules, 253*(Pt 8), Article 127571. https://doi.org/10.1016/j.ijbiomac.2023.127571

Firouz, M. S., Mohi-Alden, K., & Omid, M. (2021). A critical review on intelligent and active packaging in the food industry: Research and development. *Food Research International, 141*, Article 110113.

Francis, D. V., Dahiya, D., Gokhale, T., & Nigam, P. S. (2024). Sustainable packaging materials for fermented probiotic dairy or non-dairy food and beverage products: Challenges and innovations. *AIMS Microbiology, 10*(2), 320–339. https://doi.org/10.3934/microbiol.2024017

Gamage, A., Thiviya, P., Mani, S., Graceraj Ponnusamy, P., Manamperi, A., Evon, P., Merah, O., & Madhujith, T. (2022). Environmental properties and applications of biodegradable starch-based nanocomposites. *Polymers, 14*(21), 4578. https://doi.org/10.3390/polym14214578

Ganjeh, A. M., Saraiva, J. A., Pinto, C. A., Casal, S., & Gonçalves, I. (2024). Recent advances in nano-reinforced food packaging based on biodegradable polymers using layer-by-layer assembly: A review. *Carbohydrate Polymer Technologies and Applications, 7*. https://doi.org/10.1016/j.carpta.2023.100395

García, A. V., Álvarez-Pérez, O. B., Rojas, R., Aguilar, C. N., & Garrigós, M. C. (2020). Impact of olive extract addition on corn starch-based active edible films properties for food packaging applications. *Foods, 9*(9), 1339.

Gaviria, Y. A. R., Palencia, N. S. N., Capello, C., Trevisol, T. C., Monteiro, A. R., & Valencia, G. A. (2021). Nanostructured pH-indicator films based on cassava starch, laponite, and jambolan (Syzygium cumini) fruit manufactured by thermo-compression. *Starch - Stärke, 73*(5–6). https://doi.org/10.1002/star.202000208

Gulzar, S., Tagrida, M., Prodpran, T., Li, L., & Benjakul, S. (2023). Packaging films based on biopolymers from seafood processing wastes: Preparation, properties, and their applications for shelf-life extension of seafoods—A comprehensive review. *Comprehensive Reviews in Food Science and Food Safety, 22*(6), 4451–4483. https://doi.org/10.1111/1541-4337.13230

Hajipour, F. P., Feyzbakhsh, A., Maleknia, L., & Ahanian, I. (2025). Electrospun scaffold with bioactive polyurethane shell infused with propolis and starch-hyaluronic acid core: An advanced therapeutic platform for skin tissue engineering. *International Journal of Biological Macromolecules, 288*, Article 138745. https://doi.org/10.1016/j.ijbiomac.2024.138745

He, M., Zheng, Y., Shen, J., Shi, J., Zhang, Y., Xiao, Y., & Che, J. (2024). Chitosan/sodium alginate multilayer pH-sensitive films based on layer-by-layer self-assembly for intelligent packaging. *Journal of Renewable Materials, 12*(2), 215–233. https://doi.org/10.32604/jrm.2023.043659

Hernando, H., Marpongahtun, E. J., Nuryawan, A., Amaturrahim, S. A., Piliang, A. F. R., Yanhar, M. R., Goei, R., Soykeabkaew, N., Saputra, A. M. A., & Gea, S. (2024). Impact of glycerol on oil palm trunk starch bioplastics enhanced with citric-acid epoxidized palm oil oligomers. *Case Studies in Chemical and Environmental Engineering, 10*. https://doi.org/10.1016/j.cscee.2024.100839

Homthawornchoo, W., Kaewprachu, P., Pinijsuwan, S., Romruen, O., & Rawdkuen, S. (2022). Enhancing the UV-light barrier, thermal stability, tensile strength, and antimicrobial properties of rice starch–gelatin composite films through the incorporation of zinc oxide nanoparticles. *Polymers, 14*(12), 2505. https://doi.org/10.3390/polym14122505

Hosseini, S. F., Ghaderi, J., & Gómez-Guillén, M. C. (2022). Tailoring physico-mechanical and antimicrobial/antioxidant properties of biopolymeric films by cinnamaldehyde-loaded chitosan nanoparticles and their application in packaging of fresh rainbow trout fillets. *Food Hydrocolloids, 124*, Article 107249.

Hu, D., Xu, Y., Gao, C., Meng, L., Feng, X., Wang, Z., Shen, X., & Tang, X. (2024). Preparation and characterization of starch/PBAT film containing hydroxypropyl-β-cyclodextrin/ethyl lauroyl arginate/cinnamon essential oil microcapsules and its application in the preservation of strawberry. *International Journal of Biological Macromolecules, 259*(Pt 2), Article 129204. https://doi.org/10.1016/j.ijbiomac.2024.129204

Huang, X., Teng, Z., Xie, F., Wang, G., Li, Y., Liu, X., & Li, S. (2024). Loading of cinnamon essential oil into electrospun octenylsuccinylated starch-pullulan nanofiber mats: Electrospinnability evaluation, structural characterization, and antibacterial potential. *Food Hydrocolloids, 148*, Article 109426. https://doi.org/10.1016/j.foodhyd.2023.109426

Huang, X., Zhang, K., Li, Z., Zhang, J., Zhai, X., Zhang, N., Du, L., & Qin, Z. (2025). Exploring the integration of anthocyanins with functional materials in smart food packaging: From stabilization to application. *Foods, 14*(16), 2896.

Ikram, M., Haider, A., Imran, M., Haider, J., Naz, S., Ul-Hamid, A., & Ali, S. (2023). Assessment of catalytic, antimicrobial, and molecular docking analysis of starch-grafted polyacrylic acid doped BaO nanostructures. *International Journal of Biological Macromolecules, 230*, Article 123190.

Istiqomah, A., Prasetyo, W. E., Firdaus, M., & Kusumaningsih, T. (2022). Valorisation of lemongrass essential oils onto chitosan-starch film for sustainable active packaging: Greatly enhanced antibacterial and antioxidant activity. *International Journal of Biological Macromolecules, 210*, 669–681. https://doi.org/10.1016/j.ijbiomac.2022.04.223

Jafarzadeh, S., Salehabadi, A., Mohammadi Nafchi, A., Oladzadabbasabadi, N., & Jafari, S. M. (2021). Cheese packaging by edible coatings and biodegradable nanocomposites; improvement in shelf life, physicochemical and sensory properties. *Trends in Food Science and Technology, 116*, 218–231. https://doi.org/10.1016/j.tifs.2021.07.021

Kaur, G., Singh, P., Sharma, T., & Singh, A. (2024). Enhanced shelf-stability of cottage cheese using synergistic starch-based nanocomposite films reinforced with montmorillonite, starch nanocrystals, and antimicrobial lemongrass oil. *Journal of Food Measurement and Characterization*, 1–13.

Khoo, P. S., Ilyas, R. A., Uda, M. N. A., Hassan, S. A., Nordin, A. H., Norfarhana, A. S., Ab Hamid, N. H., Rani, M. S. A., Abral, H., Norrahim, M. N. F., Knight, V. F., Lee, C. L., & Rafiqah, S. A. (2023). Starch-based polymer materials as advanced adsorbents for sustainable water treatment: Current status, challenges, and future perspectives. *Polymers, 15*(14). https://doi.org/10.3390/polym15143114

Kumar, N., Khan, A. A., Pyngrope, D., Alanazi, A. M., Upadhyay, A., & Shukla, S. (2024). Development and characterization of novel starch (mango kernel and litchi seed) based active edible coatings and films using ultrasonication: Effects on postharvest shelf life of Khasi mandarins. *Sustainable Chemistry and Pharmacy, 39*. https://doi.org/10.1016/j.scp.2024.101610

Li, Y., Wang, F., Xu, J., Wang, T., Zhan, J., Ma, R., & Tian, Y. (2023). Improvement in the optical properties of starch coatings via chemical-physical combination strategy for fruits preservation. *Food Hydrocolloids, 137*, Article 108405.

Lin, B., Li, C., Chen, F., & Liu, C. (2021). Continuous blown film preparation of high starch content composite films with high ultraviolet aging resistance and excellent mechanical properties. *Polymers, 13*(21), 3813. https://doi.org/10.3390/polym13213813

Liu, S., Li, L., Li, B., Zhu, J., & Li, X. (2022). Size effect of carnauba wax nanoparticles on water vapor and oxygen barrier properties of starch-based film. *Carbohydrate Polymers, 296*, Article 119935.

Luchesi, B. R., Moreira, F. K. V., & Marconcini, J. M. (2024). Scalable production of hydrophobic starch/beeswax films by continuous solution casting. *Journal of Applied Polymer Science, 141*(1), Article e54730.

Lv, H., Xu, H., Xu, E., Jin, Z., Zhao, H., Yuan, C., & Cui, B. (2024). Improving structural and functional properties of starch-catechin-based green nanofiber mats for active food packaging by electrospinning and crosslinking techniques. *International Journal of Biological Macromolecules, 267*, Article 131460.

Magnaghi, L. R., Guembe-Garcia, M., Cerone, V., Perugini, P., Alberti, G., & Biesuz, R. (2024). DOE-based multi-criteria optimization of starch/gly/CMC films' composition and preparation procedure by casting deposition. *Chemometrics and Intelligent Laboratory Systems, 244*, Article 105044. https://doi.org/10.1016/j.chemolab.2023.105044

Mahajan, K., Kumar, S., Bhat, Z. F., Naqvi, Z., Mungure, T. E., & Bekhit, A. E. D. A. (2021). Functionalization of carrageenan-based edible film using Aloe vera for improved lipid oxidative and microbial stability of frozen dairy products. *Food Bioscience, 43*, Article 101336.

Maria, A. A., Salem, R. H., Salama, M. A., & Khalil, A. M. M. (2024). Antioxidant-rich biodegradable films: Incorporating date phenolic extracts into polyvinyl alcohol biofilms for strawberry preservation. *Journal of Food and Dairy Sciences, 0*(0), 203–217. https://doi.org/10.21608/jfds.2024.328102.1171

Menzel, C., González-Martínez, C., Vilaplana, F., Diretto, G., & Chiralt, A. (2020). Incorporation of natural antioxidants from rice straw into renewable starch films. *International Journal of Biological Macromolecules, 146*, 976–986. https://doi.org/10.1016/j.ijbiomac.2019.09.222

Mohseni, M. S., Khalilzadeh, M. A., Mohseni, M., Hargalani, F. Z., Getso, M. I., Raissi, V., & Raiesi, O. (2020). Green synthesis of Ag nanoparticles from pomegranate seeds extract and synthesis of Ag-Starch nanocomposite and characterization of mechanical properties of the films. *Biocatalysis and Agricultural Biotechnology, 25*. https://doi.org/10.1016/j.bcab.2020.101569

Mustafa, P., Niazi, M. B. K., Jahan, Z., Rafiq, S., Ahmad, T., Sikander, U., & Javaid, F. (2021). Improving functional properties of PVA/starch-based films as active and intelligent food packaging by incorporating propolis and anthocyanin. *Polymers and Polymer Composites, 29*(9), 1472–1484. https://doi.org/10.1177/0967391120973503

Nain, N., Katoch, G. K., Kaur, S., & Rasane, P. (2021). Recent developments in edible coatings for fresh fruits and vegetables. *Journal of Horticultural Research, 29*(2), 127–140. https://doi.org/10.2478/johr-2021-0022

Park, K., Oh, Y., Panda, P. K., & Seo, J. (2022). Effects of an acidic catalyst on the barrier and water resistance properties of crosslinked poly (vinyl alcohol) and boric acid films. *Progress in Organic Coatings, 173*. https://doi.org/10.1016/j.porgcoat.2022.107186

Pirsa, S., Mohtarami, F., & Kalantari, S. (2020). Preparation of biodegradable composite starch/tragacanth gum/nanoclay film and study of its physicochemical and mechanical properties. *Chemical Review and Letters, 3*(3), 98–103.

Prakoso, F. A. H., Indiarto, R., & Utama, G. L. (2023). Edible film casting techniques and materials and their utilization for meat-based product packaging. *Polymers, 15*(13), 2800.

Promhuad, K., Ebel, L., & Harnkarnsujarit, N. (2025). Thermoplastic starch/poly(butylene adipate-co-terephthalate) blown film with maltol and ethyl maltol preserving cake quality: Morphology and antimicrobial function. *Food Chemistry, 464*, Article 141646.

Qin, Y., Liu, Y., Zhang, X., & Liu, J. (2020). Development of active and intelligent packaging by incorporating betalains from red pitaya (Hylocereus polyrhizus) peel into starch/polyvinyl alcohol films. *Food Hydrocolloids, 100*, Article 105410. https://doi.org/10.1016/j.foodhyd.2019.105410

Rammak, T., Boonsuk, P., & Kaewtatip, K. (2021). Mechanical and barrier properties of starch blend films enhanced with kaolin for application in food packaging. *International Journal of Biological Macromolecules, 192*, 1013–1020.

Ren, M., Liu, Y., Koranteng, E., Weng, F., & Wu, Q. (2024). Special interface structure and properties for compatible packaging film from biodegradable poly (butylene adipate-co-terephthalate)/corn starch composite. *Polymer Engineering & Science, 64*(6), 2824–2840. https://doi.org/10.1002/pen.26728

Rodrigues, C., Polesca, C., Bicalho, I., Souza, V. G. L., Coelhoso, I., & Fernando, A. L. (2025). Quality preservation and shelf-life extension of prickly pear (Opuntia ficus-indica L. Mill) using edible coatings. *Foods, 14*(2). https://doi.org/10.3390/foods14020161

Satdive, A., Tayde, S., & Chatterjee, A. (2022). Flammability properties of the bionanocomposites reinforced with fire retardant filler. In *Polymer based bio-nanocomposites: Properties, durability and applications* (pp. 69–86). Singapore: Springer.

Sawitri, A., Luthfianti, H. R., Hapidin, D. A., Edikresnha, D., Palar, P. S., Surtiyeni, N., Wibowo, A., & Khairurrijal, K. (2025). Synthesis and characterization of novel electrospun nanofibers based on taro starch: Influence of solvent and isolation agent on morphology and diameter. *Polymer International, 74*(3), 217–230. https://doi.org/10.1002/pi.6709

Shapi, R. A., Othman, S. H., Basha, R. K., & Naim, M. N. (2022). Mechanical, thermal, and barrier properties of starch films incorporated with chitosan nanoparticles. *Nanotechnology Reviews, 11*(1), 1464–1477.

Sharma, M., Bains, A., Bala Dhull, S., Chawla, P., Goksen, G., & Ali, N. (2024). Extraction, characterization, and utilization of mung bean starch as an edible coating material for papaya fruit shelf-life enhancement. *Food Science and Nutrition, 12*(7), 5188–5200. https://doi.org/10.1002/fsn3.4166

Sharma, T., Kaur, G., Singh, A., Kaur, P., Dar, B. N., & Kaur, A. (2023). An emerging sustainable approach for development and characterization of gluten-based nanocomposite films reinforced with starch nanocrystals in conjugation with chitosan. *Sustainable Chemistry and Pharmacy, 36*, Article 101338. https://doi.org/10.1016/j.scp.2023.101338

Shen, Y., Zhou, J., Yang, C., Chen, Y., Yang, Y., Zhou, C., Wang, L., Xia, G., Yu, X., & Yang, H. (2022). Preparation and characterization of oregano essential oil-loaded Dioscorea zingiberensis starch film with antioxidant and antibacterial activity and its application in chicken preservation. *International Journal of Biological Macromolecules, 212*, 20–30. https://doi.org/10.1016/j.ijbiomac.2022.05.114

Sivapragasam, N., Maqsood, S., & Rupasinghe, H. V. (2024). Berry bioactive compounds immobilized in starch matrix for active and intelligent packaging: A review. *Future Foods*, Article 100397.

Sreekanth, K., Sharath, K. P., Midhun Dominic, C. D., Divya, M., & Radhakrishnan, E. K. (2024). Microbial load reduction in stored raw beef meat using chitosan/starch-based active packaging films incorporated with cellulose nanofibers and cinnamon essential oil. *Meat Science, 216*, Article 109552. https://doi.org/10.1016/j.meatsci.2024.109552

Sudheesh, C., Sunooj, K. V., Aaliya, B., Navaf, M., Parambil Akhila, P., Ahmad Mir, S., Sabu, S., Sasidharan, A., Sudheer, K. P., Sinha, S. K., Sajeevkumar, V. A., & George, J. (2023). Effect of energetic neutrals on the kithul starch retrogradation; potential utilization for improving mechanical and barrier properties of films. *Food Chemistry, 398*, Article 133881. https://doi.org/10.1016/j.foodchem.2022.133881

Taherimehr, M., Yousefnia Pasha, H., Tabatabaeekoloor, R., & Pesaranhajiabbas, E. (2021). Trends and challenges of biopolymer-based nanocomposites in food packaging. *Comprehensive Reviews in Food Science and Food Safety, 20*(6), 5321–5344.

Tamimi, N., Mohammadi Nafchi, A., Moghaddam, H. H., & Baghaie, H. (2021). The effects of nano-zinc oxide morphology on functional and antibacterial properties of tapioca starch bionanocomposite. *Food Science and Nutrition, 9*(8), 4497–4508. https://doi.org/10.1002/fsn3.2426

Umaraw, P., Munekata, P. E. S., Verma, A. K., Barba, F. J., Singh, V. P., Kumar, P., & Lorenzo, J. M. (2020). Edible films/coating with tailored properties for active packaging of meat, fish and derived products. *Trends in Food Science and Technology, 98*, 10–24. https://doi.org/10.1016/j.tifs.2020.01.032

Uzair, B., Syed, N., Kanwal, A., Samin, G., Niazi, M. B. K., Abbas, S., & Menaa, F. (2024). Synthesis of AgO/CuO/PVA/starch hydrogel by casting method and characterizations to safely overcome skin infections: A possible application in wound healing as a dressing. *Materials Today Communications, 39*, Article 109286.

Wang, W., Qu, W., Zhang, C., Fan, Q., Xue, W., Wang, J., & Zhu, J. (2024). Graphene-oxide-infused marine biopolymers: Layer-by-layer assembled nanocomposite packaging for prolonged fresh-cut apple preservation. *ACS Sustainable Chemistry & Engineering, 12*(8), 3012–3024.

Wang, W., Zhang, W., Li, L., Deng, W., Liu, M., & Hu, J. (2023a). Biodegradable starch-based packaging films incorporated with polyurethane-encapsulated essential-oil microcapsules for sustained food preservation. *International Journal of Biological Macromolecules, 235*, Article 123889.

Wang, S., Zhang, P., Li, Y., Li, J., Li, X., Yang, J., & Zhang, C. (2023b). Recent advances and future challenges of the starch-based bio-composites for engineering applications. *Carbohydrate Polymers, 307*, Article 120627.

Wardak, M. H., Nkede, F. N., Van, T. T., Meng, F., & Tanaka, F. (2024). Development of edible films and partial coating, a novel coating technique for tomato fruits, using citric acid-crosslinked starch and cellulose nanofiber. *Progress in Organic Coatings, 187*, Article 108127.

Xin, S., Xiao, L., Dong, X., Xiang, L., Wang, Y., Hu, X., Sameen, D. E., Wen, Q., & Zhu, B. (2020). Preparation of chitosan/curcumin nanoparticles based zein and potato starch composite films for Schizothorax prenati fillet preservation. *International Journal of Biological Macromolecules, 164*, 211–221. https://doi.org/10.1016/j.ijbiomac.2020.07.082

Zainal Arifin, M. A., Mohd Adzahan, N., Zainal Abedin, N. H., & Lasik-Kurdyś, M. (2023). Utilization of food waste and by-products in the fabrication of active and intelligent packaging for seafood and meat products. *Foods, 12*(3). https://doi.org/10.3390/foods12030456

Zhang, D., Chen, L., Cai, J., Dong, Q., Din, Z.u., Hu, Z. Z., Wang, G. Z., Ding, W. P., He, J. R., & Cheng, S. Y. (2021). Starch/tea polyphenols nanofibrous films for food packaging application: From facile construction to enhance mechanical, antioxidant and hydrophobic properties. *Food Chemistry, 360*. https://doi.org/10.1016/j.foodchem.2021.129922

Zhang, X., Muiruri, J. K., Yeo, J. C. C., Lin, T. T., Vijayakumar, R., Lee, C. H., Toh, J. P. W., Thitsartarn, W., & Li, Z. (2023a). Incorporating organoclays into sustainable starch/polylactide biocomposites for enhanced mechanical and thermal properties. *Journal of Applied Polymer Science, 140*(39). https://doi.org/10.1002/app.54458

Zhang, Y., Zeng, J., Jie, Z., Gao, H., Su, T., Li, Z., & Liu, F. (2024). Development and characterization of an active starch-based film as a chlorogenic acid delivery system. *International Journal of Biological Macromolecules, 255*, Article 128055.

Zhang, J., Zou, F., Tao, H., Gao, W., Guo, L., Cui, B., & Fang, Y. (2023b). Effects of different sources of cellulose on mechanical and barrier properties of thermoplastic sweet potato starch films. *Industrial Crops and Products, 194*, Article 116358.

Zhou, X., Cheng, R., Wang, B., Zeng, J., Xu, J., Li, J., Kang, L., Cheng, Z., Gao, W., & Chen, K. (2021). Biodegradable sandwich-architectured films derived from pea starch and polylactic acid with enhanced shelf-life for fruit preservation. *Carbohydrate Polymers, 251*, Article 117117. https://doi.org/10.1016/j.carbpol.2020.117117

Further reading

Alzate, P., Gerschenson, L., & Flores, S. (2021). Study of the performance of particles based on modified starches containing potassium sorbate and incorporated into biodegradable films: Physicochemical characterization and antimicrobial action. *Chemistry, 3*(2), 658–671.

Arayaphan, J., Boonsuk, P., & Chantarak, S. (2020). Enhancement of water barrier properties of cassava starch-based biodegradable films using silica particles. *Iranian Polymer Journal (English Edition), 29*(9), 749–757.

Ballesteros-Mártinez, L., Pérez-Cervera, C., & Andrade-Pizarro, R. (2020). Effect of glycerol and sorbitol concentrations on mechanical, optical, and barrier properties of sweet potato starch film. *NFS Journal, 20*, 1–9.

Cheng, M., Cui, Y., Yan, X., Zhang, R., Wang, J., & Wang, X. (2022). Effect of dual-modified cassava starches on intelligent packaging films containing red cabbage extracts. *Food Hydrocolloids, 124*, Article 107225.

Cheng, Y., Sun, C., Zhai, X., Zhang, R., Zhang, S., Sun, C., & Hou, H. (2021). Effect of lipids with different physical state on the physicochemical properties of starch/gelatin edible films prepared by extrusion blowing. *International Journal of Biological Macromolecules, 185*, 1005–1014.

Chi, K., Wang, H., & Catchmark, J. M. (2020). Sustainable starch-based barrier coatings for packaging applications. *Food Hydrocolloids, 103*, Article 105696.

El Sayed, A. M., & Alanazi, T. I. (2024). Improving the structural, optical, and electrical properties of carboxymethyl cellulose/starch/selenium oxide nanocomposites for flexible electronic devices. *Scientific Reports, 14*(1), 3398.

Ezati, P., & Rhim, J. W. (2021). Starch and agar-based color-indicator films integrated with shikonin for smart packaging application of shrimp. *ACS Food Science & Technology, 1*(10), 1963–1969.

García-Ramón, J. A., Carmona-García, R., Valera-Zaragoza, M., Aparicio-Saguilán, A., Bello-Pérez, L. A., Aguirre-Cruz, A., & Alvarez-Ramirez, J. (2021). Morphological, barrier, and mechanical properties of banana starch films reinforced with cellulose nanoparticles from plantain rachis. *International Journal of Biological Macromolecules, 187*, 35–42.

Guo, Z., Ge, X., Gou, Q., Yang, L., Han, M., Han, G., & Han, L. (2021). Changes in chilled beef packaged in starch film containing sea buckthorn pomace extract and quality changes in the film during super-chilled storage. *Meat Science, 182*, Article 108620.

Guo, Z., Han, L., Yu, Q. L., & Lin, L. (2020). Effect of a sea buckthorn pomace extract-esterified potato starch film on the quality and spoilage bacteria of beef jerky sold in supermarket. *Food Chemistry, 326*, Article 127001.

Hasan, M., Rusman, R., Khaldun, I., Ardana, L., Mudatsir, M., & Fansuri, H. (2020). Active edible sugar palm starch-chitosan films carrying extra virgin olive oil: Barrier, thermo-mechanical, antioxidant, and antimicrobial properties. *International Journal of Biological Macromolecules, 163*, 766–775.

Ilyas, R. A., Sapuan, S. M., Atiqah, A., Ibrahim, R., Abral, H., Ishak, M. R., & Ya, H. (2020). Sugar palm (*Arenga pinnata*) starch films containing sugar palm nanofibrillated cellulose as reinforcement: Water barrier properties. *Polymer Composites, 41*(2), 459–467.

Miao, Z., Lv, R., Teng, S., Cao, C., & Lu, P. (2022). Development of antioxidant active packaging films with slow release properties incorporated with tea polyphenols-loaded porous starch microcapsules. *International Journal of Biological Macromolecules, 222*, 403–412.

Othman, S. H., Nordin, N., Azman, N. A. A., Tawakkal, I. S. M. A., & Basha, R. K. (2021). Effects of nanocellulose fiber and thymol on mechanical, thermal, and barrier properties of corn starch films. *International Journal of Biological Macromolecules, 183*, 1352–1361.

Othman, S. H., Wane, B. M., Nordin, N., Noor Hasnan, N. Z., A. Talib, R., & Karyadi, J. N. W. (2021). Physical, mechanical, and water vapor barrier properties of starch/cellulose nanofiber/thymol bionanocomposite films. *Polymers, 13*(23), 4060.

Qin, Y., Yun, D., Xu, F., Chen, D., Kan, J., & Liu, J. (2021). Smart packaging films based on starch/polyvinyl alcohol and *Lycium ruthenicum* anthocyanins-loaded nano-complexes: Functionality, stability and application. *Food Hydrocolloids, 119*, Article 106850.

Sun, Z., Ning, R., Qin, M., Liang, J., Jiang, J., Sun, W., & Zi, M. (2022). Sustainable and hydrophobic polysaccharide-based mulch film with thermally stable and ultraviolet resistance performance. *Carbohydrate Polymers, 295*, Article 119865.

Tarique, J., Sapuan, S. M., & Khalina, A. (2021). Effect of glycerol plasticizer loading on the physical, mechanical, thermal, and barrier properties of arrowroot (*Maranta arundinacea*) starch biopolymers. *Scientific Reports, 11*(1), Article 13900.

Tavares, K. M., de Campos, A., Luchesi, B. R., Resende, A. A., de Oliveira, J. E., & Marconcini, J. M. (2020). Effect of carboxymethyl cellulose concentration on mechanical and water vapor barrier properties of corn starch films. *Carbohydrate Polymers, 246*, Article 116521.

Ting, L., Li, R., Luo, H., Peng, L., Wang, J., Li, S., Zhou, M., Yuan, X., Zhang, Z., & Wu, H. (2024). Eggshell powder as a bio-filler for starch and gelatin: Ternary biodegradable composite films manufactured by extrusion compression molding. *Food Hydrocolloids, 150*, Article 109632. https://doi.org/10.1016/j.foodhyd.2023.109632

Wu, C., Jiang, H., Zhao, J., Humayun, M., Wu, S., Wang, C., & Pang, J. (2022). A novel strategy to formulate edible active-intelligent packaging films for achieving dynamic visualization of product freshness. *Food Hydrocolloids, 133*, Article 107998.

Wu, X., Yan, X., Zhang, J., Wu, X., Zhang, Q., & Zhang, B. (2024). Intelligent films based on dual-modified starch and microencapsulated *Aronia melanocarpa* anthocyanins: Functionality, stability and application. *International Journal of Biological Macromolecules, 275*, Article 134076.

Yong, K. J. (2022). *Optimization study properties of corn cob/polyvinyl alcohol/glycerol biodegradable film considering the effect of chemical composition*. Doctoral dissertation, Universiti Tunku Abdul Rahman.

Zhai, X., Wang, W., Zhang, H., Dai, Y., Dong, H., & Hou, H. (2020). Effects of high starch content on the physicochemical properties of starch/PBAT nanocomposite films prepared by extrusion blowing. *Carbohydrate Polymers, 239*, Article 116231.

Chapter 24

Disruptive technologies: Impact of artificial intelligence and blockchain on starch supply chain management

Abubakar Kabir Adam[1], **Dragan Pamucar**[2,3] **and Tithli Sadhu**[1,4]

[1]*Department of Food Science and Technology, School of Agriculture, SR University, Warangal, Telangana, India;* [2]*University of Belgrade, Faculty of Organizational Sciences, Department of Operations Research and Statistics, Belgrade, Serbia;* [3]*UNEC Applied Artificial Intelligence Research Center, Azerbaijan State University of Economics (UNEC), Baku, Azerbaijan;* [4]*Department of Biochemistry, School of Agriculture, SR University, Warangal, Telangana, India*

1. Introduction

A starch supply chain faces numerous challenges, including the cost of handling, inadequate infrastructure, traceability mistrust, and environmental impacts that lead to economic and resource losses. Inefficiencies in the starch supply chain caused significant losses worldwide, making achieving sufficiency at industrial scales difficult. However, a food product supply chain involving complex and sensitive components may come up with trust, traceability, and data security requirements, which can be achieved only through innovative technologies. The widely acceptable phenomenon entrusted to address these challenges in the supply chain is disruptive technologies, including artificial intelligence (AI) and blockchain.

The concept of disruptive technologies is an emerging technology widely used by industries to increase efficiency in productivity, correct defects obtained due to conventional methods, assess the impact of the variables, assess risk, and predict the changes of the resulting analytical processes. From the supply chain perspective, disruptive technologies have shifted toward incorporating protocols to improve efficiency and reliability. Over the years, consumers have required a product that meets their demands and ensures safety, convenience, sustainability, and availability. This has called on the stakeholders of biological materials, especially food products, to use emerging ways to satisfy the ethical behavior of the consumers and standards. It remains a critical approach for key stakeholders to invest in up-to-date technologies due to the changing preferences of consumers and lifestyles. Despite the long-term use of different ways in food supply chain management, technologies are still required to outperform traditional practices. Effective supply chain assessment may be difficult with conventional practices due to inadequate historical information on the product, making it difficult to attain the current demands. However, the disruption of the process in the supply chain through advanced technologies can assist in overcoming the problems associated with and attaining sustainability practices.

Artificial intelligence (AI) is an emerging disruptive technology that performs various functions, including supply chain management (SCM) (Riahi et al., 2021). The process is instrumental in analyzing data trends to mitigate uncertainties and manage potential risks along the SCM (Budak & Sarvari, 2021). These revolutionary advancements in SCM have significantly helped to address challenges and optimize processes. Additionally, it allows users to identify patterns within the SCM, overcoming the shortcomings and uncertainties. Economic and dynamic changes in the SCM open room for the competitive use of these driven technologies, which track the data history at each stage of the product transfer. Supply chain stakeholders better understand how the supply chain process streams impact quality changes and predict better ways to optimize for enhancing economic and safety requirements. In recent scenarios, AI has been attractive in the supply chain because industries, including the food sector, collect data to explore and meet consumers' quality standards and convenience (Dedeoglu et al., 2023).

Native and Modified Starches for Food Applications. https://doi.org/10.1016/B978-0-443-27431-2.00001-8

Another paradigm disruptive technology is the use of blockchain, which creates a ledger distributed system among the various components of SCM (Ahmad et al., 2024). Its decentralized and immutable ledger system enables secure real-time data and information access to the supply chain—including farming down to the retail by all stakeholders. Concurrently, blockchain helps in regulatory compliance by ensuring records are accurate, up-to-date, and easily accessible for audits. This is increasingly important in a globalized market where food products frequently cross international borders and must adhere to regulatory standards. Blockchain technology has adopted a Peer-to-Peer (P2P) network design for transaction storage and processing. It has implemented consensus algorithms such as Proof of Stake (PoS) and Proof of Work (PoW) to verify the validity of the transactions. Blockchain platforms are classified based on their application as either public and/or private. Public blockchain platforms offer high visibility and transparency at the cost of confidentiality and privacy (Ahmad et al., 2024). Because, minimizing food waste is an evergrowing challenge in the supply chain related to inadequate transparency, traceability, tracking, and provenance of trusted data. Transparency can be introduced into food management systems by effectively tracking and tracing food products along the supply chain, which is made possible by blockchain technology. The storage of data in a unified system creates collaboration opportunities among stakeholders, and as a result, wasted food due to food loss can be identified quickly. Smart contracts can be deployed to register retailers' food demands using blockchain technology. It can also register the consent of distributors to supply food products to retailers in particular areas.

The starch supply chain (SSC) involves a complex and sensitive data encryption pattern due to the perishability and transformation of the raw materials obtained. Particularly, a large amount of the starch goes to waste, starting from farm produce procured to the final product. Lin et al. (2015) suggested that the significance of the waste can be reduced through the holistic involvement of technologies that can record real-time information. Initially, the starch product moves along the supply chain, whereby information on the source of the starch product, processing condition, and distribution is recorded to communicate with the technologies attached.

2. A major source of starch

Corn is an important source of starch that is widely used for industrial applications. The starch derived from corn was used in many bakery applications, including cakes, cookies, and fillers, due to the nature of moisture retention that prevents crystals formation in the structure of the sugar and adds a flavoring to the product (Ali et al., 2016). Secondly, starch isolated from corn helps to provide additional quality characteristics during the processing of custard paste, including enhancement of consistency and gelling characteristics (Alimi et al., 2017). Moreover, the amylose content in corn starch ranges up to 25% of the total granules. According to Lee et al. (2018) food sector utilizes the highest proportion of the total starch produced annually. In another vein, substations of corn starch based on the use of wheat flour have reduced the burden in most bakery companies and observed that better cake quality was recorded (Majzoobi et al., 2014). In addition, corn is categorized into corn starch and waxy corn starch varieties for industrial application. It was observed that the composition of amylose (20%–30%) and amylopectin (70%) varies according to the origin from which it was obtained, as commonly characterized for normal starch (Table 24.1). In contrast, the condition differs in the case of waxy starch's amylopectin (100%) (Klaochanpong et al., 2015).

Wheat (Triticum aestivum) has been widely used for various food applications since ancient times including production of cake, cookies, and many other food products obtained from it due to its chemical composition and a promising source of energy (Lu et al., 2014). However, the quality properties of the wheat products depend on the basis of the modification on the structure of the starch (Zhou et al., 2018). Though wheat is considered the most consumed food product worldwide, a negative implication for obese and diabetic patients reduces its excessive utilization (López-Barón et al., 2017).

Potatoes are typically grown in regions with a moderate climate, and the harvest is usually collected during autumn (Spooner et al., 2014). The phosphorus content in potatoes, which interlinks in potato starch, may impact its functionality. Kim et al. (2017) highlighted that starch is affected under high pressure, which results in inconsistent gelatinization. Generally, gelatinization occurs when heat and water are present. Potatoes are often associated with a high glycemic index (GI) that rapidly increases blood glucose levels (Daudt et al., 2014). Starch from both tapioca and potatoes is the key source of starch production in starch production for the food sector (Van Hung et al., 2017).

Rice (Oryza sativa L.) starch is composed of 70%–80% amylopectin (branched) and 20%–30% amylose (linear) (Lee et al., 2017). Rice's starch content makes it a crucial crop for many Asian countries, including Malaysia, Indonesia, and Thailand. A recent study highlighted the versatility of rice starch in food industries (Precha-Atsawanan et al., 2018). Different regions produce rice varieties with unique flavors and characteristics. Basmati rice has a high prospect due to its

TABLE 24.1 Source of starch and their average proportion on a dry basis.

S. no.	Source of starch	Starch content (%)
1.	Sweet potato	45.8–68
2.	Cassava	83–85
3.	Corn	70–75
4.	Sorghum	61–71
5.	Mungbean	50

aroma and quality, although it is more expensive, while Irri rice, which lacks the same aromatic properties, is more affordable (Moin et al., 2017).

Tapioca (Manihot esculenta Crantz) is another major source of starch that is used intensively for the production of different value added products, with significant production in Tamil Nadu, India (Kathiravan et al., 2017). Despite its low protein content, tapioca is a valuable source of starch and is known for forming soft and transparent gels (Ogunmuyiwa et al., 2017). Sour starch from cassava is widely used in bakery production that shows an increasing acceptance for sensitive individuals. This aligns with previous research indicating that starches from rice, potato, and tapioca as presented in Table 24.1 (Chuang et al., 2017). A recent study noted that tapioca thrives in areas with low rainfall and poor soil, making it an ideal starch source (Singh et al., 2017). Tapioca starch provides thickening and gelling properties to various food products (Trinh & Du Le, 2017). Additionally, tapioca starch have various benefits such as a lack of odor, clarity in the textural properties that is highly desirable during processing (Pfister & Zeeman, 2016).

3. Starch supply chain: Current scenario and challenges

3.1 Starch: A versatile carbohydrate with wide-ranging applications

Starch is a major carbohydrate component obtained from plants as an energy source. Starch comprising of non-branched structure of glucose units is known as amylose, whereas amylopectin features a highly branched structure. Staple crops are rich sources of starch extensively utilized in food and industrial applications. While rice starch (RS) finds limited use in diverse products due to the high economic value of milled rice, it possesses unique properties, including gluten-free and non-allergenic characteristics, small granule size, and variable amylose content, making it attractive for food applications (Majzoobi et al., 2014).

Maize starch, derived from seeds with significant starch accumulation, plays a crucial role in human and animal diets and is a key industrial raw material for bioethanol production (Niu et al., 2023). Starch's functional properties such as viscosity enhancement, gelling ability, thickening, and stabilization showcase its extensive use. It is used as a thickening agent, stabilizer, as well as texturizer in various products. Its ability to thicken solutions relies on the polysaccharide's hydrodynamic volume, with higher solubility enhancing its thickening efficiency. Associative interactions in modified starches further augment these properties. Starch is also a prominent stabilizer in food industries. Starch granules stabilize food emulsions through solid particle stabilization, forming Pickering emulsions, a novel and effective approach in emulsion technology (Anwar et al., 2020).

3.2 Potentials of starch: A case study from potato

Starch is an important component in higher plants and is the primary carbohydrate storage method. Green plants synthesize it as a means of energy storage, and it is typically formed in a granular state. This polysaccharide is found across many green plants and in nearly all tissue types, including leaves, fruits, pollen, roots, and stems. Commercially significant starches are derived from the reserve organs of various plants. Humans have long consumed starchy foods obtained from seeds, roots, and tubers, and while some starch is still consumed directly from plants, either in raw or cooked forms, with an increase in demand for commercially produced starch for incorporation into food and beverage products. The starch content in various plant tissues, such as potato tubers, maize endosperm, sweet potatoes, and the roots of cassava and yam, typically ranges from 65% to 90% of the dry weight. The properties of starch granules differ depending on the plant tissue and species, exhibiting variation in composition, size, and shape. Each tissue type has a characteristic range of these physical properties. Although starches from different sources share several common characteristics, there are notable differences between them. For example, potato starch is distinguished by its large granule size, high swelling capacity, low glass transition temperature, clear paste, and relatively neutral flavor.

Native potato starch, commonly used in food industries, has certain limitations that are not always ideal for specific applications. Starch derived from potatoes and its derivatives is valued for unique attributes such as low gelatinization temperature and high paste consistency. The food, chemical, pharmaceutical, and other industries extensively utilize potato starch. In the food sector, it is a thickener for sauces and soups and a binder in products like cake mixes, dough, biscuits, and ice cream. In the European Union, more than 40% of potato starch is allocated for non-food purposes, including manufacturing paper, bioplastics, pharmaceuticals, cosmetics, textiles, and adhesives. Its neutral flavor, clarity, and excellent binding strength enhance its desirability in these applications.

Potato starch is preferred due to the clarity of its paste, which results from its minimal lipid and protein content and neutral taste. The paper industry also values potato starch for its high molecular weight, amylose content, and solubility. Approximately 15% of the global paper starch production, estimated at 5 million tons over 3 years (2005), is sourced from potatoes. Paper mills typically purchase dry starch, which is cooked on-site before application. According to Bayer Crop Science, potato starch can enhance the durability and smoothness of textiles by providing resistance to abrasion.

Furthermore, potato starch is used in cosmetics and pharmaceuticals to produce creams, pastes (such as adhesives or sticks), and powders, reducing the reliance on chemical additives. Potato dextrins have excellent remoistening properties and favorable rheology, making them superior adhesives with optimal tack. Using potato starch in textiles improves fabric production quality due to its film-forming properties, paste penetration depth and adhesive strength. Compared to cereal starch, its superior purity makes it an ideal choice, as it contains minimal fat and protein with no color, taste, or odor. Moreover, potato starch is a key ingredient in producing biodegradable sugar-based detergents. Due to its absorbent qualities, it is more effective than other cleaning agents in removing fabric stains.

3.3 Starch extraction: A case study from yam tuber

3.3.1 Starch extraction from tubers: Methods and applications

The initial processing of tubers typically involves cleaning, peeling, and cutting into smaller pieces, followed by blending with water, an alkaline solution, or a sodium bisulfite solution to create a homogeneous mixture (Alberto Andrade et al., 2017). Similar procedures have been applied to other tuber sources. In a follow up study by Guo et al. (2024) isolated polysaccharides from Chinese yam, demonstrating notable biological activities such as immunomodulatory, anticancer, hypoglycemic, and antioxidant effects. Arueya & Ojesanmi (2019) described a method for extracting starch from water yam using a 0.2% (w/v) sodium bisulfite ($NaHSO_3$) solution. The mixture was filtered and treated with 0.2% NaOH to separate solid and liquid phases. The clear supernatant was dried at 40°C for 24 h. Hydroxypropylation of the dehydrated starch using 4%–12% propylene oxide improved its physicochemical and functional properties (Arueya & Ojesanmi, 2019). Ochoa & Osorio-Tobón (2024) explored three methods for extracting starch from purple yam residues. These yield starch with higher purity and recovery rates, promoting a circular economy by utilizing yam residues effectively (Ochoa & Osorio-Tobón, 2024). Similarly, da Costa et al. (2020) used centrifugation to extract starch, which exhibited promising water-binding capacity (WBC) and suitability as an edible film matrix (da Costa et al., 2020). Santos et al. (2022) compared aqueous and alkaline extraction techniques for purple yam starch. Alkaline extraction, employing 0.1% NaOH, efficiently dissolved mucilage without damaging starch granules, resulting in a product with higher pasting temperature and water absorption capacity (Santos et al., 2022). Conversely, the aqueous method preserved the starch granules' structural integrity, yielding starch with reduced water loss during freeze-thaw cycles and enhanced oil absorption capacity.

3.4 Key stakeholders in starch supply chains

In this context, the supply chain is conceptualized as a sequence of decision-making and execution processes designed to fulfill consumer requirements (Fig. 24.1). These processes span multiple stages, encompassing production to final consumption. The starch supply chain includes producers, suppliers, logistics providers, transporters, warehouses, retailers, and consumers, reflecting its complexity and interconnectivity.

Supply chains broadly integrate product development, marketing, operations, distribution, and customer service functions. The supply chain involves several critical stages for starch, particularly during isolation. This process is characterized by three primary phases: anatomical fragmentation, cellular disruption, and the final purification of starch.

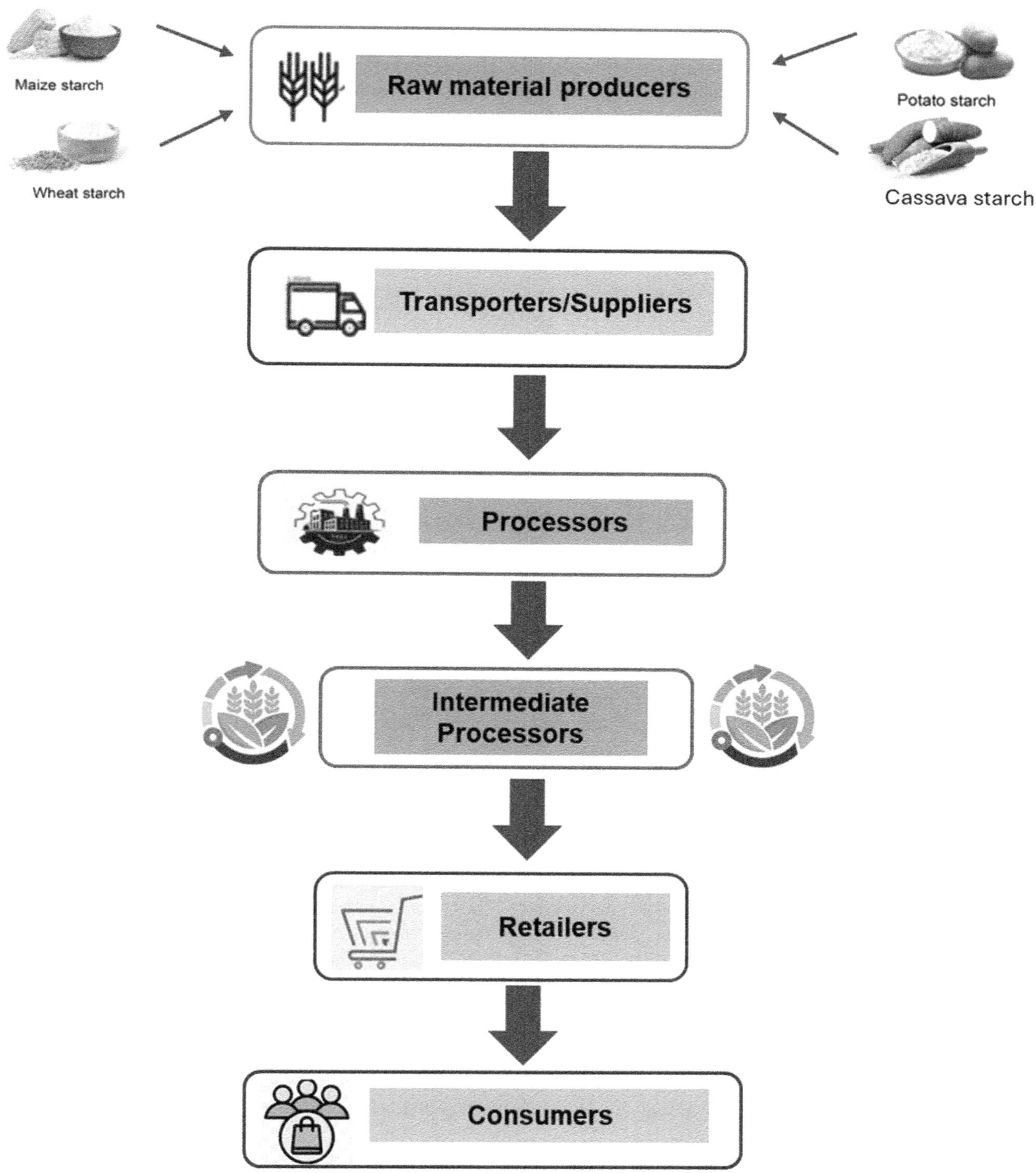

FIGURE 24.1 **Conventional starch supply chain management.** A diagram of a conventional starch supply chain management.

3.4.1 Producers

The global population has significantly increased in recent years, placing considerable pressure on starch production, including the availability of raw materials and ingredients. In addition to these challenges, the transformation of developing countries' lifestyles has altered dietary consumption patterns. The four primary starch sources corn, cassava, wheat, and potato are regularly consumed in developing countries. Beyond serving as a major energy source, starch has widespread applications in the food industry. The starch supply chain begins with producers who supply starch in its raw form, primarily sourced from cereal grains and root tubers. These producers are typically farmers engaged in the agricultural sector. Starch production businesses vary in scale among the corporations. Stakeholders in the starch industries have established measures for procuring the farms' produce for the processing and extraction of the starch component. A robust starch production sector is crucial for any country, as it directly impacts both the availability of starch for the population and the economic sustainability of the starch industry. Some actors in the supply chain, known as 'input suppliers,' provide raw materials to ensure meeting their demand. This parades the significant influence of the stage within the supply chain. Additionally, producers face numerous challenges, including unpredictable climate patterns, water scarcity, land appropriation by unethical agents, and resource deterioration from urbanization expansion. With profit margins for producers in the starch supply chain, more farmers are now focusing on cultivating crops that offer higher economic returns, which may impact the availability of essential starch products in the long term.

3.4.2 Processors

Processors are key entities within the starch supply chain, responsible for converting raw starch supplied by producers into products that align with consumer requirements, a process called starch manufacturing. This stage in the supply chain involves preparing fresh starch in a ready-to-consume format or utilizing it as a raw material to produce secondary starch-based products demanded by consumers. Starch-processing industries encompass a wide array of operations, including preserving fresh produce through methods such as freezing, pureeing, or juicing; grain milling; and the production of bakery and confectionery items, among others.

The significance of starch processing extends beyond product transformation. It underpins the starch sector's economic sustainability by fulfilling consumer demand, mitigating waste, and enhancing the shelf life of perishable starch-based raw materials. This prolongation of usability is crucial for addressing supply chain inefficiencies and ensuring consistent starch availability.

Effective collaboration with downstream supply chain entities, which facilitate the delivery of processed starch products to end-users, is critical for starch processors. Additionally, technological advancements, innovations in distribution channels, and adaptive strategies are imperative to meet evolving environmental challenges and consumer preferences.

Starch processors face significant challenges, including resource constraints, particularly water and energy, and the fluctuating availability of raw starch from producers. Addressing these issues will require adopting sustainable practices, efficient resource management, and innovative solutions to maintain resilience and competitiveness within the starch supply chain.

3.4.3 Retailers and distributors

At this stage, the starch supply chain involves two main participants: distributors and retailers. Distributors act as intermediaries connecting producers, processors, and markets. They acquire raw starch and distributes to retail outlets or secondary processors, to ensure it reaches consumers. Distributors typically purchase in large quantities and rely on networks of warehouses and distribution centers to facilitate timely delivery along the supply chain.

Their role is particularly vital in international supply chains, where they must navigate local regulations to enable seamless product movement across borders. This logistical capability makes them essential for maintaining the efficiency of global starch supply systems.

Retailers, in contrast, directly engage with consumers by presenting starch-based products in various formats, from small neighborhood stores to expansive supermarkets and hypermarkets with extensive product selections. Retailers provide consumers access to traditional and innovative starch products while competing for market share by offering better quality, pricing, and service.

Retailers employ creative business strategies to stand out in this competitive environment and explore new sales channels, including physical storefronts and online platforms. The evolving retail landscape in developing regions is marked by a shift from informal local shops to organized, large-scale retail structures such as supermarkets, necessitating corresponding adaptations in supply chain operations.

As global retail giants expand into developing markets, they drive significant changes in local supply chains. However, these shifts have raised concerns about their impact on small-scale, independent shops, prompting discussions about the balance between modernization and protecting traditional retail businesses.

3.4.4 Consumers

Consumers represent the final link in the starch supply chain, and their purchases are crucial for ensuring the economic viability of the entire system. The revenue generated by consumer spending flows upstream, sustaining various supply chain stages. Competition within the starch industry predominantly revolves around offering diverse products and added value rather than focusing on raw starch. In the UK, a competitive dynamic between traditional grocery supermarket chains and discount retailers has emerged, benefiting consumers through increased options and lower prices. However, this competition has sparked discussions about starch sustainability and wastage. Efforts to minimize starch waste at the consumer stage have become a priority for European governments and organizations within the starch sector. Simultaneously, starch safety remains a critical concern, necessitating stringent measures across the supply chain to prevent contamination. This includes addressing excessive pesticide use during production, microbial contamination during processing, and improper handling practices.

3.5 Types of starch chain

Starch supply chains can be categorized into two main types: those catering to industrial markets and those directed at end consumers. The first type involves the bulk trade of agricultural produce or commodities, while the second focuses on meeting the specific needs of individual consumers.

3.5.1 Raw commodity- and producer-focused chains

Farm produce can be procured either fresh for consumer use (such as fruits, vegetables, milk, and grains) or in bulk as raw materials for starch processing facilities. Commodities are traded using futures contracts, agreements between two parties to buy or sell a product such as starch-based items like cocoa at a predetermined price, with delivery and payment occurring later. Commodities chain encompasses palm oil, cocoa, coffee, sugar, cereals, and grains. This supply chain model typically operates with numerous sellers and relatively few buyers, functioning as a spot market where prices dictate product movement. Information exchange between trading partners is minimal in commodity systems, enabling processors to quickly secure bulk purchases, lower costs through risk hedging, and ensure product availability. However, the indirect nature of transactions between processors and producers via futures contracts creates a disconnect among the stakeholders, preventing consumer demand signals from reaching producers. Climate change, unpredictable weather patterns, global supply and demand dynamics, and trade agreements influence major commodity prices. These fluctuations create an environment conducive to futures trading.

Traders leverage diverse sourcing strategies to mitigate risks and achieve profits, as the margins in commodity trading are often narrow. Profitability relies more on high trading volumes than individual prices, with bulk transactions providing the primary financial returns.

3.6 Driven factor for the value chains

The traceability and identity of starch products are critical in consumer-driven supply chains, as this final stage directly influences consumer health and satisfaction (Trivedi et al., 2019). Starch products are processed, branded, and optimized for uniformity and high quality in retail. Commodity market products, such as coffee, are increasingly transitioning into retail as branded, gourmet items, often supported by close collaboration among stakeholders. Tracking and tracing mechanisms are vital for maintaining the integrity of the supply chain. Technological advancements are necessary to enhance efficiency and ensure compliance. Retailers often conduct rigorous scrutiny, including production site inspections, to verify adherence to buyer standards and codes, emphasizing sustainability performance. This ensures that the starch supply chain meets the expectations of both regulatory authorities and end consumers.

3.7 Factors influencing starch supply chains

The starch sector operates within a highly intricate environment shaped by industrial, technological, economic, social, and political factors. These elements collectively influence the availability of starch, the characteristics of starch-based products, and the mechanisms through which starch reaches consumers. Entities in the starch supply chain strive to enhance their operations by prioritizing quality, competitiveness, pricing efficiency, and adherence to stringent starch safety standards. The complexities within this supply chain come from several interrelated factors:

- agri-tech production;

- policymakers;
- processing and maintaining quality;
- market forces;
- logistics companies; and
- a host of other small companies are actively involved in this starch supply chain and providing secondary value.

The world is in a state of continuous evolution. Technological advancements, emerging business models, globalization, and the mobility of populations have compelled starch supply chains to reassess their fulfillment strategies and operational efficiency. Innovations in processing and transportation have enhanced the suitability of products for global distribution. At the same time, advancements in management practices and information and communication technologies (ICT) have enabled supply chains to respond more effectively to consumers' growing and sophisticated demands for starch products. Additional factors that significantly impact and shape the starch supply chain include the following:

3.8 Starch supply and value chains

Starch undergoes various transformation before it rich the final consumer. However, along these stages where raw agricultural produce are subjected to transformations aimed at improving the quality of the products are known as "value chain systems." The starch supply chain basically align with these transformational changes targeting to meets the consumers demand. Nowadays, different modules are integrated to meets the efficiency and enhance the economic viability of the system. However, this comes at the expense of reduced flexibility in selecting suppliers. These changes in the starch value chain can also affect the flexibility in the demand and purchasing strength between stakeholders.

The starch value chain encompasses all stakeholders involved in cultivating, processing, and distributing starch products for consumer requirements:

- Suppliers of the raw produce
- Starch production process
- The processing stages
- Distribution stage involving those who handle the distribution, marketing, and sale of starch
- Consumers who purchase and consume starch
- Governmental policymakers
- Regulatory agencies for quality handling

Importance of starch is not limited only to provide energy to the consumers but also helps in reducing the poverty level of the various stakeholders. While reducing starch waste and increasing starch security, the value chain must also consider accessibility, safety, and nutritional quality.

3.9 Regulations

Export and import of starch across international borders is regulated under predetermined rules that address issues such as hygiene, and labeling requirements. These regulations promote transparency in international trade. In the European Union, the starch regulation emphasizes risk management and traceability within the "farm to fork" supply chain, balancing the protection of human, animal, and plant health with environmental concerns. The underlying principle is a precautionary approach. As part of this, the European Commission introduced the "one step backwards – one step forward" strategy, which mandates that operators track the origins and destinations of products. Labeling of starch products is a central aspect of EU starch regulations.

3.10 Consumer preference

Global starch consumption patterns exhibit two distinct trends. In developed countries, there has been a growing demand for processed starch driven by the needs of a time-constrained and health impact. The demand for a convenient foods known as ready-to-eat or semi processed led to revolutionary targets by many industries to meet the demand. However, these have affected the economic stability of the starch industry as the raw materials used in the extraction increased simultaneously. In contrast, consumer behavior toward shifting their diet which is primarily on cereals and grains to one with more protein and starch is ever increasing. However, increased starch production necessitates a higher demand for animal feed and raw materials. To meet this demand, starch production has expanded into intensive farming, prioritizing crops for animal feed over those intended for human consumption, which could have long-term implications for starch security.

4. Sustainable nature of the starch supply chain management

Global starch supply chain significantly contributes to emissions of various gases including GHGs. These were generated along every stage, from starch production (and its inputs) to processing, distribution, consumption, and waste disposal. Achieving sustainability within the starch supply chain requires addressing several key factors, including:

- Carbon economy
- Energy utilization
- Ingredients usage
- Starch availability
- consumer ethical behavior

Starch supply chain have been analyzed in a holistic approaches to identify areas of non-sustainability. Mapping the supply chain also provides insights into energy usage, offering opportunities to reduce energy consumption.

4.1 Collaboration

Collaboration among stakeholders in the starch value chain is essential for enhancing efficiency and addressing issues related to safety and traceability. The interconnected roles of these stakeholders, along with their broader network, present numerous opportunities for joint efforts. Recent global concerns over starch recalls have underscored the importance of collaboration in ensuring reliability throughout the supply chain, aim to address the challenges through coordinating the major key drivers impacting the starch quality. These further helps to allow key partners to maintain a comprehensive strategies for handling the product. At the same time, partnerships between producers and processors, supported by advanced technologies, can significantly reduce post-harvest starch losses.

Strategies to Address Starch Demand Amplification and Enhance Supply Chain Performance:

- **Real-Time Information Sharing**: Minimize delays in starch conveyance by maintaining timely communication.
- **Accurate Demand Exchange**: Share precise market demand data with upstream supply chain participants using ICT and fostering collaboration.
- **Vertical Integration**: Simplify the chain by combining stages such as production and processing to eliminate intermediaries.
- **Optimized Decision-Making**: Refine order quantity practices and establish clear contracts, standards, or parameters to improve transaction efficiency.

4.2 ICT and DSS in mitigating quality degradation in starch supply chains

The starch supply chain faces significant challenges, including product quality variability, logistics inefficiencies, and raw materials' perishability. These issues often lead to substantial economic losses. One key factor exacerbating these problems is the lack of coordination among supply chain actors, which results in the "bullwhip effect." This phenomenon occurs when minor fluctuations in consumer demand lead to amplified variations in orders upstream in the supply chain (van der Vorst et al., 2007). Such coordination gaps increase inventory costs, cause stockouts, and hinder resource utilization (Lin et al., 2018). Innovative strategies that enhance communication, streamline logistics, and reduce waste are necessary to address these issues.

Advanced technologies, including Information and Communication Technology (ICT) and Decision Support Systems (DSS), offer transformative solutions for starch supply chain management. ICT facilitates frequent data exchange among stakeholders, improving coordination and responsiveness to real-time market demands (van der Vorst et al., 2007). Automated DSS, on the other hand, analyses large datasets to optimize logistics operations and provide actionable insights. These systems help increase operational efficiency, reduce costs, and tackle quality variability and perishability (Lin et al., 2018). The integration of these technologies plays a crucial role in enhancing the overall functioning of the supply chain.

The starch potato supply chain is particularly dependent on these technologies due to the perishable nature of potatoes, a key raw material for starch production. Potatoes are susceptible to quality degradation during storage, influenced by temperature, humidity, and metabolism. For instance, low storage temperatures can convert starch into reducing sugars, diminishing the product's quality. During starch campaigns, factories must process large quantities of harvested potatoes while minimizing quality loss, making predictive scheduling systems essential for determining the optimal time for harvesting, storage, and transportation (Lin et al., 2018). These systems help maximize starch yields and minimize waste.

Quality-aware DSS equipped with real-time data capabilities have proven to be effective in addressing these challenges. These systems use sensors and Radio Frequency Identification (RFID) technology to monitor environmental conditions such as temperature and humidity. This allows stakeholders to predict quality changes and make informed decisions. For example, first-order kinetic models predict starch degradation during storage, while predictive control strategies optimize schedules for harvesting, storage, and transportation (Lin et al., 2018). Simulation experiments have shown that these technologies significantly reduce waste and improve coordination within the starch supply chain.

Integrating Artificial Intelligence (AI) and the Internet of Things (IoT) further boosts the efficiency and sustainability of starch supply chain operations. AI models analyze historical and real-time data to identify patterns and predict risks, enabling stakeholders to adjust operations proactively (van der Vorst et al., 2007). IoT devices, such as smart sensors, continuously monitor environmental conditions and product quality, providing critical data for predictive scheduling. These technologies enable centralized and distributed control strategies, balancing coordination between farms and factories to ensure maximum resource utilization and the highest possible starch yields (Lin et al., 2018). Integrating these disruptive technologies not only addresses inefficiencies but also enhances the sustainability of the starch supply chain.

4.3 AI and blockchain in emerging economies for sustainable supply chains

AI technologies, including machine learning and natural language processing, have revolutionized warehouse management and inventory systems by improving operational accuracy and safety, essential for efficient supply chain operations (Marius Müller et al., 2018). AI optimizes warehouse layouts and processes by analyzing vast amounts of data, such as historical inventory records and customer demand patterns (Fig. 24.2). These capabilities enhance operational efficiency and contribute to the overall effectiveness of supply chain management (Hopp et al., 2018).

Blockchain technology further strengthens supply chain management by offering a secure and transparent information-sharing platform. Its ability to enhance traceability, promote ethical practices, and ensure sustainability makes it a critical tool for modern supply chains. Integrating blockchain improves reliability and robustness in operations, supporting organizations in achieving their sustainability goals. Moreover, the synergy between Blockchain and AI is crucial for cloud manufacturing systems, which rely on IoT for seamless connectivity and process optimization (Hopp et al., 2018; Liu et al., 2020; Sood & Tellis, 2011).

The combination of AI and Blockchain has immense potential for enhancing real-time data sharing and decision-making, especially in addressing supply chain uncertainties in emerging economies like Bangladesh. These disruptive technologies are instrumental in advancing the sustainability of supply chains, helping organizations navigate competitive and uncertain business environments through strategic adoption and prioritization (Abosuliman & Almagrabi, 2021; Benzidia et al., 2021). Ahmad et al. (2024) highlight the importance of leveraging AI and Blockchain to achieve holistic sustainability, emphasizing their transformative role in improving efficiency, transparency, and sustainability across supply chains in emerging markets.

4.4 Blockchain and AI for traceability and quality in starch supply chains

Blockchain technology enables real-time tracking of food products, ensuring transparency and traceability throughout the starch supply chain from production to consumption (Fig. 24.3). This decentralized system creates immutable records of transactions, enhancing accountability and safety within the supply chain (Kravenkit & So-In, 2022; Thangamayan et al., 2023). By leveraging blockchain, stakeholders can maintain a reliable and secure system for monitoring every stage of the supply chain, thereby building trust among producers, processors, and consumers.

Integrating blockchain with quality assurance processes in the starch supply chain allows for automated and real-time monitoring. This capability ensures consistent quality checks and enables immediate alerts in case of deviations, reducing the risks associated with quality lapses (Sri Vigna Hema & Manickavasagan, 2024). Combining Blockchain with AI further enhances decision-making and operational efficiency. AI's ability to analyze large datasets, predict trends, and identify anomalies helps optimize resource allocation and ensures the quality and safety of starch products (Patel et al., 2023).

In cases of contamination or quality issues, blockchain significantly improves the speed and accuracy of product recalls by enabling rapid identification of affected items. This feature is crucial for maintaining consumer safety and trust

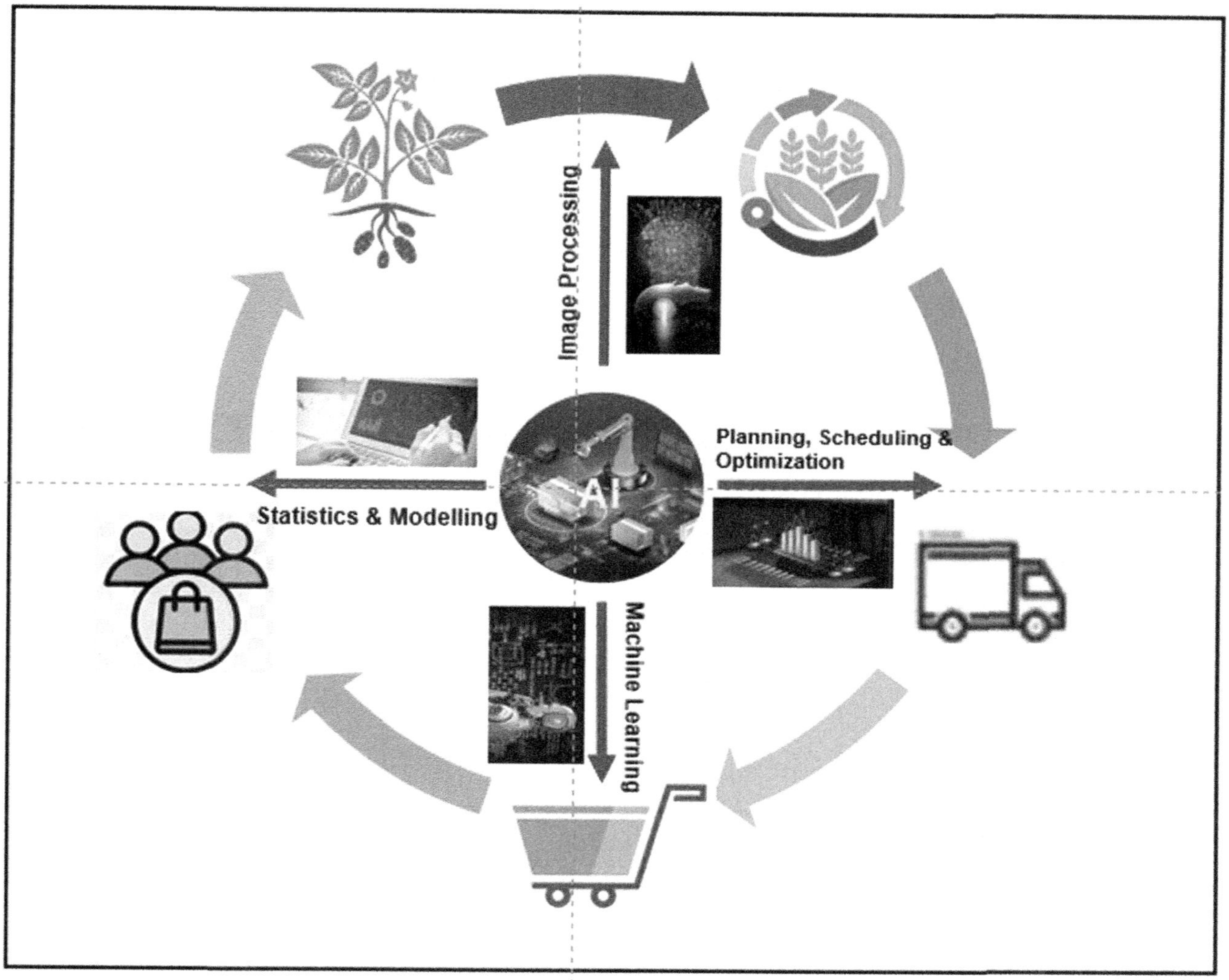

FIGURE 24.2 Starch supply chain management with AI. A diagram of starch supply chain management with AI.

(Sri Vigna Hema & Manickavasagan, 2024). Additionally, integrating disruptive technologies such as Blockchain and AI fosters better communication and collaboration among stakeholders in the starch supply chain. This enhanced coordination improves overall efficiency, benefiting all participants and driving innovation within the industry (Sri Vigna Hema & Manickavasagan, 2024).

4.5 Analyzing barriers to AI adoption in starch supply chains

Integrating AI with technologies like blockchain is crucial for creating synchronized supply networks, particularly within the starch supply chain. This combination can address the complex interactions within supply chains and improve overall efficiency (Ramamurthy et al., 2008). However, the adoption of AI in the food supply chain (FSC) faces several barriers, including technological limitations, regulatory complexities, and a shortage of skilled workers. Understanding these challenges is essential for stakeholders in the starch supply chain considering integrating disruptive technologies (Chen et al., 2020).

To analyze these barriers, a novel methodological framework is proposed that combines the NK Model with the DEMATEL technique. This framework can be adapted to assess the challenges the starch supply chain faces in adopting AI and Blockchain technologies. By identifying and understanding these obstacles, stakeholders can develop strategies to mitigate them and pave the way for successfully deploying disruptive technologies in their operations.

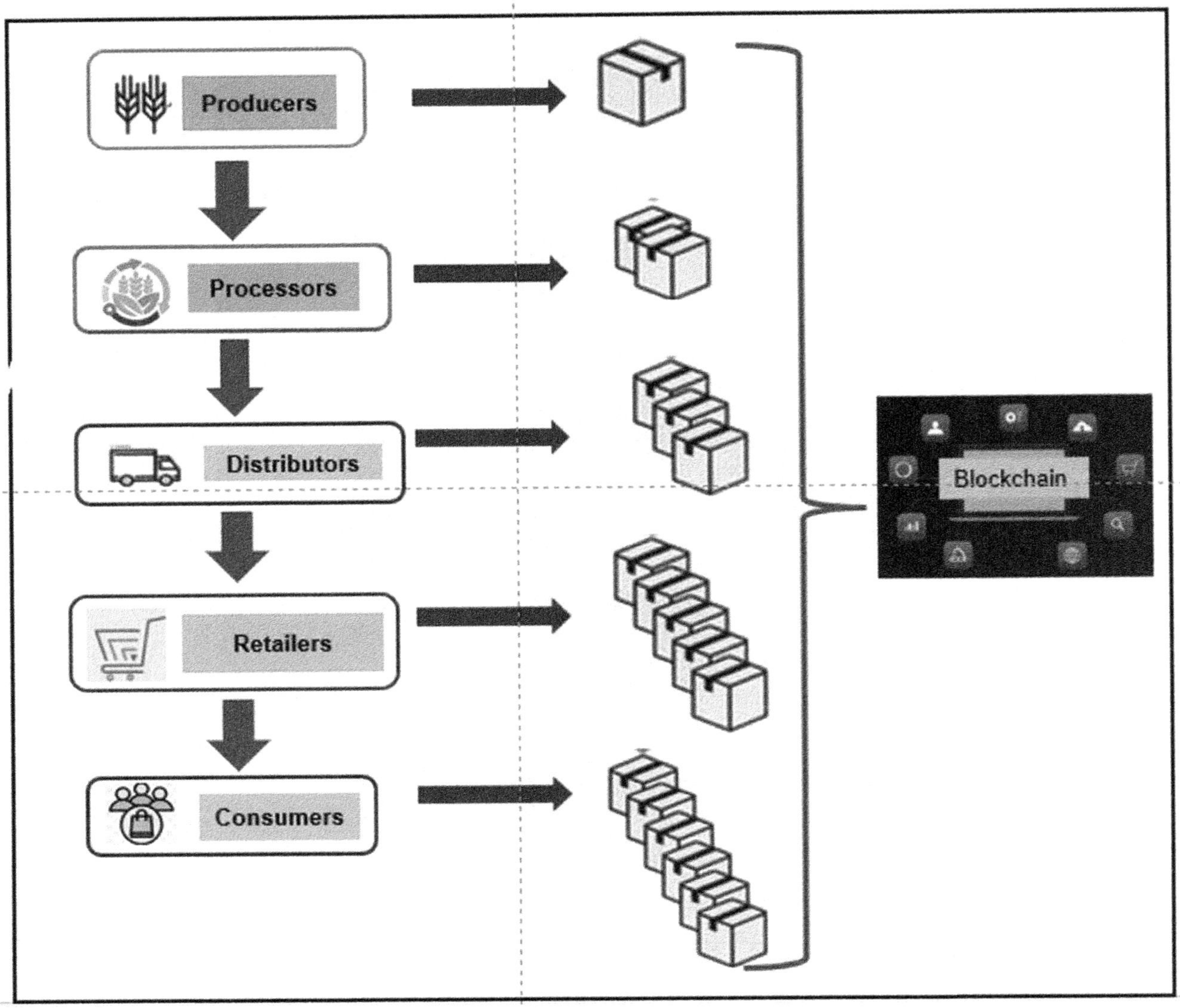

FIGURE 24.3 **Starch supply chain management integrated with blockchain.** A diagram of a blockchain process description automatically generated.

Implementing AI in supply chains is vital for improving efficiency and sustainability. Although Ghag et al.'s (2024) findings do not specifically address the starch supply chain, their insights into the challenges of AI adoption are valuable. These findings can be extrapolated to the starch supply chain, helping stakeholders understand the potential benefits and obstacles of incorporating AI and Blockchain technologies into their operations.

4.6 AI and blockchain for enhanced decision-making in starch supply chain

Blockchain technology provides a decentralized and immutable record of transactions, ensuring transparency and traceability within supply chains. This is particularly beneficial for managing starch supply chains, as it offers a reliable method to track the origin and movement of starch products from producers to consumers (Thomé et al., 2021). Additionally, blockchain can automate processes through smart contracts, enabling order management, quality control, and payment settlements without intermediaries, streamlining the entire process (Ahmad et al., 2024).

Artificial intelligence (AI) plays a pivotal and multifaceted role in the analysis and optimization of data across various stages of the starch supply chain. This includes not only traditional data sources but also information gathered from Internet of Things (IoT) devices deployed throughout the supply chain to monitor and track key variables such as temperature, humidity, product quality, and inventory levels in real time. By leveraging AI to process and analyze this vast amount of data, organizations can extract valuable insights that help streamline operations, reduce waste, predict potential disruptions, and make more informed decisions. AI's ability to analyze large datasets quickly and accurately supports more effective decision-making, which leads to increased efficiency, reduced costs, and enhanced quality control across the supply chain (Ahmad et al., 2024).

Furthermore, integrating AI with Blockchain technology significantly strengthens these processes by addressing key challenges related to data integrity and security. While AI excels at processing large, complex datasets and identifying patterns that inform decision-making, blockchain complements this by ensuring that all the data being analyzed is secure, transparent, and tamper-proof. Blockchain's decentralized and immutable ledger ensures that the data cannot be altered or manipulated without detection, which is particularly critical in industries like food production and supply chains, where data accuracy and authenticity are paramount. This integration of AI and blockchain provides organizations with a powerful combination that not only improves the quality and reliability of insights but also fosters greater trust among stakeholders. As a result, these technologies collectively improve overall decision-making, enhance operational transparency, and help organizations maintain high standards of quality and accountability (Thomé et al., 2021).

Combining AI and Blockchain can optimize operations like inventory management, quality assurance, and waste reduction in the starch supply chain (Fig. 24.4). For example, AI can forecast demand, while blockchain ensures transparency and accountability throughout the process (Ahmad et al., 2024). Moreover, blockchain's traceability feature allows stakeholders to swiftly identify and resolve issues like contamination or quality discrepancies, ensuring consumer trust in starch products (Ahmad et al., 2024). However, successfully implementing these technologies requires overcoming challenges such as data heterogeneity from IoT devices and ensuring all stakeholders are adequately equipped to use them effectively (Ahmad et al., 2024; Kravenkit & So-In, 2022).

4.7 Benefits and challenges for blockchain, AI, and IoT in supply chains

Blockchain technology offers a range of advantages to supply chains, such as enhanced transparency, immutable records, precise timestamping, decentralized control, robustness, improved privacy, and the automation of processes through smart contracts. These features facilitate secure data sharing, prevent unauthorized alterations, and foster trust and transparency among supply chain participants (Malik et al., 2018; Pal & Jadidi, 2021).

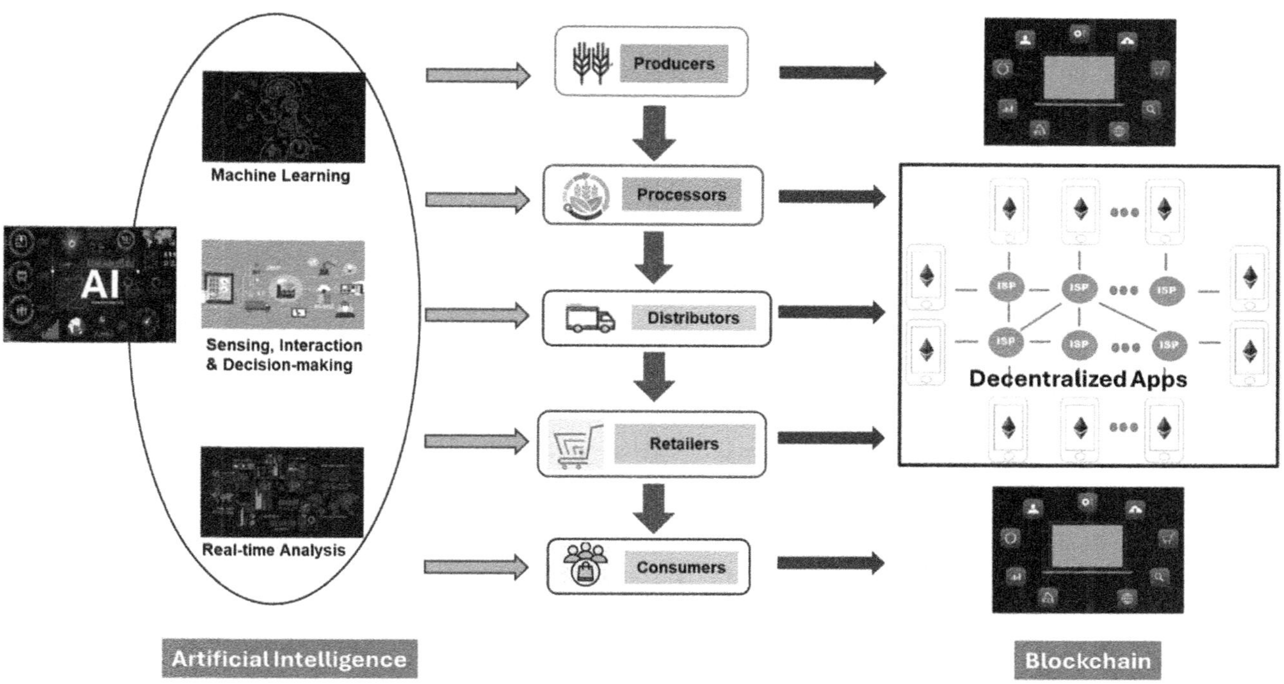

FIGURE 24.4 Integrated framework for starch supply chain by AI and Blockchain. A diagram of a software development description automatically generated with medium confidence.

Artificial intelligence (AI) transforms supply chain management by enabling predictive analytics, real-time monitoring of assets and processes, and efficient automated supervision while incorporating privacy-preserving techniques. AI applications extend across various supply chain stages, including manufacturing (e.g., production planning and quality control), warehouse management (e.g., stock planning and anomaly detection), logistics (e.g., fleet management and route optimization), retail (e.g., shelf-life prediction and counterfeit detection), and compliance management (Arinez et al., 2020; Boute & Udenio, 2022; Oosthuizen et al., 2021; Riahi et al., 2021).

Integrating Blockchain, AI, and Internet of Things (IoT) technologies offers significant benefits for food supply chains. These include improved data provenance and traceability, secure data sharing, and the prevention of unauthorized modifications (Dedeoglu et al., 2023). However, implementing these technologies has challenges, such as data quality issues, privacy concerns, usability limitations in rural areas, reluctance to share data, and security vulnerabilities (Aker et al., 2016). Future research should focus on enhancing the usability of smartphone-based sensing systems, minimizing infrastructure costs, improving interoperability across systems and standards, developing fine-grained access control mechanisms, and addressing data trust concerns (Michels et al., 2020).

5. Future perspective

The starch supply chain management is currently undergoing a significant transformation, driven by the integration of advanced technologies. Traditionally, starch is predominantly used in the food sector for applications such as stabilizers, fillings, and gelling agents, among others. However, despite its wide-ranging industrial applications, the starch supply chain faces several challenges, particularly related to the complexity of data handling. This complexity is somewhat unique compared to other food product supply chains, as it involves a more intricate process of tracking and managing raw material sources and production stages, especially due to the variety and diversity of starch content from different crops, such as grains and tubers. One of the key challenges in the starch supply chain is the need for minimal processing to ensure timely and efficient processing of starch from raw materials. Given the diverse nature of starch sources, which include various grains and tubers, the processing methods need to be carefully managed to maintain quality and meet industry standards. This complexity further exacerbates the difficulty of efficiently handling the data and optimizing the entire supply chain. To address these challenges, emerging technologies such as artificial intelligence (AI) and blockchain are being increasingly adopted in the food sector, including starch supply chain management. These disruptive technologies offer a new way of managing the flow of goods and data in the supply chain. AI, for example, can assist in predictive analytics, allowing for better forecasting of demand, efficient resource allocation, and improved decision-making. Blockchain, on the other hand, offers a secure and transparent way to track and record every stage of the starch supply chain, from production to final consumption. While the application of AI and blockchain holds great potential for improving the starch supply chain, it also presents a set of challenges. The successful implementation of these technologies requires a high level of technical knowledge and expertise, which must be shared across various stakeholders throughout the value chain. This includes producers, processors, distributors, and retailers, all of whom must understand the technologies and their implications for the supply chain. Additionally, the integration of AI and blockchain requires extensive real-time data recording, which ensures transparency and accuracy in describing the processes involved.

6. Conclusion

Starch remains a major energy source with various applications that extend across diverse sectors, heavily influenced by its origin and composition. This chapter underscores the criticality of managing starch supply chain stakeholders through an integrated and adaptive approach. A holistic framework addressing producers, processors, distributors, and end-users ensures optimal resource utilization, reduced inefficiencies, and improved sustainability. The advent of disruptive technologies, particularly Artificial Intelligence (AI) and blockchain technology, has redefined operational paradigms within the starch supply chain. AI enhances decision-making by analyzing complex datasets to predict demand, optimize storage, and streamline transportation logistics. Meanwhile, blockchain fosters transparency and traceability, offering immutable records that ensure the authenticity and quality of starch products. These technologies collectively improve supply chain efficiency, reduce losses, and mitigate risks, demonstrating tangible advancements toward sustainability goals. Moreover, integrating these technologies empowers stakeholders with real-time data, enabling proactive responses to market dynamics and environmental constraints. Blockchain's ability to enhance trust among stakeholders by providing verifiable data on sourcing and processing complements AI's capability to optimize operations. Together, they present a synergy that drives not only economic gains but also aligns with global objectives such as reducing waste and enhancing food security.

References

Abosuliman, S. S., & Almagrabi, A. O. (2021). Computer vision assisted human computer interaction for logistics management using deep learning. *Arabia Computers and Electrical Engineering, 96*. https://doi.org/10.1016/j.compeleceng.2021.107555

Ahmad, R. W., Ko, K. M., Rashid, A., & Rodrigues, J. J. P. C. (2024). Blockchain for food industry: Opportunities, requirements, case studies, and research challenges. *IEEE Access, 12*, 117363–117378. https://doi.org/10.1109/ACCESS.2024.3447918

Aker, J. C., Ghosh, I., & Burrell, J. (2016). The promise (and pitfalls) of ICT for agriculture initiatives. *Agricultural Economics, 47*(S1), 35–48. https://doi.org/10.1111/agec.12301

Alberto Andrade, L., Barbosa, N. A., & Pereira, J. (2017). Extraction and properties of starches from the non-traditional vegetables Yam and Taro. *Polímeros, 27*(2), 151–157. https://doi.org/10.1590/0104-1428.04216

Ali, A., Wani, T. A., Wani, I. A., & Masoodi, F. A. (2016). Comparative study of the physico-chemical properties of rice and corn starches grown in Indian temperate climate. *Journal of the Saudi Society of Agricultural Sciences, 15*(1), 75–82. https://doi.org/10.1016/j.jssas.2014.04.002

Alimi, B. A., Workneh, T. S., & Oyeyinka, S. A. (2017). Structural, rheological and in-vitro digestibility properties of composite corn-banana starch custard paste. *LWT - Food Science and Technology, 79*, 84–91. https://doi.org/10.1016/j.lwt.2017.01.012

Anwar, S. H., Hasni, D., Rohaya, S., Antasari, M., & Winarti, C. (2020). The role of breadfruit OSA starch and surfactant in stabilizing high-oil-load emulsions using high-pressure homogenization and low-frequency ultrasonication. *Heliyon, 6*(7), Article e04341. https://doi.org/10.1016/j.heliyon.2020.e04341

Arinez, J. F., Chang, Q., Gao, R. X., Xu, C., & Zhang, J. (2020). Artificial intelligence in advanced manufacturing: Current status and future outlook. *Journal of Manufacturing Science and Engineering, 142*(11). https://doi.org/10.1115/1.4047855

Arueya, & Ojesanmi. (2019). Evaluation of effects of increasing molar substitution of hydroxypropylene on physicochemical, functional and morphological properties of starch from water yam (*Dioscorea alata*). *Journal of Food Research, 8*(4), 58–88.

Benzidia, S., Makaoui, N., & Bentahar, O. (2021). The impact of big data analytics and artificial intelligence on green supply chain process integration and hospital environmental performance. *Technological Forecasting and Social Change, 165*, Article 120557. https://doi.org/10.1016/j.techfore.2020.120557

Boute, R. N., & Udenio, M. (2022). *AI in logistics and supply chain management. Global logistics and supply chain strategies for the 2020s: Vital skills for the next generation*. Belgium: Springer International Publishing, (pp. 49–65). https://doi.org/10.1007/978-3-030-95764-3_3

Budak, & Sarvari. (2021). Profit margin prediction in sustainable road freight transportation using machine learning. *Journal of Cleaner Production, 314*.

Chen, S., Brahma, S., Mackay, J., Cao, C., & Aliakbarian, B. (2020). The role of smart packaging system in food supply chain. *Journal of Food Science, 85*(3), 517–525. https://doi.org/10.1111/1750-3841.15046

Chuang, L., Panyoyai, N., Shanks, R. A., & Kasapis, S. (2017). Effect of salt on the glass transition of condensed tapioca starch systems. *Food Chemistry, 229*, 120–126. https://doi.org/10.1016/j.foodchem.2017.02.063

da Costa, Lima Miki, da Silva Ramos, & Teixeira-Costa. (2020). Development of biodegradable films based on purple yam starch/chitosan for food application. *Heliyon, 6*(4).

Daudt, R. M., Külkamp-Guerreiro, I. C., Cladera-Olivera, F., Thys, R. C. S., & Marczak, L. D. F. (2014). Determination of properties of pinhão starch: Analysis of its applicability as pharmaceutical excipient. *Industrial Crops and Products, 52*, 420–429. https://doi.org/10.1016/j.indcrop.2013.10.052

Dedeoglu, Malik, Ramachandran, Pal, & Jurdak. (2023). Blockchain meets edge-AI for food supply chain traceability and provenance. In smartphones for chemical analysis: From proof-of-concept to analytical applications. *Comprehensive Analytical Chemistry, 101*.

Guo, Y., Liu, F., Zhang, J., Chen, J., Chen, W., Hong, Y., Hu, J., & Liu, Q. (2024). Research progress on the structure, derivatives, pharmacological activity, and drug carrier capacity of Chinese yam polysaccharides: A review. *International Journal of Biological Macromolecules, 261*(Pt 2), Article 129853. https://doi.org/10.1016/j.ijbiomac.2024.129853

Hopp, C., Antons, D., Kaminski, J., & Oliver Salge, T. (2018). Disruptive innovation: Conceptual foundations, empirical evidence, and research opportunities in the digital age. *Journal of Product Innovation Management, 35*(3), 446–457. https://doi.org/10.1111/jpim.12448

Kathiravan, V., Saranya, S., Sangli Vikram Kumar, K., & David Rajasekar, D. (2017). Economics of processing of different tapioca based products in Salem district of Tamil Nadu, India. *International Journal of Current Microbiology and Applied Sciences, 6*(11), 1613–1619. https://doi.org/10.20546/ijcmas.2017.611.193

Kim, H.y., Oh, S. M., Bae, J. E., Yeom, J. H., Kim, B. Y., Kim, H. S., & Baik, M. Y. (2017). Preparation and characterization of amorphous granular potato starches (AGPS) and cross-linked amorphous granular potato starches (CLAGPS). *Carbohydrate Polymers, 178*, 41–47. https://doi.org/10.1016/j.carbpol.2017.09.020

Klaochanpong, N., Puttanlek, C., Rungsardthong, V., Puncha-arnon, S., & Uttapap, D. (2015). Physicochemical and structural properties of debranched waxy rice, waxy corn and waxy potato starches. *Food Hydrocolloids, 45*, 218–226. https://doi.org/10.1016/j.foodhyd.2014.11.010

Kravenkit, S., & So-In, C. (2022). Blockchain-based traceability system for product recall. *IEEE Access, 10*, 95132–95150. https://doi.org/10.1109/ACCESS.2022.3204750

López-Barón, N., Gu, Y., Vasanthan, T., & Hoover, R. (2017). Plant proteins mitigate in vitro wheat starch digestibility. *Food Hydrocolloids, 69*, 19–27. https://doi.org/10.1016/j.foodhyd.2017.01.015

Lee, S., Lee, J. H., & Chung, H. J. (2017). Impact of diverse cultivars on molecular and crystalline structures of rice starch for food processing. *Carbohydrate Polymers, 169*, 33–40. https://doi.org/10.1016/j.carbpol.2017.03.091

Lee, S. J., Lee, H. S., Kim, S. Y., & Shin, K. S. (2018). Immunostimulatory and anti-metastatic activity of polysaccharides isolated from byproducts of the corn starch industry. *Carbohydrate Polymers, 181*, 911–917. https://doi.org/10.1016/j.carbpol.2017.11.060

Lin, X., Negenborn, R. R., & Lodewijks, G. (2015). Survey on operational perishables quality control and logistics. *Lecture Notes in Computer Science, 9335*, 398–421. https://doi.org/10.1007/978-3-319-24264-4_28

Lin, Negenborn, & Lodewijks. (2018). Predictive quality-aware control for scheduling of potato starch production. *Computers and Electronics in Agriculture, 150*, 266–278.

Liu, Y., Zhang, Y., Ren, S., Yang, M., Wang, Y., & Huisingh, D. (2020). How can smart technologies contribute to sustainable product lifecycle management? *Journal of Cleaner Production, 249*, Article 119423. https://doi.org/10.1016/j.jclepro.2019.119423

Lu, Y., Luthria, D., Fuerst, E. P., Kiszonas, A. M., Yu, L., & Morris, C. F. (2014). Effect of processing on phenolic composition of dough and bread fractions made from refined and whole wheat flour of three wheat varieties. *Journal of Agricultural and Food Chemistry, 62*(43), 10431–10436. https://doi.org/10.1021/jf501941r

Majzoobi, M., Hedayati, S., Habibi, M., Ghiasi, F., & Farahnaky, A. (2014). Effects of corn resistant starch on the physicochemical properties of cake. *Journal of Agricultural Science and Technology A, 16*(3), 569–576.

Malik, S., Kanhere, S. S., & Jurdak, R. (2018). Productchain: Scalable blockchain framework to support provenance in supply chains. In *2018 IEEE 17th International Symposium on Network Computing and Applications*. Australia: Institute of Electrical and Electronics Engineers Inc. https://doi.org/10.1109/NCA.2018.8548322. http://ieeexplore.ieee.org/xpl/mostRecentIssue.jsp?punumber=8533540.

Marius Müller, J., Kiel, D., & Voigt, K.-I. (2018). What drives the implementation of industry 4.0? The role of opportunities and challenges in the context of sustainability. *Sustainability, 10*(1), 247. https://doi.org/10.3390/su10010247

Michels, M., Bonke, V., & Musshoff, O. (2020). Understanding the adoption of smartphone apps in crop protection. *Precision Agriculture, 21*(6), 1209–1226. https://doi.org/10.1007/s11119-020-09715-5

Moin, A., Ali, T. M., & Hasnain, A. (2017). Characterization and utilization of hydroxypropylated rice starches for improving textural and storage properties of rice puddings. *International Journal of Biological Macromolecules, 105*, 843–851. https://doi.org/10.1016/j.ijbiomac.2017.07.109

Niu, L., Liu, L., Zhang, J., Scali, M., Wang, W., Hu, X., & Wu, X. (2023). Genetic engineering of starch biosynthesis in maize seeds for efficient enzymatic digestion of starch during bioethanol production. *International Journal of Molecular Sciences, 24*(4), 3927. https://doi.org/10.3390/ijms24043927

Ochoa, & Osorio-Tobón. (2024). Isolation and characterization of starch from the purple yam (*Dioscorea alata*) anthocyanin extraction residue obtained by ultrasound-assisted extraction. *Waste and Biomass Valorization, 15*, 379–389.

Ogunmuyiwa, O. H., Adebowale, A. A., Sobukola, O. P., Onabanjo, O. O., Obadina, A. O., Adegunwa, M. O., Kajihausa, O. E., Sanni, L. O., & Keith, T. (2017). Production and quality evaluation of extruded snack from blends of bambara groundnut flour, cassava starch, and corn bran flour. *Journal of Food Processing and Preservation, 41*(5), Article e13183. https://doi.org/10.1111/jfpp.13183

Oosthuizen, K., Botha, E., Robertson, J., & Montecchi, M. (2021). Artificial intelligence in retail: The AI-enabled value chain. *Australasian Marketing Journal, 29*(3), 264–273. https://doi.org/10.1016/j.ausmj.2020.07.007

Pal, S., & Jadidi, Z. (2021). Analysis of security issues and countermeasures for the industrial internet of things. *Applied Sciences, 11*(20), 9393. https://doi.org/10.3390/app11209393

Patel, A. S., Brahmbhatt, M. N., Bariya, A. R., Nayak, J. B., & Singh, V. K. (2023). Blockchain technology in food safety and traceability concern to livestock products. *Heliyon, 9*(6), Article e16526. https://doi.org/10.1016/j.heliyon.2023.e16526

Pfister, B., & Zeeman, S. C. (2016). Formation of starch in plant cells. *Cellular and Molecular Life Sciences, 73*(14), 2781–2807. https://doi.org/10.1007/s00018-016-2250-x

Precha-Atsawanan, S., Puncha-arnon, S., Wandee, Y., Uttapap, D., Puttanlek, C., & Rungsardthong, V. (2018). Physicochemical properties of partially debranched waxy rice starch. *Food Hydrocolloids, 79*, 71–80. https://doi.org/10.1016/j.foodhyd.2017.12.014

Ramamurthy, K. R., Sen, A., & Sinha, A. P. (2008). An empirical investigation of the key determinants of data warehouse adoption. *Decision Support Systems, 44*(4), 817–841. https://doi.org/10.1016/j.dss.2007.10.006

Riahi, Saikouk, Gunasekaran, & Badraoui. (2021). Artificial intelligence applications in supply chain: A descriptive bibliometric analysis and future research directions. *Expert Systems with Applications, 173*.

Santos, Pires, Amante, da Cruz Rodrigues, & da Silva. (2022). Isolation and characterization of starch from purple yam (*Dioscorea trifida*). *Journal of Food Science and Technology, 59*(2), 715–723.

Singh, A., Geveke, D. J., & Yadav, M. P. (2017). Improvement of rheological, thermal and functional properties of tapioca starch by using gum arabic. *LWT, 80*, 155–162. https://doi.org/10.1016/j.lwt.2016.07.059

Sood, A., & Tellis, G. J. (2011). Demystifying disruption: A new model for understanding and predicting disruptive technologies. *Marketing Science, 30*(2), 339–354. https://doi.org/10.1287/mksc.1100.0617

Spooner, D. M., Ghislain, M., Simon, R., Jansky, S. H., & Gavrilenko, T. (2014). Systematics, diversity, genetics, and evolution of wild and cultivated potatoes. *The Botanical Review, 80*(4), 283–383. https://doi.org/10.1007/s12229-014-9146-y

Sri Vigna Hema, V., & Manickavasagan, A. (2024). Blockchain implementation for food safety in supply chain: A review. *Comprehensive Reviews in Food Science and Food Safety, 23*(5). https://doi.org/10.1111/1541-4337.70002

Thangamayan, S., Pradhan, K., Loganathan, G. B., Sitender, S., Sivamani, S., & Tesema, M. (2023). Blockchain-based secure traceable scheme for food supply chain. *Journal of Food Quality, 2023*. https://doi.org/10.1155/2023/4728840

Thomé, K. M., Cappellesso, G., Ramos, E. L. A., & Duarte, S. C.d. L. (2021). Food supply chains and short food supply chains: Coexistence conceptual framework. *Journal of Cleaner Production, 278*. https://doi.org/10.1016/j.jclepro.2020.123207

Trinh, K. S., & Du Le, H. (2017). Tapioca resistant starch production and its structural properties under annealing and plasma treatments. *Carpathian Journal of Food Science and Technology, 9*(1), 36–42.

Trivedi, S., Negi, S., & Anand, N. (2019). Role of food safety and quality in Indian food supply chain. *International Journal of Logistics Economics and Globalisation, 8*(1), 25. https://doi.org/10.1504/ijleg.2019.100208

van der Vorst, Da Silva, & Trienekens. (2007). Agro-industrial supply chain management: Concepts and applications. *Food and Agriculture Organization of the United Nations (FAO)*. FAO.

Van Hung, P., Huong, N. T. M., Phi, N. T. L., & Tien, N. N. T. (2017). Physicochemical characteristics and in vitro digestibility of potato and cassava starches under organic acid and heat-moisture treatments. *International Journal of Biological Macromolecules, 95*, 299–305. https://doi.org/10.1016/j.ijbiomac.2016.11.074

Zhou, Q., Li, X., Yang, J., Zhou, L., Cai, J., Wang, X., Dai, T., Cao, W., & Jiang, D. (2018). Spatial distribution patterns of protein and starch in wheat grain affect baking quality of bread and biscuit. *Journal of Cereal Science, 79*, 362–369. https://doi.org/10.1016/j.jcs.2017.07.017

Chapter 25

Market dynamics and consumer preferences for starch-based functional ingredients

Priyanka Balaji Sethupathy[1], **Sandeep Singh Rana**[1] **and Sujithra Sureshkumar**[2]
[1]*School of Bio Sciences & Technology (SBST), Vellore Institute of Technology, Vellore, India;* [2]*Department of Food Technology, Faculty of Engineering, Karpagam Academy of Higher Education, Coimbatore, Tamil Nadu, India*

1. Introduction

Starch-based functional ingredients have become crucial elements across multiple sectors, such as food, pharmaceuticals, and cosmetics, owing to their impressive versatility, functionality, and cost-effectiveness (Adigwe et al., 2022). Further, these ingredients perform various functions, including thickening and stabilizing formulations, as well as enhancing texture and improving the sensory experience of various food products (Alam et al., 2024; Sharma et al., 2024). Their extensive applications render them essential across various sectors, fostering ongoing innovation and industry expansion. In the food sector, starch-based functional additives are essential for enhancing texture, prolonging shelf life, and decreasing calorie content (Bertolini, 2009). They assist businesses in creating products that satisfy consumer demands for quality and nutritional value. The principal sources of these starches include corn, potato, tapioca, and rice, which are extensively cultivated and processed to produce diverse functional starch derivatives (Santana & Meireles, 2014; Yang et al., 2022; Yazid et al., 2018). Thus, the growing consumer awareness of ingredient transparency leads to an escalating need for clean-label, nonGMO, and sustainably sourced starch-based ingredients. This transition signifies an increasing inclination toward natural and minimally processed food ingredients, emphasizing the significance of plant-based starches in contemporary formulations.

In recent times, the food industries are experiencing significant transformation, influenced by changing consumer preferences and a growing awareness of health. Today's consumers are increasingly knowledgeable and selective regarding the products they choose, resulting in a significant demand for healthier, more sustainable, and allergen-free food options (Alam et al., 2025; Jadhav et al., 2024). Consequently, there has been a notable increase in the use of starch-based functional ingredients, especially in areas that promote clean-label formulations, plant-based alternatives, and products that are reduced-fat or gluten-free (Tian et al., 2022; Yang et al., 2022; Yazid et al., 2018). Accordingly, manufacturers are leveraging functional starches to meet specific dietary and lifestyle needs. For example, gluten-free baked goods featuring enhanced texture and stability are created using modified starches, increasing their attractiveness to consumers with celiac disease and gluten allergies (Gómez & Martínez, 2016; Witczak et al., 2016). Similarly, starches exhibiting fat-mimicking characteristics are incorporated into reduced-fat dairy and confectionery items to improve mouthfeel and creaminess, all whilst lessening the total fat content (Li et al., 2025; Surendra Babu et al., 2016). Starch-based hydrocolloids are used in the beverage industry to enhance suspension and viscosity in plant-based drinks, guaranteeing a uniform texture and smoothness (Reiner et al., 2010; Venkatesh et al., 2024). Further, the starch-based ingredients improve freeze-thaw stability and maintain product quality over time by playing a significant role in frozen and ready-to-eat meals (Pongsawatmanit & Srijunthongsiri, 2008; Ye et al., 2018). The growing popularity of convenience foods illustrates the critical role of functional starches in maintaining texture and preventing syneresis (water separation) in refrigerated and frozen products (Gupta et al., 2024; Rashwan et al., 2024). Hence, the food industry

Native and Modified Starches for Food Applications. https://doi.org/10.1016/B978-0-443-27431-2.00009-2

continues to explore innovative starch-based alternatives that meet consumer expectations for quality, taste, and nutritional value.

Even though starch-based functional ingredients have many uses and advantages, the sector still confronts a number of obstacles that hinder its acceptance and advancement. The regulatory environment around the use of modified starches in food items is one of the primary challenges. As the regulations governing starch modifications, clean-label claims, and permissible additives are varied across different regions of the world, mandating that manufacturers must navigate intricate compliance requirements (Gałkowska et al., 2023). Another challenge pertains to customer perception, especially concerning modified starches. Quite a few consumers express skepticism over modified starches (chemically or enzymatically), linking them to artificial additives, whereas native starches are broadly regarded as natural and lightly processed (Abbas et al., 2010; Lobato-Calleros et al., 2014). Thus, creating awareness amongst customers about the safety and advantages of these modified starches and novel processing technologies is crucial for enhancing acceptability and confidence regarding starch-based functional ingredients.

Another important element shaping the starch-based ingredients market is sustainability. In response to the increasing emphasis placed on environmentally friendly procurement and production practices by food companies, there is an increasing demand for sustainably harvested and processed starches. Further, waste reduction and sustainability enhancement within the industry are facilitated by advancements in the optimization of starch extraction methods and the up-cycling of agricultural by-products (Ashok et al., 2016). On the opportunity side, advancements in starch modification techniques, such as enzymatic treatments and physical processing, are paving the way for cleaner, more functional starches that align with clean-label trends (Maniglia et al., 2021). Additionally, the expansion of plant-based and alternative protein markets presents new avenues for starch-based ingredient applications, as these products often require texturizing and binding agents to achieve desirable sensory attributes (Abbas et al., 2010; Yazid et al., 2018).

The current chapter offers a comprehensive analysis of the current market landscape for starch-based functional ingredients, highlighting key consumer trends, challenges, and opportunities that influence innovation and adoption in the food sector. By exploring these factors, in this chapter, we have provided significant insights into the advancing position of starch-based functional ingredients in modern food formulations.

2. Significance of starch-based functional ingredients in consumer preference and market dynamics

Beyond basic nutrition, functional ingredients provide additional health benefits. These functional ingredients are nowadays becoming an essential part of basic food formulations. The categorization of starch-based functional ingredients plays a momentous role in controlling consumer preferences. A study on Slovene consumers established a deep preference for bread with functional ingredients, with a mean relative importance of 83.9%. Four consumer segments, such as uninvolved, conservative, health-conscious, and moderate, were used for latent class cluster analysis with different preferences toward selected functional ingredients (e.g., chia seeds, linseed, quinoa, and tartary buckwheat), nutrition, and other claims. The results provide a better understanding of customer preferences for functional ingredients, claims and new perspectives for bread marketing and identify the factors that affect the purchasing and consumption (Kušar et al., 2023). Another study highlighted the variety of bread as a crucial factor in their purchasing decisions by surveying 503 German consumers to recognize customer predilection and categorized respondents into segments that generally preferred bread with super food ingredients such as linseed or chia (Meyerding et al., 2018). In recent days, the market dynamics adjoining starch-based functional ingredients also imitate broader consumer values, which are increasingly supportive of environmentally sustainable and healthcentered choices. Research on Romanian consumers emphasized the functional and value perceptions. The study proposes that the functional values mean health benefits provided by added ingredients, which have a direct effect on the purchasing behavior (Corboş et al., 2024).

Due to the versatile applications and benefits in the various industries, specifically the food industry, starch-based functional ingredients play a vital role in market dynamics. Research related to modified purple yam starch was carried out in soups and sauces as thickeners and stabilizers and found that a unique color adds value to organoleptic characteristics (Shrivastava et al., 2024). Similarly, banana starch presents a substitute for chilled foods and sauces, owing to its functional properties and minimal retrogradation (Fontes et al., 2017). Incorporating pulse and buckwheat starch for a new product development enhances the product stability and sensory characteristics, by contributing a gluten-free food product, which is very well accepted by the consumer in the market (Sangokunle et al., 2020). Currently, the ingredients are selected not only for their purity and quality but also for their techno-functional properties, which provide sustainable production practices. This shift in market toward less refined starches with reduced environmental collision is increasing over time (Lie-Piang et al., 2022). In the market, resistant starch has rising awareness for its health benefits without compromising its organoleptic qualities of food products (Walsh et al., 2022). The technology recommends opportunities to progress bioavailability and functional food applications, eventually expanding the use of starch-based ingredients beyond traditional roles by enhancing innovative delivery systems, such as nano and microcapsule systems designed to enhance the stability and targeted delivery of bioactive compounds (Tian et al., 2022).

3. Market dynamics

3.1 Global market trends

The global functional ingredients market, including starch-based functional ingredients, was estimated at approximately USD 73.1 billion in 2022 and is anticipated to increase at a CAGR of 6.6% from 2023 to 2030 and according to data bridge market Meticulous Research (2025), global food starch market size was valued at USD 27.73 billion in 2024 and is projected to reach USD 46.24 billion by 2032, with a CAGR of 6.60% during the forecast period of 2025–2032 as shown in Fig. 25.1 and global food starch market overview up to 2029 is shown in Fig. 25.1. The food starch market specifically reached USD 27.1 billion in 2024 and is presumed to expand to USD 40.5 billion by 2033, growing at a CAGR of 4.33% between 2025 (IMARC Group). Within this sector, the modified starch market, a key segment, was estimated at USD 13.51 billion in 2024 and is anticipated to reach USD 20.69 billion by 2033, growing at a CAGR of 4.85% during the forecast period (Research and Markets, 2025). This expansion is piloted by expanding consumer awareness of health benefits, expanding demand for functional food and beverage products, and ongoing investments in research and development.

Regionally, the Asia Pacific area is experiencing rapid growth, driven by urbanization, evolving dietary preferences, and an expanding middle class, with substantial investments in food processing and research and development from China and India (Global Market Insights, 2025). Europe leads the sustainable and clean-label movement, with nations such as Germany, France, and the UK prioritizing plant-based and organic food markets. The Middle East and Africa, although an emerging industry, are witnessing consistent expansion as economic advancement stimulates demand for processed foods and functional ingredients (Meticulous Research, 2025). North America is a significant revenue contributor to the modified starch market, anticipated to expand at a CAGR of 3.90% from 2023 to 2031, with the U.S. at the forefront due to its sophisticated food processing sector and elevated consumer awareness (Market Research Future, 2025). In Latin America, market expansion is propelled by the rising acceptance of convenience meals and an intensified emphasis on health and well-being, with Brazil at the forefront owing to its strong agricultural foundation (Straits Research, 2025).

3.2 Factors driving market growth

The market growth of starch-based functional ingredients is driven by several key factors. The clean-label movement has led consumers to seek transparency in food labeling, favoring natural, and minimally processed ingredients. Native and organic starch varieties align with this trend, as seen in Ingredion's NOVATION functional native starches, which are labeled with familiar names like "corn starch," "tapioca starch," "rice starch," and "potato starch," enhancing consumer trust. Additionally, the rising popularity of plant-based diets has fueled innovation in meat and dairy alternatives, where starch-based ingredients play a crucial role in replicating texture and mouthfeel. This demand is particularly strong in

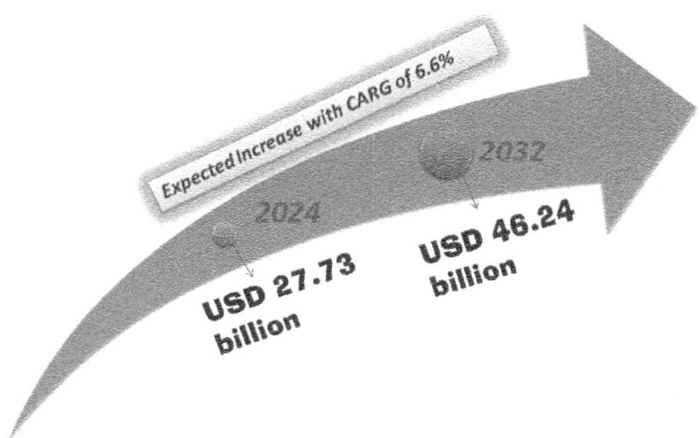

FIGURE 25.1 Global starch market projections from to 2032.

North America and Europe, where NOVATION starches are widely used in vegan-friendly formulations (NOVATION functional native starches, 2025). Furthermore, the increasing consumer preference for health-promoting ingredients is boosting the market for resistant starches, known for their prebiotic benefits and glycemic control properties. Resistant potato starch, which contains approximately 60% resistant starch in its raw form, is gaining traction in functional snacks and beverages for its digestive health benefits (Hu et al., 2024).

3.3 Challenges and barriers

The starch-based functional ingredients market faces several challenges and barriers. Supply chain and sourcing issues remain a significant concern, as the industry's reliance on agricultural raw materials makes it vulnerable to disruptions caused by climate change, geopolitical tensions, and trade restrictions. Ensuring consistent quality and supply remains a challenge for manufacturers. Additionally, the high costs of innovation and research and development (R&D) pose barriers, particularly for small and medium enterprises struggling to compete with established players. Despite these challenges, ongoing investments are driving growth in the global modified starch market. Consumer skepticism and regulatory hurdles also impact market expansion, as some modified starches face scrutiny due to concerns about chemical processing. Furthermore, navigating complex regulatory frameworks for food safety and labeling across different regions presents additional barriers. While modified starches, derived from sources like wheat, maize, and potatoes, offer functional benefits, regulatory compliance and consumer acceptance remain key challenges for industry players (Rashidinejad, 2024; Yang et al., 2022).

4. Consumer preferences and insights

4.1 Trends in clean-label and natural products

The clean-label movement is significantly influencing consumer behavior worldwide, with increasing demand for products containing simple, natural, and recognizable ingredients while avoiding artificial additives and excessive processing. Starch-based functional ingredients, especially native starches, align with these preferences due to their natural origins and versatile functionalities. A survey carried out in the USA by the International Food Information Council in 2023 to understand American consumer perception discovered that 72% of consumers prioritize clean-label products when making purchasing decisions (Public perceptions of food ingredients, 2023). This shift has led to a growing adoption of native starches like tapioca and rice starches, which have replaced synthetic thickeners in 25% of sauces and dressings launched in North America between 2020 and 2024 (IMARC Group, 2024b). Additionally, clean-label snacks incorporating tapioca starches experienced a 30% growth in sales across Europe in 2023, highlighting the increasing market preference for natural, starch-based ingredients (GlobeNewswire, 2024).

4.2 Health and nutritional preferences

The mounting importance of health and nutrition is steering demand for starch-based functional ingredients, particularly in gluten-free, resistant starch, and low-calorie product formulations. Globally, the gluten-free market is estimated at $6.7 billion in 2023 and is projected to grow at a CAGR of 9.2% through 2030, with potato, tapioca, and corn starches playing a crucial role in delivering elasticity, texture, and mouth feel in gluten-free foods like pizza dough and rice-based cereals (Grand View Research, 2023b). Resistant starches are gaining popularity due to their prebiotic benefits and blood sugar management properties, with products fortified with resistant starches, such as green banana flour, experiencing an 18% year-on-year growth in Asia–Pacific (Grand View Research, 2023c; Market Data Forecast, 2024). Meanwhile, modified starches, like acetylated distarch phosphate, are widely used in frozen meals, where demand increased by 12% in Europe in 2023 (GlobeNewswire, 2024). Additionally, prebiotic and low-calorie attributes are increasingly sought after, with resistant starches serving as dietary fibers and calorie-neutral alternatives. Innovations include prebiotic yogurt with resistant corn starch for enhanced fiber content and low-calorie bakery products using tapioca starch as a fat replacer, reducing calorie content by up to 30% (Chen et al., 2020; Karunarathna et al., 2024).

4.3 Regional and cultural differences

Regional dietary patterns and cultural preferences significantly influence the demand for starch-based functional ingredients worldwide. In North America, there is a strong focus on gluten-free, plant-based, and clean-label products, with tapioca and corn starches commonly used in nondairy alternatives. Europe emphasizes sustainability and organic ingredients, with potato starch gaining prominence due to its local availability and clean-label benefits. In the Asia–Pacific region, traditional starch sources like tapioca and rice remain in high demand, particularly in local cuisines such as bubble tea, noodles, and rice cakes, with tapioca starch imports growing by 15% in 2023. Latin America is witnessing a rise in functional snacks that incorporate cassava and maize starch, while in Africa and the Middle East, the demand for affordable, shelf-stable food products continues to grow, with maize starch playing a crucial role. Notable market trends include a 25% increase in sales of gluten-free bread fortified with potato starch in Germany and the expanding popularity of tapioca starch-based desserts and beverages in Asia–Pacific (Bonafide Research, 2023).

5. Innovations in starch-based ingredients

5.1 Advances in processing technologies

Advancements in processing technologies have significantly enhanced the functionality and sustainability of starch-based functional ingredients. Innovative modification techniques, including enzymatic treatment, physical modification (such as extrusion), and chemical modification (like cross-linking), have expanded starch applications by improving properties like freeze-thaw stability, emulsification, and resistance to acidic or high-temperature conditions (Amaraweera et al., 2021; He et al., 2023; Suri & Singh, 2023). For example, oxidized starch is now frequently employed as a thickener in acidic sauces due to its enhanced stability, while cross-linked starch plays a significant role in frozen desserts by preventing retrogradation during freezing and thawing cycles (Himashree et al., 2022; Maity et al., 2018; Steeneken et al., 2011). Additionally, enzymatically modified starch is increasingly utilized in beverages, such as transparent sports drinks, for improved solubility and clarity (Kim et al., 2023). In extraction and purification, technological advancements have led to the production of high-purity starches with a reduced environmental footprint. Innovations like ultrafiltration technology have enhanced the purity of tapioca starch, making it a preferred choice for premium confectioneries (Punia et al., 2024; Xue et al., 2025).

Recently, there has been numerous research and advances have been observed in starch modification and its application in food delivery and packaging systems, which is given in Table 25.1. Adebowale et al. (2002) investigated chemical modifications of Bambarra groundnut starch using oxidation and acetylation, which reduces gelatinization temperature, enhances swelling capacity and provides increased stability when blended with flours in bakery formulations (Shivaswamy et al., 2024). worked on mango kernel starch, in which starch was extracted by physical and chemical modifications, shows maximum water-binding capacity, amylose composition, and thermal stability, suitable for edible coatings and baked goods. Nowadays, plant-based starch processed with dual modification, which enhances the degree of substitution, is used for 3D printing of foods and packaging applications, such as biodegradable material and films, which may increase elongation strength, tensile strength and water vapor barrier properties (Lagunes-Delgado et al., 2024; Naveen & Loganathan, 2024). Twin-screw extrusion technology has been used for starch-lipid complexes for the preparation of extruded products and has been found to have more resistance to enzymatic digestion and improve the nutritional profile of the extruded products (Chao et al., 2024). Another research on nanoemulsion from flaxseed oil and faba bean starch had a vital delivery system for bioactive compounds and reported to boost stability (Sharma et al., 2024). A new possibility in nutraceutical delivery has been incorporated with starch matrices (Marinopoulou et al., 2024). examined the encapsulation of chickpea and ascorbic acid system by spray dryer. The encapsulation approach improved antioxidant stability, making it suitable for incorporating into bakery products and functional foods.

Likewise, nowadays starch-based nanocomposites have gathered momentous attention due to their biodegradability and renewable nature, making them suitable for various applications. Among the techniques, one of the effective techniques is the addition of cellulose nano-whiskers or layered silicates as a reinforcing material for creating starch-based nanocomposites. The nanocomposite materials are included in a matrix of modified potato starch, using water and sorbitol as plasticizers, which shows an increase in tensile strength and storage modulus at very high temperatures (Kvien et al., 2007). Fig. 25.2 show's innovative fabrication techniques for starch-based nanocomposites. Furthermore, different nanofillers such as nanoclays and polysaccharides are used with starch nanocomposites by homogeneously dispersing nanofillers within the starch matrix to enhance mechanical, thermal, and barrier properties that support sustainability goals in packaging applications (Abdullah & Dong, 2018). The above-mentioned advances in starch from traditional to novel products design, nutrient delivery and sustainable packaging will echo toward environmentally friendly and value-added applications in the food industries.

TABLE 25.1 Advances and innovations in starch-based technologies.

Raw material used	Method/Technology adapted	Functional property	Application	References
Bambarra groundnut starch	Oxidation and acetylation	Reduced the gelatinization temperature, increased the swelling capacity and modification enhanced stability	Blending with other flours and bakery products	Adebowale et al. (2002)
Mango kernel starch	Physical and chemical modification	Water binding capacity increased. Changes in amylose content and pasting properties are improved.Higher thermal stability.	Edible coatings and baked goods	Shivaswamy et al. (2024)
Cereal starch, tuber starch, and millet starch	Edible films	Improved tensile strength, elongation at break, and water vapor permeability	Used to package confectioneries, meat, sea foods, fruits & vegetables	Naveen & Loganathan (2024)
Different starch sources (lentil, chickpea, corn, pea, and tapioca) and a variety of antioxidants (ascorbic acid, linalool, carvacrol, and cinnamic acid)	Encapsulation – Spray dryer	Chickpea-ascorbic acid systems exhibit the highest bulk and tapped densities values and are the most hygroscopic while tapioca-cinnamon acid systems have the lowest ones.	Bakery products, antioxidants encapsulate	Marinopoulou et al. (2024)
Starch-lipid complexes	Extrusion - twin-screw extruder	More resistant to amylolysis, improves the quality and nutrition of starchy foods.	Extruded products – wheat noodles	Chao et al. (2024)
Plant based starch	Dual modification	Improves functionality, digestibility characteristics and starch degree of substitution	3D printing, biodegradable packing materials	Lagunes-Delgadoet al. (2024)
Faba bean starch and flaxseed oil	Starch extraction and Emulsification	Extracted starch yields reveals its semicrystalline structure. The optimized nanoemulsion had greater changes in stability enhancement.	stable delivery system for bioactive compounds	Sharma et al. (2024)
Corn starch & clove oil	Active packaging - cryogel	Reduction in *C.jejuni* count under micro-aerophilic conditions and cryogels also absorbed broiler meat exudates, effectivelyeliminating *C.jejuni* contamination	Reducing *C. jejuni* contamination through active packaging as absorbent pad in broiler meat packaging	Kleinübing et al. (2025)

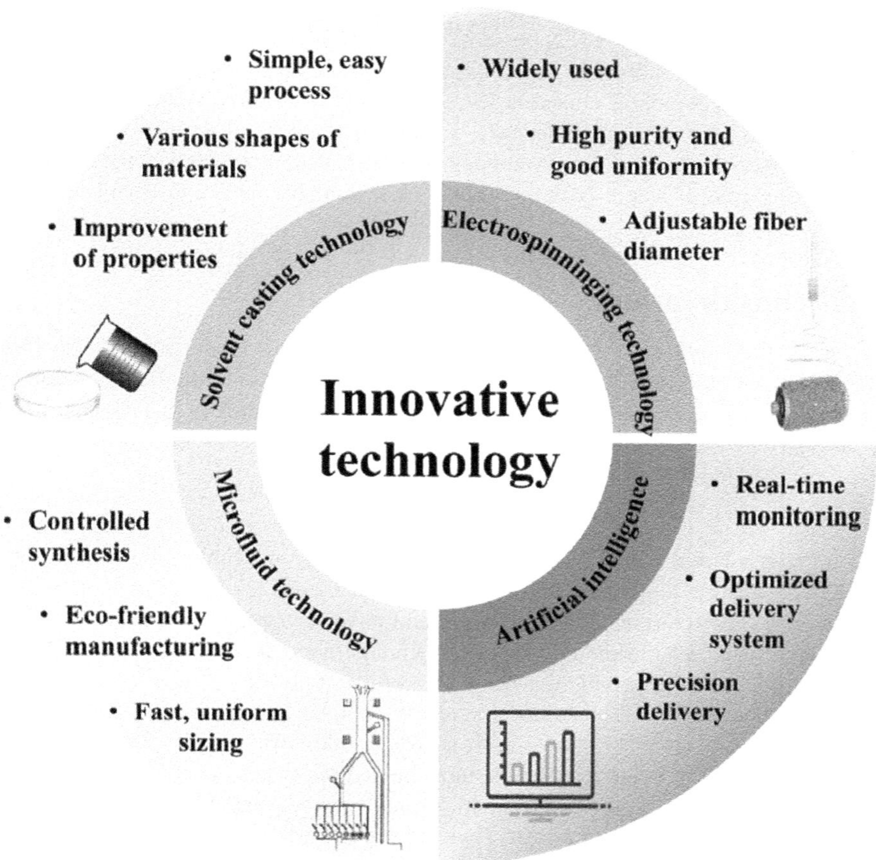

FIGURE 25.2 Innovative fabrication techniques for starch-based nanocomposites (Sun et al., 2025). *Data from open access, License not required.*

5.2 Emerging applications in the food and beverage industries

Starch-based functional ingredients are expanding their applications across both food and nonfood industries. In the food and beverage sector, these ingredients are increasingly used in functional products that improve gut health, regulate blood sugar levels, and enhance satiety (Guo et al., 2021; Himat et al., 2021; Yang et al., 2022). Resistant starches and soluble fibers play a key role, with innovations such as prebiotic smoothies incorporating resistant starch such as inulin for digestive benefits, low-glycemic breads formulated with high-amylose maize starch to reduce glycemic impact, and protein bars using tapioca starch for improved texture and shelf life (Dega & Barbhai, 2023; González-Herrera et al., 2015; Silva et al., 2013). Beyond food, starch is gaining prominence in sustainable nonfood applications. Starch-based bio-plastics, such as those developed by Novamont using potato and corn starch, offer biodegradable alternatives to petroleum-based plastics, helping to reduce environmental waste (BusinessWire, 2022). In cosmetics, tapioca starch is extensively used for its absorbent and moisturising properties in powders and creams, while in pharmaceuticals, starch capsules provide a plant-based alternative to gelatin, catering to vegetarian and vegan consumers (Arora et al., 2021). These emerging applications highlight the growing versatility and demand for starch-based ingredients across multiple industries.

5.3 Future directions in research and development

Research and development in starch-based functional ingredients are increasingly focused on enhancing sustainability and functionality through precision modification, biotechnology-driven production, and hybrid formulations. CRISPR technology is enabling the development of starches with customized amylose-to-amylopectin ratios to meet specific functional needs, while hybrid starches such as combinations of tapioca and maize are being explored for their unique performance characteristics (Fu et al., 2022). Another promising area is starch-based edible coatings, which aim to reduce packaging waste while extending the shelf life of fruits and vegetables (Sapper & Chiralt, 2018). Market projections indicate significant growth in these innovations, with the global market for starch-based bio-plastics believed to grow at a CAGR of 9.5% from 2023 to 2030, driven by rising environmental concerns (Straits Research, 2023). Additionally, functional starch ingredients in cosmetics are estimated to expand at an annual rate of 12%, particularly in Asia Pacific and Europe, where consumer demand for clean-label personal care products continues to surge (Maximize Market Research, 2023). These advancements highlight the increasing role of starch-based solutions in sustainable and high-performance applications across industries.

6. Sustainability and ethical considerations

Sustainability is a growing priority in the food industry, with consumers, regulators, and manufacturers emphasizing ethical sourcing, reduced environmental impact, and long-term viability. Starch-based functional ingredients, commonly obtained from corn, wheat, potatoes, cassava, and rice, present unique challenges and opportunities in sustainability. This section explores key considerations in raw material sourcing, environmental impact, and circular economy strategies.

6.1 Sourcing of raw materials

Sustainable sourcing of raw materials for starch-based functional ingredients begins at the agricultural level, where ethical and environmentally responsible practices ensure long-term viability (Falua et al., 2022; Gonçalves et al., 2024). Regenerative agriculture methods, including crop rotation, no-till farming, and cover cropping, enhance soil health and biodiversity while increasing carbon sequestration (Jordon et al., 2022; Khangura et al., 2023; Sher et al., 2024). Ingredion's adoption of regenerative farming for corn production has successfully reduced soil erosion and improved organic matter content (Ingredion Incorporated, 2021). Water conservation is another crucial aspect, as certain starch crops like cassava are naturally drought-resistant, whereas others, such as wheat and corn, require precision irrigation techniques (Xing & Wang, 2024). Cargill's precision irrigation models have helped cut water usage in starch processing by 25% (Cargill, 2024). Additionally, the industry is reducing chemical inputs by promoting biological fertilizers and integrated pest management (IPM), aligning with the EU's Farm to Fork strategy (Farm to Fork Strategy, 2020). The demand for organic and nonGMO starches is also rising, with the organic starch market projected to grow at a CAGR of 6.2% from 2023 to 2030, driven by applications in baby food, bakery, and plant-based products (Allied Market Research, 2024). Beyond environmental sustainability, the starch industry significantly impacts local economies, particularly in Asia, South America, and Africa. Ethical sourcing initiatives ensure fair wages, economic stability, and food security (Michel et al., 2024; Noort et al., 2022). Major corporations, including ADM and Roquette, invest in community development, infrastructure, and agricultural extension services to enhance sustainability and economic resilience in key sourcing regions (Guibert et al., 2023).

6.2 Environmental impacts of starch-based ingredients

The environmental impact of starch-based ingredients is predominantly linked with their carbon footprint and waste generation, prompting companies to adopt sustainable practices (Forfora et al., 2024; Xu et al., 2015). The carbon footprint stems from agriculture, processing, and transportation, but energy-efficient solutions are mitigating emissions (Alami et al., 2024; Thamarai et al., 2024). For instance, Roquette has implemented a global initiative known as the "30/30 energy efficiency plan," aiming to reduce specific energy consumption by 30% by 2030 across several of its sites worldwide (Roquette, 2023b). Shortening supply chains also plays a role, as transportation accounts for 18%–20% of emissions in starch ingredient supply chains, according to the FAO (FAO, 2022a,2022b). Additionally, firms like Cargill and Ingredion are advancing carbon-neutral ingredient programs, with Cargill aiming for a 30% emissions reduction in its corn starch supply chain by 2030 (Cargill, 2019; Ingredion Incorporated, 2022).

Waste management strategies focus on circular economy principles, repurposing by-products such as fibers and proteins for animal feed, bio-plastics, and bioethanol (Toplicean & Datcu, 2024; Vasileiadou, 2024). Roquette's development of pea starch-based bioplastics exemplifies this trend (Roquette, 2022, 2024). Water recycling is another priority, with facilities typically recycling around 80% of wastewater through advanced filtration (Meneses et al., 2017; Pidgeon, 2009). Furthermore, starch-based biodegradable packaging is gaining traction, with the market expected to reach $2.5 billion by 2027 due to regulations banning single-use plastics in the EU and U.S. (Dataintelo, 2023; Pro Market Reports,

2025). Regulatory frameworks like the EU Farm to Fork Strategy, U.S. Sustainable Agriculture Standard, and ISO 14,001 Environmental Management certification are shaping industry practices, promoting low-carbon farming and sustainable production (European Commission, 2020; ISO, 2015).

7. Regulatory landscape

7.1 Standards and guidelines for clean-label products

The clean-label movement has driven regulatory bodies, namely the FDA (U.S.), EFSA (Europe), and FSSAI (India), to establish guidelines ensuring ingredient transparency, minimal processing, and the avoidance of artificial additives (Balasubramaniam et al., 2023; Galanakis, 2022; Jadhav et al., 2024). While "clean label" is a market-driven term rather than a legal classification, regulations vary by region. In the U.S., the FDA mandates that "natural" claims be truthful, while organic and nonGMO labeling must meet USDA standards (Bartolomeo, 2017; Monga, 2025). The EU enforces strict additive labeling through EFSA and the FIC Regulation, encouraging the replacement of E-number additives with clean-label starches (Aryee et al., 2024; EFSA Panel on Nutrition et al., 2020; López-Rodríguez et al., 2025). In the Asia Pacific region, modified starches are regulated under China's GB standards, while Japan enforces stringent requirements for natural food labeling (Consumer Affairs Agency of Japan; Ministry of Justice, 2013). For clean-label and starch-based ingredients industry certifications, such as ISO 22,000, BRC Global Standard for Food Safety, NonGMO Project Verified, EU Organic, and USDA Organic serves as key assurance for compliance and quality.

7.2 Regional regulatory variations

Regulatory standards for clean-label and starch-based functional ingredients vary across regions, influencing market access and product development. The European Union enforces strict traceability and labeling under its "Farm to Fork" strategy, favoring native and functional starches over chemically modified alternatives (Balan et al., 2024; Pío Beltrán et al., 2022). North America adopts a more flexible, voluntary clean-label approach, with FDA guidance driving reformulations in processed food (Jadhav et al., 2024; Park & Kim, 2025). In Asia–Pacific, China maintains stricter GB standards for modified starches, while Japan emphasizes plant-based and functional starch applications Chunli et al., 2020; IMARC Group, 2024a; Liu et al., 2019). Latin America is gradually aligning with U.S. and EU standards, particularly in clean-label and nonGMO compliance (Food and Drug Law Institute, 2023). As a result, clean-label starches like native corn and tapioca starch are gaining preference, while claims such as "natural" or "additive-free" require precise regulatory adherence (Chakraborty et al., 2019; London Food Commission, 2025; Yazid et al., 2018). Companies must navigate these regional variations to ensure compliant product labeling and successful global expansion.

7.3 Impacts on product development and marketing

Regulatory compliance plays a crucial role in determining product formulation and marketing strategies in the clean-label movement (Finnegan & Krzyzaniak, 2024). Companies must invest in R&D to develop functional yet cost-effective clean-label starch ingredients, such as tapioca, rice, and potato starch, for applications in snacks, sauces, dairy, and gluten-free products. Marketing efforts must align with legal definitions of terms like "natural," "organic," and "additive-free" to ensure transparency and avoid misleading claims (Negowetti et al., 2022). Additionally, reformulation trends show a shift from modified starches (E-numbers) to functional native starches, catering to consumer demand for cleaner ingredient lists (Park & Kim, 2025). Noncompliance can lead to legal risks and erode consumer trust, making adherence to global regulatory standards essential for successful product positioning and expansion.

8. Market forecast and opportunities

8.1 Growth projections by region and segment

Globally, the starch-based functional ingredients market is set for steady growth, piloted by expanding consumer preference for natural, sustainable and clean-label products. North America and Europe will continue to dominate due to rigorous food safety regulations and high consumer awareness, while the Asia Pacific region is expected to be the fastest-growing market, supported by urbanization, rising disposable incomes, and a shift toward functional foods (Grand View Research, 2023d).

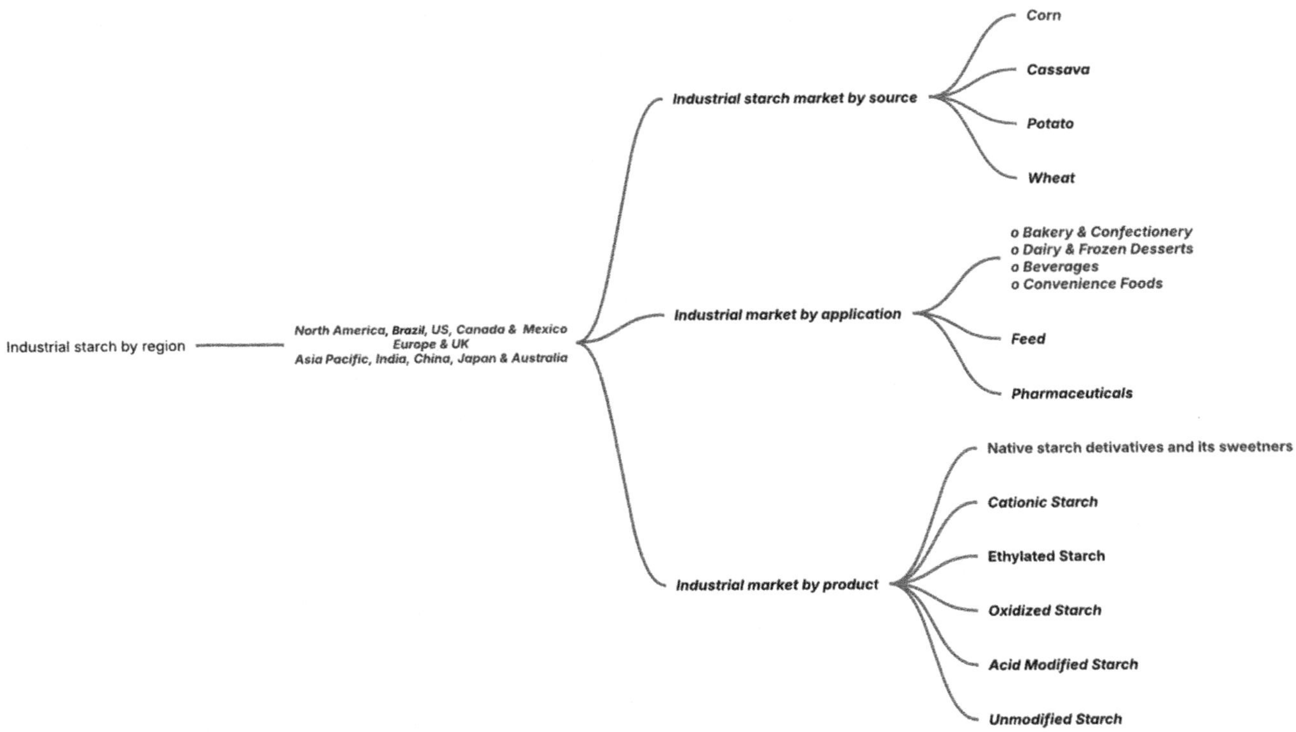

FIGURE 25.3 Market segment by industrial starch regional outlook.

Key growth segments include clean-label and organic starches (7.2% CAGR), plant-based and gluten-free applications (6.8% CAGR), and functional resistant starches for gut health (7.5% CAGR) (Custom Market Insights, 2024; Grand View Research, 2024; Precedence Research, 2024). North America leads in clean-label and plant-based innovation, Europe enforces the strictest regulatory standards, and Asia Pacific is rapidly expanding with growing functional food demand. Industrial starch regional outlook related to market segment for source, products and applications is given in Fig. 25.3 (Grand View Research, 2024).

8.2 Key drivers of future market trends

The future of the starch-based functional ingredients market will be shaped by several key trends. Consumer demand for clean-label products is driving innovation in minimally processed, natural starch alternatives (Park & Kim, 2025). The rise of plant-based and gluten-free formulations is increasing the use of starch-based texturizers and binders as replacements for gluten and animal-derived thickeners (Yang et al., 2022). Health-conscious formulations, particularly resistant starches with gut health benefits, are gaining traction (Kumar et al., 2023). Advancements in food technology will further enhance the functional properties of starches, expanding their applications (Bertolini, 2009; Sun et al., 2023). Additionally, sustainability concerns are pushing companies toward low-carbon, regenerative starch sources, while evolving regulatory frameworks will require reformulation efforts and clearer labeling to ensure compliance (Finnegan & Krzyzaniak, 2024).

8.3 Strategic opportunities for manufacturers

Manufacturers can leverage several strategic opportunities to improve progress in the starch-based functional ingredients market. Investing in research and development will enable the creation of innovative starch solutions that improve texture, stability, and nutritional benefits (Yazid et al., 2018). Expanding into emerging markets through localized strategies can help capture high-growth opportunities. Forming strategic partnerships with food and beverage companies will facilitate tailored ingredient solutions (Hu et al., 2024). The key market players globally in the starch food market are given in Fig. 25.4. Companies should prioritize clean-label starch investments, focusing on native and resistant starches, while also exploring functional and plant-based segments such as high-protein and dietary fiber-enhanced starches. Sustainability efforts, including utilizing by-products for bioplastics and biodegradable packaging, will enhance environmental

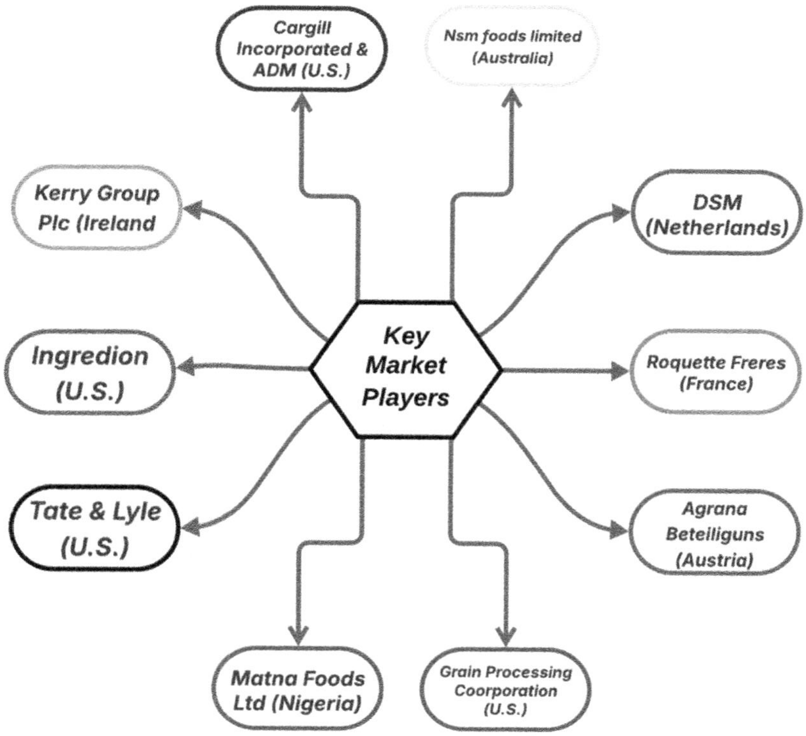

FIGURE 25.4 Key market players globally for starch food market.

responsibility (Gökçe, 2018; Gurunathan et al., 2025). Finally, improving transparency by providing clear product origin and processing details will strengthen consumer trust and regulatory compliance (Hamaker, 2021).

9. Conclusion

The market for starch-based functional ingredients is seeing significant growth, propelled by shifting consumer preferences, regulatory factors, and technical innovations. With the increasing need for clean-label, health-oriented, and sustainable food solutions, starch-based components are strategically positioned to fulfill industry requirements. Their adaptability and use render them indispensable for augmenting product texture, prolonging shelf life, and boosting nutritional quality in diverse food applications. This chapter has rigorously analyzed the principal market factors influencing the starch-based functional ingredients sector, emphasizing the evolving customer preferences that propel product innovation. The transition to clean-label and health-oriented formulations has prompted producers to investigate innovative starch modifications and sustainable sourcing methods. Nonetheless, obstacles such as regulatory limitations and consumer distrust regarding modified starches must be addressed to optimize adoption and market growth. Despite these limitations, developments in processing technologies and the enhanced requirement for plant-based, gluten-free, and reduced-fat formulations offer substantial prospects for innovation. By capitalizing on these developments and overcoming market obstacles, manufacturers might enhance the practicality and attractiveness of starch-based ingredients, thereby influencing the future of food product development. As the food industry advances, starch-based functional ingredients continue to be essential for providing high-quality, consumer-oriented food solutions that correspond with health and sustainability goals.

References

Abbas, K. A., K halil, S. K., & Hussin, A. S. M. (2010). Modified starches and their usages in selected food products: A review study. *Journal of Agricultural Science, 2*(2), 90.

Abdullah, Z. W., & Dong, Y. (2018). Recent advances and perspectives on starch nanocomposites for packaging applications. *Journal of Materials Science, 53*(22), 15319–15339. https://doi.org/10.1007/s10853-018-2613-9

Adebowale, K. O., Afolabi, T. A., & Lawal, O. S. (2002). Isolation, chemical modification and physicochemical characterisation of bambarra groundnut (voandzeia subterranean) starch and flour. *Food Chemistry, 78*(3), 305–311.

Adigwe, O. P., Egharevba, H. O., & Emeje, M. O. (2022). Starch: A veritable natural polymer. *Starch: Evolution and Recent Advances, 23.*

Alam, M. W., Kumar, J. V., Awad, M., Saravanan, P., Al-Sowayan, N. S., Rosaiah, P., & Nivetha, M. S. (2025). Emerging trends in food process engineering: Integrating sensing technologies for health, sustainability, and consumer preferences. *Journal of Food Process Engineering, 48*(1), Article e70035.

Alam, M., Pant, K., Brar, D. S., Dar, B. N., & Nanda, V. (2024). Exploring the versatility of diverse hydrocolloids to transform techno-functional, rheological, and nutritional attributes of food fillings. *Elsevier B.V., India Food Hydrocolloids, 146.* https://doi.org/10.1016/j.foodhyd.2023.109275

Alami, A. H., Olabi, A. G., Khuri, S., Aljaghoub, H., Alasad, S., Ramadan, M., & Abdelkareem, M. A. (2024). 3D printing in the food industry: Recent progress and role in achieving sustainable development goals. *Ain Shams Engineering Journal, 15*(2), Article 102386.

Allied Market Research. (2024). Organic starch market by type and application – Global opportunity analysis and industry foreca–2035. https://www.alliedmarketresearch.com/organic-starch-market-A43927.

Amaraweera, S. M., Gunathilake, C., Gunawardene, O. H. P., Fernando, N. M. L., Wanninayaka, D. B., Dassanayake, R. S., Rajapaksha, S. M., Manamperi, A., Fernando, C. A. N., Kulatunga, A. K., & Manipura, A. (2021). Development of starch-based materials using current modification techniques and their applications: A review. *Molecules, 26*(22). https://doi.org/10.3390/molecules26226880

Arora, S., Singh, D., Rajput, A., Bhatia, A., Kumar, A., Kaur, H., & Singh, B. (2021). Plant-based polysaccharides and their health functions. *Functional Foods in Health and Disease, 11*(4), 179–200.

Aryee, A. N. A., Ofori, K. F., Otchere, E. K., Dare, K. O., & Akanbi, T. O. (2024). Food- derived bioactive components and health claims: The role of regulatory agencies. In *Plant food phytochemicals and bioactive compounds in nutrition and health* (pp. 498–529). CRC Press.

Ashok, A., Mathew, M., & Rejeesh, C. R. (2016). Innovative value chain development of modified starch for a sustainable environment: A review. *Int. J. Polym. Sci. Eng, 2*(1), 20–32.

Balan, I. M., Trasca, T. I., Iancu, T., Belc, N., Radulov, I., & Tulcan, C. (2024). *Food safety in the sustainable food industry. In smart food industry: The blockchain for sustainable engineering* (pp. 218–239). CRC Press.

Balasubramaniam, V. M., Lee, J., & Serventi, L. (2023). Understanding new foods: Development of next generation of food processing, packaging, and ingredients technologies for clean label foods. In *Sustainable food innovation* (pp. 157–167). Cham: Springer International Publishing.

Bartolomeo, A. (2017). A proposal for FDA label regulations and uniform certifications for organic non-food and natural products. *Barry L. Rev., 23*, 65.

Bertolini, A. C. (2009). *Trends in starch applications starches: Characterization, properties, and applications* (pp. 1–19). Brazil: CRC Press. https://doi.org/10.1201/9781420080247-c1

Bonafide Research (2023). Global Starch Market Overview, 2023: Trends and Forecasts. Retrieved from https://www.bonafideresearch.com/press/230929661/global-starch-market.

BusinessWire (2022). Outlook on the Starch-Blended Biodegradable Polymer Global Market to 2030 - Featuring BASF, Corbion, and Novamont Among Others. Retrieved April 6, 2025, from https://www.businesswire.com/news/home/20220202005550/en/

Cargill. (2024). Cargill and goanna Ag pilot irrigation efficiency technology on cotton fields in the Mississippi Delta. https://www.cargill.co/cargill-and-goanna-ag-pilot-irrigation-efficiency-technology-on.

Cargill. (2019). Cargill Expands Climate Change Commitments. https://www.cargill.com/2019/cargill-expands-climate-change-commitments.

Chakraborty, R., Kalita, P., & Sen, S. (2019). Natural starch in biomedical and food industry: Perception and overview. *Current Drug Discovery Technologies, 16*(4), 355–367.

Chao, C., Liang, S., Sun, B., Zhang, Y., Yang, Y., & Wang, S. (2024). Effects of starch-lipid complexes on quality and starch digestibility of wheat noodles. *Starch - Stärke, 76*(11–12). https://doi.org/10.1002/star.202300155

Chen, Y., She, Y., Zhang, R., Wang, J., Zhang, X., & Gou, X. (2020). Use of starch-based fat replacers in foods as a strategy to reduce dietary intake of fat and risk of metabolic diseases. *Food Science and Nutrition, 8*(1), 16–22.

Chunli, D., Feifei, S., Yan, L., & Yanghe, L. (2020). Recent advances in modification of starch and its applications in china food industry. *The Scientific Heritage, 47-1*(47), 19–26.

Consumer Affairs Agency of Japan. (n.d.). Food labeling. Retrieved May 8, 2025, from:. Consumer Affairs Agency. https://www.caa.go.jp/en/policy/food_labeling/.

Corboş, R.-A., Bunea, O.-I., Triculescu, M., & Ioana Mişu, S. (2024). Which values matter Most to Romanian consumers? Exploring the impact of green attitudes and communication on buying behavior. *Sustainability, 16*(9), 3866. https://doi.org/10.3390/su16093866

Custom Market Insights. (2024). Global clean label starch market size, trends, share 2033.https://www.custommarketinsights.com/report/clean-label-starch-market/.

Dataintelo. (2023). Biodegradable food packaging material market – Global analysis, trends, and forecast 2023–2032. https://dataintelo.com/report/global-biodegradable-food-packaging-material-market.

Dega, V., & Barbhai, M. D. (2023). Exploring the underutilized novel foods and starches for formulation of low glycemic therapeutic foods: A review. *Frontiers in Nutrition, 10*, 1162462.

European Commission. (2020). *Farm to Fork Strategy.* https://food.ec.europa.eu/horizontal-topics/farm-fork-strategy_en.

Falua, K. J., Pokharel, A., Babaei-Ghazvini, A., Ai, Y., & Acharya, B. (2022). Valorization of starch to biobased materials: A review. *Polymers, 14*(11). https://doi.org/10.3390/polym14112215

FAO. (2022a). Food miles: The global environmental cost of food transport. https://www.fao.org/family-farming/detail/en/c/1559836/.

FAO. (2022b). Global food-miles account for nearly 20% of total food-systems emissions.https://www.fao.org/family-farming/detail/en/c/1559836/.

Farm to Fork Strategy. (2020), Available from: https://food.ec.europa.eu/horizontal-topics/farm-fork-strategy_en.

Finnegan, D. Y., & Krzyzaniak, D. S.-A. (2024). Clean labels in the food industry: Regulatory considerations and challenges in balancing consumer demand, safety and sustainability. *Food Science and Nutrition Cases, 2024.* https://doi.org/10.1079/fsncases.2024.0017

Fontes, S.d. M., Cavalcanti, M. T., Candeia, R. A., & Almeida, E. L. (2017). Characterization and study of functional properties of banana starch green variety of mysore (musa AAB - Mysore). *Food Science and Technology, 37*(2), 224–231. https://doi.org/10.1590/1678-457X.18916

Forfora, N., Azuaje, I., Kanipe, T., Gonzalez, J. A., Lendewig, M., Urdaneta, I., Venditti, R., Gonzalez, R., & Argyropoulos, D. (2024). Are starch-based materials more eco-friendly than fossil-based? A critical assessment. *Cleaner Environmental Systems, 13*, Article 100177. https://doi.org/10.1016/j.cesys.2024.100177

Fu, Y., Jiang, E., & Yao, Y. (2022). New techniques in structural tailoring of starch functionality. *Annual Review of Food Science and Technology, 13* (1), 117–143.

Galanakis, C. M. (Ed.). (2022). *The age of clean label foods*. Switzerland: Springer.

Gökçe, E. (2018). Rethinking sustainability: A research on starch based bioplastic. *Journal of Sustainable Construction Materials and Technologies, 3* (3), 249–260. https://doi.org/10.29187/jscmt.2018.28

Gómez, M., & Martínez, M. M. (2016). Changing flour functionality through physical treatments for the production of gluten-free baking goods. *Journal of Cereal Science, 67*, 68–74. https://doi.org/10.1016/j.jcs.2015.07.009

Gałkowska, D., Kapuśniak, K., & Juszczak, L. (2023). Chemically modified starches as food additives. *Molecules, 28*(22), 7543.

Food and Drug Law Institute (2023). Food labeling regulation: An overview from Brazil and Latin America. Retrieved from https://www.fdli.org/2023/11/food-labeling-regulation-an-overview-from-brazil-and-latin-america/.

Global Market Insights. (2025). Modified starch market size, growth opportun–2032. Retrieved April 1, 2025, from https://www.gminsights.com/industry-analysis/modified-starch-market.

GlobeNewswire. (2024). Native starch market analys–2032 by end-use, feedstock, and region. https://www.globenewswire.com/news-releas/03/27/2852986/28124/en/Native-Starch-Market-Analysi-2032.

Gonçalves, E. M., Silva, M., Andrade, L., & Pinheiro, J. (2024). From fields to films: Exploring starch from agriculture raw materials for biopolymers in sustainable food packaging. *Agriculture, 14*(3), 453. https://doi.org/10.3390/agriculture14030453

Grand View Research. (2023b). Gluten-free products market size, share and trends analysis report, 2023–2030.https://www.grandviewresearch.com/industry-analysis/gluten-free-products-market.

Grand View Research. (2023c). Resistant starch market size, share & trends analysis report, 2023–2030.

Grand View Research. (2023d). Starch-based functional ingredients market size, share & trends analysis report by source (corn, wheat, potato), by application (food & beverages, animal feed), by region, and segment forecasts, 2023–2030.

Grand View Research. (2024). Functional bakery ingredients market: Industry report, 2030.

G. Guibert, A. Turner, J. Frantz, J. Lewandowski, I. Hyde, Building resilience in economic development: A systems approach for sustainable growth. (2023),

González-Herrera, S. M., Herrera, R. R., López, M. G., Rutiaga, O. M., Aguilar, C. N., Esquivel, J. C. C., & Martínez, L. A. O. (2015). Inulin in food products: Prebiotic and functional ingredient. *British Food Journal, 117*(1), 371–387.

Guo, Y., Qiao, D., Zhao, S., Zhang, B., & Xie, F. (2021). Starch-based materials encapsulating food ingredients: Recent advances in fabrication methods and applications. *Carbohydrate Polymers, 270*, Article 118358. https://doi.org/10.1016/j.carbpol.2021.118358

Gupta, R. K., Pipliya, S., Kumar, S., Karunanithi, S., Rout, S., Guha, P., & Srivastav, P. P. (2024). *Starch: Application in structuring of food systems structured foods*. India: CRC Press. https://doi.org/10.1201/9781003355441-9

Gurunathan, M. K., Navasingh, R. J. H., Selvam, J. D. R., & Čep, R. (2025). Development and characterization of starch bioplastics as a sustainable alternative for packaging. *Nature Research, India Scientific Reports, 15*(1). https://doi.org/10.1038/s41598-025-00221-0

Hamaker, B. R. (2021). Current and future challenges in starch research. *Current Opinion in Food Science, 40*, 46–50. https://doi.org/10.1016/j.cofs.2021.01.003

He, R., Li, S., Zhao, G., Zhai, L., Qin, P., & Yang, L. (2023). Starch modification with molecular transformation, physicochemical characteristics, and industrial usability: A state-of-the-art review. *Polymers, 15*(13), 2935. https://doi.org/10.3390/polym15132935

Himashree, P., Sengar, A. S., & Sunil, C. K. (2022). Food thickening agents: Sources, chemistry, properties and applications-A review. *International Journal of Gastronomy and Food Science, 27*, Article 100468.

Himat, A. S., Gautam, S., Chavez Garcia, J. P., Vidrio-Sahagún, A. X., Liu, Z., Bressler, D., & Vasanthan, T. (2021). Starch-based novel ingredients for low glycemic food formulation. *Bioactive Carbohydrates and Dietary Fibre, 26*. https://doi.org/10.1016/j.bcdf.2021.100275

Hu, H., Jiang, H., Sang, S., McClements, D. J., Jiang, L., Wen, J., Jin, Z., & Qiu, C. (2024). Research advances in origin, applications, and interactions of resistant starch: Utilization for creation of healthier functional food products. *Trends in Food Science & Technology, 148*, Article 104519. https://doi.org/10.1016/j.tifs.2024.104519

IMARC Group. (2024a). Japan modified starch market: Industry trends, share, size, growth, opportunity and forecast 2025–2033.https://www.imarcgroup.com/japan-modified-starch-market.

IMARC Group. (2024b). North America native starch market repo–2033: Industry trends, share, size, growth, opportunity and forecast.https://www.imarcgroup.com/north-america-native-starch-market.

Ingredion Incorporated. (2021). Two major companies announce new AgWater Challenge commitments. https://ir.ingredionincorporated.com/news-releases/news-release-details/two-major-companies-announce-new-agwater-challenge-commitments.

Ingredion Incorporated (2022). Ingredion receives SBTi approval for its greenhouse gas emissions reduction targets. https://ir.ingredionincorporated.com/news-releases/news-release-details/ingredion-receives-sbti-approval-its-greenhouse-gas-emissions.

ISO. (2015). *ISO 14001:2015 Environmental management systems*. https://www.iso.org/iso-14001-environmental-management.html.

Jadhav, S. P., Shah, U. B., & Shelke, K. (2024). Current facts about clean label food products. In *Food intolerances* (pp. 162–200). CRC Press.

Jordon, M. W., Smith, P., Long, P. R., Bürkner, P. C., Petrokofsky, G., & Willis, K. J. (2022). Can regenerative agriculture increase national soil carbon stocks? Simulated country-scale adoption of reduced tillage, cover cropping, and ley-arable integration using RothC. *Science of the Total Environment, 825*. https://doi.org/10.1016/j.scitotenv.2022.153955

Karunarathna, S., Wickramasinghe, I., Truong, T., Brennan, C., Navaratne, S. B., & Chandrapala, J. (2024). Development of low-calorie food products with resistant starch-rich sources –a review. *Food Reviews International, 40*(2), 814–831.

Khangura, R., Ferris, D., Wagg, C., & Bowyer, J. (2023). Regenerative Agriculture—A literature review on the practices and mechanisms used to improve soil health. *Sustainability, 15*(3), 2338. https://doi.org/10.3390/su15032338

Kim, E. A., Lee, Y. R., Lee, E. H., Jeong, H. M., Kang, B. S., Kim, B. H., Park, S. J., & Shim, J. H. (2023). Development and applications of enzymatic modified starch with high water solubility providing a continuous supply of glucose. *International Journal of Biological Macromolecules, 250*. https://doi.org/10.1016/j.ijbiomac.2023.126107

Kleinübing, N. R., da Cruz, E. P., Fonseca, L. M., Ramires, T., Kröning, I. S., Crizel, R. L., … da Silva, W. P. (2025). Antimicrobial activity of corn starch cryogels incorporated with clove essential oil (Syzygium aromaticum) against antimicrobial-resistant Campylobacter jejuni: application as an absorbent pad in broiler meat packaging. *International Journal of Food Microbiology*, 111391.

Kumar, Y., Shikha, D., Guzmán-Ortiz, F. A., Sharanagat, V. S., Kumar, K., & Saxena, D. C. (2023). *Starch: Current production and consumption trends starch: Advances in modifications, technologies and applications*. India: Springer International Publishing. https://doi.org/10.1007/978-3-031-35843-2_1

Kušar, A., Pravst, I., Pivk Kupirovič, U., Grunert, K. G., Kreft, I., & Hristov, H. (2023). Consumers' preferences towards bread characteristics based on food-related lifestyles: Insights from Slovenia. *Foods, 12*(20), 3766.

Kvien, I., Sugiyama, J., Votrubec, M., & Oksman, K. (2007). Characterization of starch based nanocomposites. *Journal of Materials Science, 42*(19), 8163–8171. https://doi.org/10.1007/s10853-007-1699-2

Lagunes-Delgado, C., Agama-Acevedo, E., & Bello-Pérez, L. A. (2024). Dual starch modifications to expand its end-uses: A review. *Starch Staerke, 76* (11–12), Article 2300153.

Li, X., Xu, Z., Xu, X., zheng, K., Pan, S., Regenstein, J. M., Wang, H., & Wang, L. (2025). Carbohydrate-based fat mimetics: A focus on fat-mimicking mechanisms and a tribology perspective. *Trends in Food Science and Technology, 157*. https://doi.org/10.1016/j.tifs.2025.104885

Lie-Piang, A., Möller, A. C., Köllmann, N., Garre, A., Boom, R., & van der Padt, A. (2022). Functionality-driven food product formulation–An illustration on selecting sustainable ingredients building viscosity. *Food Research International, 152*, Article 110889.

Liu, Z., Liu, B., & Chen, G. (2019). Retrospective analysis of the development history of the Chinese food additive standards system based on the CODEX principles. *Npj Science of Food, 3*(1), 27.

London Food Commission. (2025). *Food adulteration and how to beat it*. USA: Taylor & Francis.

López-Rodríguez, R., Vermetten, J., Domínguez, L., Fernández-Ruiz, V., & Cámara, M. (2025). A critical review of synthetic novel foods within the European regulation: Proposed classification, toxicological concerns and potential health claims. *Critical Reviews in Food Science and Nutrition*, 1–21.

Maity, T., Saxena, A., & Raju, P. S. (2018). Use of hydrocolloids as cryoprotectant for frozen foods. *Critical Reviews in Food Science and Nutrition, 58* (3), 420–435.

Maniglia, B. C., Castanha, N., Le-Bail, P., Le-Bail, A., & Augusto, P. E. D. (2021). Starch modification through environmentally friendly alternatives: A review. *Critical Reviews in Food Science and Nutrition, 61*(15), 2482–2505. https://doi.org/10.1080/10408398.2020.1778633

Marinopoulou, A., Zoumaki, M., Raphaelides, S., Karageorgiou, V., & Goulas, A. (2024). Characterization of spray dried starch systems of natural antioxidant compounds. *Starch - Stärke, 76*(11–12). https://doi.org/10.1002/star.202300069

Market Data Forecast. (2024). Asia-pacific resistant starch market foreca–2029).https://www.marketdataforecast.com/market-reports/asia-pacific-resistant-starch-market.

Market Research Future. (2025). Modified starch market research report—Forecast till 2030. Retrieved April 1, 2025, from https://www.marketresearchfuture.com/reports/modified-starch-market-1038.

Maximize Market Research. (2023). Starch-based bioplastics market size, share, and growth analysis by 2030. https://www.maximizemarketresearch.com/market-report/starch-based-bioplastics-market/189415/.

Meneses, Y. E., Stratton, J., & Flores, R. A. (2017). Water reconditioning and reuse in the food processing industry: Current situation and challenges. *Trends in Food Science & Technology, 61*, 72–79.

Meticulous Research. (2025). Modified starch market size & foreca–2031. Retrieved April 1, 2025, from https://www.meticulousresearch.com/product/modified-starch-market-2718.

Meyerding, S. G., Kürzdörfer, A., & Gassler, B. (2018). Consumer preferences for superfood ingredients—The case of bread in Germany. *Sustainability, 10*(12), 4667.

Michel, M., Eldridge, A. L., Hartmann, C., Klassen, P., Ingram, J., & Meijer, G. W. (2024). Benefits and challenges of food processing in the context of food systems, value chains and sustainable development goals. *Trends in Food Science and Technology, 153*. https://doi.org/10.1016/j.tifs.2024.104703

Ministry of Justice Japan. (2013). *Food labeling act (act no. 70 of 2013)*. Japanese Law Translation. Retrieved May 8, 2025, from https://www.japaneselawtranslation.go.jp/en/laws/view/3649/en.

Monga, S. (2025). Consumer awareness and perceptions of food safety labels in packaged foods. *Indiana Journal of Humanities and Social Sciences, 6* (3), 14–17.

Naveen, R., & Loganathan, M. (2024). Role of varieties of starch in the development of edible films—A review. *Starch-Stärke, 76*(11-12), 2300138.

Negowetti, N., Ambwani, S., Karr, S., Rodgers, R. F., & Austin, S. B. (2022). Digging up the dirt on "clean" dietary labels: Public health considerations and opportunities for increased federal oversight. *International Journal of Eating Disorders, 55*(1), 39–48. https://doi.org/10.1002/eat.23585

Noort, M. W. J., Renzetti, S., Linderhof, V., du Rand, G. E., Marx-Pienaar, N. J. M. M., de Kock, H. L., Magano, N., & Taylor, J. R. N. (2022). Towards sustainable shifts to healthy diets and food security in Sub-Saharan Africa with climate-resilient crops in bread-type products: A food system analysis. *Foods, 11*(2). https://doi.org/10.3390/foods11020135

NOVATION® functional native starches. (2025), Available from: https://www.ingredion.com/na/en-us/ingredients/ingredient-product-families/novation-functional-native-starches.html.

Park, S., & Kim, Y. R. (2025). *Clean label starch handbook of starch science and technology* (pp. 336–352). South Korea: CRC Press. https://doi.org/10.1201/9781003464396-21

Pidgeon, E. F. (2009). *The application of crossflow membrane filtration to remediate wheat starch processing wastewater for reuse.*

Pío Beltrán, J., Colomer Xena, Y., Clotet, R., Gil Martín, J. C., Flotats Ripoll, X., Jarauta Bragulat, E., Pallí Guell, A., & Reguant, F. (2022). The impact of the European green deal from a sustainable global food system approach. *European Food and Feed Law Review, 17*(1), 2–38.

Pongsawatmanit, R., & Srijunthongsiri, S. (2008). Influence of xanthan gum on rheological properties and freeze–thaw stability of tapioca starch. *Journal of Food Engineering, 88*(1), 137–143. https://doi.org/10.1016/j.jfoodeng.2008.02.009

Precedence Research. (2024). Resistant starch market size to surpass USD 22.85 billion by 2034.https://www.precedenceresearch.com/resistant-starch-market.

Pro Market Reports. (2025). Starch blends biodegradable plastic market by application, by region – Global forecast 2025–2033. https://www.promarketreports.com/reports/starch-blends-biodegradable-plastic-43064.

Public perceptions of food ingredients. (2023), Available from: https://ific.org/ific-survey-public-perceptions-of-food-ingredients/.

Punia, B. S., Sunooj, K. V., Navaf, M., Phimolsiripol, Y., & Whiteside, W. S. (2024). Recent advancements in cross-linked starches for food applications-a review. *International Journal of Food Properties, 27*(1), 411–430.

Rashidinejad, A. (2024). The road ahead for functional foods: Promising opportunities amidst industry challenges. *Future Postharvest and Food, 1*(2), 266–273. https://doi.org/10.1002/fpf2.12022

Rashwan, A. K., Younis, H. A., Abdelshafy, A. M., Osman, A. I., Eletmany, M. R., Hafouda, M. A., & Chen, W. (2024). Plant starch extraction, modification, and green applications: A review. *Environmental Chemistry Letters, 22*(5), 2483–2530. https://doi.org/10.1007/s10311-024-01753-z

Reiner, S. J., Reineccius, G. A., & Peppard, T. L. (2010). A comparison of the stability of beverage cloud emulsions formulated with different gum acacia-and starch-based emulsifiers. *Journal of Food Science, 75*(5), E236–E246.

Research and Markets. (2025). Modified starch market forecast, 2024-2033.Retrieved April 1, 2025, from https://www.researchandmarkets.com/report/starch.

Roquette. (2022). Roquette launches a new line of organic pea ingredients to address consumer demand.https://www.roquette.com/-/media/media-centre/press-releases/2022/20221013.

Roquette. (2023b). Sustainability report.https://www.scribd.com/document/786111825/2023-SustainabilityReport-EnglishVersion.

Roquette. (2024). Roquette beauté launches a new biodegradable pea-based film former at in-cosmetics® glob. https://www.roquette.com/-/media/media-center/press-release-n0415-press-release-roquette-beaute-biodegradable-pea-based-pea-based-film-former.pdf.

Sangokunle, O. O., Sathe, S. K., & Singh, P. (2020). Purified starches from 18 pulses have markedly different morphology, oil absorption and water absorption capacities, swelling power, and turbidity. *Starch Staerke, 72*(11–12), Article 2000022.

Santana, Á. L., & Meireles, M. A. A. (2014). New starches are the trend for industry applications: A review. *Food and Public Health, 4*(5), 229–241.

Sapper, M., & Chiralt, A. (2018). Starch-based coatings for preservation of fruits and vegetables. *Coatings, 8*(5), 152.

Sharma, M., Patel, H., Bains, A., & Chawla, P. (2024). *Chemistry and application of starches and gums of pseudocereals pseudocereals: Production, processing, and nutrition* (pp. 106–131). India: CRC Press. https://doi.org/10.1201/9781003325277-6

Sher, A., Li, H., ullah, A., Hamid, Y., Nasir, B., & Zhang, J. (2024). Importance of regenerative agriculture: Climate, soil health, biodiversity and its socioecological impact. *Discover Sustainability, 5*(1). https://doi.org/10.1007/s43621-024-00662-z

Shivaswamy, G., Rudra, S. G., Singh, A., Kumar, R., Das, D., Goswami, A. K., & Thakur, A. (2024). Modified starch from mango pickling industry waste: Comparison of physical and chemical modification. *Starch - Stärke, 76*(11–12). https://doi.org/10.1002/star.202400019

Shrivastava, A., Gupta, R. K., & Srivastav, P. P. (2024). Exploring novel frontiers of advancements in purple yam (Dioscorea alata L.) starch extraction, modification, characterization, applications in food and other industries. *Measurement: Food, 15*, Article 100196.

Silva, E. C. D., Santos Sobrinho, V. D., & Cereda, M. P. (2013). Stability of cassava flour-based food bars. *Food Science and Technology, 33*, 192–198.

Steeneken, P. A., Woortman, A. J., & Oudhuis, A. L. (2011). Processing stability of cross-linked starches in acid sauce applications and identification of some of the molecular factors involved. *Food Hydrocolloids, 25*(3), 410–418.

Straits Research. (2023). Starch-based plastics market: Global industry analysis, share, size, growth, trends, and forecast 2023–2030. https://straitsresearch.com/report/starch-based-plastics-market/.

Straits Research. (2025). Modified starch market size, share & trends analysis report, 2025–2033. Retrieved April 1, 2025, from https://straitsresearch.com/report/modified-starch-market.

Sun, Y., Wang, Z., Ye, J., Li, Y., Wang, L., & Cao, R. (2025). Advances in starch-based nanocomposites for functional food systems: Harnessing ai and nuclear magnetic resonance technologies for tailored stability and bioactivity. *Foods, 14*(5), 773. https://doi.org/10.3390/foods14050773

Sun, C., Wei, Z., Xue, C., & Yang, L. (2023). Development, application and future trends of starch-based delivery systems for nutraceuticals: A review. *Carbohydrate Polymers, 308*, Article 120675. https://doi.org/10.1016/j.carbpol.2023.120675

Surendra Babu, A., Parimalavalli, R., Jagannadham, K., Sudhakara Rao, J., & Shalini Gaur, R. (2016). Fat mimicking properties of citric acid treated sweet potato starch. *International Journal of Food Properties, 19*(1), 139–153. https://doi.org/10.1080/10942912.2015.1020390

Suri, S., & Singh, A. (2023). Modification of starch by novel and traditional ways: Influence on the structure and functional properties. *Sustainable Food Technology, 1*(3), 348–362. https://doi.org/10.1039/d2fb00043a

Thamarai, P., Deivayanai, V. C., Saravanan, A., Vickram, A. S., & Yaashikaa, P. R. (2024). Carbon mitigation in agriculture: Pioneering technologies for a sustainable food system. *Trends in Food Science & Technology, 147*, Article 104477. https://doi.org/10.1016/j.tifs.2024.104477

Tian, S., Xue, X., Wang, X., & Chen, Z. (2022). Preparation of starch-based functional food nano-microcapsule delivery system and its controlled release characteristics. *Frontiers in Nutrition, 9*. https://doi.org/10.3389/fnut.2022.982370

Toplicean, I.-M., & Datcu, A.-D. (2024). An overview on bioeconomy in agricultural sector, biomass production, recycling methods, and circular economy considerations. *Agriculture, 14*(7), 1143. https://doi.org/10.3390/agriculture14071143

EFSA Panel on Nutrition, Novel Foods and Food Allergens (NDA), Turck, D., Castenmiller, J., Hirsch Ernst, K., Kearney, J., Knutsen, H. K., Maciuk, A., Mangelsdorf, I., McArdle, H. J., & Naska, A. (2020). Scientific opinion related to a notification from lyckeby starch AB on barley starch to be used in the manufacturing of several foods as ingredient, of the food additive modified starch and of glucose syrups pursuant to article 21(2) of regulation (EU) no 1169/2011 – For permanent exemption from labelling. *EFSA Journal, 18*(5), Article e06118. https://doi.org/10.2903/j.efsa.2020.6118

Vasileiadou, A. (2024). From organic wastes to bioenergy, biofuels, and value-added products for urban sustainability and circular economy: A review. *Urban Science, 8*(3), 121. https://doi.org/10.3390/urbansci8030121

Venkatesh, T., George, A., Sajeev, M. S., & Kothakota, A. (2024). Plant-based hydrocolloids for efficient clarification of cane juices: Rheological analysis and solidification studies. *Food & Function, 15*(4), 1977–1993.

Walsh, S. K., Lucey, A., Walter, J., Zannini, E., & Arendt, E. K. (2022). Resistant starch—An accessible fiber ingredient acceptable to the Western palate. *Comprehensive Reviews in Food Science and Food Safety, 21*(3), 2930–2955. https://doi.org/10.1111/1541-4337.12955

Witczak, M., Ziobro, R., Juszczak, L., & Korus, J. (2016). Starch and starch derivatives in gluten-free systems – A review. *Journal of Cereal Science, 67*, 46–57. https://doi.org/10.1016/j.jcs.2015.07.007

Xing, Y., & Wang, X. (2024). Precision agriculture and water conservation strategies for sustainable crop production in arid regions. *Plants, 13*(22), 3184.

Xu, Z., Sun, D. W., Zeng, X. A., Liu, D., & Pu, H. (2015). Research developments in methods to reduce the carbon footprint of the food system: A review. *Critical Reviews in Food Science and Nutrition, 55*(9), 1270–1286. https://doi.org/10.1080/10408398.2013.821593

Xue, H., Gao, Y., Wu, L., Cai, X., Liao, J., & Tan, J. (2025). Research progress in extraction, purification, structure of fruit and vegetable polysaccharides and their interaction with anthocyanins/starch. *Critical Reviews in Food Science and Nutrition, 65*(7), 1235–1260.

Yang, Z., McClements, D. J., Xu, Z., Meng, M., Li, C., Chen, L., Qiu, C., Long, J., & Jin, Z. (2022). Carbohydrate-based functional ingredients derived from starch: Current status and future prospects. *Food Hydrocolloids, 131*, Article 107729. https://doi.org/10.1016/j.foodhyd.2022.107729

Yazid, N. S. M., Abdullah, N., Muhammad, N., & Matias-Peralta, H. M. (2018). Application of starch and starch-based products in food industry. *Journal of Science and Technology, 10*(2).

Ye, J., Yang, R., Liu, C., Luo, S., Chen, J., Hu, X., & Wu, J. (2018). Improvement in freeze-thaw stability of rice starch gel by inulin and its mechanism. *Food Chemistry, 268*, 324–333. https://doi.org/10.1016/j.foodchem.2018.06.086

Index

'*Note*: Page numbers followed by "f" indicate figures and "t" indicate tables.'

D

E

F

G

H

I

K

L

M

Printed and bound by CPI Group (UK) Ltd, Croydon, CR0 4YY
14/09/2026
02216413-0001